Dictionary of Physics

Third Edition

McGraw-Hill

New York Ch … don Madrid
Mexico City … l Singapore

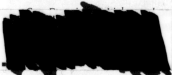

The McGraw-Hill Companies

1 2 3 4 5 6 7 8 9 0 DOC/DOC 0 9 8 7 6 5 4 3

ISBN 0-07-141048-1

This book is printed on recycled, acid-free paper containing a minimum of 50% recycled, de-inked fiber.

This book was set in Helvetica Bold and Novarese Book by the Clarinda Company, Clarinda, Iowa. It was printed and bound by RR Donnelley, The Lakeside Press.

McGraw-Hill books are available at special quantity discounts to use as premiums and sales promotions, or for use in corporate training programs. For more information, please write to the Director of Special Sales, Professional Publishing, McGraw-Hill, Two Penn Plaza, New York, NY 10121-2298. Or contact your local bookstore.

Library of Congress Cataloging-in-Publication Data

McGraw-Hill dictionary of physics — 3rd. ed.

p. cm.

"All text in this dictionary was published previously in the McGraw-Hill dictionary of scientific and technical terms, sixth edition, c2003..." — T.p. verso.

ISBN 0-07-141048-1 (alk. paper)

1. Physics—Dictionaries. I. Title: Dictionary of physics. II. Title: Physics. III. McGraw-Hill dictionary of scientific and technical terms. 6th ed.

QC5 .M424 2002
530′.03—dc21 2002033163

Contents

Preface

The *McGraw-Hill Dictionary of Physics*, Third Edition, provides a compendium of 11,000 terms that are central to the science of physics and to related fields in science and engineering. The coverage in the Third Edition is focused on the areas of acoustics, atomic physics, crystallography, electricity and electromagnetism, fluid mechanics, mechanics, nuclear physics, optics, particle physics, plasma physics, quantum mechanics, relativity, solid-state physics, spectroscopy, statistical mechanics, and thermodynamics, with new terms added and others revised as necessary.

Although there are numerous areas of specialization within physics, they all strive to understand natural phenomena in terms of elementary principles and laws. Thus, the principles of physics apply to virtually all areas of science and technology, and the terms contained in the Dictionary will also be useful in fields such as astronomy, biomedical science, chemistry, and earth science as well as in many areas of engineering.

All of the definitions are drawn from the *McGraw-Hill Dictionary of Scientific and Technical Terms*, Sixth Edition (2003). Each definition is classified according to the field with which it is primarily associated; if a definition is used in more than one area, it is identified by the general label [PHYSICS]. The pronunciation of each term is provided along with synonyms, acronyms, and abbreviations where appropriate. A guide to the use of the Dictionary is included on pages vii-viii, explaining the alphabetical organization of terms, the format of the book, cross referencing, and how synonyms, variant spellings, abbreviations, and similar information are handled. The Pronunciation Key is provided on page xi. The Appendix provides conversion tables for commonly used scientific units as well as other listings of scientific data.

It is the editors' hope that the Third Edition of the *McGraw-Hill Dictionary of Physics* will serve the needs of scientists, engineers, students, teachers, librarians, and writers for high-quality information, and that it will contribute to scientific literacy and communication.

Mark D. Licker
Publisher

Staff

Mark D. Licker, Publisher—Science

Elizabeth Geller, Managing Editor
Jonathan Weil, Senior Staff Editor
David Blumel, Staff Editor
Alyssa Rappaport, Staff Editor
Charles Wagner, Digital Content Manager
Renee Taylor, Editorial Assistant

Roger Kasunic, Vice President—Editing, Design, and Production

Joe Faulk, Editing Manager
Frank Kotowski, Jr., Senior Editing Supervisor

Ron Lane, Art Director

Thomas G. Kowalczyk, Production Manager
Pamela A. Pelton, Senior Production Supervisor

Henry F. Beechhold, Pronunciation Editor
Professor Emeritus of English
Former Chairman, Linguistics Program
The College of New Jersey
Trenton, New Jersey

How to Use the Dictionary

ALPHABETIZATION. The terms in the McGraw-Hill *Dictionary of Physics*, Third Edition, are alphabetized on a letter-by-letter basis; word spacing, hyphen, comma, solidus, and apostrophe in a term are ignored in the sequencing. For example, an ordering of terms would be:

acceptor	**air current**
Ac-Em	**A mode**
AES	**Bow's notation**
aircraft noise	**bow wave**

FORMAT. The basic format for a defining entry provides the term in boldface, the field is small capitals, and the single definition in lightface:

term [FIELD] Definition.

A field may be followed by multiple definitions, each introduced by a boldface number:

term [FIELD] **1.** Definition. **2.** Definition. **3.** Definition.

A term may have definitions in two or more fields:

term [ACOUS] Definition. [FL MECH] Definition.

A simple cross-reference entry appears as:

term *See* another term.

A cross reference may also appear in combination with definitions:

term [ACOUS] Definition. [FL MECH] *See* another term.

CROSS REFERENCING. A cross-reference entry directs the user to the defining entry. For example, the user looking up "Leduc law" finds:

Leduc law *See* Amagat-Leduc rule.

The user then turns to the "A" terms for the definition. Cross references are also made from variant spellings, acronyms, abbreviations, and symbols.

atm *See* atmosphere.
cgs system *See* centimeter-gram-second system.
CVD *See* chemical vapor deposition.
U stage *See* universal stage.

ALSO KNOWN AS . . . , etc. A definition may conclude with a mention of a synonym of the term, a variant spelling, an abbreviation for the term, or other such information, introduced by "Also known as . . . ," "Also spelled . . . ," "Abbreviated . . . ," "Symbolized . . . ," "Derived from . . . ." When a term has more than one definition, the positioning of any of these phrases conveys the extent of applicability. For example:

term [ACOUS] **1.** Definition. Also known as synonym. **2.** Definition. Symbolized T.

In the above arrangement, "Also known as . . ." applies only to the first definition; "Symbolized . . ." applies only to the second definition.

term [ACOUS] **1.** Definition. **2.** Definition. [FL MECH] Definition. Also known as synonym.

In the above arrangement, "Also known as . . ." applies only to the second field.

term [ACOUS] Also known as synonym. **1.** Definition. **2.** Definition. [FL MECH] Definition.

In the above arrangement, "Also known as . . ." applies only to both definitions in the first field.

term Also known as synonym. [ACOUS] **1.** Definition. **2.** Definition. [FL MECH] Definition.

In the above arrangement, "Also known as . . ." applies to all definitions in both fields.

Fields and Their Scope

[ACOUS] **acoustics**—The science of the production, transmission, and effects of sound.

[ATOM PHYS] **atomic physics**—The branch of physics concerned with the structures of the atom, the characteristics of the electrons and other elementary particles of which the atom is composed, the arrangement of the atom's energy states, and the processes involved in the radiation of light and x-rays.

[CRYO] **cryogenics**—The science of producing and maintaining very low temperatures, of phenomena at those temperatures, and of technical operations performed at very low temperatures.

[CRYSTAL] **crystallography**—The branch of science that deals with the geometric description of crystals, their internal arrangement, and their properties.

[ELEC] **electricity**—The science of physical phenomena involving electric charges and their effects when at rest and when in motion.

[ELECTROMAG] **electromagnetism**—The branch of physics dealing with the observations and laws relating electricity to magnetism, and with magnetism produced by an electric current.

[FL MECH] **fluid mechanics**—The science concerned with fluids, either at rest or in motion, and dealing with pressures, velocities, and accelerations in the fluid, including fluid deformation and compression or expansion.

[MECH] **mechanics**—The branch of physics which seeks to formulate general rules for predicting the behavior of a physical system under the influence of any type of interaction with its environment.

[NUC PHYS] **nuclear physics**—The study of the characteristics, behavior, and internal structure of the atomic nucleus.

[OPTICS] **optics**—The study of phenomena associated with the generation, transmission, and detection of electromagnetic radiation in the spectral range extending from the long-wave edge of the x-ray region to the short-wave edge of the radio region; and the science of light.

[PART PHYS] **particle physics**—The branch of physics concerned with understanding the properties, behavior, and structure of elementary particles, especially through study of collisions or decays involving energies of hundreds of megaelectronvolts or more.

[PHYS] **physics**—The science concerned with those aspects of nature which can be understood in terms of elementary principles and laws.

[PL PHYS] **plasma physics**—The study of highly ionized gases.

[QUANT MECH] **quantum mechanics**—The modern theory of matter, of electromagnetic radiation, and of the interaction between matter and radiation; it differs from classical physics, which it generalizes and supersedes, mainly in the realm of atomic and subatomic phenomena.

[RELAT] **relativity**—The study of the physical theory which recognizes the universal character of the propagation speed of light and the consequent dependence of space, time, and other mechanical measurements on the motion of the observer performing the measurements; the two main divisions are special theory and general theory.

[SOLID STATE] **solid-state physics**—The branch of physics centering on the physical properties of solid materials; it is usually concerned with the properties of crystalline materials only, but it is sometimes extended to include the properties of glasses or polymers.

[SPECT] **spectroscopy**—The branch of physics concerned with the production, measurement, and interpretation of electromagnetic spectra arising from either emission or absorption of radiant energy by various substances.

[STAT MECH] **statistical mechanics**—That branch of physics which endeavors to explain and predict the macroscopic properties and behavior of a system on the basis of the known characteristics and interactions of the microscopic constituents of the system, usually when the number of such constituents is very large.

[THERMO] **thermodynamics**—The branch of physics which seeks to derive, from a few basic postulates, relations between properties of substances, especially those which are affected by changes in temperature, and a description of the conversion of energy from one form to another.

Pronunciation Key

Vowels

a as in b**a**t, th**a**t
ā as in b**ai**t, cr**a**te
ä as in b**o**ther, f**a**ther
e as in b**e**t, n**e**t
ē as in b**ee**t, tr**ea**t
i as in b**i**t, sk**i**t
ī as in b**i**te, l**igh**t
ō as in b**oa**t, n**o**te
ȯ as in b**ough**t, t**au**t
u̇ as in b**oo**k, p**u**ll
ü as in b**oo**t, p**oo**l
ə as in b**u**t, sof**a**
au̇ as in cr**ow**d, p**ow**er
ȯi as in b**oi**l, sp**oi**l
yə as in form**u**la, spectac**u**lar
yü as in f**ue**l, m**u**le

Semivowels/Semiconsonants

w as in **w**ind, t**w**in
y as in **y**et, on**ion**

Stress (Accent)

ˈ precedes syllable with primary stress

ˌ precedes syllable with secondary stress

¦ precedes syllable with variable or indeterminate primary/secondary stress

Consonants

b as in **b**i**b**, dri**bb**le
ch as in **ch**arge, stre**tch**
d as in **d**og, ba**d**
f as in **f**ix, sa**f**e
g as in **g**ood, si**g**nal
h as in **h**and, be**h**ind
j as in **j**oint, di**g**it
k as in **c**ast, bri**ck**
k̲ as in Ba**ch** (used rarely)
l as in **l**oud, be**ll**
m as in **m**ild, su**mm**er
n as in **n**ew, de**n**t
n̲ indicates nasalization of preceding vowel
ŋ as in ri**ng**, si**ng**le
p as in **p**ier, sli**p**
r as in **r**ed, sca**r**
s as in **s**ign, po**s**t
sh as in **su**gar, **sh**oe
t as in **t**imid, ca**t**
th as in **th**in, brea**th**
t̲h̲ as in **th**en, brea**th**e
v as in **v**eil, wea**v**e
z as in **z**oo, crui**s**e
zh as in bei**g**e, trea**s**ure

Syllabication

· Indicates syllable boundary when following syllable is unstressed

a *See* ampere; atto-.

aΩ *See* abohm.

$(a\Omega)^{-1}$ *See* abmho.

A *See* ampere; angstrom.

Å *See* angstrom.

aA *See* abampere.

$aAcm^2$ *See* abampere centimeter squared.

aA/cm^2 *See* abampere per square centimeter.

a axis [CRYSTAL] One of the crystallographic axes used as reference in crystal description, usually oriented horizontally, front to back. { 'ā 'ak,sis }

ab- [ELECTROMAG] A prefix used to identify centimeter-gram-second electromagnetic units, as in abampere, abcoulomb, abfarad, abhenry, abmho, abohm, and abvolt. { ab }

abampere [ELEC] The unit of electric current in the electromagnetic centimeter-gram-second system; 1 abampere equals 10 amperes in the absolute meter-kilogram-second-ampere system. Abbreviated aA. Also known as Bi; biot. { ab'am·pēr }

abampere centimeter squared [ELECTROMAG] The unit of magnetic moment in the electromagnetic centimeter-gram-second system. Abbreviated $aAcm^2$. { ab'am·pēr 'sen·tə,mē·dər 'skwerd }

abampere per square centimeter [ELEC] The unit of current density in the electromagnetic centimeter-gram-second system. Abbreviated aA/cm^2. { ab'am·pēr pər 'skwer 'sen·tə,mēd·ər }

Abbe condenser [OPTICS] A variable large-aperture lens system arranged substage to image a light source into the focal plane of a microscope objective. { 'ä·bə kən'dens·ər }

Abbe number [OPTICS] A number which expresses the deviating effect of an optical glass on light of different wavelengths. { 'ä·bə ,nəm·bər }

Abbe prism [OPTICS] A system used for image erection which is composed of two double right-angle prisms and involves four reflections. { 'ä·bə 'priz·əm }

Abbe refractometer [OPTICS] An optical instrument for the measurement of the refractive index of liquids. { 'ä·bə ,rē·frak'täm·əd·ər }

Abbe's sine condition [OPTICS] A relationship which must hold to prevent aberration of a mirror or lens from producing a coma. { 'ä·bəz 'sīn kən,dish·ən }

Abbe's theory [OPTICS] The theory that for a lens to produce a true image, it must be large enough to transmit the entire diffraction pattern of the object. { 'ä·bəz 'thē·ə·rē }

abcoulomb [ELEC] The unit of electric charge in the electromagnetic centimeter-gram-second system, equal to 10 coulombs. Abbreviated aC. { ab'kü·lōm }

abcoulomb centimeter [ELEC] In the electromagnetic centimeter-gram-second system of units, the unit of electric dipole moment. Abbreviated aCcm. { ab'kü·lōm 'sen·tə,mēd·ər }

abcoulomb per cubic centimeter [ELEC] The electromagnetic centimeter-gram-second unit of volume density of charge. Abbreviated aC/cm^3. { ab'kü·lōm pər 'kyü·bik 'sen·tə,mēd·ər }

abcoulomb per square centimeter [ELEC] The electromagnetic centimeter-gram-second unit of surface density of charge, electric polarization, and displacement. Abbreviated aC/cm^2. { ab 'kü·lōm pər skwer 'sen·tə,mēd·ər }

Abelian quantum Hall state [CRYO] A quantum Hall state that contains two or more components of incompressible fluid, has a filling factor equal to p/q, where q is not divisible by p, and has a topological order that can be described as a pattern of dancing steps of the electrons and can be characterized by a symmetric matrix and a charge vector, both with integer entries. { ə'bē·lē·ən ¦kwän·təm ¦hól ,stāt }

aberration *See* optical aberration. { ,ab·ə'rā·shən }

abfarad [ELEC] A unit of capacitance in the electromagnetic centimeter-gram-second system equal to 10^9 farads. Abbreviated aF. { ab 'far·ad }

abhenry [ELEC] A unit of inductance in the electromagnetic centimeter-gram-second system of units which is equal to 10^{-9} henry. Abbreviated aH. { ab'hen·rē }

abmho [ELEC] A unit of conductance in the electromagnetic centimeter-gram-second system of units equal to 10^9 mhos. Abbreviated $(a\Omega)^{-1}$. Also known as absiemens (aS). { 'ab,mō }

Abney effect [OPTICS] A shift in the apparent hue of a light which occurs as colored light is desaturated by the addition of white light. { 'ab·nē ə'fekt }

Abney law [OPTICS] The shift in apparent hue

of spectral color that is desaturated by addition of white light is toward the red end of the spectrum if the wavelength is below 570 nanometers and toward the blue if it is above. { 'ab·nē ,lȯ }

Abney mounting [SPECT] A modification of the Rowland mounting in which only the slit is moved to observe different parts of the spectrum. { 'ab·nē ,mau̇nt·iŋ }

abnormal reflections [ELECTROMAG] Sharply defined reflections of substantial intensity at frequencies greater than the critical frequency of the ionized layer of the ionosphere. { ab'nȯr·məl re'flek·shənz }

abnormal series *See* anomalous series. { ab'nȯr·məl 'sir,ēz }

abohm [ELEC] The unit of electrical resistance in the centimeter-gram-second system; 1 abohm equals 10^{-9} ohm in the meter-kilogram-second system. Abbreviated aΩ. { a'bōm }

abohm centimeter [ELEC] The centimeter-gram-second unit of resistivity. Abbreviated aΩcm. { a'bōm 'sen·tə,mē·dər }

Abrikosov-Suhl resonance [SOLID STATE] For materials that display the Kondo effect, a long-lived scattering resonance of electronic states near the Fermi level that forms at temperatures less than the Kondo temperature; accounts for the qualitative behavior of resistivity and magnetic susceptibility as functions of temperature. Also known as Kondo resonance. { ,ab·rə'kä·sȯf 'sül 'rez·ən·əns }

absiemens *See* abmho. { ab'sē·mənz }

absolute expansion [THERMO] The true expansion of a liquid with temperature, as calculated when the expansion of the container in which the volume of the liquid is measured is taken into account; in contrast with apparent expansion. { 'ab·sə,lüt ik'span·shən }

absolute gain of an antenna [ELECTROMAG] Gain in a given direction when the reference antenna is an isotropic antenna isolated in space. Also known as isotropic gain of an antenna. { 'ab·sə,lüt ,gān əv ən an'ten·ə }

absolute humidity [PHYS] The ratio of the mass of water vapor in a sample of air to the volume of the sample. { 'ab·sə,lüt hyü'mid·ə·dē }

absolute index of refraction *See* index of refraction. { 'ab·sə,lüt 'in,deks əv ri'frak·shən }

absolute luminosity [OPTICS] The luminosity of an object expressed in units of fundamental quantities. { 'ab·sə,lüt lü·mə'näs·ə·dē }

absolute motion [PHYS] Motion of an object described by its measurement in a frame of reference that is preferred over all other frames. { 'ab·sə,lüt 'mō·shən }

absolute permeability [ELECTROMAG] The ratio of the magnetic flux density to the intensity of the magnetic field in a medium; measurement is in webers per square meter in the meter-kilogram-second system. Also known as induced capacity. { 'ab·sə,lüt pər·mē·ə'bil·ə·dē }

absolute pitch [ACOUS] The pitch of a musical tone expressed as the frequency of the sound wave of that tone. { 'ab·sə,lüt 'pich }

absolute potential vorticity *See* potential vorticity. { 'ab·sə,lüt pə'ten·shəl vȯr'tis·ə·dē }

absolute pressure [PHYS] The pressure above the absolute zero value of pressure that theoretically obtains in empty space or at the absolute zero of temperature, as distinguished from gage pressure. { 'ab·sə,lüt 'presh·ər }

absolute scale *See* absolute temperature scale. { 'ab·sə,lüt ,skāl }

absolute space-time [PHYS] A concept underlying Newtonian mechanics which postulates the existence of a preferred reference system of time and spatial coordinates; replaced in relativistic mechanics by Einstein's equivalency principle. Also known as absolute time. { 'ab·sə,lüt 'spās 'tīm }

absolute specific gravity [MECH] The ratio of the weight of a given volume of a substance in a vacuum at a given temperature to the weight of an equal volume of water in a vacuum at a given temperature. { 'ab·sə,lüt spə'sif·ək 'grav·əd·ē }

absolute standard [PHYS] A particle or object designated as a standard by assigning to it a mass of one unit; used in defining quantities in Newton's second law of motion. { 'ab·sə,lüt 'stan·dərd }

absolute system of units [PHYS] A set of units for measuring physical quantities, defined by interrelated equations in terms of arbitrary fundamental quantities of length, mass, time, and charge or current. { 'ab·sə,lüt 'sis·təm əv 'yü·nəts }

absolute temperature [THERMO] **1.** The temperature measurable in theory on the thermodynamic temperature scale. **2.** The temperature in Celsius degrees relative to the absolute zero at −273.16°C (the Kelvin scale) or in Fahrenheit degrees relative to the absolute zero at −459.69°F (the Rankine scale). { 'ab·sə,lüt 'tem·prə·chür }

absolute temperature scale [THERMO] A scale with which temperatures are measured relative to absolute zero. Also known as absolute scale. { 'ab·sə,lüt 'tem·prə·chür ,skāl }

absolute time *See* absolute space-time. { 'ab·sə,lüt 'tīm }

absolute unit [PHYS] A unit defined in terms of units of fundamental quantities such as length, time, mass, and charge or current. { 'ab·sə,lüt 'yü·nət }

absolute vacuum [PHYS] A void completely empty of matter. Also known as perfect vacuum. { 'ab·sə,lüt 'vak·yüm }

absolute velocity [PHYS] The vector sum of the velocity of a fluid parcel relative to the earth and the velocity of the parcel due to the earth's rotation; the east-west component is the only one affected. { 'ab·sə,lüt və'läs·ə·dē }

absolute viscosity [FL MECH] The tangential force per unit area of two parallel planes at unit distance apart when the space between them is filled with a fluid and one plane moves with unit velocity in its own plane relative to the other.

Also known as coefficient of viscosity. { 'ab·sə,lüt vis'käs·ə·dē }

absolute vorticity [FL MECH] The vorticity of a fluid relative to an absolute coordinate system; especially, the vorticity of the atmosphere relative to axes not rotating with the earth. { 'ab·sə,lüt vȯr'tis·ə·dē }

absolute wavemeter [ELECTROMAG] A type of wavemeter in which the frequency of an injected radio-frequency voltage is determined by measuring the length of a resonant line. { 'ab·sə,lüt 'wāv,mēd·ər }

absolute zero [THERMO] The temperature of −273.16°C, or −459.69°F, or 0 K, thought to be the temperature at which molecular motion vanishes and a body would have no heat energy. { 'ab·sə,lüt 'zir·ō }

absorb [ELECTROMAG] To take up energy from radiation. [PHYS] To take up matter or radiation. { əb'sȯrb }

absorbed charge [ELEC] Charge on a capacitor which arises only gradually when the potential difference across the capacitor is maintained, due to gradual orientation of permanent dipolar molecules. { əb'sȯrbd 'chärj }

absorptance [PHYS] The ratio of the total unabsorbed radiation to the total incident radiation; equal to one (unity) minus the transmittance. { əb'sȯrp·təns }

absorption [ELEC] The property of a dielectric in a capacitor which causes a small charging current to flow after the plates have been brought up to the final potential, and a small discharging current to flow after the plates have been short-circuited, allowed to stand for a few minutes, and short-circuited again. Also known as dielectric soak. [ELECTROMAG] Taking up of energy from radiation by the medium through which the radiation is passing. { əb'sȯrp·shən }

absorption band [PHYS] A range of wavelengths or frequencies in the electromagnetic spectrum within which radiant energy is absorbed by a substance. { əb'sȯrp·shən ,band }

absorption cell [OPTICS] A vessel with transparent walls for holding a gas or liquid whose absorptivity or absorption spectrum is to be measured. { əb'sȯrp·shən ,sel }

absorption coefficient Also known as absorption factor; absorption ratio; coefficient of absorption. [ACOUS] The ratio of the sound energy absorbed by a surface of a medium or material to the sound energy incident on the surface. [PHYS] If a flux through a material decreases with distance x in proportion to e^{-ax}, then a is called the absorption coefficient. { əb'sȯrp·shən ,kō·ə'fish·ənt }

absorption cross section [ELECTROMAG] In radar, the ratio of the amount of power removed from a beam by absorption of radio energy by a target to the power in the beam incident upon the target. { əb'sȯrp·shən ¦krȯs 'sek·shən }

absorption current [ELEC] The component of a dielectric current that is proportional to the rate of accumulation of electric charges within the dielectric. { əb'sȯrp·shən 'kər·ənt }

absorption curve [PHYS] A graph showing the curvilinear relationship of the variation in absorbed radiation as a function of wavelength. { əb'sȯrp·shən ,kərv }

absorption edge [SPECT] The wavelength corresponding to a discontinuity in the variation of the absorption coefficient of a substance with the wavelength of the radiation. Also known as absorption limit. { əb'sȯrp·shən ,ej }

absorption factor *See* absorption coefficient. { əb'sȯrp·shən ,fak·tər }

absorption index [OPTICS] The complex index of refraction may be written as $n(1 + ik)$; the coefficient k is the absorption index. Also known as index of absorption. { əb'sȯrp·shən 'in·deks }

absorption lens [OPTICS] Glass which prevents selected wavelengths from passing through it; used in eyeglasses. { əb'sȯrp·shən ,lenz }

absorption limit *See* absorption edge. { əb'sȯrp·shən 'lim·ət }

absorption line [SPECT] A minute range of wavelength or frequency in the electromagnetic spectrum within which radiant energy is absorbed by the medium through which it is passing. { əb'sȯrp·shən ,līn }

absorption peak [SPECT] A wavelength of maximum electromagnetic absorption by a chemical sample; used to identify specific elements, radicals, or compounds. { əb'sȯrp·shən ,pēk }

absorption ratio *See* absorption coefficient. { əb'sȯrp·shən ,rā·shō }

absorption spectrophotometer [SPECT] An instrument used to measure the relative intensity of absorption spectral lines and bands. Also known as difference spectrophotometer. { əb'sȯrp·shən ,spek·trə·fə'täm·ə·dər }

absorption spectroscopy [SPECT] An instrumental technique for determining the concentration and structure of a substance by measuring the intensity of electromagnetic radiation it absorbs at various wavelengths. { əb'sȯrp·shən ,spek'träs·kə·pē }

absorption spectrum [SPECT] A plot of how much radiation a sample absorbs over a range of wavelengths; the spectrum can be a plot of either absorbance or transmittance versus wavelength, frequency, or wavenumber. { əb'sȯrp·shən ,spek·trəm }

absorption unit *See* sabin. { əb'sȯrp·shən ,yü·nət }

absorptivity [THERMO] The ratio of the radiation absorbed by a surface to the total radiation incident on the surface. { əb,sȯrp'tiv·əd·ē }

abT *See* gauss.

abtesla *See* gauss. { ab'tes·lə }

abvolt [ELEC] The unit of electromotive force in the electromagnetic centimeter-gram-second system; 1 abvolt equals 10^{-8} volt in the absolute meter-kilogram-second system. Abbreviated aV. { 'ab,vōlt }

abvolt per centimeter [ELEC] In the electromagnetic centimeter-gram-second system of units, the unit of electric field strength. Abbreviated aV/cm. { 'ab,vōlt pər 'sen·tə,mēd·ər }

abwatt [ELEC] The unit of electrical power in the centimeter-gram-second system; 1 abwatt equals 1 watt in the absolute meter-kilogram-second system. { 'ab,wät }
abWb *See* maxwell.
abweber *See* maxwell. { 'ab,web·ər }
ac *See* alternating current.
aC *See* abcoulomb.
acceleration [MECH] The rate of change of velocity with respect to time. { ak,sel·ə'rā·shən }
acceleration measurement [MECH] The technique of determining the magnitude and direction of acceleration, including translational and angular acceleration. { ak,sel·ə'rā·shən 'mezh·ər·mənt }
acceleration of free fall *See* acceleration of gravity. { ak,sel·ə'rā·shən əv 'frē ,fȯl }
acceleration of gravity [MECH] The acceleration imparted to bodies by the attractive force of the earth; has an international standard value of 980.665 cm/s^2 but varies with latitude and elevation. Also known as acceleration of free fall; apparent gravity. { ak,sel·ə'rā·shən əv 'grav·ə·dē }
acceleration potential [FL MECH] The sum of the potential of the force field acting on a fluid and the ratio of the pressure to the fluid density; the negative of its gradient gives the acceleration of a point in the fluid. { ak,sel·ə'rā·shən pə'ten·shəl }
accelerator *See* particle accelerator. { ak'sel·ə,rād·ər }
accelerator mass spectrometer [SPECT] A combination of a mass spectrometer and an accelerator that can be used to measure the natural abundances of very rare radioactive isotopes. { ak¦sel·ə,rād·ər ¦mas spek'träm·əd·ər }
acceptor [SOLID STATE] An impurity element that increases the number of holes in a semiconductor crystal such as germanium or silicon; aluminum, gallium, and indium are examples. Also known as acceptor impurity; acceptor material. { ak'sep·tər }
acceptor atom [SOLID STATE] An atom of a substance added to a semiconductor crystal to increase the number of holes in the conduction band. { ak'sep·tər 'ad·əm }
acceptor impurity *See* acceptor. { ak'sep·tər im 'pyu̇r·ə·dē }
acceptor level [SOLID STATE] An energy level in a semiconductor that results from the presence of acceptor atoms. { ak'sep·tər ,lev·əl }
acceptor material *See* acceptor. { ak'sep·tər mə 'tir·ē·əl }
accessory plate [OPTICS] Thin plate of quartz, gypsum, or mica used with a petrological microscope to modify the effects of polarized light and intensify qualities in translucent minerals. { ak'ses·ə·rē ,plāt }
aCcm *See* abcoulomb centimeter.
aC/cm² *See* abcoulomb per square centimeter.
aC/cm³ *See* abcoulomb per cubic centimeter.
accommodation coefficient [STAT MECH] The ratio of the average energy actually transferred between a surface and impinging gas molecules scattered by the surface, to the average energy which would theoretically be transferred if the impinging molecules reached complete thermal equilibrium with the surface. { ə,käm·ə'dā·shən ,kō·ə'fish·ənt }
accumulator *See* storage battery. { ə'kyü·myə ,lād·ər }
accumulator battery *See* storage battery. { ə'kyü·myə,lād·ər 'bad·ə·rē }
Ac-Em *See* actinon.
a-c fracture [CRYSTAL] A type of tension fracture lying parallel to the a-c fabric plane and normal to plane b in a crystal. { 'a'sē 'frak·chər }
achromat *See* achromatic lens. { 'ak·rə,mat }
achromatic [OPTICS] Capable of transmitting light without decomposing it into its constituent colors. { ¦a·krə¦mad·ik }
achromatic color [OPTICS] A color that has no hue or saturation but only brightness, such as white, black, and various shades of gray. { ¦a·krə¦mad·ik 'kəl·ər }
achromatic condenser [OPTICS] A condenser designed to eliminate chromatic and spherical aberrations, usually through the use of four elements, two of which are achromatic lenses; used in microscopes having high magnification. { ¦a·krə¦mad·ik kən'den·sər }
achromatic fringe [OPTICS] An interference fringe of light whose position is independent of the wavelength of the light used; the first fringe of a Lloyd's mirror system and the central fringe of a Fresnel biprism system are examples. { ¦a·krə¦mad·ik 'frinj }
achromatic lens [OPTICS] A combination of two or more lenses having a focal length that is the same for two quite different wavelengths, thereby removing a major portion of chromatic aberration. Also known as achromat. { ¦a·krə¦mad·ik 'lenz }
achromatic locus [OPTICS] A region that includes those points in the chromaticity diagram which represent acceptable reference standards of illumination. Commonly referred to as white light. Also known as achromatic region. { ¦ak·rə¦mad·ik 'lō·kəs }
achromatic prism [OPTICS] A prism consisting of two or more prisms with different refractive indices combined so that light passing through the device is deviated but not dispersed. { ¦a·krə¦mad·ik 'priz·əm }
achronal set [RELAT] A set of points in a space-time with no two points of the set having timelike separation. { ¦ā,krōn·əl 'set }
ACI *See* acoustic comfort index.
Ackeret method [FL MECH] A method of studying the behavior of an airfoil in a supersonic airstream based on the hypothesis that the disturbance caused by the airfoil consists of two plane waves, at the leading and trailing edges, which propagate outward like sound waves and each makes an angle equal to the Mach angle with the direction of flow. { 'ak·ə·rət ,meth·əd }
aclastic [OPTICS] Having the property of not refracting light. { ¦ā'klas·tik }
aΩcm *See* abohm centimeter.

acoustic [ACOUS] Relating to, containing, producing, arising from, actuated by, or carrying sound. { ə'küs·tik }

acoustic absorption *See* sound absorption. { ə'küs·tik əb'sörp·shən }

acoustic absorption coefficient *See* sound absorption coefficient. { ə'küs·tik əb'sörp·shən ˌkō·əˌfish·ənt }

acoustic absorptivity *See* sound absorption coefficient. { ə'küs·tik abˌsorp'tiv·ə·tē }

acoustical [ACOUS] Having a characteristic concerning sound, of an object or quantity that in and of itself does not have properties associated with sound, such as a device, measurement, or symbol. { ə'küs·tə·kəl }

acoustical Doppler effect [ACOUS] The change in pitch of a sound observed when there is relative motion between source and observer. { ə'küs·tə·kəl 'däp·lər ə'fekt }

acoustical holography [PHYS] A technique for using sound to form visible images, in which acoustic beams form an interference pattern of an object and a beam of light interacts with this pattern and is focused to form an optical image. { ə'küs·tə·kəl hō'läg·rə·fē }

acoustic approximation [FL MECH] The approximation that leads from the nonlinear hydrodynamic equations of a gas to the linear wave equation for sound wave propagation. { ə'küs·tik əˌpräk·sə'mā·shən }

acoustic axis *See* axis of acoustic symmetry. { ə'küs·tik 'ak·səs }

acoustic Bessel bullet [ACOUS] One of a class of localized wave solutions to the three-dimensional wave equation that maintain their shape and amplitude as they propagate in space. { əˌküs·tik 'be·səl ˌbu̇l·ət }

acoustic branch [SOLID STATE] One of the parts of the dispersion relation, frequency as a function of wave number, for crystal lattice vibrations, representing vibration at low (acoustic) frequencies. { ə'küs·tik ˌbranch }

acoustic capacitance *See* acoustic compliance. { ə'küs·tik kə'pas·ə·təns }

acoustic cavitation [FL MECH] The formation of vapor-filled bubbles in a liquid during the short periodic intervals of negative pressure, or tensile stress, that accompany the passage of a sound wave. { əˌkü·stik ˌkav·ə'tā·shən }

acoustic comfort index [ACOUS] An arbitrarily designed scale to indicate the noise inside the passenger cabin of an aircraft; on this scale +100 represents ideal conditions or zero noise, 0 represents barely tolerable conditions, and −100 represents intolerable conditions. Abbreviated ACI. { ə'küs·tik 'käm·fərt 'in·deks }

acoustic compliance [ACOUS] The reciprocal of acoustic stiffness. Also known as acoustic capacitance. { ə'küs·tik kəm'plī·əns }

acoustic dispersion [ACOUS] A complex sound wave's separation into its frequency components as it passes through a medium; usually measured by the rate of change of velocity with frequency. { ə'küs·tik dis'pər·zhən }

acoustic domain [ACOUS] A concentration of crystal lattice vibrations traveling at the speed of sound; used to generate light from an array of *pn* junctions. { ə'küs·tik də'mān }

acoustic emission [ACOUS] The phenomenon of transient elastic-wave generation due to a rapid release of strain energy caused by a structural alteration in a solid material. Also known as stress-wave emission. { ə'küs·tik ē'mish·ən }

acoustic energy *See* sound energy. { ə'küs·tik 'en·ər·jē }

acoustic fatigue [MECH] The tendency of a material, such as a metal, to lose strength after acoustic stress. { ə'küs·tik fə'tēg }

acoustic grating [ACOUS] A series of rods or other suitable objects of equal size placed in a row a fixed distance apart; causes sounds with different wavelengths to be diffracted in different directions. { ə'küs·tik 'grāt·iŋ }

acoustic image [ACOUS] The geometric space figure that is made up of the acoustic foci of an acoustic lens, mirror, or other acoustic optical system and is the acoustic counterpart of an extended source of sound. Also known as image. { ə'küs·tik 'im·ij }

acoustic imaging [ACOUS] The use of ultrasound to produce real-time images of the internal structure of a metallic or biological object that is opaque to light. Also known as sonography; ultrasonic imaging; ultrasonography. { ə'küs·tik 'im·ij·iŋ }

acoustic impedance [ACOUS] The complex ratio of the sound pressure on a given surface to the sound flux through that surface, expressed in acoustic ohms. { ə'küs·tik im'pēd·əns }

acoustic inertance *See* acoustic mass. { ə'küs·tik i'nərt·əns }

acoustic intensity [ACOUS] The limit approached by the ratio of the acoustic power in a given area to the magnitude of this area as the magnitude approaches zero. { ə'küs·tik in'ten·səd·ē }

acoustic interferometer [ACOUS] A device for measuring the velocity and attenuation of sound waves in a gas or liquid by an interference method. { ə'küs·tik in·tər·fə'rä·məd·ər }

acoustic levitation [ACOUS] The use of a very intense sound wave to keep a body suspended above the device producing the sound wave. { ə'küs·tik lev·ə'tā·shən }

acoustic mass [ACOUS] The quantity which, after multiplication by 2π times the frequency, results in the acoustic reactance associated with the kinetic energy of the sound medium. Also known as acoustic inertance. { ə'küs·tik 'mas }

acoustic mass reactance [ACOUS] The part of the acoustic reactance associated with the kinetic energy of a medium. Also known as mass reactance. { ə'küs·tik 'mas rē'ak·təns }

acoustic measurement [ACOUS] The process of quantitatively determining one or more properties of sound. { ə'küs·tik 'mezh·ər·mənt }

acoustic microscope [OPTICS] An instrument which employs acoustic radiation at microwave frequencies to allow visualization of the microscopic detail exhibited in elastic properties of an object. { ə'küs·tik 'mīk·rə,skōp }

acoustic mode [SOLID STATE] The type of crystal lattice vibrations which for long wavelengths act like an acoustic wave in a continuous medium, but which for shorter wavelengths approach the Debye frequency, showing a dispersive decrease in phase velocity. { ə'küs·tik 'mōd }

acoustic noise [ACOUS] Noise in the acoustic spectrum; usually measured in decibels. { ə'küs·tik ,nȯiz }

acoustic ohm [ACOUS] The unit of acoustic impedance. Also known as acoustic reactance unit; acoustic resistance unit. { ə'küs·tik 'ōm }

acoustic particle detection [PART PHYS] A technique for detecting charged particles traversing a medium by recording the impulsive acoustic signals that result from rapid thermal expansion of the medium. { ə'küs·tik 'pärd·ə·kəl di,tek·shən }

acoustic phonon [SOLID STATE] A quantum of excitation of an acoustic mode of vibration. { ə'küs·tik 'fōn,än }

acoustic power *See* sound power. { ə'küs·tik 'pau̇·ər }

acoustic radiation [ACOUS] Infrasonic, sonic, or ultrasonic waves propagating through a solid, liquid, or gaseous medium. { ə'küs·tik ,rād·ē'ā·shən }

acoustic radiation pressure [ACOUS] A unidirectional, steady-state pressure exerted upon a surface exposed to a sound wave. { ə'küs·tik ,rād·ē'ā·shən ,presh·ər }

acoustic reactance [ACOUS] The imaginary component of the acoustic impedance. { ə'küs·tik rē'ak·təns }

acoustic reactance unit *See* acoustic ohm. { ə'küs·tik rē'ak·təns ,yü·nət }

acoustic reciprocity theorem [ACOUS] A theorem which states that in the acoustic field due to a sound source at point A, the sound pressure received at any other point B is the same as that which would be produced at A if the source were placed at B, and that this can be generalized for multiple sources and receivers. { ə'küs·tik ,res·ə'präs·əd·ē ,thir·əm }

acoustic reflection coefficient *See* acoustic reflectivity. { ə'küs·tik ri'flek·shən ,kō·ə,fish·ənt }

acoustic reflectivity [ACOUS] Ratio of the rate of flow of sound energy reflected from a surface, on the side of incidence, to the incident rate of flow. Also known as acoustic reflection coefficient; sound reflection coefficient. { ə'küs·tik ,rē·flek'tiv·əd·ē }

acoustic refraction [ACOUS] Variation of the direction of sound transmission due to spatial variation of the wave velocity in the medium. { ə'küs·tik ri'frak·shən }

acoustic resistance [ACOUS] The real component of the acoustic impedance. Also known as resistance. { ə'küs·tik ri'zis·təns }

acoustic resistance unit *See* acoustic ohm. { ə'küs·tik ri'zis·təns 'yü·nət }

acoustic resonance [ACOUS] A phenomenon exhibited by an acoustic system, such as an organ pipe or Helmholtz resonator, in which the response of the system to sound waves becomes very large when the frequency of the sound approaches a natural vibration frequency of the air in the system. { ə'küs·tik 'rez·ə·nəns }

acoustic resonator [ACOUS] An enclosure that produces sound-wave resonance at a particular frequency. { ə'küs·tik 'rez·ə,nād·ər }

acoustics [PHYS] **1.** The science of the production, transmission, and effects of sound. **2.** The characteristics of a room that determine the qualities of sound in it relevant to hearing. { ə'küs·tiks }

acoustic scattering [ACOUS] The irregular reflection, refraction, and diffraction of sound in many directions. { ə'küs·tik 'skad·ər·iŋ }

acoustic shadow [ACOUS] A region immediately behind an object placed in the path of a sound wave whose wavelength is much smaller than the object, in which the initial sound wave is cut off by the object and the sound intensity is determined by the diffraction and interference of sound waves bent around the obstacle. { ə'küs·tik 'shad·ō }

acoustic shielding [ACOUS] A sound barrier that prevents the transmission of acoustic energy. { ə'küs·tik 'shēld·iŋ }

acoustic signal processing [ACOUS] The extraction of information from signals propagated undersea, in the atmosphere, or in the solid earth in the presence of acoustic noise. { ə'küs·tik 'sig·nəl ,prä,ses·iŋ }

acoustic spectrum [ACOUS] The range of acoustic frequencies, extending from subsonic to ultrasonic frequencies, that is, approximately from zero to at least 1 megahertz. { ə'küs·tik 'spek·trəm }

acoustic stiffness [ACOUS] The product of the angular frequency and the acoustic stiffness reactance. { ə'küs·tik 'stif·nəs }

acoustic stiffness reactance [ACOUS] The part of acoustic reactance associated with the potential energy of a medium or its boundaries. Also known as stiffness reactance. { ə'küs·tik 'stif·nəs rē'ak·təns }

acoustic streaming [FL MECH] Unidirectional flow currents in a fluid that are due to the presence of sound waves. { ə'küs·tik 'strēm·iŋ }

acoustic tomography [ACOUS] An imaging or remote sensing technique in which information is collected from beams of acoustic radiation which have passed through an object, generally in the form of an image or other representation of a two-dimensional slice through the object. { ə'küs·tik tə'mäg·rə·fē }

acoustic transmission [ACOUS] The transfer of energy in the form of regular mechanical vibration through a gaseous, liquid, or solid medium. { ə'küs·tik tranz'mish·ən }

acoustic transmission coefficient *See* sound transmission coefficient. { ə'küs·tik tranz'mish·ən ,kō·ə,fish·ənt }

acoustic transmissivity *See* sound transmission coefficient. { ə'küs·tik ˌtranz·mis'iv·ə·dē }

acoustic velocity *See* speed of sound. { ə'küs·tik və'läs·ə·dē }

acoustic wave [ACOUS] **1.** An elastic nonelectromagnetic wave that has a frequency which may extend into the gigahertz range; one type is a surface acoustic wave, and the other type is a bulk or volume acoustic wave. Also known as elastic wave. **2.** *See* sound. { ə'küs·tik 'wāv }

acoustooptical filter [OPTICS] An optical filter that is tuned across the visible spectrum by acoustic waves in the frequency range of 40 to 68 megahertz. { əˌküs·tōˌäp·tə·kəl 'fil·tər }

acoustooptic interaction [OPTICS] A way to influence the propagation characteristics of an optical wave by applying a low-frequency acoustical field to the medium through which the wave passes. { əˌküs·tōˌäp·tik ˌin·tə'rak·shən }

acoustooptic modulator [OPTICS] A device utilizing acoustooptic interaction ultrasonically to vary the amplitude or the phase of a light beam. Also known as Bragg cell. { əˌküs·tōˌäp·tik 'mäd·yəˌlād·ər }

acoustooptics [OPTICS] The science that deals with interactions between acoustic waves and light. { əˌküs·tōˌäp·tiks }

a-c plane [CRYSTAL] A plane at right angles to the surface of movement in a crystal. { 'ā'sē 'plān }

acre [MECH] A unit of area, equal to 43,560 square feet, or to 4046.8564224 square meters. { 'ā·kər }

actinic [PHYS] Pertaining to electromagnetic radiation capable of initiating photochemical reactions, as in photography or the fading of pigments. { ˌak'tin·ik }

actinic achromatism [OPTICS] **1.** The design of a photographic lens system so that light sources at the wavelength of the Fraunhofer D line near 589 nanometers and the G line at 430.8 nanometers are focused at the same point and produce images of the same size. **2.** The design of an astronomical lens system so that light sources at the wavelength of the Fraunhofer F line at 486.1 nanometers and the G line at 430.8 nanometers are focused at the same point and produce images of the same size. Also known as FG achromatism. { ˌak'tin·ik ˌā'krōm·əˌtiz·əm }

actinic focus [OPTICS] The point in an optical system at which the chemically most effective rays (usually those in the ultraviolet) converge. Also known as chemical focus. { ˌak'tin·ik 'fō·kəs }

actinic glass [OPTICS] Glass that transmits more of the visible components of incident radiation and less of the infrared and ultraviolet components. { ˌak'tin·ik 'glas }

actinium emanation *See* actinon. { ˌak'tin·ē·əm ˌem·ə'nā·shən }

actinodielectric [ELEC] Of a substance, exhibiting an increase in electrical conductivity when electromagnetic radiation is incident upon it. { ˌak·tə·nōˌdī·ə'lek·trik }

actinoelectricity [ELEC] The electromotive force produced in a substance by electromagnetic radiation incident upon it. { ˌak·tə·nō·iˌlek'tris·ə·dē }

actinology [PHYS] The branch of physics dealing with electromagnetic radiation and its chemical effects. { ˌak·tə'näl·ə·jē }

actinon [NUC PHYS] A radioactive isotope of radon, symbol An, atomic number 86, atomic weight 219, belonging to the actinium series. Also known as actinium emanation (Ac-Em). { 'ak·təˌnän }

actinouranium [NUC PHYS] A naturally occurring radioactive isotope of the actinium series, emitting only alpha decay; symbol AcU; atomic number 92; mass number 235; half-life 7.1×10^8 years; isotopic symbol ^{235}U. { ˌak·tə·nōˌyu̇'rā·nē·əm }

action [MECH] An integral associated with the trajectory of a system in configuration space, equal to the sum of the integrals of the generalized momenta of the system over their canonically conjugate coordinates. Also known as phase integral. { 'ak·shən }

action at a distance theory [PHYS] A theory of the interaction of two bodies separated in space, without concern for a detailed mechanism of the propagation of effects between bodies. { 'ak·shən at ə 'dis·təns ˌthē·ə·rē }

action integral *See* action variable. { 'ak·shən 'int·ə·grəl }

action-reaction law [PHYS] The law that when one body exerts force on another, the second body exerts a collinear force on the first equal in magnitude but oppositely directed. { 'ak·shən 'rēˌak·shən ˌlȯ }

action variable [PHYS] The integral $\int p dq$ over a cycle of a dynamical system; q is some coordinate, and p the conjugate momentum. Also known as action integral. { 'ak·shən ˌver·ē·ə·bəl }

activate [ELEC] To make a cell or battery operative by addition of a liquid. [PHYS] To start activity or motion in a device or material. { 'ak·təˌvāt }

activated diffusion [SOLID STATE] Movement of atoms, ions, or lattice defects across a potential barrier in a solid. { 'ak·dəˌvād·əd di'fyü·zhən }

activation cross section [NUC PHYS] The cross section for formation of a radionuclide by a particular interaction. { ˌak·təˌvā·shən 'krȯs ˌsek·shən }

active component [ELEC] In the phasor representation of quantities in an alternating-current circuit, the component of current, voltage, or apparent power which contributes power, namely, the active current, active voltage, or active power. Also known as power component. { 'ak·tiv kəm'pō·nənt }

active current [ELEC] The component of an electric current in a branch of an alternating-current circuit that is in phase with the voltage. Also known as watt current. { 'ak·tiv 'kə·rənt }

active electric network [ELEC] Electric network containing one or more sources of energy. { 'ak·tiv ə'lek·trik 'netˌwərk }

active element [NUC PHYS] A chemical element which has one or more radioactive isotopes. { 'ak·tiv 'el·ə·mənt }

active mirror [OPTICS] A mirror whose position and shape are continually adjusted in response to changing environmental conditions in order to obtain optimum performance. { 'ak·tiv 'mir·ər }

active power [ELEC] The product of the voltage across a branch of an alternating-current circuit and the component of the electric current that is in phase with the voltage. { 'ak·tiv 'pau̇·ər }

active sound cancellation [ACOUS] Any technique in which a control sound source creates sound in a selected region equal in amplitude and opposite in phase to sound that would otherwise exist, but this sound cancellation cannot be maintained in the presence of system changes unless there is also a feedback mechanism. { ¦ak·tiv 'saund ˌkan·sə̦lā·shən }

active sound control [ACOUS] Any modification of sound fields by loudspeakers, controlled, for example, through the use of a feedforward mechanism, for reduction, equalization, or cancellation of sound. { ¦ak·tiv 'saund kən̦trōl }

active substrate [SOLID STATE] A semiconductor or ferrite material in which active elements are formed; also a mechanical support for the other elements of a semiconductor device or integrated circuit. { 'ak·tiv 'səb̦strāt }

active voltage [ELEC] In an alternating-current circuit, the component of voltage which is in phase with the current. { 'ak·tiv 'vōl·tij }

activity [NUC PHYS] The intensity of a radioactive source. Also known as radioactivity. { ˌak 'tiv·əd·ē }

AcU *See* actinouranium.

acutance [OPTICS] An objective measure of the ability of a photographic system to show a sharp edge between contiguous areas of low and high illuminance. { ə'kyü·təns }

acyclic [PHYS] Continually varying without a regularly repeated pattern. { ā'sik·lik }

acyclic motion *See* irrotational flow. { ā'sik·lik 'mō·shən }

adaptation brightness *See* adaptation luminance. { ˌa̦dap'tā·shən ˌbrīt·nəs }

adaptation illuminance *See* adaptation luminance. { ˌa̦dap 'tā·shən ə'lü·mə·nəns }

adaptation level *See* adaptation luminance. { ˌa̦dap'tā·shən ˌlev·əl }

adaptation luminance [OPTICS] The average luminance, or brightness, of objects and surfaces in the immediate vicinity of an observer estimating the visual range. Also known as adaptation brightness; adaptation illuminance; adaptation level; brightness level; field brightness; field luminance. { ˌa̦dap'tā·shən 'lü·mə·nəns }

adapter [OPTICS] An attachment to a camera that permits its use in a manner for which it was not designed. { ə'dap·tər }

adaptive optics [OPTICS] The theory and design of optical systems that measure and correct wavefront aberrations in real time, that is, simultaneous with the operation of the system. { ə'dap·tiv 'äp·tiks }

adaptive sound cancellation [ACOUS] A form of active sound cancellation that is maintained in the presence of system changes through a feedback mechanism. { ə¦dap·tiv 'saund ˌkan·sə̦lā·shən }

adaptive sound control [ACOUS] Any modification of sound fields by loudspeakers, controlled through the use of a feedback mechanism, for reduction, equalization, or cancellation of sound. { ə¦dap·tiv 'saund kən̦trōl }

Adcock antenna [ELECTROMAG] A pair of vertical antennas separated by a distance of one-half wavelength or less and connected in phase opposition to produce a radiation pattern having the shape of a figure eight. { 'ad·käk ˌan'ten·ə }

addition solid solution [CRYSTAL] Random addition of atoms or ions in the interstices within a crystal structure. { ə'di·shən 'säl·əd sə'lü·shən }

additive primary colors [OPTICS] The three colors, usually red, green, and blue, which are mixed together in an additive process. { 'ad·əd·iv 'prīm·ə·rē 'kəl·ərz }

additive process [OPTICS] The process of producing colors by mixing lights of additive primary colors in various proportions. { 'ad·əd·iv ˌprä·səs }

adhesion [ELECTROMAG] Any mutually attractive force holding together two magnetic bodies, or two oppositely charged nonconducting bodies. [MECH] The force of static friction between two bodies, or the effects of this force. [PHYS] The tendency, due to intermolecular forces, for matter to cling to other matter. { ad'hē·zhən }

adhesional work [THERMO] The work required to separate a unit area of a surface at which two substances are in contact. Also known as work of adhesion. { ad'hē·zhən·əl ˌwərk }

adhesive bond [MECH] The forces such as dipole bonds which attract adhesives and base materials to each other. { ad'hēz·iv 'bänd }

adiabatic [THERMO] Referring to any change in which there is no gain or loss of heat. { ¦ad·ē·ə¦bad·ik }

adiabatic compression [THERMO] A reduction in volume of a substance without heat flow, in or out. { ¦ad·ē·ə¦bad·ik kəm'presh·ən }

adiabatic cooling [THERMO] A process in which the temperature of a system is reduced without any heat being exchanged between the system and its surroundings. { ¦ad·ē·ə¦bad·ik 'kül·iŋ }

adiabatic demagnetization [CRYO] A method of cooling paramagnetic salts to temperatures of 10^{-3} K; the sample is cooled to the boiling point of helium in a strong magnetic field, thermally isolated, and then removed from the field to demagnetize it. Also known as Giaque-Debye method; magnetic cooling; paramagnetic cooling. { ¦ad·ē·ə¦bad·ik ˌde̦mag·nəd·ə'zā·shən }

adiabatic ellipse [FL MECH] A plot of the speed of sound as a function of the speed of flow for the adiabatic flow of a gas, which forms one quadrant of an ellipse. { ¦ad·ē·ə¦bad·ik i'lips }

adiabatic envelope [THERMO] A surface enclosing a thermodynamic system in an equilibrium which can be disturbed only by long-range forces or by motion of part of the envelope; intuitively, this means that no heat can flow through the surface. { ¦ad·ē·ə¦bad·ik 'en·və,lōp }

adiabatic expansion [THERMO] Increase in volume without heat flow, in or out. { ¦ad·ē·ə¦bad·ik ik'span·chən }

adiabatic flow [FL MECH] Movement of a fluid without heat transfer. { ¦ad·ē·ə¦bad·ik 'flō }

adiabatic invariant [PHYS] A physical quantity which may be quantized and which, to a certain degree of approximation, remains unchanged under the slow variation of any parameter. { ¦ad·ē·ə¦bad·ik in'ver·ē·ənt }

adiabatic law [PHYS] The relationship which states that, for adiabatic expansion of gases, $P\rho^{-\gamma}$ = constant, where P = pressure, ρ = density, and γ ratio of specific heats C_P/C_V. { ¦ad·ē·ə¦bad·ik 'lȯ }

adiabatic process [THERMO] Any thermodynamic procedure which takes place in a system without the exchange of heat with the surroundings. { ¦ad·ē·ə¦bad·ik prä·səs }

adiabatic recovery temperature [FL MECH] **1.** The temperature reached by a moving fluid when brought to rest through an adiabatic process. Also known as recovery temperature; stagnation temperature. **2.** The final and initial temperature in an adiabatic, Carnot cycle. { ¦ad·ē·ə¦bad·ik ri'kəv·ə·rē ,tem·prə·chər }

adiabatic vaporization [THERMO] Vaporization of a liquid with virtually no heat exchange between it and its surroundings. { ¦ad·ē·ə¦bad·ik ,vā·pər·ə'zā·shən }

adiabatic wall temperature [FL MECH] The temperature assumed by a wall in a moving fluid stream when there is no heat transfer between the wall and the stream. { ¦ad·ē·ə¦bad·ik 'wȯl ,tem·prə·chər }

adiathermanous [PHYS] Not capable of transmitting radiant heat. Also known as adiathermic. { ¦ā,dī·ə'thərm·ə·nəs }

adiathermic *See* adiathermanous. { ¦ā,dī·ə 'thərm·ik }

adjoint variable [PHYS] In classical dynamics, the canonically conjugate p_i interpreted as generalized momenta. { 'aj,ȯint 'ver·ē·ə·bəl }

adjoint wave functions [QUANT MECH] Functions in the Dirac electron theory which are formed by applying the Dirac matrix B to the Hermitian conjugates of the original wave functions. { 'aj,ȯint 'wāv ,fəŋk·shənz }

admittance [ELEC] A measure of how readily alternating current will flow in a circuit; the reciprocal of impedance, it is expressed in siemens. { əd'mit·əns }

admittance matrix [ELEC] A matrix Y whose elements are the mutual admittances between the various meshes of an electrical network, it satisfies the matrix equation I = YV, where I and V are column vectors whose elements are the currents and voltages in the meshes. { əd'mit·əns 'mā·triks }

adularescence [OPTICS] A certain type of white or bluish light seen in a gemstone (usually adularia) as it is turned. { ,aj·ə·lə'res·əns }

advanced potential [ELECTROMAG] Any electromagnetic potential arising as a solution of the classical Maxwell field equations, analogous to a retarded potential solution, but lying on the future light cone of space-time; the potential appears, at present, to have no physical interpretation. { əd'vanst pə'ten·chəl }

aeolotropic *See* anisotropic. { ¦ē·ə·lō¦träp·ik }

aeolotropy *See* anistropy. { ,ē·ə'lä·trə·pē }

aerial camera [OPTICS] A camera designed for use in aircraft and containing a mechanism to expose the film in continuous sequence at a steady rate. Also known as aerocamera. { 'e·rē·əl 'kam·rə }

aerial perspective [OPTICS] The effect produced by diffusion of light in the atmosphere whereby more distant objects have less clarity of outline and are lighter in tone. { 'e·rē·əl pər'spek·tiv }

aeriform [PHYS] Having the form or nature of air. { 'e·rə,fȯrm }

aeroacoustics [ACOUS] The science of aerodynamically generated sounds, encompassing the study of how such sounds are produced, how they propagate through various media (including propagation through ducts carrying fluid flows), how they interact with obstacles or barriers or wave-bearing surfaces, how they radiate away, how they can be managed or controlled, and how they are measured or predicted. { ,er·ō·ə'küs·tiks }

aeroballistics [MECH] The study of the interaction of projectiles or high-speed vehicles with the atmosphere. { ,e·rō·bə'lis·tiks }

aerocamera *See* aerial camera. { 'e·rō,kam·rə }

aerodiscone antenna [ELECTROMAG] Electrically small antenna for airborne applications in the very-high-frequency and ultra-high-frequency bands; it is derived from, and preserves, the desirable electrical characteristics of the discone antenna and can be designed in various physical shapes. { ,e·rō'dis,kōn an'ten·ə }

aerodynamic [FL MECH] Pertaining to forces acting upon any solid or liquid body moving relative to a gas (especially air). { ,e·rō·dī'nam·ik }

aerodynamically generated sound *See* aerodynamic sound. { ,e·rō·dī¦nam·ik·le ¦jen·ə,rād·əd 'sau̇nd }

aerodynamically rough surface [FL MECH] A surface whose irregularities are sufficiently high that the turbulent boundary layer reaches right down to the surface. { ,e·rō·dī'nam·ik·lē 'rəf 'sər·fəs }

aerodynamically smooth surface [FL MECH] A surface whose irregularities are sufficiently small to be entirely embedded in the laminar sublayer. { ,e·rō·dī'nam·ik·lē 'smüth 'sər·fəs }

aerodynamic coefficient [FL MECH] Any nondimensional coefficient relating to aerodynamic forces or moments, such as a coefficient of drag or a coefficient of lift. { ˌe·ro·dī'nam·ik ˌkō·ə'fish·ənt }

aerodynamic drag [FL MECH] A retarding force that acts upon a body moving through a gaseous fluid and that is parallel to the direction of motion of the body; it is a component of the total fluid force acting on the body. Also known as aerodynamic resistance. { ˌe·ro·dī'nam·ik 'drag }

aerodynamic force [FL MECH] The force between a body and a gaseous fluid caused by their relative motion. Also known as aerodynamic load. { ˌe·ro·dī'nam·ik 'fȯrs }

aerodynamic heating [FL MECH] The heating of a body produced by passage of air or other gases over its surface; caused by friction and by compression processes and significant chiefly at high speeds. { ˌe·ro·dī'nam·ik 'hēt·iŋ }

aerodynamic lift [FL MECH] That component of the total aerodynamic force acting on a body perpendicular to the undisturbed airflow relative to the body. Also known as lift. { ˌe·ro·dī'nam·ik 'lift }

aerodynamic load *See* aerodynamic force. { ˌe·ro·dī'nam·ik 'lōd }

aerodynamic noise *See* aerodynamic sound. { ˌe·ro·dī'nam·ik 'nȯiz }

aerodynamic phenomena [FL MECH] Acoustic, thermal, electrical, and mechanical effects, among others, that result from the flow of air over a body. { ˌe·ro·dī'nam·ik fə'näm·ə·nə }

aerodynamic resistance *See* aerodynamic drag. { ˌe·rō·dī'nam·ik ri'zis·təns }

aerodynamics [FL MECH] The science that deals with the motion of air and other gaseous fluids and with the forces acting on bodies when they move through such fluids or when such fluids move against or around the bodies. { ˌe·rō·dī'nam·iks }

aerodynamic size [PHYS] Particle size determined from inertia or settling velocity, assuming Stokes' law for the resistance to a sphere moving through a fluid. Also known as inertial size. { ˌe·ro·dī'nam·ik 'sīz }

aerodynamic sound [ACOUS] Sound that is generated by the unsteady motion of a gas and its interaction with surrounding surfaces. Also known as aerodynamically generated sound; aerodynamic noise. { ˌe·rō·dī¦nam·ik 'saủnd }

aerodynamic trail [FL MECH] A condensation trail formed by adiabatic cooling to saturation (or slight supersaturation) of air passing over the surfaces of high-speed aircraft. { ˌe·ro·dī'nam·ik 'trāl }

aerodynamic trajectory [MECH] A trajectory or part of a trajectory in which the missile or vehicle encounters sufficient air resistance to stabilize its flight or to modify its course significantly. { ˌe·ro·dī'nam·ik trə'jek·trē }

aerodynamic turbulence [FL MECH] A state of fluid flow in which the instantaneous velocities exhibit irregular and apparently random fluctuations. { ˌe·ro·dī'nam·ik 'tərb·yə·ləns }

aerodynamic wave drag [FL MECH] The force retarding an airplane, especially in supersonic flight, as a consequence of the formation of shock waves ahead of it. { ˌe·ro·dī'nam·ik 'wāv ˌdrag }

aeroelasticity [MECH] The deformation of structurally elastic bodies in response to aerodynamic loads. { ˌe·rō·iˌlas'tis·əd·ē }

aeromechanics [FL MECH] The science of air and other gases in motion or equilibrium; has two branches, aerostatics and aerodynamics. { ˌe·rō·mi'kan·iks }

aeronautical flutter [FL MECH] An aeroelastic, self-excited vibration in which the external source of energy is the airstream and which depends on the elastic, inertial, and dissipative forces of the system in addition to the aerodynamic forces. Also known as flutter. { e·rə'nȯd·ə·kəl 'fləd·ər }

aeronautics [FL MECH] The science that deals with flight through the air. { ˌe·rə'nȯd·iks }

aerosol lidar [OPTICS] A type of lidar that is designed to measure the scattering of laser light from atmospheric dust and aerosols. { 'e·rəˌsȯl 'līˌdär }

aerostatics [FL MECH] The science of the equilibrium of gases and of solid bodies immersed in them when under the influence only of natural gravitational forces. { ¦e·rō¦stad·iks }

aerothermochemistry [FL MECH] The study of gases which takes into account the effect of motion, heat, and chemical changes. { ¦e·rōˌthər·mō'kem·ə·strē }

aerothermodynamics [FL MECH] The study of aerodynamic phenomena at sufficiently high gas velocities that thermodynamic properties of the gas are important. { ˌe·rōˌthər·mō·dī'nam·iks }

aerothermoelasticity [FL MECH] The study of the response of elastic structures to the combined effects of aerodynamic heating and loading. { ¦e·rōˌthər·mōˌiˌlas'tis·əd·ē }

AES *See* Auger electron spectroscopy.

af *See* audio frequency.

aF *See* abfarad.

A15 compound *See* A15 phase. { ¦ā'fif¦tēn 'käm¦paủnd }

A15 phase [SOLID STATE] An intermetallic compound having the chemical formula A_3B, where A represents a transition element, and a crystal structure in which the B atoms are located at the corners and in the center of a cubic unit cell, while the A atoms are arranged in pairs on the cube faces. Also known as A15 compound. { ¦ā·fif'tēn ˌfāz }

afocal lens [OPTICS] A lens of zero convergent power, whose focal points are infinitely distant. { 'ā·fō·kəl 'lenz }

afocal system [OPTICS] An optical system of zero convergent power, for example, a telescope. { 'ā·fō·kəl 'sis·təm }

afterglow [ATOM PHYS] *See* phosphorescence. [PL PHYS] The transient decay of a plasma after the power has been turned off. { 'af·tərˌglō }

aggregate recoil [NUC PHYS] The ejection of atoms from the surface of a sample as a result of their being attached to one atom that is recoiling as the result of alpha-particle emission. { 'ag·rə·gət 'rē,cȯil }

aggregation *See* axisymmetrization. { ,ag·rə'gā·shən }

aging [ACOUS] The process by which the pressure disturbance from a passing aircraft is distorted as it propagates away from the aircraft, causing the signature to stretch out in duration and length, lose detail, and form shock waves. [ELEC] Allowing a permanent magnet, capacitor, meter, or other device to remain in storage for a period of time, sometimes with a voltage applied, until the characteristics of the device become essentially constant. [ELECTROMAG] Change in the magnetic properties of iron with passage of time, for example, increase in the hysteresis. { 'āj·iŋ }

agreement residual [SOLID STATE] The sum of the differences between the observed and calculated structure amplitudes of a crystal, for all observed reflections, divided by the sum of the observed amplitudes. { ə¦grē·mənt rə'zij·ə·wəl }

aH *See* abhenry.

Ah *See* ampere-hour.

Aharonov-Bohm effect [QUANT MECH] An effect manifested when a beam of electrons is split into two beams that travel in opposite directions around a region containing magnetic flux and are then recombined, whereby the intensity of the resulting beam oscillates periodically as the enclosed magnetic field is changed. { ,ä·hä'rō ,nȯf 'bām i,fekt }

A/in.² *See* ampere per square inch.

air-blower noise [ACOUS] Noise in blowers in heater and air-conditioning systems due to air turbulence. { 'er 'blō·ər }

air capacitor [ELEC] A capacitor having only air as the dielectric material between its plates. Also known as air condenser. { 'er kə'pas·əd·ər }

air compression [PHYS] The decrease of volume of a quantity of air as a result of an increase in pressure, as is accomplished by a piston moving in a cylinder. { 'er ,kəm'presh·ən }

air-core transformer [ELECTROMAG] Transformer (usually radio-frequency) having a nonmetallic core. { 'er ,kȯr tranz'fȯrm·ər }

aircraft decibel rating [ELECTROMAG] The ratio of the radar reflectivity of a specific type of aircraft to that of a selected reference aircraft, measured in decibels. { 'er,kraft 'des·ə,bəl ,rād·iŋ }

aircraft noise [ACOUS] Effective sound output of the various sources of noise associated with aircraft operation, such as propeller and engine exhaust, jet noise, and sonic boom. { 'er,kraft ,nȯiz }

air current [FL MECH] Very generally, any moving stream of air. { 'er ,kər·ənt }

air density [MECH] The mass per unit volume of air. { 'er ,den·səd·ē }

airflow [FL MECH] **1.** A flow or stream of air which may take lace in a wind tunnel or, as a relative airflow, past the wing or other parts of a moving craft. Also known as airstream. **2.** A rate of flow, measured by mass or volume per unit of time. { 'er,flō }

airflow stack effect [FL MECH] The variation of pressure with height in air flowing in a vertical duct due to a difference in temperature between the flowing air and the air outside the duct. { 'er,flō 'stak i,fekt }

airlight formula [OPTICS] A fundamental equation of visual-range theory, relating the apparent luminance of a distant black object, the apparent luminance of the background sky above the horizon, and the extinction coefficient of the air layer near the ground. { 'er,līt 'fȯr·myə·lə }

air line [SPECT] Lines in a spectrum due to the excitation of air molecules by spark discharges, and not ordinarily present in arc discharges. { 'er ,līn }

air pressure [PHYS] The force per unit area that the air exerts on any surface in contact with it, arising from the collisions of the air molecules with the surface. { 'er ,presh·ər }

air-pressure drop [FL MECH] The pressure lost in overcoming friction along an airway. { 'er ,presh·ər ,dräp }

air properties [PHYS] Characteristics of air as a gas, such as density, molecular weight, specific heats, boiling point, critical temperature, and critical pressure. { 'er ,prä·pər,tēz }

air resistance [MECH] Wind drag giving rise to forces and wear on buildings and other structures. { 'er ri'zis·təns }

air shower *See* cosmic-ray shower. { 'er ,shau̇·ər }

air-standard cycle [THERMO] A thermodynamic cycle in which the working fluid is considered to be a perfect gas with such properties of air as a volume of 12.4 cubic feet per pound at 14.7 pounds per square inch (approximately 0.7756 cubic meter per kilogram at 101.36 kilopascals) and 492°R and a ratio of specific heats of 1:4. { ¦er ¦stan·dərd 'sī·kəl }

airstream *See* airflow. { 'er,strēm }

air-velocity measurement [FL MECH] The measurement of the rate of displacement of air or gas at a specific location, as when ascertaining wind speed or airspeed of an aircraft. { ¦er və'läs·əd·ē 'mezh·ər·mənt }

air-water vapor mixture [PHYS] A mixture of dry air and water vapor, such as the atmosphere. { ¦er ¦wȯd·ər 'vā·pər ,miks·chər }

air wedge [OPTICS] A wedge-shaped film of air between two flat reflecting surfaces that produces an interference pattern consisting of a series of light and dark bands parallel to the thin edge of the wedge. { 'er ,wej }

Airy disk [OPTICS] The bright, diffuse central spot of light formed by an optical system imaging a point source of light. { ¦er·ē ¦disk }

Airy phase [ACOUS] An acoustic wave formed by an explosion in shallow water over a flat bottom. { ¦er·ē ,fāz }

Airy spirals [OPTICS] Spiral interference patterns formed by quartz cut perpendicularly to the axis in convergent circularly polarized light. { ¦er·ē ¦spī·rəlz }

Airy stress function [MECH] A biharmonic function of two variables whose second partial derivatives give the stress components of a body subject to a plane strain. { ¦er·ē 'stres ˌfəŋk·shən }

Albada finder [OPTICS] A viewfinder used with a camera held at eye level; the field of view is enclosed by a white frame that is made to appear very distant by reflection from the rear surface of the objective lens. { al'bä·də ˌfīn·dər }

albedo [OPTICS] That fraction of the total light incident on a reflecting surface, especially a celestial body, which is reflected back in all directions. { al'bēˌdō }

albite law [CRYSTAL] A rule specifying the orientation of alternating lamellae in multiple twin feldspar crystals; the twinning plane is brachypinacoid and is common in albite. { 'alˌbīt ˌlȯ }

Alcator [PL PHYS] A type of tokamak with a high toroidal field (up to 8.7 teslas) and high-density plasma (up to 3×10^{15} particles per cubic centimeter). { ¦al¦kād·ər }

alcove hologram [OPTICS] A type of hologram whose surface is bent into an arc that is concave as seen by the viewer, allowing the formation of real images that are viewable over a 180° angle. { 'alˌkōv 'häl·əˌgram }

Alexanderson antenna [ELECTROMAG] An antenna, used at low or very low frequencies, consisting of several base-loaded vertical radiators connected together at the top and fed at the bottom of one radiator. { ˌal·ig'zan·dər·sən anˌten·ə }

Alford loop [ELECTROMAG] An antenna utilizing multielements which usually are contained in the same horizontal plane and adjusted so that the antenna has approximately equal and in-phase currents uniformly distributed along each of its peripheral elements and produces a substantially circular radiation pattern in the plane of polarization; it is known for its purity of polarization. { 'ȯl·fərd ˌlüp }

Alfvén number [PHYS] The ratio of the speed of the Alfvén wave to the speed of the fluid at a point in the fluid. { äl'vān ˌnəm·bər }

Alfvén speed [PHYS] The speed of motion of the Alfvén wave, which is $v_a = B_0/\sqrt{\rho\mu}$, where B_0 is the magnetic field strength, ρ the fluid density, and μ the magnetic permeability (in meter-kilogram-second units). { äl'vān ˌspēd }

Alfvén wave [PHYS] A hydromagnetic shear wave which moves along magnetic field lines; a major accelerative mechanism of charged particles in plasma physics and astrophysics. { äl 'vān ˌwāv }

algebraic scattering theory [PHYS] An approach to the analysis of reactions between composite particles in which the fundamental role is played by the scattering matrix, which is obtained algebraically, without the use of a wave equation, by using the concept of dynamic symmetry. { ¦al·jə¦brā·ik 'skad·ə·riŋ ˌthe·ə·rē }

alignment [NUC PHYS] A population $p(m)$ of the $2I + 1$ orientational substates of a nucleus; $m = -I$ to $+I$, such that $p(m) = p(-m)$. { ə'līn·mənt }

alive *See* energized. { ə'līv }

Allard's law [OPTICS] A mathematical formula defining the relationship between the intensity of a light, atmospheric conditions, and the amount of light received at any given distance. { 'al·ərdz ˌlȯ }

allobar [NUC PHYS] A form of an element differing in its atomic weight from the naturally occurring form and hence being of different isotopic composition. [PHYS] A barometric pressure change. { 'a·loˌbär }

allochromatic crystal [CRYSTAL] A crystal having photoconductive properties due to the presence of small particles within it. { ˌa·lə·krə'mad·ik 'kris·təl }

allochromy [PHYS] Emission of electromagnetic radiation that results from incident radiation at a different wavelength, as occurs in fluorescence or the Raman effect. { ¦a·lə ¦krōm·ē }

allogyric birefringence [OPTICS] The phenomenon in active optical media whereby circularly polarized light is transmitted unchanged but the velocity of right-handed circularly polarized light is different from that of left-handed. { ˌa·lō'jī·rik ˌbī·ri'frin·jəns }

allomerism [CRYSTAL] A constancy in crystal form in spite of a variation in chemical composition. { ə'läm·əˌriz·əm }

allowable load [MECH] The maximum force that may be safely applied to a solid, or is permitted by applicable regulators. { ə'lau̇·ə·bəl 'lōd }

allowable stress [MECH] The maximum force per unit area that may be safely applied to a solid. { ə'lau̇·ə·bəl 'stres }

allowed energy bands [SOLID STATE] The restricted regions of possible electron energy levels in a solid. { ə'lau̇d 'en·ər·jē ˌbanz }

allowed transition [QUANT MECH] A transition between two states which is permitted by the selection rules and which consequently has a relatively high priority. { ə'lau̇d tranz'ish·ən }

all-sky camera [OPTICS] A camera directed vertically downward toward a horizontal convex mirror so as to photograph the entire sky simultaneously. { 'ȯl ˌskī 'kam·rə }

alnico magnet [ELECTROMAG] A permanent magnet made of alnico. { 'al·niˌkō 'mag·nət }

alpha decay [NUC PHYS] A radioactive transformation in which an alpha particle is emitted by a nuclide. { 'al·fə di'kā }

alpha emission [NUC PHYS] Ejection of alpha particles from the atom's nucleus. { 'al·fə i'mish·ən }

alpha particle [ATOM PHYS] A positively charged particle consisting of two protons and two neutrons, identical with the nucleus of the helium atom; emitted by several radioactive substances. { 'al·fə ˌpärd·ə·kəl }

alpha-particle scattering [ATOM PHYS] Deviation at various angles of a stream of alpha particles passing through a foil of material. { 'al·fə 'pärd·ə·kəl 'skad·ər·iŋ }

altazimuth telescope [OPTICS] A telescope equipped with an altazimuth mounting. { al 'taz·ə·məth 'tel·əˌskōp }

alternating current [ELEC] Electric current that

reverses direction periodically, usually many times per second. Abbreviated ac. { ¦ȯl·tərˌnād·iŋ ¦kər·ənt }

alternating-current circuit theory [ELEC] The mathematical description of conditions in an electric circuit driven by an alternating source or sources. { ¦ȯl·tərˌnād·iŋ ¦kər·ənt 'sər·kət ˌthē·ə·rē }

alternating-current Josephson effect [CRYO] The oscillating current flow resulting from the tunneling of electron pairs through a thin insulating barrier between two superconductors when a steady voltage, V, is maintained across the barrier, with a frequency equal to $2eV/h$, where e is the magnitude of the charge of the electron and h is Planck's constant. { ¦ȯl·tər·nād·iŋ ¦kər·ənt 'jō·səf·sən iˌfekt }

alternating-current Kerr effect [OPTICS] Birefringence of light passing through a crystal that is simultaneously pumped with an intense laser beam. { ¦ȯl·tərˌnād·iŋ ¦kər·ənt 'kər iˌfekt }

alternating-current resistance *See* high-frequency resistance. { ¦ȯl·tərˌnād·iŋ ¦kər·ənt ri'zis·təns }

alternating gradient [ELECTROMAG] A magnetic field in which successive magnets have gradients of opposite sign, so that the field increases with radius in one magnet and decreases with radius in the next; used in synchrotrons and cyclotrons. { 'ȯl·tər·nād·iŋ 'grād·ē·ənt }

alternating-gradient focusing [ELECTROMAG] A configuration of transverse electric or magnetic fields suitable for focusing or confining a charged-particle beam in which successive magnets or electrodes (in time or along the beam direction) have opposite polarity. { 'ȯl·tər·nād·iŋ ¦grād·ē·ənt 'fō·kəs·iŋ }

alternating stress [MECH] A stress produced in a material by forces which are such that each force alternately acts in opposite directions. { 'ȯl·tər·nād·iŋ 'stres }

alternating voltage [ELEC] Periodic voltage, the average value of which over a period is zero. { 'ȯl·tər·nād·iŋ 'vōl·tij }

alternation [PHYS] Variation, either positive or negative, of a waveform from zero to maximum and back to zero, equaling one-half of a cycle. { ˌȯl·tər'nā·shən }

A/m *See* ampere per meter.

Am² *See* ampere meter squared.

A/m² *See* ampere per square meter.

amacratic lens *See* amasthenic lens. { ¦ā·mə'krad·ik 'lenz }

Amagat density unit [PHYS] A unit of density in the Amagat system, used in the study of the behavior of gases under pressure; it is equal to the density of a gas at a pressure of 1 atmosphere and a temperature of 0°C; for an ideal gas this is 44.6148 ± 0.0004 moles per cubic meter. { 'ä·mä·gä 'den·səd·ē ˌyü·nət }

Amagat diagram [PHYS] A diagram that plots a series of isothermal curves for a gas pressure versus the gas pressure-volume product. { 'ä·mä·gä 'dī·əˌgram }

Amagat law *See* Amagat-Leduc rule. { 'ä·mä·gä ˌlȯ }

Amagat-Leduc rule [PHYS] The rule which states that the volume taken up by a gas mixture equals the sum of the volumes each gas would occupy at the temperature and pressure of the mixture. Also known as Amagat law; Leduc law. { 'ä·mä·gä lə'du̇k ˌrül }

Amagat system [PHYS] A system of units in which the unit of pressure is the atmosphere and the unit of volume is the gram-molecular volume (22.4 liters at standard conditions). { 'ä·mä·gä ˌsis·təm }

Amagat volume unit [PHYS] A unit of volume in the Amagat system, used in the study of the behavior of gases under pressure; it is equal to the volume occupied by 1 mole of a gas at a pressure of 1 atmosphere and a temperature of 0°C; for an ideal gas this is 0.02241400 ± 0.00000004 cubic meter. { 'ä·mä·gä 'väl·yəm ˌyü·nət }

amasthenic lens [OPTICS] A lens that refracts the rays of light into one focus. Also known as amacratic lens. { ¦ā·məs¦thēn·ik 'lenz }

ambient light [OPTICS] The surrounding light, such as that reaching a television picture-tube screen from light sources in a room. { 'am·bē·ənt 'līt }

ambient noise [ACOUS] The pervasive noise associated with a given environment, being usually a composite of sounds from sources both near and distant. { 'am·bē·ənt 'nȯiz }

ambient pressure [FL MECH] The pressure of the surrounding medium, such as a gas or liquid, which comes into contact with an apparatus or with a reaction. { 'am·bē·ənt 'presh·ər }

ambient temperature [PHYS] The temperature of the surrounding medium, such as gas or liquid, which comes into contact with the apparatus. { 'am·bē·ənt 'tem·prə·chər }

ambipolar diffusion [PHYS] The diffusion in a plasma of charged particles, such as electrons or ions, as a result of the almost exact local charge neutrality required. { ¦am·bē'pōl·ər di'fyü·zhən }

Amici prism [OPTICS] A compound prism, used in direct-vision spectroscopes, that disperses a beam of light into a spectrum without causing the beam as a whole to undergo any net deviation; it is made up of alternate crown and flint glass components, refracting in opposite directions. Also known as direct-vision prism. { ə'mēch·ē ˌpriz·əm }

A min *See* ampere-minute.

Am²/Js *See* ampere square meter per joule second.

ammonia-beam maser [PHYS] A gas maser using ammonia as the paramagnetic material. { ə'mōn·yə ¦bēm 'māz·ər }

A mode [ACOUS] A form of ultrasonic medical tomography that uses acoustic pulse emissions and echo reception along a single line-of-sight axial propagation path and usually displays the information on a cathode-ray oscilloscope in which the horizontal axis of the display is a linear time base, triggered at the time of the transmitted pulse, and the received echoes are manifested as vertical deflections, with vertical displacement a measure of the amplitude of the strength of the returning echo. { 'ā 'mōd }

amorphous [PHYS] Pertaining to a solid which is noncrystalline, having neither definite form nor structure. { ə'mȯr·fəs }

amorphous laser *See* glass laser. { ə'mȯr·fəs 'lāz·ər }

amorphous semiconductor [SOLID STATE] A semiconductor material which is not entirely crystalline, having only short-range order in its structure. { ə'mȯr·fəs ¦sem·ē·kən¦dək·tər }

amorphous solid [SOLID STATE] A rigid material whose structure lacks crystalline periodicity; that is, the pattern of its constituent atoms or molecules does not repeat periodically in three dimensions. { ə'mȯr·fəs 'säl·əd }

amp *See* amperage; ampere. { amp }

amperage [ELEC] The amount of electric current in amperes. Abbreviated amp. { 'am·prij }

ampere [ELEC] The unit of electric current in the rationalized meter-kilogram-second system of units; defined in terms of the force of attraction between two parallel current-carrying conductors. Abbreviated a; A; amp. { 'am ˌpir }

Ampère balance *See* current balance. { 'äm,per ˌbal·əns }

Ampère currents [ELECTROMAG] Postulated "molecular-ring" currents to explain the phenomena of magnetism as well as the apparent nonexistence of isolated magnetic poles. { 'äm,per 'kər·əns }

ampere-hour [ELEC] A unit for the quantity of electricity, obtained by integrating current flow in amperes over the time in hours for its flow; used as a measure of battery capacity. Abbreviated Ah; amp-hr. { 'am,pir ¦aù·ər }

Ampère law [ELECTROMAG] **1.** A law giving the magnetic induction at a point due to given currents in terms of the current elements and their positions relative to the point. Also known as Laplace law. **2.** A law giving the line integral over a closed path of the magnetic induction due to given currents in terms of the total current linking the path. { 'äm,per ˌlȯ }

ampere meter squared [ELECTROMAG] The SI unit of electromagnetic moment. Abbreviated Am^2. { ¦am,pir ¦mēd·ər 'skwerd }

ampere-minute [ELEC] A unit of electrical charge, equal to the charge transported in 1 minute by a current of 1 ampere, or to 60 coulombs. Abbreviated A min. { ¦am,pir ¦min·ət }

ampere per meter [ELECTROMAG] The SI unit of magnetic field strength and magnetization. Abbreviated A/m. { 'am,pir pər 'mēd·ər }

ampere per square inch [ELEC] A unit of current density, equal to the uniform current density of a current of 1 ampere flowing through an area of 1 square inch. Abbreviated A/in^2. { 'am,pir pər ˌskwer 'inch }

ampere per square meter [ELEC] The SI unit of current density. Abbreviated A/m^2. { 'am,pir pər ˌskwer 'mēd·ər }

Ampère rule [ELECTROMAG] The rule which states that the direction of the magnetic field surrounding a conductor will be clockwise when viewed from the conductor if the direction of current flow is away from the observer. { 'äm,per ˌrül }

ampere square meter per joule second [ELECTROMAG] The SI unit of gyromagnetic ratio. Abbreviated Am^2/Js. { ¦am,pir ¦skwer ¦mēd·ər pər ¦jül 'sek·ənd }

Ampère theorem [ELECTROMAG] The theorem which states that an electric current flowing in a circuit produces a magnetic field at external points equivalent to that due to a magnetic shell whose bounding edge is the conductor and whose strength is equal to the strength of the current. { 'äm,per ˌthir·əm }

ampere-turn [ELECTROMAG] A unit of magnetomotive force in the meter-kilogram-second system defined as the force of a closed loop of one turn when there is a current of 1 ampere flowing in the loop. Abbreviated amp-turn. { 'am,pir ˌtərn }

amp-hr *See* ampere-hour.

amplitude [PHYS] The maximum absolute value attained by the disturbance of a wave or by any quantity that varies periodically. { 'am·plə,tüd }

amplitude factor *See* crest factor. { 'am·plə,tüd ˌfak·tər }

amplitude level [PHYS] The natural logarithm of the ratio of two amplitudes, each measured in the same units. { 'am·plə,tüd ˌlev·əl }

amplitude modulator [PHYS] Any device which imposes amplitude modulation upon a carrier wave in accordance with a desired program. { 'am·plə,tüd 'maj·ə,lād·ər }

amplitude noise [ELECTROMAG] Effect on radar accuracy of the fluctuations in the amplitude of the signal returned by the target; these fluctuations are caused by any change in aspect if the target is not a point source. { 'am·plə,tüd ˌnȯiz }

amplitude resonance [PHYS] The frequency at which a given sinusoidal excitation produces the maximum amplitude of oscillation in a resonant system. { 'am·plə,tüd 'rez·ə·nəns }

amplitude splitting [OPTICS] A technique in which light falls on a partially reflecting surface; part of the light is transmitted, part reflected, and after further manipulation, these parts are recombined to give interference. { 'am·plə,tüd ˌsplid·iŋ }

amp-turn *See* ampere-turn. { ¦amp¦tərn }

amu *See* atomic mass unit.

amyriotic field [QUANT MECH] A quantized field that has creation and annihilation operators satisfying specified commutation rules and a vacuum state. { ə¦mir·ē,äd·ik 'fēld }

An *See* actinon.

analogous pole [SOLID STATE] The pole of a crystal that acquires a positive charge when the crystal is heated. { ə'nal·ə·gəs ˌpōl }

analog states [NUC PHYS] Certain nuclear states belonging to neighboring nuclear isobars and possessing identical structure except for the transformation of one or more neutrons into the same number of protons. Also known as isobaric analog states. { 'an·əl,äg ,stāts }

analysis line [SPECT] The spectral line used in determining the concentration of an element in spectrographic analysis. { ə'nal·ə·səs ,līn }

analytic extension [RELAT] An extension, in a real analytic manner, past a coordinate singularity of a solution to Einstein's equations of general relativity. { ,an·əl'id·ik ik'sten·chən }

analytic mechanics [MECH] The application of differential and integral calculus to classical (nonquantum) mechanics. { ,an·əl'id·ik mi 'kan·iks }

analytic regularization [QUANT MECH] A method of extracting a finite piece from an infinite result in quantum field theory, based on analytically continuing the propagators that appear in typically divergent integrals. { ,an·əl'id·ik ,reg·yə·lə·rə'zā·shən }

analyzer [OPTICS] A device, such as a Nicol prism, which passes only plane polarized light; used in the eyepiece of instruments such as the polariscope. { 'an·ə,līz·ər }

analyzing power [NUC PHYS] In a nuclear scattering process, a measure of the effect on scattering cross sections of changes in the polarization of the beam or target nuclei. { 'an·ə,liz·iŋ ,paủr }

anamorphic lens [OPTICS] A lens that produces different magnifications along lines in different directions in the image plane. { ¦an·ə¦mȯr·fik 'lenz }

anamorphic system [OPTICS] An optical system incorporating a cylindrical surface in which the image is distorted so that the angle of coverage in a direction perpendicular to the cylinder is different for the image than for the object. { ¦an·ə¦mȯr·fik 'sis·təm }

anamorphoscope [OPTICS] An optical instrument, usually consisting of a cylindrical lens or mirror, that restores an image distorted by anamorphosis to its normal proportions. { ,an·ə'mȯr·fə,skōp }

anamorphosis [OPTICS] The production of a distorted image by an optical system. { ,an·ə'mȯr·fə·səs }

anamorphote lens [OPTICS] A lens designed to produce anamorphosis. { ¦an·ə¦mȯr,fōt ,lenz }

anastigmat *See* anastigmatic lens. { a'nas·tig,mat }

anastigmatic lens [OPTICS] A compound lens corrected for astigmatism and curvature of field. Also known as anastigmat. { ¦an·ə·stig¦mad·ik 'lenz }

Anderson-Dayem bridge [CRYO] A Josephson junction in a superconducting film,formed by a constriction with length and width on the order of a few micrometers or less. { ¦an·dər·sən ¦dā·əm ,brij }

Andrade's creep law [MECH] A law which states that creep exhibits a transient state in which strain is proportional to the cube root of time and then a steady state in which strain is proportional to time. { 'an,drādz 'krēp ,lȯ }

Andrews's curves [THERMO] A series of isotherms for carbon dioxide, showing the dependence of pressure on volume at various temperatures. { 'an,drüz ,kərvz }

Andronikashvili experiment [CRYO] An experiment to determine the fractional densities of the superfluid and normal fluid components of liquid helium by measuring the period and decrement of a torsional pendulum immersed in the helium. { ,an·drə¦ni·kəsh¦vil·ē ik'sper·ə·mənt }

anelasticity [MECH] Deviation from a proportional relationship between stress and strain. { ¦an·ə·las¦tis·əd·ē }

anelectric [PHYS] Not becoming charged by friction. { ¦an·ə¦lek·trik }

aneutronic reaction [NUC PHYS] A nuclear reaction generating so few neutrons that its neutronicism is less than 0.01. { ¦ā·nü¦trän·ik rē'ak·shən }

angle-lighting luminaire [OPTICS] A luminaire whose light distribution is asymmetric with respect to a direction of specific interest. { 'aŋ·gəl ,līt·iŋ ,lü·mə'ner }

angle noise [ELECTROMAG] Tracking error introduced into radar by variations in the apparent angle of arrival of the echo from a target, because of finite target size. { 'aŋ·gəl ,nȯiz }

angle of arrival [ELECTROMAG] A measure of the direction of propagation of electromagnetic radiation upon arrival at a receiver (the term is most commonly used in radio); it is the angle between the plane of the phase front and some plane of reference, usually the horizontal, at the receiving antenna. { 'aŋ·gəl əv ə'rīv·əl }

angle of contact [FL MECH] The angle between the surface of a liquid and the surface of a partially submerged object or of the container at the line of contact. Also known as contact angle. { 'aŋ·gəl əv 'kän,takt }

angle of deviation *See* deviation. { 'aŋ·gəl əv ,dē·vē'ā·shən }

angle of divergence [OPTICS] The angular spread of a light beam from a collimating device or laser. { 'aŋ·gəl əv də'vərj·əns }

angle of fall [MECH] The vertical angle at the level point, between the line of fall and the base of the trajectory. { 'aŋ·gəl əv 'fȯl }

angle of friction *See* angle of repose. { 'aŋ·gəl əv 'frik·shən }

angle of impact [MECH] The acute angle between the tangent to the trajectory at the point of impact of a projectile and the plane tangent to the surface of the ground or target at the point of impact. { 'aŋ·gəl əv 'im,pakt }

angle of incidence [OPTICS] The angle formed by a ray arriving at a surface and the perpendicular to that surface at the point of arrival. Also known as incidence angle. { 'aŋ·gəl əv 'in·sə·dəns }

angle of lag *See* lag angle. { 'aŋ·gəl əv 'lag }

angle of lead *See* lead angle. { 'aŋ·gəl əv 'lēd }

angle of orientation [MECH] Of a projectile in

flight, the angle between the plane determined by the axis of the projectile and the tangent to the trajectory (direction of motion), and the vertical plane including the tangent to the trajectory. { 'aŋ·gəl əv ˌȯr·ē·ən'tā·shən }

angle of radiation [ELECTROMAG] Angle between the surface of the earth and the center of the beam of energy radiated upward into the sky from a transmitting antenna. Also known as angle of departure. { 'aŋ·gəl əv rād·ē'ā·shən }

angle of reflection [PHYS] The angle between the direction of propagation of a wave reflected by a surface and the line perpendicular to the surface at the point of reflection. Also known as reflection angle. { 'aŋ·gəl əv ri'flek·shən }

angle of refraction [PHYS] The angle between the direction of propagation of a wave that is refracted by a surface and the line that is perpendicular to the surface at the point of refraction. { 'aŋ·gəl əv ri'frak·shən }

angle of repose [MECH] The angle between the horizontal and the plane of contact between two bodies when the upper body is just about to slide over the lower. Also known as angle of friction. { 'aŋ·gəl əv ri'pōz }

angle of torsion [MECH] The angle through which a part of an object such as a shaft or wire is rotated from its normal position when a torque is applied. Also known as angle of twist. { 'aŋ·gəl əv 'tȯr·shən }

angle of twist *See* angle of torsion. { 'aŋ·gəl əv 'twist }

angle of view [OPTICS] The angle subtended by an image at the second nodal point of a lens. { 'aŋ·gəl əv 'vyü }

angle-resolved photoelectron spectroscopy [SPECT] A type of photoelectron spectroscopy which measures the kinetic energies of photoelectrons emitted from a solid surface and the angles at which they are emitted relative to the surface. Abbreviated ARPES. { 'aŋ·gəl ri 'zälvd ¦fōd·ō·ə'lek,trän ˌspek'träs·kə·pē }

angle variable [MECH] The dynamical variable *w* conjugate to the action variable J, defined only for periodic motion. { 'aŋ·gəl 'ver·ē·ə·bəl }

angstrom [MECH] A unit of length, 10^{-10} meter, used primarily to express wavelengths of optical spectra. Abbreviated A; å. Also known as tenthmeter. { 'aŋ·strəm }

Ångström coefficient [PHYS] The multiplying amplitude parameter inserted in Ångström's formula for the scattering of electromagnetic radiation by atmospheric dust. { 'ȯŋ·strəm ˌkō·ə'fish·ənt }

Ångström's formula [PHYS] A formula stating that the scattering coefficient for dust in the atmosphere is inversely proportional to a positive power of the wavelength of the radiation, with the power depending on the size of the dust particles. { 'ȯŋ·strəmz 'fȯrm·yə·lə }

angular acceleration [MECH] The time rate of change of angular velocity. { 'aŋ·gyə·lər akˌsel·ə'rā·shən }

angular aperture [OPTICS] The angle subtended at an axial object point of an optical instrument by the radius of the entrance pupil. { 'aŋ·gyə·lər 'ap·ə·chər }

angular correlations [NUC PHYS] A technique of nuclear experimentation for determining spins of nuclear states, the angular momentum mixtures of incoming or outgoing particles, and the multipole mixtures of emitted gamma rays, by measuring the dependence of the intensity or the cross section of a nuclear reaction on the directions of two or more radiations. { 'aŋ·gyə·lər ˌkär·ə'lā·shənz }

angular displacement [PHYS] A vector measure of the rotation of an object about an axis; the vector points along the axis according to the right-hand rule; the length of the vector is the rotation angle, in degrees or radians. { 'an·gyə·lər dis'plās·mənt }

angular distance [PHYS] The distance between two points, expressed in wavelengths at a specified frequency. { 'an·gyə·lər 'dis·təns }

angular frequency [PHYS] For any oscillation, the number of vibrations per unit time, multiplied by 2π. Also known as angular velocity; radian frequency. { 'an·gyə·lər 'frē·kwən·sē }

angular impulse [MECH] The integral of the torque applied to a body over time. { 'an·gyə·lər 'imˌpəls }

angular length [MECH] A length expressed in the unit of the length per radian or degree of a specified wave. { 'aŋ·gyə·lər 'leŋkth }

angular magnification [OPTICS] For an optical system, the ratio of the angle subtended by the image at the eye to the angle subtended by the object at the eye. { 'aŋ·gyə·lər ˌmag·nə·fə'kā·shən }

angular momentum [MECH] **1.** The cross product of a vector from a specified reference point to a particle, with the particle's linear momentum. Also known as moment of momentum. **2.** For a system of particles, the vector sum of the angular momenta (first definition) of the particles. { 'aŋ·gyə·lər mə'ment·əm }

angular momentum operator [QUANT MECH] Any vector operator satisfying communication rules of the type $[J_x, J_y] = iJ_z$. { 'aŋ·gyə·lər mə'ment·əm 'äp·əˌrād·ər }

angular rate *See* angular speed. { 'aŋ·gyə·lər ˌrāt }

angular resolution [ELECTROMAG] A measure of the ability of a radar to distinguish between two targets solely by the measurement of angles. { 'aŋ·gyə·lər ˌrez·ə'lü·shən }

angular speed [MECH] Change of direction per unit time, as of a target on a radar screen, without regard to the direction of the rotation axis; in other words, the magnitude of the angular velocity vector. Also known as angular rate. { 'aŋ·gyə·lər 'spēd }

angular travel error [MECH] The error which is introduced into a predicted angle obtained by multiplying an instantaneous angular velocity by a time of flight. { 'aŋ·gyə·lər 'trav·əl ˌer·ər }

angular velocity [MECH] The time rate of change of angular displacement. [PHYS] *See* angular frequency. { 'aŋ·gyə·lər və'läs·əd·ē }

anharmonicity [PHYS] **1.** Mechanical vibration where the restoring force acting on a system does not vary linearly with displacement from equilibrium position. **2.** Variation from a linear relationship of dipole moment with internuclear distance in the infrared portion of the electromagnetic spectrum. { ¦an,här·mə'nis·əd·ē }

anharmonic oscillator [PHYS] An oscillating system in which the restoring force opposing a displacement from the position of equilibrium is a nonlinear function of the displacement. { ¦an,här¦män·ik 'äs·ə,lād·ər }

anharmonic oscillator spectrum [SPECT] A molecular spectrum which is significantly affected by anharmonicity of the forces between atoms in the molecule. { ,an·här¦män·ik ¦äs·ə,lād·ər ¦spek·trəm }

anhysteretic remanence [ELECTROMAG] The remanence in a magnetic recording medium that results from adding an alternating current to the signal current; at low fields it is linear and thus does not exhibit hysteresis. { ,an,his·tə¦red·ik 'rem·ə·nəns }

anisotropic [PHYS] Showing different properties as to velocity of light transmission, conductivity of heat or electricity, compressibility, and so on, in different directions. Also known as aeolotropic. { ¦a,nī·sə¦träp·ik }

anisotropic magnetoresistance [SOLID STATE] A type of magnetoresistance displayed by all metallic magnetic materials, which arises because conduction electrons have more frequent collisions when they move parallel to the magnetization in the material than when they move perpendicular to it. { ,an·ə·sə¦trō·pik ,mag,ned·ō·ri'sis·təns }

anisotropy [PHYS] The characteristic of a substance for which a physical property, such as index of refraction, varies in value with the direction in or along which the measurement is made. Also known as aeolotropy; eolotropy. { ¦a,nī'sä·trə·pē }

anisotropy constant [ELECTROMAG] In a ferromagnetic material, temperature-dependent parameters relating the magnetization in various directions to the anisotropy energy. { ¦a,nī'sä·trə·pē ,kän·stənt }

anisotropy energy [ELECTROMAG] Energy stored in a ferromagnetic crystal by virtue of the work done in rotating the magnetization of a domain away from the direction of easy magnetization. { ¦a,nī'sä·trə·pē ,en·ər·jē }

anisotropy factor *See* dissymmetry factor. { ¦a,nī'sä·trə·pē ,fak·tər }

anker [MECH] A unit of capacity equal to 10 U.S. gallons (37.854 liters); used to measure liquids, especially honey, oil, vinegar, spirits, and wine. { 'aŋ·kər }

ankylosis [PHYS] The loss by a system of one or more degrees of freedom through development of one or more frictional constraints. Also spelled anchylosis. { ,aŋ·kə'lō·səs }

annealing point [THERMO] The temperature at which the viscosity of a glass is $10^{13.0}$ poises. Also known as annealing temperature; 13.0 temperature. { ə'nēl·iŋ ,pȯint }

annealing temperature *See* annealing point. { ə'nēl·iŋ ,tem·prə·chər }

annihilation [PART PHYS] A process in which an antiparticle and a particle combine and release their rest energies in other particles. { ə,nī·ə'lā·shən }

annihilation operator [QUANT MECH] An operator which reduces the occupation number of a single state by unity; for example, an annihilation operator applied to a state of one particle yields the vacuum. Also known as destruction operator. { ə,nī·ə'lā·shən ¦äp·ə,rād·ər }

annihilation radiation [PART PHYS] Electromagnetic radiation arising from the collision, and resulting annihilation, of an electron and a positron, or of any particle and its antiparticle. { ə,nī·ə'lā·shən ,rād·ē'ā·shən }

annular effect [FL MECH] A phenomenon observed in the flow of fluid in a tube when its motion is alternating rapidly, as in the propagation of sound waves, in which the mean velocity rises progressing from the center of the tube toward the walls and then falls within a thin laminar boundary layer to zero at the wall itself. { 'an·yə·lər i'fekt }

anode [ELEC] The terminal at which current enters a primary cell or storage battery; it is positive with respect to the device, and negative with respect to the external circuit. { 'a,nōd }

anodic [PHYS] Pertaining to the anode. { ə'näd·ik }

anomalon [NUC PHYS] A nuclear fragment, produced in the collision of a projectile nucleus at relativistic energy with a target nucleus at rest, that has an anomalously short mean free path, comparable to that of a uranium nucleus. { ə'näm·ə,län }

anomaloscope [OPTICS] An optical instrument for testing color vision, in which a yellow light whose intensity may be varied is matched against red and green lights whose intensity is fixed. { ə'näm·ə·lə,skōp }

anomalous Barkhausen effect [ELECTROMAG] The occurrence of large steps in the magnetization of an iron-aluminum alloy at temperature above about 400°C (750°F). { ə¦näm·ə·ləs 'bark ,haùz·ən i,fekt }

anomalous dispersion [OPTICS] Extraordinary behavior in the curve of refractive index versus wavelength which occurs in the vicinity of absorption lines or bands in the absorption spectrum of a medium. { ə'näm·ə·ləs dis'pər·zhən }

anomalous expansion [THERMO] An increase in the volume of a substance that results from a decrease in its temperature, such as is displayed by water at temperatures between 0 and 4°C (32 and 39°F). { ə'näm·ə·ləs ik'span·shən }

anomalous Hall effect [ELECTROMAG] **1.** In a current-carrying conductor in a magnetic field, development of a transverse voltage resulting from the deflection of positive charge carriers (hole states) by the Lorentz force. **2.** The Hall effect in ferromagnetic metals, which arises from the unsymmetrical scattering of conduction electrons at magnetic moments. { ə¦näm·ə·ləs 'hȯl i,fekt }

anomalous magnetic moment [PART PHYS] The difference between the observed magnetic moment and the value predicted by Dirac's theory. { ə'näm·ə·ləs mag'ned·ik 'mō·mənt }

anomalous series [ATOM PHYS] A series of spectral lines associated with atomic energy levels whose Rydberg corrections do not vary smoothly with total quantum number, generally because they involve excitation of two electrons. Also known as abnormal series. { ə'näm·ə·ləs 'sir·ēz }

anomalous skin effect [ELEC] The skin effect at very low temperatures and high frequencies at which the thickness of the conducting skin layer is less than the electron mean free path, so that the classical theory of electrical conductivity breaks down. { ə¦näm·ə·ləs 'skin i,fekt }

anomalous viscosity *See* non-Newtonian viscosity. { ə'näm·ə·ləs vis'käs·əd·ē }

anomalous Zeeman effect [SPECT] A type of splitting of spectral lines of a light source in a magnetic field which occurs for any line arising from a combination of terms of multiplicity greater than one; due to a nonclassical magnetic behavior of the electron spin. { ə'näm·ə·ləs 'zā ,män i,fekt }

antenna [ELECTROMAG] A device used for radiating or receiving radio waves. Also known as aerial; radio antenna. { an'ten·ə }

antenna amplifier [ELECTROMAG] One or more stages of wide-band electronic amplification placed within or physically close to a receiving antenna to improve signal-to-noise ratio and mutually isolate various devices receiving their feed from the antenna. { an'ten·ə 'am·plə,fī·ər }

antenna coil [ELECTROMAG] Coil through which antenna current flows. { an'ten·ə ,kȯil }

antenna coincidence [ELECTROMAG] That instance when two rotating, highly directional antennas are pointed toward each other. { an'ten·ə kō'in·səd·əns }

antenna coupler [ELECTROMAG] A radio-frequency transformer, tuned line, or other device used to transfer energy efficiently from a transmitter to a transmission line or from a transmission line to a receiver. { an'ten·ə ,kəp·lər }

antenna crosstalk [ELECTROMAG] The ratio or the logarithm of the ratio of the undesired power received by one antenna from another to the power transmitted by the other. { an' ten·ə 'krȯs,tȯk }

antenna detector [ELECTROMAG] Device consisting of an antenna and electronic equipment to warn aircraft crew members that they are being observed by radar sets. { an'ten·ə di'tek·tər }

antenna directivity diagram [ELECTROMAG] Curve representing, in polar or Cartesian coordinates, a quantity proportional to the gain of an antenna in the various directions in a particular plane or cone. { an'ten·ə di·rek'tiv·əd·ē 'dī·ə,gram }

antenna effect [ELECTROMAG] A distortion of the directional properties of a loop antenna caused by an input to the direction-finding receiver which is generated between the loop and ground, in contrast to that which is generated between the two terminals of the loop. Also known as electrostatic error; vertical component effect. { an'ten·ə i'fekt }

antenna effective area [ELECTROMAG] In any specified direction, the square of the wavelength multiplied by the power gain (or directive gain) in that direction, and divided by 4π. { an'ten·ə i'fek·tiv 'er·ē·ə }

antenna efficiency [ELECTROMAG] The ratio of the amount of power radiated into space by an antenna to the total energy received by the antenna. { an'ten·ə i,fish·ən·sē }

antenna field [ELECTROMAG] A group of antennas placed in a geometric configuration. { an'ten·ə ,fēld }

antenna gain [ELECTROMAG] A measure of the effectiveness of a directional antenna as compared to a standard nondirectional antenna. Also known as gain. { an'ten·ə ,gān }

antenna matching [ELECTROMAG] Process of adjusting impedances so that the impedance of an antenna equals the characteristic impedance of its transmission line. { an'ten·ə ,mach·iŋ }

antenna pair [ELECTROMAG] Two antennas located on a base line of accurately surveyed length, sometimes arranged so that the array may be rotated around an axis at the center of the base line; used to produce directional patterns and in direction finding. { an'ten·ə ,per }

antenna pattern *See* radiation pattern. { an'ten·ə ,pad·ərn }

antenna polarization [ELECTROMAG] The orientation of the electric field lines in the electromagnetic field radiated or received by the antenna. { an'ten·ə ,pō·lə·rə'zā·shən }

antenna power [ELECTROMAG] Radio-frequency power delivered to an antenna. { an'ten·ə ,paù·ər }

antenna power gain [ELECTROMAG] The power gain of an antenna in a given direction is 4π times the ratio of the radiation intensity in that direction to the total power delivered to the antenna. { an'ten·ə 'paù·ər ,gān }

antenna resistance [ELECTROMAG] The power supplied to an entire antenna divided by the square of the effective antenna current measured at the point where power is supplied to the antenna. { an'ten·ə ri,zis·təns }

antenna scanner [ELECTROMAG] A microwave feed horn which moves in such a way as to illuminate sequentially different reflecting elements of an antenna array and thus produce the desired field pattern. { an'ten·ə ,skan·ər }

antenna temperature [ELECTROMAG] The temperature of a blackbody enclosure which would produce the same amount of noise as the antenna if it completely surrounded the antenna and was in thermal equilibrium with it. { an'ten·ə ,tem·prə·chər }

antiatom [ATOM PHYS] An atom made up of antiprotons, antineutrons, and positrons in the same way that an ordinary atom is made up of protons, neutrons, and electrons. { 'an·tē,ad·əm }

antibaryon [ATOM PHYS] One of a class of antiparticles, including the antinucleons and the antihyperons, with strong interactions, baryon number −1, and hypercharge and charge opposite to those for the particles. { 'an·tē¦bar·ē·än }

antibonding orbital [PHYS] An atomic or molecular orbital whose energy increases as atoms are brought closer together, indicating a net repulsion rather than a net attraction and chemical bonding. { 'an·tē'bänd·iŋ 'ȯr·bə·təl }

anticoincidence [NUC PHYS] The occurrence of an event at one place without a simultaneous event at another place. { ,an·tē,kō'in·sə·dəns }

anticorona [OPTICS] A diffraction phenomenon appearing at a point before an observer with the sun or moon directly behind the observer; consists of rings of colored lights complementary to the coronal rings. Also known as Brocken bow. { ,an·tē·kə'rō·nə }

antideuteron [ATOM PHYS] The antiparticle to the deuteron, composed of an antineutron and an antiproton. { ¦an·tē¦düt·ə,rän }

antiferroelectric crystal [SOLID STATE] A crystalline substance characterized by a state of lower symmetry consisting of two interpenetrating sublattices with equal but opposite electric polarization, and a state of higher symmetry in which the sublattices are unpolarized and indistinguishable. { ¦an·tē,fer·ō·i'lek·trik 'kris·təl }

antiferromagnetic domain [SOLID STATE] A region in a solid within which equal groups of elementary atomic or molecular magnetic moments are aligned antiparallel. { ¦an·tē,fer·ō,mag'ned·ik dō'mān }

antiferromagnetic resonance [ELECTROMAG] Magnetic resonance in antiferromagnetic materials which may be observed by rotating magnetic fields in either of two opposite directions. { ¦an·tē,fer·ō,mag'ned·ik 'rez·ə·nəns }

antiferromagnetic substance [ELECTROMAG] A substance that is composed of antiferromagnetic domains. { ¦an·tē,fer·ō,mag'ned·ik 'səb·stəns }

antiferromagnetic susceptibility [ELECTROMAG] The magnetic response to an applied magnetic field of a substance whose atomic magnetic moments are aligned in antiparallel fashion. { ¦an·tē,fer·ō,mag'ned·ik sə,sep·tə'bil·əd·ē }

antiferromagnetism [SOLID STATE] A property possessed by some metals, alloys, and salts of transition elements by which the atomic magnetic moments form an ordered array which alternates or spirals so as to give no net total moment in zero applied magnetic field. { ¦an·tē,fer·ō'mag·nə,tiz·əm }

antifriction [MECH] Making friction smaller in magnitude. { ,an·tē'frik·shən }

antigravity [PHYS] The repulsion of one body by another by means of a gravitational type of force; this has never been observed. { ¦an·tē¦grav·əd·ē }

antihelium [ATOM PHYS] The antimatter counterpart of helium, whose atoms each consist of two orbiting positrons and a nucleus composed of two antiprotons and either one or two antineutrons. { ¦an·tē'hē·lē·əm }

antihydrogen [ATOM PHYS] The antimatter counterpart of hydrogen, whose atoms each consist of an orbiting positron and a nucleus that is an antiproton, antideuteron, or antitriton. { ¦an·tē'hī·drə·jən }

antihyperon [PART PHYS] An antiparticle to a hyperon, having the same mass, lifetime, and spin as the hyperon, but with charge and magnetic moment reversed in sign. { ¦an·tē¦hī·pə,rän }

antilepton [ATOM PHYS] An antiparticle of a lepton, such as an antineutrino or a positron. { ¦an·tē'lep·tän }

antilogous pole [SOLID STATE] That crystal pole which becomes electrically negative when the crystal is heated or is expanded by decompression. { an'til·ə·gəs ¦pōl }

antimatter [PHYS] Material consisting of atoms which are composed of positrons, antiprotons, and antineutrons. { 'an·tē,mad·ər }

antimolecule [ATOM PHYS] A molecule made up of antiprotons, antineutrons, and positrons in the same way that an ordinary molecule is made up of protons, neutrons, and electrons. { ¦an·tē'mäl·ə,kyül }

antimony-124 [NUC PHYS] Radioactive antimony with mass number of 124; 60-day half-life; used as tracer in solid-state and pipeline flow studies. { 'an·tə,mō·nē ,wən,twen·tē'fȯr }

antineutrino [PART PHYS] The antiparticle to the neutrino; it has zero mass, spin 1/2, and positive helicity; there are two antineutrinos, one associated with electrons and one with muons. { 'an·tē·nü¦trē·nō }

antineutron [PART PHYS] The antiparticle to the neutron; a strongly interacting baryon which has no charge, mass of 939.6 MeV, spin 1/2, and mean life of about 10^3 seconds. { 'an·tē¦nü,trän }

antinodal points *See* negative nodal points. { ,an·tē'nōd·əl ,pȯins }

antinode [PHYS] A point, line, or surface in a standing-wave system at which some characteristic of the wave has maximum amplitude. Also known as loop. { 'an·tə,nōd }

antinoise [ACOUS] Noise that is deliberately created to mimic an existing noise field in antiphase so that the two fields cancel each other, resulting in silence. Also known as antisound. { ,an·tē'nȯiz }

antinucleon [PART PHYS] An antineutron or antiproton, that is, particles having the same mass as their nucleon counterparts but opposite charge or opposite magnetic moment. { 'an·tē¦nü·klē,än }

antinucleus [NUC PHYS] A nucleus made up of antineutrons and antiprotons in the same way that an ordinary nucleus is made up of neutrons and protons. { ¦an·tē¦nü·klē·əs }

antiparallel [PHYS] Property of two displacements or other vectors which lie along parallel lines but point in opposite directions. { ¦an·tē'par·ə,lel }

antiparticle [PART PHYS] A counterpart to a particle having mass, lifetime, and spin identical to the particle but with charge and magnetic moment reversed in sign. { 'an·tē¦pärd·ə·kəl }

antiprincipal planes *See* negative principal planes. { ¦an·tē¦prin·sə·pəl 'plānz }

antiprincipal point *See* negative principal point. { ¦an·tē¦prin·sə·pəl 'pȯint }

antiproton [PART PHYS] The antiparticle to the proton; a strongly interacting baryon which is stable, carries unit negative charge, has the same mass as the proton (938.3 MeV), and has spin 1/2. { 'an·tē¦prō,tän }

antiprotonic atom [ATOM PHYS] An atom consisting of an ordinary nucleus with an orbiting antiproton. { ,an·tē·prō'tän·ik 'ad·əm }

antiquark [PART PHYS] The hypothetical antiparticle of a quark, having electric charge, baryon number, and strangeness opposite in sign to that of the corresponding quark. { 'an·tē¦kwärk }

antiresonance *See* parallel resonance. { ,an·tē'rez·ən·əns }

antiresonant circuit *See* parallel resonant circuit. { ,an·tē'rez·ən·ənt 'sər·kət }

antisound *See* antinoise. { ,an·tē'saủnd }

anti-Stokes lines [SPECT] Lines of radiated frequencies which are higher than the frequency of the exciting incident light. { ,an·tē'stōks ,līnz }

antisymmetric wave function [PHYS] A many-particle wave function which changes its sign when the coordinates of two of the particles are interchanged. { ¦an·tē·si¦me·trik 'wāv ,fəŋk·shən }

antisymmetrized wave function [QUANT MECH] A wave function of several identical fermions (such as electrons) which changes sign but in all other respects remains unaltered if two of the fermions are interchanged. { ,an·tē,sim·ə,trīzd 'wāv ,fəŋk·shən }

antitriton [ATOM PHYS] The antiparticle to the triton, composed of an antiproton and two antineutrons. { ¦an·tē'trīt·ən }

Antoine equation [PHYS] The empirical relationshipbetween temperature and vapor pressure of liquids; $\log P = B - A/(C + T)$, where A, B, C are experimental constants, T is absolute temperature, and P is vapor pressure. { 'an ,twän i,kwā·zhən }

Antonoff's rule [PHYS] The rule which states that the surface tension at the interface between two saturated liquid layers in equilibrium is equal to the difference between the individual surface tensions of similar layers when exposed to air. { an'tä,nȯfs ,rül }

anyon [QUANT MECH] A particle obeying an unconventional form of quantum statistics, which is characterized by a parameter that can take on any of a continuum of values, just two of which represent Bose-Einstein and Fermi-Dirac statistics. { 'an·ē,än }

aperiodic [PHYS] Of irregular occurrence; not periodic; not displaying resonant response. { ¦a,pir·ē¦äd·ik }

aperiodic antenna [ELECTROMAG] Antenna designed to have constant impedance over a wide range of frequencies because of the suppression of reflections within the antenna system; includes terminated wave and rhombic antennas. { ¦a,pir·ē¦äd·ik an'ten·ə }

aperiodic damping [PHYS] Condition of a system in which the amount of damping is so large that, when the system is subjected to a single disturbance, either constant or instantaneous, the system comes to a position of rest without passing through that position; while an aperiodically damped system is not strictly an oscillating system, it has such properties that it should become an oscillating system if the damping were sufficiently reduced. { ¦a,pir·ē¦äd·ik 'damp·iŋ }

aperiodic waves [ELEC] The transient current wave in a series circuit with resistance R, inductance L, and capacitance C when $R^2C = 4L$. [PHYS] Waves without a definite repetitive pattern; for example, transient waves. { ¦a,pir·ē¦äd·ik 'wāvz }

apertometer [OPTICS] An instrument designed to measure the numerical aperture of microscope objectives. { ,a·pər'täm·əd·ər }

aperture [OPTICS] The diameter of the objective of a telescope or other optical instrument, usually expressed in inches, but sometimes as the angle between lines from the principal focus to opposite ends of a diameter of the objective. { 'ap·ə,chər }

aperture aberration [OPTICS] Errors in optical imaging which occur because rays of different distances from the axis do not come to the same focus. { 'ap·ə,chər ,ab·ə'rā·shən }

aperture angle [OPTICS] The angle subtended by the radius of the entrance pupil of an optical instrument at the object. { 'ap·ə·chər ,aŋ·gəl }

aperture antenna [ELECTROMAG] Antenna in which the beam width is determined by the dimensions of a horn, lens, or reflector. { 'ap·ə,chər an'ten·ə }

aperture conductivity [ACOUS] The ratio of the density of a medium to the acoustic mass at an aperture. { 'ap·ə,chər ,kän,dək'tiv·əd·ē }

aperture illumination [ELECTROMAG] Field distribution in amplitude and phase over an aperture. { 'ap·ə,chər i,lüm·ə'nā·shən }

aperture ratio [OPTICS] The ratio of the effective diameter of a lens to its focal length. { 'ap·ə,chər 'rā·shō }

aperture slit *See* aperture slot. { 'ap·ə,chər ,slit }

aperture slot [OPTICS] A narrow rectangular opening in the optical system of a rotary camera through which light from a continuously moving document is transmitted to a film whose movement is synchronized to that of the document. Also known as aperture slit. { 'ap·ə,chər ,slät }

aperture splitting [OPTICS] A technique in which light from a single slit is divided by passing it through two other slits and is combined by a lens. { 'ap·ə·chər ,splid·iŋ }

aperture stop [OPTICS] That opening in an optical system that determines the size of the bundle of rays which traverse the system from a given point of the object to the corresponding point of the image. { 'ap·ə,chər ,stäp }

aperture synthesis [ELECTROMAG] The use of one or more pairs of instruments of relatively small aperture, acting as interferometers, to obtain the information-gathering capability of a telescope of much larger aperture. { 'ap·ə,chər ,sin·thə·səs }

A phase *See* liquid A. { 'ā ,fāz }

A_1 phase *See* liquid A. { ¦ā·səb¦wən ,fāz }

apical angle [MECH] The angle between the tangents to the curve outlining the contour of a projectile at its tip. [OPTICS] The dihedral angle between the refracting faces of a prism. Also known as refracting angle. { 'ap·i·kəl 'aŋ·gəl }

apioid [PHYS] A pear-shaped form taken by a rapidly revolving mass of liquid due to the force of gravity. { 'ap·ē,ȯid }

aplanatic lens [OPTICS] A lens corrected for spherical aberration. { ¦a·plə¦nad·ik 'lenz }

aplanatic points [OPTICS] Two points on the axis of an optical system which are located so that all the rays emanating from one converge to, or appear to diverge from, the other. { ¦a·plə¦nad·ik 'pȯins }

apochromat *See* apochromatic lens. { ,ap·ə'krō·mat }

apochromatic lens [OPTICS] A lens with corrections for chromatic and spherical aberration. { ¦ap·ə·krō¦mad·ik 'lenz }

apochromatic system [OPTICS] An optical system which is free from both spherical and chromatic aberration for two or more colors. { ¦ap·ə·krō¦mad·ik 'sis·təm }

apodization [OPTICS] The modification of the amplitude transmittance of the aperture of an optical system so as to reduce or suppress the energy of the diffraction rings relative to that of the central Airy disk. [SPECT] A mathematical transformation carried out on data received from an interferometer to alter the instrument's response function before the Fourier transformation is calculated to obtain the spectrum. { ,a·pə·də'zā·shən }

apostilb [OPTICS] A luminance unit equal to one ten-thousandth of a lambert. Also known as blandel. { 'ap·ə,stilb }

apothecaries' dram *See* dram. { ə'päth·ə,ker·ēz 'dram }

apothecaries' ounce *See* ounce. { ə'päth·ə,ker·ēz 'au̇ns }

apothecaries' pound *See* pound. { ə'päth·ə,ker·ēz 'pau̇nd }

apparent additional mass [FL MECH] A fictitious mass of fluid added to the mass of the body to represent the force required to accelerate the body through the fluid. { ə'pa·rənt ə'dish·ən·əl 'mas }

apparent candlepower [OPTICS] For an extended source of light, at a specified distance, the candlepower of a point source that would produce the same illumination as the extended source at the same distance. { ə'pa·rənt 'kan·dəl,pau̇r }

apparent depth [OPTICS] The depth of the image of an object submerged in a transparent medium; it is reduced from the real depth of the object by a factor equal to the relative index of refraction of the medium with respect to air. { ə'pa·rənt 'depth }

apparent expansion [THERMO] The expansion of a liquid with temperature, as measured in a graduated container without taking into account the container's expansion. { ə'pa·rənt ik'span·shən }

apparent force [MECH] A force introduced in a relative coordinate system in order that Newton's laws be satisfied in the system; examples are the Coriolis force and the centrifugal force incorporated in gravity. { ə'pa·rənt 'fȯrs }

apparent gravity *See* acceleration of gravity. { ə'pa·rənt 'grav·əd·ē }

apparent horizon [RELAT] The boundary of a region in space-time in which the gravitational field is so strong that the cross-sectional area of an outgoing light pulse decreases. { ə'pa·rənt hə'rīz·ən }

apparent luminance [OPTICS] Luminance, created by air light, of that portion of the visual field subtended by a dark, distant object; that is, the light scattered into the eye by particles, including air molecules, lying along the optic path from eye to object. { ə'pa·rənt 'lü·mə·nəns }

apparent motion *See* relative motion. { ə'pa·rənt 'mō·shən }

apparent power [ELEC] The product of the root-mean-square voltage and the root-mean-square current delivered in an alternating-current circuit, no account being taken of the phase difference between voltage and current. { ə'pa·rənt 'pau̇·ər }

apparent viscosity [FL MECH] The value obtained by applying the instrumental equations used in obtaining the viscosity of a Newtonian fluid to viscometer measurements of a non-Newtonian fluid. { ə¦par·ənt vi'skäs·əd·ē }

apparent visual angle [OPTICS] The angle subtended by a source at the observer's eye as calculated from the source size and distance from the eye. { ə'pa·rənt 'vizh·ə·wəl 'aŋ·gəl }

apparent volume [PHYS] The difference between the volume of a binary solution and the volume of the pure solvent at the same temperature. { ə'pa·rənt 'väl·yəm }

apparent weight [MECH] For a body immersed in a fluid (such as air), the resultant of the gravitational force and the buoyant force of the fluid acting on the body; equal in magnitude to the true weight minus the weight of the displaced fluid. { ə'pa·rənt 'wāt }

appearance potential [PHYS] The minimal potential which the electron beam in the ion source of a mass spectrometer must traverse in order to acquire enough energy to produce ions of a specified nuclide or molecular fragment. { ə'pir·əns pə'ten·chəl }

applied inverse scattering theory [PHYS] The branch of inverse scattering theory that treats the case in which the data provided are incomplete or corrupted by noise. { ə'plīd 'in,vərs ¦skad·ər·iŋ ,thē·ə·rē }

appliqué [OPTICS] A combination of lenses that provides for the same focal length at three or more wavelengths. { ¦ap·lə¦kā }

approximate absolute temperature [PHYS] A temperature scale with the ice point at 273° and boiling point of water at 373°; it is intended to approximate the Kelvin temperature scale with sufficient accuracy for many sciences, notably meteorology, and is widely used in the meteorological literature. Also known as tercentesimal thermometric scale. { ə'präk·sə·mət 'ab·sə,lüt 'tem·prə·chər }

APW method *See* augmented plane-wave method. { ¦ā¦pē'dəb·əl,yü ,meth·əd }

Arago point [OPTICS] A neutral point located about 20° directly above the antisolar point in relatively clear air and at higher elevations in turbid air. { 'a·rə,gō ,pȯint }

Arago's disk [ELECTROMAG] A device consisting of a horizontal disk of copper that can rotate about a vertical axis in an airtight box, and a horizontal bar magnet suspended above the disk but outside the box; upon rapid rotation of the disk, the bar magnet is deflected and eventually rotates in the same direction with smaller velocity. { 'a·rə,gōz ,disk }

arc *See* electric arc. { ärk }

arc discharge [ELEC] A direct-current electrical current between electrodes in a gas or vapor, having high current density and relatively low voltage drop. { 'ärk 'dis,chärj }

arc excitation [ATOM PHYS] Use of electric-arc energy to move electrons into higher energy orbits. { ¦ärk ,ek,sī'tā·shən }

arc force [MECH] The force of a plasma arc through a nozzle or opening. { 'ärk ,fȯrs }

Archimedean principle [PHYS] The principle that a body immersed in a fluid undergoes an apparent loss in weight equal to the weight of the fluid it displaces. { ¦är·kə¦mēd·ē·ən 'prin·sə·pəl }

Archimedes number [FL MECH] One of a dimensionless group of numbers denoting the ratio of gravitational force to viscous force. { ¦är·kə¦mēd,ēz 'nəm·bər }

arc spectrum [SPECT] The spectrum of a neutral atom, as opposed to that of a molecule or an ion; it is usually produced by vaporizing the substance in an electric arc; designated by the roman numeral I following the symbol for the element, for example, HeI. { 'ärk ,spek·trəm }

are [MECH] A unit of area, used mainly in agriculture, equal to 100 square meters. { är }

area redistribution [PHYS] A method of measuring the effective duration of an irregular pulse by constructing a rectangular pulse that has the same peak amplitude and the same area on a graph of amplitude versus time. { 'er·ē·ə ¦rē ,dis·trə'byü·shən }

argon laser [OPTICS] A gas laser using ionized argon; emits a 4880-angstrom line as well as infrared radiation. { 'är,gän 'lā·zər }

arm *See* branch. [PHYS] The perpendicular distance from the line along which a force is applied to a reference point. { ärm }

ARPES *See* angle-resolved photoelectron spectroscopy.

Arrhenius-Guzman equation [PHYS] The relation between the viscosity η of a liquid and the Kelvin temperature T at constant pressure: $\eta = A \exp(B/RT)$, where A and B are constants and R is the gas constant. { ar'rā·nē·əs ,güth·mən i,kwā·zhən }

Arrhenius viscosity formulas [PHYS] A series of three equations which relate the viscosity of a liquid to the temperature, the viscosity of a solution to its concentration and to the viscosity of the solvent, and the viscosity of a sol to the viscosity of the medium. { ar'rā·nē·əs vis'käs·əd·ē ,fȯr·myə·ləz }

artificial echo [ELECTROMAG] **1.** Received reflections of a transmitted pulse from an artificial target, such as an echo box, corner reflector, or other metallic reflecting surface. **2.** Delayed signal from a pulsed radio-frequency signal generator. { ¦ärd·ə¦fish·əl 'ek·ō }

aS *See* abmho.

A scale [ACOUS] A system used to filter out sound below 55 decibels; its characteristics are equal to those of the human ear. { 'ā ,skāl }

ascending branch [MECH] The portion of the trajectory between the origin and the summit on which a projectile climbs and its altitude constantly increases. { ə'send·iŋ 'branch }

asperomagnetic state [SOLID STATE] The condition of a rare-earth glass in which the spins are oriented in fixed directions, with most nearest-neighbor spins parallel or nearly parallel, so that the spin directions are distributed in one hemisphere. { a¦sper·ō,mag'ned·ik 'stāt }

aspheric surface [OPTICS] A lens or mirror surface which is altered slightly from a spherical surface in order to reduce aberrations. { ā'sfir·ik 'sər·fəs }

associated production [PART PHYS] Production of strange particles invariably in twos, never one particle alone. { ə'sō·sē,ād·əd prə'dək·shən }

astatic [PHYS] Without orientation or directional characteristics; having no tendency to change position. { ā'stad·ik }

astatic coils [ELECTROMAG] Two identical coils, connected in series and suspended from the same axis, so that a uniform, external magnetic field exerts no net torque on the system. { ā 'stad·ik 'kȯilz }

astatic pair [ELECTROMAG] A pair of parallel magnets, equal in strength and having polarities in opposite directions, and perpendicular to an axis which bisects both of them; there is no net force or torque on the pair in a uniform field. { ā'stad·ik 'per }

astatic pendulum [PHYS] A pendulum which almost never takes a position of equilibrium. { ā'stad·ik 'pen·jə·ləm }

astatic system [ELECTROMAG] A system of magnets arranged so that the net force and torque exerted on the system by a uniform magnetic field equals 0. { ā¦stad·ik ¦sis·təm }

asterism [OPTICS] A starlike optical phenomenon seen in gemstones called star stones; due to reflection of light by lustrous inclusions reduced to sharp lines of light by a domed cabochon style of cutting. [SPECT] A star-shaped pattern sometimes seen in x-ray spectrophotographs. { 'as·tə,riz·əm }

astigmat *See* astigmatic lens. { ə'stig,mat }

astigmatic difference [OPTICS] **1.** The distance between the primary and secondary foci of an astigmatic optical system. **2.** The difference between the reciprocals of the distances of the primary and secondary foci from an astigmatic thin lens or mirror. { ¦a·stig,mad·ik 'dif·rəns }

astigmatic foci [OPTICS] The two lines on which rays emanating from a point are focused by an astigmatic optical system. Also known as focal lines. { ¦a·stig,mad·ik 'fō,sī }

astigmatic interval [OPTICS] The portion of a pencil of rays in an astigmatic optical system that lies between the primary and secondary foci. Also known as conoid of Sturm; interval of Sturm. { ¦a·stig,mad·ik 'in·tər·vəl }

astigmatic lens [OPTICS] A planocylindrical, spherocylindrical, or spherotoric lens used in eyeglasses to correct astigmatism. Also known as astigmat. { ¦as·tig¦mad·ik 'lenz }

astigmatic mounting [SPECT] A mounting designed to minimize the astigmatism of a concave diffraction grating. { ¦a·stig,mad·ik 'mount·iŋ }

astigmatic surfaces [OPTICS] Two surfaces containing the astigmatic foci of points in a plane perpendicular to the optical axis of an astigmatic system. { ¦a·stig,mad·ik 'sər·fəs·əs }

astigmatism [OPTICS] The failure of an optical system, such as a lens or a mirror, to image a point as a single point; the system images the point on two line segments separated by an interval. { ə'stig·mə,tiz·əm }

astigmatizer [OPTICS] A device, as attached to a rangefinder, for drawing out a point of light into a line or band. { ə'stig·mə,tīz·ər }

astigmometer [OPTICS] An instrument which measures the amount of astigmatism in an optical system. { ,as·tig'mäm·əd·ər }

Aston whole-number rule [PHYS] The rule which states that when expressed in atomic weight units, the atomic weights of isotopes are very nearly whole numbers, and the deviations found in samples of elements are due to the presence of several isotopes with different weights. { 'as·tən ,hōl 'nəm·bər ,rül }

astroballistics [MECH] The study of phenomena arising out of the motion of a solid through a gas at speeds high enough to cause ablation; for example, the interaction of a meteoroid with the atmosphere. { ¦as·trō·bə'lis·tiks }

astron [NUC PHYS] A proposed thermonuclear device in which a deuterium plasma is confined by an axial magnetic field produced by a shell of relativistic electrons. { 'a,strän }

astronomical camera [OPTICS] A camera designed to record either point sources (stars), extended sources (nebulae, galaxies, planets, or the sun and moon), or the spectra of celestial bodies. { ,as·trə'näm·ə·kəl 'kam·rə }

astronomical photography [OPTICS] The use of the photographic process to record surface features of celestial objects, their positions and motions (for measurement), and their radiation (photometry) and spectra (spectroscopy). Also known as astrophotography. { ,as·trə'näm·ə·kəl fə'täg·rə·fē }

astronomical spectrograph [SPECT] An instrument used to photograph spectra of stars. { ,as·trə'näm·ə·kəl 'spek·trə,graf }

astronomical spectroscopy [SPECT] The use of spectrographs in conjunction with telescopes to obtain observational data on the velocities and physical conditions of astronomical objects. { ,as·trə'näm·ə·kəl ,spek'träs·kə·pē }

astronomical telescope [OPTICS] A telescope designed for viewing astronomical objects. { ,as·trə'näm·ə·kəl 'tel·ə,skōp }

astrophotography *See* astronomical photography. { ,as·trō·fə'täg·rə·fē }

asymmetrical conductivity [ELEC] A variation in the conductivity of a conductor over its cross section that is not symmetric about the conductor's central axis. { ¦ā·sə¦me·tri·kəl ,kän,dək'tiv·əd·ē }

asymmetric top [MECH] A system in which all three principal moments of inertia are different. { ¦ā·sə¦me·trik 'täp }

asymptotically flat [RELAT] A space-time is asymptotically flat if it approaches Minkowski space-time at a prescribed rate at large spatial distances. { ,ā·sim¦tät·ə·klē 'flat }

asymptotically simple [RELAT] A space-time is asymptotically simple if it satisfies certain mathematical requirements on the conformal structure of null infinity; these requirements are a definition of a type of asymptotic flatness. { ,ā·sim¦tät·ə·klē 'sim·pəl }

asymptotic freedom [PART PHYS] In some gauge theories, the property of the strong interactions of growing steadily weaker at high energies. { ā,sim'täd·ik 'frēd·əm }

asynchronous [PHYS] Not synchronous. { ā 'siŋ·krə·nəs }

at *See* technical atmosphere.

ata [MECH] A unit of absolute pressure in the metric technical system equal to 1 technical atmosphere. { 'a·tə }

athermal transformation [PHYS] A chemical or physical change not requiring a change in the temperature of the substance, as in the formation of martensite. { ¦ā'thər·məl ,tranz·fər'mā·shən }

athermancy [ELECTROMAG] Property of a substance which cannot transmit infrared radiation. { ¦ā¦thər·mən·sē }

atm *See* atmosphere.

atmolysis [FL MECH] The separation of gas mixtures by using their relative diffusibility through a porous partition. { ət'mäl·ə·səs }

atmo-meter *See* meter-atmosphere. { 'at·mō ˌmēd·ər }

atmosphere [MECH] A unit of pressure equal to 101.325 kilopascals, which is the air pressure measured at mean sea level. Abbreviated atm. Also known as standard atmosphere. { 'at·məˌsfir }

atmospheric acoustics [ACOUS] The science of sound waves in the open air. { ¦at·mə¦sfir·ik ə'kü·stiks }

atmospheric boil *See* terrestrial scintillation. { ¦at·mə¦sfir·ik 'bȯil }

atmospheric drag [FL MECH] A major perturbation of close artificial satellite orbits caused by the resistance of the atmosphere; the secular effects are decreasing eccentricity, semidiameter, and period. { ¦at·mə¦sfir·ik 'drag }

atmospheric pressure [PHYS] The pressure at any point in an atmosphere due solely to the weight of the atmospheric gases above the point concerned. Also known as barometric pressure. { ¦at·mə¦sfir·ik 'presh·ər }

atmospheric radio wave [ELECTROMAG] Radio wave that is propagated by reflection in the atmosphere; may include either the ionospheric wave or the tropospheric wave, or both. { ¦at·mə¦sfir·ik 'rād·ē·ō ˌwāv }

atmospheric shimmer *See* terrestrial scintillation. { ¦at·mə¦sfir·ik 'shim·ər }

atomic absorption coefficient [PHYS] The linear absorption coefficient divided by the number of atoms per unit volume. { ə'täm·ik əb'zȯrp·shən ˌkō·əˌfish·ənt }

atomic absorption spectroscopy [SPECT] An instrumental technique for detecting concentrations of atoms to parts per million by measuring the amount of light absorbed by atoms or ions vaporized in a flame or an electrical furnace. { ə¦tä·mik əb¦sörp·shən ˌspek'träs·kə·pē }

atomic beam [PHYS] A stream of atoms, which may or may not be ionized. { ə'täm·ik 'bēm }

atomic-beam frequency standard [PHYS] A source of precisely timed signals which are derived from an atomic-beam resonance, such as a cesium-beam cell or a hydrogen maser. { ə'täm·ik ˌbēm 'frē·kwən·sē ˌstan·dərd }

atomic-beam resonance [PHYS] Phenomenon in which an oscillating magnetic field, superimposed on a uniform magnetic field at right angles to it, causes transitions between states with different magnetic quantum numbers of the nuclei of atoms in a beam passing through the field; the transitions occur only when the frequency of the oscillating field assumes certain characteristic values. { ə'täm·ik ¦bēm ˌrez·ən·əns }

atomic charge [ATOM PHYS] The electric charge of an ion, equal to the number of electrons the atom has gained or lost in its ionization multiplied by the charge on one electron. { ə'täm·ik 'chärj }

atomic constants *See* fundamental constants. { ə'täm·ik 'kän·stəns }

atomic core [ATOM PHYS] An atom stripped of its valence electrons, so that its remaining electrons are all in closed shells. { ə¦täm·ik 'kȯr }

atomic crystal [OPTICS] A crystallike structure of atoms that occupy sites in an optical lattice. { ə¦täm·ik 'krist·əl }

atomic diamagnetism [ATOM PHYS] Diamagnetic ionic susceptibility, important in providing correction factors for measured magnetic susceptibilities; calculated theoretically by considering electron density distributions summed for each electron shell. { ə'täm·ik ˌdī·ə'mag·nə ˌtiz·əm }

atomic emission spectroscopy [SPECT] A form of atomic spectroscopy in which one observes the emission of light at discrete wavelengths by atoms which have been electronically excited by collisions with other atoms and molecules in a hot gas. { ə¦täm·ik ə¦mish·ən spek'träs·kə·pē }

atomic energy level [ATOM PHYS] A definite value of energy possible for an atom, either in the ground state or an excited condition. { ə'täm·ik 'en·ər·jē ˌlev·əl }

atomic fission *See* fission. { ə'täm·ik 'fish·ən }

atomic fluorescence spectroscopy [SPECT] A form of atomic spectroscopy in which the sample atoms are first excited by absorbing radiation from an external source containing the element to be detected, and the intensity of radiation emitted at characteristic wavelengths during transitions of these atoms back to the ground state is observed. { ə¦täm·ik flü¦res·əns spek'träs·kə·pē }

atomic form factor *See* atomic scattering factor. { ə'täm·ik 'fȯrm ˌfak·tər }

atomic fountain [ATOM PHYS] A device in which atoms in a magnetooptic trap from a thermal beam are pushed upward with a pulse of laser light, causing them to assume a ballistic trajectory; used for studying free-falling atoms. { ə¦täm·ik 'faůnt·ən }

atomic frequency [SOLID STATE] One of the vibrational frequencies of an atom in a crystal lattice. { ə¦täm·ik 'frē·kwən·sē }

atomic funnel [ATOM PHYS] A device that uses a magnetic quadrupole field and trapping and cooling laser beams to form a slowed atomic beam into a highly localized and collimated beam with a peak phase-space density over 10,000 times that of the original beam. { ə'täm·ik 'fən·əl }

atomic fusion *See* fusion. { ə'täm·ik 'fyü·zhən }

atomic gas laser [OPTICS] A gas laser, such as the helium-neon laser, in which electrons and ions accelerated between electrodes by an electric field collide and excite atoms and ions to higher energy levels; laser action occurs during subsequent decay back to lower energy levels. { ə'täm·ik 'gas ˌlā·zər }

atomic ground state [ATOM PHYS] The state of lowest energy in which an atom can exist. Also known as atomic unexcited state. { ə'täm·ik 'graůnd ˌstāt }

atomic hydrogen maser [PHYS] A maser in

which dissociated hydrogen atoms from an electric discharge source are formed into a beam that undergoes selective magnetic processing; can be used as an atomic clock. { ə'täm·ik 'hī·drə·jən 'mā·zər }

atomic magnet [ATOM PHYS] An atom which possesses a magnetic moment either in the ground state or in an excited state. { ə'täm·ik 'mag·nət }

atomic magnetic moment [ATOM PHYS] A magnetic moment, permanent or temporary, associated with an atom, measured in magnetons. { ə'täm·ik ˌmag¦ned·ik 'mō·mənt }

atomic mass [PHYS] The mass of a neutral atom usually expressed in atomic mass units. { ə'täm·ik 'mas }

atomic mass unit [PHYS] An arbitrarily defined unit in terms of which the masses of individual atoms are expressed; the standard is the unit of mass equal to one-twelfth the mass of the carbon atom, having as nucleus the isotope with mass number 12. Abbreviated amu. Also known as dalton. { ə'täm·ik 'mas 'yü·nət }

atomic number [NUC PHYS] The number of protons in an atomic nucleus. Also known as proton number. { ə'täm·ik 'nəm·bər }

atomic orbital [ATOM PHYS] The space-dependent part of a wave function describing an electron in an atom. { ə'täm·ik 'ȯr·bə·təl }

atomic paramagnetism [ELECTROMAG] The result of a permanent magnetic moment in an atom. { ə'täm·ik ˌpar·ə'mag·nəˌtiz·əm }

atomic particle [ATOM PHYS] One of the particles of which an atom is constituted, as an electron, neutron, or proton. { ə'täm·ik 'pard·ə·kəl }

atomic physics [PHYS] The science concerned with the structure of the atom, the characteristics of the elementary particles of which the atom is composed, and the processes involved in the interactions of radiant energy with matter. { ə'täm·ik 'fiz·iks }

atomic scattering factor [PHYS] A quantity which expresses the efficiency with which x-rays of a stated wavelength are scattered into a given direction by a particular atom, measured in terms of the corresponding scattering by a point electron. Also known as atomic form factor. { ə'täm·ik 'skad·ər·iŋ ˌfak·tər }

atomic second [PHYS] As defined in 1967, the duration of 9,192,631,770 periods of the radiation corresponding to the two hyperfine levels of the fundamental state of the atom of cesium-133. { ə'täm·ik 'sek·ənd }

atomic spectroscopy [SPECT] The branch of physics concerned with the production, measurement, and interpretation of spectra arising from either emission or absorption of electromagnetic radiation by atoms. { ə'täm·ik ˌspek 'träs·kə·pē }

atomic spectrum [SPECT] The spectrum of radiations due to transitions between energy levels in an atom, either absorption or emission. { ə'täm·ik 'spek·trəm }

atomic standard [PHYS] Any supposedly immutable property of an atom, such as the wavelength or frequency of a characteristic spectral line, in terms of which a unit of a physical quantity is defined. { ə'täm·ik 'stan·dərd }

atomic structure [ATOM PHYS] The arrangement of the parts of an atom, which consists of a massive, positively charged nucleus surrounded by a cloud of electrons arranged in orbits describable in terms of quantum mechanics. { ə'täm·ik 'strək·chər }

atomic susceptibility [ELECTROMAG] The magnetization of a material per atom per unit of applied field; measured in ergs per oersted per atom. { ə'täm·ik səˌsep·tə'bil·əd·ē }

atomic unexcited state *See* atomic ground state. { ə'täm·ik ˌən·ek'sīd·əd 'stāt }

atomic units *See* Hartree units. { ə'täm·ik 'yü·nəts }

atomic vibration [ATOM PHYS] Periodic, nearly harmonic changes in position of the atoms in a molecule giving rise to many properties of matter, including molecular spectra, heat capacity, and heat conduction. { ə'täm·ik ˌvī'brā·shən }

atom interferometer [PHYS] A device which measures the interference effects that result when a beam of atoms is manipulated in such a way that the de Broglie waves of the atoms are split into two components and subsequently recombined. { ˌad·əm ˌin·tə·fə'räm·əd·ər }

atom laser [PHYS] A device that generates intense coherent beams of atoms (coherent matter waves), analogous to coherent light waves emitted by a conventional laser, through a stimulated process that generally involves extraction of the beams from a Bose-Einstein condensate. { ¦ad·əm ¦lā·zər }

atom optics [PHYS] The use of laser light and nanofabricated structures to manipulate the motion of atoms in the same manner that rudimentary optical elements control light. { ¦ad·əm 'äp·tiks }

ATR *See* attenuated total reflectance.

attached shock *See* attached shock wave. { ə'tacht 'shäk }

attached shock wave [FL MECH] An oblique or conical shock wave that appears to be in contact with the leading edge of an airfoil or the nose of a body in a supersonic flow field. Also known as attached shock. { ə'tacht 'shäk ˌwāv }

attempt frequency [NUC PHYS] The frequency with which an alpha particle attempts to cross the Gamow barrier in the Gamow-Condon-Gurney theory. { ə'tempt ˌfrē·kwən·sē }

attenuated total reflectance [SPECT] A method of spectrophotometric analysis based on the reflection of energy at the interface of two media which have different refractive indices and are in optical contact with each other. Abbreviated ATR. Also known as frustrated internal reflectance; internal reflectance spectroscopy. { ə'ten·yəˌwād·əd 'tōd·əl ri'flek·təns }

attenuation [ELEC] The exponential decrease with distance in the amplitude of an electrical signal traveling along a very long uniform transmission line, due to conductor and dielectric losses. [PHYS] The reduction in level of a quantity, such as the intensity of a wave, over an interval of a variable, such as the distance from a source. { ə,ten·yə'wā·shən }

attenuation coefficient [ELECTROMAG] The space rate of attenuation of any transmitted electromagnetic radiation. { ə,ten·yə'wā·shən ,kō·ə'fish·ənt }

attenuation constant [PHYS] A rating for a line or medium through which a plane wave is being transmitted, equal to the relative rate of decrease of an amplitude of a field component, voltage, or current in the direction of propagation, in nepers per unit length. { ə,ten·yə'wā·shən ,kän·stənt }

attenuation factor *See* attenuation constant. { ə,ten·yə'wā·shən ,fak·tər }

attenuation length [PHYS] The reciprocal of the attenuation coefficient. { ə,ten·yə'wā·shən ,leŋkth }

attenuation ratio [PHYS] The magnitude of the propagation ratio. { ə,ten·yə'wā·shən ,rā·shō }

atto- [PHYS] A prefix representing 10^{-18}, which is 0.000 000 000 000 000 001, or one-millionth of a millionth of a millionth. Abbreviated a { 'ad·ō }

attractor [PHYS] A geometrical object toward which the trajectory of a dynamical system, represented by a curve in phase space, converges in the course of time. { ə'trak·tər }

atu [PHYS] A unit of underpressure or pressure below atmospheric pressure in the metric technical system; equal to 1 technical atmosphere. { ¦at¦ü }

atü [PHYS] A unit of overpressure or gage pressure in the metric technical system; equal to 1 technical atmosphere. { ¦at¦yü }

audibility [ACOUS] **1.** The state or quality of being heard. **2.** The intensity of a received audio signal, usually expressed in decibels above or below 1 milliwatt using a stated single frequency sine wave. { ,ȯd·ə'bil·əd·ē }

audibility curve [ACOUS] **1.** The limits of hearing represented graphically as an area by plotting the minimum audible intensity of a sine wave sound versus frequency. **2.** *See* equal loudness contour. { ,ȯd·ə'bil·əd·ē ,kərv }

audibility threshold [ACOUS] The sound intensity at a given frequency which is the minimum preceptible by a normal human ear under specified standard conditions. { ,ȯd·ə'bil·əd·ē ,thresh,hōld }

audible frequency *See* audible tone. { ¦ȯd·ə·bəl 'frē·kwən·sē }

audible tone [ACOUS] Sound of a frequency which the average human can hear, ranging from 30 to 16,000 hertz. Also known as audible frequency. { ¦ȯd·ə·bəl 'tōn }

audio [ACOUS] **1.** Of or pertaining to sound in the range of frequencies considered audible at reasonable listening intensities to the average young adult listener, approximately 15 to 20,000 hertz. **2.** Pertaining to equipment for the recording, transmission, reproduction, or amplification of such sound. { 'ȯd·ē·ō }

audio frequency [ACOUS] A frequency that can be detected as a sound by the average young adult, approximately 15 to 20,000 hertz. Abbreviated af. Also known as sonic frequency; sound frequency. { ¦ȯd·ē·ō ¦frē·kwən·sē }

audio-frequency range [ACOUS] The range of frequencies to which the human ear is sensitive, approximately 15 to 20,000 hertz. Also known as audio range. { 'ȯd·ē·ō ¦frē·kwən·sē ,rānj }

audiogram [ACOUS] A graph showing hearing loss, percent hearing loss, or percent hearing as a function of frequency. { 'ȯd·ē·ō,gram }

audio image [ACOUS] A sound that originates, or appears to originate, at a certain point in space. { 'ȯd·ē·ō ,im·ij }

audioimpedance measurement [ACOUS] The measurement of acoustic impedance, as in the direct assessment of the dynamic motor control of sound feedback of different parts of the ear. { ¦ȯd·ē·ō,im'pēd·əns 'mezh·ər·mənt }

audiology [ACOUS] The science of hearing. { ,ȯd·ē'äl·ə·jē }

audio masking *See* masking. { 'ȯd·ē·ō ,mask·iŋ }

audiometry [ACOUS] The study of hearing ability by means of audiometers. { ,ȯd·ē'äm·ə·trē }

audio range *See* audio-frequency range. { 'ȯd·ē·ō ,rānj }

audio signal [ACOUS] An electric signal having the frequency of a mechanical wave that can be detected as a sound by the human ear. { 'ȯd·ē·ō ,sig·nəl }

auditory perspective [ACOUS] Three-dimensional realism of sound, as produced by an actual orchestra or by a stereophonic sound system. { 'ȯd·ə,tōr·ē pər'spek·tiv }

Auger coefficient [ATOM PHYS] The ratio of the number of Auger electrons to the number of ejected x-ray photons. { ō'zhā ,kō·ə'fish·ənt }

Auger effect [ATOM PHYS] A two-electron process in which an electron makes a discrete transition from a less bound shell to a vacant electron shell and the energy gained in this process is tranferred via the electrostatic interaction to another bound electron which escapes the atom. Also known as Auger transition; internal absorption; internal photoionization. { ō'zhā i,fekt }

Auger electron [ATOM PHYS] An electron that is expelled from an atom in the Auger effect. { ō'zhā i'lek,trän }

Auger electron spectroscopy [SPECT] The energy analysis of Auger electrons produced when an excited atom relaxes by a radiationless process after ionization by a high-energy electron, ion, or x-ray beam. Abbreviated AES. { ō'zhā i'lek,trän spek'träs·kə·pē }

Auger recombination [ATOM PHYS] Recombination of an electron and a hole in which no electromagnetic radiation is emitted, and the excess energy and momentum of the recombining electron and hole are given up to another electron or hole. { ō'zhā ri,käm·bə'nā·shən }

Auger transition *See* Auger effect. { ō'zhā tran ˌzish·ən }

augmentation distance [NUC PHYS] The extrapolation distance, which is the distance between the time boundary of a nuclear reactor and its boundary calculated by extrapolation. { ˌȯg·mən'tā·shən ˌdis·təns }

augmented plane-wave method [SOLID STATE] A method of approximating the energy states of electrons in a crystal lattice; the potential is assumed to be spherically symmetrical within spheres centered at each atomic nucleus and constant in the interstitial region, wave functions (the augmented plane waves) are constructed by matching solutions of the Schrödinger equation within each sphere with plane-wave solutions in the interstitial region, and linear combinations of these wave functions are then determined by the variational method. Abbreviated APW method. { ȯg¦ment·əd 'plān ˌwāv ˌmeth·əd }

A* unit [PHYS] An atomic standard unit of length, based on the tungsten $K\alpha_1$ line, approximately 10^{-11} centimeter; used for measurements of x-ray wavelengths and of crystal dimensions. { 'āˌstär ˌyü·nət }

aural masking *See* masking. { 'ȯr·əl 'mask·iŋ }

aural signal [ACOUS] **1.** A signal that can be heard. **2.** The sound portion of a television signal. { 'ȯr·əl 'sig·nəl }

aurora *See* corona discharge. { ə'rȯr·ə }

auroral line [SPECT] A prominent green line in the spectrum of the aurora at a wavelength of 5577 angstroms, resulting from a certain forbidden transition of oxygen. { ə'rȯr·əl ˌlīn }

austausch coefficient *See* exchange coefficient. { 'au̇sˌtau̇sh ˌkō·ə'fish·ənt }

autocollimation [OPTICS] A procedure for collimating a telescope or other optical instrument with objective and crosshairs, in which the instrument is directed toward a plane mirror and the crosshairs and lens are adjusted so that the crosshairs coincide with their reflected image. { ¦ȯd·ōˌkäl·ə'mā·shən }

autocollimator [OPTICS] **1.** A device by which a single lens collimates diverging light from a slit, and then focuses the light on an exit slit after it has passed through a prism to a mirror and been reflected back through the prism. **2.** A telescope which has a graduated reticle, enabling an observer to read off the angles subtended by distant objects. **3.** A convex mirror at the focus of the principal mirror of a reflecting telescope, which causes light to leave the telescope in a parallel beam. **4.** A telescope equipped with an eyepiece designed for autocollimation. { ¦ȯd·ō'käl·əˌmād·ər }

autodetachment [ATOM PHYS] A process in which one of the excited electrons in a doubly excited state of a negative ion is ejected, while the other excited electron drop back to a lower energy level. { ˌȯd·ō·di'tach·ment }

autofocus rectifier [OPTICS] A precise, vertical photoenlarger which permits the correction of distortion in an aerial negative caused by tilt. { 'ȯd·ōˌfō·kəs 'rek·təˌfī·ər }

autogenous electrification [PHYS] The process by which net charge is built up on an object, such as an airplane, moving relative to air containing dust or ice crystals; produced by frictional effects (triboelectrification) accompanying contact between the object and the particulate matter. { ȯ'täj·ə·nəs iˌlek·trə·fə'kā·shən }

autoionization [ATOM PHYS] The radiationless transition of an electron in an atom from a discrete electronic level to an ionized continuum level of the same energy. Also known as preionization. { ¦ȯd·ōˌī·ə·nə'zā·shən }

autoluminescence [ATOM PHYS] Luminescence of a material (such as a radioactive substance) resulting from energy originating within the material itself. { ¦ȯd·ōˌlü·mə'nes·əns }

automatic exposure [OPTICS] Photoelectric exposure control by a special device that maintains an essentially constant exposure in the focal plane for a given range of field luminance. { ¦ȯd·ə¦mad·ik ik'spō·zhər }

automatic focus [OPTICS] A device in a camera or enlarger which automatically keeps the objective lens in focus through a range of magnification. { ¦ȯd·ə¦mad·ik 'fō·kəs }

autorotation [MECH] **1.** Rotation about any axis of a body that is symmetrical and exposed to a uniform airstream and maintained only by aerodynamic moments. **2.** Rotation of a stalled symmetrical airfoil parallel to the direction of the wind. { ¦ȯd·ōˌrō'tā·shən }

aV *See* abvolt.

availability [PHYS] The difference between the enthalpy per unit mass of substance and the product of entropy per unit mass multiplied by the lowest temperature available to the substance for heat discard; used in determining the ratio of actual work performed during a process by a working substance to that which theoretically should have been performed. { əˌvāl·ə'bil·ə·dē }

aV/cm *See* abvolt per centimeter.

average life *See* mean life. { 'av·rij 'līf }

Avogadro's hypothesis *See* Avogadro's law. { ¦a·və¦gäd·drōz hī'päth·ə·səs }

Avogadro's law [PHYS] The law which states that under the same conditions of pressure and temperature, equal volumes of all gases contain equal numbers of molecules; for example, 359 cubic feet at 32°F and 1 atmosphere for a perfect gas. Also known as Avogadro's hypothesis. { ¦a·və¦gäd·drōz ˌlȯ }

Avogadro's number [PHYS] The number (6.02×10^{23}) of molecules in a gram-molecular weight of a substance. { ¦a·və¦gäd·drōz ˌnəm·bər }

avogram [MECH] A unit of mass, equal to 1 gram divided by the Avogadro number. { 'a·vəˌgram }

avoirdupois pound *See* pound. { ˌav·ərd·ə'pȯiz 'pau̇nd }

avoirdupois weight [MECH] The system of units which has been commonly used in English-speaking countries for measurement of the mass of any substance except precious stones, precious metals, and drugs; it is based on the pound (approximately 453.6 grams) and includes the short ton (2000 pounds), long ton (2240 pounds), ounce (one-sixteenth pound), and dram (one-sixteenth ounce). { ˌav·ərd·ə'pȯiz 'wāt }

axes of inertia [PHYS] The three principal axes of inertia, namely, one about which the moment of inertia is a maximum, one about which the moment of inertia is a minimum, and one perpendicular to both. { 'akˌsēz əv in'ər·shə }

axial angle [CRYSTAL] **1.** The acute angle between the two optic axes of a biaxial crystal. Also known as optic angle; optic-axial angle. **2.** In air, the larger angle between the optic axes after refraction on leaving the crystal. { 'ak·sē·əl 'aŋ·gəl }

axial element [CRYSTAL] The lengths, length ratios, and angles which define a crystal's unit cell. { 'ak·sē·əl 'el·ə·mənt }

axial flow [FL MECH] Flow of fluid through an axially symmetric device such that the direction of the flow is along the axis of symmetry. Also known as axisymmetric flow. { 'ak·sē·əl 'flō }

axial jet [FL MECH] A flowing, turbulent stream which mixes with standing water in three dimensions. { 'ak·sē·əl 'jet }

axial load [MECH] A force with its resultant passing through the centroid of a particular section and being perpendicular to the plane of the section. { 'ak·sē·əl 'lōd }

axial modulus [MECH] The ratio of a simple tension stress applied to a material to the resulting strain parallel to the tension when the sides of the sample are restricted so that there is no lateral deformation. Also known as modulus of simple longitudinal extension. { ¦ak·sē·əl 'mäj·ə·ləs }

axial moment of inertia [MECH] For any object rotating about an axis, the sum of its component masses times the square of the distance to the axis. { 'ak·sē·əl 'mō·mənt əv in'ər·shə }

axial plane [CRYSTAL] **1.** A plane that includes two of the crystallographic axes. **2.** The plane of the optic axis of an optically biaxial crystal. { 'ak·sē·əl 'plān }

axial quadrupole *See* longitudinal quadrupole. { 'ak·sē·əl 'kwäd·rəˌpōl }

axial ratio [CRYSTAL] The ratio obtained by comparing the length of a crystallographic axis with one of the lateral axes taken as unity. { 'ak·sē·əl 'rā·shō }

axial vector *See* pseudovector. { 'ak·sē·əl 'vek·tər }

axiomatic S-matrix theory [PART PHYS] An approach to the study of elementary particles that seeks to formulate S-matrix theory in a rigorous manner based on a few fundamental axioms that include Lorentz invariance, unitarity, analyticity near the physical values of the energy and momentum variables, and singularities in the physical region that correspond to known particles and scattering thresholds. { ¦ak·sē·ə¦mad·ik 'es ˌmā·triks ˌthē·ə·rē }

axion [PART PHYS] A hypothetical neutral pseudoscalar boson with mass roughly of order 100 keV to 1 MeV, postulated to preserve the parity and time-reversal invariance of strong interactions, despite the effects of instantons. { 'ak·seˌän }

axis [MECH] A line about which a body rotates. { 'ak·səs }

axis of acoustic symmetry [ACOUS] An axis such that the three-dimensional directivity pattern of a transducer may be generated by rotating a two-dimensional directivity pattern around it. Also known as acoustic axis. { 'ak·səs əv ə'kü·stik 'sim·ə·trē }

axis of circulation [ELECTROMAG] The axis where the equiphase surfaces of a circulating electromagnetic wave converge. { ¦ak·səs əv ˌsər·kyə'lā·shən }

axis of rotation [MECH] A straight line passing through the points of a rotating rigid body that remain stationary, while the other points of the body move in circles about the axis. { 'ak·səs əv rō'tā·shən }

axis of symmetry [MECH] An imaginary line about which a geometrical figure is symmetric. Also known as symmetry axis. { 'ak·səs əv 'sim·ə·trē }

axis of torsion [MECH] An axis parallel to the generators of a cylinder undergoing torsion, located so that the displacement of any point on the axis lies along the axis. Also known as axis of twist. { ¦ak·səs əv 'tȯr·shən }

axis of twist *See* axis of torsion. { ¦ak·səs əv 'twist }

axisymmetric flow *See* axial flow. { ¦ak·sə·sə¦me·trik 'flō }

axisymmetrization [FL MECH] The rounding of noncircular vortices in the absence of strong external deformation. Also known as aggregation. { ¦ak·səˌsim·ə·trə'zā·shən }

Ayrton-Jones balance [ELEC] A type of balance with which force between current-carrying conductors is measured; uses single-layer solenoids as the fixed and movable coils. { ¦er·tən ¦jōnz 'bal·əns }

Ayrton-Perry winding [ELEC] Winding of two wires in parallel but opposite directions to give better cancellation of magnetic fields than is obtained with a single winding. { ¦er·tən ¦per·ē ˌwind·iŋ }

azimuthal quantum number [ATOM PHYS] The orbital angular momentum quantum number l, such that the eigenvalue of L^2 is $l(l + 1)$. { ˌaz·ə'məth·əl ¦kwän·təm ˌnəm·bər }

azimuth resolution [ELECTROMAG] Angle or distance by which two targets must be separated in azimuth to be distinguished by a radar set, when the targets are at the same range. { 'az·ə·məth ˌrez·ə'lü·shən }

B

b *See* bel.

B *See* bel; brewster.

Babcock coefficient of friction [FL MECH] An approximation to the coefficient of friction for steam flowing in a circular pipe of diameter *d* inches, given by $0.0027[1 + (3.6/d)]$. { 'bab,käk ,kō·ə'fish·ənt əv 'frik·shən }

Babinet compensator [OPTICS] A device for working with polarized light, made of two quartz prisms, assembled in a rhomb, to enable the optical retardation to be adjusted to positive or negative values. { bä·bi'nā ,käm·pen'sād·ər }

Babinet point [OPTICS] A neutral point located 15 to 20° directly above the sun. { bä·bi'nā ,pȯint }

Babinet's principle [OPTICS] The principle that the diffraction patterns produced by complementary screens are identical; two screens are said to be complementary when the opaque parts of one correspond to the transparent parts of the other. { bä·bi'nāz ,prin·sə·pəl }

backbending [NUC PHYS] A discontinuity in the rotational levels of some rare-earth nuclei around spin $20\hbar$ (where $\hbar$ is Planck's constant divided by 2π), which appears as a backbend on a graph that plots the moment of inertia versus the square of the rotational frequency. { 'bak ,bend·iŋ }

back bond [SOLID STATE] A chemical bond between an atom in the surface layer of a solid and an atom in the second layer. { 'bak ,bänd }

back-coated mirror [OPTICS] Glass with a reflective coating applied against the rear surface. { 'bak ,kōd·əd 'mir·ər }

back electromotive force *See* counterelectromotive force. { 'bak i¦lek·trō¦mōd·iv 'fȯrs }

backfire antenna [ELECTROMAG] An antenna which exhibits significant gain in a direction 180° from its principal lobe. { 'bak,fīr an'ten·ə }

backflow [FL MECH] Any flow in a direction opposite to the natural or intended direction of flow. { 'bak,flō }

back focal length [OPTICS] The distance from the rear surface of a lens to its focal plane. { 'bak ¦fō·kəl ,leŋkth }

Back-Goudsmit effect [ATOM PHYS] Breakdown of the coupling between the nuclear-spin angular momentum and the total angular momentum of the electrons in an atom at relatively small magnetic fields. { ¦bak 'gōd,smit i,fekt }

background count [PHYS] Responses of the radiation counting system to radiation coming from sources other than the source to be measured. { 'bak,graůnd ,kaůnt }

background luminance [OPTICS] In visual-range theory, the brightness of the background against which a target is viewed. { 'bak,graůnd 'lüm·i·nəns }

background mass spectrum [PHYS] The display or printed record obtained from a mass spectrometer or spectrograph before a sample has been inserted. { ¦bak,graůnd ¦mas 'spek·trəm }

background noise [ACOUS] The unwanted residual sound that is present whether or not the sound source being studied is in operation. { 'bak,graůnd ,nȯiz }

background radiation [PHYS] Radiation which is due to sources other than the source of interest in a measurement of radiation and which is detected by the measuring apparatus. { 'bak ,graůnd ,rād·ē'ā·shən }

back lobe [ELECTROMAG] The three-dimensional portion of the radiation pattern of a directional antenna that is directed away from the intended direction. { 'bak ,lōb }

back pressure [MECH] Pressure due to a force that is operating in a direction opposite to that being considered, such as that of a fluid flow. { 'bak ,presh·ər }

back-reflection photography [CRYSTAL] A method of studying crystalline structure by x-ray diffraction in which the photographic film is placed between the source of x-rays and the crystal specimen. { 'bak ri'flek·shən fə'täg·rə·fē }

backscattering Also known as back radiation; backward scattering. [ELECTROMAG] **1.** Radar echoes from a target. **2.** Undesired radiation of energy to the rear by a directional antenna. [PHYS] The deflection of radiation or nuclear particles by scattering processes through angles greater than 90° with respect to the original direction of travel. { 'bak¦skad·ə·riŋ }

back-to-front ratio [ELECTROMAG] Ratio used in connection with an antenna, metal rectifier, or any device in which signal strength or resistance in one direction is compared with that in the opposite direction. { ¦bak tə ¦frənt 'rā·shō }

backward wave [ELECTROMAG] An electromagnetic wave traveling opposite to the direction of motion of some other physical quantity in an electronic device such as a traveling-wave tube or mismatched transmission line. { 'bak·wərd ˌwāv }

baffle plate [ELECTROMAG] Metal plate inserted in a waveguide to reduce the cross-sectional area for wave conversion purposes. { 'baf·əl ˌplāt }

Bain-Ondarçuhu device [FL MECH] An inclined plane with a self-running droplet that coats its upper surface. { ¦bān ˌōn'dä·səˌhü diˌvīs }

Bairstow number [FL MECH] A term previously used for Mach number. { 'berˌstō ˌnəm·bər }

Baker-Nunn camera [OPTICS] A large camera with a Schmidt-type lens system used to track earth satellites. { ¦bāk·ər ¦nən 'kam·rə }

Baker-Schmidt telescope [OPTICS] A type of Schmidt telescope in which the light reflected from the near-spheroidal primary mirror is again reflected from a smaller, near-spheroidal secondary mirror, producing an image that is free of astigmatism and distortion. { ¦bāk·ər ¦shmit 'tel·əˌskōp }

balance [ACOUS] The condition in a stereo system wherein both speakers produce the same average sound levels. [ELEC] The state of an electrical network when it is adjusted so that voltage in one branch induces or causes no current in another branch. { 'bal·əns }

balance coil [ELEC] An iron-core solenoid with adjustable taps near the center; used to convert a two-wire circuit to a three-wire circuit, the taps furnishing a neutral terminal for the latter. { 'bal·əns ˌkȯil }

balanced bridge [ELEC] Wheatstone bridge circuit which, when in a quiescent state, has an output voltage of zero. { 'bal·ənst 'brij }

balanced circuit [ELEC] **1.** A circuit whose two sides are electrically alike and symmetrical with respect to a common reference point, usually ground. **2.** An electric circuit that has been adjusted to neutralize the mutual induction of an adjacent circuit. { 'bal·ənst 'sər·kət }

balanced voltages [ELEC] Voltages that are equal in magnitude and opposite in polarity with respect to ground. Also known as push-pull voltages. { 'bal·ənst ˌvōl·tij·əz }

ballast resistor [ELEC] A resistor that increases in resistance as current through it increases, and decreases in resistance as current decreases. Also known as barretter (British usage). { 'bal·əst ri'sis·tər }

ballistic camera [OPTICS] A ground-based camera using multiple exposures on the same plate to record the trajectory of a rocket. { bə'lis·tik ˌkam·rə }

ballistic coefficient [MECH] The numerical measure of the ability of a missile to overcome air resistance; dependent upon the mass, diameter, and form factor. { bə'lis·tik ˌkō·ə'fish·ənt }

ballistic conditions [MECH] Conditions which affect the motion of a projectile in the bore and through the atmosphere, including muzzle velocity, weight of projectile, size and shape of projectile, rotation of the earth, density of the air, temperature or elasticity of the air, and the wind. { bə'lis·tik kən'dish·əns }

ballistic curve [MECH] The curve described by the path of a bullet, a bomb, or other projectile as determined by the ballistic conditions, by the propulsive force, and by gravity. { bə'lis·tik 'kərv }

ballistic deflection [MECH] The deflection of a missile due to its ballistic characteristics. { bə'lis·tik di'flek·shən }

ballistic density [MECH] A representation of the atmospheric density encountered by a projectile in flight, expressed as a percentage of the density according to the standard artillery atmosphere. { bə'lis·tik 'den·səd·ē }

ballistic efficiency [MECH] **1.** The ability of a projectile to overcome the resistance of the air; depends chiefly on the weight, diameter, and shape of the projectile. **2.** The external efficiency of a rocket or other jet engine of a missile. { bə'lis·tik i'fish·ən·sē }

ballistic entry [MECH] Movement of a ballistic body from without to within a planetary atmosphere. { bə'lis·tik 'en·trē }

ballistic limit [MECH] The minimum velocity at which a particular armor-piercing projectile is expected to consistently and completely penetrate armor plate of given thickness and physical properties at a specified angle of obliquity. { bə'lis·tik 'lim·ət }

ballistic measurement [MECH] Any measurement in which an impulse is applied to a device such as the bob of a ballistic pendulum, or the moving part of a ballistic galvanometer, and the subsequent motion of the device is used to determine the magnitude of the impulse, and, from this magnitude, the quantity to be measured. { bə'lis·tik 'mezh·ər·mənt }

ballistics [MECH] Branch of applied mechanics which deals with the motion and behavior characteristics of missiles, that is, projectiles, bombs, rockets, guided missiles, and so forth, and of accompanying phenomena. { bə'lis·tiks }

ballistics of penetration [MECH] That part of terminal ballistics which treats of the motion of a projectile as it forces its way into targets of solid or semisolid substances, such as earth, concrete, or steel. { bə'lis·tiks əv pen·ə'trā·shən }

ballistic table [MECH] Compilation of ballistic data from which trajectory elements such as angle of fall, range to summit, time of flight, and ordinate at any time, can be obtained. { bə'lis·tik 'tā·bəl }

ballistic temperature [MECH] That temperature (in °F) which, when regarded as a surface temperature and used in conjunction with the lapse rate of the standard artillery atmosphere, would produce the same effect on a projectile as the actual temperature distribution encountered by the projectile in flight. { bə'lis·tik 'tem·prə·chər }

ballistic trajectory [MECH] The trajectory followed by a body being acted upon only by gravitational forces and resistance of the medium through which it passes. { bə'lis·tik trə'jek·tə·rē }

ballistic uniformity [MECH] The capability of a propellant, when fired under identical conditions from round to round, to impart uniform muzzle velocity and produce similar interior ballistic results. { bə'lis·tik ˌyü·nə'fȯr·məd·ē }

ballistic wave [MECH] An audible disturbance caused by compression of air ahead of a missile in flight. { bə'lis·tik ˌwāv }

ballistic wind [MECH] That constant wind which would produce the same effect upon the trajectory of a projectile as the actual wind encountered in flight. { bə'lis·tik 'wind }

balloting [MECH] A tossing or bounding movement of a projectile, within the limits of the bore diameter, while moving through the bore under the influence of the propellant gases. { 'bal·əd·iŋ }

Balmer continuum [SPECT] A continuous range of wavelengths (or wave numbers or frequencies) in the spectrum of hydrogen at wavelengths less than the Balmer limit, resulting from transitions between states with principal quantum number $n = 2$ and states in which the single electron is freed from the atom. { ¦bäl·mər kən'tin·yə·wəm }

Balmer discontinuity *See* Balmer jump. { 'bȯl·mər disˌkänt·ən'ü·əd·ē }

Balmer formula [SPECT] An equation for the wavelengths of the spectral lines of hydrogen, $1/\lambda = R[(1/m^2) - (1/n^2)]$, where λ is the wavelength, R is the Rydberg constant, and m and n are positive integers (with n larger than m) that give the principal quantum numbers of the states between which occur the transition giving rise to the line. { 'bȯl·mər ˌfȯr·myə·lə }

Balmer jump [SPECT] The sudden decrease in the intensity of the continuous spectrum of hydrogen at the Balmer limit. Also known as Balmer discontinuity. { 'bȯl·mər ˌjəmp }

Balmer limit [SPECT] The limiting wavelength toward which the lines of the Balmer series crowd and beyond which they merge into a continuum, at approximately 365 nanometers. { 'bȯl·mər ˌlim·ət }

Balmer lines [SPECT] Lines in the hydrogen spectrum, produced by transitions between $n = 2$ and $n > 2$ levels either in emission or in absorption; here n is the principal quantum number. { 'bȯl·mər ˌlīnz }

Balmer series [SPECT] The set of Balmer lines. { 'bȯl·mər ¦sir·ēz }

band [SOLID STATE] A restricted range in which the energies of electrons in solids lie, or from which they are excluded, as understood in quantum-mechanical terms. Also known as energy bands. [SPECT] *See* band spectrum. { band }

band gap [SOLID STATE] An energy difference between two allowed bands of electron energy in a metal. { 'band ˌgap }

band head [SPECT] A location on the spectrogram of a molecule at which the lines of a band pile up. { 'band ˌhed }

band-pass filter *See* Christiansen filter. { 'band ˌpas ˌfil·tər }

band scheme [SOLID STATE] The identification of energy bands of a solid with the levels of independent atoms from which they arise as the atoms are brought together to form the solid, together with the width and spacing of the bands. { 'band ˌskēm }

band sound-pressure level [ACOUS] The sound-pressure level that results from the portion of sound within a specified frequency band. Abbreviated BSPL. { ¦band 'saund ˌpresh·ər ˌlev·əl }

band spectrum [SPECT] A spectrum consisting of groups or bands of closely spaced lines in emission or absorption, characteristic of molecular gases and chemical compounds. Also known as band. { 'band ˌspek·trəm }

band theory of ferromagnetism [SOLID STATE] A theory according to which ferromagnetism is caused by electrons in the unfilled energy bands of a crystal. { ¦band ˌthē·ə·rē əv ˌfer·ō'mag·nəˌtiz·əm }

band theory of solids [SOLID STATE] A quantum-mechanical theory of the motion of electrons in solids that predicts certain restricted ranges or bands for the energies of these electrons. Also known as energy-band theory of solids. { 'band ˌthē·ə·rē əv ¦säl·ədz }

bar [MECH] A unit of pressure equal to 10^5 pascals, or 10^5 newtons per square meter, or 10^6 dynes per square centimeter. { bär }

Bardeen-Cooper-Schrieffer theory [SOLID STATE] A theory of superconductivity that describes quantum-mechanically those states of the system in which conduction electrons cooperate in their motion so as to reduce the total energy appreciably below that of other states by exploiting their effective mutual attraction; these states predominate in a superconducting material. Abbreviated BCS theory. { ¦bärˌdēn ¦kü·pər ¦shrē·fər ˌthē·ə·rē }

bare charm [PART PHYS] Charm that is carried by a quark and is not canceled by the charm of the corresponding antiquark, so that the hadron of which the quark is a constituent has net charm different from zero. { ¦ber ¦chärm }

bare value [QUANT MECH] The value which some physical property of a particle, such as its mass or charge, is supposed to have in the absence of any interactions with fields. { ¦ber 'val·yü }

barium-140 [NUC PHYS] A radioactive isotope of barium with atomic mass 140; the half-life is 12.8 days, and the decay is by negative beta-particle emission. { 'bar·ē·əm ˌwən'fȯrd·ē }

Barker method [CRYSTAL] A method utilizing a number of convenient rules which allow two observers to choose the same reference system to describe the same noncubic crystal. { 'bär·kər ˌmeth·əd }

Barkhausen effect [ELECTROMAG] The succession of abrupt changes in magnetization occurring when the magnetizing force acting on a piece of iron or other magnetic material is varied. { 'bärkˌhau̇z·ən i'fekt }

Barlow lens [OPTICS] A lens with one plane surface and one concave surface that is placed between the objective and eyepiece of a telescope to decrease the convergence of the beam from the objective and thereby increase the effective focal length. { 'bärˌlō ˌlenz }

Barlow's equation [MECH] A formula, $t = DP/2S$, used in computing the strength of cylinders subject to internal pressures, where t is the thickness of the cylinder in inches, D the outside diameter in inches, P the pressure in pounds per square inch, and S the allowable tensile strength in pounds per square inch. { 'bärˌlōz i'kwā·zhən }

bar magnet [ELECTROMAG] A bar of hard steel that has been strongly magnetized and holds its magnetism, thereby serving as a permanent magnet. { 'bär ˌmag·nət }

barn [NUC PHYS] A unit of area equal to 10^{-24} square centimeter; used in specifying nuclear cross sections. Symbolized b. { 'bärn }

Barnett effect [ELECTROMAG] The development of a slight magnetization in an initially unmagnetized iron rod when it is rotated at high speed about its axis. { 'bär·nit i'fekt }

Barnett method [ELECTROMAG] Use of the Barnett effect to determine the gyromagnetic moment of ferromagnetic material. { 'bär·nit ¦meth·əd }

baroclinic [PHYS] Of, pertaining to, or characterized by baroclinity. { ¦bar·ə¦klin·ik }

baroclinicity *See* baroclinity. { ˌbar·ə·klə'nis·əd·ē }

baroclinity [PHYS] The state of stratification in a fluid in which surfaces of constant pressure (isobaric surfaces) intersect surfaces of constant density (isosteric surfaces). Also known as baroclinicity; barocliny. { ˌbar·ə'klin·əd·ē }

barocliny *See* baroclinity. { ˌbar·ə'klin·ē }

barodynamics [MECH] The mechanics of heavy structures which may collapse under their own weight. { ˌbar·ə·dī'nam·iks }

barometric [PHYS] Loosely, pertaining to atmospheric pressure; for example, barometric gradient (meaning pressure gradient). { bar·ə'me·trik }

barometric corrections [PHYS] The corrections which must be applied to the reading of a mercury barometer in order that the observed value may be rendered accurate. Also known as barometric errors. { bar·ə'met·rik kə'rek·shənz }

barometric errors *See* barometric corrections. { bar·ə'met·rik 'er·ərz }

barometric gradient *See* pressure gradient. { bar·ə'met·rik 'grād·ē·ənt }

barometric pressure *See* atmospheric pressure. { bar·ə'met·rik 'presh·ər }

barometric surface [PHYS] A surface at each point of which the barometric pressure is the same. { bar·ə'met·rik 'sər·fəs }

baromil [MECH] The unit of length used in graduating a mercury barometer in the centimeter-gram-second system. { 'bar·əˌmil }

barotropic [PHYS] Of, pertaining to, or characterized by a condition of barotropy. { ˌbar·ə'träp·ik }

barotropic phenomenon [THERMO] The sinking of a vapor beneath the surface of a liquid when the vapor phase has the greater density. { ˌbar·ə'träp·ik fə'näm·əˌnän }

barotropy [PHYS] The state of a fluid in which surfaces of constant density (or temperature) are coincident with surfaces of constant pressure; it is the state of zero baroclinity. { bə'rä·trə·pē }

barrel [MECH] Abbreviated bbl. **1.** The unit of liquid volume equal to 31.5 gallons (approximately 119 liters). **2.** The unit of liquid volume for petroleum equal to 42 gallons (approximately 158 liters). **3.** The unit of dry volume equal to 105 quarts (approximately 116 liters). **4.** A unit of weight that varies in size according to the commodity being weighed. [OPTICS] A tapering cylindrical housing which contains the lenses of a camera and the iris diaphragm. { 'bar·əl }

barrel distortion [OPTICS] A defect in an optical system whereby lateral magnification decreases with object size; the image of a square then appears barrel-shaped. { 'bar·əl dis'tȯr·shən }

barretter [ELEC] **1.** Bolometer that consists of a fine wire or metal film having a positive temperature coefficient of resistivity, so that resistance increases with temperature; used for making power measurements in microwave devices. **2.** *See* ballast resistor. { bə'red·ər }

barrier *See* potential barrier. { 'bar·ē·ər }

barrier penetration [QUANT MECH] The passage of a particle through a potential barrier, that is, through a region of finite extent in which the particle's potential energy is greater than its total energy. { 'bar·ē·ər ˌpen·ə'trā·shən }

Bartlett force [NUC PHYS] A force between nucleons in which spin is exchanged. { 'bärt·lət ˌfȯrs }

barycentric energy [MECH] The energy of a system in its center-of-mass frame. { ˌbar·ə'sen·trik 'en·ər·jē }

barye [MECH] The pressure unit of the centimeter-gram-second system of physical units; equal to 1 dyne per square centimeter (0.001 millibar). Also known as microbar. { 'ba·rē }

baryon [PART PHYS] Any elementary particle which can be transformed into a nucleon and some number of mesons and lighter particles. Also known as heavy particle. { 'bar·ēˌän }

baryon number [PART PHYS] A quantum number equal to the number of baryons minus the number of antibaryons in a system; it is conserved at the present level of detection, but may not be exactly conserved. { 'bar·ēˌän ˌnəm·bər }

baryon octet [PART PHYS] The group of one lambda, three sigma, and two xi hyperons and two nucleons, all having spin 1/2 and positive parity, and forming a symmetrical pattern as suggested by SU_3 symmetry. { 'bar·ē,än äk'tet }

baryon resonance [PART PHYS] A cross section anomaly indicating the existence of an unstable baryon. { 'bar·ē,än 'rez·ən·əns }

baryon spectroscopy [PART PHYS] The science of the energy levels and changes of state occurring among baryon particles. { 'bar·ē,än spek'träs·kə·pē }

barytropic gas [PHYS] A gas whose pressure depends only on its density. { ¦bar·ə¦träp·ik 'gas }

basal cleavage [CRYSTAL] Cleavage parallel to the base of the crystal structure or to the lattice plane which is normal to one of the lattice axes. { 'bā·səl 'klēv·ij }

basal orientation [CRYSTAL] A crystal orientation in which the surface is parallel to the base of the lattice or to the lattice plane which is normal to one of the lattice axes. { 'bā·səl ,ȯr·ē·ən'tā·shən }

basal plane [CRYSTAL] The plane perpendicular to the long, or *c*, axis in all crystals except those of the isometric system. { 'bā·səl ¦plān }

baseball [PL PHYS] A machine used in controlled fusion research to confine a plasma; consists of a linear magnetic bottle sealed by magnetic mirrors at both ends, and has current-carrying structures, which resemble the seams of a baseball in shape, to stabilize the plasma. { 'bās,bȯl }

base-centered lattice [CRYSTAL] A space lattice in which each unit cell has lattice points at the centers of each of two opposite faces as well as at the vertices; in a monoclinic crystal, they are the faces normal to one of the lattice axes. { 'bās ,sen·tərd 'lad·əs }

base density [OPTICS] The value of the inherent optical transmission density of a film base; does not include any contribution from the emulsion layer. { 'bās 'den·səd·ē }

base drag [FL MECH] Drag owing to a base pressure lower than the ambient pressure; it is a part of the pressure drag. { 'bās ,drag }

base-loaded antenna [ELECTROMAG] Vertical antenna having an impedance in series at the base for loading the antenna to secure a desired electrical length. { 'bās ,lōd·əd an'ten·ə }

base magnification [OPTICS] The ratio of the distance between the centers of the objectives of a pair of binoculars to the distance between the centers of the eyepieces. { 'bās ,mag·nə·fə'kā·shən }

base peak [SPECT] The tallest peak in a mass spectrum; it is assigned a relative intensity value of 100, and lesser peaks are reported as a percentage of it. { 'bās ,pēk }

base-plus-fog density [OPTICS] The value of the inherent optical transmission density of a film base plus the nonimage density contributed by the developed emulsion. { 'bās ,pləs 'fäg ,den·səd·ē }

base pressure [FL MECH] The pressure exerted on the base or extreme aft end of a body, as of a cylindrical or boat-tailed body or of a blunt-trailing-edge wing in fluid flow. [MECH] A pressure used as a reference base, for example, atmospheric pressure. { 'bās ¦presh·ər }

base quantity [PHYS] One of a small number of physical quantities in a system of measurement that are defined, independent of other physical quantities, by means of a physical standard and by procedures for comparing the quantity to be measured with the standard. Also known as fundamental quantity. { 'bās 'kwän·ə·tē }

base unit [PHYS] One of a small number of units in a system of measurement that are defined, independent of other units, by means of a physical standard; equivalently, a unit of a base quantity. Also known as fundamental unit. { 'bās ,yü·nət }

basic frequency [PHYS] The frequency, in any wave, which is considered the most important; in a driven system, it would generally be the driving frequency, while in most periodic waves it would correspond to the fundamental frequency. { 'bā·sik 'frē·kwən·sē }

basic Q *See* nonloaded Q. { 'bā·sik 'kyü }

basic truss [MECH] A framework of bars arranged so that for any given loading of the bars the forces on the bars are uniquely determined by the laws of statics. { ¦bās·ik 'trəs }

basin of attraction [PHYS] The collection of all possible initial conditions of a dynamical system for which the trajectories representing that system in phase space will converge to a particular attractor. { 'bās·ən əv ə'trak·shən }

bass [ACOUS] Sounds having frequencies at the lower end of the audio range, below about 250 hertz. { bās }

Batchinsky relation [FL MECH] The relation stating that the fluidity of a liquid is proportional to the difference between the specific volume and a characteristic specific volume, approximately equal to the specific volume appearing in the van der Waals equation. { ba'chin·skē ri,lā·shən }

Bateman equations [NUC PHYS] A set of equations that give the number of atoms of each nuclide of a radioactive decay chain produced after a specified time, when a specified number of atoms of the parent nuclide are initially present. { 'bāt·mən i,kwā·zhənz }

battery [ELEC] A direct-current voltage source made up of one or more units that convert chemical, thermal, nuclear, or solar energy into electrical energy. { 'bad·ə·rē }

battery command periscope [OPTICS] An optical instrument consisting of dual telescope tubes positioned vertically on a common mounting; it provides periscopic vision for the observer, and may be used to observe artillery fire. { 'bad·ə·rē kə¦mand 'per·ə,skōp }

Baveno twin law [CRYSTAL] An uncommon twin law applicable in feldspar, in which the twin plane and composition surface are (021); a Baveno twin usually consists of two individuals. { bə'vē·nō ,twin ,lȯ }

b axis [CRYSTAL] A crystallographic axis that is oriented horizontally, right to left. { 'bē ,ak·səs }

bay [ELECTROMAG] One segment of an antenna array. { bā }

b-complete [RELAT] A criterion determining whether a space-time is free of singularities based on whether curves of finite length have an end point, where length is defined by a generalized affine parameter along the curve. { ¦bē kəm¦plēt }

BE *See* binding energy.

bead [ELECTROMAG] A glass, ceramic, or plastic insulator through which passes the inner conductor of a coaxial transmission line and by means of which the inner conductor is supported in a position coaxial with the outer conductor. { bēd }

beam [PHYS] A concentrated, nearly unidirectional flow of particles, or a like propagation of electromagnetic or acoustic waves. { bēm }

beam angle *See* beam width. { 'bēm ,aŋ·gəl }

beam antenna [ELECTROMAG] An antenna that concentrates its radiation into a narrow beam in a definite direction. { 'bēm an'ten·ə }

beam attenuator [SPECT] An attachment to the spectrophotometer that reduces reference to beam energy to accommodate undersized chemical samples. { 'bēm ə'ten·yə,wād·ər }

beam-condensing unit [SPECT] An attachment to the spectrophotometer that condenses and remagnifies the beam to provide reduced radiation at the sample. { 'bēm kən'den·siŋ ,yü·nət }

beam diameter [OPTICS] The distance between points on opposite sides of a light beam at which the power per unit area drops to $1/e$ (0.37) times its maximum value. { 'bēm dī,am·əd·ər }

beam divergence [OPTICS] The angle between points on opposite sides of a light beam at which the irradiance drops to $1/e$ (0.37) times its maximum value. [PHYS] The angular spread in the directions of the components of a beam of particles or radiation. { 'bēm də,vər·jəns }

beam drop [ELECTROMAG] Distortion of the normal rectilinear fan pattern of a detection radar in which a portion of the fan is at a lower elevation than the rest of the fan. { 'bēm ,dräp }

beam edge [PHYS] The locus of positions at which the intensity of a beam of particles or radiation is 10% of that along the axis of the beam. { 'bēm ,ej }

beam efficiency [ELECTROMAG] The fraction of the total radiated energy from an antenna contained in a single beam. { 'bēm i,fish·ən·sē }

beam expander [OPTICS] A combination of optical elements used to increase the diameter of a laser beam or other light beam. { 'bēm ik ,span·dər }

beam-foil spectroscopy [ATOM PHYS] A method of studying the structure of atoms and ions in which a beam of ions energized in a particle accelerator passes through a thin carbon foil from which the ions emerge with various numbers of electrons removed and in various excited energy levels; the light or Auger electrons emitted in the deexcitation of these levels are then observed by various spectroscopic techniques. Abbreviated BFS. { 'bēm ,fȯil spek 'träs·kə·pē }

beamguide [ELECTROMAG] A set of elements arranged and spaced so as to form and conduct a beam of electromagnetic radiation. { 'bēm,gīd }

beam resonator [OPTICS] A device which acts to confine a laser beam or other beam of electromagnetic radiation to a given region of space without continuous guidance along the beam. Also known as open resonator. { 'bēm 'rez·ən,ād·ər }

beam splitter [OPTICS] A mirror that reflects part of a beam of light falling on it and transmits part. { 'bēm ,splid·ər }

beam splitting [OPTICS] The division of a beam of light into two beams by placing a special type of mirror in the path of the beam that reflects part of the light falling on it and transmits part. { 'bēm ,splid·iŋ }

beam width [ELECTROMAG] The angle, measured in a horizontal plane, between the directions at which the intensity of an electromagnetic beam, such as a radar or radio beam, is one-half its maximum value. Also known as beam angle. { 'bēm ,width }

bearing capacity [MECH] Load per unit area which can be safely supported by the ground. { 'ber·iŋ kə'pas·əd·ē }

bearing pressure [MECH] Load on a bearing surface divided by its area. Also known as bearing stress. { 'ber·iŋ ,presh·ər }

bearing strain [MECH] The deformation of bearing parts subjected to a load. { 'ber·iŋ ,strān }

bearing strength [MECH] The maximum load that a column, wall, footing, or joint will sustain at failure, divided by the effective bearing area. { 'ber·iŋ ,strenkth }

bearing stress *See* bearing pressure. { 'ber·iŋ ,stres }

beat [PHYS] The periodic variation in amplitude of a wave that is the superposition of two simple harmonic waves of different frequencies. { bēt }

Beattie and Bridgman equation [THERMO] An equation that relates the pressure, volume, and temperature of a real gas to the gas constant. { ¦bēd·ē ən ¦brij·mən i'kwā·zhən }

beauty quark *See* bottom quark. { 'byüd·ē ,kwärk }

Becker and Kornetzki effect [PHYS] A reduction in the internal friction of a ferromagnetic substance when it is subjected to a magnetic field that is large enough to produce magnetic saturation. { 'bek·ər ən ,kȯr'nets·kē i'fekt }

Becquerel effect [ELEC] The phenomenon of a current flowing between two unequally illuminated electrodes of a certain type when they are immersed in an electrolyte. [OPTICS] *See* paramagnetic Faraday effect. { ¦bek·ə¦rel *or* be'krel i'fekt }

Becquerel rays [NUC PHYS] Formerly, radiation emitted by radioactive substances; later renamed alpha, beta, and gamma rays. { ¦bek·ə¦rel *or* be'krel ,rāz }

bedspring array *See* billboard array. { 'bed,spriŋ ə'rā }

Békésy audiometry [ACOUS] A subject-controlled auditory threshold testing procedure. { 'bā,kā·shē ȯd·ē'äm·ə·trē }

bel [PHYS] A dimensionless unit expressing the ratio of two powers or intensities, or the ratio of a power to a reference power, such that the number of bels is the common logarithm of this ratio. Symbolized b; B. { bel }

bellows [OPTICS] An accordionlike component of a camera which forms a passage between the lens and the film and allows one to vary the distance between them. { 'bel·ōz }

Bell's theorem [QUANT MECH] A theorem which states that any hidden variable that satisifies the condition of locality cannot possibly reproduce all the statistical predictions of quantum mechanics, and which places upper limits, for the predictions of any such theory, on the strength of correlations between measurements of spatially separated objects, whereas quantum mechanics predicts very strong correlations between such measurements. { 'belz ,thir·əm }

Bénard convection cells [PHYS] A regular array of hexagonal cells which sometimes appear in convection in a layer of liquid heated from below. { bā·när kən'vek·shən ,selz }

bend [ELECTROMAG] A smooth change in the direction of the longitudinal axis of a waveguide. { bend }

bending moment [MECH] Algebraic sum of all moments located between a cross section and one end of a structural member; a bending moment that bends the beam convex downward is positive, and one that bends it convex upward is negative. { 'ben·diŋ ,mō·mənt }

bending-moment diagram [MECH] A diagram showing the bending moment at every point along the length of a beam plotted as an ordinate. { 'ben·diŋ ¦mō·mənt ,dī·ə,gram }

bending stress [MECH] An internal tensile or compressive longitudinal stress developed in a beam in response to curvature induced by an external load. { 'ben·diŋ ,stres }

bend plane *See* tilt boundary. { 'bend ,plān }

Benedicks effect [PHYS] An electromotive force produced in a circuit containing one metal only, but having impurities or internal strains, in the presence of an asymmetrical temperature distribution. { 'ben·ə,diks i'fekt }

Benham top [OPTICS] A disk whose surface has black and white portions and which, when rotated at certain speeds and subjected to certain lighting, produces sensations of color. { 'ben·əm ,täp }

Benton hologram [OPTICS] A type of hologram that can be viewed in white light but lacks parallax in the vertical plane. { 'bent·ən 'häl·ə,gram }

Beranek scale [ACOUS] A scale which measures the subjective loudness of a noise; noises are arranged into six arbitrary categories: very quiet, quiet, moderately quiet, noisy, very noisy, and intolerably noisy. { bə'ran·ik ,skāl }

Bernal chart [CRYSTAL] A chart used to determine the coordinates in reciprocal space of x-ray reflections that produce the spots on an x-ray diffraction photograph of a single crystal. { bər'nal ,chärt }

Bernoulli effect [FL MECH] As a consequence of the Bernoulli theorem, the pressure of a stream of fluid is reduced as its speed of flow is increased. { ber,nü·lē i'fekt }

Bernoulli equation *See* Bernoulli theorem. { ber ,nü·lē i'kwā·zhən }

Bernoulli-Euler law [MECH] A law stating that the curvature of a beam is proportional to the bending moment. { ber,nü·lē ¦ȯil·ər ¦lȯ }

Bernoulli law *See* Bernoulli theorem. { ber,nü·lē ,lȯ }

Bernoulli theorem [FL MECH] An expression of the conservation of energy in the steady flow of an incompressible, inviscid fluid; it states that the quantity $(p/\rho) + gz + (v^{2}/2)$ is constant along any streamline, where p is the fluid pressure, v is the fluid velocity, ρ is the mass density of the fluid, g is the acceleration due to gravity, and z is the vertical height. Also known as Bernoulli equation; Bernoulli law. { ber,nü·lē 'thir·əm }

Berthelot method [THERMO] A method of measuring the latent heat of vaporization of a liquid that involves determining the temperature rise of a water bath that encloses a tube in which a given amount of vapor is condensed. { 'ber·tə,lō ,meth·əd }

Berthelot relation [PHYS] A relationship between molecular attraction constants of like and unlike species. { 'ber·tə·lō ri'lā·shən }

Bertrand lens [OPTICS] An auxiliary lens that can be inserted in the tube of a polarizing microscope to obtain interference figures. { 'ber ,tränd ,lenz }

beta [PL PHYS] The ratio of the ion energy density of a plasma to its magnetic energy diversity, or of the particle pressure to the magnetic-field pressure. { 'bād·ə }

beta decay [NUC PHYS] Radioactive transformation of a nuclide in which the atomic number increases or decreases by unity with no change in mass number; the nucleus emits or absorbs a beta particle (electron or positron). Also known as beta disintegration. { 'bād·ə di'kā }

beta-decay spectrum [NUC PHYS] The distribution in energy or momentum of the beta particles arising from a nuclear disintegration process. { 'bād·ə di'kā ,spek·trəm }

beta disintegration *See* beta decay. { 'bād·ə dis ,int·ə'grā·shən }

beta emitter [NUC PHYS] A radionuclide that disintegrates by emission of a negative or positive electron. { 'bād·ə i'mid·ər }

beta factor [PL PHYS] In plasma physics, the ratio of the plasma kinetic pressure to the magnetic pressure. { 'bād·ə ,fak·tər }

beta interaction *See* weak interaction. { 'bād·ə in·tər'ak·shən }

beta particle [NUC PHYS] An electron or positron emitted from a nucleus during beta decay. { 'bād·ə ,pard·ə·kəl }

beta ray [NUC PHYS] A stream of beta particles. { 'bād·ə ˌrā }

beta-ray spectrometer [SPECT] An instrument used to determine the energy distribution of beta particles and secondary electrons. Also known as beta spectrometer. { 'bād·ə ˌrā spek'träm·əd·ər }

beta spectrometer *See* beta-ray spectrometer. { 'bād·ə spek'träm·əd·ər }

Bethe-ansatz technique [PHYS] A method for the solution of one-dimensional many-body problems that was first applied to one-dimensional magnets and has been generalized to many-body problems with point interactions. { 'bād·ē 'anˌzats tekˌnēk }

Bethe-Salpeter equation [PART PHYS] The relativistic analog of the integral form of the two-body Schrödinger equation, the two-particle interaction kernel being the analog of the potential. { 'bāt·ə sal'pād·ər i'kwā·zhən }

Bethe-Slater curve [SOLID STATE] A graph of the exchange energy for the transition elements versus the ratio of the interatomic distance to the radius of the 3*d* shell. { ¦bet·ə 'slād·ər ˌkərv }

Betti reciprocal theorem [MECH] A theorem in the mathematical theory of elasticity which states that if an elastic body is subjected to two systems of surface and body forces, then the work that would be done by the first system acting through the displacements resulting from the second system equals the work that would be done by the second system acting through the displacements resulting from the first system. { 'bāt·tē ri'sip·rə·kəl ˌthir·əm }

Betti's method [MECH] A method of finding the solution of the equations of equilibrium of an elastic body whose surface displacements are specified; it uses the fact that the dilatation is a harmonic function to reduce the problem to the Dirichlet problem. { 'bāt·tēz ˌmeth·əd }

Beutler-Fano profile [PHYS] A function that describes a scattering cross section in the vicinity of a resonance in terms of three parameters: the resonance energy, the resonance width, and the profile index, which determines the shape of the resonance. { ¦bȯit·lər 'fan·ō ˌprōˌfīl }

BeV [PHYS] A billion (10^9) electronvolts, a unit used in the United States; the international unit is GeV (gigaelectronvolts), which has the same value. { bev }

Beverage antenna *See* wave antenna. { 'bev·rij an'ten·ə }

BFS *See* beam-foil spectroscopy.

Bhabha scattering [PART PHYS] Scattering of positrons by electrons. { 'bä·bä 'skad·ər·iŋ }

B-H curve [ELECTROMAG] A graphical curve showing the relation between magnetic induction B and magnetizing force H for a magnetic material. Also known as magnetization curve. { ¦bē¦āch ˌkərv }

Bi *See* abampere.

Bianchi classification [RELAT] A classification of possible types of spatially homogeneous space-times. { bē'aŋ·kē ˌklas·ə·fəˌkā·shən }

biaxial crystal [CRYSTAL] A crystal of low symmetry in which the index ellipsoid has three unequal axes. { bī'ak·sē·əl 'krist·əl }

biaxial indicatrix [CRYSTAL] An ellipsoid whose three axes at right angles to each other are proportional to the refractive indices of a biaxial crystal. { bī'ak·sē·əl in'dik·əˌtriks }

biaxial stress [MECH] The condition in which there are three mutually perpendicular principal stresses; two act in the same plane and one is zero. { bī'ak·sē·əl ˌstress }

Bico bi-drop *See* bi-drop. { ¦bēk·ō 'bīˌdräp }

biconcave lens *See* double-concave lens. { bī'kän ˌkāv 'lenz }

biconical antenna [ELECTROMAG] An antenna consisting of two metal cones having a common axis with their vertices coinciding or adjacent and with coaxial-cable or waveguide feed to the vertices. { bī'kän·ə·kəl an'ten·ə }

biconvex lens *See* double-convex lens. { bī'kän ˌveks 'lenz }

bidirectional antenna [ELECTROMAG] An antenna that radiates or receives most of its energy in only two directions. { ˌbī·də'rek·shən·əl an'ten·ə }

bi-drop [FL MECH] A device in which two drops of different wetting liquids are juxtaposed inside a tube, resulting in spontaneous motion of the liquid and coating of the inner surface of the tube. Also known as Bico bi-drop. { 'bīˌdräp }

Biedenharn identity [NUC PHYS] A relationship among the six-*j* symbols of Wigner. { 'bēd·ənˌhärn i'den·ə·dē }

bifilar resistor [ELEC] A resistor wound with a wire doubled back on itself to reduce the inductance. { bī'fi·lər ri'zis·tər }

bifilar transformer [ELEC] A transformer in which wires for the two windings are wound side by side to give extremely tight coupling. { bī'fi·lər tranz'fȯr·mər }

bifilar winding [ELEC] A winding consisting of two insulated wires, side by side, with currents traveling through them in opposite directions. { bī'fi·lər 'wīn·diŋ }

bifocal lens [OPTICS] **1.** A lens with two parts having different focal lengths. **2.** In particular, an eyeglass lens having one part that corrects for distant vision and one part for near vision. { bī'fō·kəl 'lenz }

bilateral antenna [ELECTROMAG] An antenna having maximum response in exactly opposite directions, 180° apart, such as a loop. { bī'lad·ə·rəl an'ten·ə }

bilateral network [ELEC] A network or circuit in which the magnitude of the current remains the same when the voltage polarity is reversed. { bī'lad·ə·rəl 'netˌwərk }

bilateral slit [SPECT] A slit for spectrometers and spectrographs that is bounded by two metal strips which can be moved symmetrically, allowing the distance between them to be adjusted with great precision. { ¦bīˌlad·ə·rəl 'slit }

billboard array [ELECTROMAG] A broadside antenna array consisting of stacked dipoles spaced one-fourth to three-fourths wavelength apart in front of a large sheet-metal reflector. Also known as bedspring array; mattress array. { 'bil,bȯrd ə'rā }

Billet split lens [OPTICS] A lens cut into two halves, along the optic axis; used in interferometry. { 'bil·ət ,split ¦lenz }

binary encounter approximation [ATOM PHYS] An approximation for predicting the probability that an incident proton will eject an inner shell electron from an atom; it uses a semiclassical treatment of momentum transfer from the incident proton to the ejected electron. { 'bīn·ə·rē in'kau̇nt·ər ə,präk·sə'mā·shən }

binary magnetic core [SOLID STATE] A ferromagnetic core that can be made to take either of two stable magnetic states. { 'bīn·ə·rē mag'ned·ik 'kȯr }

binary optics [OPTICS] A technology that uses etching technology to produce optical elements with computer-generated microscopic surface relief patterns having two or more levels. { 'bī·ner·ē 'äp·tiks }

binaural [ACOUS] Pertaining to sound that reaches the listener over two paths, to give the effect of auditory perspective. { bī'nȯr·əl }

binaural intensity effect [ACOUS] The relationship wherein, if sound of the same frequency and phase is incident at both ears, the angle between the apparent direction of the sound and the median plane of the line joining the ears is proportional to the logarithm of the ratio of the intensities of sound received at the left and right ears. { bī'nȯr·əl in'ten·səd·ē i'fekt }

binaural phase effect [ACOUS] A displacement in the apparent direction of a sound that results when a difference in phase is introduced between otherwise identical sound signals applied to the two ears; the angular displacement from the median plane is proportional to the phase difference. { bī'nȯr·əl 'fāz i'fekt }

binaural sound [ACOUS] The sound resulting from a reproduction system which has two channels, each fed into a different earphone or loudspeaker, so that a listener hears sounds coming from their original directions (with reference to the separated microphones used in recording the original sounds). { bī'nȯr·əl 'sau̇nd }

b-incomplete curve [RELAT] A curve space of finite length, where length is defined by a generalized affine parameter, that has an end point at a space-time singularity. { ¦bē ¦in·kəm,plēt 'kərv }

binding energy [PHYS] Abbreviated BE. Also known as total binding energy (TBE). **1.** The net energy required to remove a particle from a system. **2.** The net energy required to decompose a system into its constituent particles. { 'bīn·diŋ ¦en·ər·jē }

binding fraction [NUC PHYS] The ratio of the binding energy of a nucleus to the atomic mass number. { 'bīnd·iŋ ,frak·shən }

Bingham number [FL MECH] A dimensionless number used to study the flow of Bingham plastics. { 'biŋ·əm ,nəm·bər }

Bingham plastic [FL MECH] A non-Newtonian fluid exhibiting a yield stress which must be exceeded before flow starts; thereafter the rate of shear versus shear stress curve is linear. { 'biŋ·əm ,plas·tik }

binocular [OPTICS] Any optical instrument designed for use with both eyes to give enhanced views of distant objects, whose distinguishing performance feature is the depth perception obtainable. { bī'näk·yə·lər }

binocular microscope [OPTICS] A microscope having two oculars, allowing the use of both eyes at once. { bī'näk·yə·lər 'mī·krə,skōp }

binomial array antenna [ELECTROMAG] Directional antenna array for reducing minor lobes and providing maximum response in two opposite directions. { bī'nō·mē·əl ə'rā an'ten·ə }

biot [ELEC] *See* abampere. [OPTICS] A unit of rotational strength in substances exhibiting circular dichroism, equal to 10^{-40} times the corresponding centimeter-gram-second unit. { 'bī·ät }

biotar lens [OPTICS] A modern camera lens which is a modified Gauss objective with a large aperture and a field of about 24°. { 'bī·ō,tär ¦lenz }

Biot-Fourier equation [THERMO] An equation for heat conduction which states that the rate of change of temperature at any point divided by the thermal diffusivity equals the Laplacian of the temperature. { ¦byō ¦für·yā i'kwā·zhən }

Biot number [FL MECH] A dimensionless group, used in the study of mass transfer between a fluid and a solid, which gives the ratio of the mass-transfer rate at the interface to the mass-transfer rate in the interior of a solid wall of specified thickness. { 'byō ,nəm·bər }

Biot-Savart law [ELECTROMAG] A law that gives the intensity of the magnetic field due to a wire carrying a constant electric current. { ¦byō sə¦vär 'lȯ }

Biot's law [OPTICS] The law that an optically active substance rotates plane-polarized light through an angle inversely proportional in its wavelength. { 'byōz ,lȯ }

bioultrasonics [ACOUS] The study of the interaction of sound at frequencies above about 20,000 hertz with living systems. { ¦bī·ō·əl·trə'sän·iks }

biplate [OPTICS] **1.** Two plates of glass cemented together with a small angle between them, for producing a double image of a slit in interference experiments. **2.** Two half-wave plates of doubly refracting material, each cut parallel to its optical axis and cemented together with axes perpendicular; used to detect optical polarization. Also known as Bravais biplate. { 'bī,plāt }

biprism [OPTICS] A prism with apex angle only a little less than 180°, which produces a double image of a point source, giving rise to interference fringes on a nearby screen. { 'bī,priz·əm }

biprism interference [OPTICS] Light interference fringes seen on a screen near a biprism. { 'bī,priz·əm ,int·ər'fir·əns }

bipyramid [CRYSTAL] A crystal having the form of two pyramids that meet at a plane of symmetry. Also known as dipyramid. { ¦bī¦pir·ə,mid }

biquartz [OPTICS] A device consisting of two adjoining pieces of quartz of equal thickness that rotate the plane of polarization of light in opposite directions; used with a Nicol prism or other analyzer to increase the accuracy of the latter in determining the properties of polarized light. { 'bī,kwȯrts }

birefringence [OPTICS] **1.** Splitting of a light beam into two components, which travel at different velocities, by a material. **2.** For a light beam that has been split into two components by a material, the difference in the indices of refraction of the components within the material. Also known as double refraction. { ,bī·ri'frin·jəns }

birefringent filter [OPTICS] A filter consisting of alternate layers of polarizing films and plates cut from a birefringent crystal; transmits light in a series of sharp, widely spaced wavelength bands. Also known as Lyot filter; monochromatic filter. { ,bī·ri'frin·jənt 'fil·tər }

birefringent plate [OPTICS] A piece of birefringent optical material with parallel plane surfaces. { ,bī·ri'frin·jənt 'plāt }

Birge-Mieck rule [ATOM PHYS] The rule that the product of the equilibrium vibrational frequency and the square of the internuclear distance is a constant for various electronic states of a diatomic molecule. { ¦bir·gə 'mēk ,rül }

Birge-Sponer extrapolation [SPECT] A method of calculating the dissociation limit of a diatomic molecule when the convergence limit cannot be observed directly, based on the assumption that vibrational energy levels converge to a limit for a finite value of the vibrational quantum number. { ¦bir·gə 'spōn·ər ik,strap·ə'lā·shən }

Birkhoff's theorem [RELAT] A theorem which states that if a space-time containing matter or energy satisfies Einstein's equations of general relativity and is centrally symmetric, then it is necessarily static and under a coordinate transformation it becomes identical to the Schwarzschild solution. { 'bərk,hȯfs ,thir·əm }

bisectrix [CRYSTAL] A line that is the bisector of the angle between the optic axes of a biaxial crystal. { ,bī'sek,triks }

bisphenoid [CRYSTAL] A form apparently consisting of two sphenoids placed together symmetrically. { bī'sfē,noid }

bistable optical device [OPTICS] A device which can be in either of two stable states of optical transmission for a single value of the input light intensity.

bistatic reflectivity [OPTICS] The characteristic of a reflector which reflects energy along a line, or lines, different from or in addition to that of the incident ray. { 'bī,stad·ik rē,flek'tiv·əd·ē }

Bitter magnet [ELECTROMAG] A normal-conductor air-cored magnet consisting of stacked circular copper plates with a number of holes through which cooling water passes. { ¦bid·ər ¦mag·nət }

Bitter pattern [SOLID STATE] A pattern produced when a drop of a colloidal suspension of ferromagnetic particles is placed on the surface of a ferromagnetic crystal; the particles collect along domain boundaries at the surface. { 'bid·ər ,pad·ərn }

black [OPTICS] Quality of an object which uniformly absorbs large percentages of light of all visible wavelengths. { blak }

black-and-white groups *See* Shubnikov groups. { ¦blak ən ¦wīt 'grüps }

blackbody [THERMO] An ideal body which would absorb all incident radiation and reflect none. Also known as hohlraum; ideal radiator. { 'blak¦bäd·ē }

blackbody radiation [THERMO] The emission of radiant energy which would take place from a blackbody at a fixed temperature; it takes place at a rate expressed by the Stefan-Boltzmann law, with a spectral energy distribution described by Planck's equation. { 'blak¦bäd·ē ,rā·dē'ā·shən }

blackbody temperature [THERMO] The temperature of a blackbody that emits the same amount of heat radiation per unit area as a given object; measured by a total radiation pyrometer. Also known as brightness temperature. { 'blak¦bäd·ē ,tem·prə·chər }

black hole [RELAT] A region of space-time from which nothing can escape, according to classical physics; quantum corrections indicate a black hole radiates particles with a temperature inversely proportional to the mass and directly proportional to Planck's constant. { ¦blak 'hōl }

black light [OPTICS] Invisible light, such as ultraviolet rays which fall on fluorescent materials and cause them to emit visible light. { 'blak ,līt }

black-surface enclosure [THERMO] An enclosure for which the interior surfaces of the walls possess the radiation characteristics of a blackbody. { 'blak ,sər·fəs in'klozh·ər }

Blake number [FL MECH] A dimensionless number used in the study of beds of particles. { 'blāk ,nəm·bər }

blandel *See* apostilb. { blan'del }

blast [PHYS] **1.** The brief and rapid movement of air or other fluid away from a center of outward pressure, as in an explosion. **2.** The characteristic instantaneous rise in pressure, followed by a sudden decrease, that results from this movement, differentiated from less rapid pressure changes. { blast }

blast effect [PHYS] Violent air movements and pressure changes and the destruction or damage resulting therefrom, generally caused by an explosion on or above the surface of the earth. { 'blast i'fekt }

blast pressure [PHYS] The impact pressure of the air set in motion by an explosion. { 'blast ,presh·ər }

blast wave [PHYS] An air wave set in motion by an explosion. { 'blast ,wāv }

blaze-of-grating technique [OPTICS] A technique whereby the ruled grooves of a diffraction grating are given a controlled shape so that they reflect as much as 80% of the incoming light into one particular order for a given wavelength. { ˌblāz əv ˈgrād·iŋ tekˈnēk }

BLC *See* boundary-layer control.

bleaching [OPTICS] A decrease in the optical absorption of a medium, produced by radiation or by external forces. { ˈblēch·iŋ }

blink [MECH] A unit of time equal to 10^{-5} day or to 0.864 second. { bliŋk }

blink comparator [OPTICS] An optical instrument used to alternately view two pictures in the same visual field in rapid succession, to detect small differences in similar images. { ˈbliŋk kəmˈpar·əd·ər }

blink microscope [OPTICS] A blink comparator which magnifies the compared pictures. { ˈbliŋk ¦mī·krəˌskōp }

Bloch equations [SOLID STATE] Approximate equations for the rate of change of magnetization of a solid in a magnetic field due to spin relaxation and gyroscopic precession. { ˈbläk iˈkwā·zhənz }

Bloch function [SOLID STATE] A wave function for an electron in a periodic lattice, of the form $u(\mathbf{r})\exp[i\mathbf{k}\cdot\mathbf{r}]$ where $u(\mathbf{r})$ has the periodicity of the lattice. { ˈbläk ˌfəŋk·shən }

Bloch theorem [QUANT MECH] The theorem that the lowest state of a quantum-mechanical system without a magnetic field can carry no current. [SOLID STATE] The theorem that, in a periodic structure, every electronic wave function can be represented by a Bloch function. { ˈbläk ˌthir·əm }

Bloch wall [SOLID STATE] A transition layer, with a finite thickness of a few hundred lattice constants, between adjacent ferromagnetic domains. Also known as domain wall. { ˈbläk ˌwȯl }

blocking [SOLID STATE] The hindering of motion of dislocations in a solid substance by small particles of a second substance included in the solid; results in hardening of the substance. { ˈbläk·iŋ }

blondel *See* apostilb. { ˌblȯnˈdel }

Blondel-Rey law [OPTICS] A law utilized to determine the apparent point brilliance of a flashing light. { blȯn·del rā ˌlȯ }

bloom [OPTICS] Color of oil in reflected light, differing from its color in transmitted light. Also known as fluorescence. { blüm }

blowout [ELEC] The melting of an electric fuse because of excessive current. [ELECTROMAG] The extinguishing of an electric arc by deflection in a magnetic field. Also known as magnetic blowout. { ˈblōˌau̇t }

blue [OPTICS] The hue evoked in an average observer by monochromatic radiation having a wavelength in the approximate range from 455 to 492 nanometers; however, the same sensation can be produced in a variety of other ways. { blü }

blue laser [OPTICS] A laser that emits bluish-purple light efficiently at room temperature from a semiconductor diode based on multiple quantum wells of III-V nitrides such as indium gallium nitride. { ¦blü ¦lā·zər }

B meson [PART PHYS] **1.** An elementary particle with strong nuclear interactions, baryon number B = 0, spin J = 0, positive parity, negative charge parity, isotopic spin I = 1, and mass 1234 MeV, that decays into an omega meson and a pion. **2.** A meson consisting of a combination of a bottom (*b*) quark with an ordinary up (*u*) or down (*d*) quark, having a mass of approximately 5.2 GeV. { ¦bē ˈmāˌzän }

B mode [ACOUS] A form of ultrasonic medical tomography in which a two-dimensional picture is formed by scanning the line-of-sight propagation path and monitoring the position and direction of the path. { ˈbē ˌmōd }

Board of Trade unit *See* kilowatt-hour. { ¦bȯrd əv ˈtrād ˌyü·nət }

bobbin [ELECTROMAG] An insulated spool serving as a support for a coil. { ˈbäb·ən }

bobbin core [ELECTROMAG] A magnetic core having a form or bobbin on which the ferromagnetic tape is wrapped for support of the tape. { ˈbäb·ən ˌkȯr }

Bobeck effect [SOLID STATE] The contraction of magnetic strip domains in a thin magnetic film to cylindrical domains called magnetic bubbles. { ˈbäb·ək iˌfekt }

Bobillier's law [MECH] The law that, in general plane rigid motion, when *a* and *b* are the respective centers of curvature of points A and B, the angle between A*a* and the tangent to the centrode of rotation (pole tangent) and the angle between B*b* and a line from the centrode to the intersection of AB and *ab* (collineation axis) are equal and opposite. { bōˈbil·yāz ˌlȯ }

bobtail curtain antenna [ELECTROMAG] A bidirectional, vertically polarized, phased-array antenna that has two horizontal sections, each 0.5 electrical wavelength long, that connect three vertical sections, each 0.25 electrical wavelength long. { ˈbäbˌtāl ˈkərt·ən anˌten·ə }

Bodenstein number [FL MECH] A dimensionless group used in the study of diffusion in reactors. { ˈbō·dənˌstīn ˌnəm·bər }

body-centered lattice [CRYSTAL] A space lattice in which the point at the intersection of the body diagonals is identical to the points at the corners of the unit cell. { ˈbäd·ē ˌsen·tərd ˈlad·əs }

body centrode [MECH] The path traced by the instantaneous center of a rotating body relative to the body. { ¦bäd·ē ˈsenˌtrōd }

body cone [MECH] The cone in a rigid body that is swept out by the body's instantaneous axis during Poinsot motion. Also known as polhode cone. { ˈbäd·ē ˌkōn }

body force [MECH] An external force, such as gravity, which acts on all parts of a body. { ˈbäd·ē ˌfȯrs }

Bohr atom [ATOM PHYS] An atomic model having the structure postulated in the Bohr theory. { ˈbȯr ˌad·əm }

Bohr-Breit-Wigner theory *See* Breit-Wigner theory. { ¦bȯr ¦brīt ˈvig·nər ˌthē·ə·rē }

Bohr frequency condition [ATOM PHYS] The law that the frequency of the radiation emitted or absorbed during the transition of an atomic system between two stationary states equals the difference in the energies of the states divided by Planck's constant. { ¦bȯr ¦frē·kwən·sē kən,dish·ən }

Bohr magneton [ATOM PHYS] The amount $he/4\pi mc$ of magnetic moment, where h is Planck's constant, e and m are the charge and mass of the electron, and c is the speed of light. { 'bȯr 'mag·nə,tän }

Bohr orbit [ATOM PHYS] One of the electron paths about the nucleus in Bohr's model of the hydrogen atom. { 'bȯr ,ȯr·bət }

Bohr radius [ATOM PHYS] The radius of the ground-state orbit of the hydrogen atom in the Bohr theory. { 'bȯr ,rād·ē·əs }

Bohr's correspondence principle *See* correspondence principle. { 'bȯrz kär·ə'spän·dəns ,prin·sə·pəl }

Bohr-Sommerfeld theory [ATOM PHYS] A modification of the Bohr theory in which elliptical as well as circular orbits are allowed. { ¦bȯr ,zō·mər,felt ,thē·ə·rē }

Bohr theory [ATOM PHYS] A theory of atomic structure postulating an electron moving in one of certain discrete circular orbits about a nucleus with emission or absorption of electromagnetic radiation necessarily accompanied by transitions of the electron between the allowed orbits. { 'bȯr ,thē·ə·rē }

Bohr-van Leeuwen theorem [QUANT MECH] The theorem that magnetism is inexplicable in classical physics and is a quantum phenomenon. { ¦bȯr van'lā·vən ,thir·əm }

Bohr-Wheeler theory of fission [NUC PHYS] A theory accounting for the stability of a nucleus against fission by treating it as a droplet of incompressible and uniformly charged liquid endowed with surface tension. { ¦bȯr ¦wēl·ər 'thē·ə·rē əv 'fish·ən }

boil-off [THERMO] The vaporization of a liquid, such as liquid oxygen or liquid hydrogen, as its temperature reaches its boiling point under conditions of exposure, as in the tank of a rocket being readied for launch. { 'bȯil,ȯf }

Boltzmann constant [STAT MECH] The ratio of the universal gas constant to the Avogadro number. { 'bōlts·mən ,kän·stənt }

Boltzmann distribution [STAT MECH] A function giving the probability that a molecule of a gas in thermal equilibrium will have generalized position and momentum coordinates within given infinitesimal ranges of values, assuming that the molecules obey classical mechanics. { 'bōlts·mən dis·trə'byü·shən }

Boltzmann engine [THERMO] An ideal thermodynamic engine that utilizes blackbody radiation; used to derive the Stefan-Boltzmann law. { 'bōlts·mən ,en·jən }

Boltzmann entropy hypothesis [STAT MECH] The hypothesis that the entropy of a system in a given state is directly proportional to the logarithm of the probability of finding it in that state. { 'bōlts·mən 'en·trə·pē hī'päth·ə·səs }

Boltzmann factor [STAT MECH] The factor $\exp(-E/kT)$ that appears in the expression giving the probability for atoms to have an excitation energy E when at temperature T, where k is the Boltzmann constant. { 'bōlts,mən ,fak·tər }

Boltzmann H theorem [STAT MECH] The theorem that the entropy of a system never decreases; Boltzmann proved this for a classical gas of colliding particles. Also known as H theorem of Boltzmann. { 'bōlts·mən 'āch ,thir·əm }

Boltzmann statistics *See* Maxwell-Boltzmann statistics. { 'bōlts·mən stə'tis·tiks }

Boltzmann transport equation [STAT MECH] An equation used to study the nonequilibrium behavior of a collection of particles; it states that the rate of change of a function which specifies the probability of finding a particle in a unit volume of phase space is equal to the sum of terms arising from external forces, diffusion of particles, and collisions of the particles. Also known as Maxwell-Boltzmann equation. { 'bōlts·mən 'tranz,pȯrt i'kwā·zhən }

Boltzmann-Vlasov equations [PL PHYS] The equations that govern a high-temperature plasma in which the collisional mean free path is much larger than all the characteristic lengths of the system. { ¦bōlts·mən 'vlä,sȯf i'kwā·zhənz }

bomb ballistics [MECH] The special branch of ballistics concerned with bombs dropped from aircraft. { 'bäm bə'lis·tiks }

Bond albedo [OPTICS] The fraction of the total incident light that is reflected by a spherical body. { 'bänd al,bē·dō }

bonding strength [MECH] Structural effectiveness of adhesives, welds, solders, glues, or of the chemical bond formed between the metallic and ceramic components of a cermet, when subjected to stress loading, for example, shear, tension, or compression. { 'bän·diŋ ,streŋkth }

Bond number [FL MECH] A dimensionless number used in the study of atomization and the study of bubbles and drops, equal to $(\rho - \rho')L^2 g/\sigma$, where ρ is the density of a bubble or drop, ρ' is the density of the surrounding medium, L is a characteristic dimension, g is the acceleration of gravity, and σ is the surface tension of the bubble or drop. { 'bänd ,nəm·bər }

bond orientational order [PHYS] An ordering of atoms or molecules in an intermediate state of condensed matter in which the atoms or molecules are distributed at random, as in a fluid or glass, but the condensed matter is orientationally anisotropic on a macroscopic scale. { ¦bänd ,ȯr·ē·ən¦tā·shən·əl 'ȯrd·ər }

Böning effect [ELEC] The displacement of associated ions that have been bound to capturing ions in fine channels in a dielectric medium when an electric field is applied. { 'bən·iŋ i,fekt }

book capacitor [ELEC] A trimmer capacitor

consisting of two plates which are hinged at one end; capacitance is varied by changing the angle between them. { 'bu̇k kə'pas·əd·ər }

bootstrap scheme [PART PHYS] A theory of elementary particles in which the existence of each particle contributes to forces between it and other particles; these forces lead to bound systems which are the particles themselves. { 'büt ˌstrap ˌskēm }

Borda mouthpiece [FL MECH] A reentrant tube in a hydraulic reservoir, whose contraction coefficient (the ratio of the cross section of the issuing jet of liquid to that of the opening) can be calculated more simply than for other discharge openings. { ¦bȯr·də 'mau̇thˌpēs }

boresight camera [OPTICS] A camera mounted in the optical axis of a tracking radar to photograph rockets being tracked or known, fixed targets while in camera range and thus to provide a correction for the alignment of the radar. { 'bȯrˌsīt ˌkam·rə }

Born approximation [QUANT MECH] A method used for the computation of cross sections in scattering problems; the interactions are treated as perturbations of free-particle systems. { 'bȯrn ə·präk·sə'mā·shən }

Born-Haber cycle [SOLID STATE] A sequence of chemical and physical processes by means of which the cohesive energy of an ionic crystal can be deduced from experimental quantities; it leads from an initial state in which a crystal is at zero pressure and 0 K to a final state which is an infinitely dilute gas of its constituent ions, also at zero pressure and 0 K. { ¦bȯrn 'hā·bər ˌsī·kəl }

Born-Madelung model [SOLID STATE] A classical theory of cohesive energy, lattice spacing, and compressibility of ionic crystals. { ¦bȯrn 'mäd·əl·əŋ 'mäd·əl }

Born-Mayer equation [SOLID STATE] An equation for the cohesive energy of an ionic crystal which is deduced by assuming that this energy is the sum of terms arising from the Coulomb interaction and a repulsive interaction between nearest neighbors. { ¦bȯrn 'mī·ər i'kwā·zhən }

Born-von Kármán theory [SOLID STATE] A theory of specific heat which considers an acoustical spectrum for the vibrations of a system of point particles distributed like the atoms in a crystal lattice. { ¦bȯrn fən'kärˌmän ˌthē·ə·rē }

boron-10 [NUC PHYS] A nonradioactive isotope of boron with a mass number of 10; it is a good absorber for slow neutrons, simultaneously emitting high-energy alpha particles, and is used as a radiation shield in Geiger counters. { 'bȯˌrän 'ten }

Borrmann effect [PHYS] The irregular transmission of x-rays when a single crystal of high perfection is placed in a monochromatic x-ray beam in a reflecting position. { 'bȯrˌmän i'fekt }

Bosanquet's law [ELECTROMAG] The statement that, in analogy to Ohm's law for the resistance of an electric circuit, in a magnetic circuit the ratio of the magnetomotive force to the magnetic flux is a constant known as the reluctance. { 'bō·zənˌkets ˌlȯ }

Bose distribution *See* Bose-Einstein distribution. { 'bōz dis·trə'byü·shən }

Bose-Einstein condensate [CRYO] The state of matter of a gas of bosonic particles below a critical temperature such that a large number of particles occupy the ground state of the system. { ¦boz ¦īnˌstīn 'kan·dənˌsāt }

Bose-Einstein condensation [CRYO] A phase transition that occurs when a gas of bosonic particles is cooled below a critical temperature very close to absolute zero, in which a large number of the particles come to occupy the ground state of the system and form a coherent matter wave. Also known as Einstein condensation. { ¦bōz ¦īnˌstīn kän·den'sā·shən }

Bose-Einstein distribution [STAT MECH] For an assembly of independent bosons, such as photons or helium atoms of mass number 4, a function that specifies the number of particles in each of the allowed energy states. Also known as Bose distribution. { ¦bōz ¦īnˌstīn dis·trə'byü·shən }

Bose-Einstein statistics [STAT MECH] The statistical mechanics of a system of indistinguishable particles for which there is no restriction on the number of particles that may exist in the same state simultaneously. Also known as Einstein-Bose statistics. { ¦bōz ¦īnˌstīn stə'tis·tiks }

Bose gas [STAT MECH] An assemblage of noninteracting or weakly interacting bosons. { 'bōz ˌgas }

boson [STAT MECH] A particle that obeys Bose-Einstein statistics; includes photons, pi mesons, and all nuclei having an even number of particles and all particles with integer spin. { 'bōˌsän }

bottom [PART PHYS] The new quantum number associated with the bottom quark. Also known as beauty. { 'bäd·əm }

bottomonium [PART PHYS] A meson, such as the upsilon particle, that is made up of the bottom quark b and its antiquark $\bar{b}$. { ˌbäd·ə'mō·nē·əm }

bottom quark [PART PHYS] A quark with a mass of about 4.7 GeV, electric charge of $-1/3$, zero isotopic spin, strangeness and charm, and a new quantum number associated with it. Also known as b quark; beauty quark. Symbolized b. { 'bäd·əm ˌkwärk }

bougie decimale [OPTICS] Formerly, a unit of luminous intensity equal to 0.96 international standard candle. { 'bu̇ˌzhē des·ə'mäl }

boule [CRYSTAL] A pure crystal, such as silicon, having the atomic structure of a single crystal, formed synthetically by rotating a small seed crystal while pulling it slowly out of molten material in a special furnace. { bül }

boundary friction [MECH] Friction between surfaces that are neither completely dry nor completely separated by a lubricant. { 'bau̇n·drē ˌfrik·shən }

boundary-layer control [FL MECH] Control over

the development of a boundary layer by reduction of surface roughness and choice of surface contours. Abbreviated BLC. { 'bau̇n·drē ˌlā·ər kən'trōl }

boundary-layer flow [FL MECH] The flow of that portion of a viscous fluid which is in the neighborhood of a body in contact with the fluid and in motion relative to the fluid. { 'bau̇n·drē ˌlā·ər ˌflō }

boundary-layer separation [FL MECH] That point where the boundary layer no longer continues to follow the contour of the boundary because the residual momentum of the fluid (left after overcoming viscous forces) may be insufficient to allow the flow to proceed into regions of increasing pressure. Also known as flow separation. { 'bau̇n·drē ˌlā·ər sep·ə'rā·shən }

boundary-layer theory *See* film theory. { 'bau̇n·drē ˌlā·ər ˌthē·ə·rē }

boundary wavelength *See* quantum limit. { ¦bau̇n·drē 'wāvˌleŋkth }

bound charge [ELEC] Electric charge which is confined to atoms or molecules, in contrast to free charge, such as metallic conduction electrons, which is not. Also known as polarization charge. { ¦bau̇nd 'chärj }

bound electron [ATOM PHYS] An electron whose wave function is negligible except in the vicinity of an atom. { ¦bau̇nd i'lekˌträn }

bound glue state *See* glueball. { ¦bau̇nd 'glü ˌstāt }

bound level [NUC PHYS] An energy level in a nucleus so close to the ground state that it can only decay by gamma emission. { 'bau̇nd ˌlev·əl }

bound particle [PHYS] A particle which is confined to some finite region. { ¦bau̇nd 'pärd·i·kəl }

bound vector [MECH] A vector whose line of application and point of application are both prescribed, in addition to its direction. { ¦bau̇nd 'vek·tər }

bourdon [ACOUS] In a carillon, the bell with the lowest tone. { 'bu̇rd·ən }

Boussinesq approximation [FL MECH] The assumption (frequently used in the theory of convection) that the fluid is incompressible except insofar as the thermal expansion produces a buoyancy, represented by a term $g\alpha T$, where g is the acceleration of gravity, α is the coefficient of thermal expansion, and T is the perturbation temperature. { 'bü·si'nesk əˌpräk·sə'mā·shən }

Boussinesq number [FL MECH] A dimensionless number used to study wave behavior in open channels. { 'bü·si'nesk ˌnəm·bər }

Boussinesq's problem [MECH] The problem of determining the stresses and strains in an infinite elastic body, initially occupying all the space on one side of an infinite plane, and indented by a rigid punch having the form of a surface of revolution with axis of revolution perpendicular to the plane. Also known as Cerruti's problem. { 'bü·si'nesks ˌpräb·ləm }

Bowditch curve *See* Lissajous figure. { 'bau̇·dich ˌkərv }

Bow's notation [MECH] A graphical method of representing coplanar forces and stresses, using alphabetical letters, in the solution of stresses or in determining the resultant of a system of concurrent forces. { 'bōz nō'tā·shən }

bowtie antenna [ELECTROMAG] An antenna that consists of two triangular pieces of stiff wire or two triangular flat metal plates, arranged in the configuration of a bowtie, with the feed point at the gap between the apexes of the triangles. { 'bōˌtī anˌten·ə }

bow wave [FL MECH] A shock wave occurring in front of a body, such as an airfoil, or apparently attached to the forward tip of the body. { 'bau̇ ˌwāv }

box camera [OPTICS] A camera that consists of a box, an arrangement for loading and winding film, a simple lens of fixed focus, and a simple shutter with a speed of about 1/30 second. { 'bäks ˌkam·rə }

Boyle's law [PHYS] The law that the product of the volume of a gas times its pressure is a constant at fixed temperature. Also known as Mariotte's law. { 'bȯilz ˌlȯ }

Boyle's temperature [THERMO] For a given gas, the temperature at which the virial coefficient B in the equation of state $Pv = RT[1 + (B/v) + (C/v^2) + \cdots]$ vanishes. { 'bȯilz 'tem·prə·chər }

Boys camera [OPTICS] A type of camera used for the observation of lightning flashes. { 'bȯiz ˌkam·rə }

Boys' method [OPTICS] A method of measuring the refractive index of a lens, in which the curvatures of the lens surfaces are determined by positioning a light source so that reflection from a surface gives an image coincident with the object; these curvatures and the focal length are used to calculate the refractive index. { 'bȯiz ˌmeth·əd }

B phase *See* liquid B. { 'bē ˌfāz }

b quark *See* bottom quark. { 'bē ˌkwärk }

brachistochrone [MECH] The curve along which a smooth-sliding particle, under the influence of gravity alone, will fall from one point to another in the minimum time. { brə'kis·təˌkrōn }

brachyaxis [CRYSTAL] The shorter lateral axis, usually the *a* axis, of an orthorhombic or triclinic crystal. Also known as brachydiagonal. { ¦bra·kē'ak·səs }

brachydiagonal *See* brachyaxis. { ˌbrak·i·dī'ag·ə·nəl }

Brackett series [SPECT] A series of lines in the infrared spectrum of atomic hydrogen whose wave numbers are given by $R_H[(1/16) - (1/n^2)]$, where R_H is the Rydberg constant for hydrogen and n is any integer greater than 4. { 'brak·ət ˌsir·ēz }

Bragg angle [SOLID STATE] One of the characteristic angles at which x-rays reflect specularly from planes of atoms in a crystal. { 'brag ˌaŋ·gəl }

Bragg cell *See* acoustooptic modulator. { 'brag ˌsel }

Bragg curve [ATOM PHYS] **1.** A curve showing the average number of ions per unit distance along a beam of initially monoenergetic ionizing particles, usually alpha particles, passing through a gas. Also known as Bragg ionization curve. **2.** A curve showing the average specific ionization of an ionizing particle of a particular kind as a function of its kinetic energy, velocity, or residual range. { 'brag ˌkərv }

Bragg effect [OPTICS] A phenomenon observed in the recording of white-light holograms in which a photographic film irradiated from both sides displays, upon development, blackening in a series of film planes. Also known as color-filter effect. { 'brag iˌfekt }

Bragg ionization curve *See* Bragg curve. { 'brag ī·ə·nə'zā·shən ˌkərv }

Bragg-Kleeman rule *See* Bragg rule. { ¦brag 'klā·mən ˌrül }

Bragg-Pierce law [PHYS] A relationship for determining an element's atomic absorption coefficient for x-rays when the atomic number of the element and the wavelength of the x-rays are known. { ¦brag ¦pirs ˌlȯ }

Bragg reflection *See* Bragg scattering. { 'brag ri 'flek·shən }

Bragg rule [ATOM PHYS] An empirical rule according to which the mass stopping power of an element for alpha particles is inversely proportional to the square root of the atomic weight. Also known as Bragg-Kleeman rule. { 'brag ˌrül }

Bragg scattering [SOLID STATE] Scattering of x-rays or neutrons by the regularly spaced atoms in a crystal, for which constructive interference occurs only at definite angles called Bragg angles. Also known as Bragg diffraction; Bragg reflection. { 'brag ˌskad·ər·iŋ }

Bragg's equation *See* Bragg's law. { 'bragz i'kwā·zhən }

Bragg's law [SOLID STATE] A statement of the conditions under which a crystal will reflect a beam of x-rays with maximum intensity. Also known as Bragg's equation; Bravais' law. { 'bragz ˌlȯ }

branch [ELEC] A portion of a network consisting of one or more two-terminal elements in series. Also known as arm. [NUC PHYS] A product resulting from one mode of decay of a radioactive nuclide that has two or more modes of decay. { branch }

branching [NUC PHYS] The occurrence of two or more modes by which a radionuclide can undergo radioactive decay. Also known as multiple decay; multiple disintegration. { 'branch·iŋ }

branching fraction [NUC PHYS] That fraction of the total number of atoms involved which follows a particular branch of the disintegration scheme; usually expressed as a percentage. { 'branch·iŋ 'frak·shən }

branching ratio [NUC PHYS] The ratio of the number of parent atoms or particles decaying by one mode to the number decaying by another mode; the ratio of two specified branching fractions. { 'branch·iŋ 'rā·shō }

branch point [ELEC] A terminal in an electrical network that is common to more than two elements or parts of elements of the network. Also known as junction point; node. { 'branch ˌpȯint }

Brans-Dicke theory [RELAT] A theory of gravitation in which the gravitational field is described by the tensor field of general relativity and by a new scalar field, which is determined by the distribution of mass-energy in the universe and replaces the gravitational constant. { ¦bränz ¦dik ˌthē·ə·rē }

Bravais biplate *See* biplate. { brə¦vā 'bīˌplāt }

Bravais indices [CRYSTAL] A modification of the Miller indices; frequently used for hexagonal and trigonal crystalline systems; they refer to four axes: the *c* axis and three others at 120° angles in the basal plane. { brə'vā 'in·dəˌsēz }

Bravais lattice [CRYSTAL] One of the 14 possible arrangements of lattice points in space such that the arrangement of points about any chosen point is identical with that about any other point. { brə'vā 'lad·əs }

Bravais' law *See* Bragg's law. { brə'vāz ˌlȯ }

bra vector [QUANT MECH] A vector describing the state of a dynamic system in Hilbert space; the dual of a ket vector. { 'brä ˌvek·tər }

Brayton cycle [THERMO] A thermodynamic cycle consisting of two constant-pressure processes interspersed with two constant-entropy processes. Also known as complete-expansion diesel cycle; Joule cycle. { 'brāt·ən ˌsī·kəl }

breakaway [FL MECH] Boundary-layer separation in which the boundary layer does not become reattached to the surface. { 'brāk·əˌwā }

breakdown [ELEC] A large, usually abrupt rise in electric current in the presence of a small increase in voltage; can occur in a confined gas between two electrodes, a gas tube, the atmosphere (as lightning), an electrical insulator, and a reverse-biased semiconductor diode. Also known as electrical breakdown. { 'brākˌdau̇n }

breakdown potential *See* breakdown voltage. { 'brākˌdau̇n pə'ten·shəl }

breakdown voltage [ELEC] **1.** The voltage measured at a specified current in the electrical breakdown region of a semiconductor diode. Also known as Zener voltage. **2.** The voltage at which an electrical breakdown occurs in a dielectric. **3.** The voltage at which an electrical breakdown occurs in a gas. Also known as breakdown potential; sparking potential; sparking voltage. { 'brākˌdau̇n ˌvōl·tij }

breaking load [MECH] The stress which, when steadily applied to a structural member, is just sufficient to break or rupture it. Also known as ultimate load. { 'brāk·iŋ ˌlōd }

breaking strength [MECH] The ability of a material to resist breaking or rupture from a tension force. { 'brāk·iŋ ˌstreŋkth }

breaking stress [MECH] The stress required to fracture a material whether by compression, tension, or shear. { 'brāk·iŋ ˌstres }

Breit-Wigner formula [NUC PHYS] A formula which relates the cross section of a particular nuclear reaction with the energy of the incident particle, when the energy is near that required to form a discrete resonance level of the component nucleus. { ¦brīt 'vig·nər ,fór·myə·lə }

Breit-Wigner theory [NUC PHYS] A theory of nuclear reactions from which the Breit-Wigner formula is derived. Also known as Bohr-Breit-Wigner theory. { ¦brīt 'vig·nər ,thē·ə·rē }

bremsstrahlung [ELECTROMAG] Radiation that is emitted by an electron accelerated in its collision with the nucleus of an atom. { 'brem ,shträ·ləŋ }

brewster [OPTICS] A unit of stress optical coefficient of a material; it is equal to the stress optical coefficient of a material in which a stress of 1 bar produces a relative retardation between the components of a linearly polarized light beam of 1 angstrom when the light passes through a thickness of 1 millimeter in a direction perpendicular to the stress. Abbreviated B. { 'brü·stər }

Brewster fringes [OPTICS] Interference fringes observed when white light is viewed through two plane parallel plates of nearly equal thickness. { 'brü·stər ,frin·jəz }

Brewster point [OPTICS] A neutral point located 15 to 20° directly below the sun. { 'brü·stər ,pȯint }

Brewster's angle [OPTICS] The angle of incidence of light reflected from a dielectric surface at which the reflectivity for light whose electrical vector is in the plane of incidence becomes zero; given by Brewster's law. Also known as polarizing angle. { ¦brü·stərz ¦aŋ·gəl }

Brewster's law [OPTICS] The law that the index of refraction for a material is equal to the tangent of the polarizing angle for the material. { ¦brü·stərz ¦lȯ }

Brewster stereoscope [OPTICS] A type of stereoscope that uses prisms to enable the eyes to form a fused image of two pictures whose separation is greater than the interocular distance. { ¦brü·stər 'ster·ē·ə,skōp }

Brewster window [OPTICS] A special glass window used at opposite ends of some gas lasers to transmit one polarization of the laser output beam without loss. { 'brü·stər ,win,dō }

bridge [ELEC] **1.** An electrical instrument having four or more branches, by means of which one or more of the electrical constants of an unknown component may be measured. **2.** An electrical shunt path. { brij }

bridge circuit [ELEC] An electrical network consisting basically of four impedances connected in series to form a rectangle, with one pair of diagonally opposite corners connected to an input device and the other pair to an output device. { 'brij ,sər·kət }

bridge vibration [MECH] Mechanical vibration of a bridge superstructure due to natural and human-produced excitations. { 'brij vī'brā·shən }

Bridgman anvil [PHYS] A device for producing high static pressures using two large massive opposed pistons bearing on a small thin sample confined by a gasket material. { 'brij·mən 'an·vəl }

Bridgman effect [SOLID STATE] The phenomenon that when an electric current passes through an anisotropic crystal, there is an absorption or liberation of heat due to the nonuniformity in current distribution. { 'brij·mən i'fekt }

Bridgman relation [SOLID STATE] $P = QT\Sigma$ in a metal or semiconductor, where P is the Ettingshausen coefficient, Q the Nernst-Ettingshausen coefficient, T the temperature, and Σ the thermal conductivity in a transverse magnetic field. { 'brij·mən ri'lā·shən }

Bridgman technique [SOLID STATE] A method of growing single crystals in which a vertical cylinder that tapers conically to a point at the bottom and contains the substance to be crystallized in molten form is slowly lowered into a cold zone, resulting in crystallization beginning at the tip. { 'brij·mən tek'nēk }

brig [PHYS] A unit to express the ratio of two quantities, as a logarithm to the base 10; that is, a ratio of 10^x is equal to x brig; it is analogous to the bel, but the latter is restricted to power ratios. Also known as dex. { brig }

bright [OPTICS] Attribute of an area that appears to emit a large amount of light. { brīt }

bright-field [OPTICS] Having a brightly lighted background. { 'brīt ,fēld }

bright-line spectrum [SPECT] An emission spectrum made up of bright lines on a dark background. { 'brīt ,līn 'spek·trəm }

brightness [OPTICS] **1.** The characteristic of light that gives a visual sensation of more or less light. **2.** *See* luminance. { 'brīt·nəs }

brightness level *See* adaptation luminance. { 'brīt·nəs ,lev·əl }

brightness temperature *See* blackbody temperature. { 'brīt·nəs ,tem·prə·chər }

bril [OPTICS] A unit of subjective luminance; 100 brils is the luminance level that corresponds to a luminance of 1 millilambert, and a doubling of luminance level corresponds to an increase of 1 bril. { bril }

Brillouin function [SOLID STATE] A function of x with index (or parameter) n that appears in the quantum-mechanical theories of paramagnetism and ferromagnetism and is expressed as $[(2n + 1)/2n] \coth [(2n + 1)x/2n] - (1/2n) \coth (x/2n)$. { brē·wan ¦fəŋk·shən }

Brillouin scattering [SOLID STATE] Light scattering by acoustic phonons. { brē·wan ¦skad·ər·iŋ }

Brillouin zone [SOLID STATE] A fundamental region of wave vectors in the theory of the propagation of waves through a crystal lattice; any wave vector outside this region is equivalent to some vector inside it. { brē·wan ¦zōn }

Brinkmann number [FL MECH] A dimensionless number used to study viscous flow. { 'briŋk·män ,nəm·bər }

British absolute system of units [PHYS] A measurement system based on the foot, the second, and the pound mass; force unit is the poundal. Also known as foot-pound-second system of units (fps system of units). { 'brid·ish 'ab·sə,lüt ,sis·təm əv 'yü·nəts }

British engineering system of units *See* British gravitational system of units. { 'brid·ish en·jə'nir·iŋ 'sis·təm əv 'yü·nəts }

British gravitational system of units [PHYS] A measurement system based on the foot, the second, and the slug mass; 1 slug weighs 32.174 pounds at sea level and 45° latitude, and equals 14.594 kilograms. Also known as British engineering system of units; engineer's system of units. { 'brid·ish grav·ə'tā·shən·əl 'sis·təm əv 'yü·nəts }

British imperial pound [MECH] The British standard of mass, of which a standard is preserved by the government. { 'brid·ish im'pir·ē·əl 'paünd }

British thermal unit [THERMO] Abbreviated Btu. **1.** A unit of heat energy equal to the heat needed to raise the temperature of 1 pound of air-free water from 60° to 61°F at a constant pressure of 1 standard atmosphere; it is found experimentally to be equal to 1054.5 joules. Also known as sixty degrees Fahrenheit British thermal unit ($Btu_{60/61}$). **2.** A unit of heat energy that is equal to 1/180 of the heat needed to raise 1 pound of air-free water from 32°F (0°C) to 212°F (100°C) at a constant pressure of 1 standard atmosphere; it is found experimentally to be equal to 1055.79 joules. Also known as mean British thermal unit (Btu_{mean}). **3.** A unit of heat energy whose magnitude is such that 1 British thermal unit per pound equals 2326 joules per kilogram; it is equal to exactly 1055.05585262 joules. Also known as international table British thermal unit (Btu_{IT}). { 'brid·ish 'thər·məl ,yü·nət }

brittleness [MECH] That property of a material manifested by fracture without appreciable prior plastic deformation. { 'brid·əl·nəs }

brittle temperature [THERMO] The temperature point below which a material, especially metal, is brittle; that is, the critical normal stress for fracture is reached before the critical shear stress for plastic deformation. { 'brid·əl ,tem·prə·chər }

broadband antenna [ELECTROMAG] An antenna that functions satisfactorily over a wide range of frequencies, such as for all 12 very-high-frequency television channels. { 'bròd,band an'ten·ə }

broad beam [PHYS] In measurements of the attenuation of a beam of ionizing radiation, a beam in which much of the scattered radiation reaches the detector, along with the unscattered radiation. { 'bròd ,bēm }

broadening of spectral lines [SPECT] A widening of spectral lines by collision or pressure broadening, or possibly by Doppler effect. { 'bròd·ən·iŋ əv ¦spek·trəl 'līn }

broadside [ELECTROMAG] Perpendicular to an axis or plane. { 'bròd,sīd }

broadside array [ELECTROMAG] An antenna array whose direction of maximum radiation is perpendicular to the line or plane of the array. { 'bròd,sīd ə'rā }

broadside-on position [ELECTROMAG] The position of a point which lies on a line through the center of a magnet, perpendicular to the magnetic axis. Also known as Gauss B position. { ¦bròd,sīd ¦òn pə,zish·ən }

Brocken bow *See* anticorona. { 'bräk·ən ,bō }

broken-back transit [OPTICS] A type of transit telescope in which the light path is broken by insertion of a right-angled prism at the intersection of the optical and rotational axes. Also known as prism transit. { 'brō·kən ,bak 'tranz·ət }

Brooks variable inductometer [ELEC] An inductometer providing a nearly linear scale and consisting of two movable coils, side by side in a plane, sandwiched between two pairs of fixed coils. { ¦brüks 'ver·ē·ə·bəl ,in,dək'täm·əd·ər }

broomy flow [FL MECH] A swirling flow of a fluid in a pipe after passing through a constricted section or after a sudden change of direction. { ¦brü·mē ¦flō }

Brownian movement [STAT MECH] Random movements of small particles suspended in a fluid, caused by the statistical pressure fluctuations over the particle. { 'braün·ē·ən ,müv·mənt }

brush discharge [ELEC] A luminous electric discharge that starts from a conductor when its potential exceeds a certain value but remains too low for the formation of an actual spark. { ¦brəsh ¦dis,chärj }

BSPL *See* band sound pressure level.

Btu *See* British thermal unit.

bu *See* bushel.

bubble [PHYS] **1.** A small, approximately spherical body of fluid within another fluid or solid. **2.** A thin, approximately spherical film of liquid inflated with air or other gas. [SOLID STATE] *See* magnetic bubble. { 'bəb·əl }

bubble cavitation [FL MECH] **1.** Formation of vapor- or gas-filled cavities in liquids by mechanical forces. **2.** The formation of vapor-filled cavities in the interior of liquids in motion when the pressure is reduced without change in ambient temperature. { 'bəb·əl kav·ə'tā·shən }

bubble raft [SOLID STATE] A visual demonstration for the structure of dislocations in metal lattices, showing slip propagation; it consists of many identical bubbles floating on a liquid surface in something like a crystalline array. { 'bəb·əl ,raft }

Buckingham's π theorem [PHYS] The theorem that if there are n physical quantities, $x_1, x_2, \ldots, x_n$, which can be expressed in terms of m fundamental quantities and if there exists one and only one mathematical expression connecting them which remains formally true no matter how the units of the fundamental quantities are changed, namely $\phi(x_1, x_2, \ldots, x_n) = 0$, then the relation ϕ can be expressed by a relation of the form $F(\pi_1, \pi_2, \ldots, \pi_{n-m}) = 0$, where the π's are $n - m$ independent dimensionless products of $x_1, x_2, \ldots, x_n$. Also known as pi theorem. { 'bək·iŋ·əmz 'pī ,thir·əm }

buckling [MECH] Bending of a sheet, plate, or

column supporting a compressive load. { 'bək·liŋ }

buckling stress [MECH] Force exerted by the crippling load. { 'bək·liŋ ˌstres }

Buerger precession method [CRYSTAL] The recording on film of a single level of the reciprocal lattice of an individual crystal, by means of x-ray diffraction, for the purpose of determining unit cell dimensions and space groups. { 'bər·gər prē'sesh·ən ˌmeth·əd }

buffer [ELEC] An electric circuit or component that prevents undesirable electrical interaction between two circuits or components. { 'bəf·ər }

buffeting flutter [FL MECH] A phenomenon that can occur when one wire or cable in a strong wind is in the wake of another, whereby flow separated from the front wire induces the rear wire into oscillations with large amplitude. Also known as wake-induced galloping; wire-induced flutter. { 'bəf·əd·iŋ ˌfləd·ər }

built-in antenna [ELECTROMAG] An antenna that is located inside the cabinet of a radio or television receiver. { 'biltˌin an'ten·ə }

bulk acoustic wave [ACOUS] An acoustic wave that travels through a piezoelectric material, as in a quartz delay line. Also known as volume acoustic wave. { ¦bəlk ə'kü·stik 'wāv }

bulk flow *See* convection. { ¦bəlk 'flō }

bulk lifetime [SOLID STATE] The average time that elapses between the formation and recombination of minority charge carriers in the bulk material of a semiconductor. { ¦bəlk 'līfˌtīm }

bulk modulus *See* bulk modulus of elasticity. { ¦bəlk 'mäj·ə·ləs }

bulk modulus of elasticity [MECH] The ratio of the compressive or tensile force applied to a substance per unit surface area to the change in volume of the substance per unit volume. Also known as bulk modulus; compression modulus; hydrostatic modulus; modulus of compression; modulus of volume elasticity. { ¦bəlk 'mäj·ˌ·ləs əv iˌlas'tis·əd·ē }

bulk rheology [MECH] The branch of rheology wherein study of the behavior of matter neglects effects due to the surface of a system. { ¦bəlk rē'äl·ə·jē }

bulk strain [MECH] The ratio of the change in the volume of a body that occurs when the body is placed under pressure, to the original volume of the body. { 'balk ˌstrān }

bulk strength [MECH] The strength per unit volume of a solid. { ¦bəlk 'streŋkth }

bullet drop [MECH] The vertical drop of a bullet. { 'bu̇l·ət ˌdräp }

Bulygen number [THERMO] A dimensionless number used in the study of heat transfer during evaporation. { 'bül·ə·jən ˌnəm·bər }

Bunn chart [CRYSTAL] A chart for classifying x-ray diffraction powder photographs of substances whose crystals have tetragonal or hexagonal symmetry. { 'bən ˌchärt }

Bunsen disk [OPTICS] The screen generally used in a grease-spot photometer, with a circular translucent spot at the center. { 'bən·sən ˌdisk }

Bunsen-Kirchhoff law [SPECT] The law that every element has a characteristic emission spectrum of bright lines and absorption spectrum of dark lines. { ¦bən·sən 'kirˌkóf ˌló }

buoyancy [FL MECH] The resultant vertical force exerted on a body by a static fluid in which it is submerged or floating. { 'bói·ən·sē }

buoyancy parameter [FL MECH] The Grashof number divided by the square of the Reynolds number. { 'bói·ən·sē pə'ram·əd·ər }

buoyant density [PHYS] A technique that uses the sedimentation equilibrium in a density gradient to characterize a solute. { 'bói·ənt 'den·səd·ē }

buoyant force [FL MECH] The force exerted vertically upward by a fluid on a body wholly or partly immersed in it; its magnitude is equal to the weight of the fluid displaced by the body. { 'bói·ənt 'fórs }

burble [FL MECH] **1.** A separation or breakdown of the laminar flow past a body. **2.** The eddying or turbulent flow resulting from this occurrence. { 'bər·bəl }

burble angle *See* burble point. { 'bər·bəl ˌaŋ·gəl }

burble point [FL MECH] A point reached in an increasing angle of attack at which burble begins. Also known as burble angle. { 'bər·bəl ˌpóint }

Burgers vector [CRYSTAL] A translation vector of a crystal lattice representing the displacement of the material to create a dislocation. { 'bər·gərz ˌvek·tər }

burning glass [OPTICS] A converging lens used to produce intense heat by converging the rays of the sun on a small area. { 'bər·niŋ ˌglas }

Burstein effect [SPECT] The shift of the absorption edge in the spectrum of a semiconductor to higher energies at high carrier densities in the semiconductor. { 'bərˌstīn iˌfekt }

bursting strength [MECH] A measure of the ability of a material to withstand pressure without rupture; it is the hydraulic pressure required to burst a vessel of given thickness. { 'bər·stiŋ ˌstreŋkth }

burst pressure [MECH] The maximum inside pressure that a process vessel can safely withstand. { 'bərst ˌpresh·ər }

burst wave [FL MECH] Wave of compressed air caused by a bursting projectile or bomb; a detonation wave; it may produce extensive local damage. { 'bərst ˌwāv }

bushel [MECH] Abbreviated bu. **1.** A unit of volume (dry measure) used in the United States, equal to 2150.42 cubic inches or approximately 35.239 liters. **2.** A unit of volume (liquid and dry measure) used in Britain, equal to 2219.36 cubic inches or 8 imperial gallons (approximately 36.369 liters). { 'bu̇sh·əl }

butterfly capacitor [ELEC] A variable capacitor having stator and rotor plates shaped like butterfly wings, with the stator plates having an outer ring to provide an inductance so that both capacitance and inductance may be varied, thereby giving a wide tuning range. { 'bəd·ər,flī kə'pas·əd·ər }

butterfly effect [PHYS] In a chaotic system, the ability of miniscule changes in initial conditions (such as the flap of a butterfly's wings) to have far-reaching, large-scale effects on the development of the system (such as the course of weather a continent away). { 'bəd·ər,flī i,fekt }

buzz [FL MECH] In supersonic diffuser aerodynamics, a nonsteady shock motion and airflow associated with the shock system ahead of the inlet. { bəz }

C

c *See* calorie; charmed quark.

C *See* capacitance; capacitor; coulomb.

^{45}Ca *See* calcium-45.

Cabibbo theory [PART PHYS] A theory describing baryon beta-decay processes, according to which the amplitude for such processes is given by $G\{\cos \Theta[V(\Delta s = 0) + A(\Delta s = 0)] + \sin \Theta\{V(\Delta s = +1) + A(\Delta s = +1)]\}$, where Θ is the Cabibbo angle, Δs is the change in strangeness for the baryon, G is a universal beta-decay amplitude, and V and A are vector and axial vector amplitudes, respectively; it is experimentally determined that $\sin \Theta \approx 0.25$, so that $\cos \Theta = 0.97$. { ka'bi·bō ˌthē·ə·rē }

cage [CRYSTAL] A void occurring in a crystal structure capable of trapping one or more foreign atoms. { kāj }

cage antenna [ELECTROMAG] Broad-band dipole antenna in which each pole consists of a cage of wires whose overall shape resembles that of a cylinder or a cone. { 'kāj an'ten·ə }

cal *See* calorie.

Cal *See* kilocalorie.

calcium-45 [NUC PHYS] A radioisotope of calcium having a mass number of 45, often used as a radioactive tracer in studying calcium metabolism in humans and other organisms; half-life is 165 days. Designated ^{45}Ca. { 'kal·sē·əm ˌfórd·ē'fīv }

calcium reversal lines [SPECT] Narrow calcium emission lines that appear as bright lines in the center of broad calcium absorption bands in the spectra of certain stars. { 'kal·se·əm ri'vər·səl ˌlīnz }

Callendar's equation [THERMO] **1.** An equation of state for steam whose temperature is well above the boiling point at the existing pressure, but is less than the critical temperature: $(V - b) = (RT/p) - (a/T^n)$, where V is the volume, R is the gas constant, T is the temperature, p is the pressure, n equals 10/3, and a and b are constants. **2.** A very accurate equation relating temperature and resistance of platinum, according to which the temperature is the sum of a linear function of the resistance of platinum and a small correction term, which is a quadratic function of temperature. { 'kal·ən·dərz i'kwā·zhən }

calorescence [PHYS] The production of visible light by infrared radiation; the transformation is indirect, the light being produced by heat and not by any direct change of wavelength. { ˌkal·ə'res·əns }

calorie [THERMO] Abbreviated cal; often designated c. **1.** A unit of heat energy, equal to 4.1868 joules. Also known as International Table calorie (IT calorie). **2.** A unit of energy, equal to the heat required to raise the temperature of 1 gram of water from 14.5° to 15.5°C at a constant pressure of 1 standard atmosphere; equal to 4.1855 ± 0.0005 joules. Also known as fifteen-degrees calorie; gram-calorie (g-cal); small calorie. **3.** A unit of heat energy equal to 4.184 joules; used in thermochemistry. Also known as thermochemical calorie. { 'kal·ə·rē }

Calzecchi-Onesti effect [ELEC] A change in the conductivity of a loosely aggregated metallic powder caused by an applied electric field. { ˌkält'se·kē ˌō'nes·tē i'fekt }

camera [OPTICS] A light-tight enclosure containing an aperture (usually provided with an optical lens or system of lenses) through which the light from an object passes and forms an image, oftenon a light-sensitive material, inside. { 'kam·rə }

camera lucida [OPTICS] An instrument having a peculiarly shaped prism or a system of mirrors, and often a microscope, which causes a virtual image of an object to be produced on a plane surface, enabling the image's outline to be traced. { ¦kam·rə 'lü·səd·ə }

camera obscura [OPTICS] A primitive camera in which the real image of an object can be observed or traced on the wall of the enclosure opposite the aperture, rather than being recorded photographically. { ¦kam·rə əb'skyůr·ə }

Campbell bridge [ELEC] **1.** A bridge designed for comparison of mutual inductances. **2.** A circuit for measuring frequencies by adjusting a mutual inductance, until the current across a detector is zero. { 'kam·əl ˌbrij }

Campbell's formula [ELECTROMAG] A formula which relates the propagation constant of a loaded transmission line to the propagation constant and characteristic impedance of an unloaded line and the impedance of each loading coil. { 'kam·əlz ¦fór·myə·lə }

canal ray [ATOM PHYS] The name given in early gaseous discharge experiments to the particles passing through a hole or canal in the cathode;

the ray comprises positive ions of the gas being used in the discharge. { kə'nal ˌrā }

candela [OPTICS] A unit of luminous intensity, defined as 1/60 of the luminous intensity per square centimeter of a blackbody radiator operating at the temperature of freezing platinum. Formerly known as candle. Also known as new candle. { kan'del·ə }

candle *See* candela. { 'kan·dəl }

candlepower [OPTICS] Luminous intensity expressed in candelas. Abbreviated cp. { 'kan·dəlˌpaü·ər }

canonical distribution [STAT MECH] The density of members of the canonical ensemble in phase space. { kə'nän·ə·kəl ˌdis·trə'byü·shən }

canonical ensemble [STAT MECH] A hypothetical collection of systems of particles used to describe an actual individual system which is in thermal contact with a heat reservoir but is not allowed to exchange particles with its environment. { kə'nän·ə·kəl än'säm·bəl }

canonical equations of motion *See* Hamilton's equations of motion. { kə'nän·ə·kəl i'kwā·zhənz əv 'mō·shən }

canonically conjugate variables [MECH] A generalized coordinate and its conjugate momentum. { kə'nän·ə·klē ¦kan·jə·gət 'ver·ē·ə·bəlz }

canonical momentum *See* conjugate momentum. { kə'nän·ə·kəl mə'ment·əm }

canonical transformation [MECH] A transformation which occurs among the coordinates and momenta describing the state of a classical dynamical system and which leaves the form of Hamilton's equations of motion unchanged. Also known as contact transformation. { kə'nän·ə·kəl ˌtranz·fər'mā·shən }

cantilever vibration [MECH] Transverse oscillatory motion of a body fixed at one end. { 'kant·əlˌē·vər vī'brā·shən }

canting [MECH] Displacing the free end of a beam which is fixed at one end by subjecting it to a sideways force which is just short of that required to cause fracture. { 'kant·iŋ }

capacitance [ELEC] The ratio of the charge on one of the conductors of a capacitor (there being an equal and opposite charge on the other conductor) to the potential difference between the conductors. Symbolized C. Formerly known as capacity. { kə'pas·ə·təns }

capacitance box [ELEC] An assembly of capacitors and switches which permits adjustment of the capacitance existing at the terminals in nominally uniform steps, from a minimum value near zero to the maximum which exists when all the capacitors are connected in parallel. { kə'pas·ə·təns ˌbäks }

capacitance bridge [ELEC] A bridge for comparing two capacitances, such as a Schering bridge. { kə'pas·ə·təns ˌbrij }

capacitance hat [ELECTROMAG] A network of wires that is placed at the top of an antenna either to increase its bandwidth or to lower its resonant frequency. { kə'pas·əd·əns ˌhat }

capacitance standard *See* standard capacitor. { kə'pas·ə·təns ˌstan·dərd }

capacitive coupling [ELEC] Use of a capacitor to transfer energy from one circuit to another. { kə'pas·ə·təns ˌkəp·liŋ }

capacitive diaphragm [ELECTROMAG] A resonant window used in a waveguide to provide the equivalent of capacitive reactance at the frequency being transmitted. { kə'pas·əd·iv 'dī·əˌfram }

capacitive divider [ELEC] Two or more capacitors placed in series across a source, making available a portion of the source voltage across each capacitor; the voltage across each capacitor will be inversely proportional to its capacitance. { kə¦pas·əd·iv di'vīd·ər }

capacitive load [ELECTROMAG] A load in which the capacitive reactance exceeds the inductive reactance; the load draws a leading current. { kə¦pas·əd·iv ¦lōd }

capacitive loading [ELECTROMAG] **1.** Raising the resonant frequency of an antenna by connecting a fixed capacitor or capacitors in series with it. **2.** Lowering the resonant frequency of an antenna by installing a capacitance hat. { kə'pas·əd·iv 'lōd·iŋ }

capacitive post [ELECTROMAG] Metal post or screw extending across a waveguide at right angles to the E field, to provide capacitive susceptance in parallel with the waveguide for tuning or matching purposes. { kə¦pas·əd·iv ¦pōst }

capacitive reactance [ELECTROMAG] Reactance due to the capacitance of a capacitor or circuit, equal to the inverse of the product of the capacitance and the angular frequency. { kə¦pas·əd·iv rē'ak·təns }

capacitive window [ELECTROMAG] Conducting diaphragm extending into a waveguide from one or both sidewalls, producing the effect of a capacitive susceptance in parallel with the waveguide. { kə¦pas·əd·iv 'win·dō }

capacitor [ELEC] A device which consists essentially of two conductors (such as parallel metal plates) insulated from each other by a dielectric and which introduces capacitance into a circuit, stores electrical energy, blocks the flow of direct current, and permits the flow of alternating current to a degree dependent on the capacitor's capacitance and the current frequency. Symbolized C. Also known as condenser; electric condenser. { kə'pas·əd·ər }

capacitor antenna [ELECTROMAG] Antenna consisting of two conductors or systems of conductors, the essential characteristic of which is its capacitance. Also known as condenser antenna. { kə'pas·əd·ər an'ten·ə }

capacitor bank [ELEC] A number of capacitors connected in series or in parallel. { kə'pas·əd·ər ˌbaŋk }

capacitor color code [ELEC] A method of marking the value on a capacitor by means of dots or bands of colors as specified in the Electronic Industry Association color code. { kə'pas·əd·ər 'kəl·ər ˌkōd }

capacity *See* capacitance. { kə'pas·əd·ē }

cape foot [MECH] A unit of length equal to 1.033 feet or to 0.3148584 meter. { 'kāp ˌfùt }

capillarity [FL MECH] The action by which the surface of a liquid where it contacts a solid is elevated or depressed, because of the relative attraction of the molecules of the liquid for each other and for those of the solid. Also known as capillary action. { ˌkap·əˈlar·əd·ē }

capillary action *See* capillarity. { ˈkap·əˌler·ē ˈak·shən }

capillary attraction [FL MECH] The force of adhesion existing between a solid and a liquid in capillarity. { ˈkap·əˌler·ē əˈtrak·shən }

capillary curve [FL MECH] The curve along which the surface of a liquid intersects a vertical plane perpendicular to a vertical glass plane surface. { ˈkap·əˌler·ē ˌkərv }

capillary depression [FL MECH] The depression of the meniscus of a liquid contained in a tube where the liquid does not wet the walls of the container, as in a mercury barometer; the meniscus has a convex shape, resulting in a depression. { ˈkap·əˌler·ē diˈpresh·ən }

capillary number [FL MECH] A dimensionless number associated with a liquid that compares the intensity of liquid viscosity and surface tension, equal to $\mu V/\sigma$, where μ is the viscosity, σ is the surface tension, and V is a fluid velocity such as the deposition velocity on a solid that is drawn out of the liquid. { ˈkap·ə·ler·ē ˌnəm·bər }

capillary pressure [FL MECH] **1.** The difference of pressure across the interface of two immiscible fluid phases. **2.** The pressure or adhesive force exerted by water in an enclosed space as a result of surface tension. { ˈkap·əˌler·ē ˌpresh·ər }

capillary ripple *See* capillary wave. { ˈkap·əˌler·ē ˌrip·əl }

capillary rise [FL MECH] The rise of a liquid in a capillary tube times the radius of the tube. { ˈkap·əˌler·ē ˌrīz }

capillary wave [FL MECH] **1.** A wave occurring at the interface between two fluids, such as the interface between air and water on oceans and lakes, in which the principal restoring force is controlled by surface tension. **2.** A water wave of less than 1.7 centimeters. Also known as capillary ripple; ripple. { ˈkap·əˌler·ē ˌwāv }

capture [PHYS] A process in which an atomic or nuclear system acquires an additional particle; for example, the capture of electrons by positive ions, or capture of neutrons by nuclei. { ˈkap·chər }

capture cross section [NUC PHYS] The cross section that is effective for radiative capture. { ˈkap·chər ¦krȯs ˈsek·shən }

capture gamma rays [NUC PHYS] The gamma rays emitted in radiative capture. { ¦kap·chər ˈgam·ə ˌrāz }

Carathéodory's principle [THERMO] An expression of the second law of thermodynamics which says that in the neighborhood of any equilibrium state of a system, there are states which are not accessible by a reversible or irreversible adiabatic process. Also known as principle of inaccessibility. { ˌkär·əˌtā·əˈdȯr·ēz ˈprin·sə·pəl }

carbon-12 [NUC PHYS] A stable isotope of carbon with mass number of 12, forming about 98.9% of natural carbon; used as the basis of the newer scale of atomic masses, having an atomic mass of exactly 12u (relative nuclidic mass unit) by definition. { ˈkär·bən ˈtwelv }

carbon-13 [NUC PHYS] A heavy isotope of carbon having a mass number of 13. { ˈkär·bən ˈthərˌtēn }

carbon-14 [NUC PHYS] A naturally occurring radioisotope of carbon having a mass number of 14 and half-life of 5780 years; used in radiocarbon dating and in the elucidation of the metabolic path of carbon in photosynthesis. Also known as radiocarbon. { ˈkär·bən ˈfȯrˌtēn }

carbon arc [ELEC] An electric arc between two electrodes, at least one of which is made of carbon; used in welding and high-intensity lamps, such as in searchlights and photography lamps. { ¦kär·bən ¦ärk }

carbon burning [NUC PHYS] The synthesis of nuclei in stars through reactions involving the fusion of two carbon-12 nuclei at temperatures of about 5×10^8 K. { ¦kär·bən ¦bərn·iŋ }

carbon cycle *See* carbon-nitrogen cycle. { ˈkär·bən ˌsī·kəl }

carbon dioxide gas laser [PHYS] A powerful, continuously operating laser in the infrared that can emit several hundred watts of power at a wavelength of 10.6 micrometers. { ¦kär·bən dīˈäkˌsīd ˈgas ˌlā·zər }

carbon-film resistor [ELEC] A resistor made by depositing a thin carbon film on a ceramic form. { ˈkär·bən ˌfilm riˈzis·tər }

carbon monoxide laser [OPTICS] A molecular gas laser in which the active laser molecule is carbon monoxide, and the strongest wavelengths are 4.9 to 5.7 micrometers. Also known as CO laser. { ¦kär·bən məˈnäkˌsīd ˈlā·zər }

carbon-nitrogen cycle [NUC PHYS] A series of thermonuclear reactions, with release of energy, which presumably occurs in stars that are more massive than the sun; the net accomplishment is the synthesis of four hydrogen atoms into a helium atom, the emission of two positrons and much energy, and restoration of a carbon-12 atom with which the cycle began. Also known as carbon cycle; nitrogen cycle. { ¦kär·bən ¦nī·trə·jən ˌsī·kəl }

carbon-nitrogen-oxygen bicycle [NUC PHYS] The pair of carbon-nitrogen-oxygen cycles formed from the original carbon-nitrogen cycle and an alternative cycle that results when proton capture by a nitrogen-15 nucleus results in the formation of an oxygen-16 nucleus. { ¦kär·bən ¦nī·trə·jən ¦äk·sə·jən ˈbīˌsī·kəl }

carbon-nitrogen-oxygen cycles [NUC PHYS] A group of nuclear reactions involving the interaction of protons with carbon, nitrogen, and oxygen nuclei; completion of any of the cycles results in the consumption of four protons and the production of a helium-4 nucleus, two positrons, two neutrinos, and energy. Abbreviated CNO cycles. { ¦kär·bən ¦nī·trə·jən ¦äk·sə·jən ˌsī·kəlz }

carbon pile [ELEC] A variable resistor consisting of a stack of carbon disks mounted between a fixed metal plate and a movable one that serve as the terminals of the resistor; the resistance value is reduced by applying pressure to the movable plate. { 'kär·bən ,pīl }

cardinal point [OPTICS] Any one of six points in an optical system, namely, the two principal points, two nodal points, and two focal points. Also known as Gauss point. { 'kärd·nəl ¦pȯint }

cardioid condenser [OPTICS] A substage condenser that cuts off the direct light and allows only the light diffracted or dispersed from the object to enter the microscope; used in dark-field microscopes. { 'kärd·ē,ȯid kən'den·sər }

Carlsbad law [CRYSTAL] A feldspar twin law in which the twinning axis is the *c* axis, the operation is rotation of 180°, and the contact surface is parallel to the side pinacoid. { 'kärlz,bad ,lȯ }

Carlsbad turn [CRYSTAL] A twin crystal in the monoclinic system with the vertical axis as the turning axis. { 'kärlz,bad ,tərn }

Carnot-Clausius equation [THERMO] For any system executing a closed cycle of reversible changes, the integral over the cycle of the infinitesimal amount of heat transferred to the system divided by its temperature equals 0. Also known as Clausius theorem. { kär¦nōt 'klȯz·ē·əs i,kwā·zhən }

Carnot cycle [THERMO] A hypothetical cycle consisting of four reversible processes in succession: an isothermal expansion and heat addition, an isentropic expansion, an isothermal compression and heat rejection process, and an isentropic compression. { kär'nō ,si·kəl }

Carnot efficiency [THERMO] The efficiency of a Carnot engine receiving heat at a temperature absolute T_1 and giving it up at a lower temperature absolute T_2; equal to $(T_1 - T_2)/T_1$. { kär'nō i'fish·ən·sē }

Carnot number [THERMO] A property of two heat sinks, equal to the Carnot efficiency of an engine operating between them. { kär'nō ,nəm·bər }

Carnot's theorem [THERMO] **1.** The theorem that all Carnot engines operating between two given temperatures have the same efficiency, and no cyclic heat engine operating between two given temperatures is more efficient than a Carnot engine. **2.** The theorem that any system has two properties, the thermodynamic temperature T and the entropy S, such that the amount of heat exchanged in an infinitesimal reversible process is given by $dQ = TdS$; the thermodynamic temperature is a strictly increasing function of the empirical temperature measured on an arbitrary scale. { kär'noz 'thir·əm }

carrier density [SOLID STATE] The density of electrons and holes in a semiconductor. { 'kar·ē·ər ,den·səd·ē }

carrier mobility [SOLID STATE] The average drift velocity of carriers per unit electric field in a homogeneous semiconductor; the mobility of electrons is usually different from that of holes. { 'kar·ē·ər mō'bil·əd·ē }

Carter chart [ELECTROMAG] An Argand diagram of the complex reflection coefficient of a waveguide junction on which are drawn lines of constant magnitude and phase of the impedance. { 'kärd·ər ,chärt }

Carter's theorem [RELAT] Theorem proving that the only stationary, charged black hole solutions to the equations of general relativity are the Kerr-Newman solutions. { 'kärd·ərz ,thir·əm }

Carvallo paradox [OPTICS] The absurdity that since light is composed from infinitely long wave trains of various frequencies, a spectrograph should show the spectrum of a source both before and after it is illuminated. { kär'väl·ō ,par·ə,däks }

cascade [ELEC] An electric-power circuit arrangement in which circuit breakers of reduced interrupting ratings are used in the branches, the circuit breakers being assisted in their protection function by other circuit breakers which operate almost instantaneously. Also known as backup arrangement. [PHYS] The emission of a series of photons by a quantum system, such as an atomic nucleus or a laser, in an excited state, accompanying transitions of the system to successively lower excited states, until the system reaches the ground state. { ka'skād }

cascade gamma emission [NUC PHYS] The emission by a nucleus of two or more gamma rays in succession. { ka'skād 'gam·ə i'mish·ən }

cascade hyperon *See* xi hyperon. { ka'skād ,hī·pə,rän }

cascade liquefaction [CRYO] A method of liquefying gases in which a gas with a high critical temperature is liquefied by increasing its pressure; evaporation of this liquid cools a second liquid so that it can also be liquefied by compression, and so on. { ka'skād lik·wə'fak·shən }

cascade particle *See* xi hyperon; xi-minus particle. { ka'skād 'pard·ə·kəl }

cascade shower [PART PHYS] A cosmic-ray shower of electrons, positrons, and gamma rays which grows by pair production and bremsstrahlung events. { ka'skād ,shaủ·ər }

Casimir-du Pré theory [SOLID STATE] A theory of spin-lattice relaxation which treats the lattice and spin systems as distinct thermodynamic systems in thermal contact with one another. { 'kaz·ə,mir dyü'prā ,thē·ə·rē }

Casimir effect [QUANT MECH] An attractive force between two parallel, conducting plates in empty space that arises from zero-point quantum fluctuations of the vacuum electromagnetic field and is proportional to $1/d^4$, where d is the plate separation. { 'kaz·ə,mir i,fekt }

Cassegrain antenna [ELECTROMAG] A microwave antenna in which the feed radiator is mounted at or near the surface of the main reflector and aimed at a mirror at the focus; energy from the feed first illuminates the mirror, then spreads outward to illuminate the main reflector. { kas·gran an'ten·ə }

Cassegrain focus [OPTICS] The principal focus

of a Cassegrain telescope, located just behind the primary mirror. { kas·gran 'fō·kəs }

Cassegrain-Newtonian telescope *See* Newtonian-Cassegrain telescope. { kas·gran nü'tōn·ē·ən 'tel·ə,skōp }

Cassegrain telescope [OPTICS] A reflecting telescope in which a small hyperboloidal mirror reflects the convergent beam from the paraboloidal primary mirror through a hole in the primary mirror to an eyepiece in back of the primary mirror. { kas·gran 'tel·ə'skōp }

cast [OPTICS] A change in a color because of the adding of a different hue. { kast }

Castigliano's principle *See* Castigliano's theorem. { ,kas·til'yä·nōz ,prin·sə·pəl }

Castigliano's theorem [MECH] The theorem that the component in a given direction of the deflection of the point of application of an external force on an elastic body is equal to the partial derivative of the work of deformation with respect to the component of the force in that direction. Also known as Castigliano's principle. { ,kas·til'yä·nōz ,thir·əm }

casting strain [MECH] Any strain that results from the cooling of a casting, causing casting stress. { 'kast·iŋ ,strān }

casting stress [MECH] Any stress that develops in a casting due to geometry and casting shrinkage. { 'kast·iŋ ,stres }

catadioptric [OPTICS] Involving both reflection and refraction of light. { ,kad·ə,dī'äp·trik }

cathode [ELEC] The terminal at which current leaves a primary cell or storage battery; it is negative with respect to the device, and positive with respect to the external circuit. Symbolized K. { 'kath,ōd }

catoptric light [OPTICS] Light reflected from a mirror, for example, light from a filament, concentrated into a parallel beam by means of a reflector. { kə'täp·trik 'līt }

Cauchy data [RELAT] The Cauchy data for a hyperbolic partial differential equation consist of the value of the field and its time derivative on some spacelike surface. { kō·shē ,dad·ə }

Cauchy dispersion formula [OPTICS] A semiempirical formula for the index of refraction n of a medium as a function of wavelength λ, according to which $n = A + (B/\lambda^2)$, where A and B are constants. { kō·shē dis'pər·zhən ,fór·myə·lə }

Cauchy horizon [RELAT] Boundary of the region that can be predicted by Cauchy data set on a spacelike surface (partial Cauchy surface). { kō·shē hə,rīz·ən }

Cauchy number [FL MECH] A dimensionless number used in the study of compressible flow, equal to the density of a fluid times the square of its velocity divided by its bulk modulus. Also known as Hooke number. { kō·shē ,nəm·bər }

Cauchy relations [SOLID STATE] A set of six relations between the compliance constants of a solid which should be satisfied provided the forces between atoms in the solid depend only on the distances between them and act along the lines joining them, and provided that each atom is a center of symmetry in the lattice. { kō·shē ri'lā·shənz }

Cauchy surface [RELAT] A surface S in a space-time M is a (global) Cauchy surface if every non-spacelike curve in M intersects S exactly once; that is, the Cauchy development of S equals M. { kō·shē ,sər·fəs }

causal boundary [RELAT] A boundary attached to a space-time that depends only on the causal structure; it does not distinguish between boundary points at finite distances (singularities) or those at infinity. Also known as C boundary. { ¦kóz·əl 'baún·drē }

causal curve [RELAT] A curve in space-time that is nowhere spacelike. { ¦kóz·əl 'kərv }

causal future [RELAT] The causal future relative to a set of points S in a space-time M is the set of points in M which can be reached from S by future-directed timelike or null curves. { ¦kóz·əl 'fyü·chər }

causality [MECH] In classical mechanics, the principle that the specification of the dynamical variables of a system at a given time, and of the external forces acting on the system, completely determines the values of dynamical variables at later times. Also known as determinism. [PHYS] **1.** The principle that an event cannot precede its cause; in a relativistic theory, an event cannot have an effect outside its future light cone. **2.** In relativistic quantum field theory, the principle that the field operators at different space-time points commute (for boson fields; anticommute in the case of fermion fields) if the separation of the points is spacelike. [QUANT MECH] The principle that the specification of the dynamical state of a system at a given time, and of the interaction of the system with its environment, determines the dynamical state of the system at later times, from which a probability distribution for the observation of any dynamical variable may be determined. Also known as determinism. { kó'zal·əd·ē }

causality condition [RELAT] The condition of a space-time requiring there be no closed non-spacelike curves. { kó'zal·əd·ē kən,dish·ən }

causally simple [RELAT] A set of points U in a space-time is said to be causally simple if the causal past and causal future of every compact subset of U is closed in U. { ¦kóz·əl·ē 'sim·pəl }

causal past [RELAT] The causal past relative to a set of points S in a space-time M is the set of points in M which can be reached from S by past-directed timelike or null curves. { ¦kóz·əl 'past }

caustic [OPTICS] A curve or surface which is tangent to the rays of an initially parallel beam after reflection or refraction in an optical system. [PHYS] A curve or surface which is tangent to adjacent orthogonals to waves that have been reflected or refracted from a curved surface. { 'kó·stik }

cavitation [FL MECH] Formation of gas- or vapor-filled cavities within liquids by mechanical forces; broadly includes bubble formation when water is brought to a boil and effervescence of carbonated drinks; specifically, the formation of vapor-filled cavities in the interior or on the solid boundaries of vaporized liquids in motion where the pressure is reduced to a critical value without a change in ambient temperature. { ˌkav·əˈtā·shən }

cavitation noise [ACOUS] Noise resulting from the formation of vapor- or gas-filled cavities in liquids by mechanical forces, as occurs near a propeller. { ˌkav·əˈtā·shən ˌnȯiz }

cavitation number [FL MECH] The excess of the local static pressure head over the vapor pressure head divided by the velocity head. { ˌkav·əˈtā·shən ˌnəm·bər }

caviton [PL PHYS] A region of a plasma having reduced mass density and enhanced wave energy density. { ˈkav·əˌtän }

cavity *See* cavity resonator. { ˈkav·əd·ē }

cavity coupling [ELECTROMAG] The extraction of electromagnetic energy from a resonant cavity, either waveguide or coaxial, using loops, probes, or apertures. { ˈkav·əd·ē ˌkəp·liŋ }

cavity filter [ELECTROMAG] A microwave filter that uses quarter-wavelength-coupled cavities inserted in waveguides or coaxial lines to provide band-pass or other response characteristics at frequencies in the gigahertz range. { ˈkav·əd·ē ˌfil·tər }

cavity radiator [THERMO] A heated enclosure with a small opening which allows some radiation to escape or enter; the escaping radiation approximates that of a blackbody. { ˈkav·əd·ē ˈrād·ēˌād·ər }

cavity resonance [ELECTROMAG] The resonant oscillation of the electromagnetic field in a cavity. { ˈkav·əd·ē ˈrez·ən·əns }

cavity resonator [ELECTROMAG] A space totally enclosed by a metallic conductor and excited in such a way that it becomes a source of electromagnetic oscillations. Also known as cavity; microwave cavity; microwave resonance cavity; resonant cavity; resonant chamber; resonant element; rhumbatron; tuned cavity; waveguide resonator. { ˈkav·əd·ē ˈrez·ənˌād·ər }

cavity ringdown laser absorption spectroscopy [SPECT] A direct absorption technique used for measuring short-lived species and for trace-gas analysis in which the rate of decay of light, injected with a pulsed laser and trapped in a cavity formed by two highly reflective mirrors, is measured, allowing the calculation of the amount of light absorbed by the sample. Abbreviated CRLAS. { ¦kav·əd·ē ¦riŋˌdau̇n ¦lā·zər əb¦sörp·shən ˌspekˈträs·kə·pē }

cavity tuning [ELECTROMAG] Use of an adjustable cavity resonator as a tuned circuit in an oscillator or amplifier, with tuning usually achieved by moving a metal plunger in or out of the cavity to change the volume, and hence the resonant frequency of the cavity. { ˈkav·əd·ē ˌtün·iŋ }

c axis [CRYSTAL] A vertically oriented crystal axis, usually the principal axis; the unique symmetry axis in tetragonal and hexagonal crystals. { ˈsē ˌak·səs }

C-band waveguide [ELECTROMAG] A rectangular waveguide, with dimensions 3.48 by 1.58 centimeters, which is used to excite only the dominant mode (TE_{01}) for wavelengths in the range 3.7–5.1 centimeters. { ˈsē ˌband ˈwāvˌgīd }

C boundary *See* causal boundary. { ˈsē ˌbau̇n·drē }

C core [ELECTROMAG] A spirally wound magnetic core that is formed to a desired rectangular shape before being cut into two C-shaped pieces and placed around a transformer or magnetic amplifier coil. { ˈsē ˌkȯr }

CD *See* circular dichroism.

celerity *See* phase velocity. { səˈler·əd·ē }

cell [ELEC] A single unit of a battery. { sel }

cellular horn *See* multicellular horn. { ˈsel·yə·lər ˈhȯrn }

celo [MECH] A unit of acceleration equal to the acceleration of a body whose velocity changes uniformly by 1 foot (0.3048 meter) per second in 1 second. { ˈse·lō }

Celor lens system [OPTICS] An anastigmatic lens system consisting of two air-spaced achromatic doublet lenses, one on each side of the stop. Also known as Gauss lens system; Gauss objective lens. { ˈse·lȯr ˈlenz ˌsis·təm }

Celsius degree [THERMO] Unit of temperature interval or difference equal to the kelvin. { ˈsel·sē·əs diˈgrē }

Celsius temperature scale [THERMO] Temperature scale in which the temperature Θ_c in degrees Celsius (°C) is related to the temperature T_k in kelvins by the formula $\Theta_c = T_k - 273.15$; the freezing point of water at standard atmospheric pressure is very nearly 0°C and the corresponding boiling point is very nearly 100°C. Formerly known as centigrade temperature scale. { ˈsel·se·əs ˈtem·prə·chər ˌskāl }

cemented lens *See* compound lens. { siˈmen·təd ˈlenz }

cent [ACOUS] The interval between two sounds whose basic frequency ratio is the twelve-hundredth root of 2; the interval, in cents, between any two frequencies is 1200 times the logarithm to the base 2 of the frequency ratio. { sent }

centare *See* centiare. { ˈsenˌtär }

center [OPTICS] To adjust the components of an optical system so that their centers of curvature lie on a common optical axis. Also known as square-on. { ˈsen·tər }

centered lattice [CRYSTAL] A crystal lattice in which the axes have been chosen according to the rules for the crystal system, and in which there are lattice points at the centers of certain planes as well as at the corners. { ˈsen·tərd ˈlad·əs }

center loading [ELECTROMAG] Alteration of the resonant frequency of a transmitting antenna by inserting an inductance or capacitance about halfway between the feed point and the end of the antenna. { ˈsen·tər ˈlōd·iŋ }

center of attraction [MECH] A point toward which a force on a body or particle (such as gravitational or electrostatic force) is always directed; the magnitude of the force depends only on the distance of the body or particle from this point. { 'sen·tər əv ə'trak·shən }

center of buoyancy [MECH] The point through which acts the resultant force exerted on a body by a static fluid in which it is submerged or floating; located at the centroid of displaced volume. { 'sen·tər əv 'bȯi·ən·sē }

center of force [MECH] The point toward or from which a central force acts. { 'sen·tər əv 'fȯrs }

center of gravity [MECH] A fixed point in a material body through which the resultant force of gravitational attraction acts. { 'sen·tər əv 'grav·əd·ē }

center of inertia *See* center of mass. { 'sen·tər əv i'nər·shə }

center of inversion [CRYSTAL] A point in a crystal lattice such that the lattice is left invariant by an inversion in the point. { 'sen·tər əv in'vər·zhən }

center of mass [MECH] That point of a material body or system of bodies which moves as though the system's total mass existed at the point and all external forces were applied at the point. Also known as center of inertia; centroid. { 'sen·tər əv 'mas }

center-of-mass coordinate system [MECH] A reference frame which moves with the velocity of the center of mass, so that the center of mass is at rest in this system, and the total momentum of the system is zero. Also known as center of momentum coordinate system. { 'sen·tər əv 'mas kō'ȯrd·nət ,sis·təm }

center-of-momentum coordinate system *See* center-of-mass coordinate system. { 'sen·tər əv mə'men·təm kō'ȯrd·nət ,sis·təm }

center of oscillation [MECH] Point in a physical pendulum, on the line through the point of suspension and the center of mass, which moves as if all the mass of the pendulum were concentrated there. { 'sen·tər əv ,äs·ə'lā·shən }

center of percussion [MECH] If a rigid body, free to move in a plane, is struck a blow at a point O, and the line of force is perpendicular to the line from O to the center of mass, then the initial motion of the body is a rotation about the center of percussion relative to O; it can be shown to coincide with the center of oscillation relative to O. { 'sen·tər əv pər'kəsh·ən }

center of pressure [FL MECH] For a body immersed in a fluid, the point through which the resultant of the forces on the surface of the body due to hydrostatic pressure acts. { 'sen·tər əv 'presh·ər }

center of suspension [MECH] The intersection of the axis of rotation of a pendulum with a plane perpendicular to the axis that passes through the center of mass. { 'sen·tər əv sə'spen·shən }

center of twist [MECH] A point on a line parallel to the axis of a beam through which any transverse force must be applied to avoid twisting of the section. Also known as shear center. { 'sen·tər əv 'twist }

center tap [ELEC] A terminal at the electrical midpoint of a resistor, coil, or other device. Abbreviated CT. { 'sen·tər ,tap }

centiare [MECH] Unit of area equal to 1 square meter. Also spelled centare. { 'sen·tē,är }

centibar [MECH] A unit of pressure equal to 0.01 bar or to 1000 pascals. { 'sent·ə,bär }

centigrade heat unit [THERMO] A unit of heat energy, equal to 0.01 of the quantity of heat needed to raise 1 pound of air-free water from 0 to 100°C at a constant pressure of 1 standard atmosphere; equal to 1900.44 joules. Symbolized CHU; (more correctly) CHU_{mean}. { 'sent·ə,grād 'hēt ,yü·nət }

centigrade temperature scale *See* Celsius temperature scale. { 'sent·ə,grād 'tem·prə·chər ,skāl }

centigram [MECH] Unit of mass equal to 0.01 gram or 10^{-5} kilogram. Abbreviated cg. { 'sent·ə,gram }

centihg *See* centimeter of mercury. { 'sen,tig *or* ¦sent·ē,āch'jē }

centiliter [MECH] A unit of volume equal to 0.01 liter or to 10^{-5} cubic meter. { 'sent·ə,lēd·ər }

centimeter [MECH] A unit of length equal to 0.01 meter. Abbreviated cm. { 'sent·ə,mēd·ər }

centimeter-candle *See* phot. { 'sent·ə,mēd·ər 'kand·əl }

centimeter-gram-second system [PHYS] An absolute system of metric units in which the centimeter, gram mass, and the second are the basic units. Abbreviated cgs system. { ¦sent·ə,mēd·ər ¦gram 'sek·ənd ,sis·təm }

centimeter of mercury [MECH] A unit of pressure equal to the pressure that would support a column of mercury 1 centimeter high, having a density of 13.5951 grams per cubic centimeter, when the acceleration of gravity is equal to its standard value (980.665 centimeters per second per second); it is equal to 1333.22387415 pascals; it differs from the dekatorr by less than 1 part in 7,000,000. Abbreviated cmHg. Also known as centihg. { 'sent·ə,mēd·ər əv 'mər·kyə·rē }

centipoise [FL MECH] A unit of viscosity which is equal to 0.01 poise. Abbreviated cp. { 'sent·ə,pȯiz }

centistoke [FL MECH] A cgs unit of kinematic viscosity in customary use, equal to the kinematic viscosity of a fluid having a dynamic viscosity of 1 centipoise and a density of 1 gram per cubic centimeter. Abbreviated cs. { 'sent·ə,stōk }

central field approximation [PHYS] The approximation that the electrons in an atom or the nucleons in a nucleus move in the potential of a central force which is the same for all the particles. { 'sen·trəl ,fēld ə,präk·sə'mā·shən }

central force [MECH] A force whose line of action is always directed toward a fixed point; the force may attract or repel. { 'sen·trəl 'fȯrs }

central orbit [MECH] The path followed by a body moving under the action of a central force. { 'sen·trəl 'ȯr·bət }

centrifugal [MECH] Acting or moving in a direction away from the axis of rotation or the center of a circle along which a body is moving. { ,sen 'trif·i·gəl }

centrifugal barrier [MECH] A steep rise, located around the center of force, in the effective potential governing the radial motion of a particle of nonvanishing angular momentum in a central force field, which results from the centrifugal force and prevents the particle from reaching the center of force, or causes its Schrödinger wave function to vanish there in a quantum-mechanical system. { ,sen'trif·i·gəl 'bar·ē·ər }

centrifugal distortion [PHYS] Tendency of a molecule to stretch slightly as its speed of rotation increases. { ,sen'trif·i·gəl di'stȯr·shən }

centrifugal force [MECH] **1.** An outward pseudo-force, in a reference frame that is rotating with respect to an inertial reference frame, which is equal and opposite to the centripetal force that must act on a particle stationary in the rotating frame. **2.** The reaction force to a centripetal force. { ,sen'trif·i·gəl 'fȯrs }

centrifugal moment [MECH] The product of the magnitude of centrifugal force acting on a body and the distance to the center of rotation. { ,sen'trif·i·gəl 'mō·mənt }

centrifugal stretching [PHYS] Stretching of the bonds of a rotating molecule caused by centrifugal force, resulting in an increase in the molecule's moment of inertia and a modification of its energy levels. { ,sen'trif·i·gəl 'strech·iŋ }

centrifuge microscope [OPTICS] An instrument which permits magnification and observation of living cells being centrifuged; image of the material magnified by the objective which rotates near the periphery of the centrifuge head is brought to the axis of rotation where it is observed in a stationary ocular. { 'sen·trə,fyüj 'mī·krə,skōp }

centripetal [MECH] Acting or moving in a direction toward the axis of rotation or the center of a circle along which a body is moving. { ,sen'trip·əd·əl }

centripetal acceleration [MECH] The radial component of the acceleration of a particle or object moving around a circle, which can be shown to be directed toward the center of the circle. Also known as radial acceleration. { ,sen'trip·əd·əl ik,sel·ə'rā·shən }

centripetal force [MECH] The radial force required to keep a particle or object moving in a circular path, which can be shown to be directed toward the center of the circle. { ,sen'trip·əd·əl 'fȯrs }

centrobaric [MECH] **1.** Pertaining to the center of gravity, or to some method of locating it. **2.** Possessing a center of gravity. { ¦sen·trō ¦bar·ik }

centrode [MECH] The path traced by the instantaneous center of a plane figure when it undergoes plane motion. { 'sen,trōd }

centroid *See* center of mass. { 'sen,trȯid }

centrosymmetry [PHYS] Property of a body or system which is unchanged under space inversion through a specified point. { ¦sen·trō'sim·ə·trē }

cepstrum [ACOUS] The Fourier transform of the logarithm of a speech power spectrum; used to separate vocal tract information from pitch excitation in voiced speech. { 'sep·trəm }

ceramagnet [ELECTROMAG] A ferrimagnet composed of the hard magnetic material $BaO \cdot 6Fe_2O_3$. { 'se·rə,mag·nət }

ceramic capacitor [ELEC] A capacitor whose dielectric is a ceramic material such as steatite or barium titanate, the composition of which can be varied to give a wide range of temperature coefficients. { sə'ram·ik kə'pas·əd·ər }

ceramic magnet [ELECTROMAG] A permanent magnet made from pressed and sintered mixtures of ceramic and magnetic powders. Also known as ferromagnetic ceramic. { sə'ram·ik 'mag·nət }

Cerenkov radiation [ELECTROMAG] Light emitted by a high-speed charged particle when the particle passes through a transparent, nonconducting material at a speed greater than the speed of light in the material. { chə'reŋ·kəf rād·ē'ā·shən }

cerium-140 [NUC PHYS] An isotope of cerium with atomic mass number of 140, 88.48% of the known amount of the naturally occurring element. { 'sir·ē·əm ,wən'fȯrd·ē }

cerium-142 [NUC PHYS] A radioactive isotope of cerium with atomic mass number of 142; emits α-particles and has a half-life of 5×10^{15} years. { 'sir·ē·əm ,wən,fȯrd·ē'tü }

cerium-144 [NUC PHYS] A radioactive isotope of the element cerium with atomic mass number of 144; a beta emitter with a half-life of 285 days. { 'sir·ē·əm ,wən,fȯrd·ē'fȯr }

cermet resistor [ELEC] A metal-glaze resistor, consisting of a mixture of finely powdered precious metals and insulating materials fired onto a ceramic substrate. { 'sər,met ri'zis·tər }

Cerruti's problem *See* Boussinesq's problem. { se'rü·dēz ,präb·ləm }

cesium fountain [ATOM PHYS] A device for performing highly accurate frequency measurements, in which cesium atoms are cooled and trapped by pairs of counterpropagating tuned laser beams and are thrown upward by shifting the frequency of the vertical lasers; they are further cooled and placed in one of the hyperfine states, and then pass through a microwave field region before falling into a detection region. { ¦sēz·ē·əm 'faunt·ən }

cesium-134 [NUC PHYS] An isotope of cesium, atomic mass number of 134; emits negative beta particles and has a half-life of 2.19 years; used in photoelectric cells and in ion propulsion systems under development. { 'sē·zē·əm ,wən,thərd·ē'fȯr }

cesium-137 [NUC PHYS] An isotope of cesium with atomic mass number of 137; emits negative beta particles and has a half-life of 30 years; offers promise as an encapsulated radiation source for therapeutic and other purposes. Also known as radiocesium. { 'sē·zē·əm ,wən,thərd·ē'sev·ən }

CFI *See* computational flow imaging.

cfs *See* cusec.

cg *See* centigram.

cgs system *See* centimeter-gram-second system. { ¦sē¦gē¦es 'sis·təm }

Chadwick-Goldhaber effect *See* photodisintegration. { 'chad·wik ˌgōlt'häb·ər i'fekt }

chain decay *See* series disintegration. { 'chān di'kā }

chain disintegration *See* series disintegration. { 'chān disˌint·ə·'grā·shən }

chain fission yield [NUC PHYS] The sum of the independent fission yields for all isobars of a particular mass number. { ¦chān 'fish·ən ˌyēld }

chain structure [SOLID STATE] A crystalline structure in which forces between atoms in one direction are greater than those in other directions, so that the atoms are concentrated in chains. { 'chān ˌstrək·chər }

chaldron [MECH] **1.** A unit of volume in common use in the United Kingdom, equal to 36 bushels, or 288 gallons, or approximately 1.30927 cubic meters. **2.** A unit of volume, formerly used for measuring solid substances in the United States, equal to 36 bushels, or approximately 1.26861 cubic meters. { 'chȯl·drən }

channeling [PHYS] The steering of energetic charged particles by the atomic rows or atomic planes of a crystalline solid. { 'chan·əl·iŋ }

channeling radiation [PHYS] The radiation emitted by energetic charged particles that pass through a solid. { 'chan·əl·iŋ ˌrād·ē'ā·shən }

channel spin [NUC PHYS] The vector sum of the spins of the particles involved in a nuclear reaction, either before or after the reaction takes place. { 'chan·əl ˌspin }

channel width [NUC PHYS] The part of the total energy width of a nuclear energy level that corresponds to a particular mode of decay. { 'chan·əl ˌwidth }

chaos *See* chaotic behavior. { 'kāˌäs }

chaotic advection [FL MECH] Fluid motion in which the trajectories of particles initially quite close diverge rapidly, even though the flow carrying the particles may be simple. { kā'äd·ik ad·'vek·shən }

chaotic behavior [MECH] The behavior of a system whose final state depends so sensitively on the system's precise initial state that the behavior is in effect unpredictable and cannot be distinguished from a random process, even though it is strictly determinate in a mathematical sense. Also known as chaos. { kā'äd·ik bi'hā·vyər }

Chaplygin-Kármán-Tsien relation [FL MECH] The relation that in the case of isentropic flow of an ideal gas with negligible viscosity and thermal conductivity, the sum of the pressure and a constant times the reciprocal of the density of the fluid is constant along a streamline; a useful, although physically impossible, approximation. { chə'plē·gən ˌkärˌmän 'tsyen ri'lā·shən }

Chapman-Enskog approximations [STAT MECH] Approximations to a solution of the Boltzmann transport equation in the Chapman-Enskog theory. { ¦chap·mən 'enˌskȯg əˌpräk·sə'mā·shənz }

Chapman-Enskog solution [STAT MECH] The solution of the Boltzmann transport equation according to the Chapman-Enskog theory. { ¦chap·mən 'enˌskȯg sə'lü·shən }

Chapman-Enskog theory [STAT MECH] A method of solving the Boltzmann transport equation by successive approximations, essentially in powers of the mean free path. Also known as Enskog theory. { ¦chap·mən 'enˌskȯg ˌthē·ə·rē }

Chapman equation [STAT MECH] The relationship that the viscosity of a gas equals $(0.499)mv/[\sqrt{2}\,\pi\sigma^2\,(1 + C/T)]$, where m is the mass of a molecule, v its average speed, σ its collision diameter, C the Sutherland constant, and T the absolute temperature (Kelvin scale). { 'chap·mən iˌkwā·zhən }

Chapman-Jouguet plane [MECH] A hypothetical, infinite plane, behind the initial shock front, in which it is variously assumed that reaction (and energy release) has effectively been completed, that reaction product gases have reached thermodynamic equilibrium, and that reaction gases, streaming backward out of the detonation, have reached such a condition that a forward-moving sound wave located at this precise plane would remain a fixed distance behind the initial shock. { ¦chap·mən zhü¦gwā ˌplān }

characteristic acoustic impedance [ACOUS] The product of the density and the speed of sound in a medium; it is analogous to the characteristic impedance of an infinitely long transmission line. Also known as intrinsic impedance. { ˌkar·ik·tə'ris·tik ə'kü·stik im'pēd·əns }

characteristic equation [PHYS] An equation relating a set of variables, such as pressure, volume, and temperature, whose values determine a substance's physical condition. [PL PHYS] An equation whose solutions give the frequencies and modes of those perturbations of a hydromagnetic system which decay or grow exponentially in time, and indicate regions of stability of such a system. { ˌkar·ik·tə'ris·tik i'kwā·zhən }

characteristic function [PHYS] A function, such as the point characteristic function or the principal function, which is the integral of some property of an optical or mechanical system over time or over the path followed by the system, and whose value for a path actually followed by a system is a maximum or a minimum with respect to nearby paths with the same end points. { ˌkar·ik·tə'ris·tik 'fəŋk·shən }

characteristic length [MECH] A convenient reference length (usually constant) of a given configuration, such as overall length of an aircraft, the maximum diameter or radius of a body of revolution, or a chord or span of a lifting surface. { ˌkar·ik·tə'ris·tik 'leŋkth }

characteristic loss spectroscopy [SPECT] A branch of electron spectroscopy in which a solid surface is bombarded with monochromatic electrons, and backscattered particles which have lost an amount of energy equal to the core-level binding energy are detected. Abbreviated CLS. { ˌkar·ik·tə'ris·tik 'lȯs ˌspek'träs·kə·pē }

characteristic radiation [ATOM PHYS] Radiation

originating in an atom following removal of an electron, whose wavelength depends only on the element concerned and the energy levels involved. { ˌkar·ik·tə'ris·tik ˌrād·ē'ā·shən }

characteristic temperature *See* Debye temperature. { ˌkar·ik·tə'ris·tik 'tem·prə·chər }

characteristic x-rays [ATOM PHYS] Electromagnetic radiation emitted as a result of rearrangements of the electrons in the inner shells of atoms; the spectrum consists of lines whose wavelengths depend only on the element concerned and the energy levels involved. { ˌkar·ik·tə'ris·tik 'eks,rāz }

charge [ELEC] **1.** A basic property of elementary particles of matter; the charge of an object may be a positive or negative number or zero; only integral multiples of the proton charge occur, and the charge of a body is the algebraic sum of the charges of its constituents; the value of the charge may be inferred from the Coulomb force between charged objects. Also known as electric charge, quantity of electricity. **2.** To convert electrical energy to chemical energy in a secondary battery. **3.** To feed electrical energy to a capacitor or other device that can store it. { chärj }

charge carrier [SOLID STATE] A mobile conduction electron or mobile hole in a semiconductor. Also known as carrier. { 'chärj ˌkar·ē·ər }

charge collector [ELEC] The structure within a battery electrode that provides a path for the electric current to or from the active material. Also known as current collector. { 'chärj kəˌlek·tər }

charge conjugation conservation [PART PHYS] The principle that the laws of motion are left unchanged by the charge conjugation operation; it is violated by the weak interactions, but no other violations have as yet been established. { 'chärj ˌkän·jə¦gā·shən ˌkän·sər'vā·shən }

charge conjugation operation [PART PHYS] The operation of changing every particle into its antiparticle. { 'chärj ˌkän·jə¦gā·shən ˌäp·ə'rā·shən }

charge conjugation parity *See* charge parity. { 'chärj ˌkän·jə¦gā·shən 'par·əd·ē }

charge conservation *See* conservation of charge. { 'chärj ˌkän·sər'vā·shən }

charged-current interaction [PART PHYS] A weak interaction in which the charges of the interacting fermions are changed; easily observed processes such as beta decay are of this type. { ¦chärjd ¦kər·ənt in·tər'ak·shən }

charge density [ELEC] The charge per unit area on a surface or per unit volume in space. { 'chärj ˌden·səd·ē }

charge-density wave [SOLID STATE] The ground state of a metal in which the conduction-electron charge density is sinusoidally modulated in space. { 'chärj ˌden·səd·ē ˌwāv }

charged particle [PART PHYS] A particle whose charge is not zero; the charge of a particle is added to its designation as a superscript, with particles of charge +1 and −1 (in terms of the charge of the proton) denoted by + and − respectively; for example, π^+, Σ^-. { 'chärjd 'pärd·ə·kəl }

charge exchange [PHYS] The transfer of electric charge from one particle to another during a collision between the two particles. { 'chärj iksˌchānj }

charge independence [NUC PHYS] The principle that the nuclear (strong) force between a neutron and a proton is identical to the force between two protons or two neutrons in the same orbital and spin state. [PART PHYS] As a generalization of the nuclear physics definition, the principle that the strong interactions of particles are unchanged if a particle is replaced by another particle of the same isotopic spin multiplet. { ¦chärj in·də'pen·dəns }

charge invariance [NUC PHYS] The principle that interactions between nucleons are left unchanged by rotations in isotopic spin space. { 'chärj in'ver·ē·əns }

charge-mass ratio [ELEC] The ratio of the electric charge of a particle to its mass. { ˌchärj ˌmas 'rā·shō }

charge multiplet *See* isospin multiplet. { ¦chärj 'məl·tə·plət }

charge neutrality [PL PHYS] The near equality in the density of positive and negative charges throughout a volume, which is characteristic of a plasma. [SOLID STATE] The condition in which electrons and holes are present in equal numbers in a semiconductor. { ¦chärj nü'tral·əd·ē }

charge parity [PART PHYS] The eigenvalue of the charge conjugation operation; it exists only for a system which goes into itself under this operation. Also known as charge conjugation parity. { 'chärj ˌpar·əd·ē }

charge quantization [ELEC] The principle that the electric charge of an object must equal an integral multiple of a universal basic charge. { 'chärj ˌkwan·tə'zā·shən }

charge-state process [SOLID STATE] A process involving the motion of preexisting crystal defects in a solid, following a change in the charges of the defects. { 'chärj ˌstāt ˌpräs·əs }

charging current [ELEC] The current that flows into a capacitor when a voltage is first applied. { 'chär·jiŋ ˌkər·ənt }

Charles' law [PHYS] The law that at constant pressure the volume of a fixed mass or quantity of gas varies directly with the absolute temperature; a close approximation. Also known as Gay-Lussac's first law. { 'chärlz ˌlȯ }

charm [PART PHYS] A quantum number which has been proposed to account for an apparent lack of symmetry in the behavior of hadrons relative to that of leptons, to explain why certain reactions of elementary particles do not occur, and to account for the longevity of the J-1 and J-2 particles. { chärm }

charmed particle [PART PHYS] A particle whose total charm is not equal to zero. { ¦chärmd 'pärd·ə·kəl }

charmed quark [PART PHYS] A quark with an electric charge of +2/3, baryon number of 1/3, zero strangeness, and charm of +1. Symbolized c. { ¦chärmd 'kwärk }

charmonium [PART PHYS] A meson, such as the J/ψ particle, that is made up of the charmed quark c and its antiparticle $\bar{c}$. { chär'mō·nē·əm }

Charpak-Massonet current distribution system [PART PHYS] An electronic data readout method used in spark chambers to locate a single spark, as determined by observing how the spark current divides between the two available paths to the ground. { ¦chär,päk ,mas·ō'nā 'kər·ənt dis·trə'byü·shən ,sis·təm }

cheese antenna [ELECTROMAG] An antenna having a parabolic reflector between two metal plates, dimensioned to permit propagation of more than one mode in the desired direction of polarization. { 'chēz an'ten·ə }

chelate laser [OPTICS] A liquid laser that uses a rare-earth chelate (a metalloorganic compound), with initial excitation taking place within the organic part of the liquid molecule and then transferring to the metallic ions to give lasing action. Also known as rare-earth chelate laser. { 'kē,lāt ,lā·zər }

chemical crystallography [CRYSTAL] The geometric description, and study, of the internal arrangement of atoms in crystals formed from chemical compounds. { 'kem·i·kəl kris·tə'läg·rə·fē }

chemical film dielectric [ELEC] An extremely thin layer of material on one or both electrodes of an electrolytic capacitor, which conducts electricity in only one direction and thereby constitutes the insulating element of the capacitor. { 'kem·i·kəl ,film ,dī·ə'lek·trik }

chemical focus *See* actinic focus. { 'kem·i·kəl 'fō·kəs }

chemical laser *See* chemically pumped laser. { 'kem·i·kəl 'lā·zər }

chemically pumped laser [OPTICS] A laser in which pumping is achieved by using a chemical action rather than electrical energy to produce the required pulses of light. Also known as chemical laser. { 'kem·ik·lē ¦pəmt 'lā·zər }

chemical vapor deposition [SOLID STATE] The growth of thin solid films on a crystalline substrate as the result of thermochemical vapor-phase reactions. Abbreviated CVD. { 'kem·i·kəl ¦vā·pər ,dep·ə'zish·ən }

Chevalier lens [OPTICS] A type of magnifying lens composed of an achromatic negative lens combined with a distant collecting front lens; a magnifying power up to 10X with an object distance up to 3 inches (7.62 centimeters) can be obtained. { shə'val·yā ,lenz }

Chevrel phase [SOLID STATE] One of a series of ternary molybdenum chalcogenide compounds with unusual superconducting properties and the general formula $M_xMo_6X_8$, where M represents any one of a large number of metallic elements, x has values between 1 and 4, and X is a chalcogen (sulfur, selenium, or tellurium). { she'vrel ,fāz }

Chézy formula [FL MECH] For the velocity V of open-channel flow which is steady and uniform, $V = \sqrt{8g/f \cdot mS}$, *where* f is the Darcy-Weisbach friction coefficient, m the hydraulic radius, S the energy dissipation per unit length, and g the acceleration of gravity. { 'shā·zē ,fȯr·myə·lə }

chief ray [OPTICS] A ray in a pencil that passes through the intersection of the axis of an optical system with the plane of the aperture stop. { ¦chēf 'rā }

chi meson [PART PHYS] A meson resonance of mass 958 MeV/c^2, designated χ_0, which has 0 isospin and charge, negative parity, positive G parity, and spin probably equal to 0. Also known as eta-prime meson (η'). Also denoted η'_A (958). { ,kī 'mā,zän }

chimney effect [FL MECH] The tendency of air or gas in a vertical passage to rise when it is heated because its density is lower than that of the surrounding air or gas. { 'chim,nē i'fekt }

chirality [PART PHYS] The characteristic of particles of spin 1/2 $\hbar$ that are allowed to have only one spin state with respect to an axis of quantization parallel to the particle's momentum; if the particle's spin is always parallel to its momentum, it has positive chirality; antiparallel, negative chirality. [PHYS] The characteristic of an object that cannot be superimposed upon its mirror image. { kī'ral·əd·ē }

chiral symmetry group [PART PHYS] A group of symmetry transformations that act differently on the left- and right-handed parts of fermion fields. { 'kī·rəl 'sim·ə·trē ,grüp }

chiral twinning *See* optical twinning. { 'kī·rəl 'twin·iŋ }

Chireix antenna [ELECTROMAG] A phased array composed of two or more coplanar square loops, connected in series. Also known as Chireix-Mesny antenna. { ki'rāks an,ten·ə }

Chireix-Mesny antenna *See* Chireix antenna. { ki'rāks ,mez,nē an,ten·ə }

Chladni's figures [MECH] Figures produced by sprinkling sand or similar material on a horizontal plate and then vibrating the plate while holding it rigid at its center or along its periphery; indicate the nodal lines of vibration. { 'klad,nēz ,fig·yərz }

choke [ELEC] An inductance used in a circuit to present a high impedance to frequencies above a specified frequency range without appreciably limiting the flow of direct current. Also known as choke coil. [ELECTROMAG] A groove or other discontinuity in a waveguide surface so shaped and dimensioned as to impede the passage of guided waves within a limited frequency range. { chōk }

choke coil *See* choke. { 'chōk ,kȯil }

choke coupling [ELECTROMAG] Coupling between two parts of a waveguide system that are not in direct mechanical contact with each other. { 'chōk ,kəp·liŋ }

choked flow [FL MECH] Flow in a duct or passage such that the flow upstream of a certain critical section cannot be increased by a reduction of downstream pressure. { ¦chōkt 'flō }

choke flange [ELECTROMAG] A waveguide flange having in its mating surface a slot (choke) so shaped and dimensioned as to restrict leakage of microwave energy within a limited frequency range. { 'chōk ˌflanj }

choke joint [ELECTROMAG] A connection between two waveguides that uses two mating choke flanges to provide effective electrical continuity without metallic continuity at the inner walls of the waveguide. { 'chōk ˌjȯint }

choke piston [ELECTROMAG] A piston in which there is no metallic contact with the walls of the waveguide at the edges of the reflecting surface; the short circuit for high-frequency currents is achieved by a choke system. Also known as noncontacting piston; noncontacting plunger. { 'chōk ˌpis·tən }

choking [FL MECH] The condition prevailing in compressible fluid flow when the upper limit of mass flow is reached, or when the speed of sound is reached in a duct. { 'chōk·iŋ }

choking Mach number [FL MECH] The Mach number at some reference point in a duct or passage (for example, at the inlet) at which the flow in the passage becomes choked. { ¦chōk·iŋ 'mäk ˌnəm·bər }

chopper [PHYS] A device for interrupting an electric current, beam of light, or beam of infrared radiation at regular intervals, to permit amplification of the associated electrical quantity or signal by an alternating-current amplifier; also used to interrupt a continuous stream of neutrons to measure velocity. { 'chäp·ər }

chopping [PHYS] The act of interrupting an electric current, beam of light, beam of infrared radiation, or stream of neutrons at regular intervals. { 'chäp·iŋ }

chord [ACOUS] A combination of two or more tones. { kȯrd }

Christiansen filter [OPTICS] A type of color filter, a solid-in-liquid suspension, which scatters all incident energy except that of a narrow frequency range out of the direct beam. Also known as band-pass filter. { 'kris·chən·sən 'fil·tər }

chroma [OPTICS] **1.** The dimension of the Munsell system of color that corresponds most closely to saturation, which is the degree of vividness of a hue. Also known as Munsell chroma. **2.** *See* color saturation. { 'krō·mə }

chromascope [OPTICS] An instrument used to determine the optical effects of color. { 'krō·məˌskōp }

chromatic [OPTICS] Relating to color. { krō 'mad·ik }

chromatic aberration [OPTICS] An optical lens defect causing color fringes, because the lens material brings different colors of light to focus at different points. Also known as color aberration. { krō'mad·ik ab·ə'rā·shən }

chromatic diagram *See* chromaticity diagram. { krō'mad·ik 'dī·əˌgram }

chromatic difference of magnification [OPTICS] Variation in the size of the image produced by an optical system with the wavelength (or, equivalently, color) of light. Also known as lateral chromatic aberration. { krō¦mad·ik ¦dif·rəns əv ˌmag·nə·fə'kā·shən }

chromaticity [OPTICS] The color quality of light that can be defined by its chromaticity coordinates; depends only on hue and saturation of a color, and not on its luminance (brightness). { ˌkrō·mə'tis·əd·ē }

chromaticity coordinates [OPTICS] The fractional amounts of the x, y, and z primary colors, specified by the International Committee on Illumination, in a color sample; more precisely, $x = X/(X + Y + Z)$, $y = Y/(X + Y + Z)$, $z = Z/(X + Y + Z)$, where X, Y, and Z are the integrals over wavelength λ of the product of the amount of light emerging from the sample per unit wavelength, and the tristimulus values, $\bar{x}(\lambda)$, $\bar{y}(\lambda)$, and $\bar{z}(\lambda)$, respectively. { ˌkrō·mə'tis·əd·ē kō'ȯrd·ən·ˌāts }

chromaticity diagram [OPTICS] A triangular graph for specifying colors, whose ordinate is the y chromaticity coordinate and whose abscissa is the x chromaticity coordinate; the apexes of the triangle represent primary colors. Also known as chromatic diagram. { ˌkrō·mə'tis·əd·ē dī·əˌgram }

chromatic parallax [OPTICS] A type of optical parallax that arises from the dependence of the position of the focal plane on the wavelength of light. { krō'mad·ik 'par·əˌlaks }

chromatic resolving power [OPTICS] The difference between two equally strong spectral lines that can barely be separated by a spectroscopic instrument, divided into the average wavelength of these two lines; for prisms and gratings Rayleigh's criteria are used, and the term is defined as the width of the emergent beam times the angular dispersion. { krō'mad·ik rə'zälv·iŋ ˌpau̇·ər }

chromatics [OPTICS] **1.** The branch of optics concerned with the properties of colors. **2.** The part of colorimetry concerned with hue and saturation. { krō'mad·iks }

chromatic sensitivity [OPTICS] The smallest change in wavelength of light that produces a change in hue which is just large enough to be detected by human vision. { krō'mad·ik sen·sə'tiv·əd·ē }

chromatoscope [OPTICS] An instrument in which light beams are used to mix color stimuli. { krō'mad·əˌskōp }

chrominance [OPTICS] The difference between any color and a specified reference color of equal brightness; in color television, this reference color is white having coordinates $x = 0.310$ and $y = 0.316$ on the chromaticity diagram. { 'krō·mə·nəns }

chromium-51 [NUC PHYS] A radioactive isotope with atomic mass 51 made by neutron bombardment of chromium; radiates gamma rays. { 'krō·mē·əm ˌfif·tē'wən }

chromodynamics [PART PHYS] A theory of the interaction between quarks carrying color in which the quarks exchange gluons in a manner analogous to the exchange of photons between charged particles in electrodynamics. { ¦krō·mō·dī'nam·iks }

chromoscope [OPTICS] An instrument for analyzing color values and intensities. { 'krō·mə,skōp }

chronological future [RELAT] The chronological future relative to a set of points S in a space-time M is the set of points in M which can be reached from S by future-directed timelike curves. { ,krän·ə¦läj·ə·kəl 'fyü·chər }

chronological past [RELAT] The chronological past relative to a set of points S in a space-time M is the set of points in M which can be reached from S by past-directed timelike curves. { ,krän·ə¦läj·ə·kəl 'past }

chronon [PHYS] A hypothetical quantum of time, given approximately by the time taken for light to traverse the classical electron radius, on the order of 10^{-23} second. { 'krän·ən }

CHU *See* centigrade heat unit.

CHU_{mean} *See* centigrade heat unit.

circle diagram [ELEC] A diagram which gives a graphical solution of equations for a transmission line, giving the input impedance of the line as a function of load impedance and electrical length of the line. { ¦sər·kəl ¦dī·ə,gram }

circle of confusion [OPTICS] The blurred circular image of a point object which is formed by a camera lens, even with the best focusing. Also known as circle of least confusion. { 'sər·kəl əv kən'fyü·zhən }

circle of least confusion *See* circle of confusion. { 'sər·kəl əv 'lēst kən'fyü·zhən }

circuit [ELEC] *See* electric circuit. [ELECTROMAG] A complete wire, radio, or carrier communications channel. { 'sər·kət }

circuital field *See* rotational field. { sə¦kyü·əd·əl 'fēld }

circuit design [ELEC] The art of specifying the components and interconnections of an electrical network. { 'sər·kət də'zīn }

circuit diagram [ELEC] A drawing, using standardized symbols, of the arrangement and interconnections of the conductors and components of an electrical or electronic device or installation. Also known as schematic circuit diagram; wiring diagram. { 'sər·kət ,dī·ə,gram }

circuit element *See* component. { 'sər·kət ¦el·ə·mənt }

circuit loading [ELEC] Power drawn from a circuit by an electric measuring instrument, which may alter appreciably the quantity being measured. { 'sər·kət ,lōd·iŋ }

circuitry [ELEC] The complete combination of circuits used in an electrical or electronic system or piece of equipment. { 'sər·kə·trē }

circuit theory [ELEC] The mathematical analysis of conditions and relationships in an electric circuit. Also known as electric circuit theory. { 'sər·kət ,thē·ə·rē }

circular antenna [ELECTROMAG] A folded dipole that is bent into a circle, so the transmission line and the abutting folded ends are at opposite ends of a diameter. { 'sər·kyə·lər an 'ten·ə }

circular birefringence [OPTICS] The phenomenon in which an optically active substance transmits right circularly polarized light with a different velocity from left circularly polarized light. { 'sər·kyə·lər ,bī·rə'frin·jəns }

circular coil [ELECTROMAG] In eddy-current nondestructive tests, a type of test coil which surrounds an object. { 'sər·kyə·lər ¦kȯil }

circular current [ELEC] An electric current moving in a circular path. { 'sər·kyə·lər 'kər·ənt }

circular dichroism [OPTICS] A change from planar to elliptic polarization when an initially plane-polarized light wave traverses an optically active medium. Abbreviated CD. { 'sər·kyə·lər 'dī·krō,iz·əm }

circular electric wave [ELECTROMAG] A transverse electric wave for which the lines of electric force form concentric circles. { 'sər·kyə·lər i¦lek·trik 'wāv }

circular flow method [FL MECH] A method to determine viscosities of Newtonian fluids by measuring the torque from viscous drag of sample material between a closely spaced rotating plate-stationary cone assembly. { 'sər·kyə·lər ,flō 'meth·əd }

circular horn [ELECTROMAG] A circular-waveguide section that flares outward into the shape of a horn, to serve as a feed for a microwave reflector or lens. { 'sər·kyə·lər 'hȯrn }

circular inch [MECH] The area of a circle 1 inch (25.4 millimeters) in diameter. { 'sər·kyə·lər 'inch }

circular magnetic wave [ELECTROMAG] A transverse magnetic wave for which the lines of magnetic force form concentric circles. { 'sər·kyə·lər mag'ned·ik 'wāv }

circular magnetostriction *See* Wiedemann effect. { 'sər·kyə·lər mag'ned·ə,strik·shən }

circular mil [MECH] A unit equal to the area of a circle whose diameter is 1 mil (0.001 inch); used chiefly in specifying cross-sectional areas of round conductors. Abbreviated cir mil. { 'sər·kyə·lər 'mil }

circular motion [MECH] **1.** Motion of a particle in a circular path. **2.** Motion of a rigid body in which all its particles move in circles about a common axis, fixed with respect to the body, with a common angular velocity. { 'sər·kyə·lər 'mō·shən }

circular polarization [PHYS] Attribute of a transverse wave (either of electromagnetic radiation, or in an elastic medium) whose electric or displacement vector is of constant amplitude and, at a fixed point in space, rotates in a plane perpendicular to the propagation direction with constant angular velocity. { 'sər·kyə·lər ,pō·lə·rə'zā·shən }

circular polarized loop vee [ELECTROMAG] Airborne communications antenna with an omnidirectional radiation pattern to provide optimum near-horizon communications coverage. { 'sər·kyə·lər 'pō·lə,rīzd 'lüp ,vē }

circular velocity [MECH] At any specific distance from the primary, the orbital velocity required to maintain a constant-radius orbit. { 'sər·kyə·lər və'läs·əd·ē }

circular waveguide [ELECTROMAG] A waveguide whose cross-sectional area is circular. { 'sər·kyə·lər 'wāv,gīd }

circulating electromagnetic wave [ELECTROMAG] An electromagnetic wave whose equiphase surfaces are half-planes originating at a common axis. { ¦sər·kyə,lād·iŋ i,lek·trō·mag¦ned·ik 'wāv }

circulation [FL MECH] The flow or motion of fluid in or through a given area or volume. { ,sər·kyə·'lā·shən }

circulator [ELECTROMAG] A waveguide component having a number of terminals so arranged that energy entering one terminal is transmitted to the next adjacent terminal in a particular direction. Also known as microwave circulator. { ,sər·kyə·'lād·ər }

circumaural cushion [ACOUS] An earphone cushion that completely surrounds the auricle. { ,sər·kəm,ȯr·əl 'ku̇sh·ən }

circumhorizontal arc [OPTICS] A halo phenomenon consisting of a colored arc, red on its upper margin; it extends for about 90° parallel to the horizon and lies about 46° below the sun. { ¦sər·kəm,här·ə'zänt·əl 'ärk }

circumzenithal arc [OPTICS] A brilliant rainbow-colored arc of about a quarter of a circle with its center at the zenith and about 46° above the sun, produced by refraction and dispersion of the sun's light striking the top of prismatic ice crystals in the atmosphere, and usually lasting only a few minutes. { ¦sər·kəm¦zē·nə·thəl 'ärk }

cir mil *See* circular mil.

Clapeyron-Clausius equation *See* Clausius-Clapeyron equation. { kla·pā·rōn ¦klȯz·ē·əs i,kwā·zhən }

Clapeyron equation *See* Clausius-Clapeyron equation. { kla·pā·rōn i'kwā·zhən }

Clapeyron's theorem [MECH] The theorem that the strain energy of a deformed body is equal to one-half the sum over three perpendicular directions of the displacement component times the corresponding force component, including deforming loads and body forces, but not the six constraining forces required to hold the body in equilibrium. { kla·pā·rōnz ,thir·əm }

classical approximation [QUANT MECH] The approximation that Planck's constant may be considered infinitely small; the laws of quantum mechanics must then reduce to those of classical mechanics. { 'klas·ə·kəl ə,präk·sə'mā·shən }

classical attenuation [ACOUS] Sound absorption through mechanisms that do not involve molecular relaxation, namely, shear viscosity, heat conduction, heat radiation, and diffusion. Also known as thermoviscous attenuation. { ,klas·i·kəl ə,ten·yə'wā·shən }

classical conductivity theory [STAT MECH] A theory which treats the system of electrons in a metal as a gas and uses the Boltzmann transport equation to calculate conductivity. Also known as Lorentz conductivity theory. { 'klas·ə·kəl ,kän·dək'tiv·əd·ē 'thē·ə·rē }

classical electron radius [ELECTROMAG] The quantity expressed as e^2/m_ec^2, where e is the electron's charge in electrostatic units, m_e its mass, and c the speed of light; equal to approximately 2.82×10^{-13} centimeter. { 'klas·ə·kəl i'lek ,trän 'rād·ē·əs }

classical field theory [PHYS] The study of distributions of energy, matter, and other physical quantities under circumstances where their discrete nature is unimportant, and they may be regarded as (in general, complex) continuous functions of position. Also known as c-number theory; continuum mechanics; continuum physics. { 'klas·ə·kəl 'fēld ,thē·ə·rē }

classical mechanics [MECH] Mechanics based on Newton's laws of motion. { 'klas·ə·kəl mə'kan·iks }

classical physics [PHYS] The branch of physics that is based on the assumption of Newtonian mechanics and excludes relativity and quantum mechanics. { 'klas·ə·kəl 'fiz·iks }

classical wave equation *See* wave equation. { 'klas·ə·kəl 'wāv i'kwā·zhən }

classons [PART PHYS] Massless bosons which are quanta of the two classical fields, gravitational and electromagnetic. { 'kla,sänz }

Claude process [CRYO] A method of liquefying air or other gases in stages, in which the gas is cooled by doing work in an expansion engine and then undergoing the Joule-Thomson effect as it passes through an expansion valve. { 'klōd ,präs·əs }

clausius [THERMO] A unit of entropy equal to the increase in entropy associated with the absorption of 1000 international table calories of heat at a temperature of 1 K, or to 4186.8 joules per kelvin. { 'klȯz·ē·əs }

Clausius-Clapeyron equation [THERMO] An equation governing phase transitions of a substance, $dp/dT = \Delta H/(T\Delta V)$, in which p is the pressure, T is the temperature at which the phase transition occurs, ΔH is the change in heat content (enthalpy), and ΔV is the change in volume during the transition. Also known as Clapeyron-Clausius equation; Clapeyron equation. { klȯz·ē·əs kla·pā,rōn i,kwā·zhən }

Clausius equation [THERMO] An equation of state in reference to gases which applies a correction to the van der Waals equation: $\{P + (n^2a/[T(V + c)^2])\}\ (V - nb) = nRT$, where P is the pressure, T the temperature, V the volume of the gas, n the number of moles in the gas, R the gas constant, a depends only on temperature, b is a constant, and c is a function of a and b. { 'klȯz·ē·əs i'kwā·zhən }

Clausius inequality [THERMO] The principle that for any system executing a cyclical process, the integral over the cycle of the infinitesimal amount of heat transferred to the system divided by its temperature is equal to or less than zero. Also known as Clausius theorem; inequality of Clausius. { 'klȯz·ē·əs in·i'kwäl·əd·ē }

Clausius law [THERMO] The law that an ideal

gas's specific heat at constant volume does not depend on the temperature. { 'klȯz·ē·əs ˌlō }

Clausius-Mosotti equation [ELEC] An expression for the polarizability γ of an individual molecule in a medium which has the relative dielectric constant ϵ and has N molecules per unit volume: $\gamma = (3/4\pi N)[(\epsilon - 1)/(\epsilon + 2)]$ (Gaussian units). { ¦klȯz·ē·əs mə'zäd·ē i'kwā·zhən }

Clausius-Mosotti-Lorentz-Lorenz equation [ELECTROMAG] The equation that results from replacing the real relative dielectric constant in the Clausius-Mosotti equation, or the real index of refraction in the Lorentz-Lorenz equation, with its complex counterpart. { ¦klȯz·ē·əs mə'zäd·ē 'lȯ·rens'lȯ·rens i'kwā·zhən }

Clausius number [THERMO] A dimensionless number used in the study of heat conduction in forced fluid flow, equal to $V^3L\rho/k\Delta T$, where V is the fluid velocity, ρ is its density, L is a characteristic dimension, k is the thermal conductivity, and ΔT is the temperature difference. { 'klȯz·ē·əs ˌnəm·bər }

Clausius range [STAT MECH] The condition in which the mean free path of molecules in a gas is much smaller than the dimensions of the container. { 'klȯz·ē·əs 'rānj }

Clausius' statement [THERMO] A formulation of the second law of thermodynamics, stating it is not possible that, at the end of a cycle of changes, heat has been transferred from a colder to a hotter body without producing some other effect. { 'klȯz·ē·əs 'stāt·mənt }

Clausius theorem *See* Clausius inequality. { 'klȯz·ē·əs 'thir·əm }

Clausius virial theorem [STAT MECH] The theorem that in a system of particles whose positions and velocities are bounded, the total kinetic energy of the system averaged over a long period of time equals the virial of the system. Also known as virial theorem. { 'klȯz·ē·əs 'vir·ē·əl 'thir·əm }

cleavage [CRYSTAL] Splitting, or the tendency to split, along planes determined by crystal structure and always parallel to a possible face. { 'klēv·ij }

cleavage crystal [CRYSTAL] A crystal fragment bounded by cleavage faces giving it a regular form. { 'klēv·ij ˌkris·təl }

cleavage fracture [CRYSTAL] **1.** Manner of breaking a crystalline substance along the cleavage plane. **2.** The appearance of such a broken surface. { 'klēv·ij ˌfrak·chər }

cleavage plane [CRYSTAL] Plane along which a crystalline substance may be split. { 'klēv·ij ˌplān }

Clebsch-Gordan coefficient *See* vector coupling coefficient. { ¦klepsh ¦gȯrd·ən ˌkō·ə'fish·ənt }

clinoaxis [CRYSTAL] The inclined lateral axis that makes an oblique angle with the vertical axis in the monoclinic system. Also known as clinodiagonal. { ¦klī·nō'ak·səs }

clinodiagonal *See* clinoaxis. { ¦klī·nō·dī'ag·ən·əl }

clinohedral class [CRYSTAL] A rare class of crystals in the monoclinic system having a plane of symmetry but no axis of symmetry. Also known as domatic class. { ¦klī·nō¦hē·drəl ˌklas }

clinopinacoid [CRYSTAL] A form of monoclinic crystal whose faces are parallel to the inclined and vertical axes. { ¦klī·nə'pin·əˌkȯid }

clock paradox [RELAT] The apparent contradiction between the principle of relativity, which asserts the equivalence of different observers, and the prediction, also part of the theory of relativity, that the clock of an observer who passes back and forth will be slower than the clock of an observer at rest. Also known as twin paradox. { 'kläk ˌpar·əˌdäks }

close coupling [ELEC] **1.** The coupling obtained when the primary and secondary windings of a radio-frequency or intermediate-frequency transformer are close together. **2.** A degree of coupling that is greater than critical coupling. Also known as tight coupling. { ¦klōs 'kəp·liŋ }

close-coupling method [ATOM PHYS] A method of approximating the wave function describing a process in which an initially free electron impinges on an isolated atom; in this method, the unknown solution of the Schrödinger equation is expanded in terms of the known wave functions of the target atom. { ˌklōs 'kəp·liŋ ˌmeth·əd }

closed circuit [ELEC] A complete path for current. { ¦klōzd 'sər·kət }

closed cycle [THERMO] A thermodynamic cycle in which the thermodynamic fluid does not enter or leave the system, but is used over and over again. { ˌklōzd 'sī·kəl }

closed magnetic circuit [ELECTROMAG] A complete circulating path for magnetic flux around a core of ferromagnetic material. { ¦klōzd mag'ned·ik 'sər·kət }

closed pair [MECH] A pair of bodies that are subject to constraints which prevent any relative motion between them. { ¦klōzd 'per }

closed shell [PHYS] An atomic or nuclear shell containing the maximum number of electrons or nucleons allowed by the Pauli exclusion principle. { ¦klōzd 'shel }

closed system [THERMO] A system which is isolated so that it cannot exchange matter or energy with its surroundings and can therefore attain a state of thermodynamic equilibrium. Also known as isolated system. { ¦klōzd 'sis·təm }

closed trapped surface [RELAT] A compact spacelike two-surface in space-time such that outgoing null rays perpendicular to the surface are not expanding. { ¦klōzd ¦trapt 'sər·fəs }

close-packed crystal [CRYSTAL] A crystal structure in which the lattice points are centers of spheres of equal radius arranged so that the volume of the interstices between the spheres is minimal. { ¦klōs ¦pakt 'kris·təl }

closing line [MECH] The vector required to complete a polygon consisting of a set of vectors whose sum is zero (such as the forces acting on a body in equilibrium). { 'klōz·iŋ ˌlīn }

closure domain [SOLID STATE] A small ferromagnetic domain whose position and orientation ensure that the flux lines between adjacent larger domains close on themselves. Also known as flux-closure domain. { 'klō·zhər dō,mān }

cloud [NUC PHYS] The nucleons that are in the nucleus of an atom but not in closed shells. { klaůd }

cloud attenuation [ELECTROMAG] The attenuation of microwave radiation by clouds (for the centimeter-wavelength band, clouds produce Rayleigh scattering); due largely to scattering, rather than absorption, for both ice and water clouds. { 'klaůd ə,ten·yə'wā·shən }

cloudy-crystal-ball model [NUC PHYS] An optical analogy used in explaining scattering of nucleons by nuclei, in which the nucleus is thought of as a sphere of nuclear matter which partially refracts and partially absorbs the incident nucleon (de Broglie) wave. Also known as optical model. { 'klaůd·ē ¦kris·təl ¦bȯl ,mäd·əl }

cloverleaf antenna [ELECTROMAG] Antenna having radiating units shaped like a four-leaf clover. { 'klō·vər,lēf an 'ten·ə }

CLS *See* characteristic loss spectroscopy.

cluster aggregation [PHYS] A mathematical model of a coagulation process in which a collection of particles all move randomly at once, and two particles, or a particle and a previously formed cluster, stick together whenever they come within a certain fixed distance of each other. { 'kləs·tər ,ag·rə'gā·shən }

cluster expansion [STAT MECH] A virial expansion in which the virial coefficients (of inverse powers of the volume of the gas in question) are obtained from integrals, over positions of a small number of molecules, of functions involving intermolecular potentials. { 'kləs·tər ik'span·shən }

cluster radioactivity [NUC PHYS] A process in which a nucleus emits a fragment that is heavier than an alpha particle but lighter than a fission fragment, such as carbon-14 or neon-24. { 'kləs·tər 'rād·ē·ō ak,tiv·əd·ē }

cm *See* centimeter.

cmHg *See* centimeter of mercury.

C mode [ACOUS] A form of acoustic tomography that provides a two-dimensional image display at constant time delay, and presumably at constant distance from the ultrasonic transducer. { 'sē ,mōd }

CNO cycles *See* carbon-nitrogen-oxygen cycles. { ¦sē¦en¦ō ,sī·kəlz }

c-number theory *See* classical field theory. { 'sē ,nəm·bər 'thē·ə·rē }

^{60}Co *See* cobalt-60.

coalescence [PHYS] The uniting by growth in one body, as particles, gas, or a liquid. { ,kō·ə'les·əns }

Coanda effect [FL MECH] The tendency of a gas or liquid coming out of a jet to travel close to the wall contour even if the wall's direction of curvature is away from the jet's axis; a factor in the operation of a fluidic element. { kō'an·də i'fekt }

coated lens [OPTICS] A lens whose surfaces have been coated with a thin, transparent film having an index of refraction that minimizes light loss by reflection. { 'kōd·əd 'lenz }

coax *See* coaxial cable. { 'kō,aks }

coaxial [MECH] Sharing the same axes. { kō 'ak·sē·əl }

coaxial antenna [ELECTROMAG] An antenna consisting of a quarter-wave extension of the inner conductor of a coaxial line and a radiating sleeve that is in effect formed by folding back the outer conductor of the coaxial line for a length of approximately a quarter wavelength. { kō'ak·sē·əl an'ten·ə }

coaxial attenuator [ELECTROMAG] An attenuator that has a coaxial construction and terminations suitable for use with coaxial cable. { kō'ak·sē·əl ə'ten·yə,wād·ər }

coaxial cable [ELECTROMAG] A transmission line in which one conductor is centered inside and insulated from an outer metal tube that serves as the second conductor. Also known as coax; coaxial line; coaxial transmission line; concentric cable; concentric line; concentric transmission line. { kō'ak·sē·əl 'kā·bəl }

coaxial capacitor *See* cylindrical capacitor. { kō'ak·sē·əl kə'pas·əd·ər }

coaxial cavity [ELECTROMAG] A cylindrical resonating cavity having a central conductor in contact with its pistons or other reflecting devices. { kō'ak·sē·əl 'kav·əd·ē }

coaxial connector [ELECTROMAG] An electric connector between a coaxial cable and an equipment circuit, so constructed as to maintain the conductor configuration, through the separable connection, and the characteristic impedance of the coaxial cable. { kō'ak·sē·əl kə'nek·tər }

coaxial filter [ELECTROMAG] A section of coaxial line having reentrant elements that provide the inductance and capacitance of a filter section. { kō'ak·sē·əl 'fil·tər }

coaxial hybrid [ELECTROMAG] A hybrid junction of coaxial transmission lines. { kō'ak·sē·əl 'hī,brəd }

coaxial isolator [ELECTROMAG] An isolator used in a coaxial cable to provide a higher loss for energy flow in one direction than in the opposite direction; all types use a permanent magnetic field in combination with ferrite and dielectric materials. { kō'ak·sē·əl 'ī·sə,lād·ər }

coaxial line *See* coaxial cable. { kō'ak·sē·əl 'līn }

coaxial-line resonator [ELECTROMAG] A resonator consisting of a length of coaxial line short-circuited at one or both ends. { kō'ak·sē·əl ,līn 'rez·ən,ād·ər }

coaxially fed linear array [ELECTROMAG] A beacon antenna having a uniform azimuth pattern. { kō'ak·sē·ə·lē ,fed 'lin·ē·ər ə'rā }

coaxial stub [ELECTROMAG] A length of nondissipative cylindrical waveguide or coaxial cable branched from the side of a waveguide to produce some desired change in its characteristics. { kō'ak·sē·əl 'stəb }

coaxial transmission line *See* coaxial cable. { kō'ak·sē·əl tranz'mish·ən ˌlīn }

cobalt-60 [NUC PHYS] A radioisotope of cobalt, symbol ^{60}Co, having a mass number of 60; emits gamma rays and has many medical and industrial uses; the most commonly used isotope for encapsulated radiation sources. { 'kōˌbȯlt 'siks·tē }

Coddington lens [OPTICS] A magnifier consisting of a glass sphere with a deep groove cut around a great circle to serve as a stop. { 'käd·iŋ·tən ˌlenz }

Coddington shape factor *See* shape factor. { 'käd·iŋ·tən 'shāp ˌfak·tər }

coded passive reflector antenna [ELECTROMAG] An object intended to reflect Hertzian waves and having variable reflecting properties according to a predetermined code for the purpose of producing an indication on a radar receiver. { 'kōd·əd 'pas·iv ri'flek·tər anˌten·ə }

coefficient of absorption *See* absorption coefficient. { ¦kō·ə'fish·ənt əv əb'sȯrp·shən }

coefficient of capacitance [ELEC] One of the coefficients which appears in the linear equations giving the charges on a set of conductors in terms of the potentials of the conductors; a coefficient is equal to the ratio of the charge on a given conductor to the potential of the same conductor when the potentials of all the other conductors are 0. { ¦kō·ə'fish·ənt əv kə'pas·ə·təns }

coefficient of compressibility [MECH] The decrease in volume per unit volume of a substance resulting from a unit increase in pressure; it is the reciprocal of the bulk modulus. { ¦kō·ə'fish·ənt əv kəmˌpres·ə'bil·əd·ē }

coefficient of condensation [STAT MECH] The ratio of the number of molecules condensed on the surface of a solid or liquid in equilibrium with its vapor phase to the total number of vapor molecules striking the surface. { ¦kō·ə'fish·ənt əv ˌkän·dən'sā·shən }

coefficient of conductivity *See* thermal conductivity. { ¦kō·ə'fish·ənt əv ˌkän·dək'tiv·əd·ē }

coefficient of contraction [FL MECH] The ratio of the minimum cross-sectional area of a jet of liquid discharging from an orifice to the area of the orifice. Also known as contraction coefficient. { ¦kō·ə'fish·ənt əv kən'trak·shən }

coefficient of coupling *See* coupling constant. { ¦kō·ə'fish·ənt əv 'kəp·liŋ }

coefficient of cubical expansion [THERMO] The increment in volume of a unit volume of solid, liquid, or gas for a rise of temperature of 1° at constant pressure. Also known as coefficient of expansion; coefficient of thermal expansion; coefficient of volumetric expansion; expansion coefficient; expansivity. { ¦kō·ə'fish·ənt əv 'kyüb·ə·kəl ik'span·shən }

coefficient of discharge *See* discharge coefficient. { ¦kō·ə'fish·ənt əv 'disˌchärj }

coefficient of eddy diffusion *See* eddy diffusivity. { ¦kō·ə'fish·ənt əv 'ed·ē də'fyü·zhən }

coefficient of eddy viscosity [FL MECH] The portion of the kinematic viscosity of a turbulent fluid that is associated with its eddy viscosity. Also known as coefficient of turbulence. { ¦kō·ə'fish·ənt əv 'ed·ē viˌskäs·əd·ē }

coefficient of elasticity *See* modulus of elasticity. { ¦kō·ə'fish·ənt əv iˌlas'tis·əd·ē }

coefficient of expansion *See* coefficient of cubical expansion. { ¦kō·ə'fish·ənt əv ik'span·shən }

coefficient of friction [MECH] The ratio of the frictional force between two bodies in contact, parallel to the surface of contact, to the force, normal to the surface of contact, with which the bodies press against each other. Also known as friction coefficient. { ¦kō·ə'fish·ənt əv 'frik·shən }

coefficient of friction of rest *See* coefficient of static friction. { ¦kō·ə'fish·ənt əv 'frik·shən əv 'rest }

coefficient of induction [ELEC] One of the coefficients which appears in the linear equations giving the charges on a set of conductors in terms of the potentials of the conductors; a coefficient is equal to the ratio of the charge on a given conductor to the potential on another conductor, when the potentials of all the other conductors equal 0. { ¦kō·ə'fish·ənt əv in'dək·shən }

coefficient of kinematic viscosity *See* kinematic viscosity. { ¦kō·ə'fish·ənt əv ¦kin·ə¦mad·ik vis'käs·əd·ē }

coefficient of kinetic friction [MECH] The ratio of the frictional force, parallel to the surface of contact, that opposes the motion of a body which is sliding or rolling over another, to the force, normal to the surface of contact, with which the bodies press against each other. { ¦kō·ə'fish·ənt əv kə'ned·ik 'frik·shən }

coefficient of linear expansion [THERMO] The increment of length of a solid in a unit of length for a rise in temperature of 1° at constant pressure. Also known as linear expansivity. { ¦kō·ə'fish·ənt əv 'lin·ē·ər ik'span·shən }

coefficient of performance [THERMO] In a refrigeration cycle, the ratio of the heat energy extracted by the heat engine at the low temperature to the work supplied to operate the cycle; when used as a heating device, it is the ratio of the heat delivered in the high-temperature coils to the work supplied. { ¦kō·ə'fish·ənt əv pər'fȯr·məns }

coefficient of permeability *See* permeability coefficient. { ¦kō·ə'fish·ənt əv ˌpər·mē·ə'bil·əd·ē }

coefficient of potential [ELEC] One of the coefficients which appears in the linear equations giving the potentials of a set of conductors in terms of the charges on the conductors. { ¦kō·ə'fish·ənt əv pə'ten·chəl }

coefficient of reflection *See* reflection coefficient. { ¦kō·ə'fish·ənt əv ri'flek·shən }

coefficient of resistance [FL MECH] The ratio of the loss of head of fluid, issuing from an orifice or passing over a weir, to the remaining head. { ¦kō·ə'fish·ənt əv ri'zis·təns }

coefficient of restitution [MECH] The constant e, which is the ratio of the relative velocity of two elastic spheres after direct impact to that before impact; e can vary from 0 to 1, with 1 equivalent to an elastic collision and 0 equivalent to a perfectly elastic collision. Also known as restitution coefficient. { ¦kō·ə'fish·ənt əv ˌres·tə'tü·shən }

coefficient of rigidity *See* modulus of elasticity in shear. { ¦kō·ə'fish·ənt əv rə'jid·əd·ē }

coefficient of rolling friction [MECH] The ratio of the frictional force, parallel to the surface of contact, opposing the motion of a body rolling over another, to the force, normal to the surface of contact, with which the bodies press against each other. { ¦kō·ə'fish·ənt əv 'rōl·iŋ 'frik·shən }

coefficient of sliding friction [MECH] The ratio of the frictional force, parallel to the surface of contact, opposing the motion of a body sliding over another, to the force, normal to the surface of contact, with which the bodies press against each other. { ¦kō·ə'fish·ənt əv 'slīd·iŋ 'frik·shən }

coefficient of static friction [MECH] The ratio of the maximum possible frictional force, parallel to the surface of contact, which acts to prevent two bodies in contact, and at rest with respect to each other, from sliding or rolling over each other, to the force, normal to the surface of contact, with which the bodies press against each other. Also known as coefficient of friction of rest. { ¦kō·ə'fish·ənt əv 'stad·ik 'frik·shən }

coefficient of strain [MECH] For a substance undergoing a one-dimensional strain, the ratio of the distance along the strain axis between two points in the body, to the distance between the same points when the body is undeformed. { ¦kō·ə'fish·ənt əv 'strān }

coefficient of superficial expansion [THERMO] The increment in area of a solid surface per unit of area for a rise in temperature of 1° at constant pressure. Also known as superficial expansivity. { ¦kō·ə'fish·ənt əv ˌsü·pər'fish·əl ik'span·chən }

coefficient of thermal expansion *See* coefficient of cubical expansion. { ¦kō·ə'fish·ənt əv 'thər·məl ik'span·shən }

coefficient of turbulence *See* coefficient of eddy viscosity. { ¦kō·ə'fish·ənt əv 'tər·byə·ləns }

coefficient of velocity *See* velocity coefficient. { ¦kō·ə'fish·ənt əv və'läs·əd·ē }

coefficient of viscosity *See* absolute viscosity. { ¦kō·ə'fish·ənt əv vis'käs·əd·ē }

coefficient of volumetric expansion *See* coefficient of cubical expansion. { ¦kō·ə'fish·ənt əv ¦väl·yə¦me·trik ik'span·chən }

coercive force [ELECTROMAG] The magnetic field H which must be applied to a magnetic material in a symmetrical, cyclicly magnetized fashion, to make the magnetic induction B vanish. Also known as magnetic coercive force. { kō'ər·siv 'fȯrs }

coercivity [ELECTROMAG] The coercive force of a magnetic material in a hysteresis loop whose maximum induction approximates the saturation induction. { ˌkō·ər'siv·əd·ē }

cogging [ELECTROMAG] Variations in torque and speed of an electric motor due to variations in magnetic flux as rotor poles move past stator poles. { 'käg·iŋ }

coherence [PHYS] **1.** The existence of a correlation between the phases of two or more waves, so that interference effects may be produced between them, or of a correlation between the phases of part of a single wave. **2.** Property of moving in unison, such as is characteristic of the particles in a synchrotron. { kō'hir·əns }

coherence area [OPTICS] A quantitative measure of the spatial coherence of a light beam, equal to the largest cross-sectional area such that light passing through any two pinholes placed in this area will produce interference fringes. { kō'hir·əns ˌer·ē·ə }

coherence distance *See* coherence length. { kō'hir·əns ˌdis·təns }

coherence length [PHYS] For a beam of particles, the typical length of a wave packet along the beam; the more monochromatic the beam, the greater its coherence length. [SOLID STATE] A measure of the distance through which the effect of any local disturbance is spread out in a superconducting material. Also known as coherence distance. { kō'hir·əns ˌleŋkth }

coherence time [PHYS] The average time required for the relative phase of two waves, or the phase of a single wave, to fluctuate appreciably. { kō'hir·əns ˌtīm }

coherent light [OPTICS] Radiant electromagnetic energy of the same, or almost the same, wavelength, and with definite phase relationships between different points in the field. { kō'hir·ənt 'līt }

coherent radiation [PHYS] Radiation in which there are definite phase relationships between different points in a cross section of the beam. { kō'hir·ənt ˌrād·ē'ā·shən }

coherent scattering [PHYS] Scattering in which there is a definite phase relationship between incoming and scattered particles or photons. { kō'hir·ənt 'skad·ə·riŋ }

coherent source [PHYS] A source in which there is a constant phase difference between waves emitted from different parts of the source. { kō'hir·ənt 'sȯrs }

coherent units [PHYS] A system of units, such as the International System, in which the units of derived quantities are formed as products or quotients of units of the base quantities according to the algebraic relations linking these quantities. { kō'hir·ənt 'yü·nəts }

cohesion [PHYS] The tendency of parts of a body of like composition to hold together, as a result of intermolecular attractive forces. { kō'hē·zhən }

cohesional work [PHYS] The work per unit area required to separate a column of liquid into two parts. { kō'hēzh·ən·əl 'wərk }

cohesive energy [SOLID STATE] The difference between the energy per atom of a system of free atoms at rest far apart from each other, and the energy of the solid. { kō'hē·siv 'en·ər·jē }

cohesive strength [MECH] **1.** Strength corresponding to cohesive forces between atoms.

2. Hypothetically, the stress causing tensile fracture without plastic deformation. { kō'hē·siv 'streŋkth }

coil [ELECTROMAG] A number of turns of wire used to introduce inductance into an electric circuit, to produce magnetic flux, or to react mechanically to a changing magnetic flux; in high-frequency circuits a coil may be only a fraction of a turn. Also known as electric coil; inductance; inductance coil; inductor. { kȯil }

coil antenna [ELECTROMAG] An antenna that consists of one or more complete turns of wire. { 'kȯil an'ten·ə }

coil form [ELECTROMAG] The tubing or spool of insulating material on which a coil is wound. { 'kȯil ,fȯrm }

coincidence boundary [CRYSTAL] A grain boundary separating crystal lattices which are rotated with respect to each other by an angle with a special value, resulting in a periodic grain boundary structure and an extension of a sublattice of the original lattice across the boundary. { kō'in·sə·dəns ,bau̇n·drē }

coincidence effect *See* track adaptation effect. { kō'in·sə·dəns i,fekt }

coincidence rangefinder [OPTICS] An optical rangefinder in which one-eyed viewing through a single eyepiece provides the basis for manipulation of the rangefinder adjustment to cause two images of the target or parts of each, viewed over different paths, to match or coincide. { kō'in·sə·dəns 'rānj,fīnd·ər }

CO laser *See* carbon monoxide laser. { ¦sē¦ō ,lā·zər }

Colburn analogy [FL MECH] Dimensionless Reynolds equation for fluid-flow resistance modified to be analogous to the Colburn *j* factor heat-transfer equation. { 'kōl·bərn ə'nal·ə·jē }

Colburn j factor equation [THERMO] Dimensionless heat-transfer equation to calculate the natural convection movement of heat from vertical surfaces or horizontal cylinders to fluids (gases or liquids) flowing past these surfaces. { 'kol·bərn 'jā ,fak·tər i'kwā·zhən }

cold [ELEC] Pertaining to electrical circuits that are disconnected from voltage supplies and at ground potential; opposed to hot, pertaining to carrying an electrical charge. { kōld }

cold-conductor effect [SOLID STATE] A sudden increase in resistivity, up to seven orders of magnitude, as the temperature increases over a narrow range; observed in certain semiconducting materials, particularly ferroelectric titanate ceramics. { ¦kōld kən'dək·tər i,fekt }

cold flow [MECH] Creep in polymer plastics. { 'kōld ,flō }

cold gas approximation [PL PHYS] An approximation according to which the sound speed is much smaller than the Alfvén speed or the gas pressure is much smaller than the magnetic pressure. { 'kōld ¦gas ə,präk·sə,mā·shən }

cold light [PHYS] **1.** Light emitted in luminescence. **2.** Visible light which is accompanied by little or no infrared radiation, and therefore has little heating effect. { 'kōld ,līt }

cold neutron [SOLID STATE] A very-low-energy neutron in a reactor, used for research into solid-state physics because it has a wavelength of the order of crystal lattice spacings and can therefore be diffracted by crystals. { 'kōld 'nü,trän }

cold stress [MECH] Forces tending to deform steel, cement, and other materials, resulting from low temperatures. { 'kōld ,stres }

Colebrook equation [FL MECH] An empirical equation for the flow of liquids in ducts, relating the friction factor to the Reynolds number and the relative roughness of the duct. { 'kōl,bru̇k ik'wā·zhən }

Cole-Cole plot [ELEC] For a substance displaying orientation polarization, a graph of the imaginary part versus the real part of the complex relative permittivity that is a circular arc, with its center below the abscissa. { 'kōl 'kōl ,plät }

Cole-Davidson plot [ELEC] For a substance displaying orientation polarization, a graph of the real part versus the imaginary part of the complex relative permittivity that is a skewed arc which approximates a straight line at the high-frequency end and a circular arc at the low-frequency end. { 'kōl 'dā·vəd·sən ,plät }

colidar *See* ladar. { 'kä·lə,där }

collapse properties [MECH] Strength and dimensional attributes of piping, tubing, or process vessels, related to the ability to resist collapse from exterior pressure or internal vacuum. { kə'laps ,präp·ərd·ēz }

collapsing pressure [MECH] The minimum external pressure which causes a thin-walled body or structure to collapse. { kə'lap·siŋ ,presh·ər }

collar vortex *See* vortex ring. { 'käl·ər ,vȯr,teks }

collateral series [NUC PHYS] A radioactive decay series, initiated by transmutation, that eventually joins into one of the four radioactive decay series encountered in natural radioactivity. { kə¦lad·ə·rəl 'sir,ēz }

collecting power [OPTICS] The power of a lens to make parallel rays converge or reduce the divergence of divergent rays. { kə'lek·tiŋ ,pau̇·ər }

collective electron theory [SOLID STATE] A theory of ferromagnetism in which electrons responsible for ferromagnetism are supposed to move more or less freely throughout a crystal, and to align with one another as the result of an exchange interaction. { kə¦lek·tiv i'lek,trän ,thē·ə·rē }

collective mode [PHYS] A weakly damped and therefore long-lived coherent motion of a large fraction of the particles in a system. { kə'lek·tiv ,mōd }

collective motion [NUC PHYS] Motion of nucleons in a nucleus correlated so that their overall space pattern is essentially constant or undergoes changes which are slow compared to the motions of individual nucleons. { kə'lek·tiv 'mō·shən }

collective paramagnetism [ELECTROMAG] Magnetization of a collection of extremely small ferromagnetic particles, each containing only one magnetic domain, that resembles paramagnetism of a collection of atoms or molecules. Also known as superparamagnetism. { kə'lek·tiv ˌpar·ə'mag·nəˌtiz·əm }

collective transition [NUC PHYS] A nuclear transition from one state of collective motion to another. { kə'lek·tiv tranz'ish·ən }

collider *See* colliding-beam accelerator. { kə'līd·ər }

colliding-beam accelerator [PART PHYS] A particle accelerator in which two beams of high-energy particles are allowed to collide head-on, resulting in high center-of-mass energies. Also known as collider. { kə'līd·iŋ ¦bēm ək'sel·əˌrād·ər }

colliding-pulse-ring dye laser [OPTICS] A laser consisting of a series of mirrors that form a ring cavity containing an optically pumped saturable gain dye and a saturable absorber dye. { kə'līd·iŋ ¦pəls ˌriŋ 'dī 'lā·zər }

collimate [PHYS] To render parallel to a certain line or direction; paths of electrons in a flooding beam, or paths of various rays of a scanning beam are collimated to cause them to become more nearly parallel as they approach the storage assembly of a storage tube. { 'käl·əˌmāt }

collimated beam [PHYS] A beam of radiation or matter whose rays or particles are nearly parallel so that the beam does not converge or diverge appreciably. { 'käl·əˌmād·əd 'bēm }

collimating lens [OPTICS] A lens on a collimator used to focus light from a source near one of its focal points into a parallel beam. { 'käl·əˌmād·iŋ ˌlenz }

collimator [OPTICS] An instrument which produces parallel rays of light. [PHYS] A device for confining the elements of a beam within an assigned solid angle. { 'käl·əˌmād·ər }

collinear array *See* linear array. { kə'lin·ē·ər ə'rā }

collinear transformation [OPTICS] The mapping of object space into image space produced by an ideal optical image-forming system, in which a unique image point corresponds to each object point and every straight line in the object space has as its corresponding image a unique straight line. { kə'lin·ē·ər ˌtranz·fər'mā·shən }

Collins helium liquefier [CRYO] A machine which uses the Joule-Thomson effect and work done by helium gas in expansion against a movable piston to liquefy helium. { 'käl·ənz 'hē·lē·əm 'lik·wəˌfī·ər }

collision [PHYS] An interaction resulting from the close approach of two or more bodies, particles, or systems of particles, and confined to a relatively short time interval during which the motion of at least one of the particles or systems changes abruptly. { kə'lizh·ən }

collision broadening *See* collision line-broadening. { kə'lizh·ən ˌbrȯd·ən·iŋ }

collision density [PHYS] The number of collisions of a specified type per unit volume per unit time. { kə'lizh·ən ˌden·səd·ē }

collision excitation [ATOM PHYS] The excitation of a gas by collisions of moving charged particles. { kə'lizh·ən ˌekˌsī'tā·shən }

collision frequency [PHYS] The average number of collisions undergone by a particle traveling through a material, such as an electron traveling through a gas, in a unit time. { kə'lizh·ən ˌfrē·kwən·sē }

collision ionization [ATOM PHYS] The ionization of atoms or molecules of a gas or vapor by collision with other particles. { kə'lizh·ən ī·ən·ə'zā·shən }

collisionless Boltzmann equation *See* Vlasov equation. { kə'lizh·ən·ləs 'bōltsˌmän i'kwā·zhən }

collisionless plasma [PL PHYS] A plasma in which particles interact through the mutually induced space-charge field, and collisions are assumed to be negligible. { kə'lizh·ən·ləs 'plaz·mə }

collision line-broadening [SPECT] Spreading of a spectral line due to interruption of the radiation process when the radiator collides with another particle. Also known as collision broadening. { kə'lizh·ən 'līn ˌbrȯd·ən·iŋ }

collision matrix *See* scattering matrix. { kə'lizh·ən ˌmāˌtriks }

collision of the first kind [PHYS] An inelastic collision in which some of the kinetic energy of translational motion is converted to internal energy of the colliding systems. Also known as endoergic collision. { kə'lizh·ən əv t͟hə 'fərst ˌkīnd }

collision of the second kind [PHYS] An inelastic collision in which some of the internal energy of the colliding systems is converted to kinetic energy of translation. Also known as exoergic collision. { kə'lizh·ən əv t͟hə 'sek·ənd ˌkīnd }

collision probability [PHYS] The ratio of the cross section for a given type of collision between two particles to the total cross section for all types of collision between the particles. { kə'lizh·ən ˌpräb·əˌbil·əd·ē }

collision-radiative recombination [ATOM PHYS] The capture of an electron by an ion in a gas, accompanied by the emission of one or more photons. { ¦kə'lizh·ən ¦rād·ē·ād·iv riˌkäm·bə'nā·shən }

collision theory [QUANT MECH] Theory to describe collisions of simple or complex particles, the derivation of collision cross sections from postulated interactions and the study of properties of collision amplitudes which follow from invariance principles such as conservation of probability and time-reversal invariance. { kə'lizh·ən ˌthē·ə·rē }

color [OPTICS] A general term that refers to the wavelength composition of light, with particular reference to its visual appearance. [PART PHYS] A hypothetical quantum number carried by quarks, so that each type of quark comes in three varieties which are identical in all measurable qualities but which differ in this additional property; this quantity determines the coupling of quarks to the gluon field. { 'kəl·ər }

color attribute [OPTICS] Any of the visual qualities of hue, saturation, or brightness. { 'kəl·ər ˌa·trəˌbyüt }

color center [SOLID STATE] A point lattice defect which produces optical absorption bands in an otherwise transparent crystal. { 'kəl·ər ˌsen·tər }

color circle [OPTICS] An arrangement of hues about the circumference of a circle in the order in which they appear in the electromagnetic spectrum, with pairs of complementary colors at opposite ends of diameters. { 'kəl·ər ˌsər·kəl }

color code [ELEC] A system of colors used to indicate the electrical value of a component or to identify terminals and leads. { 'kəl·ər ˌkōd }

color correction [OPTICS] The construction of an optical system so that the image positions of an object are the same for two or more wavelengths, and chromatic aberration is thus minimized. { 'kəl·ər kə'rek·shən }

color disk [OPTICS] A rotating circular disk having three filter sections to produce the individual red, green, and blue pictures in a field-sequential color television system. { 'kəl·ər ˌdisk }

color Doppler flow image [ACOUS] An acoustic image in which Doppler information is encoded in color, with red and blue representing fluid flow toward and away from the transducer. { ¦kəl·ər 'däp·lər ˌflō ˌim·ij }

color Doppler optical coherence tomography [OPTICS] An augmentation of optical coherence tomography which uses the interferometric phase information ignored in conventional optical coherence tomography to achieve simultaneous blood flow mapping and spatially resolved imaging. Also known as optical Doppler tomography. { ¦kəl·ər ¦däp·lər ¦äp·tə·kəl kō¦hēr·əns tə'mä·grə·fē }

color emissivity *See* monochromatic emissivity. { ¦kəl·ər ˌe·mi'siv·əd·ē }

color equation [OPTICS] An algebraic equation that expresses a specified color as an additive mixture of primary colors. { 'kəl·ər i'kwā·zhən }

color filter [OPTICS] An optical element that partially absorbs incident light, consisting of a pane of glass or other partially transparent material, or of films separated by narrow layers; the absorption may be either selective or nonselective with respect to wavelength. Also known as light filter. { 'kəl·ər ˌfil·tər }

color-filter effect *See* Bragg effect. { ¦kəl·ər ¦fil·tər iˌfekt }

color force [PART PHYS] The force that acts between quarks to bind them in hadrons and is thought to be the basis of all nuclear forces. { 'kəl·ər ˌfȯrs }

colorimeter [OPTICS] An instrument that measures color by determining the intensities of the three primary colors that will give that color. { ˌkəl·ə'rim·əd·ər }

colorimetric photometer [OPTICS] A photometer that can measure light intensities in several spectral regions, using color filters placed in the path of the light. { ˌkəl·ə·rə'me·trik fō'täm·əd·ər }

colorimetry [OPTICS] Any technique by which an unknown color is evaluated in terms of standard colors; the technique may be visual, photoelectric, or indirect by means of spectrophotometry; used in chemistry and physics. { ˌkəl·ə'rim·ə·trē }

color medium [OPTICS] Any colored, transparent material that is placed in front of a lighting unit to color the light transmitted. { 'kəl·ər ˌmēd·ē·əm }

color rendering [OPTICS] For a light source, the extent of the agreement between the perceived color of a surface illuminated by the source and that of the same surface illuminated by a reference source under specified viewing conditions, measured and expressed in terms of the chromaticity coordinates of the source and the luminance of the source in agreed spectral bands. { ¦kəl·ər 'ren·dər·iŋ }

color saturation [OPTICS] The degree to which a color is mixed with white; high saturation means little white, low saturation means much white. Also known as chroma; saturation. { 'kəl·ər sach·ə'rā·shən }

color solid [OPTICS] A three-dimensional diagram which represents the relationship of three attributes of surface color: hue, saturation, and brightness. { 'kəl·ər ˌsäl·əd }

color SU_3 [PART PHYS] A unitary symmetry based on the equivalence of the three differently colored quarks of a given flavor, which form a fundamental multiplet. { 'kəl·ər ˌesˌyü 'thrē }

color system [OPTICS] Any three-component coordinate system used to represent the attributes of colors. { 'kəl·ər ˌsis·təm }

color temperature [STAT MECH] Of a solid surface, that temperature of a blackbody from which the radiant energy has essentially the same spectral distribution as that from the surface. { ¦kəl·ər ¦tem·prə·chər }

color-translating microscope [OPTICS] A type of compound microscope that employs three different wavelengths of light to reveal details produced by ultraviolet or other nonvisible radiation. { ¦kəl·ər tranz'lād·iŋ 'mī·krə'skōp }

color triangle [OPTICS] A triangle on a chromaticity diagram that represents the range of chromaticities that can be obtained as additive mixtures of three prescribed primary colors represented by the corners of the triangle. { 'kəl·ər 'trīˌaŋ·gəl }

colossal magnetoresistance [SOLID STATE] A very large magnetoresistance associated with magnetic phase transitions in certain homogeneous materials, particularly a class of rare-earth perovskite manganites. { kə¦läs·əl magˌned·ō·ri'zis·təns }

columnar ionization [PHYS] Ionization of atoms in a region confined to one or more paths of very small cross-sectional area. { kə'ləm·nər ˌī·ə·nə'zā·shən }

coma [OPTICS] A manifestation of errors in an optical system, so that a point has an asymmetrical image (that is, appears as a pear-shaped spot). { 'kō·mə }

coma lobe [ELECTROMAG] Side lobe that occurs in the radiation pattern of a microwave antenna when the reflector alone is tilted back and forth to sweep the beam through space because the feed is no longer always at the center of the reflector; used to eliminate the need for a rotary joint in the feed waveguide. { 'kō·mə ,lōb }

comatic circle [OPTICS] A circle formed in the focal plane by rays from an off-axis point passing through a given zone of a lens that displays coma. { kō¦mad·ik 'sər·kəl }

comb antenna [ELECTROMAG] A broad-band antenna for vertically polarized signals, in which half of a fishbone antenna is erected vertically and fed against ground by a coaxial line. { 'kōm an,ten·ə }

combination principle *See* Ritz's combination principle. { ,käm·bə'nā·shən ,prin·sə·pəl }

combination tone [ACOUS] A subjective tone produced by simultaneously sounding two pure tones whose frequencies differ by a large amount. { ,käm·bə'nā·shən 'tōn }

combination vibration [SPECT] A vibration of a polyatomic molecule involving the simultaneous excitation of two or more normal vibrations. { ,käm·bə'nā·shən vī'brā·shən }

combined flexure [MECH] The flexure of a beam under a combination of transverse and longitudinal loads. { kəm'bīnd 'flek·shər }

combined stresses [MECH] Bending or twisting stresses in a structural member combined with direct tension or compression. { kəm'bīnd 'stres·əz }

combining glass [OPTICS] A glass screen designed to reflect display imagery to the viewer, usually at selected wavelengths of light, while being sufficiently transmissive for the viewer to see the scene beyond. { kəm'bīn·iŋ ,glas }

comma [ACOUS] The difference between the larger and smaller whole tones in the just scale, corresponding to a frequency ratio of 81/80. { 'käm·ə }

common branch [ELEC] A branch of an electrical network which is common to two or more meshes. Also known as mutual branch. { 'käm·ən 'branch }

common impedance coupling [ELECTROMAG] The interaction of two circuits by means of an inductance or capacitance in a branch which is common to both circuits. { ¦käm·ən im'ped·əns ,kəp·liŋ }

commutation rules [QUANT MECH] The specification of the commutators of operators corresponding to the dynamical variables of a system, which are equal to $i\hbar$ times the Poisson brackets of the classical variables to which the operators correspond. { ,käm·yə'tā·shən ,rülz }

commutator [ELECTROMAG] That part of a direct-current motor or generator which serves the dual function, in combination with brushes, of providing an electrical connection between the rotating armature winding and the stationary terminals, and of permitting reversal of the current in the armature windings. [QUANT MECH] The commutator of a and b is $[a,b] = ab - ba$. { 'käm·yə,tād·ər }

comovement effect [ATOM PHYS] The effect on atomic energy levels of the movement of the atomic nucleus, together with the electrons, about their common center of mass. { ¦kō'müv·mənt i,fekt }

comparative lifetime [NUC PHYS] The product of the mean life of a nucleus that undergoes beta decay and the probability per unit time that beta decay would occur if the matrix element between the initial and final states of this transition were unity. { kəm'par·əd·iv 'līf,tīm }

comparator method [THERMO] A method of determining the coefficient of linear expansion of a substance in which one measures the distance that each of two traveling microscopes must be moved in order to remain centered on scratches on a rod-shaped specimen when the temperature of the specimen is raised by a measured amount. { kəm'par·əd·ər ,meth·əd }

comparison lamp [OPTICS] An incandescent lamp whose luminous intensity is constant (although not necessarily known), and which is compared against other lamps in a photometer. { kəm'par·ə·sən ,lamp }

comparison microscope [OPTICS] **1.** An arrangement of two microscopes connected by a special receiving ocular so that the field of one microscope is seen at one side of a dividing line and the field of the other microscope at the opposite side. **2.** A projection type of microscope in which the image is compared with a template or known pattern. { kəm'par·ə·sən 'mī·krə,skōp }

comparison spectrum [SPECT] A line spectrum whose wavelengths are accurately known, and which is matched with another spectrum to determine the wavelengths of the latter. { kəm'par·ə·sən ,spek·trəm }

compatibility conditions [MECH] A set of six differential relations between the strain components of an elastic solid which must be satisfied in order for these components to correspond to a continuous and single-valued displacement of the solid. { kəm,pad·ə'bil·əd·ē kən,dish·ənz }

compensating eyepiece [OPTICS] A type of Huygens eyepiece in which the eye lens is achromatized to compensate for the color errors of the objective. { 'käm·pən,sād·iŋ 'ī,pēs }

compensating impurity [SOLID STATE] A semiconductor impurity that is of the opposite electrical type to a given impurity, and that reduces the concentration of charge carriers (electrons or holes) that resulted from the given impurity. { 'käm·pən,sād·iŋ im'pyu̇r·əd·ē }

compensating plate [OPTICS] The first of two plates in a Brace compensator, which covers the entire field of view. { 'käm·pən,sād·iŋ ,plāt }

compensation effect *See* Jaccarino-Peter effect. { ,käm·pən'sā·shən i,fekt }

compensator [OPTICS] A device, usually consisting of two quartz wedges, for determining the phase difference between the two components of elliptically polarized light. { 'käm·pən,sād·ər }

complementarity [QUANT MECH] The principle that nature has complementary aspects, particle and wave; the two aspects are related by $p = h/\lambda$ and $E = h\nu$, where p and E are the momentum and energy of the particle, λ and ν are the length and frequency of the wave, and h is Planck's constant. { ˌkäm·plə·mənˈtar·əd·ē }

complementary colors [OPTICS] Two colors which lie on opposite sides of the white point in the chromaticity diagram so that an additive mixture of the two, in appropriate proportions, can be made to yield an achromatic mixture. { ˌkäm·pləˈmen·trē ˈkəl·ərz }

complementary variables *See* conjugate variables. { ˌkäm·pləˈmen·trē ˈver·ē·ə·bəlz }

complementary wave [ELECTROMAG] Wave brought into existence at the ends of a coaxial cable, or two-conductor transmission lines, or any discontinuity along the line. { ˌkäm·pləˈmen·trē ˈwāv }

complementary wavelength [OPTICS] The wavelength of light that, when combined with a sample color in suitable proportions, matches a reference standard light. { ˌkäm·pləˈmen·trē ˈwāvˌleŋkth }

complete degeneracy [QUANT MECH] The condition in which all the states of interest have the same energy. { kəmˈplēt diˈjen·ə·rə·sē }

complete electron shell [ATOM PHYS] An inner electron shell of an atom that contains its maximun number of electrons. { kəmˈplēt iˈlekˌträn ˈshel }

complete-expansion diesel cycle *See* Brayton cycle. { kəmˈplēt ikˈspan·shən ˈdē·zəl ˌsi·kəl }

completely inelastic collision *See* perfectly inelastic collision. { kəmˈplēt·lē in·əˈlas·tik kəˈlizh·ən }

complex degree of coherence [PHYS] A measure of the coherence between two waves, equal to the cross-correlation function between the normalized amplitude of one wave and the complex conjugate of the normalized amplitude of the other. { ˈkämˌpleks dəˈgrē əv ˌkōˈhir·əns }

complex impedance *See* electrical impedance; impedance. { ˈkämˌpleks imˈpēd·əns }

complex notation [PHYS] The representation of a physical quantity by a complex number whose real component equals the instantaneous value of the physical quantity, a sinusoidally varying quantity thus being represented by a point rotating in a circle centered at the origin of the complex plane with uniform speed. { ˈkämˌpleks nōˈtā·shən }

complex permeability [ELECTROMAG] A property, designated by μ^*, of a magnetic material, equal to $\mu_0 (L/L_0)$, where L is the complex inductance of an inductance coil in which the magnetic material forms the core when the coil is connected to a sinusoidal voltage source, and L_0 is the vacuum inductance of the coil. { ˈkämˌpleks ˌpər·mē·əˈbil·əd·ē }

complex permittivity [ELEC] A property of a dielectric, equal to $\epsilon_0(C/C_0)$, where C is the complex capacitance of a capacitor in which the dielectric is the insulating material when the capacitor is connected to a sinusoidal voltage source, and C_0 is the vacuum capacitance of the capacitor. { ˈkämˌpleks ˌpər·məˈtiv·əd·ē }

complex potential [FL MECH] An analytic function in ideal aerodynamics whose real part is the velocity potential and whose imaginary part is the stream function. [NUC PHYS] A generalization of the potential in the Schrödinger equation describing the scattering of a nucleon by a nucleus in the cloudy crystal-ball model. { ˈkämˌpleks pəˈten·chəl }

complex tone [ACOUS] A sound wave produced by the combination of simple sinusoidal components of different frequencies. { ˈkämˌpleks ˈtōn }

complex velocity [FL MECH] In ideal aerodynamic flow, the derivative of the complex potential with respect to $z = x + iy$, where x and y are the chosen coordinates. { ˈkämˌpleks vəˈläs·əd·ē }

complex wave [PHYS] A waveform which varies from instant to instant, but can be resolved into a number of sine-wave components, each of a different frequency and probably of a different amplitude. { ˈkämˌpleks ˈwāv }

compliance [MECH] The displacement of a linear mechanical system under a unit force. { kəmˈplī·əns }

compliance constant [MECH] Any one of the coefficients of the relations in the generalized Hooke's law used to express strain components as linear functions of the stress components. Also known as elastic constant. { kəmˈplī·əns ˌkän·stənt }

component [ELEC] Any electric device, such as a coil, resistor, capacitor, generator, line, or electron tube, having distinct electrical characteristics and having terminals at which it may be connected to other components to form a circuit. Also known as circuit element; element. { kəmˈpō·nənt }

component symbol [ELEC] A graphical design used to represent a component in a circuit diagram. { kəmˈpō·nənt ˌsim·bəl }

composite balance [ELEC] An electric balance made by modifying the Kelvin balance to measure amperage, voltage, or wattage. { kəmˈpäz·ət ˈbal·əns }

composition [MECH] The determination of a force whose effect is the same as that of two or more given forces acting simultaneously; all forces are considered acting at the same point. { ˌkäm·pəˈzish·ən }

composition face *See* composition surface. { ˌkäm·pəˈzish·ən ˌfās }

composition-of-velocities law [MECH] A law relating the velocities of an object in two references frames which are moving relative to each other with a specified velocity. { ˌkäm·pəˈzish·ən əv vəˈläs·əd·ēz ˌlȯ }

composition plane [CRYSTAL] A planar composition surface in a crystal uniting two individuals of a contact twin. { ˌkäm·pəˈzish·ənˌplān }

composition surface [CRYSTAL] The surface uniting individuals of a crystal twin; may or may not be planar. Also known as composition face. { ˌkäm·pə'zish·ən ˌsər·fəs }

compound elastic scattering [NUC PHYS] Scattering in which the final state is the same as the initial state, but there is an intermediate state with the colliding systems amalgamating to form a compound system. { 'käm,paůnd i'las·dik 'skad·ə·riŋ }

compound lens [OPTICS] **1.** A combination of two or more lenses in which the second surface of one lens has the same radius as the first surface of the following lens, and the two lenses are cemented together. Also known as cemented lens. **2.** Any optical system consisting of more than one element, even when they are not in contact. { 'käm,paůnd 'lenz }

compound magnet [ELEC] A permanent magnet that is constructed from a number of thin magnets having the same shape. { 'käm,paůnd 'mag·nət }

compound microscope [OPTICS] A microscope which utilizes two lenses or lens systems; one lens forms an enlarged image of the object, and the second magnifies the image formed by the first. { 'käm,paůnd 'mī·krə,skōp }

compound nucleus [NUC PHYS] An intermediate state in a nuclear reaction in which the incident particle combines with the target nucleus and its energy is shared among all the nucleons of the system. { 'käm,paůnd 'nü·klē·əs }

compound pendulum *See* pendulum. { 'käm ,paůnd 'pen·jə·ləm }

compound twins [CRYSTAL] Individuals of one mineral group united in accordance with two or more different twin laws. { 'käm,paůnd 'twinz }

compound wave [FL MECH] A plane wave of finite amplitude in which neither the sum of the velocity potential and the component of velocity in the direction of wave motion, nor the difference of these two quantities, is constant. { 'kä m,paůnd 'wāv }

compressadensity function [MECH] A function used in the acoustic levitation technique to determine either the density or the adiabatic compressibility of a submicroliter droplet suspended in another liquid, if the other property is known. { kəm,pres·ə'den·səd·ē ,fəŋk·shən }

compressed air [MECH] Air whose density is increased by subjecting it to a pressure greater than atmospheric pressure. { kəm'prest 'er }

compressibility [MECH] The property of a substance capable of being reduced in volume by application of pressure; quantitively, the reciprocal of the bulk modulus. { kəm,pres·ə'bil·əd·ē }

compressibility burble [FL MECH] A region of disturbed flow, produced by and rearward of a shock wave. { kəm,pres·ə'bil·əd·ē ,bər·bəl }

compressibility correction [FL MECH] The correction of the calibrated airspeed caused by compressibility error. { kəm,pres·ə'bil·əd·ē kə'rek·shən }

compressibility error [FL MECH] The error in the readings of a differential-pressure-type airspeed indicator due to compression of the air on the forward part of the pitot tube component moving at high speeds. { kəm,pres·ə'bil·əd·ē ,er·ər }

compressibility factor [THERMO] The product of the pressure and the volume of a gas, divided by the product of the temperature of the gas and the gas constant; this factor may be inserted in the ideal gas law to take into account the departure of true gases from ideal gas behavior. Also known as deviation factor; gas-deviation factor; supercompressibility factor. { kəm,pres·ə'bil·əd·ē ,fak·tər }

compressible flow [FL MECH] Flow in which the fluid density varies. { kəm'pres·ə·bəl 'flō }

compressible-flow principle [FL MECH] The principle that when flow velocity is large, it is necessary to consider that the fluid is compressible rather than to assume that it has a constant density. { kəm¦pres·ə·bal ¦flō 'prin·sə·pəl }

compression [MECH] Reduction in the volume of a substance due to pressure; for example in building, the type of stress which causes shortening of the fibers of a wooden member. { kəm'presh·ən }

compressional wave [PHYS] A disturbance traveling in an elastic medium; characterized by changes in volume and by particle motion parallel with the direction of wave movement. Also known as dilatational wave; irrotational wave; pressure wave; P wave. { kəm'presh·ən·əl ,wāv }

compression modulus *See* bulk modulus of elasticity. { kəm'presh·ən ,mäj·ə·ləs }

compression strength [MECH] Property of a material to resist rupture under compression. { kəm'presh·ən ,streŋkth }

compression wave [FL MECH] A wave in a fluid in which a compression is propagated. { kəm'presh·ən ,wāv }

compressive strength [MECH] The maximum compressive stress a material can withstand without failure. { kəm'pres·iv 'streŋkth }

compressive stress [MECH] A stress which causes an elastic body to shorten in the direction of the applied force. { kəm'pres·iv 'stres }

Compton absorption [QUANT MECH] The absorption of an x-ray or gamma-ray photon in Compton scattering, accompanied by the emission of another photon of lower energy. { 'käm·tən əb'sȯrp·shən }

Compton cross section [QUANT MECH] The differential cross section for the elastic scattering of photons by electrons. { 'käm·tən 'krȯs ,sek·shən }

Compton-Debye effect *See* Compton effect. { ¦käm·tən də'be·ə i'fekt }

Compton effect [QUANT MECH] The increase in wavelength of electromagnetic radiation in the x-ray and gamma-ray region on being scattered by material objects; the scattering is due to the interaction of the photons with electrons that are effectively free. Also known as Compton-Debye effect. { 'käm·tən i'fekt }

Compton electron *See* Compton recoil electron. { 'käm·tən i'lek,trän }

Compton equation [QUANT MECH] The equation for the change in wavelength $\Delta\lambda$ of radiation scattered by electrons in the Compton effect, $\Delta\lambda = \lambda_c(1 - \cos\theta)$, where λ_c is the Compton wavelength of the electron, and θ is the angle between the directions of incident and scattered radiation. { 'käm·tən i,kwā·zhən }

Compton incoherent scattering [NUC PHYS] Scattering of gamma rays by individual nucleons in a nucleus or electrons in an atom when the energy of the gamma rays is large enough so that binding effects may be neglected. { 'käm·tən in·kō'hir·ənt 'skad·ə·riŋ }

Compton process *See* Compton scattering. { 'käm·tən ,präs·əs }

Compton recoil electron [QUANT MECH] An electron set in motion by its interaction with a photon in Compton scattering. Also known as Compton electron. { 'käm·tən ri'kȯil i'lek,trän }

Compton recoil particle [QUANT MECH] Any particle that has acquired its momentum in a scattering process similar to Compton scattering. { 'käm·tən ri'kȯil ,pard·ə·kəl }

Compton scattering [QUANT MECH] The elastic scattering of photons by electrons. Also known as Compton process; gamma-ray scattering. { ¦käm·tən ¦skad·ə·riŋ }

Compton shift [QUANT MECH] The change in wavelength of scattered radiation due to the Compton effect. { 'käm·tən ,shift }

Compton wavelength [QUANT MECH] A convenient unit of length that is characteristic of an elementary particle, equal to Planck's constant divided by the product of the particle's mass and the speed of light. { 'käm·tən 'wāv,leŋkth }

computational aeroacoustics [ACOUS] The study of the problems of aeroacoustics using computational techniques. { ,käm·pyü¦tā·shən·əl ,er·ō·ə'küs·tiks }

computational flow imaging [FL MECH] A technology for generating digital images of theoretic fluid dynamic phenomena in optical formats that mimic real observations of the corresponding real flow fields. Abbreviated CFI. { ,kam·pyə,tā·shən·əl 'flō ,im·ij·iŋ }

computational fluid dynamics [FL MECH] A field of study concerned with the use of high-speed digital computers to numerically solve the complete nonlinear partial differential equations governing viscous fluid flows. { ,käm·pyə'tā·shən·əl 'flü·əd dī'nam·iks }

concave grating [SPECT] A reflection grating which both collimates and focuses the light falling upon it, made by spacing straight grooves equally along the chord of a concave spherical or paraboloid mirror surface. Also known as Rowland grating. { 'kän,kāv 'grād·iŋ }

concave spherical mirror [OPTICS] A round mirror having a concavely curved surface, in the form of a portion of a sphere. { 'kän,kāv ¦sfer·ə·kəl 'mir·ər }

concentrated load [MECH] A force that is negligible because of a small contact area; a beam supported on a girder represents a concentrated load on the girder. { 'kän·sən,trād·əd 'lōd }

concentric cable *See* coaxial cable. { kən'sen·trik 'kā·bəl }

concentric lens [OPTICS] A lens whose two spherical surfaces have the same center. { kən'sen·trik 'lenz }

concentric line *See* coaxial cable. { kən'sen·trik 'līn }

concentric resonator [OPTICS] A beam resonator that consists of a pair of spherical mirrors that have the same axis of rotational symmetry and are positioned so that their centers of curvature coincide on this axis. { kən'sen·trik 'rez·ən,ād·ər }

concentric transmission line *See* coaxial cable. { kən'sen·trik tranz'mish·ən ,līn }

condensation [ACOUS] A measure of the increase in the instantaneous density at a given point owing to a sound wave, namely $(\rho - \rho_0)/\rho_0$, where ρ is the density and ρ_0 is the constant mean density at the point. [CRYO] *See* Bose-Einstein condensation. [ELEC] An increase of electric charge on a capacitor conductor. [MECH] An increase in density. [OPTICS] Focusing or collimation of light. { ,kän·dən'sā·shən }

condensation shock wave [FL MECH] A sheet of discontinuity associated with a sudden condensation and fog formation in a field of flow; it occurs, for example, on a wing where a rapid drop in pressure causes the temperature to drop considerably below the dew point. { ,kän·dən'sā·shən 'shäk ,wāv }

condensed matter [PHYS] Matter in the liquid or solid state. { kən¦denst 'mad·ər }

condensed-matter physics [PHYS] The branch of physics concerned with the study of very large numbers of strongly interacting particles, including the study of the solid and liquid states, dense plasmas, liquid crystals, glasses, polymers, and gels. { kən¦denst ¦mad·ər 'fiz·iks }

condenser [ELEC] *See* capacitor. [OPTICS] A system of lenses or mirrors in an optical projection system, which gathers as much of the light from the source as possible and directs it through the projection lens. { kən'den·sər }

condenser antenna *See* capacitor antenna. { kən'den·sər an'ten·ə }

condensing flow [FL MECH] The flow and simultaneous condensation (partial or complete) of vapor through a cooled pipe or other closed conduit or container. { kən'dens·iŋ ,flō }

conditionally periodic motion [MECH] Motion of a system in which each of the coordinates undergoes simple periodic motion, but the associated frequencies are not all rational fractions of each other so that the complete motion is not simply periodic. { kən'dish·ən·əl·ē ,pir·ē¦ad·ik ,mō·shən }

Condon-Shortley-Wigner phase convention [QUANT MECH] Convention relating the phases of states having the same eigenvalue of $J^2 = J_x^2 + J_y^2 + J_z^2$, and different eigenvalues of J_z, where $\mathbf{J}$ is the total angular momentum, according to which the matrix elements of $\mathbf{J}_+ = J_x + iJ_y$ and $\mathbf{J}_- = J_x - iJ_y$ between such states are real. { ¦kän·dən ¦shȯrt·lē ¦wig·nər 'fās kən,ven·shən }

conductance [ELEC] The real part of the admittance of a circuit; when the impedance contains no reactance, as in a direct-current circuit, it is the reciprocal of resistance, and is thus a measure of the ability of the circuit to conduct electricity. Also known as electrical conductance. Designated G. [FL MECH] For a component of a vacuum system, the amount of a gas that flows through divided by the pressure difference across the component. [THERMO] *See* thermal conductance. { kən'dək·təns }

conductance-variation method [ELEC] A technique for measuring low admittances; measurements in a parallel-resonance circuit with the terminals open-circuited, with the unknown admittance connected, and then with the unknown admittance replaced by a known conductance standard are made; from them the unknown can be calculated. { kən'dək·təns ver·ē'ā·shən ,meth·əd }

conduction [ELEC] The passage of electric charge, which can occur by a variety of processes, such as the passage of electrons or ionized atoms. Also known as electrical conduction. [PHYS] Transmission of energy by a medium which does not involve movement of the medium itself. { kən'dək·shən }

conduction band [SOLID STATE] An energy band in which electrons can move freely in a solid, producing net transport of charge. { kən'dək·shən ,band }

conduction current [SOLID STATE] A current due to a flow of conduction electrons through a body. { kən'dək·shən ,kər·ənt }

conduction electron [SOLID STATE] An electron in the conduction band of a solid, where it is free to move under the influence of an electric field. Also known as outer-shell electron; valence electron. { kən'dək·shən i'lek,trän }

conduction field [ELECTROMAG] Energy surrounding a conductor when an electric current is passed through the conductor, which, because of the difference in phase between the electrical field and magnetic field set up in the conductor, cannot be detached from the conductor. { kən'dək·shən ,fēld }

conductive coupling [ELEC] Electric connection of two electric circuits by their sharing the same resistor. { kən'dək·tiv 'kəp·liŋ }

conductivity [ELEC] The ratio of the electric current density to the electric field in a material. Also known as electrical conductivity; specific conductance. { ,kän,dək'tiv·əd·ē }

conductivity bridge [ELEC] A modified Kelvin bridge for measuring very low resistances. { ,kän,dək'tiv·əd·ē ,brij }

conductivity cell [ELEC] A glass vessel with two electrodes at a definite distance apart and filled with a solution whose conductivity is to be measured. { ,kän,dək'tiv·əd·ē ,sel }

conductivity ellipsoid [ELEC] For an anisotropic material, an ellipsoid whose axes are the eigenvectors of the conductivity tensor. { ,kän ,dək¦tiv·əd·ē i'lip,sȯid }

conductivity tensor [ELEC] A tensor which, when multiplied by the electric field vector according to the rules of matrix multiplication, gives the current density vector. { ,kän,dək'tiv·əd·ē ,ten·sər }

conductivity theory [STAT MECH] Theory which treats the system of electrons in a metal as a gas and uses the Boltzmann transport equation to calculate conductivity. { ,kän,dək'tiv·əd·ē ,thē·ə·rē }

conductor [ELEC] A wire, cable, or other body or medium that is suitable for carrying electric current. Also known as electric conductor. { kən'dək·tər }

conductor skin effect *See* skin effect. { kən'dək·tər ,skin i'fekt }

cone antenna *See* conical antenna. { 'kōn an'ten·ə }

cone flow *See* conical flow. { 'kōn ,flō }

cone of friction [MECH] A cone in which the resultant force exerted by one flat horizontal surface on another must be located when both surfaces are at rest, as determined by the coefficient of static friction. { ¦kōn əv 'frik·shən }

cone of nulls [ELECTROMAG] In antenna practice, a conical surface formed by directions of negligible radiation. { 'kōn əv 'nəlz }

configuration [ELEC] A group of components interconnected to perform a desired circuit function. [MECH] The positions of all the particles in a system. { kən,fig·yə'rā·shən }

configurational free energy [STAT MECH] The free energy of a solid lattice associated with the interaction between neighboring atoms, and with external electric and magnetic fields. { kən,fig·yə'rā·shən·əl ¦frē 'en·ər·jē }

configuration-interaction method [ATOM PHYS] A method of accounting for the electron correlation in an atom or molecule in which the total many-body wave function is expanded in a basis set and the coefficients in this expansion are determined by minimization of the expectation value of the energy. { kən,fig·yə¦rā·shən ,in·tər'ak·shən ,meth·əd }

confinement [PART PHYS] A property of quantum chromodynamics whereby isolated quarks cannot exist, nor can any other isolated particles that carry color charges, such as gluons. [PL PHYS] Restriction of a hot plasma to a given volume as long as possible, by such means as magnetic mirrors and pinch effect. { kən'fīn·mənt }

confocal resonator [ELECTROMAG] A wavemeter for millimeter wavelengths, consisting of two spherical mirrors facing each other; changing the spacing between the mirrors affects propagation of electromagnetic energy between them, permitting direct measurement of free-space wavelength. [OPTICS] A beam resonator that consists of a pair of spherical mirrors which have the same axis of rotational symmetry and are positioned so that their focal points coincide on this axis. { kän'fō·kəl 'rez·ən,ād·ər }

conformal diagram *See* Penrose diagram. { kən'fȯr·məl 'dī·ə,gram }

conformal optics [OPTICS] The design of optical systems whereby the shape of the outer surfaces is chosen to optimize the interaction with the environment in which the optical system is being used. { kən¦fȯrm·əl 'äp·tiks }

conformal reflection chart [ELECTROMAG] An Argand diagram for plotting the complex reflection coefficient of a waveguide junction and its image, the two being related by a conformal transformation. { kən'fȯr·məl ri'flek·shən ,chärt }

congruent melting point [THERMO] A point on a temperature composition plot of a nonstoichiometric compound at which the one solid phase and one liquid phase are adjacent. { kən'grü·ənt 'melt·iŋ ,pȯint }

conical antenna [ELECTROMAG] A wide-band antenna in which the driven element is conical in shape. Also known as cone antenna. { 'kän·ə·kəl an'ten·ə }

conical flow [FL MECH] Steady supersonic flow of a perfect, inviscid gas past a conical solid body in a region of the flow field where the principal physical quantities such as velocity, pressure, and density are constant on rays passing through a fixed point. Also known as cone flow. { 'kän·ə·kəl 'flō }

conical helimagnet [SOLID STATE] A helimagnet in which the directions of atomic magnetic moments all make the same angle (greater than 0° and less than 90°) with a specified axis of the crystal,moments of atoms in successive basal planes are separated by equal azimuthal angles, and all moments have the same magnitude. { 'kän·ə·kəl 'hel·ə,mag·nət }

conical horn [ACOUS] A horn having a circular cross section and straight sides. { 'kän·ə·kəl 'hȯrn }

conical-horn antenna [ELECTROMAG] A horn antenna having a circular cross section and straight sides. { 'kän·ə·kəl ,hȯrn an'ten·ə }

conical monopole antenna [ELECTROMAG] A variation of a biconical antenna in which the lower cone is replaced by a ground plane and the upper cone is usually bent inward at the top. { 'kän·ə·kəl 'män·ə,pōl an'ten·ə }

conical pendulum [MECH] A weight suspended from a cord or light rod and made to rotate in a horizontal circle about a vertical axis with a constant angular velocity. { 'kän·ə·kəl 'pen·jə·ləm }

conical refraction [OPTICS] Phenomenon in which a ray incident on the surface of a biaxial crystal at a certain direction splits into a family of rays which lie along a cone. { 'kän·ə·kəl ri'frak·shən }

conjugate branches [ELEC] Any two branches of an electrical network such that a change in the electromotive force in either does not result in a change in current in the other. Also known as conjugate conductors. { 'kän·jə·gət 'bran·chəz }

conjugate conductors *See* conjugate branches. { 'kän·jə·gət kən'dək·tərz }

conjugate impedances [ELEC] Impedances having resistance components that are equal, and reactance components that are equal in magnitude but opposite in sign. { 'kän·jə·gət im'pēd·ən·səz }

conjugate momentum [MECH] If q_j ($j = 1,2, \ldots$) are generalized coordinates of a classical dynamical system, and L is its Lagrangian, the momentum conjugate to q_j is $p_j = \partial L/\partial q_j$. Also known as canonical momentum; generalized momentum. { 'kän·jə·gət mə'men·təm }

conjugate particles [PART PHYS] A particle and its antiparticle. { 'kän·jə·gət 'pärd·ə·kəlz }

conjugate points [OPTICS] Any pair of points such that all rays from one are imaged on the other within the limits of validity of Gaussian optics. Also known as conjugate foci. { 'kän·jə·gət 'pȯins }

conjugate variables [QUANT MECH] A pair of physical variables describing a quantum-mechanical system such that their commutator is a nonzero constant; either of them, but not both, can be precisely specified at the same time. Also known as complementary variables. { 'kän·jə·gət 'ver·ē·ə·bəlz }

conoid of Sturm *See* astigmatic interval. { ¦kä ,nȯid əv 'stərm }

conoscope [OPTICS] An instrument, essentially a wide-angle microscope, used for study and observation of interference figures and related phenomena of specially cut crystal plates, especially for measuring the axial angle. Also known as hodoscope. { kän·ə,skōp }

consequent poles [ELECTROMAG] Pairs of magnetic poles in a magnetized body that are in excess of the usual single pair. { 'kän·sə·kwənt ¦pōlz }

conservation law [PHYS] A law which states that some physical quantity associated with an isolated system is constant. { ,kän·sər'vā·shən ,lȯ }

conservation of angular momentum [MECH] The principle that, when a physical system is subject only to internal forces that bodies in the system exert on each other, the total angular momentum of the system remains constant, provided that both spin and orbital angular momentum are taken into account. { ,kän·sər'vā·shən əv 'aŋ·gyə·lər mə'men·təm }

conservation of areas [MECH] A principle governing the motion of a body moving under the action of a central force, according to which a line joining the body with the center of force sweeps out equal areas in equal times. { ,kän·sər'vā·shən əv 'er·ē·əz }

conservation of charge [ELEC] A law which states that the total charge of an isolated system is constant; no violation of this law has been discovered. Also known as charge conservation. { ,kän·sər'vā·shən əv 'chärj }

conservation of condensation [FL MECH] The

principle that the rapid rise of pressure associated with the spherical wave propagating outward from an explosion must be followed by a region of diminished pressure. { ˌkän·sərˈvā·shən əv ˌkän·dənˈsā·shən }

conservation of energy [PHYS] The principle that energy cannot be created or destroyed, although it can be changed from one form to another; no violation of this principle has been found. Also known as energy conservation. { ˌkän·sərˈvā·shən əv ˈen·ər·jē }

conservation of mass [PHYS] The notion that mass can neither be created nor destroyed; it is violated by many microscopic phenomena. { ˌkän·sərˈvā·shən əv ˈmas }

conservation of matter [PHYS] The notion that matter can be neither created nor destroyed; it is violated by microscopic phenomena. { ˌkän·sərˈvā·shən əv ˈmad·ər }

conservation of momentum [MECH] The principle that, when a system of masses is subject only to internal forces that masses of the system exert on one another, the total vector momentum of the system is constant; no violation of this principle has been found. Also known as momentum conservation. { ˌkän·sərˈvā·shən əv məˈmən·təm }

conservation of parity [QUANT MECH] The law that, if the wave function describing the initial state of a system has even (odd) parity, the wave function describing the final state has even (odd) parity; it is violated by the weak interactions. Also known as parity conservation. { ˌkän·sərˈvā·shən əv ˈpar·əd·ē }

conservation of probability [QUANT MECH] The requirement that the sum of the probabilities of finding a system in each of its possible states is constant. { ˌkän·sərˈvā·shən əv ˌpräb·əˈbil·əd·ē }

conservation of vorticity [FL MECH] **1.** The principle that the vertical component of the absolute vorticity of each particle in an inviscid, autobarotropic fluid flowing horizontally remains constant. **2.** The hypothesis that the vorticity of fluid particles remains constant during the turbulent mixing of the fluid. { ˌkän·sərˈvā·shən əv ˌvȯrˈtis·əd·ē }

conservative force field [MECH] A field of force in which the work done on a particle in moving it from one point to another depends only on the particle's initial and final positions. { kənˈsər·və·tiv ˈfȯrs ˌfēld }

conservative property [THERMO] A property of a system whose value remains constant during a series of events. { kənˈsər·və·tiv ˈpräp·ərd·ē }

conservative scattering [ELECTROMAG] Scattering of radiation without accompanying absorption. { kənˈsər·vəd·iv ˈskad·ə·riŋ }

conservative system [PHYS] A system in which there is no dissipation of energy so that the total energy remains constant with time. { kənˈsər·vəd·iv ˈsis·təm }

conserved quantity [PHYS] A quantity that remains unchanged with time during the evolution of a dynamical system. { kənˈsərvd ˈkwän·əd·ē }

conserved vector current [PART PHYS] The hypothesis that the weak hadronic vector current is identical to the conserved isotopic-spin current. Abbreviated CVC. { kənˈservd ˈvek·tər ˌkər·ənt }

consolute temperature [THERMO] The upper temperature of immiscibility for a two-component liquid system. Also known as upper consolute temperature; upper critical solution temperature. { ˈkan·səˌlüt ˈtem·prə·chər }

consonance [ACOUS] The interval between two tones whose frequencies are in a ratio approximately equal to the quotient of two whole numbers, each equal to or less than 6, or to such a quotient multiplied or divided by some power of 2. { ˈkän·sə·nəns }

constant-angle fringes *See* Haidinger fringes. { ¦kän·stənt ˈaŋ·gəl ˌfrin·jəz }

constant-bandwidth analyzer [ACOUS] A tunable sound analyzer which has a fixed pass band that is swept through the frequency range of interest. Also known as constant-bandwidth filter. { ¦kän·stənt ˈbandˌwidth ˌan·əˌlīz·ər }

constant-bandwidth filter *See* constant-bandwidth analyzer. { ¦kän·stəntˈbandˌwidth ˈfil·tər }

constant-current dc potentiometer [ELEC] A potentiometer in which the unknown electromotive force is balanced by a constant current times the resistance of a calibrated resistor or slidewire. Also known as Poggendorff's first method. { ¦kän·stənt ˈkər·ənt ¦dē¦sē pəˌten·chēˈam·əd·ər }

constant-deviation fringes *See* Haidinger fringes. { ¦kän·stənt ˌdē·vēˈā·shən ˌfrin·jəz }

constant-deviation prism [OPTICS] A prism whose deviation is constant and does not depend on the index of refraction or wavelength. { ¦kän·stənt ˌdē·vē¦ā·shən ˈpriz·əm }

constant-deviation spectrometer [SPECT] A spectrometer in which the collimator and telescope are held fixed and the observed wavelength is varied by rotating the prism or diffraction grating. { ¦kän·stənt ˌdē·vē¦ā·shən spek ˈträm·əd·ər }

constant field *See* stationary field. { ˈkän·stənt ˈfēld }

constant-k lens [ELECTROMAG] A microwave lens that is constructed as a solid dielectric sphere; a plane electromagnetic wave brought to a focus at one point on the sphere emerges from the opposite side of the sphere as a parallel beam. { ¦kän·stənt ¦kā ˈlenz }

constant of gravitation *See* gravitational constant. { ˈkän·stənt əv grav·əˈtā·shən }

constant of motion [MECH] A dynamical variable of a system which remains constant in time. { ˈkän·stənt əv ˈmō·shən }

constant-resistance dc potentiometer [ELEC] A potentiometer in which the ratio of an unknown and a known potential are set equal to the ratio of two known constant resistances. Also known as Poggendorff's second method. { ¦kän·stənt riˈzis·təns ¦dē¦sē pəˌten·chēˈäm·əd·ər }

constitutive equations [ELECTROMAG] The equations $D = \epsilon E$ and $B = \mu H$, which relate the electric displacement D with the electric field intensity E, and the magnetic induction B with the magnetic field intensity H. { 'kän·stə,tüd·iv i'kwā·zhənz }

constraint [MECH] A restriction on the natural degrees of freedom of a system; the number of constraints is the difference between the number of natural degrees of freedom and the number of actual degrees of freedom. { kən'strānt }

constringence *See* nu value. { kən'strin·jəns }

constructive interference [PHYS] Phenomenon in which the phases of waves arriving at a specified point over two or more paths of different lengths are such that the square of the resultant amplitude is greater than the sum of the squares of the component amplitudes. { kən'strək·div ,in·tər'fir·əns }

contact [ELEC] *See* electric contact. [FL MECH] The surface between two immiscible fluids contained in a reservoir. { 'kän,takt }

contact angle *See* angle of contact. { 'kän,takt ,aŋ·gəl }

contact arc [ELEC] A spark that occurs immediately after the breaking of an electric contact carrying a current. { 'kän,takt ,ärk }

contact drop [ELEC] The voltage drop across the terminals of an electric contact. { 'kän ,takt ,dräp }

contact electricity [ELEC] An electric charge at the surface of contact of two different materials. { 'kän,takt i,lek'tris·əd·ē }

contact electromotive force *See* contact potential difference. { 'kän,takt i¦lek·trə'mōd·iv 'fōrs }

contact lens [OPTICS] **1.** A thin lens fitted over the cornea to correct defects of vision. **2.** A similar lens or prism used with a gonioscope in eye examinations. { 'kän,takt ,lenz }

contact piston [ELECTROMAG] A waveguide piston that makes contact with the walls of the waveguide. Also known as contact plunger. { 'kän,takt ,pis·tən }

contact plunger *See* contact piston. { 'kän,takt ,plən·jər }

contact potential *See* contact potential difference. { 'kän,takt pə'ten·chəl }

contact potential difference [ELEC] The potential difference that exists across the space between two electrically connected materials. Also known as contact electromotive force; contact potential; Volta effect. { 'kän,takt pə'ten·chəl 'dif·rəns }

contact resistance [ELEC] The resistance in ohms between the contacts of a relay, switch, or other device when the contacts are touching each other. { 'kän,takt ri'zis·təns }

contact sparking [ELEC] The formation of a spark or arc at the contact points when a circuit is opened while it is carrying a current. { 'kän,takt ,spärk·iŋ }

contact twin [CRYSTAL] Twinned crystals whose members are symmetrically arranged about a twin plane. { 'kän,takt ,twin }

continuity [ELEC] Continuous effective contact of all components of an electric circuit to give it high conductance by providing low resistance. { ,känt·ən'ü·əd·ē }

continuity equation [PHYS] An equation obeyed by any conserved, indestructible quantity such as mass, electric charge, thermal energy, electrical energy, or quantum-mechanical probability, which is essentially a statement that the rate of increase of the quantity in any region equals the total current flowing into the region. Also known as equation of continuity. { ,känt·ən'ü·əd·ē i'kwā·zhən }

continuity of state [THERMO] Property of a transition between two states of matter, as between gas and liquid, during which there are no abrupt changes in physical properties. { ,känt·ən'ü·əd·ē əv 'stāt }

continuity test [ELEC] An electrical test used to determine the presence and location of a broken connection. { ,känt·ən'ü·əd·ē ,test }

continuous radiation [ELECTROMAG] Electromagnetic radiation that includes all the wavelengths in some interval. Also known as white radiation. { kən¦tin·yə·wəs ,rād·ē'ā·shən }

continuous spectrum [SPECT] A radiation spectrum which is continuously distributed over a frequency region without being broken up into lines or bands. { kən¦tin·yə·wəs 'spek·trəm }

continuous wave [ELECTROMAG] A radio or radar wave whose successive sinusoidal oscillations are identical under steady-state conditions. Abbreviated CW. Also known as type A wave. { kən¦tin·yə·wəs 'wāv }

continuous-wave gas laser [OPTICS] A laser having a quartz envelope filled with a mixture of helium and neon at low pressure, with Brewster-angle mirrors at opposite ends and an external optical system. { kən¦tin·yə·wəs ¦wāv 'gas ,lā·zər }

continuous-wave laser [OPTICS] A laser in which the beam of coherent light is generated continuously, as required for communication and certain other applications. Abbreviated CW laser. { kən¦tin·yə·wəs ¦wāv 'lā·zər }

continuous x-rays [ELECTROMAG] The electromagnetic radiation, having a continuous spectral distribution, that is produced when high-velocity electrons strike a target. { kən¦tin·yə·wəs 'eks ,rāz }

continuum mechanics *See* classical field theory. { kən'tin·yə·wəm mə'kan·iks }

continuum physics *See* classical field theory. { kən'tin·yə·wəm 'fiz·iks }

contour [PHYS] A curve drawn up on a two-dimensional diagram through points which satisfy $f(x,y) = c$, where c is a constant and f is some function, such as the field strength for a transmitter. { 'kän,tu̇r }

contraction [MECH] The action or process of becoming smaller or pressed together, as a gas on cooling. { kən'trak·shən }

contraction coefficient *See* coefficient of contraction. { kən'trak·shən ,kō·i'fish·ənt }

contraction loss [FL MECH] In fluid flow, the loss in mechanical energy in a stream flowing through a closed duct or pipe when there is a sudden contraction of the cross-sectional area of the passage. { kən'trak·shən ,lo̊s }

contracurrent system *See* katoptric system. { ¦kän·trə¦kər·ənt ˌsis·təm }

contrast sensitivity *See* threshold contrast. { 'kän ˌtrast sen·sə'tiv·əd·ē }

contrast threshold *See* threshold contrast. { 'kän ˌtrast ˌthresh,hōld }

convection [FL MECH] Diffusion in which the fluid as a whole is moving in the direction of diffusion. Also known as bulk flow. [PHYS] Transmission of energy or mass by a medium involving movement of the medium itself. { kən'vek·shən }

convection coefficient *See* film coefficient. { kən'vek·shən ˌkō·i'fish·ənt }

convection modulus [FL MECH] An intrinsic property of a fluid which is important in determining the Nusselt number, equal to the acceleration of gravity times the volume coefficient of thermal expansion divided by the product of the kinematic viscosity and the thermal diffusivity. { kən'vek·shən ˌmäj·ə·ləs }

conventional current [ELEC] The concept of current as the transfer of positive charge, so that its direction of flow is opposite to that of electrons which are negatively charged. { kən'ven·chən·əl 'kər·ənt }

convergence [NEUROSCI] The coming together of a group of afferent nerves upon a motoneuron of the ventral horn of the spinal cord. [PHYS] The intersection of light beams or particles within a small region, or the narrowing of a single beam so that it passes through a small region. { kən'vər·jəns }

convergence circuit [ELECTROMAG] An auxiliary deflection system in a color television receiver which maintains convergence, having separate convergence coils for electromagnetic controls of the positions of the three beams in a convergence yoke around the neck of the kinescope. { kən'vər·jəns ˌsər·kət }

convergence limit [SPECT] **1.** The short-wavelength limit of a set of spectral lines that obey a Rydberg series formula; equivalently, the long-wavelength limit of the continuous spectrum corresponding to ionization from or recombination to a given state. **2.** The wavelength at which the difference between successive vibrational bands in a molecular spectrum decreases to 0. { kən'vər·jəns ˌlim·ət }

convergence ratio [OPTICS] The ratio of the tangent of the angle between a meridional ray and the optical axis after it passes through an optical system to the tangent of the angle between the ray and the axis before it passes through the system. { kən'vər·jəns ˌrā·shō }

convergence zone [ACOUS] A sound transmission channel produced in sea water by a combination of pressure and temperature changes in the depth range between 2500 and 15,000 feet (750 and 4500 meters); utilized by sonar systems. { kən'vər·jəns ˌzōn }

converging lens [OPTICS] A lens that has a positive focal length, and therefore causes rays of light parallel to its axis to converge. Also known as positive lens. { kən'vər·jiŋ ˌlenz }

converging mirror [OPTICS] A concave mirror that causes rays of light parallel to its axis to converge. Also known as positive mirror. { kən'vərj·iŋ ˌmir·ər }

conversion [NUC PHYS] Nuclear transformation of a fertile substance into a fissile substance. [PHYS] Change in a quantity's numerical value as a result of using a different unit of measurement. { kən'vər·zhən }

conversion coefficient [NUC PHYS] Also known as conversion fraction; internal conversion coefficient. **1.** The ratio of the number of conversion electrons emitted per unit time to the number of photons emitted per unit time in the de-excitation of a nucleus between two given states. **2.** In older literature, the ratio of the number of conversion electrons emitted per unit time to the number of conversion electrons plus the number of photons emitted per unit time in the de-excitation of a nucleus between two given states. { kən'vər·zhən ˌkō·i'fish·ənt }

conversion electron [NUC PHYS] An electron which receives energy directly from a nucleus in an internal conversion process and is thereby expelled from the atom. { kən'vər·zhən i'lek,trän }

conversion fraction *See* conversion coefficient. { kən'vər·zhən ˌfrak·shən }

conversion length [PHYS] The average distance traveled by an energetic photon in a given medium before it is converted into an electron and a positron through pair production. { kən'vər·zhən ˌleŋkth }

Conwell-Weisskopf equation [SOLID STATE] An equation for the mobility of electrons in a semiconductor in the presence of donor or acceptor impurities, in terms of the dielectric constant of the medium, the temperature, the concentration of ionized donors (or acceptors), and the average distance between them. { ¦kän,wel ¦vīs,kȯpf i'kwā·zhən }

Cooke objective [OPTICS] A three-lens objective consisting of one biconcave lens, the dispersive component, between two biconvex lens, the collective components; used in astronomical cameras. { 'kuk əb'jek·div }

Coolidge tube [ELECTROMAG] An x-ray tube in which the needed electrons are produced by a hot cathode. { 'kül·ij ˌtüb }

cooling correction [THERMO] A correction that must be employed in calorimetry to allow for heat transfer between a body and its surroundings. Also known as radiation correction. { 'kül·iŋ kə'rek·shən }

cooling curve [THERMO] A curve obtained by plotting time against temperature for a solid-liquid mixture cooling under constant conditions. { 'kül·iŋ ˌkərv }

cooling method [THERMO] A method of determining the specific heat of a liquid in which the times taken by the liquid and an equal volume of water in an identical vessel to cool through the same range of temperature are compared. { 'kül·iŋ ˌmeth·əd }

cooling stress [MECH] Stress resulting from

uneven contraction during cooling of metals and ceramics due to uneven temperature distribution. { 'kül·iŋ ,stres }

cooperative phenomenon [SOLID STATE] A process that involves a simultaneous collective interaction among many atoms or electrons in a crystal, such as ferromagnetism, superconductivity, and order-disorder transformations. { kō'äp·rəd·iv fə'näm·ə,nän }

Cooper pairs [SOLID STATE] Pairs of bound electrons which occur in a superconducting medium according to the Bardeen-Cooper-Schrieffer theory. { 'kü·pər ,perz }

coordination lattice [CRYSTAL] The crystal structure of a coordination compound. { kō,ȯrd·ən'ā·shən ,lad·əs }

coordination number [PHYS] The number of nearest neighbors of a point in a space lattice, of an atom or an ion in a solid, or of an anion or cation in a solution. { kō,ȯrd·ən'ā·shən ,nəm·bər }

coplanar forces [MECH] Forces that act in a single plane; thus the forces are parallel to the plane and their points of application are in the plane. { kō'plān·ər ,fȯrs·əz }

copper-64 [NUC PHYS] Radioactive isotope of copper with mass number of 64; derived from pile-irradiation of metallic copper; used as a research aid to study diffusion, corrosion, and friction wear in metals and alloys. { 'käp·ər ,sik·stē'fȯr }

copper vapor laser [OPTICS] A high-power laser that emits intense pulses of very short duration (typically 30 nanoseconds) at a rate of 5000–50,000 pulses per second, at wavelengths of 510.5 nanometers (green) and 578.2 nanometers (yellow). { 'käp·ər ,vā·pər ,lās·ər }

Corbino disk [ELECTROMAG] A variable-resistance device utilizing the effect of a magnetic field on the flow of carriers from the center to the circumference of a disk made of semiconducting or conducting material. { kȯr'bē·nō ,disk }

Corbino effect [ELECTROMAG] The production of an electric current around the circumference of a disk when a magnetic field perpendicular to the disk acts on a radial current in the disk. { kȯr'bēn·ō i,fekt }

core [ATOM PHYS] The electrons in the filled shells of an atom. [ELECTROMAG] *See* magnetic core. [NUC PHYS] The nucleons in the filled shells of a nucleus. { kȯr }

core electron [ATOM PHYS] An electron in a filled shell of an atom. { 'kȯr i,lek,trän }

core-form transformer [ELECTROMAG] A transformer in which half of the turns of the primary winding and half of those of the secondary are on each of two legs. { ¦kȯr ,fȯrm tranz'fȯr·mər }

core loss [ELECTROMAG] The rate of energy conversion into heat in a magnetic material due to the presence of an alternating or pulsating magnetic field. Also known as excitation loss; iron loss. { 'kȯr ,lȯs }

core state [PHYS] An energy state corresponding to an energy level in a filled shell of an atom or nucleus.

Coriolis acceleration [MECH] **1.** An acceleration which, when added to the acceleration of an object relative to a rotating coordinate system and to its centripetal acceleration, gives the acceleration of the object relative to a fixed coordinate system. **2.** A vector which is equal in magnitude and opposite in direction to that of the first definition. { kȯr·ē'ō·ləs ik,sel·ə'rā·shən }

Coriolis deflection *See* Coriolis effect. { kȯr·ē'ō·ləs di'flek·shən }

Coriolis effect [MECH] **1.** Also known as Coriolis deflection. **2.** The deflection relative to the earth's surface of any object moving above the earth, caused by the Coriolis force; an object moving horizontally is deflected to the right in the Northern Hemisphere, to the left in the Southern. **3.** The effect of the Coriolis force in any rotating system. { kȯr·ē'ō·ləs i'fekt }

Coriolis force [MECH] A velocity-dependent pseudoforce in a reference frame which is rotating with respect to an inertial reference frame; it is equal and opposite to the product of the mass of the particle on which the force acts and its Coriolis acceleration. { kȯr·ē'ō·ləs ,fȯrs }

Coriolis operator [SPECT] An operator which gives a large contribution to the energy of an axially symmetric molecule arising from the interaction between vibration and rotation when two vibrations have equal or nearly equal frequencies. { kȯr·ē'ō·ləs ,äp·ə,rād·ər }

Coriolis resonance interactions [SPECT] Perturbation of two vibrations of a polyatomic molecule, having nearly equal frequencies, on each other, due to the energy contribution of the Coriolis operator. { kȯr·ē'ō·ləs 'rez·ən·əns ,in·tər,ak·shənz }

corkscrew rule [ELECTROMAG] The rule that the direction of the current and that of the resulting magnetic field are related to each other as the forward travel of a corkscrew and the direction in which it is rotated. { 'kȯrk,skrü ,rül }

corner reflector [ELECTROMAG] An antenna consisting of two conducting surfaces intersecting at an angle that is usually 90°, with a dipole or other antenna located on the bisector of the angle. [OPTICS] A reflector which returns a laser beam in the direction of its source, consisting of perpendicular reflecting surfaces; used to make precise determinations of distances in surveying. { 'kȯr·nər ri'flek·tər }

Cornu-Hartmann formula *See* Hartmann dispersion formula. { ¦kȯr·nü 'härt·män ,fȯr·myə·lə }

Cornu quartz prism [OPTICS] A prism constructed of two 30° quartz prisms, left- and right-handed, used in conjunction with left- and right-handed lenses, so that the rotation of polarization occurring in one half of the optical path is exactly compensated by the reverse rotation in the other; used in a quartz spectrograph. { 'kȯr·nü ,kwȯrts 'priz·əm }

corona *See* corona discharge. { kə'rō·nə }

corona current [ELEC] The current of electricity equivalent to the rate of charge transferred to the air from an object experiencing corona discharge. { kə'rō·nə ¦kər·ənt }

corona discharge [ELEC] A discharge of electricity appearing as a bluish-purple glow on the surface of and adjacent to a conductor when the voltage gradient exceeds a certain critical value; due to ionization of the surrounding air by the high voltage. Also known as aurora; corona; electric corona. { kə'rō·nə 'dis,chärj }

corona stabilization [ELEC] The increase in the breakdown voltage of a gas separating two electrodes, where the electric field is very high at one pointed electrode and low at the other, due to the reduction of electric field around the pointed electrode by corona discharge. { kə'rō·nə ,stā·bə·lə'zā·shən }

corona start voltage [ELEC] The voltage difference at which corona discharge is initiated in a given system. { kə'rō·nə 'stärt ,vōl·tij }

corona tube [ELEC] A gas-discharge voltage-reference tube employing a corona discharge. { kə'rō·nə ,tüb }

corpuscle [NEUROSCI] An encapsulated sensory-nerve end organ. [OPTICS] A particle of light in the corpuscular theory, corresponding to the photon in the quantum theory. { 'kȯr·pəs·əl }

corpuscular radiation [PHYS] Radiation consisting of subatomic particles, such as electrons, protons, deuterons, and neutrons, as distinguished from electromagnetic radiation. { kȯr'pəs·kyə·lər ,rād·ē'ā·shən }

corpuscular theory of light [OPTICS] Theory that light consists of a stream of particles; now considered a limiting case of the quantum theory. Also known as Newton's theory of light. { kȯr'pəs·kyə·lər ,thē·ə·rē əv 'līt }

correcting plate *See* corrector plate. { kə'rek·tiŋ ,plāt }

corrector plate [OPTICS] A thin lens or system of lenses used to correct the spherical aberration of a spherical lens or the coma of a parabolic lens; used particularly in telescopes such as the Schmidt telescope. Also known as correcting plate. { kə'rek·tər ,plāt }

correlation [ATOM PHYS] *See* electron correlation. [PHYS] They tendency of two or more systems that independently exhibit simple behavior to show complex and novel behavior together because of their interaction. { ,kär·ə'lā·shən }

correlation array *See* multiplicative acoustic array. { ,kär·ə'lā·shən ə,rā }

correlation energy [ATOM PHYS] The difference between the experimentally measured energy of a particular energy level of an atom or molecule and the energy calculated in the Hartree-Fock approximation. [SOLID STATE] The modification of the Coulomb energy of a crystal that results from the tendency of electrons to stay apart from each other. { ,kär·ə'lā·shən ,en·ər·jē }

correspondence principle [QUANT MECH] The principle that quantum mechanics has a classical limit in which it is equivalent to classical mechanics. Also known as Bohr's correspondence principle. { ,kär·ə'spän·dəns ,prin·sə·pəl }

corrugated conical-horn antenna [ELECTROMAG] A horn antenna that has a circular cross section and a series of equally spaced ridges protruding from otherwise straight sides. { ¦kär·ə,gād·əd ,kän·ə·kəl ,hȯrn an'ten·ə }

corrugated lens [OPTICS] A lens having circular sections cut out from the surface to reduce its weight without lowering its focal power. { 'kär·ə,gād·əd 'lenz }

cosecant antenna [ELECTROMAG] An antenna that gives a beam whose amplitude varies as the cosecant of the angle of depression below the horizontal; used in navigation radar. { kō'sē ,kant an'ten·ə }

cosecant-squared antenna [ELECTROMAG] An antenna that has a cosecant-squared pattern. { kō'sē,kant ¦skwerd an'ten·ə }

cosecant-squared pattern [ELECTROMAG] A ground radar-antenna radiation pattern that sends less power to nearby objects than to those farther away in the same sector; the field intensity varies as the square of the cosecant of the elevation angle. { kō'sē,kant ¦skwerd 'pad·ərn }

cosine emission law [OPTICS] The law that the energy emitted by a radiating surface in any direction is proportional to the cosine of the angle which that direction makes with the normal. { 'kō,sīn i'mish·ən ,lȯ }

cosine pulse [PHYS] A pulse whose amplitude varies during some time interval in proportion to the cosine function over the range from $-\pi/2$ to $\pi/2$, and vanishes outside this time interval. { 'kō,sīn ,pəls }

cosine-squared pulse [PHYS] A pulse whose amplitude varies during some time interval in proportion to the square of the cosine function over the range from $-\pi/2$ to $\pi/2$, and vanishes outside this time interval. { 'kō,sīn ¦skwerd ,pəls }

cosmic radiation *See* cosmic rays. { 'käz·mik ,rād·ē'ā·shən }

cosmic rays [NUC PHYS] Electrons and the nuclei of atoms, largely hydrogen, that impinge upon the earth from all directions of space with nearly the speed of light. Also known as cosmic radiation; primary cosmic rays. { 'käz·mik 'rāz }

cosmic-ray shower [NUC PHYS] The simultaneous appearance of a number of downward-directed ionizing particles, with or without accompanying photons, caused by a single cosmic ray. Also known as air shower; shower. { 'käz·mik ,rā 'shau̇·ər }

cosmological constant [RELAT] The multiplicative constant for a term proportional to the metric in Einstein's equation relating the curvature of space to the energy-momentum tensor. { ¦käz·mə¦läj·ə·kəl 'kän·stənt }

cosmological term [RELAT] A term proportional to the metric tensor in Einstein's field equations for special relativity. { ¦käz·mə¦läj·ə·kəl 'tərm }

cospectrum [PHYS] **1.** The spectral decomposition of the in-phase components of the covariance of two functions of time. **2.** The real part of the cross spectrum of two functions. { ¦kō 'spek·trəm }

Coster-Kronig transition [ATOM PHYS] An Auger transition in which the vacant electron level is filled by an electron from a higher subshell of the same shell. { 'kas·tər 'krō,nig tran,zi·shən }

Cotton-Mouton birefringence *See* Cotton-Mouton effect. { ¦kät·ən ¦mü·ton ,bī·ri'frin·jəns }

Cotton-Mouton constant [OPTICS] A constant giving the strength of the Cotton-Mouton effect in a liquid; when multiplied by the path length and the square of the magnetic field, it gives the phase difference between the components of light parallel and perpendicular to the field. { ¦kät·ən ¦mü·ton ,kän·stənt }

Cotton-Mouton effect [OPTICS] The double refraction (birefringence) of light in a liquid in a magnetic field at right angles to the direction of light propagation. Also known as Cotton-Mouton birefringence. { ¦kät·ən ¦mü·ton i¦fekt }

Cottrell atmosphere [SOLID STATE] A cluster of impurity atoms surrounding a dislocation in a crystal. { kä¦trel 'at·mə,sfir }

Cottrell hardening [SOLID STATE] Hardening of a material caused by locking of its dislocations when impurity atoms whose size differs from that of the solvent cluster around them. { 'kä·trəl ,härd·ən·iŋ }

coudé focus [OPTICS] Focus achieved with a coudé telescope. { kü'dā ,fō·kəs }

coudé-Newtonian-Cassegrain telescope [OPTICS] A reflecting telescope designed so that observations can be made at the coudé, Newtonian, or Cassegrain focus. { kü'dā nü'tōn·ē·ən kas·gran 'tel·ə,skōp }

coudé spectrograph [SPECT] A stationary spectrograph that is attached to the tube of a coudé telescope. { kü'dā 'spek·trə,graf }

coudé spectroscopy [SPECT] The production and investigation of astronomical spectra using a coudé spectrograph. { kü'dā spek'träs·kə·pē }

coudé telescope [OPTICS] An instrument in which light is reflected along the polar axis to come to focus at a fixed place where it is viewed through a fixed eyepiece or where a spectrograph can be mounted. { kü'dā 'tel·ə,skōp }

Couette flow [FL MECH] Low-speed, steady motion of a viscous fluid between two infinite plates moving parallel to each other. { kü'et ,flō }

Couette-Taylor flow [FL MECH] The flow of a fluid within the annular space between two concentric cylinders when one or both of the cylinders rotate. { kü'et 'tā·lər ,flō }

coul *See* coulomb.

coulomb [ELEC] A unit of electric charge, defined as the amount of electric charge that crosses a surface in 1 second when a steady current of 1 absolute ampere is flowing across the surface; this is the absolute coulomb and has been the legal standard of quantity of electricity since 1950; the previous standard was the international coulomb, equal to 0.999835 absolute coulomb. Abbreviated coul. Symbolized C. { 'kü,läm }

Coulomb attraction [ELEC] The electrostatic force of attraction exerted by one charged particle on another charged particle of opposite sign. Also known as electrostatic attraction. { 'kü ,läm ə'trak·shən }

Coulomb barrier [NUC PHYS] **1.** The Coulomb repulsion which tends to keep positively charged bombarding particles out of the nucleus. **2.** Specifically, the Coulomb potential associated with this force. { 'kü,läm ,bar·ē·ər }

Coulomb crystal [ATOM PHYS] A crystalline array that is formed from laser-cooled ions stored in an electromagnetic trap and in which the relative positions of the ions are approximately fixed and are determined by the balance between the confining forces of the trap and the Coulomb repulsion of the ions. Also known as ion crystal. [CRYO] A structure formed by electrons trapped at a liquid helium surface at sufficiently high electron densities and low temperatures, in which the electrons occupy the points of a two-dimensional hexagonal lattice. { 'kü,läm ,krist·əl }

Coulomb energy [PHYS] The part of the binding energy of a system of particles, such as an atomic nucleus of a solid, which is associated with electrostatic forces between the particles. { 'kü,läm ,en·ər·jē }

Coulomb excitation [NUC PHYS] Inelastic scattering of a positively charged particle by a nucleus and excitation of the nucleus, caused by the interaction of the nucleus with the rapidly changing electric field of the bombarding particle. { 'kü,läm ,ek,sī'tā·shən }

Coulomb explosion [PHYS] A process in which a molecule moving with high velocity strikes a solid and the electrons that bond the molecule are torn off rapidly in violent collisions with the electrons of the solid; as a result, the molecule is transformed into a cluster of charged atomic constituents that then separate under the influence of their mutual Coulomb repulsion. { 'kü ,läm ik,splō·zhən }

Coulomb field [ELEC] The electric field created by a stationary charged particle. { 'kü,läm ,fēld }

Coulomb force [ELEC] The electrostatic force of attraction or repulsion exerted by one charged particle on another, in accordance with Coulomb's law. { 'kü,läm ,fȯrs }

Coulomb friction [MECH] Friction occurring between dry surfaces. { 'kü,läm ,frik·shən }

Coulomb gage [ELECTROMAG] A gage in which the divergence of the magnetic vector potential is equal to 0. { 'kü,läm ,gāj }

Coulomb interactions [ELEC] Interactions of charged particles associated with the Coulomb forces they exert on one another. Also known as electrostatic interactions. { 'kü,läm in·tər'ak·shənz }

Coulomb potential [ELEC] A scalar point function equal to the work per unit charge done against the Coulomb force in transferring a particle bearing an infinitesimal positive charge from infinity to a point in the field of a specific charge distribution. { kü'läm pə'ten·chəl }

Coulomb repulsion [ELEC] The electrostatic force of repulsion exerted by one charged particle on another charged particle of the same sign. Also known as electrostatic repulsion. { kü'läm ri'pəl·shən }

Coulomb scattering [PHYS] A collision of two charged particles in which the Coulomb force is the dominant interaction. { kü'läm ˌskad·ə·riŋ }

Coulomb's law [ELEC] The law that the attraction or repulsion between two electric charges acts along the line between them, is proportional to the product of their magnitudes, and is inversely proportional to the square of the distance between them. Also known as law of electrostatic attraction. { 'kü'lämz ˌlȯ }

Coulomb's theorem [ELEC] The proposition that the intensity of an electric field near the surface of a conductor is equal to the surface charge density on the nearby conductor surface divided by the absolute permittivity of the surrounding medium. { 'küˌlämz ˌthir·əm }

counterelectromotive force [ELECTROMAG] The voltage developed in an inductive circuit by a changing current; the polarity of the induced voltage is at each instant opposite that of the applied voltage. Also known as back electromotive force. { ¦kaünt·ər·iˌlek·trō'mōd·iv 'fȯrs }

counterpoise method *See* substitution weighing. { 'kaün·tərˌpȯiz ˌmeth·əd }

counterpropagating beams [OPTICS] Two light beams that are propagating through a medium in precisely opposite directions. { ˌkaün·tərˌpräp·əˌgād·iŋ 'bēmz }

counter terms [QUANT MECH] Additional terms added to a Lagrangian in quantum field theory in order to absorb the typical divergences that occur in a perturbation expansion of the theory. { 'kaün·tər ˌtərmz }

counter voltage [ELEC] The reverse voltage that appears across an inductor when current through the inductor is shut off. { 'kaünt·ər ˌvōl·tij }

counting rate [PHYS] The average rate of occurrence of events as observed by means of a counting system. { 'kaünt·iŋ ˌrāt }

couple [ELEC] To connect two circuits so signals are transferred from one to the other. [MECH] A system of two parallel forces of equal magnitude and opposite sense. { 'kəp·əl }

coupled antenna [ELECTROMAG] An antenna electromagnetically coupled to another. { 'kəp·əld an'ten·ə }

coupled circuits [ELEC] Two or more electric circuits so arranged that energy can transfer electrically or magnetically from one to another. { 'kəp·əld 'sər·kəts }

coupled field vectors [ELECTROMAG] The electric-and magnetic-field vectors, which depend upon each other according to Maxwell's field equations. { 'kəp·əld 'fēld ˌvek·tərz }

coupled harmonic oscillators [PHYS] Linear oscillators with an interaction, often also linear or weak. { 'kəp·əld har'män·ik 'äs·əˌlād·ərz }

coupled modes [ACOUS] Modes of acoustic transmission along a duct having a discontinuity, so that the reflected and transmitted waves contain modes other than the incident ones. { 'kəp·əld 'mōdz }

coupled oscillators [ELECTROMAG] A set of alternating-current circuits which interact with each other, for example, through mutual inductances or capacitances. [MECH] A set of particles subject to elastic restoring forces and also to elastic interactions with each other. { 'kəp·əld 'äs·əˌlād·ərz }

coupled systems [PHYS] Mechanical, electrical, or other systems which are connected in such a way that they interact and exchange energy with each other. { 'kəp·əld 'sis·təmz }

coupled wave [FL MECH] A surface wave which is being continuously generated by another wave having the same phase velocity. Also known as C wave. { 'kəp·əld 'wāv }

coupler [ELEC] A component used to transfer energy from one circuit to another. [ELECTROMAG] **1.** A passage which joins two cavities or waveguides, allowing them to exchange energy. **2.** A passage which joins the ends of two waveguides, whose cross section changes continuously from that of one to that of the other. { 'kəp·lər }

coupling [ELEC] **1.** A mutual relation between two circuits that permits energy transfer from one to another, through a wire, resistor, transformer, capacitor, or other device. **2.** A hardware device used to make a temporary connection between two wires. { 'kəp·liŋ }

coupling aperture [ELECTROMAG] An aperture in the wall of a waveguide or cavity resonator, designed to transfer energy to or from an external circuit. Also known as coupling hole; coupling slot. { 'kəp·liŋ ˌap·ə·chər }

coupling coefficient *See* coupling constant. { 'kəp·liŋ ˌkō·i'fish·ənt }

coupling constant [PART PHYS] A measure of the strength of a type of interaction between particles, such as the strong interaction between mesons and nucleons, and the weak interaction between four fermions; analogous to the electric charge, which is the coupling constant between charged particles and electromagnetic radiation. [PHYS] **1.** A measure of the strength of the coupling between two systems, especially electric circuits; maximum coupling is 1 and no coupling is 0. Also known as coefficient of coupling; coupling coefficient. **2.** A measure of the dependence of one physical quantity on another. { 'kəp·liŋ 'kän·stənt }

coupling hole *See* coupling aperture. { 'kəp·liŋ ˌhōl }

coupling loop [ELECTROMAG] A conducting loop projecting into a waveguide or cavity resonator, designed to transfer energy to or from an external circuit. { 'kəp·liŋ ˌlüp }

coupling probe [ELECTROMAG] A probe projecting into a waveguide or cavity resonator, designed to transfer energy to or from an external circuit. { 'kəp·liŋ ˌprōb }

coupling slot *See* coupling aperture. { 'kəp·liŋ ˌslät }

Courant condition [FL MECH] A condition on numerical hydrodynamics calculations requiring that the time interval employed be no greater than that required for a sound wave to cross a spatial cell. { 'ku̇rˌänt kənˌdish·ən }

covalent crystal [CRYSTAL] A crystal held together by covalent bonds. Also known as valence crystal. { kō'vā·lənt 'krist·əl }

covariant [RELAT] A scalar, vector, or higher-order tensor. { kō'ver·ē·ənt }

covariant equation [PHYS] An equation which has the same form in all inertial frames of reference; that is, its form is unchanged by Lorentz transformations. { kō'ver·ē·ənt i'kwā·zhən }

covariant theory [PHYS] A theory in which the equations have the same form in any inertial reference frame, the frames being related to each other by Lorentz transformations. { kō'ver·ē·ənt 'thē·ə·rē }

covering power [OPTICS] The field of view over which a camera lens can produce a sharp image, frequently expressed as an angle. { 'kəv·riŋ ˌpau̇·ər }

cp *See* candlepower; centipoise.

CP invariance [PART PHYS] The principle that the laws of physics are left unchanged by a combination of the operations of charge conjugation C and space inversion P; a small violation of this principle has been observed in the decay of neutral K - mesons. { ¦sē¦pē in'ver·ē·əns }

cpm *See* cycle per minute.

cps *See* hertz.

CPT theorem [PART PHYS] A theorem which states that a Lorentz invariant field theory is invariant to the product of charge conjugation C, space inversion P, and time reversal T. { ¦sē ¦pē¦tē 'thir·əm }

creation operator [QUANT MECH] An operator which increases the occupation number of a single state by unity and leaves all the other occupation numbers unchanged. { krē'ā·shən ˌäp·əˌrād·ər }

creep [MECH] A time-dependent strain of solids caused by stress. { krēp }

creepage [ELEC] The conduction of electricity across the surface of a dielectric. { 'krē·pij }

creep buckling [MECH] Buckling that may occur when a compressive load is maintained on a member over a long period, leading to creep which eventually reduces the member's bending stiffness. { 'krēp ˌbək·liŋ }

creeping flow [FL MECH] Fluid flow in which the velocity of flow is very small. { 'krē·piŋ ˌflō }

creep limit [MECH] The maximum stress a given material can withstand in a given time without exceeding a specified quantity of creep. { 'krēp ˌlim·ət }

creep recovery [MECH] Strain developed in a period of time after release of load in a creep test. { 'krēp ri'kəv·ə·rē }

creep rupture strength [MECH] The stress which, at a given temperature, will cause a material to rupture in a given time. { 'krēp 'rəp·chər ˌstreŋkth }

creep strength [MECH] The stress which, at a given temperature, will result in a creep rate of 1% deformation within 100,000 hours. { 'krēp ˌstreŋkth }

crest factor [PHYS] The ratio of the peak value to the effective value of any periodic quantity such as a sinusoidal alternating current. Also known as amplitude factor; peak factor. { 'krest ˌfak·tər }

crest value *See* peak value. { 'krest ˌval·yü }

crinal [MECH] A unit of force equal to 0.1 newton. { 'krīn·əl }

crispation number [PHYS] A dimensionless number used in the study of convection currents, equal to the product of a fluid's dynamic viscosity and its thermal diffusivity, divided by the product of its undisturbed surface tension and a layer thickness. { kri'spā·shən ˌnəm·bər }

crith [MECH] A unit of mass, used for gases, equal to the mass of 1 liter of hydrogen at standard pressure and temperature; it is found experimentally to equal 8.9885×10^{-5} kilogram. { krith }

critical absorption wavelength [SPECT] The wavelength, characteristic of a given electron energy level in an atom of a specified element, at which an absorption discontinuity occurs. { 'krid·ə·kəl əb'sȯrp·shən 'wāvˌleŋkth }

critical angle [PHYS] An angle associated with total reflection of electromagnetic or acoustic radiation back into a medium from the boundary with another medium in which the radiation has a higher phase velocity; it is the smallest angle with the normal to the boundary at which total reflection occurs. { 'krid·ə·kəl ¦aŋ·gəl }

critical angle refractometer [OPTICS] A refractometer, such as the Abbe or Pulfrich refractometer, in which the index of refraction of a medium A is measured by observing its critical angle with respect to another medium B with a known index of refraction, or by measuring the critical angle of B with respect to A. { 'krid·ə·kəl ¦aŋ·gəl ˌrē ˌfrak'täm·əd·ər }

critical coupling [ELEC] The degree of coupling that provides maximum transfer of signal energy from one radio-frequency resonant circuit to another when both are tuned to the same frequency. Also known as optimum coupling. { 'krid·ə·kəl 'kəp·liŋ }

critical current [SOLID STATE] The current in a superconductive material above which the material is normal and below which the material is superconducting, at a specified temperature and in the absence of external magnetic fields. { 'krid·ə·kəl 'kər·ənt }

critical damping [PHYS] Damping in a linear system on the threshold between oscillatory and exponential behavior. { 'krid·ə·kəl 'dam·piŋ }

critical density [THERMO] The density of a substance at the liquid-vapor critical point. { 'krid·ə·kəl 'den·səd·ē }

critical exponent [THERMO] A parameter n that

characterizes the temperature dependence of a thermodynamic property of a substance near its critical point; the temperature dependence has the form $|T - T_c|^n$, where T is the temperature and T_c is the critical temperature. { 'krid·ə·kəl ik'spō·nənt }

critical field [SOLID STATE] The magnetic field strength below which magnetic flux is excluded from a type I superconductor. Symbolized H_c. { 'krid·ə·kəl 'fēld }

critical flicker frequency [OPTICS] That frequency of an intermittent light source at which the light appears half the time as flickering and half the time as continuous. { 'krid·ə·kəl 'flik·ər ˌfrē·kwən·sē }

critical flow [FL MECH] The rate of flow of a fluid equivalent to the speed of sound in that fluid. { 'krid·ə·kəl 'flō }

critical frequency [ELECTROMAG] The limiting frequency below which a radio wave will be reflected by an ionospheric layer at vertical incidence at a given time. { 'krid·ə·kəl 'frē·kwən·sē }

critical isotherm [THERMO] A curve showing the relationship between the pressure and volume of a gas at its critical temperature. { 'krid·ə·kəl 'ī·səˌthərm }

critical magnetic field [SOLID STATE] The field below which a superconductive material is superconducting and above which the material is normal, at a specified temperature and in the absence of current. { 'krid·ə·kəl mag'ned·ik 'fēld }

critical magnetic scattering [SOLID STATE] Intense scattering of low-energy neutrons by a ferromagnetic crystal at temperatures near the Curie point. { 'krid·ə·kəl mag'ned·ik 'skad·ər·iŋ }

critical opalescence [OPTICS] Extreme opalescence resulting from strong density fluctuations in a medium near a critical point. { 'krid·ə·kəl ˌōp·ə'les·əns }

critical potential [ATOM PHYS] The energy needed to raise an electron to a higher energy level in an atom (resonance potential) or to remove it from the atom (ionization potential). [ELEC] A potential which results in sudden change in magnitude of the current. { 'krid·ə·kəl pə'ten·chəl }

critical pressure [FL MECH] For a nozzle whose cross section at each point is such that a fluid in isentropic flow just fills it, the pressure at the section of minimum area of the nozzle; if the nozzle is cut off at this point with no diverging section, decrease in the discharge pressure below the critical pressure (at constant admission pressure) does not result in increased flow. [THERMO] The pressure of the liquid-vapor critical point. { 'krid·ə·kəl 'presh·ər }

critical-pressure ratio [FL MECH] The ratio of the critical pressure of a nozzle to the admission pressure of the nozzle (equals 0.53 for gases). { 'krid·ə·kəl 'presh·ər ˌrā·shō }

critical Reynolds number [FL MECH] The Reynolds number at which there is a transition from laminar to turbulent flow. { 'krid·ə·kəl 'ren·əlz ˌnəm·bər }

critical scattering [PHYS] Intense scattering of some form of radiation by a substance at a temperature near a second-order transition, as in critical opalescence or critical magnetic scattering. { 'krid·ə·kəl 'skad·ər·iŋ }

critical shear stress [SOLID STATE] The shear stress needed to cause slip in a given direction along a given crystallographic plane of a single crystal. { 'krid·ə·kəl 'shirˌstres }

critical speed *See* critical velocity. { 'krid·ə·kəl 'spēd }

critical velocity [CRYO] The velocity of a superfluid in very narrow channels (on the order of 10^{-5} centimeter), which is nearly constant. Also known as critical speed. [FL MECH] **1.** The speed of flow equal to the local speed of sound. Also known as critical speed. **2.** The speed of fluid flow through a given conduit above which it becomes turbulent. { 'krid·ə·kəl və'läs·əd·ē }

critical volume [PHYS] The volume occupied by one mole of a substance at the liquid-vapor critical point, that is, at the critical temperature and pressure. { 'krid·ə·kəl 'väl·yəm }

critical zone [FL MECH] In fluid flow, the area on a graph of the Reynolds number versus friction factor indicating unstable flow (Reynolds number 2000 to 4000) between laminar flow and the transition to turbulent flow. { 'krid·ə·kəl 'zōn }

CRLAS *See* cavity ringdown laser absorption spectroscopy.

CR law [ELEC] A law which states that when a constant electromotive force is applied to a circuit consisting of a resistor and capacitor connected in series, the time taken for the potential on the plates of the capacitor to rise to any given fraction of its final value depends only on the product of capacitance and resistance. { ¦sē¦är ˌlȯ }

Crocco's equation [FL MECH] A relationship, expressed as $\mathbf{v} \times \omega = -T \operatorname{grad} S$, between vorticity and entropy gradient for the steady flow of an inviscid compressible fluid; **v** is the fluid velocity vector, ω (= curl **v**) is the vorticity vector, T is the fluid temperature, and S is the entropy per unit mass of the fluid. { 'kräˌkōz i'kwā·zhən }

crocodile [ELEC] A unit of potential difference or electromotive force, equal to 10^6 volts; used informally at some nuclear physics laboratories. { 'kräk·əˌdīl }

Crookes radiometer [PHYS] A radiometer used to demonstrate that radiant energy from the sun can produce motion; a miniature four-vane windmill is mounted in a glass-envelope vacuum tube, with each vane polished on one side and black on the other. { 'krůks ˌrād·ē'äm·əd·ər }

cross antenna [ELECTROMAG] An array of two or more horizontal antennas connected to a single feed line and arranged in the pattern of a cross. { 'krȯs anˌten·ə }

crosscurrent [FL MECH] A current that flows across or opposite to another current. { 'krȯsˌkər·ənt }

crossed cylinder [OPTICS] **1.** A thin lens whose surfaces are portions of circular cylinders whose axes cross at right angles or obliquely. **2.** A weak lens whose effect is equivalent to that of lenses with convex and concave cylindrical surfaces of equal curvature crossed at right angles. { ¦kröst 'sil·ən·dər }

crossed lens [OPTICS] A lens designed with radii of curvature which give minimum spherical aberration for parallel incident rays. { 'kröst 'lenz }

crossed prisms [OPTICS] A pair of Nicol prisms whose principal planes are perpendicular to each other, so that light passing through one is extinguished by the other. { 'kröst 'priz·əmz }

cross effect [PHYS] Any phenomenon in which two or more transport effects are coupled, such as thermal and electrical conductivity, or thermal conductivity and diffusion. { 'krös i,fekt }

cross flux [ELECTROMAG] A component of magnetic flux perpendicular to that produced by the field magnets in an electrical rotating machine. { 'krös ,fləks }

crossing symmetry [PART PHYS] The amplitude for a process that involves creation of a particle with four-momentum P_μ is equal to the amplitude for a process which is the same except it involves destruction of the antiparticle with four-momentum $-P_\mu$. { 'krös·iŋ ,sim·ə·trē }

cross-magnetizing effect [ELECTROMAG] The distortion in the flux-density distribution in the air gap of an electric rotating machine caused by armature reaction. { ¦krös 'mag·nə,tīz·iŋ i'fekt }

crossover [ELEC] A point at which two conductors cross, with appropriate insulation between them to prevent contact. { 'krös,ō·vər }

cross-polarization [ELECTROMAG] The component of the electric field vector normal to the desired polarization component. { ¦krös ,pō·lə·rə'zā·shən }

cross section [PHYS] An area characteristic of a collision reaction between atomic or nuclear particles or systems, such that the number of reactions which occur equals the product of the number of target particles or systems and the number of incident particles or systems which would pass through this area if their velocities were perpendicular to it. Also known as collision cross section. { 'krös ,sek·shən }

cross section per atom [NUC PHYS] The microscopic cross section for a given nuclear reaction referred to the natural element, even though the reaction involves only one of the natural isotopes. { 'krös ,sek·shən pər 'ad·əm }

cross spectrum [PHYS] The complex vector sum of the cospectrum and quadrature spectrum. { ¦krös ¦spek·trəm }

Crova wavelength [STAT MECH] The wavelength in the spectrum of a radiator whose intensity divided by the intensity of the total radiation equals the derivative of the intensity of the wavelength with respect to temperature divided by the derivative of the total intensity with respect to temperature. { 'krō·və 'wāv,leŋkth }

crushing strain [MECH] Compression which causes the failure of a material. { 'krəsh·iŋ ,strān }

crushing strength [MECH] The compressive stress required to cause a solid to fail by fracture; in essence, it is the resistance of the solid to vertical pressure placed upon it. { 'krəsh·iŋ ,streŋkth }

cryogen *See* cryogenic fluid. { 'krī·ə·jən }

cryogenic coil [CRYO] A high-purity coil refrigerated to very low temperatures to reduce effective coil resistivity. { ,krī·ə'jen·ik 'köil }

cryogenic conductor *See* superconductor. { ,krī·ə'jen·ik kən'dək·tər }

cryogenic device [CRYO] A device whose operation depends on superconductivity as produced by temperatures near absolute zero. Also known as superconducting device. { ,krī·ə'jen·ik di'vīs }

cryogenic fluid [CRYO] A liquid which boils at temperatures of less than about 110 K at atmospheric pressure, such as hydrogen, helium, nitrogen, oxygen, air, or methane. Also known as cryogen; cryogenic liquid. { ,krī·ə'jen·ik 'flü·əd }

cryogenic liquid *See* cryogenic fluid. { ,krī·ə'jen·ik 'lik·wəd }

cryogenic pump [CRYO] A high-speed vacuum pump that can produce an extremely low vacuum and has a low power consumption; to reduce the pressure, gases are condensed on surfaces within an enclosure at extremely low temperatures, usually attained by using liquid helium or liquid or gaseous hydrogen. Also known as cryopump. { ,krī·ə'jen·ik 'pəmp }

cryogenics [PHYS] The production and maintenance of very low temperatures, and the study of phenomena at these temperatures. { ,krī·ə'jen·iks }

cryogenic temperature [CRYO] A temperature within a few degrees of absolute zero. { ,krī·ə'jen·ik 'tem·prə·chər }

cryomagnetic [CRYO] Pertaining to production of very low temperatures by adiabatic demagnetization of paramagnetic salts. { ¦krī·ō·mag'ned·ik }

cryophysics [CRYO] Physics as restricted to phenomena occurring at very low temperatures, approaching absolute zero. { ¦krī·ō'fiz·iks }

cryopump *See* cryogenic pump. { 'krī·ō,pəmp }

crystal [CRYSTAL] A homogeneous solid made up of an element, chemical compound or isomorphous mixture throughout which the atoms or molecules are arranged in a regularly repeating pattern. { 'krist·əl }

crystal axis [CRYSTAL] A reference axis used for the vectoral properties of a crystal. { ¦krist·əl 'ak·səs }

crystal base [CRYSTAL] The contents of a primitive cell of a crystal. { ¦krist·əl 'bās }

crystal chemistry [CRYSTAL] The study of the crystalline structure and properties of a mineral or other solid. { 'krist·əl 'kem·ə·strē }

crystal class [CRYSTAL] One of 32 categories of crystals according to the inversions, rotations about an axis, reflections, and combinations of these which leaves the crystal invariant. Also known as symmetry class. { 'krist·əl 'klas }

crystal defect [CRYSTAL] Any departure from crystal symmetry caused by free surfaces, disorder, impurities, vacancies and interstitials, dislocations, lattice vibrations, and grain boundaries. Also known as lattice defect. { 'krist·əl 'dē,fekt }

crystal diffraction [SOLID STATE] Diffraction by a crystal of beams of x-rays, neutrons, or electrons whose wavelengths (or de Broglie wavelengths) are comparable with the interatomic spacing of the crystal. { 'krist·əl di'frak·shən }

crystal dynamics *See* lattice dynamics. { ¦krist·əl də'nam·iks }

crystal face [CRYSTAL] One of the outward planar surfaces which define a crystal and reflect its internal structure. Also known as face. { ¦krist·əl ¦fās }

crystal form [CRYSTAL] A collection of crystal faces generated by operating on a single face with a subgroup of the symmetry elements of the crystal class. { ¦krist·əl 'fȯrm }

crystal gliding [CRYSTAL] Slip along a crystal plane due to plastic deformation; often produces crystal twins. Also known as translation gliding. { 'krist·əl ,glīd·iŋ }

crystal grating [SPECT] A diffraction grating for gamma rays or x-rays which uses the equally spaced lattice planes of a crystal. { 'krist·əl ,grād·iŋ }

crystal growth [CRYSTAL] The growth of a crystal, which involves diffusion of the molecules of the crystallizing substance to the surface of the crystal, diffusion of these molecules over the crystal surface to special sites on the surface, incorporation of molecules into the surface at these sites, and diffusion of heat away from the surface. { 'krist·əl ,grōth }

crystal habit [CRYSTAL] The size and shape of the crystals in a crystalline solid. Also known as habit. { 'krist·əl ,hab·ət }

crystal indices *See* Miller indices. { 'krist·əl 'in·də,sēz }

crystal laser [OPTICS] A laser that uses a pure crystal of ruby or other material for generating a coherent beam of output light. { ¦krist·əl 'lā·zər }

crystal lattice [CRYSTAL] A lattice from which the structure of a crystal may be obtained by associating with every lattice point an assembly of atoms identical in composition, arrangement, and orientation. { ¦krist·əl 'lad·əs }

crystalline [CRYSTAL] Of, pertaining to, resembling, or composed of crystals. { 'kris·tə·lən }

crystalline anisotropy [SOLID STATE] The tendency of crystals to have different properties in different directions; for example, a ferromagnet will spontaneously magnetize along certain crystallographic axes. { 'kris·tə·lən an·ə'sä·trə·pē }

crystalline double refraction [OPTICS] The splitting which a wavefront experiences when a wave disturbance propagates through an anisotropic crystal. { 'kris·tə·lən 'dəb·əl ri'frak·shən }

crystalline field [SOLID STATE] The internal electric field in a solid due to localized charges, especially ions, inside. { 'kris·tə·lən 'fēld }

crystalline laser [OPTICS] A solid laser in which the lasing material is a pure crystal like ruby or a doped crystal like neodymium-doped ruby or neodymium-doped yttrium aluminum garnet. { 'kris·tə·lən 'lā·zər }

crystallinity [CRYSTAL] The quality or state of being crystalline. { ,kris·tə'lin·əd·ē }

crystallization [CRYSTAL] The formation of crystalline substances from solutions or melts. { ,kris·tə·lə'zā·shən }

crystallogram [CRYSTAL] A photograph of the x-ray diffraction pattern of a crystal. { 'kris·tə·lō,gram }

crystallographic axis [CRYSTAL] One of three lines (sometimes four, in the case of a hexagonal crystal), passing through a common point, that are chosen to have definite relation to the symmetry properties of a crystal, and are used as a reference in describing crystal symmetry and structure. { ¦kris·tə·lō¦graf·ik 'ak·səs }

crystallography [PHYS] The branch of science that deals with the geometric description of crystals and their internal arrangement. { ,kris·tə'läg·rə·fē }

crystallomagnetic [SOLID STATE] Pertaining to magnetic properties of crystals. { ,kris·tə·lō·mag'ned·ik }

crystal momentum [SOLID STATE] The product of Planck's constant and the wave vector associated with an elementary excitation in a crystal (the magnitude of the wave vector being taken as the reciprocal of the wavelength). { ¦krist·əl mə'men·təm }

crystal monochromator [SPECT] A spectrometer in which a collimated beam of slow neutrons from a reactor is incident on a single crystal of copper, lead, or other element mounted on a divided circle. { ¦krist·əl ,män·ə'krō,mād·ər }

crystal optics [OPTICS] The study of the propagation of light, and associated phenomena, in crystalline solids. { ¦krist·əl 'äp·tiks }

crystal photoeffect [SOLID STATE] An electromotive force induced by illumination of natural cuprite crystals or transparent zinc sulfide, and having a direction dependent on that of the incident light beam. { 'krist·əl 'fōd·ō·i,fekt }

crystal plane [CRYSTAL] One of a set of parallel, equally spaced planes in a crystal structure, each of which contains an infinite periodic array of lattice points. { ¦krist·əl 'plān }

crystal projection [CRYSTAL] Any method of displaying the positions of the poles of a crystal by projecting them on a plane. { ¦krist·əl prə'jek·shən }

crystal pulling [CRYSTAL] A method of crystal growing in which the developing crystal is gradually withdrawn from a melt. { 'krist·əl ,pu̇l·iŋ }

crystal shutter [ELECTROMAG] Mechanical waveguide or coaxial-cable shorting switch that, when closed, prevents undesired radio-frequency energy from reaching and damaging a crystal detector. { ¦krist·əl 'shəd·ər }

crystal structure [CRYSTAL] The arrangement of

atoms or ions in a crystalline solid. { ¦krist·əl 'strək·chər }

crystal symmetry [CRYSTAL] The existence of nontrivial operations, consisting of inversions, rotations around an axis, reflections, and combinations of these, which bring a crystal into a position indistinguishable from its original position. { ¦krist·əl 'sim·ə·trē }

crystal system [CRYSTAL] One of seven categories (cubic, hexagonal, tetragonal, trigonal, orthorhombic, monoclinic, and triclinic) into which a crystal may be classified according to the shape of the unit cell of its Bravais lattice, or according to the dominant symmetry elements of its crystal class. { ¦krist·əl 'sis·təm }

crystal twin *See* twin crystal. { 'krist·əl ˌtwin }

crystal whisker [CRYSTAL] A single crystal that has grown in a filamentary form. Also known as whisker. { ¦krist·əl 'wis·kər }

cs *See* centistoke.

cu *See* cubic.

cube-surface coil [ELECTROMAG] A system of five equally spaced square coils that produces a region of uniform magnetic field over a large volume which is easily accessible from outside the coils. { 'kyüb ˌsər·fəs ˌkȯil }

cubic [MECH] Denoting a unit of volume, so that if x is a unit of length, a cubic x is the volume of a cube whose sides have length $1x$; for example, a cubic meter, or a meter cubed, is the volume of a cube whose sides have a length of 1 meter. Abbreviated cu. { 'kyü·bik }

cubical antenna [ELECTROMAG] An antenna array, the elements of which are positioned to form a cube. { 'kyü·bə·kəl an'ten·ə }

cubical dilation [MECH] The isotropic part of the strain tensor describing the deformation of an elastic solid, equal to the fractional increase in volume. { 'kyü·bə·kəl di'lā·shən }

cubical expansion [PHYS] The increase in volume of a substance with a change in temperature or pressure. { 'kyü·bə·kəl ik'span·shən }

cubic cleavage [CRYSTAL] Isometric crystal cleavage occuring parallel to the faces of a cube. { 'kyü·bik 'klē·vij }

cubic crystal [CRYSTAL] A crystal whose lattice has a unit cell with perpendicular axes of equal length. { 'kyü·bik 'krist·əl }

cubic foot per minute [MECH] A unit of volume flow rate, equal to a uniform flow of 1 cubic foot in 1 minute; equal to 1/60 cusec. Abbreviated cfm. { ¦kyü·bik ¦fu̇t pər 'min·ət }

cubic foot per second *See* cusec. { ¦kyü·bik ¦fu̇t pər 'sek·ənd }

cubic measure [MECH] A unit or set of units to measure volume. { 'kyü·bik 'mezh·ər }

cubic packing [CRYSTAL] The spacing pattern of uniform solid spheres in a clastic sediment or crystal lattice in which the unit cell is a cube. { 'kyü·bik 'pak·iŋ }

cubic plane [CRYSTAL] A plane that is at right angles to any one of the three crystallographic axes of the cubic system. { 'kyü·bik 'plān }

cubic system *See* isometric system. { 'kyü·bik ˌsis·təm }

cumec [MECH] A unit of volume flow rate equal to 1 cubic meter per second. { 'kyüˌmek }

cumulative excitation [ATOM PHYS] Process by which the atom is raised from one excited state to a higher state by collision, for example, with an electron. { 'kyü·myə·ləd·iv ek·sə'tā·shən }

cumulative ionization [ATOM PHYS] Ionization of an excited atom in the metastable state by means of cumulative excitation. { 'kyü·myə·ləd·iv ˌī·ən·ə'zā·shən }

cup core [ELECTROMAG] A core that encloses a coil to provide magnetic shielding; usually has a powdered iron center post through the coil. { 'kəp ˌkȯr }

Curie constant [ELECTROMAG] The electric or magnetic susceptibility at some temperature times the difference of the temperature and the Curie temperature, which is a constant at temperatures above the Curie temperature according to the Curie-Weiss law. { 'kyu̇r·ē ¦kän·stənt }

Curie point *See* Curie temperature. { 'kyu̇r·ē ˌpȯint }

Curie principle [THERMO] The principle that a macroscopic cause never has more elements of symmetry than the effect it produces; for example, a scalar cause cannot produce a vectorial effect. { 'kyu̇r·ē ˌprin·sə·pəl }

Curie scale of temperature [THERMO] A temperature scale based on the susceptibility of a paramagnetic substance, assuming that it obeys Curie's law; used at temperatures below about 1 kelvin. { ¦kyu̇r·ē ¦skāl əv 'tem·prə·chər }

Curie's law [ELECTROMAG] The law that the magnetic susceptibilities of most paramagnetic substances are inversely proportional to their absolute temperatures. { 'kyu̇rˌēz ˌlȯ }

Curie temperature [ELECTROMAG] The temperature marking the transition between ferromagnetism and paramagnetism, or between the ferroelectric phase and paraelectric phase. Also known as Curie point. { 'kyu̇r·ē ˌtem·prə·chər }

Curie-Weiss law [ELECTROMAG] A relation between magnetic or electric susceptibilities and the absolute temperatures which is followed by ferromagnets, antiferromagnets, nonpolar ferroelectrics, antiferroelectrics, and some paramagnets. { ¦kyu̇r·ē ¦vīs ˌlȯ }

curium-242 [NUC PHYS] An isotope of curium, mass number 242; half-life is 165.5 days for α-particle emission; 7.2 × 10^6 years for spontaneous fission. { 'kyu̇r·ē·əm ¦tü¦fȯrd·ē¦tü }

curium-244 [NUC PHYS] An isotope of curium, mass number 244; half-life is 16.6 years for α-particle emission; 1.4 × 10^7 years for spontaneous fission; potential use as compact thermoelectric power source. { 'kyu̇r·ē·əm ¦tü¦fȯrd·ē¦fȯr }

current [ELEC] The net transfer of electric charge per unit time; a specialization of the physics definition. Also known as electric current. [PHYS] **1.** The rate of flow of any conserved, indestructible quantity across a surface per unit time. **2.** *See* current density. { 'kər·ənt }

current algebra [PART PHYS] The application of

algebraic relationships among currents derived from approximate symmetries, such as broken SU_3 symmetry, to the study of hadrons. { 'kər·ənt ˌal·jə·brə }

current antinode [ELEC] A point at which current is a maximum along a transmission line, antenna, or other circuit element having standing waves. Also known as current loop. { 'kər·ənt 'an·tə,nōd }

current balance [ELEC] An apparatus with which force is measured between current-carrying conductors, with the purpose of assigning the value of the ampere. Also known as ampere balance. { 'kər·ənt ˌbal·əns }

current collector *See* charge collector. { 'kər·ənt kə,lek·tər }

current comparator [ELEC] An instrument for determining the ratio of two direct or alternating currents, based on Ampère's laws, in which the two currents are passed through a toroid by two windings of known numbers of turns and the ampere-turn unbalance is measured by a detection winding. { 'kə·rənt kəm,par·əd·ər }

current density [ELEC] The current per unit cross-sectional area of a conductor; a specialization of the physics definition. Also known as electric current density. [PHYS] A vector quantity whose component perpendicular to any surface equals the rate of flow of some conserved, indestructible quantity across that surface per unit area per unit time. Also known as current. { 'kər·ənt ˌden·səd·ē }

current divider [ELEC] A device used to deliver a desired fraction of a total current to a circuit. { 'kər·ənt di,vīd·ər }

current drain [ELEC] The current taken from a voltage source by a load. Also known as drain. { 'kər·ənt ˌdrān }

current function *See* Lagrange stream function. { 'kər·ənt ˌfəŋk·shən }

current intensity [ELEC] The magnitude of an electric current. Also known as current strength. { 'kər·ənt in'ten· səd·ē }

current loop *See* current antinode. { 'kər·ənt ˌlüp }

current measurement [ELEC] The measurement of the flow of electric current. { 'kər·ənt ˌmezh·ər·mənt }

current node [ELEC] A point at which current is zero along a transmission line, antenna, or other circuit element having standing waves. { 'kər·ənt ˌnōd }

current ratio [ELECTROMAG] In a waveguide, the ratio of maximum to minimum current. { 'kər·ənt ˌrā·shō }

current strength *See* current intensity. { 'kər·ənt ˌstreŋkth }

current tap *See* multiple lamp holder; plug adapter lamp holder. { 'kər·ənt ˌtap }

current transformer [ELEC] An instrument transformer intended to have its primary winding connected in series with a circuit carrying the current to be measured or controlled; the current is measured across the secondary winding. { 'kər·ənt tranz'fȯr·mər }

current-transformer phase angle [ELEC] Angle between the primary current vector and the secondary current vector reversed; it is conveniently considered as positive when the reversed secondary current vector leads the primary current vector. { 'kər·ənt tranz'fȯr·mər 'fāz ˌaŋ·gəl }

current-voltage dual [ELEC] A circuit which is equivalent to a specified circuit when one replaces quantities with dual quantities; current and voltage impedance and admittance, and meshes and nodes are examples of dual quantities. { ¦kər·ənt ¦vōl·tij ¦dül }

curtain array [ELECTROMAG] An antenna array consisting of vertical wire elements stretched between two suspension cables. { 'kərt·ən ə'rā }

curtain rhombic antenna [ELECTROMAG] A multiple-wire rhombic antenna having a constant input impedance over a wide frequency range; two or more conductors join at the feed and terminating ends but are spaced apart vertically from 1 to 5 feet (30 to 150 centimeters) at the side poles. { 'kərt·ən 'räm·bik an'ten·ə }

curvature of field [OPTICS] Error in the image of a plane object formed on a flat screen by an optical system when the best image lies on a curved surface. Also known as field curvature. { 'kər·və·chər əv 'fēld }

curvature of space [RELAT] **1.** The deviation of a spacelike three-dimensional subspace of curved space-time from euclidean geometry. **2.** The Gaussian curvature of a spacelike three-dimensional subspace of curved space-time. { 'kər·və·chər əv 'spās }

curved space-time [RELAT] A four-dimensional Riemannian space, in which there are no straight lines but only curves, which is a generalization of the Minkowski universe in the general theory of relativity. { ¦kərvd 'spās 'tīm }

curve resistance [MECH] The force opposing the motion of a railway train along a track due to track curvature. { 'kərv ri'zis·təns }

curvilinear motion [MECH] Motion along a curved path. { 'kər·və'lin·ē·ər 'mō·shən }

cusec [MECH] A unit of volume flow rate, used primarily to describe pumps, equal to a uniform flow of 1 cubic foot in 1 second. Also known as cubic foot per second (cfs). { 'kyü,sek }

cushion effect *See* Poisson effect. { 'kəsh·ən i,fekt }

cusped magnetic field [ELECTROMAG] A magnetic field created by adjacent parallel coils that carry current in opposite directions; used in fusion research, to contain a plasma of high-energy deuterium ions. { 'kəspt mag'ned·ik 'fēld }

cut [CRYSTAL] A section of a crystal having two parallel major surfaces; cuts are specified by their orientation with respect to the axes of the natural crystal, such as X cut, Y cut, BT cut, and AT cut. { kət }

Cutler feed [ELECTROMAG] A resonant cavity that transfers radio-frequency energy from the end of a waveguide to the reflector of a radar spinner assembly. { 'kət·lər ˌfēd }

cutoff [PHYS] Technique used when the contribution to the value of a physical quantity given by integration over a certain variable is absurd (in particular, when the contribution is infinite); involves cutting off the integral at some limit. { 'kət,ȯf }

cutoff attenuator [ELECTROMAG] Variable length of waveguide used below its cutoff frequency to introduce variable nondissipative attenuation. { 'kət,ȯf ə'ten·yə,wād·ər }

cutoff wavelength [ELECTROMAG] **1.** The ratio of the velocity of electromagnetic waves in free space to the cutoff frequency in a uniconductor waveguide. **2.** The wavelength corresponding to the cutoff frequency. { 'kət,ȯf 'wāv,leŋkth }

cut-set [ELEC] A set of branches of a network such that the cutting of all the branches of the set increases the number of separate parts of the network, but the cutting of all the branches except one does not. { 'kət ,set }

CVC *See* conserved vector current.

CVD *See* chemical vapor deposition.

CW *See* continuous wave.

C wave *See* coupled wave. { 'sē ,wāv }

CW laser *See* continuous-wave laser. { ¦sē¦dəb·əl,yü 'lā·zər }

cyanometer [OPTICS] An instrument designed to measure or estimate the degree of blueness of light, as of the sky. { sī·ə'näm·əd·ər }

cyanometry [OPTICS] The study and measurement of the blueness of light. { ,sī·ə'nam·ə·trē }

cybotaxis [PHYS] A transient molecular orientation in a liquid evidenced by x-ray diffraction effects. { ,si·bə'tak·səs }

cycle [FL MECH] A system of phases through which the working substance passes in an engine, compressor, pump, turbine, power plant, or refrigeration system. { 'sī·kəl }

cycle per minute [PHYS] A unit of frequency of action, equal to 1/60 hertz. Abbreviated cpm. { 'sī·kəl pər 'min·ət }

cycle per second *See* hertz. { 'sī·kəl pər 'sek·ənd }

cyclic coordinate [MECH] A generalized coordinate on which the Lagrangian of a system does not depend explicitly. Also known as ignorable coordinate. { 'sīk·lik kō'ȯrd·ən·ət }

cyclic currents *See* mesh currents. { 'sīk·lik ¦kər·ənts }

cyclic magnetization [ELECTROMAG] A magnetizing force varying between two specific limits long enough so that the magnetic induction has the same value for corresponding points in successive cycles. { 'sīk·lik mag·nə·tə'zā·shən }

cyclic permeability *See* normal permeability. { 'sik·lik ,pər·mē·ə'bil·əd·ē }

cyclic twinning [CRYSTAL] Repeated twinning of three or more individuals in accordance with the same twinning law but without parallel twinning axes. { 'sīk·lik 'twin·iŋ }

cycloidal mass spectrometer [SPECT] Small mass spectrometer of limited mass range fitted with a special-type analyzer that generates a cycloidal-path beam of the sample mass. { sī'klȯid·əl 'mas spek'träm·əd·ər }

cycloidal pendulum [MECH] A modification of a simple pendulum in which a weight is suspended from a cord which is slung between two pieces of metal shaped in the form of cycloids; as the bob swings, the cord wraps and unwraps on the cycloids; the pendulum has a period that is independent of the amplitude of the swing. { sī'klȯid·əl 'pen·jə·ləm }

cycloidal wave [FL MECH] A very steep, symmetrical wave in the form of a cycloid whose crest forms an angle of 120°. { sī'klȯid·əl ,wāv }

cyclostrophic flow [FL MECH] A form of gradient flow in which the centripetal acceleration exactly balances the horizontal pressure force. { ¦sī·klō¦strä·fik 'flō }

cyclotron emission *See* cyclotron radiation. { 'sī·klə,trän i'mish·ən }

cyclotron frequency [ELECTROMAG] The angular frequency of the motion of a charged particle in a uniform magnetic field in a plane perpendicular to the field. Also known as gyrofrequency. { 'sī·klə,trän ,frē·kwen·sē }

cyclotron radiation [ELECTROMAG] The electromagnetic radiation emitted by charged particles as they orbit in a magnetic field, at a speed which is not close to the speed of light. Also known as cyclotron emission. { 'sī·klə,trän ,rād·ē'ā·shən }

cyclotron resonance [PHYS] Resonance absorption of energy from an alternating-current electric field by electrons or ions in a uniform magnetic field when the frequency of the electric field equals the cyclotron frequency, or the cyclotron frequency corresponds to the effective mass of electrons in a solid. Also known as diamagnetic resonance. { 'sī·klə,trän 'rez·ən·əns }

cyclotron-resonance heating [PL PHYS] A modification of magnetic pumping that involves compressing and expanding plasma at a frequency approximating the cyclotron frequency of the ions in the plasma; the goal is temperatures above several million degrees. { 'sī·klə,trän 'rez·ən·əns 'hēd·iŋ }

cyclotron wave [ELECTROMAG] A wave associated with the electron beam of a traveling-wave tube. { 'sī·klə,trän ,wāv }

cylindrical antenna [ELECTROMAG] An antenna in which hollow cylinders serve as radiating elements. { sə'lin·drə·kəl an'ten·ə }

cylindrical capacitor [ELEC] A capacitor made of two concentric metal cylinders of the same length, with dielectric filling the space between the cylinders. Also known as coaxial capacitor. { sə'lin·drə·kəl kə'pas·əd·ər }

cylindrical cavity [ELECTROMAG] A cavity resonator in the shape of a right circular cylinder. { sə'lin·drə·kəl 'kav·əd·ē }

cylindrical lens [OPTICS] A lens one or both of whose surfaces are a portion of a circular cylinder. { si'lin·drə·kəl 'lenz }

cylindrical pinch *See* pinch effect. { sə'lin·drə·kəl 'pinch }

cylindrical reflector [ELECTROMAG] A reflector that is a portion of a cylinder; this cylinder is usually parabolic. { sə'lin·drə·kəl ri'flek·tər }

cylindrical wave [ELECTROMAG] A wave whose

equiphase surfaces form a family of coaxial cylinders. { sə'lin·drə·kəl 'wāv }

Czerny-Turner spectrograph [SPECT] A spectrograph used chiefly in laboratory work, which has a plane reflection grating and spherical reflectors for the collimator and camera. { ¦cher·nē¦tərn·ər 'spek·trə,graf }

Czochralski process [CRYSTAL] A method of producing large single crystals by inserting a small seed crystal of germanium, silicon, or other semiconductor material into a crucible filled with similar molten material, then slowly pulling the seed up from the melt while rotating it. { chə'kräl·skē ,präs·əs }

D

D *See* diopter.

Dagor lens [OPTICS] An anastigmatic lens consisting of two lens systems that are nearly symmetrical with respect to the stop, each system containing three or more lenses. { 'dā‚gȯr ‚lenz }

d'Alembert's paradox [FL MECH] The paradox that no forces act on a body moving at constant velocity in a straight line through a large mass of incompressible, inviscid fluid which was initially at rest, or in uniform motion. { ¦dal·əm¦bərz 'par·ə‚däks }

d'Alembert's principle [MECH] The principle that the resultant of the external forces and the kinetic reaction acting on a body equals zero. { ¦dal·əm¦bərz ‚prin·sə·pəl }

d'Alembert's solution [PHYS] A general solution to the linearized small-amplitude one-dimensional wave equation, consisting of two traveling waves of arbitrary shape which travel in opposite directions with a constant wave speed and with no change in shape or amplitude. { ‚da·ləm'berz sə‚ü·shən }

d'Alembert's wave equation *See* wave equation. { ¦dal·əm¦bərz 'wāv i¦kwā·zhən }

Dalitz pair [PART PHYS] The electron and positron resulting from the decay of a neutral pion to these particles and a photon. { 'dä·lits ‚per }

Dalitz plot [PART PHYS] Pictorial representation for data on the distribution of certain three-particle configurations that result from elementary-particle decay processes or high-energy nuclear reactions. { 'dä·lits ‚plät }

dalton *See* atomic mass unit. { 'dȯl·tən }

Dalton's law [PHYS] The law that the pressure of a gas mixture is equal to the sum of the partial pressures of the gases composing it. Also known as law of partial pressures. { 'dȯl·tənz ‚lȯ }

Dalton's temperature scale [THERMO] A scale for measuring temperature such that the absolute temperature T is given in terms of the temperature on the Dalton scale τ by $T = 273.15(373.15/273.15)^{\tau/100}$. { 'dȯl·tənz 'tem·prə·chər ‚skāl }

damaging stress [MECH] The minimum unit stress for a given material and use that will cause damage to the member and make it unfit for its expected length of service. { ¦dam·ə·jiŋ 'stres }

Damköhler number I [PHYS] A dimensionless number, equal to the ratio of the time it takes a fluid to flow some characteristic distance, to the time it takes some chemical reaction or other physical process to be completed. Symbolized *Da* I. Also known as Damköhler's ratio. { 'däm‚kər·lər ¦nəm·bər ¦wən }

Damköhler number II [PHYS] A measure of the ratio of the rate of a chemical reaction to the rate of molecular diffusion, equal to the square of a characteristic length divided by the product of the diffusivity and the time it takes for a chemical reaction or other physical process to be completed. Symbolized *Da* II. { 'däm‚kər·lər ¦nəm·bər ¦tü }

Damköhler number III [PHYS] A measure of the ratio of the heat liberated by a chemical reaction to the bulk transport of heat in a fluid, equal to the time it takes the fluid to travel a characteristic length, divided by the product of the fluid temperature and the time it would take for the chemical reaction to raise this temperature one unit if all the heat liberated by it were immediately absorbed by the fluid. Symbolized *Da* III. { 'däm‚kər·lər ¦nəm·bər ¦thrē }

Damköhler number IV [PHYS] A measure of the ratio of the heat liberated by a chemical reaction to the conductive heat transfer, equal to a characteristic length times the heat liberated per unit volume per unit time divided by the product of the thermal conductivity and the temperature. Symbolized *Da* IV. { 'däm‚kər·lər ¦nəm·bər ¦fȯr }

Damköhler number V *See* Reynolds number. { 'däm‚kər·lər ¦nəm·bər ¦fīv }

Damköhler's ratio *See* Damköhler number I. { 'däm‚kər·lərz ‚rā·shō }

damp [PHYS] To gradually diminish the amplitude of a vibration or oscillation. { damp }

damped harmonic motion [PHYS] **1.** Also known as damped oscillation; damped vibration. **2.** The linear motion of a particle subject both to an elastic restoring force proportional to its displacement and to a frictional force in the direction opposite to its motion and proportional to its speed. **3.** A similar variation in a quantity analogous to the displacement of a particle, such as the charge on a capacitor in a simple series circuit containing a resistance. { ¦dampt har ¦män·ik 'mō·shən }

damped oscillation [PHYS] **1.** Any oscillation in which the amplitude of the oscillating quantity decreases with time. Also known as damped vibration. **2.** *See* damped harmonic motion. { ¦dampt ˌäs·əˈlā·shən }

damped vibration *See* damped harmonic motion; damped oscillation. { ¦dampt vīˈbrā·shən }

damped wave [PHYS] **1.** A wave whose amplitude drops exponentially with distance because of energy losses which are proportional to the square of the amplitude. **2.** A wave in which the amplitudes of successive cycles progressively diminish at the source. { ¦dampt ˌwāv }

damping [PHYS] **1.** The dissipation of energy in motion of any type, especially oscillatory motion and the consequent reduction or decay of the motion. **2.** The extent of such dissipation and decay. { ˈdam·piŋ }

damping capacity [MECH] A material's capability in absorbing vibrations. { ˈdam·piŋ kəˈpas·əd·ē }

damping coefficient *See* damping factor; resistance. { ˈdam·piŋ ˌkō·iˌfish·ənt }

damping constant *See* resistance. { ˈdam·piŋ ˌkän·stənt }

damping factor [PHYS] **1.** The ratio of the logarithmic decrement of any underdamped harmonic motion to its period. Also known as damping coefficient. **2.** *See* decrement. { ˈdam·piŋ ˌfak·tər }

damping magnet [ELECTROMAG] A permanent magnet used in conjunction with a disk or other moving conductor to produce a force that opposes motion of the conductor and thereby provides damping. { ˈdam·piŋ ˌmag·nət }

damping ratio [PHYS] The ratio of the actual resistance in damped harmonic motion to that necessary to produce critical damping. Also known as relative damping ratio. { ˈdam·piŋ ˌrā·shō }

damping resistor [ELEC] **1.** A resistor that is placed across a parallel resonant circuit or in series with a series resonant circuit to decrease the Q factor and thereby eliminate ringing. **2.** A noninductive resistor placed across an analog meter to increase damping. { ˈdam·piŋ riˌzis·tər }

dangling bond [SOLID STATE] A chemical bond associated with an atom in the surface layer of a solid that does not join the atom with a second atom but extends in the direction of the solid's exterior. { ¦daŋ·gliŋ ˈbänd }

daraf [ELEC] The unit of elastance, equal to the reciprocal of 1 farad. { ˈdaˌraf }

darcy [PHYS] A unit of permeability, equivalent to the passage of 1 cubic centimeter of fluid of 1 centipoise viscosity flowing in 1 second under a pressure of 1 atmosphere through a porous medium having a cross-sectional area of 1 square centimeter and a length of 1 centimeter. { ˈdär·sē }

Darcy number 1 [FL MECH] A dimensionless group, equal to four times the Fanning friction factor. Symbolized Da_1. Also known as Darcy-Weisbach coefficient; resistance coefficient 2. { ˈdär·sē ˌnəm·bər ˈwən }

Darcy number 2 [FL MECH] A dimensionless group used in the study of the flow of fluids in porous media, equal to the fluid velocity times the flow path divided by the permeability of the medium. Symbolized Da_2. { ˈdär·sē ˌnəm·bər ˌtü }

Darcy's law [FL MECH] The law that the rate at which a fluid flows through a permeable substance per unit area is equal to the permeability, which is a property only of the substance through which the fluid is flowing, times the pressure drop per unit length of flow, divided by the viscosity of the fluid. { ˈdär·sēz ˌlȯ }

Darcy-Weisbach coefficient *See* Darcy number 1. { ¦där·sē ˈvīsˌbäk ˌkō·iˌfish·ənt }

Darcy-Weisbach equation [FL MECH] An equation for the loss of head due to friction h_f during turbulent flow of a fluid through a duct of any shape; in the case of a circular pipe, $h_f = f(L/d)(V^2/2g)$, where L and d are the length and diameter of the pipe, V is the fluid velocity, g the acceleration of gravity, and f a dimensionless number called Darcy number 1. { ¦där·sē ˈvīsˌbäk iˌkwā·zhən }

dark-field illumination [OPTICS] A method of microscope illumination in which the illuminating beam is a hollow cone of light formed by an opaque stop at the center of the condenser large enough to prevent direct light from entering the objective; the specimen is placed at the concentration of the light cone, and is seen with light scattered or diffracted by it. { ˈdärk ˌfēld əˌlüm·əˈnā·shən }

dark-line spectrum [SPECT] The absorption spectrum that results when white light passes through a substance, consisting of dark lines against a bright background. { ˈdärk¦līn ˈspek·trəm }

darkroom filter [OPTICS] An optical component of glass, gelatin, or other material used to alter the radiation emitted by the darkroom light source so that only specific wavelengths are transmitted. { ˈdärkˌrüm ˌfil·tər }

Darwin curve [CRYSTAL] A plot of the intensity of diffracted x-rays from a perfect crystal as a function of angle. { ˈdär·win ˌkərv }

dasymeter [PHYS] A thin glass globe used to measure the density of gas by weighing the globe in the gas. { daˈsim·əd·ər }

daughter [NUC PHYS] The immediate product of radioactive decay of an element, such as uranium. Also known as decay product; radioactive decay product. { ˈdȯd·ər }

Dauphine law [CRYSTAL] A twin law in which the twinned parts are related by a rotation of 180° around the *c* axis. { dȯˌfēn ˌlȯ }

Davis correction [FL MECH] Empirical relation of flow-line diameters used to correct data calculated from the Atherton equation (friction loss in annular passages). { ˈdā·vəs kəˈrek·shən }

Davis-Gibson color filter [OPTICS] A two-component filter for converting the spectral energy distribution of an incandescent light source to that of white light. { ¦dā·vəs ¦gib·sən ˈkəl·ər ˌfil·tər }

Davisson-Germer experiment [QUANT MECH]

The first experiment to demonstrate electron diffraction, in which a beam of electrons was directed at the surface of a nickel crystal, and the distribution of electrons scattered back from the crystal was measured by a Faraday cylinder. { ¦da·və·sən ¦ger·mər ik,sper·ə·mənt }

Dawes' limit [OPTICS] The resolving power of a telescope, limited by diffraction effects, is 4.5/*a* seconds of arc, where *a* is the aperture in inches. { 'do̊z ,lĭm·ət }

dB *See* decibel.

dBf *See* decibels above 1 femtowatt.

dBic [ELECTROMAG] The directive gain of a circularly polarized antenna, expressed as the ratio, in decibels, of the antenna's directivity to that of an isotropic antenna with the same polarization characteristic. Derived from decibels over isotropic. { 'dē,bik }

dBk *See* decibels above 1 kilowatt.

dBm *See* decibels above 1 milliwatt.

dBp *See* decibels above 1 picowatt.

D-brane [PART PHYS] In superstring theory, a point, curve, surface, or higher-dimensional surface to which the end points of strings can be attached; its existence is implied by string duality. { 'dē,brān }

dBV *See* decibels above 1 volt.

dBW *See* decibels above 1 watt.

dc *See* direct current.

D center *See* R center. { 'dē ,sen·tər }

d constant [SOLID STATE] The ratio of the induced strain in a piezoelectric material to the applied electric field that produces this strain. { 'dē ,kän·stənt }

dead [ELEC] Free from any electric connection to a source of potential difference from electric charge; not having a potential different from that of earth; the term is used only with reference to current-carrying parts which are sometimes alive or charged. { ded }

dead band [ELEC] The portion of a potentiometer element that is shortened by a tap; when the wiper traverses this area, there is no change in output. { 'ded ,band }

deadbeat [MECH] Coming to rest without vibration or oscillation, as when the pointer of a meter moves to a new position without overshooting. Also known as deadbeat response. { 'ded,bēt }

deadbeat response *See* deadbeat. { 'ded,bēt ri'späns }

dead end [ACOUS] The end of a sound studio that has the greater sound-absorbing characteristics. [ELEC] The portion of a tapped coil through which no current is flowing at a particular switch position. { 'ded ,end }

dead ground [ELEC] A low-resistance connection between the ground and an electric circuit. { ¦ded 'graůnd }

dead load *See* static load. { 'ded ,lōd }

dead short [ELEC] A short-circuit path that has extremely low resistance. { ¦ded'sho̊rt }

dead space [THERMO] A space filled with gas whose temperature differs from that of the main body of gas, such as the gas in the capillary tube of a constant-volume gas thermometer. { 'ded ,spās }

Dean number [FL MECH] A dimensionless number giving the ratio of the viscous force acting on a fluid flowing in a curved pipe to the centrifugal force; equal to the Reynolds number times the square root of the ratio of the radius of the pipe to its radius of curvature. Symbolized N_D. { 'dēn ,nəm·bər }

Deborah number [MECH] A dimensionless number used in rheology, equal to the relaxation time for some process divided by the time it is observed. Symbolized D. { də'bo̊r·ə ,nəm·bər }

de Broglie equation *See* de Broglie relation. { də¦brō¦glē i'kwā·zhən }

de Broglie relation [QUANT MECH] The relation in which the de Broglie wave associated with a free particle of matter, and the electromagnetic wave in a vacuum associated with a photon, has a wavelength equal to Planck's constant divided by the particle's momentum and a frequency equal to the particle's energy divided by Planck's constant. Also known as de Broglie equation. { də¦brō¦glē ri'lā·shən }

de Broglie's theory [QUANT MECH] The theory that particles of matter have wavelike properties which can give rise to interference effects, and electrons in an atom are associated with standing waves on a Bohr orbit. { də¦brō¦glēz ,thē·ə·rē }

de Broglie wave [QUANT MECH] The quantum-mechanical wave associated with a particle of matter. Also known as matter wave. { də¦brō ¦glē ,wāv }

de Broglie wavelength [QUANT MECH] The wavelength of the wave associated with a particle as given by the de Broglie relation. { də¦brō ¦glē 'wāv,leŋkth }

debye [ELEC] A unit of electric dipole moment, equal to 10^{-18} Franklin centimeter. { də'bī }

Debye effect [ELECTROMAG] Selective absorption of electromagnetic waves by a dielectric, due to molecular dipoles. { də'bī i'fekt }

Debye equation [SOLID STATE] The equation for the Debye specific heat, which satisfies the Dulong and Petit law at high temperatures and the Debye T^3 law at low temperatures. { də'bī i'kwā·zhən }

Debye equation for polarization [STAT MECH] The Langevin-Debye formula for the polarization of a dielectric material, relating the total polarization for *n* molecules to the permanent moment of the specific molecule and its polarizability. { də'bī i,kwā·zhən fər pō·lə·rə'zā·shən }

Debye frequency [SOLID STATE] The maximum allowable frequency in the computation of the Debye specific heat. { də'bī ,frē·kwən·sē }

Debye-Hückel screening radius *See* Debye shielding length. { de¦bi 'hik·əl 'skrēn·iŋ,rā·dē·əs }

Debye-Jauncey scattering [SOLID STATE] Incoherent background scattering of x-rays from a crystal in directions between those of the Bragg reflections. { də¦bī 'jo̊n·sē ,skad·ə·riŋ }

Debye length *See* Debye shielding length. { də'bī ˌleŋkth }

Debye potentials [ELECTROMAG] Two scalar potentials, designated Π_e and Π_m, in terms of which one can express the electric and magnetic fields resulting from radiation or scattering of electromagnetic waves by a distribution of localized sources in a homogeneous isotropic medium. { də'bī pə'ten·chəlz }

Debye-Scherrer method [SOLID STATE] An x-ray diffraction method in which the sample, consisting of a powder stuck to a thin fiber or contained in a thin-walled silica tube, is rotated in a monochromatic beam of x-rays, and the diffraction pattern is recorded on a cylindrical film whose axis is parallel to the axis of rotation of the sample. { də¦bī 'sher·ər ˌmeth·əd }

Debye-Sears ultrasonic cell [ACOUS] A process in ultrasonic imaging for which the acoustic wavefronts act as optical gratings to diffract the light on either side of the central spot. { də¦bī 'sirz ¦əl·trə¦sän·ik 'sel }

Debye shielding length [PL PHYS] A characteristic distance in a plasma beyond which the electric field of a charged particle is shielded by particles having charges of the opposite sign. Also known as Debye-Hückel screening radius; Debye length; shielding distance. { də'bī 'shēld·iŋ ˌleŋkth }

Debye specific heat [SOLID STATE] The specific heat of a solid under the assumption that the energy of the lattice arises entirely from acoustic lattice vibration modes which all have the same sound velocity, and that frequencies are cut off at a maximum such that the total number of modes equals the number of degrees of freedom of the solid. { də'bī spəˌsif·ik 'hēt }

Debye temperature [SOLID STATE] The temperature θ arising in the computation of the Debye specific heat, defined by $k\theta = h\nu$, where k is the Boltzmann constant, h is Planck's constant, and ν is the Debye frequency. Also known as characteristic temperature. { də'bī 'tem·prə·chər }

Debye theory [ELEC] The classical theory of the orientation polarization of polar molecules in which the molecules have a single relaxation time, and the plot of the imaginary part of the complex relative permittivity against the real part is a semicircle. { də'bī ˌthē·ə·rē }

Debye T^3 law [SOLID STATE] The law that the specific heat of a solid at constant volume varies as the cube of the absolute temperature T at temperatures which are small with respect to the Debye temperature. { də'bī ˌtē'kyübd ˌlò }

Debye-Waller factor [SOLID STATE] A reduction factor for the intensity of coherent (Bragg) scattering of x-rays, neutrons, or electrons by a crystal, arising from thermal motion of the atoms in the lattice. { də¦bī 'väl·ər ˌfak·tər }

decade box [ELEC] An assembly of precision resistors, coils, or capacitors whose individual values vary in submultiples and multiples of 10; by appropriately setting a 10-position selector switch for each section, the decade box can be set to any desired value within its range. { de 'kād ˌbäks }

decaliter [MECH] A unit of volume, equal to 10 liters, or to 0.01 cubic meter. { 'dek·əˌlēd·ər }

decameter [MECH] A unit of length in the metric system equal to 10 meters. { 'dek·əˌmēd·ər }

decametric wave [ELECTROMAG] British term for a radio wave ranging from 10 to 100 meters long. { 'dek·əˌme·trik ˌwāv }

decastere [MECH] A unit of volume, equal to 10 cubic meters. { 'dek·əˌstir }

decay [NUC PHYS] *See* radioactive decay. [PHYS] Gradual reduction in the magnitude of a quantity, as of current, magnetic flux, a stored charge, or phosphorescence. { di'kā }

decay chain *See* radioactive series. { di'kā ˌchān }

decay coefficient *See* decay constant. { di'kā ˌkō·iˌfish·ənt }

decay constant [PHYS] The constant c in the equation $I = I_0 e^{-ct}$, for the time dependence of rate of decay of a radioactive species; here, I is the number of disintegrations per unit time. Also known as decay coefficient; disintegration constant; radioactive decay constant; transformation constant. { di'kā ˌkän·stənt }

decay curve [NUC PHYS] A graph showing how the activity of a radioactive sample varies with time; alternatively, it may show the amount of radioactive material remaining at any time. { di'kā ˌkərv }

decay family *See* radioactive series. { di'kā ˌfam·lē }

decay gammas [NUC PHYS] The characteristic gamma rays emitted during the decay of most radioisotopes. { di'kā ˌgam·əz }

decay mode [NUC PHYS] A possible type of decay of a radionuclide or elementary particle. { di'kā ˌmōd }

decay product *See* daughter. { di'kā ˌprä·dəkt }

decay rate [NUC PHYS] The time rate of disintegration of radioactive material, generally accompanied by emission of particles or gamma radiation. { di'kā ˌrāt }

decay series *See* radioactive series. { di'kā ˌsir·ēz }

decay time [PHYS] The time taken by a quantity to decay to a stated fraction of its initial value; the fraction is commonly $1/e$. Also known as storage time (deprecated). { di'kā ˌtīm }

deceleration [MECH] The rate of decrease of speed of a motion. { dēˌsel·ə'rā·shən }

decentered lens [OPTICS] A lens whose optical center does not coincide with the geometrical center of the rim of the lens; has the effect of a lens combined with a weak prism. { dē'sent·ərd 'lenz }

deciare [MECH] A unit of area, equal to 0.1 are or 10 square meters. { 'des·ēˌer }

decibar [MECH] A metric unit of pressure equal to one-tenth bar. { 'des·əˌbär }

decibel [PHYS] A unit for describing the ratio of two powers or intensities, or the ratio of a power to a reference power; in the measurement of sound intensity, the pressure of the reference sound is usually taken as 2×10^{-4} dyne per square centimeter; equal to one-tenth bel; if P_1 and P_2 are two amounts of power, the first is said to be *n* decibels greater, where $n = 10 \log_{10} (P_1/P_2)$. Abbreviated dB. { 'des·ə,bel }

decibels above 1 femtowatt [ELEC] A power level equal to 10 times the common logarithm of the ratio of the given power in watts to 1 femtowatt (10^{-15} watt). Abbreviated dBf. { 'des·ə·bəlz ə¦bəv ¦wən 'fem·tō,wät }

decibels above 1 kilowatt [ELEC] A measure of power equal to 10 times the common logarithm of the ratio of a given power to 1000 watts. Abbreviated dBk. { 'des·ə·bəlz ə¦bəv ¦wən 'kil·ə,wät }

decibels above 1 milliwatt [ELEC] A measure of power equal to 10 times the common logarithm of the ratio of a given power to 0.001 watt; a negative value, such as −2.7 dBm, means decibels below 1 milliwatt. Abbreviated dBm. { 'des·ə·bəlz ə¦bəv ¦wən 'mil·i,wät }

decibels above 1 picowatt [ELEC] A measure of power equal to 10 times the common logarithm of the ratio of a given power to 1 picowatt. Abbreviated dBp. { 'des·ə·bəlz ə¦bəv ¦wən 'pē·kō,wät }

decibels above 1 volt [ELEC] A measure of voltage equal to 20 times the common logarithm of the ratio of a given voltage to 1 volt. Abbreviated dBV. { 'des·ə·bəlz ə¦bəv ¦wən 'vōlt }

decibels above 1 watt [ELEC] A measure of power equal to 10 times the common logarithm of the ratio of a given power to 1 watt. Abbreviated dBW. { 'des·ə·bəlz ə¦bəv ¦wən 'wät }

decibels over isotropic *See* dBic. { 'des·ə·bəlz ō·vər ,ī·sə'träp·ik }

decigram [MECH] A unit of mass, equal to 0.1 gram. { 'des·ə,gram }

deciliter [MECH] A unit of volume, equal to 0.1 liter, or 10^{-4} cubic meter. { 'des·ə,lēd·ər }

decimeter [MECH] A metric unit of length equal to one-tenth meter. { 'des·ə,mēd·ər }

decimetric wave [ELECTROMAG] An electromagnetic wave having a wavelength between 0.1 and 1 meter, corresponding to a frequency between 300 and 3000 megahertz. { ¦des·ə¦me·trik 'wāv }

decineper [PHYS] One-tenth of a neper. { 'des·ə,nep·ər }

Deck effect [PART PHYS] The simulation of resonances in multiple-particle production processes in high-energy scattering, wherein the true resonance pertains to a subset of the particles produced. { 'dek i,fekt }

decoherence [QUANT MECH] The process whereby the quantum-mechanical state of any macroscopic system is rapidly correlated with that of its environment in such a way that no measurement on the system alone (without a simultaneous measurement of the complete state of the environment) can demonstrate any interference between two quantum states of the system. { ,dē·kō'hir·əns }

decoupling [ELEC] Preventing transfer or feedback of energy from one circuit to another. { dē'kəp·liŋ }

decoupling network [ELEC] Any combination of resistors, coils, and capacitors placed in power supply leads or other leads that are common to two or more circuits, to prevent unwanted interstage coupling. { dē'kəp·liŋ ,net,wərk }

decrement [PHYS] The ratio of the amplitudes of an underdamped harmonic motion during two successive oscillations. Also known as damping factor; numerical decrement. { 'dek·rə·mənt }

deenergize [ELEC] To disconnect from the source of power. { dē'en·ər,jīz }

deep inelastic collision [NUC PHYS] A nuclear reaction in which two nuclei interact strongly, dissipating sizable amounts of energy and exchanging energy and nucleons, while their surfaces overlap for a brief period corresponding to a partial rotation of the intermediate dinuclear complex. Also known as deep inelastic transfer; incomplete fusion; quasi-fission; relaxed peak process; strongly damped collision. { 'dēp ,in·ə'las·tik kə'lizh·ən }

deep inelastic transfer *See* deep inelastic collision. { 'dēp in·ə'las·tik 'tranz·fər }

deerhorn antenna [ELECTROMAG] A dipole antenna whose ends are swept back to reduce wind resistance when mounted on an airplane. { 'dir,hȯrn an'ten·ə }

defect chemistry [SOLID STATE] The study of the dynamic properties of crystal defects under particular conditions, such as raising of the temperature or exposure to electromagnetic particle radiation. { 'dē,fekt ,kem·ə·strē }

defect cluster [CRYSTAL] A macroscopic cluster of crystal defects which can arise from attraction among defects. { 'dē,fekt ,kləs·tər }

defect conduction [SOLID STATE] Electric conduction in a semiconductor by holes in the valence band. { 'dē,fekt kən'dək·shən }

defect motion [CRYSTAL] Movement of a point defect from one lattice point to another. { 'dē ,fekt 'mō·shən }

defect scattering [SOLID STATE] Scattering of particles or electromagnetic radiation by crystal defects. { di'fekt skad·ər·iŋ }

defect structure [SOLID STATE] A crystal structure in which some atomic positions are occupied by atoms other than those that would be found in a perfect crystal, or are unoccupied. { di'fekt ,strək·chər }

definition [OPTICS] Lens image clarity or discernible detail. { ,def·ə'nish·ən }

deflecting torque [MECH] An instrument's moment, resulting from the quantity measured, that acts to cause the pointer's deflection. { di'flek·diŋ ,tȯrk }

deflection curve [MECH] The curve, generally downward, described by a shot deviating from its true course. { di'flek·shən ,kərv }

deformation [MECH] Any alteration of shape or dimensions of a body caused by stresses, thermal expansion or contraction, chemical or metallurgical transformations, or shrinkage and expansions due to moisture change. { ,def·ər'mā·shən }

deformation curve [MECH] A curve showing the relationship between the stress or load on a structure, structural member, or a specimen and the strain or deformation that results. Also known as stress-strain curve. { ,def·ər'mā·shən ,kərv }

deformation ellipsoid *See* strain ellipsoid. { ,def·ər'mā·shən ə'lip,sȯid }

deformation energy [NUC PHYS] The energy which must be supplied to an initially spherical nucleus to give it a certain deformation in the Bohr-Wheeler theory. { ,def·ər'mā·shən ,en·ər·jē }

deformation potential [SOLID STATE] The effective electric potential experienced by free electrons in a semiconductor or metal resulting from a local deformation in the crystal lattice. { ,def·ər'mā·shən pə,ten·chəl }

defrost [THERMO] To thaw out from a frozen state. { dē'frȯst }

degauss [ELECTROMAG] To neutralize (demagnetize) a magnetic field of, for example, a ship hull or television tube; a direct current of the correct value is sent through a cable around the ship hull; a current-carrying coil is brought up to and then removed from the television tube. Also known as deperm. { dē'gau̇s }

degaussing cable [ELECTROMAG] A single-conductor or multiple-conductor cable used on ships for degaussing. { dē'gau̇s·iŋ ,kābəl }

degaussing coil [ELECTROMAG] A plastic-encased coil, about 1 foot (0.3 meter) in diameter, that can be plugged into a 120-volt alternating-current wall outlet and moved slowly toward and away from a color television picture tube to demagnetize adjacent parts. { dē'gau̇s·iŋ ,kȯil }

degaussing control [ELECTROMAG] A control that automatically varies the current in the degaussing cable as a ship changes heading or rolls and pitches. { dē'gau̇s·iŋ kən,trōl }

degeneracy [PHYS] The condition in which two or more modes of a vibrating system have the same frequency; a special case of the mathematics definition. [QUANT MECH] The condition in which two or more stationary states of the same system have the same energy even though their wave functions are not the same; a special case of the mathematics definition. { di'jen·ə·rə·sē }

degeneracy pressure [STAT MECH] The pressure exerted by a degenerate electron or neutron gas. { di'jen·ə·rə·sē ,presh·ər }

degenerate conduction band [SOLID STATE] A band in which two or more orthogonal quantum states exist that have the same energy, the same spin, and zero mean velocity. { di'jen·ə·rət kən'dək·shən ,band }

degenerate electron gas [STAT MECH] An electron gas that is far below its Fermi temperature and is therefore described in first approximation by the Fermi distribution; most of the electrons completely fill the lower energy levels and are unable to take part in physical processes until excited out of these levels. { di'jen·ə·rət i'lek ,trän ,gas }

degenerate four-wave mixing [OPTICS] A method of achieving optical phase conjugation in which two strong counterpropagating pump beams, having the same frequency, set up a standing wave in a clear material whose index of refraction varies linearly with intensity, thereby providing the conditions in which a third beam, at the same frequency, incident upon the material from any direction, results in a fourth beam which precisely retraces the third one. { di'jen·ə·rə·rət 'fȯr ,wāv 'mik·siŋ }

degenerate matter [PHYS] Matter that has been stripped of its orbital electrons, so the nuclei are packed close together. { di'jen·ə·rət 'mad·ər }

degenerate semiconductor [SOLID STATE] A semiconductor in which the number of electrons in the conduction band approaches that of a metal. { di'jen·ə·rət 'sem·i·kən,dək·tər }

degeneration [STAT MECH] A phenomenon which occurs in gases at very low temperatures when the molecular heat drops to less than 3/2 the gas constant. { di,jen·ə'rā·shən }

degradation [PHYS] Loss of energy of a particle, such as a neutron or photon, through a collision. [THERMO] The conversion of energy into forms that are increasingly difficult to convert into work, resulting from the general tendency of entropy to increase. { ,deg·rə'dā·shən }

degree [FL MECH] One of the units in any of various scales of specific gravity, such as the Baumé scale. [THERMO] One of the units of temperature or temperature difference in any of various temperature scales, such as the Celsius, Fahrenheit, and Kelvin temperature scales (the Kelvin degree is now known as the kelvin). { di'grē }

degree Engler [FL MECH] A measure of viscosity; the ratio of the time of flow of 200 milliliters of the liquid through a viscometer devised by Engler, to the time for the flow of the same volume of water. { di'grē 'eŋ·glər }

degree of freedom [MECH] **1.** Any one of the number of ways in which the space configuration of a mechanical system may change. **2.** Of a gyro, the number of orthogonal axes about which the spin axis is free to rotate, the spin axis freedom not being counted; this is not a universal convention; for example, the free gyro is frequently referred to as a three-degree-of-freedom gyro, the spin axis being counted. { di'grē əv 'frē·dəm }

de Haas-van Alphen effect [SOLID STATE] An effect occurring in many complex metals at low temperatures, consisting of a periodic variation in the diamagnetic susceptibility of conduction electrons with changes in the component of the applied magnetic field at right angles to the principal axis of the crystal. { də¦häs ,van'äl·fən i,fekt }

dekapoise [FL MECH] A unit of absolute viscosity, equal to 10 poises. { 'dek·ə,pȯiz }

Delaborne prism [OPTICS] A special compound prism which, when rotated about an axis parallel to the reflecting face and lying in a plane perpendicular to the refracting faces, rotates the image through twice the angle. Also known as Dove prism. { del·ə'bȯrn ˌpriz·əm }

Delaunay orbit element [MECH] In the *n*-body problem, certain functions of variable elements of an ellipse with a fixed focus along which one of the bodies travels; these functions have rates of change satisfying simple equations. { də·lō·nā 'ȯr·bət ˌel·ə·mənt }

delayed alpha particle [NUC PHYS] An alpha particle emitted by an excited nucleus that was formed an appreciable time after a beta disintegration process. { di'lād 'al·fə ˌpard·ə·kəl }

delayed neutron [NUC PHYS] A neutron emitted spontaneously from a nucleus as a consequence of excitation left from a preceding radioactive decay event; in particular, a delayed fission neutron. { di'lād 'nü,trän }

delayed neutron fraction [NUC PHYS] The ratio of the mean number of delayed fission neutrons per fission to the mean total number of neutrons (prompt plus delayed) per fission. { di'lād ¦nü ˌträn ˌfrak·shən }

delayed proton [NUC PHYS] A proton emitted spontaneously from a nucleus as a consequence of excitation left from a previous radioactive decay event. { di'lād 'prō,tän }

Delbrück scattering [NUC PHYS] Elastic scattering of gamma rays by a nucleus caused by virtual electron-positron pair production. { 'del·brik ˌskad·ə·riŋ }

d electron [ATOM PHYS] An atomic electron that has an orbital angular momentum of 2 in the central field approximation. { 'dē i,lek,trän }

delocalized state [QUANT MECH] A state of motion in which a charge carrier is spread over a whole molecule or crystal. { dē'lō·kə,līzd 'stāt }

delta baryon [PART PHYS] **1.** Any excited baryon state belonging to a multiplet having a total isospin of 3/2, a hypercharge of +1, a spin of 3/2, positive parity, and an approximate mass of 1232 MeV. Designated $\Delta(1232)$. **2.** Any excited baryon state belonging to any multiplet having a total isospin of 3/2 and a hypercharge of +1. { 'del·tə 'bar·ē,än }

delta connection [ELEC] A combination of three components connected in series to form a triangle like the Greek letter delta. Also known as mesh connection. { 'del·tə kə'nek·shən }

delta current [ELEC] Electricity going through a delta connection. { 'del·tə ˌkər·ənt }

delta E effect [ELECTROMAG] Magnetization of a ferromagnetic substance that is caused by elastic tension. { ¦del·tə 'ē i,fekt }

delta particle [ATOM PHYS] An electron or proton ejected by recoil when a rapidly moving alpha particle or other primary ionizing particle passes through matter. { 'del·tə ˌpärd·ə·kəl }

delta-Y transformation *See* Y-delta transformation. { 'del·tə ˌwī ˌtranz·fər'mā·shən }

deltohedron [CRYSTAL] A polyhedron which has 12 quadrilateral faces, and is the form of a crystal belonging to the cubic system and having hemihedral symmetry. Also known as deltoid dodecahedron; tetragonal tristetrahedron. { ˌdel·tə'hē·drən }

deltoid dodecahedron *See* deltohedron. { 'del ˌtȯid dō,dek·ə'hē·drən }

demagnetization [ELECTROMAG] **1.** The process of reducing or removing the magnetism of a ferromagnetic material. **2.** The reduction of magnetic induction by the internal field of a magnet. { dē,mag·nəd·ə'zā·shən }

demagnetization coefficient *See* demagnetizing factor. { dē,mag·nə·tə'zā·shən ˌkō·ə,fish·ənt }

demagnetization curve [ELECTROMAG] Graph of magnetic induction B versus magnetic field H in a ferromagnetic material, as the magnetic field is reduced to 0 from its saturation value. { dē ˌmag·nə·tə'zā·shən ˌkərv }

demagnetizing factor [ELECTROMAG] The ratio of the negative of the demagnetizing field to the magnetization of a sample. Also known as demagnetization coefficient. { dē'mag·nə,tīz·iŋ ˌfak·tər }

demagnetizing field [ELECTROMAG] An additional magnetic field that is produced in a magnetic material subject to an applied magnetic field, due to the magnetic material itself. { dē 'mag·nə,tīz·iŋ ˌfēld }

demolitional measurement [QUANT MECH] A measurement that alters the value of the physical observable being measured. { ˌdem·ə'lish·ən·əl 'mezh·ər·mənt }

demon of Maxwell [THERMO] Hypothetical creature who controls a trapdoor over a microscopic hole in an adiabatic wall between two vessels filled with gas at the same temperature, so as to supposedly decrease the entropy of the gas as a whole and thus violate the second law of thermodynamics. Also known as Maxwell's demon. { 'dē·mən əv 'maks,wel }

dendrite [NEUROSCI] The part of a neuron that carries the unidirectional nerve impulse toward the cell body. Also known as dendron. [CRYSTAL] A crystal having a treelike structure. { 'den,drīt }

Denisyuk hologram [OPTICS] A type of hologram that can be viewed in ordinary white light through use of the depth dimension of the emulsion. { 'den·ə·syu̇k 'häl·ə,gram }

dense-air refrigeration cycle *See* reverse Brayton cycle. { ¦dens ¦er ri,frij·ə'rā·shən ˌsī·kəl }

density [MECH] The mass of a given substance per unit volume. [OPTICS] **1.** The degree of opacity of a translucent material. **2.** The common logarithm of opacity. [PHYS] The total amount of a quantity, such as energy, per unit of space. { 'den· səd·ē }

density matrix [QUANT MECH] A matrix ρ_{mn} describing an ensemble of quantum-mechanical systems in a representation based on an orthonormal set of functions ϕ_n; for any operator G with representation G_{mn}, the ensemble average of the expectation value of G is the trace of ρG. { 'den·səd·ē 'mā·triks }

density of states [SOLID STATE] A function of energy E equal to the number of quantum states in the energy range between E and E + dE divided by the product of dE and the volume of the substance. { 'den·səd·ē əv 'stăts }

density wave [PHYS] A sound wave or other type of material wave which causes the density of the matter through which it passes to alternately rise above and drop below its mean value. { 'den·səd·ē ˌwāv }

depolarization [ELEC] The removal or prevention of polarization in a substance (for example, through the use of a depolarizer in an electric cell) or of polarization arising from the field due to the charges induced on the surface of a dielectric when an external field is applied. [OPTICS] The resolution of polarized light in an optical depolarizer. { dēˌpō·lə·rə'zā·shən }

depolarization factor [ELEC] The ratio of the internal electric field induced by the charges on the surface of a dielectric when an external field is applied to the polarization of the dielectric. { dēˌpō·lə·rə'zā·shən ˌfak·tər }

depth magnification [OPTICS] The ratio of the distance between two nearby points of the axis on the image side of an optical system to the distance between their conjugate points on the object side. { 'depth ˌmag·nə·fə'kā·shən }

depth of field [OPTICS] The range of distances over which a camera gives satisfactory definition, when its lens is in the best focus for a certain specific distance. { 'depth əv 'fēld }

depth of focus [OPTICS] The range of image distances corresponding to the range of object distances included in depth of field. { ¦depth əv 'fō·kəs }

derived quantity [PHYS] A physical quantity which, in a specified system of measurement, is defined by operations based on other physical quantities. { də'rīvd 'kwän·əd·ē }

derived unit [PHYS] A unit that is formed, in a specified system of measurement, by combining base units and other derived units according to the algebraic relations linking the corresponding quantities. { də'rīvd 'yü·nət }

Deryagin number [PHYS] A dimensionless group equal to the ratio of the thickness of a film coating a liquid to the capillary length of the liquid. Symbolized *De*. { ˌder·ē'ag·ən ˌnəm·bər }

desaturated color [OPTICS] A color that is neither a pure spectral color nor a purple formed from a mixture of deep red and violet. { dē ¦sach·əˌrād·əd 'kəl·ər }

DeSauty's bridge [ELEC] A four-arm bridge used to compare two capacitances; two adjacent arms contain capacitors in series with resistors, while the other two arms contain resistors only. Also known as Wien-DeSauty bridge. { də'sōd·ēz ˌbrij }

Descartes laws of refraction *See* Snell laws of refraction. { dā'kärt 'lȯz əv ri'frak·shən }

Descartes ray [OPTICS] A ray of light incident on a sphere of transparent material, such as a water droplet, which after one internal reflection leaves the drop at the smallest possible angle of deviation from the direction of the incident ray; these rays make the primary rainbow. { dā 'kärt ˌrā }

descending branch [MECH] That portion of a trajectory which is between the summit and the point where the trajectory terminates, either by impact or air burst, and along which the projectile falls, with altitude constantly decreasing. Also known as descent trajectory. { di'sen·diŋ 'branch }

descent trajectory *See* descending branch. { di 'sent trə'jek·tə·rē }

de Sitter space [RELAT] A constant-curvature, vacuum solution to Einstein's equations of general relativity with cosmological term. { də'sid·ər ˌspās }

despun antenna [ELECTROMAG] Satellite directional antenna pointed continuously at earth by electrically or mechanically despinning the antenna at the same rate that the satellite is spinning for stabilization. { dē'spən an'ten·ə }

Destriau effect [SOLID STATE] Sustained emission of light by suitable phosphor powders that are embedded in an insulator and subjected only to the action of an alternating electric field. { 'des·trē·au̇ iˌfekt }

destruction operator *See* annihilation operator. { di'strək·shən 'äp·əˌrād·ər }

destructive interference [OPTICS] The interaction of superimposed light from two different sources when the phase relationship is such as to reduce or cancel the resultant intensity to less than the sum of the individual lights. { di'strək·tiv ˌin·tər'fir·əns }

detached shock wave [FL MECH] A shock wave not in contact with the body which originates it. { di'tacht 'shäk ˌwāv }

detailed balance [STAT MECH] The hypothesis that when a system is in equilibrium any process occurs with the same frequency as the reverse process. { də'tāld 'bal·əns }

determinate structure [MECH] A structure in which the equations of statics alone are sufficient to determine the stresses and reactions. { də'tər·mə·nət 'strək·chər }

determinism *See* causality. { də'tər·məˌniz·əm }

deterministic equation [PHYS] An equation that governs the motion of a dynamical system and does not contain terms corresponding to random forces. { dəˌtər·mə'nis·tik i'kwā·zhən }

detonating rate [MECH] The velocity at which the explosion wave passes through a cylindrical charge. { 'det·ənˌād·iŋ ˌrāt }

detonation wave [FL MECH] A shock wave that accompanies detonation and has a shock front followed by a region of decreasing pressure in which the reaction occurs. { ˌdet·ən'ā·shən ˌwāv }

detuning stub [ELECTROMAG] Quarter-wave stub used to match a coaxial line to a sleeve-stub antenna; the stub detunes the outside of the coaxial feed line while tuning the antenna itself. { dē'tün·iŋ 'stəb }

deuterium cycle *See* proton-proton chain. { dü'tir·ē·əm ˌsī·kəl }

deuteron [NUC PHYS] The nucleus of a deuterium atom, consisting of a neutron and a proton. Designated d. Also known as deuton. { 'düd·ə,rän }

deuteron capture [NUC PHYS] The absorption of a deuteron by a nucleus, giving rise to a compound nucleus which subsequently decays. { 'düd·ə,rän ,kap·chər }

deuton *See* deuteron. { 'dü,tän }

deviation [OPTICS] The angle between the incident ray on an object or optical system and the emergent ray, following reflection, refraction, or diffraction. Also known as angle of deviation. { ,dēv·ē'ā·shən }

deviation factor *See* compressibility factor. { ,dēv·ē'ā·shən ,fak·tər }

deviatonic stress [MECH] The portion of the total stress that differs from an isostatic hydrostatic pressure; it is equal to the difference between the total stress and the spherical stress. { ,dev·ē·ə'tän·ik 'stres }

Dewar flask [PHYS] A vessel having double walls, the space between being evacuated to prevent the transfer of heat and the surfaces facing the vacuum being heat-reflective; used to hold liquid gases and to study low-temperature phenomena. { 'dü·ər ,flask }

dewcap [OPTICS] An open tube attached to the end of a refracting telescope to prevent moisture from condensing on the objective. { 'dü,kap }

dex *See* brig. { deks }

dextro *See* dextrorotatory. { 'dek,strō }

dextrorotatory [OPTICS] Rotating clockwise the plane of polarization of a wave traveling through a medium in a clockwise direction, as seen by an eye observing the light. Abbreviated dextro. { ,dek·strə'rōd·ə,tȯr·ē }

diabatic [THERMO] A thermodynamic change of state of a system in which there is a transfer of heat across the boundaries of the system. Also known as nonadiabatic. { ¦dī·ə¦bad·ik }

diad axis [CRYSTAL] A rotation axis whose multiplicity is equal to 2. { 'dī,ad ,aks·əs }

diadochy [CRYSTAL] Replacement or ability to be replaced of one atom or ion by another in a crystal lattice. { dī'ad·ə·kē }

diafocal point [OPTICS] For a ray of light refracted by a lens, a point on the ray which lies on a plane passing through the axis of the lens which is parallel to the ray on the opposite side of the lens. { ¦dī·ə¦fō·kəl 'pȯint }

diagonal [OPTICS] A plane mirror or prism face mounted near the eyepiece of a telescope at an angle to the light path, to redirect the light for convenience of observation or to reduce the intensity of the image of the sun so that it can be observed directly. { dī'ag·ən·əl }

diagonal horn antenna [ELECTROMAG] Horn antenna in which all cross sections are square and the electric vector is parallel to one of the diagonals; the radiation pattern in the far field has almost perfect circular symmetry. { dī'ag·ən·əl 'hȯrn an'ten·ə }

diamagnet [ELECTROMAG] A substance which is diamagnetic, such as the alkali and alkaline earth metals, the halogens, and the noble gases. { ¦dī·ə'mag·nət }

diamagnetic [ELECTROMAG] Having a magnetic permeability less than 1; materials with this property are repelled by a magnet and tend to position themselves at right angles to magnetic lines of force. { ¦dī·ə·mag'ned·ik }

diamagnetic Faraday effect [OPTICS] Faraday effect at frequencies near an absorption line which is split due to the splitting of the upper level only. { ¦dī·ə·mag¦ned·ik 'far·ə,dā i,fekt }

diamagnetic resonance *See* cyclotron resonance. { ¦dī·ə·mag¦ned·ik 'rez·ən·əns }

diamagnetic susceptibility [ELECTROMAG] The susceptibility of a diamagnetic material, which is always negative and usually on the order of -10^{-5} cm^3/mole. { ¦dī·ə·mag¦ned·ik sə,sep·tə'bil·əd·ē }

diamagnetism [ELECTROMAG] The property of a material which is repelled by magnets. { ¦dī·ə'mag·nə,tiz·əm }

diamond antenna *See* rhombic antenna. { 'dī,mənd an'ten·ə }

diamonds [FL MECH] The pattern of shock waves often visible in a rocket exhaust which resembles a series of diamond shapes placed end to end. { 'dī·mənz }

diamond structure [CRYSTAL] A crystal structure in which each atom is the center of a tetrahedron formed by its nearest neighbors. { ,dī·mənd ,strək·chər }

diamond-turned optics [OPTICS] Optical elements that have been machined to a specular finish on metal-working lathes whose precision is so great that these elements can be used in the infrared without further optical working, and small elements with simple shapes can be used even in the visible region. { 'dī·mənd ¦tərnd 'äp·tiks }

diaphragm [ELECTROMAG] *See* iris. [OPTICS] Any opening in an optical system which controls the cross section of a beam of light passing through it, to control light intensity, reduce aberration, or increase depth of focus. Also known as lens stop. [PHYS] **1.** A separating wall or membrane, especially one which transmits some substances and forces but not others. **2.** In general, any opening, sometimes adjustable in size, which is used to control the flow of a substance or radiation. { 'dī·ə,fram }

diaphragm setting [OPTICS] The position of a camera's diaphragm after opening or closing it. { 'dī·ə,fram ,sed·iŋ }

diasporometer [OPTICS] Oppositely rotating wedges used in optical rangefinders to obtain deviation of the axis of the image. { ¦dī·ə·spȯr'äm·əd·ər }

diathermanous [PHYS] Capable of transmitting radiant heat. Also known as diathermic. { ¦dī·ə¦thər·mə·nəs }

diathermic *See* diathermanous. { ¦dī·ə¦thər·mik }

diathermous envelope [THERMO] A surface enclosing a thermodynamic system in equilibrium that is not an adiabatic envelope; intuitively, this means that heat can flow through the surface. { ¦dī·ə¦thər·məs 'en·və,lōp }

diatonic scale [ACOUS] A musical scale in which the octave is divided into intervals of two different sizes, five of one and two of the other, with adjustments in tuning systems other than equal temperament. { ¦dī·ə¦tän·ik 'skāl }

dichotic listening *See* dichotic presentation. { di ¦käd·ik 'lis·ən·iŋ }

dichotic presentation [ACOUS] The simultaneous reception of one message through one ear and another message through the other ear. Also known as dichotic listening. { di¦käd·ik ,prē·zin'tā·shən }

dichotomic variable [QUANT MECH] A variable with a range consisting of two values; used, for example, to describe a particle with spin 1/2. { ¦dī·kə¦täm·ik 'ver·ē·ə·bəl }

dichroic mirror [OPTICS] A glass surface coated with a special metal film that reflects certain colors of light while allowing others to pass through. { dī'krō·ik 'mir·ər }

dichroism [OPTICS] In certain anisotropic materials, the property of having different absorption coefficients for light polarized in different directions. { 'dī·krō,iz·əm }

dielectric absorption [ELEC] The persistence of electric polarization in certain dielectrics after removal of the electric field. *See* dielectric loss. { ,dī·ə'lek·trik əb'sörp·shən }

dielectric antenna [ELECTROMAG] An antenna in which a dielectric is the major component used to produce a desired radiation pattern. { ,dī·ə'lek·trik an'ten·ə }

dielectric circuit [ELEC] Any electric circuit which has capacitors. { ,di·ə'lek·trik 'sər·kət }

dielectric constant [ELEC] **1.** For an isotropic medium, the ratio of the capacitance of a capacitor filled with a given dielectric to that of the same capacitor having only a vacuum as dielectric. **2.** More generally, $1 + \gamma\chi$, where γ is 4π in Gaussian and cgs electrostatic units or 1 in rationalized mks units, and χ is the electric susceptibility tensor. Also known as relative dielectric constant; relative permittivity; specific inductive capacity (SIC). { ,dī·ə'lek·trik 'kän·stənt }

dielectric crystal [ELEC] A crystal which is electrically nonconducting. { ,dī·ə'lek·trik 'krist·əl }

dielectric current [ELEC] The current flowing at any instant through a surface of a dielectric that is located in a changing electric field. { ,dī·ə'lek·trik 'kər·ənt }

dielectric displacement *See* electric displacement. { ,dī·ə'lek·trik di'splās·mənt }

dielectric ellipsoid [ELEC] For an anisotropic medium in which the dielectric constant is a tensor quantity **K**, the locus of points **r** satisfying $\mathbf{r}\cdot\mathbf{K}\cdot\mathbf{r} = 1$. { ,dī·ə'lek·trik ə'lip,söid }

dielectric field [ELEC] The average total electric field acting upon a molecule or group of molecules inside a dielectric. Also known as internal dielectric field. { ,dī·ə'lek·trik 'fēld }

dielectric film [ELEC] A film possessing dielectric properties; used as the central layer of a capacitor. { ,dī·ə'lek·trik 'film }

dielectric flux density *See* electric displacement. { ,dī·ə'lek·trik 'fləks ,den·səd·ē }

dielectric gas [ELEC] A gas having a high dielectric constant, such as sulfur hexafluoride. { ,dī·ə'lek·trik 'gas }

dielectric heating [ELEC] Heating of a nominally electrical insulating material due to its own electrical (dielectric) losses, when the material is placed in a varying electrostatic field. { ,dī·ə'lek·trik 'hēd·iŋ }

dielectric hysteresis *See* ferroelectric hysteresis. { ,dī·ə'lek·trik hi·stə'rē·səs }

dielectric imperfection levels [SOLID STATE] Energy levels that occur in the forbidden zone between the valence and conduction bands of a dielectric crystal, because of imperfections in the crystal. { ,dī·ə'lek·trik ,im·pər'fek·shən ,lev·əlz }

dielectric leakage [ELEC] A very small steady current that flows through a dielectric subject to a steady electric field. { ,dī·ə'lek·trik 'lēk·ij }

dielectric lens [ELECTROMAG] A lens made of dielectric material so that it refracts radio waves in the same manner that an optical lens refracts light waves; used with microwave antennas. { ,dī·ə'lek·trik 'lenz }

dielectric-lens antenna [ELECTROMAG] An aperture antenna in which the beam width is determined by the dimensions of a dielectric lens through which the beam passes. { ,dī·ə¦lek·trik ¦lenz an'ten·ə }

dielectric loss [ELECTROMAG] The electric energy that is converted into heat in a dielectric subjected to a varying electric field. Also known as dielectric absorption. { ,dī·ə'lek·trik 'lös }

dielectric loss angle [ELEC] Difference between 90° and the dielectric phase angle. { ,dī·ə¦lek·trik ¦lös ,aŋ·gəl }

dielectric loss factor [ELEC] Product of the dielectric constant of a material and the tangent of its dielectric loss angle. { ,dī·ə¦lek·trik ¦los ,fak·tər }

dielectric matching plate [ELECTROMAG] In waveguide technique, a dielectric plate used as an impedance transformer for matching purposes. { ,dī·ə'lek·trik 'mach·iŋ ,plāt }

dielectric phase angle [ELEC] Angular difference in phase between the sinusoidal alternating potential difference applied to a dielectric and the component of the resulting alternating current having the same period as the potential difference. { ,dī·ə'lek·trik 'fāz ,aŋ·gəl }

dielectric polarization *See* polarization. { ,dī·ə'lek·trik ,pō·lə·rə'zā·shən }

dielectric power factor [ELEC] Cosine of the dielectric phase angle (or sine of the dielectric loss angle). { ,dī·ə'lek·trik 'paůr ,fak·tər }

dielectric-rod antenna [ELECTROMAG] A surface-wave antenna in which an end-fire radiation pattern is produced by propagation of a surface wave on a tapered dielectric rod. { ,dī·ə¦lek·trik ¦räd an'ten·ə }

dielectric shielding [ELEC] The reduction of an electric field in some region by interposing a dielectric substance, such as polystyrene, glass, or mica. { ˌdī·ə'lek·trik 'shēld·iŋ }

dielectric soak *See* absorption. { 'dī·ə'lek·trik 'sōk }

dielectric strength [ELEC] The maximum electrical potential gradient that a material can withstand without rupture; usually specified in volts per millimeter of thickness. Also known as electric strength. { ˌdī·ə'lek·trik 'streŋkth }

dielectric susceptibility *See* electric susceptibility. { ˌdī·ə'lek·trik səˌsep·tə'bil·əd·ē }

dielectric waveguide [ELEC] A waveguide consisting of a dielectric cylinder surrounded by air. { ˌdī·ə'lek·trik 'wāvˌgīd }

dielectric wedge [ELECTROMAG] A wedge-shaped piece of dielectric used in a waveguide to match its impedance to that of another waveguide. { dī·ə'lek·trik 'wej }

dielectric wire [ELECTROMAG] A dielectric waveguide used to transmit ultra-high-frequency radio waves short distances between parts of a circuit. { dī·ə'lek·trik 'wīr }

dielectronic recombination [ATOM PHYS] The combination of an electron with a positive-ion in a gas, so that the energy released is taken up by two electrons of the resulting atom. { di·əˌlek'trän·ik ˌrēˌkäm·bə'nā· shən }

diesel cycle [THERMO] An internal combustion engine cycle in which the heat of compression ignites the fuel. { 'dē·zəl' ˌsī·kəl }

Dieterici equation of state [THERMO] An empirical equation of state for gases, $pe^{a/RT}(v - b) = RT$, where p is the pressure, T is the absolute temperature, v is the molar volume, R is the gas constant, and a and b are constants characteristic of the substance under consideration. { dē·də'rē·chē i'kwā·zhen əv 'stāt }

difference number *See* neutron excess. { 'dif·rəns ˌnəm·bər }

difference spectrophotometer *See* absorption spectrophotometer. { 'dif·rəns ˌspek·trə·fə 'täm·əd·ər }

difference tone [ACOUS] A combination tone whose frequency equals the difference of the frequencies of the pure tones producing it. { 'dif·rəns ˌtōn }

differential calorimetry [THERMO] Technique for measurement of and comparison (differential) of process heats (reaction, absorption, hydrolysis, and so on) for a specimen and a reference material. { ˌdif·ə'ren·chəl ˌkal·ə'rim·ə·trē }

differential capacitor [ELEC] A two-section variable capacitor having one rotor and two stators so arranged that as capacitance is reduced in one section it is increased in the other. { ˌdif·ə'ren·chəl kə'pas·əd·ər }

differential cross section [PHYS] The cross section for a collision process resulting in the emission of particles or photons at a specified angle relative to the direction of the incident particles, per unit angle or per unit solid angle. { ˌdif·ə'ren·chəl ¦krȯs ˌsek·shən }

differential effects [MECH] The effects upon the elements of the trajectory due to variations from standard conditions. { ˌdif·ə'ren·chəl i'feks }

differential electromagnet [ELEC] An electromagnet having part of its winding opposed to the other part, so that the force exerted by the magnet can be adjusted. { ˌdif·ə'ren·chəl iˌlek·trō'mag·nət }

differential frequency circuit [ELEC] A circuit that provides a continuous output frequency equal to the absolute difference between two continuous input frequencies. { ˌdif·ə'ren·chəl ¦frē·kwən·sē ¦sər·kət }

differential heat of solution [THERMO] The partial derivative of the total heat of solution with respect to the molal concentration of one component of the solution, when the concentration of the other component or components, the pressure, and the temperature are held constant. { ˌdif·ə'ren·chəl 'hēt əv sə'lü·shən }

differential interferometer *See* lateral shear interferometer. { ˌdif·ə'ren·chəl ˌin·tər·fə'räm·əd·ər }

differential permeability [ELECTROMAG] The slope of the magnetization curve for a magnetic material. { ˌdif·ə'ren·chəl ˌpər·mē·ə'bil·əd·ē }

differential pressure [PHYS] The difference in pressure between two points of a system, such as between the well bottom and wellhead or between the two sides of an orifice. { ˌdif·ə'ren·chəl 'presh·ər }

differential spectrophotometry [SPECT] Spectrophotometric analysis of a sample when a solution of the major component of the sample is placed in the reference cell; the recorded spectrum represents the difference between the sample and the reference cell. { ˌdif·ə'ren·chəl ˌspek·trō·fə'täm·ə·trē }

differential thermal analysis [THERMO] A method of determining the temperature at which thermal reactions occur in a material undergoing continuous heating to elevated temperatures; also involves a determination of the nature and intensity of such reactions. { ˌdif·ə'ren·chəl 'thər·məl ə'nal·ə·səs }

differential thermogravimetric analysis [THERMO] Thermal analysis in which the rate of material weight change upon heating versus temperature is plotted; used to simplify reading of weight-versus-temperature thermogram peaks that occur close together. { ˌdif·ə'ren·chəl ¦thər·mō ˌgrav·ə¦me·trik ə'nal·ə·səs }

differentiating circuit [ELEC] A circuit whose output voltage is proportional to the rate of change of the input voltage. Also known as differentiating network. { ˌdif·ə¦ren·chēˌād·iŋ ¦sər·kət }

differentiating network *See* differentiating circuit. { ˌdif·ə¦ren·chēˌād·iŋ 'netˌwərk }

diffluence [FL MECH] A region of fluid flow in which the fluid is diverging from the direction of flow. { 'diˌflü·əns }

diffracted wave [PHYS] A wave whose front has been changed in direction by an obstacle or other nonhomogeneity in a medium, other than by reflection or refraction. { di'frak·təd 'wāv }

diffraction [PHYS] Any redistribution in space of the intensity of waves that results from the presence of an object causing variations of either the amplitude or phase of the waves; found in all types of wave phenomena. { di'frak·shən }

diffraction analysis [PHYS] The study of the atomic structure of solids, liquids, or gases by means of diffraction of x-rays or particles, such as neutrons or electrons. { di'frak·shən ə,nal·ə·səs }

diffraction grating [SPECT] An optical device consisting of an assembly of narrow slits or grooves which produce a large number of beams that can interfere to produce spectra. Also known as grating. { di'frak·shən ,grād·iŋ }

diffraction instrument *See* diffractometer. { di 'frak·shən ,in·strə·mənt }

diffraction-limited [OPTICS] Capable of producing images whose separations are as small as the theoretical limit imposed by diffraction effects. { di'frak·shən 'lim·əd·əd }

diffraction loss [PHYS] That part of the reduction in power of a propagating wave or beam that results from diffraction. { di'frak·shən ,lȯs }

diffraction pattern [PHYS] Pattern produced on a screen or plate by waves which have undergone diffraction. { di'frak·shən ,pad·ərn }

diffraction propagation [ELECTROMAG] Propagation of electromagnetic waves around objects or over the horizon by diffraction. { di'frak·shən präp·ə'gā·shən }

diffraction ring [OPTICS] Circular light pattern which appears to surround particles in a microscope field. { di'frak·shən ,riŋ }

diffraction scattering [PHYS] Elastic scattering that occurs when inelastic processes remove particles from a beam. { di'frak·shən ,skad·ə·riŋ }

diffraction spectrum [SPECT] Parallel light and dark or colored bands of light produced by diffraction. { di'frak·shən ,spek·trəm }

diffraction symmetry [CRYSTAL] Any symmetry in a crystal lattice which causes the systematic annihilation of certain beams in x-ray diffraction. { di'frak·shən ,sim·ə·trē }

diffraction velocimeter [OPTICS] A velocity-measuring instrument that uses a continuous-wave laser to send a beam of coherent light at objects moving at right angles to the beam; the needlelike diffraction lobes reflected by the moving objects sweep past the optical grating in the receiver, thereby generating in a photomultiplier a series of impulses from which velocity can be determined and read out. Also known as optical diffraction velocimeter. { di'frak·shən ,vel·ə'sim·əd·ər }

diffraction zone [ELECTROMAG] The portion of a radio propagation path which lies outside a line-of-sight path. { ,di'frak·shən ,zōn }

diffractometer [PHYS] An instrument used to study the structure of matter by means of the diffraction of x-rays, electrons, neutrons, or other waves. Also known as diffraction instrument. { ,di,frak'täm·əd·ər }

diffractometry [CRYSTAL] The science of determining crystal structures by studying the diffraction of beams of x-rays or other waves. { ,di ,frak'täm·ə·trē }

diffuse-cutting filter [OPTICS] A color filter that gradually changes in absorption with wavelength. { də¦fyüs ¦kəd·iŋ ,fil·tər }

diffuse illumination [OPTICS] Lighting so arranged that the object is illuminated from many directions or sources. { də'fyüs i,lüm·ə'nā·shən }

diffuse radiation [PHYS] Radiant energy propagating in many different directions through a given small volume of space. { də'fyüs ,rād·ē'ā·shən }

diffuse reflection [PHYS] Reflection of light, sound, or radio waves from a surface in all directions according to the cosine law. { də'fyüs ri'flek·shən }

diffuse reflection model [PHYS] A model for the behavior of gas molecules striking the surface of a solid body, in which the molecules are absorbed and reemitted with a Maxwellian velocity distribution corresponding to a temperature intermediate between that of the surface and that of the incoming flow of gas. { də¦fyüs ri'flek·shən ,mäd·əl }

diffuse reflector [OPTICS] Any surface whose irregularities are so large compared to the wavelength of the incident radiation that the reflected rays are sent back in a multiplicity of directions. { də'fyüs ri'flek·tər }

diffuse series [SPECT] A series occurring in the spectra of many atoms having one, two, or three electrons in the outer shell, in which the total orbital angular momentum quantum number changes from 2 to 1. { də'fyüs 'sir·ēz }

diffuse sound [ACOUS] Sound that has uniform energy density in a given region so that all directions of energy flux at all parts of the region are equally probable. { də'fyüs 'sau̇nd }

diffuse spectrum [SPECT] Any spectrum having lines which are very broad even when there is no possibility of line broadening by collisions. { də'fyüs 'spek·trəm }

diffuse transmission [PHYS] Transmission of electromagnetic or acoustic radiation in all directions by a transmitting body. { də'fyüs tranz 'mish·ən }

diffuse transmission density [OPTICS] The value of the photographic transmission density obtained when light flux impinges normally on the sample and all the transmitted flux is collected and measured. { də¦fyüs tranz'mish·ən ,den·səd·ē }

diffusing disk *See* diffusion disk. { də'fyüz,iŋ ,disk }

diffusiometer [PHYS] An instrument which measures diffusion in liquids. { də,fyüz·ē'äm·əd·ər }

diffusion [ACOUS] The degree of variation in the propagation directions of sound waves over the volume of a sound field. [OPTICS] **1.** The distribution of incident light by reflection. **2.** Transmission of light through a translucent material. [PHYS] **1.** The spontaneous movement and scattering of particles (atoms and molecules), of liquids, gases, and solids. **2.** In particular, the macroscopic motion of the components of a system of fluids that is driven by differences in concentration. [SOLID STATE] **1.** The actual transport of mass, in the form of discrete atoms, through the lattice of a crystalline solid. **2.** The movement of carriers in a semiconductor. { də'fyü·zhən }

diffusion area [PHYS] One-sixth of the mean-square displacement between the appearance and disappearance of a subatomic particle of a given type. { də'fyü·zhən ˌer·ē·ə }

diffusion coefficient [PHYS] The weight of a material, in grams, diffusing across an area of 1 square centimeter in 1 second in a unit concentration gradient. Also known as diffusivity. { də'fyü·zhən ˌkō·i'fish·ənt }

diffusion constant [SOLID STATE] The diffusion current density in a hologeneous semiconductor divided by the charge carrier concentration gradient. { də'fyü·zhən ˌkän·stənt }

diffusion diameter [STAT MECH] For a gas, the diameter of identical hard spheres that display the same diffusion as that observed for the molecules of the actual gas when their motion is treated classically. { di'fyü·zhən dīˌam·əd·ər }

diffusion disk [OPTICS] A piece of transparent material that is marked or embossed, and is used with a camera lens to give the image a hazy softened quality. Also known as diffusing disk. { də'fyü·zhən ˌdisk }

diffusion equation [PHYS] **1.** An equation for diffusion which states that the rate of change of the density of the diffusing substance, at a fixed point in space, equals the sum of the diffusion coefficient times the Laplacian of the density, the amount of the quantity generated per unit volume per unit time, and the negative of the quantity absorbed per unit volume per unit time. **2.** More generally, any equation which states that the rate of change of some quantity, at a fixed point in space, equals a positive constant times the Laplacian of that quantity. { də'fyü·zhən i'kwā·zhən }

diffusion gradient [PHYS] The graphed distance of penetration (diffusion) versus concentration of the material (or effect) diffusing through a second material; applies to heat, liquids, solids, or gases. { də'fyü·zhən ˌgrād·ē·ənt }

diffusion length [PHYS] The average distance traveled by a particle, such as a minority carrier in a semiconductor or a thermal neutron in a nuclear reactor, from the point at which it is formed to the point at which it is absorbed. { də'fyü·zhən ˌleŋkth }

diffusion-limited aggregation [PHYS] A mathematical model for particle aggregation processes, such as the growth of a metal deposit on an electrochemical cell, in which particles move according to a random walk process until they arrive at a certain fixed distance from the current aggregate, where they stick to it. { də'fyü·zhən ¦lim·əd·əd ag·rə'gā·shən }

diffusion number [FL MECH] A dimensionless number used in the study of mass transfer, equal to the diffusivity of a solute through a stationary solution contained in the solid, times a characteristic time, divided by the square of the distance from the midpoint of the solid to the surface. Symbolized β. { də'fyü·zhən ˌnəm·bər }

diffusion theory [ELEC] The theory that in semiconductors, where there is a variation of carrier concentration, a motion of the carriers is produced by diffusion in addition to the drift determined by the mobility and the electric field. { də'fyü·zhən ˌthē·ə·rē }

diffusion velocity [FL MECH] **1.** The relative mean molecular velocity of a selected gas undergoing diffusion in a gaseous atmosphere, commonly taken as a nitrogen (N_2) atmosphere; a molecular phenomenon that depends upon the gaseous concentration as well as upon the pressure and temperature gradients present. **2.** The velocity or speed with which a turbulent diffusion process proceeds as evidenced by the motion of individual eddies. { də'fyü·zhən və'läs·əd·ē }

diffusivity [PHYS] *See* diffusion coefficient. [THERMO] The quantity of heat passing normally through a unit area per unit time divided by the product of specific heat, density, and temperature gradient. Also known as thermal diffusivity; thermometric conductivity. { dif·yü'ziv·əd·ē }

dihedral reflector [OPTICS] A corner reflector having two sides meeting at a line. { dī'hē·drəl ri'flek·tər }

dihexagonal [CRYSTAL] Of crystals, having a symmetrical form with 12 sides. { dī·hek'sag·ən·əl }

dihexagonal-dipyramidal [CRYSTAL] Characterized by the class of crystals in the hexagonal system in which any section perpendicular to the sixfold axis is dihexagonal. { dī·hek'sag·ən·əl dī·pir·ə'mid·əl }

dihexahedron [CRYSTAL] A type of crystal that has 12 faces, such as a double six-sided pyramid. { dī¦hek·sə¦hē·drən }

dilatant fluid [FL MECH] A fluid whose apparent viscosity increases simultaneously with an increase in shear rate. Also known as inverted pseudoplastic fluid. { dī¦lāt·ənt 'flü·əd }

dilatation [PHYS] The increase in volume per unit volume of any continuous substance, caused by deformation. { ˌdil·ə'tā·shən }

dilatational wave *See* compressional wave. { ˌdil·ə'tā·shən·əl 'wāv }

dilatometry [PHYS] The measurement of changes in the volume of a liquid or dimensions of a solid which occur in phenomena such as allotropic transformations, thermal expansion, compression, creep, or magnetostriction. { ˌdil·ə'täm·ə·trē }

dilaton [PART PHYS] A hypothetical elementary particle having zero mass and zero spin, which is introduced in constructing a scale invariant theory involving massive particles. { 'dī·lə,tän }

dilepton event [PART PHYS] The inelastic scattering of a neutrino or antineutrino from a nucleus in which there are two leptons among the products of the collision. { dī'lep,tän ə'vent }

dilution [OPTICS] Reducing the intensity of a color by adding white. { də'lü·shən }

dilution factor [ELECTROMAG] The energy density of a radiation field divided by the equilibrium value for radiation of the same color temperature. { də'lü·shən ,fak·tər }

DIM *See* nonthermal decimetric emission.

dimensional analysis [PHYS] A technique that involves the study of dimensions of physical quantities, used primarily as a tool for obtaining information about physical systems too complicated for full mathematical solutions to be feasible. { də'men·chən·əl ə'nal·ə·səs }

dimensional constant [PHYS] A physical quantity whose numerical value depends on the units chosen for fundamental quantities but not on the system being considered. { də'men·chən·əl 'kän·stənt }

dimensional formula [PHYS] The expression of a derived quantity as a product of powers of the fundamental quantities. { də'men·chən·əl 'fȯr·myə·lə }

dimensional regularization [QUANT MECH] A method of extracting a finite piece from an infinite result in quantum field theory based on analytically continuing a typically divergent integral in its number of space-time dimensions. { di'men·chən·əl ,reg·yəl·ər·ə'zā·shən }

dimensionless group [PHYS] Any combination of dimensional or dimensionless quantities possessing zero overall dimensions; an example is the Reynolds number. { də'men·chən·ləs 'grüp }

dimensions [PHYS] The product of powers of fundamental quantities (or of convenient derived quantities) which are used to define a physical quantity; the fundamental quantities are often mass, length, and time. { də'men·chənz }

dimple crystal [CRYO] A periodic pattern of hexagons that forms on a liquid helium surface when electrons, forming an electric charge sheet, are trapped at the surface, and an intense electric field is applied; the hexagons are typically about 2 millimeters across and the interiors of the hexagons are depressions about 10 micrometers deep. { 'dim·pəl ,krist·əl }

dimuon event [PART PHYS] An inelastic collision of a neutrino or antineutrino with a nucleus in which there are two muons among the products of the collision. { dī'myü,än i'vent }

dineutron [NUC PHYS] **1.** A hypothetical bound state of two neutrons, which probably does not exist. **2.** A combination of two neutrons which has a transitory existence in certain nuclear reactions. { dī'nü,trän }

dioctahedral [CRYSTAL] Pertaining to a crystal structure in which only two of the three available octahedrally coordinated positions are occupied. { ,dī,äk·tə'hē·drəl }

diode laser *See* semiconductor laser. { 'dī,ōd ,lāz·ər }

diode theory [ELEC] The theory that in a semiconductor, when the barrier thickness is comparable to or smaller than the mean free path of the carriers, then the carriers cross the barrier without being scattered, much as in a vacuum tube diode. { 'dī,ōd ,thē·ə·rē }

diopter [OPTICS] A measure of the power of a lens or a prism, equal to the reciprocal of its focal length in meters. Abbreviated D. { dī'äp·tər }

dioptometer [OPTICS] An instrument for determining ocular refraction. { dī,äp'täm·əd·ər }

dioptric [OPTICS] **1.** Serving in or effecting refraction. **2.** Produced by means of refraction. { dī'äp·trik }

dioptrics [OPTICS] The branch of optics that treats of the refraction of light, especially by the transparent medium of the eye, and by lenses. { dī'äp·triks }

diploid [CRYSTAL] A crystal form in the isometric system having 24 similar quadrilateral faces arranged in pairs. { 'di,plȯid }

dipole [ELECTROMAG] Any object or system that is oppositely charged at two points, or poles, such as a magnet or a polar molecule; more precisely, the limit as either charge goes to infinity, the separation distance to zero, while the product remains constant. Also known as doublet; electric doublet. { 'dī,pōl }

dipole antenna [ELECTROMAG] An antenna approximately one-half wavelength long, split at its electrical center for connection to a transmission line whose radiation pattern has a maximum at right angles to the antenna. Also known as doublet antenna; half-wave dipole. { 'dī,pōl an'ten·ə }

dipole-dipole interaction [ATOM PHYS] The interaction of two atoms, molecules, or nuclei by means of their electric or magnetic dipole moments. The interaction energy depends on the strength and relative orientation of the two dipoles, as well as on the distance between the centers and the orientation of the radius vector connecting the centers with respect to the dipole vectors. { ¦dī,pōl ¦dī,pōl ,in·tər 'ak·shən }

dipole disk feed [ELECTROMAG] Antenna, consisting of a dipole near a disk, used to reflect energy to the disk. { 'dī,pōl 'disk ,fēd }

dipole moment *See* electric dipole moment; magnetic dipole moment. { 'dī,pōl ,mō·mənt }

dipole polarization *See* orientation polarization. { 'dī,pōl ,pō·lə·rə'zā·shən }

dipole radiation [ELECTROMAG] The electromagnetic radiation generated by an oscillating electric or magnetic dipole. { 'dī,pōl ,rād·ē'ā·shən }

dipole relaxation [ELEC] The process, occupying a certain period of time after a change in the applied electric field, in which the orientation polarization of a substance reaches equilibrium. { 'dī,pōl ,rē,lak'sā·shən }

dipole sound field [ACOUS] A sound field generated by an oscillating dipole source. { 'dī,pōl 'saủnd ,fēld }

dipole transition [ATOM PHYS] A transition of an atom or nucleus from one energy state to another in which dipole radiation is emitted or absorbed. { 'dī,pōl tran'zish·ən }

dipping refractometer *See* immersion refractometer. { 'dip·iŋ ,rē,frak'täm·əd·ər }

dip pole *See* magnetic pole. { 'dip ,pōl }

diproton [NUC PHYS] A hypothetical bound state of two protons, which probably does not exist. { dī'prō,tän }

dipyramid *See* bipyramid. { dī'pir·ə,mid }

Dirac charge [QUANT MECH] A fundamental magnetic charge g such that only integral multiples of g are consistent with quantum mechanics. { di'rak ,chärj }

Dirac covariants [QUANT MECH] Quantities which behave as a scalar, a pseudoscalar, a vector, an axial vector, or a second-rank tensor under Lorentz transformations, and whose elements consist of basis elements of the Dirac gamma algebra multiplied by the Dirac wave function on the right and its adjoint on the left. { di'rak kō'ver·ē·əns }

Dirac electron theory *See* Dirac theory. { di'rak i'lek,trän ,thē·ə·rē }

Dirac equation [QUANT MECH] A relativistic wave equation for an electron in an electromagnetic field, in which the wave function has four components corresponding to four internal states specified by a two-valued spin coordinate and an energy coordinate which can have a positive or negative value. { di'rak i'kwā·zhən }

Dirac fields [QUANT MECH] Operators, arising in the second quantization of the Dirac theory, which correspond to the Dirac wave functions in the original theory. { di'rak ,fēlz }

Dirac gamma algebra [QUANT MECH] An algebra whose basis consists of 16 linearly independent 4×4 matrices constructed from products of the four basic Dirac matrices. { di'rak ¦gam·ə 'al·jə·brə }

Dirac h *See* h-bar. { di'rak 'āch }

Dirac hole theory [QUANT MECH] The theory that the continuum of negative energy states that are solutions to the Dirac equation are filled with electrons, and the vacancies in this continuum (holes) are manifested as positrons with energy and momentum that are the negative of those of the state. { di'rak 'hōl ,thē·ə·rē }

Dirac matrix [QUANT MECH] Any one of four matrices, designated γ_μ ($\mu = 1, 2, 3, 4$), each having four rows and four columns and satisfying $\gamma_\mu\gamma_\nu + \gamma_\nu\gamma_\mu = \delta_{\mu\nu}$, where $\delta_{\mu\nu}$ is the Kronecker delta function, which matrices operate on the four-component wave function in the Dirac equation. Also known as gamma matrix. { di'rak 'mā,triks }

Dirac moment [QUANT MECH] Magnetic moment of the electron according to the Dirac theory, equal to $e\hbar/2mc$, where e and m are the charge and mass of the positron respectively, $\hbar$ is Planck's constant divided by 2π, and c is the speed of light. { di'rak 'mō·mənt }

Dirac monopole [QUANT MECH] A hypothetical magnetic monopole whose magnetic charge is an integral multiple of $\hbar c/(2e)$, where $\hbar$ is Planck's constant divided by 2π, c is the speed of light, and e is the charge of the electron. { di'rak 'män·ə,pōl }

Dirac particle [PART PHYS] A particle behaving according to the Dirac theory, which describes the behavior of electrons and muons except for radiative corrections, and is envisaged as describing a central core of a hadron of spin $^1/_2\,\hbar$ which remains when the effects of nuclear forces are removed. { di'rak ,pärd·ə·kəl }

Dirac quantization [QUANT MECH] The condition, arising from conservation of angular momentum, that for any electric charge q and magnetic monopole with magnetic charge m, one has $2qm = n\hbar c$, where n is an integer, $\hbar$ is Planck's constant divided by 2π, and c is the speed of light (gaussian units). { di'rak ,kwän·tə'zā·shən }

Dirac sea [QUANT MECH] The continuum of negative energy states that are solutions of the Dirac equation, and that are filled with electrons according to Dirac hole theory. { di'rak ,sē }

Dirac theory [QUANT MECH] Theory of the electron based on the Dirac equation, which accounts for its spin angular momentum and gives its magnetic moment and its behavior in an electromagnetic field (except for higher-order corrections). Also known as Dirac electron theory. { di'rak 'thē·ə·rē }

Dirac wave function [QUANT MECH] A function appropriate for describing a spin $^1/_2$ particle and antiparticle; it is a column matrix with four entries, each of which is a function of the space and time coordinates; the four-components form two first-rank Lorentz spinors. { di'rak 'wāv ,fəŋk·shən }

direct-aperture antenna [ELECTROMAG] An antenna whose conductor or dielectric is a surface or solid, such as a horn, mirror, or lens. { də¦rekt ¦ap·ə·chər an'ten·ə }

direct-band-gap semiconductor [SOLID STATE] A semiconductor material in which the state of minimum energy in the conduction band and the state of maximum energy in the valence band have the same momentum, so that optical transitions between free electrons and holes are allowed. { də'rekt 'band ,gap 'sem·i·kən,dək·tər }

direct coupling [ELEC] Coupling of two circuits by means of a non-frequency-sensitive device, such as a wire, resistor, or battery, so both direct and alternating current can flow through the coupling path. { də¦rekt 'kəp·liŋ }

direct current [ELEC] Electric current which flows in one direction only, as opposed to alternating current. Abbreviated dc. { də¦rekt 'kə·rənt }

direct-current circuit [ELEC] Any combination of dc voltage or current sources, such as generators and batteries, in conjunction with transmission lines, resistors, and power converters such as motors. { də¦rekt ¦kə·rənt 'sər·kət }

direct-current circuit theory [ELEC] An analysis of relationships within a dc circuit. { də¦rekt ¦kə·rənt 'sər·kət ,thē·ə·rē }

direct-current continuity [ELEC] Property of a circuit in which there is an established pathway for conduction of current from a direct-current source. { də¦rekt ¦kə·rənt ˌkänt·ən'ü·əd·ē }

direct-current Josephson effect [CRYO] The current flow resulting from the tunneling of electron pairs through a thin insulating barrier between two superconductors in the absence of a voltage drop across the barrier. { di¦rekt ¦kər·ənt 'jō·sef·sən iˌfekt }

direct-current voltage *See* direct voltage. { də¦rekt ¦kə·rənt 'vōl·tij }

direct-current working volts [ELEC] The maximum continuously applied dc voltage for which a capacitor is rated. Abbreviated dcwV. { də¦rekt ¦kə·rənt 'wərk·iŋ ˌvōlts }

direct electromotive force [ELEC] Unidirectional electromotive force in which the changes in values are either zero or so small that they may be neglected. { də'rekt iˌlek·trō'mōd·iv 'fȯrs }

direct-gap semiconductor [SOLID STATE] A semiconductor in which the minimum of the conduction band occurs at the same wave vector as the maximum of the valence band, and recombination radiation consequently occurs with relatively large intensity. { də¦rekt ¦gap 'sem·i·kənˌdək·tər }

directional antenna [ELECTROMAG] An antenna that radiates or receives radio waves more effectively in some directions than others. { də'rek·shən·əl an'ten·ə }

directional beam [ELECTROMAG] A radio or radar wave that is concentrated in a given direction. { də'rek·shən·əl 'bēm }

directional gyro [MECH] A two-degrees-of-freedom gyro with a provision for maintaining its spin axis approximately horizontal. { də'rek·shən·əl 'jī·rō }

directional pattern *See* radiation pattern. { də'rek·shən·əl 'pad·ərn }

direction of propagation [PHYS] **1.** The normal to a surface of constant phase, in a propagating wave. **2.** The direction of the group velocity. **3.** The direction of time-average energy flow. (In a homogeneous isotropic medium, these three directions coincide.) { də'rek·shən əv ˌpräp·ə'gā·shən }

directive gain [ELECTROMAG] Of an antenna in a given direction, 4π times the ratio of the radiation intensity in that direction to the total power radiated by the antenna. { də'rek·tiv ˌgān }

directivity [ELECTROMAG] **1.** The value of the directive gain of an antenna in the direction of its maximum value. **2.** The ratio of the power measured at the forward-wave sampling terminals of a directional coupler, with only a forward wave present in the transmission line, to the power measured at the same terminals when the direction of the forward wave in the line is reversed; the ratio is usually expressed in decibels. { dəˌrek'tiv·əd·ə }

direct nuclear reaction [NUC PHYS] A nuclear reaction which is completed in the time required for the incident particle to transverse the target nucleus, so that it does not combine with the nucleus as a whole but interacts only with the surface or with some individual constituent. { də¦rekt ¦nü·klē·ər rē'ak·shən }

direct numerical simulation [FL MECH] Simulation of fluid flow that is carried out through numerical integration of the Navier-Stokes equation in a two-stage process: a temporal approximation involving the splitting of the equation into linear and nonlinear parts and the approximation of time derivatives with finite differences, and a spatial approximation involving the replacement of spatial derivatives with finite difference, finite element, spectral, or hybrid numerical approximations. Abbreviated DNS. { də ¦rekt nü¦mer·ə·kəl ˌsim·yə'lā·shən }

director [ELECTROMAG] A parasitic element placed a fraction of a wavelength ahead of a dipole receiving antenna to increase the gain of the array in the direction of the major lobe. { də'rek·tər }

direct piezoelectricity [SOLID STATE] Name sometimes given to the piezoelectric effect in which an electric charge is developed on a crystal by the application of mechanical stress. { də'rekt pē¦ā·zōˌiˌlek'tris·əd·ē }

direct reflection *See* specular reflection. { də¦rekt ri'flek·shən }

direct scattering theory *See* scattering theory. { də'rekt 'skad·ə·riŋ ˌthē·ə·rē }

direct simulation Monte Carlo [FL MECH] An iterated two-step procedure for simulation of fluid flow, in which the calculation of the outcomes of collisions of paired particles based on intermolecular potentials alternates with the partitioning of the phase space into cells and the random selection of 10–100 particle pairs as candidates for collision, consistent with collision frequency predictions based on kinetic theory. { də¦rekt ˌsim·yə¦lā·shən ¦män·tə kär·lō }

direct sound [ACOUS] The portion of the room impulse response consisting of the combination of the true direct sound, which has traveled directly from the sound source to the listener, and the various reflections within the first 20 milliseconds after it. { di¦rekt 'sau̇nd }

direct viewfinder [OPTICS] A viewfinder in which the user views the subject directly through a lens or sight. { də¦rekt 'vyüˌfīnd·ər }

direct-vision nephoscope [OPTICS] A type of nephoscope in which the cloud motion is observed by looking directly into the instrument. { də¦rekt ¦vizh·ən 'nef·əˌskōp }

direct-vision prism *See* Amici prism. { də¦rekt ¦vizh·ən 'priz·əm }

direct-vision spectroscope [SPECT] A spectroscope that allows the observer to look in the direction of the light source by means of an Amici prism. { də¦rekt ¦vizh·ən 'spek·trəˌskōp }

direct voltage [ELEC] A voltage that forces electrons to move through a circuit in the same direction continuously, thereby producing a direct current. Also known as direct-current voltage. { də¦rekt 'vōl·tij }

discharge [ELEC] To remove a charge from a

battery, capacitor, or other electric-energy storage device. [FL MECH] The flow rate of a fluid at a given instant expressed as volume per unit of time. { 'dis,chärj }

discharge coefficient [FL MECH] In a nozzle or other constriction, the ratio of the mass flow rate at the discharge end of the nozzle to that of an ideal nozzle which expands an identical working fluid from the same initial conditions to the same exit pressure. Also known as coefficient of discharge. { 'dis,chärj ,kō·i'fish·ənt }

discone antenna [ELECTROMAG] A biconical antenna in which one of the cones is spread out to 180° to form a disk; the center conductor of the coaxial line terminates at the center of the disk, and the cable shield terminates at the vertex of the cone. { 'dis,kōn an'ten·ə }

discontinuity [ELECTROMAG] An abrupt change in the shape of a waveguide. Also known as waveguide discontinuity. [PHYS] A break in the continuity of a medium or material at which a reflection of wave energy can occur. { dis,känt·ən'ü·əd·ē }

discord *See* dissonance. { 'di,skȯrd }

discrete spectrum [SPECT] A spectrum in which the component wavelengths constitute a discrete sequence of values rather than a continuum of values. { di'skrēt 'spek·trəm }

dish *See* parabolic reflector. { dish }

disinclination [CRYSTAL] A type of crystal imperfection in which one part of the crystal is rotated and therefore displaced relative to the rest of the crystal; observed in liquid crystals and protein coats of viruses. { dis,in·klə'nā·shən }

disintegration [NUC PHYS] Any transformation of a nucleus, whether spontaneous or induced by irradiation, in which particles or photons are emitted. { dis,in·tə'grā·shən }

disintegration chain *See* radioactive series. { dis,in·tə'grā·shən ,chān }

disintegration constant *See* decay constant. { dis,in·tə'grā·shən ,kän·stənt }

disintegration energy [NUC PHYS] The energy released, or the negative of the energy absorbed, during a nuclear or particle reaction. Designated Q. Also known as Q value; reaction energy. { dis,in·tə'grā·shən ,en·ər·jē }

disintegration family *See* radioactive series. { dis,in·tə'grā·shən ,fam·lē }

disintegration series *See* radioactive series. { dis,in·tə'grā·shən ,sir·ēz }

disk camera [OPTICS] A camera that uses a disk of color negative film enclosed in a light-tight plastic pack and containing 15 rectangular frames measuring approximately 8 by 10 millimeters. { 'disk ,kam·rə }

disk capacitor [ELEC] A small, flat, circular capacitor that usually has a ceramic dielectric. { 'disk kə,pas·əd·ər }

disk telescope [OPTICS] A telescope designed for observations of the brilliant solar disk; examples are the tower telescope and the horizontal fixed telescope. { ¦disk 'tel·ə,skōp }

dislocation [CRYSTAL] A defect occurring along certain lines in the crystal structure and present as a closed ring or a line anchored at its ends to other dislocations, grain boundaries, the surface, or other structural feature. Also known as line defect. { ,dis·lō'kā·shən }

dislocation line [CRYSTAL] A curve running along the center of a dislocation. { ,dis·lō'kā·shən ,līn }

disorder [CRYSTAL] Departures from regularity in the occupation of lattice sites in a crystal containing more than one element. { dis'ȯrd·ər }

disordered crystalline alloy [SOLID STATE] A mixture of two elements in which the atoms of the mixture are found at more or less random positions on a crystal lattice. { dis'ȯrd·ərd ¦krist·əl·ən 'al,ȯi }

dispersing prism [OPTICS] An optical prism which deviates light of different wavelengths by different amounts and can therefore be used to separate white light into its monochromatic parts. { də'spərs·iŋ ,priz·əm }

dispersion [ELECTROMAG] Scattering of microwave radiation by an obstruction. [PHYS] **1.** The separation of a complex of electromagnetic or sound waves into its various frequency components. **2.** Quantitatively, the rate of change of refractive index with wavelength or frequency at a given wavelength or frequency. **3.** The rate of change of deviation with wavelength or frequency. **4.** In general, any process separating radiation into components having different frequencies, energies, velocities, or other characteristics, such as the sorting of electrons according to velocity in a magnetic field. { də'spər·zhən }

dispersion equation *See* dispersion formula. { də'spər·zhən i'kwā·zhən }

dispersion formula [PHYS] Any formula which gives the refractive index as a function of wavelength of electromagnetic radiation. Also known as dispersion equation. { də'spər·zhən ,fōr·myə·lə }

dispersion relation [NUC PHYS] A relation between the cross section for a given effect and the de Broglie wavelength of the incident particle, which is similar to a classical dispersion formula. [PHYS] An integral formula relating the real and imaginary parts of some function of frequency or energy, such as a refractive index or scattering amplitude, based on the causality principle and the Cauchy integral formula. [PL PHYS] A relation between the radian frequency and the wave vector of a wave motion or instability in a plasma. { də'spər·zhən ri,lā·shən }

dispersion strengthening [SOLID STATE] The reduction of plastic deformation of a solid by the presence of a uniform dispersion of another substance which inhibits the motion of plastic dislocations. { də'spər·zhən ,streŋk·thən·iŋ }

dispersive medium [ELECTROMAG] A medium in which the phase velocity of an electromagnetic wave is a function of frequency. { də'spər·siv 'mē·dē·əm }

dispersive power [OPTICS] A measure of the power of a medium to separate different colors of light, equal to $(n_2 - n_1)/(n - 1)$, where n_1 and n_2 are the indices of refraction at two specified widely differing wavelengths, and n is the index of refraction for the average of these wavelengths, or for the D line of sodium. { də'spər·siv ˌpaů·ər }

disphenoid [CRYSTAL] **1.** A crystal form with four similar triangular faces combined in a wedge shape; can be tetragonal or orthorhombic. **2.** A crystal form with eight scalene triangles combined in pairs. { dī'sfēˌnȯid }

displacement [ELEC] *See* electric displacement. [FL MECH] **1.** The weight of fluid which is displaced by a floating body, equal to the weight of the body and its contents; the displacement of a ship is generally measured in long tons (1 long ton = 2240 pounds). **2.** The volume of fluid which is displaced by a floating body. [MECH] **1.** The linear distance from the initial to the final position of an object moved from one place to another, regardless of the length of path followed. **2.** The distance of an oscillating particle from its equilibrium position. { dis'plās·mənt }

displacement current [ELECTROMAG] The rate of change of the electric displacement vector, which must be added to the current density to extend Ampère's law to the case of time-varying fields (meter-kilogram-second units). Also known as Maxwell's displacement current. { dis'plās·mənt ˌkə·rənt }

displacement law *See* radioactive displacement law; Wien's displacement law. { dis'plās·mənt ˌlȯ }

disruptive discharge [ELEC] A sudden and large increase in current through an insulating medium due to complete failure of the medium under electrostatic stress. { dis¦rəp·tiv 'disˌchärj }

dissecting microscope [OPTICS] Either of two types of optical microscope used to magnify materials undergoing dissection. { də'sek·tiŋ 'mī·krəˌskōp }

dissipation [PHYS] Any loss of energy, generally by conversion into heat; quantitatively, the rate at which this loss occurs. Also known as energy dissipation. { ˌdis·ə'pā·shən }

dissipation coefficient *See* scattering coefficient. { ˌdis·ə'pā·shən ˌkō·i'fish·ənt }

dissipation factor [ELEC] The inverse of Q, the storage factor. { ˌdis·ə'pā·shən ˌfak·tər }

dissipation function *See* Rayleigh's dissipation function; viscous dissipation function. { ˌdis·ə'pā·shən ˌfəŋk·shən }

dissipation line [ELECTROMAG] A length of stainless steel or Nichrome wire used as a noninductive terminating impedance for a rhombic transmitting antenna when several kilowatts of power must be dissipated. { ˌdis·ə'pā·shən ˌlīn }

dissipation trail [FL MECH] A clear rift left behind an aircraft as it flies in a thin cloud layer; the opposite of a condensation trail. Also known as distrail. { ˌdis·ə'pā·shən ˌtrāl }

dissipative tunneling [SOLID STATE] Quantum-mechanical tunneling of individual electrons, rather than pairs, across a thin insulating layer separating two superconducting metals when there is a voltage across this layer, resulting in partial disruption of cooperative motion. { ˌdis·ə'pād·iv 'tən·əl·iŋ }

dissociation limit [SPECT] The wavelength, in a series of vibrational bands in a molecular spectrum, corresponding to the point at which the molecule dissociates into its constituent atoms; it corresponds to the convergence limit. { də'sō·sē'ā·shən ˌlim·ət }

dissociative recombination [ATOM PHYS] The combination of an electron with a positive molecular ion in a gas followed by dissociation of the molecule in which the resulting atoms carry off the excess energy. { də¦sō·shəd·iv ˌrē·käm·bə'nā·shən }

dissonance [ACOUS] An unpleasant combination of harmonics heard when certain musical tones are played simultaneously. Also known as discord. { 'dis·ə·nəns }

dissymmetry factor [OPTICS] A quantity which expresses the strength of circular dichroism, equal to the difference in the absorption indices for left and right circularly polarized light divided by the absorption index for ordinary light of the same wavelength. Also known as anisotropy factor. { di'sim·ə·trē ˌfak·tər }

distance [MECH] The spatial separation of two points, measured by the length of a hypothetical line joining them. { 'dis·təns }

distant field [ELECTROMAG] The electromagnetic field at a distance of five wavelengths or more from a transmitter, where the radial electric field becomes negligible. { ¦dis·tənt ¦fēld }

distortion [OPTICS] A type of aberration in which there is variation in magnification with the distance from the axis of an optical system, so that images are not geometrically similar to their objects. { di'stȯr·shən }

distrail *See* dissipation trail. { 'disˌtrāl }

distributed capacitance [ELEC] Capacitance that exists between the turns in a coil or choke, or between adjacent conductors or circuits, as distinguished from the capacitance concentrated in a capacitor. { di'strib·yəd·əd kə'pas·əd·əns }

distributed constant [ELECTROMAG] A circuit parameter that exists along the entire length of a transmission line. Also known as distributed parameter. { di'strib·yəd·əd 'kän·stənt }

distributed inductance [ELECTROMAG] The inductance that exists along the entire length of a conductor, as distinguished from inductance concentrated in a coil. { di'strib·yəd·əd in 'dək·təns }

distributed parameter *See* distributed constant. { di'strib·yəd·əd pə'ram·əd·ər }

distribution coefficient [OPTICS] One of the tristimulus values of monochromatic radiations having equal power, usually denoted by x, y, z. { ˌdis·trə'byü·shən ˌko·i'fish·ənt }

distribution law [STAT MECH] A law which gives

a density function specifying the probability of finding a particle in a unit volume of phase space, or the number of particles in each of the states which a particle may occupy, or the number of particles per unit volume of phase space. { ˌdis·trəˈbyü·shən ˌlō }

distribution photometer *See* light-distribution photometer. { ˌdis·trəˈbyü·shən fōˈtäm·əd·ər }

Dittus-Boelter equation [FL MECH] An equation used to calculate the surface coefficient of heat transfer for fluids in turbulent flow inside clean, round pipes. { ¦did·əs ¦bel·tər iˌkwā·zhən }

divariant system [THERMO] A system composed of only one phase, so that two variables, such as pressure and temperature, are sufficient to define its thermodynamic state. { di¦ver·ē·ənt ˈsis·təm }

divergence [FL MECH] The ratio of the area of any section of fluid emerging from a nozzle to the area of the throat of the nozzle. [PHYS] The spreading apart of a beam of particles of light. { dəˈvər·jəns }

divergent I-beam technique [PHYS] A method of x-ray diffraction analysis in which a divergent beam of x-rays is used to produce Kossel lines. { dəˈvər·jent ˈīˌbēm ˈtekˈnēk }

diverging lens [OPTICS] A lens whose focal length is negative, so that light incident parallel to its axis diverges after passing through it. Also known as negative lens. { dəˈvərj·iŋ ˌlenz }

diverging meniscus lens *See* negative meniscus lens. { dəˈvərj·iŋ məˈnis·kəs ˌlenz }

diverging mirror [OPTICS] A convex mirror that causes rays of light parallel to its axis to diverge. Also known as negative mirror. { dəˈvərj·iŋ ˌmir·ər }

djalmaite *See* microlite. { ˈjal·məˌīt }

D line [SPECT] The yellow line that is the first line of the major series of the sodium spectrum; the doublet in the Fraunhofer lines whose almost equal components have wavelengths of 5895.93 and 5889.96 angstroms respectively. { ˈdē ˌlīn }

D meson [PART PHYS] Collective name for four charmed mesons that form two isotopic spin doublets, have masses of approximately 1865 MeV, and are pseudoscalar particles. { ¦dē ˈmāˌsän }

DNS *See* direct numerical simulation.

Dobson spectrophotometer [SPECT] A photoelectric spectrophotometer used in the determination of the ozone content of the atmosphere; compares the solar energy at two wavelengths in the absorption band of ozone by permitting the radiation of each to fall alternately upon a photocell. { ˈdäb·sən ˌspek·trō·fəˈtäm·əd·ər }

domain [SOLID STATE] A region in a solid within which elementary atomic or molecular magnetic or electric moments are uniformly arrayed. { dōˈmān }

domain growth [SOLID STATE] A stage in the process of magnetization in which there is a growth of those magnetic domains in a ferromagnet oriented most nearly in the direction of an applied magnetic field. { dōˈmān ˌgrōth }

domain rotation [SOLID STATE] The stage in the magnetization process in which there is rotation of the direction of magnetization of magnetic domains in a ferromagnet toward the direction of a magnetic applied field and against anisotropy forces. { dōˈmān rōˈtā·shən }

domain theory [SOLID STATE] A theory of the behavior of ferromagnetic and ferroelectric crystals according to which changes in the bulk magnetization and polarization arise from changes in size and orientation of domains that are each polarized to saturation but which point in different directions. { dōˈmān ˌthē·ə·rē }

domain wall *See* Bloch wall. { dōˈmān ˌwȯl }

domatic class *See* clinohedral class. { dōˈmad·ik ˈklas }

dome [CRYSTAL] An open crystal form consisting of two faces astride a symmetry plane. { dōm }

dominant energy condition [RELAT] The condition used in general relativity theory that all observers see a nonnegative energy density and a nonnegative energy flux. { ˈdäm·ə·nənt ˈen·ər·jē kənˌdish·ən }

dominant mode *See* fundamental mode. { ˈdäm·ə·nənt ˈmōd }

dominant wave [ELECTROMAG] The electromagnetic wave that has the lowest cutoff frequency in a given uniconductor waveguide. { ˈdäm·ə·nənt ˈwāv }

dominant wavelength [OPTICS] The single wavelength of light that, when combined in suitable proportions with a reference standard light, matches the color of a given sample. { ˈdäm·ə·nənt ˈwāvˌleŋkth }

Donders reduced eye [OPTICS] An optical model used to simplify calculations of the size and position of images produced by the human eye, consisting of a convex refracting surface that separates air in front from water, with a refractive index of 4/3, behind, and having an anterior focal distance of 15 millimeters and a posterior focal distance of 20 millimeters. { ˈdän·dərz ri¦düst ˈī }

donkey power [PHYS] A unit of power equal to 250 watts; it is approximately $^1/_3$ horsepower. { ˈdəŋ·kē ˌpau̇·ər }

Donohue equation [THERMO] Equation used to determine the heat-transfer film coefficient for a fluid on the outside of a baffled shell-and-tube heat exchanger. { ˈdän·ə·hü iˌkwā·zhən }

donor [SOLID STATE] An impurity that is added to a pure semiconductor material to increase the number of free electrons. Also known as donor impurity; electron donor. { ˈdō·nər }

donor impurity *See* donor. { ˈdō·nər imˌpyu̇r·əd·ē }

donor level [SOLID STATE] An intermediate energy level close to the conduction band in the energy diagram of an extrinsic semiconductor. { ˈdō·nər ˌlev·əl }

doorknob capacitor [ELEC] A high-voltage, plastic-encased capacitor resembling a doorknob in size and shape. { ˈdȯrˌnäb kəˌpas·əd·ər }

Doppler-averaged cross section [PHYS] A

cross section averaged over the energy of the incident particles and weighted to take into account the Doppler shifts associated with the thermal motions of the target particles. { ¦däp·lər ¦av·rijd 'krȯs ˌsek·shən }

Doppler broadening [SPECT] Frequency spreading that occurs in single-frequency radiation when the radiating atoms, molecules, or nuclei do not all have the same velocity and may each give rise to a different Doppler shift. { 'däp·lər ˌbrȯd·ən·iŋ }

Doppler effect [PHYS] The change in the observed frequency of an acoustic or electromagnetic wave due to relative motion of source and observer. { 'däp·lər iˌfekt }

Doppler-free spectroscopy [SPECT] Any of several techniques which make use of the intensity and monochromatic nature of a laser beam to overcome the Doppler broadening of spectral lines and measure their wavelengths with extremely high accuracy. { däp·lər ˌfrē spek'träs·kə·pē }

Doppler-free two-photon spectroscopy [SPECT] A version of Doppler free spectroscopy in which the wavelength of a transition induced by the simultaneous absorption of two photons is measured by placing a sample in the path of a laser beam reflected on itself, so that the Doppler shifts of the incident and reflected beams cancel. { däp·lər ˌfrē ¦tü ¦fōˌtän spek'träs·kə·pē }

Doppler frequency *See* Doppler shift. { däp·lər ˌfrē·kwən·sē }

Doppler lidar [OPTICS] A type of lidar that measures the frequency shift of the scattered laser pulse; this frequency shift is proportional to the velocity of the scatterer along the propagation path. { 'dop·lər 'līˌdär }

Doppler shift [PHYS] The amount of the change in the observed frequency of a wave due to Doppler effect, usually expressed in hertz. Also known as Doppler frequency. { 'däp·lər ˌshift }

Doppler spectroscopy [SPECT] A technique for measuring the speed with which an object is moving toward or away from the observer by measuring the amount that light from the object is shifted to a higher or lower frequency by the Doppler effect. { ˌdäp·lər spek'träs·kə·pē }

doroid [ELECTROMAG] A coil resembling half a toroid, using a removable core segment to simplify the winding process. { 'dȯˌrȯid }

dotter [OPTICS] A worker who uses a centering machine to locate the optical center and optical axis of a lens, for the guidance of workers who will cut, edge, trim, and mount the lens. Also known as spotter. { 'däd·ər }

double-beam spectrophotometer [SPECT] An instrument that uses a photoelectric circuit to measure the difference in absorption when two closely related wavelengths of light are passed through the same medium. { ¦dəb·əl ¦bēm spek·trō·fə'täm·əd·ər }

double beta decay [NUC PHYS] A nuclear transformation in which the atomic number changes by 2 and the mass number does not change; either two electrons are emitted or two orbital electrons are captured. { ¦dəb·əl ¦bād·ə di'kā }

double bridge *See* Kelvin bridge. { ¦dəb·əl 'brij }

double Compton scattering [QUANT MECH] A process in which a photon collides with a free electron and two photons are given off. { ¦dəb·əl 'käm·tən ˌskad·ər·iŋ }

double-concave lens [OPTICS] A lens having surfaces that are adjacent portions of nonintersecting spheres whose centers lie on opposite sides of the plane of the lens. Also known as biconcave lens. { ¦dəb·əl ¦känˌkāv 'lenz }

double-convex lens [OPTICS] A lens having surfaces that are adjacent portions of intersecting spheres whose centers lie on opposite sides of the plane of the lens. Also known as biconvex lens. { ¦dəb·əl ¦känˌveks 'lenz }

double diffusion [FL MECH] A type of convective transport in fluids that depends on the difference in diffusion rates of at least two density-affecting components. Also, it is necessary to have an unstable or top-heavy distribution of one component. For example, in oceanography, the two components are heat and dissolved salts, and the unstable component is the slower-diffusing salt. { ˌdəb·əl də'fyü·zhən }

double-doublet antenna [ELECTROMAG] Two half-wave doublet antennas criss-crossed at their center, one being shorter than the other to give broader frequency coverage. { ¦dəb·əl ¦dəb·lət an'ten·ə }

double electron excitation [ATOM PHYS] An excited state of an atom in which two electrons are excited rather than one. { ¦dəb·əl i¦lekˌträn ˌek·sī'tā·shən }

double-ended Q machine [PL PHYS] A Q machine in which the plasma is generated at a hot tungsten plate at one end and the plasma column is reflected at the other end by a second hot tungsten plate. { ¦dəb·əl ¦end·əd 'kyü məˌshēn }

double-exposure holographic interferometry [OPTICS] The study of the interference fringes generated by the superposition of two holograms of the same object, one with the object in an undeformed state and the second after a small deformation. { ¦dəb·əl ik'spō·zhər ¦häl·ə¦graf·ik ˌin·tər·fə'räm·ə·trē }

double group [QUANT MECH] A type of group useful in studying systems of half-integral spin; it is formed by modifying a finite point group by introducing an element which is a rotation through an angle of 2π about an arbitrary axis and which is not the unit element but gives the unit element when applied twice. { ¦dəb·əl 'grüp }

double-hump fission barrier [NUC PHYS] Two separated maxima in a plot of potential energy against nuclear deformation of an actinide nucleus, which inhibit spontaneous fission of the nucleus and give rise to isomeric states in the valley between the two maxima. { ¦dəb·əl ˌhəmp 'fish·ən ˌbar·ē·ər }

double-integrating gyro [MECH] A single-degree-of-freedom gyro having essentially no restraint of its spin axis about the output axis. { ¦dəb·əl ¦in·tə,grād·iŋ 'jī·rō }

double mirror [OPTICS] Two plane mirrors inclined at an angle to each other. { ¦dəb·əl 'mir·ər }

double pendulum [MECH] Two masses, one suspended from a fixed point by a weightless string or rod of fixed length, and the other similarly suspended from the first; often the system is constrained to remain in a vertical plane. { ¦dəb·əl 'pen·jə·ləm }

double-quantum stimulated-emission device [OPTICS] A laser in which the crystal contains two species of fluorescent ions whose fluorescence frequencies are so related that, when the flash lamp coils produce pumping action, the ions of one species contribute photons to the fluorescence of the other species, which is the active ion. { ¦dəb·əl 'kwänt·əm ¦stim·yə,lād·əd ē'mish·ən də,vīs }

double quantum transition [ATOM PHYS] A radiative transition between atomic or molecular states in which two or more photons are simultaneously emitted or absorbed. { ¦dəb·əl 'kwänt·əm tran'sish·ən }

double refraction *See* birefringence. { ¦dəb·əl ri'frak·shən }

double-shield enclosure [ELEC] Type of shielded enclosure or room in which the inner wall is partially isolated electrically from the outer wall. { ¦dəb·əl ,shēld in'klō·zhər }

double-slit interference *See* Young's two-slit interference. { 'dəb·əl ¦slit ,in·tər'fir·əns }

double-stub tuner [ELECTROMAG] Impedance-matching device, consisting of two stubs, usually fixed three-eighths of a wavelength apart, in parallel with the main transmission lines. { ¦dəb·əl ,stəb 'tün·ər }

doublet [ATOM PHYS] Two stationary states which have the same orbital and spin angular momentum but which have different total angular momenta, and therefore have slightly different energies due to spin-orbit coupling. [ELECTROMAG] *See* dipole. [FL MECH] A source and a sink separated by an infinitesimal distance, each having an infinitely large strength so that the product of this strength and the separation is finite. [OPTICS] A lens made up of two components, especially an achromat. [PART PHYS] Two elementary particles which have slightly differing masses and the same baryon number, spin, parity, and charge conjugation parity (if self-conjugate), but have different charges. [SPECT] Two closely separated spectral lines arising from a transition between a single state and a pair of states forming a doublet as described in the atomic physics definition. { 'dəb·lət }

doublet antenna *See* dipole antenna. { 'dəb·lət an'ten·ə }

doublet flow [FL MECH] The motion of a fluid in the vicinity of a doublet; can be superposed with uniform flow to yield flow around a cylinder or a sphere. { 'dəb·lət ,flō }

double weighing [MECH] A method of weighing to allow for differences in lengths of the balance arms, in which object and weights are balanced twice, the second time with their positions interchanged. Also known as Gauss method of weighing. { ¦dəb·əl 'wā·iŋ }

Dove prism *See* Delaborne prism. { 'dəv ,priz·əm }

down-Doppler [ACOUS] The sonar situation wherein the target is moving away from the transducer, so that the frequency of the echo is less than the frequency of the reverberations received immediately after the end of the outgoing ping; opposite of up-Doppler. { 'dau̇n ,däp·lər }

downdraft [PHYS] A current of air or other gas that travels downward, as during a thunderstorm or in a mine shaft. { 'dau̇n,draft }

down quark [PART PHYS] A quark with an electric charge of $-\frac{1}{3}$, baryon number of $\frac{1}{3}$, and 0 strangeness and charm. { 'dau̇n ,kwärk }

downwash [FL MECH] The downward deflection of air, relative to the direction of motion of an airfoil. { 'dau̇n,wäsh }

dr *See* dram.

drachm *See* dram. { dram }

draft Also spelled draught. [FL MECH] **1.** An air current in a confined space, such as that in a cooling tower or chimney. **2.** The difference between atmospheric pressure and some lower pressure in a confined space that causes air to flow, such as exists in the furnace or gas passages of a steam-generating unit or in a chimney. { draft }

draft differential [FL MECH] The difference in static pressure between two locations of gas flow. { 'draft dif·ə'ren·chəl }

drag [FL MECH] Resistance caused by friction in the direction opposite to that of the motion of the center of gravity of a moving body in a fluid. { drag }

drag coefficient [FL MECH] A characteristic of a body in a flowing inviscous fluid, equal to the ratio of twice the force on the body in the direction of flow to the product of the density of the fluid, the square of the flow velocity, and the effective cross-sectional area of the body. { 'drag ,kō·i'fish·ənt }

drag force [PL PHYS] A force on an electrically conducting fluid arising from inelastic collisions of electrons and ions and proportional to the fluid velocity. { 'drag ,fȯrs }

dragging of inertial frames [RELAT] A relativistic effect whereby, loosely speaking, a spinning body drags space around with it; this causes a gyroscope in orbit about the earth to display a precession distinct from the geodetic precession. Also known as frame dragging; Lense-Thirring effect. { 'drag·iŋ əv ə¦nər·shəl 'frāmz }

drain *See* current drain. { drān }

dram [MECH] **1.** A unit of mass, used in the apothecaries' system of mass units, equal to $\frac{1}{8}$ apothecaries' ounce or 60 grains or 3.8879346 grams. Also known as apothecaries' dram (dram ap); drachm (British). **2.** A unit of mass, formerly used in the United Kingdom, equal to $^1/_{16}$ ounce (avoirdupois) or approximately 1.77185 grams. Abbreviated dr. { dram }

dram ap *See* dram. { 'dram ˌap }

drift [SOLID STATE] The movement of current carriers in a semiconductor under the influence of an applied voltage. { drift }

drift current [PL PHYS] A current of free charged particles in perpendicular electric and magnetic fields that results from an average motion of the particles in a direction perpendicular to both fields. { 'drift ˌkə·rənt }

drift mobility [SOLID STATE] The average drift velocity of carriers per unit electric field in a homogeneous semiconductor. Also known as mobility. { 'drift mō'bil·əd·ē }

drift speed [ELEC] Average speed at which electrons or ions progress through a medium. { 'drift ˌspēd }

drift velocity [SOLID STATE] The average velocity of a carrier that is moving under the influence of an electric field in a semiconductor, conductor, or electron tube. { 'drift və'läs·əd·ē }

drift wave [PL PHYS] An oscillation in a magnetically confined plasma which arises in the presence of density gradients, for example, at the plasma's surface, and which resembles the waves that propagate at the interface of two fluids of different density in a gravity field. { 'drift ˌwāv }

drip line [NUC PHYS] The boundary, on a chart of the nuclides, beyond which a nucleon (proton or neutron) is no longer bound to the nucleus. { 'drip ˌlīn }

driven array [ELECTROMAG] An antenna array consisting of a number of driven elements, usually half-wave dipoles, fed in phase or out of phase from a common source. { ¦driv·ən ə'rā }

driven element [ELECTROMAG] An antenna element that is directly connected to the transmission line. { ¦driv·ən 'el·ə·mənt }

driver element [ELECTROMAG] Antenna array element that receives power directly from the transmitter. { 'drī·vər ˌel·ə·mənt }

driving resistance [MECH] The force exerted by soil on a pile being driven into it. { 'drīv·iŋ ri'zis·təns }

drop [FL MECH] The quantity of liquid that coalesces into a single globule; sizes vary according to physical conditions and the properties of the fluid itself. { dräp }

droplet condensation [THERMO] The formation of numerous discrete droplets of liquid on a wall in contact with a vapor, when the wall is cooled below the local vapor saturation temperature and the liquid does not wet the wall. { ¦dräp·lət ˌkän·dən'sā·shən }

drop model of nucleus *See* liquid-drop model. { 'dräp ˌmäd·əl əv 'nü·klē·əs }

drop weight [FL MECH] The weight of the largest drop that can hang from the end of a tube of given radius. { 'dräp ˌwāt }

drop-weight method [FL MECH] A method of measuring surface tension by measuring the weight of a slowly increasing drop of the liquid hanging from the end of a tube, just before it is detached from the tube. { 'dräp ˌwāt ˌmeth·əd }

dropwise condensation [THERMO] Condensation of a vapor on a surface in which the condensate forms into drops. { ¦dräpˌwīz ˌkän·dən'sā·shən }

Drude equation [OPTICS] An equation which states that the rotation of the plane of polarization of plane-polarized light passing through an optically active substance is inversely proportional to the difference between the square of the wavelength of the light and the square of a constant wavelength. { 'drüd iˌkwā·zhən }

Drude's theory of conduction [SOLID STATE] A theory which treats the electrons in a metal as a gas of classical particles. { ¦drüdz ¦thē·ə·rē əv kən'dək·shən }

dry battery [ELEC] A battery made up of a series, parallel, or series-parallel arrangement of dry cells in a single housing to provide desired voltage and current values. { ¦drī 'bad·ə·rē }

dry-bulb temperature [PHYS] The actual air temperature as measured by a dry-bulb thermometer. { ¦drī ˌbəlb 'tem·prə·chər }

dry cell [ELEC] A voltage-generating cell having an immobilized electrolyte. { 'drī ˌsel }

dry-charged battery [ELEC] A storage battery in which the electrolyte is drained from the battery for storage, and which is filled with electrolyte and charged for a few minutes to prepare for use. { ¦drī ˌchärjd 'bad·ə·rē }

dry contact [ELEC] A contact that does not break or make current. { ¦drī 'känˌtakt }

dry electrolytic capacitor [ELEC] An electrolytic capacitor in which the electrolyte is a paste rather than a liquid; the dielectric is a thin film of gas formed on one of the plates by chemical action. { ¦drī i¦lek·trə¦lid·ik kə'pas·əd·ər }

dry friction [MECH] Resistance between two dry solid surfaces, that is, surfaces free from contaminating films or fluids. { ¦drī 'frik·shən }

dry measure [MECH] A measure of volume for commodities that are dry. { ¦drī ¦mezh·ər }

dry pint *See* pint. { ¦drī ¦pīnt }

dry pt *See* pint.

dry steam [PHYS] Steam with no liquid phase dispersed in it. { 'drī 'stēm }

duality principle Also known as principle of duality. [ELEC] The principle that for any theorem in electrical circuit analysis there is a dual theorem in which one replaces quantities with dual quantities; current and voltage, impedance and admittance, and meshes and nodes are examples of dual quantities. [ELECTROMAG] The principle that one can obtain new solutions of Maxwell's equations from known solutions by replacing **E** with **H**, **H** with $-$**E**, ϵ with μ, and μ with ϵ. [QUANT MECH] *See* wave-particle duality. { dü'al·əd·ē ˌprin·sə·pəl }

dual laser [OPTICS] A gas laser having Brewster windows and concave mirrors at opposite ends, the mirrors having different reflectivities so as to produce two different visible or infrared wavelengths from a helium-neon laser beam. { 'dü·əl 'lā·zər }

dual network [ELEC] A network which has the same number of terminal pairs as a given network, and whose open-circuit impedance network is the same as the short-circuit admittance matrix of the given network, and vice versa. { 'dü·əl 'net ˌwərk }

dual radioactive decay [NUC PHYS] Property exhibited by a nucleus which has two or more independent and alternative modes of decay. { ¦dü·əl ˌrād·ē·ō¦ak·tiv di'kā }

Duane-Hunt law [QUANT MECH] The law that the frequency of x-rays resulting from electrons striking a target cannot exceed eV/h, where e is the charge of the electron, V is the exciting voltage, and h is Planck's constant. { ¦dwān ¦hənt ˌlȯ }

Duane-Hunt limit [QUANT MECH] The upper limit on the frequency of radiation from an x-ray tube given by the Duane-Hunt law. { ¦dwān ¦hənt 'lim·ət }

Duchemin's formula [PHYS] An expression for normal wind pressure per square foot on an inclined surface, $N = F[(2 \sin a)/(1 + \sin^2 a)]$, where F = normal wind force in pounds per square foot on a vertical surface, and a = angle of inclination of inclined surface. { dü·shmanz ˌfȯr·myə·lə }

ductile fracture *See* fibrous fracture. { ¦dək·təl 'frak·chər }

Dufour effect [THERMO] Energy flux due to a mass gradient occurring as a coupled effect of irreversible processes. { ¦dü·fȯr i'fekt }

Dufour number [THERMO] A dimensionless number used in studying thermodiffusion, equal to the increase in enthalpy of a unit mass during isothermal mass transfer divided by the enthalpy of a unit mass of mixture. Symbol Du_2. { ¦dü·fȯr ˌnəm·bər }

Duhem-Margules equation [THERMO] An equation showing the relationship between the two constituents of a liquid-vapor system and their partial vapor pressures: $\frac{d \ln p_A}{d \ln x_A} = \frac{d \ln p_B}{d \ln x_B}$ where x_A *and* x_B are the mole fractions of the two constituents, and p_A and p_B are the partial vapor pressures. { dü'em 'mär·gyə·lēz iˌkwā·zhən }

Dulong number *See* Eckert number. { də'lȯŋ ˌnəm·bər }

Dulong-Petit law [THERMO] The law that the product of the specific heat per gram and the atomic weight of many solid elements at room temperature has almost the same value, about 6.3 calories (264 joules) per degree Celsius. { də'lȯŋ pə'tē ˌlȯ }

duolateral coil *See* honeycomb coil. { ˌdü·ō'lad·ə·rəl 'kȯil }

duplet lens system [OPTICS] A system of lenses in which there are two groups of lenses separated by a space, and successive lenses in each group are in contact. { ¦düp·lət 'lenz ˌsis·təm }

Dupré equation [THERMO] The work W_{LS} done by adhesion at a gas-solid-liquid interface, expressed in terms of the surface tensions γ of the three phases, is $W_{LS} = \gamma_{GS} + \gamma_{GL} - \gamma_{LS}$. { dü'prā iˌkwā·zhən }

duration [MECH] A basic concept of kinetics which is expressed quantitatively by time measured by a clock or comparable mechanism. { də'rā·shən }

dust [PHYS] A loose term applied to solid particles predominantly larger than colloidal size and capable of temporary gas suspension. { dəst }

dust extinction [OPTICS] The contribution to total extinction of light made by scattering and absorption by dust particles in the path of a light beam. { 'dəst ik'stiŋk·shən }

dwt *See* pennyweight.

dye laser [OPTICS] A type of tunable laser in which the active material is a dye such as acridine red or esculin, with very large molecules, and laser action takes place between the first excited and ground electronic states, each of which comprises a broad vibrational-rotational continuum. { 'dī ˌlā·zər }

dynamical friction [PHYS] **1.** The drag force between electrons and ions drifitng with respect to each other. **2.** Sliding friction, in contrast to static friction. { dī¦nam·ə·kəl 'frik·shən }

dynamical similarity [MECH] Two flow fields are dynamically similar if one can be transformed into the other by a change of length and velocity scales. All dimensionless numbers of the flows must be the same. { dī¦nam·ə·kəl sim·ə'lar·əd·ē }

dynamical symmetry [PHYS] A type of symmetry of the Hamiltonian of a physical system that can specify detailed properties of the system. { dīˌnam·ə·kəl 'sim·ə·trē }

dynamical variable [MECH] One of the quantities used to describe a system in classical mechanics, such as the coordinates of a particle, the components of its velocity, the momentum, or functions of these quantities. { dī¦nam·ə·kəl 'ver·ē·ə·bəl }

dynamic analogies [PHYS] Analogies that make it possible to convert the differential equations for mechanical and acoustical systems to equivalent electrical equations that can be represented by electric networks and solved by circuit theory. { dī¦nam·ik ə'nal·ə·jēz }

dynamic balance [MECH] The condition which exists in a rotating body when the axis about which it is forced to rotate, or to which reference is made, is parallel with a principal axis of inertia; no products of inertia about the center of gravity of the body exist in relation to the selected rotational axis. { dī¦nam·ik 'bal·əns }

dynamic boundary condition [FL MECH] The condition that the pressure must be continuous across an internal boundary or free surface in a fluid. { dī¦nam·ik 'bau̇n·drē kənˌdish·ən }

dynamic braking [MECH] A technique of electric braking in which the retarding force is supplied by the same machine that originally was the driving motor. { dī¦nam·ik 'brāk·iŋ }

dynamic condenser electrometer [ELEC] A sensitive voltage-measuring instrument in which an object carrying charge resulting from the voltage is moved back and forth in an electrostatic field and the resulting alternating-current signal is observed. { dī¦nam·ik kən¦den·sər i,lek'träm·əd·ər }

dynamic creep [MECH] Creep resulting from fluctuations in a load or temperature. { dī¦nam·ik 'krēp }

dynamic equilibrium Also known as kinetic equilibrium. [MECH] The condition of any mechanical system when the kinetic reaction is regarded as a force, so that the resultant force on the system is zero according to d'Alembert's principle. [PHYS] A condition in which several processes act simultaneously to maintain a system in an overall state that does not change with time. { dī¦nam·ik ē·kwə'lib·rē·əm }

dynamic fluidity [FL MECH] The reciprocal of the dynamic viscosity. { dī¦nam·ik flü'id·əd·ē }

dynamic height [PHYS] The amount of work done when a water particle of unit mass is moved vertically from one level to another. Also known as geodynamic height. { dī¦nam·ik 'hīt }

dynamic impedance [ELEC] The impedance of a circuit having an inductance and a capacitance in parallel at the frequency at which this impedance has a maximum value. Also known as rejector impedance. { dī¦nam·ik im'ped·əns }

dynamic instability *See* inertial instability. { dī¦nam·ik ,in·stə'bil·əd·ē }

dynamic meter [PHYS] The standard unit of dynamic height expressed as 10 square meters per second per second. { dī¦nam·ik 'mēd·ər }

dynamic nuclear polarization [NUC PHYS] The creation of assemblies of nuclei whose spin axes are not oriented at random, and which are in a steady state that is not a state of thermal equilibrium. { dī'nam·ik 'nü·klē·ər ,pō·lə·rə'zā·shən }

dynamic pressure [FL MECH] **1.** The pressure that a moving fluid would have if it were brought to rest by isentropic flow against a pressure gradient. Also known as impact pressure; stagnation pressure; total pressure. **2.** The difference between the quantity in the first definition and the static pressure. { dī¦nam·ik 'presh·ər }

dynamic resistance [ELEC] A device's electrical resistance when it is in operation. { dī¦nam·ik ri'zis·təns }

dynamics [MECH] That branch of mechanics which deals with the motion of a system of material particles under the influence of forces, especially those which originate outside the system under consideration. { dī¦nam·iks }

dynamic scattering device [OPTICS] A type of numerical display device in which a voltage is applied to a cell containing a nematic liquid crystal of negative dielectric anisotropy; opposing influences from the electric field and electrical conduction produce turbulence, causing the cell to become milk white. { dī¦nam·ik 'skad·ə·riŋ di,vīs }

dynamic stability [MECH] The characteristic of a body, such as an aircraft, rocket, or ship, that causes it, when disturbed from an original state of steady motion in an upright position, to damp the oscillations set up by restoring moments and gradually return to its original state. Also known as stability. { dī¦nam·ik stə'bil·əd·ē }

dynamic symmetry [PHYS] A symmetry law related, not to the geometric structure of the constituents of matter, but to the laws which govern the dynamic behavior of these constituents. { dī¦nam·ik 'sim·ə·trē }

dynamic temperature difference [PHYS] The difference between the temperature of a static medium and the surface temperature at the stagnation point of a heat-insulated body immersed in a flowing medium of the same composition. { dī¦nam·ik 'tem·prə·chər ,dif·rəns }

dynamic viscosity *See* absolute viscosity. { dī¦nam·ik vis'käs·əd·ē }

dynamoelectric [PHYS] Pertaining to the conversion of mechanical energy to electric energy, or vice versa. { ¦dī·nə,mō·i'lek·trik }

dyne [MECH] The unit of force in the centimeter-gram-second system of units, equal to the force which imparts an acceleration of 1 cm/s^2 to a 1 gram mass. { dīn }

dyne-centimeter *See* erg. { ¦dīn 'sen·tə,mēd·ər }

dyne-cm *See* erg.

Dyson microscope [OPTICS] A type of interference microscope, now obsolete, in which a light ray is split into two parallel beams and then recombined by reflections from surfaces of parallel plates, and one of the beams passes through the object under observation. { 'dī·sən ,mī·krə,skōp }

E

E *See* electric-field vector.

eagle mounting [SPECT] A mounting for a diffraction grating, based on the principle of the Rowland circle, in which the diffracted ray is returned along nearly the same direction as the incident beam. { 'ē·gəl ˌmau̇n·tiŋ }

Earnshaw's theorem [ELEC] The theorem that a charge cannot be held in stable equilibrium by an electrostatic field. { 'ərnˌshȯz ˌthir·əm }

earth *See* ground. { ərth }

earth current [ELEC] Return, fault, leakage, or stray current passing through the earth from electrical equipment. Also known as ground current. { 'ərth ˌkə·rənt }

earthed system *See* grounded system. { 'ərtht ˌsis·təm }

easy glide [SOLID STATE] A large increase in plastic deformation of a single crystal accompanying a small increase in stress as the result of the passage of many thousands of dislocations through the crystal along a single glide system. { 'ē·zē ˌglīd }

E beam [PHYS] An intense burst of fast electrons from a small particle accelerator, used to excite a pulsed gas laser in which the gas pressure is too high to permit an electric discharge. { 'ē ˌbēm }

E bend [ELECTROMAG] A smooth change in the direction of the axis of a waveguide, throughout which the axis remains in a plane parallel to the direction of polarization. Also known as E-plane bend. { 'ē ˌbend }

ebullition [PHYS] The process or state of a liquid bubbling up or boiling. { ˌeb·ə'li·shən }

eccentricity [MECH] The distance of the geometric center of a revolving body from the axis of rotation. { ˌek·sən'tris·əd·ē }

E center *See* R center. { 'ē ˌsen·tər }

echelette grating [SPECT] A diffraction grating with coarse groove spacing, designed for the infrared region; has grooves with comparatively flat sides and concentrates most of the radiation by reflection into a small angular coverage. { ¦esh·ə¦let *or* ¦āsh¦let ˌgrād·iŋ }

echelle grating [SPECT] A diffraction grating designed for use in high orders and at angles of illumination greater than 45° to obtain high dispersion and resolving power by the use of high orders of interference. { ā'shel ˌgrād·iŋ }

echelle spectrograph [SPECT] A spectrograph that employs gratings intended to be used in very high orders (greater than 10), and is equipped with a second dispersal element (another grating or a prism) at right angles to the first in order to separate the successive spectral strips from each other. { eˌshel 'spek·trəˌgraf }

echelon grating [SPECT] A diffraction grating which consists of about 20 plane-parallel plates about 1 centimeter thick, cut from one sheet, each plate extending beyond the next by about 1 millimeter, and which has a resolving power on the order of 10^6. { 'esh·əˌlän ˌgrād·iŋ }

echo [PHYS] A wave packet that has been reflected or otherwise returned with sufficient delay and magnitude to be perceived as a signal distinct from that directly transmitted. { 'ek·ō }

echo area [ELECTROMAG] In radar, the area of a fictitious perfect reflector of electromagnetic waves that would reflect the same amount of energy back to the radar as the actual target. Also known as radar cross section; target cross section. { 'ekˌō ˌer·ē·ə }

echo chamber [ACOUS] A reverberant room or enclosure used in a studio to add echo effects to sounds for radio or television programs. { 'ek·ō ˌchām·bər }

Eckert number [PHYS] A dimensionless group used in the study of compressible flow around a body, equal to the square of the fluid velocity far from the body divided by the product of the specific heat of the fluid at constant temperature and the difference between the temperatures of the fluid and the body. Symbolized N_E. Also known as Dulong number. { 'ek·ərt ˌnəm·bər }

E core [ELECTROMAG] A transformer core made from E-shaped laminations and used in conjunction with I-shaped laminations. { 'ē ˌkȯr }

eddy [FL MECH] A vortexlike motion of a fluid running contrary to the main current. { 'ed·ē }

eddy coefficient *See* exchange coefficient. { 'ed·ē ˌkō·iˌfish·ənt }

eddy conduction *See* eddy heat conduction. { 'ed·ē kənˌdək·shən }

eddy conductivity [THERMO] The exchange coefficient for eddy heat conduction. { 'ed·ē ˌkän ˌdək'tiv·əd·ē }

eddy current [ELECTROMAG] An electric current induced within the body of a conductor when that conductor either moves through a nonuniform magnetic field or is in a region where there is a change in magnetic flux. Also known as Foucault current. { 'ed·ē ,kə·rənt }

eddy-current loss [ELECTROMAG] Energy loss due to undesired eddy currents circulating in a magnetic core. { 'ed·ē ,kə·rənt ,lȯs }

eddy-current test [ELECTROMAG] A nondestructive test in which the change of impedance of a test coil brought close to a conducting specimen indicates the eddy currents induced by the coil, and thereby indicates certain properties or defects of the specimen. { 'ed·ē ,kə·rənt ,test }

eddy diffusion [FL MECH] Diffusion which occurs in turbulent flow, by the rapid process of mixing of the swirling eddies of fluid. Also known as turbulent diffusion. { 'ed·ē də,fyü·zhən }

eddy-diffusion coefficient *See* eddy diffusivity. { 'ed·ē də,fyü·zhən ,kō·i,fish·ənt }

eddy diffusivity [FL MECH] The exchange coefficient for the diffusion of a conservative property by eddies in a turbulent flow. Also known as coefficient of eddy diffusion; eddy-diffusion coefficient. { 'ed·ē di,fyü'siv·əd·ē }

eddy flux [FL MECH] The rate of transport (or flux) of fluid properties such as momentum, mass heat, or suspended matter by means of eddies in a turbulent motion; the rate of turbulent exchange. Also known as moisture flux; turbulent flux. { 'ed·ē ,fləks }

eddy heat conduction [THERMO] The transfer of heat by means of eddies in turbulent flow, treated analogously to molecular conduction. Also known as eddy heat flux; eddy conduction. { 'ed·ē 'hēt kən'dək·shən }

eddy heat flux *See* eddy heat conduction. { 'ed·ē 'hēt ,fləks }

eddy kinetic energy [FL MECH] The kinetic energy of that component of fluid flow which represents a departure from the average kinetic energy of the fluid, the mode of averaging depending on the particular problem. Also known as turbulence energy. { 'ed·ē kə'ned·ik,en·ər·jē }

eddy resistance [FL MECH] Resistance or drag of a ship resulting from eddies that are shed from the hull or appendages of the ship and carry away energy. { 'ed·ē ri'zis·təns }

eddy spectrum [FL MECH] **1.** The distribution of frequencies of rotation of eddies in a turbulent flow, or of those eddies having some range of sizes. **2.** The distribution of kinetic energy among eddies with various frequencies or sizes. { 'ed·ē 'spek·trəm }

eddy stress *See* Reynolds stress. { 'ed·ē ,stres }

eddy velocity [FL MECH] The difference between the mean velocity of fluid flow and the instantaneous velocity at a point. Also known as fluctuation velocity. { 'ed·ē və,läs·əd·ē }

eddy viscosity [FL MECH] The turbulent transfer of momentum by eddies giving rise to an internal fluid friction, in a manner analogous to the action of molecular viscosity in laminar flow, but taking place on a much larger scale. { ,ed·ē vi'skäs·əd·ē }

eddy viscosity model [FL MECH] A model of the Reynolds stresses in turbulent flow which is based on the idea that turbulent mixing, analogous to molecular mixing, is governed by an effective viscosity (the eddy viscosity) which is not a property of the fluid but a consequence of the local state of turbulence. { 'ed·ē vi'skäs·əd·ē ,mäd·əl }

edge dislocation [CRYSTAL] A dislocation which may be regarded as the result of inserting an extra plane of atoms, terminating along the line of the dislocation. Also known as Taylor-Orowan dislocation. { 'ej ,dis·lō,kā·shən }

edge effect [ELEC] An outward-curving distortion of lines of force near the edges of two parallel metal plates that form a capacitor. { 'ej i,fekt }

edge excitation [CRYO] An excitation of a droplet of incompressible quantum Hall liquid in which a surface wave propagates along the edge of the droplet in the same direction that electrons drift along the edge, as determined by the direction of the magnetic field. { 'ej ,ek·sī'tā·shən }

edge focusing [ELECTROMAG] Axial focusing of a stream of ions which occurs when it crosses a fringe magnetic field obliquely; used in mass spectrometers and cyclotrons. { 'ej ,fō·kəs·iŋ }

edge tones [ACOUS] Tones produced when an air jet of sufficient speed is split by a sharp edge. { 'ej ,tōnz }

Edser and Butler's bands [OPTICS] Dark bands at intervals of equal frequency in a spectrum of light which has passed through a thin plate of transparent material with parallel sides. { ¦ed·sər ən 'bət·lərz ,banz }

EDXD *See* energy-dispersive x-ray diffraction.

EELS *See* electron energy loss spectroscopy.

effective ampere [ELEC] The amount of alternating current flowing through a resistance that produces heat at the same average rate as 1 ampere of direct current flowing in the same resistance. { ə¦fek·tiv 'am,pir }

effective antenna length [ELECTROMAG] Electrical length of an antenna, as distinguished from its physical length. { ə¦fek·tiv an'ten·ə ,leŋkth }

effective aperture [OPTICS] The diameter of the image of the aperture stop of an optical system, as viewed from the object. { ə¦fek·tiv 'ap·ə·chər }

effective area [ELECTROMAG] Of an antenna in any specified direction, the square of the wavelength multiplied by the power gain (or directive gain) in that direction and divided by 4π (12.57). { ə¦fek·tiv 'er·ē·ə }

effective capacitance [ELEC] Total capacitance existing between any two given points of an electric circuit. { ə¦fek·tiv kə'pas·əd·əns }

effective current [ELEC] The value of alternating current that will give the same heating effect as the corresponding value of direct current. Also known as root-mean-square current. { ə¦fek·tiv 'kə·rənt }

effective energy [OPTICS] The energy of a quantum of a beam of monochromatic radiation that is absorbed or scattered by a given medium to the same extent as a given beam of polychromatic radiation. { ə¦fek·tiv 'en·ər·jē }

effective field intensity [ELECTROMAG] Root-mean-square value of the inverse distance fields at a distance of 1 mile (1.6 kilometers) from the transmitting antenna in all directions in the horizontal plane. { ə¦fek·tiv 'fēld in,ten·səd·ē }

effective force *See* inertial force. { ə¦fek·tiv 'fȯrs }

effective gun bore line [MECH] The line which a projectile should follow when the muzzle velocity of the antiaircraft gun is vectorially added to the aircraft velocity. { ə¦fek·tiv 'gən ¦bȯr ,līn }

effective height [ELECTROMAG] The height of the center of radiation of a transmitting antenna above the effective ground level. { ə¦fek·tiv 'hīt }

effective launcher line [MECH] The line along which the aircraft rocket would go if it were not affected by gravity. { ə¦fek·tiv 'lȯn·chər ,līn }

effective magnetic length [ELECTROMAG] The distance between the effective magnetic poles of a magnet. Also known as equivalent magnetic length. { ə¦fek·tiv mag¦ned·ik 'leŋkth }

effective mass [SOLID STATE] A parameter with the dimensions of mass that is assigned to electrons in a solid; in the presence of an external electromagnetic field the electrons behave in many respects as if they were free, but with a mass equal to this parameter rather than the true mass. { ə¦fek·tiv 'mas }

effective radiated power [ELECTROMAG] The product of antenna input power and antenna power gain, expressed in kilowatts. Abbreviated ERP. { ə¦fek·tiv ,rād·ē,ād·əd 'pau̇·ər }

effective resistance *See* high-frequency resistance. { ə¦fek·tiv ri'zis·təns }

effective sound pressure [ACOUS] The root-mean-square value of the instantaneous sound pressure at a point during a complete cycle, expressed in dynes per square centimeter. Also known as root-mean-square sound pressure; sound pressure. { ə¦fek·tiv 'sau̇nd ,presh·ər }

effective temperature [STAT MECH] That temperature which can be inserted in the Boltzmann distribution formula to describe the relative populations of two energy levels which may or may not be in thermal equilibrium. { ə¦fek·tiv 'tem·prə·chər }

effective value *See* root-mean-square value. { ə¦fek·tiv 'val·yü }

efficiency Abbreviated eff. [PHYS] The ratio, usually expressed as a percentage, of the useful power output to the power input of a device. [THERMO] The ratio of the work done by a heat engine to the heat energy absorbed by it. Also known as thermal efficiency. { ə'fish·ən·sē }

e-folding length [PHYS] The distance over which the amplitude of any exponentially varying quantity increases or decreases by a factor of e (2.718 . . .). { ¦ē ¦fōld·iŋ ,leŋkth }

e-folding time [PHYS] The time required for the amplitude of an oscillation to increase or decrease by a factor of e (2.718. . .). { 'ē ,fōld·iŋ ,tīm }

Egerton's effusion method [THERMO] A method of determining vapor pressures of solids at high temperatures, in which one measures the mass lost by effusion from a sample placed in a tightly sealed silica pot with a small hole; the pot rests at the bottom of a tube that is evacuated for several hours, and is maintained at a high temperature by a heated block of metal surrounding it. { ¦ej·ər·tənz ə'fyü·zhən ,meth·əd }

Ehrenfest's adiabatic law [QUANT MECH] The law that, if the Hamiltonian of a system undergoes an infinitely slow change, and if the system is initially in an eigenstate of the Hamiltonian, then at the end of the change it will be in the eigenstate of the new Hamiltonian that derives from the original state by continuity, provided certain conditions are met. Also known as Ehrenfest's theorem. { 'er·ən,fests ,ad·ē·ə¦bad·ik ¦lȯ }

Ehrenfest's equations [THERMO] Equations which state that for the phase curve P(T) of a second-order phase transition the derivative of pressure P with respect to temperature T is equal to $(Cfp - Cip)/TV(\gamma^f - \gamma^i) = (\gamma^f - \gamma^i)/(K^f - K^i)$, where i and f refer to the two phases, γ is the coefficient of volume expansion, K is the compressibility, C_p is the specific heat at constant pressure, and V is the volume. { 'er·ən,fests i,kwā·zhənz }

Ehrenfest's theorem [QUANT MECH] **1.** The theorem that a quantum-mechanical wave packet obeys the equations of motion of the corresponding classical particle when the position, momentum, and force acting on the particle are replaced by the expectation values of these quantities. **2.** *See* Ehrenfest's adiabatic law. { 'er·ən,fests ,thir·əm }

Ehrenhaft effect [ELECTROMAG] A helical motion of fine particles along the lines of force of a magnetic field during exposure to light, resulting from radiometer effects. { 'er·ən,haft i,fekt }

eht *See* extra-high tension.

E-H T junction [ELECTROMAG] In microwave waveguides, a combination of E- and H-plane T junctions forming a junction at a common point of intersection with the main waveguide. { ¦ē ¦āch 'tē ,jəŋk·shən }

E-H tuner [ELECTROMAG] Tunable E-H T junction having two arms terminated in adjustable plungers used for impedance transformation. { ¦ē ¦āch 'tün·ər }

eigenfrequency [PHYS] One of the frequencies at which an oscillatory system can vibrate. { 'ī·gən,frē·kwən·sē }

eigenstate [QUANT MECH] **1.** A dynamical state whose state vector (or wave function) is an eigenvector (or eigenfunction) of an operator corresponding to a specified physical quantity. **2.** *See* energy state. { 'ī·gən,stāt }

eightfold way [PART PHYS] The classification of hadrons composed of up, down, and strange quarks by SU_3 symmetry; in particular, the eight-dimensional representation of the group SU_3 over its center, which gives rise to unitary octets. { 'āt¦fōld 'wā }

eikonal equation [PHYS] An equation for propagation of electromagnetic or acoustic waves in a nonhomogeneous medium; it is valid only when the variation of the properties of the medium is small over the distance of a wavelength. { ī'kōn·əl i,kwā·zhən }

eikonometer [OPTICS] A scale used to measure sizes of objects viewed through a microscope, usually attached to the eye-piece so that it is seen superimposed on the image. { ī·kə'näm·əd·ər }

einstein [PHYS] A unit of light energy used in photochemistry, equal to Avogadro's number times the energy of one photon of light of the frequency in question. { 'īn,stīn }

Einstein-Bohr equation [QUANT MECH] In a system undergoing a transition between two states so that it emits or absorbs radiation, that equation indicating that the radiation frequency equals the difference in energy between the two states divided by Planck's constant. { 'īn,stīn 'bȯr i,kwā·zhən }

Einstein-Bose statistics *See* Bose-Einstein statistics. { 'īn,stīn 'bōz stə,tis·tiks }

Einstein characteristic temperature [SOLID STATE] A temperature, characteristic of a substance, that appears in Einstein's equation for specific heat; it is equal to the product of Planck's constant and the Einstein frequency divided by Boltzmann's constant. { ¦īn,stīn ,kar·ik·tə¦ris·tik 'tem·prə·chər }

Einstein condensation *See* Bose-Einstein condensation. { 'īn,stīn ,kän·dən'sā·shən }

Einstein-de Haas effect [ELECTROMAG] A freely suspended body consisting of a ferromagnetic material acquires a rotation when its magnetization changes. { 'īn,stīn də'häs i,fekt }

Einstein-de Haas method [ELECTROMAG] Method of measuring the gyromagnetic ratio of a ferromagnetic substance; one measures the angular displacement induced in a ferromagnetic cylinder suspended from a torsion fiber when magnetization of the object is reversed, and the magnetization change is measured with a magnetometer. { 'īn,stīn də'häs ,meth·əd }

Einstein-de Sitter model [RELAT] A model of the universe in which ordinary Euclidean geometry holds good, the distribution of matter extends infinitely at all times, and the universe expands from an infinitely condensed state at such a rate that the density is inversely proportional to the square of the time elapsed since the beginning of the expansion. { 'īn,stīn də'sid·ər ,mäd·əl }

Einstein diffusion equation [STAT MECH] An equation which gives the mean square displacement caused by Brownian movement of spherical, colloidal particles in a gas or liquid. { 'īn,stīn də'fyü·zhən i,kwā·zhən }

Einstein displacement *See* Einstein shift. { 'īn,stīn di,splās·mənt }

Einstein elevator [RELAT] A windowless elevator freely falling in its shaft, inside of which conditions resemble interstellar space; used to elucidate the principle of equivalence. { 'īn,stīn 'el·ə,vād·ər }

Einstein equations [STAT MECH] Equations for the density and pressure of a Bose-Einstein gas in terms of power series in a parameter which appears in the Bose-Einstein distribution law. { 'īn,stīn i,kwā·zhənz }

Einstein frequency [SOLID STATE] Single frequency with which each atom vibrates independently of other atoms, in a model of lattice vibrations; equal to the frequency observed in infrared absorption studies. { 'īn,stīn ,frē·kwən·sē }

Einstein frequency condition [SOLID STATE] The assumption that all vibrations of a crystal lattice are harmonic with the same characteristic frequency. { 'īn,stīn 'frē·kwən·se kən,dish·ən }

Einstein mass-energy relation [RELAT] The relation in which the energy of a system is equivalent to its mass times the square of the speed of light. { 'īn,stīn ¦mas 'en·ər·jē ri,lā·shən }

Einstein number [PL PHYS] A dimensionless number used in magnetofluid dynamics, equal to the ratio of the velocity of a fluid to the speed of light. { 'īn,stīn ,nəm·bər }

Einstein partition function [STAT MECH] The partition function for a solid, based on the Einstein frequency condition. { 'īn,stīn par'tish·ən ,fəŋk·shən }

Einstein photoelectric law [QUANT MECH] The law that the energy of an electron emitted from a system in the photoelectric effect is $h\nu - W$, where h is Planck's constant, ν is the frequency of the incident radiation, and W is the energy needed to remove the electron from the system; if $h\nu$ is less than W, no electrons are emitted. { 'īn,stīn ¦fōd·ō·i¦lek·trik ,lȯ }

Einstein-Planck law [QUANT MECH] The law that the energy of a photon is given by Planck's constant times the frequency. [RELAT] The equation of motion of a charged particle in an electromagnetic field, according to which its rate of change of momentum is equal to the Lorentz force, where the magnitude of the momentum is $mv/(1 - v^2/c^2)^{1/2}$, where m and v are the particle's mass and velocity, and c is the speed of light. { ¦īn,stīn ¦pläŋk ,lȯ }

Einstein-Podolsky-Rosen experiment [QUANT MECH] A Gedanken experiment which was introduced to argue that quantum mechanics is not a complete theory, involving polarization measurements on two photons emitted in opposite directions in an atomic cascade. Abbreviated EPR experiment. { 'īn,stīn pə'däl·skē 'rōz·ənik·,sper·ə·mənt }

Einstein radius [RELAT] The radius of a ring-shaped region through which light is bent by a gravitational lens directly between the light source and the observer. { 'īn,stīn ,rād·ē·əs }

Einstein relation [PHYS] The relation in which the mobility of charges in an ionic solution or semiconductor is equal to the magnitude of the charge times the diffusion coefficient divided by the product of the Boltzmann constant and the absolute temperature. { 'īn,stīn ri,lā·shən }

Einstein-Rosen waves [RELAT] Gravitational waves produced by oscillating ponderable matter, along an infinitely long cylindrical axis, in an exact solution of Einstein's field equations. { ¦īn,stīn ¦rōz·ən ,wāvz }

Einstein's absorption coefficient [ATOM PHYS] The proportionality constant governing the absorption of electromagnetic radiation by atoms, equal to the number of quanta absorbed per second divided by the product of the energy of radiation per unit volume per unit wave number and the number of atoms in the ground state. { 'īn,stīnz əb'sȯrp·shən ,kō·i,fish·ənt }

Einstein's coefficient of spontaneous emission [ATOM PHYS] Proportionality constant governing the rate at which atoms or molecules pass spontaneously from an upper energy state to a lower one by emission of radiation, equal to the number of such transitions per second divided by the number of atoms in the upper state. { ¦īn,stīnz ,kō·ə¦fish·ənt əv spän¦tā·nē·əs ə'mish·ən }

Einstein's coefficient of stimulated emission [ATOM PHYS] Proportionality constant governing the rate at which atoms or molecules pass from an upper energy state to a lower one by stimulated emission of radiation, equal to the number of such transitions per second divided by the product of energy of the radiation inducing the transition per unit volume per unit wave number and the number of atoms in the upper state. { ¦īn,stīnz ,kō·ə¦fish·ənt əv ¦stim·yə,lād·əd ə'mish·ən }

Einstein's equation for specific heat [SOLID STATE] The earliest equation based on quantum mechanics for the specific heat of a solid; uses the assumption that each atom oscillates with the same frequency. { 'īn,stīnz i'kwā·zhən fər spə,si·fik 'hēt }

Einstein's equivalency principle *See* equivalence principle. { 'īn,stīnz i'kwiv·ə·lən·sē ,prin·sə·pəl }

Einstein's field equations [RELAT] Those equations relevant to the relationship in which the Einstein tensor equals -8π times the energy momentum tensor times the gravitational constant divided by the square of the speed of light. Also known as Einstein's law of gravitation. { 'īn,stīnz 'fēld i,kwā·zhənz }

Einstein shift [RELAT] A shift toward longer wavelengths of spectral lines emitted by atoms in strong gravitational fields. Also known as Einstein displacement. { 'īn,stīn ,shift }

Einstein's law of gravitation *See* Einstein's field equations. { 'īn,stīnz ,lȯ əv ,grav·ə'tā·shən }

Einstein's principle of relativity [RELAT] The principle that all the laws of physics must assume the same mathematical form in any inertial frame of reference; thus, it is impossible to determine the absolute motion of a system by any means. { 'īn,stīnz ¦prin·sə·pəl əv rel·ə'tiv·əd·ē }

Einstein static universe [RELAT] A nonvacuum, globally static solution to Einstein's equations of general relativity with cosmological term. { 'īn,stīn 'stad·ik ,yü·nə,vərs }

Einstein's unified field theories [RELAT] A series of theories attempting to express a general unifying principle underlying electromagnetism and gravity. { 'īn,stīnz ,yü·nə,fīd 'fēld ,thē·ə·rēz }

Einstein tensor [RELAT] The tensor expressed as $E_{\mu}\nu = R_{\mu}\nu - {}^1/_2(g_{\mu}\nu R - 2\Lambda)$, where $R_{\mu}\nu$ is the contracted curvature tensor, R is the curvature of space-time, $g_{\mu}\nu$ is the metric tensor, and Λ is the cosmological constant. { 'īn,stīn 'ten·sər }

Einstein universe [RELAT] A model of the universe which is a four-dimensional cylindrical surface in a five-dimensional space. { 'īn,stīn 'yü·nə,vərs }

Ekman sucking [FL MECH] A boundary-layer phenomenon in which fluid near the bottom of a spinning vessel is drawn toward the edge of the vessel along the bottom. { 'ek·mən ,sək·iŋ }

elastance [ELEC] The reciprocal of capacitance. { i'las·təns }

elastica [MECH] The elastic curve formed by a uniform rod that is originally straight, then is bent in a principal plane by applying forces, and couples only at its ends. { i'las·tə·kə }

elastic [MECH] Capable of sustaining deformation without permanent loss of size or shape. { i'las·tik }

elastic aftereffect [MECH] The delay of certain substances in regaining their original shape after being deformed within their elastic limits. Also known as elastic lag. { i'las·tik 'af·tər·i,fekt }

elastic axis [MECH] The lengthwise line of a beam along which transverse loads must be applied in order to produce bending only, with no torsion of the beam at any section. { i'las·tik 'ak·səs }

elastic body [MECH] A solid body for which the additional deformation produced by an increment of stress completely disappears when the increment is removed. Also known as elastic solid. { i'las·tik 'bäd·ē }

elastic buckling [MECH] An abrupt increase in the lateral deflection of a column at a critical load while the stresses acting on the column are wholly elastic. { i'las·tik 'bək·liŋ }

elastic center [MECH] That point of a beam in the plane of the section lying midway between the flexural center and the center of twist in that section. { i'las·tik 'sen·tər }

elastic collision [MECH] A collision in which the sum of the kinetic energies of translation of the participating systems is the same after the collision as before. { i'las·tik kə'lizh·ən }

elastic constant *See* compliance constant; stiffness constant. { i'las·tik 'kän·stənt }

elastic cross section [PHYS] The cross section for an elastic collision between two particles or systems. { i'las·tik 'krȯs ,sek·shən }

elastic curve [MECH] The curved shape of the longitudinal centroidal surface of a beam when the transverse loads acting on it produced wholly elastic stresses. { i'las·tik 'kərv }

elastic deformation [MECH] Reversible alteration of the form or dimensions of a solid body under stress or strain. { i'las·tik ,dē·fər'mā·shən }

elastic equilibrium [MECH] The condition of an elastic body in which each volume element of the body is in equilibrium under the combined effect of elastic stresses and externally applied body forces. { i¦las·tik ˌē·kwə'lib·rē·əm }

elastic failure [MECH] Failure of a body to recover its original size and shape after a stress is removed. { i'las·tik 'fāl·yər }

elastic flow [MECH] Return of a material to its original shape following deformation. { i'las·tik 'flō }

elastic force [MECH] A force arising from the deformation of a solid body which depends only on the body's instantaneous deformation and not on its previous history, and which is conservative. { i'las·tik 'fȯrs }

elastic hysteresis [MECH] Phenomenon exhibited by some solids in which the deformation of the solid depends not only on the stress applied to the solid but also on the previous history of this stress; analogous to magnetic hysteresis, with magnetic field strength and magnetic induction replaced by stress and strain respectively. { i'las·tik ˌhis·tə'rē·səs }

elasticity [MECH] **1.** The property whereby a solid material changes its shape and size under action of opposing forces, but recovers its original configuration when the forces are removed. **2.** The existence of forces which tend to restore to its original position any part of a medium (solid or fluid) which has been displaced. { iˌlas'tis·əd·ē }

elasticity modulus *See* modulus of elasticity. { iˌlas'tis·əd·ē ˌmäj·ə·ləs }

elasticity number 1 [FL MECH] A dimensionless number which is a measure of the ratio of elastic forces to inertial forces on a viscoelastic fluid flowing in a pipe, and is equal to the product of the fluid's relaxation time and its dynamic viscosity, divided by the product of the fluid's density and the square of the radius of the pipe. Symbolized N_{El}. { iˌlas'tis·əd·ē ¦nəm·bər ¦wən }

elasticity number 2 [FL MECH] A dimensionless number used in studying the effect of elasticity on a flow process, equal to the fluid's density times its specific heat at constant pressure, divided by the product of its coefficient of bulk expansion and its bulk modulus. Symbolized $\bar{K}_E$. { iˌlas'tis·əd·ē ¦nəm·bər ¦tü }

elastic lag *See* elastic aftereffect. { i'las·tik 'lag }

elastic limit [MECH] The maximum stress a solid can sustain without undergoing permanent deformation. { iˌlas'tis·tik 'lim·ət }

elastic-limit effect [SOLID STATE] A phenomenon in which a material acquires a sharp elastic limit because of the introduction of foreign atoms. { i¦las·tik ¦lim·ət iˌfekt }

elastic modulus *See* modulus of elasticity. { iˌlas·tik 'mäj·ə·ləs }

elasticoviscosity [FL MECH] That property of a fluid whose rate of deformation under stress is the sum of a part corresponding to a viscous Newtonian fluid and a part obeying Hooke's law. { i¦las·tə·kō·vis'käs·əd·ē }

elastic potential energy [MECH] Capacity that a body has to do work by virtue of its deformation. { i'las·tik pə¦ten·chəl ¦en·ər·jē }

elastic ratio [MECH] The ratio of the elastic limit to the ultimate strength of a solid. { i'las·tik 'rā·shō }

elastic recovery [MECH] That fraction of a given deformation of a solid which behaves elastically. { i'las·tik ri'kəv·ə·rē }

elastic scattering [MECH] Scattering due to an elastic collision. { i'las·tik 'skad·ə·riŋ }

elastic solid *See* elastic body. { i'las·tik 'säl·əd }

elastic strain energy [MECH] The work done in deforming a solid within its elastic limit. { i'las·tik 'strān ˌen·ər·jē }

elastic theory [MECH] Theory of the relations between the forces acting on a body and the resulting changes in dimensions. { i'las·tik 'thē·ə·rē }

elastic vibration [MECH] Oscillatory motion of a solid body which is sustained by elastic forces and the inertia of the body. { i'las·tik vī'brā·shen }

elastic wave [ACOUS] *See* acoustic wave. [PHYS] A wave propagated by a medium having inertia and elasticity (the existence of forces which tend to restore any part of a medium to its original position), in which displaced particles transfer momentum to adjoining particles, and are themselves restored to their original position. { i'las·tik 'wāv }

elastodynamics [MECH] The study of the mechanical properties of elastic waves. { i¦la·stō·dī¦nam·iks }

elastoplasticity [MECH] State of a substance subjected to a stress greater than its elastic limit but not so great as to cause it to rupture, in which it exhibits both elastic and plastic properties. { i¦las·tō·plə'stis·əd·ē }

elastoresistance [ELEC] The change in a material's electrical resistance as it undergoes a stress within its elastic limit. { i¦las·tō·ri'zis·təns }

elbow [ELECTROMAG] In a waveguide, a bend of comparatively short radius, normally 90°, and sometimes for acute angles down to 15°. { 'elˌbō }

ELDOR *See* electron electron double resonance. { 'elˌdȯr *or* ¦ē¦el¦dē¦ō'är }

electret [ELEC] A solid dielectric possessing persistent electric polarization, by virtue of a long time constant for decay of a charge instability. { i'lekˌtret }

electric [ELEC] Containing, producing, arising from, or actuated by electricity; often used interchangeably with electrical. { i'lek·trik }

electrical [ELEC] Related to or associated with electricity, but not containing it or having its properties or characteristics; often used interchangeably with electric. { ə'lek·trə·kəl }

electrical analog [PHYS] An electric circuit whose behavior may be described by the same mathematical equations as some physical system under study. { ə'lek·trə·kəl 'an·əˌläg }

electrical angle [ELEC] An angle that specifies a particular instant in an alternating-current cycle or expresses the phase difference between two alternating quantities; usually expressed in electrical degrees. { ə'lek·trə·kəl 'aŋ·gəl }

electrical axis [SOLID STATE] The *x* axis in a quartz crystal; there are three such axes in a crystal, each parallel to one pair of opposite sides of the hexagon; all pass through and are perpendicular to the optical, or *z*, axis. { ə'lek·trə·kəl 'ak·səs }

electrical breakdown *See* breakdown. { ə'lek·trə·kəl 'brāk,daủn }

electrical center [ELEC] Point approximately midway between the ends of an inductor or resistor that divides the inductor or resistor into two equal electrical values. { ə'lek·trə·kəl 'sen·tər }

electrical circuit theory *See* circuit theory. { ə'lek·trə·kəl 'sər·kət ,thē·ə·rē }

electrical conductance *See* conductance. { ə'lek·trə·kəl kən'dək·təns }

electrical conduction *See* conduction. { ə'lek·trə·kəl kən'dək·shən }

electrical conductivity *See* conductivity. { ə'lek·trə·kəl ,kän,dək'tiv·əd·ē }

electrical degree [ELEC] A unit equal to 1/360 cycle of an alternating quantity. { i'lek·trə·kəl də'grē }

electrical distance [ELECTROMAG] The distance between two points, expressed in terms of the duration of travel of an electromagnetic wave in free space between the two points. { i'lek·trə·kəl 'dis·təns }

electrical fault *See* fault. { i'lek·trə·kəl 'fȯlt }

electrical impedance Also known as impedance. [ELEC] **1.** The total opposition that a circuit presents to an alternating current, equal to the complex ratio of the voltage to the current in complex notation. Also known as complex impedance. **2.** The ratio of the maximum voltage in an alternating-current circuit to the maximum current; equal to the magnitude of the quantity in the first definition. { i'lek·trə·kəl im'pēd·əns }

electrical insulation *See* insulation. { i'lek·trə·kəl 'in·sə'lā·shən }

electrical insulator *See* insulator. { i'lek·trə·kəl 'in·sə,lād·ər }

electrical length [ELECTROMAG] The length of a conductor expressed in wavelengths, radians, or degrees. { i'lek·trə·kəl 'leŋkth }

electrically connected [ELEC] Connected by means of a conducting path, or through a capacitor, as distinguished from connection merely through electromagnetic induction. { i'lek·trə·klē kə'nek·təd }

electrical measurement [ELEC] The measurement of any one of the many quantities by which electricity is characterized. { i'lek·trə·kəl 'mezh·ər·mənt }

electrical potential energy [ELEC] Energy possessed by electric charges by virtue of their position in an electrostatic field. { i'lek·trə·kəl pə'ten·chəl'en·ər·jē }

electrical properties [ELEC] Properties of a substance which determine its response to an electric field, such as its dielectric constant or conductivity. { i'lek·trə·kəl 'präp·ərd·ēz }

electrical resistance *See* resistance. { i'lek·trə·kəl ri'zis·təns }

electrical resistivity [ELEC] The electrical resistance offered by a material to the flow of current, times the cross-sectional area of current flow and per unit length of current path; the reciprocal of the conductivity. Also known as resistivity; specific resistance. { i'lek·trə·kəl ,rē·zis'tiv·əd·ē }

electrical resistor *See* resistor. { i'lek·trə·kəl ri'zis·tər }

electrical symbol [ELEC] A simple geometrical symbol used to represent a component of a circuit in a schematic circuit diagram. { i'lek·trə·kəl 'sim·bəl }

electrical unit [ELEC] A standard in terms of which some electrical quantity is evaluated. { i'lek·trə·kəl 'yü·nət }

electric arc [ELEC] A discharge of electricity through a gas, normally characterized by a voltage drop approximately equal to the ionization potential of the gas. Also known as arc. { i¦lek·trik 'ärk }

electric charge *See* charge. { i¦lek·trik 'chärj }

electric chopper [ELECTROMAG] A chopper in which an electromagnet driven by a source of alternating current sets into vibration a reed carrying a moving contact that alternately touches two fixed contacts in a signal circuit, thus periodically interrupting the signal. { i¦lek·trik 'chäp·ər }

electric circuit [ELEC] Also known as circuit. **1.** A path or group of interconnected paths capable of carrying electric currents. **2.** An arrangement of one or more complete, closed paths for electron flow. { i¦lek·trik 'sər·kət }

electric circuit theory *See* circuit theory. { i¦lek·trik 'sər·kət ,thē·ə·rē }

electric coil *See* coil. { i¦lek·trik 'kȯil }

electric condenser *See* capacitor. { i¦lek·trik kən'den·sər }

electric conductor *See* conductor. { i¦lek·trik kən'dək·tər }

electric connection [ELEC] A direct wire path for current between two points in a circuit. { i¦lek·trik kə'nek·shən }

electric constant [ELEC] The permittivity of empty space, equal to 1 in centimeter-gram-second electrostatic units and to $10^7/4\pi c^2$ farads per meter or, numerically, to 8.854×10^{-12} farad per meter in International System units, where c is the speed of light in meters per second. Symbolized ϵ_0. { i¦lek·trik 'kän·stənt }

electric contact [ELEC] A physical contact that permits current flow between conducting parts. Also known as contact. { i¦lek·trik 'kän,takt }

electric corona *See* corona discharge. { i¦lek·trik kə'rō·nə }

electric current *See* current. { i¦lek·trik 'kə·rənt }

electric current density *See* current density. { i¦lek·trik ¦kə·rənt ,den·səd·ē }

electric dipole [ELEC] A localized distribution of positive and negative electricity, without net charge, whose mean positions of positive and negative charges do not coincide. { i¦lek·trik 'dī,pōl }

electric dipole moment [ELEC] A quantity characteristic of a charge distribution, equal to the vector sum over the electric charges of the product of the charge and the position vector of the charge. { i¦lek·trik 'dī,pōl ,mō·mənt }

electric dipole transition [ATOM PHYS] A transition of an atom or nucleus from one energy state to another, in which electric dipole radiation is emitted or absorbed. { i¦lek·trik 'dī,pōl tran 'zish·ən }

electric displacement [ELEC] The electric field intensity multiplied by the permittivity. Symbolized D. Also known as dielectric displacement; dielectric flux density; displacement; electric displacement density; electric flux density; electric induction. { i'lek·trik dis'plās·mənt }

electric displacement density *See* electric displacement. { i'lek·trik dis'plās·mənt ,den·səd·ē }

electric doublet *See* dipole. { i¦lek·trik 'dəb·lət }

electric energy [ELECTROMAG] **1.** Energy of electric charges by virtue of their position in an electric field. **2.** Energy of electric currents by virtue of their position in a magnetic field. { i¦lek·trik 'en·ər·jē }

electric energy measurement [ELEC] The measurement of the integral, with respect to time, of the power in an electric circuit. { i¦lek·trik 'en·ər·jē ,mezh·ər·mənt }

electric field [ELEC] **1.** One of the fundamental fields in nature, causing a charged body to be attracted to or repelled by other charged bodies; associated with an electromagnetic wave or a changing magnetic field. **2.** Specifically, the electric force per unit test charge. { i¦lek·trik 'fēld }

electric-field effect *See* Stark effect. { i¦lek·trik ¦fēld i'fekt }

electric-field intensity *See* electric-field vector. { i¦lek·trik ¦fēld in'ten·səd·ē }

electric-field strength *See* electric-field vector. { i¦lek·trik ¦fēld 'streŋkth }

electric-field vector [ELEC] The force on a stationary positive charge per unit charge at a point in an electric field. Designated **E**. Also known as electric-field intensity; electric-field strength; electric vector. { i¦lek·trik ¦fēld 'vek·tər }

electric flux [ELEC] **1.** The integral over a surface of the component of the electric displacement perpendicular to the surface; equal to the number of electric lines of force crossing the surface. **2.** The electric lines of force in a region. { i¦lek·trik 'fləks }

electric flux density *See* electric displacement. { i¦lek·trik 'fləks ,den·səd·ē }

electric flux line *See* electric line of force. { i¦lek·trik 'fləks ,līn }

electric hysteresis *See* ferroelectric hysteresis. { i¦lek·trik ,his·tə'rē·səs }

electric image [ELEC] A fictitious charge used in finding the electric field set up by fixed electric charges in the neighborhood of a conductor; the conductor, with its distribution of induced surface charges, is replaced by one or more of these fictitious charges. Also known as image. { i¦lek·trik 'im·ij }

electric induction *See* electric displacement. { i¦lek·trik in'dək·shən }

electricity [PHYS] Physical phenomenon involving electric charges and their effects when at rest and when in motion. { i,lek'tris·əd·ē }

electric line of force [ELEC] An imaginary line drawn so that each segment of the line is parallel to the direction of the electric field or of the electric displacement at that point, and the density of the set of lines is proportional to the electric field or electrical displacement. Also known as electric flux line. { i¦lek·trik ¦līn əv 'fȯrs }

electric-magnetic duality [PART PHYS] The property of a physical theory in which the electric and magnetic charges exchange roles when the coupling between electric charges and electric fields, which is actually small, is made to be large, with electric charges becoming fuzzy, heavy, and strongly coupled, while the magnetic charges become pointlike, light, and weakly coupled. { i¦lek·trik mag¦ned·ik dü'al·əd·ē }

electric moment [ELEC] One of a series of quantities characterizing an electric charge distribution; an l-th moment is given by integrating the product of the charge density, the l-th power of the distance from the origin, and a spherical harmonic Y^*_{lm} over the charge distribution. { i¦lek·trik 'mō·mənt }

electric monopole [ELEC] A distribution of electric charge which is concentrated at a point or is spherically symmetric. { i¦lek·trik 'män·ə,pōl }

electric multipole [ELECTROMAG] One of a series of types of static or oscillating charge distributions; the multipole of order 1 is a point charge or a spherically symmetric distribution, and the electric and magnetic fields produced by an electric multipole of order 2^n are equivalent to those of two electric multipoles of order 2^{n-1} of equal strengths, but opposite sign, separated from each other by a short distance. { i¦lek·trik 'məl·tə,pōl }

electric multipole field [ELECTROMAG] The electric and magnetic fields generated by a static or oscillating electric multipole. { i¦lek·trik 'məl·tə,pōl ,fēld }

electric octupole moment [ELEC] A quantity characterizing an electric charge distribution; obtained by integrating the product of the charge density, the third power of the distance from the origin, and a spherical harmonic Y^*_{3m} over the charge distribution. { i¦lek·trik 'äk·tə,pōl 'mō·mənt }

electric polarizability [ELEC] Induced dipole moment of an atom or molecule in a unit electric field. { i¦lek·trik ,pō·lə,rī·zə'bil·əd·ē }

electric polarization *See* polarization. { i¦lek·trik ,pō·lə·rə'zā·shən }

electric potential [ELEC] The work which must be done against electric forces to bring a unit charge from a reference point to the point in question; the reference point is located at an infinite distance, or, for practical purposes, at the surface of the earth or some other large conductor. Also known as electrostatic potential;

potential. Abbreviated V. { i¦lek·trik pə'ten·chəl }

electric power [ELEC] The rate at which electric energy is converted to other forms of energy, equal to the product of the current and the voltage drop. { i¦lek·trik 'pau̇·ər }

electric probe [PL PHYS] A device used to measure electron temperatures, electron and ion densities, space and wall potentials, and random electron currents in a plasma; consists substantially of one or two small collecting electrodes to which various potentials are applied, with the corresponding collection currents being measured. Also known as electrostatic probe. { i¦lek·trik 'prōb }

electric quadrupole [ELEC] A charge distribution that produces an electric field equivalent to that produced by two electric dipoles whose dipole moments have the same magnitude but point in opposite directions and which are separated from each other by a small distance. { i¦lek·trik 'kwä·drə,pōl }

electric quadrupole moment [ELEC] A quantity characterizing an electric charge distribution, obtained by integrating the product of the charge density, the second power of the distance from the origin, and a spherical harmonic Y^*_{2m} over the charge distribution. { i¦lek·trik 'kwä·drə,pōl ,mō·mənt }

electric quadrupole transition [ATOM PHYS] A transition of an atom or molecule from one energy state to another, in which electric quadrupole radiation is emitted or absorbed. { i¦lek·trik 'kwä·drə,pōl tran'zish·ən }

electric reactor *See* reactor. { i¦lek·trik rē'ak·tər }

electric shielding [ELECTROMAG] Any means of avoiding pickup of undesired signals or noise, suppressing radiation of undesired signals, or confining wanted signals to desired paths or regions, such as electrostatic shielding or electromagnetic shielding. Also known as screening; shielding. { i¦lek·trik 'shēld·iŋ }

electric shock tube [PL PHYS] A gas-filled tube used in plasma physics to ionize a gas suddenly; a capacitor bank charged to a high voltage is discharged into the gas at one tube end to ionize and heat the gas, producing a shock wave that may be studied as it travels down the tube. { i¦lek·trik 'shäk ,tüb }

electric shunt *See* shunt. { i¦lek·trik 'shənt }

electric solenoid *See* solenoid. { i¦lek·trik 'sō·lə,nȯid }

electric spark *See* spark. { i¦lek·trik 'spärk }

electric strength *See* dielectric strength. { i¦lek·trik 'streŋkth }

electric susceptibility [ELEC] A dimensionless parameter measuring the ease of polarization of a dielectric, equal (in meter-kilogram-second units) to the ratio of the polarization to the product of the electric field strength and the vacuum permittivity. Also known as dielectric susceptibility. { i¦lek·trik sə,sep·tə'bil·əd·ē }

electric transient [ELEC] A temporary component of current and voltage in an electric circuit which has been disturbed. { i¦lek·trik 'tran·zhənt }

electric twinning [SOLID STATE] A defect occurring in natural quartz crystals, in which adjacent regions of quartz have their electric axes oppositely poled. { i¦lek·trik 'twin·iŋ }

electric vector *See* electric-field vector. { i¦lek·trik 'vek·tər }

electric wave [ELECTROMAG] An electromagnetic wave, especially one whose wavelength is at least a few centimeters. Also known as Hertzian wave. { i¦lek·trik 'wāv }

electric wire *See* wire. { i¦lek·trik 'wīr }

electrization [ELEC] The electric polarization divided by the permittivity of empty space. { i,lek·trə'zā·shən }

electrocaloric effect [SOLID STATE] A temperature change in certain crystals caused by alteration of the permanent polarization by application of an external electric field. { i¦lek·trō·kə¦lȯr·ik i'fekt }

electrocapillarity [PHYS] A change in the surface tension of a liquid caused by an electric field at the surface. { i,lek·trō,kap·ə'lar·əd·ē }

electrochemical thermodynamics [THERMO] The application of the laws of thermodynamics to electrochemical systems. { i,lek·trō'kem·ə·kəl ,thərm·ō·dī'nam·iks }

electrode [ELEC] **1.** An electric conductor through which an electric current enters or leaves a medium, whether it be an electrolytic solution, solid, molten mass, gas, or vacuum. **2.** One of the terminals used in dielectric heating or diathermy for applying the high-frequency electric field to the material being heated. { i'lek,trōd }

electrode couple [ELEC] The pair of electrodes in an electric cell, between which there is a potential difference. { i'lek,trōd ,kə·pəl }

electrodisintegration [NUC PHYS] The breakup of a nucleus into two or more fragments as a result of bombardment by electrons. { i¦lek·trō·dis,int·ə'grā·shən }

electrodynamics [ELECTROMAG] The study of the relations between electrical, magnetic, and mechanical phenomena. { i,lek·trō·dī'nam·iks }

electroendosmosis [PHYS] The production of an endosmosis effect by an electrical potential; that is, the use of electricity to cause diffusion of a liquid through an organic membrane. { i,lek·trō¦en·däs'mō·səs }

electrofluid [FL MECH] Newtonian (or shear-thinning) fluid whose rheological or flow properties are changed into those of a viscoplastic type by the addition of electric-field modulation. { i,lek·trō'flü·əd }

electrogasdynamics [PHYS] Conversion of the kinetic energy of a moving gas to electricity, for such applications as high-voltage electric power generation, air-pollution control, and paint-spraying. { i¦lek·trō,gas·dī'nam·iks }

electrohydrodynamic ionization mass spectroscopy [SPECT] A technique for analysis of nonvolatile molecules in which the nonvolatile material is dissolved in a volatile solvent with a high dielectric constant such as glycerol, and high electric-field gradients at the surface of droplets of the liquid solution induce ion emission. { i¦lek·trō¦hī·drō·dī'nam·ik ˌī·ə·nə'zā·shən ¦mas spek'träs·kə·pē }

electrokinetic potential *See* zeta potential. { i¦lek·trō·kə'ned·ik pə'ten·chəl }

electrokinetics [ELECTROMAG] The study of the motion of electric charges, especially of steady currents in electric circuits, and of the motion of electrified particles in electric or magnetic fields. { i¦lek·trō·kə'ned·iks }

electrolytic capacitor [ELEC] A capacitor consisting of two electrodes separated by an electrolyte; a dielectric film, usually a thin layer of gas, is formed on the surface of one electrode. Also known as electrolytic condenser. { i'lek·trəˌlid·ik kə'pas·əd·ər }

electrolytic condenser *See* electrolytic capacitor. { i'lek·trəˌlid·ik kən'den·sər }

electromagnet [ELECTROMAG] A magnet consisting of a coil wound around a soft iron or steel core; the core is strongly magnetized when current flows through the coil, and is almost completely demagnetized when the current is interrupted. { i¦lek·trō'mag·nət }

electromagnetic [PHYS] Pertaining to phenomena in which electricity and magnetism are related. { i¦lek·trō·mag'ned·ik }

electromagnetic amplifying lens [ELECTROMAG] Large numbers of waveguides symmetrically arranged with respect to an excitation medium in order to become excited with equal amplitude and phase to provide a net gain in energy. { i¦lek·trō·mag'ned·ik 'am·pləˌfī·iŋ ˌlenz }

electromagnetic complex [ELECTROMAG] Electromagnetic configuration of an installation, including all significant radiators of energy. { i¦lek·trō·mag'ned·ik 'kämˌpleks }

electromagnetic constant *See* speed of light. { i¦lek·trō·mag'ned·ik 'kän·stənt }

electromagnetic coupling [ELECTROMAG] Coupling that exists between circuits when they are mutually affected by the same electromagnetic field. { i¦lek·trō·mag'ned·ik 'kəp·liŋ }

electromagnetic damping [ELEC] Retardation of motion that results from the reaction between eddy currents in a moving conductor and the magnetic field in which it is moving. { i¦lek·trō·mag'ned·ik 'damp·iŋ }

electromagnetic energy [ELECTROMAG] The energy associated with electric or magnetic fields. { i¦lek·trō·mag'ned·ik 'en·ər·jē }

electromagnetic field [ELECTROMAG] An electric or magnetic field, or a combination of the two, as in an electromagnetic wave. { i¦lek·trō·mag'ned·ik 'fēld }

electromagnetic field equations *See* Maxwell field equations. { i¦lek·trō·mag'ned·ik 'fēld iˌkwā·zhənz }

electromagnetic field tensor [ELECTROMAG] An antisymmetric, second-rank Lorentz tensor, whose elements are proportional to the electric and magnetic fields; the Maxwell field equations can be expressed in a simple form in terms of this tensor. { i¦lek·trō·mag'ned·ik 'fēld ˌten·sər }

electromagnetic horn *See* horn antenna. { i¦lek·trō·mag'ned·ik 'hȯrn }

electromagnetic induction [ELECTROMAG] The production of an electromotive force either by motion of a conductor through a magnetic field so as to cut across the magnetic flux or by a change in the magnetic flux that threads a conductor. Also known as induction. { i¦lek·trō·mag'ned·ik in'dək·shən }

electromagnetic inertia [ELECTROMAG] **1.** Characteristic delay of a current in an electric circuit in reaching its maximum value, or in returning to zero, after the source voltage has been removed or applied. **2.** The property of a circuit whereby variation of the current in the circuit gives rise to a voltage in the circuit. { i¦lek·trō·mag'ned·ik i'nər·shə }

electromagnetic interaction [PART PHYS] The interaction of elementary particles that results from the coupling of charge to the electromagnetic field. { i¦lek·trō·mag'ned·ik ˌint·ə'rak·shən }

electromagnetic mass [ELECTROMAG] The contribution to the mass of an object from its electric and magnetic field energy. { i¦lek·trō·mag'ned·ik 'mas }

electromagnetic mirror [ELECTROMAG] Surface or region capable of reflecting radio waves, such as one of the ionized layers in the upper atmosphere. { i¦lek·trō·mag'ned·ik 'mir·ər }

electromagnetic moment [ELECTROMAG] The magnetic moment of a current-carrying coil, equal to the product of the current, the number of turns, and the area of the coil. { i¦lek·trō·mag¦ned·ik 'mō·mənt }

electromagnetic momentum [ELECTROMAG] The momentum transported by electromagnetic radiation; its volume density equals the Poynting vector divided by the square of the speed of light. { i¦lek·trō·mag'ned·ik mə'men·təm }

electromagnetic oscillograph [ELECTROMAG] An oscillograph in which the recording mechanism is controlled by a moving-coil galvanometer, such as a direct-writing recorder or a light-beam oscillograph. { i¦lek·trō·mag'ned·ik ä'sil·əˌgraf }

electromagnetic potential [ELECTROMAG] Collective name for a scalar potential, which reduces to the electrostatic potential in a time-independent system, and the vector potential for the magnetic field; the electric and magnetic fields can be written in terms of these potentials. { i¦lek·trō·mag'ned·ik pə'ten·chəl }

electromagnetic properties [ELECTROMAG] The response of materials or equipment to electromagnetic fields, and their ability to produce such fields. { i¦lek·trō·mag'ned·ik 'präp·ərd·ēz }

electromagnetic radiation [ELECTROMAG] Electromagnetic waves and, especially, the associated electromagnetic energy. { i¦lek·trō·mag'ned·ik ˌrād·ē'ā·shən }

electromagnetic scattering [PHYS] The process in which energy is removed from a beam of electromagnetic radiation and reemitted without appreciable changes in wavelength. { i¦lek·trō·mag'ned·ik 'skad·ə·riŋ }

electromagnetic separator [ELECTROMAG] Device in which ions of varying mass are separated by a combination of electric and magnetic fields. { i¦lek·trō·mag'ned·ik 'sep·əˌrād·ər }

electromagnetic shielding [ELECTROMAG] Means, similar to electrostatic or magnetostatic shielding, for suppressing changing magnetic fields or electromagnetic radiation at a device. { i¦lek·trō·mag'ned·ik 'shēld·iŋ }

electromagnetic shock wave [ELECTROMAG] Electromagnetic wave of great intensity which results when waves with different intensities propagate with different velocities in a nonlinear optical medium, and faster-traveling waves from a pulse of light catch up with preceding, slower traveling waves. { i¦lek·trō·mag'ned·ik 'shäk ˌwāv }

electromagnetic spectrum [ELECTROMAG] The total range of wavelengths or frequencies of electromagnetic radiation, extending from the longest radio waves to the shortest known cosmic rays. { i¦lek·trō·mag'ned·ik 'spek·trəm }

electromagnetic system of units [ELECTROMAG] A centimeter-gram-second system of electric and magnetic units in which the unit of current is defined as the current which, if maintained in two straight parallel wires having infinite length and being 1 centimeter apart in vacuum, would produce between these conductors a force of 2 dynes per centimeter of length; other units are derived from this definition by assigning unit coefficients in equations relating electric and magnetic quantities. Also known as electromagnetic units (emu). { i¦lek·trō·mag'ned·ik ¦sis·təm əv 'yü·nəts }

electromagnetic theory of light [ELECTROMAG] Theory according to which light is an electromagnetic wave whose electric and magnetic fields obey Maxwell's equations. { i¦lek·trō·mag'ned·ik ¦thē·ə·rē əv 'līt }

electromagnetic units *See* electromagnetic system of units. { i¦lek·trō·mag'ned·ik 'yü·nəts }

electromagnetic wave [ELECTROMAG] A disturbance which propagates outward from any electric charge which oscillates or is accelerated; far from the charge it consists of vibrating electric and magnetic fields which move at the speed of light and are at right angles to each other and to the direction of motion. { i¦lek·trō·mag'ned·ik 'wāv }

electromagnetic-wave filter [ELECTROMAG] Any device to transmit electromagnetic waves of desired frequencies while substantially attenuating all other frequencies. { i¦lek·trō·mag'ned·ik ¦wāv ˌfil·tər }

electromagnetism [PHYS] **1.** Branch of physics relating electricity to magnetism. **2.** Magnetism produced by an electric current rather than by a permanent magnet. { i¦lek·trō'mag·nəˌtiz·əm }

electromechanical coupling coefficient [SOLID STATE] The ratio of the mutual elastodielectric energy density in a piezoelectric material to the square root of the product of the stored elastic and dielectric energy densities. { iˌlek·trō·mi ¦kan·ə·kəl 'kəp·liŋ ˌkō·əˌfish·ənt }

electromodulation [SPECT] Modulation spectroscopy in which changes in transmission or reflection spectra induced by a perturbing electric field are measured. { i¦lek·trōˌmäj·ə'lā·shən }

electron [PHYS] **1.** A stable elementary particle which is the negatively charged constituent of ordinary matter, having a mass of about 9.11×10^{-28} gram (equivalent to 0.511 MeV), a charge of about -1.602×10^{-19} coulomb, and a spin of 1/2. Also known as negative electron; negatron. **2.** Collective name for the electron, as in the first definition, and the positron. { i'lekˌträn }

electron acceptor *See* acceptor. { i'lekˌträn ak'-sep·tər }

electron-acoustic microscopy [PHYS] A technique for producing images that show variations in an object's thermal and elastic properties; an electron beam generates ultrasonic waves in the specimen which are detected by a piezoelectric transducer whose output controls the brightness of a spot sweeping a cathode-ray tube in synchronism with the electron beam. { i'lekˌträn ə'küs·tik mī'kräs·kə·pē }

electron affinity [ATOM PHYS] The work needed in removing an electron from a negative ion, thus restoring the neutrality of an atom or molecule. { i'lekˌträn ə'fin·əd·ē }

electron attachment [ATOM PHYS] The combination of an electron with a neutral atom or molecule to form a negative ion. *See* electron capture. { i'lekˌträn ə'tach·mənt }

electron-beam laser [OPTICS] A semiconductor laser in which the electron beam that provides pumping action in a thin plate of cadmium sulfide or other material is swept electrically in two dimensions by a deflection yoke, much as in a cathode-ray tube. { i'lekˌträn ˌbēm 'lā·zər }

electron capture [ATOM PHYS] The process in which an atom or ion passing through a material medium either loses or gains one or more orbital electrons. [NUC PHYS] A radioactive transformation of nuclide in which a bound electron merges with its nucleus. Also known as electron attachment. { i'lekˌträn 'kap·chər }

electron charge [PHYS] The charge carried by an electron, equal to about -1.602×10^{-19} coulomb, or -4.803×10^{-10} statcoulomb. { i'lek ˌträn ˌchärj }

electron cloud [ATOM PHYS] Picture of an electron state in which the charge is thought of as being smeared out, with the resulting charge density distribution corresponding to the probability distribution function associated with the Schrödinger wave function. { i'lekˌträn ˌklau̇d }

electron conduction [ELEC] Conduction of electricity resulting from motion of electrons, rather than from ions in a gas or solution, or holes in a solid. [THERMO] The transport of energy in highly ionized matter primarily by electrons of relatively high temperature moving in one direction and electrons of lower temperature moving in the other. { i'lek,trän kən,dək·shən }

electron configuration [ATOM PHYS] The orbital and spin arrangement of an atom's electrons, specifying the quantum numbers of the atom's electrons in a given state. { i'lek,trän kən,fig·yə'rā·shən }

electron correlation [ATOM PHYS] The difference between the actual wave function of an atomic system and the wave function in the Hartree-Fock approximation. Also known as correlation. { i¦lek,trän ,kä·rə'ā·shən }

electron cyclotron resonance [PHYS] Resonance absorption of energy from a radio-frequency or microwave-frequency electromagnetic field by electrons in a uniform magnetic field when the frequency of the electromagnetic field equals the cyclotron frequency of the electrons. { i¦lek,trän ¦sī·klə,trän 'rez·ə·nəns }

electron cyclotron wave [PL PHYS] A wave in a plasma which propagates parallel to the magnetic field produced by currents outside the plasma at frequencies less than that of the electron cyclotron resonance, and which is circularly polarized, rotating in the same sense as electrons in the plasma; responsible for whistlers. Also known as whistler wave. { i'lek,trän 'sī·klə,trän ,wāv }

electron density [PHYS] **1.** The number of electrons in a unit volume. **2.** When quantum-mechanical effects are significant, the total probability of finding an electron in a unit volume. { i'lek,trän 'den·səd·ē }

electron diffraction [PHYS] The phenomenon associated with the interference processes which occur when electrons are scattered by atoms in crystals to form diffraction patterns. { i'lek,trän di'frak·shən }

electron diffraction analysis [PHYS] Examination of solid surfaces by observing the diffraction of a stream of electrons by the surface. { i'lek ,trän di'frak·shən ə,nal·ə·səs }

electron diffraction camera [OPTICS] A camera used to obtain a photographic record of the position and intensity of the diffracted beams produced when a specimen is irradiated by a beam of electrons. { i'lek,trän di'frak·shən ,kam·rə }

electron diffractograph [PHYS] A device, allied to the electron microscope, in which a beam of electrons strikes the sample, showing crystal pattern and other physical attributes on the resulting diffraction pattern; used for chemical analysis, atomic structure determination, and so on. { i'lek,trän di'frak·tə,graf }

electron dipole moment *See* electron magnetic moment. { i'lek,trän 'dī,pōl ,mō·mənt }

electron distribution [PHYS] A function which gives the number of electrons per unit volume of phase space. { i'lek,trän dis·trə'byü·shən }

electron donor *See* donor. { i'lek,trän ,dō·nər }

electronegative [ELEC] **1.** Carrying a negative electric charge. **2.** Capable of acting as the negative electrode in an electric cell. { i¦lek·trō 'neg·əd·iv }

electron-electron double resonance [SPECT] A type of electron paramagnetic resonance (EPR) spectroscopy in which a material is irradiated at two different microwave frequencies, and the changes in the EPR spectrum resulting from sweeping either the second frequency or the magnetic field are monitored through detection at the first frequency. Abbreviated ELDOR. { i,lek,trän i¦lek,tran ,dəb·əl 'rez·ən·əns }

electron energy level [ATOM PHYS] A quantum-mechanical concept for energy levels of electrons about the nucleus; electron energies are functions of each particular atomic species. { i'lek ,trän 'en·ər·jē ,lev·əl }

electron energy loss spectroscopy [SPECT] A technique for studying atoms, molecules, or solids in which a substance is bombarded with monochromatic electrons, and the energies of scattered electrons are measured to determine the distribution of energy loss. Abbreviated EELS. { i'lek,trän 'en·ər·jē ,lȯs spek'träs·kə·pē }

electron flow [ELEC] A current produced by the movement of free electrons toward a positive terminal; the direction of electron flow is opposite to that of current. { i'lek,trän ,flō }

electron gas [PHYS] A concentration of electrons whose behavior, is, in first approximation, not governed by forces. { i'lek,trän ,gas }

electron hole *See* hole. { i'lek,trän ¦hōl }

electron-hole droplets [SOLID STATE] A form of electronic excitation observed in germanium and silicon at sufficiently low cryogenic temperatures; it is associated with a liquid-gas phase transition of the charge carriers, and consists of regions of conducting electron-hole Fermi liquid coexisting with regions of insulating exciton gas. { i'lek,trän ¦hōl 'dräp·ləts }

electron-hole recombination [SOLID STATE] The process in which an electron, which has been excited from the valence band to the conduction band of a semiconductor, falls back into an empty state in the valence band, which is known as a hole. { i'lek,trän 'hōl rē,käm·bə'nā·shən }

electronic absorption spectrum [SPECT] Spectrum resulting from absorption of electromagnetic radiation by atoms, ions, and molecules due to excitations of their electrons. { i,lek 'trän·ik əb'sȯrp·shən ,spek·trəm }

electronic angular momentum [ATOM PHYS] The total angular momentum associated with the orbital motion of the spins of all the electrons of an atom. { i,lek'trän·ik 'aŋ·gyə·lər mə'ment·əm }

electronic band spectrum [SPECT] Bands of spectral lines associated with a change of electronic state of a molecule; each band corresponds to certain vibrational energies in the initial and final states and consists of numerous rotational lines. { i,lek'trän·ik 'band ,spek·trəm }

electronic emission spectrum [SPECT] Spectrum resulting from emission of electromagnetic radiation by atoms, ions, and molecules following excitations of their electrons. { i,lek'trän·ik i'mish·ən ,spek·trəm }

electronic magnetic moment [ATOM PHYS] The total magnetic dipole moment associated with the orbital motion of all the electrons of an atom and the electron spins; opposed to nuclear magnetic moment. { i,lek'trän·ik mag'ned·ik 'mō·mənt }

electronic polarization [ELEC] Polarization arising from the displacement of electrons with respect to the nuclei with which they are associated, upon application of an external electric field. { i,lek'trän·ik ,pō·lə·rə'zā·shən }

electronic pumping *See* pumping. { i,lek'trän·ik 'pəmp·iŋ }

electronics [PHYS] Study, control, and application of the conduction of electricity through gases or vacuum or through semiconducting or conducting materials. { i,lek'trän·iks }

electronic specific heat [SOLID STATE] Contribution to the specific heat of a metal from the motion of conduction electrons. { i,lek'trän·ik spə¦sif·ik 'hēt }

electronic spectrum [SPECT] Spectrum resulting from emission or absorption of electromagnetic radiation during changes in the electron configuration of atoms, ions, or molecules, as opposed to vibrational, rotational, fine-structure, or hyperfine spectra. { i,lek'trän·ik 'spek·trəm }

electronic state [QUANT MECH] The physical state of electrons of a system, as specified, for example, by a Schrödinger-Pauli wave function of the positions and spin orientations of all the electrons. { i,lek'trän·ik 'stāt }

electronic structure [PHYS] The arrangement of electrons in an atom, molecule, or solid, specified by their wave functions, energy levels, or quantum numbers. { i,lek'trän·ik 'strək·chər }

electronic work function [SOLID STATE] The energy required to raise an electron with the Fermi energy in a solid to the energy level of an electron at rest in vacuum outside the solid. { i,lek'trän·ik 'wərk ,fəŋk·shən }

electron lepton number [PART PHYS] The number of electrons and electron-associated neutrinos minus the number of positrons and electron-associated antineutrinos. { i'lek,trän 'lep,tän ,nəm·bər }

electron magnetic moment [ATOM PHYS] The magnetic dipole moment which an electron possesses by virtue of its spin. Also known as electron dipole moment. { i'lek,trän ¦mag·ned·ik 'mō·mənt }

electron mass [PHYS] The mass of an electron, equal to about 9.11×10^{-28} gram, equivalent to 0.511 MeV. Also known as electron rest mass. { i,lek'trän 'mas }

electron microprobe [PHYS] An x-ray machine in which electrons emitted from a hot-filament source are accelerated electrostatically, then focused to an extremely small point on the surface of a specimen by an electromagnetic lens; nondestructive analysis of the specimen can then be made by measuring the backscattered electrons, the specimen current, the resulting x-radiation, or any other resulting phenomenon. Also known as electron probe. { i'lek,trän 'mī·krō ,prōb }

electron mobility [SOLID STATE] The drift mobility of electrons in a semiconductor, being the electron velocity divided by the applied electric field. { i'lek,trän mō'bil·əd·ē }

electron multiplicity [ATOM PHYS] In an atom with Russell-Saunders coupling, the quantity 2S + 1, where S is the total spin quantum number. { i'lek,trän məl·tə'plis·əd·ē }

electron nuclear double resonance [SPECT] A type of electron paramagnetic resonance (EPR) spectroscopy permitting greatly enhanced resolution, in which a material is simultaneously irradiated at one of its EPR frequencies and by a second oscillatory field whose frequency is swept over the range of nuclear frequencies. Abbreviated ENDOR. { i'lek,trän ¦nü·klē·ər ¦dəb·əl 'rez·ən·əns }

electron number [ATOM PHYS] The number of electrons in an ion or atom. { i'lek,trän ,nəm·bər }

electron orbit [PHYS] The path described by an electron. { i'lek,trän 'ȯr·bət }

electron paramagnetic resonance [PHYS] Magnetic resonance arising from the magnetic moment of unpaired electrons in a paramagnetic substance or in a paramagnetic center in a diamagnetic substance. Abbreviated EPR. Also known as electron spin resonance (ESR); paramagnetic resonance. { i'lek,trän ¦par·ə·mag ¦ned·ik 'rez·ən·əns }

electron paramagnetism [PHYS] Paramagnetism in a substance whose atoms or molecules possess a net electronic magnetic moment; arises because of the tendency of a magnetic field to orient the electronic magnetic moments parallel to itself. { i'lek,trän ,par·ə'mag·nə,tiz·əm }

electron-positron pair [PHYS] An electron and a positron produced at the same time in the interaction of a photon with a high-intensity electric field. { i¦lek,trän 'päz·ə,trän ,per }

electron probe *See* electron microprobe. { i'lek ,trän ,prōb }

electron radius [PHYS] The classical value *r* of $2.8179403 \times 10^{-13}$ centimeter for the radius of an electron; obtained by equating mc^2 for the electron to e^2/r, where e and m are the charge and mass of the electron respectively; any classical model for an electron will have approximately this radius. { i'lek,trän 'rād·ē·əs }

electron rest mass *See* electron mass. { i'lek,trän 'rest ,mas }

electron shell [ATOM PHYS] **1.** The collection of all the electron states in an atom which have a given principal quantum number. **2.** The collection of all the electron states in an atom which have a given principal quantum number and a given orbital angular momentum quantum number. { i'lek‚trän 'shel }

electron spectroscopy [SPECT] The study of the energy spectra of photoelectrons or Auger electrons emitted from a substance upon bombardment by electromagnetic radiation, electrons, or ions; used to investigate atomic, molecular, or solid-state structure, and in chemical analysis. { i'lek‚trän spek'träs·kə·pē }

electron spectroscopy for chemical analysis *See* x-ray photoelectron spectroscopy. { i'lek‚trän spek'träs·kə·pē fər 'kem·i·kəl ə'nal·ə·səs }

electron spectrum [SPECT] Visual display, photograph, or graphical plot of the intensity of electrons emitted from a substance bombarded by x-rays or other radiation as a function of the kinetic energy of the electrons. { i'lek‚trän 'spek·trəm }

electron spin [QUANT MECH] That property of an electron which gives rise to its angular momentum about an axis within the electron. { i'lek‚trän 'spin }

electron spin echo [SOLID STATE] A net magnetization of a material that is sometimes observed at a particular time following the application of two or more short, intense pulses of microwave radiation; in the simplest case, two pulses separated by a time interval t are followed by a net magnetization at time t' after the second pulse. { i‚lek‚trän 'spin ‚ek·ō }

electron spin echo envelope modulation [SPECT] **1.** The variation in the intensity of an electron spin echo as the time interval between the two microwave pulses producing the echo is incremented in small steps in the case of a two-pulse echo, or time intervals between suitable pulses are incremented for multiple-pulse echoes. **2.** A type of electron paramagnetic resonance spectroscopy in which this variation is mathematically transformed, using the Fourier transform, to yield the spectrum of nuclear frequencies. Abbreviated ESEEM. { i‚lek‚trän ¦spin ‚ek·ō ¦en·və‚lōp ‚mäj·ə'lā·shən }

electron spin density [PHYS] The vector sum of the spin angular momenta of electrons at each point in a substance per unit volume. { i'lek ‚trän 'spin ‚den·səd·ē }

electron spin resonance *See* electron paramagnetic resonance. { i'lek‚trän 'spin ‚rez·ən·əns }

electron temperature [PL PHYS] The temperature at which ideal gas molecules would have an average kinetic energy equal to that of electrons in a plasma under consideration. { i'lek ‚trän 'tem·prə·chər }

electron transfer [PHYS] The passage of an electron from one constituent of a system to another. { i'lek‚trän 'trans·fər }

electron transition [QUANT MECH] Change of an electron from one state to another, accompanied by emission or absorption of electromagnetic radiation. { i'lek‚trän tran'zish·ən }

electron trap [SOLID STATE] A defect or chemical impurity in a semiconductor or insulator which captures mobile electrons in a special way. { i'lek‚trän ‚trap }

electron tunneling [QUANT MECH] The passage of electrons through a potential barrier which they would not be able to cross according to classical mechanics, such as a thin insulating barrier between two superconductors. { i'lek ‚trän 'tən·əl·iŋ }

electronvolt [PHYS] A unit of energy which is equal to the energy acquired by an electron when it passes through a potential difference of 1 volt in a vacuum; it is equal to $(1.60217646 \pm 0.00000006) \times 10^{-19}$ volt. Abbreviated eV. { i'lek‚trän ‚vōlt }

electron wave [QUANT MECH] The de Broglie wave or probability amplitude wave of an electron. { i'lek‚trän ‚wāv }

electron wave function [QUANT MECH] Function of the spin orientation and position of one or more electrons, specifying the dynamical state of the electrons; the square of the function's modulus gives the probability per unit volume of finding electrons at a given position. { i'lek ‚trän ‚wāv ‚fəŋk·shən }

electron wavelength [QUANT MECH] The de Broglie wavelength of an electron, given by Planck's constant divided by the momentum. { i'lek‚trän 'wāv‚leŋkth }

electrooptical birefringence *See* electrooptical Kerr effect. { i‚lek·trō'äp·tə·kəl bī·ri'frin·jəns }

electrooptical Kerr effect [OPTICS] Birefringence induced by an electric field. Also known as electrooptical birefringence; Kerr effect. { i‚lek·trō'äp·tə·kəl 'kər i‚fekt }

electrooptic material [OPTICS] A material in which the indices of refraction are changed by an applied electric field. { i‚lek·trō'äp·tik mə 'tir·ē·əl }

electrooptics [OPTICS] The study of the influence of an electric field on optical phenomena, as in the electrooptical Kerr effect and the Stark effect. Also known as optoelectronics. { i‚lek·trō'äp·tiks }

electrophoretic display [OPTICS] A liquid crystal display in which a light-absorbing dye has been added to the liquid to improve both color and luminance contrast. { i‚lek·trō·fə'red·ik di'splā }

electrophorus [ELEC] A device used to produce electric charges; it consists of a hard-rubber disk, which is negatively charged by rubbing with fur, and a metal plate, held by an insulating handle, which is placed on the disk; the plate is then touched with a grounded conductor, so that negative charge is removed and the plate has net positive charge. { i‚lek'trä·fə·rəs }

electrophotophoresis [PHYS] Helical motion of small particles suspended in a gas along the direction of an electric field when exposed to a beam of light. { i‚lek·trō‚fōd·ə·fə'rē·səs }

electropositive [ELEC] **1.** Carrying a positive electric charge. **2.** Capable of acting as the positive electrode in an electric cell. { i‚lek·trə'päz·əd·iv }

electroreflectance [SPECT] Electromodulation in which reflection spectra are studied. Abbreviated ER. { i¦lek·trō·ri'flek·təns }

electrostatic [ELEC] Pertaining to electricity at rest, such as an electric charge on an object. { i,lek·trə'stad·ik }

electrostatic attraction *See* Coulomb attraction. { i,lek·trə'stad·ik ə'trak·shən }

electrostatic energy [ELEC] The potential energy which a collection of electric charges possesses by virtue of their positions relative to each other. { i,lek·trə'stad·ik 'en·ər·jē }

electrostatic error *See* antenna effect. { i,lek·trə'stad·ik 'er·ər }

electrostatic field [ELEC] A time-independent electric field, such as that produced by stationary charges. { i,lek·trə'stad·ik 'fēld }

electrostatic force [ELEC] Force on a charged particle due to an electrostatic field, equal to the electric field vector times the charge of the particle. { i,lek·trə'stad·ik ' fȯrs }

electrostatic induction [ELEC] The process of charging an object electrically by bringing it near another charged object, then touching it to ground. Also known as induction. { i,lek·trə'stad·ik in'dək·shən }

electrostatic interactions *See* Coulomb interactions. { i,lek·trə'stad·ik int·ə'rak·shənz }

electrostatic ion-cyclotron wave [PL PHYS] A longitudinal wave that propagates in a magnetically confined plasma at a very large angle to the magnetic field, with a frequency somewhat in excess of the ion-cyclotron frequency. { i¦lek·trə,stad·ik ¦ī·ən 'sī·klə,trän ,wāv }

electrostatic potential *See* electric potential. { i'lek·trə,stad·ik pə'ten·chəl }

electrostatic probe *See* electric probe. { i'lek·trə,stad·ik 'prōb }

electrostatic repulsion *See* Coulomb repulsion. { i'lek·trə,stad·ik ri'pəl·shən }

electrostatics [ELEC] The study of electric charges at rest, their electric fields, and potentials. { i,lek·trə'stad·iks }

electrostatic shielding [ELEC] The placing of a grounded metal screen, sheet, or enclosure around a device or between two devices to prevent electric fields from interacting. { i'lek·trə,stad·ik 'shēld·iŋ }

electrostatic stress [ELEC] An electrostatic field acting on an insulator, which produces polarization in the insulator and causes electrical breakdown if raised beyond a certain intensity. { i'lek·trə,stad·ik 'stres }

electrostatic tape camera [OPTICS] A camera in which images are stored electrostatically on a plastic tape; designed for use in satellites, where the stored image is not damaged by Van Allen or other radiation. { i'lek·trə,stad·ik 'tāp ,kam·rə }

electrostatic units [ELEC] A centimeter-gram-second system of electric and magnetic units in which the unit of charge is that charge which exerts a force of 1 dyne on another unit charge when separated from it by a distance of 1 centimeter in vacuum; other units are derived from this definition by assigning unit coefficients in equations relating electric and magnetic quantities. Abbreviated esu. { i'lek·trə,stad·ik 'yü·nəts }

electrostatic wave [PL PHYS] Wave motion of a plasma whose restoring forces are primarily electrostatic. { i'lek·trə,stad·ik 'wāv }

electrostriction [MECH] A form of elastic deformation of a dielectric induced by an electric field, associated with those components of strain which are independent of reversal of field direction, in contrast to the piezoelectric effect. Also known as electrostrictive strain. { i¦lek·trō 'strik·shən }

electrostrictive strain *See* electrostriction. { i¦lek·trō'strik·tiv 'strān }

electrothermal [PHYS] **1.** Pertaining to both heat and electricity. **2.** In particular, pertaining to conversion of electrical energy into heat energy. { i¦lek·trō'thər·məl }

electroviscous effect [FL MECH] Change in a liquid's viscosity induced by a strong electrostatic field. { i¦lek·trō'vis·kəs i'fekt }

electroweak interaction [PART PHYS] The unification of the electromagnetic and weak interactions described by the Weinberg-Salam theory. { i'lek·trō,wēk ,in·tər'ak·shən }

element [ELEC] **1.** A part of an electron tube, semiconductor device, or antenna array that contributes directly to the electrical performance. **2.** *See* component. [ELECTROMAG] Radiator, active or parasitic, that is a part of an antenna. { 'el·ə·mənt }

elementary charge [PHYS] An electric charge such that the electric charge of any body is an integral multiple of it, equal to the electron charge. { ,el·ə'men·trē ,chärj }

elementary excitation [QUANT MECH] The quantum of energy of some vibration or wave, such as a photon, phonon, plasmon, magnon, polaron, or exciton. { ,el·ə'men·trē ,ek,sī'tā·shən }

elementary particle [PART PHYS] A particle which, in the present state of knowledge, cannot be described as compound, and is thus one of the fundamental constituents of all matter. Also known as fundamental particle; particle; subnuclear particle. { ,el·ə'men·trē 'pärd·i·kəl }

elements [MECH] The various features of a trajectory such as the angle of departure, maximum ordinate, angle of fall, and so on. { 'el·ə·mənts }

elevation head [FL MECH] The energy per unit mass possessed by a fluid as a result of its height above some reference level. Also known as potential head. { ,el·ə'vā·shən ,hed }

E lines [ELEC] Contour lines of constant electrostatic field strength referred to some reference base. { 'ē ,līnz }

ellipsoidal of wave normals *See* index ellipsoid. { ə,lip'sȯid·əl əv ¦wāv 'nȯr·məlz }

ellipsoidal reflector [OPTICS] A concave ellipsoidal surface from which light is specularly reflected; used in a light projector to focus rays from a light source at the near focal point onto the opposite focal point of the ellipse. { ə¦lip 'sȯid·əl ri'flek·tər }

ellipsoid of wave normals *See* index ellipsoid. { ə'lip¦sȯid əv ¦wāv 'nȯr·məlz }

ellipsometer [OPTICS] An instrument for determining the degree of ellipticity of polarized light; used to measure the thickness of very thin transparent films by observing light reflected from the film. { ə¦lip'säm·əd·ər }

ellipsometry [OPTICS] A technique for determining the properties of a material from the polarization characteristics of linearly polarized incident light reflected from its surface. { ə¦lip 'säm·ə·trē }

elliptical orbit [MECH] The path of a body moving along an ellipse, such as that described by either of two bodies revolving under their mutual gravitational attraction but otherwise undisturbed. { ə'lip·tə·kəl 'ȯr·bət }

elliptical polarization [ELECTROMAG] Polarization of an electromagnetic wave in which the electric field vector at any point in space describes an ellipse in a plane perpendicular to the propagation direction. { ə'lip·tə·kəl ¦pō·lə·rə'zā·shən }

elongation [MECH] The fractional increase in a material's length due to stress in tension or to thermal expansion. { ē¦lȯŋ'gā·shən }

ELR scale *See* equal listener response scale. { ¦ē¦el¦är ¦skāl }

Elster-Geitel effect [PHYS] The phenomenon in which a heated conductor acquires a positive or negative electric charge in the presence of a gas, while in a vacuum it always acquires a negative charge. { ¦el·stər ¦gīt·əl i¦fekt }

emagram [THERMO] A graph of the logarithm of the pressure of a substance versus its temperature, when it is held at constant volume; in meteorological investigations, the potential temperature is often the parameter. { 'em·ə¦gram }

emanation *See* radioactive emanation. { ¦em·ə'nā·shən }

embrittlement [MECH] Reduction or loss of ductility or toughness in a metal or plastic with little change in other mechanical properties. { ¦em'brid·əl·mənt }

emission [ELECTROMAG] Any radiation of energy by means of electromagnetic waves, as from a radio transmitter. { i'mish·ən }

emission lines [SPECT] Spectral lines resulting from emission of electromagnetic radiation by atoms, ions, or molecules during changes from excited states to states of lower energy. { i'mish·ən ¦līnz }

emission spectrometer [SPECT] A spectrometer that measures percent concentrations of preselected elements in samples of metals and other materials; when the sample is vaporized by an electric spark or arc, the characteristic wavelengths of light emitted by each element are measured with a diffraction grating and an array of photodetectors. { i'mish·ən spek'träm·əd·ər }

emission spectrum [SPECT] Electromagnetic spectrum produced when radiations from any emitting source, excited by any of various forms of energy, are dispersed. { i'mish·ən ¦spek·trəm }

emissive power *See* emittance. { i¦mis·iv 'pau̇·ər }

emissivity [THERMO] The ratio of the radiation emitted by a surface to the radiation emitted by a perfect blackbody radiator at the same temperature. Also known as thermal emissivity. { ¦ē·mə'siv·əd·ē }

emittance [THERMO] The power radiated per unit area of a radiating surface. Also known as emissive power; radiating power. { i'mit·əns }

E mode *See* transverse magnetic mode. { 'ē ¦mōd }

emu *See* electromagnetic system of units. { 'ē¦myü }

end correction [ACOUS] A correction that must be made to the assumption that an antinode exists at an open end of a pipe in which air is vibrating, in order to take into account the radiation of sound waves from the pipe. { 'end kə¦rek·shən }

end effect [ELECTROMAG] The effect of capacitance at the ends of an antenna; it requires that the actual length of a half-wave antenna be about 5% less than a half wavelength. { 'end i¦fekt }

end-fire antenna *See* end-fire array. { 'end ¦fīr an'ten·ə }

end-fire array [ELECTROMAG] A linear array whose direction of maximum radiation is along the axis of the array; it may be either unidirectional or bidirectional; the elements of the array are parallel and in the same plane, as in a fishbone antenna. Also known as end-fire antenna. { 'end ¦fīr ə'rā }

end loss [ELECTROMAG] The difference between the actual and the effective lengths of a radiating antenna element. { 'end ¦lȯs }

endoergic collision *See* collision of the first kind. { ¦en·dō¦ər·jik kə'lizh·ən }

end-on position [ELECTROMAG] The position of a point which lies on the magnetic axis of a magnet. Also known as Gauss A position. { ¦end 'ȯn pə¦zish·ən }

ENDOR *See* electron nuclear double resonance. { 'en¦dȯr }

endothermic [NUC PHYS] Petaining to a nuclear or particle reaction in which some of the kinetic energy of the initial particles is converted to mass energy. { ¦en·də'thər·mik }

end product [PHYS] The final product of a chemical or nuclear reaction or process. { 'end ¦prä·dəkt }

end radiation *See* quantum limit. { 'end ¦rād·ē¦ā·shən }

endurance limit *See* fatigue limit. { in'du̇r·əns ¦lim·ət }

endurance ratio *See* fatigue ratio. { in'du̇r·əns ¦rā·shō }

endurance strength *See* fatigue strength. { in 'du̇r·əns ¦streŋkth }

energetics [PHYS] The study of energy and of its transformation from one form to another. { ¦en·ər'jed·iks }

energized [ELEC] Electrically connected to a

voltage source. Also known as alive; hot; live. { 'en·ər,jīzd }

energy [PHYS] The capacity for doing work. { 'en·ər·jē }

energy absorption [PHYS] Conversion of mechanical or radiant energy into the internal potential energy or heat energy of a system. { 'en·ər·jē ab,sórp·shən }

energy balance [PHYS] The arithmetic balancing of energy inputs versus outputs for an object, reactor, or other processing system; it is positive if energy is released, and negative if it is absorbed. { 'en·ər·jē ,bal·əns }

energy band *See* band. { ¦en·ər·jē ,band }

energy-band theory of solids *See* band theory of solids. { 'en·ər·jē ,band ¦thē·ə·rē əv 'säl·ədz }

energy conservation *See* conservation of energy. { 'en·ər·jē ,kän·sər'vā·shən }

energy conversion [PHYS] The process of changing energy from one form to another. { 'en·ər·jē kən'vər·zhən }

energy density [PHYS] The energy per unit volume of a medium; in the case of an electric or magnetic field, the energy needed to set up the field is thought of as residing in the field. { 'en·ər·jē ,den·səd·ē }

energy diagram *See* energy-level diagram. { 'en·ər·jē ,dī·ə,gram }

energy-dispersive x-ray diffraction [PHYS] A technique in which an energy spectrum is obtained of the x-rays scattered from a polychromatic x-ray beam through a fixed angle by a polycrystalline sample. Abbreviated EDXD. { ¦en·ər·jē di¦spər·səv 'eks,rā di'frak·shən }

energy dissipation *See* dissipation. { 'en·ər·jē dis·ə'pā·shən }

energy eigenstate *See* energy state. { 'en·ər·jē'ī·gən,stāt }

energy ellipsoid *See* momental ellipsoid. { ¦en·ər·jē i'lip,sóid }

energy flux [PHYS] A vector quantity whose component perpendicular to any surface equals the energy transported across that surface by some medium per unit area per unit time. { 'en·ər·jē ,fləks }

energy gap [SOLID STATE] A range of forbidden energies in the band theory of solids. { 'en·ər·jē ,gap }

energy gradient [PHYS] Any change in energy over time or space. { 'en·ər·jē ,grād·ē·ənt }

energy head [FL MECH] The elevation of the hydraulic grade line at any section of a waterway plus the velocity head of the mean velocity of the water in that section. { 'en·ər·jē ,hed }

energy integral [MECH] A constant of integration resulting from integration of Newton's second law of motion in the case of a conservative force; equal to the sum of the kinetic energy of the particle and the potential energy of the force acting on it. { 'en·ər·jē 'in·tə·grəl }

energy level [QUANT MECH] An allowed energy of a physical system; there may be several allowed states at one level. { 'en·ər·jē ,lev·əl }

energy-level diagram [QUANT MECH] A diagram in which the energy levels of a quantized system are indicated by distances of horizontal lines from a zero energy level. Also known as energy diagram; level scheme. { 'en·ər·jē ,lev·əl dī·ə,gram }

energy momentum tensor [PHYS] A tensor whose 16 elements give the energy density, momentum density, and stresses in a distribution of matter or radiation. { 'en·ər·jē mə'men·təm ,ten·sər }

energy of a charge [ELEC] Charge energy measured in ergs according to the equation $E = QV$, where Q is the charge and V is the potential in electrostatic units. { 'en·ər·jē əv ə 'chärj }

energy of rotation [PHYS] Kinetic energy of a mass with moment of inertia I rotating with angular velocity ω about the axis, expressed as $E = {}^1/_2 I\omega^2$. { 'en·ər·jē əv rō'tā·shən }

energy operator [QUANT MECH] The operator corresponding to the energy or Hamiltonian of a classical system. Also known as Hamiltonian operator. { 'en·ər·jē ,äp·ə,rād·ər }

energy product curve [ELECTROMAG] Curve obtained by plotting the product of the values of magnetic induction B and demagnetizing force H for each point on the demagnetization curve of a permanent magnet material; usually shown with the demagnetization curve. { 'en·ər·jē 'prä,dəkt ,kərv }

energy spectrum [PHYS] Any plot, display, or photographic record of the intensity of some type of radiation as a function of its energy. { 'en·ər·jē ,spek·trəm }

energy spread [QUANT MECH] The width in energy of a wave packet or metastable state. { 'en·ər·jē ,spred }

energy state [QUANT MECH] An eigenstate of the energy (Hamiltonian) operator, so that the energy has a definite stationary value. Also known as eigenstate; energy eigenstate; quantum state; stationary state. { 'en·ər·jē ,stāt }

energy winds [PHYS] A group of winds which contain the bulk of recoverable kinetic energy for each month. { 'en·ər·jē 'winz }

engine cycle [THERMO] Any series of thermodynamic phases constituting a cycle for the conversion of heat into work; examples are the Otto cycle, Stirling cycle, and Diesel cycle. { 'en·jən ,sī·kəl }

engineer's system of units *See* British gravitational system of units. { ,en·jə'nirz ¦sis·təm əv 'yü·nəts }

enhanced line *See* enhanced spectral line. { en 'hanst 'līn }

enhanced spectral line [SPECT] A spectral line of a very hot source, such as a spark, whose intensity is much greater than that of a line in a flame or arc spectrum. Also known as enhanced line. { en'hanst 'spek·trəl ,līn }

enlargement loss [FL MECH] Energy loss by friction in a flowing fluid when it moves into a cross-sectional area of sudden enlargement. { en'lärj·mənt ,lós }

enlarger [OPTICS] An optical projector used to

project an enlarged image of a photograph's negative onto photosensitized film or paper. Also known as photoenlarger. { en'lär·jər }

ensemble [STAT MECH] A collection of systems of particles used to describe an individual system; time averages of quantities describing the individual system are found by averaging over the systems in the ensemble at a fixed time. { än'säm·bəl }

Enskog theory *See* Chapman-Enskog theory. { 'en,skäg ,thē·ə·rē }

ensonify [ACOUS] To fill the ocean or any fluid medium with acoustic radiation, which is then observed and analyzed to study the medium or to locate or image objects within it. Also spelled insonify. { en'sän·i·fī }

enthalpy [THERMO] The sum of the internal energy of a system plus the product of the system's volume multiplied by the pressure exerted on the system by its surroundings. Also known as heat content; sensible heat; total heat. { en 'thal·pē }

enthalpy-entropy chart [THERMO] A graph of the enthalpy of a substance versus its entropy at various values of temperature, pressure, or specific volume; useful in making calculations about a machine or process in which this substance is the working medium. { en¦thal·pē 'en·trə·pē ,chärt }

enthalpy of vaporization *See* heat of vaporization. { en'thal·pē əv ,vā·pə·rə'zā·shən }

enthalpy-pressure chart *See* pressure-enthalpy chart. { en¦thal·pē 'presh·ər ,chärt }

entrained fluid [FL MECH] Fluid in the form of mist, fog, or droplets that is carried out of a column or vessel by a rising gas or vapor stream. { en'trānd 'flü·əd }

entrance loss [FL MECH] Energy loss by friction in a flowing fluid when it moves into a cross-sectional area of sudden contraction, as at the entrance of a pipe or a suddenly reduced area of a duct. { 'en·trəns ,lȯs }

entrance pupil [OPTICS] The image of the aperture stop of an optical system formed in the object space by rays emanating from a point on the optical axis in the image space. { 'en·trəns ,pyü·pəl }

entrance slit [SPECT] Narrow slit through which passes the light entering a spectrometer. { 'en·trəns ,slit }

entropy [STAT MECH] Measure of the disorder of a system, equal to the Boltzmann constant times the natural logarithm of the number of microscopic states corresponding to the thermodynamic state of the system; this statistical-mechanical definition can be shown to be equivalent to the thermodynamic definition. [THERMO] Function of the state of a thermodynamic system whose change in any differential reversible process is equal to the heat absorbed by the system from its surroundings divided by the absolute temperature of the system. Also known as thermal charge. { 'en·trə·pē }

entry ballistics [MECH] That branch of ballistics which pertains to the entry of a missile, spacecraft, or other object from outer space into and through an atmosphere. { 'en·trē bə,lis·tiks }

envelope soliton [PHYS] A rapidly oscillating wave that propagates with a characteristic constant shape, and can be pictured as cut off by a smoothly modulating envelope. { 'en·və,lōp 'säl·ə,tän }

envelopmental sound [ACOUS] The portion of the room impulse response consisting of sound that arrives at a listener's location between 20 and 150 milliseconds after the first direct sound, and which has been reflected against walls and ceilings relatively few times. { in,vel·əp,ment·əl 'saủnd }

environment [PHYS] The aggregate of all the conditions and the influences that determine the behavior of a physical system. { in'vī·ərn·mənt *or* in'vī·rən·ment }

environmental fluid mechanics [FL MECH] The study of the flows of air and water, of the species carried by them (especially pollution), and of their interactions with geological, biological, social, and engineering systems in the vicinity of a planet's surface. { in,vī·ərn,ment·əl ,flü·əd mi'kan·iks }

environmental stress cracking [MECH] The susceptibility of a material to crack or craze in the presence of surface-active agents or other factors. { in¦vī·ərn¦mənt·əl 'stres ,krak·iŋ }

eolian sounds [ACOUS] Sounds produced by eddying motions of air in the lee of obstacles, such as wires, twigs, and even the ear itself, when wind blows over those obstacles. { ē'ōl·yən 'saủnz }

eolotropy *See* anistropy. { ,ē·ə'lä·trə·pē }

eon [MECH] A unit of time, equal to 10^9 years. { 'ē,än }

Eötvös constant [PHYS] A constant that appears in an expression for the behavior of the surface tension γ of a liquid as the temperature T drops to a critical temperature T_c at which the surface tension disappears, equal to $\gamma(M/\rho)^{2/3}/(T_c - T)$, where M is the molecular weight and ρ the density of the liquid. { 'ət·vəsh ,kän·stənt }

Eötvös effect [MECH] An apparent decrease (or increase) in the weight of a body moving from west to east (or east to west) because of its greater (or smaller) centrifugal acceleration. { 'ət·vəsh i,fekt }

Eötvös experiment [RELAT] An experiment which tests the equality of inertial mass and gravitational mass by balancing on a given body the earth's gravitational attraction against the kinetic reaction arising from the rotation of the earth. { 'ət·vəsh ik,sper·ə·mənt }

Eötvös number *See* Bond number. { 'ət·vəsh ,nəm·bər }

Eötvös rule [THERMO] The rule that the rate of change of molar surface energy with temperature is a constant for all liquids; deviations are encountered in practice. { 'ət·vəsh ,rül }

epidiascope [OPTICS] **1.** An optical projection system for forming an enlarged real image of a flat opaque object, in which light is reflected from the object and then from a mirror before being focused by a projection lens. Also known as episcope. **2.** An optical projection system which can easily be altered to project either transparent or opaque objects. { ˌep·əˈdī·əˌskōp }

episcope *See* epidiascope. { ˈep·əˌskōp }

episcotister [OPTICS] A device for reducing the intensity of light by a known fraction, consisting of a rapidly rotating disk with transparent and opaque sectors. { ˌep·ə·skōˈtis·tər }

epitaxial layer [SOLID STATE] A semiconductor layer having the same crystalline orientation as the substrate on which it is grown. { ˌep·əˈtak·sē·əl ˌlā·ər }

epitaxy [CRYSTAL] Growth of one crystal on the surface of another crystal in which the growth of the deposited crystal is oriented by the lattice structure of the substrate. { ˈep·əˌtak·sē }

E-plane antenna [ELECTROMAG] An antenna which lies in a plane parallel to the electric field vector of the radiation that it emits. { ˈē ˌplān anˌten·ə }

E-plane bend *See* E bend. { ˈē ˌplān ˌbend }

E-plane T junction [ELECTROMAG] Waveguide T junction in which the change in structure occurs in the plane of the electric field. Also known as series T junction. { ˈē ˌplān ˈtē ˌjəŋk·shən }

epoch *See* time. { ˈep·ək }

EPR *See* electron paramagnetic resonance.

EPR experiment *See* Einstein-Podolsky-Rosen experiment. { ˌēˌpēˈär ikˈsper·ə·mənt }

epsilon meson [PART PHYS] Neutral, scalar, meson resonance having positive charge conjugation parity and G-parity, a mass of about 730 MeV, and a width of about 600 MeV; decays to two pions. { ˈep·səˌlän ˈmāˌsän }

epsilon structure [SOLID STATE] The hexagonal close-packed structure of the ϵ-phase of an electron compound. { ˈep·səˌlän ˌstrək·chər }

equal-arm balance [MECH] A simple balance in which the distances from the point of support of the balance-arm beam to the two pans at the end of the beam are equal. { ¦ē·kwal ¦ärm ˈbal·əns }

equal-energy source [PHYS] Electromagnetic or sound source of energy which emits the same amount of energy for each frequency of the spectrum. { ¦ē·kwal ¦en·ər·jē ˌsȯrs }

equal listener response scale [ACOUS] An arbitrary scale of noisiness which measures the average response of a listener to a noise when allowance is made for the apparent increase of intensity of a noise as its frequency increases. Abbreviated ELR scale. { ¦ē·kwəl ¦lis·nər ri ˈspäns ˌskāl }

equal loudness contour [ACOUS] A curve on a graph of sound intensity in decibels versus frequency at each point along which sound appears to be equally loud to a listener. Also known as Fletcher-Munson contour. { ¦ē·kwəl ˈlau̇d·nəs ˌkänˌtu̇r }

equally tempered scale [ACOUS] A musical scale formed by dividing the octave into 12 equal intervals and selecting from the resulting notes; thus, the frequency ratio between any two successive notes is exactly $2^{1/2}$ or $2^{1/6}$. Also known as equitempered scale. { ¦ē·kwə·lē ¦tem·pərd ˈskāl }

equation of continuity *See* continuity equation. { iˈkwā·zhən əv ˌkänt·ənˈü·əd·ē }

equation of motion [FL MECH] One of a set of hydrodynamical equations representing the application of Newton's second law of motion to a fluid system; the total acceleration on an individual fluid particle is equated to the sum of the forces acting on the particle within the fluid. [MECH] **1.** Equation which specifies the coordinates of particles as functions of time. **2.** A differential equation, or one of several such equations, from which the coordinates of particles as functions of time can be obtained if the initial positions and velocities of the particles are known. [QUANT MECH] A differential equation which enables one to predict the statistical distribution of the results of any measurement upon a system at any time if the initial dynamical state of the system is known. { iˈkwā·zhən əv ˈmō·shən }

equation of piezotropy [THERMO] An equation obeyed by certain fluids which states that the time rate of change of the fluid's density equals the product of a function of the thermodynamic variables and the time rate of change of the pressure. { iˈkwā·zhən əv pē·əˈzä·trə·pē }

equatorial plane [MECH] A plane perpendicular to the axis of rotation of a rotating body and equidistant from the intersections of this axis with the body's surface, provided that the body is symmetric about the axis of rotation and is symmetric under reflection through this plane. [OPTICS] *See* sagittal plane. { ˌe·kwəˈtȯr·ē·əl ˈplān }

equiangular spiral antenna [ELECTROMAG] A frequency-independent broad-band antenna, cut from sheet metal, that radiates a very broad, circularly polarized beam on both sides of its surface; this bidirectional radiation pattern is its chief limitation. { ¦ē·kwē¦aŋ·gyə·lər ¦spī·rəl anˈten·ə }

equilibrant [MECH] A single force which cancels the vector sum of a given system of forces acting on a rigid body and whose torque cancels the sum of the torques of the system. { iˈkwil·ə·brənt }

equilibrium [MECH] Condition in which a particle, or all the constituent particles of a body, are at rest or in unaccelerated motion in an inertial reference frame. Also known as static equilibrium. [PHYS] Condition in which no change occurs in the state of a system as long as its surroundings are unaltered. [STAT MECH] Condition in which the distribution function of a system is time-independent. { ˌē·kwəˈlib·rē·əm }

equilibrium vapor pressure [PHYS] The vapor pressure of a system in which two or more phases of water coexist in equilibrium. { ˌē·kwəˈlib·rē·əm ˈvā·pər ˌpresh·ər }

equipartition law [STAT MECH] In a classical

ideal gas, the average kinetic energy per molecule associated with any degree of freedom which occurs as a quadratic term in the expression for the mechanical energy, is equal to half of Boltzmann's constant times the absolute temperature. { ¦e·kwə·pär'tish·ən ,lō }

equiphase wave surface [PHYS] Any surface in a wave over which the field vectors at the same instant are in the same phase or 180° out of phase. { 'e·kwə,fāz 'wāv ,sər·fəs }

equipollent [MECH] Of two systems of forces, having the same vector sum and the same total torque about an arbitrary point. { ¦e·kwə¦päl·ənt }

equipotential surface [ELEC] A surface on which the electric potential is the same at every point. [MECH] A surface which is always normal to the lines of force of a field and on which the potential is everywhere the same. { ¦e·kwə·pə'ten·chəl 'sər·fəs }

equisignal surface [ELECTROMAG] Surface around an antenna formed by all points at which, for transmission, the field strength (usually measured in volts per meter) is constant. { ¦e·kwə¦sig·nəl ,sər·fəs }

equitempered scale *See* equally tempered scale. { ¦e·kwə¦tem·pərd 'skāl }

equivalence principle [RELAT] In general relativity, the principle that the observable local effects of a gravitational field are indistinguishable from those arising from acceleration of the frame of reference. Also known as Einstein's equivalency principle; principle of equivalence. { i'kwiv·ə·ləns ,prin·sə·pəl }

equivalent absorption area [ACOUS] Area of perfectly absorbing surface that will absorb sound energy at the same rate as the given object under the same conditions; the acoustic unit of equivalent absorption is the sabin. { i'kwiv·ə·lənt əb'sȯrp·shən ,er·ē·ə }

equivalent bending moment [MECH] A bending moment which, acting alone, would produce in a circular shaft a normal stress of the same magnitude as the maximum normal stress produced by a given bending moment and a given twisting moment acting simultaneously. { i'kwiv·ə·lənt 'bend·iŋ ,mō·mənt }

equivalent blackbody temperature [THERMO] For a surface, the temperature of a blackbody which emits the same amount of radiation per unit area as does the surface. { i'kwiv·ə·lənt 'blak,bäd·ē ,tem·prə·chər }

equivalent circuit [ELEC] A circuit whose behavior is identical to that of a more complex circuit or device over a stated range of operating conditions. { i'kwiv·ə·lənt 'sər·kət }

equivalent electrons [ATOM PHYS] Electrons in an atom which have the same principal and orbital quantum numbers, but not necessarily the same magnetic orbital and magnetic spin quantum numbers. { i'kwiv·ə·lənt i'lek,tränz }

equivalent evaporation [FL MECH] The amount of water, usually in pounds per hour, evaporated from a temperature of 212°F (100°C) to saturated steam at the same temperature. { i'kwiv·ə·lənt i,vap·ə'rā·shən }

equivalent focal length [OPTICS] The focal length of a thin lens which forms images that most nearly duplicate those of a given compound lens, thick lens, or system of lenses. { i'kwiv·ə·lənt 'fō·kəl ,leŋkth }

equivalent footcandle *See* footlambert. { i'kwiv·ə·lənt 'fu̇t,kan·dəl }

equivalent magnetic length *See* effective magnetic length. { i¦kwiv·ə·lənt mag¦ned·ik 'leŋkth }

equivalent nitrogen pressure [MECH] The pressure that would be indicated by a device if the gas inside it were replaced by nitrogen of equivalent molecular density. { i'kwiv·ə·lənt 'nī·trə·jən ,presh·ər }

equivalent resistance [ELEC] Concentrated or lumped resistance that would cause the same power loss as the actual small resistance values distributed throughout a circuit. { i'kwiv·ə·lənt ri'zis·təns }

equivalent sine wave [PHYS] A sine wave whose root-mean-square value and period are the same as that of a given periodic wave. { i¦kwiv·ə·lənt 'sīn ,wāv }

equivalent temperature [THERMO] A term used in British engineering for that temperature of a uniform enclosure in which, in still air, a sizable blackbody at 75°F (23.9°C) would lose heat at the same rate as in the environment. { i'kwiv·ə·lənt 'tem·prə·chər }

equivalent twisting moment [MECH] A twisting moment which, if acting alone, would produce in a circular shaft a shear stress of the same magnitude as the shear stress produced by a given twisting moment and a given bending moment acting simultaneously. { i'kwiv·ə·lənt 'twist·iŋ ,mō·mənt }

equivalent viscous damping [MECH] An assumed value of viscous damping used in analyzing a vibratory motion, such that the dissipation of energy per cycle at resonance is the same for the assumed or the actual damping force. { i'kwiv·ə·lənt ¦vis·kəs 'damp·iŋ }

equivalent width [PHYS] A measure of the total absorption of radiant energy as indicated by an absorption line or absorption band. { i'kwiv·ə·lənt 'width }

erect image [OPTICS] An image in which directions are the same as those in the object, in contrast to an inverted image. { i'rekt 'im·ij }

erecting lens [OPTICS] An eyepiece sometimes used in Kepler telescopes that consists of four lenses and provides an erect image, which is more convenient for viewing terrestrial objects than the inverted image provided by simpler eyepieces. { i'rek·tiŋ ,lenz }

erecting prism [OPTICS] A system of prisms that converts the inverted image formed by most types of astronomical telescopes into an erect image. Also known as inverting prism. { i'rek·tiŋ ,priz·əm }

erection stress [MECH] The internal forces exerted on a structural member during construction. { i'rek·shən ,stres }

erg [PHYS] A unit of energy or work in the centimeter-gram-second system of units, equal to the work done by a force of magnitude of 1 dyne when the point at which the force is applied is displaced 1 centimeter in the direction of the force. Also known as dyne centimeter (dyne-cm). { ərg }

ergodic theory [STAT MECH] Mathematical theory which attempts to show that the various possible microscopic states of a system are equally probable, and that the system is therefore ergodic. { ər'gäd·ik 'thē·ə·rē }

ergon [QUANT MECH] A quantum of energy; for any oscillator it is equal to the product of the oscillator's frequency and Planck's constant. { 'ər,gän }

ergoregion *See* ergosphere. { 'ər·gō,rē·jən }

ergosphere [RELAT] The region outside the event horizon but inside the stationary limit of a Kerr black hole, within which no object can appear stationary to a distant observer. Also known as ergoregion. { 'ər·gə,sfir }

Ericsson cycle [THERMO] An ideal thermodynamic cycle consisting of two isobaric processes interspersed with processes which are, in effect, isothermal, but each of which consists of an infinite number of alternating isentropic and isobaric processes. { 'er·ik·sən ,sī·kəl }

eriometer [OPTICS] A device used to measure diameters of small particles or fibers by observing the diameter of the diffraction pattern produced by them in light coming from a small hole in a metal plate. { ,er·ē'äm·əd·ər }

ERP *See* effective radiated power.

ESCA *See* x-ray photoelectron spectroscopy.

Esclangon effect [OPTICS] Bending of a reflected light ray caused by movement of the mirror in a direction making an acute angle with its surface. { e'skläŋ·gən i'fekt }

ESEEM *See* electron spin echo envelope modulation. { 'ē,sēm *or* ¦ē¦es¦ē¦ē'em }

E-set [ACOUS] The set of 10 English letters and digits that share the E sound and therefore tend to be more easily confused by speech recognition systems than other elements of an alphadigit vocabulary: the letters B, C, D, E, G, P, T, V, and Z, and the digit 3. { 'ē,set }

Eshelby twist [SOLID STATE] A torsional deformation of a crystal whisker resulting from a screw dislocation along the whisker axis. { 'esh·əl·bē 'twist }

ESR *See* electron paramagnetic resonance.

established flow [FL MECH] The flow when the boundary layer of a fluid flowing in a duct completely fills the duct; that is, when the effect of the wall shearing stress extends completely across the duct. { i'stab·lisht 'flō }

esu *See* electrostatic units.

etalon [OPTICS] **1.** Two adjustable parallel mirrors mounted so that either one may serve as one of the mirrors in a Michelson interferometer; used to measure distances in terms of wavelengths of spectral lines. **2.** An instrument similar to the Fabry-Pérot interferometer, except that the distance between the plates is fixed. Also known as Fabry-Pérot etalon. { 'ed·əl,än }

eta meson [PART PHYS] Neutral pseudoscalar meson having zero isotopic spin and hypercharge, positive charge parity and G parity, and a mass of about 549 MeV; decays via electromagnetic interactions. { 'ād·ə 'mā,sän }

eta-prime meson *See* chi meson. { ¦ād·ə ,prīm ,mā,sän }

ether [ELECTROMAG] The medium postulated to carry electromagnetic waves, similar to the way a gas carries sound waves. { 'e·thər }

ether drag [ELECTROMAG] The hypothesis, advanced unsuccessfully to account for results of the Michelson-Morley experiment, that ether is dragged along with matter. { 'ē·thər ,drag }

ether drift [ELECTROMAG] Hypothetical motion of the ether relative to the earth. { 'ē·thər ,drift }

ether thermoscope [PHYS] A device for detecting radiant heat; consists of an evacuated U-shaped tube, with ether at the bottom of the tube and a bulb at each end, one bulb being blackened. { 'ē·thər 'thər·mə,skōp }

E transformer [ELECTROMAG] A transformer consisting of two coils wound around a laminated iron core in the shape of an E, with the primary and secondaries occupying the center and outside legs respectively. { 'ē tranz,fór·mər }

Ettingshausen coefficient [PHYS] A measure of the strength of the Ettingshausen effect, equal to the ratio of the temperature gradient to the product of the current density and magnetic field strength which produce this gradient. { 'ed·iŋz ,haủz·ən ,kō·i,fish·ənt }

Ettingshausen effect [PHYS] The phenomenon that, when a metal strip is placed with its plane perpendicular to a magnetic field and an electric current is sent longitudinally through the strip, corresponding points on opposite edges of the strip have different temperatures. { 'ed·iŋz ,haủz·ən i,fekt }

Ettingshausen-Nernst coefficient [PHYS] A measure of the strength of the Ettingshausen-Nernst effect, equal to the ratio of the electric field to the product of the temperature gradient and magnetic field strength which produce this field. { ¦ed·iŋz,haủz·ən ¦nərnst,kō·i,fish·ənt }

Ettingshausen-Nernst effect [PHYS] The phenomenon that, when a conductor or semiconductor is subjected to a temperature gradient and to a magnetic field perpendicular to the temperature gradient, an electric field arises perpendicular to both the temperature gradient and the magnetic field. { ¦ed·iŋz,haủz·ən ¦nərnst i,fekt }

Euclidean quantum field theory [QUANT MECH] A relativistic quantum field theory in which time is replaced by a purely formal imaginary time, resulting in replacement of Lorentz covariance by euclidean group covariance. { yü'klid·ē·ən ¦kwänt·əm 'fēld ,thē·ə·rē }

Euler angles [MECH] Three angular parameters

that specify the orientation of a body with respect to reference axes. { 'ȯi·lər ˌaŋ·gəlz }

Euler equation [MECH] Expression for the energy removed from a gas stream by a rotating blade system (as a gas turbine), independent of the blade system (as a radial- or axial-flow system). { 'ȯi·lər iˌkwā·zhən }

Euler equations of motion [MECH] A set of three differential equations expressing relations between the force moments, angular velocities, and angular accelerations of a rotating rigid body. { 'ȯi·lər i¦kwā·zhənz əv 'mō·shən }

Euler force [MECH] The greatest load that a long, slender column can carry without buckling, according to the Euler formula for long columns. { 'ȯi·lər ˌfȯrs }

Euler formula for long columns [MECH] A formula which gives the greatest axial load that a long, slender column can carry without buckling, in terms of its length, Young's modulus, and the moment of inertia about an axis along the center of the column. { 'ȯi·lər ¦fȯr·myə·lə fər ˌlȯŋ 'käl·əmz }

Eulerian coordinates [FL MECH] Any system of coordinates in which properties of a fluid are assigned to points in space at each given time, without attempt to identify individual fluid parcels from one time to the next; a sequence of synoptic charts is a Eulerian representation of the data. { ȯi'ler·ē·ən kō'ȯrd·ənˌəts }

Eulerian correlation [FL MECH] The correlation between the properties of a flow at various points in space at a single instant of time. Also known as synoptic correlation. { ȯi'ler·ē·ən ˌkä·rə'lā·shən }

Eulerian description *See* Euler method. { ȯi¦ler·ē·ən di'skrip·shən }

Eulerian equation [FL MECH] A mathematical representation of the motions of a fluid in which the behavior and the properties of the fluid are described at fixed points in a coordinate system. { ȯi'ler·ē·ən iˌkwā·zhən }

Euler method [MECH] A method of studying fluid motion and the mechanics of deformable bodies in which one considers volume elements at fixed locations in space, across which material flows; the Euler method is in contrast to the Lagrangian method. { 'oi·lər ˌmeth·əd }

Euler number 1 [FL MECH] A dimensionless number used in the study of fluid friction in conduits, equal to the pressure drop due to friction divided by the product of the fluid density and the square of the fluid velocity. { 'ȯi·lər ˌnəm·bər 'wən }

Euler number 2 [FL MECH] A dimensionless number equal to two times the Fanning friction factor. { 'ȯi·lər ˌnəm·bər 'tü }

Euler-Rodrigues parameter [MECH] One of four numbers which may be used to specify the orientation of a rigid body; they are components of a quaternion. { ¦ȯi·lər rə'drē·gəs pəˌram·əd·ər }

Euler's expansion [FL MECH] The transformation of a derivative (*d/dt*) describing the behavior of a moving particle with respect to time, into a local derivative ($\delta/\delta t$) and three additional terms that describe the changing motion of a fluid as it passes through a fixed point. { 'ȯi·lərz ik ¦span·shən }

EUV radiation *See* vacuum ultraviolet radiation. { ¦ē¦yü¦vē ˌrād·ē'ā·shən }

eV *See* electronvolt.

evaporation [PHYS] Conversion of a liquid to the vapor state by the addition of latent heat. { iˌvap·ə'rā·shən }

E vector [ELECTROMAG] Vector representing the electric field of an electromagnetic wave. { 'ē ˌvek·tər }

even-even nucleus [NUC PHYS] A nucleus which has an even number of neutrons and an even number of protons. { ¦ē·vən ¦ē·vən ˌnü·klē·əs }

even harmonic [PHYS] A harmonic that is an even multiple of the fundamental frequency. { ¦ē·vən härˌmän·ik }

even-odd nucleus [NUC PHYS] A nucleus which has an even number of protons and an odd number of neutrons. { ¦ē·vən ¦äd 'nü·klē·əs }

event [PHYS] A point in space-time. { i'vent }

event horizon [RELAT] The boundary of a region of space-time from which it is not possible to escape to infinity. Symbolized I^+. { i'vent həˌrīz·ən }

Everett-Wheeler interpretation [QUANT MECH] An interpretation of quantum mechanics which holds (at least according to some expositions) that the subjective impression of having observed one and only one outcome of a given experiment is an illusion and that there exist other parallel universes, said to be equally real, in which the unknown outcomes are realized. Also known as many-world interpretation, relative-state interpretation. { ¦ev·rit 'wēl·ər ˌin·tər·prəˌtā·shən }

Evjen method [SOLID STATE] Method of calculating lattice sums in which groups of charges whose total charge is zero are taken together, so that the contribution of each group is small and the series rapidly converges. { 'ev·yən ˌmeth·əd }

Ewald-Kornfeld method [SOLID STATE] An extension of the Ewald method to calculate Coulomb energies of dipole arrays. { ¦ē·valt ¦kȯrn ˌfeld ˌmeth·əd }

Ewald method [SOLID STATE] Method of calculating lattice sums in which certain mathematical techniques are employed to make series converge rapidly. { 'ē·valt ˌmeth·əd }

Ewald sphere [SOLID STATE] A sphere superimposed on the reciprocal lattice of a crystal, used to determine the directions in which an x-ray or other beam will be reflected by a crystal lattice. { 'ē·valt ˌsfir }

E wave *See* transverse magnetic wave. { 'ē ˌwāv }

Ewing theory of ferromagnetism [SOLID STATE] Theory of ferromagnetic phenomena which assumes each atom is a permanent magnet which can turn freely about its center under the influence of applied fields and other magnets. { 'yü·iŋ ˌthē·ə·rē əv ˌfe·rō'mag·nəˌtiz·əm }

EXAFS *See* extended x-ray absorption fine structure.

exceptional space [QUANT MECH] A space used to describe a system with a finite number of degrees of freedom in a generalization of quantum mechanics; this generalization is achieved by reformulating quantum mechanics in terms of a Jordan algebra of observables and states, and then generalizing this to the exceptional Jordan algebra realized by the algebra of 3 × 3 Hermitian matrices over the Cayley numbers. { ek'sep·shən·əl 'spās }

excess conduction [SOLID STATE] Electrical conduction by excess electrons in a semiconductor. { 'ek,ses kən'dək·shən }

excess electron [SOLID STATE] Electron introduced into a semiconductor by a donor impurity and available for conduction. { 'ek,ses i'lek,trän }

exchange [QUANT MECH] **1.** Operation of exchanging the space and spin coordinates in a Schrödinger-Pauli wave function representing two identical particles; this operation must leave the wave function unchanged, except possibly for sign. **2.** Process of exchanging a real or virtual particle between two other particles. { iks,chānj }

exchange anisotropy [ELECTROMAG] Phenomenon observed in certain mixtures of magnetic materials under certain conditions, in which magnetization is favored in some direction (rather than merely along some axis); thought to be caused by exchange coupling across the interface between compounds when one is ferromagnetic and one is antiferromagnetic. { iks'chānj ,an·ə'sä·trə·pē }

exchange broadening [SPECT] The broadening of a spectral line by some type of chemical or spin exchange process which limits the lifetime of the absorbing or emitting species and produces the broadening via the Heisenberg uncertainty principle. { iks'chānj 'bròd·ən·iŋ }

exchange coefficient [FL MECH] A coefficient of eddy flux in turbulent flow, defined in analogy to those coefficients of the kinetic theory of gases. Also known as austausch coefficient; eddy coefficient; interchange coefficient. { iks'chānj ,kō·i,fish·ənt }

exchange current [ELEC] The magnitude of the current which flows through a galvanic cell when it is operating in a reversible manner. { iks'chānj ,kə·rənt }

exchange degeneracy [PART PHYS] Coincidence of two Regge trajectories for particles having the same quantum numbers (except for parity, charge parity, and G parity) where one would have expected separate trajectories for alternate Regge recurrences. [QUANT MECH] An exchange process that leads back to the original configuration. Also known as exchange symmetry. { iks'chānj dē'jen·ə·rə·sē }

exchange force [QUANT MECH] The force arising in an exchange interaction. { iks'chānj ,fòrs }

exchange integral [QUANT MECH] Integral over the coordinates of two identical particles which can be thought of as the interaction between a given state and a second state in which the coordinates of the particles are exchanged. { iks'chānj 'in·tə·grəl }

exchange interaction [QUANT MECH] **1.** An interaction represented by a potential involving exchange of space or spin coordinates, or both, of the particles involved; can be visualized physically in terms of exchange of particles. **2.** Any interaction which can be looked upon as due to exchange of particles. { iks'chānj ,int·ə'rak·shən }

exchange narrowing [SPECT] The phenomenon in which, when a spectral line is split and thereby broadened by some variable perturbation, the broadening may be narrowed by a dynamic process that exchanges different values of the perturbation. { iks'chānj 'nar·ə·wiŋ }

exchange operator [QUANT MECH] An operator which exchanges the spatial coordinates of the particles in a wave function, or their spins, or both positions and spins. { iks'chānj ,äp·ə,rād·ər }

exchange symmetry *See* exchange degeneracy. { iks'chānj 'sim·ə·trē }

excimer laser [OPTICS] A laser containing a noble gas, such as helium or neon, which is based on a transition between an excited state in which a metastable bond exists between two gas atoms and a rapidly dissociating ground state. { 'ek·sə·mər ,lā·zər }

excitation [ATOM PHYS] A process in which an atom or molecule gains energy from electromagnetic radiation or by collision, raising it to an excited state. [ELEC] The application of voltage to field coils to produce a magnetic field, as required for the operation of an excited-field loudspeaker or a generator. [NEUROSCI] A change in the electrical state of a neuron leading to an action potential. [QUANT MECH] The addition of energy to a particle or system of particles at ground state to produce an excited state. { ,ek,sī'tā·shən }

excitation curve [NUC PHYS] A curve showing the relative yield of a specified nuclear reaction as a function of the energy of the incident particles or photons. Also known as excitation function. { ,ek,sī'tā·shən ,kərv }

excitation energy [QUANT MECH] The minimum energy required to change a system from its ground state to a particular excited state. { ,ek,sī'tā·shən ,en·ər·jē }

excitation function [ATOM PHYS] The cross section for an incident electron to excite an atom to a particular excited state expressed as a function of the electron energy. [NUC PHYS] *See* excitation curve. { ,ek,sī'tā·shən ,fəŋk·shən }

excitation index [SPECT] In emission spectroscopy, the ratio of intensities of a pair of extremely nonhomologous spectra lines; used to provide a sensitive indication of variation in excitation conditions. { ,ek,sī'tā·shən ,in,deks }

excitation loss *See* core loss. { ,ek,sī'tā·shən ,lòs }

excitation potential [QUANT MECH] Electric potential which gives the excitation energy when multiplied by the magnitude of the electron charge. { ˌekˌsīˈtā·shən pəˌten·chəl }

excitation spectrum [SPECT] The graph of luminous efficiency per unit energy of the exciting light absorbed by a photoluminescent body versus the frequency of the exciting light. { ˌek ˌsīˈtā·shən ˌspek·trəm }

excitation volume [PHYS] In electron-probe microanalysis, the volume of the x-ray source used to penetrate and diffuse into the target sample. { ˌekˌsīˈtā·shən ˌväl·yəm }

excited state [QUANT MECH] A stationary state of higher energy than the lowest stationary state or ground state of a particle or system of particles. { ekˈsīd·əd ˈstāt }

excited-state effect [SOLID STATE] The motion of a crystal defect through a process in which the defect is first raised into an excited state and then decays, together with the surroundings, into a state in which motion of the defect readily occurs. { ekˈsīd·əd ˈstāt iˌfekt }

excited-state maser [PHYS] A maser whose amplifying transition has a terminal level that is not appreciably populated at thermal equilibrium for the ambient temperature. { ekˈsīd·əd ˈstāt ˈmā·zər }

exciter [ELEC] **1.** A small auxiliary generator that provides field current for an alternating-current generator. **2.** *See* exciter lamp. [ELECTROMAG] **1.** The portion of a directional transmitting antenna system that is directly connected to the transmitter. **2.** A loop or probe extending into a resonant cavity or waveguide. { ekˈsīd·ər }

exciting line [SPECT] The frequency of electromagnetic radiation, that is, the spectral line from a noncontinuous source, which is absorbed by a system in connection with some particular process. { ekˈsīd·iŋ ˌlīn }

exciton [SOLID STATE] An excited state of an insulator or semiconductor which allows energy to be transported without transport of electric charge; may be thought of as an electron and a hole in a bound state. { ˈek·səˌtän }

exclusion principle [QUANT MECH] The principle that no two fermions of the same kind may simultaneously occupy the same quantum state. Also known as Pauli exclusion principle. { ik ˈsklü·zhən ˌprin·sə·pəl }

exergy [THERMO] The portion of the total energy of a system that is available for conversion to useful work; in particular, the quantity of work that can be performed by a fluid relative to a reference condition, usually the surrounding ambient condition. { ˈeks·ər·jē }

exhaust velocity [FL MECH] The velocity of gaseous or other particles in the exhaust stream of the nozzle of a reaction engine, relative to the nozzle. { igˈzȯst vəˈläs·əd·ē }

exit pupil [OPTICS] The image of the aperture stop of an optical system formed in the image space by rays emanating from a point on the optical axis in the object space. { ˈeg·zət ˌpyü·pəl }

exline correction [FL MECH] Calculation of fluid-flow friction loss through annular sections with a correction for the flow eccentricity in the laminar-flow range. { ˈeksˌlīn kəˈrek·shən }

exoelectrons [PHYS] Electrons emitted from the surfaces of metals and certain ceramics after these surfaces have been freshly formed by a process such as abrasion or fracture; electrons obtain energy required for emission from processes such as establishment of surface films and rearrangement of disturbed atoms. { ¦ek·sō·iˈlekˌtränz }

exoergic *See* exothermic. { ¦ek·sō¦wər·jik }

exoergic collision *See* collision of the second kind. { ¦ek·sə¦wər·jik kəˈlizh·ən }

exogenous electrification [ELEC] The separation of electric charge in a conductor placed in a preexisting electric field, especially applied to the charge separation observed on metal-covered aircraft, resulting from induction effects, and by itself does not create any net total charge on the conductor. { ˌekˈsäj·ə·nəs iˌlek·trə·fəˈkā·shən }

exothermic [PHYS] Indicating liberation of heat. Also known as exoergic. { ¦ek·sōˈthər·mik }

exotic atom [ATOM PHYS] A system in which either the proton that forms the nucleus of a hydrogen atom is replaced by another particle (such as a muon, to form muonium, or a positron, to form positronium), one electron in an ordinary atom is replaced by another particle (such as a muon, pion, or antiproton), or both substitutions are made (as in antihydrogen). { ik¦säd·ik ˈad·əm }

exotic nucleus [NUC PHYS] An atomic nucleus in which the ratio of neutron number to proton number is much larger or much smaller than that of naturally occurring nuclei. { igˈzäd·ik ˈnü·klē·əs }

expanding universe [RELAT] A model of the universe describing the process defined in the astronomy definition, in which the universe is nonstatic, homogeneous, and isotropic; based on Einstein's field equations with a nonvanishing cosmical constant. { ik¦spand·iŋ ˈyü·nə·vərs }

expansion [PHYS] Process in which the volume of a constant mass of a substance increases. { ikˈspan·shən }

expansion coefficient *See* coefficient of cubical expansion. { ikˈspan·shən kō·əˈfish·ənt }

expansion ellipsoid [SOLID STATE] An ellipsoid whose axes have lengths which are proportional to the coefficient of linear expansion in the corresponding direction in a crystal. { ikˈspan·shən əˈlipˌsȯid }

expansion ratio [FL MECH] For the calculation of the mass flow of a gas out of a nozzle or other expanding duct, the ratio of the nozzle exit section area to the nozzle throat area, or the ratio of final to initial volume. { ikˈspan·shən ˌrā·shō }

expansion wave [FL MECH] A pressure wave or shock wave that decreases the density of air as the air passes through it. { ik'span·shən ˌwāv }

expansivity *See* coefficient of cubical expansion. { ˌek,span'siv·əd·ē }

expectation value [QUANT MECH] The average of the results of a large number of measurements of a quantity made on a system in a given state; in case the measurement disturbs the state, the state is reprepared before each measurement. { ˌek,spek'tā·shən ˌval·yü }

explicit symmetry breaking [PHYS] A phenomenon in which a system is not quite, but almost, the same for two configurations related by exact symmetry. { ik'splis·ət 'sim·ə·trē ˌbrāk·iŋ }

exploring coil [ELECTROMAG] A small coil used to measure a magnetic field or to detect changes produced in a magnetic field by a hidden object; the coil is connected to an indicating instrument either directly or through an amplifier. Also known as magnetic test coil; search coil. { ik 'splȯr·iŋ ˌkȯil }

explosion method [THERMO] Method of measuring the specific heat of a gas at constant volume by enclosing the gas with an explosive mixture, whose heat of reaction is known, in a chamber closed with a corrugated steel membrane which acts as a manometer, and by deducing the maximum temperature reached on ignition of the mixture from the pressure change. { ik 'splō·zhən ˌmeth·əd }

exponential decay [PHYS] The decrease of some physical quantity according to the exponential law $N(t)=N_0e^{t/\tau}$, where τ is a constant called the decay time. { ˌek·spə'n en·chəl di'kā }

exponential law [PHYS] The principle that growth or decay of some physical quantity is at a rate such that its value at a certain time or place is the initial value times e raised to a power equal to a constant times some convenient coordinate, such as the elapsed time or the distance traveled by a wave; there is growth if the constant is positive, decay if it is negative. { ˌek·spə'nen·chəl 'lȯ }

exponential pulse [PHYS] Variation of some quantity with time similar to the displacement of a critically damped harmonic oscillator which is initially given an impulse in its equilibrium position. { ˌek·spə'nen·chəl 'pəls }

exposure *See* light exposure; radiant exposure. { ik'spō·zhər }

exposure meter [OPTICS] An instrument used to measure the intensity of light reflected from an object, for the purpose of determining proper camera exposure. { ik'spō·zhər ˌmēd·ər }

exposure time [PHYS] The amount of time a material is illuminated or irradiated. { ik'spō·zhər ˌtīm }

extended close-coupling method [ATOM PHYS] A method of extending the close-coupling method to the case of ionizing collisions of an electron with a hydrogen atom by replacing the true continuum of ionized hydrogenic target states with a finite number of discrete, normalized, positive-energy pseudostates, while treating the incident electron with conventional, two-body scattering boundary conditions. { ik ¦stend·əd ˌklōs 'kəp·liŋ ˌmeth·əd }

extended dislocation [CRYSTAL] A dislocation in a close-packed structure consisting of a strip of stacking fault edged by two lines across which slip through a fraction of a lattice constant, into one of the alternative stacking positions, has occurred. { ik'stend·əd ˌdis,lō'kā·shən }

extended source [OPTICS] A source of radiation that can be resolved by the eye or a specified instrument into a geometrical image. { ik 'stend·əd 'sȯrs }

extended state [QUANT MECH] A state of motion in which an electron may be found anywhere within a region of a material of linear extent equal to that of the material itself. { ik¦stend·əd 'stāt }

extended x-ray absorption fine structure [PHYS] A variation in the x-ray absorption of a substance as a function of energy, at energies just above that required for photons to liberate core electrons into the continuum; it is due to interference between the outgoing photoelectron waves and electron waves backscattered from atoms adjacent to the absorbing atoms. Abbreviated EXAFS. { ik¦stend·əd 'eks,rā əb'sȯrp·shən ¦fīn 'strək·chər }

extensibility [MECH] The amount to which a material can be stretched or distorted without breaking. { ik,sten·sə'bil·əd·ē }

exterior ballistics [MECH] The science concerned with behavior of a projectile after leaving the muzzle of the firing weapon. { ek'stir·ē·ər bə'lis·tiks }

exterior complex scaling [ATOM PHYS] A mathematical transformation, which has been used to simplify the boundary conditions on the wave functions in an electron-atom collision, in which the variable r, representing the distance of the electron from the nucleus, is replaced, at values of r greater than some constant R, by $R + (r - R)C$, where C is a complex number with unit modulus and positive imaginary part. { ik¦stir·ē·ər ¦käm,pleks ˌskāl·iŋ }

external force [MECH] A force exerted on a system or on some of its components by an agency outside the system. { ek¦stərn·əl 'fȯrs }

external line [QUANT MECH] A component of a Feynman graph (in the diagrammatic presentation of perturbative quantum field theory) describing an incoming or outgoing particle in a scattering. { ek¦stərn·əl 'līn }

external wave [FL MECH] **1.** A wave in fluid motion having its maximum amplitude at an external boundary such as a free surface. **2.** Any surface wave on the free surface of a homogeneous incompressible fluid is an external wave. { ek¦stərn·əl 'wāv }

external work [THERMO] The work done by a system in expanding against forces exerted from outside. { ek¦stərn·əl 'wərk }

extinction [OPTICS] Phenomenon in which plane polarized light is almost completely absorbed by a polarizer whose axis is perpendicular to the plane of polarization. { ek'stiŋk·shən }

extinction meter [OPTICS] An exposure meter in which light intensity is measured by gradually attenuating the light by a known fraction until a selected design is just visible or disappears. { ek'stiŋk·shən ˌmēd·ər }

extraordinary index [OPTICS] The index of refraction of the extraordinary wave propagating in a direction perpendicular to the optical axis of a uniaxial crystal. { ik'strȯr·dənˌer·ē 'inˌdeks }

extraordinary ray [OPTICS] One of two rays into which a ray incident on an anisotropic uniaxial crystal is split; its deviation at the crystal's surface depends on the orientation of the crystal, and it is deviated even in the case of normal incidence. { ik'strȯr·dənˌer·ē 'rā }

extraordinary wave [OPTICS] Component of electromagnetic radiation propagating in an anisotropic uniaxial crystal whose electric displacement vector lies in the plane containing the optical axis and the direction normal to the wavefront; it gives rise to the extraordinary ray. Also known as extraordinary compoment. { ik'strȯr·dənˌer·ē 'wāv }

extraterrestrial noise [ELECTROMAG] Cosmic and solar noise; radio disturbances from sources other than those related to the earth. { ¦ek·strə·tə'res·trē·əl 'nȯiz }

extreme narrowing approximation [SPECT] A mathematical approximation in the theory of spectral-line shapes to the effect that the exchange narrowing of a perturbation is complete. { ek'strēm 'nar·ə·wiŋ əˌpräk·sə'mā·shən }

extreme relativistic limit [PHYS] Limit which a formula describing a particle's behavior approaches when the speed of the particle approaches the speed of light. { ek¦strēm ˌrel·ə·tə'vis·tik 'lim·ət }

extreme ultraviolet radiation *See* vacuum ultraviolet radiation. { ek¦strēm ˌəl·trə'vī·lət ˌrād·ē'ā·shən }

eyeglasses [OPTICS] Optical devices containing corrective lenses for defects of vision or for special purposes. { 'īˌglas·əs }

eye lens [OPTICS] The lens in a two-lens eyepiece which is nearer to the eye. { 'ī ˌlenz }

eyepiece [OPTICS] A lens or optical system which offers to the eye the image originating from another system (the objective) at a suitable viewing distance. Also known as ocular. { 'īˌpēs }

eyepoint [OPTICS] That point on the axis of a lens at which the brightest and sharpest visual image is obtained. { 'īˌpȯint }

Eykman formula [OPTICS] An empirical formula which relates the molal refraction of a liquid at a given optical frequency to its index of refraction, density, and molecular weight. { 'īk·mən ˌfȯr·myə·lə }

Eyring formula [FL MECH] A formula, based on the Eyring theory of rate processes, which relates shear stress acting on a liquid and the resulting rate of shear. { 'ī·riŋ ˌfȯr·myə·lə }

Eyring molecular system [FL MECH] Theory to account for liquid properties; assumes that each liquid molecule can move freely within a certain free volume. Also known as Eyring theory. { 'ī·riŋ mə'lek·yə·lər ˌsis·təm }

Eyring theory *See* Eyring molecular system. { 'ī·riŋ ˌthē·ə·rē }

F

f *See* Fanning friction factor..

F *See* farad.

fA *See* femtoampere.

Faber flaw [SOLID STATE] A deformation in a superconducting material that acts as a nucleation center for the growth of a superconducting region. { 'fā·bər ˌflȯ }

Fabry-Barot method [OPTICS] Method of determining the index of refraction of a prism in which the prism is set up so that the incident beam is perpendicular to the emergent face, and the index of refraction is calculated from the angle of the prism and the angle of deviation. { fä'brē bə'rō ˌmeth·əd }

Fabry-Pérot etalon *See* etalon. { fä'brē pə'rō 'ed·əlˌän }

Fabry-Pérot filter [OPTICS] An optical interference filter, similar to the Fabry-Pérot interferometer except that the space between the partially reflecting surfaces is only a few thousand angstroms. { fä'brē pə'rō 'fil·tər }

Fabry-Pérot fringes [OPTICS] Series of rings observed when a monochromatic light source is viewed through a Fabry-Pérot interferometer. { fä'brē pə'rō 'frin·jəz }

Fabry-Pérot interferometer [OPTICS] An interferometer having two parallel glass plates (whose separation of a few centimeters may be varied), silvered on their inner surfaces so that the incoming wave is multiply reflected between them and ultimately transmitted. { fä'brē pə'rō ˌint·ə·fə'räm·əd·ər }

face *See* crystal face. { fās }

face-centered cubic lattice [CRYSTAL] A lattice whose unit cells are cubes, with lattice points at the center of each face of the cube, as well as at the vertices. Abbreviated fcc lattice. { 'fās ˌsen·tərd ¦kyüb·ik 'lad·əs }

face-centered orthorhombic lattice [CRYSTAL] An orthorhombic lattice which has lattice points at the center of each face of a unit cell, as well as at the vertices. { 'fās ˌsen·tərd ¦ȯr·thō¦räm·bik 'lad·əs }

factor of safety [MECH] **1.** The ratio between the breaking load on a member, appliance, or hoisting rope and the safe permissible load on it. Also known as safety factor. **2.** *See* factor of stress intensity. { 'fak·tər əv 'sāf·tē }

factor of stress concentration [MECH] Any irregularity producing localized stress in a structural member subject to load. Also known as fatigue-strength reduction factor. { 'fak·tər əv 'stres ˌkäns·ənˌtrā·shən }

factor of stress intensity [MECH] The ratio of the maximum stress to which a structural member can be subjected, to the maximum stress to which it is likely to be subjected. Also known as factor of safety. { 'fak·tər əv 'stres inˌten·səd·ē }

fade chart [ELECTROMAG] Graph on which the null areas of an air-search radar antenna are plotted as an aid to estimating target altitude. { 'fād ˌchärt }

Fahrenheit scale [THERMO] A temperature scale; the temperature in degrees Fahrenheit (°F) is the sum of 32 plus 9/5 the temperature in degrees Celsius; water at 1 atmosphere (101,325 pascals) pressure freezes very near 32°F and boils very near 212°F. { 'far·ənˌhīt ˌskāl }

failure [MECH] Condition caused by collapse, break, or bending, so that a structure or structural element can no longer fulfill its purpose. { 'fāl·yər }

falling body [MECH] A body whose motion is accelerated toward the center of the earth by the force of gravity, other forces acting on it being negligible by comparison. { 'fȯl·iŋ 'bäd·ē }

falling-drop method [PHYS] Technique for measurement of liquid densities in which the time of fall of a drop of the sample liquid through a reference liquid is measured. { 'fȯl·iŋ ˌdräp ˌmeth·əd }

falling film [FL MECH] A theoretical liquid film that moves downward in even flow on a vertical surface in laminar flow; the concept is used for heat-and mass-transfer calculations. { 'fȯl·iŋ ˌfilm }

fall time [ELEC] Measure of time required for a circuit to change its output from a high level to a low level. { 'fȯl ˌtīm }

fall velocity *See* settling velocity. { 'fȯl vəˌlä·səd·ē }

false color [OPTICS] Color assigned to frequency bands that are normally invisible to the human eye (such as infrared radiation) in an image in order to enhance contrasts or to display those colors. { ¦fȯls ¦kəl·ər }

false pyroelectricity *See* tertiary pyroelectricity. { ¦fȯls ¦pī· rō·iˌlek'tri·səd·ē }

false white rainbow *See* fogbow. { ¦fȯls ¦wīt 'rān,bō }

fan [ELECTROMAG] Volume of space periodically energized by a radar beam (or beams) repeatedly traversing an established pattern. { fan }

fan antenna [ELECTROMAG] An array of folded dipoles of different length forming a wide-band ultra-high-frequency or very-high-frequency antenna. { 'fan an,ten·ə }

fan beam [ELECTROMAG] **1.** A radio beam having an elliptically shaped cross section in which the ratio of the major to the minor axis usually exceeds 3 to 1; the beam is broad in the vertical plane and narrow in the horizontal plane. **2.** A radar beam having the shape of a fan. { 'fan ,bēm }

fanned-beam antenna [ELECTROMAG] Unidirectional antenna so designed that transverse cross sections of the major lobe are approximately elliptical. { ¦fand ¦bēm an,ten·ə }

fanning beam [ELECTROMAG] Narrow antenna beam which is repeatedly scanned over a limited arc. { 'fan·iŋ ,bēm }

Fanning friction factor [FL MECH] A dimensionless number used in studying fluid friction in pipes, equal to the pipe diameter times the drop in pressure in the fluid due to friction as it passes through the pipe, divided by the product of the pipe length and the kinetic energy of the fluid per unit volume. Symbolized *f*. { 'fan·iŋ 'frik·shən ,fak·tər }

Fanning's equation [FL MECH] The equation expressing that frictional pressure drop of fluid flowing in a pipe is a function of the Reynolds number, rate of flow, acceleration due to gravity, and length and diameter of the pipe. { 'fan·iŋz i,kwā·zhən }

Fanno flow [FL MECH] An ideal flow used to study the flow of fluids in long pipes; the flow obeys the same simplifying assumptions as Rayleigh flow except that the assumption there is no friction is replaced by the requirement the flow be adiabatic. { 'fan·ō ,flō }

Fano effect [ATOM PHYS] The spin polarization of photoelectrons from alkali atoms that is produced upon the atoms' absorption of circularly polarized light. { 'fan·ō i,fekt }

farad [ELEC] The unit of capacitance in the meter-kilogram-second system, equal to the capacitance of a capacitor which has a potential difference of 1 volt between its plates when the charge on one of its plates is 1 coulomb, there being an equal and opposite charge on the other plate. Symbolized F. { 'fa,rad }

faraday [PHYS] The electric charge required to liberate 1 gram-equivalent of a substance by electrolysis; experimentally equal to 96,485.3415 ± 0.0039 coulombs. Also known as Faraday constant. { 'far·ə,dā }

Faraday birefringence [OPTICS] Difference in the indices of refraction of left and right circularly polarized light passing through matter parallel to an applied magnetic field; it is responsible for the Faraday effect. { 'far·ə,dā ,bī·ri'frin·jəns }

Faraday cage *See* Faraday shield. { 'far·ə,dā ,kāj }

Faraday constant *See* faraday. { 'far·ə,dā ,kän·stənt }

Faraday cylinder [ELEC] **1.** A closed, or nearly closed, hollow conductor, usually grounded, within which apparatus is placed to shield it from electrical fields. **2.** A nearly closed, insulated, hollow conductor, usually shielded by a second grounded cylinder, used to collect and detect a beam of charged particles. { 'far·ə,dā ,sil·ən·dər }

Faraday disk machine [ELECTROMAG] A device for demonstrating electromagnetic induction, consisting of a copper disk in which a radial electromotive force is induced when the disk is rotated between the poles of a magnet. Also known as Faraday generator. { 'far·ə,dā 'disk mə,shēn }

Faraday effect [OPTICS] Rotation of polarization of a beam of linearly polarized light when it passes through matter in the direction of an applied magnetic field; it is the result of Faraday birefringence. Also known as Faraday rotation; Kundt effect; magnetic rotation. { 'far·ə,dā i'fekt }

Faraday generator *See* Faraday disk machine. { 'far·ə,dā 'jen·ə,rād·ər }

Faraday ice bucket experiment [ELEC] Experiment in which one lowers a charged metal body into a pail and observes the effect on an electroscope attached to the pail, with and without contact between body and pail; the experiment shows that charge resides on a conductor's outside surface. { 'far·ə,dā 'īs ,bək·ət ik,spər·ə·mənt }

Faraday rotation *See* Faraday effect. { 'far·ə,dā rō'tā·shən }

Faraday rotation experiment [ELECTROMAG] An experiment in which a wire dipping in a pool of mercury surrounding a magnet rotates around the magnet when a current passes through it, demonstrating the effect of a magnetic field on a current-carrying conductor. { ¦far·ə,dā rō'tā·shən ik,sper·ə·mənt }

Faraday rotation isolator *See* ferrite isolator. { 'far·ə,dā rō'tā·shən 'īs·əl,ād·ər }

Faraday screen *See* Faraday shield. { 'far·ə,dā ,skrēn }

Faraday shield [ELEC] Electrostatic shield composed of wire mesh or a series of parallel wires, usually connected at one end to another conductor which is grounded. Also known as Faraday cage; Faraday screen. { 'far·ə,dā ,shēld }

Faraday's law of electromagnetic induction [ELECTROMAG] The law that the electromotive force induced in a circuit by a changing magnetic field is equal to the negative of the rate of change of the magnetic flux linking the circuit. Also known as law of electromagnetic induction. { 'far·ə,dāz 'lȯ əv i¦lek·trō,mag¦ned·ik in'dək·shən }

Faraday tube [ELEC] A tube of force for electric

displacement which is of such size that the integral over any surface across the tube of the component of electric displacement perpendicular to that surface is unity. { 'far·ə,dā ,tüb }

faradic current Also spelled faradaic current. [ELEC] An intermittent and nonsymmetrical alternating current like that obtained from the secondary winding of an induction coil. { fə'rad·ik ,kə·rənt }

far field *See* Fraunhofer region. { ¦fär ¦fēld }

far-infrared radiation [ELECTROMAG] Infrared radiation the wavelengths of which are the longest of those in the infrared region, about 50–1000 micrometers; requires diffraction gratings for spectroscopic analysis. { ¦fär in·frə'red ,rād·ē'ā·shən }

far point [OPTICS] The farthest point from an eye at which an object is distinctly seen; for a normal eye it is theoretically at infinity. Also known as punctum remotum. { ¦fär ¦pȯint }

far region *See* Fraunhofer region. { ¦fär ¦rē·jən }

far-ultraviolet radiation [ELECTROMAG] Ultraviolet radiation in the wavelength range of 200–300 nanometers; germicidal effects are greatest in this range. Abbreviated FUV radiation. { 'fär ,əl·trə'vī·lət ,rād·ē'ā·shən }

far zone *See* Fraunhofer region. { ¦fär ¦zōn }

fast axis [OPTICS] The direction of the electrical displacement vector of light propagating in an anisotropic crystal with the greatest possible phase velocity corresponding to a specified direction of propagation. { ¦fast 'ak·səs }

fast carbon-nitrogen-oxygen cycle *See* hot carbon-nitrogen-oxygen cycle. { ¦fast ¦kär·bən ¦nī·trə·jən ¦äks·ə·jən ,sī·kəl }

fast fission [NUC PHYS] Fission caused by fast neutrons. { ¦fast 'fish·ən }

fast-neutron spectrometry [NUC PHYS] Neutron spectrometry in which nuclear reactions are produced by or yield fast neutrons; such reactions are more varied than in the slow-neutron case. { ¦fast 'nü,trän spek'träm·ə·trē }

fast time constant [ELEC] An electric circuit which combines resistance and capacitance to give a short time constant for capacitor discharge through the resistor. { 'fast 'tīm ,kän·stənt }

fast-vibration direction [OPTICS] The direction of the electric field vector of the ray of light that travels with the greatest velocity in an anisotropic crystal and therefore corresponds to the minimum refractive index. { ¦fast vī'brā·shən də,rek·shən }

Fata Morgana [OPTICS] A complex mirage characterized by multiple distortions of images, generally in the vertical, so that such objects as cliffs or cottages are distorted and magnified into fantastic castles. { 'fäd·ə ,mȯr'gän·ə }

fatigue [MECH] Failure of a material by cracking resulting from repeated or cyclic stress. { fə'tēg }

fatigue life [MECH] The number of applied repeated stress cycles a material can endure before failure. { fə'tēg ,līf }

fatigue limit [MECH] The maximum stress that a material can endure for an infinite number of stress cycles without breaking. Also known as endurance limit. { fə'tēg ,lim·ət }

fatigue ratio [MECH] The ratio of the fatigue limit or fatigue strength to the static tensile strength. Also known as endurance ratio. { fə'tēg ,rā·shō }

fatigue strength [MECH] The maximum stress a material can endure for a given number of stress cycles without breaking. Also known as endurance strength. { fə'tēg ,streŋkth }

fatigue-strength reduction factor *See* factor of stress concentration. { fə'tēg ,streŋkth ri'dək·shən ,fak·tər }

fault [ELEC] A defect, such as an open circuit, short circuit, or ground, in a circuit, component, or line. Also known as electrical fault; faulting. { fȯlt }

faulting *See* fault. { 'fȯl·tiŋ }

F band [SOLID STATE] The optical absorption band arising from F centers. { 'ef ,band }

fcc lattice *See* face-centered cubic lattice. { ¦ef ¦sē¦sē 'lad·əs }

F center [SOLID STATE] A color center consisting of an electron trapped by a negative ion vacancy in an ionic crystal, such as an alkali halide or an alkaline-earth fluoride or oxide. { ¦ef ¦sen·tər }

F′ center [SOLID STATE] A color center that gives rise to a broad absorption band at longer wavelengths than the band of the F center; probably an F center that has trapped an additional electron. { ¦ef,prīm ,sen·tər }

Fechner color [OPTICS] A sensation of color caused by achromatic stimuli at intervals in time. { 'fek·nər ,kəl·ər }

Fedorov stage *See* universal stage. { fyȯ·dȯr·ȯf ,stāj }

feed [ELECTROMAG] The part of a radar antenna that is connected to or mounted on the end of the transmission line and serves to radiate radio-frequency electromagnetic energy to the reflector or receive energy therefrom. { fēd }

feed horn [ELECTROMAG] A device located at the focus of a receiving paraboloidal antenna that acts as a receiver of radio waves which the antenna collects, focuses, and couples to transmission lines to the amplifier. { 'fēd ,hȯrn }

f electron [ATOM PHYS] An atomic electron that has an orbital angular momentum quantum number of 3 in the central field approximation. { 'ef i,lek,trän }

femtoampere [ELEC] A unit of current equal to 10^{-15} ampere. Abbreviated fA. { ¦fem·tō¦am·pir }

femtometer [MECH] A unit of length, equal to 10^{-15} meter; used particularly in measuring nuclear distances. Abbreviated fm. Also known as fermi. { 'fem·tō,mēd·ər }

femtovolt [ELEC] A unit of voltage equal to 10^{-15} volt. Abbreviated fV. { 'fem·tō,vōlt }

Fermat's principle [OPTICS] The principle that an electromagnetic wave will take a path that involves the least travel time when propagating between two points. Also known as least-time principle; stationary time principle. { fer'mäz 'prin·sə·pəl }

fermi *See* femtometer. { 'fer·mē }

Fermi beta-decay theory [NUC PHYS] Theory in which a nucleon source current interacts with an electron-neutrino field to produce beta decay, in a manner analogous to the interaction of an electric current with an electromagnetic field during the emission of a photon of electromagnetic radiation. { 'fer·mē ¦bād·ə di¦kā ,thē·ə·rē }

Fermi constant [NUC PHYS] A universal constant, introduced in beta-disintegration theory, that expresses the strength of the interaction between the transforming nucleon and the electron-neutrino field. { 'fer·mē ,kän·stənt }

Fermi derivative [RELAT] A generalization of covariant differentiation along a curve that reduces to covariant differentiation when the curve is geodesic; an orthonormal tetrad constructed at each point along a timelike curve such that the Fermi derivative of the tetrad along the curve is zero has (1) its timelike basis vector equal to the curve's unit tangent vector and (2) its spatial basis vectors nonrotating along the curve. { 'fer·mē də,riv·əd·iv }

Fermi-Dirac distribution function [STAT MECH] A function specifying the probability that a member of an assembly of independent fermions, such as electrons in a semiconductor or metal, will occupy a certain energy state when thermal equilibrium exists. { ¦fer·mē di¦rak ,dis·trə'byü·shən ,fəŋk·shən }

Fermi-Dirac gas *See* Fermi gas. { ¦fer·mē di¦rak ,gas }

Fermi-Dirac statistics [STAT MECH] The statistics of an assembly of identical half-integer spin particles; such particles have wave functions antisymmetrical with respect to particle interchange and satisfy the Pauli exclusion principle. { ¦fer·mē di¦rak stə'tis·tiks }

Fermi distribution [SOLID STATE] Distribution of energies of electrons in a semiconductor or metal as given by the Fermi-Dirac distribution function; nearly all energy levels below the Fermi level are filled, and nearly all above this level are empty. { 'fer·mē ,dis·trə,byü·shən }

Fermi energy [STAT MECH] **1.** The average energy of electrons in a metal, equal to three-fifths of the Fermi level. **2.** *See* Fermi level. { 'fer·mē ,en·ər·jē }

Fermi gas [STAT MECH] An assembly of independent particles that obey Fermi-Dirac statistics, and therefore obey the Pauli exclusion principle; this concept is used in the free-electron theory of metals and in one model of the behavior of the nucleons in a nucleus. Also known as Fermi-Dirac gas. { 'fer·mē ,gas }

Fermi hole [SOLID STATE] A region surrounding an electron in a solid in which the energy band theory predicts that the probability of finding other electrons is less than the average over the volume of the solid. { 'fer·mē ,hōl }

Fermi interaction [PART PHYS] The direct interaction between four Dirac fields at a single point in space-time, postulated in conventional theories of the weak interactions.

Fermi level [STAT MECH] The energy level at which the Fermi-Dirac distribution function of an assembly of fermions is equal to one-half. Also known as Fermi energy. { 'fer·mē ,lev·əl }

Fermi liquid [CRYO] A liquid of particles which have Fermi-Dirac statistics; an example is the liquid phase of helium-3, in which the atoms belong to the isotope with mass number 3. { 'fer·mē ,lik·wəd }

fermion [QUANT MECH] A particle, such as the electron, proton, or neutron, which obeys the rule that the wave function of several identical particles changes sign when the coordinates of any pair are interchanged; it therefore obeys the Pauli exclusion principle. { 'fer·mē,än }

fermion field [QUANT MECH] An operator defined at each point in space-time that creates or annihilates a particular type of fermion and its antiparticle. { 'fer·mē,än ,fēld }

Fermi plot *See* Kurie plot. { 'fer·mē ,plät }

Fermi-propagated [RELAT] A vector field is said to be Fermi-propagated along a curve γ when it is constructed so that its Fermi derivative along γ is 0. { 'fer·mē ,präp·ə,gād·əd }

Fermi selection rules [NUC PHYS] Selection rules for beta decay in a Fermi transition; that is, there is no change in total angular momentum or parity of the nucleus in an allowed transition. { 'fer·mē si'lek·shən ,rülz }

Fermi's golden rules [QUANT MECH] The equations giving the first-order (rule number 2) and second-order (rule number 1) contributions to the transition probability per unit time induced by a perturbation Hamiltonian, in terms of matrix elements of the perturbation Hamiltonian. { ¦fer·mēz ¦gōld·ən 'rülz }

Fermi sphere [STAT MECH] The Fermi surface of an assembly of fermions in the approximation that the fermions are free particles. { 'fer·mē ,sfir }

Fermi surface [SOLID STATE] A constant-energy surface in the space containing the wave vectors of states of members of an assembly of independent fermions, such as electrons in a semiconductor or metal, whose energy is that of the Fermi level. { 'fer·mē ,sər·fəs }

Fermi temperature [STAT MECH] The energy of the Fermi level of an assembly of fermions divided by Boltzmann's constant, which appears as a parameter in the Fermi-Dirac distribution function. { 'fer·mē ,tem·prə·chər }

Fermi transition [NUC PHYS] Beta decay subject to Fermi selection rules. { 'fer·mē tran'zish·ən }

Ferranti effect [ELEC] A rise in voltage occurring at the end of a long transmission line when its load is disconnected. { fə'ran·tē i,fekt }

ferrimagnet *See* ferrimagnetic material. { 'fe·ri ,mag·nət }

ferrimagnetic limiter [ELECTROMAG] Power limiter used in microwave systems to replace transmit-receive tubes; uses ferrimagnetic material (such as a piece of ferrite or garnet) that exhibits nonlinear properties. { ˌfe·ri·mag'ned·ik 'lim·əd·ər }

ferrimagnetic material [SOLID STATE] A material displaying ferrimagnetism; the ferrites are the principal example. Also known as ferrimagnet. { ˌfe·ri·mag'ned·ik mə'tir·ē·əl }

ferrimagnetic resonance [PHYS] Magnetic resonance of a ferrimagnetic material. { ˌfe·ri·mag'ned·ik 'rez·ən·əns }

ferrimagnetism [SOLID STATE] A type of magnetism in which the magnetic moments of neighboring ions tend to align nonparallel, usually antiparallel, to each other, but the moments are of different magnitudes, so there is an appreciable resultant magnetization. Also known as Néel ferromagnetism. { ˌfe·ri'mag·nəˌtiz·əm }

ferrite [SOLID STATE] Any ferrimagnetic material having high electrical resistivity which has a spinel crystal structure and the chemical formula XFe_2O_4, where X represents any divalent metal ion whose size is such that it will fit into the crystal structure. { 'feˌrīt }

ferrite attenuator *See* ferrite limiter. { 'feˌrīt ə'ten·yəˌwād·ər }

ferrite circulator [ELECTROMAG] A combination of two dual-mode transducers and a 45° ferrite rotator, used with rectangular waveguides to control and switch microwave energy. Also known as ferrite phase-differential circulator. { 'feˌrīt 'sər·kyəˌlād·ər }

ferrite isolator [ELECTROMAG] A device consisting of a ferrite rod, centered on the axis of a short length of circular waveguide, located between rectangular-waveguide sections displaced 45° with respect to each other, which passes energy traveling through the waveguide in one direction while absorbing energy from the opposite direction. Also known as Faraday rotation isolator. { 'feˌrīt 'ī·səˌlād·ər }

ferrite limiter [ELECTROMAG] A passive, low-power microwave limiter having an insertion loss of less than 1 decibel when operating in its linear range, with minimum phase distortion; the input signal is coupled to a single-crystal sample of either yttrium iron garnet or lithium ferrite, which is biased to resonance by a magnetic field. Also known as ferrite attenuator. { 'feˌrīt 'lim·əd·ər }

ferrite phase-differential circulator *See* ferrite circulator. { 'feˌrīt ¦fāz dif·ə¦ren·chəl 'sər·kyəˌlād·ər }

ferrite-rod antenna [ELECTROMAG] An antenna consisting of a coil wound on a rod of ferrite; used in place of a loop antenna in radio receivers. Also known as ferrod; loopstick antenna. { 'feˌrīt ¦räd an'ten·ə }

ferrite rotator [ELECTROMAG] A gyrator consisting of a ferrite cylinder surrounded by a ring-type permanent magnet, inserted in a waveguide to rotate the plane of polarization of the electromagnetic wave passing through the waveguide. { 'feˌrīt 'rōˌtād·ər }

ferrite switch [ELECTROMAG] A ferrite device that blocks the flow of energy through a waveguide by rotating the electric field vector 90°; the switch is energized by sending direct current through its magnetizing coil; the rotated electromagnetic wave is then reflected from a reactive mismatch or absorbed in a resistive card. { 'feˌrīt 'swich }

ferrod *See* ferrite-rod antenna. { 'feˌräd }

ferroelectric [SOLID STATE] A crystalline substance displaying ferroelectricity, such as barium titanate, potassium dihydrogen phosphate, and Rochelle salt; used in ceramic capacitors, acoustic transducers, and dielectric amplifiers. Also known as Rochelle electric; Seignette electric. { ¦fe·rō·i'lek·trik }

ferroelectric Barkhausen effect [SOLID STATE] A series of abrupt changes in the dielectric polarization of a ferroelectric material that occurs when the external electric field acting on the material is varied. { ¦fe·rō·i¦lek·trik 'bärkˌhau̇z·ən iˌfekt }

ferroelectric crystal [SOLID STATE] A crystal of a ferroelectric material. { ¦fe·rō·i'lek·trik 'krist·əl }

ferroelectric domain [SOLID STATE] A region of a ferroelectric material within which the spontaneous polarization is constant. { ¦fe·rō·i'lek·trik də'mān }

ferroelectric hysteresis [ELEC] The dependence of the polarization of ferroelectric materials not only on the applied electric field but also on their previous history; analogous to magnetic hysteresis in ferromagnetic materials. Also known as dielectric hysteresis; electric hysteresis. { fe·rō·i'lek·trik ˌhis·tə'rē·səs }

ferroelectric hysteresis loop [ELEC] Graph of polarization or electric displacement versus applied electric field of a material displaying ferroelectric hysteresis. { ¦fe·rō·i'lek·trik ˌhis·tə'rē·səs ˌlüp }

ferroelectricity [SOLID STATE] Spontaneous electric polarization in a crystal; analogous to ferromagnetism. { ¦fe·rō·i'lek·tris·əd·ē }

ferroelectric shutter [OPTICS] A shutter consisting of a slab of ferroelectric crystal located between polarizers whose planes are at right angles; opens to pass light when activated by a pulse of up to 100 volts. { ¦fe·rō·i'lek·trik 'shəd·ər }

ferrohydrodynamics [PHYS] The study of the motion of strongly magnetizable fluids subjected to magnetic fields. { ˌfer·ōˌhī·drə·di'nam·iks }

ferromagnetic ceramic *See* ceramic magnet. { ¦fe·rō·mag¦ned·ik sə'ram·ik }

ferromagnetic crystal [SOLID STATE] A crystal of a ferromagnetic material. Also known as polar crystal. { ¦fe·rō·mag¦ned·ik 'krist·əl }

ferromagnetic domain [SOLID STATE] A region of a ferromagnetic material within which atomic or molecular magnetic moments are aligned parallel. Also known as magnetic domain. { ¦fe·rō·mag¦ned·ik də'mān }

ferromagnetic film *See* magnetic thin film. { ¦fe·rō·mag¦ned·ik 'film }

ferromagnetic material [SOLID STATE] A material displaying ferromagnetism, such as the various forms of iron, steel, cobalt, nickel, and their alloys. { ¦fe·rō·mag¦ned·ik mə'tir·ē·əl }

ferromagnetic resonance [SOLID STATE] Magnetic resonance of a ferromagnetic material. { ¦fe·rō·mag¦ned·ik 'rez·ən·əns }

ferromagnetic tape [ELECTROMAG] A tape made of magnetic material for use in winding closed magnetic cores of toroids and transformers. { ¦fe·rō·mag¦ned·ik 'tāp }

ferromagnetism [SOLID STATE] A property, exhibited by certain metals, alloys, and compounds of the transition (iron group) rare-earth and actinide elements, in which the internal magnetic moments spontaneously organize in a common direction; gives rise to a permeability considerably greater than that of vacuum, and to magnetic hysteresis. { ¦fe·rō'mag·nə,tiz·əm }

Féry spectrograph [SPECT] A spectrograph whose only optical element consists of a back-reflecting prism with cylindrically curved faces. { ¦fār·ē 'spek·trə,graf }

Feshbach resonance [QUANT MECH] A sharp resonance or peak which is seen when the cross section of an atomic or nuclear scattering process is plotted as a function of energy. It is associated with an energy threshold above which the scattering process can lead to a new result (such as excitation or ionization of one of the colliding objects), and it lies at an energy slightly below this threshold. { 'fesh,bäk ,rez·ən·əns }

Feynman diagram [QUANT MECH] A diagram which gives an intuitive picture of a term in a perturbation expansion of a scattering matrix element or other physical quantity associated with interactions of particles; each line represents a particle, each vertex an interaction. { 'fīn·mən ,dī·ə,gram }

Feynman integral [QUANT MECH] A term in a perturbation expansion of a scattering matrix element; it is an integral over the Minkowski space of various particles (or over the corresponding momentum space) of the product of propagators of these particles and quantities representing interactions between the particles. { 'fīn·mən 'int·ə·grəl }

Feynman propagator [QUANT MECH] A factor $(p + m)/(p^2 - m^2 + i\epsilon)$ in a transition amplitude corresponding to a line that connects two vertices in a Feynman diagram, and that represents a virtual particle. { 'fīn·mən 'präp·ə,gād·ər }

Feynman's rules [QUANT MECH] Rules for carrying out perturbation expansions in quantum field theory codified by Feynman diagrams. { 'fīn·mənz ,rülz }

Feynman's superfluidity theory [CRYO] Microscopic theory of superfluid helium which accounts for the spectrum of elementary excitations assumed by Landau's superfluidity theory. { 'fīn·mənz ,sü·pər,flü'id·əd·ē ,thē·ə·rē }

FG achromatism *See* actinic achromatism. { ¦ef¦jē ,ā'krō·mə,tiz·əm }

fiber [OPTICS] A transparent threadlike object made of glass or clear plastic, used to conduct light along selected paths. { 'fī·bər }

fiber bundle [OPTICS] A flexible bundle of glass or other transparent fibers, parallel to each other, used in fiber optics to transmit a complete image from one end of the bundle to the other. { 'fī·bər ,bən·dəl }

fiber diagram [SOLID STATE] The x-ray diffraction pattern of a collection of crystallites that have one crystallographic axis approximately parallel to a common direction but are otherwise randomly oriented. { 'fī·bər ,dī·ə,gram }

fiber-optic imaging [OPTICS] The formation of optical images by transmission through precisely aligned bundles of optical fibers; each fiber transmits one element of the image. { ¦fī·bər ¦äp·tik 'im·ij·iŋ }

fiber optics [OPTICS] The technique of transmitting light through long, thin, flexible fibers of glass, plastic, or other transparent materials; bundles of parallel fibers can be used to transmit complete images. { 'fī·bər ,äp·tiks }

fiberscope [OPTICS] An arrangement of parallel glass fibers with an objective lens on one end and an eyepiece at the other; the assembly can be bent as required to view objects that are inaccessible for direct viewing. { 'fī·bər,skōp }

fiber stress [MECH] **1.** The tensile or compressive stress on the fibers of a fiber metal or other fibrous material, especially when fiber orientation is parallel with the neutral axis. **2.** Local stress through a small area (a point or line) on a section where the stress is not uniform, as in a beam under bending load. { 'fī·bər ,stres }

fibrous fracture [MECH] Failure of a material resulting from a ductile crack; broken surfaces are dull and silky. Also known as ductile fracture. { 'fī·brəs 'frak·chər }

Fick's law [PHYS] The law that the rate of diffusion of matter across a plane is proportional to the negative of the rate of change of the concentration of the diffusing substance in the direction perpendicular to the plane. { 'fiks ,lȯ }

FID *See* free induction decay.

fiducial point [OPTICS] A mark, or one of several marks, visible in the field of view of an optical instrument, used as a reference or for measurement. Also known as fiduciary point. { fə'dü·shəl ,pȯint }

fiducial temperature [THERMO] Any of the temperatures assigned to a number of reproducible equilibrium states on the International Practical Temperature Scale; standard instruments are calibrated at these temperatures. { fə'dü·shəl 'tem·prə·chər }

fiduciary point *See* fiducial point. { fə'dü·shē,er·ē ,pȯint }

field [ELEC] That part of an electric motor or generator which produces the magnetic flux which reacts with the armature, producing the desired machine action. [OPTICS] *See* field of view. [PHYS] **1.** An entity which acts as an intermediary in interactions between particles, which is distributed over part or all of space, and whose properties are functions of space coordinates and, except for static fields, of time; examples include gravitational field, sound field, and the strain tensor of an elastic medium. **2.** The quantum-mechanical analog of this entity, in which the function of space and time is replaced by an operator at each point in space-time. { fēld }

field brightness *See* adaptation luminance. { 'fēld ˌbrīt·nəs }

field curvature *See* curvature of field. { 'fēld ˌkər·və·chər }

field desorption [SOLID STATE] A technique which tears atoms from a surface by an electric field applied at a sharp dip to produce very well-ordered, clean, plane surfaces of many crystallographic orientations. { ¦fēld dē'sȯrp·shən }

field-desorption mass spectroscopy [SPECT] A technique for analysis of nonvolatile molecules in which a sample is deposited on a thin tungsten wire containing sharp microneedles of carbon on the surface; a voltage is applied to the wire, thus producing high electric-field gradients at the points of the needles, and moderate heating then causes desorption from the surface or molecular ions, which are focused into a mass spectrometer. { ¦fēld dē'sȯrp·shən ¦mas spek 'trä·skə·pē }

field-effect display [OPTICS] A type of numerical display device in which a liquid-crystal cell is sandwiched between polarizers; the cell is treated so that it normally rotates light 90°, but ceases to rotate light when an electric field is applied to it, altering the transmission of the device. { 'fēld iˌfekt di'splā }

field flattener [OPTICS] A thin planoconvex lens placed in front of the photographic plate in some telescopes that have a curved focal plane so as to focus light on the flat plate. { 'fēld ˌflat·ən·ər }

field gradient [PHYS] **1.** A vector obtained by applying the del operator to a scalar field. **2.** A tensor obtained by dyadic multiplication of the del operator with a vector field. { 'fēld ˌgrād·ē·ənt }

field intensity *See* field strength. { 'fēld inˌten·səd·ē }

field lens [OPTICS] The lens in a two-lens eyepiece which is farther from the eye. { 'fēld ˌlenz }

field luminance *See* adaptation luminance. { 'fēld ˌlü·mə·nəns }

field of view [OPTICS] The area or solid angle which can be viewed through an optical instrument. Also known as field. { 'fēld əv 'vyü }

field operator [QUANT MECH] An operator function of space and time for the annihilation or creation of a particle. { 'fēld ˌäp·əˌrād·ər }

field pattern *See* radiation pattern. { 'fēld ˌpad·ərn }

field quenching [SOLID STATE] Decrease in the emission of light of a phosphor excited by ultraviolet radiation, x-rays, alpha particles, or cathode rays when an electric field is simultaneously applied. { 'fēld ˌkwench·iŋ }

field shift [NUC PHYS] The portion of the mass shift produced by the change in the size and shape of the nuclear charge distribution when neutrons are added to the nucleus. Also known as volume shift. { 'fēld ˌshift }

field stop [OPTICS] An opening, usually circular, in an opaque screen, whose edges determine the limits of the field of view of an optical instrument. { 'fēld ˌstäp }

field strength [PHYS] A vector characterizing a field. Also known as field intensity. { 'fēld ˌstreŋkth }

field theory [PHYS] A theory in which the basic quantities are fields; classically the equations governing the fields may be given; in quantum field theory the commutation rules satisfied by the field operators also are specified. { 'fēld ˌthē·ə·rē }

field waveguide [ELECTROMAG] A single wire, threaded or coated with dielectric, which guides an electromagnetic field. Also known as G string. { 'fēld 'wāvˌgīd }

Fierz interference [NUC PHYS] Interference between the axial vector and tensor parts of the weak interaction of nucleon and lepton (electron-neutrino) fields in beta decay; measurements of the beta-particle energy spectrum indicate that it vanishes. { 'firts ˌint·ə'fir·əns }

fifteen-degrees calorie *See* calorie. { ¦fif·tēn di ¦grēz ¦kal·ə·rē }

fifth sound [CRYO] A temperature oscillation which propagates in helium II contained in a porous material such as a tightly packed powder, where the normal component is immobilized by its viscosity. { ¦fifth ¦saůnd }

figuring [OPTICS] Grinding or polishing of surfaces of optical components to remove aberrations. { 'fig·yər·iŋ }

filled band [SOLID STATE] An energy band, each of whose energy levels is occupied by an electron. { ¦fild 'band }

filled shell [PHYS] A set of energy levels in an atom or nucleus which have approximately the same energy and which are all occupied. { ˌfild 'shel }

filling factor [CRYO] The ratio of the electron density of a quantum Hall liquid to the density of magnetic flux quanta. { 'fil·iŋ ˌfak·tər }

film boiling [THERMO] Boiling in which a continuous film of vapor forms at the hot surface of the container holding the boiling liquid, reducing heat transfer across the surface. { 'film ˌbȯil·iŋ }

film coefficient [THERMO] For a fluid confined in a vessel, the rate of flow of heat out of the fluid, per unit area of vessel wall divided by the difference between the temperature in the interior of the fluid and the temperature at the surface of the wall. Also known as convection coefficient. { 'film ˌkō·iˌfish·ənt }

film condensation [THERMO] The formation of a continuous film of liquid on a wall in contact with a vapor, when the wall is cooled below the local vapor saturation temperature and the liquid wets the cold surface. { 'film ˌkän·dənˌsā·shən }

film cooling [THERMO] The cooling of a body or surface, such as the inner surface of a rocket combustion chamber, by maintaining a thin fluid layer over the affected area. { 'film ˌkül·iŋ }

film pressure [PHYS] The difference between

the surface tension of a pure liquid and the surface tension of the liquid with a unimolecular layer of a given substance adsorbed on it. Also known as surface pressure. { 'film ˌpresh·ər }

film reader [OPTICS] A device for projecting or displaying microfilm so that an operator can read the data on the film; usually provided with equipment for moving or holding the film. { 'film ˌrēd·ər }

film resistor [ELEC] A fixed resistor in which the resistance element is a thin layer of conductive material on an insulated form; the conductive material does not contain binders or insulating material. { 'film riˌzis·tər }

film theory [PHYS] A theory of the transfer of material or heat across a phase boundary, where one or both of the phases are flowing fluids, the main controlling factor being resistance to heat conduction or mass diffusion through a relatively stagnant film of the fluid next to the surface. Also known as boundary-layer theory. { 'film ˌthē·ə·rē }

filter [OPTICS] An optical element that partially absorbs incident electromagnetic radiation in the visible, ultraviolet, or infrared spectra, consisting of a pane of glass or other partially transparent material, or of films separated by narrow layers; the absorption may be either selective or nonselective with respect to wavelength. Also known as optical filter. { 'fil·tər }

filter factor [OPTICS] The number of times the exposure must be increased when a filter is used on a camera, because the filter absorbs some of the light. { 'fil·tər ˌfak·tər }

filter slot [ELECTROMAG] Choke in the form of a slot designed to suppress unwanted modes in a waveguide. { 'fil·tər ˌslät }

filter spectrophotometer [SPECT] Spectrophotographic analyzer of spectral radiations in which a filter is used to isolate narrow portions of the spectrum. { 'fil·tər spek·trə·fə'täm·əd·ər }

finder [OPTICS] A small telescope having a wide-angle lens and low power, which is attached to a larger telescope and points in the same direction; used to locate objects that are to be viewed in the larger telescope. { 'fīnd·ər }

fine structure [ATOM PHYS] The splitting of spectral lines in atomic and molecular spectra caused by the spin angular momentum of the electrons and the coupling of the spin to the orbital angular momentum. { ¦fīn 'strək·chər }

fine-structure constant [PHYS] A fundamental dimensionless constant, equal to $e^2/(4\pi\epsilon_0\hbar c)$ in International System (SI) units and to $e^2/(\hbar c)$ in centimeter-gram-second (cgs) electrostatic units, where e is the elementary charge, $\hbar$ is Planck's constant divided by 2π, c is the speed of light, and ϵ_0 is the electric constant; numerically, it is equal to 0.007 297 352 533 ± 0.000 000 000 027 or to 1/(137.035 999 76 ± 0.000 000 50); symbolized α. Also known as Sommerfeld fine-structure constant. { 'fīn ˌstrək·chər 'kän·stənt }

finite elasticity theory *See* finite strain theory. { ¦fīˌnīt iˌlas'tis·əd·ē ˌthē·ə·rē }

finite strain theory [MECH] A theory of elasticity, appropriate for high compressions, in which it is not assumed that strains are infinitesimally small. Also known as finite elasticity theory. { 'fīˌnīt 'strān ˌthē·ə·rē }

finsen unit [ELECTROMAG] A unit of intensity of ultraviolet radiation, equal to the intensity of ultraviolet radiation at a specified wavelength whose energy flux is 100,000 watts per square meter; the wavelength usually specified is 296.7 nanometers. Abbreviated FU. { 'fin·sən ˌyü·nət }

fin waveguide [ELECTROMAG] Waveguide containing a thin longitudinal metal fin that serves to increase the wavelength range over which the waveguide will transmit signals efficiently; usually used with circular waveguides. { ¦fin 'wāvˌgīd }

firmoviscosity [MECH] Property of a substance in which the stress is equal to the sum of a term proportional to the substance's deformation, and a term proportional to its rate of deformation. { ¦fər·mō·vis¦käs·əd·ē }

first-class current [PART PHYS] A weak-interaction current whose charge symmetry (or G parity) properties are the same as those of currents which arise in the Fermi theory of beta decay. { ¦fərst ˌklas 'kə·rənt }

first Fresnel zone [ELECTROMAG] Circular portion of a wavefront transverse to the line between an emitter and a more distant point, where the resultant disturbance is being observed, whose center is the intersection of the front with the direct ray, and whose radius is such that the shortest path from the emitter through the periphery to the receiving point is one-half wavelength longer than the direct ray. { ¦fərst frə'nel ˌzōn }

first harmonic *See* fundamental. { ¦fərst här 'män·ik }

first law of motion *See* Newton's first law. { 'fərst ˌlȯ əv 'mō·shən }

first law of thermodynamics [THERMO] The law that heat is a form of energy, and the total amount of energy of all kinds in an isolated system is constant; it is an application of the principle of conservation of energy. { 'fərst ˌlȯ əv ˌthər·mō·dī'nam·iks }

first-order spectrum [SPECT] A spectrum, produced by a diffraction grating, in which the difference in path length of light from adjacent slits is one wavelength. { ¦fərst ˌȯrd·ər 'spek·trəm }

first-order theory [OPTICS] *See* Gaussian optics. [PHYS] A theory which takes into account only the most important terms, such as the term proportional to the independent variable in the series expansion of a function appearing in the theory. { ¦fərst ˌȯrd·ər 'thē·ə·rē }

first-order transition [THERMO] A change in state of aggregation of a system accompanied by a discontinuous change in enthalpy, entropy, and volume at a single temperature and pressure. { ¦fərst ˌȯrd·ər trans'zish·ən }

first radiation constant [STAT MECH] A constant appearing in the Planck radiation formula; its value depends on the form of the formula used; in the formula for power emitted by a blackbody per unit area per unit wavelength interval, it is 2π times Planck's constant, times the square of the speed of light, or approximately 3.74177×10^{-16} watt (meter)^2. Symbolized c_1; C_1. { 'fərst ˌrād·ē'ā·shən ˌkän·stənt }

first sound [CRYO] Ordinary sound in helium II, in which pressure and density variations are propagated; in contrast to second sound. { 'fərst ˌsau̇nd }

Fischer-Hinnen method [ELEC] Method of analysis of a complex waveform which has like loops above and below the time axis, in which the amplitude and phase of the *n*-th harmonic is determined from the ordinates of the resultant wave at a series of times which divide the half wave into $2n$ equal time intervals. { ¦fish·ər¦hin·ən ˌmeth·əd }

fish-bone antenna [ELECTROMAG] **1.** Antenna consisting of a series of coplanar elements arranged in collinear pairs, loosely coupled to a balanced transmission line. **2.** Directional antenna in the form of a plane array of doublets arranged transversely along both sides of a transmission line. { 'fish ˌbōn anˌten·ə }

fish-eye lens [OPTICS] A photographic lens that has a highly curved protruding front, enabling it to cover an angle of about 180°; provides a circular image with barrel distortion. { ¦fish ¦ī 'lenz }

fishpole antenna *See* whip antenna. { 'fishˌpōl anˌten·ə }

fission [NUC PHYS] The division of an atomic nucleus into parts of comparable mass; usually restricted to heavier nuclei such as isotopes of uranium, plutonium, and thorium. Also known as atomic fission; nuclear fission. { 'fish·ən }

fission barrier [NUC PHYS] One or more maxima in the plot of potential energy against nuclear deformation of a heavy nucleus, which inhibits spontaneous fission of the nucleus. { 'fish·ən ˌbar·ē·ər }

fission cross section [NUC PHYS] The cross section for a bombarding neutron, gamma ray, or other particle to induce fission of a nucleus. { ¦fish·ən 'kro̍ˌsek·shən }

fission isomer [NUC PHYS] A highly deformed nuclear state lying in the second well of a double-hump fission barrier. { ¦fish·ən 'ī·sə·mər }

fission neutron [NUC PHYS] A neutron emitted as a result of nuclear fission. { ¦fish·ən 'nüˌträn }

fission product [NUC PHYS] Any radioactive or stable nuclide resulting from fission, including both primary fission fragments and their radioactive decay products. { 'fish·ən ˌpräd·əkt }

fission spectrum [NUC PHYS] The energy distribution of neutrons arising from fission. { 'fish·ən ˌspek·trəm }

fission threshold [NUC PHYS] The minimum kinetic energy of a bombarding neutron required to induce fission of a nucleus. { 'fish·ən ˌthreshˌhōld }

fission yield [NUC PHYS] The percent of fissions that gives a particular nuclide or group of isobars. { 'fish·ən ˌyēld }

FitzGerald-Lorentz contraction [RELAT] The contraction of a moving body in the direction of its motion when its speed is comparable to the speed of light. Also known as Lorentz contraction; Lorentz-FitzGerald contraction. { fits¦jer·əld lə'rens kənˌtrak·shən }

five-fourths power law [THERMO] The proposition that the rate of heat loss from a body by free convection is proportional to the five-fourths power of the difference between the temperature of the body and that of its surroundings. { ¦fīv ¦fo̍rths 'pau̇·ər ˌlo̍ }

fixed capacitor [ELEC] A capacitor having a definite capacitance value that cannot be adjusted. { ¦fikst kə'pas·əd·ər }

fixed end [MECH] An end of a structure, such as a beam, that is clamped in place so that both its position and orientation are fixed. { 'fikst ˌend }

fixed end moment *See* fixing moment. { 'fikst ˌend 'mō·mənt }

fixed-focus lens [OPTICS] A lens whose focus is invariable, as on inexpensive cameras with no mechanism for adjusting focus but so designed that all objects from a few feet away to infinity are tolerably in focus. { 'fikst ˌfō·kəs 'lenz }

fixed-point attractor [PHYS] An attractor that consists of a single point in phase space and describes a stationary state of a system. { 'fikst ¦po̍int ə'trak·tər }

fixed resistor [ELEC] A resistor that has no provision for varying its resistance value. { ¦fikst ri'zis·tər }

fixing moment [MECH] The bending moment at the end support of a beam necessary to fix it and prevent rotation. Also known as fixed end moment. { 'fik·siŋ ˌmō·mənt }

Fizeau effect [OPTICS] The change in the speed of light in a material medium that results from the motion of the medium relative to the source and to the observer. { 'fē'zō iˌfekt }

Fizeau fringes [OPTICS] **1.** Interference fringes of monochromatic light from interference in a geometrical situation other than plane parallel plates. Also known as fringes of equal thickness. **2.** Interference fringes in light from a Fizeau interferometer. { fē'zō ˌfrin·jəz }

Fizeau interferometer [OPTICS] Interferometer in which light from a point source is collimated and multiply reflected between a plane mirror and the partially silvered inner surface of a parallel plane plate, and is viewed in reflection. { fē'zō ˌin·tər·fə'räm·əd·ər }

Fizeau toothed wheel [OPTICS] Rapidly rotating toothed wheel which was used to measure the speed of light by adjusting the rotation speed until light passing through one tooth opening and reflected from a distant mirror would pass through the next tooth opening on return. { fē'zō ¦tütht 'wēl }

flame emission spectroscopy [SPECT] A flame photometry technique in which the solution containing the sample to be analyzed is optically excited in an oxyhydrogen or oxyacetylene flame. { 'flām i,mish·ən spek'träs·kə·pē }

flame excitation [SPECT] Use of a high-temperature flame (such as oxyacetylene) to excite spectra emission lines from alkali and alkaline-earth elements and metals. { ¦flām ,ek·sī'tā·shən }

flame laser [OPTICS] A molecular gas laser in which gases such as carbon disulfide and oxygen are mixed at low pressures and ignited; the flame is then self-sustaining and produces carbon monoxide laser emission. { 'flām ,lā·zər }

flame photometer [SPECT] One of several types of instruments used in flame photometry, such as the emission flame photometer and the atomic absorption spectrophotometer, in each of which a solution of the chemical being analyzed is vaporized; the spectral lines resulting from the light source going through the vapors enters a monochromator that selects the band or bands of interest. { 'flām ,fə'täm·əd·ər }

flame photometry [SPECT] A branch of spectrochemical analysis in which samples in solution are excited to produce line emission spectra by introduction into a flame. { 'flām fə'täm·ə·tre }

flame spectrometry [SPECT] A procedure used to measure the spectra or to determine wavelengths emitted by flame-excited substances. { ¦flām spek'träm·ə·trē }

flame spectrophotometry [SPECT] A method used to determine the intensity of radiations of various wavelengths in a spectrum emitted by a chemical inserted into a flame. { ¦flām ¦spek·trə·fə'täm·ə·trē }

flame spectrum [SPECT] An emission spectrum obtained by evaporating substances in a nonluminous flame. { 'flām ,spek·trəm }

flap attenuator [ELECTROMAG] A waveguide attenuator in which a contoured sheet of dissipative material is moved into the guide through a nonradiating slot to provide a desired amount of power absorption. Also known as vane attenuator. { ¦flap ə'ten·yə,wād·ər }

flare *See* horn antenna. { fler }

flare spot [OPTICS] A small, diffuse, brightly illuminated region produced by multiple reflections of light from the various surfaces of an optical system. { 'fler ,spät }

flash factor [OPTICS] In photography using a photoflash lamp, a number dependent on the lamp and the film speed, equal to the product of the distance of the lamp from the subject and the correct *f*-number for that distance. { 'flash ,fak·tər }

flashing [OPTICS] The apparent filling of a curved mirror or lens with light when viewed from a distance, as a result of the production of a parallel beam by a light source at the focus. { 'flash·iŋ }

flash magnetization [ELECTROMAG] Magnetization of a ferromagnetic object by a current impulse of short duration. { ¦flash ,mag·nə·tə'zā·shən }

flashover [ELEC] An electric discharge around or over the surface of an insulator. { 'flash,ō·vər }

flash spectroscopy [SPECT] The study of the electronic states of molecules after they absorb energy from an intense, brief light flash. { ¦flash spek'träs·kə·pē }

flat [ACOUS] A musical note that is a half step lower than a specified note. { flat }

flat line [ELECTROMAG] A radio-frequency transmission line, or part thereof, having essentially 1-to-1 standing wave ratio. { 'flat ,līn }

flat space-time [RELAT] Space-time in which the Riemann-Christoffel tensor vanishes; geometry is then equivalent to that of the Minkowski universe used in special relativity. { 'flat ¦spās ¦tīm }

flat spin [MECH] Motion of a projectile with a slow spin and a very large angle of yaw, happening most frequently in fin-stabilized projectiles with some spin-producing moment, when the period of revolution of the projectile coincides with the period of its oscillation; sometimes observed in bombs and in unstable spinning projectiles. { ¦flat 'spin }

flat-top antenna [ELECTROMAG] An antenna having two or more lengths of wire parallel to each other and in a plane parallel to the ground, each fed at or near its midpoint. { 'flat ,täp an,ten·ə }

flat-top boom [ACOUS] A sonic boom whose pressure signature is shaped to reduce the perceived amplitude of the shocks by allowing plateaus following the bow shock and preceding the tail shock. { ¦flat ¦täp 'büm }

flat trajectory [MECH] A trajectory which is relatively flat, that is, described by a projectile of relatively high velocity. { ¦flat trə'jek·trē }

flavor [PART PHYS] A label used to distinguish different types of leptons (the electron, electron neutrino, muon, muon neutrino, and possibly others) and different color triplets of quarks (the up, down, strange, and charmed quarks, and possibly others). { 'flā·vər }

fl dr *See* fluid dram.

Fleming-Kennelly law [ELECTROMAG] The reluctivity of a ferromagnetic substance varies linearly with magnetic field strength at points near magnetic saturation. { ¦flem·iŋ 'ken·ə·lē ,lȯ }

Fleming's rule *See* left-hand rule; right-hand rule. { 'flem·iŋz ,rül }

Fletcher-Munson contour *See* equal loudness contour. { ¦flech·ər ¦mən·sən 'kän,tür }

flexibility [MECH] The quality or state of being able to be flexed or bent repeatedly. { ,flek·sə'bil·əd·ē }

flexible coupling [ELECTROMAG] A coupling designed to allow a limited angular movement between the axes of two waveguides. { ,flek·sə·bəl 'kəp·liŋ }

flexible waveguide [ELECTROMAG] A waveguide that can be bent or twisted without appreciably changing its electrical properties. { ,flek·sə·bəl 'wāv,gīd }

flexural modulus [MECH] A measure of the resistance of a beam of specified material and cross section to bending, equal to the product of Young's modulus for the material and the square of the radius of gyration of the beam about its neutral axis. { 'flek·shə·rəl 'mäj·ə·ləs }

flexural rigidity [MECH] The ratio of the sideward force applied to one end of a beam to the resulting displacement of this end, when the other end is clamped. { 'flek·shə·rəl ri'jid·əd·ē }

flexural strength [MECH] Strength of a material in blending, that is, resistance to fracture. { 'flek·shə·rəl 'streŋkth }

flexure [MECH] **1.** The deformation of any beam subjected to a load. **2.** Any deformation of an elastic body in which the points originally lying on any straight line are displaced to form a plane curve. { 'flek·shər }

flexure theory [MECH] Theory of the deformation of a prismatic beam having a length at least 10 times its depth and consisting of a material obeying Hooke's law, in response to stresses within the elastic limit. { 'flek·shər ˌthē·ə·rē }

flicker [OPTICS] A visual sensation produced by periodic fluctuations in light at rates ranging from a few cycles per second to a few tens of cycles per second. { 'flik·ər }

flicker photometer [OPTICS] A photometer in which a single field of view is alternately illuminated by the light sources to be compared, and the rate of alternation is such that color flicker is absent but brightness flicker is not; disappearance of flicker signifies equality of luminance. { 'flik·ər fə'täm·əd·ər }

F line [SPECT] A green-blue line in the spectrum of hydrogen, at a wavelength of 486.133 nanometers. { 'ef ˌlīn }

flip coil [ELECTROMAG] A small coil used to measure the strength of a magnetic field; it is placed in the field, connected to a ballistic galvanometer or other instrument, and suddenly flipped over 180°; alternatively, the coil may be held stationary and the magnetic field reversed. { 'flip ˌkȯil }

flip-flop amplifier *See* wall-attachment amplifier. { 'flipˌfläp ˌam·pləˌfī·ər }

flip-over process *See* Umklapp process. { 'flipˌō·vər ˌpräs·əs }

floating neutral [ELEC] Neutral conductor whose voltage to ground is free to vary when circuit conditions change. { ¦flōd·iŋ 'nü·trəl }

floating reticle [OPTICS] A reticle the image of which is movable within the field of view. { ¦flōd·iŋ 'red·ə·kəl }

florentium *See* promethium-147. { flȯ'ren·chəm }

flotation analysis [PHYS] Technique to measure liquid density in which a float of known density is adjusted with weights to match that of the liquid. { flō'tā·shən əˌnal·ə·səs }

flow [FL MECH] The forward continuous movement of a fluid, such as gases, vapors, or liquids, through closed or open channels or conduits. [PHYS] The movement of electric charges, gases, liquids, or other materials or quantities. { flō }

flowability [FL MECH] Capability of a liquid or loose particulate solid to move by flow. { ˌflō·ə'bil·əd·ē }

flow coefficient [FL MECH] An experimentally determined proportionality constant, relating the actual velocity of fluid flow in a pipe, duct, or open channel to the theoretical velocity expected under certain assumptions. { ¦flō ˌkō·i'fish·ənt }

flow cone [FL MECH] One of a collection of elements of conical shape that may be attached to an aerodynamic surface to visualize the fluid flow over the surface. { 'flō ˌkōn }

flow curve [FL MECH] A graph of the total shear of a fluid as a function of time. [MECH] The stress-strain curve of a plastic material. { 'flō ˌkərv }

flow distribution *See* flow field. { 'flō ˌdis·trə ˌbyü·shən }

flow equation [FL MECH] Equation for the calculation of fluid (gas, vapor, liquid) flow through conduits or channels; consists of an interrelation of fluid properties (such as density or viscosity), environmental conditions (such as temperature or pressure), and conduit or channel geometry and conditions (such as diameter, cross-sectional shape, or surface roughness). { 'flō iˌkwā·zhən }

flow field [FL MECH] The velocity and the density of a fluid as functions of position and time. Also known as flow distribution. { 'flō ˌfēld }

flow figure *See* strain figure. { 'flō ˌfig·yər }

flow-induced vibration [FL MECH] Structural and mechanical oscillations of structures immersed in or conveying fluid flow as a result of an interaction between the fluid-dynamic forces and the inertia, damping, and elastic forces in the structures. { ¦flō inˌdüst vī'brā·shən }

flowing-temperature factor [THERMO] Calculation correction factor for gases flowing at temperatures other than that for which a flow equation is valid, that is, other than 60°F (15.5°C). { ¦flō·iŋ 'tem·prə·chər ˌfak·tər }

flow net [FL MECH] A diagram used in studying the flow of a fluid through a permeable substance (such as water through a soil structure) having two nests of curves, one representing the flow lines, which follow the path of the fluid, and the other the equipotential lines, which connect points of equal head. { 'flō ˌnet }

flow noise [ACOUS] **1.** Pressure variations associated with a turbulent flow field that do not propagate away from the turbulent source but are sensed as sound by a receiver in direct contact or close to the turbulent flow. Also known as near-field noise. **2.** More generally, any noise generated by turbulent fluid flow. { 'flō ˌnȯiz }

flow pattern [FL MECH] Pattern of two-phase flow in a conduit or channel pipe, taking into consideration the ratio of gas to liquid and conditions of flow resistance and liquid holdup. { 'flō ˌpad·ərn }

flow rate [FL MECH] Also known as rate of flow. **1.** Time required for a given quantity of flowable material to flow a measured distance.

2. Weight or volume of flowable material flowing per unit time. { 'flō ˌrāt }

flow resistance [FL MECH] **1.** Any factor within a conduit or channel that impedes the flow of fluid, such as surface roughness or sudden bends, contractions, or expansions. **2.** *See* viscosity. { 'flō riˌzis·təns }

flow separation *See* boundary-layer separation. { 'flō ˌsep·əˌrā·shən }

flow stress [MECH] The stress along one axis at a given value of strain that is required to produce plastic deformation. { 'flō ˌstres }

fl oz *See* fluid ounce.

fluctuating current [ELEC] Direct current that changes in value but not at a steady rate. { ¦fləkchəˌwād·iŋ 'kə·rənt }

fluctuation theory [OPTICS] The theory proposed by M. von Smoluchowski and A. Einstein which states that the scattering of light occurs in pure water because random molecular motion causes density variations which effect changes in the refraction of light. { ˌflək·chə'wā·shən ˌthē·ə·rē }

fluctuation velocity *See* eddy velocity. { ˌflək·chə'wā·shən vəˌläs·əd·ē }

fluence [ELECTROMAG] The total energy per unit area carried by a pulse of electromagnetic radiation. [PHYS] A measure of time-integrated particle flux, expressed in particles per square centimeter. { 'flü·əns }

fluid [PHYS] An aggregate of matter in which the molecules are able to flow past each other without limit and without fracture planes forming. { 'flü·əd }

fluid density [FL MECH] The mass of a fluid per unit volume. { ¦flü·əd ¦den·səd·ē }

fluid dram [MECH] Abbreviated fl dr. **1.** A unit of volume used in the United States for measurement of liquid substances, equal to 1/8 fluid ounce, or $3.6966911953125 \times 10^{-6}$ cubic meter. **2.** A unit of volume used in the United Kingdom for measurement of liquid substances and occasionally of solid substances, equal to 1/8 fluid ounce or $3.5516328125 \times 10^{-6}$ cubic meter. { ¦flü·əd 'dram }

fluid dynamics [FL MECH] The science of fluids in motion. { ¦flü·əd dī'nam·iks }

fluid friction [FL MECH] Conversion of mechanical energy in fluid flow into heat energy. { ¦flü·əd 'frik·shən }

fluidity [FL MECH] The reciprocal of viscosity; expresses the ability of a substance to flow. { flü'id·əd·ē }

fluid mechanics [MECH] The science concerned with fluids, either at rest or in motion, and dealing with pressures, velocities, and accelerations in the fluid, including fluid deformation and compression or expansion. { ¦flü·əd mə'kan·iks }

fluid ounce [MECH] Abbreviated fl oz. **1.** A unit of volume that is used in the United States for measurement of liquid substances, equal to 1/16 liquid pint, or 231/128 cubic inches, or $2.95735295625 \times 10^{-5}$ cubic meter. **2.** A unit of volume used in the United Kingdom for measurement of liquid substances, and occasionally of solid substances, equal to 1/20 pint or $2.84130625 \times 10^{-5}$ cubic meter. { ¦flü·əd 'au̇ns }

fluid resistance [FL MECH] The force exerted by a gas or liquid opposing the motion of a body through it. Also known as resistance. { ¦flü·əd ri'zis·təns }

fluid statics [FL MECH] The determination of pressure intensities and forces exerted by fluids at rest. { ¦flü·əd 'stad·iks }

fluid stress [MECH] Stress associated with plastic deformation in a solid material. { ¦flü·əd 'stres }

fluid ton [MECH] A unit of volume equal to 32 cubic feet or approximately 0.90614 cubic meter; used for many hydrometallurgical, hydraulic, and other industrial purposes. { ¦flü·əd 'tən }

fluophor *See* luminophor. { 'flü·əˌfȯr }

fluorescence [ATOM PHYS] **1.** Emission of electromagnetic radiation that is caused by the flow of some form of energy into the emitting body and which ceases abruptly when the excitation ceases. **2.** Emission of electromagnetic radiation that is caused by the flow of some form of energy into the emitting body and whose decay, when the excitation ceases, is temperature-independent. [NUC PHYS] Gamma radiation scattered by nuclei which are excited to and radiate from an excited state. [OPTICS] *See* bloom. { flu̇'res·əns }

fluorescence digital imaging microscope [OPTICS] An instrument for observing images formed by emitted fluorescent light; consists of an electronic camera or a laser-photomultiplier scanning system, an analog-to-digital converter, and a computer that manipulates the digitized signal and converts it back to analog form for display. { flu̇'res·əns 'dij·əd·əl ¦im·ij·iŋ 'mī·krəˌskōp }

fluorescence microscope [OPTICS] A variation of the compound laboratory light microscope which is arranged to admit ultraviolet, violet, and sometimes blue radiations to a specimen, which then fluoresces. { flu̇'res·əns 'mī·krəˌskōp }

fluorescence spectra [SPECT] Emission spectra of fluorescence in which an atom or molecule is excited by absorbing light and then emits light of characteristic frequencies. { flu̇'res·əns ˌspek·trə }

fluorescence x-rays [ATOM PHYS] Characteristic x-rays emitted as the result of the absorption of x-rays of higher frequency. { flu̇¦res·əns 'eksˌrāz }

fluorescence yield [ATOM PHYS] The probability that an atom in an excited state will emit an x-ray photon in its first transition rather than an Auger electron. { flu̇'res·əns ˌyēld }

fluorescent minituft method [FL MECH] An adaptation of the tuft method in which the tuft material is treated with a fluorescent dye and the image is recorded with fluorescence photography, allowing reduction of the tuft size, so that there is less interference by the tufts with the fluid flow. { flə¦res·ənt 'min·ēˌtəft ˌmeth·əd }

flutter [ACOUS] Distortion that occurs in sound reproduction as a result of undesired speed variations during the recording, duplicating, or reproducing process. [ELECTROMAG] A fast-changing variation in received signal strength, such as may be caused by antenna movements in a high wind or interaction with a signal or another frequency. [FL MECH] *See* aeronautical flutter. { 'fləd·ər }

flutter echo [ACOUS] A multiple echo in which the reflections rapidly follow each other. [ELECTROMAG] A radar echo consisting of a rapid succession of reflected pulses resulting from a single transmitted pulse. { 'fləd·ər ˌek·ō }

flux [ELECTROMAG] The electric or magnetic lines of force in a region. [PHYS] **1.** The integral over a given surface of the component of a vector field (for example, the magnetic flux density, electric displacement, or gravitational field) perpendicular to the surface; by definition, it is proportional to the number of lines of force crossing the surface. **2.** The amount of some quantity flowing across a given area (often a unit area perpendicular to the flow) per unit time; the quantity may be, for example, mass or volume of fluid, electromagnetic energy, or number of particles. { fləks }

fluxball [ELECTROMAG] A type of magnetic test coil in which the wire is wound into the form of a solid spherical winding by combining a series of coaxial cylindrical windings of different lengths; it gives accurate values of the magnetic flux density (or its variation) at its center, even in a nonuniform magnetic field. { 'fləks,bȯl }

flux-closure domain *See* closure domain. { 'fləks ˌklō·zhər dō,mān }

flux density [PHYS] Any vector field whose flux is a significant physical quantity; examples are magnetic flux density, electric displacement, gravitational field, and the Poynting vector. { 'fləks ˌden·səd·ē }

flux density threshold *See* threshold illuminance. { 'fləks ˌden·səd·ē 'thresh,hōld }

flux-gate magnetometer [ELECTROMAG] A magnetometer in which the degree of saturation of the core by an external magnetic field is used as a measure of the strength of the earth's magnetic field; the essential element is the flux gate. { 'fləks ¦gāt ˌmag·nə'täm·əd·ər }

flux jumping *See* Meissner effect. { 'fləks ˌjəmp·iŋ }

flux lattice [SOLID STATE] The regular array of fluxoids in a type II superconductor in the mixed state, when the superconductor is sufficiently pure and free of defects. { 'fləks ˌlad·əs }

flux-lattice melting [SOLID STATE] A phenomenon in high-temperature superconductors in the mixed state, in which the regular ordering of fluxoids breaks down above a certain temperature. { 'fləks,lad·əs ˌmelt·iŋ }

flux leakage [ELECTROMAG] Magnetic flux that does not pass through an air gap or other part of a magnetic circuit where it is required. { 'fləks ˌlēk·ij }

flux line *See* fluxoid; line of force. { 'fləks ˌlīn }

flux-line pinning [SOLID STATE] The introduction, in a type II superconductor, of microscopic crystalline defects that suppress the motion of flux lines which would otherwise occur in the presence of a magnetic field. { 'fləks ˌlīn ˌpin·iŋ }

flux linkage [ELECTROMAG] The product of the number of turns in a coil and the magnetic flux passing through the coil. Also known as linkage. { 'fləks ˌliŋk·ij }

flux of energy [PHYS] The energy which passes through a surface per unit area per unit time. { 'fləks əv 'en·ər·jē }

fluxoid [SOLID STATE] One of the microscopic filaments of magnetic flux that penetrates a type II superconductor in the mixed state, consisting of a normal core in which the magnetic field is large, surrounded by a superconducting region in which flows a vortex of persistent supercurrent which maintains the field in the core. Also known as flux line; fluxon; vortex. { 'fluk,sȯid }

fluxon *See* fluxoid. { 'flək,sän }

flux path [ELECTROMAG] A path which is followed by magnetic lines of force and in which the magnetic flux density is significant. { 'fləks ˌpath }

flux pump [CRYO] A cryogenic direct-current generator that converts a small alternating-current input to a large direct-current output when cooled to about 4 K; the output current builds up in a series of steps, much like the action of a pump. { 'fləks ˌpəmp }

flux refraction [ELECTROMAG] The abrupt change in direction of magnetic flux lines at the boundary between two media having different permeabilities, or of the electric flux lines at the boundary between two media having different dielectric constants, when these lines are oblique to the boundary. { 'fləks ri,frak·shən }

flying-spot microscope [OPTICS] A microscope in which a minute spot of light, produced in the lens system, passes through a specimen while sweeping over it systematically, and falls on a photocell; the image is produced on a cathode-ray tube that is scanned in synchronization with the spot. { ¦flī·iŋ ˌspät 'mī·krə,skōp }

fm *See* femtometer.

F meson [PART PHYS] Former name of charged meson that carries both strangeness and charm and has a mass of approximately 1.97 GeV, spin 0, and negative parity; now known as D_s meson. { 'ef ˌmā,sän }

f number [OPTICS] A lens rating obtained by dividing the lens's focal length by its effective maximum diameter; the larger the *f* number, the less exposure is given. Also known as focal ratio. Also known as stop number. { 'ef ˌnəm·bər }

foam [FL MECH] A collection of bubbles on the surface of a liquid, often stabilized by organic contaminants, as found at sea or along shore. Also known as froth. { fōm }

foaminess [PHYS] The volume of foam produced in a liquid, in cubic centimeters, produced by passing air through it divided by the rate of flow of air, in cubic centimeters per second. { 'fō·mē·nəs }

focal distance *See* focal length. { 'fō·kəl ˌdis·təns }

focal length [OPTICS] The distance from the focal point of a lens or curved mirror to the principal point; for a thin lens it is approximately the distance from the focal point to the lens. Also known as focal distance. { 'fō·kəl ˌleŋkth }

focal lines *See* astigmatic foci. { 'fō·kəl ˌlīnz }

focal plane [OPTICS] A plane perpendicular to the axis of an optical system and passing through the focal point of the system. { 'fō·kəl ˌplān }

focal-plane shutter [OPTICS] A camera shutter consisting of a blind containing a slot; the blind is pulled rapidly across the film, exposing it through the slot. { 'fō·kəl ˌplān 'shəd·ər }

focal point [OPTICS] The point to which rays that are initially parallel to the axis of a lens, mirror, or other optical system are converged or from which they appear to diverge. Also known as principal focus. { 'fō·kəl ˌpȯint }

focal power [OPTICS] A measure of the ability of a lens, mirror, prism, or optical system to converge a parallel beam of light; equals the reciprocal of the focal length. Also known as power. { 'fō·kəl ˌpaů·ər }

focal ratio *See* f number. { ¦fō·kəl 'rā·shō }

Foch space [QUANT MECH] An infinite-dimensional vector space in which the state of a quantum-mechanical system with a variable number of particles is represented by an infinite number of wave functions, each of which corresponds to a fixed number of particles. { 'fōsh ˌspās }

focus [OPTICS] **1.** The point or small region at which rays converge or from which they appear to diverge. **2.** To move an optical lens toward or away from a screen or film to obtain the sharpest possible image of a desired object. { 'fō·kəs }

focus control [OPTICS] A device to adjust a lens system to produce a sharp image. { 'fō·kəs kən,trōl }

focused collision sequence [PHYS] A cascade of interatomic collisions, initiated by the bombardment of a crystal with energetic particles, that propagates in a particular direction along a closely packed row of atoms in the crystal. { ¦fō·kəst kə'lizh·ən ˌsē·kwəns }

focusing glass [OPTICS] A magnifying glass designed to enlarge the image thrown on the ground glass of the viewfinder of a camera, to help achieve exact focusing. { 'fō·kəs·iŋ ˌglas }

focusing scale [OPTICS] A graduated scale to indicate appropriate lens-to-image plane positions for given lens-to-object plane distances. { 'fō·kəs·iŋ ˌskāl }

focus wave mode [PHYS] A localized wave solution of the three-dimensional wave equation whose overall characteristics depend on a free parameter such that it resembles a transverse plane wave at one extreme and a narrow spatially transverse pulse at the other extreme. { ¦fō·kəs 'wāv ˌmōd }

fogbow [OPTICS] A faintly colored circular arc similar to a rainbow but formed on fog layers containing drops whose diameters are of the order of 100 micrometers or less. Also known as false white rainbow; mistbow; white rainbow. { 'fäg,bō }

foil electret [ELEC] A thin film of strongly insulating material capable of trapping charge carriers, such as polyfluoroethylenepropylene, that is electrically charged to produce an external electric field; in the conventional design, charge carriers of one sign are injected into one surface, and a compensation charge of opposite sign forms on the opposite surface or an adjacent electrode. { ¦fȯil i'lek·trət }

Fokker-Planck equation [STAT MECH] An equation for the distribution function of a gas, analogous to the Boltzmann equation but applying where the forces are long-range and the collisions are not binary. { ¦fō·kər ¦pläŋk i'kwā·zhən }

folded dipole *See* folded-dipole antenna. { ¦fōld·əd 'dī,pōl }

folded-dipole antenna [ELECTROMAG] A dipole antenna whose outer ends are folded back and joined together at the center; the impedance is about 300 ohms, as compared to 70 ohms for a single-wire dipole; widely used with television and frequency-modulation receivers. Also known as folded dipole. { ¦fōld·əd 'dī,pōl an'ten·ə }

Foldy effect [ATOM PHYS] The interaction of the electric field that is produced by a neutron and results from the nonvanishing charge distribution in the neutron's interior with the electrons in an atom. { 'fōl·dē i,fekt }

fondant *See* flux. { 'fän·dənt }

foot [MECH] The unit of length in the British systems of units, equal to exactly 0.3048 meter. Abbreviated ft. { fůt }

footcandle [OPTICS] A unit of illumination, equal to the illumination of a surface, 1 square foot in area, on which there is a luminous flux of 1 lumen uniformly distributed, or equal to the illumination of a surface all points of which are at a distance of 1 foot from a uniform point source of 1 candela; equal to approximately 10.7639 lux. Abbreviated ftc. { 'fůt ˌkand·əl }

footlambert [OPTICS] A unit of luminance (photometric brightness), equal to $1/\pi$ candela per square foot, or to the uniform luminance of a perfectly diffusing surface emitting or reflecting light at the rate of 1 lumen per square foot; equal to approximately 3.42625 nit. Abbreviated ft-L. Also known as equivalent footcandle. { 'fůt ¦lam·bərt }

foot-pound [MECH] **1.** Unit of energy or work in the English gravitational system, equal to the work done by 1 pound of force when the point at which the force is applied is displaced 1 foot in the direction of the force; equal to approximately 1.355818 joule. Abbreviated ft-lb; ft-lbf. **2.** Unit of torque in the English gravitational system, equal to the torque produced by 1 pound of force acting at a perpendicular distance of 1 foot from an axis of rotation. Also known as pound-foot. Abbreviated lbf-ft. { 'fůt ¦paůnd }

foot-poundal [MECH] **1.** A unit of energy or work in the English absolute system, equal to the work done by a force of magnitude 1 poundal when the point at which the force is applied is displaced 1 foot in the direction of the force; equal to approximately 0.04214011 joule. Abbreviated ft-pdl. **2.** A unit of torque in the English absolute system, equal to the torque produced by a force of magnitude 1 poundal acting at a perpendicular distance of 1 foot from the axis of rotation. Also known as poundal-foot. Abbreviated pdl-ft. { 'fu̇t ¦pau̇nd·əl }

foot-pound-second system of units *See* British absolute system of units. { ¦fu̇t ¦pau̇nd ¦sek·ənd ˌsis·təm əv 'yü·nəts }

Forbes bar [THERMO] A metal bar which has one end immersed in a crucible of molten metal and thermometers placed in holes at intervals along the bar; measurement of temperatures along the bar together with measurement of cooling of a short piece of the bar enables calculation of the thermal conductivity of the metal. { 'fȯrbz ˌbär }

forbidden band [SOLID STATE] A range of unallowed energy levels for an electron in a solid. { fər¦bid·ən 'band }

forbidden line [ATOM PHYS] A spectral line associated with a transition forbidden by selection rules; optically this might be a magnetic dipole or electric quadrupole transition. { fər¦bid·ən 'līn }

forbidden transition [QUANT MECH] A transition between two states of a quantum-mechnical system which is considerably less probable than a competing allowed transition. { fər¦bid·ən tran'zish·ən }

force [MECH] That influence on a body which causes it to accelerate; quantitatively it is a vector, equal to the body's time rate of change of momentum. { fȯrs }

force constant [MECH] The ratio of the force to the deformation of a system whose deformation is proportional to the applied force. { 'fȯrs ˌkän·stənt }

forced convection [THERMO] Heat convection in which fluid motion is maintained by some external agency. { ¦fȯrst kən'vek·shən }

forced oscillation [MECH] An oscillation produced in a simple oscillator or equivalent mechanical system by an external periodic driving force. Also known as forced vibration. { ¦fȯrst ˌäs·ə'lā·shən }

forced vibration *See* forced oscillation. { ¦fȯrst vī'brā·shən }

forced wave [FL MECH] Any wave which is required to fit irregularities at the boundary of a system or satisfy some impressed force within the system; the forced wave will not in general be a characteristic mode of oscillation of the system. { ¦fȯrst 'wāv }

force polygon [MECH] A closed polygon whose sides are vectors representing the forces acting on a body in equilibrium. { ¦fȯrs 'päl·əˌgän }

formant [ACOUS] A set of resonances of a musical instrument or voice mechanism that form partials of sounds produced by the instrument, independent of the fundamental frequency, and give these sounds their quality. { 'fȯr·mənt }

form birefringence [OPTICS] Birefringence of a liquid caused by the orientation of rod-shaped particles in the liquid whose thickness and separation are much smaller than a wavelength of light. { ¦fȯrm ˌbī·ri'frin·jəns }

form cutter *See* formed cutter. { 'fȯrm ˌkəd·ər }

form drag [FL MECH] **1.** The drag from all causes resulting from the particular shape of a body relative to its direction of motion, as of fuselage, wing, or nacelle. **2.** At supersonic speed, the drag caused by losses due to shock waves, exclusive of losses due to skin friction. { 'fȯrm ˌdrag }

formed cutter [MECH] A cutting tool shaped to make surfaces with irregular geometry. Also known as form cutter. { ¦fȯrmd 'kəd·ər }

form factor [ELEC] **1.** The ratio of the effective value of a periodic function, such as an alternating current, to its average absolute value. **2.** A factor that takes the shape of a coil into account when computing its inductance. Also known as shape factor. [MECH] The theoretical stress concentration factor for a given shape, for a perfectly elastic material. [PHYS] A function which describes the internal structure of a particle, allowing calculations to be made even though the structure is unknown. [QUANT MECH] An expression used in studying the scattering of electrons or radiation from atoms, nuclei, or elementary particles, which gives the deviation from point particle scattering due to the distribution of charge and current in the target. { 'fȯrm ˌfak·tər }

fors *See* G; gram-force. { fȯrs }

Fortrat parabola [SPECT] Graph of wave numbers of lines in a molecular spectral band versus the serial number of the successive lines. { ˌfȯrträ pə'rab·ə·lə }

forward scattering [PHYS] **1.** Scattering in which there is no change in the direction of motion of the scattered particles. **2.** Scattering in which the angle between the initial and final directions of motion of the scattered particles is less than 90°. { ¦fȯr·wərd 'skad·ə·riŋ }

forward-scatter propagation *See* scatter propagation. { ¦fȯr·wərd ¦skad·ər präp·ə'gā·shən }

Foucault current *See* eddy current. { fü'kō ¦kə·rənt }

Foucault knife-edge test [OPTICS] Test of a lens or a concave mirror in which a pinhole source is placed at twice the focal length behind the lens or at the mirror's center of curvature, the eye is placed at the image of the pinhole, and defects in the lens or mirror result in irregular darkening of the image when a knife edge is moved across the image immediately in front of the eye. { fü'kō 'nīf ˌej ˌtest }

Foucault mirror [OPTICS] Experiment for measuring the speed of light in which light is reflected from a rapidly rotating mirror to a distant mirror and back, and the speed of light is deduced from the displacement of the beam after its second reflection from the rotating mirror, the angular speed of the rotating mirror, and the distance the light travels. { fü'kō ¦mir·ər }

Foucault pendulum [MECH] A swinging weight supported by a long wire, so that the wire's upper support restrains the wire only in the vertical direction, and the weight is set swinging with no lateral or circular motion; the plane of the pendulum gradually changes, demonstrating the rotation of the earth on its axis. { fü'kō 'pen·jə·ləm }

fountain effect [FL MECH] The effect occurring when two containers of superfluid helium are connected by a capillary tube and one of them is heated, so that helium flows through the tube in the direction of higher temperature. { 'faunt·ən i,fekt }

four-current density [RELAT] A four-vector whose three space components are those of the ordinary current density and whose time component is the charge density. { ¦fȯr ¦kə·rənt 'den·səd·ē }

four-force [RELAT] A four-vector equal to the product of the rest mass of a particle and the rate of change of its four-momentum with respect to its proper time. { ¦fȯr ¦fȯrs }

fourier *See* thermal ohm. { fu̇r·ē,ā }

Fourier heat equation *See* Fourier law of heat conduction; heat equation. { ,fu̇r·ē,ā 'hēt i,kwā·zhən }

Fourier law of heat conduction [THERMO] The law that the rate of heat flow through a substance is proportional to the area normal to the direction of flow and to the negative of the rate of change of temperature with distance along the direction of flow. Also known as Fourier heat equation. { 'fu̇r·ē,ā ,lȯ əv 'hēt kən,dək·shən }

Fourier number [FL MECH] A dimensionless number used in unsteady-state flow problems, equal to the product of the dynamic viscosity and a characteristic time divided by the product of the fluid density and the square of a characteristic length. Symbolized Fo_f. [PHYS] A dimensionless number used in the study of unsteady-state mass transfer, equal to the product of the diffusion coefficient and a characteristic time divided by the square of a characteristic length. Symbolized N_{Fom}. [THERMO] A dimensionless number used in the study of unsteady-state heat transfer, equal to the product of the thermal conductivity and a characteristic time, divided by the product of the density, the specific heat at constant pressure, and the distance from the midpoint of the body through which heat is passing to the surface. Symbolized N_{Foh}. { ,fu̇r·ē,ā ,nəm·bər }

Fourier spectrum [PHYS] A plot of the magnitude and phase of the Fourier transform of a function. { ,fu̇r·ē,ā ,spek·trəm }

Fourier transform spectroscopy [SPECT] A spectroscopic technique in which all pertinent wavelengths simultaneously irradiate the sample for a short period of time, and the absorption spectrum is found by mathematical manipulation of the Fourier transform so obtained. { ¦fu̇r·ē¦ā 'tranz,fȯrm spek'träs·kə·pē }

four laws of black hole mechanics [RELAT] Four laws of general relativity theory describing black holes, which are closely analogous to the four laws of classical thermodynamics. { ¦fȯr ¦lȯz əv ¦blak ¦hōl mi'kan·iks }

four-level laser [PHYS] A laser in which the lowest level for a laser transition is an excited state rather than the ground level. { ¦fȯr ¦lev·əl 'lā·zər }

fourth dimension [RELAT] Time in the theory of relativity, in which space and time are conceived as particular aspects of a four-dimensional world. { ¦fȯrth də'men·chən }

fourth-power law *See* Stefan-Boltzmann law. { ¦fȯrth ¦pau̇·ər 'lȯ }

fourth sound [CRYO] A pressure wave which propagates in helium II contained in a porous material such as a tightly packed powder, and which results entirely from motion of the superfluid component, the normal component being immobilized by its viscosity. { ¦fȯrth 'sau̇nd }

four-vector [RELAT] A set of four quantities which transform under a Lorentz transformation in the same way as the three space coordinates and the time coordinate of an event. Also known as Lorentz four-vector. { 'fȯr ,vek·tər }

four-vector potential [ELECTROMAG] A four-vector whose space components are the magnetic vector potential and whose time component is the electric scalar potential. { 'fȯr ,vek·tər pə'ten·chəl }

four-velocity [RELAT] A four-vector whose components are the rates of change of the space and time coordinates of a particle with respect to the particle's proper time. { 'fȯr və,läs·əd·ē }

four-wire line [ELECTROMAG] A transmission line in which four conductors lie at the corners of a rectangle, and each conductor is in phase with the conductor at the opposite corner and out of phase with the conductors at adjacent corners. { 'fȯr ,wīr 'līn }

Fowler-DuBridge theory [SOLID STATE] Theory of photoelectric emission from a metal based on the Sommerfeld model, which takes into account the thermal agitation of electrons in the metal and predicts the photoelectric yield and the energy spectrum of photoelectrons as functions of temperature and the frequency of incident radiation. { ¦fau̇l·ər dü'brij ,thē·ə·rē }

Fowler function [SOLID STATE] A mathematical function used in the Fowler-DuBridge theory to calculate the photoelectric yield. { 'fau̇lər ,faŋk·shən }

fps system of units *See* British absolute system of units. { ¦ef ¦pē¦es 'sis·təm əv 'yü·nəts }

fractal persistence length [PHYS] A length L characterizing a solid that has a fractal material distribution at shorter length scales, such that the solid is homogeneous at length scales larger than L. { ¦frak·təl pər'sis·təns ,leŋkth }

fractional sine wave [PHYS] A pulse train whose waveform is a truncated sine wave. { ¦frak·shən·əl 'sīn ,wāv }

fracton [PHYS] **1.** A quantum corresponding to

a vibrational excitation of random fractal structure, which is localized in space. **2.** A vibrational quantum of a fractal structure, either random or ordered, that may be localized or delocalized. **3.** A quantum of a localized vibrational excitation of a disordered structure, whether or not it is fractal. { 'frak‚tän }

fracton dimension [PHYS] A parameter $\bar{\bar{d}}$ associated with a fractal material distribution of fractal dimension D such that the average frequency ω of fractons of size l is given by a power law stating that $\omega^{\bar{\bar{d}}}$ is proportional to l^{-D}, and the density of vibrational states is proportional to $\omega^{\bar{\bar{d}}-1}$. Also known as spectral dimension. { 'frak·tən di‚men·shən }

fracture strength *See* fracture stress. { 'frak·shər ‚streŋkth }

fracture stress [MECH] The minimum tensile stress that will cause fracture. Also known as fracture strength. { 'frak·shər ‚stres }

fracture wear [MECH] The wear on individual abrasive grains on the surface of a grinding wheel caused by fracture. { 'frak·shər ‚wer }

frame dragging *See* dragging of inertial frames. { 'frām ‚drag·iŋ }

frame of reference [PHYS] A coordinate system for the purpose of assigning positions and times to events. Also known as reference frame. { ¦frām əv 'ref·rəns }

framework structure [SOLID STATE] A crystalline structure in which there are strong interatomic bonds which are not confined to a single plane, in contrast to a layer structure. { 'frām‚wərk ‚strək·chər }

framing camera [OPTICS] A motion picture camera that automatically controls the position of successive still photographs on the film so that, when the film is subsequently projected, the image will appear steady on the screen. { 'frām·iŋ ‚kam·rə }

Francis formula [FL MECH] An equation for the calculation of water flow rate over a rectangular weir in terms of length and head. { 'fran·səs ‚fȯr·myə·lə }

frangible [MECH] Breakable, fragile, or brittle. { 'fran·jə·bəl }

franklin *See* statcoulomb. { 'fraŋk·lən }

franklin centimeter [ELEC] A unit of electric dipole moment, equal to the dipole moment of a charge distribution consisting of positive and negative charges of 1 statcoulomb separated by a distance of 1 centimeter. { 'fraŋk·lən 'sent·ə‚mēd·ər }

Frank partial dislocation [CRYSTAL] A partial dislocation whose Burger's vector is not parallel to the fault plane, so that it can only diffuse and not glide, in contrast to a Schockley partial dislocation. { 'fränk ¦pär·shəl ‚dis·lō'kā·shən }

Franz-Keldysh effect [OPTICS] A shift to longer wavelength in the spectrum transmitted by a semiconductor when a strong electric field is applied. { ¦fränts 'kel·dəsh i‚fekt }

fraunhofer [SPECT] A unit for measurement of the reduced width of a spectrum line such that a spectrum line's reduced width in fraunhofers equals 10^6 times its equivalent width divided by its wavelength. { 'frau̇n‚hōf·ər }

Fraunhofer diffraction [OPTICS] Diffraction of a beam of parallel light observed at an effectively infinite distance from the diffracting object, usually with the aid of lenses which collimate the light before diffraction and focus it at the point of observation. { 'frau̇n‚hōf·ər di‚frak·shən }

Fraunhofer lines [SPECT] The dark lines constituting the Fraunhofer spectrum. { 'frau̇n‚hōf·ər ‚līnz }

Fraunhofer region [ELECTROMAG] The region far from an antenna compared to the dimensions of the antenna and the wavelength of the radiation. Also known as far field; far region; far zone; radiation zone. { 'frau̇n‚hōf·ər ‚rē·jən }

Fraunhofer spectrum [SPECT] The absorption lines in sunlight, due to the cooler outer layers of the sun's atmosphere. { 'frau̇n‚hōf·ər ‚spek·trəm }

free atom [ATOM PHYS] An atom, as in a gas, whose properties, such as spectrum and magnetic moment, are not significantly affected by other atoms, ions, or molecules nearby. { ¦frē 'ad·əm }

free charge [ELEC] Electric charge which is not bound to a definite site in a solid, in contrast to the polarization charge. { ¦frē 'chärj }

free convection *See* natural convection. { ¦frē kən'vek·shən }

free convection number *See* Grashof number. { ¦frē kən'vek·shən ‚nəm·bər }

free electromagnetic field [ELECTROMAG] An electromagnetic field in empty space that does not interact with matter. { 'frē i‚lek·trō·mag‚ned·ik 'fēld }

free electron [PHYS] An electron that is not constrained to remain in a particular atom, and is therefore able to move in matter or in a vacuum when acted on by external electric or magnetic fields. { ¦frē i'lek‚trän }

free-electron laser [OPTICS] A device in which a beam of relativistic electrons passes through a static periodic magnetic field to amplify a superimposed coherent optical wave and thereby produce a powerful beam of coherent light. { 'frē i¦lek‚trän 'lā·zər }

free-electron paramagnetism [ELECTROMAG] Paramagnetism of certain metals that results from the magnetic moments of nearly free electrons in their conduction bands. Also known as Pauli paramagnetism. { ¦frē i¦lek‚trän ‚par·ə'mag·nə‚tiz·əm }

free-electron theory of metals [SOLID STATE] A model of a metal in which the free electrons, that is, those giving rise to the conductivity, are regarded as moving in a potential (due to the metal ions in the lattice and to all the remaining free electrons) which is approximated as constant everywhere inside the metal. Also known as Sommerfeld model; Sommerfeld theory. { ¦frē i'lek‚trän ¦thē·ə·rē əv 'med·əlz }

free energy [THERMO] **1.** The internal energy of a system minus the product of its temperature and its entropy. Also known as Helmholtz free energy; Helmholtz function; Helmholtz potential; thermodynamic potential at constant volume; work function. **2.** *See* Gibbs free energy. { ¦frē 'en·ər·jē }

free enthalpy *See* Gibbs free energy. { ¦frē 'en ˌthal·pē }

free fall [MECH] The ideal falling motion of a body acted upon only by the pull of the earth's gravitational field. { 'frē ˌfȯl }

free field [ACOUS] An isotropic, homogeneous sound field that is free from all bounding surfaces. [PHYS] A field in empty space not interacting with other fields or sources. { 'frē ˌfēld }

free flight [MECH] Unconstrained or unassisted flight. { 'frē ˌflīt }

free-flight angle [MECH] The angle between the horizontal and a line in the direction of motion of a flying body, especially a rocket, at the beginning of free flight. { 'frē ˌflīt ˌaŋ·gəl }

free-flight trajectory [MECH] The path of a body in free fall. { 'frē ˌflīt trə'jek·trē }

free-free absorption *See* inverse bremsstrahlung. { 'frē 'frē ab'sȯrp·shən }

free gas [PHYS] Any gas at any pressure not in solution, or mechanically held in the liquid hydrocarbon phase. { ¦frē ¦gas }

free hole [SOLID STATE] Any hole which is not bound to an impurity or to an exciton. { ¦frē ¦hōl }

free induction decay [SPECT] A type of electron paramagnetic resonance spectroscopy in which a material is exposed to a short high-power pulse (as short as 2 nanoseconds) of microwave radiation, and the response of the material is Fourier transformed into the normal spectrum. Abbreviated FID. { ¦frē in'dək·shən diˌkā }

free molecule flow [PHYS] Flow of a gas in which the mean free path of the molecules is long compared to a characteristic dimension of the flow field, such as the diameter of a tube through which gas is flowing. Also known as Knudsen flow. { ¦frē 'mäl·əˌkyül ˌflō }

free oscillation [PHYS] The oscillation of a physical system with no externally applied stimuli. Also known as free vibration. { ¦frē ˌäs·ə'lā·shən }

free progressive wave [PHYS] A wave in a medium or in vacuum, free from boundary effects. Also known as free wave. { 'frē prə'gres·iv 'wāv }

free quark [PART PHYS] A hypothetical quark that is not bound together with other quarks within a hadron, and whose charge, mass, and other properties can therefore be measured individually. { 'frē 'kwärk }

free space [PHYS] A region of space in which there are no particles of matter and no electromagnetic or gravitational fields other than those whose behavior is under consideration. { ¦frē 'spās }

free-space field intensity [ELECTROMAG] Radio field intensity that would exist at a point in a uniform medium in the absence of waves reflected from the earth or other objects. { 'frē ˌspās 'fēld inˌten·səd·ē }

free-space loss [ELECTROMAG] The theoretical radiation loss, depending only on frequency and distance, that would occur if all variable factors were disregarded when transmitting energy between two antennas. { 'frē ˌspās ˌlȯs }

free-space propagation [ELECTROMAG] Propagation of electromagnetic radiation over a straight-line path in a vacuum or ideal atmosphere, sufficiently removed from all objects that affect the wave in any way. { 'frē ˌspās ˌpräp·ə'gā·shən }

free-space radar equation [ELECTROMAG] Equation that governs a radar signal characteristic when it is propagated between a radar set and a reflecting object or target in otherwise empty space. { 'frē ˌspās 'rāˌdär iˌkwā·zhən }

free-space radiation pattern [ELECTROMAG] Radiation pattern that an antenna would have if it were in free space where there is nothing to reflect, refract, or absorb the radiated waves. { 'frē ˌspās rād·ē'ā·shən ˌpad·ərn }

free-space wave [ELECTROMAG] An electromagnetic wave propagating in a vacuum, free from boundary effects. { 'frē ˌspās 'wāv }

free streamline [FL MECH] A streamline separating fluid in motion from fluid at rest. { ¦frē 'strēmˌlīn }

free surface [FL MECH] A boundary between two homogeneous fluids. { ¦frē 'sər·fəs }

free vector [MECH] A vector whose direction in space is prescribed but whose point of application and line of application are not prescribed. { ¦frē 'vek·tər }

free vibration *See* free oscillation. { ¦frē vī'brā·shən }

free volume [STAT MECH] In a lattice theory of a dense gas or liquid, the volume of the cage in which a given molecule is free to wander when its nearest neighbors are fixed at their lattice positions. { ¦frē 'väl·yəm }

free vortex [FL MECH] Two-dimensional fluid flow in which the fluid moves in concentric circles at speeds inversely proportional to the radii of the circles. { ¦frē 'vȯrˌteks }

free wave *See* free progressive wave. { ¦frē ¦wāv }

freeze etching [CRYO] A method using cryogenics to prepare specimens for study with a microscope. { 'frēz ˌech·iŋ }

french [MECH] A unit of length used to measure small diameters, especially those of fiber optic bundles, equal to 1/3 millimeter. { french }

Frenkel defect [SOLID STATE] A crystal defect consisting of a vacancy and an interstitial which arise when an atom is plucked out of a normal lattice site and forced into an interstitial position. Also known as Frenkel pair. { 'freŋk·əl 'dēˌfekt }

Frenkel exciton [SOLID STATE] A tightly bound exciton in which the electron and the hole are usually on the same atom, although the pair can travel anywhere in the crystal. { 'freŋk·əl 'ek·səˌtän }

Frenkel-Halsey-Hill isotherm equation [PHYS] An equation for the volume v of a gas adsorbed on a surface at a given temperature, $\ln (p/p_o) = k/v^s$, where p is the pressure of the gas, p_o is the vapor pressure, and *k* and *s* are constants. { ¦freŋk·əl ¦hōlz·ē ¦hil 'ī·sō,thərm i,kwā·zhən }

Frenkel pair *See* Frenkel defect. { 'freŋk·əl ,per }

frequency [PHYS] The number of cycles completed by a periodic quantity in a unit time. { 'frē·kwən·sē }

frequency band [PHYS] A continuous range of frequencies extending between two limiting frequencies. { 'frē·kwən·sē ,band }

frequency-domain reflectometer [ELECTROMAG] A tuned reflectometer used for measuring reflection coefficients and impedance of waveguides over a wide frequency range, by sweeping a band of frequencies and analyzing the reflected returns. { 'frē·kwən·sē də,mān ,rē,flek'täm·əd·ər }

frequency mixing [OPTICS] The combination of two or more electromagnetic waves in a nonlinear medium to form another wave whose frequency is a sum or difference of the frequencies of the incident waves. { 'frē·kwən·sē ,mik·siŋ }

frequency-modulated laser [OPTICS] A helium-neon or other laser in which an ultrasonic modulation cell is used to impress a frequency-modulated video signal on the output beam of the laser. { 'frē·kwən·sē ,mäj·ə,lād·əd 'lā·zər }

frequency-modulation laser [OPTICS] Conventional laser containing a phase modulator inside its Fabry-Pérot cavity; characterized by the lack of noise resulting from the random fluctuation in the phase in the various modes. { 'frē·kwən·sē ,mäj·ə,lā·shən 'lā·zər }

frequency scan antenna [ELECTROMAG] A radar antenna similar to a phased array antenna in which one dimensional scanning is accomplished through frequency variation. { 'frē·kwən·sē ,skan an'ten·ə }

frequency spectrum [PHYS] A plot of the distribution of the intensity of some type of electromagnetic or acoustic radiation as a function of frequency. { 'frē·kwən·sē ,spek·trəm }

fresnel [PHYS] A unit of frequency, equal to 10^{12} hertz. { frā'nel }

Fresnel-Arago laws [OPTICS] The three laws stating that two rays of polarized light interfere in the same way as ordinary light if they are polarized in the same plane, but do not interfere if they are polarized at right angles; two rays polarized from ordinary light at right angles do not interfere in the ordinary sense when they are brought into the same plane of polarization; and two rays polarized at right angles from plane polarized light, and then brought into the same polarization plane, interfere. { frā'nel ä·rä'gō ,lȯz }

Fresnel biprism [OPTICS] A very flat triangular prism which has two very acute angles and one very obtuse angle; used to observe the interference of light from a slit passing through the two halves of the prism. { frā'nel 'bī,priz·əm }

Fresnel diffraction [OPTICS] Diffraction in which the source of light or the observing screen are at a finite distance from the aperture or obstacle. { frā'nel di'frak·shən }

Fresnel drag coefficient [OPTICS] The quantity $1 - (1/n^2)$, where n is the index of diffraction of a transparent medium, believed by Fresnel to be the ratio of the velocity with which ether was dragged along in the medium to the velocity of the medium itself. { frā'nel 'drag ,kō·i,fish·ənt }

Fresnel ellipsoid [OPTICS] An ellipsoid whose three perpendicular axes are proportional to the principal values of the wave velocity of light in an anisotropic medium. Also known as ray ellipsoid. { frā'nel i'lip,sȯid }

Fresnel equations [OPTICS] Equations which give the intensity of each of the two polarization components of light which is reflected or transmitted at the boundary between two media with different indices of refraction. { frā'nel i,kwā·zhənz }

Fresnel fringe [OPTICS] One of a series of light and dark bands that appear near the edge of a shadow in Fresnel diffraction. { frā'nel 'frinj }

Fresnel-Huygens principle *See* Huygens-Fresnel principle. { frā¦nel 'hī·gənz ,prin·sə·pəl }

Fresnel lens [OPTICS] A thin lens constructed with stepped setbacks so as to have the optical properties of a much thicker lens. { frā'nel 'lenz }

Fresnel mirrors [OPTICS] Two plane mirrors which are inclined to each other on the order of a degree and used to observe the interference of light which originates from a slit and is reflected from both mirrors. { frā'nel 'mir·ərz }

Fresnel ovaloid [OPTICS] For an anisotropic crystal, an ovaloid whose central section normal to the propagation direction of an electromagnetic wave gives the axes of polarization of the displacement vector and the associated wave velocities. { frā'nel 'ōv·ə,lȯid }

Fresnel reflection formula [OPTICS] The Fresnel equations for light reflected from a boundary. { frā'nel ri'flek·shən ,fȯr·myə·lə }

Fresnel region [ELECTROMAG] The region between the near field of an antenna (close to the antenna compared to a wavelength) and the Fraunhofer region. { frā'nel ,rē·jən }

Fresnel rhomb [OPTICS] A glass rhomb which has an acute angle of about 52°; light which is incident normal to the end of the rhomb undergoes two internal reflections, and if it is initially linearly polarized at an angle of 45° to the plane of incidence, it emerges circularly polarized. { frā'nel 'räm }

Fresnel theory of double refraction [OPTICS] The theory which explains double refraction of a crystal in terms of nonspherical wave surfaces. { frā'nel ,thē·ə·rē əv ,dəb·əl ri'frak·shən }

Fresnel zones [ELECTROMAG] Circular portions of a wavefront transverse to a line between an emitter and a point where the disturbance is being observed; the nth zone includes all paths whose lengths are between $n-1$ and n half-wavelengths longer than the line-of-sight path. Also known as half-period zones. { frā'nel ,zōnz }

Freundlich isotherm equation [PHYS] Equation which states that the volume of gas adsorbed on a surface at a given temperature is proportional to the pressure of the gas raised to a constant power. { 'frȯind·lik̲ 'ī·sō,thərm i,kwā·zhən }

friction [MECH] A force which opposes the relative motion of two bodies whenever such motion exists or whenever there exist other forces which tend to produce such motion. { 'frik·shən }

frictional electricity [ELEC] The electric charges produced on two different objects, such as silk and glass or catskin and ebonite, by rubbing them together. Also known as triboelectricity. { 'frik·shən·əl i,lek'tri·səd·ē }

frictional grip [MECH] The adhesion between the wheels of a locomotive and the rails of the railroad track. { ¦frik·shən·əl 'grip }

frictional secondary flow *See* secondary flow. { ¦frik·shən·əl ¦sek·ən,der·ē 'flō }

friction coefficient *See* coefficient of friction. { 'frik·shən ,kō·i'fish·ənt }

friction damping [MECH] The conversion of the mechanical vibrational energy of solids into heat energy by causing one dry member to slide on another. { 'frik·shən ,damp·iŋ }

friction factor [FL MECH] Any of several dimensionless numbers used in studying fluid friction in pipes, equal to the Fanning friction factor times some dimensionless constant. { 'frik·shən ,fak·tər }

friction flow [FL MECH] Fluid flow in which a significant amount of mechanical energy is dissipated into heat by action of viscosity. { 'frik·shən ,flō }

friction head [FL MECH] The head lost by the flow in a stream or conduit due to frictional disturbances set up by the moving fluid and its containing conduit and by intermolecular friction. { 'frik·shən ,hed }

frictionless flow *See* inviscid flow. { ¦frik·shən·ləs 'flō }

friction loss [MECH] Mechanical energy lost because of mechanical friction between moving parts of a machine. { 'frik·shən ,lōs }

friction torque [MECH] The torque which is produced by frictional forces and opposes rotational motion, such as that associated with journal or sleeve bearings in machines. { 'frik·shən ,tȯrk }

Friedel's law [CRYSTAL] The law that x-ray or electron diffraction measurements cannot determine whether or not a crystal has a center of symmetry. { frē'delz ,lȯ }

Friedman solution [RELAT] A solution of Einstein's equations of general relativity with flat spatial sections describing a cosmological model. { 'frēd·mən sə,lü·shən }

frigorie [THERMO] A unit of rate of extraction of heat used in refrigeration, equal to 1000 fifteen-degree calories per hour, or 1.16264 ± 0.00014 watts. { 'frig·ə·rē }

fringe [OPTICS] One of the light or dark bands produced by interference or diffraction of light. { frinj }

fringe magnetic field [ELECTROMAG] The part of the magnetic field of a horseshoe magnet that extends outside the space between its poles. { 'frinj mag¦ned·ik 'fēld }

fringes of equal thickness *See* Fizeau fringes. { 'frin·jəz əv ¦ē·kwəl 'thik·nəs }

fringe value [OPTICS] A quantity used in photoelastic work, equal to the stress which must be applied to a material, in pounds per square inch (1 pound per square inch equals approximately 6.89476 kilopascals), to produce a relative retardation of 1 wavelength between the components of a linearly polarized light beam when the light passes through a thickness of 1 inch (2.54 centimeters) in a direction perpendicular to the stress. { 'frinj ,val·yü }

Froissart bound [PART PHYS] A limit on the rate at which the cross section of a completely absorptive collision between hadrons can increase with energy, so that the interaction radius cannot increase more rapidly than the logarithm of the energy. { frwä'sär ,bau̇nd }

front pinacoid [CRYSTAL] The {100} pinacoid in an orthorhombic, monoclinic, or triclinic crystal. Also known as macropinacoid; orthopinacoid. { ¦frənt 'pin·ə,kȯid }

front-to-back ratio [ELECTROMAG] Ratio of the effectiveness of a directional antenna, loudspeaker, or microphone toward the front and toward the rear. [SOLID STATE] Ratio of resistance of a crystal to current flowing in the normal direction to current flowing in the opposite direction. { ¦frənt tə ¦bak 'rā·shō }

froth *See* foam. { frȯth }

Froude number 1 [FL MECH] A dimensionless number used in studying the motion of a body floating on a fluid with production of surface waves and eddies; equal to the ratio of the square of the relative speed to the product of the acceleration of gravity and a characteristic length of the body. Symbolized N_{Fr1}. { ¦früd ,nəm·bər 'wən }

Froude number 2 [FL MECH] A dimensionless number, equal to the ratio of the speed of flow of a fluid in an open channel to the speed of very small gravity waves, the latter being equal to the square root of the product of the acceleration of gravity and a characteristic length. Symbolized N_{Fr2}. { ¦früd ,nəm·bər 'tü }

frozen flux [PL PHYS] The lines of force of a frozen-in field. { ¦frōz·ən 'fləks }

frozen-in field [PL PHYS] A magnetic field in a plasma which has negligible electrical resistance; it can be shown that the lines of force of this field are constrained to move with the material. { ¦frōz·ən 'in ,fēld }

frustrated internal reflectance *See* attenuated total reflectance. { 'frəs,trād·əd in,tərn·əl ri'flek·təns }

frustration [SOLID STATE] In spin glasses, a phenomenon in which individual magnetic moments receive competing ordering instructions via different routes, because of the variation of the interaction between pairs of atomic moments with separation. { frəs'trā·shən }

f stop [OPTICS] An aperture setting for a camera lens; indicated by the f number. { 'ef ,stäp }

f-sum rule [ATOM PHYS] The rule that the sum

of the *f* values (or oscillator strengths) of absorption transitions of an atom in a given state, minus the sum of the *f* values of the emission transitions in that state, equals the number of electrons which take part in these transitions. Also known as Thomas-Reiche-Kuhn sum rule. { 'ef ˌsəm ˌrül }

ftc *See* footcandle.

ft-L *See* footlambert.

ft-lb *See* foot-pound.

ft-lbf *See* foot-pound.

ft-pdl *See* foot-poundal.

fT value [NUC PHYS] The product of the half-period T of a nucleus that undergoes beta decay and an integral *f* that depends on the beta-decay energy and the type of transition; decays of a particular degree of forbiddenness have similar values of this product. { ¦ef¦tē ˌval·yü }

FU *See* finsen unit.

fugacity [THERMO] A function used as an analog of the partial pressure in applying thermodynamics to real systems; at a constant temperature it is proportional to the exponential of the ratio of the chemical potential of a constituent of a system divided by the product of the gas constant and the temperature, and it approaches the partial pressure as the total pressure of the gas approaches zero. { fyü'gas·əd·ē }

fugacity coefficient [THERMO] The ratio of the fugacity of a gas to its pressure. { fyü'gas·əd·ē ˌkō·əˌfish·ənt }

Fulcher bands [SPECT] A group of bands in the spectrum of molecular hydrogen that are preferentially excited by a low-voltage discharge. { 'fəl·chər ˌbanz }

fulcrum [MECH] The rigid point of support about which a lever pivots. { 'fu̇l·krəm }

full load [ELEC] The greatest load that a circuit or piece of equipment is designed to carry under specified conditions. { ¦fu̇l 'lōd }

full-load current [ELEC] The greatest current that a circuit or piece of equipment is designed to carry under specified conditions. { ¦fu̇l ˌlōd 'kə·rənt }

full width at half maximum [PHYS] The difference between the energies or frequencies on either side of a spectral line or resonance curve at which the line absorption or emission or the resonant quantity reaches half its maximum intensity. Abbreviated FWHM. { 'fu̇l ˌwidth at 'haf 'mak·sə·məm }

funal *See* sthène. { 'fyün·əl }

fundamental [PHYS] The lowest frequency component of a complex wave. Also known as first harmonic; fundamental component. { ¦fən·də¦ment·əl }

fundamental component *See* fundamental. { ¦fən·də¦ment·əl kəm'pō·nənt }

fundamental constants [PHYS] The physical constants which play a fundamental role in the basic theories of physics, including the speed of light, electronic charge, electronic mass, Planck's constant, and the fine-structure constant. Also known as atomic constants; universal constants. { ˌfən·də'ment·əl 'kän·stəns }

fundamental frequency [PHYS] **1.** The lowest frequency at which a system vibrates freely. **2.** The lowest frequency in a complex wave. { ¦fən·də¦ment·əl 'frē·kwən·sē }

fundamental interaction [PART PHYS] One of the fundamental forces that act between the elementary particles of matter. { ¦fən·də¦ment·əl ˌin·tər'ak·shən }

fundamental interval [THERMO] **1.** The value arbitrarily assigned to the difference in temperature between two fixed points (such as the ice point and steam point) on a temperature scale, in order to define the scale. **2.** The difference between the values recorded by a thermometer at two fixed points; for example, the difference between the resistances recorded by a resistance thermometer at the ice point and steam point. { ¦fən·də¦men·təl 'int·ər·vəl }

fundamental mode [ELECTROMAG] The waveguide mode having the lowest critical frequency. Also known as dominant mode; principal mode. [PHYS] The normal mode of vibration having the lowest frequency. { ¦fən·də¦ment·əl 'mōd }

fundamental particle *See* elementary particle. { ¦fən·də¦ment·əl 'pärd·ə·kəl }

fundamental quantity *See* base quantity. { ¦fən·də¦ment·əl 'kwän·əd·ē }

fundamental series [SPECT] A series occurring in the line spectra of many atoms and ions having one, two, or three electrons in the outer shell, in which the total orbital angular momentum quantum number changes from 3 to 2. { ¦fən·də¦ment·əl 'sir·ēz }

fundamental tone [ACOUS] The component tone of lowest pitch in a complex tone. { ¦fən·də¦ment·əl 'tōn }

fundamental unit *See* base unit. { ¦fən·də¦ment·əl 'yü·nət }

fundamental wavelength [PHYS] Of an oscillatory device, that wavelength corresponding to its fundamental frequency. { ¦fən·də¦ment·əl 'wāvˌleŋkth }

funicular polygon [MECH] **1.** The figure formed by a light string hung between two points from which weights are suspended at various points. **2.** A force diagram for such a string, in which the forces (weights and tensions) acting on points of the string from which weights are suspended are represented by a series of adjacent triangles. { fə'nik·yə·lər 'päl·əˌgän }

furlong [MECH] A unit of length, equal to $^{1}/_{8}$ mile, 660 feet, or 201.168 meters. { 'fərˌlȯŋ }

Furry theorem [QUANT MECH] In quantum electrodynamics, the theorem that the contribution of a Feynman diagram, consisting of a closed polygon of fermion lines connected to an odd number of photon lines, vanishes. { 'fər·ē ˌthir·əm }

fusibility [THERMO] The quality or degree of being capable of being liquefied by heat. { ˌfyü·zə'bil·əd·ē }

fusion [NUC PHYS] Combination of two light nuclei to form a heavier nucleus (and perhaps other reaction products) with release of some binding energy. Also known as atomic fusion; nuclear fusion. { 'fyü·zhən }

future [RELAT] For an event in space-time, those events that can be reached by a signal that is emitted at the original event and moves at a speed less than or equal to the speed of light in a vacuum. { 'fyü·chər }

future asymptotically predictable [RELAT] A mathematical restriction on the global nature of an asymptotically flat space-time such that Cauchy data set on a spacelike surface (partial Cauchy surface) S will determine the evolution of the space-time to the future of S; naked singularities to the future of S are thereby ruled out. { ¦fyü·chər ‚ā‚sim¦täd·ə·klē prə'dik·tə·bəl }

future Cauchy development [RELAT] The set of points *p* relative to a surface S in a space-time such that every past-directed inextendible timelike or null curve through *p* intersects S. Symbolized $D^+(s)$. Also known as domain of dependence. { ¦fyü·chər 'kō·shē di‚vel·əp·mənt }

future horismos [RELAT] The set of points *p* relative to a surface S that can be causally affected by events in S; that is, the set of points in the future of S which can be reached from S by future-directed timelike curves. { ¦fyü·chər hə'riz·mōs }

future light cone [RELAT] The set of all points in space-time that are reached by signals traveling at the speed of light from a specified point. { 'fyü·chər 'līt ‚kōn }

future trapped set [RELAT] A set of points in a space-time such that no two points of the set have timelike separation and the future horismos is compact. { ¦fyü·chər ¦trapt 'set }

FUV radiation *See* far-ultraviolet radiation. { ¦ef ¦yü¦vē ‚rād·ē'ā·shən }

fuzzy-structure acoustics [ACOUS] A class of conceptual viewpoints in the study of vibrations of large structure in which precise, computationally intensive models of the overall structure are replaced by nonprecise analytical models. { ‚fəz·ē ‚strək·chər ə'küs·tiks }

fV *See* femtovolt.

f value *See* oscillator strength. { 'ef ‚väl·yü }

FWHM *See* full width at half maximum.

G

g *See* gram.

G [ELEC] *See* conductance. [MECH] A unit of acceleration equal to the standard acceleration of gravity, 9.80665 meters per second per second, or approximately 32.1740 feet per second per second. Also known as fors; grav.

gain *See* antenna gain. { gān }

gal [MECH] **1.** The unit of acceleration in the centimeter-gram-second system, equal to 1 centimeter per second squared; commonly used in geodetic measurement. Formerly known as galileo. Symbolized Gal. **2.** *See* gallon. { gal }

Gal *See* gal. { gal }

Galilean glass *See* Galilean telescope. { ˌgal·əˌlē·ən 'glas }

Galilean telescope [OPTICS] A refracting telescope whose objective is a converging (convex) lens and whose eyepiece is a diverging (concave) lens; it forms erect images. Also known as Galilean glass. { ˌgal·əˌlē·ən 'tel·əˌskōp }

Galilean transformation [MECH] A mathematical transformation used to relate the space and time variables of two uniformly moving (inertial) reference systems in nonrelativistic kinematics. { ˌgal·əˌlē·ən ˌtranz·fər'mā·shən }

galileo *See* gal. { ˌgal·ə'lē·ō }

Galileo number [FL MECH] A dimensionless number used in studying the circulation of viscous liquids, equal to the cube of a characteristic dimension, times the acceleration of gravity, times the square of the liquid's density, divided by the square of its viscosity. Symbol N_{Ga}. { ˌgal·ə'lē·ō ˌnəm·bər }

Galileo's law of inertia *See* Newton's first law. { ˌgal·ə'lē·ōz ˌlȯ əv i'nər·shə }

Galitzin pendulum [MECH] A massive horizontal pendulum that is used to measure variations in the direction of the force of gravity with time, and thus serves as the basis of a seismograph. { gäˌlit·sən 'pen·jə·ləm }

gallium arsenide laser [OPTICS] A laser that emits light at right angles to a junction region in gallium arsenide, at a wavelength of 9000 angstroms (900 nanometers); can be modulated directly at microwave frequencies; cryogenic cooling is required. { ˌgal·ē·əm ˌärs·ənˌīd 'lā·zər }

gallium arsenide semiconductor [SOLID STATE] A semiconductor having a forbidden-band gap of 1.4 electronvolts and a maximum operating temperature of 400°C when used in a transistor. { ˌgal·ē·əm ˌärs·ənˌīd 'sem·i·kənˌdək·tər }

gallium phosphide semiconductor [SOLID STATE] A semiconductor having a forbidden-band gap of 2.4 electronvolts and a maximum operating temperature of 870°C when used in a transistor. { ˌgal·ē·əm ˌfäsˌfīd 'sem·i·kənˌdək·tər }

gallon [MECH] Abbreviated gal. **1.** A unit of volume used in the United States for measurement of liquid substances, equal to 231 cubic inches, or to 3.785 411 784 × 10^{-3} cubic meter, or to 3.785 411 784 liters; equal to 128 fluid ounces. **2.** A unit of volume used in the United Kingdom for measurement of liquid and solid substances, usually the former; equal to 4.54609 × 10^{-3} cubic meter, or to 4.54609 liters; equal to 160 fluid ounces. Also known as imperial gallon. { 'gal·ən }

galloping [FL MECH] Large-amplitude oscillations of a wire or cable in a strong wind, which may become destructive. { 'gal·əp·iŋ }

galvanic [ELEC] Pertaining to electricity flowing as a result of chemical action. { gal'van·ik }

galvanic battery [ELEC] A galvanic cell, or two or more such cells electrically connected to produce energy. { gal'van·ik 'bad·ə·rē }

galvanic cell [ELEC] An electrolytic cell that is capable of producing electric energy by electrochemical action. { gal'van·ik 'sel }

galvanic couple [ELEC] A pair of unlike substances, such as metals, which generate a voltage when brought in contact with an electrolyte. { gal'van·ik 'kəp·əl }

galvanic current [ELEC] A steady direct current. { gal'van·ik 'kə·rənt }

galvanoluminescence [PHYS] Light emission which may occur when electrodes of certain metals, such as aluminum or tantalum, are immersed in suitable electrolytes and current is passed between them. { ˌgal·və·nōˌlü·mə'nes·əns }

galvanomagnetic effect [ELECTROMAG] One of the electrical or thermal phenomena occurring when a current-carrying conductor or semiconductor is placed in a magnetic field; examples are the Hall effect, Ettingshausen effect, transverse magnetoresistance, and Nernst effect. Also known as magnetogalvanic effect. { ˌgal·və·nō ˌmagˌned·ik i'fekt }

gamma [ELECTROMAG] A unit of magnetic field

strength, equal to 10 microoersteds, or 0.00001 oersted. [MECH] A unit of mass equal to 10^{-6} gram or 10^{-9} kilogram. { 'gam·ə }

gamma cross section [NUC PHYS] The cross section for absorption or scattering of gamma rays by a nucleus or atom. { ¦gam·ə 'krȯ ˌsek·shən }

gamma decay *See* gamma emission. { 'gam·ə diˌkā }

gamma emission [NUC PHYS] A quantum transition between two energy levels of a nucleus in which a gamma ray is emitted. Also known as gamma decay. { ¦gam·ə i'mish·ən }

gamma flux density [NUC PHYS] The number of gamma rays passing through a unit area in a unit time. { ¦gam·ə 'fləks ˌden·səd·ē }

gamma matrix *See* Dirac matrix. { 'gam·ə ˌmā·triks }

gamma ray [NUC PHYS] A high-energy photon, especially as emitted by a nucleus in a transition between two energy levels. { 'gam·ə ˌrā }

gamma-ray laser [PHYS] A hypothetical device which would generate coherent radiation in the range 0.005–0.5 nanometer by inducing isomeric radiative transitions between isomeric nuclear states. Also known as graser. { 'gam·ə ˌrā 'lā·zər }

gamma-ray scattering *See* Compton scattering. { 'gam·ə ˌrā 'skad·ə·riŋ }

gamma-ray spectrum [SPECT] The set of wavelengths or energies of gamma rays emitted by a given source. { 'gam·ə ˌrā ˌspek·trəm }

gamma-ray transformation [NUC PHYS] A radioactive decay in which gamma rays are emitted. { 'gam·ə ˌrā ˌtranz·fərˌmā·shən }

gamma structure [SOLID STATE] A Hume-Rothery designation for structurally analogous phases or intermetallic phases having 21 valence electrons to 13 atoms, analogous to the γ-brass structure. { 'gam·ə ˌstrək·chər }

gamma transition *See* glass transition. { 'gam·ə tran'zish·ən }

Gamow barrier [NUC PHYS] The potential barrier which retards the escape of alpha particles from the nucleus according to the Gamow-Condon-Gurney theory. { 'gaˌmȯf ˌbar·ē·ər }

Gamow-Condon-Gurney theory [NUC PHYS] An early quantum-mechanical theory of alpha-particle decay according to which the alpha particle penetrates a potential barrier near the surface of the nucleus by a tunneling process. { ¦gaˌmȯf ¦känd·ən 'gər·nē ˌthē·ə·rē }

Gamow-Teller interaction [NUC PHYS] Interaction between a nucleon source current and a lepton field which has an axial vector or tensor form. { ¦gaˌmof 'tel·ər ˌinˌtər'ak·shən }

Gamow-Teller selection rules [NUC PHYS] Selection rules for beta decay caused by the Gamow-Teller interaction; that is, in an allowed transition there is no parity change of the nuclear state, and the spin of the nucleus can either remain unchanged or change by ±1; transitions from spin 0 to spin 0 are excluded, however. { ¦gaˌmof 'tel·ər si'lek·shən ˌrülz }

gap [ELEC] The spacing between two electric contacts. [ELECTROMAG] A break in a closed magnetic circuit, containing only air or filled with a nonmagnetic material. { gap }

gap filling [ELECTROMAG] Electrical or mechanical rearrangement of an antenna array, or the use of a supplementary array, to produce lobes where gaps previously occurred. { 'gap ˌfil·iŋ }

Garvey-Kelson mass relations [NUC PHYS] A set of equations relating the masses of atomic nuclei with slightly different numbers of neutrons and protons. { 'gär·vē 'kel·sən 'mas ri'lā·shənz }

gas [PHYS] A phase of matter in which the substance expands readily to fill any containing vessel; characterized by relatively low density. { gas }

gas capacitor [ELEC] A capacitor consisting of two or more electrodes separated by a gas, other than air, that serves as a dielectric. { ¦gas kə'pas·əd·ər }

gas cell [ELEC] Cell in which the action depends on the absorption of gases by the electrodes. { 'gas ˌsel }

gas-cell frequency standard [ATOM PHYS] An atomic frequency standard in which the frequency-determining element is a gas cell containing rubidium, cesium, or sodium vapor. { 'gas ˌsel 'frē·kwən·sē ˌstan·dərd }

gas constant [THERMO] The constant of proportionality appearing in the equation of state of an ideal gas, equal to the pressure of the gas times its molar volume divided by its temperature. Also known as gas-law constant; universal gas constant. { 'gas ˌkän·stənt }

gas cycle [THERMO] A sequence in which a gaseous fluid undergoes a series of thermodynamic phases, ultimately returning to its original state. { 'gas ˌsī·kəl }

gas-deviation factor *See* compressibility factor. { ¦gas ˌdē·vē'ā·shən ˌfak·tər }

gas-discharge laser [OPTICS] A gas laser in which optical pumping is caused by nonequilibrium processes in a gas discharge. { ¦gas 'dis ˌchärj ˌlā·zər }

gas dynamic laser [OPTICS] A gas laser that converts thermal energy directly into coherent radiation at an efficiency high enough to offer promise of wireless power transmission. { ¦gas dī¦nam·ik 'lā·zər }

gas dynamics [PHYS] The study of the motion of gases, and of its causes, which takes into account thermal effects generated by the motion. { ¦gas dī'nam·iks }

gas kinematics [FL MECH] The motion of a gas considered by itself, without regard for the causes of motion. { ¦gas ˌkin·ə'mad·iks }

gas laser [OPTICS] A laser in which the active medium is a discharge in a gas contained in a glass or quartz tube with a Brewster-angle window at each end; the gas can be excited by a high-frequency oscillator or direct-current flow between electrodes inside the tube; the function of the discharge is to pump the medium, to obtain population inversion. { 'gas ˌlā·zər }

gas law [THERMO] Any law relating the pressure, volume, and temperature of a gas. { 'gas ˌlȯ }

gas-law constant *See* gas constant. { 'gas ˌlȯ ˌkän·stənt }

gas lens [OPTICS] An optical lens formed by a flow of gas which gives rise to gradients of refractive index that bring about the focusing of light. { 'gas ˌlenz }

gas maser [PHYS] A maser in which the microwave electromagnetic radiation interacts with the molecules of a gas such as ammonia; used chiefly in highly stable oscillator applications, as in atomic clocks. { 'gas ˌmā·zər }

gas mechanics [FL MECH] The action of forces on gases. { 'gas məˌkan·iks }

gas slippage [FL MECH] Phenomenon of gas bypassing liquids that occurs when the diameter of capillary openings approaches the mean free path of the gas; occurs not only in capillary tubing, but in porous oil-reservoir formations. { ¦gas ¦slip·ij }

gas viscosity [FL MECH] The internal fluid function of a gas. { 'gas vi'skäs·əd·ē }

gauge [ELECTROMAG] One of the family of possible choices for the electric scalar potential and magnetic vector potential, given the electric and magnetic fields. { gāj }

gauge boson [PHYS] A massless spin-1 particle, such as the photon and gluons, whose existence is required by gauge invariance in a gauge theory; such particles can acquire mass through spontaneous symmetry breaking, as in the case of intermediate vector bosons. Also known as gauge particle. { 'gāj ˌbōˌsän }

gauge-fixing term [QUANT MECH] A term added to a Lagrangian in quantum field theory that breaks gauge invariance. { 'gāj ¦fiks·iŋ ˌtərm }

gauge group [PHYS] The group of gauge transformations in a gauge theory. { 'gāj ˌgrüp }

gauge invariance [ELECTROMAG] The invariance of electric and magnetic fields and electrodynamic interactions under gauge transformations. [PHYS] The invariance of any field theory under gauge transformations. [QUANT MECH] An invariance of a Lagrangian based on an internal gauge group, such as U(1) for electromagnetism or U(1) × SU(2) for the Weinberg-Salam unified model of weak and electromagnetic interactions. { 'gāj in'ver·ē·əns }

gauge particle *See* gauge boson. { 'gāj ˌpärd·ə·kəl }

gauge theory [PHYS] Any field theory in which, as the result of the conservation of some quantity, it is possible to perform a transformation in which the phase of the fields is altered by a function of space and time without altering any measurable physical quantity, so that the fields obtained by any such transformation give a valid description of a given physical situation. { 'gāj ˌthē·ə·rē }

gauge transformation [ELECTROMAG] The addition of the gradient of some function of space and time to the magnetic vector potential, and the addition of the negative of the partial derivative of the same function with respect to time, divided by the speed of light, to the electric scalar potential; this procedure gives different potentials but leaves the electric and magnetic fields unchanged. [PHYS] An alteration of the phase of the fields of a gauge theory as a function of space and time which does not alter the value of any measurable physical quantity. { 'gaj tranz·fər'mā·shən }

gauss [ELECTROMAG] Unit of magnetic induction in the electromagnetic and Gaussian systems of units, equal to 1 maxwell per square centimeter, or 10^{-4} weber per square meter. Also known as abtesla (abt). { gau̇s }

Gauss A position *See* end-on position. { ¦gau̇s ¦ā pəˌzish·ən }

Gauss B position *See* broadside-on position. { ¦gau̇s ¦bē pəˌzish·ən }

Gauss eyepiece [OPTICS] A Ramsden eyepiece which has a thin glass plate between the two lenses, making an angle of 45° with the optical axis; used to set a telescope perpendicular to a plane reflecting surface. { 'gau̇s 'īˌpēs }

Gaussian beam [ELECTROMAG] A beam of electromagnetic radiation whose wave front is approximately spherical at any point along the beam and whose transverse field intensity over any wave front is a Gaussian function of the distance from the axis of the beam. { 'gau̇s·ē·ən 'bēm }

Gaussian optics [OPTICS] An approximation which describes rays which are very close to the axis of an optical system and are nearly parallel to this axis, so that only the linear terms of Taylor series for the distance of a point from the axis or the angle which a ray makes with the axis need be considered. Also known as first-order theory. { ¦gau̇·sē·ən 'äp·tiks }

Gaussian pulse [PHYS] A pulse for which the graph of intensity as a function of time is a Gaussian curve. { ¦gau̇·sē·ən 'pəls }

Gaussian system [ELECTROMAG] A combination of the electrostatic and electromagnetic systems of units (esu and emu), in which electrostatic quantities are expressed in esu and magnetic and electromagnetic quantities in emu, with appropriate use of the conversion constant c (the speed of light) between the two systems. Also known as Gaussian units. { ¦gau̇·sē·ən ˌsis·təm }

Gaussian units *See* Gaussian system. { ¦gau̇·sē·ən ˌyü·nəts }

Gauss image point [OPTICS] A point through which pass all paraxial rays from a specified point source in an optical system. { ¦gau̇s 'im·ij ˌpȯint }

Gauss' law of flux [ELEC] The law that the total electric flux which passes out from a closed surface equals (in rationalized units) the total charge within the surface. { 'gau̇s ˌlȯ əv 'fləks }

Gauss lens system *See* Celor lens system. { 'gau̇s 'lenz ˌsis·təm }

Gauss method of weighing *See* double weighing. { ¦gau̇s ¦meth·əd əv 'wā·iŋ }

Gauss objective lens *See* Celor lens system. { 'gaůs əb¦jek·tiv 'lenz }

Gauss positions [ELECTROMAG] The Gauss A and B positions; that is, a point on the axis of a bar magnet is in Gauss A position, and a point on the magnetic equator of the magnet is in Gauss B position, with respect to the magnet. { 'gaůs pə,zish·ənz }

Gauss' principle of least constraint [MECH] The principle that the motion of a system of interconnected material points subjected to any influence is such as to minimize the constraint on the system; here the constraint, during an infinitesimal period of time, is the sum over the points of the product of the mass of the point times the square of its deviation from the position it would have occupied at the end of the time period if it had not been connected to other points. { 'gaůs 'prin·sə·pəl əv ¦lēst kən'strānt }

gauze tones *See* howling tones. { 'gȯz ,tōnz }

Gay-Lussac's first law *See* Charles' law. { ,gā·lü ,säks 'fərst ,lȯ }

Gay-Lussac's second law [THERMO] The law that the internal energy of an ideal gas is independent of its volume. { ,gā·lü,säks 'sek·ənd ,lȯ }

g-cal *See* calorie. { 'jē,kal }

G center *See* N center. { 'jē ,sen·tər }

g-cm *See* gram-centimeter.

g-completeness *See* geodesic completeness. { ¦jē kəm'plēt·nəs }

g constant [SOLID STATE] The ratio of the induced electric field in a piezoelectric material to the applied force that produces this field. { 'jē ,kän·stənt }

Gedanken experiment [PHYS] A hypothetical ("thought") experiment which is possible in principle and is analyzed (but not performed) to test some hypothesis. Also known as thought experiment. { ge'däŋk·ən ik,sper·ə·mənt }

geepound *See* slug. { 'jē,paůnd }

Geiger-Nutall rule [NUC PHYS] The rule that the logarithm of the decay constant of an alpha emitter is linearly related to the logarithm of the range of the alpha particles emitted by it. { ¦gī·gər 'nəd,ȯl ,rül }

Gell-Mann-Nishijima scheme [PART PHYS] A classification of elementary particles according to hypercharge, total isotopic spin, and its third component (which distinguishes between members of an isospin multiplet). { 'gel¦män ,nē·shē·jē·mə ,skēm }

Gell-Mann-Okubo mass formula [PART PHYS] A formula, based on SU(3) symmetry, giving the masses of the members of a unitary multiplet of mesons or baryons in terms of their total isotopic spin, hypercharge, and coefficients characteristic of the multiplet considered. { 'gel¦män ,ō·kü·bō ¦mas ¦fȯr·myə·lə }

Gell-Mann relation [PART PHYS] A relation, derived from the Gell-Mann-Okubo mass formula, between the masses of the pi, eta, and K mesons of a meson octet. { 'gel¦män ri,lā·shən }

gel strength [FL MECH] Of a colloid, the ability or a measure of its ability to form gels. { 'jel ,streŋkth }

gemmho [ELEC] A unit of conductance, equal to 10^{-6} mho, being the conductance of a substance which has a resistance of 10^6 ohms. { 'je,mō }

Gemolite [OPTICS] A binocular magnifier with dark-field illumination, used to distinguish natural from synthetic gem materials. { 'jem·ə,līt }

generalized coordinates [MECH] A set of variables used to specify the position and orientation of a system, in principle defined in terms of cartesian coordinates of the system's particles and of the time in some convenient manner; the number of such coordinates equals the number of degrees of freedom of the system Also known as Lagrangian coordinates. { 'jen·rə,līzd kō'ȯrd·ən·əts }

generalized force [MECH] The generalized force corresponding to a generalized coordinate is the ratio of the virtual work done in an infinitesimal virtual displacement, which alters that coordinate and no other, to the change in the coordinate. { 'jen·rə,līzd 'fȯrs }

generalized momentum *See* conjugate momentum. { 'jen·rə,līzd mə'ment·əm }

generalized velocity [MECH] The derivative with respect to time of one of the generalized coordinates of a particle. Also known as Lagrangian generalized velocity. { 'jen·rə,līzd və'läs·əd·ē }

general relativistic collapse [RELAT] Process in which a star undergoing gravitational collapse cannot release its kinetic energy to the outside universe, but continues to collapse into a general relativistic singularity of infinite density. { ¦jen·rəl ,rel·ə·tə¦vis·tik kə'laps }

general relativity [RELAT] The theory of Einstein which generalizes special relativity to noninertial frames of reference and incorporates gravitation, and in which events take place in a curved space. { ¦jen·rəl ,rel·ə'tiv·əd·ē }

generating flow [FL MECH] For a liquid allowed to flow smoothly into a duct, the flow while the boundary layer, which starts at the entrance and grows until it fills the duct, is growing. { 'jen·ə,rād·iŋ ,flō }

geoacoustics [ACOUS] Study of the acoustic properties of rock, mainly to study possible use of the rock system as a carrier of seismic signals in a communications system. { ¦jē·ō·ə'küs·tiks }

geodesic completeness [RELAT] Property of a space-time wherein all timelike and null geodesics can be extended to arbitrary values of their affine parameter. Also known as g-completeness. { ¦jē·ə¦des·ik kəm'plēt·nəs }

geodesic coordinates [RELAT] Coordinates in the neighborhood of a point P such that the gradient of the metric tensor is zero at P. { ¦jē·ə¦des·ik kō'ȯrd·ən·əts }

geodesic incompleteness [RELAT] Property of a space-time wherein there exists at least one timelike or null geodesic that cannot be extended to arbitrarily large values of its affine parameter; such a space-time contains a singularity. Also known as g-incompleteness. { ¦jē·ə¦des·ik ¸iŋ·kəm'plēt·nəs }

geodesic motion [RELAT] Motion of a particle along a geodesic path in the four dimensional space-time continuum; according to general relativity, this is the motion which occurs in the absence of nongravitational forces. { ¦jē·ə¦des·ik 'mō·shən }

geodetic precession [RELAT] The precession of a gyroscope in orbit about the earth due to the curvature of space of the Schwarzschild metric. { ¦jē·ə¦ded·ik prē'sesh·ən }

geographical mile [MECH] The length of 1 minute of arc of the Equator, or 6087.08 feet (1855.34 meters), which approximates the length of the nautical mile. { ¦jē·ə¦graf·ə·kəl 'mīl }

geometrical acoustics *See* ray acoustics. { ¦jē·ə¦me·trə·kəl ə'kü·stiks }

geometric albedo [OPTICS] The ratio of the light flux received from an object to that which would be received from a perfectly reflecting, perfectly diffusing disk of the same size at the same distance at zero phase angle. { ¦jē·ə¦me·trik al'bē·dō }

geometrical optics [OPTICS] The geometry of paths of light rays and their imagery through optical systems. { ¦jē·ə¦me·trə·kəl 'äp·tiks }

geometrical similarity [FL MECH] Property of two fluid flows for which a simple alteration of scales of length and velocity transforms one into the other. { ¦jē·ə¦me·trə·kəl ¸sim·ə'lar·əd·ē }

geometric phase [PHYS] A unifying mathematical concept that describes the relation between the history of internal states of a system and the system's resulting orientation in space. { ¸jē·ə¸me·trik 'fāz }

geometric shadow [PHYS] That region which a given type of radiation would not reach, because of the presence of an object, if the effects of diffraction and interference could be neglected. { ¦jē·ə¦me·trik 'shad·ō }

geometrodynamics [RELAT] A theory involving only geometry which attempts to combine gravitational and electromagnetic theory; characterized by a multiply connected space-time manifold containing structures, descriptively called wormholes, associated with electric charge. { ¦jē·ō¸me·trə·dī'nam·iks }

geon [PHYS] A hypothetical electromagnetic field that is held together by its own gravitational attraction. { 'jē¸än }

geonium [ATOM PHYS] A microscopic system consisting of a single electron in a Penning trap, forming a kind of synthetic atom. Also known as monoelectron oscillator. { ¸jē'ō·nē·əm }

geopotential [PHYS] The potential energy of a unit mass relative to sea level, numerically equal to the work that would be done in lifting the unit mass from sea level to the height at which the mass is located, against the force of gravity. { ¦jē·ō·pə'ten·chəl }

Gerstner wave [FL MECH] A rotational gravity wave of finite amplitude. { 'gerst·nər ¸wāv }

GeV *See* gigaelectronvolt.

gf *See* gram-force.

g factor *See* Landé g factor.

g force [PHYS] A force such that a body subjected to it would have the acceleration of gravity at sea level; used as a unit of measurement for bodies undergoing the stress of acceleration. { 'jē ¸fȯrs }

gf-value [ATOM PHYS] The product of the oscillator strength f of an atomic transition and the statistical weight g of the lower level. Also known as weighted oscillator strength. { ¸jē'ef ¸val·yü }

Ghirardi-Rimini-Weber-Pearle theory *See* GRWP theory. { ¦gi¸rär·dē ¦rim·i·nē ¦vā·bər 'pərl ¸thē·ə·rē }

ghost crystal *See* phantom crystal. { 'gōst ¸krist·əl }

ghost image [OPTICS] An undesired image appearing at the image plane of an optical system; it may be a false image of the object or an out-of-focus image of a bright source of light in the field of the optical system. [SPECT] A false image of a spectral line produced by irregularities in the ruling of a diffraction grating. { 'gōst ¸im·ij }

ghost mode [ELECTROMAG] Waveguide mode having a trapped field associated with an imperfection in the wall of the waveguide; a ghost mode can cause trouble in a waveguide operating close to the cutoff frequency of a propagation mode. { 'gōst ¸mōd }

giant magnetoresistance [SOLID STATE] A very large decrease in electrical resistance upon application of a magnetic field in certain structures composed of alternating layers of magnetic and nonmagnetic metals. { ¦jī·ənt ¸mag·nēd·ō·ri'zis·təns }

giant nuclear resonance [NUC PHYS] A systematic excitation of the atomic nucleus which occurs with great strength in a concentrated energy region. { 'jī·ənt 'nü·klē·ər 'rez·ən·əns }

giant pulse laser *See* Q-switched laser. { 'jī·ənt 'pəls ¸lā·zər }

Giaque-Debye method *See* adiabatic demagnetization. { ¦zhyäk də'bī ¸meth·əd }

Giaque's temperature scale [THERMO] The internationally accepted scale of absolute temperature, in which the triple point of water is defined to have a temperature of 273.16 K. { ¦zhyäks 'tem·prə·chər ¸skāl }

gibbs [PHYS] A unit of amount of adsorption, equal to a surface concentration of 10^{-6} mole per square meter. { gibz }

Gibbs elasticity [PHYS] The elasticity of a film of liquid, equal to twice the product of the surface area and the derivative of the surface tension with respect to surface area. { 'gibz i¸las'tis·əd·ē }

Gibbs free energy [THERMO] The thermodynamic function $G = H - TS$, where H is enthalpy, T absolute temperature, and S entropy. Also known as free energy; free enthalpy; Gibbs function. { 'gibz ¦frē 'en·ər·jē }

Gibbs function *See* Gibbs free energy. { 'gibz ¸fəŋk·shən }

Gibbs-Helmholtz equation [THERMO] **1.** Either

of two thermodynamic relations that are useful in calculating the internal energy U or enthalpy H of a system; they may be written $U = F - T(\partial F/\partial T)_V$ and $H = G - T(\partial G/\partial T)_P$, where F is the free energy, G is the Gibbs free energy, T is the absolute temperature, V is the volume, and P is the pressure. **2.** Any of the similar equations for changes in thermodynamic potentials during an isothermal process. { 'gibz 'helm,hōlts i,kwā·zhən }

Gibbs paradox [STAT MECH] The paradox in which there is an increase in entropy when two separate volumes of gases of the same kind, at the same temperature and pressure, are mixed. { 'gibz 'par·ə,däks }

Gibbs system [STAT MECH] **1.** A hypothetical replica of a physical system. **2.** A set of such replicas forming an ensemble. { 'gibz ,sis·təm }

gigaelectronvolt [PHYS] A unit of energy, used primarily in high-energy physics, equal to 10^9 electronvolts or approximately 1.602×10^{-10} joule. Abbreviated GeV. { ¦gig·ə·i'lek,trän ,vōlt }

gigawatt [ELEC] One billion watts, or 10^9 watts. Abbreviated GW. { 'gig·ə,wät }

gigohm [ELEC] One thousand megohms, or 10^9 ohms. { 'gig,ōm }

gilbert [ELECTROMAG] The unit of magnetomotive force in the electromagnetic system, equal to the magnetomotive force of a closed loop of one turn in which there is a current of $1/(4\pi)$ abamp. { 'gil·bərt }

gill [MECH] **1.** A unit of volume used in the United States for the measurement of liquid substances, equal to 1/4 U.S. liquid pint, or to $1.1829411825 \times 10^{-4}$ cubic meter. **2.** A unit of volume used in the United Kingdom for the measurement of liquid substances, and occasionally of solid substances, equal to 1/4 U.K. pint, or to approximately $1.420653125 \times 10^{-4}$ cubic meter. { gil }

g-incompleteness *See* geodesic incompleteness. { ¦jē ,iŋ·kəm'plēt·nəs }

Ginzburg-Landau theory [CRYO] A phenomenological theory of superconductivity which accounts for the coherence length; the ordered state of a superconductor is described by a complex order parameter which is similar to a Schrödinger wave function, but describes all the condensed superelectrons, rather than a single charged particle. Also known as Landau-Ginzburg theory. { 'ginz·bərg 'lan·daů ,thē·ə·rē }

Ginzburg-London superconductivity theory [SOLID STATE] A modification of the London superconductivity theory to take into account the boundary energy. { 'ginz·bərg 'lən·dən 'sü·pər,kän,dək'tiv·əd·ē ,thē·ə·rē }

Giorgi system *See* meter-kilogram-second-ampere system. { ,gyȯr·gē ,sis·təm }

Gladstone-Dale constant [OPTICS] The ratio $(n - 1)/\rho$, where n is the index of refraction of a gas and ρ is its density. { 'glad·stōn 'dāl ,kän·stənt }

Gladstone-Dale law [OPTICS] A law for the variation of the index of refraction n of a substance, according to which $n - 1$ is proportional to its density. { 'glad·stōn 'dāl ,lȯ }

glancing angle [PHYS] The angle between a surface and a beam of particles or radiation incident upon it; it is the complement of the angle of incidence. { 'glans·iŋ ,aŋ·gəl }

glare [OPTICS] **1.** Discomfort produced in an observer by one or more visible sources of light. Also known as discomfort glare. **2.** Visual disability caused by visible sources or areas of luminance which are in an observer's field of view but do not assist in viewing. Also known as disability glare. **3.** Dazzling brightness of the atmosphere, caused by excessive reflection and scattering of light by particles in the line of sight. { gler }

glass capacitor [ELEC] A capacitor whose dielectric material is glass. { ¦glas kə'pas·əd·ər }

glass laser [OPTICS] A solid laser in which glass serves as the host for laser ions of such materials as erbium, holmium, neodymium, and ytterbium. Also known as amorphous laser. { ¦glas 'lā·zər }

glass transition [PHYS] The transition that occurs when a liquid is cooled to an amorphous or glassy solid. { 'glas tran,zish·ən }

glass-plate capacitor [ELEC] High-voltage capacitor in which the metal plates are separated by sheets of glass serving as the dielectric, with the complete assembly generally immersed in oil. { 'glas ,plāt kə'pas·əd·ər }

glass resistor [ELEC] A glass tube with a helical carbon resistance element painted on it. { ¦glas ri'zis·tər }

glassy state *See* vitreous state. { ¦glas·ē 'stāt }

glassy transition *See* glass transition. { ¦glas·ē tran'zish·ən }

glide *See* slip. { glīd }

glide plane [CRYSTAL] A lattice plane in a crystal on which translation or twin gliding occurs. Also known as slip plane. { 'glīd ,plān }

G line [ELECTROMAG] A single dielectric-coated, round wire used for transmitting microwave energy. { 'jē,līn }

glint [OPTICS] A small region designed to strongly reflect light from a target. { glint }

glissile dislocation *See* Shockley partial dislocation. { 'glis·əl ,dis·lō'kā·shən }

glitter [OPTICS] The spots of light reflected from a point source by the surface of the sea or wave facets, that is, specular reflection. { 'glid·ər }

g load [PHYS] The numerical ratio of any applied force to the gravitational force at the earth's surface. { 'jē ,lōd }

globally hyperbolic [RELAT] Property of a space-time M that satisfies certain causality conditions ensuring that the solution to the wave equation for a delta function source at a point p in M is unique and vanishes outside the causal future of p. { ¦glō·bə·lē ,hī·pər'bäl·ik }

global symmetry [PHYS] A type of symmetry of the Hamiltonian of a physical system, specifying that the system must obey some general conservation law. { ,glō·bəl 'sim·ə·trē }

glory [OPTICS] A set of concentric, colored rings

of light around the shadow cast by an observer or his head onto a cloud or fog bank. { 'glȯ·rē }

gloss [OPTICS] The ratio of the light specularly reflected from a surface to the total light reflected. { gläs }

glossy [OPTICS] Pertaining to a surface from which much more light is specularly reflected than is diffusely reflected. { 'gläs·ē }

glueball [PART PHYS] A hadron consisting entirely of gluons, without any quarks. Also known as bound glue state; gluonia. { 'glü,bȯl }

glug [MECH] A unit of mass, equal to the mass which is accelerated by 1 centimeter per second per second by a force of 1 gram-force, or to 980.665 grams. { gləg }

gluon [PART PHYS] One of eight hypothetical massless particles with spin quantum number and negative parity that mediate strong interactions between quarks. { 'glü,än }

gluonia *See* glueball. { glü'ō·nyə }

gm *See* gram.

gnomonic projection [CRYSTAL] A projection for displaying the poles of a crystal in which the poles are projected radially from the center of a reference sphere onto a plane tangent to the sphere. { nō'män·ik prə'jek·shən }

Gödel's universe [RELAT] An exact solution of the nonvacuum equations of general relativity with matter in the form of dust; there are closed timelike lines in this solution. { ¦gərd·əlz 'yü·nə,vərs }

Goertler parameter [FL MECH] A dimensionless number used in studying boundary-layer flow on curved surfaces, equal to the Reynolds number, where the characteristic length is the boundary-layer momentum thickness, times the square root of this thickness, divided by the square root of the surface's radius of curvature. { 'gərt·lər pə'ram·əd·ər }

gold-198 [NUC PHYS] The radioisotope of gold, atomic mass-number 198 and half-life 2.7 days; used in medical treatment of tumors by injecting it in colloidal form directly into tumor tissue. { ¦gōld ¦wən¦nīn·tē'āt }

Goldberg-Mohn friction [FL MECH] A force proportional to the velocity of a current and the density of the medium; used as a first approximation in estimating frictional effects in the atmosphere and the ocean. { 'gōlt·berk 'mōn 'frik·shən }

Goldhaber triangle [PART PHYS] A plot describing a high-energy reaction leading to four or more particles; its coordinates are the invariant masses of two intermediate-state quasi-particle composites, and its kinematical limits form a right-angled isosceles triangle; resonances in the quasi-particle composites appear as horizontal and vertical bands. { 'gōlt,hä·bər ,trī,aŋ·gəl }

gold-leaf electroscope [ELEC] An electroscope in which two narrow strips of gold foil or leaf suspended in a glass jar spread apart when charged; the angle between the strips is related to the charge. { 'gōld ,lēf i'lek·trə,skōp }

gold point [THERMO] The temperature of the freezing point of gold at a pressure of 1 standard atmosphere (101,325 pascals); used to define the International Temperature Scale of 1940, on which it is assigned a value of 1337.33 K or 1064.18°C. { 'gold ,pȯint }

Goldschmidt's law [SOLID STATE] The law that crystal structure is determined by the ratios of the numbers of the constituents, the ratios of their sizes, and their polarization properties. { 'gōl,shmits ,lȯ }

Goldstone bosons [PHYS] Particles with zero mass and zero spin which accompany spontaneous breaking of exact fundamental symmetries. { 'gōl,stōn 'bō,sänz }

goniophotometer [OPTICS] A photometer designed to measure the intensity of light reflected from a surface at various angles. { ¦gō·nē·ō·fə'täm·əd·ər }

good geometry [NUC PHYS] An arrangement of source and detecting equipment such that little error is produced by the finite sizes of the source and the detector aperture. { ¦gu̇d jē'äm·ə·trē }

Goos-Hähnchen effect [OPTICS] A shift of a few wavelengths in the position of a light beam that undergoes total internal reflection from a surface. { ¦gos 'hench·ən i,fekt }

G parity [PART PHYS] The eigenvalue of a system under the operation of inversion in isotopic spin space; it is conserved by the strong interactions. Also known as isotopic parity. { 'jē ¦par·əd·ē }

graded index lens *See* gradient index lens. { 'grād·əd ¦in,deks ,lenz }

graded refractive index rod lens [OPTICS] A type of gradient index lens used in optical fiber components, consisting of a rod with a refractive index that has its maximum value on the axis, and decreases approximately as the square of the distance. Abbreviated GRIN-rod lens. { 'grād·əd ri'frak·tiv 'in,deks ¦räd ,lenz }

gradient coupling [PART PHYS] A hypothetical interaction of particles in which the interaction Hamilton depends explicitly on first derivatives of wave functions associated with the particles with respect to position and time. { 'grād·ē·ənt ,kəp·liŋ }

gradient index lens [OPTICS] An optical element within which the refractive index is a smooth, but not constant, function of position and, as a result, the ray paths are curved. Also known as graded index lens. { 'grād·ē·ənt ¦in ,deks ,lenz }

Graetz number [THERMO] A dimensionless number used in the study of streamline flow, equal to the mass flow rate of a fluid times its specific heat at constant pressure divided by the product of its thermal conductivity and a characteristic length. Also spelled Grätz number. Symbolized N_{Gz}. { 'grets ,nəm·bər }

Graetz problem [FL MECH] The problem of determining the steady-state temperature field in a fluid flowing in a circular tube when the wall of the tube is held at a uniform temperature and the fluid enters the tube at a different uniform temperature. { 'grets ,präb·ləm }

Graham's law of diffusion [FL MECH] The law that the rate of diffusion of a gas is inversely proportional to the square root of its density. { 'grā·əmz ¦lȯ əv di'fyü·zhən }

grain [MECH] A unit of mass in the United States and United Kingdom, common to the avoirdupois, apothecaries', and troy systems, equal to 1/7000 of a pound, or to 6.479891×10^{-5} kilogram. Abbreviated gr. { grān }

grain boundary [SOLID STATE] An interface between individual crystals in a polycrystalline solid. { 'grān ,bau̇n·drē }

gram [MECH] The unit of mass in the centimeter-gram-second system of units, equal to 0.001 kilogram. Abbreviated g; gm. { gram }

gram-calorie *See* calorie. { 'gram ¦kal·ə·rē }

gram-centimeter [MECH] A unit of energy in the centimeter-gram-second gravitational system, equal to the work done by a force of magnitude 1 gram force when the point at which the force is applied is displaced 1 centimeter in the direction of the force. Abbreviated g-cm. { 'gram 'sent·ə,mēd·ər }

gram-force [MECH] A unit of force in the centimeter-gram-second gravitational system, equal to the gravitational force on a 1-gram mass at a specified location. Abbreviated gf. Also known as fors; gram-weight; pond. { 'gram ,fȯrs }

gram-weight *See* gram-force. { 'gram¦wāt }

grand canonical ensemble [STAT MECH] A collection of systems of particles used to describe an individual system which is allowed to exchange both energy and particles with its environment. { 'grand kə¦nän·ə·kəl än'säm·bəl }

grand unified field theory [PART PHYS] A theory in which the strong, electromagnetic, and weak interactions become aspects of one interaction. { 'grand ¦yü·nə,fīd 'fēld ,thē·ə·rē }

graphical statics [MECH] A method of determining forces acting on a rigid body in equilibrium, in which forces are represented on a diagram by straight lines whose lengths are proportional to the magnitudes of the forces. { ¦graf·ə·kəl 'stad·iks }

graser *See* gamma-ray laser. { 'grā·zər }

Grashof formula [FL MECH] A formula, $m = 0.0165A_2p_1^{0.97}$, used to express the discharge m of saturated steam in pounds per second, where A_2 is the area of the orifice in square feet, and p_1 is reservoir pressure in pounds per square foot. { 'gräs,hȯf ,fȯr·myə·lə }

Grashof number [FL MECH] A dimensionless number used in the study of free convection of a fluid caused by a hot body, equal to the product of the fluid's coefficient of thermal expansion, the temperature difference between the hot body and the fluid, the cube of a typical dimension of the body and the square of the fluid's density, divided by the square of the fluid's dynamic viscosity. Also known as free convection number. { 'gräs,hȯf ,nəm·bər }

graticule [OPTICS] A scale at the focal plane of an optical instrument to aid in the measurement of objects. { 'grad·ə,kyül }

grating [ELECTROMAG] **1.** An arrangement of fine, parallel wires used in waveguides to pass only a certain type of wave. **2.** An arrangement of crossed metal ribs or wires that acts as a reflector for a microwave antenna and offers minimum wind resistance. [SPECT] *See* diffraction grating. { 'grād·iŋ }

grating constant [OPTICS] The distance between consecutive diffraction centers of a ultrasonic wave which is producing a light diffraction spectrum. [SPECT] The distance between consecutive grooves of a diffraction grating. { 'grād·iŋ ,kän·stənt }

grating spectrograph [SPECT] A grating spectroscope provided with a photographic camera or other device for recording the spectrum. { 'grād·iŋ 'spek·trə,graf }

grating spectroscope [SPECT] A spectroscope which employs a transmission or reflection grating to disperse light, and usually also has a slit, a mirror or lenses to collimate the light sent through the slit and to focus the light dispersed by the grating into spectrum lines, and an eyepiece for viewing the spectrum. { 'grād·iŋ 'spek·trə,skōp }

Grätz number *See* Graetz number. { 'grets ,nəm·bər }

grav *See* G. { grav }

gravitation [PHYS] The mutual attraction between all masses in the universe. Also known as gravitational attraction. { ,grav·ə'tā·shən }

gravitational acceleration [PHYS] The acceleration imparted to a body by the attraction of the earth; approximately equal to 980.7 cm/s², or 32.2 ft/s². { ,grav·ə'tā·shən·əl ak,sel·ə'rā·shən }

gravitational attraction *See* gravitation. { ,grav·ə'tā·shən·əl ə'trak·shən }

gravitational bremsstrahlung [RELAT] The emission of gravitational radiation by two massive objects that pass each other at a high relative velocity and deflect each other slightly. Also known as relativistic bremsstrahlung. { ,grav·ə'tā·shən·əl 'brem,shträ·ləŋ }

gravitational constant [MECH] The constant of proportionality in Newton's law of gravitation, equal to the gravitational force between any two particles times the square of the distance between them, divided by the product of their masses. Also known as constant of gravitation. { ,grav·ə'tā·shən·əl 'kän·stənt }

gravitational displacement [MECH] The gravitational field strength times the gravitational constant. Also known as gravitational flux density. { ,grav·ə'tā·shən·əl dis'plās·mənt }

gravitational energy *See* gravitational potential energy. { ,grav·ə'tā·shən·əl 'en·ər·jē }

gravitational field [MECH] The field in a region in space in which a test particle would experience a gravitational force; quantitatively, the gravitational force per unit mass on the particle at a particular point. { ,grav·ə'tā·shən·əl 'fēld }

gravitational-field theory [RELAT] A theory in which gravity is treated as a field, as opposed to a theory in which the force acts instantaneously at a distance. { ,grav·ə'tā·shən·əl 'fēld ,thē·ə·rē }

gravitational flux density *See* gravitational displacement. { ˌgrav·əˈtā·shən·əl ˈfləks ˌden·səd·ē }

gravitational force [MECH] The force on a particle due to its gravitational attraction to other particles. { ˌgrav·əˈtā·shən·əl ˈfȯrs }

gravitational geon [PHYS] A hypothetical gravitational field that is held together by its own gravitational attraction. { ˌgrav·ə¦tā·shən·əl ˈjēˌän }

gravitational instability [MECH] Instability of a dynamic system in which gravity is the restoring force. { ˌgrav·əˈtā·shən·əl ˌin·stəˈbil·əd·ē }

gravitational mass [PHYS] The mass of a particle as it determines the force it experiences in a gravitational field; equal to inertial mass according to the equivalence principle. { ˌgrav·əˈtā·shən·əl ˈmas }

gravitational potential [MECH] The amount of work which must be done against gravitational forces to move a particle of unit mass to a specified position from a reference position, usually a point at infinity. { ˌgrav·əˈtā·shən·əl pəˈten·chəl }

gravitational potential energy [MECH] The energy that a system of particles has by virtue of their positions, equal to the work that must be done against gravitational forces to assemble the particles from some reference configuration, such as mutually infinite separation. Also known as gravitational energy. { ˌgrav·əˈtā·shən·əl pə¦ten·chəl ˈen·ər·jē }

gravitational pressure *See* hydrostatic pressure. { ˌgrav·əˈtā·shən·əl ˈpresh·ər }

gravitational radiation *See* gravitational wave. { ˌgrav·əˈtā·shən·əl ˌrād·ēˈā·shən }

gravitational radius *See* Schwarzschild radius. { ˌgrav·əˈtā·shən·əl ˈrād·ē·əs }

gravitational redshift [RELAT] A displacement of spectral lines toward the red when the gravitational potential at the observer of the light is greater than at its source. { ˌgrav·əˈtā·shən·əl ˈred ˌshift }

gravitational repulsion [PHYS] Hypothetical repulsion of matter and antimatter; however, experimental results indicate that matter and antimatter attract according to the same laws as matter and matter. { ˌgrav·əˈtā·shən·əl riˈpəl·shən }

gravitational systems of units [MECH] Systems in which length, force, and time are regarded as fundamental, and the unit of force is the gravitational force on a standard body at a specified location on the earth's surface. { ˌgrav·əˈtā·shən·əl ¦sis·təmz əv ˈyü·nəts }

gravitational wave [RELAT] A propagating gravitational field predicted by general relativity, which is produced by some change in the distribution of matter; it travels at the speed of light, exerting forces on masses in its path. Also known as gravitational radiation. { ˌgrav·əˈtā·shən·əl ˈwāv }

gravitino [PART PHYS] A hypothetical counterpart of the graviton; postulated to exist in supersymmetry theories. { ˌgrav·əˈtēn·ō }

gravitoelectric field [RELAT] In general relativity, those components of the gravitational field that are analogous to the electric-field components of the electromagnetic field and normally provide the dominant contribution. { ¦grav·ə·tō·i¦lek·trik ˈfēld }

gravitomagnetic field [RELAT] In general relativity, those components of the gravitational field that are analogous to the magnetic-field components of the electromagnetic field. { ¦grav·ə·tō·mag¦ned·ik ˈfēld }

graviton [PHYS] A theoretically deduced particle postulated as the quantum of the gravitational field, having a rest mass and charge of zero and a spin of 2. { ˈgrav·əˌtän }

gravity [MECH] The gravitational attraction at the surface of a planet or other celestial body. { ˈgrav·əd·ē }

gravity gradiometry [PHYS] The study and measurement of variations in the acceleration due to gravity. { ¦grav·əd·ē ˌgrād·ēˈäm·ə·trē }

gravity pendulum *See* pendulum. { ˈgrav·əd·ē ˈpen·jə·ləm }

gravity vector [MECH] The force of gravity per unit mass at a given point. Symbolized **g**. { ˈgrav·əd·ē ˌvek·tər }

gravity wave [FL MECH] **1.** A wave at a gas-liquid interface which depends primarily upon gravitational forces, surface tension and viscosity being of secondary importance. **2.** A wave in a fluid medium in which restoring forces are provided primarily by buoyancy (that is, gravity) rather than by compression. { ˈgrav·əd·ē ˌwāv }

graybody [THERMO] An energy radiator which has a blackbody energy distribution, reduced by a constant factor, throughout the radiation spectrum or within a certain wavelength interval. Also known as nonselective radiator. { ˈgrā ˌbäd·ē }

gray filter *See* neutral-density filter. { ˈgra ˌfil·tər }

gray scale [OPTICS] A series of achromatic tones having varying proportions of white and black, to give a full range of grays between white and black; a gray scale is usually divided into 10 steps; however, electronic scanners can typically differentiate 16 to 256 levels. { ˈgrā ˌskāl }

grazing angle [PHYS] A very small glancing angle. { ˈgraz·iŋ ˌaŋ·gəl }

grazing incidence [PHYS] Incidence at a small glancing angle. { ˈgraz·iŋ ˌin·sə·dəns }

grease spot photometer [OPTICS] A photometer in which the light sources whose intensities are to be compared illuminate a thin sheet of opaque paper with a translucent spot at the center. { ˈgrēs ˌspät fəˈtäm·ədər }

green [OPTICS] The hue evoked in an average observer by monochromatic radiation having a wavelength in the approximate range from 492 to 577 nanometers; however, the same sensation can be produced in a variety of other ways. { grēn }

green laser [OPTICS] A gas laser using mercury and argon to generate a green line at 5225 angstroms, corresponding to the wavelength that is most readily transmitted through seawater. { ¦grēn ¦lā·zər }

Gregorian telescope [OPTICS] A reflecting telescope having a paraboloidal mirror with a hole in the center and a small secondary (concave ellipsoidal) mirror placed beyond the focus of the primary mirror; light is reflected to the secondary mirror and back to an eyepiece at the hole; the telescope produces an erect image but a small field of view. { grə'gȯr·ē·ən 'tel·ə,skōp }

Greninger chart [CRYSTAL] A chart that enables angular relations between planes and zones in a crystal to be read directly from an x-ray diffraction photograph. { 'gren·iŋ·ər ,chärt }

grid spectrometer [SPECT] A grating spectrometer in which a large increase in light flux without loss of resolution is achieved by replacing entrance and exit slits with grids consisting of opaque and transparent areas, patterned to have large transmittance only when the entrance grid image coincides with that of the exit grid. { 'grid spek'träm·əd·ər }

Griebe-Schiebe method [SOLID STATE] A method of observing the piezoelectric behavior of small crystals, in which the crystals are placed between two electrodes connected to the resonant circuit of an oscillator, and tuning of the resonant circuit results in jumps in the oscillator frequency which produce clicks in headphones or a loudspeaker attached to the plate circuit of the oscillator. { 'grē·bə 'shē·bə ,meth·əd }

Griebhard's rings [ELEC] A method of producing lines of constant color on a copper sheet, coinciding with the equipotential lines of an electric field. { 'grēb·härts ,riŋz }

Griffith's criterion [MECH] A criterion for the fracture of a brittle material under biaxial stress, based on the theory that the strength of such a material is limited by small cracks. { 'grif·əths krī,tir·ē·ən }

Griffiths' method [THERMO] A method of measuring the mechanical equivalent of heat in which the temperature rise of a known mass of water is compared with the electrical energy needed to produce this rise. { 'grif·əths ,meth·əd }

grinding stress [MECH] Residual tensile or compressive stress, or a combination of both, on the surface of a material due to grinding. { 'grīn·diŋ ,stres }

GRIN-rod lens *See* graded refractive index rod lens. { 'grin ¦räd ,lenz }

grism [SPECT] A combination of a diffraction grating and a prism, wherein the grating spreads light into colors and the prism moves the spectrum's position to the point in an image where the observed object appears. { 'griz·əm }

gross ton *See* ton. { ¦grōs ¦tən }

ground [ELEC] **1.** A conducting path, intentional or accidental, between an electric circuit or equipment and the earth, or some conducting body serving in place of the earth. Abbreviated gnd. Also known as earth (British usage); earth connection. **2.** To connect electrical equipment to the earth or to some conducting body which serves in place of the earth. { graůnd }

ground absorption [ELECTROMAG] Loss of energy in transmission of radio waves, due to dissipation in the ground. { 'graůnd əb,sȯrp·shən }

ground conductivity [ELEC] The effective conductivity of the ground, used in calculating the attenuation of radio waves. { 'graůnd ,kän·dək¦tiv·əd·ē }

ground current *See* earth current. { 'graůnd ,kə·rənt }

ground dielectric constant [ELEC] Dielectric constant of the earth at a given location. { 'graůnd di·ə¦lek·trik 'kän·stənt }

grounded system [ELEC] Any conducting apparatus connected to ground. Also known as earthed system. { ¦graůnd·əd 'sis·təm }

ground equalizer inductors [ELECTROMAG] Coils, having relatively low inductance, inserted in the circuit to one or more of the grounding points of an antenna to distribute the current to the various points in any desired manner. { ¦graůnd 'ē·kwə,līz·ər in,dək·tərz }

ground glass [OPTICS] A sheet of matte-surfaced glass on the back of a view camera or process camera so that the image of the subject can be focused on it; it is exactly in the film plane. { ¦graůnd 'glas }

grounding [ELEC] Intentional electrical connection to a reference conducting plane, which may be earth, but which more generally consists of a specific array of interconnected electrical conductors referred to as the grounding conductor. { 'graůnd·iŋ }

grounding conductor [ELEC] An array of interconnected electric conductors at a uniform potential, to which electrical connections are made for the purpose of grounding. { 'graůnd·iŋ kən,dək·tər }

ground-plane antenna [ELECTROMAG] Vertical antenna combined with a grounded horizontal disk, turnstile element, or similar ground-plane simulation; such antennas may be mounted several wavelengths above the ground, and provide a low radiation angle. { 'graůnd ,plān an'ten·ə }

ground potential [ELEC] Zero potential with respect to the ground or earth. { 'graůnd pə,ten·chəl }

ground recharge [ELEC] The flow of electrons from the ground, in reference to lightning effects. { 'graůnd ¦rē,chärj }

ground-reflected wave [ELECTROMAG] Component of the ground wave that is reflected from the ground. { 'graůnd ri¦flek·təd 'wāv }

ground resistance [ELEC] Opposition of the earth to the flow of current through it; its value depends on the nature and moisture content of the soil, on the material, composition, and nature of connections to the earth, and on the electrolytic action present. { 'graůnd ri,zis·təns }

ground state [QUANT MECH] The stationary state of lowest energy of a particle or a system of particles. { 'graůnd ,stāt }

ground-state maser [PHYS] A maser whose amplifying transition has a terminal level that is appreciably populated at thermal equilibrium for the ambient temperature. { 'graůnd ,stāt 'mā·zər }

ground system [ELECTROMAG] The portion of an antenna that is closely associated with an extensive conducting surface, which may be the earth itself. { 'graůnd ,sis·təm }

group frequency [ELECTROMAG] Frequency corresponding to group velocity of propagated waves in a transmission line or waveguide. { 'grüp ¦frē·kwən·sē }

group velocity [PHYS] The velocity of the envelope of a group of interfering waves having slightly different frequencies and phase velocities. { ¦grüp və'läs·əd·ē }

growth spiral [CRYSTAL] A structure on a crystal surface, observed after growth, consisting of a growth step winding downward and outward in an Archimedean spiral which may be distorted by the crystal structure. { 'grōth ,spī·rəl }

growth step [CRYSTAL] A ledge on a crystal surface, one or more lattice spacings high, where crystal growth can take place. { 'grōth ,step }

Grüneisen constant [SOLID STATE] Three times the bulk modulus of a solid times its linear expansion coefficient, divided by its specific heat per unit volume; it is reasonably constant for most cubic crystals. Also known as Grüneisen gamma. { 'grü·nīz·ən ,kän·stənt }

Grüneisen gamma *See* Grüneisen constant. { 'grü·nīz·ən ,gam·ə }

Grüneisen relation [SOLID STATE] The relation stating that the electrical resistivity of a very pure metal is proportional to a mathematical function which depends on the ratio of the temperature to a characteristic temperature. { 'grü·nīz·ən ri,lā·shən }

GRWP theory [QUANT MECH] A theory that attempts to resolve the quantum measurement paradox by postulating the existence of new laws whose corrections to quantum mechanics become significant over time periods of t_0/N, where t_0 is a characteristic time of the order of the age of the universe and N is the number of particles in the system in question. Derived from Ghirardi-Rimini-Weber-Pearle theory. { ¦jē¦är¦dəb·əl,yü 'pē ,thē·ə·rē }

G string *See* field waveguide. { 'jē ,striŋ }

guard ring [ELEC] A ring-shaped auxiliary electrode surrounding one of the plates of a parallel-plate capacitor to reduce edge effects. [THERMO] A device used in heat flow experiments to ensure an even distribution of heat, consisting of a ring that surrounds the specimen and is made of a similar material. { 'gärd ,riŋ }

guided wave [ELECTROMAG] A wave whose energy is concentrated near a boundary or between substantially parallel boundaries separating materials of different properties and whose direction of propagation is effectively parallel to these boundaries; waveguides transmit guided waves. { 'gīd·əd 'wāv }

guide wavelength [ELECTROMAG] Wavelength of electromagnetic energy conducted in a waveguide; guide wavelength for all air-filled guides is always longer than the corresponding free-space wavelength. { 'gīd ¦wāv,leŋkth }

guiding center [ELECTROMAG] A slowly moving point about which a charged particle rapidly revolves; this is used in an approximation for the motion of a charged particle in slowly varying electric and magnetic fields. { 'gīd·iŋ ¦sent·ər }

guiding telescope [OPTICS] A telescope that is mounted so that it remains parallel to a photographic telescope and is used by a person observing through it to supplement the clock motion in keeping the image of a celestial body motionless on a photographic plate. { 'gīd·iŋ ¦tel·ə,skōp }

Guillemin effect [ELECTROMAG] The tendency of a bent magnetorestrictive rod to straighten in a magnetic field parallel to its length. { gē·yə'ma i,fekt }

Gukhman number [THERMO] A dimensionless number used in studying convective heat transfer in evaporation, equal to $(t_0 - t_m)/T_0$, where t_0 is the temperature of a hot gas stream, t_m is the temperature of a moist surface over which it is flowing, and T_0 is the absolute temperature of the gas stream. Symbolized *Gu*; N_{Gu}. { 'gu̇k·mən ,nəm·bər }

gun camera [OPTICS] **1.** A camera used in gunnery training that records the image of the target at which it is aimed on a strip of motion-picture film. **2.** A camera synchronized to a gun to record the results of firing, with the film revealing if the camera was correctly sighted on the target. { 'gən ¦kam·rə }

gun reaction [MECH] The force exerted on the gun mount by the rearward movement of the gun resulting from the forward motion of the projectile and hot gases. Also known as recoil. { 'gən rē,ak·shən }

Gurevich effect [SOLID STATE] An effect observed in electric conductors in which phonon-electron collisions are important, in the presence of a temperature gradient, in which phonons carrying a thermal current tend to drag the electrons with them from hot to cold. Also known as phonon-drag effect. { 'gu̇r·ə·vich i,fekt }

gust load [MECH] The wind load on an antenna due to gusts. { 'gəst ,lōd }

GW *See* gigawatt.

gyration tensor [SOLID STATE] A tensor characteristic of an optically active crystal, whose product with a unit vector in the direction of propagation of a light ray gives the gyration vector. { ji'rā·shən ¦ten·sər }

gyration vector [OPTICS] For light propagating in an optically active medium, a vector whose cross product with the time derivative of the electric displacement vector gives a negative contribution to the electric field. { ji'rā·shən ¦vek·tər }

gyrator [ELECTROMAG] A waveguide component that uses a ferrite section to give zero phase shift for one direction of propagation and 180°

phase shift for the other direction; in other words, it causes a reversal of signal polarity for one direction of propagation but not for the other direction. Also known as microwave gyrator. { 'jī,rād·ər }

gyrodynamics [MECH] The study of rotating bodies, especially those subject to precession. { ,jī·rō·dī'nam·iks }

gyrofrequency *See* cyclotron frequency. { 'jī·rō ,frē·kwən·sē }

gyromagnetic effect [ELECTROMAG] The rotation induced in a body by a change in its magnetization, or the magnetization resulting from a rotation. { ¦jī·rō·mag'ned·ik i'fekt }

gyromagnetic radius *See* Larmor radius. { ¦jī·rō·mag'ned·ik 'rād·ē·əs }

gyromagnetic ratio [PHYS] **1.** The ratio of the magnetic dipole moment to the angular momentum for a classical, atomic, or nuclear system. **2.** Occasionally, the reciprocal of the quantity in the first definition. { ¦jī·rō·mag'ned·ik 'rā·shō }

gyromagnetics [ELECTROMAG] The study of the relation between the angular momentum and the magnetization of a substance as exhibited in the gyromagnetic effect. { ¦jī·rō·mag'ned·iks }

gyroscopic precession [MECH] The turning of the axis of spin of a gyroscope as a result of an external torque acting on the gyroscope; the axis always turns toward the direction of the torque. { ,jī·rə'skäp·ik prē'sesh·ən }

gyroscopics [MECH] The branch of mechanics concerned with gyroscopes and their use in stabilization and control of ships, aircraft, projectiles, and other objects. { ,jī·rə'skäp·iks }

H

h *See* Planck's constant.

H *See* henry.

ha *See* hectare.

Haas effect [ACOUS] A phenomenon whereby sound produced by the second of two loudspeakers cannot be detected if it is delayed relative to sound from the other loudspeaker by a time interval between 1 and 30 milliseconds and has an intensity less that 10 decibels above that of the primary sound. { häs iˌfekt }

habit *See* crystal habit. { 'hab·ət }

habit plane [CRYSTAL] The crystallographic plane or system of planes along which certain phenomena such as twinning occur. { 'hab·ət ˌplān }

hadron [PART PHYS] An elementary particle which has strong interactions. { 'hadˌrän }

hadronic atom [ATOM PHYS] An atom consisting of a negatively charged, strongly interacting particle orbiting around an ordinary nucleus. { ha'drän·ik 'ad·əm }

Hagen-Poiseuille law [FL MECH] In the case of laminar flow of fluid through a circular pipe, the loss of head due to fluid friction is 32 times the product of the fluid's viscosity, the pipe length, and the fluid velocity, divided by the product of the acceleration of gravity, the fluid density, and the square of the pipe diameter. { 'häg·ən pwä·zȯi ˌlȯ }

Hagen-Rubens relation [OPTICS] An equation for the reflectivity of a solid surface in terms of the frequency of radiation of the conductivity of the solid; it applies at wavelengths long enough that the product of the frequency and the relaxation time is much less than unity. { ¦häg·ən ¦rü·bənz ri'lā·shən }

Hahn technique [SOLID STATE] A method of studying changes in solids under various treatments that involves incorporating small amounts of radium into the solid and measuring the emanating power. { 'hän tekˌnēk }

Haidinger brushes [OPTICS] Faint yellow, brushlike patterns that are observed when a bright surface is viewed through a polarizer such as a rotating Nicol prism or sheet of Polaroid film; believed to be caused by birefringence of fibers at the fovea of the eye. { 'hī·diŋ·ər ˌbrəsh·əz }

Haidinger fringes [OPTICS] Interference fringes produced by nearly normal incidence of light on thick, flat plates. Also known as constant-angle fringes; constant-deviation fringes. { 'hī·diŋ·ər ˌfriŋ·jəz }

halation [OPTICS] A halo on a photographic image of a bright object caused by light reflected from the back of the film or plate. { hā'lā·shən }

Halbach array [ELECTROMAG] An array of permanent magnets that produces a strong, concentrated, spatially periodic magnetic field; used on the moving object in the Inductrack magnetic levitation system. { 'hälˌbäk̲ əˌrā }

half-period zones *See* Fresnel zones. { 'haf ˌpir·ē·əd ˌzōnz }

half-power beamwidth [ELECTROMAG] The angle across the main lobe of an antenna pattern between the two directions at which the antenna's sensitivity is half its maximum value at the center of the lobe. Abbreviated HPBW. { 'haf ¦pau̇·ər 'bēmˌwidth }

half-shade plate [OPTICS] A half-wave plate that is placed near the polarizer of a polariscope, between it and the analyzer. { 'haf ˌshād ˌplāt }

half-silvered surface [OPTICS] A surface covered with metallic film of a thickness such that approximately half the light falling on it at normal incidence is reflected and half is transmitted. { 'haf ¦sil·vərd ˌsər·fəs }

half step *See* semitone. { 'haf ˌstep }

half thickness [PHYS] The thickness of a sheet of material which reduces the intensity of a beam of radiation passing through it to one-half its initial value. Also known as half-value layer; half-value thickness. { 'haf ¦thik·nəs }

half-value layer *See* half thickness. { 'haf ¦val·yü ˌlā·ər }

half-value thickness *See* half thickness. { 'haf ¦val·yü ˌthik·nəs }

half-wave [ELEC] Pertaining to half of one cycle of a wave. [ELECTROMAG] Having an electrical length of a half wavelength. { 'haf ¦wāv }

half-wave antenna [ELECTROMAG] An antenna whose electrical length is half the wavelength being transmitted or received. { 'haf ¦wāv an'ten·ə }

half-wave dipole *See* dipole antenna. { 'haf ¦wāv 'dīˌpōl }

half-wavelength [ELECTROMAG] The distance corresponding to an electrical length of half a wavelength at the operating frequency of a transmission line, antenna element, or other device. { 'haf ¦wāv,leŋkth }

half-wave plate [OPTICS] A thin section of a doubly refracting crystal, of a thickness such that the ordinary and extraordinary components of normally incident light emerge from it with a phase difference corresponding to an odd number of half wavelengths. { 'haf ¦wāv 'plāt }

half-wave transmission line [ELECTROMAG] Transmission line which has an electrical length equal to one-half the wavelength of the signal being transmitted or received. { 'haf ¦wāv tranz'mish·ən ,līn }

Hall accelerator [PL PHYS] A plasma accelerator based on the Hall effect. { 'hȯl ak'sel·ə,rād·ər }

Hall angle [ELECTROMAG] The electric field, resulting from the Hall effect, perpendicular to a current, divided by the electric field generating the current. { 'hȯl ,aŋ·gəl }

Hall coefficient [ELECTROMAG] A measure of the Hall effect, equal to the transverse electric field (Hall field) divided by the product of the current density and the magnetic induction. Also known as Hall constant. { 'hȯl ,kō·i'fish·ənt }

Hall constant *See* Hall coefficient. { 'hȯl ,kän·stənt }

Hall effect [ELECTROMAG] The development of a transverse electric field in a current-carrying conductor placed in a magnetic field; ordinarily the conductor is positioned so that the magnetic field is perpendicular to the direction of current flow and the electric field is perpendicular to both. { 'hȯl i,fekt }

Hall-effect isolator [ELECTROMAG] An isolator that makes use of the Hall effect in a semiconductor plate mounted in a magnetic field, to provide greater loss in one direction of signal travel through a waveguide than in the other direction. { 'hȯl i,fekt 'ī·sə,lād·ər }

Hall generator [ELECTROMAG] A generator using the Hall effect to give an output voltage proportional to magnetic field strength. { 'hȯl 'jen·ə,rād·ər }

Hall mobility [SOLID STATE] The product of conductivity and the Hall constant for a conductor or semiconductor; a measure of the mobility of the electrons or holes in a semiconductor. { 'hȯl mō'bil·əd·ē }

Hallwachs effect [PHYS] The ability of ultraviolet radiation to discharge a negatively charged body in a vacuum. { 'häl,väks i,fekt }

halo [OPTICS] A ring around the photographic image of a bright source caused by light scattering in any one of a number of possible ways. { 'hā·lō }

halo of 22° [OPTICS] A halo phenomenon in the form of a prismatically colored circle of 22° angular radius around the sun or moon, exhibiting coloration from red on the inside to blue on the outside. { 'hā·lō əv ¦twen·tē'tü di'grēz }

halo of 46° [OPTICS] A halo phenomenon in the form of a prismatically colored circle or incomplete arc thereof, centered on the sun or moon and having an angular radius of about 46°; the coloration is red on the inner edge to blue on the outer edge. { 'hā·lō əv ¦fȯrd·ē'siks di'grēz }

halo of 90° *See* Hevelian halo. { 'hā·lō əv 'nīn·tē di'grēz }

halo of Hevelius *See* Hevelian halo. { 'hā·lō əv hə'vel·yəs }

Hamiltonian function [MECH] A function of the generalized coordinates and momenta of a system, equal in value to the sum over the coordinates of the product of the generalized momentum corresponding to the coordinate, and the coordinate's time derivative, minus the Lagrangian of the system; it is numerically equal to the total energy if the Lagrangian does not depend on time explicitly; the equations of motion of the system are determined by the functional dependence of the Hamiltonian on the generalized coordinates and momenta. { ,ham·əl'tō·nē·ən ¦fəŋk·shən }

Hamiltonian operator *See* energy operator. { ,ham·əl'tō·nē·ən ¦äp·ə,rād·ər }

Hamilton-Jacobi theory [MECH] A theory that provides a means for discussing the motion of a dynamic system in terms of a single partial differential equation of the first order, the Hamilton-Jacobi equation. { 'ham·əl·tən jə'kō·bē ,thē·ə·rē }

Hamilton's equations of motion [MECH] A set of first-order, highly symmetrical equations describing the motion of a classical dynamical system, namely $\dot{q}_j = \partial H/\partial p_j$, $\dot{p}_j = -\partial H/\partial q_j$; here q_j ($j = 1, 2, \ldots$) are generalized coordinates of the system, p_j is the momentum conjugate to q_j, and H is the Hamiltonian. Also known as canonical equations of motion. { 'ham·əl·tənz i¦kwā·zhənz əv 'mō·shən }

Hamilton's principle [MECH] A variational principle which states that the path of a conservative system in configuration space between two configurations is such that the integral of the Lagrangian function over time is a minimum or maximum relative to nearby paths between the same end points and taking the same time. { 'ham·əl·tənz ¦prin·sə·pəl }

Hampson process [CRYO] A process for liquefying gases which resembles the Linde process except that the Joule-Thomson expansion reduces the gas pressure to approximately atmospheric pressure. { 'ham·sən ,prä·səs }

handedness [PHYS] A division of objects, such as coordinate systems, screws, and circularly polarized light beams, into two classes (right and left), which distinguishes an object from a mirror image but not from a rotated object. { 'han·dəd·nəs }

hand lens *See* simple microscope. { 'hand ,lenz }

hand rule *See* right-hand rule. { 'hand ,rül }

hand viewer [OPTICS] An optical magnifying device small enough for hand use. { 'hand ¦vyü·ər }

Hanle effect [OPTICS] A reduction in the polarization of light that is emitted from atoms excited by linearly polarized light when a magnetic field is applied in the direction of observation. { 'hän·lē i,fekt }

hardness [ELECTROMAG] That quality which determines the penetrating ability of x-rays; the shorter the wavelength, the harder and more penetrating the rays. { 'härd·nəs }

hard radiation [PHYS] Radiation whose particles or photons have a high energy and, as a result, readily penetrate all kinds of materials, including metals. { 'härd ,rād·ē'ā·shən }

hard superconductor [CRYO] A superconductor that requires a strong magnetic field, over 1000 oersteds (79,577 amperes per meter), to destroy superconductivity; niobium and vanadium are examples. { 'härd 'sü·pər·kən,dək·tər }

Harker-Kasper inequalities [SOLID STATE] Inequalities used in the analysis of crystal structure by x-ray diffraction which relate the structure factors and help to determine their phase factors. { 'härk·ər 'kas·pər ,in·i'kwäl·əd·ēz }

Harkin's rule [PHYS] An empirical rule for the calculation of the nuclear abundances of an element's isotopes stating that isotopes with an odd mass number are less abundant than their even-mass-number neighbors. { 'här·kənz ,rül }

harmonic [ACOUS] One of a series of sounds, each of which has a frequency which is an integral multiple of some fundamental frequency. [PHYS] A sinusoidal component of a periodic wave, having a frequency that is an integral multiple of the fundamental frequency. Also known as harmonic component. { här'män·ik }

harmonic analysis [PHYS] Any method of identifying and evaluating the harmonics that make up a complex waveform of sound pressure, voltage, current, or some other varying quantity. { här'man·ik ə'nal·ə·səs }

harmonic antenna [ELECTROMAG] An antenna whose electrical length is an integral multiple of a half-wavelength at the operating frequency of the transmitter or receiver. { här'män·ik an'ten·ə }

harmonic component *See* harmonic. { här'män·ik kəm'pō·nənt }

harmonic content [PHYS] The components remaining after the fundamental frequency has been removed from a complex wave. { här 'män·ik 'kän·tent }

harmonic echo [ACOUS] An echo that appears to be higher in pitch than the original sound, due to enhancement of harmonics in the original complex tone. { här¦män·ik 'ek·ō }

harmonic fields [ELECTROMAG] The sinusoidal Fourier components of a magnetic or other field confined to a finite region of space; their half-wavelengths are integral divisors of the length of the space in which the field is confined. { här 'män·ik 'fēlz }

harmonic frequency [PHYS] An integral multiple of the fundamental frequency of a periodic wave. { här'män·ik 'frē·kwən·sē }

harmonic loss [ELECTROMAG] Energy loss in a generator due to space harmonics of the magnetomotive force produced by armature current, especially losses resulting from the fifth and seventh harmonics. { här'män·ik 'lȯs }

harmonic motion [MECH] A periodic motion that is a sinusoidal function of time, that is, motion along a line given by the equation $x = a\cos(kt + \theta)$, where t is the time parameter, and a, k, and θ are constants. Also known as harmonic vibration; simple harmonic motion (SHM). { här'män·ik 'mō·shən }

harmonic oscillator [MECH] Any physical system that is bound to a position of stable equilibrium by a restoring force or torque proportional to the linear or angular displacement from this position. [PHYS] Anything which has equations of motion that are the same as the system in the mechanics definition. Also known as linear oscillator; simple oscillator. { här'män·ik 'äs·ə,lād·ər }

harmonic synthesizer [MECH] A machine which combines elementary harmonic constituents into a single periodic function; a tide-predicting machine is an example. { här'män·ik 'sin·thə,sīz·ər }

harmonic vibration *See* harmonic motion. { här 'män·ik vī'brā·shən }

harmonic vibration-rotation band [SPECT] A vibration-rotation band of a molecule in which the harmonic oscillator approximation holds for the vibrational levels, so that the vibrational levels are equally spaced. { här'män·ik vī'brā·shən rō'tā·shən ,band }

harmonic wave [PHYS] A transverse waveform obtained by mapping onto a time base the periodic up and down excursions of simple harmonic motion. { här'män·ik 'wāv }

Hartmann dispersion formula [OPTICS] A semiempirical formula relating the index of refraction n and wavelengths λ; $n = n_0 + a/(\lambda - \lambda_0)$, where n_0, a, and λ_0 are empirical constants. Also known as Cornu-Hartmann formula. { 'härt·män di'sper·zhən ,fȯr·myə·lə }

Hartmann flow [PL PHYS] The steady flow of an electrically conducting fluid between two parallel plates when there is a uniform applied magnetic field normal to the plates. { 'härt·män ,flō }

Hartmann number [PL PHYS] A dimensionless number which gives a measure of the relative importance of drag forces resulting from magnetic induction and viscous forces in Hartmann flow, and determines the velocity profile for such flow. { 'härt·män ,nəm·bər }

Hartmann test [OPTICS] A test for telescope mirrors in which the mirror is covered with a screen with regularly spaced holes, and a photographic plate is placed near the focus; for a perfect mirror, this results in regularly spaced dots on the plate. [SPECT] A test for spectrometers in which light is passed through different parts of the entrance slit; any resulting changes of the spectrum indicate a fault in the instrument. { 'härt·män ,test }

hartree [ATOM PHYS] A unit of energy used in studies of atomic spectra and structure, equal to $2R_\infty hc$ or $\alpha^2 mc^2$, where R_∞ is the Rydberg constant, h is Planck's constant, c is the speed of light, α is the fine-structure constant, and m is the mass of the electron; numerically, it is approximately 27.21 electronvolts or 4.360×10^{-18} joule. Also known as Hartree energy. { 'här·trē }

Hartree energy *See* hartree.

Hartree-Fock approximation [QUANT MECH] A refinement of the Hartree method in which one uses determinants of single-particle wave functions rather than products, thereby introducing exchange terms into the Hamiltonian. { 'här·trē ,fäk ə,präk·sə,mā·shən }

Hartree method [QUANT MECH] An iterative variational method of finding an approximate wave function for a system of many electrons, in which one attempts to find a product of single-particle wave functions, each one of which is a solution of the Schrödinger equation with the field deduced from the charge density distribution due to all the other electrons. Also known as self-consistent field method. { 'här·trē ,meth·əd }

Hartree units [ATOM PHYS] A system of units in which the unit of angular momentum is Planck's constant divided by 2π, the unit of mass is the mass of the electron, and the unit of charge is the charge of the electron. Also known as atomic units. { 'här·trē ,yü·nəts }

Haüy law [CRYSTAL] The law that for a given crystal there is a set of ratios such that the ratios of the intercepts of any crystal plane on the crystal axes are rational fractions of these ratios. { ä'wē ,lȯ }

Havelock's law [OPTICS] The law that in a substance displaying the Kerr effect, $n_p - n = 2(n_s - n)$, where n is the index of refraction in the absence of an electric field, and n_p and n_s are the indices of refraction of light whose magnetic vector is parallel and perpendicular to the applied electric field. { 'hav,läks ,lȯ }

Hay bridge [ELEC] A four-arm alternating-current bridge used to measure inductance in terms of capacitance, resistance, and frequency; bridge balance depends on frequency. { 'hā ,brij }

haze [OPTICS] The degree of cloudiness in a solution, cured plastic material, or coating material. { hāz }

h-bar [QUANT MECH] A fundamental constant equal to $h/2\pi$, where h is Planck's constant. Symbolized $\hbar$. Also known as Dirac h; h-line. { 'āch ,bär }

H bend *See* H-plane bend. { 'āch ,bend }

hcp structure *See* hexagonal close-packed structure. { ¦āch¦sē¦pē 'strək·chər }

head *See* pressure head. { hed }

head loss [FL MECH] The drop in the sum of pressure head, velocity head, and potential head between two points along the path of a flowing fluid, due to causes such as fluid friction. { 'hed ,lȯs }

head-up display [OPTICS] A device that enables an aircraft pilot to view the instrument panel while looking out the cockpit window, by projecting an image of the panel in the direction of the window and forming the image at infinity. { 'hed ,əp di'splā }

heat [THERMO] Energy in transit due to a temperature difference between the source from which the energy is coming and a sink toward which the energy is going; other types of energy in transit are called work. { hēt }

heat balance [THERMO] The equilibrium which is known to exist when all sources of heat gain and loss for a given region or body are accounted for. { 'hēt ,bal·əns }

heat budget [THERMO] The statement of the total inflow and outflow of heat for a planet, spacecraft, biological organism, or other entity. { 'hēt ,bəj·ət }

heat capacity [THERMO] The quantity of heat required to raise a system one degree in temperature in a specified way, usually at constant pressure or constant volume. Also known as thermal capacity. { 'hēt kə,pas·əd·ē }

heat conduction [THERMO] The flow of thermal energy through a substance from a higher-to a lower-temperature region. { 'hēt kən,dək·shən }

heat conductivity *See* thermal conductivity. { 'hēt ,kän·dək'tiv·əd·ē }

heat content *See* enthalpy. { 'hēt ¦kän·tent }

heat convection [THERMO] The transfer of thermal energy by actual physical movement from one location to another of a substance in which thermal energy is stored. Also known as thermal convection. { 'hēt kən¦vek·shən }

heat cycle *See* thermodynamic cycle. { 'hēt ,sī·kəl }

heat death [THERMO] The condition of any isolated system when its entropy reaches a maximum, in which matter is totally disordered and at a uniform temperature, and no energy is available for doing work. { 'hēt ,deth }

heat dissipation *See* heat loss. { 'hēt ,dis·ə¦pā·shən }

heat energy *See* internal energy. { 'hēt ,en·ər·jē }

heat engine [THERMO] A thermodynamic system which undergoes a cyclic process during which a positive amount of work is done by the system; some heat flows into the system and a smaller amount flows out in each cycle. { 'hēt ,en·jən }

heat equation [THERMO] A parabolic second-order differential equation for the temperature of a substance in a region where no heat source exists: $\partial t/\partial \tau = (k/\rho c)(\partial^2 t/\partial x^2 + \partial^2 t/\partial y^2 + \partial t^2/\partial z^2)$, where x, y, and z are space coordinates, τ is the time, $t(x,y,z,\tau)$ is the temperature, k is the thermal conductivity of the body, ρ is its density, and c is its specific heat; this equation is fundamental to the study of heat flow in bodies. Also known as Fourier heat equation; heat flow equation. { 'hēt i,kwā·zhən }

heat filter [OPTICS] Special glass in condenser lens systems to keep heat from film. { 'hēt ,fil·tər }

heat flow [THERMO] Heat thought of as energy flowing from one substance to another; quantitatively, the amount of heat transferred in a unit time. Also known as heat transmission. { 'hēt ˌflō }

heat flow equation *See* heat equation. { 'hēt ¦flō iˌkwā·zhən }

heat flux [THERMO] The amount of heat transferred across a surface of unit area in a unit time. Also known as thermal flux. { 'hēt ˌfləks }

heat loss [PHYS] Energy or power transmitted out of a system in the form of heat. Also known as heat dissipation. { 'hēt ˌlȯs }

heat of ablation [THERMO] A measure of the effective heat capacity of an ablating material, numerically the heating rate input divided by the mass loss rate which results from ablation. { 'hēt əv ə'blā·shən }

heat of adsorption [THERMO] The increase in enthalpy when 1 mole of a substance is adsorbed upon another at constant pressure. { 'hēt əv ad'sȯrp·shən }

heat of aggregation [THERMO] The increase in enthalpy when an aggregate of matter, such as a crystal, is formed at constant pressure. { 'hēt əv ˌag·rə'gā·shən }

heat of compression [THERMO] Heat generated when air is compressed. { 'hēt əv kəm'presh·ən }

heat of condensation [THERMO] The increase in enthalpy accompanying the conversion of 1 mole of vapor into liquid at constant pressure and temperature. { 'hēt əv ˌkänd·ən'sā·shən }

heat of cooling [THERMO] Increase in enthalpy during cooling of a system at constant pressure, resulting from an internal change such as an allotropic transformation. { 'hēt əv 'kül·iŋ }

heat of crystallization [THERMO] The increase in enthalpy when 1 mole of a substance is transformed into its crystalline state at constant pressure. { 'hēt əv ˌkrist·əl·ə'zā·shən }

heat of evaporation *See* heat of vaporization. { 'hēt əv iˌvap·ə'rā·shən }

heat of fusion [THERMO] The increase in enthalpy accompanying the conversion of 1 mole, or a unit mass, of a solid to a liquid at its melting point at constant pressure and temperature. Also known as latent heat of fusion. { 'hēt əv 'fyü·zhən }

heat of mixing [THERMO] The difference between the enthalpy of a mixture and the sum of the enthalpies of its components at the same pressure and temperature. { 'hēt əv 'mik·siŋ }

heat of solidification [THERMO] The increase in enthalpy when 1 mole of a solid is formed from a liquid or, less commonly, a gas at constant pressure and temperature. { 'hēt əv səˌlid·ə·fə'kā·shən }

heat of sublimation [THERMO] The increase in enthalpy accompanying the conversion of 1 mole, or unit mass, of a solid to a vapor at constant pressure and temperature. Also known as latent heat of sublimation. { 'hēt əv ˌsəb·lə'mā·shən }

heat of transformation [THERMO] The increase in enthalpy of a substance when it undergoes some phase change at constant pressure and temperature. { 'hēt əv ˌtranz·fər'mā·shən }

heat of vaporization [THERMO] The quantity of energy required to evaporate 1 mole, or a unit mass, of a liquid, at constant pressure and temperature. Also known as enthalpy of vaporization; heat of evaporation; latent heat of vaporization. { 'hēt əv ˌvā·pə·rə'zā·shən }

heat of wetting [THERMO] **1.** The heat of adsorption of water on a substance. **2.** The additional heat required, above the heat of vaporization of free water, to evaporate water from a substance in which it has been absorbed. { 'hēt əv 'wed·iŋ }

heat quantity [THERMO] A measured amount of heat; units are the small calorie, normal calorie, mean calorie, and large calorie. { 'hēt ¦kwän·əd·ē }

heat radiation [THERMO] The energy radiated by solids, liquids, and gases in the form of electromagnetic waves as a result of their temperature. Also known as thermal radiation. { 'hēt ˌrād·ē'ā·shən }

heat release [THERMO] The quantity of heat released by a furnace or other heating mechanism per second, divided by its volume. { 'hēt riˌlēs }

heatsink [ELEC] A mass of metal that is added to a device for the purpose of absorbing and dissipating heat; used with power transistors and many types of metallic rectifiers. Also known as dissipator. [THERMO] Any (gas, solid, or liquid) region where heat is absorbed. { 'hētˌsiŋk }

heat source [THERMO] Any device or natural body that supplies heat. { 'hēt ˌsȯrs }

heat transfer [THERMO] The movement of heat from one body to another (gas, liquid, solid, or combinations thereof) by means of radiation, convection, or conduction. { 'hēt ¦tranz·fər }

heat-transfer coefficient [THERMO] The amount of heat which passes through a unit area of a medium or system in a unit time when the temperature difference between the boundaries of the system is 1 degree. { 'hēt ¦tranz·fər ˌkō·i'fish·ənt }

heat transmission *See* heat flow. { 'hēt tranz ˌmish·ən }

heat transport [THERMO] Process by which heat is carried past a fixed point or across a fixed plane, as in a warm current. { 'hēt ¦tranzˌpȯrt }

heat wave [ELECTROMAG] Infrared radiation, much higher in frequency than radio waves. { 'hēt ˌwāv }

Heaviside-Lorentz system [ELECTROMAG] A system of electrical units which is the same as the Gaussian system except that the units of charge and current are smaller by a factor of $1/\sqrt{4\pi}$, and those of electric and magnetic field are larger by a factor by $\sqrt{4}$. Also known as Lorentz-Heaviside system. { 'hev·ēˌsīd lȯ'rents ˌsis·təm }

heavy-fermion superconductor [SOLID STATE] A superconductor in which the superconducting electrons have unusually large effective masses, more than 100 times the mass of a free electron. { ¦hev·ē 'fər·mēˌän 'sü·pər·kənˌdək·tər }

heavy-fermion system [SOLID STATE] A lanthanide-based or actinide-based intermetallic compound in which the low-energy excitations (quasiparticles) of the conduction electron system have effective masses at low temperatures that are several hundred times the free-electron mass. { ˌhev·ē 'fər·mēˌän ˌsis·təm }

heavy hydrogen [NUC PHYS] Hydrogen consisting of isotopes whose mass number is greater than one, namely deuterium or tritium. { 'hev·ē 'hī·drə·jən }

heavy oxygen *See* oxygen-18. { 'hev·ē 'äk·sə·jən }

heavy particle *See* baryon. { 'hev·ē 'pärd·ə·kəl }

hectare [MECH] A unit of area in the metric system equal to 100 ares or 10,000 square meters. Abbreviated ha. { 'hekˌtar }

hectogram [MECH] A unit of mass equal to 100 grams. Abbreviated hg. { 'hek·təˌgram }

hectoliter [MECH] A metric unit of volume equal to 100 liters or to 0.1 cubic meter. Abbreviated hl. { 'hek·təˌlēd·ər }

hectometer [MECH] A unit of length equal to 100 meters. Abbreviated hm. { 'hek·təˌmēd·ər }

Hedvall effect I [SOLID STATE] A discontinuous change in the temperature dependence of the chemical reaction rate of certain substances at the Curie temperture. { 'hed·vȯl iˌfekt 'wən }

Hedvall effect II [SOLID STATE] A discontinuous change in the activation energy of certain substances at the Curie temperature. { 'hed·vȯl iˌfekt 'tü }

HEED *See* high-energy electron diffraction.

Hefner candle [OPTICS] A luminous intensity standard, formerly used in Germany, equal to 0.9 international candle; produced by a Hefner lamp burning under standard conditions. Abbreviated HK. Also known as Hefnerkerze. { 'hef·nər ¦kand·əl }

Hefnerkerze *See* Hefner candle. { 'hef·nərˌkert·sə }

heiligenschein [OPTICS] A diffuse white ring surrounding the shadow cast by the observer's head upon a dew-covered lawn when the solar elevation is low and, therefore, the distance from observer to shadow is great. { 'hī·lə·gənˌshīn }

Heisenberg algebra [QUANT MECH] The Lie algebra formed by the operators of position and momentum. { 'hīz·ən·bərg ˌal·jə·brə }

Heisenberg equation of motion [QUANT MECH] An equation which gives the rate of change of an operator corresponding to a physical quantity in the Heisenberg picture. { 'hīz·ən·bərg i¦kwā·zhən əv 'mō·shən }

Heisenberg exchange coupling [SOLID STATE] The exchange forces between electrons in neighboring atoms which give rise to ferromagnetism in the Heisenberg theory. { 'hīz·ən·bərg iks'chānj ˌkəp·liŋ }

Heisenberg force [NUC PHYS] A force between two nucleons derivable from a potential with an operator which exchanges both the positions and the spins of the particles. { 'hīz·ən·bərg ˌfȯrs }

Heisenberg picture [QUANT MECH] A mode of description of a system in which dynamic states are represented by stationary vectors and physical quantities are represented by operators which evolve in the course of time. Also known as Heisenberg representation. { 'hīz·ən·bərg ˌpik·chər }

Heisenberg representation *See* Heisenberg picture. { 'hīz·ən·bərg ˌre·prəˌzen'tā·shən }

Heisenberg theory of ferromagnetism [SOLID STATE] A theory in which exchange forces between electrons in neighboring atoms are shown to depend on relative orientations of electron spins, and ferromagnetism is explained by the assumption that parallel spins are favored so that all the spins in a lattice have a tendency to point in the same direction. { 'hīz·ən·bərg 'thē·ə·rē əv ˌfer·ō'mag·nəˌtiz·əm }

Heisenberg uncertainty principle *See* uncertainty principle. { 'hīz·ən·bərg ən'sərt·ən·tē ˌprin·sə·pəl }

Heisenberg uncertainty relation *See* uncertainty relation. { 'hīz·ən·bərg ən'sərt·ən·tē riˌlā·shən }

helical angle [MECH] In the study of torsion, the angular displacement of a longitudinal element, originally straight on the surface of an untwisted bar, which becomes helical after twisting. { 'hel·ə·kəl 'aŋ·gəl }

helical antenna [ELECTROMAG] An antenna having the form of a helix. Also known as helix antenna. { 'hel·ə·kəl an'ten·ə }

helical potentiometer [ELEC] A multiturn precision potentiometer in which a number of complete turns of the control knob are required to move the contact arm from one end of the helically wound resistance element to the other end. { 'hel·ə·kəl pəˌten·chē'äm·əd·ər }

helical resonator [ELECTROMAG] A cavity resonator with a helical inner conductor. { 'hel·ə·kəl 'rez·ənˌād·ər }

helicity [QUANT MECH] The component of the spin of a particle along its momentum. { he'lis·əd·ē }

helicon [ELECTROMAG] A low-frequency, circularly polarized electromagnetic wave that is propagated in a metal in the presence of an external magnetic field. { 'hēl·əˌkän }

helimagnet [SOLID STATE] A metal, alloy, or salt that possesses helimagnetism. { 'hel·əˌmag·nət }

helimagnetism [SOLID STATE] A property possessed by some metals, alloys, and salts of transition elements or rare earths, in which the atomic magnetic moments, at sufficiently low temperatures, are arranged in ferromagnetic planes, the direction of the magnetism varying in a uniform way from plane to plane. { ˌhel·ə'mag·nəˌtiz·əm }

heliometer [OPTICS] A split-lens telescope used to measure the sun's diameter as well as small distances between stars or other celestial bodies. { ˌhē·lē'äm·əd·ər }

helion [NUC PHYS] The nucleus of a helium-3 atom, consisting of two protons and one neutron. { 'hē·lēˌän }

helioscope [OPTICS] A telescope for observing the sun that protects the observer's eyes from the sun's glare. { 'hē·lē·ə,skōp }

helium I [CRYO] The phase of liquid helium-4 which is stable at temperatures above the lambda point (about 2.2 K) and has the properties of a normal liquid, except low density. { 'hē·lē·əm 'wən }

helium II [CRYO] The phase of liquid helium-4 which is stable at temperatures between absolute zero and the lambda point (about 2.2 K), and has many remarkable properties such as vanishing viscosity, extremely high heat conductivity, and the fountain effect. { 'hē·lē·əm 'tü }

helium-3 [NUC PHYS] The isotope of helium with mass number 3, constituting approximately 1.3 parts per million of naturally occurring helium. { 'hē·lē·əm 'thrē }

helium-4 [NUC PHYS] The isotope of helium with mass number 4, constituting nearly all naturally occurring helium. { 'hē·lē·əm 'fȯr }

helium burning [NUC PHYS] The synthesis of elements in stars through the fusion of three alpha particles to form a carbon-12 nucleus, followed by further captures of alpha particles. { 'hē·lē·əm ,bərn·iŋ }

helium-cadmium laser [OPTICS] A metal-vapor ion laser in which cadmium vapor, produced by heat or other means, migrates through a high-voltage glow discharge in helium, generating a continuous laser beam at wavelengths in the ultraviolet and blue parts of the spectrum (about 0.3 to 0.5 micrometer). { 'hē·lē·əm 'kad·mē·əm 'lā·zər }

helium film [CRYO] A superfluid film that covers any surface in contact with helium II. Also known as Rollin film. { 'hē·lē·əm ,film }

helium-like ion [ATOM PHYS] An atom from which all the electrons except two have been removed. { 'hē·lē·əm ,līk ,i·ən }

helium liquefier [CRYO] Any one of several machines which liquefy helium by causing it to undergo adiabatic expansion and to do external work. { 'hē·lē·əm 'lik·wə,fī·ər }

helium magnetometer [PHYS] A device for measuring magnetic fields by observing the Zeeman effect in the lowest triplet level of helium atoms subjected to the field. { 'hē·lē·əm mag·nə'täm·əd·ər }

helium-3 maser [PHYS] A gas maser in which the gas used is helium-3. { 'hē·lē·əm ¦thrē 'mā·zər }

helium-neon laser [OPTICS] An atomic gas laser in which a combination of helium and neon gases is used. { 'hē·lē·əm ¦nē,än 'lā·zər }

helium spectrometer [SPECT] A small mass spectrometer used to detect the presence of helium in a vacuum system; for leak detection, a jet of helium is applied to suspected leaks in the outer surface of the system. { 'hē·lē·əm spek'träm·əd·ər }

helix antenna *See* helical antenna. { 'hē,liks an'ten·ə }

Hellman-Feynman theorem [QUANT MECH] A theorem which states that in the Born-Oppenheimer approximation the forces on nuclei in molecules or solids are those which would arise electrostatically if the electron probability density were treated as a static distribution of negative electric charge. { ¦hel·mən ¦fīn·mən ,thir·əm }

helmholtz [ELEC] A unit of dipole moment per unit area, equal to 1 Debye unit per square angstrom, or approximately 3.335×10^{-10} coulomb per meter. { 'helm,hōlts }

Helmholtz coils [ELECTROMAG] A pair of flat, circular coils having equal numbers of turns and equal diameters, arranged with a common axis, and connected in series; used to obtain a magnetic field more nearly uniform than that of a single coil. { 'helm,hōlts ,kȯilz }

Helmholtz double layer [PHYS] An electrical double layer of positive and negative charges one molecule thick which occurs at a surface where two bodies of different materials are in contact, or at the surface of a metal or other substance capable of existing in solution as ions and immersed in a dissociating solvent. { 'helm,hōlts ¦dəb·əl 'lā·ər }

Helmholtz equation [OPTICS] An equation which relates the linear and angular magnifications of a spherical refracting interface. Also known as Lagrange-Helmholtz equation. { 'helm,hōlts i,kwā·zhən }

Helmholtz flow [FL MECH] Flow with free streamlines or vortex sheets. { 'helm,hōlts ,flō }

Helmholtz free energy *See* free energy. { 'helm ,hōlts ¦frē 'en·ər·jē }

Helmholtz function *See* free energy. { 'helm,hōlts ,fəŋk·shən }

Helmholtz instability [FL MECH] The hydrodynamic instability arising from a shear, or discontinuity, in current speed at the interface between two fluids in two-dimensional motion; the perturbation gains kinetic energy at the expense of that of the basic currents. Also known as shearing instability. { 'helm,hōlts ,in·stə'bil·əd·ē }

Helmholtz-Keteller formula [OPTICS] A dispersion formula in which the difference between the square of the index of refraction and unity is set equal to a sum of terms each of which is associated with a resonant wavelength of the medium. { 'helm,hōlts 'ked·əl·ər ,fȯr·myə·lə }

Helmholtz potential *See* free energy. { 'helm,hōlts pə¦ten·chəl }

Helmholtz's theorem [ELEC] *See* Thévenin's theorem. [FL MECH] The theorem that in the isentropic flow of a nonviscous fluid which is not subject to body forces, individual vortices always consist of the same fluid particles. { 'helm,hōlt·səz ,thir·əm }

Helmholtz wave [FL MECH] An unstable wave in a system of two homogeneous fluids with a velocity discontinuity at the interface. { 'helm ,hōlts ,wāv }

hemihedral symmetry [CRYSTAL] The possession by a crystal of only half of the elements of symmetry which are possible in the crystal system to which it belongs. { ¦he·mē¦hē·drəl 'sim·ə·trē }

hemiholohedral [CRYSTAL] Of hemihedral form but with half of the octants having the full number of planes. { ¦he·mē,hō·lə'hē·drəl }

hemimorphic crystal [CRYSTAL] A crystal with no transverse plane of symmetry and no center of symmetry; composed of forms belonging to only one end of the axis of symmetry. { ¦he·mē¦mȯr·fik 'krist·əl }

hemiprism [CRYSTAL] A pinacoid that cuts two crystallographic axes. { 'he·mē,priz·əm }

hemispherical candlepower [OPTICS] Luminous intensity of a hemispherical light source. { ,he·mē'sfir·ə·kəl 'kand·əl,pau̇·ər }

hemitropic [CRYSTAL] Pertaining to a twinned structure in which, if one part were rotated 180°, the two parts would be parallel. { ¦he·mē ¦träp·ik }

henry [ELECTROMAG] The mks unit of self and mutual inductance, equal to the self-inductance of a circuit or the mutual inductance between two circuits if there is an induced electromotive force of 1 volt when the current is changing at the rate of 1 ampere per second. Symbolized H. { 'hen·rē }

hereditary mechanics [MECH] A field of mechanics in which quantities, such as stress, depend not only on other quantities, such as strain, at the same instant but also on integrals involving the values of such quantities at previous times. { hə'red·ə,ter·ē mi'kan·iks }

Hermann-Mauguin symbols [CRYSTAL] Symbols representing the 32 symmetry classes, consisting of series of numbers giving the multiplicity of symmetry axes in descending order, with other symbols indicating inversion axes and mirror planes. { 'her·män 'mō,gan ,sim·bəlz }

herpolhode [MECH] The curve traced out on the invariable plane by the point of contact between the plane and the inertia ellipsoid of a rotating rigid body not subject to external torque. { ¦hər·pəl'hōd }

herpolhode cone *See* space cone. { ¦hər·pəl'hōd ,kōn }

Herschel-Cassegrain telescope [OPTICS] A modification of a Cassegrain telescope in which the primary paraboloidal mirror is slightly inclined to the optical axis, and both the secondary hyperboloidal mirror and the eyepiece are located off the axis, so that it is not necessary to pierce the primary. { 'hər·shəl 'kas·gran ,tel·ə,skōp }

Herschel-Quincke tube [ACOUS] A device for demonstrating the interference of sound in which sound waves from a common source travel through two tubes of different lengths and recombine, producing reinforcement or cancellation of sound depending on the difference in path length; used to demonstrate the interference of sound and as a wave filter. Also known as Quincke tube. { ¦hər·shəl 'kwiŋ·kə ,tüb }

hertz [PHYS] Unit of frequency; a periodic oscillation has a frequency of *n* hertz if in 1 second it goes through *n* cycles. Also known as cycle per second (cps). Symbolized Hz. { hərts }

Hertz antenna [ELECTROMAG] An ungrounded half-wave antenna. { 'hərts an¦ten·ə }

Hertz effect [ELECTROMAG] A dependence of the attenuation of a linearly polarized electromagnetic wave passing through a grating of metal rods on the angle between the electric vector and the rod direction, with the attenuation being a minimum when the two are perpendicular. { 'hərts i,fekt }

Hertzian oscillator [ELECTROMAG] **1.** A generator of electric dipole radiation; consists of two capacitors joined by a conducting rod having a small spark gap; an oscillatory discharge occurs when the two halves of the oscillator are raised to a sufficiently high potential difference. **2.** A dumbbell-shaped conductor in which electrons oscillate from one end to the other, producing electric dipole radiation. { 'hərt·sē·ən 'äs·ə,lād·ər }

Hertzian wave *See* electric wave. { 'hərt·sē·ən 'wāv }

Hertz's law [MECH] A law which gives the radius of contact between a sphere of elastic material and a surface in terms of the sphere's radius, the normal force exerted on the sphere, and Young's modulus for the material of the sphere. { 'hərt·səs ,lȯ }

Hertz vector *See* polarization potential. { 'hərts ,vek·tər }

heterochromatic photometry [OPTICS] The branch of photometry concerned with comparing the illuminating powers of light sources with different colors. { ,hed·ə·rō·krə¦mad·ik fō'täm·ə·trē }

heterodesmic [CRYSTAL] Pertaining to those atoms bonded in more than one way in crystals. { ,hed·ə·rō'dez·mik }

heterogeneous fluid [FL MECH] A fluid within which the density varies from point to point; for most purposes the atmosphere must be treated as heterogeneous, particularly with regard to the decrease of density with height. { ,hed·ə·rə'jē·nē·əs 'flü·əd }

heterogeneous radiation [PHYS] Radiation having a number of different frequencies, different particles, or different particle energies. { ,hed·ə·rə'jē·nē·əs ,rād·ē'ā·shən }

heterogeneous strain [MECH] A strain in which the components of the displacement of a point in the body cannot be expressed as linear functions of the original coordinates. { ,hed·ə·rə¦jē·nē·əs 'strān }

heteromorphic transformation [THERMO] A change in the values of the thermodynamic variables of a system in which one or more of the component substances also undergo a change of state. { ,hed·ə·rə¦mȯr·fik ,tranz·fər'mā·shən }

heterostatic [ELEC] Pertaining to the measurement of one electrostatic potential by means of a different potential. { ¦hed·ə·rō'stad·ik }

heterostatic connection [ELEC] An arrangement of a quadrant electrometer in which the vane is maintained at a high potential with respect to one of the quadrant pairs and the deflection of the vane is linearly proportional to the unknown voltage applied across the quadrant pairs. { ¦hed·ə·rə͵stad·ik kə'nek·shən }

heterotic superstring theory [PART PHYS] The most promising version of superstring theory, with only closed strings, 10 dimensions, and 496 massless gauge bosons. { ¦hed·ə¦räd·ik 'sü·pər͵striŋ ͵thē·ə·rē }

Heulinger equations [PHYS] Equations which relate the values of various quantities, such as the Hall coefficient, thermoelectric power, electrical resistivity, and thermal conductivity, in isothermal and adiabatic thermoelectric and thermomagnetic effects. { 'hȯi·liŋ·ər i͵kwā·zhənz }

Hevelian halo [OPTICS] A faint, white halo with an angular radius of 90°, centered on the sun or moon, and only occasionally seen; it is a member of the class of halos reported but not yet fully explained. Also known as halo of Hevelius; halo of 90°. { he'vāl·yən 'hā·lō }

Hevelius's parhelia [OPTICS] Bright spots infrequently observed on the parhelic circle halfway around from the sun to the anthelion; these two brighter areas on the parhelic circle are probably a result of superposition of luminosity of the parhelic circle and the Hevelian halo. { he'vel·yəs·əz pär'hēl·yə }

hexad axis [CRYSTAL] A rotation axis whose multiplicity is equal to 6. { 'hek·sad ͵ax·səs }

hexagonal close-packed structure [CRYSTAL] Close-packed crystal structure characterized by the regular alternation of two layers; the atoms in each layer lie at the vertices of a series of equilateral triangles, and the atoms in one layer lie directly above the centers of the triangles in neighboring layers. Abbreviated hcp structure. { hek'sag·ə·nəl ¦klōs ¦pakt 'strək·chər }

hexagonal lattice [CRYSTAL] A Bravais lattice whose unit cells are right prisms with hexagonal bases and whose lattice points are located at the vertices of the unit cell and at the centers of the bases. { hek'sag·ə·nəl 'lad·əs }

hexagonal system [CRYSTAL] A crystal system that has three equal axes intersecting at 120° and lying in one plane; a fourth, unequal axis is perpendicular to the other three. { hek'sag·ə·nəl 'sis·təm }

hexatic phase [PHYS] A phase of matter that is intermediate between the normal solid and the isotropic liquid phases, and corresponds to a two-dimensional fluid with sixfold orientational order but no positional order. { hek¦sad·ik 'fāz }

hexoctahedron [CRYSTAL] A cubic crystal form that has 48 equal triangular faces, each of which cuts the three crystallographic axes at different distances. { ͵hek'säk·tə͵hē·drən }

hextetrahedron [CRYSTAL] A 24-faced form of crystal in the tetrahedral group of the isometric system. { ͵heks͵te·trə'hē·drən }

hfs *See* hyperfine structure.

HFS *See* type II superconductor.

hg *See* hectogram.

hidden variables [QUANT MECH] Hypothetical additional variables or parameters which would supplement quantum mechanics, making it possible to unambiguously predict the result of a single measurement on a single microscopic system. { 'hid·ən 'ver·ē·ə·bəlz }

hidden-variable theory of the first kind [QUANT MECH] A theory postulating the existence of hidden variables and constructed so as to be self-consistent and to reproduce all the statistical predictions of quantum mechanics when the hidden variables are in an equilibrium distribution. { 'hid·ən ¦ver·ē·ə·bəl ͵thē·ə·rē əv thə 'fərst ͵kīnd }

hidden-variable theory of the second kind [QUANT MECH] A theory, postulating the existence of hidden variables, that predicts deviations from the statistical predictions of quantum mechanics, even for the equilibrium situations; such theories are required to satisfy a locality condition. Also known as local hidden-variable theory. { 'hid·ən ¦ver·ē·ə·bəl ͵thē·ə·rē əv thə 'sek·ənd 'kīnd }

Higgs bosons [PART PHYS] Massive scalar mesons whose existence is predicted by certain unified gage theories of the weak and electromagnetic interactions; they are not eliminated by the Higgs mechanism. { 'higz ¦bō͵sänz }

Higgs mechanism [PART PHYS] The feature of the spontaneously broken gage symmetries that the Goldstone bosons do not appear as physical particles, but instead constitute the zero helicity states of gage vector bosons of nonzero mass (such as the intermediate vector boson). [QUANT MECH] A mathematical procedure in which particles in a field theory gain or lose mass due to spontaneous breakdown of symmetry. { 'higz ͵mek·ə͵niz·əm }

high-energy electron diffraction [PHYS] The diffraction of electrons with high energies, usually in the range of 30,000–70,000 electronvolts, mainly to study the structure of atoms and molecules in gases and liquids. Abbreviated HEED. { 'hī ͵en·ər·jē i'lek͵trn di͵frak·shən }

high-energy neutron-proton scattering [NUC PHYS] A collision of a neutron having an energy greater than 440 megaelectronvolts with a proton; the collision is inelastic because of pion production. { ¦hī ¦en·ər·jē ¦nü͵trän 'prō͵tän ͵skad·ər·iŋ }

high-energy particle [PART PHYS] An elementary particle having an energy of hundreds of megaelectronvolts or more. { 'hī ͵en·ər·jē 'pärd·ə·kəl }

high-energy physics *See* particle physics. { 'hī ͵en·ər·jē 'fiz·iks }

high-energy proton-proton scattering [NUC PHYS] A collision of a proton having an energy greater than 440 megaelectronvolts with another proton; the collision is inelastic because of pion production. { ¦hī ¦en·ər·jē ¦prō͵tän 'prō͵tän ͵skad·ər·iŋ }

high-energy scattering [PART PHYS] Collisions of particles having energies of hundreds of megaelectronvolts or more, sufficient to produce new particles. { 'hī ͵en·ər·jē 'skad·ə·riŋ }

higher mode [ELECTROMAG] A waveguide mode whose frequency is higher than the lowest one. { 'hī·ər ˌmōd }

high-field superconductor *See* type II superconductor. { hī ˌfēld 'sü·pər·kən,dək·tər }

high-frequency resistance [ELEC] The total resistance offered by a device in an alternating-current circuit, including the direct-current resistance and the resistance due to eddy current, hysteresis, dielectric, and corona losses. Also known as alternating-current resistance; effective resistance; radio-frequency resistance. { 'hī ¦frē·kwən·sē ri'zis·təns }

high heat [THERMO] Heat absorbed by the cooling medium in a calorimeter when products of combustion are cooled to the initial atmospheric (ambient) temperature. { 'hī ¦hēt }

high-K capacitor [ELEC] A capacitor whose dielectric material is a ferroelectric having a high dielectric constant, up to about 6000. { 'hī ˌkā kə'pas·əd·ər }

high-pressure phenomena [PHYS] Natural conditions and processes occurring at high pressures, and their duplication in the laboratory. { 'hī ¦presh·ər fə'näm·ə·nä }

high-pressure physics [PHYS] The study of the effects of high pressure on the properties of matter. { 'hī ¦presh·ər 'fiz·iks }

high-Q cavity [ELECTROMAG] A cavity resonator which has a large Q factor, and thus has a small energy loss. Also known as high-Q resonator. { 'hī ˌkyü 'kav·əd·ē }

high-Q resonator *See* high-Q cavity. { 'hī ˌkyü 'rez·ən,ād·ər }

high-resolution electron energy loss spectroscopy [SPECT] A type of electron energy loss spectroscopy in which electron scattering is performed by using a monoenergetic beam and electron energy analyzers to achieve a resolution of 5 to 10 millielectronvolts. Abbreviated HREELS. { 'hī ˌrez·ə'lü·shən i'lek,trän 'en·ər·jē ˌlȯs spek'träs·kə·pē }

high-temperature phenomena [PHYS] Phenomena occurring at temperatures above about 500 K. { 'hī ˌtem·prə·chər fə'näm·ə·nə }

high-temperature superconductor [SOLID STATE] A ceramic material, consisting of an oxide of a rare-earth element, barium, and copper, which displays superconductivity at temperatures of 90 K (−298°F) or more. { 'hī 'tem·prə·chər 'sü·pər·kən,dək·tər }

high vacuum [PHYS] A vacuum with a pressure between 1×10^{-3} and 1×10^{-6} mmHg (0.1333224 and 0.0001333 pascal). { 'hī ¦vak·yüm }

Hildebrand function [THERMO] The heat of vaporization of a compound as a function of the molal concentration of the vapor; it is nearly the same for many compounds. { 'hil·də,brand ˌfəŋk·shən }

HK *See* Hefner candle.

hl *See* hectoliter.

h-line *See* h-bar. { 'āch ˌlīn }

hm *See* hectometer.

H mode *See* transverse electric mode. { 'āch,mōd }

hodograph [PHYS] **1.** The curve traced out in the course of time by the tip of a vector representing some physical quantity. **2.** In particular, the path traced out by the velocity vector of a given particle. { 'häd·ə,graf }

hodograph method [FL MECH] A method for studying two-dimensional steady fluid flow in which the independent variables are taken as the components of the velocity with respect to cartesian or polar coordinates, rather than the coordinates themselves. { 'häd·ə,graf ˌmeth·əd }

hodoscope *See* conoscope. { 'häd·ə,skōp }

hoghorn antenna *See* horn antenna. { 'häg,hȯrn an'ten·ə }

hohlraum *See* blackbody. { 'hōl,raüm }

holding magnet *See* lifting magnet. { 'hōl·diŋ ˌmag·nət }

hole [SOLID STATE] A vacant electron energy state near the top of an energy band in a solid; behaves as though it were a positively charged particle. Also known as electron hole. { hōl }

hole burning [PHYS] Saturation of attenuation or gain that is confined to a narrow range of frequencies (hole) within an inhomogeneously broadened transition when the saturating radiation is confined to frequencies within this range. { 'hōl ˌbərn·iŋ }

hole-burning spectroscopy [SPECT] A method of observing extremely narrow line widths in certain ions and molecules embedded in crystalline solids, in which broadening produced by crystal-site-dependent statistical field variations is overcome by having a monochromatic laser temporarily remove ions or molecules at selected crystal sites from their absorption levels, and observing the resulting dip in the absorption profile with a second laser beam. { 'hōl ˌbərn·iŋ spek 'träs·kə·pē }

hole theory [QUANT MECH] A theory about the significance of negative energy states in the Dirac theory which leads to the prediction of the existence of the positron and, by extension, to that of other antiparticles. { 'hōl ˌthē·ə·rē }

hollow atom [ATOM PHYS] An atom whose electrons are in highly excited states, leaving the states of lower energy (in which the electron is, on average, closer to the nucleus) vacant. { ¦häl·ō 'ad·əm }

hollow-pipe waveguide [ELECTROMAG] A waveguide consisting of a hollow metal pipe; electromagnetic waves are transmitted through the interior and electric currents flow on the inner surfaces. { 'häl·ō ˌpīp 'wāv,gīd }

holoaxial [CRYSTAL] Having all possible axes of symmetry. { ¦häl·ō'ak·sē·əl }

hologram [OPTICS] The special photographic plate used in holography; when this negative is developed and illuminated from behind by a coherent gas-laser beam, it produces a three-dimensional image in space. Also known as hologram interferometer. { 'häl·ə,gram }

hologram interferometer *See* hologram. { 'häl·ə,gram ,in·tər·fə'räm·əd·ər }

holographic interferometry [OPTICS] The study of the formation and interpretation of the fringe pattern which appears when a wave, generated at some earlier time and stored in a hologram, is later reconstructed and caused to interfere with a comparison wave. { ,häl·ə'graf·ik ,in·tər·fə'räm·ə·trē }

holographic multiplexing *See* quasi-acoustical holography. { ¦häl·ə¦graf·ik 'məl·tə,pleks·iŋ }

holographic optical element [OPTICS] A hologram that is used to control transmitted light beams, rather than to display images, based on the principles of diffraction. { ¦häl·ə¦graf·ik 'äp·tə·kəl 'el·ə·mənt }

holography [PHYS] A technique for recording, and later reconstructing, the amplitude and phase distributions of a wave disturbance; widely used as a method of three-dimensional optical image formation, and also with acoustical and radio waves; in optical image formation, the technique is accomplished by recording on a photographic plate the pattern of interference between coherent light reflected from the object of interest, and light that comes directly from the same source or is reflected from a mirror. { hə'läg·rə·fē }

holohedral [CRYSTAL] Pertaining to a crystal structure having the highest symmetry in each crystal class. Also known as holosymmetric; holosystemic. { ¦häl·ō¦hē·drəl }

holohedron [CRYSTAL] A crystal form of the holohedral class, having all the faces needed for complete symmetry. { ¦häl·ō'hē·drən }

holomicrography [OPTICS] The use of holography to produce three-dimensional images with various types of microscopes. { ¦häl·ō·mī'kräg·rə·fē }

holonomic constraints [MECH] An integrable set of differential equations which describe the restrictions on the motion of a system; a function relating several variables, in the form $f(x_1, \ldots, x_n) = 0$, in optimization or physical problems. { ¦häl·ə¦näm·ik kən'strāns }

holonomic system [MECH] A system in which the constraints are such that the original coordinates can be expressed in terms of independent coordinates and possibly also the time. { ¦häl·ə¦näm·ik 'sis·təm }

holophotal [OPTICS] **1.** Pertaining to a holophote. **2.** Reflecting all the light from a source in one direction. { ¦häl·ə¦fōd·əl }

holophote [OPTICS] An optical system consisting of lenses or reflectors that collect a large amount of the light from a source (such as the lamp of a lighthouse) and send it in a desired direction. { 'häl·ə,fōt }

holosymmetric *See* holohedral. { ¦häl·ō·si'me·trik }

holosystemic *See* holohedral. { ¦häl·ō·si'stem·ik }

Holzer's method [MECH] A method of determining the shapes and frequencies of the torsional modes of vibration of a system, in which one imagines the system to consist of a number of flywheels on a massless flexible shaft and, starting with a trial frequency and motion for one flywheel, determines the torques and motions of successive flywheels. { 'hōt·sərz ,meth·əd }

homenergic flow [THERMO] Fluid flow in which the sum of kinetic energy, potential energy, and enthalpy per unit mass is the same at all locations in the fluid and at all times. { 'häm·ə,nər·jik 'flō }

homentropic flow [FL MECH] Fluid flow in which the entropy per unit mass is the same at all locations in the fluid and at all times. { 'häm·ən,träp·ik 'flō }

homeomorph [CRYSTAL] A crystal that displays a form similar to that of a crystal with a different chemical composition. { 'hō·mē·ə,mórf }

homing antenna [ELECTROMAG] A directional antenna array used in flying directly to a target that is emitting or reflecting radio or radar waves. { 'hōm·iŋ an,ten·ə }

homocentric [OPTICS] Pertaining to rays which have the same focal point, or which are parallel. Also known as stigmatic. { ,häm·ə'sen·trik }

homodesmic [CRYSTAL] Of a crystal, having atoms bonded in a single way. { ,hä·mə'dez·mik }

homogeneity [PHYS] Quality of a substance whose properties are independent of position. { ,hō·mə·jə'nē·əd·ē }

homogeneous line-broadening [OPTICS] An increase beyond the natural linewidth of an absorption or emission line which results from a disturbance (such as collisions or lattice vibrations) that is the same for all the source emitters. { ¦hō·mə¦jē·nē·əs 'līn ,bród·ən·iŋ }

homogeneous radiation [PHYS] Radiation having an extremely narrow band of frequencies, or a beam of monoenergetic particles of a single type, so that all components of the radiation are alike. { ,hä·mə'jē·nē·əs ,rād·ē'ā·shən }

homogeneous strain [MECH] A strain in which the components of the displacement of any point in the body are linear functions of the original coordinates. { ¦hō·mə,jē·nē·əs 'strān }

homometric pair [CRYSTAL] A pair of crystal structures whose x-ray diffraction patterns are identical. { ¦hä·mə¦me·trik 'per }

homomorphous transformation [THERMO] A change in the values of the thermodynamic variables of a system in which none of the component substances undergoes a change of state. { ,hō·mə¦mór·fəs ,tranz·fər'mā·shən }

homopolar [ELEC] **1.** Electrically symmetrical. **2.** Having equal distribution of charge. { ¦hä·mə'pō·lər }

homopolar crystal [SOLID STATE] A crystal in which the bonds are all covalent. { ¦hä·mə'pō·lər 'krist·əl }

honeycomb coil [ELECTROMAG] A coil wound in a crisscross manner to reduce distributed capacitance. Also known as duolateral coil; lattice-wound coil. { 'hən·ē,kōm ,kóil }

Hookean deformation [MECH] Deformation of a substance which is proportional to the force applied to it. { 'hu̇k·ē·ən ,dəf·ər'mā·shən }

Hookean solid [MECH] An ideal solid which obeys Hooke's law exactly for all values of stress, however large. { 'hu̇k·ē·ən 'säl·əd }

Hooke number *See* Cauchy number. { 'hu̇k ˌnəm·bər }

Hooke's law [MECH] The law that the stress of a solid is directly proportional to the strain applied to it. { 'hu̇ks ˌlȯ }

Hope's apparatus [THERMO] An apparatus consisting of a vessel containing water, a freezing mixture in a tray surrounding the vessel, and thermometers inserted in the water at points above and below the freezing mixture; used to show that the maximum density of water lies at about 4°C. { 'hōps ˌap·əˌrad·əs }

Hopkinson effect [ELECTROMAG] A phenomenon in which the permeability of a ferromagnetic material at low field strengths, measured as a function of temperature, reaches a maximum at a temperature a little below the Curie temperature. { 'häp·kən·sən iˌfekt }

Hopkinson's coefficient [ELECTROMAG] The average magnetic flux per turn of an induction coil divided by the average flux per turn of another coil linked with it. { 'häp·kən·sənz ˌkō·i'fish·ənt }

horizontal field-strength diagram [ELECTROMAG] Representation of the field strength at a constant distance from an antenna and in a horizontal plane; unless otherwise specified, this plane is that passing through the antenna. { ˌhär·ə'zänt·əl 'fēld ˌstreŋkth ˌdī·əˌgram }

horizontal pendulum [MECH] A pendulum that moves in a horizontal plane, such as a compass needle turning on its pivot. { ˌhär·ə'zänt·əl 'pen·jə·ləm }

horizontal vee [ELECTROMAG] An antenna consisting of two linear radiators in the form of the letter V, lying in a horizontal plane. { ˌhär·ə'zänt·əl 'vē }

horn *See* horn antenna. { hȯrn }

horn antenna [ELECTROMAG] A microwave antenna produced by flaring out the end of a circular or rectangular waveguide into the shape of a horn, for radiating radio waves directly into space. Also known as electromagnetic horn; flare (British usage); hoghorn antenna (British usage); horn; horn radiator. { 'hȯrn an'ten·ə }

horn equation [ACOUS] A second-order partial differential equation for the velocity potential as a function of time and of distance along an acoustic horn. { 'hȯrn iˌkwā·zhən }

horn gap [ELEC] Type of spark gap which is provided with divergent electrodes. { 'hȯrn ˌgap }

horn radiator *See* horn antenna. { 'hȯrn 'rād·ēˌād·ər }

horsepower [MECH] The unit of power in the British engineering system, equal to 550 foot-pounds per second, approximately 745.7 watts. Abbreviated hp. { 'hȯrsˌpau̇·ər }

horseshoe magnet [ELECTROMAG] A permanent magnet or electromagnet in which the core is horseshoe-shaped or U-shaped, to bring the two poles near each other. { 'hȯrˌshü ˌmag·nət }

hot [ELEC] *See* energized. [PHYS] Having or charged with high energy, such as high thermal energy or a high level of radioactivity. { hät }

hot carbon-nitrogen-oxygen cycle [NUC PHYS] A modification of the carbon-nitrogen cycle that occurs at the high temperatures and high densities encountered in stellar explosions such as novae and supernovae, in which a nitrogen-13 nucleus captures a proton to form oxygen-14 before it can undergo beta decay to carbon-13. Also known as fast carbon-nitrogen-oxygen cycle. { 'hät ¦kär·bən ¦nī·trə·jən ¦äk·sə·jən ˌsī·kəl }

hot nucleus [NUC PHYS] An excited nucleus whose energy is shared among its many degrees of freedom. { ¦hät 'nü·klē·əs }

hot spot [PHYS] A localized region with temperature higher than the surroundings. { 'hät ˌspät }

hot strength *See* tensile strength. { 'hät ˌstreŋkth }

hour [MECH] A unit of time equal to 3600 seconds. Abbreviated h; hr. { au̇r }

howling tones [ACOUS] The sounds produced by a howling tube. Also known as gauze tones. { ¦hau̇l·iŋ ¦tōnz }

howling tube [ACOUS] A vertical open tube with a piece of gauze in the lower half which is placed over a flame to make the tube produce powerful sound waves with many overtones. { ¦hau̇l·iŋ ¦tüb }

hp *See* horsepower.

HPBW *See* half-power beamwidth.

H plane [ELECTROMAG] The plane of an antenna in which lies the magnetic field vector of linearly polarized radiation. { ¦āch ˌplān }

H-plane bend [ELECTROMAG] A rectangular waveguide bend in which the longitudinal axis of the waveguide remains in a plane parallel to the plane of the magnetic field vector throughout the bend. Also known as H bend. { ¦āch ˌplān ˌbend }

H-plane T junction [ELECTROMAG] Waveguide T junction in which the change in structure occurs in the plane of the magnetic field. Also known as shunt T junction. { ¦āch ˌplān 'tē ˌjəŋk·shən }

hr *See* hour.

HREELS *See* high-resolution electron energy loss spectroscopy. { 'āch ˌrēlz }

H theorem of Boltzmann *See* Boltzmann H theorem. { 'āch ˌthir·əm əv 'bōltsˌmän }

Hubble Space Telescope [OPTICS] An astronomical reflecting telescope with a mirror 94.5 inches (2.4 meters) in diameter; placed in orbit above the earth's atmosphere in April 1990. { ¦həb·əl 'spās ˌtel·əˌskōp }

Hübner rhomb [OPTICS] A glass rhombohedron used in photometry to compare two illuminated surfaces. { 'hyib·nər ˌräm }

hue [OPTICS] The name of a color, such as red, yellow, green, blue, or purple, as perceived subjectively. { hyü }

Hughes effect [ELECTROMAG] An asymmetry in the hysteresis curves of laminated cores made of certain magnetic materials such as permalloy or mu metal in alternating magnetic fields. { 'yüz i,fekt }

Hugoniot function [PHYS] A function specifying the locus of states which are possible immediately after the passage of a shock front; gives the state's pressure as a function of its specific volume. { yü'gōn·yō ,fəŋk·shən }

Hume-Rothery rule [SOLID STATE] The rule that the ratio of the number of valence electrons to the number of atoms in a given phase of an electron compound depends only on the phase, and not on the elements making up the compounds. { 'hyüm 'rȯth·ə·rē ,rül }

Humphreys series [SPECT] A series of lines in the infrared spectrum of atomic hydrogen whose wave numbers are given by $R_H(1/36) - (1/n^2)$, where R_H is the Rydberg constant for hydrogen, and n is any number greater than 6. { 'həm·frēz ,sir·ēz }

Humphries equation [THERMO] An equation which gives the ratio of specific heats at constant pressure and constant volume in moist air as a function of water vapor pressure. { 'həm·frēz i,kwā·zhən }

Hund coupling cases [ATOM PHYS] Five ways of combining the electron-spin angular momentum, electron-orbital angular momentum, and nuclear-rotation angular momentum to form the total angular momentum of a molecule. { 'hənd ¦kəp·liŋ ,kā·səz }

Hund rules [ATOM PHYS] Two rules giving the order in energy of atomic states formed by equivalent electrons: of the terms given by equivalent electrons, the ones with greatest multiplicity have the least energy, and of these the one with greatest orbital angular momentum is lowest; the state of a multiplet with lowest energy is that in which the total angular momentum is the least possible, if the shell is less than half-filled, and the greatest possible, if more than half filled. { 'hənd ,rülz }

Huttig equation [THERMO] An equation which states that the ratio of the volume of gas adsorbed on the surface of a nonporous solid at a given pressure and temperature to the volume of gas required to cover the surface completely with a unimolecular layer equals $(1 + r)\, c^r/(1 + c^r)$, where r is the ratio of the equilibrium gas pressure to the saturated vapor pressure of the adsorbate at the temperature of adsorption, and c is the product of a constant and the exponential of $(q - q_l)/RT$, where q is the heat of adsorption into a first layer molecule, q_l is the heat of liquefaction of the adsorbate, T is the temperature, and R is the gas constant. { 'həd·ik i,kwā·zhən }

Huygens eyepiece [OPTICS] An eyepiece in which there are two plano-convex lenses, and the plane sides of both lenses face the eye. { 'hī·gənz ¦ī,pēs }

Huygens-Fresnel principle [OPTICS] A modification of Huygens' principle according to which the amplitude of secondary waves falls off in proportion to the cosine of the angle between the normals to the original and secondary waves, and the secondary waves interfere with each other according to the principle of superposition. Also known as Fresnel-Huygens principle. { ¦hī·gənz frā'nel ,prin·sə·pəl }

Huygens' principle [OPTICS] The principle that each point on a light wavefront may be regarded as a source of secondary waves, the envelope of these secondary waves determining the position of the wavefront at a later time. { 'hī·gənz ,prin·sə·pəl }

Huygens wavelet [OPTICS] A secondary wave as used in Huygens' principle. { 'hī·gənz ,wāv·lət }

H vector [ELECTROMAG] A vector that is the magnetic field. For a plane wave in free space, it is perpendicular to the E vector and to the direction of propagation. { 'āch ,vek·tər }

H wave *See* transverse electric wave. { 'āch ,wāv }

hybrid electromagnetic wave [ELECTROMAG] Wave which has both transverse and longitudinal components of displacement. { 'hī·brəd i¦lek·trō·mag¦ned·ik 'wāv }

hybrid magnet [CRYO] A type of superconducting magnet consisting of a large-bore NbTi (niobium-titanium) external coil, which provides an external field of about 5 teslas, and an inner Nb_3Sn (niobium-tin) coil which provides additional field strength. [ELECTROMAG] An air-cooled magnet consisting of a large-volume superconducting magnet surrounding a water-cooled normal-conductor magnet that operates at the highest field. { 'hī·brəd 'mag·nət }

hybrid tee [ELECTROMAG] A microwave hybrid junction composed of an E-H tee with internal matching elements; it is reflectionless for a wave propagating into the junction from any arm when the other three arms are match-terminated. Also known as magic tee. { 'hī·brəd 'tē }

hybrid wave function [QUANT MECH] A linear combination of wave functions of one problem used as an approximation to the wave function in another problem; for example, a linear combination of atomic orbitals used to represent a molecular bond. { 'hī·brəd 'wāv ,fəŋk·shən }

hydraulic analog table [FL MECH] An experimental facility based on the hydraulic analogy; the water flows over a smooth horizontal surface and is bounded by vertical walls geometrically similar to the boundaries of the corresponding compressible gas flow; flow patterns are easily observed, and boundary changes may be made rapidly and inexpensively during exploratory studies. { hī'drȯ·lik 'an·ə,läg ,tā·bəl }

hydraulic analogy [FL MECH] The analogy between the flow of a shallow liquid and the flow of a compressible gas; various phenomena such as shock waves occur in both systems; the analogy requires neglect of vertical accelerations in the liquid, and restrictions on the ratio of specific heats for the gas. { hī'drȯ·lik ə'nal·ə·jē }

hydraulic conductivity *See* permeability coefficient. { hī'drȯ·lik ,kän,dək'tiv·əd·ē }

hydraulic friction [FL MECH] Resistance to flow which is exerted on the surface of contact between a stream and its conduit and which induces a loss of energy. { hī'drȯ·lik 'frik·shən }

hydraulic grade line [FL MECH] **1.** In a closed channel, a line joining the elevations that water would reach under atmospheric pressure. **2.** The free water surface in an open channel. { hī'drȯ·lik 'grād ˌlīn }

hydraulic gradient [FL MECH] With regard to an aquifer, the rate of change of pressure head per unit of distance of flow at a given point and in a given direction. { hī'drȯ·lik 'grād·ē·ənt }

hydraulic jump [FL MECH] A steady-state, finite-amplitude disturbance in a channel, in which water passes turbulently from a region of (uniform) low depth and high velocity to a region of (uniform) high depth and low velocity; when applied to hydraulic jumps, the usual hydraulic formulas governing the relations of velocity and depth do not conserve energy. { hī'drȯ·lik 'jəmp }

hydraulic loss [FL MECH] The loss in fluid power due to flow friction within the system. { hī'drȯ·lik 'lȯs }

hydraulic radius [FL MECH] The ratio of the cross-sectional area of a conduit in which a fluid is flowing to the inner perimeter of the conduit. { hi'drȯ·lik 'rād·ē·əs }

hydraulics [FL MECH] The branch of engineering that focuses on the practical problems of collecting, storing, measuring, transporting, controlling, and using water and other liquids. { hī'drȯ·liks }

hydroacoustics *See* underwater acoustics. { ˌhī·drō·ə'kü·stiks }

hydrodynamic equations [FL MECH] Three equations which express the net acceleration of a unit water particle as the sum of the partial accelerations due to pressure gradient force, frictional force, earth's deflecting force, gravitational force, and other factors. { ˌhī·drō·dī'nam·ik i'kwā·zhənz }

hydrodynamic pressure [FL MECH] The difference between the pressure of a fluid and the hydrostatic pressure; this concept is useful chiefly in problems of the steady flow of an incompressible fluid in which the hydrostatic pressure is constant for a given elevation (as when the fluid is bounded above by a rigid plate), so that the external force field (gravity) may be eliminated from the problem. { ˌhī·drō·dī'nam·ik 'presh·ər }

hydrodynamics [FL MECH] The study of the motion of a fluid and of the interactions of the fluid with its boundaries, especially in the incompressible inviscid case. { ˌhī·drō·dī'nam·iks }

hydroelasticity [FL MECH] **1.** Theory of elasticity of a fluid. **2.** The interaction between the flow of water or other liquid and the elastic behavior of a body immersed in it. { ˌhī·drō‚ē·las'tis·əd·ē }

hydrogen-bubble method [FL MECH] A method of flow visualization in which hydrogen bubbles are generated by electrolysis of water using wire electrodes, and are made to form either time lines (by applying a pulsating voltage to a bare cathode) or streamlines (by applying a continuous voltage to a kink in an insulated cathode). { ¦hī·drə·jən 'bəb·əl ˌmeth·əd }

hydrogen cyanide laser [OPTICS] A gas laser using hydrogen cyanide, which emits infrared radiation at wavelengths of 311 and 337 micrometers. { 'hī·drə·jən ¦sī·ə‚nīd 'lā·zər }

hydrogenic ion [ATOM PHYS] An atom from which all but one of the electrons have been removed. { ¦hī·drə¦jen·ik 'ī‚än }

hydrogen laser [OPTICS] A molecular gas laser in which hydrogen is used to generate coherent wavelengths near 0.6 micrometer in the vacuum ultraviolet region. { 'hī·drə·jən 'lā·zər }

hydrogen-like atom [ATOM PHYS] An atom from which all the electrons except one have been removed. { 'hī·drə·jən ˌlīk ˌad·əm }

hydrogen line [SPECT] A spectral line emitted by neutral hydrogen having a frequency of 1420 megahertz and a wavelength of 21 centimeters; radiation from this line is used in radio astronomy to study the amount and velocity of hydrogen in the Galaxy. { 'hī·drə·jən ˌlīn }

hydrogen maser [PHYS] A maser in which hydrogen gas is the basis for providing an output signal with a high degree of stability and spectral purity. { 'hī·drə·jən 'mā·zər }

hydrokinematics [FL MECH] The study of the motion of a liquid apart from the cause of motion. { 'hī·drə‚kin·ə'mad·iks }

hydrokinetics [FL MECH] The study of the forces produced by a liquid as a consequence of its motion. { 'hī·drə·kə'ned·iks }

hydromagnetic instability *See* magnetohydrodynamic instability. { ¦hī·drō·mag¦ned·ik ˌin·stə'bil·əd·ē }

hydromagnetics *See* magnetohydrodynamics. { ¦hī·drō·mag¦ned·iks }

hydromagnetic stability *See* magnetohydrodynamic stability. { ¦hī·drō·mag¦ned·ik stə'bil·əd·ē }

hydromagnetic wave *See* magnetohydrodynamic wave. { ¦hī·drō·mag¦ned·ik 'wāv }

hydromechanics [FL MECH] The study of liquids, traditionally water, as a medium for the transmission of forces. { ¦hī·drō·mi'kan·iks }

hydrometry [FL MECH] The science and technology of measuring specific gravities, particularly of liquids. { hī'dräm·ə·trē }

hydrophotometer [OPTICS] An instrument for measuring the attenuation coefficient of collimated light in sea water, in which light from a collimated source is directed through a column of sea water and is measured by a photocell or other electronic device at the other end of the column. { ˌhī·drə·fə'täm·əd·ər }

hydroscope [OPTICS] An instrument designed to observe objects an appreciable distance below the surface of water, consisting of a series of mirrors enclosed in a steel tube. { 'hī·drə‚skōp }

hydrostatic analogy [PHYS] An analogy between the relations among current, potential difference, and resistance in an electric circuit and the relations among corresponding quantities describing water flowing under a hydrostatic head. { ˌhī·drə¦stad·ik ə'nal·ə·jē }

hydrostatic balance [MECH] An equal-arm balance in which an object is weighed first in air and then in a beaker of water to determine its specific gravity. { ˌhī·drə¦stad·ik i'kwā·zhən }

hydrostatic equation [PHYS] The form assumed by the vertical component of the vector equation of motion when all Coriolis force, earth curvature, frictional, and vertical acceleration terms are considered negligible compared with those involving the vertical pressure force and the force of gravity. { ˌhī·drə'stad·ik i'kwā·zhən }

hydrostatic equilibrium [PHYS] The state of a fluid whose surfaces of constant mass (or density) coincide and are horizontal throughout; complete balance exists between the force of gravity and the pressure force; the relation between the pressure and the geometic height is given by the hydrostatic equation. { ˌhī·drə'stad·ik ˌē·kwə'lib·rē·əm }

hydrostatic modulus *See* bulk modulus of elasticity. { ˌhī·drə'stad·ik 'mäj·ə·ləs }

hydrostatic pressure [FL MECH] **1.** The pressure at a point in a fluid at rest due to the weight of the fluid above it. Also known as gravitational pressure. **2.** The negative of the stress normal to a surface in a fluid. { ˌhī·drə'stad·ik 'presh·ər }

hydrostatics [FL MECH] The study of liquids at rest and the forces exerted on them or by them. { ˌhī·drə'stad·iks }

hydrostatic strength [MECH] The ability of a body to withstand hydrostatic stress. { ˌhī·drə'stad·ik 'streŋkth }

hydrostatic stress [MECH] The condition in which there are equal compressive stresses or equal tensile stresses in all directions, and no shear stresses on any plane. { ˌhī·drə'stad·ik 'stres }

hydrostatic weighing [FL MECH] A method of determining the density of a sample in which the sample is weighed in air, and then weighed in a liquid of known density; the volume of the sample is equal to the loss of weight in the liquid divided by the density of the liquid. { ˌhī·drə'stad·ik 'wā·iŋ }

hyl *See* metric-technical unit of mass.

Hylleraas coordinates [ATOM PHYS] Coordinates for two particles used in studying the helium atom; they comprise the distance between the two particles, the sum of the distances of the particles from the origin, and the difference of the distances of the particles from the origin. { 'hil·əˌräs kōˌȯrd·ən·əts }

hyperbolic antenna [ELECTROMAG] A radiator whose reflector in cross section describes a half hyperbola. { ¦hī·pər¦bäl·ik an'ten·ə }

hyperbolic distance [ELECTROMAG] A function of pairs of points within a unit circle, where the interior of this circle is a conformal or projective representation of a hyperbolic space used in transmission line theory and waveguide analysis. { ¦hī·pər¦bäl·ik 'dis·təns }

hyperbolic point [FL MECH] A singular point in a streamline field which constitutes the intersection of a convergence line and a divergence line; it is analogous to a col in the field of a single-valued scalar quantity. Also known as neutral point. { ¦hī·pər¦bäl·ik 'pȯint }

hypercharge [PART PHYS] A quantum number conserved by strong interactions, equal to twice the average of the charges of the members of an isospin multiplet. { 'hī·pərˌchärj }

hyperfine enhanced nuclear cooling [CRYO] A version of adiabatic demagnetization in which a sample containing magnetic ions, embedded in a suitable crystal which quenches the hyperfine fields, is cooled in a moderate external field which reinduces the hyperfine fields, and is then thermally isolated and removed from the external field. { 'hī·pərˌfīn in'hanst ¦nü·klē·ər 'kül·iŋ }

hyperfine structure [SPECT] A splitting of spectral lines due to the spin of the atomic nucleus or to the occurrence of a mixture of isotopes in the element. Abbreviated hfs. { 'hī·pərˌfīn 'strək·chər }

hyperfocal distance [OPTICS] The distance from the camera lens to the nearest object in acceptable focus when the lens is focused on infinity. { ¦hī·pər¦fō·kəl 'dis·təns }

hyperfrequency waves [ELECTROMAG] Microwaves having wavelengths in the range from 1 centimeter to 1 meter. { ˌhī·pər'frē·kwən·sē 'wāvz }

hypernucleus [NUC PHYS] A nucleus that consists of protons, neutrons, and one or more strange particles, such as lambda particles. { ¦hī·pər'nü·klē·əs }

hyperon [PART PHYS] **1.** An elementary particle which has baryon number B = + 1, that is, which can be transformed into a nucleon and some number of mesons or lighter particles, and which has nonzero strangeness number. **2.** A hyperon (as in the first definition) which is semistable (the lifetime is much longer than 10^{-22} second). { 'hī·pəˌrän }

hyper-Raman effect [OPTICS] The phenomenon whereby, when light is scattered from an intense laser beam with frequency ν_0, the scattered light has components not only with frequency $2\nu_0$ but also with frequencies $2\nu_0 \pm \nu_m$, where ν_m is the frequency of a transition in the scattering molecules. { ˌhī·pər'räˌmän iˌfekt }

hypersonic [ACOUS] Pertaining to frequencies above 500 megahertz. [FL MECH] Pertaining to hypersonic speeds, or air currents moving at hypersonic speeds. { ¦hī·pər'sän·ik }

hypersonic flow [FL MECH] Flow of a fluid over a body at hypersonic speeds, and in which shock waves start at a finite distance from the surface of the body. { ¦hī·pər'sän·ik 'flō }

hypersonic inlet [FL MECH] An entrance or orifice for admission of fluids at hypersonic speeds. { ¦hī·pər'sän·ik 'inˌlet }

hypersonic nozzle [FL MECH] A supersonic nozzle designed to accelerate a fluid to hypersonic speeds. { ¦hī·pər'sän·ik 'näz·əl }

hypersonics [ACOUS] Production and utilization of sound waves of frequencies above 500 megahertz. { ¦hī·pər'sän·iks }

hypersonic speed [FL MECH] A speed of an object greater than about five times the speed of sound in the fluid through which the object is moving. { ¦hī·pər'sän·ik 'spēd }

hypervelocity [MECH] **1.** Muzzle velocity of an artillery projectile of 3500 feet per second (1067 meters per second) or more. **2.** Muzzle velocity of a small-arms projectile of 5000 feet per second (1524 meters per second) or more. **3.** Muzzle velocity of a tank-cannon projectile in excess of 3350 feet per second (1021 meters per second). { ˌhī·pər·və'läs·əd·ē }

hypobaric [PHYS] Having less weight or pressure. { ¦hī·pō¦bar·ik }

hysteresis [ELECTROMAG] *See* magnetic hysteresis. [PHYS] The dependence of the state of a system on its previous history, generally in the form of a lagging of a physical effect behind its cause. { ˌhis·tə'rē·səs }

hysteresis coefficient [PHYS] A constant, characteristic of a particular material, in a formula for hysteresis loss. { ˌhis·tə'rē·səs ˌkō·i'fish·ənt }

hysteresis damping [MECH] Damping of a vibration due to energy lost through mechanical hysteresis. { ˌhis·tə'rē·səs 'dam·piŋ }

hysteresis error [PHYS] The maximum separation due to hysteresis between upscale-going and downscale-going indications of a measured variable. { ˌhis·tə'rē·səs ˌer·ər }

hysteresis heating [PHYS] **1.** Supply of heat to a material through hysteresis loss. **2.** In particular, supply of a controlled amount of heat to a thermally isolated paramagnetic sample at temperatures below 1 kelvin by taking it through a magnetic hysteresis loop. { ˌhis·tə'rē·səs ˌhēd·iŋ }

hysteresis loop [PHYS] The closed curve followed by a material displaying hysteresis (such as a ferromagnet or ferroelectric) on a graph of a driven variable (such as magnetic flux density or electric polarization) versus the driving variable (such as magnetic field or electric field). { ˌhis·tə'rē·səs ˌlüp }

hysteresis loss [PHYS] The energy converted to heat in a material because of magnetic or other hysteresis, accompanying cyclic variation of the magnetic field or other driving variable. { ˌhis·tə'rē·səs ˌlȯs }

hysteretic damping [MECH] Damping of a vibrating system in which the retarding force is proportional to the velocity and inversely proportional to the frequency of the vibration. { ˌhis·tə'red·ik ¦damp·iŋ }

Hz *See* hertz.

HZE particles [NUC PHYS] A component of cosmic radiation consisting of energetic heavy nuclei (atomic number 3 or greater); so named for their high atomic number (Z) and energy (E). { ¦āch¦zē'ē ˌpärd·i·kəlz }

I

IA *See* international angstrom.

ice line [THERMO] A graph of the freezing point of water as a function of pressure. { 'īs ˌlīn }

iconocenter [ELECTROMAG] The image of the reflection coefficient of a matched load as plotted on an Argand diagram. { ī'kän·əˌsen·tər }

iconometer [OPTICS] **1.** An instrument used to find the size of an object of known distance or the distance of an object of known size by measurement of the image of it produced by a lens whose focal length is known. **2.** A direct viewfinder with a metal frame. { ˌī·kə'näm·əd·ər }

icositetrahedron *See* trapezohedron. { ī¦kä·səˌte·trə'hē·drən }

ICP-AES *See* inductively coupled plasma-atomic emission spectroscopy.

ICT *See* International Critical Tables.

ideal aerodynamics [FL MECH] A branch of aerodynamics that deals with simplifying assumptions that help explain some airflow problems and provide approximate answers. Also known as ideal fluid dynamics. { ī'dēl ˌer·ō·dī'nam·iks }

ideal crystal *See* perfect crystal. { ¦īˌdēl 'krist·əl }

ideal dielectric [ELEC] Dielectric in which all the energy required to establish an electric field in the dielectric is returned to the source when the field is removed. Also known as perfect dielectric. { ī'dēl ˌdī·i'lek·trik }

ideal exhaust velocity [FL MECH] The theoretical maximum velocity, relative to the nozzle, of the gas flow as it passes from a given nozzle inlet temperature and pressure to a given ambient pressure, when the combustion gas has a given mean molecular weight. { ī'dēl ig'zȯst vəˌläs·əd·ē }

ideal flow [FL MECH] **1.** Fluid flow which is incompressible, two-dimensional, irrotational, steady, and nonviscous. **2.** *See* inviscid flow. { ī'dēl 'flō }

ideal fluid [FL MECH] **1.** A fluid which has ideal flow. **2.** *See* inviscid fluid. { ī'dēl 'flü·əd }

ideal fluid dynamics *See* ideal aerodynamics. { ī'dēl ¦flü·əd dī'nam·iks }

ideal gas [THERMO] Also known as perfect gas. **1.** A gas whose molecules are infinitely small and exert no force on each other. **2.** A gas that obeys Boyle's law (the product of the pressure and volume is constant at constant temperature) and Joule's law (the internal energy is a function of the temperature alone). { ī'dēl 'gas }

ideal gas law [THERMO] The equation of state of an ideal gas which is a good approximation to real gases at sufficiently high temperatures and low pressures; that is, $PV = RT$, where P is the pressure, V is the volume per mole of gas, T is the temperature, and R is the gas constant. { ī'dēl 'gas ˌlȯ }

ideal propeller [FL MECH] A propeller which is considered as acting alone on an inviscid, incompressible fluid stream. { ī'dēl prə'pel·ər }

ideal radiator *See* blackbody. { ī'dēl 'rād·ēˌād·ər }

ideal transformer [ELEC] A hypothetical transformer that neither stores nor dissipates energy, has unity coefficient of coupling, and has pure inductances of infinitely great value. { ī'dēl tranz'fȯr·mər }

idiostatic connection [ELEC] An arrangement of a quadrant electrometer in which the vane is electrically connected to one of the quadrant pairs and the deflection of the vane is proportional to the square of the unknown voltage applied across the quadrant pairs. { ˌid·ē·əˌstad·ik kə'nek·shən }

idle component *See* reactive component. { 'īd·əl kəm'pō·nənt }

idle current *See* reactive current. { 'īd·əl ˌkə·rənt }

ignition temperature [PL PHYS] The lowest temperature at which the fusion energy generated in a plasma exceeds the energy lost through bremsstrahlung radiation. { ig'nish·ən ˌtem·prə·chər }

ignorable coordinate *See* cyclic coordinate. { ig'nȯr·ə·bəl kō'ȯrd·ən·ət }

illinium *See* promethium-147. { ə'lin·ē·əm }

illuminance [OPTICS] The density of the luminous flux on a surface. Also known as illumination; luminous flux density. { ə'lü·mə·nəns }

illumination [ELECTROMAG] **1.** The geometric distribution of power reaching various parts of a dish reflector in an antenna system. **2.** The power distribution to elements of an antenna array. [OPTICS] **1.** The science of the application of lighting. **2.** *See* illuminance. { əˌlü·məˌnā·shən }

illumination distribution [OPTICS] The manner in which light is dispersed on a surface. { əˌlü·məˌnā·shən ˌdi·strə'byü·shən }

illuminometer [OPTICS] A portable photometer

which is used in the field or outside the laboratory and yields results of lower accuracy than a laboratory photometer. { ə,lü·mə'näm·əd·ər }

image [ACOUS] *See* acoustic image. [ELEC] *See* electric image. [ELECTROMAG] The input reflection coefficient corresponding to the reflection coefficient of a specified load when the load is placed on one side of a waveguide junction and a slotted line is placed on the other. [OPTICS] An optical counterpart of a self-luminous or illuminated object formed by the light rays that traverse an optical system; each point of the object has a corresponding point in the image from which rays diverge or appear to diverge. [PHYS] Any reproduction of an object produced by means of focusing light, sound, electron radiation, or other emanations coming from the object or reflected by the object. { 'im·ij }

image antenna [ELECTROMAG] A fictitious electrical counterpart of an actual antenna, acting mathematically as if it existed in the ground directly under the real antenna and served as the direct source of the wave that is reflected from the ground by the actual antenna. { 'im·ij an¦ten·ə }

image converter [OPTICS] A converter that uses a fiber optic bundle to change the form of an image, for more convenient recording and display or for the coding of secret messages. { 'im·ij kən,vərd·ər }

image effect [ELECTROMAG] Effect produced on the field of an antenna due to the presence of the earth; electromagnetic waves are reflected from the earth's surface, and these reflections often are accounted for by an image antenna at an equal distance below the earth's surface. { 'im·ij i,fekt }

image force [ELEC] The electrostatic force on a charge in the neighborhood of a conductor, which may be thought of as the attraction to the charge's electric image. { 'im·ij ,fòrs }

image plane [OPTICS] The plane in which an image produced by an optical system is formed; if the object plane is perpendicular to the optical axis, the image plane will ordinarily also be perpendicular to the axis. { 'im·ij ,plān }

image potential [ELEC] The potential set up by an electric image. { 'im·ij pə,ten·chəl }

image space [OPTICS] The region of space where real or virtual images are formed by an optical system. { 'im·ij ,spās }

image surface [OPTICS] A surface on which lie images of points on a given plane perpendicular to the axis of an optical system. { 'im·ij ,sər·fəs }

imaging [PHYS] The formation of images of objects. { 'im·i·jiŋ }

immersion lens *See* immersion objective. { ə'mər·zhən ,lenz }

immersion objective [OPTICS] A high-power microscope objective designed to work with the space between the objective and the cover glass over the object filled with an oil whose index of refraction is nearly the same as that of the objective and the cover glass, in order to reduce reflection losses and increase the index of refraction of the object space. Also known as immersion lens. { ə'mər·zhən əb,jek·tiv }

immersion refractometer [OPTICS] Device to measure refractive indices by immersing the prism portion in the sample being checked. Also known as dipping refractometer. { ə'mər·zhən ,rē,frak'täm·əd·ər }

immittance [ELEC] A term used to denote both impedance and admittance, as commonly applied to transmission lines, networks, and certain types of measuring instruments. { i'mit·əns }

immittance bridge [ELECTROMAG] A modification of an admittance bridge which compares the output current of a four-terminal device with admittance standards in a T configuration in order to measure transfer admittance by a null method. { i'mit·əns ,brij }

impact [MECH] A forceful collision between two bodies which is sufficient to cause an appreciable change in the momentum of the system on which it acts. Also known as impulsive force. { 'im,pakt }

impact energy [MECH] The energy necessary to fracture a material. Also known as impact strength. { 'im,pakt ,en·ər·jē }

impact force *See* set forward force. { 'im,pakt ,fòrs }

impact law [PHYS] The relationship of fluid density, particle density, and fluid viscosity in the settling velocity of large particles in a given liquid: settling velocity is directly proportional to the square root of the particle diameter. { 'im ,pakt ,lò }

impact loss [FL MECH] Loss of head in a flowing stream due to the impact of water particles upon themselves or some bounding surface. { 'im ,pakt ,lòs }

impact parameter [NUC PHYS] In a nuclear collision, the perpendicular distance from the target nucleus to the initial line of motion of the incident particle. { 'im,pakt pə,ram·əd·ər }

impact pressure *See* dynamic pressure. { 'im,pakt ,presh·ər }

impact strength [MECH] **1.** Ability of a material to resist shock loading. **2.** *See* impact energy. { 'im,pakt ,streŋkth }

impact stress [MECH] Force per unit area imposed on a material by a suddenly applied force. { 'im,pakt ,stres }

impact velocity [MECH] The velocity of a projectile or missile at the instant of impact. Also known as striking velocity. { 'im,pakt və'läs·əd·ē }

impedance [ELEC] *See* electrical impedance. [PHYS] **1.** The ratio of a sinusoidally varying quantity to a second quantity which measures the response of a physical system to the first, both being considered in complex notation; examples are electrical impedance, acoustic impedance, and mechanical impedance. Also known as complex impedance. **2.** The ratio of the greatest magnitude of a sinusoidally varying quantity to the greatest magnitude of a second quantity which measures the response of a physical system to the first; equal to the magnitude of the quantity in the first definition. { im'pēd·əns }

impedance bridge [ELEC] A device similar to a Wheatstone bridge, used to compare impedances which may contain inductance, capacitance, and resistance. { im'pēd·əns ˌbrij }

impedance coil [ELEC] A coil of wire designed to provide impedance in an electric circuit. { im'pēd·əns ˌkȯil }

impedance component [ELEC] **1.** Resistance or reactance. **2.** A device such as a resistor, inductor, or capacitor designed to provide impedance in an electric circuit. { im'pēd·əns kəmˌpō·nənt }

impedance drop [ELEC] The total voltage drop across a component or conductor of an alternating-current circuit, equal to the phasor sum of the resistance drop and the reactance drop. { im'pēd·əns ˌdräp }

impedance match [ELEC] The condition in which the external impedance of a connected load is equal to the internal impedance of the source or to the surge impedance of a transmission line, thereby giving maximum transfer of energy from source to load, minimum reflection, and minimum distortion. { im'pēd·əns ˌmach }

impedance-matching network [ELEC] A network of two or more resistors, coils, or capacitors used to couple two circuits in such a manner that the impedance of each circuit will be equal to the impedance into which it locks. Also known as line-building-out network. { im'pēd·əns ¦mach·iŋ ˌnetˌwərk }

impedance matrix [ELEC] A matrix Z whose elements are the mutual impedances between the various meshes of an electrical network; satisfies the matrix equation $V = ZI$, where V and I are column vectors whose elements are the voltages and currents in the meshes. { im'pēd·əns ˌmā·triks }

impedometer [ELECTROMAG] An instrument used to measure impedances in waveguides. { ˌim·pə'däm·əd·ər }

imperfect crystal [CRYSTAL] A crystal in which the regular, periodic structure is interrupted by various defects. { im¦pərˌfekt 'krist·əl }

imperfect gas *See* real gas. { im'pər·fikt 'gas }

imperial gallon *See* gallon. { im'pir·ē·əl 'gal·ən }

imperial pint *See* pint. { im'pir·ē·əl 'pīnt }

implosion [PHYS] A bursting inward, as in the inward collapse of an evacuated container (such as the glass envelope of a cathode-ray tube) or the compression of fissionable material by ordinary explosives in a nuclear weapon. { im 'plō·zhən }

imprisoned incompleteness [RELAT] The property of incomplete geodesics in a space-time being confined to a compact neighborhood. { im ¦priz·ənd ˌiŋ·kəm'plēt·nəs }

impulse [MECH] The integral of a force over an interval of time. [PHYS] A pulse which lasts for so short a time that its duration can be thought of as infinitesimal. { 'imˌpəls }

impulse approximation [PHYS] An approximation for studying the collision of an incident particle with a bound target particle, in which the binding forces on the target particle during the collision are ignored. { ¦imˌpəls ə¦präk·səˌmā·shən }

impulse generator [ELEC] An apparatus which produces very short surges of high-voltage or high-current power by discharging capacitors in parallel or in series. Also known as pulse generator. { 'imˌpəls ˌjen·əˌrād·ər }

impulse solenoid [ELECTROMAG] A solenoid that operates on pulse power, at speeds up to several hundred strokes per second. { 'imˌpəls ¦sō·ləˌnȯid }

impulse voltage [ELEC] A unidirectional voltage that rapidly rises to a peak value and then drops to zero more or less rapidly. Also known as pulse voltage. { 'imˌpəls ˌvōl·tij }

impulsive force *See* impact. { im'pəl·siv 'fȯrs }

impulsive sound [ACOUS] A sound that lasts for a short period of time and includes frequencies over a large portion of the acoustic spectrum, such as a hammer blow or hand clap. { im'pəl·siv ˌsau̇nd }

impulsive sound equation [ACOUS] An equation which states that the total sound energy produced by a short burst of sound in a room is an exponentially decreasing function of time, whose decay constant depends on the speed of sound, the sound absorption coefficient, and the volume and surface area of the room. { im'pəl·siv 'sau̇nd iˌkwā·zhən }

impurity [SOLID STATE] A substance that, when diffused into semiconductor metal in small amounts, either provides free electrons to the metal or accepts electrons from it. { im'pyu̇r·əd·ē }

impurity band [SOLID STATE] The impurity levels in a semiconductor, occupying a certain range of energies. { im'pyu̇r·əd·ē ˌband }

impurity level [SOLID STATE] An energy level in the band gap of a semiconductor that results from the presence of an impurity atom. { im 'pyu̇r·əd·ē ˌlev·əl }

impurity scattering [SOLID STATE] Scattering of electrons by holes or phonons in the crystal. { im'pyu̇r·əd·ē ¦skad·ə·riŋ }

impurity semiconductor [SOLID STATE] A semiconductor whose properties are due to impurity levels produced by foreign atoms. { im'pyu̇r·əd·ē ¦sem·i·kən¦dək·tər }

in. *See* inch.

incandescence [OPTICS] The emission of visible radiation by a hot body. { ˌin·kən'des·əns }

inch [MECH] A unit of length in common use in the United States and the United Kingdom, equal to $^{1}/_{12}$ foot or 2.54 centimeters. Abbreviated in. { inch }

inch of mercury [MECH] The pressure exerted by a 1-inch-high (2.54-centimeter) column of mercury that has a density of 13.5951 grams per cubic centimeter when the acceleration of gravity has the standard value of 9.80665 m/s^2 or approximately 32.17398 ft/s^2; equal to 3386.388640341 pascals; used as a unit in the measurement of atmospheric pressure. { 'inch əv 'mər·kyə·rē }

incidence angle *See* angle of incidence. { 'in·səd·əns ,aŋ·gəl }

incidence plane *See* plane of incidence. { 'in·səd·əns ,plān }

incident field intensity [ELECTROMAG] Field strength of a sky wave without including the effects of earth reflections at the receiving location. { 'in·sə·dənt 'fēld in,ten·səd·ē }

incident light [OPTICS] The direct light that falls on a surface. { 'in·sə·dənt ¦līt }

incident wave [PHYS] A wave that impinges on a discontinuity, particle, or body, or on a medium having different propagation characteristics. { 'in·sə·dənt ¦wāv }

inclined extinction [OPTICS] Extinction in which the vibration directions are inclined to a crystal axis or direction of cleavage. Also known as oblique extinction. { in'klīnd ik'stiŋk·shən }

inclined plane [MECH] A plane surface at an angle to some force or reference line. { 'in,klīnd 'plān }

inclusion [CRYSTAL] **1.** A crystal or fragment of a crystal found in another crystal. **2.** A small cavity filled with gas or liquid in a crystal. { in'klü·zhən }

incoherent light [OPTICS] Electromagnetic radiant energy not all of the same phase, and possibly also consisting of various wavelengths. { ,in·kō'hir·ənt 'līt }

incoherent scattering [PHYS] Scattering of particles or photons in which the scattering elements act independently of one another, so that there are no definite phase relationships among the different parts of the scattered beam. { ,in·kō'hir·ənt 'skad·ə·riŋ }

incoherent waves [PHYS] Waves having no fixed phase relationship. { ,in·kō'hir·ənt 'wāvz }

incomplete fusion *See* deep inelastic collision. { ,in·kəm'plēt 'fyü·zhən }

incompressibility [MECH] Quality of a substance which maintains its original volume under increased pressure. { ¦in·kəm,pres·ə'bil·əd·ē }

incompressibility condition [FL MECH] The condition prevailing when the time rate of change of the density of a fluid is zero; this is a valid assumption for most problems in dynamic oceanography. { ¦in·kəm,pres·ə'bil·əd·ē kən,dish·ən }

incompressible flow [FL MECH] Fluid motion without any change in density. { ¦in·kəm'pres·ə·bal 'flō }

incompressible fluid [FL MECH] A fluid which is not reduced in volume by an increase in pressure. { ¦in·kəm'pres·ə·bal 'flü·əd }

incremental hysteresis loss [ELECTROMAG] Hysteresis loss when a magnetic material is subjected to a pulsating magnetizing force. { ,iŋ·krə'ment·əl ,his·tə'rē·səs ,lȯs }

incremental induction [ELECTROMAG] The quantity lying between the highest and lowest value of a magnetic induction at a point in a polarized material, when subjected to a small cycle of magnetization. { ,iŋ·krə'ment·əl in'dək·shən }

incremental permeability [ELECTROMAG] The ratio of a small cyclic change in magnetic induction to the corresponding cyclic change in magnetizing force when the average magnetic induction is greater than zero. { ,iŋ·krə'ment·əl ,pər·mē·ə'bil·əd·ē }

independent-particle model [ATOM PHYS] A model of an atomic system in which the electrons are assumed to move independently of each other in the average field generated by the nucleus and the other electrons. { ,in·də¦pen·dənt 'pard·i·kəl ,mäd·əl }

indeterminacy principle *See* uncertainty principle. { ,in·də'tərm·ə·nə·sē ,prin·sə·pəl }

index [PHYS] A numerical quantity, usually dimensionless, denoting the magnitude of some physical effect, such as the refractive index. { 'in,deks }

index ellipsoid [OPTICS] An ellipsoid whose three perpendicular axes are proportional in length to the principal values of the index of refraction of light in an anisotropic medium and point in the direction of the corresponding electric vector. Also known as ellipsoid of wave normals; indicatrix; optical indicatrix; polarizability ellipsoid; reciprocal ellipsoid. { 'in,deks ə 'lip,sȯid }

index liquid [OPTICS] A liquid whose index of refraction is known, used to find the index of refraction of powdered substances with a microscope. { 'in,deks ,lik·wəd }

index of absorption *See* absorption index. { 'in ,deks əv əb'sȯrp·shən }

index of refraction [OPTICS] The ratio of the phase velocity of light in a vacuum to that in a specified medium. Also known as absolute index of refraction; absolute refractive constant; refractive constant; refractive index. { 'in,deks əv ri'frak·shən }

indicatrix *See* index ellipsoid. { in'dik·ə,triks }

indifferent equilibrium *See* neutral equilibrium. { in'dif·ərnt ,ē·kwə'lib·rē·əm }

indirect-band-gap semiconductor [SOLID STATE] A semiconductor material in which the state of minimun energy in the conduction band and the state of maximum energy in the valence band have different momenta, and consequently optical transitions between free electrons and holes are forbidden. { ,in·də'rekt 'band ,gap 'sem·i·kən¦dək·tər }

indirect wave [PHYS] Any radio wave which arrives by an indirect path, having undergone an abrupt change of direction by refraction or reflection. { ,in·də'rekt 'wāv }

induced anisotropy [SOLID STATE] A type of uniaxial anisotropy in a magnetic material produced by annealing the magnetic material in a magnetic field. { in'düst ,an·ə'sä·trə·pē }

induced capacity *See* absolute permeability. { in 'düst kə'pas·əd·ē }

induced current [ELECTROMAG] A current produced in a conductor by a time-varying magnetic field, as in induction heating. { in'düst 'kə·rənt }

induced dipole [ELEC] An electric dipole produced by application of an electric field. { in 'düst 'dī,pōl }

induced drag [FL MECH] That part of the drag caused by the downflow or downwash of the airstream passing over the wing of an aircraft, equal to the lift times the tangent of the induced angle of attack. { in'düst 'drag }

induced electromotive force [ELECTROMAG] An electromotive force resulting from the motion of a conductor through a magnetic field, or from a change in the magnetic flux that threads a conductor. { in'düst i,lek·trə¦mōd·iv 'fȯrs }

induced emission *See* stimulated emission. { in 'düst i'mish·ən }

induced fission [NUC PHYS] Fission which takes place only when a nucleus is bombarded with neutrons, gamma rays, or other carriers of energy. { in'düst 'fish·ən }

induced magnetism [ELECTROMAG] The magnetism acquired by magnetic material while it is in a magnetic field. { in'düst 'mag·nə,tiz·əm }

induced moment [ELEC] The average electric dipole moment per molecule which is produced by the action of an electric field on a dielectric substance. { in'düst 'mō·mənt }

induced potential *See* induced voltage. { in'düst pə'ten·chəl }

induced voltage [ELECTROMAG] A voltage produced by electromagnetic or electrostatic induction. Also known as induced potential. { in 'düst 'vōl·tij }

inductance [ELECTROMAG] **1.** That property of an electric circuit or of two neighboring circuits whereby an electromotive force is generated (by the process of electromagnetic induction) in one circuit by a change of current in itself or in the other. **2.** Quantitatively, the ratio of the emf (electromotive force) to the rate of change of the current. { in'dək·təns }

inductance bridge [ELECTROMAG] **1.** A device, similar to a Wheatstone bridge, for comparing inductances. **2.** A four-coil alternating-current bridge circuit used for transmitting a mechanical movement to a remote location over a three-wire circuit; half of the bridge is at each location. { in'dək·təns ,brij }

inductance measurement [ELECTROMAG] The determination of the self-inductance of a circuit or the mutual inductance of two circuits. { in'dək·təns ,mezh·ər·mənt }

inductance meter [ELECTROMAG] A device which measures the self-inductance of a circuit or the mutual inductance of two circuits. { in'dək·təns ,mēd·ər }

inductance standards [ELECTROMAG] Two equal, multilayer coils, wound on toroidal cores of nonmagnetic materials, connected in series and located so that their interactions with external fields tend to cancel one another. { in'dək·təns ,stan·dərdz }

induction *See* electrostatic induction; electromagnetic induction. { in'dək·shən }

induction charging [ELEC] Production of electric charge on a body by means of electrostatic induction. { in'dək·shən ,chär·jiŋ }

induction coil [ELECTROMAG] A device for producing high-voltage alternating current or high-voltage pulses from low-voltage direct current, in which interruption of direct current in a primary coil, containing relatively few turns of wire, induces a high voltage in a secondary coil, containing many turns of wire wound over the primary. { in'dək·shən ,kȯil }

induction disk relay [ELECTROMAG] A unit widely used in regulating and protective relays, in which alternating current applied to a coil produces torque to rotate a disk. { in'dək·shən ¦disk 'rē,lā }

induction field [ELECTROMAG] A component of an electromagnetic field associated with an alternating current in a loop, coil, or antenna which carries energy alternately away from and back into the source, with no net loss, and which is responsible for self-inductance in a coil or mutual inductance with neighboring coils. { in'dək·shən ,fēld }

induction noise [ACOUS] The noise caused by the periodic inrush of intake air into the cylinder of an automobile engine or an air compressor as the intake valve opens and the piston moves on the intake stroke. { in'dək·shən ,nȯiz }

inductive capacities [ELECTROMAG] The permeability and permittivity of a substance. { in¦dək·tiv kə'pas·əd·ēz }

inductive charge [ELEC] The charge that exists on an object as a result of its being near another charged object. { in'dək·tiv 'chärj }

inductive circuit [ELEC] A circuit containing a higher value of inductive reactance than capacitive reactance. { in'dək·tiv 'sər·kət }

inductive coupler [ELEC] A mutual inductance that provides electrical coupling between two circuits; used in radio equipment. { in'dək·tiv 'kəp·lər }

inductive coupling [ELEC] Coupling of two circuits by means of the mutual inductance provided by a transformer. Also known as transformer coupling. { in'dək·tiv 'kəp·liŋ }

inductive divider [ELECTROMAG] A device for incorporating a desired fraction of an inductance into a circuit, usually consisting of an autotransformer with an intermediate tap. { in'dək·tiv di'vīd·ər }

inductive load [ELEC] A load that is predominantly inductive, so that the alternating load current lags behind the alternating voltage of the load. Also known as lagging load. { in'dək·tiv 'lōd }

inductively coupled plasma-atomic emission spectroscopy [SPECT] A type of atomic spectroscopy in which the light emitted by atoms and ions in an inductively coupled plasma is observed. Abbreviated ICP-AES. { in'dək·tiv·lē ¦kəp·əld ¦plaz·mə ə¦täm·ik i¦mish·ən spek'träs·kə·pē }

inductively coupled plasma discharge [PL PHYS] A high-temperature (8000–10,000 K) discharge generated by inducing a magnetic field in a flowing conducting gas, usually argon or argon and nitrogen, by means of a water-cooled copper coil which surrounds tubes through which the gas flows. { in'dək·tiv·lē ¦kəp·əld 'plaz·mə 'dis,chärj }

inductive post [ELECTROMAG] Metal post or screw extending across a waveguide parallel to the E field, to add inductive susceptance in parallel with the waveguide for tuning or matching purposes. { in'dək·tiv 'pōst }

inductive reactance [ELEC] Reactance due to the inductance of a coil or circuit. { in'dək·tiv rē'ak·təns }

inductive spacing [ELECTROMAG] Spacing of parallel transmission lines so that there is transfer of energy by mutual inductance. { in'dək·tiv 'spās·iŋ }

inductive surge [ELECTROMAG] A surge in voltage caused by sudden interruption of current in an inductive circuit. { in'dək·tiv 'sərj }

inductive susceptance [ELEC] In a circuit containing almost no resistance, the part of the susceptance due to inductance. { in'dək·tiv sə 'sep·təns }

inductive waveform [ELEC] A graph or trace of the effect of current buildup across an inductive network; proportional to the exponential of the product of a negative constant and the time. { in'dək·tiv 'wāv,fȯrm }

inductive window [ELECTROMAG] Conducting diaphragm extending into a waveguide from one or both sidewalls of the waveguide, to give the effect of an inductive susceptance in parallel with the waveguide. { in'dək·tiv 'win,dō }

inductometer [ELECTROMAG] A coil of wire of known inductance; the inductance may be fixed as in the case of primary standards, adjustable by means of switches, or continuously variable by means of a movable-coil construction. { ,in ,dək'täm·əd·ər }

inelastic [MECH] Not capable of sustaining a deformation without permanent change in size or shape. { ,in·ə'las·tik }

inelastic buckling [MECH] Sudden increase of deflection or twist in a column when compressive stress reaches the elastic limit but before elastic buckling develops. { ,in·ə'las·tik 'bək·liŋ }

inelastic collision [MECH] A collision in which the total kinetic energy of the colliding particles is not the same after the collision as before it. { ,in·ə'las·tik kə'lizh·ən }

inelastic cross section [PHYS] The cross section for an inelastic collision. { ,in·ə'las·tik 'krȯs ,sek·shən }

inelastic neutron scattering *See* slow-neutron spectroscopy. { ,in·ə'las·tik 'nü,trän ,skad·ə·riŋ }

inelastic scattering [PHYS] Scattering that results from inelastic collisions. { ,in·ə'las·tik 'skad·ə·riŋ }

inelastic stress [MECH] A force acting on a solid which produces a deformation such that the original shape and size of the solid are not restored after removal of the force. { ,in·ə'las·tik 'stres }

inequality of Clausius *See* Clausius inequality. { ,in·i'kwäl·əd·ē əv 'klaů·zē·əs }

inertia [MECH] That property of matter which manifests itself as a resistance to any change in the momentum of a body. { i'nər·shə }

inertia ellipsoid [MECH] An ellipsoid used in describing the motion of a rigid body; it is fixed in the body, and the distance from its center to its surface in any direction is inversely proportional to the square root of the moment of inertia about the corresponding axis. Also known as Poinsot ellipsoid. { i'nər·shə i'lip,sȯid }

inertial coordinate system *See* inertial reference frame. { i'nər·shəl kō'ȯrd·ən,ət ,sis·təm }

inertial flow [FL MECH] Flow in which no external forces are exerted on a fluid. { i'nər·shəl 'flō }

inertial force [MECH] The fictitious force acting on a body as a result of using a noninertial frame of reference; examples are the centrifugal and Coriolis forces that appear in rotating coordinate systems. Also known as effective force. { i'nər·shəl 'fȯrs }

inertial instability [FL MECH] **1.** Generally, instability in which the only form of energy transferred between the steady state and the disturbance in the fluid is kinetic energy. **2.** The hydrodynamic instability arising in a rotating fluid mass when the velocity distribution is such that the kinetic energy of a disturbance grows at the expense of kinetic energy of the rotation. Also known as dynamic instability. { i'nər·shəl ,in·stə'bil·əd·ē }

inertial mass [MECH] The mass of an object as determined by Newton's second law, in contrast to the mass as determined by the proportionality to the gravitational force. { i'nər·shəl 'mas }

inertial reference frame [MECH] A coordinate system in which a body moves with constant velocity as long as no force is acting on it. Also known as inertial coordinate system. { i'nər·shəl 'ref·rəns ,frām }

inertial size *See* aerodynamic size. { i'nər·shəl 'sīz }

inertia matrix [MECH] A matrix **M** used to express the kinetic energy T of a mechanical system during small displacements from an equilibrium position, by means of the equation $T = 1/2\dot{q}^T \mathbf{M} \dot{q}$, where $\dot{q}$ is the vector whose components are the derivatives of the generalized coordinates of the system with respect to time, and $\dot{q}^T$ is the transpose of $\dot{q}$. { i'nər·shə ,mā·triks }

inertia of energy [RELAT] The principle that the inertial properties of matter both determine and are determined by its total energy content. { i'nər·shə əv 'en·ər·jē }

inertia tensor [MECH] A tensor associated with a rigid body whose product with the body's rotation vector yields the body's angular momentum. { i'nər·shə ,ten·sər }

inertia wave [FL MECH] **1.** Any wave motion in

which no form of energy other than kinetic energy is present; in this general sense, Helmholtz waves, barotropic disturbances, Rossby waves, and so forth, are inertia waves. **2.** More restrictedly, a wave motion in which the source of kinetic energy of the disturbance is the rotation of the fluid about some given axis; in the atmosphere a westerly wind system is such a source, the inertia waves here being, in general, stable. { i'nər·shə ,wāv }

inextensional deformation [MECH] A bending of a surface that leaves unchanged the length of any line drawn on the surface and the curvature of the surface at each point. { in,ek'sten·chən·əl ,def·ər'mā·shən }

inferior mirage [OPTICS] A spurious image of an object formed below the true position of that object by abnormal refraction conditions along the line of sight; one of the most common types of mirage, and the opposite of a superior mirage. { in'fir·ē·ər mə'räzh }

infinity method [OPTICS] Method of adjusting two lines of sight to make them parallel; lines are adjusted on an object at great distance, for example, a star. { in'fin·əd·ē ,meth·əd }

influence line [MECH] A graph of the shear, stress, bending moment, or other effect of a movable load on a structural member versus the position of the load. { 'in,flü·əns ,līn }

infragravity wave [FL MECH] A gravity wave whose period ranges from 30 seconds to 5 minutes. { ,in·frə'grav·əd·ē ,wāv }

infrared [ELECTROMAG] Pertaining to infrared radiation. { ¦in·frə¦red }

infrared absorption [ELECTROMAG] The taking up of energy from infrared radiation by a medium through which the radiation is passing. { ¦in·frə¦red əb'sȯrp·shən }

infrared binoculars [OPTICS] An instrument for viewing an enlarged infrared image with both eyes; it has two infrared telescopes whose lens systems are similar to those of ordinary binoculars. { ¦in·frə¦red bə'näk·yə·lərz }

infrared catastrophe [QUANT MECH] The logarithmic divergence in the cross section (which one would expect to be finite) for the emission of low-energy photons in bremsstrahlung and in the double Compton effect, according to quantum electrodynamics; the difficulty is resolved by taking radiative corrections to elastic scattering into account. Also known as infrared problem. { ¦in·frə¦red kə'ts·trə·fē }

infrared dome *See* irdome. { ¦in·frə¦red 'dōm }

infrared emission [PHYS] The act of emitting infrared waves. { ¦in·frə¦red i'mish·ən }

infrared filter [OPTICS] A substance or device which is highly transparent to infrared radiation at certain wavelengths while absorbing other types of electromagnetic radiation. { ¦in·frə¦red 'fil·tər }

infrared laser [PHYS] A laser which emits infrared radiation, especially in the near- and intermediate-infrared regions. { ¦in·frə¦red 'lā·zər }

infrared maser [PHYS] A laser which emits infrared radiation, especially in the far-infrared region, or which is pumped with radiation at infrared frequencies and emits radiation at millimeter wavelengths. { ¦in·frə¦red 'mā·zər }

infrared microscope [OPTICS] A type of reflecting microscope which uses radiation of wavelengths greater than 700 nanometers and is used to reveal detail in materials that are opaque to light, such as molybdenum, wood, corals, and many red-dyed materials. { ¦in·frə¦red 'mī·krə,skōp }

infrared optical material [ELECTROMAG] A material which is transparent to infrared radiation. { ¦in·frə¦red 'äp·tə·kəl mə,tir·ē·əl }

infrared phosphor [SOLID STATE] A phosphor which, when exposed to infrared radiation during or even after decay of luminescence resulting from its usual or dominant activator, emits light having the same spectrum as that of the dominant activator; sulfide and selenide phosphors are the most important examples. { ¦in·frə¦red 'fäs·fər }

infrared problem *See* infrared catastrophe. { ¦in·frə¦red 'präb·ləm }

infrared radiation [ELECTROMAG] Electromagnetic radiation whose wavelengths lie in the range from 0.75 or 0.8 micrometer (the long-wavelength limit of visible red light) to 1000 micrometers (the shortest microwaves). { ¦in·frə¦red ,rād·ē'ā·shən }

infrared searchlight [OPTICS] A device for illuminating a scene with infrared radiation so that it may be viewed through an infrared image converter tube, although it is invisible to the unaided eye. { ¦in·frə¦red 'sərch,līt }

infrared spectrometer [SPECT] An instrument used to identify and measure the concentration of chemical compounds (gases, nonaqueous liquids, and solids) with electromagnetic radiation from 800 nanometers to 1 millimeter. { ¦in·frə¦red spek'träm·əd·ər }

infrared spectrophotometry [SPECT] Spectrophotometry in the infrared region, usually for the purpose of chemical analysis through measurement of absorption spectra associated with rotational and vibrational energy levels of molecules. { ¦in·frə¦red ¦spek·trə·fə'täm·ə·trē }

infrared spectroscopy [SPECT] The study of the properties of material systems by means of their interaction with infrared radiation; ordinarily the radiation is dispersed into a spectrum after passing through the material. { ¦in·frə¦red spek'träs·kə·pē }

infrared spectrum [ELECTROMAG] **1.** The range of wavelengths of infrared radiation. **2.** A display or graph of the intensity of infrared radiation emitted or absorbed by a material as a function of wavelength or some related parameter. { ¦in·frə¦red ¦spek·trəm }

infrared telescope [OPTICS] An instrument that converts an invisible infrared image into a visible image and enlarges this image, consisting of an infrared image converter tube, an objective lens for imaging the scene to be viewed onto the photocathode of the tube, and an ocular for viewing the phosphor screen of the tube. { ¦in·frə¦red 'tel·ə,skōp }

infrasonic [ACOUS] Pertaining to signals, equipment, or phenomena involving frequencies below the range of human hearing, hence below about 15 hertz. Also known as subsonic (deprecated usage). { ¦in·frə¦sän·ik }

infrasound [ACOUS] Vibrations of the air at frequencies too low to be perceived as sound by the human ear, below about 15 hertz. { 'in·frə,saủnd }

Ingen-Hausz apparatus [THERMO] An apparatus for comparing the thermal conductivities of different conductors; specimens consisting of long wax-coated rods of equal length are placed with one end in a tank of boiling water covered with a radiation shield, and the lengths along the rods from which the wax melts are compared. { ¦iŋ·gən 'haủs ,ap·ə,rad·əs }

inhomogeneous line-broadening [OPTICS] An increase beyond the natural linewidth in the width of an absorption or emission line that results from a disturbance (such as strains or imperfections) that can differ from one source emitter to another. { in,hä·mə'jē·nē·əs 'līn ,brȯd·ən·iŋ }

initial free space [MECH] In interior ballistics, the portion of the effective chamber capacity not displaced by propellant. { i'nish·əl ¦frē 'spās }

initial permeability [ELECTROMAG] The limit of the normal permeability as the magnetic induction and magnetic field strength approach 0. { i¦nish·əl ,pər·mē·ə'bil·əd·ē }

initial shot start pressure [MECH] In interior ballistics, the pressure required to start the motion of the projectile from its initial loaded position; in fixed ammunition, it includes pressure required to separate projectile and cartridge case and to start engraving the rotating band. { i'nish·əl 'shät ¦stärt ,presh·ər }

initial-value problem [FL MECH] A dynamical problem whose solution determines the state of a system at all times subsequent to a given time at which the state of the system is specified by given initial conditions; the initial-value problem is contrasted with the steady-state problem, in which the state of the system remains unchanged in time. Also known as transient problem. { i'nish·əl ¦val·yü ,präb·ləm }

initial velocity [PHYS] The velocity of anything at the beginning of a specific phase of its motion. { i'nish·əl və'läs·əd·ē }

initial yaw [MECH] The yaw of a projectile the instant it leaves the muzzle of a gun. { i'nish·əl 'yȯ }

injection laser [OPTICS] A laser in which a forward-biased gallium arsenide diode converts direct-current input power directly into coherent light, without optical pumping. { in'jek·shən ¦lā·zər }

inner bremsstrahlung [NUC PHYS] The emission of a photon during beta decay or electron capture by a nucleus. Also known as internal bremsstrahlung. { ¦in·ər 'brem,shträ·ləŋ }

inner potential [SOLID STATE] The average value of the electrostatic potential, taken over the volume of a crystal. { ¦in·ər pə'ten·chəl }

inner quantum number [ATOM PHYS] A quantum number J which gives an atom's total angular momentum, excluding the nuclear spin. { ¦in·ər 'kwänt·əm ,nəm·bər }

inorganic liquid laser [OPTICS] A liquid laser in which an inorganic liquid such as neodymium-selenium oxychloride or neodymium-doped phosphorus chloride is used as the active material. Also known as neodymium liquid laser. { ¦in·ȯr¦gan·ik ¦lik·wəd 'lā·zər }

in phase [PHYS] Having waveforms that are of the same frequency and that pass through corresponding values at the same instant. { 'in ,fāz }

in-phase component [ELEC] The component of the phasor representing an alternating current which is parallel to the phasor representing voltage. { 'in ,fāz kəm'pō·nənt }

input admittance [ELEC] The admittance measured across the input terminals of a four-terminal network with the output terminals short-circuited. { 'in,pu̇t əd,mit·əns }

input impedance [ELEC] The impedance across the input terminals of a four-terminal network when the output terminals are short-circuited. { 'in,pu̇t im,pēd·əns }

insonify *See* ensonify. { in'sän·ə·fī }

instability [PHYS] A property of the steady state of a system such that certain disturbances or perturbations introduced into the steady state will increase in magnitude, the maximum perturbation amplitude always remaining larger than the initial amplitude. { ,in·stə'bil·əd·ē }

instantaneous axis [MECH] The axis about which a rigid body is carrying out a pure rotation at a given instant in time. { ¦in·stən¦tā·nē·əs 'ak·səs }

instantaneous center [MECH] A point about which a rigid body is rotating at a given instant in time. Also known as instant center. { ¦in·stən¦tā·nē·əs 'sen·tər }

instantaneous condition [PHYS] The condition of a system at a particular instant in time. { ¦in·stən¦tā·nē·əs kən'dish·ən }

instantaneous field of view [OPTICS] The solid angle within which radiation is detected by an imaging system employing some form of scanning mechanism, at a given instant of time. { ¦in·stən¦tā·nē·əs ¦fēld əv 'vyü }

instantaneous power [ELEC] The product of the instantaneous voltage and the instantaneous current for a circuit or component. { ¦in·stən¦tā·nē·əs 'pau̇·ər }

instantaneous recovery [MECH] The immediate reduction in the strain of a solid when a stress is removed or reduced, in contrast to creep recovery. { ¦in·stən¦tā·nē·əs ri'kəv·ə·rē }

instantaneous strain [MECH] The immediate deformation of a solid upon initial application of a stress, in contrast to creep strain. { ¦in·stən¦tā·nē·əs 'strān }

instantaneous value [PHYS] The value of a sinusoidal or otherwise varying quantity at a particular instant. { ¦in·stən¦tā·nē·əs 'val·yü }

instant center *See* instantaneous center. { 'in·stənt 'sen·tər }

instanton [PART PHYS] A hypothetical pseudoparticle which provides solutions to equations describing the gauge fields of quantum chromodynamics, and represents large vacuum fluctuations in these fields that would exert forces on quarks. { 'in·stən,tän }

instrument transformer [ELEC] A transformer that transfers primary current, voltage, or phase values to the secondary circuit with sufficient accuracy to permit connecting an instrument to the secondary rather than the primary; used so only low currents or low voltages are brought to the instrument. { 'in·strə·mənt tranz,fȯr·mər }

insulated [ELEC] Separated from other conducting surfaces by a nonconducting material. { 'in·sə,lād·əd }

insulated conductor [ELEC] A conductor surrounded by insulation to prevent current leakage or short circuits. Also known as insulated wire. { 'in·sə,lād·əd kən'dək·tər }

insulated wire *See* insulated conductor. { 'in·sə,lād·əd 'wīr }

insulating strength [ELEC] Measure of the ability of an insulating material to withstand electric stress without breakdown; it is defined as the voltage per unit thickness necessary to initiate a disruptive discharge; usually measured in volts per centimeter. { 'in·sə,lād·iŋ ,streŋkth }

insulation [ELEC] A material having high electrical resistivity and therefore suitable for separating adjacent conductors in an electric circuit or preventing possible future contact between conductors. Also known as electrical insulation. { ,in·sə'lā·shən }

insulation resistance [ELEC] The electrical resistance between two conductors separated by an insulating material. { ,in·sə'lā·shən ri¦zis·təns }

insulator [ELEC] A device having high electrical resistance and used for supporting or separating conductors to prevent undesired flow of current from them to other objects. Also known as electrical insulator. [SOLID STATE] A substance in which the normal energy band is full and is separated from the first excitation band by a forbidden band that can be penetrated only by an electron having an energy of several electronvolts, sufficient to disrupt the substance. { 'in·sə,lād·ər }

integer spin [QUANT MECH] Property of a particle whose spin angular momentum is a whole number times Planck's constant divided by 2π; bosons have this property; in contrast, fermions have half-integer spin. { 'int·ə·jər ,spin }

integrable system [MECH] A dynamical system whose motion is governed by an integrable differential equation. { ¦int·i·grə·bəl ¦sis·təm }

integral photography [OPTICS] A type of three-dimensional photography in which the photographic medium is placed at the focal plane of a microlens array, and the developed image is viewed through the same lens array, allowing the object to be reconstructed in full parallax. { ,int·i·grəl fə'täg·rə·fē }

integral hologram [OPTICS] A type of hologram that is automatically synthesized from a large collection of photographs, each taken from a slightly different position. { 'int·ə·grəl 'häl·ə,gram }

integrated optics [OPTICS] A thin-film device containing tiny lenses, prisms, and switches to transmit very thin laser beams, and serving the same purposes as the manipulation of electrons in thin-film devices of integrated electronics. { 'in·tə,grād·əd 'äp·tiks }

integrated radiation [OPTICS] The integral of the radiance over the duration of exposure. { 'in·tə,grād·əd ,rād·ē'ā·shən }

integrated reflection [PHYS] The intensity of a beam of x-rays or neutrons reflected from a given atomic plane of a crystal, integrated over a small range of angles about the general direction of the beam. { 'int·ə,grād·əd ri'flek·shən }

integrating-sphere photometer [OPTICS] An instrument for measuring the total luminous flux of a lamp or luminaire; the source is placed inside a sphere whose inside surface has a diffusely reflecting white finish, and the light reflected from this surface onto a window is measured by an ordinary photometer. Also known as sphere photometer. { 'int·ə,grād·iŋ ,sfir fə'täm·əd·ər }

intensity [PHYS] **1.** The strength or amount of a quantity, as of electric field, current, magnetization, radiation, or radioactivity. **2.** The power transmitted by a light or sound wave across a unit area perpendicular to the wave. { in'ten·səd·ē }

intensity level [PHYS] The logarithm of the ratio of two intensities, powers or energies, usually expressed in decibels. { in'ten·səd·ē ,lev·əl }

intensity of magnetization *See* intrinsic induction. { in'ten·səd·ē əv ,mag·nəd·ə'zā·shən }

interacting boson model [NUC PHYS] A model of nuclear structure which assumes that the particles making up a complex nucleus are bosons, each of which corresponds to a pair of correlated nucleons. { ,in·tər¦ak·tiŋ 'bō,sän ,mäd·əl }

interaction [FL MECH] With respect to wave components, the nonlinear action by which properties of fluid flow (such as momentum, energy, vorticity), are transferred from one portion of the wave spectrum to another, or viewed in another manner, between eddies of different size-scales. [PHYS] A process in which two or more bodies exert mutual forces on each other. { ¦in·tə¦rak·shən }

interaction picture [QUANT MECH] A mode of description of a system in which the time dependence is carried partly by the operators and partly by the state vectors, the time dependence of the state vectors being due entirely to that part of the Hamiltonian arising from interactions between particles. Also known as interaction representation. { ¦in·tə¦rak·shən ,pik·chər }

interaction representation *See* interaction picture. { ¦in·tə¦rak·shən ˌrep·ri·zen'tā·shən }

intercept [CRYSTAL] One of the distances cut off a crystal's reference axis by planes. { ¦in·tər¦sept }

interchangeable lens [OPTICS] A lens which can be used in place of another, generally of different magnification. { ¦in·tər¦chānj·ə·bəl 'lenz }

interchange coefficient *See* exchange coefficient. { 'in·tərˌchānj ˌkō·i'fish·ənt }

interface resistance [THERMO] **1.** Impairment of heat flow caused by the imperfect contact between two materials at an interface. **2.** Quantitatively, the temperature difference across the interface divided by the heat flux through it. { 'in·tərˌfās ri'zis·təns }

interfacial angle [CRYSTAL] The angle between two crystal faces. { 'in·tərˌfā·shəl ¦aŋ·gəl }

interfacial energy [PHYS] The free energy of the surfaces at an interface, resulting from differences in the tendencies of each phase to attract its own molecules; equal to the surface tension. Also known as surface energy. { 'in·tərˌfā·shəl ¦en·ər·jē }

interfacial force *See* surface tension. { 'in·tərˌfā·shəl ¦fȯrs }

interfacial polarization [ELEC] *See* space-charge polarization. [OPTICS] Polarization of light by reflection from the surface of a dielectric at Brewster's angle. { 'in·tərˌfā·shəl ˌpō·lə·rə'zā·shən }

interfacial tension *See* surface tension. { 'in·tərˌfā·shəl 'ten·chən }

interference [PHYS] The variation with distance or time of the amplitude of a wave which results from the superposition (algebraic or vector addition) of two or more waves having the same, or nearly the same, frequency. Also known as wave interference. { ˌin·tər'fir·əns }

interference colors [OPTICS] Colors formed by interference of a beam of light passed through a thin section of a mineral placed in a polarizing microscope. { ˌin·tər'fir·əns ˌkəl·ərz }

interference figure [OPTICS] A pattern of light and dark areas observed with a conoscope when a birefringent crystal is placed in a convergent beam of linearly polarized light. { ˌin·tər'fir·əns ˌfig·yər }

interference filter [OPTICS] An optical filter in which the wavelengths that are not transmitted are removed by interference phenomena rather then by absorbtion or scattering. { ˌin·tər'fir·əns ¦fil·tər }

interference fringes [OPTICS] A series of light and dark bands produced by interference of light waves. { ˌin·tər'fir·əns ¦frin·jəz }

interference microscope [OPTICS] A microscope used for visualizing and measuring differences in phase or optical paths in transparent or reflecting specimens; it differs from the phase contrast microscope in that the incident and diffracted waves are not separated, but interference is produced between the transmitted wave and another wave which originates from the same source. { ˌin·tər'fir·əns ¦mī·krəˌskōp }

interference pattern [PHYS] Resulting space distribution of pressure, particle density, particle velocity, energy density, or energy flux when progressive waves of the same frequency and kind are superimposed. { ˌin·tər'fir·əns ¦pad·ərn }

interference spectrum [SPECT] A spectrum that results from interference of light, as in a very thin film. { ˌin·tər'fir·əns ¦spek·trəm }

interferogram [SPECT] A graph of the variation of the output signal from an interferometer as the condition for interference within the interferometer is varied. { ˌin·tə'fir·əˌgram }

interferometer [OPTICS] An instrument in which light from a source is split into two or more beams, which are subsequently reunited after traveling over different paths and display interference. { ˌin·tə·fə'räm·əd·ər }

interferometry [OPTICS] The design and use of optical inferometers; uses include precise measurement of wavelength, measurement of very small distances and thicknesses, study of hyperfine structure of spectral lines, precise measurement of indices of refraction, and determination of separations of binary stars and diameters of very large stars. { ˌin·tə·fə'räm·ə·trē }

interior ballistics [MECH] The science concerned with the combustion of powder, development of pressure, and movement of a projectile in the bore of a gun. { in'tir·ē·ər bə'lis·tiks }

intermediate-energy neutron-proton scattering [NUC PHYS] An elastic collision of a neutron having an energy from 10 to 440 megaelectronvolts with a proton (usually the nucleus of a hydrogen atom). { ˌin·tər¦mē·dē·ət 'en·ər·jē ¦nüˌträn ¦prōˌtän ˌskad·ər·iŋ }

intermediate-energy proton-proton scattering [NUC PHYS] An elastic collision of a proton having an energy from 10 to 440 megaelectronvolts with another proton (usually the nucleus of a hydrogen atom). { ˌin·tər¦mē·dē·ət 'en·ər·jē ¦prōˌtän ¦prōˌtän ˌskad·ər·iŋ }

intermediate-infrared radiation [ELECTROMAG] Infrared radiation having a wavelength between about 2.5 micrometers and about 50 micrometers; this range includes most molecular vibrations. Also known as mid-infrared radiation. { ˌin·tər'mēd·ē·ət ¦in·frə¦red ˌrād·ē'ā·shən }

intermediate state [CRYO] A state of partial superconductivity that occurs when a magnetic field slightly less than the critical field is applied to a superconducting material below its critical temperature. [QUANT MECH] A state through which a system may pass during transition from an initial state to a final state. { ˌin·tər'mēd·ē·ət ¦stāt }

intermediate vector boson [PART PHYS] One of the three fundamental particles that transmits the weak nuclear force in the same manner that the photon transmits the electromagnetic force. { ˌin·tər'mēd·ē·ət ¦vek·tər 'bōˌsän }

intermittency [PHYS] The alternation in time of a dynamical system between nearly periodic and chaotic behavior. { ˌin·tər'mit·ən·sē }

intermittent current [ELEC] A unidirectional current that flows and ceases to flow at irregular or regular intervals. { ¦in·tər¦mit·ənt 'kə·rənt }

internal absorptance [ELECTROMAG] The value of absorptance, corrected to eliminate the effects of scattering and of reflection from the surfaces of the substance; that is, the ratio of the radiant power absorbed between the entry and exit surfaces of the substance to the radiant power leaving the entry surface. { in'tərn·əl əb'sȯrp·təns }

internal absorption *See* Auger effect. { in'tərn·əl əp'sȯrp·shən }

internal bremmsstrahlung *See* inner bremsstrahlung. { in'tərn·əl 'brem,shträ·lůŋ }

internal conversion [NUC PHYS] A nuclear deexcitation process in which energy is transmitted directly from an excited nucleus to an orbital electron, causing ejection of that electron from the atom. { in'tərn·əl kən'vər·zhən }

internal conversion coefficient *See* conversion coefficient. { in'tərn·əl kən'vər·zhən ˌkō·i'fish·ənt }

internal dielectric field *See* dielectric field. { in'tərn·əl ˌdī·ə'lek·trik 'fēld }

internal energy [THERMO] A characteristic property of the state of a thermodynamic system, introduced in the first law of thermodynamics; it includes intrinsic energies of individual molecules, kinetic energies of internal motions, and contributions from interactions between molecules, but excludes the potential or kinetic energy of the system as a whole; it is sometimes erroneously referred to as heat energy. { in 'tərn·əl 'en·ər·jē }

internal force [MECH] A force exerted by one part of a system on another. { in'tərn·əl 'fōrs }

internal friction [FL MECH] *See* viscosity. [MECH] **1.** Conversion of mechanical strain energy to heat within a material subjected to fluctuating stress. **2.** In a powder, the friction that is developed by the particles sliding over each other; it is greater than the friction of the mass of solid that comprises the individual particles. { in'tərn·əl 'frik·shən }

internal line [QUANT MECH] A component of a Feynman graph (in the diagrammatic presentation of perturbative quantum field theory) describing the propagation of a virtual particle whose momentum is integrated over all possible values. { in'tərn·əl 'līn }

internal photoelectric effect [SOLID STATE] A process in which the absorption of a photon in a semiconductor results in the excitation of an electron from the valence band to the conduction band. { in'tərn·əl ˌfōd·ō·ə'lek·trik i,fekt }

internal photoionization *See* Auger effect. { in'tərn·əl ˌfōd·ō,ī·ə·nə'zā·shən }

internal pressure *See* intrinsic pressure. { in'tərn·əl 'presh·ər }

internal reflectance spectroscopy *See* attenuated total reflectance. { in'tərn·əl ri¦flek·təns spek 'träs·kə·pē }

internal reflection [OPTICS] The reflection of electromagnetic radiation in a given medium from the boundary with a less dense medium. { in,tərn·əl ri'flek·shən }

internal resistance [ELEC] The resistance within a voltage source, such as an electric cell or generator. { in'tərn·əl ri'zis·təns }

internal standard [SPECT] The principal line in spectrum analysis by the logarithmic sector method, a quantitative spectroscopy procedure. { in'tərn·əl 'stan·dərd }

internal stress [MECH] A stress system within a solid that is not dependent on external forces. Also known as residual stress. { in'tərn·əl 'stres }

internal transmittance [ELECTROMAG] The value of transmittance, corrected to eliminate the effects of scattering and of reflection from the surfaces of the substance; that is, the ratio of the radiant power reaching the exit surface of the substance to the radiant power leaving the entry surface. { in'tərn·əl ˌtranz'mit·əns }

internal wave [FL MECH] A wave motion of a stably stratified fluid in which the maximum vertical motion takes place below the surface of the fluid. { in'tərn·əl 'wāv }

internal work [THERMO] The work done in separating the particles composing a system against their forces of mutual attraction. { in'tərn·əl 'wərk }

international ampere [ELEC] The current that, when flowing through a solution of silver nitrate in water, deposits silver at a rate of 0.001118 gram per second; it has been superseded by the ampere as a unit of current, and is equal to approximately 0.999850 ampere. { ¦in·tər¦nash·ən·əl 'am,pir }

international angstrom [PHYS] A unit of length, equal to 1/6438.4696 of the wavelength of the red cadmium line in dry air at standard atmospheric pressure, at a temperature of 15°C containing 0.03% by volume of carbon dioxide; equal to 1.0000002 angstroms. Abbreviated IA. { ¦in·tər¦nash·ən·əl 'aŋg·strəm }

international candle [OPTICS] A unit of luminous intensity, now replaced by the candela; as defined in the United States, it was a specified fraction of the average luminous intensity radiated in a horizontal direction by a group of 45 carbon-filament lamps preserved at the National Bureau of Standards when the lamps were operated at a specified voltage. Also known as standard candle. { ¦in·tər¦nash·ən·əl 'kand·əl }

International Critical Tables [PHYS] A seven-volume series of tables of numerical data in physics, chemistry, and technology, published in 1926–1930, prepared by experts who gave the "best" value which could be derived from all the data available at the time. Abbreviated ICT. { ¦in·tər¦nash·ən·əl 'krid·ə·kəl ˌtā·bəlz }

international henry [ELECTROMAG] A unit of electrical inductance which has been superseded by the henry, and is equal to 1.00049 henry. Also known as quadrant; secohm. { ¦in·tər¦nash·ən·əl 'hen·rē }

international ohm [ELEC] A unit of resistance, equal to that of a column of mercury of uniform cross section that has a length of 160.3 centimeters and a mass of 14.4521 grams at the temperature of melting ice; it has been superseded by the ohm, and is equal to 1.00049 ohms. { ¦in·tər¦nash·ən·əl 'ōm }

international practical temperature scale [THERMO] Temperature scale based on six points: the water triple point, the boiling points of oxygen, water, sulfur, and the solidification points of silver and gold; designated as °C, degrees Celsius, or t_{int}; replaced in 1990 by the international temperature scale. { ¦in·tər¦nash·ən·əl ¦prak·tə·kəl 'tem·prə·chər ˌskāl }

international system of electrical units [ELEC] System of electrical units based on agreed fundamental units for the ohm, ampere, centimeter, and second, in use between 1893 and 1947, inclusive; in 1948, the Giorgi, or meter-kilogram-second-absolute system, was adopted for international use. { ¦in·tər¦nash·ən·əl ¦sistəm əv i¦lek·trə·kəl 'yü·nəts }

International System of Units [PHYS] A system of physical units in which the fundamental quantities are length, time, mass, electric current, temperature, luminous intensity, and amount of substance, and the corresponding units are the meter, second, kilogram, ampere, kelvin, candela, and mole; it has been given official status and recommended for universal use by the General Conference on Weights and Measures. Also known (in French) as Système International d'Unités. Abbreviated SI (in all languages). { ¦in·tər¦nash·ən·əl ¦sis·təm əv 'yü·nəts }

international table British thermal unit *See* British thermal unit. { ¦in·tər¦nash·ən·əl ¦tā·bəl ¦brid·ish 'thər·məl ˌyü·nət }

international table calorie *See* calorie. { ¦in·tər¦nash·ən·əl ¦tā·bəl 'kal·ə·rē }

international temperature scale [THERMO] A standard temperature scale, adopted in 1990, that approximates the thermodynamic scale, based on assigned temperature values of 17 thermodynamic equilibrium fixed points and prescribed thermometers for interpolation between them. Abbreviated ITS-90. { ¦in·tər¦nash·ən·əl 'tem·prə·chər ˌskāl }

international volt [ELEC] A unit of potential difference or electromotive force, equal to 1/1.01858 of the electromotive force of a Weston cell at 20°C; it has been superseded by the volt, and is equal to 1.00034 volts. { ¦in·tər¦nash·ən·əl 'vōlt }

interpenetration twin [CRYSTAL] Two or more individual crystals so twinned that they appear to have grown through one another. Also known as penetration twin. { ¦in·tərˌpen·ə'trā·shən ¦twin }

interplanar spacing [CRYSTAL] The perpendicular distance between successive parallel planes of atoms in a crystal. { ¦in·tərˌplā·nər 'spās·iŋ }

interspace [PHYS] An interval of space or time. { 'in·tərˌspās }

interstice [SOLID STATE] A space or volume between atoms of a lattice, or between groups of atoms or grains of a solid structure. { in'tərs·təs }

interstitial [CRYSTAL] A crystal defect in which an atom occupies a position between the regular lattice positions of a crystal. { ¦in·tər¦stish·əl }

interstitial atom [CRYSTAL] A displaced atom which is forced into a nonequilibrium site within a crystal lattice. { ¦in·tər¦stish·əl 'ad·əm }

interstitial compound [SOLID STATE] A binary compound in which atoms of one element (usually a light, nonmetallic element) occupy spaces between atoms of the crystal lattice formed by the other element (usually a heavy, metallic element). { ¦in·tər¦stish·əl 'kämˌpau̇nd }

interstitial impurity [SOLID STATE] An atom which is not normally found in a solid, and which is located at a position in the lattice structure where atoms or ions normally do not exist. { ¦in·tər¦stish·əl im'pyu̇r·əd·ē }

interval [ACOUS] The spacing in pitch or frequency between two sounds; the frequency interval is the ratio of the frequencies or the logarithm of this ratio. [PHYS] The time separating two events, or the distance between two objects. [RELAT] **1.** In special relativity, the Lorentz invariant quantity $c^2(\Delta t)^2 - (\Delta x)^2 - (\Delta y)^2 - (\Delta z)^2$, where c is the speed of light, Δt is the difference in the time coordinates of two specified events, and Δx, Δy, and Δz are differences in their x, y, and z coordinates, respectively. **2.** In general relativity, a generalization of this concept, namely the sum over the indices μ and ν of $g_\mu \nu dx_{\mu\nu} dx_\nu$, where $dx_{\mu\nu}$ and dx_ν are the differences in the x_μ and x_ν coordinates of two specified neighboring events, and $g_\mu \nu$ is an element of the metric tensor. { 'in·tər·vəl }

interval of Sturm *See* astigmatic interval. { ¦in·tər·vəl əv 'stərm }

intrabeam viewing [OPTICS] The viewing condition in which the eye is exposed to all or part of a laser beam. { 'in·trəˌbēm 'vyü·iŋ }

intracavity absorption spectroscopy [SPECT] A highly sensitive technique in which an absorbing sample is placed inside the resonator of a broadband dye laser, and absorption lines are detected as dips in the laser emission spectrum. { ¦in·trə'kav·əd·ē əb¦sȯrp·shən spek'träs·kə·pē }

intranuclear cascade model [NUC PHYS] A model of nuclear collisions that assumes a series of independent nucleon-nucleon collisions between particles that act like billiard balls. { ¦in·trə'nü·klē·ər kas'kād ˌmäd·əl }

intraoptical light sighting system [OPTICS] A target sighting system used with infrared pyrometers in which a visible, pulsating light is directed through the infrared optics, and gives an indication of the exact field of view, not just the centerpoint. { ˌin·trə'äp·tə·kəl 'līt ¦sīd·iŋ ˌsis·təm }

intrinsic conductivity [SOLID STATE] The conductivity of a semiconductor or metal in which impurities and structural defects are absent or have a very low concentration. { in'trin·sik ˌkän ˌdək'tiv·əd·ē }

intrinsic contact potential difference [ELEC] True potential difference between two perfectly clean metals in contact. { in'trin·sik ¦kän,takt pə¦ten·chəl 'dif·ərns }

intrinsic electric strength [ELEC] The extremely high dielectric strength displayed by a substance at low temperatures. { in¦trin·sik i¦lek·trik ,streŋkth }

intrinsic flux density *See* intrinsic induction. { in 'trin·sik 'fləks ,den·səd·ē }

intrinsic induction [ELECTROMAG] The vector difference between the magnetic flux density at a given point and the magnetic flux density which would exist there, for the same magnetic field strength, if the point were in a vacuum. Symbol Bi. Also known as intensity of magnetization; intrinsic flux density; magnetic polarization. { in'trin·sik in'dək·shən }

intrinsic mobility [SOLID STATE] The mobility of the electrons in an intrinsic semiconductor. { in'trin·sik mō'bil·əd·ē }

intrinsic parity [PART PHYS] A quantum number, equal to +1 or −1, which is assigned to particles so that the product of the intrinsic parities of the particles composing a system times the parity of the system's wave function yields the total parity. { in'trin·sik 'par·əd·ē }

intrinsic photoconductivity [SOLID STATE] Photoconductivity associated with excitation of charge carriers across the band gap of a material. { in'trin·sik ¦fōd·ō,kän,dək'tiv·əd·ē }

intrinsic photoemission [SOLID STATE] Photoemission which can occur in an ideally pure and perfect crystal, in contrast to other types of photoemission which are associated with crystal defects. { in'trin·sik ¦fōd·ō·i'mish·ən }

intrinsic pressure [PHYS] Pressure in a fluid resulting from inward forces on molecules near the fluid surface, caused by attraction between molecules. Also known as internal pressure. { in'trin·sik 'presh·ər }

intrinsic property [SOLID STATE] A property of a substance that is not seriously affected by impurities or imperfections in the crystal structure. { in'trin·sik 'präp·ərd·ē }

intrinsic semiconductor [SOLID STATE] A semiconductor in which the concentration of charge carriers is characteristic of the material itself rather than of the content of impurities and structural defects of the crystal. Also known as *i*-type semiconductor. { in'trin·sik ¦sem·i·kən¦dək·tər }

intrinsic temperature range [SOLID STATE] In a semiconductor, the temperature range in which its electrical properties are essentially not modified by impurities or imperfections within the crystal. { in'trin·sik 'tem·prə·chər ,rānj }

intrinsic tracer [NUC PHYS] An isotope that is present naturally in a form suitable for tracing a given element through chemical and physical processes. { in'trin·sik 'trā·sər }

in vacuo [PHYS] In a vacuum. { in 'vak·yə·wō }

invariable line [MECH] A line which is parallel to the angular momentum vector of a body executing Poinsot motion, and which passes through the fixed point in the body about which there is no torque. { in'ver·ē·ə·bəl 'līn }

invariable plane [MECH] A plane which is perpendicular to the angular momentum vector of a rotating rigid body not subject to external torque, and which is always tangent to its inertia ellipsoid. { in'ver·ē·ə·bəl 'plān }

invariance [OPTICS] Any property of a light beam that remains constant when the light is reflected or refracted at one or more surfaces. [PHYS] The property of a physical quantity or physical law of being unchanged by certain transformations or operations, such as reflection of spatial coordinates, time reversal, charge conjugation, rotations, or Lorentz transformations. Also known as symmetry. { in'ver·ē·əns }

invariance principle [PHYS] Any principle which states that a physical quantity or physical law possesses invariance under certain transformations. Also known as symmetry law; symmetry principle. [RELAT] In general relativity, the principle that the laws of motion are the same in all frames of reference, whether accelerated or not. { in'ver·ē·əns ,prin·sə·pəl }

invariant plane [ATOM PHYS] The plane perpendicular to the total angular momentum (orbital plus spin) of an atom. { in'ver·ē·ənt 'plān }

inverse beta decay [NUC PHYS] A reaction providing evidence for the existence of the neutrino, in which an antineutrino (or neutrino) collides with a proton (or neutron) to produce a neutron (or proton) and a positron (or electron). { 'in,vərs 'bād·ə di,kā }

inverse bremsstrahlung [ATOM PHYS] The absorption by an electron of a photon in a strong electric field such as that surrounding an atomic nucleus. Also known as free-free absorption. { 'in,vərs 'brem,shträ·lúŋ }

inverse Compton effect [QUANT MECH] A process in which relativistic particles give up some of their energy to long-wavelength radiation, converting it to shorter-wavelength radiation. { 'in,vərs 'kam·tən i,fekt }

inverse magnetostriction *See* magnetic tension effect. { 'in,vərs mag,ned·ō'strik·shən }

inverse network [ELEC] Two two-terminal networks are said to be inverse when the product of their impedances is independent of frequency within the range of interest. { 'in,vərs 'net,wərk }

inverse piezoelectric effect [SOLID STATE] The contraction or expansion of a piezoelectric crystal under the influence of an electric field, as in crystal headphones; also occurs at *pn* junctions in some semiconductor materials. { 'in,vərs pē¦ā·zō·i¦lek·trik i,fekt }

inverse scattering theory [PHYS] The discipline that determines the nature of the scattering object, or an interaction potential energy, in a scattering process or collision, from knowledge of the amplitudes of the scattered fields. { 'in,vərs 'skad·ə·riŋ ,thē·ə·rē }

inverse-square law [PHYS] Any law in which a physical quantity varies with distance from a source inversely as the square of that distance. { 'in,vərs ¦skwer ,lȯ }

inverse Stark effect [SPECT] The Stark effect as

observed with absorption lines, in contrast to emission lines. { 'in,vərs 'stärk i,fekt }

inverse Zeeman effect [SPECT] A splitting of the absorption lines of atoms or molecules in a static magnetic field; it is the Zeeman effect observed with absorption lines. { 'in,vərs 'zē·mən i,fekt }

inversion [CRYSTAL] A change from one crystal polymorph to another. Also known as transformation. [ELEC] The solution of certain problems in electrostatics through the use of the transformation in Kelvin's inversion theorem. [OPTICS] The formation of an inverted image by an optical system. [PHYS] The simultaneous reflection of all three directions in space, so that each coordinate is replaced by the negative of itself. Also known as space inversion. [SOLID STATE] The production of a layer at the surface of a semiconductor which is of opposite type from that of the bulk of the semiconductor, usually as the result of an applied electric field. [THERMO] A reversal of the usual direction of a variation or process, such as the change in sign of the expansion coefficient of water at 4°C, or a change in sign in the Joule-Thomson coefficient at a certain temperature. { in'vər·zhən }

inversion axis *See* rotation-inversion axis. { in'vər·zhən ,ak·səs }

inversion ratio [PHYS] The negative of the ratio in a maser medium of the difference in populations between two nondegenerate energy states under a condition of population inversion, to the population difference at equilibrium. { in'vər·zhən ,rā·shō }

inversion spectrum [SPECT] Lines in the microwave spectra of certain molecules (such as ammonia) which result from the quantum-mechanical analog of an oscillation of the molecule between two configurations which are mirror images of each other. { in'vər·zhən ,spek·trəm }

inversion symmetry [PHYS] The principle that the laws of physics are unchanged by the operation of inversion; it is violated by the weak interactions. { in'vər·zhən 'sim·ə·trē }

inversion temperature [THERMO] The temperature at which the Joule-Thomson effect of a gas changes sign. { in'vər·zhən ,tem·prə·chər }

inverted image [OPTICS] An image in which up and down, as well as left and right, are interchanged; that is, an image that results from rotating the object 180° about a line from the object to the observer; such images are formed by most astronomical telescopes. Also known as reversed image. { in'vərd·əd 'im·ij }

inverted L antenna [ELECTROMAG] An antenna consisting of one or more horizontal wires to which a connection is made by means of a vertical wire at one end. { in'vərd·əd ¦el an,ten·ə }

inverted microscope [OPTICS] A microscope in which the body of the microscope, including the objective and the ocular, are below the stage, the illumination for transmitted light is above the stage, and with opaque materials, the vertical illuminator is used under the stage near the objective. { in'vərd·əd 'mī·krə,skōp }

inverted pseudoplastic fluid *See* dilatant fluid. { in¦vərd·əd ,süd·ō¦plas·tik 'flü·əd }

inverted vee [ELECTROMAG] **1.** A directional antenna consisting of a conductor which has the form of an inverted V, and which is fed at one end and connected to ground through an appropriate termination at the other. **2.** A center-fed horizontal dipole antenna whose arms have ends bent downward 45°. { in'vərd·əd 'vē }

inverting prism *See* erecting prism. { in'vərd·iŋ 'priz·əm }

inverting telescope [OPTICS] A telescope that inverts the usual telescopic image, allowing the object to be seen right side up. { in'vərd·iŋ 'tel·ə,skōp }

inviscid flow [FL MECH] Flow of an inviscid fluid. Also known as frictionless flow; ideal flow; nonviscous flow. { in'vis·əd 'flō }

inviscid fluid [FL MECH] A fluid which has no viscosity; it therefore can support no shearing stress, and flows without energy dissipation. Also known as ideal fluid; nonviscous fluid; perfect fluid. { in'vis·əd 'flü·əd }

invisible image [PHYS] An image having a form which cannot be perceived by unaided vision, such as a latent image on a photographic emulsion. { in'viz·ə·bəl 'im·ij }

Io *See* ionium. { 'ī,ō }

iodine-131 [NUC PHYS] A radioactive, artificial isotope of iodine, mass number 131; its half-life is 8 days with beta and gamma radiation; used in medical and industrial radioactive tracer work; moderately radiotoxic. { 'ī·ə,dīn ¦wən¦thərd·ē'wən }

Ioffe bars [PL PHYS] Heavy current-carrying bars that are used to increase plasma stability in some types of controlled fusion reactor. { 'yäf·ē ,bärz }

Ioffe effect [SOLID STATE] An effect in which the simultaneous exposure of an ionic crystal such as rock salt to a mechanical stress and a solvent results in an increase in its plasticity. { 'yäf·ē i,fekt }

ion-acoustic wave [PL PHYS] A longitudinal compression wave in the ion density of a plasma which can occur at high electron temperatures and low frequencies, caused by a combination of ion inertia and electron pressure. { ¦ī,än ə¦kü·stik 'wāv }

ion backscattering [SOLID STATE] Large-angle elastic scattering of monoenergetic ions in a beam directed at a metallized film on silicon or some other thin multilayer system. { 'ī,än 'bak,skad·ə·riŋ }

ion concentration *See* ion density. { 'ī,än ,kän·sən'trā·shən }

ion crystal *See* Coulomb crystal. { 'ī,än ,krist·əl }

ion current [PHYS] The electric current resulting from motion of ions. { 'ī,än ,kə·rənt }

ion cyclotron frequency [ELECTROMAG] The angular frequency of the motion of an ion in a uniform magnetic field in a plane perpendicular to the field. { 'ī,än 'si·klə,trän ,frē·kwən·sē }

ion-cyclotron-resonance mass spectrometer [SPECT] A device for detecting and measuring the mass distribution of ions orbiting in an applied magnetic field, either by applying a constant radio-frequency signal and varying the magnetic field to bring ion frequencies equal to the applied radio frequency sequentially into resonance, or by rapidly varying the radio frequency and applying Fourier transform techniques. { 'ī,än 'sī·klə,trän 'rez·ən·əns 'mas spek'träm·əd·ər }

ion density [PHYS] The number of ions per unit volume. Also known as ion concentration. { 'ī,än ,den·səd·ē }

ion emission [PHYS] The ejection of ions from the surface of a substance into the surrounding space. { 'ī,än i,mish·ən }

ionic charge [PHYS] **1.** The total charge of an ion. **2.** The charge of an electron; the charge of any ion is equal to this electron charge in magnitude, or is an integral multiple of it. { ī'än·ik 'chärj }

ionic conduction [SOLID STATE] Electrical conduction of a solid due to the displacement of ions within the crystal lattice. { ī'än·ik kən'dək·shən }

ionic conductivity [SOLID STATE] The portion of the electrical conductivity of a solid that results from ionic conduction. { i'än·ik ,kän,dək'tiv·əd·ē }

ionic crystal [CRYSTAL] A crystal in which the lattice-site occupants are charged ions held together primarily by their electrostatic interaction. { ī'än·ik 'krist·əl }

ionic lattice [CRYSTAL] The lattice of an ionic crystal. { ī'än·ik 'lad·əs }

ionic mobility [PHYS] The ratio of the average drift velocity of an ion in a liquid or gas to the electric field. { ī'än·ik mō'bil·əd·ē }

ionic semiconductor [SOLID STATE] A solid whose electrical conductivity is due primarily to the movement of ions rather than that of electrons and holes. { ī'än·ik ¦sem·i·kən¦dək·tər }

ionic solid [SOLID STATE] A solid made up of ions held together primarily by their electrostatic interaction. { i'än·ik 'säl·əd }

ion irradiation [PHYS] Bombardment of a substance by high-velocity ions. { 'ī,än i,rād·ē'ā·shən }

ionium [NUC PHYS] A naturally occurring radioisotope, symbol Io, of thorium, atomic weight 230. { ī'ō·nē·əm }

ionization cross section [PHYS] The cross section for a particle or photon to undergo a collision with an atom, thus removing or adding one or more electrons to the atom. { ,ī·ə·nə'zā·shən 'krȯs ¦sek·shən }

ionization energy [ATOM PHYS] The amount of energy needed to remove an electron from a given kind of atom or molecule to an infinite distance; usually expressed in electron volts, and numerically equal to the ionization potential in volts. { ,ī·ə·nə'zā·shən ¦en·ər·jē }

ionization potential [ATOM PHYS] The energy per unit charge needed to remove an electron from a given kind of atom or molecule to an infinite distance; usually expressed in volts. Also known as ion potential. { ,ī·ə·nə'zā·shən pə'ten·chəl }

ionization temperature [STAT MECH] The temperature at which the average kinetic energy of gas molecules having a Maxwell distribution equals the ionization energy. { ,ī·ə·nə'zā·shən ,tem·prə·chər }

ionized gas [PHYS] A gas, some of whose atoms or molecules have undergone ionization. { 'ī·ə,nīzd 'gas }

ionizing event [PHYS] Any occurrence in which an ion or group of ions is produced; for example, by passage of charged particles through matter. { 'ī·ə,niz·iŋ i,vent }

ion kinetic energy spectrometry [SPECT] A spectrometric technique that uses a beam of ions of high kinetic energy passing through a field-free reaction chamber from which ionic products are collected and energy analyzed; it is a generalization of metastable ion studies in which both unimolecular and bimolecular reactions are considered. { 'ī,än ki¦ned·ik 'en·ər·jē spek'träm·ə·trē }

ion laser [OPTICS] A gas laser in which stimulated emission takes place between two energy levels of an ion; gases used include argon, krypton, neon, and xenon; examples include helium-cadmium lasers and metal vapor lasers. { 'ī,än ,lā·zər }

ion migration [ELEC] Movement of ions produced in an electrolyte, semiconductor, and so on, by the application of an electric potential between electrodes. { 'ī,än mī'grā·shən }

ion potential *See* ionization potential. { 'ī,än pə,ten·chəl }

ion scattering spectroscopy [SPECT] A spectroscopic technique in which a low-energy (about 1000 electronvolts) beam of inert-gas ions in directed at a surface, and the energies and scattering angles of the scattered ions are used to identify surface atoms. Abbreviated ISS. { ¦ī,än ¦skad·ə·riŋ spek'träs·kə·pē }

ion-solid interaction [SOLID STATE] An atomic process that occurs as a result of the collision of energetic ions, atoms, or molecules with condensed matter. { 'ī,än ¦säl·əd ,in·tər'ak·shən }

irdome [OPTICS] A dome used to protect an infrared detector and its optical elements, generally made from quartz, silicon, germanium, sapphire, calcium aluminate, or other material having high transparency to infrared radiation. Derived from infrared dome. { ¦ī'är,dōm }

IR drop *See* resistance drop. { ¦ī¦är 'dräp }

iridescence [OPTICS] A rainbow color effect exhibited in various bodies as a result of interference in a thin film (as of soap bubbles or mother of pearl) or of diffraction of light reflected from a ribbed surface (as of the plumage of some birds). { ,ir·ə'des·əns }

iridium-192 [NUC PHYS] Radioactive isotope of iridium with a 75-day half-life; β and γ radiation; used in cancer treatment and for radiography of light metal castings. { i'rid·ē·əm ¦wən¦nīn·tē'tü }

iris [ELECTROMAG] A conducting plate

mounted across a waveguide to introduce impedance; when only a single mode can be supported, an iris acts substantially as a shunt admittance and may be used for matching the waveguide impedance to that of a load. Also known as diaphragm; waveguide window. [OPTICS] A circular mechanical device, whose diameter can be varied continuously, which controls the amount of light reaching the film of a camera. Also known as iris diaphragm. { 'ī·rəs }

iron-55 [NUC PHYS] Radioactive isotope of iron, symbol ^{55}Fe, with a 2.91-year half-life; highly toxic. { 'ī·ərn ¦fif·tē'fīv }

iron-59 [NUC PHYS] Radioactive isotope of iron, symbol ^{59}Fe, 46.3-day half-life; β and γ radiation; highly toxic; used to study metallic welds, corrosion mechanisms, engine wear, and bodily functions. { 'ī·ərn ¦fif·tē'nīn }

iron core [ELECTROMAG] A core made of solid or laminated iron, or some other magnetic material which may contain very little iron. { 'ī·ərn 'kōr }

iron-core choke *See* iron-core coil. { 'ī·ərn ¦kór 'chōk }

iron-core coil [ELECTROMAG] A coil in which solid or laminated iron or other magnetic material forms part or all of the magnetic circuit linking its winding. Also known as iron-core choke; magnet coil. { 'ī·ərn ¦kór 'kóil }

iron-core transformer [ELECTROMAG] A transformer in which laminations of iron or other magnetic material make up part or all of the path for magnetic lines of force that link the transformer windings. { 'ī·ərn ¦kór tranz'fōr·mər }

iron-dust core [ELECTROMAG] A core made by mixing finely powdered magnetic material with an insulating binder and molding under pressure to form a rod-shaped core that can be moved into or out of a coil or transformer to vary the inductance or degree of coupling for tuning purposes. { 'ī·ərn ˌdəst ˌkōr }

iron loss *See* core loss. { 'ī·ərn ˌlós }

irradiance *See* radiant flux density. { i'rād·ē·əns }

irradiation [OPTICS] An optical illusion which makes bright objects appear larger than they really are. { iˌrād·ē'ā·shən }

irrationality of dispersion [OPTICS] The effect whereby spectra produced by prisms of different types of glass are not geometrically similar. { iˌrash·ə¦nal·əd·ē əv di'spər·zhən }

irreversible energy loss [THERMO] Energy transformation process in which the resultant condition lacks the driving potential needed to reverse the process; the measure of this loss is expressed by the entropy increase of the system. { ˌi·ri'vər·sə·bəl 'en·ər·je ˌlós }

irreversible process [THERMO] A process which cannot be reversed by an infinitesimal change in external conditions. { ˌi·ri'vər·sə·bəl 'prä·səs }

irreversible thermodynamics *See* nonequilibrium thermodynamics. { ˌi·ri'vər·sə·bəl ¦thər·mə·dī'nam·iks }

irrotational flow [FL MECH] Fluid flow in which the curl of the velocity function is zero everywhere, so that the circulation of the velocity about any closed curve vanishes. Also known as acyclic motion; irrotational motion. { ¦ir·ə'tā·shən·əl 'flō }

irrotational motion *See* irrotational flow. { ¦ir·ə'tā·shən·əl 'mō·shən }

irrotational wave *See* compressional wave. { ¦ir·ə'tā·shən·əl 'wāv }

isenergic flow [THERMO] Fluid flow in which the sum of the kinetic energy, potential energy, and enthalpy of any part of the fluid does not change as that part is carried along with the fluid. { ¦ī·səˌnər·jik 'flō }

isenthalpic expansion [THERMO] Expansion which takes place without any change in enthalpy. { ¦īs·ən¦thal·mik ik'span·chən }

isenthalpic process [THERMO] A process that is carried out at constant enthalpy. { ˌī·sən¦thal·pik 'präˌses }

isentrope [THERMO] A line of equal or constant entropy. { 'īs·ənˌtrōp }

isentropic [THERMO] Having constant entropy; at constant entropy. { ¦īs·ən¦träp·ik }

isentropic compression [THERMO] Compression which occurs without any change in entropy. { ¦īs·ən¦träp·ik kəm'presh·ən }

isentropic expansion [THERMO] Expansion which occurs without any change in entropy. { ¦īs·ən'träp·ik ik'span·chən }

isentropic flow [THERMO] Fluid flow in which the entropy of any part of the fluid does not change as that part is carried along with the fluid. { ¦īs·ən'träp·ik 'flō }

isentropic process [THERMO] A change that takes place without any increase or decrease in entropy, such as a process which is both reversible and adiabatic. { ¦īs·ən'träp·ik 'prä·ses }

Ising coupling [SOLID STATE] A model of coupling between two atoms in a lattice, used to study ferromagnetism, in which the spin component of each atom along some axis is taken to be +1 or −1, and the energy of interaction is proportional to the negative of the product of the spin components along this axis. { 'ī·ziŋ ˌkəp·liŋ }

Ising model [SOLID STATE] A crude model of a ferromagnetic material or an analogous system, used to study phase transitions, in which atoms in a one-, two-, or three-dimensional lattice interact via Ising coupling between nearest neighbors, and the spin components of the atoms are coupled to a uniform magnetic field. { 'ī·ziŋ ˌmäd·əl }

isobar [NUC PHYS] One of two or more nuclides having the same number of nucleons in their nuclei but differing in their atomic numbers and chemical properties. [PHYS] **1.** A line connecting points of equal pressure along a given surface in a physical system. **2.** A line connecting points of equal pressure on a graph plotting thermodynamic variables. { 'ī·səˌbär }

isobaric [THERMO] Of equal or constant pressure, with respect to either space or time. { ¦ī·sə¦bär·ik }

isobaric analog states *See* analog states. { ¦ī·sə¦bär·ik 'an·ə,läg ,stāts }

isobaric process [THERMO] A thermodynamic process of a gas in which the heat transfer to or from the gaseous system causes a volume change at constant pressure. { ¦ī·sə¦bär·ik 'prä·səs }

isobaric spin *See* isotopic spin. { ¦ī·sə¦bär·ik 'spin }

isocandle diagram [OPTICS] A diagram showing the distribution of light from a lighting system in various directions by means of contours connecting directions of equal luminous intensity, projected in a suitable manner. { ¦ī·sə¦kan·dəl 'dī·ə,gram }

isochor *See* isochore. { 'ī·sə,kȯr }

isochoric [PHYS] Taking place without change in volume. Also known as isovolumic. { ,ī·sə'kȯr·ik }

isochromatic [OPTICS] **1.** Pertaining to a variation of certain quantities related to light (such as density of the medium through which the light is passing, index of refraction), in which the color or wavelength of the light is held constant. **2.** Pertaining to lines connecting points of the same color. { ¦ī·sō·krə'mad·ik }

isochromatic fringe pattern [OPTICS] A pattern of bands, each of uniform color, observed when a plate is placed in a polariscope and subjected to stress, making it birefringent. { ¦ī·sō·krə'mad·ik 'frinj ,pad·ərn }

isochrone [PHYS] A line on a chart connecting all points having the same time of occurrence of particular phenomena or of a particular value of a quantity. { 'ī·sə,krōn }

isochronism [MECH] The property of having a uniform rate of operation or periodicity, for example, of a pendulum or watch balance. { ī'sä·krə,niz·əm }

isochronous [PHYS] Having a fixed frequency or period. { ī'sä·krə·nəs }

isochronous circuits [ELEC] Circuits having the same resonant frequency. { ī'sä·krə·nəs 'sər·kəts }

isocirculator [ELECTROMAG] A circulator that has an absorber in one of its terminals and thereby acts as an isolator. { ,ī·sō'sər·kyə,lād·ər }

isoclinic line [SOLID STATE] A line joining points in a plate at which the principal stresses have parallel directions. { ¦ī·sə¦klin·ik 'līn }

isodesmic structure [SOLID STATE] An ionic crystal structure in which all bonds are of the same strength, so that no distinct groups of atoms are formed. { ¦ī·sə¦dez·mik 'strək·chər }

isodiapheres [NUC PHYS] Nuclides which have the same difference in the number of neutrons and protons. { ,ī·sə'dī·ə,firz }

isodynamic [MECH] Pertaining to equality of two or more forces or to constancy of a force. { ¦ī·sō·dī'nam·ik }

isoelectric [ELEC] Pertaining to a constant electric potential. { ¦ī·sō·i'lek·trik }

isoelectronic [ATOM PHYS] Pertaining to atoms having the same number of electrons outside the nucleus of the atom. { ¦ī·sō·i,lek'trän·ik }

isoelectronic sequence [SPECT] A set of spectra produced by different chemical elements ionized so that their atoms or ions contain the same number of electrons. { ¦ī·sō·i,lek'trän·ik 'sē·kwəns }

isofootcandle *See* isolux. { ¦ī·sō'fu̇t,kand·əl }

isogyre [OPTICS] A dark band in an interference figure located at those points that correspond to directions of transmission through the crystal plate in which the polarization of the incident light is not affected by passing through the plate. { ¦ī·sə'jīr }

isolate [ELEC] To disconnect a circuit or piece of equipment from an electric supply system. { 'ī·sə,lāt }

isolux [OPTICS] A curve or surface connecting points at which light intensity is the same. Also known as isofootcandle; isophot. { 'ī·sə,ləks }

isomer [NUC PHYS] One of two or more nuclides having the same mass number and atomic number, but existing for measurable times in different quantum states with different energies and radioactive properties. Also known as nuclear isomer. { 'ī·sə·mər }

isomeric transition [NUC PHYS] A radioactive transition from one nuclear isomer to another of lower energy. { ¦ī·sə¦mer·ik tran'zish·ən }

isomerism [NUC PHYS] The occurrence of nuclear isomers. { ī'säm·ə,riz·əm }

isometric *See* isochore. { ¦ī·sə'me·trik }

isometric process [THERMO] A constant-volume, frictionless thermodynamic process in which the system is confined by mechanically rigid boundaries. { ¦ī·sə'me·trik 'prä·səs }

isometric system [CRYSTAL] The crystal system in which the forms are referred to three equal, mutually perpendicular axes. Also known as cubic system. { ¦ī·sə'me·trik 'sis·təm }

isophot *See* isolux. { 'ī·sə,fät }

isophotometer [OPTICS] A direct-recording photometer that automatically scans and measures optical density of all points in a film transparency or plate, and plots the measured density values in a quantitative two-dimensional isodensity tracing of the scanned areas. { ¦ī·sō·fə'täm·əd·ər }

isopiestic [PHYS] Denoting equal or constant pressure. { ¦ī·sə'pī·es·tik }

isopycnic [PHYS] Of equal or constant density, with respect to either space or time. { ¦ī·sō¦pik·nik }

isospin *See* isotopic spin. { 'ī·sə,spin }

isospin multiplet [PART PHYS] A collection of elementary particles which have approximately the same mass and the same quantum numbers except for charge, but have a sequence of charge values, $(Y/2) - I, (Y/2) - I + 1, \ldots, (Y/2) + I$ times the proton charge, where Y is an integer known as the hypercharge, and I is an integer or half-integer known as the isospin; examples are the pions ($Y = 0$, $I = 1$) and the nucleons ($Y = 1$, $I = 1/2$). Also known as charge multiplet; particle multiplet. { 'ī·sə,spin 'məl·tə·plət }

isostatics [MECH] In photoelasticity studies of stress analyses, those curves, the tangents to which represent the progressive change in principal-plane directions. Also known as stress trajectories. Also known as stress lines. { ¦ī·sə'stad·iks }

isostatic surface [MECH] A surface in a three-dimensional elastic body such that at each point of the surface one of the principal planes of stress at that point is tangent to the surface. { ¦ī·sə'stad·ik 'sər·fəs }

isosteric [PHYS] Of equal or constant specific volume with respect to either time or space. { ¦ī·sə¦ster·ik }

isostructural [CRYSTAL] Pertaining to crystalline materials that have corresponding atomic positions, and have a considerable tendency for ionic substitution. { ¦ī·sō'strək·chə·rəl }

isotherm [THERMO] A curve or formula showing the relationship between two variables, such as pressure and volume, when the temperature is held constant. Also known as isothermal. { 'ī·sə,thərm }

isothermal [THERMO] **1.** Having constant temperature; at constant temperature. **2.** *See* isotherm. { ¦ī·sə¦thər·məl }

isothermal calorimeter [THERMO] A calorimeter in which the heat received by a reservoir, containing a liquid in equilibrium with its solid at the melting point or with its vapor at the boiling point, is determined by the change in volume of the liquid. { ¦ī·sə¦thər·məl ,kal·ə'rim·əd·ər }

isothermal compression [THERMO] Compression at constant temperature. { ¦ī·sə¦thər·məl kəm'presh·ən }

isothermal equilibrium [THERMO] The condition in which two or more systems are at the same temperature, so that no heat flows between them. { ¦ī·sə¦thər·məl ,ē·kwə'lib·rē·əm }

isothermal expansion [THERMO] Expansion of a substance while its temperature is held constant. { ¦ī·sə¦thər·məl ik'span·chən }

isothermal flow [THERMO] Flow of a gas in which its temperature does not change. { ¦ī·sə¦thər·məl 'flō }

isothermal layer [THERMO] A layer of fluid, all points of which have the same temperature. { ¦ī·sə¦thər·məl 'lā·ər }

isothermal magnetization [THERMO] Magnetization of a substance held at constant temperature; used in combination with adiabatic demagnetization to produce temperatures close to absolute zero. { ¦ī·sə¦thər·məl ,mag·nə·tə'zā·shən }

isothermal process [THERMO] Any constant-temperature process, such as expansion or compression of a gas, accompanied by heat addition or removal from the system at a rate just adequate to maintain the constant temperature. { ¦ī·sə¦thər·məl 'prä·səs }

isothermal transformation [THERMO] Any transformation of a substance which takes place at a constant temperature. { ¦ī·sə¦thər·məl ,tranz·fər'mā·shən }

isotone [NUC PHYS] One of several nuclides having the same number of neutrons in their nuclei but differing in the number of protons. { 'ī·sə,tōn }

isotope [NUC PHYS] One of two or more atoms having the same atomic number but different mass number. { 'ī·sə,tōp }

isotope abundance [NUC PHYS] The ratio of the number of atoms of a particular isotope in a sample of an element to the number of atoms of a specified isotope, or to the total number of atoms of the element. { 'ī·sə,tōp ə¦bən·dəns }

isotope effect [SOLID STATE] Variation of the transition temperatures of the isotopes of a superconducting element in inverse proportion to the square root of the atomic mass. { 'ī·sə,tōp i,fekt }

isotope shift [SPECT] A displacement in the spectral lines due to the different isotopes of an element. { 'ī·sə,tōp ,shift }

isotopic element [NUC PHYS] An element which has more than one naturally occurring isotope. { ¦ī·sə¦täp·ik 'el·ə·mənt }

isotopic incoherence [PHYS] Incoherence in the scattering of neutrons from a crystal lattice due to differences in scattering lengths of different isotopes of the same element. { ¦ī·sə¦täp·ik ,iŋ·kō'hir·əns }

isotopic number *See* neutron excess. { ¦ī·sə¦täp·ik 'nəm·bər }

isotopic parity *See* G parity. { ¦ī·sə¦täp·ik 'par·əd·ē }

isotopic spin [NUC PHYS] A quantum-mechanical variable, resembling the angular momentum vector in algebraic structure whose third component distinguished between members of groups of elementary particles, such as the nucleons, which apparently behave in the same way with respect to strong nuclear forces, but have different charges. Also known as isobaric spin; isospin; i-spin. { ¦ī·sə¦täp·ik 'spin }

isotropic [PHYS] Having identical properties in all directions. { ¦ī·sə¦trä·pik }

isotropic antenna *See* unipole. { ¦ī·sə¦trä·pik an'ten·ə }

isotropic dielectric [ELEC] A dielectric whose polarization always has a direction that is parallel to the applied electric field, and a magnitude which does not depend on the direction of the electric field. { ¦ī·sə¦trä·pik ,dī·ə'lek·trik }

isotropic fluid [FL MECH] A fluid whose properties are not dependent on the direction along which they are measured. { ¦ī·sə¦trä·pik 'flü·əd }

isotropic flux [PHYS] Radiation, or a flow of particles or matter, which reaches a location from all directions with equal intensity. { ¦ī·sə¦trä·pik 'fləks }

isotropic gain of an antenna *See* absolute gain of an antenna. { ¦ī·sə¦trä·pik 'gān əv ən an'ten·ə }

isotropic material [PHYS] A material whose properties are not dependent on the direction along which they are measured. { ¦ī·sə¦trä·pik mə'tir·ē·əl }

isotropic noise [ELECTROMAG] Random noise radiation which reaches a location from all directions with equal intensity. { ¦ī·sə¦trä·pik 'nȯiz }

isotropic plasma [PL PHYS] A plasma whose properties, such as pressure, are not dependent on the direction along which they are measured. { ¦ī·sə¦trä·pik 'plaz·mə }

isotropic radiation [ELECTROMAG] Radiation which is emitted by a source in all directions with equal intensity, or which reaches a location from all directions with equal intensity. { ¦ī·sə¦trä·pik ˌrād·ē'ā·shən }

isotropic radiator [PHYS] An energy source that radiates uniformly in all directions. { ¦ī·sə¦trä·pik 'rād·ēˌād·ər }

isotropic turbulence [FL MECH] Turbulence whose properties, especially statistical correlations, do not depend on direction. { ¦ī·sə¦trä·pik 'tər·byə·ləns }

isotropy [PHYS] The quality of a property which does not depend on the direction along which it is measured, or of a medium or entity whose properties do not depend on the direction along which they are measured. { ī'sä·trə·pē }

isotypic [CRYSTAL] Pertaining to a crystalline substance whose chemical formula is analogous to, and whose structure is like, that of another specified compound. { ¦ī·sə¦tip·ik }

isovolumic *See* isochoric. { ¦ī·sō¦väl·yə·mik }

i-spin *See* isotopic spin. { 'ī ˌspin }

Israel's theorem [RELAT] A theorem of general relativity essentially proving that the Schwarzschild solution is the unique solution of Einstein's equations describing nonrotating black holes in empty space and that the Reissner-Nordstrom solution is the unique solution describing nonrotating charged black holes. { 'iz·rē·əlz ˌthir·əm }

ISS *See* ion scattering spectroscopy.

IT calorie *See* calorie. { ¦ī'tē ˌkal·ə·rē }

ITS-90 *See* international temperature scale.

i-type semiconductor *See* intrinsic semiconductor. { 'ī ˌtīp ¦sem·i·kən'dək·tər }

Ixion [PHYS] An experimental magnetic-mirror device used for research on controlled fusion; involves study of plasma rotation in a magnetic-mirror confinement system using crossed electric and magnetic fields. { 'ik·sē·ən }

J

J *See* joule.

Jaccarino-Peter effect [SOLID STATE] The production of superconductivity in certain ferromagnetic metals through the application of an external magnetic field that compensates for the polarization of the conduction electrons. Also known as compensation effect. { ˌjak·əˈrē·nō ˈpēd·ər iˌfekt }

Jaeger method [FL MECH] A method of determining surface tension of a liquid in which one measures the pressure required to cause air to flow from a capillary tube immersed in the liquid. { ˈyā·gər ˌmeth·əd }

Jaeger-Steinwehr method [THERMO] A refinement of the Griffiths method for determining the mechanical equivalent of heat, in which a large mass of water, efficiently stirred, is used, the temperature rise of the water is small, and the temperature of the surroundings is carefully controlled. { ˈyā·gər ˈshtīn·ver ˌmeth·əd }

Jamin effect [FL MECH] Resistance to flow of a column of liquid divided by air bubbles in a capillary tube, even when subjected to a substantial pressure difference between the ends of the tube. { jəˈmēn iˌfekt }

Jamin refractometer [OPTICS] An instrument for measuring the index of refraction of a gas in which two light beams from a common source are each passed through an evacuated tube and recombined, and the displacement of interference fringes is noted as gas is slowly admitted into one of the tubes. { jəˈmēn ˌrēˌfrakˈtäm·əd·ər }

J antenna [ELECTROMAG] Antenna having a configuration resembling a J, consisting of a half-wave antenna end-fed by a parallel-wire quarter-wave section. { ˈjā antˌen·ə }

jar [ELEC] A unit of capacitance equal to 1000 statfarads, or approximately 1.11265×10^{-9} farad; it is approximately equal to the capacitance of a Leyden jar; this unit is now obsolete. { jär }

Jeans viscosity equation [THERMO] An equation which states that the viscosity of a gas is proportional to the temperature raised to a constant power, which is different for different gases. { ˈjēnz viˈskäs·əd·ē iˌkwā·zhən }

jellium model [SOLID STATE] A model of electron-electron interactions in a metal in which the positive charge associated with the ion cores immersed in the sea of conduction electrons is replaced by a uniform positive background charge terminating along a plane that represents the surface of the metal. { ˈjel·ē·əm ˌmäd·əl }

jerk [MECH] **1.** The rate of change of acceleration; it is the third derivative of position with respect to time. **2.** A unit of rate of change of acceleration, equal to 1 foot (30.48 centimeters) per second squared per second. { jərk }

jet [FL MECH] A strong, well-defined stream of compressible fluid, either gas or liquid, issuing from an orifice or nozzle or moving in a contracted duct. [PART PHYS] A group of particles issuing in approximately the same direction from a high-energy collision of elementary particles, believed to consist of decay products of a member of a quark-antiquark pair created in the collision. { jet }

jet tones [ACOUS] Unsteady tones produced when a stream of air issues into still air from an orifice. { ˈjet ˌtōnz }

J factor [THERMO] A dimensionless equation used for the calculation of free convection heat transmission through fluid films. { ˈjā ˌfak·tər }

j-j coupling [ATOM PHYS] A process for building up many-electron wave functions; the spin and orbital functions of each particle are combined to form eigenfunctions of the particle's total angular momentum, and then the wave functions of all the particles are combined to form eigenfunctions of the total angular momentum of the system; this coupling is used when the spin-orbit interaction is strong compared to the electrostatic interaction. { ¦jā¦jā ˌkəp·liŋ }

jog [CRYSTAL] A shift in a dislocation from one crystal plane to another. { jäg }

Johann crystal geometry [CRYSTAL] The focusing shape of a diffracting crystal for x-ray dispersion used in electron-probe microanalysis; less stringent than Johannson crystal geometry. { ˈyōˌhän ¦krist·əl jēˈäm·ə·trē }

Johannson crystal geometry [CRYSTAL] The full-focusing shape of a diffracting crystal for x-ray dispersion used in electron-probe microanalyzers; more stringent than Johann crystal geometry. { jōˈhan·sən ¦krist·əl jēˈäm·ə·trē }

Johnson and Lark-Horowitz formula [SOLID

STATE] A formula according to which the resistivity of a metal or degenerate semiconductor resulting from impurities which scatter the electrons is proportional to the cube root of the density of impurities. { 'jän·sən ən ¦lärk 'här·ə,witz ,fȯr·myə·lə }

Johnson-Rahbeck effect [PHYS] An increase in frictional force between two electrodes in contact with a semiconductor that arises when a potential difference is applied between the electrodes. { 'jän·sən 'rä,bek i,fekt }

Joly photometer [OPTICS] A photometer consisting of two equal paraffin wax or opal glass blocks separated by a thin opaque sheet; the positions of two light sources under comparison are adjusted until the two blocks appear equally bright. Also known as wax-block photometer. { ¦jäl·ē fō'täm·əd·ər }

Jordan lag [ELECTROMAG] A type of magnetic viscosity in which the angular lag of the magnetic induction behind a sinusoidally varying magnetic field strength, and also the energy loss per cycle, is independent of frequency. { 'jȯrd·ən ,lag }

Jordan-Wigner commutation rules [QUANT MECH] Rules obtained by replacing commutators of creation and destruction operators by anticommutators; applicable to fermion fields in a quantized field theory. { 'jȯrd·ən 'wig·nər ,käm·yə'tā·shən rülz }

Josephson constant [PHYS] **1.** The quantity $K_J = 2e/h$, which appears in the equations for the alternating-current Josephson effect, where e is the magnitude of the charge of the electron and h is Planck's constant. **2.** The conventional value of this quantity adopted by international agreement on January 1, 1990, to establish a standard for the volt, K_{J-90} = 493,597.9 gigahertz per volt. { 'jō·sef·sən ,kän·stənt }

Josephson current [CRYO] The current across a Josephson junction in the absence of voltage across the junction, resulting from the Josephson effect. { 'jō·səf·sən ,kə·rənt }

Josephson effect [CRYO] The tunneling of electron pairs through a thin insulating barrier between two superconducting materials. Also known as Josephson tunneling. { 'jō·səf·sən i,fekt }

Josephson equation [CRYO] An equation according to which the Josephson current is a sinusoidally varying function of the applied magnetic field. { 'jō·səf·sən i,kwā·zhən }

Josephson junction [CRYO] A thin insulator separating two superconducting materials; it displays the Josephson effect. { 'jō·səf·sən ,jəŋk·shən }

Josephson penetration depth [CRYO] A measure of the distance that a magnetic field extends into a Josephson junction. { 'jō·səf·sən 'pen·ə'trā·shən ,depth }

Josephson tunneling *See* Josephson effect. { 'jō·səf·sən ,tən·əl·iŋ }

Joukowski profile [FL MECH] An airfoil profile with a cuspshaped trailing edge, resulting from the Joukowski transformation of a circle which passes through the point $z = a$ and which is located so that the point $z = -a$ does not lie outside the circle. { yü'kəf·skē ,prō·fīl }

Joukowski transformation [FL MECH] A conformal mapping used to transform circles into airfoil profiles for the purpose of studying fluid flow past the airfoil profiles; it assigns to each complex number z the number $w = z + (a^2/z)$. { yü'kəf·skē ,tranz·fər,mā·shən }

joule [MECH] The unit of energy or work in the meter-kilogram-second system of units, equal to the work done by a force of 1 newton magnitude when the point at which the force is applied is displaced 1 meter in the direction of the force. Symbolized J. Also known as newton-meter of energy. { jül *or* jaủl }

Joule and Playfairs' experiment [THERMO] An experiment in which the temperature of the maximum density of water is measured by taking the mean of the temperatures of water in two columns whose densities are determined to be equal from the absence of correction currents in a connecting trough. { ¦jül and 'plā,fārz ik,sper·ə·mənt }

Joule-Clausius velocity [STAT MECH] A quantity used in the description of the kinetic behavior of a gas, equal to the square root of the ratio of the pressure of the gas to one-third of its density. { 'jül 'klaủz·ē·əs və,läs·əd·ē }

Joule cycle *See* Brayton cycle. { 'jül ,sī·kəl }

Joule effect [PHYS] **1.** The heating effect produced by the flow of current through a resistance. **2.** A change in the length of a ferromagnetic substance which occurs parallel to an applied magnetic field. Also known as Joule magnetorestriction; longitudinal magnetorestriction. { 'jül i,fekt }

Joule equivalent [THERMO] The numerical relation between quantities of mechanical energy and heat; the present accepted value is 1 fifteen-degrees calorie equals 4.1855 ± 0.0005 joules. Also known as mechanical equivalent of heat. { 'jül i,kwiv·ə·lənt }

Joule experiment [THERMO] **1.** An experiment to detect intermolecular forces in a gas, in which one measures the heat absorbed when gas in a small vessel is allowed to expand into a second vessel which has been evacuated. **2.** An experiment to measure the mechanical equivalent of heat, in which falling weights cause paddles to rotate in a closed container of water whose temperature rise is measured by a thermometer. { 'jül ik,sper·ə·mənt }

Joule heat [ELEC] The heat which is evolved when current flows through a medium having electrical resistance, as given by Joule's law. { 'jül ,hēt }

Joule-Kelvin effect *See* Joule-Thomson effect. { 'jül 'kel·vən i,fekt }

Joule magnetorestriction *See* Joule effect. { 'jül mag¦ned·ō·ri'strik·shən }

Joule's law [ELEC] The law that when electricity flows through a substance, the rate of evolution of heat in watts equals the resistance of the substance in ohms times the square of the current in amperes. [THERMO] The law that at constant temperature the internal energy of a gas tends to a finite limit, independent of volume, as the pressure tends to zero. { 'jülz ,lȯ }

Joule-Thomson coefficient [THERMO] The ratio of the temperature change to the pressure change of a gas undergoing isenthalpic expansion. { 'jül 'täm·sən ,kō·ə,fish·ənt }

Joule-Thomson effect [THERMO] A change of temperature in a gas undergoing Joule-Thomson expansion. Also known as Joule-Kelvin effect. { 'jül 'täm·sən i,fekt }

Joule-Thomson expansion [THERMO] The adiabatic, irreversible expansion of a fluid flowing through a porous plug or partially opened valve. Also known as Joule-Thomson process. { 'jül 'täm·sən ik,span·chən }

Joule-Thomson inversion temperature [THERMO] A temperature at which the Joule-Thomson coefficient of a given gas changes sign. { ¦jül ¦täm·sən in'vər·zhən ,tem·prə·chər }

Joule-Thomson process *See* Joule-Thomson expansion. { 'jül 'täm·sən ,prä·səs }

Joule-Thomson valve [CRYO] A valve through which a gas is allowed to expand adiabatically, resulting in lowering of its temperature; used in production of liquid hydrogen and helium. { 'jül 'täm·sən ,valv }

J particle [PART PHYS] A neutral meson which has a mass of 3095 megaelectronvolts, spin quantum number 1, and negative parity and charge parity; it has an anomalously long lifetime of approximately 10^{-20} second (corresponding to a width of approximately 70 kiloelectronvolts). Also known as psi particle (symbolized ψ). { 'jā ,pärd·ə·kəl }

junction [ELEC] *See* major node. [ELECTROMAG] A fitting used to join a branch waveguide at an angle to a main waveguide, as in a tee junction. Also known as waveguide junction. { 'jəŋk·shən }

junction laser [OPTICS] A laser in which a junction in a semiconductor serves as the source of the coherent laser beam. { 'jəŋk·shən ¦lā·zər }

junction magnetoresistance *See* tunneling magnetoresistance. { ¦jəŋk·shən mag,ned·ō·ri'zis·təns }

junior [OPTICS] A 1000- or 2000-watt Fresnel spotlight. { 'jün·yər }

Jurin rule [FL MECH] The rule that a height to which a liquid rises in a capillary tube is twice the liquid's surface tension times the cosine of its contact angle with the capillary, divided by the product of the liquid's weight density and the internal radius of the tube. { 'jůr·ən ,rül }

just scale [ACOUS] A diatonic scale rendered in the just tuning system. { 'jəst 'skāl }

just ton *See* ton. { 'jəst 'tən }

just tuning [ACOUS] A tuning system generated by octave rearrangements of the notes of three consecutive triads, each having the frequency ratio 4:5:6, with the highest note of one triad serving as the lowest note of the next triad. { 'jəst 'tün·iŋ }

Kellogg equation [THERMO] An equation of state for a gas, of the form

$$p = RT\rho + \sum_{n=2}^{\infty} [b_n T - a_n - (c_n/T^2)]\rho^n$$

where p is the pressure, T the absolute temperature, ρ the density, R the gas constant, and a_n, b_n, and c_n are constants. { 'kel,äg i,kwā·zhən }

kelvin [ELEC] A name formerly given to the kilowatt-hour. Also known as thermal volt. [THERMO] A unit of absolute temperature equal to 1/273.16 of the absolute temperature of the triple point of water. Symbolized K. Formerly known as degree Kelvin. { 'kel·vən }

Kelvin absolute temperature scale [THERMO] A temperature scale in which the ratio of the temperatures of two reservoirs is equal to the ratio of the amount of heat absorbed from one of them by a heat engine operating in a Carnot cycle to the amount of heat rejected by this engine to the other reservoir; the temperature of the triple point of water is defined as 273.16 K. Also known as Kelvin temperature scale. { 'kel·vən ¦ab·sə,lüt 'tem·prə·chər ,skāl }

Kelvin balance [ELECTROMAG] An ammeter in which the force between two coils in series that carry the current to be measured, one coil being attached to one arm of a balance, is balanced against a known weight at the other end of the balance arm. { 'kel·vən ¦bal·əns }

Kelvin body [MECH] An ideal body whose shearing (tangential) stress is the sum of a term proportional to its deformation and a term proportional to the rate of change of its deformation with time. Also known as Voigt body. { 'kel·vən ,bäd·ē }

Kelvin bridge [ELEC] A specialized version of the Wheatstone bridge network designed to eliminate, or greatly reduce, the effect of lead and contact resistance, and thus permit accurate measurement of low resistance. Also known as double bridge; Kelvin network; Thomson bridge. { 'kel·vən ,brij }

Kelvin equation [THERMO] An equation giving the increase in vapor pressure of a substance which accompanies an increase in curvature of its surface; the equation describes the greater rate of evaporation of a small liquid droplet as compared to that of a larger one, and the greater solubility of small solid particles as compared to that of larger particles. { 'kel·vən i,kwā·zhən }

Kelvin guard-ring capacitor [ELEC] A capacitor with parallel circular plates, one of which has a guard ring separated from the plate by a narrow gap; it is used as a standard, whose capacitance can be accurately calculated from its dimensions. { 'kel·vən 'gärd ,riŋ kə,pas·əd·ər }

Kelvin-Helmholtz instability [FL MECH] An instability that occurs at the interface between two fluid layers if their relative motion is sufficiently large, and eventually results in the disruption of the interface. { ¦kel·vin 'helm,hōls ,in·stə'bil·əd·ē }

Kelvin network *See* Kelvin bridge. { 'kel·vən ,net,wərk }

Kelvin relations *See* Thomson relations. { 'kel·vən ri,lā·shənz }

Kelvin scale [THERMO] The basic scale used for temperature definition; the triple point of water (comprising ice, liquid, and vapor) is defined as 273.16 K; given two reservoirs, a reversible heat engine is built operating in a cycle between them, and the ratio of their temperatures is defined to be equal to the ratio of the heats transferred. { 'kel·vən ,skāl }

Kelvin's circulation theorem [FL MECH] The theorem that, if the external forces acting on an inviscid fluid are conservative and if the fluid density is a function of the pressure only, then the circulation along a closed curve which moves with the fluid does not change with time. { 'kel·vənz ,sər·kyə'lā·shən ,thir·əm }

Kelvin's formula *See* Thomson formula. { 'kel·vənz ,fór·myə·lə }

Kelvin skin effect *See* skin effect. { 'kel·vən 'skin i,fekt }

Kelvin's minimum-energy theorem [FL MECH] The theorem that the irrotational motion of an incompressible, inviscid fluid occupying a simply connected region has less kinetic energy than any other fluid motion consistent with the boundary condition of zero relative velocity normal to the boundaries of the region. { 'kel·vənz ¦min·ə·məm 'en·ər·jē ,thir·əm }

Kelvin's statement of the second law of thermodynamics [THERMO] The statement that it is not possible that, at the end of a cycle of changes, heat has been extracted from a reservoir and an equal amount of work has been produced without producing some other effect. { 'kel·vənz 'stāt·mənt əv thə 'sek·ənd ,ló əv ,thər·mō·dī'nam·iks }

Kelvin temperature scale [THERMO] **1.** An International Temperature Scale which agrees with the Kelvin absolute temperature scale within the limits of experimental determination. **2.** *See* Kelvin absolute temperature scale. { 'kel·vən 'tem·prə·chər ,skāl }

Kennard packet [QUANT MECH] A wave packet for which the product of the root-mean-square deviations of position and momentum from their respective mean values is as small as possible, being equal to Planck's constant divided by 4π. { 'ken·ərd ,pak·ət }

Kennedy and Pancu circle [MECH] For a harmonic oscillator subject to hysteretic damping and subjected to a sinusoidally varying force, a plot of the in-phase and quadrature components of the displacement of the oscillator as the frequency of the applied vibration is varied. { ¦ken·ə·dē ən 'pän·chü ,sər·kəl }

Keplerian telescope [OPTICS] A telescope that forms a real intermediate image in the focal plane and can be used for introducing a reticle or a scale into the focal plane. { ke'plir·ē·ən 'tel·ə,skōp }

kernel [ATOM PHYS] An atom that has been stripped of its valence electrons, or a positively charged nucleus lacking the outermost orbital electrons. { 'kərn·əl }

K

K *See* cathode.

kA *See* kiloampere.

K-A decay [NUC PHYS] Radioactive decay of potassium-40 (^{40}K) to argon-40 (^{40}A), as the nucleus of potassium captures an orbital electron and then decays to argon-40; the ratio of ^{40}K to ^{40}A is used to determine the age of rock (K-A age). { 'kā'ā di,kā }

Kaiser effect [ACOUS] An effect observed in most metals, in which acoustic emissions are not observed during the reloading of a material until the stress exceeds its previous high value. { 'kī·zər i,fekt }

kaleidoscope [OPTICS] An optical toy consisting of a tube containing two plane mirrors placed at an angle of 60° and mounted so that a symmetrical pattern produced by multiple reflection is observed through a peephole at one end when objects (such as pieces of colored glass) at the other end are suitably illuminated. { kə'līd·ə,skōp }

Kaluza theory [RELAT] An attempted unified field theory in which the four-dimensional world that one observes is taken to be a projection of a five-dimensional continuum. { kə'lü·zə ,thē·ə·rē }

KAM theorem *See* Kolmogorov-Arnold-Moser theorem. { 'kam ,thir·əm }

kaon *See* K meson. { 'kā,än }

kaonic atom [ATOM PHYS] An atom consisting of a negatively charged kaon orbiting around an ordinary nucleus. { kā'än·ik 'ad·əm }

Kapitza resistance [CRYO] A thermal resistance to the flow of heat across the interface between liquid helium and a solid. { 'kä·pit·sə ri¦zis·təns }

Kármán constant [FL MECH] A dimensionless number formed from the velocity of turbulent flow parallel to a plane wall, the distance from the wall, the shear stress, and the density of the fluid; for a wide range of flow patterns it has a constant value. { 'kär,män ,kän·stənt }

Kármán-Tsien method [FL MECH] A method of approximating equations for two-dimensional compressible flow which yields a simple rule for estimating compressibility effects of subsonic flow. { ¦kär,män 'tsyen ,meth·əd }

Kármán vortex street [FL MECH] A double row of line vortices in a fluid which, under certain conditions, is shed in the wake of cylindrical bodies when the relative fluid velocity is perpendicular to the axis of the cylinder. { 'kär,män 'vȯr,teks ,strēt }

Kater's reversible pendulum [MECH] A gravity pendulum designed to measure the acceleration of gravity and consisting of a body with two knife-edge supports on opposite sides of the center of mass. { 'kā·dərz ri¦vər·sə·bəl 'pen·jə·ləm }

katoptric system [OPTICS] An optical system such that, when the object is displaced in a direction parallel to the axis, the image is displaced in the opposite direction (in contrast to a dioptric system). Also known as contracurrent system. { kə'täp·trik ,sis·təm }

kayser [SPECT] A unit of reciprocal length, especially wave number, equal to the reciprocal of 1 centimeter. Also known as rydberg. { 'kī·zər }

kb *See* kilobar.

K band [SOLID STATE] An optical absorption band which appears together with an F-band and has a lower intensity and shorter wavelength than the latter. { 'kā ,band }

kc *See* kilohertz.

kcal *See* kilocalorie.

K capture [NUC PHYS] A type of beta interaction in which a nucleus captures an electron from the K shell of atomic electrons (the shell nearest the nucleus) and emits a neutrino. { 'kā ,kap·chər }

KdV soliton *See* Korteweg-de Vries soliton. { 'kā·dē,vē 'säl·ə,tän }

keeper [ELECTROMAG] A bar of iron or steel placed across the poles of a permanent magnet to complete the magnetic circuit when the magnet is not in use, to avoid the self-demagnetizing effect of leakage lines. Also known as magnet keeper. { 'kēp·ər }

Keldysh theory [ATOM PHYS] A theory of multiphoton ionization, in which an atom is ionized by rapid absorption of a sufficient number of photons; it predicts that the ionization rate depends primarily upon the ratio of the mean binding electric field to the peak strength of the incident electromagnetic field, and upon the ratio of the binding energy to the energy of photons in the field. { 'kel·dish ,thē·ə·rē }

K electron [ATOM PHYS] An electron in the K shell. { 'kā i'lek,trän }

Kellner eyepiece [OPTICS] A Ramsden eyepiece with an achromatic eye lens. { 'kel·nər 'ī,pēs }

Kerr cell [OPTICS] A glass cell containing a dielectric liquid that exhibits the Kerr effect, such as nitrobenzene, in which is inserted the two plates of a capacitor, used to observe the Kerr effect on light passing through the cell. { 'kər ˌsel }

Kerr constant [OPTICS] A measure of the strength of the Kerr effect in a substance, equal to the difference between the extraordinary and ordinary indices of refraction divided by the product of the light's wavelength and the square of the electric field. { 'kər 'kän·stənt }

Kerr effect *See* electrooptical Kerr effect. { 'kər iˌfekt }

Kerr magnetooptical effect *See* magnetooptic Kerr effect. { 'kər magˌned·ō'äp·tə·kəl iˌfekt }

Kerr-Newman solution [RELAT] A solution to Einstein's field equations that describes a rotating, charged black hole. { ¦kər 'nü·mən səˌlü·shən }

Kerr solution [RELAT] A solution to Einstein's field equations that describes a rotating, uncharged, axisymmetric black hole. { 'kər səˌlü·shən }

Ketteler formula [ELECTROMAG] The case of Sellmeier's equation where only two characteristic frequencies are involved. { 'ket·lər ˌfȯr·myə·lə }

ket vector [QUANT MECH] A vector in Hilbert space specifying the state of a system (opposed to bra vector); represented by the symbol |>, with a letter or one or more indices inserted to distinguish it from other vectors. { 'ketˌvek·tər }

keV *See* kiloelectronvolt.

Keyes equation [THERMO] An equation of state of a gas which is designed to correct the van der Waals equation for the effect of surrounding molecules on the term representing the volume of a molecule. { 'kēz iˌkwā·zhən }

kg *See* kilogram; kilogram force.

kg-cal *See* kilocalorie.

kgf *See* kilogram force.

kgf-m *See* meter-kilogram.

kg-wt *See* kilogram force.

kHz *See* kilohertz.

Kikuchi lines [CRYSTAL] A pattern consisting of pairs of white and dark parallel lines, obtained when an electron beam is scattered (diffracted) by a crystalline solid; the pattern gives information on the structure of the crystal. { kē'kü·chē ˌlīnz }

killer [SOLID STATE] An impurity that inhibits luminescence in a solid. { 'kil·ər }

kiloampere [ELEC] A metric unit of current flow equal to 1000 amperes. Abbreviated kA. { 'ki·lō'amˌpir }

kilobar [MECH] A unit of pressure equal to 1000 bars (100 megapascals). Abbreviated kb. { 'kil·əˌbär }

kilocalorie [THERMO] A unit of heat energy equal to 1000 calories. Abbreviated kcal. Also known as kilogram-calorie (kg-cal); large calorie (Cal). { 'kil·əˌkal·ə·rē }

kilocycle *See* kilohertz. { 'kil·əˌsī·kəl }

kiloelectronvolt [PHYS] A nit of energy, equal to 1000 electronvolts. Abbreviated keV. { ¦ki·lō·i'lekˌtränˌvōlt }

kilogram [MECH] **1.** The unit of mass in the meter-kilogram-second system, equal to the mass of the international prototype kilogram stored at Sèvres, France. Abbreviated kg. **2.** *See* kilogram force. { 'kil·əˌgram }

kilogram-calorie *See* kilocalorie. { 'kil·əˌgram 'kal·ə·rē }

kilogram force [MECH] A unit of force equal to the weight of a 1-kilogram mass at a point on the earth's surface where the acceleration of gravity is 9.80665 m/s^2. Abbreviated kgf. Also known as kilogram (kg); kilogram weight (kg-wt). { 'kil·əˌgram 'fȯrs }

kilogram-meter *See* meter-kilogram. { 'kil·əˌgram 'mēd·ər }

kilogram weight *See* kilogram force. { 'kil·əˌgram 'wāt }

kilohertz [PHYS] A unit of frequency equal to 1000 hertz. Abbreviated kHz. Also known as kilocycle (kc). { 'kil·əˌhərts }

kilohm [ELEC] A unit of electrical resistance equal to 1000 ohms. Abbreviated K; kohm. { 'kilˌōm }

kilojoule [PHYS] A unit of energy or work equal to 1000 joules. Abbreviated kJ. { 'kil·əˌjül }

kiloliter [MECH] A unit of volume equal to 1000 liters or to 1 cubic meter. Abbreviated kl. { 'kil·əˌlēd·ər }

kilometer [MECH] A unit of length equal to 1000 meters. Abbreviated km. { 'kil·əˌmēd·ər }

kiloton [PHYS] A unit used in specifying the yield of a fission or fusion bomb, equal to the explosive power of 1000 metric tons of trinitrotoluene (TNT). Abbreviated kt. { 'kil·əˌtən }

kilovar [ELEC] A unit equal to 1000 volt-amperes reactive. Abbreviated kvar. { 'kil·əˌvär }

kilovolt [ELEC] A unit of potential difference equal to 1000 volts. Abbreviated kV. { 'kil·əˌvōlt }

kilovolt-ampere [ELEC] A unit of apparent power in an alternating-current circuit, equal to 1000 volt-amperes. Abbreviated kVA. { 'kil·əˌvōlt 'amˌpir }

kilowatt [PHYS] A unit of power equal to 1000 watts. Abbreviated kW. { 'kil·əˌwät }

kilowatt-hour [ELEC] A unit of energy or work equal to 1000 watt-hours. Abbreviated kWh; kW-hr. Also known as Board of Trade Unit. { 'kil·əˌwät ˌau̇r }

kinematically admissible motion [MECH] Any motion of a mechanical system which is geometrically compatible with the constraints. { ˌkin·ə¦mad·ə·klē id¦mis·ə·bəl 'mō·shən }

kinematic boundary condition [FL MECH] The condition that the component of fluid velocity perpendicular to a solid boundary must vanish on the boundary itself; when the boundary is a fluid surface, the condition applies to the vector difference of velocities across the interface. { ¦kin·ə¦mad·ik 'bau̇n·drē kənˌdish·ən }

kinematic fluidity [FL MECH] The reciprocal of the kinematic viscosity. { ¦kin·ə¦mad·ik flü'id·əd·ē }

kinematics [MECH] The study of the motion of a system of material particles without reference to the forces which act on the system. { ¦kin·ə¦mad·iks }

kinematic similarity [FL MECH] A relationship between fluid-flow systems in which corresponding fluid velocities and velocity gradients are in the same ratios at corresponding locations. { ¦kin·ə¦mad·ik ˌsim·ə'lar·əd·ē }

kinematic viscosity [FL MECH] The absolute viscosity of a fluid divided by its density. Also known as coefficient of kinematic viscosity. { ¦kin·ə¦mad·ik vi'skäs·əd·ē }

kinetic energy [MECH] The energy which a body possesses because of its motion; in classical mechanics, equal to one-half of the body's mass times the square of its speed. { kə'ned·ik 'en·ər·jē }

kinetic equilibrium *See* dynamic equilibrium. { kə'ned·ik ˌē·kwə'lib·rē·əm }

kinetic friction [MECH] The friction between two surfaces which are sliding over each other. { kə'ned·ik 'frik·shən }

kinetic momentum [MECH] The momentum which a particle possesses because of its motion; in classical mechanics, equal to the particle's mass times its velocity. { kə'ned·ik mə'men·təm }

kinetic potential *See* Lagrangian. { kə'ned·ik pə'ten·chəl }

kinetic pressure [FL MECH] The kinetic energy per unit volume of a fluid, equal to one-half the product of its density and the square of its velocity. { kə'ned·ik 'presh·ər }

kinetic reaction [MECH] The negative of the mass of a body multiplied by its acceleration. { kə'ned·ik rē'ak·shən }

kinetics [MECH] The dynamics of material bodies. { kə'ned·iks }

kinetic stress [STAT MECH] A stress which arises, in a theory taking the motions of individual molecules into account, from the existence of a velocity distribution of molecules, an example is the pressure of an ideal gas. { kə'ned·ik 'stres }

kinetic theory [STAT MECH] A theory which attempts to explain the behavior of physical systems on the assumption that they are composed of large numbers of atoms or molecules in vigorous motion; it is further assumed that energy and momentum are conserved in collisions of these particles, and that statistical methods can be applied to deduce the particles' average behavior. Also known as molecular theory. { kə'ned·ik 'thē·ə·rē }

Kingdon trap [ELEC] A thin charged wire for confining charged particles; ions are attracted toward the wire, but their angular momentum causes them to spiral around the wire in trajectories that have a low probability of hitting the wire. { 'kin·dən trap }

kink instability [PL PHYS] A type of hydromagnetic instability in which the ionized gas and its magnetic confining field tend to form a loop or kink, which then grows steadily larger. Also known as sausage instability. { 'kiŋk ˌin·stə'bil·əd·ē }

kip [MECH] A 1000-pound (453.6-kilogram) load. { kip }

Kirchhoff formula [THERMO] A formula for the dependence of vapor pressure p on temperature T, valid over limited temperature ranges; it may be written $\log p = A - (B/T) - C \log T$, where A, B, and C are constants. { 'kərk,hōf ˌfȯr·myə·lə }

Kirchhoff's current law [ELEC] The law that at any given instant the sum of the instantaneous values of all the currents flowing toward a point is equal to the sum of instantaneous values of all the currents flowing away from the point. Also known as Kirchhoff's first law. { 'kərk,hōfs 'kə·rənt ˌlȯ }

Kirchhoff's equations [THERMO] Equations which state that the partial derivative of the change of enthalpy (or of internal energy) during a reaction, with respect to temperature, at constant pressure (or volume) equals the change in heat capacity at constant pressure (or volume). { 'kərk,hōfs i,kwā·zhənz }

Kirchhoff's first law *See* Kirchhoff's current law. { 'kərk,hōfs 'fərst ˌlȯ }

Kirchhoff's law [ELEC] Either of the two fundamental laws dealing with the relation of currents at a junction and voltages around closed loops in an electric network; they are known as Kirchhoff's current law and Kirchhoff's voltage law. [THERMO] The law that the ratio of the emissivity of a heat radiator to the absorptivity of the same radiator is the same for all bodies, depending on frequency and temperature alone, and is equal to the emissivity of a blackbody. Also known as Kirchhoff's principle. { 'kərk ˌhōfs ˌlȯ }

Kirchhoff's principle *See* Kirchhoff's law. { 'kərk,hōfs ˌprin·sə·pəl }

Kirchhoff's second law *See* Kirchhoff's voltage law. { 'kərk,hōfs 'sek·ənd ˌlȯ }

Kirchhoff's voltage law [ELEC] The law that at each instant of time the algebraic sum of the voltage rises around a closed loop in a network is equal to the algebraic sum of the voltage drops, both being taken in the same direction around the loop. Also known as Kirchhoff's second law. { 'kərk,hōfs 'vōl·tij ˌlȯ }

Kirchhoff theory [OPTICS] A theory of diffraction of light which gives a mathematical formulation of Huygens' principle, based on the wave equation and Green's theorem, and enables quantitative determination of the amplitude and phase at any point to a very close approximation. { 'kərk,hōf ˌthē·ə·rē }

Kirchhoff vapor pressure formula [THERMO] An approximate formula for the variation of vapor pressure p with temperature T, valid over a limited temperature range; it is $\ln p = A - B/T - C \ln T$, where A, B, and C are constants. { ¦kirch,hōf 'vā·pər ˌpre·shər ˌfȯr·myə·lə }

Kirkwood-Brinkely's theory [MECH] In terminal ballistics, a theory formulating the scaling laws from which the effect of blast at high altitudes may be inferred, based upon observed results at ground level. { 'kərk,wu̇d 'briŋk·lēz ,thē·ə·rē }

kJ *See* kilojoule.

kl *See* kiloliter.

Klein-Gordon equation [QUANT MECH] A wave equation describing a spinless particle which is consistent with the special theory of relativity. Also known as Schrödinger-Klein-Gordon equation. { 'klīn 'go̊rd·ən i,kwā·zhən }

Klein-Nishina formula [QUANT MECH] A formula, based on the Dirac electron theory without radiative correction, for the differential cross section for scattering of a photon by an unbound electron. { 'klīn ni'shē·nə ,fo̊r·myə·lə }

Klein paradox [QUANT MECH] The paradox whereby, according to the Dirac electron theory, an electron can penetrate into a potential barrier which is greater than twice the rest energy of the electron (about 1 MeV) by making a transition from a positive energy state to a negative energy state, provided the potential change occurs over a distance on the order of a Compton wavelength or less. { 'klīn ,par·ə,däks }

K/L ratio [NUC PHYS] The ratio of the number of internal conversion electrons emitted from the K shell of an atom during de-excitation of a nucleus to the number of such electrons emitted from the L shell. { 'kā'el ,rā·shō }

km *See* kilometer.

K meson [PART PHYS] **1.** Collective name for four pseudo-scalar mesons, having masses of about 495 MeV (megaelectronvolts) and decaying via weak interactions: K^+, K^-, $K_S{}^0$, and $K_L{}^0$; they consist of two isotopic spin doublets, the (K^+, K^0) doublet and its antiparticle doublet (K^-, $\bar{K}^0$), having hypercharge or strangeness of +1 and −1 respectively, where K^0 and $\bar{K}^0$ are certain combinations of $K_L{}^0$ and $K_S{}^0$ states. Also known as kaon. **2.** Collective name for any meson resonance belonging to an isotopic doublet with hypercharge +1 or −1, denoted $K_{JP}(m)$ or $\bar{K}_{JP}(m)$ respectively, where m is the mass in megaelectronvolts, and J and P are the spin and parity. { 'kā ¦mā,sän }

knife-edge refraction [ELECTROMAG] Radio propagation effect in which the atmospheric attenuation of a signal is reduced when the signal passes over and is diffracted by a sharp obstacle such as a mountain ridge. { 'nīf ,ej ri'frak·shən }

Knight shift [PHYS] A shift of the nuclear magnetic resonance frequency in a metal to higher values than that of the same nucleus in a diamagnetic compound. { 'nīt ,shift }

knock-on atom [SOLID STATE] An atom which is knocked out of its equilibrium position in a crystal lattice by an energetic bombarding particle, and is displaced many atomic distances away into an interstitial position, leaving behind a vacant lattice site. { 'näk,o̊n ,ad·əm }

knot [PHYS] A speed unit of 1 nautical mile (1.852 kilometers) per hour, equal to approximately 0.51444 meters per second. { nät }

Knudsen cosine law [PHYS] A law which states that the probability of a gas molecule leaving a solid surface in a given direction within a solid angle $d\omega$ is proportional to cos θ $d\omega$, where θ is the angle between the direction and the normal to the surface. { kə'nüd·sən 'kō,sīn ,lo̊ }

Knudsen flow *See* free molecule flow. { kə'nüd·sən ,flō }

Knudsen number [FL MECH] The ratio of the mean free path length of the molecules of a fluid to a characteristic length; used to describe the flow of low-density gases. { kə'nüd·sən ,nəm·bər }

Knudsen's equation [PHYS] An equation for the amount of gas that flows through a tube in free molecule flow,

$$q\sqrt{2\pi}\Delta p d^3/6l\sqrt{\rho}$$

where q is the volume of gas measured at unit pressure that flows through the tube per second, Δp is the difference between the pressures at the ends of the tube, d is the inside diameter of the tube, l is the length of the tube, and ρ is the density of the gas at unit pressure. { kə'nüd·sənz i,kwā·shən }

Kohler illumination [OPTICS] A method of illumination for the optical microscope used with coiled filaments or other sources of irregular form or brightness; an image of the filament large enough to fill the iris opening is focused on the condenser which is focused so that the image of the iris diaphragm on the lamp is in focus with the specimen, and the lamp iris is opened only enough to fill the field of view; the iris of the microscope is opened only enough to illuminate the back aperture of the objective; no ground glass is used. { 'kō·lər i,lü·mə'nā·shən }

kohm *See* kilohm. { ¦kā¦ōm }

Kohn effect [SOLID STATE] A sharp change in the phonon dispersion curve of a material when the wave-number vector of the phonons corresponds to the diameter of the Fermi sphere, because of the production of standing waves. { 'kōn i,fekt }

Kolmogorov-Arnold-Moser theorem [PHYS] A theorem that oscillatory motions in conservative dynamical systems persist when small perturbations are added to the system. Abbreviated KAM theorem. { ,ko̊l·mə'go̊·ro̊f 'ar·nəld 'mō·zər ,thir·əm }

Kolmogorov inertial subrange [FL MECH] The middle portion of the turbulence spectrum, between the low-wave-number (long-wave) part and the high-wave-number (short-wave) part. { ,ko̊l·mə'go̊·ro̊f i'nər·shəl 'səb,ranj }

Kolosov-Muskhelishvili formulas [MECH] Formulas which express plane strain and plane stress in terms of two holomorphic functions of the complex variable $z = x + iy$, where x and y are plane coordinates. { ¦kōl·ə,so̊f ,mu̇sh'kel·ish,vil·ē ,fo̊r·myə·ləz }

Kondo resonance *See* Abrikosov-Suhl resonance. { 'kän·dō ,rez·ən·əns }

konig [OPTICS] The X tristimulus value. { 'kō·nig }

Korteweg-de Vries soliton [PHYS] A soliton composed of a single oscillation, in contrast to an envelope soliton. Abbreviated KdV soliton. { 'kȯrd·ə,vek də'vrēz 'säl·ə,tän }

Kossel effect [PHYS] The production of a series of cones of reflected x-rays by characteristic x-rays generated by atoms in a single crystal. { 'kȯs·əl i,fekt }

Kossel lines [PHYS] Conic sections recorded on a flat film from the cones generated in the Kossel effect. { 'kȯs·əl ,līnz }

Kossel-Sommerfeld law [SPECT] The law that the arc spectra of the atom and ions belonging to an isoelectronic sequence resemble each other, especially in their multiplet structure. { 'käs·əl 'zȯm·ər,felt ,lȯ }

Kosterlitz-Thouless transition [CRYO] The transition from the superfluid to the normal state in thin films of helium-3, which proceeds through the unbinding of vortices in the phase of the order parameter having opposite directions of rotation. { 'käs·tər,lits tü'les tran,zish·ən }

Kozeny-Carmen equation [FL MECH] Equation for streamline flow of fluids through a powdered bed. { ,kō,zā·nē 'kär·mən i,kwā·zhən }

Kramers-Kronig relation [OPTICS] A relation between the real and imaginary parts of the index of refraction of a substance, based on the causality principle and Cauchy's theorem. { 'krā·mərz 'krō·nig ri,lā·shən }

Kramer's theorem [SOLID STATE] The theorem that the states of a system consisting of an odd number of electrons in an external electrostatic field are at least twofold degenerate. { 'krā·mərz ,thir·əm }

Krigar-Menzel law [MECH] A generalization of the second Young-Helmholtz law which states that when a string is bowed at a point which is at a distance of p/q times the string's length from one of the ends, where p and q are relative primes, then the string moves back and forth with two constant velocities, one of which is $q - 1$ times as large as the other. { ¦krē·gər 'menz·əl ,lȯ }

Kronig-Penney model [SOLID STATE] An idealized one-dimensional model of a crystal in which the potential energy of an electron is an infinite sequence of periodically spaced square wells. { 'krō·nig 'pen·ē ,mäd·əl }

Kruskal coordinates [RELAT] Coordinate system used in general relativity to describe in a nonsingular manner the geometry of a nonrotating black hole in empty space. { ¦krüs·kəl kō'ȯrd·ən·əts }

Kruskal diagram [RELAT] A space-time diagram displaying the Schwarzschild metric in a form that eliminates the formal singularity that appears at the Schwarzschild radius in the form in which the metric is usually written. { 'krüs·kəl ,dī·ə,gram }

krypton-86 [NUC PHYS] An isotope of krypton, atomic mass 86; used in measurement of the standard meter. { 'krip·tän ¦ād·ē'siks }

K shell [ATOM PHYS] The innermost shell of electrons surrounding the atomic nucleus, having electrons characterized by the principal quantum number 1. { 'kā ,shel }

k-space *See* wave-vector space. { 'kā ,spās }

kt *See* kiloton.

Kubelka-Munk model [OPTICS] A widely used theoretical model of reflectance; the model supposes that some light passing through a homogeneous sample is scattered and absorbed so that the light is attenuated in both directions. { kü'bel·kə 'məŋk ,mäd·əl }

Kumakhov optics [OPTICS] Systems composed of bundles of glass fibers that behave as waveguides to propagate x-rays or neutrons by multiple internal refllections, and that can be bent or tapered to concentrate, collimate, or focus the radiation. Also known as polycapillary optics. { kü'mä,kȯf ,äp·tiks }

Kundt effect [OPTICS] **1.** The occurrence of a very large magnetic rotation when polarized light passes through very thin films of pure ferromagnetic materials. **2.** *See* Faraday effect. { 'ku̇nt i,fekt }

Kundt rule [SPECT] The rule that the optical absorption bands of a solution are displaced toward the red when its refractive index increases because of changes in composition or other causes. { 'ku̇nt ,rül }

Kundt's constant [OPTICS] A measure of the strength of the Faraday effect in a material, equal to the ratio of the Faraday rotation to the product of the path length and the magnetization of the material; it depends only on the temperature for any magnetic material. { 'ku̇ns ,kän·stənt }

Kundt tube [ACOUS] A tube used to measure the speed of sound; it is filled with air or other gas and contains a light powder which becomes lumped at nodes, giving the length of standing waves generated in the tube. { 'ku̇nt ,tüb }

Kurie plot [NUC PHYS] Graph used in studying beta decay, in which the square root of the number of beta particles whose momenta (or energy) lie within a certain narrow range, divided by a function worked out by Fermi, is plotted against beta-particle energy; it is a straight line for allowed transitions and some forbidden transitions, in accord with the Fermi beta-decay theory. Also known as Fermi plot. { 'kyu̇r·ē ,plät }

Kutta-Joukowski airfoil [FL MECH] A class of airfoils that may be produced by mapping circles with the complex variable transform $w = z + (c^2/z)$. { 'ku̇d·ə jü'kȯv·skē 'er,fȯil }

Kutta-Joukowski condition [FL MECH] A boundary condition or fluid flow about an airfoil which requires that the circulation of the flow be such that a streamline leaves the trailing edge of the airfoil smoothly, or, equivalently, that the fluid velocity at the trailing edge be finite. { ¦ku̇d·ə jü'kȯv,skē kən,dish·ən }

Kutta-Joukowski equation [FL MECH] An equation which states that the lift force exerted on a body by an ideal fluid, per unit length of body perpendicular to the flow, is equal to the product of the mass density of the fluid, the linear velocity of the fluid relative to the body, and the fluid circulation. Also known as Kutta-Joukowski theorem. { 'ku̇d·ə jü'kȯv·skēi,kwā·zhən }

Kutta-Joukowski theorem *See* Kutta-Joukowski equation. { 'ku̇d·ə jü'kȯv·skē ,thir·əm }

kV *See* kilovolt.

kVA *See* kilovolt-ampere.

kvar *See* kilovar. { 'kā,vär }

kW *See* kilowatt.

kWh *See* kilowatt-hour.

kW-hr *See* kilowatt-hour.

L

l *See* liter.

L *See* lambert; liter.

laboratory coordinate system [MECH] A reference frame attached to the laboratory of the observer, in contrast to the center-of-mass system. { 'lab·rə,tȯr·ē kō'ȯrd·ən,ət ,sis·təm }

ladar [OPTICS] A missile-tracking system that uses a visible light beam in place of a microwave radar beam to obtain measurements of speed, altitude, direction, and range of missiles. Derived from laser detecting and ranging. Also known as colidar; laser radar. { 'lā,där }

Ladenburg f value *See* oscillator strength. { 'läd·ən,bərg 'ef ,val·yü }

lag [PHYS] **1.** The difference in time between two events or values considered together. **2.** *See* lag angle. { lag }

lag angle [PHYS] The negative of phase difference between a sinusoidally varying quantity and a reference quantity which varies sinusoidally at the same frequency, when this phase difference is negative. Also known as angle of lag; lag. { 'lag ,aŋ·gəl }

lag coefficient *See* time constant. { 'lag ,kō·i,fish·ənt }

lagging current [ELEC] An alternating current that reaches its maximum value up to 90° behind the voltage that produces it. { 'lag·iŋ ,kə·rənt }

lagging load *See* inductive load. { 'lag·iŋ ,lōd }

Lagrange bracket [MECH] Given two functions of coordinates and momenta in a system, their Lagrange bracket is an expression measuring how coordinates and momenta change jointly with respect to the two functions. { lə'gränj ,brak·ət }

Lagrange function *See* Lagrangian. { lə'gränj ,fəŋk·shən }

Lagrange-Hamilton theory [MECH] The formalized study of continuous systems in terms of field variables where a Lagrangian density function and Hamiltonian density function are introduced to produce equations of motion. { lə'gränj 'ham·əl·tən ,thē·ə·rē }

Lagrange's equations [MECH] Equations of motion of a mechanical system for which a classical (non-quantum-mechanical) description is suitable, and which relate the kinetic energy of the system to the generalized coordinates, the generalized forces, and the time. Also known as Lagrangian equations of motion. { lə'grän·jəz i,kwā·zhənz }

Lagrange stream function [FL MECH] A scalar function of position used to describe steady, incompressible two-dimensional flow; constant values of this function give the streamlines, and the rate of flow between a pair of streamlines is equal to the difference between the values of this function on the streamlines. Also known as current function; stream function. { lə'gränj 'strēm ,fəŋk·shən }

Lagrangian [MECH] **1.** The difference between the kinetic energy and the potential energy of a system of particles, expressed as a function of generalized coordinates and velocities from which Lagrange's equations can be derived. Also known as kinetic potential; Lagrange function. **2.** For a dynamical system of fields, a function which plays the same role as the Lagrangian of a system of particles; its integral over a time interval is a maximum or a minimum with respect to infinitesimal variations of the fields, provided the initial and final fields are held fixed. { lə'grän·jē·ən }

Lagrangian coordinates *See* generalized coordinates. { lə'grän·jē·ən ko'ȯrd·ən·əts }

Lagrangian density [MECH] For a dynamical system of fields or continuous media, a function of the fields, of their time and space derivatives, and the coordinates and time, whose integral over space is the Lagrangian. { lə'grän·jē·ən 'den· səd·ē }

Lagrangian description *See* Lagrangian method. { lə ¦grän·jē·ən di'skrip·shən }

Lagrangian equations of motion *See* Lagrange's equations. { lə'grän·jē·ən i¦kwā·zhənz əv 'mō·shən }

Lagrangian function [MECH] The function which measures the difference between the kinetic and potential energy of a dynamical system. { lə'grän·jē·ən ,fəŋk·shən }

Lagrangian generalized velocity *See* generalized velocity. { lə'grän·jē·ən ¦jen·rə,līzd və'läs·əd·ē }

Lagrangian method [FL MECH] A method of studying fluid motion and the mechanics of deformable bodies in which one considers volume elements which are carried along with the fluid or body, and across whose boundaries material does not flow; in contrast to Euler method.

Also known as Lagrangian description. { lə'grän·jē·ən ˌmeth·əd }

lambda [MECH] A unit of volume equal to 10^{-6} liter or 10^{-9} cubic meter. { 'lam·də }

lambda hyperon [PART PHYS] **1.** A quasi-stable baryon, forming an isotopic singlet, having zero charge and hypercharge, a spin of 1/2, positive parity and mass of approximately 1115.5 megaelectronvolts. Designated Λ. Also known as lambda particle. **2.** Any baryon resonance having zero hypercharge and total isotopic spin; designated $\Lambda_{JP}(m)$, where m is the mass of the baryon in megaelectronvolts, and J and P are its spin and parity (if known). { 'lam·də 'hī·pəˌrän }

lambda leak [CRYO] A leak of liquid helium II through small holes where normal liquids cannot pass. Also known as superleak. { 'lam·də ˌlēk }

lambda particle *See* lambda hyperon. { 'lam·də ˌpard·ə·kəl }

lambda point [CRYO] The temperature (2.1780 K), at atmospheric pressure, at which the transformation between the liquids helium I and helium II takes place; a special case of the thermodynamics definition. [THERMO] A temperature at which the specific heat of a substance has a sharply peaked maximum, observed in many second-order transitions. { 'lam·də ˌpȯint }

lambert [OPTICS] A unit of luminance (photometric brightness) that is equal to $1/\pi$ candela per square centimeter, or to the uniform luminance of a perfectly diffusing surface emitting or reflecting light at the rate of 1 lumen per square centimeter. Abbreviated L. { 'lam·bərt }

Lambert's law [OPTICS] **1.** The law that the illumination of a surface by a light ray varies as the cosine of the angle of incidence between the normal to the surface and the incident ray. **2.** The law that the luminous intensity in a given direction radiated or reflected by a perfectly diffusing plane surface varies as the cosine of the angle between that direction and the normal to the surface. { 'lam·bərts ˌlȯ }

Lambert surface [THERMO] An ideal, perfectly diffusing surface for which the intensity of reflected radiation is independent of direction. { 'lam·bərt ˌsər·fəs }

Lamb shift [ATOM PHYS] A small shift in the energy levels of a hydrogen atom, and of hydrogenlike ions, from those predicted by the Dirac electron theory, in accord with principles of quantum electrodynamics. { 'lam ¦shift }

Lamb wave [ACOUS] *See* plate wave. [ELECTROMAG] Electromagnetic wave propagated over the surface of a solid whose thickness is comparable to the wavelength of the wave. { 'lam ˌwāv }

Lamé constants [MECH] Two constants which relate stress to strain in an isotropic, elastic material. { lä'mā ˌkän·stəns }

lamellar crystal [CRYSTAL] A polycrystalline substance whose grains are in the form of thin sheets. { lə'mel·ər 'krist·əl }

laminar boundary layer [FL MECH] A thin layer over the surface of a body immersed in a fluid, in which the fluid velocity relative to the surface increases rapidly with distance from the surface and the flow is laminar. { 'lam·ə·nər 'bau̇n·drē ˌlā·ər }

laminar flow [FL MECH] Streamline flow of an incompressible, viscous Newtonian fluid; all particles of the fluid move in distinct and separate lines. { 'lam·ə·nər 'flō }

laminar sublayer [FL MECH] The laminar boundary layer underlying a turbulent boundary layer. { 'lam·ə·nər 'səbˌlā·ər }

laminated core [ELECTROMAG] An iron core for a coil transformer, armature, or other electromagnetic device, built up from laminations stamped from sheet iron or steel and more or less insulated from each other by surface oxides and sometimes also by application of varnish. { 'lam·əˌnād·əd 'kȯr }

Lami's theorem [MECH] When three forces act on a particle in equilibrium, the magnitude of each is proportional to the sine of the angle between the other two. { la'mēz ˌthir·əm }

Lanchester's rule [MECH] The rule that a torque applied to a rotating body along an axis perpendicular to the rotation axis will produce precession in a direction such that, if the body is viewed along a line of sight coincident with the torque axis, then a point on the body's circumference, which initially crosses the line of sight, will appear to describe an ellipse whose sense is that of the torque. { 'lanˌches·tərz ˌrülz }

Landau damping [PL PHYS] Damping of a plasma oscillation wave which occurs in situations where the particles of the plasma are able to increase their average energy at the expense of the wave, and thus to damp it out, even in cases where the dissipative effects of collisions are unimportant. { 'lanˌdau̇ ˌdam·piŋ }

Landau-Ginzburg theory *See* Ginzburg-Landau theory. { 'lanˌdau̇ 'ginz·bərg ˌthē·ə·rē }

Landau levels [SOLID STATE] Energy levels of conduction electrons which occur in a metal subjected to a magnetic field at very low temperatures and which are quantized because of the quantization of the electron motion perpendicular to the field. { 'lanˌdau̇ ˌlev·əlz }

Landau-Levich-Derjaguin picture [FL MECH] A theory of fluid coating at low velocities, according to which the thickness of the film that forms when a solid is drawn out of a bath results from a balance between (1) the effects of viscosity, which causes a macroscopic entrainment of liquid by the solid, and (2) surface tension, which resists the film entrainment, so that the film thickness is proportional to the capillary number raised to the 2/3 power. { ¦lanˌdu̇ ¦lev·ich 'der·zhəˌgēn ˌpik·chər }

Landé Γ-permanence rule [ATOM PHYS] The rule that the sum of the shifts of energy levels produced by the spin-orbit interaction, over a series of states having the same spin and orbital angular momentum quantum numbers (or the same total angular momentum quantum numbers for individual electrons) but different total angular momenta, and having the same total magnetic quantum number, is independent of the strength of an applied magnetic field. { län'dā ¦gam·ə 'pər·mə·nəns ˌrül }

Landé g factor [ATOM PHYS] Also known as g factor. **1.** The negative ratio of the magnetic moment of an electron or atom, in units of the Bohr magneton, to its angular momentum, in units of Planck's constant divided by 2π. **2.** The ratio of the difference in energy between two energy levels which differ only in magnetic quantum number to the product of the Bohr magneton, the applied magnetic field, and the difference between the magnetic quantum numbers of the levels; identical to the first definition for free atoms. Also known as Landé splitting factor; spectroscopic splitting factor. [NUC PHYS] The ratio of the magnetic moment of a nucleon, in units of the nuclear magneton, to its angular momentum in units of Planck's constant divided by 2π. { län'dā 'jē ˌfak·tər }

Landé interval rule [ATOM PHYS] The rule that when the spin-orbit interaction is weak enough to be treated as a perturbation, an energy level having definite spin angular momentum and orbital angular momentum is split into levels of differing total angular momentum, so that the interval between successive levels is proportional to the larger of their total angular momentum values. { län'dā 'int·ər·vəl ˌrül }

Landé splitting factor *See* Landé g factor. { län'dā 'splid·iŋ ˌfak·tər }

Landholt fringe [OPTICS] A black fringe that crosses the darkened field which is produced when a brilliant source of light is viewed through two Nicol prisms oriented with their principal axes at right angles to one another. { 'land ˌhōlt ˌfrinj }

land measure [MECH] **1.** Units of area used in measuring land. **2.** Any system for measuring land. { 'land ˌmezh·ər }

land mile *See* mile. { 'lan ¦mīl }

Langevin-Debye formula [STAT MECH] A formula for the polarizability of a dielectric material or the paramagnetic susceptibility of a magnetic material, in which these quantities are the sum of a temperature-independent contribution and a contribution arising from the partial orientation of permanent electric or magnetic dipole moments which varies inversely with the temperature. Also known as Langevin-Debye law. { länzh·van də'bī ˌfȯr·myə·lə }

Langevin-Debye law *See* Langevin-Debye formula. { länzh·van də'bī ˌlȯ }

Langevin function [ELECTROMAG] A mathematical function, $L(x)$, which occurs in the expressions for the paramagnetic susceptibility of a classical (non-quantum-mechanical) collection of magnetic dipoles, and for the polarizability of molecules having a permanent electric dipole moment; given by $L(x) = \coth x - 1/x$. { länzh·van ˌfəŋk·shən }

Langevin radiation pressure [ACOUS] A measure of acoustic radiation pressure, equal to the difference between the mean pressure on an absorbing or reflecting wall and that in the same acoustic medium, at rest, behind the wall. { länzh·van ˌrād·ē'ā·shən ˌpresh·ər }

Langevin theory of diamagnetism [ELECTROMAG] A theory based on the idea that diamagnetism results from electronic currents caused by Larmor precession of electrons inside atoms. { länzh·van ¦thē·ə·rē əv ˌdī·ə'mag·nəˌtiz·əm }

Langevin theory of paramagnetism [ELECTROMAG] A theory which treats a substance as a classical (non-quantum-mechanical) collection of permanent magnetic dipoles with no interactions between them, having a Boltzmann distribution with respect to energy of interaction with an applied field. { länzh·van ¦thē·ə·rē əv ˌpar·ə'mag·nəˌtiz·əm }

langley [PHYS] A unit of energy per unit area commonly employed in radiation theory; equal to 1 gram-calorie per square centimeter. { 'laŋ·lē }

Langmuir effect [SOLID STATE] The ionization of atoms of low ionization potential that come into contact with a hot metal with a high work function. { 'laŋˌmyu̇r iˌfekt }

Langmuir plasma frequency [PL PHYS] The frequency of nonpropagating oscillations in a plasma; in rationalized mks units, it is $(ne^2/\epsilon_0 m)^{1/2}$, where e and m are the charge and mass of the oscillating electrons or ions, n is their number density, and ϵ_0 is the permittivity of empty space. Also known as plasma frequency. { 'laŋˌmyu̇r 'plaz·mə ˌfrē·kwən·sē }

Langmuir probe [PL PHYS] A device for measuring the temperature and electron density of a plasma, consisting of an electrode in contact with the plasma whose potential is varied while the resulting collection currents are measured. { 'laŋˌmyu̇r ˌprōb }

Langmuir wave [PL PHYS] A longitudinal, electrostatic wave that propagates in a plasma, because of variations in the plasma's electron density. { 'laŋˌmyu̇r ˌwāv }

L antenna [ELECTROMAG] An antenna that consists of an elevated horizontal wire having a vertical down-lead connected at one end. { 'el anˌten·ə }

lanthanide contraction [ATOM PHYS] A phenomenon encountered in the rare-earth elements; the radii of the atoms of the members of the series decrease slightly as the atomic numbers increase; starting with element 58 in the periodic table, the balancing electron fills in an inner incomplete $4f$ shell as the charge on the nucleus increases. { 'lan·thəˌnīd kənˌtrak·shən }

Laplace irrotational motion [FL MECH] Irrotational flow of an inviscid, incompressible fluid. { lə'pläs ˌir·ō'tā·shən·əl ˌmō·shən }

Laplace law *See* Ampère law. { lə'pläs ˌlȯ }

Laplace's equation [ACOUS] An equation for the speed c of sound in a gas; it may be written $c = \sqrt{\gamma p/\rho}$, where p is the pressure, ρ is the density, and γ is the ratio of specific heats. { lə'pläs·əz iˌkwā·zhən }

Laplacian speed of sound [FL MECH] The phase speed of a sound wave in a compressible fluid under the assumption that the expansions and compressions are adiabatic. { lə'pläs·ē·ən ¦spēd əv 'saủnd }

Laporte selection rule [ATOM PHYS] The rule that an electric dipole transition can occur only between states of opposite parity. { lə'pȯrt si'lek·shən ˌrül }

large calorie *See* kilocalorie. { 'lärj 'kal·ə·rē }

large dyne *See* newton. { 'lärj 'dīn }

large-eddy simulation [FL MECH] A technique for the prediction of complex turbulent flows in which the contribution of the large, energy-containing scales of motion is computed directly, and only the effect of the smallest scales of turbulence is modeled. { ¦lärj ¦ed·ē ˌsim·yə'lā·shən }

large number hypothesis [PHYS] The hypothesis that there is a physical basis for the approximate equality of two numbers on the order of 10^{40}: the ratio of the electrostatic to the gravitational force between a proton and an electron in a hydrogen atom, and the ratio of the age of the universe to the time required for light to cross an elementary particle diameter. { 'lärj ˌnəm·bər hī'päth·ə·səs }

large polaron [SOLID STATE] An electron in a crystal lattice together with the surrounding lattice deformation, for the case in which the deformation extends over many lattice sites so that the lattice can be treated as a continuum. { 'lärj 'pō·ləˌrän }

Larmor formula [ELECTROMAG] The rate at which energy is radiated by a nonrelativistic, accelerated charge is $2q^2a^2/3c^3$, where q is the particle's charge in esu (electrostatic units), a is its acceleration, and c is the speed of light. { 'lär·mȯr ˌfȯr·myə·lə }

Larmor frequency [ELECTROMAG] The angular frequency of the Larmor precession, equal in esu (electrostatic units) to the negative of a particle's charge times the magnetic induction divided by the product of twice the particle's mass and the speed of light. { 'lär·mȯr ˌfrē·kwən·sē }

Larmor orbit [ELECTROMAG] The motion of a charged particle in a uniform magnetic field, which is a superposition of uniform circular motion in a plane perpendicular to the field, and uniform motion parallel to the field. { 'lär·mȯr ˌȯr·bət }

Larmor precession [ELECTROMAG] A common rotation superposed upon the motion of a system of charged particles, all having the same ration of charge to mass, by a magnetic field. { 'lär·mȯr prēˌsesh·ən }

Larmor radius [ELECTROMAG] For a charged particle moving transversely in a uniform magnetic field, the radius of curvature of the projection of its path on a plane perpendicular to the field. Also known as gyromagnetic radius. { 'lär·mȯr ˌrād·ē·əs }

Larmor's theorem [ELECTROMAG] The theorem that for a system of charged particles, all having the same ratio of charge to mass, moving in a central field of force, the motion in a uniform magnetic induction B is, to first order in B, the same as a possible motion in the absence of B except for the superposition of a common precession of angular frequency equal to the Larmor frequency. { 'lär·mȯrz ˌthir·əm }

Larson-Miller parameter [MECH] The effects of time and temperature on creep, being defined empirically as $P = T\,(C + \log t) \times 10^{-3}$, where T = test temperature in degrees Rankine (degrees Fahrenheit + 460) and t = test time in hours; the constant C depends upon the material but is frequently taken to be 20. { 'lärs·ən 'mil·ər pə'ram·əd·ər }

laser [OPTICS] An active electron device that converts input power into a very narrow, intense beam of coherent visible or infrared light; the input power excites the atoms of an optical resonator to a higher energy level, and the resonator forces the excited atoms to radiate in phase. Derived from light amplification by stimulated emission of radiation. { 'lā·zər }

laser beam [OPTICS] A narrow beam of coherent, powerful, and nearly monochromatic electromagnetic radiation emitted by a laser. { 'lā·zər ˌbēm }

laser camera [OPTICS] An airborne camera system for night photography in which a laser beam is split into two beams; one beam, which is almost invisible, scans the ground, while the second beam is modulated by a detector of light reflected from the ground area being scanned, and is in turn swept back and forth over a moving film by the same scanner. { 'lā·zər ˌkam·rə }

laser cooling [ATOM PHYS] A method of slowing atoms in an atomic beam to very low velocities, by directing a beam of properly tuned laser light opposite to the atomic beam, and compensating for the Doppler shift of the slowing atoms by varying the laser frequency or Zeeman-shifting the atomic levels with a varying magnetic field. { 'lā·zər ˌkül·iŋ }

laser detecting and ranging *See* ladar. { 'lā·zər di¦tek·tiŋ ən 'rān·jiŋ }

laser diode *See* semiconductor laser. { 'lā·zər ¦dīˌōd }

laser Doppler velocimeter [OPTICS] A type of laser velocimeter used for determining the velocity of a fluid flow from the Doppler shift in the frequency of laser light scattered from particles in the fluid. { 'lā·zər 'däp·lər ˌvel·ə'sim·əd·ər }

laser drill [OPTICS] A drill in which concentrated light from a ruby laser generates intense heat for drilling holes as small as 0.0001 inch (2.5 micrometers) in diameter in tungsten, gemstones, and other hard materials. { 'lā·zər ˌdril }

laser extensometer [OPTICS] A device which uses interference of laser beams to measure small changes in distance; it can operate between points as much as 0.6 mile (1 kilometer) apart, and has been used to measure effects produced by earth tides. { 'lā·zər ˌekˌsten'säm·əd·ər }

laser heterodyne spectroscopy [SPECT] A high-resolution spectroscopic technique, used in astronomical and atmospheric observations, in which the signal to be measured is mixed with a laser signal in a solid-state diode, producing a difference-frequency signal in the radio-frequency range. { 'lā·zər ¦hed·ə·rə,dīn spek'träs·kə·pē }

laser-induced fluorescence imaging [FL MECH] A flow visualization method in which fluorescent tracers are excited by a laser beam. { ¦lā·zər in,düst flu̇r¦es·ənt 'im·ij·iŋ }

laser-induced nuclear polarization [NUC PHYS] A technique for making the spin vectors of an ensemble of nuclei point preferentially in one direction by means of an optical pumping process using either circularly or linearly polarized laser light. Abbreviated LINUP. { 'lā·zər in ,düst ¦nü·klē·ər ,pō·lə·rə'zā·shən }

laser infrared radar *See* lidar. { 'lā·zər ¦in·frə¦red 'rā,där }

laser interferometer [OPTICS] An interferometer which uses a laser as a light source; because of the monochromaticity and high intrinsic brilliance of laser light, it can operate with path differences in the interfering beams of hundreds of meters, in contrast to a maximum of about 20 centimeters (8 inches) for classical interferometers. { 'lā·zər ,in·tər·fə'räm·əd·ər }

laser radar *See* ladar. { 'lā·zər ,rā,där }

laser rangefinder [OPTICS] A portable rangefinder using a battery-powered ruby laser in combination with an optical telescope to aim a laser beam and a photomultiplier for picking up the laser beam reflected from the target. { 'lā·zər 'rānj,find·ər }

laser-solid interaction [SOLID STATE] Interaction of laser light with a solid, especially the thermal effects of absorption of a high-intensity laser beam. { 'lā·zər 'säl·əd ,in·tər'ak·shən }

laser spectroscopy [SPECT] A branch of spectroscopy in which a laser is used as an intense, monochromatic light source; in particular, it includes saturation spectroscopy, as well as the application of laser sources to Raman spectroscopy and other techniques. { 'lā·zər spek'träs·kə,pē }

laser spectrum [PHYS] The spectrum that includes all optical wavelengths, ranging from infrared through visible light to ultraviolet, in which coherent radiation can be produced by various types of lasers. { 'lā·zər ¦spek·trəm }

laser trap [OPTICS] A device for confining atoms, molecules, and larger neutral particles up to 10 micrometers in diameter, consisting of a focused laser beam tuned to a frequency below an atomic resonance, which attracts the particles toward regions of high laser intensity. { 'lā·zər ,trap }

laser tweezers [OPTICS] A laser trap used to hold microscopic organisms and their organelles and move them through the objective of an optical microscope without apparent damage. Also known as optical tweezers. { 'lā·zər ,twē·zərz }

laser velocimeter [OPTICS] Any velocity measuring instrument that makes use of a laser. { 'lā·zər ,vel·ə'sim·əd·ər }

lasing [OPTICS] Generation of visible or infrared light waves having very nearly a single frequency by pumping or exciting electrons into high-energy states in a laser. { 'lāz·iŋ }

latent heat [THERMO] The amount of heat absorbed or evolved by 1 mole, or a unit mass, of a substance during a change of state (such as fusion, sublimation or vaporization) at constant temperature and pressure. { 'lāt·əŋt 'hēt }

latent heat of fusion *See* heat of fusion. { 'lāt·ənt ¦hēt əv 'fyü·zhən }

latent heat of sublimation *See* heat of sublimation. { 'lāt·ənt ¦hēt əv ,səb·lə'mā·shən }

latent heat of vaporization *See* heat of vaporization. { 'lāt·ənt ¦hēt əv ,vā·pə·rə'zā·shən }

lateral aberration [OPTICS] **1.** The distance from the axis of an optical system at which a ray intersects a plane perpendicular to the axis through the focus of paraxial rays. **2.** The difference between the reciprocals of the image distances for paraxial and rim rays. **3.** For chromatic aberration, the difference in sizes of the images of an object for two different colors. { 'lad·ə·rəl ,ab·ə'rā·shən }

lateral chromatic aberration *See* chromatic difference of magnification. { ¦lad·ə·rəl krō¦mad·ik ,ab·ə'rā·shən }

lateral inversion [OPTICS] The effect produced by a mirror in reversing images from left to right. Also known as perversion. { 'lad·ə·rəl in'vər·zhən }

lateral magnification [OPTICS] The ratio of some linear dimension, perpendicular to the optical axis, of an image formed by an optical system, to the corresponding linear dimension of the object. Also known as magnification. { 'lad·ə·rəl ,mag·nə·fə'kā·shən }

lateral mirage [OPTICS] A very rare type of mirage in which the apparent position of an object appears displaced to one side of its true position. { 'lad·ə·rəl mə'räzh }

lateral quadrupole [ACOUS] A sound source resulting from a variation of a component of the velocity of matter in a direction perpendicular to the velocity component. [ELECTROMAG] An electric or magnetic quadrupole which produces a field equivalent to that of two equal and opposite electric or magnetic dipoles separated by a small distance perpendicular to the direction of the dipoles. { 'lad·ə·rəl 'kwäd·rə,pōl }

lateral shear interferometer [OPTICS] An interferometer in which a wavefront is interfered with a shifted version of itself, resulting in fringes along which the slope or derivative of the wavefront is constant. Also known as differential interferometer. { 'lad·ə·rəl ¦shir ,in·ter·fə'räm·əd·ər }

lattice [CRYSTAL] A regular periodic arrangement of points in three-dimensional space; it consists of all those points P for which the vector from a given fixed point to P has the form $n_1\mathbf{a} + n_2\mathbf{b} + n_3\mathbf{c}$, where n_1, n_2, and n_3 are integers, and **a**, **b**, and **c** are fixed, linearly independent vectors. Also known as periodic lattice; space lattice. { 'lad·əs }

lattice Boltzmann method [STAT MECH] A numerical method of solving the Boltzmann transport equation, using a finite difference approximation on a discrete phase space in which both discrete particle velocities and discrete spatial locations are assumed. Abbreviated LBM. { ¦lad·əs 'bōlts,män ,meth·əd }

lattice constant [CRYSTAL] A parameter defining the unit cell of a crystal lattice, that is, the length of one of the edges of the cell or an angle between edges. Also known as lattice parameter. { 'lad·əs ,kän·stənt }

lattice defect *See* crystal defect. { 'lad·əs ,dē,fekt }

lattice dynamics [SOLID STATE] The study of the thermal vibrations of a crystal lattice. Also known as crystal dynamics. { 'lad·əs dī,nam·iks }

lattice energy [SOLID STATE] The energy required to separate ions in an ionic crystal an infinite distance from each other. { 'lad·əs ,en·ər·jē }

lattice field theory *See* lattice-gauge theory. { 'lad·ən ,fēld ,thē·ə·rē }

lattice-gauge theory [PART PHYS] A formulation of the theory of hadron structure in which the continuum of space-time is replaced by a discrete set of points or sites, and quarks move through this structure by sequential hops between neighboring sites, interacting with gauge gluons which are represented by fields located on bonds connecting the lattice sites. Also known as lattice field theory. { 'lad·əs ,gāj ,thē·ə·rē }

lattice network [ELEC] A network that is composed of four branches connected in series to form a mesh; two nonadjacent junction points serve as input terminals, and the remaining two junction points serve as output terminals. { 'lad·əs 'net,wərk }

lattice parameter *See* lattice constant. { 'lad·əs pə,ram·əd·ər }

lattice polarization [SOLID STATE] Electric polarization of a solid due to displacement of ions from equilibrium positions in the lattice. { 'lad·əs pō·lə·rə'zā·shən }

lattice scattering [SOLID STATE] Scattering of electrons by collisions with vibrating atoms in a crystal lattice, reducing the mobility of charge carriers in the crystal and thereby affecting its conductivity. { 'lad·əs ,skad·ə·riŋ }

lattice vibration [SOLID STATE] A periodic oscillation of the atoms in a crystal lattice about their equilibrium positions. { 'lad·əs vī'brā·shən }

lattice wave [SOLID STATE] A disturbance propagated through a crystal lattice in which atoms oscillate about their equilibrium positions. { 'lad·əs ,wāv }

lattice-wound coil *See* honeycomb coil. { 'lad·əs ,waủnd ,kȯil }

Laue camera [CRYSTAL] The apparatus used in the Laue method; the x-ray beam usually enters through a hole in the x-ray film, which records beams bent through an angle of nearly 180° by the crystal; less commonly, the film is placed beyond the crystal. { 'laủ·ə ,kam·rə }

Laue condition [CRYSTAL] **1.** The condition for a vector to lie in a Laue plane: its scalar product with a specified vector in the reciprocal lattice must be one-half of the scalar product of the latter vector with itself. **2.** *See* Laue equations. { 'laủ·ə kən,dish·ən }

Laue equations [CRYSTAL] Three equations which must be satisfied for an x-ray beam of specified wavelength to be diffracted through a specified angle by a crystal; they state that the scaler products of each of the crystallographic axial vectors with the difference between unit vectors in the directions of the incident and scattered beams, are integral multiples of the wavelength. Also known as Laue condition. { 'laủ·ə i,kwā·zhənz }

Laue method [CRYSTAL] A method of studying crystalline structures by x-ray diffraction, in which a finely collimated beam of polychromatic x-rays falls on a single crystal whose orientation can be set as desired, and diffracted beams are recorded on a photographic film. { 'laủ·ə ,meth·əd }

Laue pattern [CRYSTAL] The characteristic photographic record obtained in the Laue method. { 'laủ·ə ,pad·ərn }

Laue plane [CRYSTAL] A plane which is the perpendicular bisector of a vector in the reciprocal lattice; such planes form the boundaries of Brillouin zones. { 'laủ·ə ,plān }

Laue theory [CRYSTAL] A theory of diffraction of x-rays by crystals, based on the Laue equations. { 'laủ·ə ,thē·ə·rē }

Laughlin state [CRYO] The simplest type of quantum Hall state, which contains only one component of incompressible fluid and has a filling factor equal to $1/m$, where m is an integer. { 'läk·lin ,stāt }

launching [ELECTROMAG] The process of transferring energy from a coaxial cable or transmission line to a waveguide. { 'lȯn·chiŋ }

laurence [OPTICS] A shimmering seen over a hot surface on a calm, cloudless day, caused by the unequal refraction of light by innumerable convective air columns of different temperatures and densities. { 'lȯr·əns }

Laurent half-shade plate [OPTICS] A device used to determine the direction of polarization of plane polarized light; it consists of a quartz plate of special thickness that covers half of the plane polarized beam, followed by a plane polarization analyzer. { lȯ'rän 'haf ,shād ,plāt }

Lauritsen electroscope [ELEC] A rugged and sensitive electroscope in which a metallized quartz fiber is the sensitive element. { 'laủ·rət·sən i'lek·trə,skōp }

law of action and reaction *See* Newton's third law. { 'lȯ əv 'ak·shən ən 'rē,ak·shən }

law of constant angles [CRYSTAL] The law that the angles between the faces of a crystal remain constant as the crystal grows. { 'lȯ əv ¦kän·stənt 'aŋ·gəlz }

law of corresponding times [MECH] The principle that the times for corresponding motions of dynamically similar systems are proportional to

L/V and also to $\sqrt{L/G}$, where L is a typical dimension of the system, V a typical velocity, and G a typical force per unit mass. { ¦lȯ əv ˌkär·ə ¦spänd·iŋ 'tīmz }

law of electric charges [ELEC] The law that like charges repel, and unlike charges attract. { 'lȯ əv i¦lek·trik 'chärj·əz }

law of electromagnetic induction *See* Faraday's law of electromagnetic induction. { 'lȯ əv i¦lek·trō·mag¦ned·ik in'dək·shən }

law of electrostatic attraction *See* Coulomb's law. { 'lȯ əv i¦lek·trə¦stad·ik ə'trak·shən }

law of flotation [FL MECH] The principle that an object floating in a fluid displaces its own weight of fluid. { 'lȯ əv flō'tā·shən }

law of gravitation *See* Newton's law of gravitation. { 'lȯ əv ˌgrav·ə'tā·shən }

law of magnetism [ELECTROMAG] The law that like poles repel, and unlike poles attract. { 'lȯ əv 'mag·nəˌtiz·əm }

law of parallel solenoids [PHYS] The law that under stationary conditions isopycnals and isobars must be parallel at all levels, and isobars and isopycnals at one level must be parallel to those at all other levels. { 'lȯ əv ¦par·əˌlel 'sō·ləˌnȯidz }

law of partial pressures *See* Dalton's law. { 'lȯ əv ¦pär·shəl 'presh·ərz }

law of reflection *See* reflection law. { 'lȯ əv ri'flek·shən }

laws of refraction *See* Snell laws of refraction. { 'lȯz əv ri'frak·shən }

Lawson criterion [PL PHYS] The requirement for the energy produced by fusion in a plasma to exceed that required to produce the confined plasma; it states that for a mixture of deuterium and tritium in the temperature range from 1×10^8 to 5×10^8 degrees Celsius, the product of the ionic density and the confinement time must be about 10^{14} seconds per cubic centimeter. { 'lȯs·ən krīˌtir·ē·ən }

layer lattice *See* layer structure. { 'lā·ər ˌlad·əs }

layer structure [CRYSTAL] A crystalline structure found in substances such as graphites and clays, in which the atoms are largely concentrated in a set of parallel planes, with the regions between the planes comparatively vacant. Also known as layer lattice. { 'lā·ər ˌstrək·chər }

lazy H antenna [ELECTROMAG] An antenna array in which two or more dipoles are stacked one above the other to obtain greater directivity. { ¦lā·zē 'āch anˌten·ə }

lb *See* pound.

lb ap *See* pound.

lb apoth *See* pound.

lbf *See* pound.

lbf-ft *See* foot-pound.

lb t *See* pound.

lb tr *See* pound.

LBM *See* lattice Boltzmann method.

LCAO *See* linear combination of atomic orbitals.

L capture [NUC PHYS] A type of generalized beta interaction in which a nucleus captures an electron from the L shell of atomic electrons (the shell second closest to the nucleus). { 'el ˌkap·chər }

LC ratio [ELEC] The inductance of a circuit in henrys divided by capacitance in farads. { ¦el¦sē ˌrā·shō }

lead [ELEC] A wire used to connect two points in a circuit. *See* lead angle. { led }

lead angle [PHYS] The phase difference between a sinusoidally varying quantity and a reference quantity which varies sinusoidally at the same frequency, when this phase difference is positive. Also known as angle of lead; lead; phase lead. { 'lēd ˌaŋ·gəl }

lead-I-lead junction [SOLID STATE] A Josephson junction consisting of two pieces of lead separated by a thin insulating barrier of lead oxide. Abbreviated Pb-I-Pb junction. { 'led ¦ī 'led ˌjəŋk·shən }

leading current [ELEC] An alternating current that reaches its maximum value up to 90° ahead of the voltage that produces it. { 'lēd·iŋ ¦kə·rənt }

leading edge [PHYS] The major portion of the rise of a pulse. { 'lēd·iŋ 'ej }

leading load [ELEC] Load that is predominately capacitive, so that its current leads the voltage applied to the load. { 'lēd·iŋ ¦lōd }

lead-208 [NUC PHYS] Lead isotope, atomic mass number of 208, which is formed by the radioactive decay of thorium. { 'led ¦tü¦ō'āt }

league [MECH] A unit of length equal to 3 miles or 4828.032 meters. { lēg }

leakage coefficient *See* leakage factor. { 'lēk·ij ˌkō·ə¦fish·ənt }

leakage conductance [ELEC] The conductance of the path over which leakage current flows; it is normally a low value. { 'lēk·ij kən¦dək·təns }

leakage current [ELEC] **1.** Undesirable flow of current through or over the surface of an insulating material or insulator. **2.** The flow of direct current through a poor dielectric in a capacitor. { 'lēk·ij ˌkə·rənt }

leakage factor [ELECTROMAG] The total magnetic flux in an electric rotating machine or transformer divided by the useful flux that passes through the armature or secondary winding. Also known as leakage coefficient. { 'lēk·ij ˌfak·tər }

leakage flux [ELECTROMAG] Magnetic lines of force that go beyond their intended path and do not serve their intended purpose. { 'lēk·ij ˌfləks }

leakage inductance [ELECTROMAG] Self-inductance due to leakage flux in a transformer. { 'lēk·ij in¦dək·təns }

leakage radiation [ELECTROMAG] In a radio transmitting system, radiation from anything other than the intended radiating system. { 'lēk·ij ˌrād·ē'ā·shən }

leakage reactance [ELECTROMAG] Inductive reactance due to leakage flux that links only the primary winding of a transformer. { 'lēk·ij rē¦ak·təns }

leakage resistance [ELEC] The resistance of

the path over which leakage current flows; it is normally high. { 'lēk·ij ri¦zis·təns }

leaky [ELEC] Pertaining to a condition in which the leakage resistance has dropped so much below its normal value that excessive leakage current flows; usually applied to a capacitor. [CELL MOL] Pertaining to a protein coded for by a mutant gene that shows subnormal activity. { 'lēk·ē }

leaky-wave antenna [ELECTROMAG] A wide-band microwave antenna that radiates a narrow beam whose direction varies with frequency; it is fundamentally a perforated waveguide, thin enough to permit flush mounting for aircraft and missile radar applications. { 'lēk·ē ¦wāv an'ten·ə }

least-action principle *See* principle of least action. { ¦lēst 'ak·shən ,prin·sə·pəl }

least-energy principle [MECH] The principle that the potential energy of a system in stable equilibrium is a minimum relative to that of nearby configurations. { ¦lēst 'en·ər·jē ,prin·sə·pəl }

least-time principle *See* Fermat's principle. { ¦lēst 'tīm ,prin·sə·pəl }

least-work theory [MECH] A theory of statically indeterminate structures based on the fact that when a stress is applied to such a structure the individual parts of it are deflected so that the energy stored in the elastic members is minimized. { ¦lēst 'wərk ,thē·ə·rē }

Le Chatelier's principle [PHYS] The principle that when an external force is applied to a system at equilibrium, the system adjusts so as to minimize the effect of the applied force. { lə'shäd·əl,yāz ,prin·sə·pəl }

Lecher line *See* Lecher wires. { 'lek·ər ,līn }

Lecher wires [ELECTROMAG] Two parallel wires that are several wavelengths long and a small fraction of a wavelength apart, used to measure the wavelength of a microwave source that is connected to one end of the wires; a shorting bar which slides along the wires is used to determine the position of standing-wave nodes. Also known as Lecher line; Lecher wire wavemeter. { 'lek·ər ,wīrz }

Lecher wire wavemeter *See* Lecher wires. { 'lek·ər ¦wīr 'wāv,mēd·ər }

Leduc current [ELEC] An asymmetrical alternating current obtained from, or similar to that obtained from, the secondary winding of an induction coil; used in electrobiology. { lə'dük ,kə·rənt }

Leduc effect *See* Righi-Leduc effect. { lə'dük i,fekt }

Leduc law *See* Amagat-Leduc rule. { lə'dük ,lȯ }

LEED *See* low-energy electron diffraction.

lee eddies [FL MECH] The small, irregular motions or eddies produced immediately in the rear of an obstacle in a turbulent fluid. { 'lē ,ed·ēz }

Lee's disk [THERMO] A device for determining the thermal conductivity of poor conductors in which a thin, cylindrical slice of the substance under study is sandwiched between two copper disks, a heating coil is placed between one of these disks and a third copper disk, and the temperatures of the three copper disks are measured. { 'lēz ,disk }

Leeson disk [OPTICS] The screen sometimes used in a grease-spot photometer, in which the translucent spot at the center is star-shaped to provide a fine line of demarcation. { 'lē·sən ,disk }

lee wave [FL MECH] Any wave disturbance which is caused by, and is therefore stationary with respect to, some barrier in the fluid flow. { 'lē ,wāv }

left-handed [CRYSTAL] Having a crystal structure with a mirror-image relationship to a right-handed structure. { 'left ¦hand·əd }

left-hand polarization [ELECTROMAG] In elementary-particle discussions, circular or elliptical polarization of an electromagnetic wave in which the electric field vector at a fixed point in space rotates in the left-hand sense about the direction of propagation; in optics, the opposite convention is used; in facing the source of the beam, the electric vector is observed to rotate counterclockwise. { 'left ¦hand ,pō·lə·rə'zā·shən }

left-hand rule [ELECTROMAG] **1.** For a current-carrying wire, the rule that if the fingers of the left hand are placed around the wire so that the thumb points in the direction of electron flow, the fingers will be pointing in the direction of the magnetic field produced by the wire. **2.** For a current-carrying wire in a magnetic field, such as a wire on the armature of a motor, the rule that if the thumb, first, and second fingers of the left hand are extended at right angles to one another, with the first finger representing the direction of magnetic lines of force and the second finger representing the direction of current flow, the thumb will be pointing in the direction of force on the wire. Also known as Fleming's rule. { 'left ¦hand ,rül }

Legendre contact transformation *See* Legendre transformation. { lə'zhän·drə 'kän,tak ,tranz·fər,mā·shən }

Legendre transformation [FL MECH] The basis for a version of the hodograph method for compressible flow in which a replacement is made not only of the independent variables but also of the dependent variables, that is, of the velocity potential and the stream function. { lə'zhän·drə ,tranz·fər'mā·shən }

Leidenfrost point [THERMO] The lowest temperature at which a hot body submerged in a pool of boiling water is completely blanketed by a vapor film; there is a minimum in the heat flux from the body to the water at this temperature. { 'līd·ən,frȯst ,pȯint }

Leidenfrost's phenomenon [THERMO] A phenomenon in which a liquid dropped on a surface that is above a critical temperature becomes insulated from the surface by a layer of vapor, and does not wet the surface as a result. { 'līd·ən,frȯsts fə,nam·ə,nän }

L electron [ATOM PHYS] An electron in the L shell. { 'el i,lek,trän }

length [MECH] Extension in space. { leŋkth }

lengthened dipole [ELECTROMAG] An antenna element with lumped inductance to compensate an end loss. { 'leŋk·thənd 'dī,pōl }

lens [ELECTROMAG] *See* magnetic lens. [OPTICS] A curved piece of ground and polished or molded material, usually glass, used for the refraction of light, its two surfaces having the same axis; or two or more such surfaces cemented together. Also known as optical lens. { lenz }

lens antenna [ELECTROMAG] A microwave antenna in which a dielectric lens is placed in front of the dipole or horn radiator to concentrate the radiated energy into a narrow beam or to focus received energy on the receiving dipole or horn. { 'lenz an,ten·ə }

lens coating [OPTICS] A transparent substance coated on an optical surface to derive maximum light transmission. { 'lenz ,kōd·iŋ }

lens element [OPTICS] Separate component lens of a multielement lens. { 'lenz ,el·ə·mənt }

lens equation [OPTICS] Any equation which relates the distance of a point object from some well-defined reference point in an optical system to the distance of its image from a similar point. { 'lenz i,kwā·zhən }

Lense-Thirring effect *See* dragging of inertial frames. { 'len·zə 'tir·iŋ i,fekt }

lens shim [OPTICS] Thin piece of material used to position and focus a lens. { 'lenz ,shim }

lens stop *See* diaphragm. { 'lenz ,stäp }

lenticular [OPTICS] Of or pertaining to a lens. { len'tik·yə·lər }

lentor *See* stoke. { 'len,tȯr }

Lenz's law [ELECTROMAG] The law that whenever there is an induced electromotive force (emf) in a conductor, it is always in such a direction that the current it would produce would oppose the change which causes the induced emf. { 'lenz·əz ,lȯ }

leo [MECH] A unit of acceleration, equal to 10 meters per second per second; it has rarely been employed. { 'lē·ō }

LEPD *See* low-energy positron diffraction.

lepton [PART PHYS] A fermion having a mass smaller than the proton mass; leptons interact with electromagnetic and gravitational fields, but beyond this they interact only through weak interactions. { 'lep,tän }

lepton conservation [PART PHYS] The principle that the number of electrons and *e*-neutrinos minus the number of positrons and *e*-antineutrinos is unchanged in any interaction; similarly, the number of negatively charged muons and μ-neutrinos minus the number of positively charged muons and μ-antineutrinos is unchanged. { 'lep,tän ,kän·sər,vā·shən }

leptonic decay [PART PHYS] Decay of an elementary particle in which at least some of the products are leptons. { lep'tän·ik di'kā }

lepton number [PART PHYS] A conserved quantum number, equal to the number of leptons minus the number of antileptons in a system. { 'lep,tän ,nəm·bər }

leptoquark [PART PHYS] A hypothetical elementary particle; a colored technipion with a mass of the order of 100–300 GeV, which would decay into a lepton plus a quark. { 'lep·tō,kwärk }

leptoquark boson [PART PHYS] A charged-vector gauge boson postulated in grand unified theories of color and electroweak forces, with a mass of the order of 10^{15} GeV, that can change quarks to leptons or quarks to antiquarks, and is responsible for proton decay. { 'lep·tō,kwärk 'bō,sän }

Leslie cube [THERMO] A metal box, with faces having different surface finishes, in which water is heated and next to which a thermopile is placed in order to compare the heat emission properties of different surfaces. { 'lez·lē ,kyüb }

leucitohedron *See* trapezohedron. { ,lü·sə·tō'hē·drən }

level measurement [MECH] The determination of the linear vertical distance between a reference point or datum plane and the surface of a liquid or the top of a pile of divided solid. { 'lev·əl 'mezh·ər·mənt }

level point *See* point of fall. { 'lev·əl ,pȯint }

level scheme *See* energy-level diagram. { 'lev·əl ,skēm }

level width [QUANT MECH] A measure of the spread in energy of an unstable state, equal to the difference between the energies at which intensity of emission or absorption of photons or particles, or the cross section for a reaction, is one-half its maximum value. { 'lev·əl 'width }

leverage [MECH] The multiplication of force or motion achieved by a lever. { 'lev·rij }

Leverett function [FL MECH] A dimensionless number used in studying two-phase flow in porous mediums, written as $(\xi/e)^{1/2}(p/\sigma)$, where ξ is the permeability of a medium (as defined by Darcy's law), e is the medium's porosity, σ is the surface tension between two liquids flowing through it, and p is the capillary pressure. { 'lev·rət ,fəŋk·shən }

levitation [PHYS] The use of a force that does not involve physical contact to balance gravity, such as that associated with an electric or magnetic field, or electromagnetic or acoustic radiation. { ,lev·ə'tā·shən }

levorotation [OPTICS] Rotation of the plane of polarization of plane polarized light in a counterclockwise direction, as seen by an observer facing in the direction of light propagation. Also known as levulorotation. { ¦lē·və·rō'tā·shən }

levulorotation *See* levorotation. { ¦lē·vyə·lō·rō'tā·shən }

Lewis number [PHYS] **1.** A dimensionless number used in studies of combined heat and mass transfer, equal to the thermal diffusivity divided by the diffusion coefficient. Symbolized Le; N_{Le}. **2.** Sometimes, the reciprocal of this quantity. { 'lü·əs ,nəm·bər }

Leyden jar [ELEC] An early type of capacitor, consisting simply of metal foil sheets on the inner and outer surfaces of a glass jar. { 'līd·ən ,jär }

LiBeB process *See* l-process. { ¦el,ī¦bē¦ē'bē ,prä·səs }

libration [PHYS] Any oscillatory rotational motion, such as that of the moon, or of a molecule in a solid which does not have enough energy to make full rotations. { lī'brā·shən }

Lichenberger figures *See* Lichtenberg figures. { 'lī·kən,bər·gər ,fig·yərz }

Lichtenberg figures [ELEC] Patterns produced on a photographic emulsion, or in fine powder spread over the surface of a solid dielectric, by an electric discharge produced by a high transient voltage. Also known as Lichenberger figures. { 'lik·tən·bərg ,fig·yərz }

lidar [OPTICS] An instrument in which a laser generates intense infrared pulses in beam widths as small as 30 seconds of arc; beam reflections and scattering effects of clouds, smog layers, and some atmospheric discontinuities are measured by radar techniques; it can also be used for tracking weather balloons, smoke puffs, and rocket trails. Derived from laser infrared radar. { 'lī,där }

Liebmann effect [OPTICS] The effect whereby it is more difficult to visually distinguish contrasting forms when they have the same luminance and different chromaticities than when they have different luminances and the same chromaticity. { 'lēb,mən i,fekt }

Liénard-Wiechert potentials [ELECTROMAG] The retarded and advanced electromagnetic scalar and vector potentials produced by a moving point charge, expressed in terms of the (retarded or advanced) position and velocity of the charge. { 'lē,närt 'vē·kərt pə,ten·chəlz }

lifetime *See* mean life. { 'līf,tīm }

lift *See* aerodynamic lift. { lift }

lifting magnet [ELECTROMAG] A type of electromagnet in which a material to be held or moved is initially placed in contact with the magnet, in contrast to a traction magnet. Also known as holding magnet. { 'lift·iŋ ,mag·nət }

light [OPTICS] **1.** Electromagnetic radiation with wavelengths capable of causing the sensation of vision, ranging approximately from 400 (extreme violet) to 770 nanometers (extreme red). Also known as light radiation; visible radiation. **2.** More generally, electromagnetic radiation of any wavelength; thus, the term is sometimes applied to infrared and ultraviolet radiation. { līt }

light absorption [OPTICS] The process in which energy of light radiation is transferred to a medium through which it is passing. { 'līt əb,sorp·shən }

light amplification by stimulated emission of radiation *See* laser. { 'līt ,am·plə·fə'kā·shən bī ¦stim·yə,lād·əd i¦mish·ən əv ,rād·ē'ā·shən }

light cone [RELAT] The set of all points in spacetime that are reached by signals traveling at the speed of light from a specified point, or from which signals traveling at the speed of light reach that point. Also known as null cone. { 'līt ,kōn }

light-distribution photometer [OPTICS] A device which measures the luminous intensity of a light source in various directions; the light source is fixed, and a mirror system is rotated about an axis passing through the centers of the light source and a photocell so that the light emitted by the source in any direction perpendicular to this axis is reflected to the photocell. Also known as distribution photometer. { 'līt dis·trə,byü·shən fə'täm·əd·ər }

light exposure [OPTICS] A measure of the total amount of light falling on a surface; equal to the integral over time of the luminance of the surface. Also known as exposure. { 'līt ik ,spō·zhər }

light filter *See* color filter. { 'līt ,fil·tər }

light guide *See* optical fiber. { 'līt ,gīd }

light hydrogen *See* protium. { 'līt 'hī·drə·jən }

light intensity *See* luminous intensity. { 'līt in,ten·səd·ē }

light microscope *See* optical microscope. { 'līt ,mī·krə,skōp }

light microsecond [ELECTROMAG] Distance a light wave travels in free space in one-millionth of a second. { 'līt ,mī·krə,sek·ənd }

light pipe [OPTICS] A solid, transparent plastic rod that transmits light from one end to the other even when bent. { 'līt ,pīp }

light projector [OPTICS] A device designed to produce controlled beams of light that can be projected over considerable distances. { 'līt prə,jek·tər }

light quantum *See* photon. { 'līt ,kwän·təm }

light radiation *See* light. { 'līt ,rād·ē,ā·shən }

light ray [OPTICS] A beam of light having a small cross section. { 'līt ,rā }

light scattering [OPTICS] The process in which energy is removed from a beam of light radiation and reemitted without appreciable change in wavelength. { 'līt ,skad·ə·riŋ }

light-sheet method [FL MECH] A method of flow visualization in which a light beam is broadened by a cylindrical lens to illuminate a plane sheet of fluid; the light scattered from tracer particles in the illuminated plane exhibits the flow pattern. { 'līt ,shēt ,meth·əd }

light source [OPTICS] A lamp used to supply radiant energy, as for an optical microscope, projector, or photoelectric control system. { 'līt ,sȯrs }

light transmission [OPTICS] The process in which light travels through a medium without being absorbed or scattered. { 'līt tranz,mish·ən }

light watt [OPTICS] A unit of luminous power equal to the luminous power of light of a single wavelength λ whose radiant power is $1/V^{\lambda}$ watts, where V^{λ} is the value of the luminosity function at λ. { 'līt ,wät }

liminal contrast *See* threshold contrast. { 'lim·ə·nəl 'kän,trast }

limiting friction *See* static friction. { 'lim·əd·iŋ ,frik·shən }

limiting ray [ACOUS] Any ray which is tangent to a plane at which the velocity of propagation of sound has a maximum value, either at aboundary of the medium of propagation or at a level where the velocity gradient changes sign. { 'lim·əd·iŋ 'rā }

limit of resolution [OPTICS] The minimum distance or angular separation between two point objects which allows them to be resolved according to the Rayleigh criterion. { 'lim·ət əv ˌrez·ə'lü·shən }

limit velocity [MECH] In armor and projectile testing, the lowest possible velocity at which any one of the complete penetrations is obtained; since the limit velocity is difficult to obtain, a more easily obtainable value, designated as the ballistic limit, is usually employed. { 'lim·ət və'läs·əd·ē }

Lindeck potentiometer [ELEC] A potentiometer in which an unknown potential difference is balanced against a known potential difference derived from a fixed resistance carrying a variable current; the converse of most potentiometers. { 'lin,dek pə,ten·chē'äm·əd·ər }

Lindemann theory [SOLID STATE] A theory of the melting point of solids according to which solids melt when the amplitude of oscillation of the atoms becomes so great that neighboring atoms collide. { 'lin·də·mən ˌthē·ə·rē }

Linde process [CRYO] A cyclic process for liquefying gases in which compressed gas is cooled by Joule-Thomson expansion through a valve to a pressure of about 40 atmospheres (4 megapascals), further cools the incoming gas in a heat exchanger, and is compressed for the next cycle. { 'lin·də ˌprä,ses }

Linde's rule [SOLID STATE] The rule that the increase in electrical resistivity of a monovalent metal produced by a substitutional impurity per atomic percent impurity is equal to $a + b(v - 1)^2$, where a and b are constants for a given solvent metal and a given row of the periodic table for the impurity, and v is the valence of the impurity. { 'lin·dəz ˌrül }

lineage structure [CRYSTAL] An imperfection structure characterizing a crystal, parts of which have slight differences in orientation. { 'lin·ē·ij ˌstrək·chər }

linear array [ELECTROMAG] An antenna array in which the dipole or other half-wave elements are arranged end to end on the same straight line. Also known as collinear array. { 'lin·ē·ər ə'rā }

linear birefringence [OPTICS] Birefringence effects which are proportional to applied stresses. { 'lin·ē·ər ˌbī·ri'frin·jəns }

linear circuit See linear network. { 'lin·ē·ər 'sər·kət }

linear collision cascade [SOLID STATE] A sputtering event in which the bombarding projectile collides directly with a small number of target atoms, which collide with others, and the sharing of energy then proceeds through many generations before one or more target atoms are ejected; the density of atoms in motion remains sufficiently small so that collisions between atoms can be ignored. { 'lin·ē·ər kə'lizh·ən ˌkas,kād }

linear combination of atomic orbitals [PHYS] A method of constructing approximate wave functions for molecular orbitals or for electrons in solids, by taking sums of atomic orbitals of the component atoms, each centered on an atom in the structure and multiplied by a coefficient, and then varying these coefficients to minimize the energy of the wave function. Abbreviated LCAO. { ¦lin·ē·ər ˌkäm·bə¦nā·shən əv ə¦täm·ik 'ȯrb·əd·əlz }

linear conductor antenna [ELECTROMAG] An antenna consisting of one or more wires which all lie along a straight line. { 'lin·ē·ər kən'dək·tər an,ten·ə }

linear density [PHYS] The quantity of anything distributed along a line per unit length of line. { 'lin·ē·ər 'den·səd·ē }

linear expansion [PHYS] Expansion of a body in one direction. { 'lin·ē·ər ik'span·chən }

linear expansity See coefficient of linear expansion. { 'lin·ē·ər ik'span·səd·ē }

linearity [PHYS] The relationship that exists between two quantities when a change in one of them produces a directly proportional change in the other. { ˌlin·ē'ar·əd·ē }

linearized theory of fluid flow [FL MECH] An approximate method for solving aerodynamic problems; it treats the flow of an inviscid gas past a body whose geometry and motion are such that the disturbance velocities caused by its introduction into some previously known flow are small compared with the speed of sound; as a result, the equations of motion can be approximated by retaining only those terms which are linear in disturbance or perturbation velocities, pressures, densities, and so forth. { ¦lin·ē·ə,rīzd ˌthē·ə·rē əv 'flü·əd ˌflō }

linear momentum See momentum. { 'lin·ē·ər mə'men·təm }

linear motion See rectilinear motion. { 'lin·ē·ər 'mō·shən }

linear network [ELEC] A network in which the parameters of resistance, inductance, and capacitance are constant with respect to current or voltage, and in which the voltage or current of sources is independent of or directly proportional to other voltages and currents, or their derivatives, in the network. Also known as linear circuit. { 'lin·ē·ər 'net,wərk }

linear polarization [OPTICS] Polarization of an electromagnetic wave in which the electric vector at a fixed point in space remains pointing in a fixed direction, although varying in magnitude. Also known as plane polarization. { 'lin·ē·ər ˌpō·lə·rə'zā·shən }

linear Stark effect [ATOM PHYS] A splitting of spectral lines of hydrogenlike atoms placed in an electric field; each energy level of principal quantum number n is split into $2n$-1 equidistant levels of separation proportional to the field strength. { 'lin·ē·ər 'stärk i,fekt }

linear strain [MECH] The ratio of the change in the length of a body to its initial length. Also known as longitudinal strain. { 'lin·ē·ər ¦strān }

linear velocity See velocity. { 'lin·ē·ər və'läs·əd·ē }

line broadening [SPECT] An increase in the range of wavelengths over which the characteristic absorption or emission of a spectral line takes place, due to a number of causes such as collision broadening and Doppler broadening. { 'līn ˌbrȯd·ən·iŋ }

line-building-out network *See* impedance-matching network. { 'līn 'bild·iŋ ¦au̇t ˌnetˌwərk }

line defect *See* dislocation. { 'līn diˌfekt }

line flux [ELECTROMAG] A local inductive field of a telephone or power line. { 'līn ˌfləks }

line impedance [ELECTROMAG] The impedance measured across the terminals of a transmission line. { 'līn imˌpēd·əns }

line influence [ELECTROMAG] The effect of a local inductive field around a telephone line. { 'līn ¦in·flü·əns }

line lengthener [ELECTROMAG] Device for altering the electrical length of a waveguide or transmission line without altering other electrical characteristics, or the physical length. { 'līn ˌleŋk·thə·nər }

line of collimation [OPTICS] In a surveying telescope, the imaginary line through the optical center of the object glass and the cross-hair intersection in the diaphragm. { 'līn əv ˌkäl·ə'mā·shən }

line of electrostatic induction [ELEC] A unit of electric flux equal to the electric flux associated with a charge of 1 statcoulomb. { 'līn əv iˌlek·trəˌstad·ik in'dək·shən }

line of fall [MECH] The line tangent to the ballistic trajectory at the level point. { 'līn əv ¦fȯl }

line of flight [MECH] The line of movement, or the intended line of movement, of an aircraft, guided missile, or projectile in the air. { 'līn əv ¦flīt }

line of flux *See* line of force. { 'līn əv ¦fləks }

line of force [PHYS] An imaginary line in a field of force (such as an electric, magnetic, or gravitational field) whose tangent at any point gives the direction of the field at that point; the lines are spaced so that the number through a unit area perpendicular to the field represents the intensity of the field. Also known as flux line; line of flux. { 'līn əv ¦fȯrs }

line of impact [MECH] A line tangent to the trajectory of a missile at the point of impact. { 'līn əv 'imˌpakt }

line of magnetic induction *See* maxwell. { 'līn əv mag¦ned·ik in'dək·shən }

line of sight [ELECTROMAG] The straight line for a transmitting radar antenna in the direction of the beam. { 'līn əv 'sīt }

line-of-sight velocity *See* radial velocity. { 'līn əv 'sīt və'läs·əd·ē }

line of thrust [MECH] Locus of the points through which the resultant forces pass in an arch or retaining wall. { 'līn əv 'thrəst }

line pair [SPECT] In spectrographic analysis, a particular spectral line and the internal standard line with which it is compared to determine the concentration of a substance. { 'līn ˌper }

line source [OPTICS] An idealized source of light consisting of an infinitely long line from which light is emitted with uniform intensity. { 'līn ˌsȯrs }

line spectrum [SPECT] **1.** A spectrum of radiation in which the quantity being studied, such as frequency or energy, takes on discrete values. **2.** Conventionally, the spectra of atoms, ions, and certain molecules in the gaseous phase at low pressures; distinguished from band spectra of molecules, which consist of a pattern of closely spaced spectral lines which could not be resolved by early spectroscopes. { 'līn ˌspek·trəm }

line strength [ATOM PHYS] The intensity of a spectrum line. { 'līn ˌstreŋkth }

line stretcher [ELECTROMAG] Section of waveguide or rigid coaxial line whose physical length is variable to provide impedance matching. { 'līn ˌstrech·ər }

line-turn *See* Maxwell-turn. { 'līn ˌtərn }

line vortex [FL MECH] A type of fluid motion in which fluid flows approximately in circles about a line, at speeds inversely proportional to the distance from the line, so that there is an infinite concentration of vorticity on the line, and vorticity vanishes elsewhere. { 'līn 'vȯrˌteks }

linewidth [ATOM PHYS] A measure of the width of the band of frequencies of radiation emitted or absorbed in an atomic or molecular transition, given by the difference between the upper and lower frequencies at which the intensity of radiation reaches half its maximum value. { 'līnˌwidth }

linkage *See* flux linkage. { 'liŋ·kij }

link circuit [ELECTROMAG] Closed loop used for coupling purposes; it generally consists of two coils, each having a few turns of wire, connected by a twisted pair of wires or by other means, with each coil placed over, near, or in one of the two coils that are to be coupled. { 'liŋk ˌsər·kət }

link coupling [ELECTROMAG] Modification of inductive coupling where the two coils are connected together by a short length of transmission line, with each coil inductively coupled to the coil of a separate tuned circuit. { 'liŋk ˌkəp·liŋ }

Linnik interference microscope [OPTICS] A type of interference microscope used for studying the surface structure of reflecting specimens; light from a source is divided by a semireflecting mirror into two beams, one of which is focused through an objective onto the specimen surface, the other onto a comparison surface; after reflection from the respective surfaces, the beams are reunited by the mirror. { 'lin·ik ˌin·tər¦fir·əns 'mī·krəˌskōp }

LINUP *See* laser-induced nuclear polarization. { 'līnˌəp }

Liouville equation [STAT MECH] An equation which states that the density of points representing an ensemble of systems in phase space which are in the neighborhood of some given system does not change with time. { 'lyüˌvēl iˌkwā·zhən }

Lippich prism [OPTICS] A Nicol prism which is placed in the eyepiece of a polarimeter, covering half the field of view, to identify the character of polarized light emerging from the instrument. { 'lip·ik ˌpriz·əm }

Lippmann effect [PHYS] A change in surface tension that results from a potential difference across the interface between two immiscible liquid conductors. { 'lip·mən iˌfekt }

Lippmann fringes [OPTICS] Interference fringes in standing electromagnetic waves generated when light is reflected by a mercury coating at the back of a special fine-grained photographic emulsion; originally used in color photography. { 'lip·mən ˌfrin·jəz }

liq pt *See* pint.

liquefaction [PHYS] A change in the phase of a substance to the liquid state; usually, a change from the gaseous to the liquid state, especially of a substance which is a gas at normal pressure and temperature. { ˌlik·wə'fak·shən }

liquid [PHYS] A state of matter intermediate between that of crystalline substances and gases in which a substance has the capacity to flow under extremely small shear stresses and conforms to the shape of a confining vessel, but is relatively incompressible, lacks the capacity to expand without limit, and can possess a free surface. { 'lik·wəd }

liquid A [CRYO] A phase of superfluid helium-3 in which the helium-3 pairs only occur in those two of the three possible nuclear spin states in which the nuclear spins are parallel, and these pairs couple coherently to give macroscopic orbital and spin angular momenta and anisotropic superfluid properties. Also known as A phase. { 'lik·wəd 'ā }

liquid A_1 [CRYO] A phase of liquid helium-3 intermediate between liquid A and liquid B that appears only in the presence of a magnetic field and then only in a narrow portion of the pressure-temperature diagram, and in which only pairs of one of the three possible nuclear spin states are superfluid. Also known as A_1 phase. { 'lik·wəd 'ā 'wən }

liquid air [PHYS] Air in the liquid state obtained as a faintly bluish, transparent, mobile, intensely cold liquid by compressing purified air and cooling it to a temperature below the boiling points of its principal components, nitrogen and oxygen; used chiefly as a refrigerant. { 'lik·wəd 'er }

liquid B [CRYO] A phase of superfluid helium-3 in which pairs of all three possible nuclear spin states are coupled to give superfluid properties that are isotropic except in the more subtle aspects of the spin configuration. Also known as B phase. { 'lik·wəd 'bē }

liquid-bubble tracer [FL MECH] A method of observing the motion of a liquid by following tiny particles of an immiscible liquid of the same density as the moving liquid. { 'lik·wəd ¦bəb·əl ˌtrā·sər }

liquid degeneracy [STAT MECH] A process in which a liquid cooled below a certain temperature loses the entropy associated with disordered motion of its molecules, without becoming a solid. { ¦lik·wəd di'jen·ə·rə·sē }

liquid-dielectric capacitor [ELEC] A capacitor in which the plate assemblies are mounted in a tank filled with a suitable oil or liquid dielectric. { 'lik·wəd ˌdī·ə¦lek·trik kə'pas·əd·ər }

liquid-drop model [NUC PHYS] A model of the nucleus in which it is compared to a drop of incompressible liquid, and the nucleons are analogous to molecules in the liquid; used to study binding energies, fission, collective motion, decay, and reactions. Also known as drop model. { 'lik·wəd ¦dräp ˌmäd·əl }

liquid flow [FL MECH] The flow or movement of materials in the liquid phase. { 'lik·wəd 'flō }

liquid fluorine [CRYO] Cold, liquefied fluorine gas; used as a cryogenic propellant. { 'lik·wəd 'flu̇rˌēn }

liquid gas [PHYS] A gas in the liquid state. { 'lik·wəd 'gas }

liquid helium [CRYO] The state of helium which exists at atmospheric pressure at temperatures below −268.95°C (4.2 K), and for temperatures near absolute zero at pressures up to about 25 atmospheres (2.53 megapascals); has two phases, helium I and helium II. { 'lik·wəd 'hē·lē·əm }

liquid holdup [FL MECH] A condition in two-phase flow through a vertical pipe; when gas flows at a greater linear velocity than the liquid, slippage takes place and liquid holdup occurs. { 'lik·wəd 'hōlˌdəp }

liquid hydrogen [CRYO] Hydrogen that exists as a liquid at atmospheric pressure, at −252.7°C (20.46 K); used for high-impulse rocket fuels. { 'lik·wəd 'hī·drə·jən }

liquid laser [OPTICS] A laser whose active material is dissolved in a liquid contained in a transparent cylindrical shell; rare-earth ions in suitable dissolved molecules and organic dye solutions are used. { 'lik·wəd 'lā·zər }

liquid measure [MECH] A system of units used to measure the volumes of liquid substances in the United States; the units are the fluid dram, fluid ounce, gill, pint, quart, and gallon. { 'lik·wəd ¦mezh·ər }

liquid methane [CRYO] Methane that has been cooled to at least −161°C; used for cryogenic applications and for tankship transport of methane. { 'lik·wəd 'methˌān }

liquid nitrogen [CRYO] Nitrogen that exists as a liquid at atmospheric pressure, at −195°C (77.4 K); used in research work, cryogenics, and cryosurgery. { 'lik·wəd 'nī·trə·jən }

liquid oxygen [CRYO] Oxygen that exists as a liquid at atmospheric pressure, at −182.97°C (90.18 K); a pale-blue, transparent, mobile liquid. { 'lik·wəd 'äk·sə·jən }

liquid-phase epitaxy [SOLID STATE] A process for growing thin epitaxial layers on a crystalline substrate in which the substrate is sequentially brought into contact with solutions that are at the desired composition and may be supersaturated or cooled to achieve growth. Abbreviated LPE. { 'lik·wəd ¦fāz 'ep·əˌtak·sē }

liquid pint *See* pint. { 'lik·wəd 'pīnt }

liquidus line [THERMO] For a two-component system, a curve on a graph of temperature versus concentration which connects temperatures at which fusion is completed as the temperature is raised. { 'lik·wəd·əs ˌlīn }

Lissajous figure [PHYS] The path of a particle moving in a plane when the components of its position along two perpendicular axes each undergo simple harmonic motions and the ratio of their frequencies is a rational number. Also known as Bowditch curve. { ¦lē·sə¦zhü ˌfig·yər }

liter [MECH] A unit of volume or capacity, equal to 1 decimeter cubed, or 0.001 cubic meter, or 1000 cubic centimeters. Abbreviated l; L. { lēd·ər }

liter-atmosphere [PHYS] A unit of energy equal to the work done on a piston by a fluid at a pressure of 1 standard atmosphere (101,325 pascals) when the piston sweeps out a volume of 1 liter; equal to 101.325 joules. { 'lēd·ə·r ¦at·məˌsfir }

Littrow grating spectrograph [SPECT] A spectrograph having a plane grating at an angle to the axis of the instrument, and a lens in front of the grating which both collimates and focuses the light. { 'liˌtrō ¦grād·iŋ 'spek·trəˌgraf }

Littrow mounting [SPECT] The arrangement of the grating and other components of a Littrow grating spectrograph, which is analogous to that of a Littrow quartz spectrograph. { 'liˌtrō ˌmau̇nt·iŋ }

Littrow prism [OPTICS] A prism having angles of 30, 60, and 90°, silvered on the side opposite the 60° angle; a lens used with it can serve both as a telescope and as a collimator. { 'liˌtrō ˌpriz·əm }

Littrow quartz spectrograph [SPECT] A spectrograph in which dispersion is accomplished by a Littrow quartz prism with a rear reflecting surface that reverses the light; a lens in front of the prism acts as both collimator and focusing lens. { 'liˌtrō ¦kwȯrts 'spek·trəˌgraf }

litzendraht wire *See* litz wire. { 'lits·ənˌdrät ˌwīr }

litz wire [ELEC] Wire consisting of a number of separately insulated strands woven together so each strand successively takes up all possible positions in the cross section of the entire conductor, to reduce skin effect and thereby reduce radio-frequency resistance. Derived from litzendraht wire. { 'lits ˌwīr }

live *See* energized. { līv }

live end [ACOUS] The end of a radio studio that gives almost complete reflection of sound waves. { 'līv ¦end }

live load [MECH] A moving load or a load of variable force acting upon a structure, in addition to its own weight. { 'līv 'lōd }

live room [ACOUS] A room having a minimum of sound-absorbing material. { 'līv ¦rüm }

livre [MECH] A unit of mass, used in France, equal to 0.5 kilogram. { 'lēv·rə }

Lloyd's mirror interference [OPTICS] The interference pattern produced when part of the light from a slit falls directly on a screen, and part is reflected from a mirror whose surface makes a small angle with the incident beam. { 'loidz ¦mir·ər ˌin·tər'fir·əns }

L/M [NUC PHYS] The ratio of the number of internal conversion electrons emitted from the L shell in the de-excitation of a nucleus to the number of such electrons emitted from the M shell.

lm-hr *See* lumen-hour.

lm-sec *See* lumen-second.

lm/w *See* lumen per watt.

load [ELEC] **1.** A device that consumes electric power. **2.** The amount of electric power that is drawn from a power line, generator, or other power source. **3.** The material to be heated by an induction heater or dielectric heater. Also known as work. [MECH] **1.** The weight that is supported by a structure. **2.** Mechanical force that is applied to a body. **3.** The burden placed on any machine, measured by units such as horsepower, kilowatts, or tons. { lōd }

loaded Q [ELEC] The Q factor of an impedance which is connected or coupled under working conditions. Also known as working Q. [ELECTROMAG] The Q factor of a specific mode of resonance of a microwave tube or resonant cavity when there is external coupling to that mode. { 'lōd·əd kyü }

load factor [ELEC] The ratio of average electric load to peak load, usually calculated over a 1-hour period. [MECH] The ratio of load to the maximum rated load. { 'lōd ˌfak·tər }

loading [ELEC] The addition of inductance to a transmission line to improve its transmission characteristics throughout a given frequency band. Also known as electrical loading. [FL MECH] **1.** The relative concentration of particles in a flowing fluid. **2.** In particular, the ratio of particle mass flow to fluid mass flow. { 'lōd·iŋ }

loading coil [ELECTROMAG] **1.** An iron-core coil connected into a telephone line or cable at regular intervals to lessen the effect of line capacitance and reduce distortion. Also known as Pupin coil; telephone loading coil. **2.** A coil inserted in series with a radio antenna to increase its electrical length and thereby lower the resonant frequency. { 'lōd·iŋ ˌkȯil }

loading disk [ELECTROMAG] Circular metal piece mounted at the top of a vertical antenna to increase its natural wavelength. { 'lōd·iŋ ˌdisk }

loading noise [ACOUS] The component of propeller noise that is related to the lift and drag forces acting on the propeller blade. { 'lōd·iŋ ˌnȯiz }

load isolator [ELECTROMAG] Waveguide or coaxial device that provides a good energy path from a signal source to a load, but provides a poor energy path for reflections from a mismatched load back to the signal source. { 'lōd ˌī·səˌlād·ər }

load stress [MECH] Stress that results from a pressure or gravitational load. { 'lōd ˌstres }

lobe [ELECTROMAG] A part of the radiation pattern of a directional antenna representing an area of stronger radio-signal transmission. Also known as radiation lobe. { lōb }

lobe-half-power width [ELECTROMAG] In a plane containing the direction of the maximum energy of a lobe, the angle between the two directions in that plane about the maximum in which the radiation intensity is one-half the maximum value of the lobe. { ¦lōb ¦haf ¦pau̇·ər ˌwidth }

lobe penetration [ELECTROMAG] Penetration of the radar coverage of a station which is not limited by pulse repetition frequency, scope limitations, or the screening angle at the azimuth of penetration. { 'lōb ˌpen·ə'trā·shən }

lobing [ELECTROMAG] Formation of maxima and minima at various angles of the vertical plane antenna pattern by the reflection of energy from the surface surrounding the radar antenna; these reflections reinforce the main beam at some angles and detract from it at other angles, producing fingers of energy. { 'lōb·iŋ }

local buckling [MECH] Buckling of thin elements of a column section in a series of waves or wrinkles. { 'lō·kəl 'bək·liŋ }

local cell [ELEC] A galvanic cell resulting from differences in potential between adjacent areas on the surface of a metal immersed in an electrolyte. { 'lō·kəl 'sel }

local coefficient of heat transfer [THERMO] The heat transfer coefficient at a particular point on a surface, equal to the amount of heat transferred to an infinitesimal area of the surface at the point by a fluid passing over it, divided by the product of this area and the difference between the temperatures of the surface and the fluid. { 'lō·kəl ˌkō·i'fish·ənt əv 'hēt ˌtranz·fər }

local derivative [FL MECH] The rate of change of a quantity f with respect to time at a fixed point of a fluid, $\partial f/\partial t$; it is related to the individual derivative df/dt through the expression $\partial f/\partial t = df/dt - \mathrm{V} \cdot \nabla f$, where f is a thermodynamic property $f(x,y,z,t)$ of the fluid, V the vector velocity of the fluid, and ∇ the del operator. { 'lō·kəl də'riv·əd·iv }

local hidden-variable theory *See* hidden-variable theory of the second kind. { 'lō·kəl ¦hid·ən 'ver·ē·ə·bəl ˌthē·ə·rē }

local invariance [PHYS] The property of physical laws which remain unchanged under a specified set of symmetry transformations even when these transformations are chosen independently at every point of space and time. { 'lō·kəl in'ver·ē·əns }

locality [PHYS] The condition that two events at spatially separated locations are entirely independent of each other, provided that the time interval between the events is less than that required for a light signal to travel from one location to the other. { lō'kal·əd·ē }

localization discrimination suppression [ACOUS] The ability of human hearing to process changes in the location of a sound source without significant disruption from reflections. { ˌlōk·ə·lə¦zā·shən diˌskrim·ənā·shən səˌpresh·ən }

localization dominance [ACOUS] In human hearing, the effect whereby the perceived location of sound from a source combined with reflected sounds is dominated by the sound source. { ˌlōk·əl·ə¦zā·shən 'däm·ə·nəns }

localized state [QUANT MECH] A state of motion in which an electron may be found anywhere within a region of a material of linear extent smaller than that of the material. { 'lō·kəˌlīzd 'stāt }

localized vector [MECH] A vector whose line of application or point of application is prescribed, in addition to its direction. { 'lō·kəˌlīzd 'vek·tər }

localized wave solution [PHYS] A solution to the multidimensional wave equation in which the energy is concentrated in certain regions of space and time. { ˌlōk·əˌlīzd 'wāv səˌlü·shən }

local structural discontinuity [MECH] The effect of intensified stress on a small portion of a structure. { 'lō·kəl 'strək·chə·rəl disˌkänt·ən'ü·əd·ē }

logarithmic decrement [PHYS] The natural logarithm of the ratio of the amplitude of one oscillation to that of the next which has the same polarity, when no external forces are applied to maintain the oscillation. { 'läg·əˌrith·mik 'dek·rə·mənt }

logarithmic potential [PHYS] A potential function that is proportional to the logarithm of some coordinate; for example, a straight, electrically charged cylinder of circular cross section and effectively infinite length gives rise to an electrostatic potential that is the sum of a constant and a term proportional to the logarithm of the distance from the cylinder's axis. { 'läg·əˌrith·mik pə'ten·chəl }

logarithmic profile of velocity [FL MECH] The mean velocity parallel to a boundary of a fluid in turbulent motion as a function of distance from the boundary, on the assumption that the shearing stress is independent of distance from the boundary, and the mixing length is proportional either to the distance from the boundary or to the ratio of the first derivative of the profile of velocity itself to the second derivative. { 'läg·əˌrith·mik 'prōˌfīl əv və'läs·əd·ē }

log-mean temperature difference [THERMO] The log-mean temperature difference $T_{LM} = (T_2 - T_1)/\ln T_2/T_1$, where T_2 and T_1 are the absolute (K or °R) temperatures of the two extremes being averaged; used in heat transfer calculations in which one fluid is cooled or heated by a second held separate by pipes or process vessel walls. { 'läg ¦mēn 'tem·prə·chər ˌdif·rəns }

log-periodic antenna [ELECTROMAG] A broadband antenna which consists of a sheet of metal with two wedge-shaped cutouts, each with teeth cut into its radii along circular arcs; characteristics are repeated at a number of frequencies that are equally spaced on a logarithmic scale. { 'läg ˌpir·ē¦ad·ik an'ten·ə }

Lombard effect [ACOUS] The change in a talker's articulation effort when he or she speaks in a noisy environment; for example, trying to raise the voice or to make the voice better understood by the listener. { 'lumˌbärd iˌfekt }

London equations [SOLID STATE] Equations for the time derivative and the curl of the current in a superconductor in terms of the electric and magnetic field vectors respectively, derived in the London superconductivity theory. { 'lən·dən i¦kwā·zhənz }

London penetration depth [SOLID STATE] A measure of the depth which electric and magnetic fields can penetrate beneath the surface of a superconductor from which they are otherwise excluded, according to the London superconductivity theory. { 'lən·dən ,pen·ə'trā·shən ,depth }

London superconductivity theory [SOLID STATE] An extension of the two-fluid model of superconductivity, in which it is assumed that superfluid electrons behave as if the only force acting on them arises from applied electric fields, and that the curl of the superfluid current vanishes in the absence of a magnetic field. { 'lən·dən ¦sü·pər,kän,dək'tiv·əd·ē ,thē·ə·rē }

London superfluidity theory [CRYO] A theory, based on the fact that helium-4 obeys Bose-Einstein statistics, in which helium-4 is treated as an ideal Bose-Einstein gas, and its superfluid component is equated with the finite fraction of the atoms of such a gas which are in the ground state at very low temperatures. { 'lən·dən ,sü·pər,flü'id·əd·ē ,thē·ə·rē }

long-conductor antenna *See* long-wire antenna. { 'lȯŋ kən¦dək·tər an,ten·ə }

long discharge [ELEC] **1.** A capacitor or other electrical charge accumulator which takes a long time to leak off. **2.** A gaseous electrical discharge in which the length of the discharge channel is very long compared with its diameter; lightning discharges are natural examples of long discharges. Also known as long spark. { 'lȯŋ 'dis,chärj }

longitudinal aberration [OPTICS] **1.** The distance along the optical axis from the focus of paraxial rays to the point where rays coming from the outer edges of its lens or reflecting surface intersect this axis. **2.** In chromatic aberration, the distance along the optical axis between the foci of two standard colors. { ,län·jə'tüd·ən·əl ,ab·ə'rā·shən }

longitudinal acceleration [MECH] The component of the linear acceleration of an aircraft, missile, or particle parallel to its longitudinal, or X, axis. { ,län·jə'tüd·ən·əl ak,sel·ə'rā·shən }

longitudinal magnetoresistance [ELECTROMAG] The change of electrical resistance produced in a current-carrying metal or semiconductor upon application of a magnetic field parallel to the current flow. { ,län·jə'tüd·ən·əl mag¦ned·ō·ri 'zis·təns }

longitudinal magnetorestriction *See* Joule effect. { ,län·jə'tüd·ən·əl mag¦ned·ō·ri'strik·shən }

longitudinal mass [RELAT] The ratio of a force acting on a relativistic particle in the direction of its velocity to the resulting acceleration; equal to $m_0(1 - v^2/c^2)^{-3/2}$, where m_0 is the particle's rest mass, v is its speed, and c is the speed of light. { ,län·jə'tüd·ən·əl 'mas }

longitudinal quadrupole [ACOUS] A sound source resulting from a variation of a component of the velocity of matter in a direction parallel to the velocity component. [ELECTROMAG] An electric or magnetic quadrupole which produces a field equivalent to that of two equal and opposite electric or magnetic dipoles separated by a small distance parallel to the direction of the dipoles. Also known as axial quadrupole. { ,län·jə'tüd·ən·əl 'kwäd·rə,pōl }

longitudinal strain *See* linear strain. { ,län·jə'tüd·ən·əl 'strān }

longitudinal vibration [MECH] A continuing periodic change in the displacement of elements of a rod-shaped object in the direction of the long axis of the rod. { ,län·jə'tüd·ən·əl vī'brā·shən }

longitudinal wave [PHYS] A wave in which the direction of some vector characteristic of the wave, for example, the displacement of particles of the transmitting medium, is along the direction of propagation. { ,län·jə'tüd·ən·əl 'wāv }

long-range order [SOLID STATE] A tendency for some property of atoms in a lattice (such as spin orientation or type of atom) to follow a pattern which is repeated every few unit cells. { 'lȯŋ ,rānj 'ȯr·dər }

long spark *See* long discharge. { 'lȯŋ ¦spärk }

long ton *See* ton. { 'lȯŋ 'təŋ }

long-wavelength infrared radiation [ELECTROMAG] Infrared radiation having a wavelength greater than 8 micrometers. { 'lȯŋ 'wāv,leŋkth ¦in·frə¦red ,rād·ē'ā·shən }

long-wire antenna [ELECTROMAG] An antenna whose length is a number of times greater than its operating wavelength, so as to give a directional radiation pattern. Also known as long-conductor antenna. { 'lȯŋ ,wīr an'ten·ə }

looming [OPTICS] A form of mirage in which images of objects normally hidden below the horizon are seen in the sky, sometimes upside down; a common occurrence in the Far North. { 'lüm·iŋ }

Loomis-Wood diagram [SPECT] A graph used to assign lines in a molecular spectrum to the various branches of rotational bands when these branches overlap, in which the difference between observed wave numbers and wave numbers extrapolated from a few lines that apparently belong to one branch are plotted against arbitrary running numbers for that branch. { 'lü·məs 'wu̇d ,dī·ə,gram }

loop [ELEC] **1.** A closed path or circuit over which a signal can circulate, as in a feedback control system. **2.** Commercially, the portion of a connection from central office to subscriber in a telephone system. [ELECTROMAG] *See* coupling loop; loop antenna. [PHYS] **1.** A closed curve on a graph, such as a hysteresis loop. **2.** That part of a standing wave where the vertical motion is greatest and the horizontal velocities are least. { lüp }

loop antenna [ELECTROMAG] A directional-type antenna consisting of one or more complete turns of a conductor, usually tuned to resonance by a variable capacitor connected to the terminals of the loop. Also known as loop. { 'lüp an,ten·ə }

loop coupling [ELECTROMAG] A method of

transferring energy between a waveguide and an external circuit, by inserting a conducting loop into the waveguide, oriented so that electric lines of flux pass through it. { 'lüp ˌkəp·liŋ }

loop flow *See* parallel flow. { 'lüp ˌflō }

loopstick antenna *See* ferrite-rod antenna. { 'lüp ˌstikanˌten·ə }

loose coupling [ELEC] Coupling of a degree less than the critical coupling. { 'lüs 'kəp·liŋ }

Lorentz-Boltzmann equation [STAT MECH] An approximation to the Boltzmann transport equation for states that are near equilibrium, which shows that the Maxwell-Boltzmann distribution applies at equilibrium. { 'lȯrˌens ˌbōlts·mən iˌkwā·zhən }

Lorentz conductivity theory *See* classical conductivity theory. { 'lȯrˌens ˌkänˌdək'tiv·əd·ē ˌthē·ə·rē }

Lorentz contraction *See* FitzGerald-Lorentz contraction. { 'lȯrˌens kən'trak·shən }

Lorentz electron [ELECTROMAG] A model of the electron as a damped harmonic oscillator; used to explain the variation of the real and imaginary parts of the index of refraction of a substance with frequency. { 'lȯrˌens i'lekˌträn }

Lorentz equation [ELECTROMAG] The equation of motion for a charged particle, which sets the rate of change of its momentum equal to the Lorentz force. { 'lȯrˌens i'kwā·zhən }

Lorentz factor [RELAT] An important parameter in special relativity, equal to $1\sqrt{1-(v/c)^2}$, where c is the speed of light and v is the constant relative velocity of two frames of reference. { 'lȯrˌens ˌfak·tər }

Lorentz-FitzGerald contraction *See* FitzGerald-Lorentz contraction. { 'lȯrˌens fits'jer·əld kən ˌtrak·shən }

Lorentz force [ELECTROMAG] The force on a charged particle moving in electric and magnetic fields, equal to the particle's charge times the sum of the electric field and the cross product of the particle's velocity with the magnetic flux density. { 'lȯrˌens ˌfȯrs }

Lorentz-force density [ELECTROMAG] The force per unit volume on a charge density and current density, assuming that these densities arise from large numbers of charged particles experiencing a Lorentz force. { 'lȯrˌens ¦fȯrs ˌden·səd·ē }

Lorentz four-vector *See* four-vector. { 'lȯrˌens 'fȯr ˌvek·tər }

Lorentz frame [RELAT] Any of the family of inertial coordinate systems, with three space coordinates and one time coordinate, used in the special theory of relativity; each frame is in uniform motion with respect to all the other Lorentz frames, and the interval between any two events is the same in all frames. { 'lȯrˌens ˌfrām }

Lorentz gage [ELECTROMAG] Any gage in which the sum of the divergence of the vector potential and the partial derivative of the scalar potential divided by the speed of light (in Gaussian units) vanishes identically; it is always possible to find a gage satisfying this condition. { 'lȯrˌens ˌgāj }

Lorentz-Heaviside system *See* Heaviside-Lorentz system. { 'lȯrˌens 'hev·ēˌsīd ˌsis·təm }

Lorentz invariance [RELAT] The property, possessed by the laws of physics and of certain physical quantities, of being the same in any Lorentz frame, and thus unchanged by a Lorentz transformation. { 'lȯrˌens inˌver·ē·əns }

Lorentz line-splitting theory [ATOM PHYS] A theory predicting that when a light source is placed in a strong magnetic field, its spectral lines are each split into three components, one of them retaining the zero-field frequency, and the other two shifted upward and downward in frequency by the Larmor frequency (the normal Zeeman effect). { 'lȯrˌens 'līn ˌsplid·iŋ ˌthē·ə·rē }

Lorentz local field [ELEC] In a theory of electric polarization, the average electric field due to the polarization at a molecular site that is calculated under the assumption that the field due to polarization by molecules inside a small sphere centered at the site may be neglected. Also known as Mossotti field. { 'lȯrˌens ¦lō·kəl 'fēld }

Lorentz-Lorenz equation [OPTICS] The equation that results from replacing the relative dielectric constant with the square of the index of refraction in the Clausius-Mossotti equation. { 'lȯrˌens 'lȯrˌens iˌkwā·shən }

Lorentz-Lorenz molar refraction *See* molar refraction. { 'lȯrˌens 'lȯrˌens ¦mō·lər ri'frak·shən }

Lorentz matrix [RELAT] A matrix whose product with a vector whose components are the space and time coordinates of an event yields a vector whose components are new coordinates derived from the original ones by a Lorentz transformation. { 'lȯrˌens ˌmā·triks }

Lorentz number [PL PHYS] The ratio of the velocity of a fluid to the velocity of light. Symbolized N_{Lo}. [SOLID STATE] The thermal conductivity of a metal divided by the product of its temperature and its electrical conductivity, according to the Wiedemann-Franz law. { 'lȯrˌens ˌnəm·bər }

Lorentz polarization factor [OPTICS] A geometric factor in the equation for the intensity of x-rays or other radiation diffracted through a given angle by a crystalline substance. { 'lȯrˌens ˌpō·lə·rə'zā·shən ˌfak·tər }

Lorentz relation *See* Wiedemann-Franz law. { 'lȯr ˌens riˌlā·shən }

Lorentz theory of light sources [ATOM PHYS] A theory according to which light is emitted by vibrations of electrons, which are damped harmonic oscillators attached to atoms. { 'lȯrˌens 'thē·ə·rē əv 'līt ˌsȯrs·əz }

Lorentz transformation [RELAT] Any of the family of mathematical transformations used in the special theory of relativity to relate the space and time variables of different Lorentz frames. { 'lȯrˌens ˌtranz·fərˌmā·shən }

Lorentz unit [SPECT] A unit of reciprocal length used to measure the difference, in wave numbers, between a (zero field) spectrum line and its Zeeman components; equal to $eH/4\pi mc^2$, where H is the magnetic field strength, c is the speed of light, and e and m are the charge and mass of the electron respectively (gaussian units). { 'lȯr,ens ,yü·nət }

Lorenz attractor [PHYS] The strange attractor for the solution of a system of three coupled, nonlinear, first-order differential equations that are encountered in the study of Rayleigh-Bénard convection; it is highly layered and has a fractal dimension of 2.06. Also know as Lorenz butterfly. { 'lȯr,ens ə,trak·tər }

Lorenz butterfly *See* Lorenz attractor. { ,lȯr·ənz 'bəd·ər,flī }

Loschmidt number [PHYS] The number of molecules in 1 cubic centimeter of an ideal gas at 1 atmosphere pressure and 0°C, equal to approximately 2.687×10^{19}. Symbolized n_0. { 'lō,shmit ,nəm·bər }

loss angle [ELECTROMAG] A measure of the power loss in an inductor or a capacitor, equal to the amount by which the angle between the phasors denoting voltage and current across the inductor or capacitor differs from 90°. { 'lȯs ,aŋ·gəl }

loss cone [PL PHYS] A cone in the velocity space of particles in a plasma confined by magnetic mirrors; particles with velocities in the cone are not trapped by the mirrors and are lost out of the system. { 'lȯs ,kōn }

loss-cone instability [PL PHYS] An instability in a plasma confined between magnetic mirrors. { 'lȯs ,kōn ,in·stə'bil·əd·ē }

loss current [ELEC] The current which passes through a capacitor as a result of the conductivity of the dielectric and results in power loss in the capacitor. [ELECTROMAG] The component of the current across an inductor which is in phase with the voltage (in phasor notation) and is associated with power losses in the inductor. { 'lȯs ,kə·rənt }

loss factor [ELEC] The power factor of a material multiplied by its dielectric constant; determines the amount of heat generated in a material. { 'lȯs ,fak·tər }

lossless junction [ELECTROMAG] A waveguide junction in which all the power incident on the junction is reflected from it. { 'lȯs·ləs ,jəŋk·shən }

lossless material [PHYS] An ideal material that dissipates none of the energy of electromagnetic or acoustic waves passing through it. { 'lȯs·ləs mə'tir·ē·əl }

loss of head [FL MECH] Energy decrease between two points in a hydraulic system due to such causes as friction, bends, obstructions, or expansions. { 'lȯs əv 'hed }

lossy attenuator [ELECTROMAG] In waveguide technique, a length of waveguide deliberately introducing a transmission loss by the use of some dissipative material. { 'lȯs·ē ə'ten·yə,wād·ər }

lossy material [PHYS] A material that dissipates energy of electromagnetic or acoustic energy passing through it. { 'lȯs·ē mə'tir·ē·əl }

loudness [ACOUS] The magnitude of the physiological sensation produced by a sound, which varies directly with the physical intensity of sound but also depends on frequency of sound and waveform. { 'lau̇d·nəs }

loudness level [ACOUS] The level of a sound, in phons, equal to the sound pressure level in decibels, relative to 0.0002 microbar, of a pure 1000-hertz tone that is judged to be equally loud by listeners. { 'lau̇d·nəs ,lev·əl }

loudness recruitment [ACOUS] An abnormal increase in perceived loudness as a sound is intensified. { 'lau̇d·nəs ri,krüt·mənt }

loudness unit [ACOUS] A unit of loudness equal to the loudness of a sound having a loudness level of 0 phon; the loudness unit has been replaced by the sone. { 'lau̇d·nəs ,yü·nət }

Lovibond tintometer [OPTICS] A colorimeter which compares a solution or object under examination with a series of slides of each of three colors. { 'lō·və,bänd tin'täm·əd·ər }

low-angle scattering *See* small-angle scattering. { 'lō ,aŋ·gəl 'skad·ər·iŋ }

low-energy electron diffraction [SOLID STATE] A technique for studying the atomic structure of single crystal surfaces, in which electrons of uniform energy in the approximate range 5–500 electronvolts are scattered from a surface, and those scattered electrons that have lost no energy are selected and accelerated to a fluorescent screen where the diffraction pattern from the surface can be observed. Abbreviated LEED. { 'lō ,en·ər·jē i,lek,trän di'frak·shən }

low-energy physics [PHYS] That part of physics which studies microscopic phenomena involving energies of several million electronvolts or less, such as the arrangement of electrons in an atom or a solid, and the arrangement of protons and neutrons within the atomic nucleus, and the nature of forces between these particles. { 'lō ,en·ər·jē ,fiz·iks }

low-energy positron diffraction [SOLID STATE] A technique for studying the atomic structure of solid surfaces in which a narrow beam of low-energy monoenergetic positrons is made to strike a solid surface, and the diffracted beams in certain directions that are permitted by the regular array of surface atoms are observed. Abbreviated LEPD. { 'lō ,en·ər·jē 'päz·ə,trän di,frak·shən }

lower critical field [SOLID STATE] The magnetic field strength below which magnetic flux is completely excluded from type II superconductor and above which it penetrates the superconductor as microscopic filaments called fluxoids. Symbolized H_{c1}. { ¦lō·ər ¦krid·i·kəl 'fēld }

lower heating value *See* low heat value. { 'lō·ər 'hēd·iŋ ,val·yü }

lower pitch limit [ACOUS] Minimum frequency, for a sinusoidal sound wave, that will produce a pitch sensation. { 'lō·ər 'pich ,lim·ət }

low-frequency antenna [ELECTROMAG] An antenna designed to transmit or receive radiation at frequencies of less than about 300 kilohertz. { 'lō ,frē·kwən·sē an'ten·ə }

low-frequency propagation [ELECTROMAG]

Propagation of radio waves at frequencies between 30 and 300 kilohertz. { 'lō ˌfrē·kwən·sē ˌpräp·ə'gā·shən }

low-frequency spectrum [SPECT] Spectrum of atoms and molecules in the microwave region, arising from such causes as the coupling of electronic and nuclear angular momenta, and the Lamb shift. { 'lō ˌfrē·kwən·sē 'spek·trəm }

low heat value [THERMO] The heat value of a combustion process assuming that none of the water vapor resulting from the process is condensed out, so that its latent heat is not available. Also known as lower heating value; net heating value. { 'lō 'hēt ˌval·yü }

low-loss [ELEC] Having a small dissipation of electric or electromagnetic power. { 'lō ¦lȯs }

low-pressure fluid flow [FL MECH] Flow of fluids below atmospheric pressures, particularly gases and vapors following ideal gas laws, in pipes, fittings, and other common configurations. { 'lō ¦presh·ər 'flü·əd ˌflō }

low-reflection film [OPTICS] A transparent film covering a glass surface, designed so that a small proportion of the light incident will be reflected and a correspondingly large proportion transmitted into the glass. { 'lō riˌflek·shən 'film }

low-temperature physics [CRYO] A study of the properties of gross matter at low temperatures, especially at temperatures so low that the quantum character of the substance becomes observable in effects such as superconductivity, superfluid liquid helium, magnetic cooling, and nuclear orientation. { 'lō ˌtem·prə·chər 'fiz·iks }

low-temperature production [CRYO] Production of temperatures from about 80 K down to about 10^{-6} K by techniques such as isentropic expansion of gases, refrigeration cycles, and adiabatic demagnetization. { 'lō ˌtem·prə·chər prə'dək·shən }

low-temperature thermometry [CRYO] The assignment of numbers on the Kelvin absolute temperature scale to achievable and reproducible low-temperature states, and the choice and calibration of suitable instruments for the practical measurement of low temperatures, such as thermocouples, and resistance, vapor-pressure, gas, and magnetic thermometers. { 'lō ˌtem·prə·chər thər'mäm·ə·trē }

low velocity [MECH] Muzzle velocity of an artillery projectile of 2499 feet (762 meters) per second or less. { 'lō və'läs·əd·ē }

LPE *See* liquid-phase epitaxy.

l-process [NUC PHYS] The synthesis of certain light nuclides through the breakup of heavier nuclides, probably by cosmic-ray bombardment of the interstellar medium. Also known as LiBeB process. { 'el ˌprä·səs }

LS coupling *See* Russell-Saunders coupling. { ¦el¦es ˌkəp·liŋ }

L shell [ATOM PHYS] The second shell of electrons surrounding the nucleus of an atom, having electrons whose principal quantum number is 2. { ¦el ˌshel }

Ludwig-Soret effect [THERMO] A phenomenon in which a temperature gradient in a mixture of substances gives rise to a concentration gradient. { ¦lüdˌvik sə'rā iˌfekt }

lumberg [OPTICS] A unit of luminous energy equal to the luminous energy corresponding to a radiant energy of 1/K ergs, where K is the luminous efficiency in lumens per watt. Formerly known as lumerg. { 'lümˌbərg }

lumen [OPTICS] The unit of luminous flux, equal to the luminous flux emitted within a unit solid angle (1 steradian) from a point source having a uniform intensity of 1 candela, or to the luminous flux received on a unit surface, all points of which are at a unit distance from such a source. Symbolized lm. { 'lü·mən }

lumen-hour [OPTICS] A unit of quantity of light (luminous energy), equal to the quantity of light radiated or received for a period of 1 hour by a flux of 1 lumen. Abbreviated lm-hr. { 'lü·mən ¦au̇r }

lumen per watt [OPTICS] The unit of luminosity factor and of luminous efficacy. Abbreviated lm/w. { 'lü·mən pər 'wät }

lumen-second [OPTICS] A unit of quantity of light (luminous energy), equal to the quantity of light radiated or received for a period of 1 second by a flux of 1 lumen. Abbreviated lm-sec. { 'lü·mən ¦sek·ənd }

lumerg *See* lumberg. { 'lüˌmərg }

luminance [OPTICS] The ratio of the luminous intensity in a given direction of an infinitesimal element of a surface containing the point under consideration, to the orthogonally projected area of the element on a plane perpendicular to the given direction. Formerly known as brightness. { 'lü·mə·nəns }

luminance factor [OPTICS] The ratio of the luminance of a body when illuminated and observed under certain conditions to that of a perfect diffuser under the same conditions. { 'lü·mə·nəns ˌfak·tər }

luminescence [PHYS] Light emission that cannot be attributed merely to the temperature of the emitting body, but results from such causes as chemical reactions at ordinary temperatures, electron bombardment, electromagnetic radiation, and electric fields. { ˌlü·mə'nes·əns }

luminescent [PHYS] Capable of exhibiting luminescence. { ˌlü·mə'nes·ənt }

luminescent center [SOLID STATE] A point-lattice defect in a transparent crystal that exhibits luminescence. { ˌlü·mə'nes·ənt 'sen·tər }

luminophor [PHYS] A luminescent material that converts part of the absorbed primary energy into emitted luminescent radiation. Also known as fluophor; fluor; phosphor. { lü'min·əˌfȯr }

luminosity *See* luminosity factor. { ˌlü·mə'näs·əd·ē }

luminosity curve *See* luminosity function. { ˌlü·mə'näs·əd·ē ˌkərv }

luminosity factor [OPTICS] The ratio of luminous flux in lumens emitted by a source at a particular wavelength to the corresponding radiant flux in watts at the same wavelength; thus this is a measure of the visual sensitivity of the eye. Also known as luminosity. { ˌlü·mə'näs·əd·ē ˌfak·tər }

luminosity function [OPTICS] A standard measure of the response of an eye to monochromatic light at various wavelengths; the function is normalized to unity at its maximum value. Also known as luminosity curve; spectral luminous efficiency; visibility function. { ˌlü·mə'näs·əd·ē ˌfəŋk·shən }

luminous coefficient [OPTICS] A measure of the fraction of the radiant power of a light source which contributes to its luminous properties, equal to the average of the luminosity function at various wavelengths, weighted according to the spectral intensity of the source. Also known as luminous efficiency. { 'lü·mə·nəs ˌkō·i'fish·ənt }

luminous efficacy [OPTICS] **1.** The ratio of the total luminous flux in lumens emitted by a light source over all wavelengths to the total radiant flux in watts. Formerly known as luminous efficiency. **2.** The ratio of the total luminous flux emitted by a light source to the power input of the source; expressed in lumens per watt. { 'lü·mə·nəs ˌef·ə·kə·sē }

luminous efficiency *See* luminous coefficient; luminous efficacy. { 'lü·mə·nəs i'fish·ən·sē }

luminous emittance [OPTICS] The emittance of visible radiation weighted to take into account the different response of the human eye to different wavelengths of light; in photometry, luminous emittance is always used as a property of a self-luminous source, and therefore should be distinguished from luminance. Also known as luminous exitance. { 'lü·mə·nəs i'mit·əns }

luminous energy [OPTICS] The total radiant energy emitted by a source, evaluated according to its capacity to produce visual sensation; measured in lumen-hours or lumen-seconds. { 'lü·mə·nəs 'en·ər·jē }

luminous exitance *See* luminous emittance. { 'lü·mə·nəs 'ek·səd·əns }

luminous flux [OPTICS] The time rate of flow of radiant energy, evaluated according to its capacity to produce visual sensations; measured in lumens. { 'lü·mə·nəs 'fləks }

luminous flux density *See* illuminance. { 'lü·mə·nəs 'fləks 'den·səd·ē }

luminous intensity [OPTICS] The luminous flux incident on a small surface which lies in a specified direction from a light source and is normal to this direction, divided by the solid angle (in steradians) which the surface subtends at the source of light. Also known as light intensity. { 'lü·mə·nəs in'ten·səd·ē }

luminous quantities [OPTICS] Physical quantities used in photometry, such as luminous intensity and luminance, which are based on the response of the human eye, and are thus weighted to take into account the difference in response at different wavelengths of light. { 'lü·mə·nəs 'kwän·əd·ēz }

Lummer-Brodhun sight box [OPTICS] A device, having a series of prisms, for viewing simultaneously the two sides of a white diffuse plaster screen illuminated by light sources whose luminous intensities are being compared. { 'lu̇m·ər 'brȯdˌhu̇n 'sīt ˌbäks }

Lummer-Gehrcke plate [OPTICS] An interferometer consisting of a glass or quartz plate with parallel surfaces and sizable thickness in which multiple reflections take place. { 'lu̇m·ər 'ger·kə ˌplāt }

lumped constant [ELEC] A single constant that is electrically equivalent to the total of that type of distributed constant existing in a coil or circuit. Also known as lumped parameter. { 'ləmpt 'kän·stənt }

lumped-constant network [ELEC] An analytical tool in which distributed constants (inductance, capacitance, and resistance) are represented as hypothetical components. { 'ləmpt ¦kän·stənt 'netˌwərk }

lumped discontinuity [ELECTROMAG] An analytical tool in the study of microwave circuits in which the effective values of inductance, capacitance, and resistance representing a discontinuity in a waveguide are shown as discrete components of equivalent value. { 'ləmpt ˌdisˌkänt·ən'ü·əd·ē }

lumped element [ELECTROMAG] A section of a transmission line designed so that electric or magnetic energy is concentrated in it at specified frequencies, and inductance or capacitance may therefore be regarded as concentrated in it, rather than distributed over the length of the line. { 'ləmpt 'el·ə·mənt }

lumped impedance [ELECTROMAG] An impedance concentrated in a single component rather than distributed throughout the length of a transmission line. { 'ləmpt im'pēd·əns }

lumped parameter *See* lumped constant. { 'ləmpt pə'ram·əd·ər }

lunar rainbow *See* moonbow. { 'lü·nər 'rānˌbō }

Luneberg lens [ELECTROMAG] A type of antenna consisting of a dielectric sphere whose index of refraction varies with distance from the center of the sphere so that a beam of parallel rays falling on the lens is focused at a point on the lens surface diametrically opposite from the direction of incidence, and, conversely, energy emanating from a point on the surface is focused into a plane wave. Accurately spelled Luneburg lens. { 'lü·nəˌbərg ˌlenz }

Luneburg lens *See* Luneberg lens. { 'lü·nəˌbərg ˌlenz }

lusec [PHYS] A unit used for the measurement of power of evacuation of a vacuum pump, equal to the power associated with a leak rate of 1 liter per second at a pressure of 1 millitorr, or to approximately 1.33322×10^{-4} watt. { 'lüˌsek }

luster [OPTICS] The appearance of a surface dependent on reflected light; types include metallic, vitreous, resinous, adamantine, silky, pearly, greasy, dull, and earthy; applied to minerals, textiles, and many other materials. { 'ləs·tər }

lux [OPTICS] A unit of illumination, equal to the illumination on a surface 1 square meter in area on which there is a luminous flux of 1 lumen uniformly distributed, or the illumination on a surface all points of which are at a distance of 1 meter from a uniform point source of 1 candela. Symbolized lx. Also known as meter-candle. { ləks }

luxon *See* troland. { 'lək,sän }

lx *See* lux.

Lyapunov exponent [PHYS] One of a number of coefficients that describe the rates at which nearby trajectories in phase space converge or diverge, and that provide estimates of how long the behavior of a mechanical system is predictable before chaotic behavior sets in. { li·pu̇'nȯf ik,spō·nənt }

Lyddane-Sachs-Teller relation [SOLID STATE] For an infinite ionic crystal, the relation $\epsilon(0)/\epsilon(_n) = \omega_L{}^2/\omega_T{}^2$, where $\epsilon(0)$ is the crystal's static dielectric constant, $\epsilon(_n)$ is the dielectric constant at a frequency at which electronic polarizability is effective but ionic polarizability is not, ω_L is the frequency of longitudinal optical phonons with zero wave vectors, and ω_T is the frequency of transverse optical phonons with large wave vector. { lə'dān 'saks 'tel·ər ri,lā·shən }

Lyman-alpha radiation [SPECT] Radiation emitted by hydrogen associated with the spectral line in the Lyman series whose wavelength is 121.5 nanometers. { 'lī·mən 'al·fə ,rād·ē'ā·shən }

Lyman band [SPECT] A band in the ultraviolet spectrum of molecular hydrogen, extending from 125 to 161 nanometers. { 'lī·mən ,band }

Lyman continuum [SPECT] A continuous range of wavelengths (or wave numbers or frequencies) in the spectrum of hydrogen at wavelengths less than the Lyman limit, resulting from transitions between the ground state of hydrogen and states in which the single electron is freed from the atom. { 'lī·mən kən'tin·yə·wəm }

Lyman ghost [SPECT] A false line observed in a spectroscope as a result of a combination of periodicities in the ruling. { 'lī·mən ,gōst }

Lyman limit [SPECT] The lower limit of wavelengths of spectral lines in the Lyman series (912 angstrom units), or the corresponding upper limit in frequency, energy of quanta, or wave number (equal to the Rydberg constant for hydrogen). { 'lī·mən ,lim·ət }

Lyman series [SPECT] A group of lines in the ultraviolet spectrum of hydrogen covering the wavelengths of 121.5–91.2 nanometers. { 'lī·mən ,sir·ēz }

Lyot filter *See* birefringent filter. { 'lyō ,fil·tər }

M

m *See* meter.

mA *See* milliampere.

Macaluso-Corbino effect [PHYS] Anomalously large Faraday rotation exhibited by a medium at wavelengths in the neighborhood of an absorption line. { ˌmäk·əˈlü·sō kȯrˈbē·nō iˌfekt }

Macbeth illuminometer [OPTICS] A type of portable visual photometer in which the light to be measured is balanced by a Lummer-Brodhun sight box against a comparison lamp, whose apparent brightness can be varied by moving it along a tube; a control box supplies a calibrated current to the comparison lamp, and calibrated optical filters can be placed in the light paths to correct for color differences in the comparison and measured sources and to extend the range of the instrument. { məkˈbeth əˌlüm·əˈnäm·əd·ər }

MacCullagh's formula [PHYS] A formula for the potential due to a distribution of mass or charge at an external point: the potential V at a point P resulting from a distribution of mass or positive charge centered about a point O is $V = (kM/r) + (k/2r^3)(A + B + C - 3I) + O(1/r^4)$, where r is the distance from O to P, k is the gravitational or electrostatic constant, M is the total mass or charge, A, B, and C are the principal moments about O, I is the moment about OP, and $O(1/r^4)$ is a quantity that falls off at least as rapidly as $1/r^4$. { məˈkəl·əz ˌfȯr·myə·lə }

Mach angle [FL MECH] The vertex half angle of the Mach cone generated by a body in supersonic flight. { ˈmäk ˌaŋ·gəl }

Mach cone [FL MECH] **1.** The cone-shaped shock wave theoretically emanating from an infinitesimally small particle moving at supersonic speed through a fluid medium; it is the locus of the Mach lines. **2.** The cone-shaped shock wave generated by a sharp-pointed body, as at the nose of a high-speed aircraft. { ˈmäk ˌkōn }

Mach disk [FL MECH] A structure visible on a schlieren photograph of a supersonic air jet exhausting from a nozzle at low pressure into higher-pressure air at rest; it is formed by the focusing and strengthening of oblique shock waves emanating from the edges of the nozzle as they approach the jet axis. { ˈmäk ˌdisk }

Mach front *See* Mach stem. { ˈmäk ˌfrənt }

Mach line [FL MECH] **1.** A line representing a Mach wave. **2.** *See* Mach wave. { ˈmäk ˌlīn }

Mach number [FL MECH] The ratio of the speed of a body or of a point on a body with respect to the surrounding air or other fluid, or the ratio of the speed of a fluid, to the speed of sound in the medium. Symbolized N_{Ma}. Also known as relative Mach number. { ˈmäk ˌnəm·bər }

Mach principle [RELAT] The principle that the motion of a particle is only meaningful when referred to the rest of the matter in the universe; this motion is thus determined by the distribution of this matter and is not an intrinsic property of an absolute space. { ˈmäk ˌprin·sə·pəl }

Mach reflection [FL MECH] The reflection of a shock wave from a rigid wall in which the shock strength of the reflected wave and the angle of reflection both have the smaller of the two values which are theoretically possible. { ˈmäk ri ˌflek·shən }

Mach refractometer *See* Mach-Zehnder interferometer. { ˈmäk ˌrēˌfrakˈtäm·əd·ər }

Mach stem [FL MECH] A shock wave or front formed above the surface of the earth by the fusion of direct and reflected shock waves resulting from an airburst bomb. Also known as Mach front. { ˈmäk ˌstem }

Mach wave [FL MECH] Also known as Mach line. **1.** A shock wave theoretically occurring along a common line of intersection of all the pressure disturbances emanating from an infinitesimally small particle moving at supersonic speed through a fluid medium, with such a wave considered to exert no changes in the condition of the fluid passing through it. **2.** A very weak shock wave appearing, for example, at the nose of a very sharp body, where the fluid undergoes no substantial change in direction. { ˈmäk ˌwāv }

Mach-Zehnder interferometer [OPTICS] A variation of the Michelson interferometer used mainly in measuring the spatial variation of the index of refraction of a gas; the device has two semitransparent mirrors and two wholly reflecting mirrors at alternate corners of a rectangle, and half the beam travels along each side of the rectangle. Also known as Mach refractometer. { ˈmäk ˈtsān·dər ˌin·tər·fəˈräm·əd·ər }

macle [CRYSTAL] A twinned crystal. { ˈmak·əl }

Macleod equation [FL MECH] An equation which states that the fourth root of the surface tension of a liquid is proportional to the difference between the densities of the liquid and of its vapor. { mə'klaüd i,kwā·zhən }

MacMichael degree [FL MECH] An arbitrary unit used in measuring viscosity with a type of Couette viscometer; its size depends on the stiffness of the suspension of the inner cylinder of the viscometer. { mik'mī·kəl di,grē }

macrodome [CRYSTAL] Dome of a crystal in which planes are parallel to the longer lateral axis. { 'mak·rə,dōm }

macro lens [OPTICS] A camera lens designed to focus at very short distances and to form an image as large as the subject. { 'ma·krō ,lenz }

macrometer [OPTICS] Instrument that has two mirrors and a focusing telescope with which the ranges of distant objects can be found. { ma 'kräm·əd·ər }

macropinacoid *See* front pinacoid. { ¦mak·rō 'pin·ə,kòid }

macrorheology [MECH] A branch of rheology in which materials are treated as homogeneous or quasi-homogeneous, and processes are treated as isothermal. { ¦mak·rō·rē'äl·ə·jē }

macroscopic cross section [PHYS] The sum of the cross sections of an atom in a substance. { ¦ma·krə¦skäp·ik 'kròs ,sek·shən }

macroscopic property *See* thermodynamic property. { ¦mak·rə¦skäp·ik 'präp·ərd·ē }

macroscopic state [STAT MECH] Any state of a system as described by actual or hypothetical observations of its macroscopic statistical properties. Also known as macrostate. { ¦mak·rə¦skäp·ik 'stāt }

macroscopic theory [PHYS] A theory concerning only phenomena observable with the naked eye or with an ordinary light microscope, and not with the behavior of atoms, molecules, or their constituents which may underlie these phenomena. { ¦mak·rə¦skäp·ik 'thē·ə·rē }

macrosonics [ACOUS] The technology of sound at signal amplitudes so large that linear approximations are not valid, as in the use of ultrasonics for cleaning or drilling. { ¦mak·rō¦sän·iks }

macrostate *See* macroscopic state. { 'mak·rō,stāt }

MAD *See* magnetic anomaly detector; multiwavelength anomalous dispersion. { ¦em¦ā'dē *or* mad }

Madelung constant [SOLID STATE] A dimensionless constant which determines the electrostatic energy of a three-dimensional periodic crystal lattice consisting of a large number of positive and negative point charges when the number and magnitude of the charges and the nearest-neighbor distance between them is specified. { 'mä·də,lùŋ ,kän·stənt }

Maggi-Righi-Leduc effect [PHYS] A phenomenon in which the thermal conductivity of a conductor changes when it is placed in a magnetic field. { 'mä·jē 'rē·gē lə'dük i,fekt }

magic numbers [NUC PHYS] The integers 8, 20, 28, 50, 82, 126; nuclei in which the number of protons, neutrons, or both is magic have a stability and binding energy which is greater than average, and have other special properties. { 'maj·ik 'nəm·bərz }

magic tee *See* hybrid tee. { 'maj·ik 'tē }

maglev *See* magnetic levitation. { 'mag,lev }

magn [ELECTROMAG] A unit of absolute permeability equal to 1 henry per meter; it was proposed by the former Soviet Union, but has not won general acceptance. { 'mäg·ən }

magnet [ELECTROMAG] A piece of ferromagnetic or ferrimagnetic material whose domains are sufficiently aligned so that it produces a net magnetic field outside itself and can experience a net torque when placed in an external magnetic field. { 'mag·nət }

magnet coil *See* iron-core coil. { 'mag·nət ,kòil }

magnetic [ELECTROMAG] Pertaining to magnetism or a magnet. { mag'ned·ik }

magnetic anisotropy [ELECTROMAG] The dependence of the magnetic properties of some materials on direction. { mag'ned·ik ,an·ə'sä·trə·pē }

magnetic axis [ELECTROMAG] A line through the center of a magnet such that the torque exerted on the magnet by a magnetic field in the direction of this line equals 0. [PL PHYS] The single line of force that closes on itself after one revolution in a magnetic field with a rotational transform. { mag'ned·ik 'ak·səs }

magnetic bias [ELECTROMAG] A steady magnetic field applied to the magnetic circuit of a relay or other magnetic device. { mag'ned·ik 'bī·əs }

magnetic blowout [ELECTROMAG] **1.** A permanent magnet or electromagnet used to produce a magnetic field that lengthens the arc between opening contacts of a switch or circuit breaker, thereby helping to extinguish the arc. **2.** *See* blowout. { mag'ned·ik 'blō,aüt }

magnetic bottle [PL PHYS] A magnetic field used to confine or contain a plasma in controlled fusion experiments. { mag'ned·ik 'bäd·əl }

magnetic bubble [SOLID STATE] A cylindrical stable (nonvolatile) region of magnetization produced in a thin-film magnetic material by an external magnetic field; direction of magnetization is perpendicular to the plane of the material. Also known as bubble. { mag'ned·ik 'bəb·əl }

magnetic circuit [ELECTROMAG] A group of magnetic flux lines each forming a closed path, especially when this circuit is regarded as analogous to an electric circuit because of the similarity of its magnetic field equations to direct-current circuit equations. { mag'ned·ik 'sər·kət }

magnetic coercive force *See* coercive force. { mag'ned·ik kō'ər·siv 'fòrs }

magnetic confinement [PL PHYS] The containment of a plasma within a region of space by the forces of magnetic fields on the charged particles in the gas. { mag'ned·ik kən'fīn·mənt }

magnetic constant [ELECTROMAG] The absolute permeability of empty space, equal to 1 electromagnetic unit in the centimeter-gram-second system, and to $4\pi \times 10^{-7}$ henry per meter or, numerically, to 1.25664×10^{-6} henry per meter in the International System of units. Symbolized μ_0. { mag'ned·ik 'kän·stənt }

magnetic cooling *See* adiabatic demagnetization. { mag'ned·ik 'kül·iŋ }

magnetic core [ELECTROMAG] A quantity of ferrous material placed in a coil or transformer to provide a better path than air for magnetic flux, thereby increasing the inductance of the coil and increasing the coupling between the windings of a transformer. Also known as core. { mag'ned·ik 'kȯr }

magnetic coupling [ELECTROMAG] For a pair of particles or systems, the effect of the magnetic field created by one system on the magnetic moment or angular momentum of the other. { mag'ned·ik 'kəp·liŋ }

magnetic Curie temperature [SOLID STATE] The temperature below which a magnetic material exhibits ferromagnetism, and above which ferromagnetism is destroyed and the material is paramagnetic. { mag'ned·ik 'kyu̇r·ē ,tem·prə·chər }

magnetic damping [ELECTROMAG] Damping of a mechanical motion by means of the reaction between a magnetic field and the current generated by the motion of a coil through the magnetic field. { mag'ned·ik 'dam·piŋ }

magnetic diffusivity [ELECTROMAG] A measure of the tendency of a magnetic field to diffuse through a conducting medium at rest; it is equal to the partial derivative of the magnetic field strength with respect to time divided by the Laplacian of the magnetic field, or to the reciprocal of $4\pi\mu\sigma$, where μ is the magnetic permeability and σ is the conductivity in electromagnetic units. { mag'ned·ik ,di,fyü'siv·əd·ē }

magnetic dipole [ELECTROMAG] An object, such as a permanent magnet, current loop, or particle with angular momentum, which experiences a torque in a magnetic field, and itself gives rise to a magnetic field, as if it consisted of two magnetic poles of opposite sign separated by a small distance. { mag'ned·ik 'dī,pōl }

magnetic dipole antenna [ELECTROMAG] Simple loop antenna capable of radiating an electromagnetic wave in response to a circulation of electric current in the loop. { mag'ned·ik 'dī,pōl an,ten·ə }

magnetic dipole density *See* magnetization. { mag'ned·ik 'dī,pōl ,den·səd·ē }

magnetic dipole moment [ELECTROMAG] A vector associated with a magnet, current loop, particle, or such, whose cross product with the magnetic induction (or alternatively, the magnetic field strength) of a magnetic field is equal to the torque exerted on the system by the field. Also known as dipole moment; magnetic moment. { mag'ned·ik 'dī,pōl ,mō·mənt }

magnetic displacement *See* magnetic flux density. { mag'ned·ik di'splās·mənt }

magnetic domain *See* ferromagnetic domain. { mag'ned·ik də'mān }

magnetic double refraction [OPTICS] The double refraction of light passing through certain substances when the substance is placed in a transverse magnetic field. { mag'ned·ik 'dəbəl ri,frak·shən }

magnetic energy [ELECTROMAG] The energy required to set up a magnetic field. { mag'ned·ik 'en·ər·jē }

magnetic ferroelectric [SOLID STATE] A substance which possesses both magnetic ordering and spontaneous electric polarization. { mag'ned·ik ¦fer·ō·i'lek·trik }

magnetic field [ELECTROMAG] **1.** One of the elementary fields in nature; it is found in the vicinity of a magnetic body or current-carrying medium and, along with electric field, in a light wave; charges moving through a magnetic field experience the Lorentz force. **2.** *See* magnetic field strength. { mag'ned·ik 'fēld }

magnetic field intensity *See* magnetic field strength. { mag'ned·ik 'fēld in,ten·səd·ē }

magnetic field strength [ELECTROMAG] An auxiliary vector field, used in describing magnetic phenomena, whose curl, in the case of static charges and currents, equals (in meter-kilogram-second units) the free current density vector, independent of the magnetic permeability of the material. Also known as magnetic field; magnetic field intensity; magnetic force; magnetic intensity; magnetizing force. { mag'ned·ik 'fēld ,streŋkth }

magnetic film *See* magnetic thin film. { mag'ned·ik 'film }

magnetic flux [ELECTROMAG] **1.** The integral over a specified surface of the component of magnetic induction perpendicular to the surface. **2.** *See* magnetic lines of force. { mag'ned·ik 'fləks }

magnetic flux density [ELECTROMAG] A vector quantity that is used as a quantitative measure of magnetic field; the force on a charged particle moving in the field is equal to the particle's charge times the cross product of the particle's velocity with the magnetic flux density (SI units). Also known as magnetic displacement; magnetic induction; magnetic vector. { mag'ned·ik 'fləks ,den·səd·ē }

magnetic flux quantum [ELEC] A fundamental unit of magnetic flux, the total magnetic flux in a fluxoid in a type II superconductor, equal to $h/(2e)$, where h is Planck's constant and e is the magnitude of the electron charge, or approximately 2.07×10^{-15} weber. { mag,ned·ik 'fləks ,kwän·təm }

magnetic focusing [ELECTROMAG] Focusing a beam of electrons or other charged particles by using the action of a magnetic field. { mag'ned·ik 'fō·kə·siŋ }

magnetic force *See* magnetic field strength. { mag'ned·ik 'fȯrs }

magnetic force parameter [PL PHYS] A dimensionless number used in magnetofluid dynamics, equal to the product of the square of the magnetic permeability, the square of the magnetic field strength, the electrical conductivity, and a characteristic length, divided by the product of the mass density and the fluid velocity.

Symbolized N. { mag'ned·ik ¦fȯrs pə'ram·əd·ər }

magnetic gap [ELECTROMAG] The space between a magnet's pole faces. { mag'ned·ik 'gap }

magnetic groups *See* Shubnikov groups. { mag 'ned·ik 'grüps }

magnetic hysteresis [ELECTROMAG] Lagging of changes in the magnetization of a substance behind changes in the magnetic field as the magnetic field is varied. Also known as hysteresis. { mag'ned·ik ,his·tə'rē·səs }

magnetic induction *See* magnetic flux density. { mag'ned·ik in'dək·shən }

magnetic intensity *See* magnetic field strength. { mag'ned·ik in'ten·səd·ē }

magnetic leakage [ELECTROMAG] Passage of magnetic flux outside the path along which it can do useful work. { mag'ned·ik 'lēk·ij }

magnetic lens [ELECTROMAG] A magnetic field with axial symmetry, capable of converging beams of charged particles of uniform velocity and of forming images of objects placed in the path of such beams; the field may be produced by solenoids, electromagnets, or permanent magnets. Also known as lens. { mag'ned·ik 'lenz }

magnetic levitation [ELECTROMAG] Contactless, frictionless support of objects through the controlled use of magnetic forces to balance gravitational forces. Abbreviated maglev. { mag'ned·ik ,lev·ə'tā·shən }

magnetic lines of flux *See* magnetic lines of force. { mag'ned·ik 'līnz əv 'fləks }

magnetic lines of force [ELECTROMAG] Lines used to represent the magnetic induction in a magnetic field, selected so that they are parallel to the magnetic induction at each point, and so that the number of lines per unit area of a surface perpendicular to the induction is equal to the induction. Also known as magnetic flux; magnetic lines of flux. { mag'ned·ik 'līnz əv 'fȯrs }

magnetic Mach number [PL PHYS] A dimensionless number equal to the ratio of the velocity of a fluid to the velocity of Alfvén waves in the fluid. Symbolized M_{Ma}. { mag'ned·ik 'mäk ,nəm·bər }

magnetic material [ELECTROMAG] A material exhibiting ferromagnetism. { mag'ned·ik mə'tir·ē·əl }

magnetic mirror [PL PHYS] A magnetic field used in controlled-fusion experiments to reflect charged particles into the central region of a magnetic bottle; reflection occurs in the region where the magnetic field increases abruptly in strength. { mag'ned·ik 'mir·ər }

magnetic moment *See* magnetic dipole moment. { mag'ned·ik 'mō·mənt }

magnetic monopole [ELECTROMAG] A hypothetical particle carrying magnetic charge; it would be a source for a magnetic field in the same way that a charged particle is a source for an electric field. Also known as monopole. { mag'ned·ik 'män·ə,pōl }

magnetic multipole [ELECTROMAG] One of a series of types of static or oscillating distributions of magnetization, which is a magnetic multipole of order 2; the electric and magnetic fields produced by a magnetic multipole of order 2^n are equivalent to those of two magnetic multipoles of order 2^{n-1} of equal strength but opposite sign, separated from each other by a short distance. { mag'ned·ik 'məl·tə,pōl }

magnetic multipole field [ELECTROMAG] The electric and magnetic fields generated by a static or oscillating magnetic multipole. { mag'ned·ik ¦məl·tə,pōl ,fēld }

magnetic needle [ELECTROMAG] **1.** A bar magnet or collection of bar magnets which is hung so as to show the direction of the magnetic field. **2.** In particular, a slender bar magnet, pointed at both ends, that is pivoted or freely suspended in a magnetic compass. { mag'ned·ik 'nēd·əl }

magnetic nuclear resonance *See* nuclear magnetic resonance. { mag'ned·ik ¦nü·klē·ər 'rez·ən·əns }

magnetic number [PL PHYS] A dimensionless number used in magnetofluid dynamics, equal to the square root of the magnetic force parameter. Symbolized R_M. { mag'ned·ik 'nəm·bər }

magnetic octupole moment [ELECTROMAG] A quantity characterizing a distribution of magnetization; obtained by integrating the product of the divergence of the magnetization, the third power of the distance from the origin, and a spherical harmonic Y^*3_m over the magnetization distribution. { mag'ned·ik ¦äk·tə,pōl ,mō·mənt }

magnetic Oseen number [PL PHYS] A dimensionless number used in magnetofluid dynamics, equal to 1/2 $(1 - N_{AL}{}^2)R_M$, where N_{AL} is the Alfvén number, and R_M is the magnetic number. Symbolized *k*. { mag'ned·ik ü'sān ,nəm·bər }

magnetic pendulum [ELECTROMAG] A bar magnet which is hung by a thread or balanced on a pivot so that it oscillates in a horizontal plane when disturbed and released in a magnetic field having a horizontal component. { mag'ned·ik 'pen·jə·ləm }

magnetic permeability *See* permeability. { mag'ned·ik ,pər·mē·ə'bil·əd·ē }

magnetic pinch *See* pinch effect. { mag'ned·ik 'pinch }

magnetic polarization *See* intrinsic induction. { mag'ned·ik ,pō·lə·rə'zā·shən }

magnetic pole [ELECTROMAG] **1.** One of two regions located at the ends of a magnet that generate and respond to magnetic fields in much the same way that electric charges generate and respond to electric fields. **2.** A particle which generates and responds to magnetic fields in exactly the same way that electric charges generate and respond to electric fields; the particle probably does not have physical reality, but it is often convenient to imagine that a magnetic dipole consists of two magnetic poles of opposite sign, separated by a small distance. { mag'ned·ik 'pōl }

magnetic pole strength [ELECTROMAG] The magnitude of a (fictional) magnetic pole, equal to the force exerted on the pole divided by the magnetic induction (or, alternatively, by the magnetic field strength). Also known as pole strength. { mag'ned·ik 'pōl ˌstreŋkth }

magnetic potential *See* magnetic scalar potential. { mag'ned·ik pə'ten·chəl }

magnetic pressure [PL PHYS] A function, proportional to the square of the magnetic induction, such that the force exerted by a magnetic field on an electrically conducting fluid (excluding the force associated with curvature of magnetic flux lines) is the same as the force that would be exerted by a hydrostatic pressure equal to this function. { mag'ned·ik 'presh·ər }

magnetic probe [ELECTROMAG] A small coil inserted in a magnetic field to measure changes in field strength. { mag'ned·ik 'prōb }

magnetic pumping [ELECTROMAG] A method of moving a conducting liquid by applying a magnetic field which varies with time. [PL PHYS] A method of heating a plasma to a high ion temperature by applying an oscillating electromagnetic field. { mag'ned·ik 'pəmp·iŋ }

magnetic quadrupole lens [ELECTROMAG] A magnetic field generated by four magnetic poles of alternating sign arranged in a circle; used to focus beams of charged particles in devices such as electron microscopes and particle accelerators. { mag'ned·ik ¦kwä·drəˌpōl ˌlenz }

magnetic quantum number [ATOM PHYS] The eigenvalue of the component of an angular momentum operator in a specified direction, such as that of an applied magnetic field, in units of Planck's constant divided by 2π. { mag'ned·ik ¦kwän·təm ˌnəm·bər }

magnetic refrigerator [CRYO] A device for keeping substances cooled to about 0.2 K, in which a working substance consisting of a paramagnetic salt undergoes a cycle of processes which approximates a Carnot cycle between a high-temperature reservoir consisting of a liquid-helium bath at 1.2 K and a low-temperature reservoir consisting of the substance to be cooled, and isentropic cooling of the working substance is accomplished by demagnetization. { mag'ned·ik ri'frij·əˌrād·ər }

magnetic relaxation [PHYS] The approach of a magnetic system to an equilibrium or steady-state condition, over a period of time. { mag'ned·ik ˌrēˌlak'sā·shən }

magnetic reluctance *See* reluctance. { mag'ned·ik ri'lək·təns }

magnetic reluctivity *See* reluctivity. { mag'ned·ik ˌrēˌlək'tiv·əd·ē }

magnetic resonance [PHYS] A phenomenon exhibited by the magnetic spin systems of certain atoms whereby the spin systems absorb energy at specific (resonant) frequencies when subjected to magnetic fields alternating at frequencies which are in synchronism with natural frequencies of the system. Also known as spin resonance. { mag'ned·ik ¦rez·ən·əns }

magnetic Reynolds number [PL PHYS] A dimensionless number used to compare the transport of magnetic lines of force in a conducting fluid to the leakage of such lines from the fluid, equal to a characteristic length of the fluid times the fluid velocity, divided by the magnetic diffusivity. Symbolized R_M. { mag'ned·ik 'ren·əlz ˌnəm·bər }

magnetic rigidity [ELECTROMAG] A measure of the momentum of a particle moving perpendicular to a magnetic field, equal to the magnetic induction times the particle's radius of curvature. [PL PHYS] The existence of restoring forces which resist displacements of a conducting fluid when a magnetic field is present. { mag'ned·ik ri'jid·əd·ē }

magnetic rotation [OPTICS] **1.** In a weak magnetic field, the rotation, of the plane of polarization of fluorescent light emitted perpendicular to the field and perpendicular to the propagation direction of the incident light. **2.** *See* Faraday effect. { mag'ned·ik rō'tā·shən }

magnetics [ELECTROMAG] The study of magnetic phenomena, comprising magnetostatics and electromagnetism. { mag'ned·iks }

magnetic saturation [ELECTROMAG] The condition in which, after a magnetic field strength becomes sufficiently large, further increase in the magnetic field strength produces no additional magnetization in a magnetic material. Also known as saturation. { mag'ned·ik ˌsach·ə'rā·shən }

magnetic scalar potential [ELECTROMAG] The work which must be done against a magnetic field to bring a magnetic pole of unit strength from a reference point (usually at infinity) to the point in question. Also known as magnetic potential. { mag'ned·ik ¦skāl·ər pə'ten·chəl }

magnetic scanning [SPECT] The magnetic field sorting of ions into their respective spectrums for analysis by mass spectroscopy; accomplished by varying the magnetic field strength while the electrostatic field is held constant. { mag'ned·ik 'skan·iŋ }

magnetic scattering [PHYS] Scattering of neutrons as a result of the interaction of the magnetic moment of the neutron with the magnetic moments of atoms or other particles. { mag'ned·ik 'skad·ə·riŋ }

magnetic separatrix [ELECTROMAG] A surface that forms the boundary between an internal region of closed magnetic surfaces and an external region of open field lines. { mag'ned·ik 'sep·rəˌtriks }

magnetic shell [ELECTROMAG] Two layers of magnetic charge of opposite sign, separated by an infinitesimal distance. { mag'ned·ik 'shel }

magnetic shielding *See* magnetostatic shielding. { mag'ned·ik 'shēld·iŋ }

magnetic shunt [ELECTROMAG] Piece of iron, usually adjustable as to position, used to divert a portion of the magnetic lines of force passing through an air gap in an instrument or other device. { mag'ned·ik 'shənt }

magnetic strain energy [SOLID STATE] The potential energy of a magnetic domain, subject to both a tensile stress and a magnetic field, associated with the domain's magnetostriction expansion. { mag'ned·ik 'strān ,en·ər·jē }

magnetic stress [PL PHYS] The force which acts across a surface in a conducting fluid because of curving or stretching of magnetic flux lines. { mag'ned·ik 'stres }

magnetic stress tensor [PL PHYS] A second-rank tensor, proportional to the dyad product of the magnetic induction with itself, whose divergence gives that part of the force of a magnetic field on a unit volume of conducting fluid which is due to curvature or stretching of magnetic flux lines. { mag'ned·ik 'stres ,ten·sər }

magnetic superconductor [SOLID STATE] A superconductor which is not magnetic in the ordinary sense, but which contains elements with large magnetic moments or large spin. { mag'ned·ik 'sü·pər·kən,dək·tər }

magnetic susceptibility [ELECTROMAG] The ratio of the magnetization of a material to the magnetic field strength; it is a tensor when these two quantities are not parallel; otherwise it is a simple number. Also known as susceptibility. { mag'ned·ik sə,sep·tə'bil·əd·ē }

magnetic tension effect [ELECTROMAG] The ability of stresses on a ferromagnetic material to alter its remanence. Also known as inverse magnetostriction. { mag¦ned·ik 'ten·shən i,fekt }

magnetic test coil *See* exploring coil. { mag'ned·ik 'test ,kȯil }

magnetic thermometer [SOLID STATE] A sample of a paramagnetic salt whose magnetic susceptibility is measured and whose temperature is then calculated from the inverse relationship between the two quantities; useful at temperatures below about 1 K. { mag'ned·ik thər'mäm·əd·ər }

magnetic thin film [SOLID STATE] A sheet or cylinder of magnetic material less than 5 micrometers thick, usually possessing uniaxial magnetic anisotropy; used mainly in computer storage and logic elements. Also known as ferromagnetic film; magnetic film. { mag'ned·ik 'thin ¦film }

magnetic vector *See* magnetic flux density. { mag'ned·ik 'vek·tər }

magnetic vector potential *See* vector potential. { mag'ned·ik ¦vek·tər pə,ten·chəl }

magnetic viscosity [ELECTROMAG] The existence of a time delay between a change in the magnetic field applied to a ferromagnetic material and the resulting change in magnetic induction which is too great to be explained by the existence of eddy currents. [PL PHYS] The effect, possessed by a magnetic field in the absence of sizable mechanical forces or electric fields, of damping motions of a conducting fluid perpendicular to the field similar to ordinary viscosity. { mag'ned·ik vis'käs·əd·ē }

magnetic wave [SOLID STATE] The spread of magnetization from a small portion of a substance where an abrupt change in the magnetic field has taken place. { mag'ned·ik 'wāv }

magnetic wave device [ELECTROMAG] A device that depends on magnetoelastic or magnetostatic wave propagation through or on the surface of a magnetic or dielectric material. { mag'ned·ik ¦wāv di,vīs }

magnetic well [PL PHYS] A configuration of magnetic fields used to contain a plasma in controlled fusion experiments, in which the plasma is confined in a central region surrounded by fields which keep it from escaping in any direction. { mag'ned·ik 'wel }

magnetic x-ray scattering [ELECTROMAG] A process in which the electric and magnetic fields of incident x-rays interact with electronic magnetic moments, giving rise to magnetic reradiation. { mag'ned·ik 'eks,rā ,skad·ə·riŋ }

magnetism [PHYS] Phenomena involving magnetic fields and their effects upon materials. { 'mag·nə,tiz·əm }

magnetization [ELECTROMAG] **1.** The property and in particular, the extent of being magnetized; quantitatively, the magnetic moment per unit volume of a substance. Also known as magnetic dipole density; magnetization intensity. **2.** The process of magnetizing a magnetic material. { ,mag·nəd·ə'zā·shən }

magnetization curve *See* B-H curve; normal magnetization curve. { ,mag·nəd·ə'zā·shən ,kərv }

magnetization intensity *See* magnetization. { ,mag·nəd·ə'zā·shən in'ten·səd·ē }

magnetizing force *See* magnetic field strength. { 'mag·nə,tīz·iŋ ,fȯrs }

magnet keeper *See* keeper. { 'mag·nət ,kēp·ər }

magnetoacoustics [PHYS] The study of the effects of magnetic fields on acoustical phenomena, such as various oscillations in the attenuation of ultrasonic sound waves by a crystal placed in a magnetic field at a very low temperature, as the magnetic field strength or sound frequency is varied. { mag¦nēd·ō·ə'kü·stiks }

magnetoaerodynamics [PL PHYS] Study of the properties and characteristics of, and the forces exerted by, highly ionized air and other gases; applied principally to study of reentering ballistic missiles and spacecraft. { mag¦nēd·ō¦er·ō·dī'nam·iks }

magnetocaloric effect [THERMO] The reversible change of temperature accompanying the change of magnetization of a ferromagnetic material. { mag¦nēd·ō·kə'lȯr·ik i,fekt }

magnetoelastic coupling [SOLID STATE] The interaction between the magnetization and the strain of a magnetic material. { mag¦nēd·ō·i'las·tik 'kəp·liŋ }

magnetoelasticity [SOLID STATE] Phenomenon in which an elastic strain alters the magnetization of a ferromagnetic material. { mag¦nēd·ō,i ,las'tis·əd·ē }

magnetoelectric effect [SOLID STATE] A linear coupling between magnetization and polarization found in certain magnetic ferroelectrics, such as $BaMnF_4$ at low temperatures. { mag¦nēd·ō·i'lek·trik i,fekt }

magnetoelectricity [ELECTROMAG] Magnetic techniques for generating voltages, such as in an ordinary generator. [SOLID STATE] The appearance of an electric field in certain substances, such as chromic oxide (Cr_2O_3), when they are subjected to a static magnetic field. { mag¦nēd·ō·i,lek'tris·əd·ē }

magnetofluid dynamics [PHYS] **1.** The study of the motion of an electrically conducting metal, such as mercury, in the presence of electric and magnetic fields. **2.** *See* magnetohydrodynamics. { ¦mag·nəd·ō'flü·əd dī'nam·iks }

magnetogalvanic effect *See* galvanomagnetic effect. { mag,nēd·ō·gal'van·ik i,fekt }

magnetogas dynamics [PL PHYS] The science of motion in a plasma under the influence of mechanical, electric, and magnetic forces. { mag'nēd·ō,gas dī'nam·iks }

magnetograph [ELECTROMAG] A set of three variometers attached to a suitable recording unit, which records the components of the magnetic field vector in each of three perpendicular directions. { mag'ned·ə,graf }

magnetohydrodynamic electromagnetic pulse [ELECTROMAG] A relatively slow pulse of electromagnetic radiation generated by a nuclear explosion, and comprising magnetic bubble electromagnetic pulse and atmospheric heave electromagnetic pulse. { mag¦nēd·ō,hī·drō·dī'nam·ik i¦lek·trō·mag¦ned·ik 'pəls }

magnetohydrodynamic instability [PL PHYS] An instability of a plasma in which the plasma expands while moving into a region of weaker magnetic field, until it is expelled from the field. Also known as hydromagnetic instability. { mag¦nēd·ō,hī·drə·dī'nām·ik ,in·stə'bil·əd·ē }

magnetohydrodynamics [PHYS] The study of the dynamics or motion of an electrically conducting fluid, such as an ionized gas or liquid metal, interacting with a magnetic field. Abbreviated MHD. Also known as hydromagnetics; magnetofluid dynamics. { mag¦nēd·ō,hī·drə·dī'nam·iks }

magnetohydrodynamic stability [PL PHYS] The condition of a plasma in which fluctuations in density, pressure, velocity, or the distribution of particles in phase space, die out rather than increase. Also known as hydromagnetic stability. { mag¦nēd·ō,hī·drə·dī'nam·ik stə'bil·əd·ē }

magnetohydrodynamic turbulence [PL PHYS] Motion of a plasma in which velocities and pressures fluctuate irregularly. { mag¦nēd·ō,hī·drə·dī'nam·ik 'tər·byə·ləns }

magnetohydrodynamic wave [PHYS] Wave motion in an electrically conducting fluid, such as plasma or liquid metal, in a strong magnetic field at a frequency much less than that of the ion cyclotron frequency. Also known as hydromagnetic wave. { mag¦nēd·ō,hī·drə·dī'nam·ik 'wāv }

magnetomechanical factor [PHYS] The gyromagnetic ratio of an atom or substance (magnetic dipole moment divided by angular momentum) divided by the quantity $e/2mc$, where e and m are the charge (in esu, or electrostatic units) and mass of the electron respectively, and c is the speed of light. Also known as g factor. { mag¦nēd·ō·mi'kan·ə·kəl 'fak·tər }

magnetomechanics [PHYS] The study of the effects which the magnetization of a material and its strain have on each other. { mag¦nēd·ō·mi 'kan·iks }

magnetomotive force [ELECTROMAG] The work that would be required to carry a magnetic pole of unit strength once around a magnetic circuit. Abbreviated mmf. { mag¦nēd·ō'mōd·iv 'fōrs }

magneton [PHYS] A unit of magnetic moment used for atomic, molecular, or nuclear magnets, such as the Bohr magneton, Weiss magneton, or nuclear magneton. { 'mag·nə,tän }

magneton number [PHYS] The ratio of the magnetic moment per atom, ion, or molecule of a paramagnetic or ferromagnetic material to the Bohr magneton. { 'mag·nə,tän ,nəm·bər }

magnetooptical modulator [ELECTROMAG] An arrangement for modulating a beam of light by passing it through a single crystal of yttrium iron garnet, which provides intensity modulation by using a magnetic field to produce optical rotation. { mag¦nēd·ō¦äp·tə·kəl 'mäj·ə,lād·ər }

magnetooptical shutter [OPTICS] A device in which light passes through crossed Nicol prisms and a glass cell containing a liquid displaying the Faraday effect between the prisms; light can pass through the system only when a magnetic field is applied to the cell at an angle of 45° to the polarization planes of both prisms. { mag¦nēd·ō¦äp·tə·kəl 'shəd·ər }

magnetooptic Kerr effect [OPTICS] Changes produced in the optical properties of a reflecting surface of a ferromagnetic substance when the substance is magnetized; this applies especially to the elliptical polarization of reflected light, when the ordinary rules of metallic reflection would give only plane polarized light. Also known as Kerr magnetooptical effect. { mag ¦nēd·ō¦äp·tik 'kər i,fekt }

magnetooptic material [OPTICS] A material whose optical properties are changed by an applied magnetic field. { mag¦nēd·ō¦äp·tik mə'tir·ē·əl }

magnetooptics [OPTICS] The study of the effect of a magnetic field on light passing through a substance in the field. { mag¦nēd·ō¦äp·tiks }

magnetoplasmadynamics [ELECTROMAG] The generation of electric current by shooting a beam of ionized gas through a magnetic field, to give the same effect as moving copper bars near a magnet. { mag¦nēd·ə,plaz·mə·dī'nam·iks }

magnetoresistance [ELECTROMAG] The change in electrical resistance produced in a current-carrying conductor or semiconductor on application of a magnetic field. { mag¦nēd·ō·ri'zis·təns }

magnetoresistivity [ELECTROMAG] The change in resistivity produced in a current-carrying conductor or semiconductor on application of a magnetic field. { mag,nēd·ō·ri,zis'tiv·əd·ē }

magnetostatic [ELECTROMAG] Pertaining to magnetic properties that do not depend upon the motion of magnetic fields. { mag¦nēd·ə¦stad·ik }

magnetostatic mode [SOLID STATE] A spin wave in a magnetic material whose wavelength is greater than about one-tenth the size of the sample. { mag¦nēd·ə¦stad·ik 'mōd }

magnetostatics [ELECTROMAG] The study of magnetic fields that remain constant with time. Also known as static magnetism. { mag¦nēd·ō ¦stad·iks }

magnetostatic shielding [ELECTROMAG] The use of an enclosure made of a high-permeability magnetic material to prevent a static magnetic field outside the enclosure from reaching objects inside it, or to confine a magnetic field within the enclosure. Also known as magnetic shielding. { mag¦nēd·ō¦stad·ik 'shēld·iŋ }

magnetostriction [ELECTROMAG] The dependence of the state of strain (dimensions) of a ferromagnetic sample on the direction and extent of its magnetization. { mag,nēd·ō'strik·shən }

magnetostriction transducer [ELECTROMAG] A transducer used with sonar equipment to change an alternating current to sound energy at the same frequency and to form the sound energy into a beam; its operation depends on the interaction between the magnetization and the deformation of a material having magnetostrictive properties. { mag,nēd·ō'strik·shən tranz'dü·sər }

magnetostrictive delay line [ELECTROMAG] A delay line made of nickel or other magnetostrictive material, in which the amount of delay is determined by a shock wave traveling through the length of the line at the speed of sound. { mag¦nēd·ō¦strik·tiv di'lā ,līn }

magnetostrictive resonator [SOLID STATE] Ferromagnetic rod so designed that it can be excited magnetically into resonant vibration at one or more definite and known frequencies. { mag¦nēd·ō¦strik·tiv 'rez·ən,ād·ər }

magnetostrictor [ELECTROMAG] A device for converting electric oscillations to mechanical oscillations by employing the property of magnetostriction. { mag¦nēd·ō¦strik·tər }

magnet power [ELECTROMAG] The electric power supplied to the coils of an electromagnet. { 'mag·nət ,paù·ər }

magnification [OPTICS] **1.** A measure of the effectiveness of an optical system in enlarging or reducing an image; the magnification may be lateral, longitudinal, or angular. **2.** *See* lateral magnification. { ,mag·nə·fə'kā·shən }

magnifier *See* simple microscope. { 'mag·nə,fī·ər }

magnifying glass [OPTICS] **1.** Any device that uses a simple lens which enlarges the object being viewed. **2.** *See* simple microscope. { 'mag·nə,fī·iŋ ,glas }

magnifying power [OPTICS] The ratio of the tangent of the angle subtended at the eye by an image formed by an optical system, to the tangent of the angle subtended at the eye by the corresponding object at a distance for convenient viewing. { 'mag·nə,fī·iŋ ,paù·ər }

magnon [SOLID STATE] A quasi-particle which is introduced to describe small departures from complete ordering of electronic spins in ferro-, ferri-, antiferro-, and helimagnetic arrays. Also known as quantized spin wave. { 'mag,nän }

Magnus effect [FL MECH] A force on a rotating cylinder in a fluid flowing perpendicular to the axis of the cylinder; the force is perpendicular to both flow direction and cylinder axis. Also known as Magnus force. { 'mäg·nəs i,fekt }

Magnus force *See* Magnus effect. { 'mäg·nəs ,fòrs }

Magnus moment [FL MECH] A torque associated with the Magnus effect, such as moments about the pitch and yaw axes of a missile or aircraft due to rotation about the roll axis. { 'mäg·nəs ,mō·mənt }

main lobe *See* major lobe. { 'mān 'lōb }

Majorana effect [OPTICS] The effect in which a transverse magnetic field acting on a colloidal solution, such as a sol of iron oxide, produces optical anisotropy, resulting in magnetic birefringence. { mā·jə'ran·ə i,fekt }

Majorana force [NUC PHYS] A force between two nucleons postulated to explain various phenomena, which can be derived from a potential containing an operator which exchanges the nucleons' positions but not their spins. { ,mā·jə'ran·ə ,fòrs }

Majorana neutrino [PART PHYS] A particle described by a wave function that satisfies the Dirac equation with mass equal to zero, and that is self-charge-conjugate. { ,mā·jə'ran·ə nü'trē·nō }

major diatonic scale [ACOUS] A diatonic scale in which the relative sizes of the sequence of intervals are approximately 2,2,1,2,2,2,1. { 'mā·jər ¦dī·ə¦tan·ik 'skāl }

major lobe [ELECTROMAG] Antenna lobe indicating the direction of maximum radiation or reception. Also known as main lobe. { 'mā·jər 'lōb }

major node [ELEC] A point in an electrical network at which three or more elements are connected together. Also known as junction. { 'mā·jər 'nōd }

Maksutov-Schmidt telescope *See* meniscus-Schmidt telescope. { mak'sü,tòf 'shmit ,tel·ə,skōp }

Maksutov system [OPTICS] A catadioptric telescope optical system capable of covering a large field (60° and more); used to survey large areas of the sky. { mak'sü,tòf ,sis·təm }

Maldacena duality [PART PHYS] A form of weak-strong duality between black holes in superstring theory and ordinary quantum field theory with many fields. { ¦mäl·də,sān·ə dü'al·əd·ē }

Malter effect [SOLID STATE] A phenomenon in which a metal with a nonconducting surface film has a large coefficient of secondary electron emission; this is particularly notable in aluminum whose surface has been oxidized and then coated with cesium oxide. { 'mäl·tər i,fekt }

Malus cosine-squared law [OPTICS] The law that if a beam of plane polarized light passes through a Nicol prism, the intensity of light emerging from the prism is proportional to the square of the cosine between the plane of polarization of the incident light and the plane of polarization of the prism. { 'mä·ləs ¦kō,sīn 'skwerd ,lȯ }

Malus' law of rays [OPTICS] The law that an orthotomic system of rays is still orthotomic after the rays have been reflected and refracted any number of times. { 'mä·ləs ¦lȯ əv 'rāz }

Mandelstam plane [PART PHYS] A method of plotting energy versus scattering angle of three reactions, each having two particles both before and after scattering, which can be derived from each other by the crossing principle; the three reactions are on an equal footing, and poles in the scattering amplitude representing exchanged particles lie along straight lines. { 'mänd·əl,shtäm ,plān }

Mandelstam representation [PART PHYS] For a reaction in which there are two particles both before and after scattering: an expression, containing several integrals, for a function related to the scattering amplitude; the arguments of the function are the center-of-mass energy and scattering angle, extended to complex values; the function is conjectured to be analytic in these variables except for certain cuts and to have values along these cuts which give the scattering amplitude of the reaction, and of the two reactions derivable from it by the crossing principle. { 'mänd·əl,shtäm ,rep·ri,zən'tā·shən }

Mangin mirror [OPTICS] A negative meniscus lens whose shallower surface is silvered to act as a spherical mirror while the other surface corrects for spherical aberration of the reflecting surface; used in searchlights and aircraft gunsights. { män,zhan ,mir·ər }

manifest covariance [RELAT] Property of an expression composed of Lorentz invariant numbers and operators, four-vectors, and tensors in such a way that its Lorentz covariance is immediately obvious. { 'man·ə,fest kō'ver·ē·əns }

manifold of states [ATOM PHYS] A set of states sufficient to form a representation of an operator or a Lie group of operators. { 'man·ə,fōld əv 'stāts }

Manning equation [FL MECH] An equation used to compute the velocity of uniform flow in an open channel. { 'man·iŋ i,kwā·zhən }

manocryometer [THERMO] An instrument for measuring the change of a substance's melting point with change in pressure; the height of a mercury column in a U-shaped capillary supported by an equilibrium between liquid and solid in an adjoining bulb is measured, and the whole apparatus is in a thermostat. { ,man·ō ,krī'äm·əd·ər }

manometric capsule [ACOUS] A device for studying air vibrations in a pipe or resonator, consisting of a rubber membrane which is stretched over a hole in the pipe, or over the end of a flange attached to such a hole, and apparatus for measuring vibrations of the membrane. { ,man·ə¦me·trik 'kap·səl }

many-body force [PHYS] A force exerted on a particle, in the presence of two or more other particles, which differs from the vector sum of the forces which would be exerted on it if each of the other particles were present alone. { 'men·ē 'bäd·ē ,fȯrs }

many-body problem [MECH] The problem of predicting the motions of three or more objects obeying Newton's laws of motion and attracting each other according to Newton's law of gravitation. Also known as *n*-body problem. { 'men·ē 'bäd·ē ,präb·ləm }

many-body theory [PHYS] A scheme for calculating physical quantities for systems with large numbers of particles, without finding details of each particle's motion, often at temperatures close to absolute zero. { 'men·ē 'bäd·ē ,thē·ə·rē }

many-worlds interpretation *See* Everett-Wheeler interpretation. { ¦men·ē 'wərlz ,in·tər·prə,tā·shən }

Marconi antenna [ELECTROMAG] Antenna system of which the ground is an essential part, as distinguished from a Hertz antenna. { mär'kō·nē an'ten·ə }

marginal dimensionality [STAT MECH] The largest number of spatial dimensions for which nonlinear effects are important in calculating the behavior of a substance near a critical point. { 'mär·jən·əl di,men·shə'nal·əd·ē }

marginally outer trapped surface [RELAT] A spacelike, two-dimensional surface in a spacetime such that outgoing null rays perpendicular to the surface are neither diverging nor converging. { ¦mär·jen·əl·ē ¦au̇d·ər ¦trapt 'sər·fəs }

Margoulis number *See* Stanton number. { mär'gü·ləs ,nəm·bər }

marine rainbow [OPTICS] A rainbow seen in ocean spray. Also known as sea rainbow. { mə'rēn 'rān·bō }

Mariotte's law *See* Boyle's law. { ¦mar·ē¦äts ,lȯ }

Martens wedge [OPTICS] A wedge-shaped piece of quartz used to rotate the plane of polarization of linearly polarized light. { 'märt·ənz ,wej }

Marx circuit [ELEC] An electric circuit used in an impulse generator in which capacitors are charged in parallel through charging resistors, and then connected in series and discharged through the test piece by the simultaneous sparkover of spark gaps. { 'märks ,sər·kət }

Marx effect [SOLID STATE] The effect wherein the energy of photoelectrons emitted from an illuminated surface is decreased when the surface is simultaneously illuminated by light of lower frequency than that causing the emission. { 'märks i,fekt }

maser [PHYS] A device for coherent amplification or generation of electromagnetic waves in which an ensemble of atoms or molecules, raised to an unstable energy state, is stimulated by an electromagnetic wave to radiate excess energy at the same frequency and phase as the stimulating wave. Derived from microwave amplification by stimulated emission of radiation. Also known as paramagnetic amplifier. { 'mā·zər }

masking [ACOUS] The amount by which the threshold of audibility of a sound is raised by the presence of another sound; the unit customarily used is the decibel. Also known as audio masking; aural masking. { 'mask·iŋ }

mass [MECH] A quantitative measure of a body's resistance to being accelerated; equal to the inverse of the ratio of the body's acceleration to the acceleration of a standard mass under otherwise identical conditions. { mas }

mass absorption coefficient [PHYS] The linear absorption coefficient divided by the density of the medium. { 'mas əb'sȯrp·shən ˌkō·iˌfish·ənt }

mass-analyzed ion kinetic energy spectrometry [SPECT] A type of ion kinetic energy spectrometry in which the ionic products undergo mass analysis followed by energy analysis. Abbreviated MIKES. { 'mas ¦an·əˌlīzd 'īˌän kə¦ned·ik 'en·ər·jē spek'träm·ə·trē }

mass defect [NUC PHYS] The difference between the mass of an atom and the sum of the masses of its individual components in the free (unbound) state. { 'mas 'dēˌfekt }

mass divergence [FL MECH] The divergence of the momentum field, a measure of the rate of net flux of mass out of a unit volume of a system; in symbols, $\nabla \cdot \rho \mathbf{V}$, where ρ is the fluid density, $\mathbf{V}$ the velocity vector, and ∇ the del operator. { 'mas də'vər·jəns }

mass-energy conservation [RELAT] The principle that energy cannot be created or destroyed; however, one form of energy is that which a particle has because of its rest mass, equal to this mass times the square of the speed of light. { 'mas 'en·ər·jē ˌkän·sərˌvā·shən }

mass-energy relation [RELAT] The relation whereby the total energy content of a body is equal to its inertial mass times the square of the speed of light. { 'mas 'en·ər·jē riˌlā·shən }

Massey formula [ATOM PHYS] A formula for the probability that an excited atom approaching the surface of a metal will emit secondary electrons. { 'mas·ē ˌfȯr·myə·lə }

mass flow [FL MECH] The mass of a fluid in motion which crosses a given area in a unit time. { 'mas 'flō }

mass formula [NUC PHYS] An equation giving the atomic mass of a nuclide as a function of its atomic number and mass number. { 'mas ˌfȯr·myə·lə }

Massieu function [THERMO] The negative of the Helmholtz free energy divided by the temperature. { ma'syü ˌfəŋk·shən }

mass number [NUC PHYS] The sum of the numbers of protons and neutrons in the nucleus of an atom or nuclide. Also known as nuclear number; nucleon number. { 'mas ˌnəm·bər }

mass operator [QUANT MECH] An operator which is added to the Lagrangian in a quantized field theory in order to eliminate certain infinite quantities, and whose sum with the mechanical mass gives the observed mass. { 'mas 'äp·əˌrād·ər }

mass reactance *See* acoustic mass reactance. { 'mas rē'ak·təns }

mass renormalization [QUANT MECH] The mathematical operation of adding the mass which a particle possesses because of its self interaction, to its mechanical mass in order to obtain its measured mass. { 'mas rēˌnȯr·mə·lə'zā·shən }

mass resistivity [ELEC] The product of the electrical resistance of a conductor and its mass, divided by the square of its length; the product of the electrical resistivity and the density. { 'mas ˌrēˌzis'tiv·əd·ē }

mass shift [NUC PHYS] The portion of the isotope shift which results from the difference between the nuclear masses of different isotopes. { 'mas 'shift }

mass spectrum [PART PHYS] A plot of masses of elementary particles, including unstable states. Also known as particle spectrum. [PHYS] A display, record, or plot of the distribution in mass, or in mass-to-charge ratio, of ionized atoms, molecules, or molecular fragments. { 'mas 'spek·trəm }

mass-transfer rate [PHYS] The measurement of the movement of matter as a function of time. { ¦mas 'tranz·fər ˌrāt }

mass transport [FL MECH] **1.** Carrying of loose materials in a moving medium such as water or air. **2.** The movement of fluid, especially water, from one place to another. { 'mas 'tranzˌpȯrt }

mass units [MECH] Units of measurement having to do with masses of materials, such as pounds or grams. { 'mas ˌyü·nəts }

mass velocity [FL MECH] The weight flow rate of a fluid divided by the cross-sectional area of the enclosing chamber or conduit; for example, $lb/(h \cdot ft^2)$. { 'mas vəˌläs·əd·ē }

master equation [ATOM PHYS] An equation which determines the rate of change of the population of an energy level in terms of the populations of other levels and transition probabilities. { 'mas·tər i'kwā·zhən }

matched-field processing [ACOUS] A signal-processing technique that has a variety of applications in underwater acoustics and is based on the comparison of measured data with predictions for the data that are calculated from a model of underwater sound propagation. { ¦macht ˌfēld 'präˌses·iŋ }

matched impedance [ELEC] An impedance of a load which is equal to the impedance of a generator, so that maximum power is delivered to the load. { 'macht im'pēd·əns }

matched transmission line [ELEC] Transmission line terminated with a load equivalent to its characteristic impedance. { 'macht tranz'mish·ən ˌlīn }

matching [ELEC] Connecting two circuits or parts together with a coupling device in such a way that the maximum transfer of energy occurs between the two circuits, and the impedance of either circuit will be terminated in its image. { 'mach·iŋ }

matching diaphragm [ELECTROMAG] Diaphragm consisting of a slit in a thin sheet of metal, placed transversely across a waveguide for matching purposes; the orientation of the slit with respect to the long dimension of the waveguide determines whether the diaphragm acts as a capacitive or inductive reactance. { 'mach·iŋ 'dī·ə,fram }

matching section [ELECTROMAG] A section of transmission line, a quarter or half wavelength long, inserted between a transmission line and a load to obtain impedance matching. { 'mach·iŋ ¦sek·shən }

matching stub [ELECTROMAG] Device placed on a radio-frequency transmission line which varies the impedance of the line; the impedance of the line can be adjusted in this manner. { 'mach·iŋ ,stəb }

materialization [PHYS] The direct conversion of energy into mass, as in pair production. { mə,tir·ē·ə·lə'zā·shən }

material particle [MECH] An object which has rest-mass and an observable position in space, but has no geometrical extension, being confined to a single point. Also known as particle. { mə'tir·ē·əl 'pärd·ə·kəl }

mathematical physics [PHYS] The study of the mathematical systems which represent physical phenomena; particular areas are, for example, quantum and statistical mechanics and field theory. { ¦math·ə¦mad·ə·kəl 'fiz·iks }

m-atm *See* meter-atmosphere.

matrix element [QUANT MECH] The scalar product of a member of a complete, orthogonal set of vectors, representing states, with a vector which results from applying a specified operator to another member of this set. { 'mā·triks ,el·ə·mənt }

matrix isolation [SPECT] A spectroscopic technique in which reactive species can be characterized by maintaining them in a very cold, inert environment while they are examined by an absorption, electron-spin resonance, or laser excitation spectroscope. { 'mā·triks ,ī·sə'lā·shən }

matrix mechanics [QUANT MECH] The theory of quantum mechanics developed by using the Heisenberg picture and representing operators by their matrix elements between eigenfunctions of the Hamiltonian operator; Heisenberg's original formulation of quantum mechanics. { 'mā·triks mə'kan·iks }

matrix spectrophotometry [SPECT] Spectrophotometric analysis in which the specimen is irradiated in sequence at more than one wavelength, with the visible spectrum evaluated for the energy leaving for each wavelength of irradiation. { 'mā·triks ,spek·trō·fə'täm·ə·trē }

matter [PHYS] The substance composing bodies perceptible to the senses; includes any entity possessing mass when at rest. { 'mad·ər }

matter wave *See* de Broglie wave. { 'mad·ər ,wāv }

Matteuci effect [PHYS] A phenomenon in which an electric potential difference appears between the ends of a ferromagnet that is twisted in a magnetic field. { mad·ə'ü·chē i,fekt }

Matthias' rules [SOLID STATE] Several empirical rules giving the dependence of the transition temperatures of superconducting metals and alloys on the position of the metals in the periodic table and in the composition of the alloys. { mə'thī·əs ,rülz }

Matthiessen sinker method [THERMO] A method of determining the thermal expansion coefficient of a liquid, in which the apparent weight of a sinker when immersed in the liquid is measured for two different temperatures of the liquid. { ¦math·ə·sən 'siŋ·kər ,meth·əd }

Matthiessen's rule [SOLID STATE] An empirical rule which states that the total resistivity of a crystalline metallic specimen is the sum of the resistivity due to thermal agitation of the metal ions of the lattice and the resistivity due to imperfections in the crystal. { 'math·ə·sənz ,rül }

mattress array *See* billboard array. { 'ma·trəs ə'rā }

Maupertius' principle [MECH] The principle of least action is sufficient to determine the motion of a mechanical system. { mō'pər·shəs ,prin·sə·pəl }

maximal analytic extension [RELAT] An extension, in a real analytic manner, past all coordinate singularities of a solution to Einstein's equations of general relativity. { ¦mak·sə·məl ,an·ə¦lid·ik ik'sten·chən }

maximum ordinate [MECH] Difference in altitude between the origin and highest point of the trajectory of a projectile. { 'mak·sə·məm 'ȯrd·ən·ət }

maximum sound pressure [ACOUS] For any given cycle of a periodic wave, the maximum absolute value of the instantaneous sound pressure occurring during that cycle. { 'mak·sə·məm 'sau̇nd ,presh·ər }

maximum unambiguous range [ELECTROMAG] The range beyond which the echo from a pulsed radar signal returns after generation of the next pulse, and can thus be mistaken as a short-range echo of the next cycle. { 'mak·sə·məm ,ən·am ¦big·yə·wəs 'rānj }

maxwell [ELECTROMAG] A centimeter-gram-second electromagnetic unit of magnetic flux, equal to the magnetic flux which produces an electromotive force of 1 abvolt in a circuit of one turn linking the flux, as the flux is reduced to zero in 1 second at a uniform rate; equal to 10^{-8} weber. Abbreviated Mx. Also known as abweber (abWb); line of magnetic induction. { 'mak,swel }

Maxwell body *See* Maxwell liquid. { 'mak,swel ,bäd·ē }

Maxwell-Boltzmann density function *See* Maxwell-Boltzmann distribution. { 'mak,swel 'bōlts·mən 'den·səd·ē ,fəŋk·shən }

Maxwell-Boltzmann distribution [STAT MECH] Any function giving the probability (or some function proportional to it) that a molecule of a gas in thermal equilibrium will have values of certain variables within given infinitesimal ranges, assuming that the gas molecules obey classical mechanics, and possibly making other assumptions; examples are the Maxwell distribution and the Boltzmann distribution. Also known as Maxwell-Boltzmann density function. { 'mak,swel 'bōlts·mən ,di·strə,byü·shən }

Maxwell-Boltzmann equation *See* Boltzmann transport equation. { 'mak,swel 'bōlts·mən i,kwā·zhən }

Maxwell-Boltzmann statistics [STAT MECH] The classical statistics of identical particles, as opposed to the Bose-Einstein or Fermi-Dirac statistics. Also known as Boltzmann statistics. { 'mak,swel 'bōlts·mən stə'tis·tiks }

Maxwell bridge [ELEC] A four-arm alternating-current bridge used to measure inductance (or capacitance) in terms of resistance and capacitance (or inductance); bridge balance is independent of frequency. Also known as Maxwell-Wien bridge; Wien-Maxwell bridge. { 'mak ,swel ,brij }

Maxwell distribution [STAT MECH] A function giving the number of molecules of a gas in thermal equilibrium whose velocities lie within a given, infinitesimal range of values, assuming that the molecules obey classical mechanics, and do not interact. Also known as Maxwellian distribution. { 'mak,swel ,di·strə,byü·shən }

Maxwell effect [OPTICS] Double refraction of a viscous liquid having anisotropic molecules, which results from components of the velocity gradient perpendicular to the fluid velocity itself. { 'mak,swel i,fekt }

Maxwell equal-area rule [THERMO] At temperatures for which the theoretical isothermal of a substance, on a graph of pressure against volume, has a portion with positive slope (as occurs in a substance with liquid and gas phases obeying the van der Waals equation), a horizontal line drawn at the equilibrium vapor pressure and connecting two parts of the isothermal with negative slope has the property that the area between the horizontal and the part of the isothermal above it is equal to the area between the horizontal and the part of the isothermal below it. { 'mak,swel ¦ē·kwəl 'er·ē·ə ,rül }

Maxwell equations *See* Maxwell field equations. { 'mak,swel i,kwā·zhənz }

Maxwell field equations [ELECTROMAG] Four differential equations which relate the electric and magnetic fields to electric charges and currents, and form the basis of the theory of electromagnetic waves. Also known as electromagnetic field equations; Maxwell equations. { 'mak,swel 'fēld i,kwā·zhənz }

Maxwellian distribution *See* Maxwell distribution. { mak,swel·ē·ən ,di·strə'byü·shən }

Maxwellian distribution law [STAT MECH] Equation relating the statistical distribution of speeds and energies of molecules of a pure gas at a uniform temperature where there are no convection currents. { mak,swel·ē·ən ,di·strə'byü·shən ,lȯ }

Maxwellian equilibrium [STAT MECH] Thermal equilibrium of a gas, or of some group of particles, in which the velocity distribution of the particles is the Maxwell distribution corresponding to the temperature of the object with which they are in equilibrium. { mak,swel·ē·ən ,ē·kwə'lib·rē·əm }

Maxwellian gas [STAT MECH] A gas whose molecules have the Maxwell distribution of velocities. { mak,swel·ē·ən 'gas }

Maxwellian view [OPTICS] A method of using an optical instrument in which a real image of a light source is focused on the pupil of the eye, instead of using an eyepiece. { mak,swel·ē·ən 'vyü }

Maxwell liquid [FL MECH] A liquid whose rate of deformation is the sum of a term proportional to the shearing stress acting on it and a term proportional to the rate of change of this stress. Also known as Maxwell body. { 'mak,swel ,lik·wəd }

Maxwell primaries [OPTICS] The primary colors in a system of colorimetry devised by J.C. Maxwell; they are cyan, green, and magenta. { 'mak ,swel 'prī,mer·ēz }

Maxwell relation [ELECTROMAG] According to Maxwell's electromagnetic theory, that relation wherein the dielectric constant of a substance equals the square of its index of refraction. [THERMO] One of four equations for a system in thermal equilibrium, each of which equates two partial derivatives, involving the pressure, volume, temperature, and entropy of the system. { 'mak,swel ri'lā·shən }

Maxwell's coefficient of diffusion [FL MECH] A number in an equation for the difference between mean velocities of two gases which are allowed to mix, which determines the contribution to this quantity of the concentration gradient. { 'mak,swelz ,kō·i,fish·ənt əv di'fyü·zhən }

Maxwell's cyclic currents *See* mesh currents. { 'mak,swelz 'sī·klik 'kə·rəns }

Maxwell's demon *See* demon of Maxwell. { 'mak ,swelz 'dē·mən }

Maxwell's displacement current *See* displacement current. { 'mak,swelz di'splās·mənt ,kə·rənt }

Maxwell's electromagnetic theory [ELECTROMAG] A mathematical theory of electric and magnetic fields which predicts the propagation of electromagnetic radiation, and is valid for electromagnetic phenomena where effects on an atomic scale can be neglected. { 'mak,swelz i,lek·trō·mag'ned·ik 'thē·ə·rē }

Maxwell's law [ELECTROMAG] A movable portion of a circuit will always move in such a direction as to give maximum magnetic flux linkages through the circuit. { 'mak,swelz 'lȯ }

Maxwell's stress functions [MECH] Three functions of position, ϕ_1, ϕ_2, and ϕ_3, in terms of which the elements of the stress tensor σ of a body may be expressed, if the body is in equilibrium and is not subjected to body forces; the elements of the stress tensor are given by $\sigma_{11}=\partial^2\phi_2/\partial x_3^2+\partial^2\phi_3/\partial x_2^2$, $\sigma_{23}=-\partial^2\phi_1/\partial x_2\partial x_3$, and cyclic permutations of these equations. { 'mak ,swelz 'stres ,fəŋk·shənz }

Maxwell's stress tensor [ELECTROMAG] A second-rank tensor whose product with a unit vector normal to a surface gives the force per unit area transmitted across the surface by an electromagnetic field. { 'mak,swelz 'stres ,ten·sər }

Maxwell's theorem [MECH] If a load applied at one point A of an elastic structure results in a given deflection at another point B, then the same load applied at B will result in the same deflection at A. { 'mak,swelz 'thir·əm }

Maxwell's theory of light [OPTICS] An application of Maxwell's electromagnetic theory in which light is treated as a propagating electromagnetic wave. { 'mak,swelz ,thē·ə·rē əv 'līt }

Maxwell triangle [OPTICS] Color-matching chromaticity values plotted on an *x,y* diagram. Also known as *x,y* chromaticity diagram. { 'mak,swel 'trī,aŋ·gəl }

Maxwell-turn [ELECTROMAG] A centimeter-gram-second electromagnetic unit of flux linkage, equal to the flux linkage of a coil consisting of one complete loop of wire through which passes a magnetic flux of one maxwell. Also known as line-turn. { 'mak,swel ,tərn }

Maxwell-Wagner mechanism [ELEC] A capacitor consisting of two parallel metal plates with two layers of material between them, one with vanishing conductivity, the other with finite conductivity and vanishing electric susceptibility. { 'mak,swel 'wag·nər 'mek·ə,niz·əm }

Maxwell-Wien bridge *See* Maxwell bridge. { 'mak ,swel 'wēn 'brij }

mayer [THERMO] A unit of heat capacity equal to the heat capacity of a substance whose temperature is raised 1° Celsius by 1 joule. { 'mī·ər }

Mayer condensation theory [STAT MECH] A theory of the condensation and critical state of a system of chemically saturated molecules, in which the system is assumed to consist of independent clusters of molecules. { 'mī·ər ,kän·dən'sā·shən ,thē·ə·rē }

Mayer's formula [THERMO] A formula which states that the difference between the specific heat of a gas at constant pressure and its specific heat at constant volume is equal to the gas constant divided by the molecular weight of the gas. { 'mī·ərz ,fȯr·myə·lə }

mb *See* millibar; millibarn.

MBE *See* molecular-beam epitaxy.

mc *See* millihertz.

Mc *See* megahertz.

M center [SOLID STATE] A color center consisting of an F center combined with two ion vacancies. { 'em ,sen·tər }

MCHF approximation *See* multiconfiguration Hartree-Fock approximation. { ¦em¦sē¦āch¦ef ə ,präk·sə'mā·shən }

McLeod gage [FL MECH] A type of instrument used to measure vacuum by measuring the height of a column of mercury supported by the gas whose pressure is to be measured, when this gas is trapped and compressed into a capillary tube. { mə'klau̇d ,gāj }

McMath telescope [OPTICS] A unique 60-inch (1.5-meter) solar telescope at Kitt Peak, Arizona, that has an unconventional configuration; the sun's light is reflected from an 80-inch (2.0-meter) mirror into a long, fixed tube. { mik'math 'tel·ə,skōp }

md *See* millidarcy.

mean British thermal unit *See* British thermal unit. { 'mēn ¦brid·ish 'thər·məl ,yü·nət }

mean calorie [THERMO] One-hundredth of the heat needed to raise 1 gram of water from 0 to 100°C. { 'mēn 'kal·ə·rē }

mean free path [ACOUS] For sound waves in an enclosure, the average distance sound travels between successive reflections in the enclosure. [PHYS] The average distance traveled between two similar events, such as elastic collisions of molecules in a gas, of electrons or phonons in a crystal, or of neutrons in a moderator. { 'mēn ¦frē 'path }

mean life [PHYS] The average time during which a system, such as an atom, nucleus, or elementary particle, exists in a specified form; for a radionuclide or an excited state of an atom or nucleus, it is the reciprocal of the decay constant. Also known as average life; lifetime. { 'mēn 'līf }

mean normal stress [MECH] In a system stressed multiaxially, the algebraic mean of the three principal stresses. { 'mēn ¦nȯrm·əl 'stres }

mean specific heat [THERMO] The average over a specified range of temperature of the specific heat of a substance. { 'mēn spə'sif·ik 'hēt }

mean spherical intensity [OPTICS] The luminous intensity of a light source averaged over all directions. { 'mēn 'sfer·ə·kəl in'ten·səd·ē }

mean-square velocity [PHYS] The average value of the square of the velocities of a group of particles, such as the molecules of a gas. { ¦mēn ¦skwər və'läs·əd·ē }

mean stress [MECH] **1.** The algebraic mean of the maximum and minimum values of a periodically varying stress. **2.** *See* octahedral normal stress. { 'mēn 'stres }

mean-tone scale [ACOUS] A musical scale formed by giving the major third a ratio of exactly 5:4 and adapting other intervals to equalize them. { 'mēn ¦tōn ,skāl }

mean trajectory [MECH] The trajectory of a missile that passes through the center of impact or center of burst. { 'mēn trə'jek·trē }

mean velocity [PHYS] The average value of the velocities of a group of particles, such as the molecules of a gas. { 'mēn və'läs·əd·ē }

measured spectrum *See* spectrogram. { 'mezh·ərd 'spek·trəm }

mechanical birefringence [OPTICS] A change in the double refraction of a solid material when it is subjected to stress. Also known as stress birefringence. { mi'kan·ə·kəl ,bī·ri'frin·jəns }

mechanical equivalent of heat [THERMO] The amount of mechanical energy equivalent to a unit of heat. { mi'kan·ə·kəl i'kwiv·ə·lənt əv 'hēt }

mechanical equivalent of light [OPTICS] The ratio of the radiant power emitted by a monochromatic light source whose wavelength is that at which the sensitivity of phototopic vision is greatest (about 555 nanometers), to its luminous flux measured in lumens. { mi'kan·ə·kəl i'kwiv·ə·lənt əv 'līt }

mechanical hysteresis [MECH] The dependence of the strain of a material not only on the instantaneous value of the stress but also on the previous history of the stress; for example, the elongation is less at a given value of tension when the tension is increasing than when it is decreasing. { mi'kan·ə·kəl ,his·tə'rē·səs }

mechanical impedance [MECH] The complex ratio of a phasor representing a sinusoidally varying force applied to a system to a phasor representing the velocity of a point in the system. { mi'kan·ə·kəl im'pēd·əns }

mechanical mass [QUANT MECH] The part of a particle's mass which is supposed to exist in the absence of any interaction of the particle with itself through a field. { mi'kan·ə·kəl 'mas }

mechanical ohm [MECH] A unit of mechanical resistance, reactance, and impedance, equal to a force of 1 dyne divided by a velocity of 1 centimeter per second. { mi'kan·ə·kəl 'ōm }

mechanical property [MECH] A property that involves a relationship between stress and strain or a reaction to an applied force. { mi'kan·ə·kəl 'präp·ərd·ē }

mechanical reactance [MECH] The imaginary part of mechanical impedance. { mi'kan·ə·kəl rē'ak·təns }

mechanical resistance *See* resistance. { mi'kan·ə·kəl ri'zis·təns }

mechanical rotational impedance *See* rotational impedance. { mi'kan·ə·kəl rō'tā·shən·əl im'pēd·əns }

mechanical rotational reactance *See* rotational reactance. { mi'kan·ə·kəl rō'tā·shən·əl rē'ak·təns }

mechanical rotational resistance *See* rotational resistance. { mi'kan·ə·kəl rō'tā·shən·əl ri'zis·təns }

mechanical units [MECH] Units of length, time, and mass, and of physical quantities derivable from them. { mi'kan·ə·kəl ,yü·nəts }

mechanical vibration [MECH] The continuing motion, often repetitive and periodic, of parts of machines and structures. { mi'kan·ə·kəl vī'brā·shən }

mechanics [PHYS] **1.** In the original sense, the study of the behavior of physical systems under the action of forces. **2.** More broadly, the branch of physics which seeks to formulate general rules for predicting the behavior of a physical system under the influence of any type of interaction with its environment. { mi'kan·iks }

mechanocaloric effect [CRYO] An effect resulting from the fact that a temperature gradient in helium II is invariably accompanied by a pressure gradient, and conversely; examples are the fountain effect, and the heating of liquid helium left behind in a container when part of it leaks out through a small orifice. { ¦mek·ə·nō·kə'lór·ik i,fekt }

mechanomotive force [MECH] The root-mean-square value of a periodically varying force. { ¦mek·ə·nō¦mōd·iv ,fórs }

medium [PHYS] That entity in which objects exist and phenomena take place; examples are free space and various fluids and solids. { 'mē·dē·əm }

megacycle *See* megahertz. { 'meg·ə,sī·kəl }

megaelectronvolt [PHYS] A unit of energy commonly used in nuclear and particle physics, equal to the energy acquired by an electron in falling through a potential of 1,000,000 volts. Abbreviated MeV. { ¦meg·ə·i'lek,trän,vōlt }

megagauss [ELECTROMAG] A unit of magnetic induction equal to 10^6 gauss or 100 tesla. { 'meg·ə,gaůs }

megagauss physics [PHYS] The production, measurement, and application of megagauss fields, as produced by discharge of capacitor banks or explosive flux-compression techniques. { 'meg·ə,gaůs 'fiz·iks }

megahertz [PHYS] Unit of frequency, equal to 1,000,000 hertz. Abbreviated MHz. Also known as megacycle (Mc). { 'meg·ə,hərts }

megaphone [ACOUS] A conical or rectangular horn used to amplify or direct the sound of a speaker's voice. { 'meg·ə,fōn }

megasecond [MECH] A unit of time, equal to 1,000,000 seconds. Abbreviated Ms; Msec. { 'meg·ə,sek·ənd }

megaton [PHYS] The energy released by 1,000,000 metric tons of chemical high explosive calculated at a rate of 1000 calories per gram, or a total of 4.18×10^{15} joules; used principally in expressing the energy released by a nuclear bomb. Abbreviated MT. { 'meg·ə,tən }

megavolt [ELEC] A unit of potential difference or emf (electromotive force), equal to 1,000,000 volts. Abbreviated MV. { 'meg·ə,vōlt }

megawatt [MECH] A unit of power, equal to 1,000,000 watts. Abbreviated MW. { 'meg·ə,wät }

megohm [ELEC] A unit of resistance, equal to 1,000,000 ohms. { 'me,gōm }

megohmmeter [ELEC] An instrument which is used for measuring the high resistance of electrical materials of the order of 20,000 megohms at 1000 volts; one direct-reading type employs a permanent magnet and a moving coil. { 'me ,gōmē,mēd·ər }

Meinzer unit *See* permeability coefficient. { 'mīnt·sər ,yü·nət }

Meissner effect [SOLID STATE] The expulsion of magnetic flux from the interior of a piece of superconducting material as the material undergoes the transition to the superconducting phase. Also known as flux jumping; Meissner-Ochsenfeld effect. { 'mīs·nər i,fekt }

Meissner-Ochsenfeld effect *See* Meissner effect. { 'mīs·nər 'äk·sən,feld i,fekt }

mel [ACOUS] A unit of pitch, equal to one-thousandth of the pitch of a simple tone whose frequency is 1000 hertz and whose loudness is 40 decibels above a listener's threshold. { mel }

Melde's experiment [MECH] An experiment to study transverse vibrations in a long, horizontal thread when one end of the thread is attached to a prong of a vibrating tuning fork, while the other passes over a pulley and has weights suspended from it to control the tension in the thread. { 'mel·dēz ik,sper·ə·mənt }

M electron [ATOM PHYS] An electron whose principal quantum number is 3. { 'em i,lek,trän }

melt fracture [MECH] Melt flow instability through a die during plastics molding, leading to helicular, rippled surface irregularities on the finished product. { 'melt ,frak·chər }

melting point [THERMO] **1.** The temperature at which a solid of a pure substance changes to a liquid. Abbreviated mp. **2.** For a solution of two or more components, the temperature at which the first trace of liquid appears as the solution is heated. { 'melt·iŋ ,pȯint }

melt instability [MECH] Instability of the plastic melt flow through a die. { 'melt ,in·stə'bil·əd·ē }

melt strength [MECH] Strength of a molten plastic. { 'melt ,streŋkth }

membrane analogy [MECH] A formal identity between the differential equation and boundary conditions for a stress function for torsion of an elastic prismatic bar, and those for the deflection of a uniformly stretched membrane with the same boundary as the cross section of the bar, subjected to a uniform pressure. { 'mem,brān ə,nal·ə·jē }

membrane stress [MECH] Stress which is equivalent to the average stress across the cross section involved and normal to the reference plane. { 'mem,brān ,stres }

meniscus [FL MECH] The free surface of a liquid which is near the walls of a vessel and which is curved because of surface tension. { mə'nis·kəs }

meniscus lens [OPTICS] A lens with one convex surface and one concave surface. { mə'nis·kəs 'lenz }

meniscus-Schmidt telescope [OPTICS] A variant of the Schmidt system in which the corrector plate is replaced by a weaker corrector plate followed by a meniscus lens. Also known as Maksutov-Schmidt telescope; Schmidt-Maksutov telescope. { mə'nis·kəs 'shmit 'tel·ə,skōp }

meridian telescope [OPTICS] Any telescope used to make observations in the plane of the meridian, such as a transit telescope or zenith telescope. { mə'rid·ē·ən 'tel·ə,skōp }

meridional focus *See* primary focus. { mə'rid·ē·ən·əl 'fō·kəs }

meridional plane [OPTICS] A plane containing the axis of an optical system. Also known as tangential plane. { mə'rid·ē·ən·əl 'plān }

meridional ray [OPTICS] A ray that lies within a plane which also contains the axis of an optical system. { mə'rid·ē·ən·əl 'rā }

merohedral [CRYSTAL] Of a crystal class in a system, having a general form with only one-half, one-fourth, or one-eighth the number of equivalent faces of the corresponding form in the holohedral class of the same system. Also known as merosymmetric. { ¦mer·ə¦hē·drəl }

merosymmetric *See* merohedral. { ¦mer·ə·si¦me·trik }

Merrington effect [FL MECH] The pronounced expansion of a non-Newtonian fluid when it emerges from a nozzle so that the diameter of the emerging stream can be several times the nozzle diameter. { 'mer·iŋ·tən i,fekt }

Mersenne's law [MECH] The fundamental frequency of a vibrating string is proportional to the square root of the tension and inversely proportional both to the length and the square root of the mass per unit length. { mər'senz ,lȯ }

Merton grating [OPTICS] A type of diffraction grating which is produced by a process in which a helical thread is cut on a cylinder, and errors are smoothed by cutting a second thread further along the same cylinder with a Merton nut. { 'mərt·ən ,grād·iŋ }

mesh [ELEC] A set of branches forming a closed path in a network so that if any one branch is omitted from the set, the remaining branches of the set do not form a closed path. Also known as loop. { mesh }

mesh analysis [ELEC] A method of electrical circuit analysis in which the mesh currents are taken as independent variables and the potential differences around a mesh are equated to 0. { 'mesh ə'nal·ə·səs }

mesh connection *See* delta connection. { 'mesh kə,nek·shən }

mesh currents [ELEC] The currents which are considered to circulate around the meshes of an electric network, so that the current in any branch of the network is the algebraic sum of the mesh currents of the meshes to which that branch belongs. Also known as cyclic currents; Maxwell's cyclic currents. { 'mesh kə·rəns }

mesh impedance [ELEC] The ratio of the voltage to the current in a mesh when all other meshes are open. Also known as self-impedance. { 'mesh im'pēd·əns }

mesic atom [PART PHYS] An atom in which one of the electrons is replaced by a negative muon or meson orbiting close to or within the nucleus. Also known as mesonic atom. { 'me·zik 'ad·əm }

mesic molecule [PART PHYS] A molecule in which one of the electrons is replaced by a negative muon or meson orbiting close to or within one of the nuclei. Also known as mesonic molecule. { 'me·zik 'mäl·ə,kyül }

meson [PART PHYS] Any elementary (noncomposite) particle with strong nuclear interactions and baryon number equal to zero. { 'me,sän }

meson capture [PART PHYS] Process in which an atomic nucleus acquires a negative muon or meson which circles it in a tightly bound orbit until it decays. { 'me,sän ,kap·chər }

mesonic atom *See* mesic atom. { me'zän·ik 'ad·əm }

mesonic molecule *See* mesic molecule. { me'zän·ik 'mäl·ə,kyül }

mesonic x-ray [PART PHYS] An x-ray emitted by a mesic atom when the muon or meson makes a transition from one bound state to another. { me'zän·ik 'eks,rā }

meson resonance [PART PHYS] Any elementary particle with a baryon number of zero which decays through strong interactions, and therefore has an extremely short lifetime on the order of 10^{-23} second. { 'me,zän 'rez·ən·əns }

mesoscopic [PHYS] Pertaining to a size regime, intermediate between the microscopic and the macroscopic, that is characteristic of a region where a large number of particles can interact in a quantum-mechanically correlated fashion. { ¦mez·ə¦skäp·ik }

mesoscopic physics [PHYS] A subdiscipline of condensed-matter physics that focuses on the properties of solids in a size range intermediate between bulk matter and individual atoms or molecules. { ,mez·ə,skäp·ik 'fiz·iks }

metacenter [FL MECH] The intersection of a vertical line through the center of buoyancy of a floating body, slightly displaced from its equilibrium position, with a line connecting the center of gravity and the equilibrium center of buoyancy; the floating body is stable if the metacenter lies above the center of gravity. { 'med·ə,sen·tər }

metal antenna [ELECTROMAG] An antenna which has a relatively small metal surface, in contrast to a slot antenna. { 'med·əl an'ten·ə }

metal-film resistor [ELEC] A resistor in which the resistive element is a thin film of metal or alloy, deposited on an insulating substrate of an integrated circuit. { 'med·əl ¦film ri'zis·tər }

metal-insulator semiconductor [SOLID STATE] Semiconductor construction in which an insulating layer, generally a fraction of a micrometer thick, is deposited on the semiconducting substrate before the pattern of metal contacts is applied. Abbreviated MIS. { 'med·əl ¦in·sə,lād·ər 'sem·i·kən'dək·tər }

metal-insulator transition [SOLID STATE] The change of certain low-dimensional conductors from metals to insulators as the temperature is lowered through a certain value, due to the lattice distortion and band gap accompanying the onset of a charge-density wave. { 'med·əl ¦in·sə,lād·ər tran'zish·ən }

metallic [OPTICS] Having a brilliant mineral luster characteristic of metals. { mə'tal·ik }

metallic circuit [ELEC] Wire circuit of which the ground or earth forms no part. { mə'tal·ik 'sər·kət }

metallic insulator [ELECTROMAG] Section of transmission line used as a mechanical support device; the section is an odd number of quarter-wavelengths long at the frequency of interest, and the input impedance becomes high enough so that the section effectively acts as an insulator. { mə'tal·ik 'in·sə,lād·ər }

metallized capacitor [ELEC] A capacitor in which a film of metal is deposited directly on the dielectric to serve in place of a separate foil strip; has self-healing characteristics. { 'med·əl,īzd kə'pas·əd·ər }

metallized-paper capacitor [ELEC] A modification of a paper capacitor in which metal foils are replaced by extremely thin films of metal deposited on the paper; if a breakdown occurs, these films burn away in the area of the breakdown. { 'med·əl,īzd ¦pāp·ər kə'pas·əd·ər }

metallized resistor [ELEC] A resistor made by depositing a thin film of high-resistance metal on the surface of a glass or ceramic rod or tube. { 'med·əl,īzd ri'zis·tər }

metallograph [OPTICS] An optical microscope equipped with a camera for both visual observation and photography of the structure and constitution of a metal or alloy. { mə'tal·ə,graf }

metal-nitride-oxide semiconductor [SOLID STATE] A semiconductor structure that has a double insulating layer; typically, a layer of silicon dioxide (SiO_2) is nearest the silicon substrate, with a layer of silicon nitride (Si_3N_4) over it. Abbreviated MNOS. { 'med·əl ¦nī,trīd ¦äk,sīd 'sem·i·kən,dək·tər }

metal-organic chemical vapor deposition [SOLID STATE] A technique for growing thin layers of compound semiconductors in which metal organic compounds, having the formula MR_x, where M is a group III metal and R is an organic radical, are decomposed near the surface of a heated substrate wafer, in the presence of a hydride of a group V element. Abbreviated MOCVD. { 'med·əl ȯr'gan·ik 'kem·ə·kəl 'vā·pər ,dep·ə'zish·ən }

metal oxide resistor [ELEC] A metal-film resistor in which an oxide of a metal such as tin is deposited as a film onto an insulating substrate. { 'med·əl ¦äk,sīd ri'zis·tər }

metal oxide semiconductor [SOLID STATE] A metal insulator semiconductor structure in which the insulating layer is an oxide of the substrate material; for a silicon substrate, the insulating layer is silicon dioxide (SiO_2). Abbreviated MOS. { 'med·əl ¦äk,sīd 'sem·i·kən,dək·tər }

metal rolling *See* rolling. { 'med·əl ,rōl·iŋ }

metal vapor laser [OPTICS] An ion laser based on vaporization of a solid or liquid metal, such as cadmium, calcium, copper, lead, manganese, selenium, strontium, and tin, vaporized with a buffer gas such as helium. { 'med·əl ,vā·pər 'lā·zər }

metarheology [MECH] A branch of rheology whose approach is intermediate between those of macrorheology and microrheology; certain processes that are not isothermal are taken into consideration, such as kinetic elasticity, surface tension, and rate processes. { ,med·ə·rē'äl·ə·jē }

metastable equilibrium [PHYS] A condition in which a system returns to equilibrium after small (but not large) displacements; it may be represented by a ball resting in a small depression on top of a hill. { ¦med·ə'stā·bəl ˌē·kwə'lib·rē·əm }

metastable state [QUANT MECH] An excited stationary energy state whose lifetime is unusually long. { ¦med·ə'stā·bəl ¦stāt }

metastasis [PHYS] A transition of an electron or nucleon from one bound state to another in an atom or molecule, or the capture of an electron by a nucleus. { mə'tas·tə·səs }

meteorological optics [OPTICS] A branch of atmospheric physics or physical meteorology in which optical phenomena occurring in the atmosphere are described and explained. Also known as atmospheric optics. { ˌmed·ē·ə·rə'läj·ə·kəl 'äp·tiks }

meter [MECH] The international standard unit of length, equal to the length of the path traveled by light in vacuum during a time interval of 1/299,792,458 of a second. Abbreviated m. { 'mēd·ər }

meter-atmosphere [PHYS] The depth of an equivalent atmosphere of a given gas, in meter-atmospheres, is equal to the depth in meters that the atmosphere would have if it were composed entirely of the gas in question and in the same amount as exists in the actual atmosphere, and had a uniform temperature and pressure of 0°C and 1 standard atmosphere. Abbreviated m-atm. Also known as atmo-meter. { 'mēd·ər 'at·məˌsfir }

meter bridge [ELEC] A uniform resistance wire 1 meter in length, mounted above a scale marked in millimeters, with terminals added to make the device usable as either part of a Wheatstone bridge or of a potentiometer. { 'mēd·ər ˌbrij }

meter-candle *See* lux. { 'mēd·ər 'kan·dəl }

meter-kilogram [MECH] **1.** A unit of energy or work in a meter-kilogram-second gravitational system, equal to the work done by a kilogram-force when the point at which the force is applied is displaced 1 meter in the direction of the force; equal to 9.80665 joules. Abbreviated m-kgf. Also known as meter kilogram-force. **2.** A unit of torque, equal to the torque produced by a kilogram-force acting at a perpendicular distance of 1 meter from the axis of rotation. Also known as kilogram-meter (kgf-m). { 'mēd·ər 'kil·ə ˌgram }

meter kilogram-force *See* meter-kilogram. { 'mēd·ər 'kil·əˌgram 'fȯrs }

meter-kilogram-second-ampere system [PHYS] A system of electrical and mechanical units in which length, mass, time, and electric current are the fundamental quantities, and the units of these quantities are the meter, the kilogram, the second, and the ampere respectively. Abbreviated mksa system. Also known as Giorgi system; practical system. { 'mēd·ər 'kil·əˌgram 'sek·ənd 'amˌpir ˌsis·təm }

meter-kilogram-second system [MECH] A metric system of units in which length, mass, and time are fundamental quantities, and the units of these quantities are the meter, the kilogram, and the second respectively. Abbreviated mks system. { 'mēd·ər 'kil·əˌgram 'sek·ənd ˌsis·təm }

meter sizing factor [FL MECH] A dimensionless number used in calculating the rate of flow of fluid through a pipe from the readings of a flowmeter that measures the drop in pressure when the fluid is forced to flow through a circular orifice; it is equal to $K(d/D)^2$, where K is the flow coefficient, d is the orifice bore diameter, and D is the internal diameter of the pipe. { 'mēd·ər 'sīz·iŋ ˌfak·tər }

meter-ton-second system [MECH] A modification of the meter-kilogram-second system in which the metric ton (1000 kilograms) replaces the kilogram as the unit of mass. { 'mēd·ər 'tən 'sek·ənd ˌsis·təm }

method of images [ELEC] In electrostatics, a method of determining the electric fields and potentials set up by charges in the vicinity of a conductor, in which the conductor and its induced surface charges are replaced by one or more fictitious charges. [PHYS] Any method of solving magnetostatic, hydrodynamic, and other problems involving boundary conditions at the interface between two media, in which fictitious objects, such as magnetic dipoles and sources and sinks of fluid, are introduced to satisfy the boundary conditions; these methods are generalizations of the method in electrostatics. { 'meth·əd əv 'im·ij·əz }

method of mixtures [THERMO] A method of determining the heat of fusion of a substance whose specific heat is known, in which a known amount of the solid is combined with a known amount of the liquid in a calorimeter, and the decrease in the liquid temperature during melting of the solid is measured. { 'meth·əd əv 'miks·chərz }

metric centner [MECH] **1.** A unit of mass equal to 50 kilograms. **2.** A unit of mass equal to 100 kilograms. Also known as quintal. { 'me·trik 'sent·nər }

metric grain [MECH] A unit of mass, equal to 50 milligrams; used in commercial transactions in precious stones. { 'me·trik 'grān }

metric horsepower [PHYS] A unit of power, equal to 75 meter kilograms-force per second; equal to 735.49875 watts. { 'me·trik ˌhȯrs ˌpau̇·ər }

metric line *See* millimeter. { 'me·trik 'līn }

metric ounce *See* mounce. { 'me·trik 'au̇ns }

metric slug *See* metric-technical unit of mass. { 'me·trik 'sləg }

metric system [MECH] A system of units used in scientific work throughout the world and employed in general commercial transactions and engineering applications; its units of length, time, and mass are the meter, second, and kilogram respectively, or decimal multiples and submultiples thereof. { 'me·trik ˌsis·təm }

metric-technical unit of mass [MECH] A unit of mass, equal to the mass which is accelerated by 1 meter per second per second by a force of 1 kilogram-force; it is equal to 9.80665 kilograms.

Abbreviated TME. Also known as hyl; metric slug. { 'me·trik ¦tek·ni·kəl ¦yü·nət əv 'mas }

metric ton *See* tonne. { 'me·trik 'tən }

metric waves [ELECTROMAG] Radio waves having wavelengths between 1 and 10 meters, corresponding to frequencies between 30 and 300 megahertz (the very-high-frequency band). { 'me·trik 'wāvz }

metrology [PHYS] The science of measurement. { mə'träl·ə·jē }

MeV *See* megaelectronvolt.

Meyer atomic volume curve [ATOM PHYS] A graph of the atomic volumes of the elements versus their atomic numbers; it reveals a periodicity, with peaks at the alkali elements and valleys at the transition elements. { 'mī·ər ə¦täm·ik ˌväl·yəm 'kərv }

mF *See* millifarad.

mg *See* milligram.

mG *See* milligauss.

mGal *See* milligal.

mH *See* millihenry.

MHD *See* magnetohydrodynamics.

mho *See* siemens. { mō }

mHz *See* millihertz.

MHz *See* megahertz.

mi *See* mile.

mica capacitor [ELEC] A capacitor whose dielectric consists of thin rectangular sheets of mica and whose electrodes are either thin sheets of metal foil stacked alternately with mica sheets, or thin deposits of silver applied to one surface of each mica sheet. { 'mī·kə kə'pas·əd·ər }

Michel parameter [PART PHYS] A number appearing in an equation for the momentum spectrum of muon decay, which depends on the nature of the weak interactions; the number is equal to 3/4 in any two-component neutrino theory before radiative corrections are taken into account. { mi'shel pəˌram·əd·ər }

Michelson interferometer [OPTICS] An interferometer in which light strikes a partially reflecting plate at an angle of 45°, the light beams reflected and transmitted by the plate are both reflected back to the plate by mirrors, and the beams are recombined at the plate, interfering constructively or destructively depending on the distances from the plate to the two mirrors. { 'mī·kəl·sən ˌin·tər·fə'räm·əd·ər }

Michelson-Morley experiment [OPTICS] An experiment which uses a Michelson interferometer to determine the difference between the speeds of light in two perpendicular directions. { 'mī·kəl·sən 'mȯr·lē ikˌsper·ə·mənt }

Michelson stellar interferometer [OPTICS] An instrument for measuring angular diameters of astronomical objects, in which a system of mirrors directs two parallel beams of light into a telescope, and angular diameter is determined from the maximum distance between the beams at which interference fringes are observable. { 'mī·kəl·sən ¦stel·ər ˌin·tər·fə'räm·əd·ər }

microampere [ELEC] A unit of current equal to one-millionth of an ampere. Abbreviated μA. { ¦mī·krō'amˌpir }

microangstrom [MECH] A unit of length equal to one-millionth of an angstrom, or 10^{-16} meter. Abbreviated μA. { ¦mī·krō'aŋ·strəm }

microbar *See* barye. { 'mī·krəˌbär }

microbarom [ACOUS] An infrasound wave that originates with surface waves on the seas or oceans, having a period of 4–7 seconds and a sound pressure level of about 85 decibels. { 'mī·krōˌbar·əm }

microbeam [ELECTROMAG] An x-ray beam with submicrometer dimensions. { 'mī·krōˌbēm }

microbridge [CRYO] A Josephson junction formed by configuration of thin superconducting films. { 'mī·krōˌbrij }

microcanonical ensemble [STAT MECH] A collection of systems describing a single isolated system of specified energy; its members are uniformly distributed over a part of phase space whose energies lie within an infinitesimal range. { ¦mī·krō·kə'nän·ə·kəl än'säm·bəl }

microcoulomb [ELEC] A unit of electric charge equal to one-millionth of a coulomb. Abbreviated μC. { ¦mī·krō'küˌläm }

microcrystalline [CRYSTAL] Composed of or containing crystals that are visible only under the microscope. { ¦mī·krō'krist·əl·ən }

microdensitometer [SPECT] A high-sensitivity densitometer used in spectroscopy to detect spectrum lines too faint on a negative to be seen by the human eye. { ¦mī·krōˌden·sə'täm·əd·ər }

microdisk laser [OPTICS] A very small semiconductor laser that consists of a quantum well structure formed into a disk, such that total internal reflection of photons traveling around the perimeter of the disk results in high-Q whispering-gallery resonances. { 'mī·krōˌdisk ˌlā·zər }

microfarad [ELEC] A unit of capacitance equal to one-millionth of a farad. Abbreviated μF. { ¦mī·krō'far·əd }

microfluid [FL MECH] A fluid in which the effects of local motion of contained material particles on properties and behavior of the fluid are not disregarded. { 'mī·krōˌflü·əd }

microgram [MECH] A unit of mass equal to one-millionth of a gram. Abbreviated μg. { 'mī·krəˌgram }

microgravity [MECH] A state of very weak gravity, such that the gravitational acceleration experienced by an observer inside the system in question is of the order of one-millionth of that on earth. { ˌmī·krō'grav·əd·ē }

microhm [ELEC] A unit of resistance, reactance, and impedance, equal to 10^{-6} ohm. { 'mī·krōm }

microholography *See* x-ray holography. { ¦mī·krō·hō'läg·rə·fē }

microhysteresis effect [SOLID STATE] Hysteresis that results from the motion of domain walls lagging behind an applied magnetic or elastic stress when these walls are held up by dislocations and other imperfections in the material. { ¦mī·krōˌhis·tə'rē·səs iˌfekt }

microinterferometer [OPTICS] Functional combination of a microscope with an interferometer;

used to study thin films, platings, or transparent coatings. { ¦mī·krō,int·ə·fə'räm·əd·ər }

microlaser *See* single-atom laser. { 'mī·krō,lā·zər }

microlens array [OPTICS] An array of very small lenses with diameters between 20 micrometers and 1 millimeter; used in a variety of applications, including integral photography, photocopying, facsimile, and high-speed parallel switching networks. { 'mī·krō,lenz ə,rā }

microlite [CRYSTAL] A microscopic crystal which polarizes light. Also known as microlith. { 'mī·krə,līt }

microlith *See* microlite. { 'mī·krə,lith }

micromaser *See* single-atom laser. { 'mī·krə,māz·ər }

micrometer [MECH] A unit of length equal to one-millionth of a meter. Abbreviated μm. Also known as micron (μ). { mī'kräm·əd·ər }

micrometer of mercury *See* micron. { mī'kräm·əd·ər əv 'mər·kyə·rē }

micromicrowatt *See* picowatt. { ¦mī·krō¦mī·krō'wät }

micron [MECH] **1.** A unit of pressure equal to the pressure exerted by a column of mercury 1 micrometer high, having a density of 13.5951 grams per cubic centimeter, under the standard acceleration of gravity; equal to 0.133322387415 pascal; it differs from the millitorr by less than one part in seven million. Also known as micrometer of mercury. **2.** *See* micrometer. { 'mī,krän }

microoptics [OPTICS] A technology that utilizes optical elements that range in diameter from 20 micrometers to 1 millimeter. { ,mī·krō'äp·tiks }

microprism [OPTICS] A usually circular area in the focusing screen of a camera viewfinder that is made up of tiny prisms and causes the image in the viewfinder to blur if the subject is out of focus. { 'mī·krə,priz·əm }

microprobe [SPECT] An instrument for chemical microanalysis of a sample, in which a beam of electrons is focused on an area less than a micrometer in diameter, and the characteristic x-rays emitted as a result are dispersed and analyzed in a crystal spectrometer to provide a qualitative and quantitative evaluation of chemical composition. Also known as x-ray microprobe. { 'mī·krə,prōb }

microprobe spectrometry [SPECT] Microanalysis of a sample, using a microprobe. { 'mī·krə,prōb spek'träm·ə·trē }

microradiogram [PHYS] A two-dimensional x-ray image of a sample, produced by one type of x-ray microscope used in microradiography; all levels of the sample object are imaged into essentially a single focal plane for subsequent microphotographic enlargement. { ¦mī·krō'rād·ē·ə,gram }

micro-reciprocal-degree *See* mired. { ¦mī·krō ri'sip·rə·kəl di'grē }

microrefractometry [OPTICS] The measurement of refractive indices of microscopic objects; this is often done by immersing an object in a series of mediums of graded refractive index until one is found that makes the object invisible in a phase-contrast microscope. { ¦mī·krō,rē,frak'täm·ə·trē }

microrheology [MECH] A branch of rheology in which the heterogeneous nature of dispersed systems is taken into account. { ¦mī·krō·rē'äl·ə·jē }

microscope [OPTICS] An instrument through which minute objects are enlarged by means of a lens or lens system; principal types include optical, electron, and x-ray. { 'mī·krə,skōp }

microscope stage [OPTICS] The platform on which specimens are placed for microscopic examination. { 'mī·krə,skōp ,stāj }

microscopic *See* microscopical. { ¦mī·krə¦skäp·ik }

microscopical [OPTICS] Also known as microscopic. **1.** Of or pertaining to the microscope. **2.** Visible only under a microscope. { ,mī·krə'skäp·ə·kəl }

microscopic reversibility [STAT MECH] A principle which requires that in a system at equilibrium any molecular process and its reverse take place at the same average rate. Also known as reversibility principle. { ¦mī·krə¦skäp·ik ri,vər·sə'bil·əd·ē }

microscopic state [STAT MECH] The state of a system as specified by the actual properties of each individual, elemental component, in the ultimate detail permitted by the uncertainty principle. Also known as microstate. { ¦mī·krə¦skäp·ik 'stāt }

microscopic theory [PHYS] A theory concerned with the interactions of atoms, molecules, or their constituents, involving distances on the order of 10^{-10} meter or less, which underlie observable phenomena. { ¦mī·krə¦skäp·ik 'thē·ə·rē }

microscopy [OPTICS] The interpretive application of microscope magnification to the study of materials that cannot be properly seen by the unaided eye. { mī'kräs·kə·pē }

microsecond [MECH] A unit of time equal to one-millionth of a second. Abbreviated μs. { ¦mī·krə,sek·ənd }

microspectrograph [SPECT] A microspectroscope provided with a photographic camera or other device for recording the spectrum. { ¦mī·krō'spek·trə,graf }

microspectrophotometer [SPECT] A split-beam or double-beam spectrophotometer including a microscope for the localization of the object under study, and capable of carrying out spectral analyses within the dimensions of a single cell. { ¦mī·kro¦spek·trə·fə'täm·əd·ər }

microspectroscope [SPECT] An instrument for analyzing the spectra of microscopic objects, such as living cells, in which light passing through the sample is focused by a compound microscope system, and both this light and the light which has passed through a reference sample are dispersed by a prism spectroscope, so that the spectra of both can be viewed simultaneously. { ¦mī·krō'spek·trə,skōp }

microstate *See* microscopic state. { 'mī·krə,stāt }

microstrip [ELECTROMAG] A strip transmission line that consists basically of a thin-film strip in intimate contact with one side of a flat dielectric substrate, with a similar thin-film ground-plane conductor on the other side of the substrate. { 'mī·krə,strip }

microvolt [ELEC] A unit of potential difference equal to one-millionth of a volt. Abbreviated μV. { 'mī·krə,vōlt }

microvolts per meter [ELECTROMAG] Field strength of antenna which is the ratio of the antenna voltage in microvolts to the antenna length in meters, as measured at a given point. { 'mī·krə,vols pər 'mēd·ər }

microwatt [MECH] A unit of power equal to one-millionth of a watt. Abbreviated μW. { 'mī·krə,wät }

microwave [ELECTROMAG] An electromagnetic wave which has a wavelength between about 0.3 and 30 centimeters, corresponding to frequencies of 1–100 gigahertz; however, there are no sharp boundaries distinguishing microwaves from infrared and radio waves. { 'mī·krə,wāv }

microwave acoustics [ACOUS] The production and study of elastic vibrations in materials at microwave frequencies, on the order of 10^9 to 10^{11} hertz, such as in single-crystal delay lines used in radar systems. { 'mī·krə,wāv ə'küs·tiks }

microwave amplification by stimulated emission of radiation *See* maser. { 'mī·krə,wāv ,am·plə·fə'kā·shən bī 'stim·yə,lād·əd i'mish·ən əv ,rād·ē'ā·shən }

microwave antenna [ELECTROMAG] A combination of an open-end waveguide and a parabolic reflector or horn, used for receiving and transmitting microwave signal beams at microwave repeater stations. { 'mī·krə,wāv an'ten·ə }

microwave attenuator [ELECTROMAG] A device that causes the field intensity of microwaves in a waveguide to decrease by absorbing part of the incident power; usually consists of a piece of lossy material in the waveguide along the direction of the electric field vector. { 'mī·krə,wāv ə'ten·yə,wād·ər }

microwave bridge [ELECTROMAG] A microwave circuit equivalent to an ordinary electrical bridge and used to measure impedance; consists of six waveguide sections arranged to form a multiple junction. { 'mī·krə,wāv ,brij }

microwave cavity *See* cavity resonator. { 'mī·krə,wāv ,kav·ə·dē }

microwave circuit [ELECTROMAG] Any particular grouping of physical elements, including waveguides, attenuators, phase changers, detectors, wavemeters, and various types of junctions, which are arranged or connected together to produce certain desired effects on the behavior of microwaves. { 'mī·krə,wāv ,sər·kət }

microwave circulator *See* circulator. { 'mī·krə,wāv 'sər·kyə,lād·ər }

microwave filter [ELECTROMAG] A device which passes microwaves of certain frequencies in a transmission line or waveguide while rejecting or absorbing other frequencies; consists of resonant cavity sections or other elements. { 'mī·krə,wāv ,fil·tər }

microwave frequency [PHYS] A frequency on the order of 10^9–10^{11} hertz. { 'mī·krə,wāv ,frē·kwən·sē }

microwave gyrator *See* gyrator. { 'mī·krə,wāv 'jī,rād·ər }

microwave heating [ELECTROMAG] Heating of food by means of electromagnetic energy in or just below the microwave spectrum for cooking, dehydration, sterilization, thawing, and other purposes. { 'mī·krə,wāv 'hēd·iŋ }

microwave maser [PHYS] A maser which emits microwave radiation. { 'mī·krə,wāv 'mā·zər }

microwave optics [ELECTROMAG] The study of those properties of microwaves which are analogous to the properties of light waves in optics. { 'mī·krə,wāv 'äp·tiks }

microwave pumping [ELECTROMAG] The use of microwaves to produce large departures from thermal equilibrium in the relative populations of selected quantized states of different energy in atomic, molecular, or nuclear systems. { 'mī·krə,wāv 'pəmp·iŋ }

microwave reflectometer [ELECTROMAG] A pair of single-detector couplers on opposite sides of a waveguide, one of which is positioned to monitor transmitted power, and the other to measure power reflected from a single discontinuity in the line. { 'mī·krə,wāv ,rē,flek'täm·əd·ər }

microwave refractometer [ELECTROMAG] An instrument that measures the index of refraction of the atmosphere by measuring the travel time of microwave signals through each of two precision microwave transmission cavities, one of which is hermetically sealed to serve as a reference. { 'mī·krə,wāv ,rē,frak'täm·əd·ər }

microwave resonance cavity *See* cavity resonator. { 'mī·krə,wāv 'rez·ən·əns ,kav·əd·ē }

microwave spectrometer [SPECT] An instrument which makes a graphical record of the intensity of microwave radiation emitted or absorbed by a substance as a function of frequency, wavelength, or some related variable. { 'mī·krə,wāv spek 'träm·əd·ər }

microwave spectroscope [SPECT] An instrument used to observe the intensity of microwave radiation emitted or absorbed by a substance as a function of frequency, wavelength, or some related variable. { 'mī·krə,wāv 'spek·trə,skōp }

microwave spectroscopy [SPECT] The methods and techniques of observing and the theory for interpreting the selective absorption and emission of microwaves at various frequencies by solids, liquids, and gases. { 'mī·krə,wāv spek'träs·kə·pē }

microwave spectrum [ELECTROMAG] The range of wavelengths or frequencies of electromagnetic radiation that are designated microwaves. [SPECT] A display, photograph, or plot of the intensity of microwave radiation emitted or absorbed by a substance as a function of frequency, wavelength, or some related variable. { 'mī·krə,wāv 'spek·trəm }

microwave transmission line [ELECTROMAG] A material structure forming a continuous path from one place to another and capable of directing the transmission of electromagnetic energy along this path. { 'mī·krə,wāv tranz'mish·ən ,līn }

microwave waveguide *See* waveguide. { 'mī·krə,wāv 'wāv,gīd }

microwave wavemeter [ELECTROMAG] Any device for measuring the free-space wavelengths (or frequencies) of microwaves; usually made of a cavity resonator whose dimensions can be varied until resonance with the microwaves is achieved. { 'mī·krə,wāv 'wāv,mēd·ər }

mid-infrared radiation *See* intermediate-infrared radiation. { ,mid¦in·frə,red ,rad·ē'ā·shən }

Mie-Grüneisen equation [THERMO] An equation of state particularly useful at high pressure, which states that the volume of a system times the difference between the pressure and the pressure at absolute zero equals the product of a number which depends only on the volume times the difference between the internal energy and the internal energy at absolute zero. { 'mē 'grü,nīz·ən i,kwā·zhən }

Mie scattering [OPTICS] The scattering of light by a sphere of dielectric material. { 'mē ,skad·ə·riŋ }

Mie's double plate [ELEC] A device consisting of two small metal disks with insulating handles; they are held in contact in an electric field and then separated, and the charge on one of the disks is then measured to determine the electric displacement. { 'mēz ¦dəb·əl 'plāt }

migma plasma [PL PHYS] A hybrid physical state between a colliding beam and a plasma, which is generated by accelerating ions to energies of several megaelectronvolts and causing them to travel in self-colliding orbits in the presence of thermal, ambient electrons. { 'mig·mə ,plaz·mə }

migration [SOLID STATE] **1.** The movement of charges through a semiconductor material by diffusion or drift of charge carriers or ionized atoms. **2.** The movement of crystal defects through a semiconductor crystal under the influence of high temperature, strain, or a continuously applied electric field. { mī'grā·shən }

MIKES *See* mass-analyzed ion kinetic energy spectrometry. { mīks }

mil [MECH] **1.** A unit of length, equal to 0.001 inch, or to 2.54×10^{-5} meter. Also known as milli-inch; thou. **2.** *See* milliliter. { mil }

mile [MECH] A unit of length in common use in the United States, equal to 5280 feet, or 1609.344 meters. Abbreviated mi. Also known as land mile; statute mile. { mīl }

Miller indices [CRYSTAL] Three integers identifying a type of crystal plane; the intercepts of a plane on the three crystallographic axes are expressed as fractions of the crystal parameters; the reciprocals of these fractions, reduced to integral proportions, are the Miller indices. Also known as crystal indices. { 'mil·ər 'in·də,sēz }

Miller law [CRYSTAL] If the edges formed by the intersections of three faces of a crystal are taken as the three reference axes, then the three quantities formed by dividing the intercept of a fourth face with one of these axes by the intercept of a fifth face with the same axis are proportional to small whole numbers, rarely exceeding 6. Also known as law of rational intercepts. { 'mil·ər ,lȯ }

milliampere [ELEC] A unit of current equal to one-thousandth of an ampere. Abbreviated mA. { ¦mil·ē'am,pir }

millibar [MECH] A unit of pressure equal to one-thousandth of a bar. Abbreviated mb. Also known as vac. { 'mil·ə,bär }

millibarn [NUC PHYS] A unit of cross section equal to one-thousandth of a barn. Abbreviated mb. { 'mil·ə,bärn }

millicycle *See* millihertz. { 'mil·ə,sī·kəl }

millidarcy [PHYS] A unit of fluid permeability equal to one-thousandth of a darcy. Abbreviated md. { ¦mil·ə¦där·sē }

millier *See* tonne. { mil'yā }

millifarad [ELEC] A unit of capacitance equal to one-thousandth of a farad. Abbreviated mF. { ¦mil·ē'far·əd }

milligal [MECH] A unit of acceleration commonly used in geodetic measurements, equal to 10^{-3} galileo, or 10^{-5} meter per second per second. Abbreviated mGal. { 'mil·ə,gal }

milligauss [ELECTROMAG] A unit of magnetic flux density equal to one-thousandth of a gauss. Abbreviated mG. { 'mil·ə,gau̇s }

milligram [MECH] A unit of mass equal to one-thousandth of a gram. Abbreviated mg. { 'mil·ə,gram }

millihenry [ELECTROMAG] A unit of inductance equal to one-thousandth of a henry. Abbreviated mH. { 'mil·ə,hen·rē }

millihertz [PHYS] A unit of frequency equal to one-thousandth of a hertz. Abbreviated mHz. Also known as millicycle (mc). { 'mil·ə,hərts }

millihg *See* millimeter of mercury.

milli-inch *See* mil. { 'mil·ē,inch }

Millikan oil-drop experiment [ATOM PHYS] A method of determining the charge on an electron, in which one measures the terminal velocities of rise and fall of oil droplets in an electric field after the droplets have picked up charge from ionization in the surrounding gas produced by an x-ray beam. { 'mil·ə·kən 'ȯil,dräp ik ,sper·əmənt }

milliliter [MECH] A unit of volume equal to 10^{-3} liter or 10^{-6} cubic meter. Abbreviated ml. Also known as mil. { 'mil·ə,lēd·ər }

milli-mass-unit [PHYS] One-thousandth of an atomic mass unit. Abbreviated mmu. { ¦mil·ə ¦mas 'yü·nət }

millimeter [MECH] A unit of length equal to one-thousandth of a meter. Abbreviated mm. Also known as metric line; strich. { 'mil·ə,mēd·ər }

millimeter of mercury [MECH] A unit of pressure, equal to the pressure exerted by a column of mercury 1 millimeter high with a density of 13.5951 grams per cubic centimeter under the standard acceleration of gravity; equal to 133.322387415 pascals; it differs from the torr by less than 1 part in 7,000,000. Abbreviated mmHg. Also known as millihg. { 'mil·ə,mēd·ər əv 'mər·kyə·rē }

millimeter of water [MECH] A unit of pressure, equal to the pressure exerted by a column of water 1 millimeter high with a density of 1 gram per cubic centimeter under the standard acceleration of gravity; equal to 9.80665 pascals. Abbreviated mmH_2O. { 'mil·ə,mēd·ər əv 'wȯdər }

millimeter wave [ELECTROMAG] An electromagnetic wave having a wavelength between 1 millimeter and 1 centimeter, corresponding to frequencies between 30 and 300 gigahertz. Also known as millimetric wave. { 'mil·ə,mēd·ər 'wāv }

millimetric wave *See* millimeter wave. { ¦mil·ə¦me·trik 'wāv }

millimicron *See* nanometer. { 'mil·ə,mī·krȯn }

Millington reverberation formula [ACOUS] A formula that states that the reverberation time of a chamber in seconds is 0.05 times its volume in cubic feet, divided by the sum over the surfaces of the chamber of the product of the surface's area in square feet by the natural logarithm of 1 minus its absorption coefficient. { 'mil·iŋ·tən ri,vər·bə'rā·shən ,fȯr·myə·lə }

million electronvolts *See* megaelectronvolt. { 'mil·yən i'lek,trän,vōlts }

millisecond [MECH] A unit of time equal to one-thousandth of a second. Abbreviated ms; msec. { 'mil·ə,sek·ənd }

millivolt [ELEC] A unit of potential difference or emf equal to one-thousandth of a volt. Abbreviated mV. { 'mil·ə,vōlt }

milliwatt [MECH] A unit of power equal to one-thousandth of a watt. Abbreviated mW. { 'mil·ə,wät }

Mills cross [ELECTROMAG] An antenna array that consists of two antennas oriented perpendicular to each other and that produces a narrow pencil beam. { 'milz 'krȯs }

mimetic [CRYSTAL] Pertaining to a crystal that is twinned or malformed but whose crystal symmetry appears to be of a higher grade than it actually is. { mə'med·ik }

min *See* minim. { min }

minim [MECH] A unit of volume in the apothecaries' measure; equals 1/60 fluidram (approximately 0.061612 cubic centimeter) or about 1 drop (of water). Abbreviated min. { 'min·əm }

minimization principle [PHYS] A principle requiring that the final state of a system is determined by the attainment of the minimum possible value of a certain quantity. { ,min·ə·mə'zā·shən ,prin·sə·pəl }

minimum deviation [OPTICS] For a prism, the smallest possible angle between the incident and refracted rays; this angle is realized when refraction is symmetrical. { 'min·ə·məm ,dē,vē'ā·shən }

minimum ionizing speed [ATOM PHYS] The smallest speed at which a charged particle passing through a gas can ionize an atom or molecule. { 'min·ə·məm 'ī·ə,niz·iŋ ,spēd }

minimum resolvable temperature difference [THERMO] The change in equivalent blackbody temperature that corresponds to a change in radiance which will produce a just barely resolvable change in the output of an infrared imaging device, taking into account the characteristics of the device, the display, and the observer. Abbreviated MRTD. { 'min·ə·məm ri'zäl·və·bəl 'tem·prə·chər ,dif·rəns }

minimum-shock boom [ACOUS] A sonic boom whose pressure signature is shaped to reduce the perceived amplitude of the shocks by rounding the shape of the signature near the maximum. { ¦min·ə·məm 'shäk ,büm }

Minkowski electrodynamics [ELECTROMAG] An electromagnetic theory, compatible with the special theory of relativity, which takes into account the presence of matter with electric and magnetic polarization. { miŋ'kȯf·skē i¦lek·trō·dī¦nam·iks }

Minkowski metric [RELAT] The metric tensor of the Minkowski space-time used in special relativity; it is a 4 × 4 matrix whose nonzero entries lie on the diagonal, with one entry (corresponding to the time coordinate) equal to 1, and three entries (corresponding to space coordinates) equal to -1; sometimes, the negative of this matrix is used. { miŋ'kȯf·skē 'me·trik }

Minkowski space-time [RELAT] The space-time of special relativity; it is completely flat and contains no gravitating matter. Also known as Minkowski universe. { miŋ'kȯf·skē 'spās 'tīm }

Minkowski universe *See* Minkowski space-time. { miŋ'kȯf·skē 'yü·nə,vərs }

minor bend [ELECTROMAG] Rectangular waveguide bent so that throughout the length of a bend a longitudinal axis of the guide lies in one plane which is parallel to the narrow side of the waveguide. { 'mīn·ər 'bend }

minor diatonic scale [ACOUS] A diatonic scale in which the relative sizes of the sequence of intervals are approximately 2,1,2,2,2,2,1. { 'mīn·ər ¦dī·ə¦tän·ik 'skāl }

minority carrier [SOLID STATE] The type of carrier, electron, or hole that constitutes less than half the total number of carriers in a semiconductor. { mə'när·əd·ē 'kar·ē·ər }

minor lobe [ELECTROMAG] Any lobe except the major lobe of an antenna radiation pattern. Also known as secondary lobe; side lobe. { 'mīn·ər ¦lōb }

minute [MECH] A unit of time, equal to 60 seconds. { 'min·ət }

mirage [OPTICS] Any one of a variety of unusual images of distant objects seen as a result of the bending of light rays in the atmosphere during abnormal vertical distribution of air density. { mə'räzh }

mired [THERMO] A unit used to measure the

reciprocal of color temperature, equal to the reciprocal of a color temperature of 10^6 kelvins. Derived from micro-reciprocal-degree. { mīrd }

mirror [OPTICS] A surface which specularly reflects a large fraction of incident light. { 'mir·ər }

mirror coating [OPTICS] A thin film of highly reflective material spread over a correctly shaped glass surface to produce a mirror; aluminum is usually used in the visible region. Also known as reflective coating. { 'mir·ər ˌkōd·iŋ }

mirror image [OPTICS] A form that is identical to another except that it is reversed, as if viewed in a mirror. { 'mir·ər 'im·ij }

mirror interference [OPTICS] Interference occurring between two beams, one or both of which are reflected from a mirror at a small angle. { 'mir·ər ˌin·tər'fir·əns }

mirror interferometer [OPTICS] Any interferometer which makes use of mirror interference. { 'mir·ər ˌin·tər·fə'räm·əd·ər }

mirror machine [PL PHYS] A device which confines plasma in a tube with magnetic mirrors at each end to prevent it from escaping. { 'mir·ər məˌshēn }

mirror nuclei [NUC PHYS] A pair of atomic nuclei, each of which would be transformed into the other by changing all its neutrons into protons, and vice versa. { 'mir·ər 'nü·klēˌī }

mirror optics [OPTICS] The science and technology of mirrors which, by means of reflecting rays of light, either revert optical bundles or focus them to form images. { 'mir·ər 'äp·tiks }

mirror plane of symmetry *See* plane of mirror symmetry. { 'mir·ər 'plān əv 'sim·ə·trē }

mirror reflection *See* specular reflection. { 'mir·ər ri'flek·shən }

MIR technique *See* multiple isomorphous replacement technique. { ¦em¦ī'är tekˌnēk }

MIS *See* metal-insulator semiconductor.

mismatch [ELEC] The condition in which the impedance of a source does not match or equal the impedance of the connected load or transmission line. { 'misˌmach }

mismatch factor *See* reflection factor. { 'misˌmach ˌfak·tər }

mismatch slotted line [ELECTROMAG] A slotted line linking two waveguides which is not properly designed to minimize the power reflected or transmitted by it. { 'misˌmach 'släd·əd 'līn }

missile attitude [MECH] The position of a missile as determined by the inclination of its axes (roll, pitch, and yaw) in relation to another object, as to the earth. { 'mis·əl ˌad·əˌtüd }

missing mass spectrometer [PART PHYS] An apparatus which measures the momentum of the recoil protons in a reaction such as $\pi^- + p \rightarrow p + (MM)^-$, in order to determine the distribution of masses of the MM system, without any detailed observations on this system. { ˌmis·iŋ ¦mas spek'träm·əd·ər }

mist [FL MECH] Fine liquid droplets suspended in or falling through a moving or stationary gas atmosphere. { mist }

mistbow *See* fogbow. { 'mistˌbō }

mistuning [MECH] The difference between the square of the natural frequency of vibration of a vibrating system, without the effect of damping, and the square of the frequency of an external, oscillating force. { mis'tün·iŋ }

MIT bag model [PART PHYS] A model describing quark confinement in hadrons in which a hadron is viewed as a bubble of gas in a uniform, isotropic, perfect fluid, with the thermodynamic pressure of the gas replaced by the quantum pressure of quarks. Derived from Massachusetts Institute of Technology bag model. { ¦em¦ī¦tē 'bag ˌmäd·əl }

mix crystal *See* mixed crystal. { 'miks ˌkrist·əl }

mixed crystal [CRYSTAL] A crystal whose lattice sites are occupied at random by different ions or molecules of two different compounds. Also known as mix crystal. { 'mikst 'krist·əl }

mixed reflection *See* spread reflection. { 'mikst ri'flek·shən }

mixer [OPTICS] A nonlinear device in which two light beams are combined to form new beams having frequencies equal to the sum or the difference of the input wavelengths. { 'mik·sər }

mixing length [PHYS] A mean length of travel, characteristic of a particular motion, over which an eddy maintains its identity; it is analogous to the mean free path of a molecule; physically, the idea implies that mixing occurs by discontinuous steps, that fluctuations which arise as eddies with different characteristics wander about, and that the mixing is done almost entirely by the small eddies. { 'mik·siŋ ˌleŋkth }

m-kgf *See* meter-kilogram.

mksa system *See* meter-kilogram-second-ampere system. { ¦em¦kā¦es'ā ˌsis·təm }

mks system *See* meter-kilogram-second system. { ¦em¦kā'es ˌsis·təm }

ml *See* milliliter.

mm *See* millimeter.

mmf *See* magnetomotive force.

mmHg *See* millimeter of mercury.

mmH$_2$O *See* millimeter of water.

M mode [ACOUS] A modification of the A mode of ultrasonic medical tomography used to display the movement of time-varying echo-producing structures by intensity-modulating the trace as it is swept slowly across the oscilloscope screen in a direction at right angles to the fast time-base sweep. { 'em ˌmōd }

MMT *See* multi-mirror telescope.

mmu *See* milli-mass-unit.

MNOS *See* metal-nitride-oxide semiconductor. { 'emˌnòs }

mobility [FL MECH] The reciprocal of the plastic viscosity of a Bingham plastic. [PHYS] Freedom of particles to move, either in random motion or under the influence of fields or forces. *See* drift mobility. { mō'bil·əd·ē }

mobility tensor [PL PHYS] A second-rank tensor whose product with the electric field vector for a plane wave in a plasma gives a vector equal to the average velocity of electrons or ions; components of both vectors are in phasor notation. { mō'bil·əd·ē ˌten·sər }

Möbius resistor [ELEC] A nonreactive resistor made by placing strips of aluminum or other metallic tape on opposite sides of a length of dielectric ribbon, twisting the strip assembly half a turn, joining the ends of the metallic tape, then soldering leads to opposite surfaces of the resulting loop. { 'mər·bē·əs riˌzis·tər }

MOCVD *See* metal-organic chemical vapor deposition.

mode [ELECTROMAG] A form of propagation of guided waves that is characterized by a particular field pattern in a plane transverse to the direction of propagation. Also known as transmission mode. [PHYS] A state of an oscillating system that corresponds to a particular field pattern and one of the possible resonant frequencies of the system. { mōd }

mode filter [ELECTROMAG] A waveguide filter designed to separate waves of the same frequency but of different transmission modes. { 'mōd ˌfil·tər }

mode-locked laser [OPTICS] A laser designed so that several modes of oscillation with closely spaced wavelengths, in which the laser would normally oscillate, are synchronized so that a pulse of light, lasting for as little as a picosecond, is generated. { 'mōd ˌläkt 'lā·zər }

mode of oscillation *See* mode of vibration. { 'mōd əv ˌäs·ə'lā·shən }

mode of vibration [MECH] A characteristic manner in which a system which does not dissipate energy and whose motions are restricted by boundary conditions can oscillate, having a characteristic pattern of motion and one of a discrete set of frequencies. Also known as mode of oscillation. { 'mōd əv vī'brā·shən }

modulated Raman scattering [SPECT] Application of modulation spectroscopy to the study of Raman scattering; in particular, use of external perturbations to lower the symmetry of certain crystals and permit symmetry-forbidden modes, and the use of wavelength modulation to analyze second-order Raman spectra. { 'mäj·əˌlād·əd 'rä·mən ˌskad·ə·riŋ }

modulation-doped structure [SOLID STATE] An epitaxially grown crystal structure in which successive semiconductor layers contain different types of electrical dopants. { ˌmäj·ə'lā·shən ¦dōpt 'strək·chər }

modulation spectroscopy [SPECT] A branch of spectroscopy concerned with the measurement and interpretation of changes in transmission or reflection spectra induced (usually) by externally applied perturbation, such as temperature or pressure change, or an electric or magnetic field. { ˌmäj·ə'lā·shən spek'träs·kə·pē }

modulator crystal [OPTICS] Crystal which is used to modulate a polarized light beam by the use of the Pockel's effect; useful as a modulator in laser systems. { 'mäj·əˌlād·ər ˌkrist·əl }

modulus of compression *See* bulk modulus of elasticity. { 'mäj·ə·ləs əv kəm'presh·ən }

modulus of decay [MECH] The time required for the amplitude of oscillation of an underdamped harmonic oscillator to drop to $1/e$ of its initial value; the reciprocal of the damping factor. { ¦mäj·ə·ləs əv di'kā }

modulus of deformation [MECH] The modulus of elasticity of a material that deforms other than according to Hooke's law. { 'mäj·ə·ləs əv ˌdē ˌfȯr'mā·shən }

modulus of elasticity [MECH] The ratio of the increment of some specified form of stress to the increment of some specified form of strain, such as Young's modulus, the bulk modulus, or the shear modulus. Also known as coefficient of elasticity; elasticity modulus; elastic modulus. { 'mäj·ə·ləs əv iˌlas'tis·əd·ē }

modulus of elasticity in shear [MECH] A measure of a material's resistance to shearing stress, equal to the shearing stress divided by the resultant angle of deformation expressed in radians. Also known as coefficient of rigidity; modulus of rigidity; rigidity modulus; shear modulus. { 'mäj·ə·ləs əv iˌlas'tis·əd·ē in 'shir }

modulus of resilience [MECH] The maximum mechanical energy stored per unit volume of material when it is stressed to its elastic limit. { 'mäj·ə·ləs əv ri'zil·yəns }

modulus of rigidity *See* modulus of elasticity in shear. { 'mäj·ə·ləs əv ri'jid·əd·ē }

modulus of rupture in bending [MECH] The maximum stress per unit area that a specimen can withstand without breaking when it is bent, as calculated from the breaking load under the assumption that the specimen is elastic until rupture takes place. { 'mäj·ə·ləs əv 'rəp·chər in 'bend·iŋ }

modulus of rupture in torsion [MECH] The maximum stress per unit area that a specimen can withstand without breaking when its ends are twisted, as calculated from the breaking load under the assumption that the specimen is elastic until rupture takes place. { 'mäj·ə·ləs əv 'rəp·chər in 'tȯr·shən }

modulus of simple longitudinal extension *See* axial modulus. { ¦mäj·ə·ləs əv ¦sim·pəl ˌlän·jə¦tüd·ən·əl ik'sten·chən }

modulus of torsion *See* torsional modulus. { 'mäj·ə·ləs əv 'tȯr·shən }

modulus of volume elasticity *See* bulk modulus of elasticity. { 'mäj·ə·ləs əv 'väl·yəm iˌlas'tis·əd·ē }

mohm [MECH] A unit of mechanical mobility, equal to the reciprocal of 1 mechanical ohm. { mōm }

Mohr's circle [MECH] A graphical construction making it possible to determine the stresses in a cross section if the principal stresses are known. { 'mȯrz 'sər·kəl }

moiré effect [OPTICS] The effect whereby, when one family of curves is superposed on another family of curves so that the curves cross at angles of less than about 45°, a new family of curves appears which pass through intersections of the original curves. { mȯ'rā iˌfekt }

moiré fringes [OPTICS] The bands which appear in the moiré effect. { mȯ'rā 'frin·jəz }

moisture content [MECH] The quantity of water in a mass of soil, sewage, sludge, or screenings; expressed in percentage by weight of water in the mass. { 'mȯis·chər ˌkän·tent }

moisture flux *See* eddy flux. { 'mȯis·chər ˌfləks }

moisture-vapor transmission [FL MECH] The rate at which water vapor permeates a porous film (such as plastic or paper) or a wall. { 'mȯis·chər ¦vā·pər tranzˌmish·ən }

molar dispersion [OPTICS] In refractometry, the difference in molar refraction (refractive index) of a compound at two different light-beam wavelengths. { 'mō·lər di'spər·shən }

molar magnetic rotation [OPTICS] A measure of the strength of the Faraday effect in a substance, equal to $M\alpha\rho'/(M'\alpha'\rho)$,where α is the angle of rotation, M is the molecular weight of the substance, ρ is its density, and α', M', and ρ' are corresponding quantities for water. { ¦mō·lər mag¦ned·ik rō'tā·shən }

molar refraction [OPTICS] Equation for the refractive index of a compound modified by the compound's molecular weight and density. Also known as the Lorentz-Lorenz molar refraction. { 'mō·lər ri'frak·shən }

molded capacitor [ELEC] Capacitor, usually mica, that has been encased in a molded plastic insulating material. { 'mōl·dəd kə'pas·əd·ər }

molecular beam [PHYS] A beam of neutral molecules whose directions of motion lie within a very small solid angle. { mə'lek·yə·lər 'bēm }

molecular-beam apparatus [PHYS] A device in which a molecular beam in a vacuum is subjected to magnetic fields, oscillating fields, or other influences, and a detector measures the resulting intensity of the beam at some location; used primarily in radio-frequency spectroscopy. { mə'lek·yə·lər¦bēm ˌap·ə'rad·əs }

molecular-beam epitaxy [SOLID STATE] A technique of growing single crystals in which beams of atoms or molecules are made to strike a single-crystalline substrate in a vacuum, giving rise to crystals whose crystallographic orientation is related to that of the substrate. Abbreviated MBE. { mə'lek·yə·lər¦bēm 'ep·əˌtak·sē }

molecular binding [SOLID STATE] The force which holds a molecule at some site on the surface of a crystal. { mə'lek·yə·lər 'bind·iŋ }

molecular crystal [CRYSTAL] A solid consisting of a lattice array of molecules such as hydrogen, methane, or more complex organic compounds, bound by weak van der Waals forces, and therefore retaining much of their individuality. { mə'lek·yə·lər 'krist·əl }

molecular diffusion [FL MECH] The transfer of mass between adjacent layers of fluid in laminar flow. { mə'lek·yə·lər di'fyü·zhən }

molecular effusion [FL MECH] Mass-transfer flow mechanism of free-molecule transfer through pores or orifices. { mə'lek·yə·lər i'fyü·zhən }

molecular field theory *See* Weiss theory. { mə'lek·yə·lər 'fēld ˌthē·ə·rē }

molecular flow [FL MECH] Gas-flow phenomenon at low pressures or in small channels when the mean free path is of the same order of magnitude as the channel diameter; a gas molecule thus migrates along the channel independent of other gas molecules present. { mə'lek·yə·lər 'flō }

molecular gas laser [OPTICS] Any gas laser in which the gas consists of molecules rather than atoms; such a laser can be operated on a large number of rotational-vibrational lines, and, at a sufficiently high pressure, these lines overlap and a wide gain region is obtained. Also known as molecular laser. { mə'lek·yə·lər ¦gas 'lā·zər }

molecular heat [THERMO] The heat capacity per mole of a substance. { mə'lek·yə·lər 'hēt }

molecular heat diffusion [THERMO] Transfer of heat through the motion of molecules. { mə'lek·yə·lər ¦hēt diˌfyü·shən }

molecular laser *See* molecular gas laser. { mə'lek·yə·lər 'lā·zər }

molecular optics [OPTICS] The study of the propagation of light and associated phenomena, such as refraction, absorption, and scattering, through collections of molecules in gases, liquids, and solids. { mə'lek·yə·lər 'äp·tiks }

molecular physics [PHYS] The study of the behavior and structure of molecules, including the quantum-mechanical explanation of several kinds of chemical binding between atoms in a molecule, directed valence, the polarizability of molecules, the quantization of vibrational, rotational, and electronic motions of molecules, and the phenomena arising from intermolecular forces. { mə'lek·yə·lər 'fiz·iks }

molecular rotation [OPTICS] In a solution of an optically active compound, the specific rotation (angular rotation of polarized light) multiplied by the compound's molecular weight. { mə'lek·yə·lər rō'tā·shən }

molecular spectroscopy [SPECT] The production, measurement, and interpretation of molecular spectra. { mə'lek·yə·lər spek'träs·kə·pē }

molecular spectrum [SPECT] The intensity of electromagnetic radiation emitted or absorbed by a collection of molecules as a function of frequency, wave number, or some related quantity. { mə'lek·yə·lər 'spek·trəm }

molecular theory *See* kinetic theory. { mə'lek·yə·lər 'thē·ə·rē }

Molenbroeck-Chaplygin transformation [FL MECH] A version of the hodograph method for compressible flow in which only the independent variables are replaced and no change is made in the dependent variables, that is, the velocity potential and stream function. { ¦mō·lən·brük chap'lē·gən ˌtranz·fərˌmā·shən }

Moller scattering [QUANT MECH] Scattering of electrons by electrons. { 'mȯl·ər ˌskad·ə·riŋ }

Mollier diagram [THERMO] Graph of enthalpy versus entropy of a vapor on which isobars, isothermals, and lines of equal dryness are plotted. { mȯl'yā ˌdī·əˌgram }

moment [MECH] Static moment of some quantity, except in the term "moment of inertia." { 'mō·mənt }

momental ellipsoid [MECH] An inertia ellipsoid whose size is specified to be such that the tip of the angular velocity vector of a freely rotating object, with origin at the center of the ellipsoid, always lies on the ellipsoid's surface. Also known as energy ellipsoid. { mō'ment·əl ə'lip,sȯid }

moment diagram [MECH] A graph of the bending moment at a section of a beam versus the distance of the section along the beam. { 'mō·mənt ,dī·ə,gram }

moment of force *See* torque. { 'mō·mənt əv 'fȯrs }

moment of inertia [MECH] The sum of the products formed by multiplying the mass (or sometimes, the area) of each element of a figure by the square of its distance from a specified line. Also known as rotational inertia. { 'mō·mənt əv i'nər·shə }

moment of momentum *See* angular momentum. { 'mō·mənt əv mō'ment·əm }

momentum [MECH] Also known as linear momentum; vector momentum. **1.** For a single nonrelativistic particle, the product of the mass and the velocity of a particle. **2.** For a single relativistic particle, $m\mathbf{v}/(1 - v^2/c^2)^{1/2}$, where m is the rest-mass, $\mathbf{v}$ the velocity, and c the speed of light. **3.** For a system of particles, the vector sum of the momenta (as in the first or second definition) of the particles. { mō'ment·əm }

momentum conservation *See* conservation of momentum. { mōm'ment·əm ,kän·sər'vā·shən }

momentum density [PHYS] The momentum per unit volume of any given field. { mō'ment·əm 'den·səd·ē }

momentum-transport hypothesis [FL MECH] The hypothesis that the principle of conservation of momentum is valid in turbulent eddy transfer. { mō'ment·əm ¦tranz,pȯrt hī,päth·ə·səs }

momentum wave function [QUANT MECH] A function of the momenta of a system of particles and of time which results from taking Fourier transforms, over the coordinates of all the particles, of the Schrödinger wave function; the absolute value squared is proportional to the probability that the particles will have given momenta at a given time. { mō'ment·əm 'wāv ,fəŋk·shən }

monochromatic [OPTICS] Pertaining to the color of a surface which radiates light having an extremely small range of wavelengths. [PHYS] Consisting of electromagnetic radiation having an extremely small range of wavelengths, or particles having an extremely small range of energies. { ¦män·ə·krə'mad·ik }

monochromatic emissivity [THERMO] The ratio of the energy radiated by a body in a very narrow band of wavelengths to the energy radiated by a blackbody in the same band at the same temperature. Also known as color emissivity. { ,män·ə·krə'mad·ik ,ē·mi'siv·əd·ē }

monochromatic filter *See* birefringent filter. { män·ə·krə'mad·ik 'fil·tər }

monochromatic interference [OPTICS] Interference between beams coming from a source of monochromatic light. { män·ə·krə'mad·ik ,in·tər'fir·əns }

monochromatic light [OPTICS] Light of one color, having wavelengths confined to an extremely narrow range. { män·ə·krə'mad·ik 'līt }

monochromatic radiation [ELECTROMAG] Electromagnetic radiation having wavelengths confined to an extremely narrow range. { män·ə·krə'mad·ik ,rād·ē'ā·shən }

monochromatic temperature scale [THERMO] A temperature scale based upon the amount of power radiated from a blackbody at a single wavelength. { män·ə·krə'mad·ik 'tem·prə·chər ,skāl }

monochromator [SPECT] A spectrograph in which a detector is replaced by a second slit, placed in the focal plane, to isolate a particular narrow band of wavelengths for refocusing on a detector or experimental object. { ¦män·ə¦krō ,mād·ər }

monochrome [OPTICS] Having only one chromaticity. { 'män·ə,krōm }

monoclinic system [CRYSTAL] One of the six crystal systems characterized by a single, twofold symmetry axis or a single symmetry plane. { ¦män·ə'klin·ik ,sis·təm }

monoelectron oscillator *See* geonium. { ¦män·ō·i'lek,trän 'äs·ə,lād·ər }

monoenergetic gamma rays [PHYS] A beam of gamma rays whose energies are confined to an extremely narrow range. { ¦män·ō,en·ər'jed·ik 'gam·ə ,rāz }

monoenergetic radiation [PHYS] Radiation consisting of photons or particles whose energies are confined to an extremely narrow range. { ,mä·nō,en·ər,jed·ik ,rād·ē'ā·shən }

monopole *See* magnetic monopole. { 'män·ə,pōl }

monopole antenna [ELECTROMAG] An antenna, usually in the form of a vertical tube or helical whip, on which the current distribution forms a standing wave, and which acts as one part of a dipole whose other part is formed by its electrical image in the ground or in an effective ground plane. Also known as spike antenna. { 'män·ə,pōl an'ten·ə }

monotrophic [CRYSTAL] Of crystal pairs, having one of the pair always metastable with respect to the other. { ¦män·ə¦träf·ik }

monotropic [PHYS] Pertaining to an element which may exist in two or more forms, but in which one form is the stable modification at all temperatures and pressures. { ¦män·ə¦träp·ik }

monotropy coefficient [FL MECH] A coefficient v related to the ratio of velocity coefficients, A_y/A_x, in an equation developed by P. Raethjen for the velocity profile in a fluid. { mə'nä·trə·pē ,kō·i,fish·ənt }

Montonen-Olive conjecture [PART PHYS] The conjecture that unified theories of elementary particle interactions with supersymmetry have the property that, when the coupling constant between electric charges and electric fields becomes large, the electric charges become fuzzy, heavy, and strongly coupled while the magnetic charges become point-like, light, and weakly coupled, which is the reverse of the situation when the coupling constant is small. { ˌmän·təˌnən 'äl·əv kənˌjek·chər }

Moody friction factor [FL MECH] Modification of the friction factor-Reynolds number-fluid flow relationship into which a roughness factor has been incorporated. { 'müd·ē 'frik·shən ˌfak·tər }

moonbow [OPTICS] A rainbow formed by light from the moon; the colors in a moonbow are usually very difficult to detect. Also known as lunar rainbow. { 'münˌbō }

moon illusion [OPTICS] An optical illusion whereby the moon appears larger when it is close to the horizon than when it is higher up. { 'mün iˌlü·zhən }

Morera's stress functions [MECH] Three functions of position, ψ_1, ψ_2, and ψ_3, in terms of which the elements of the stress tensor σ of a body may be expressed, if the body is in equilibrium and is not subjected to body forces; the elements of the stress tensor are given by $\sigma_{11} = -2\partial^2\psi_1/\partial x_2 \partial x_3$, $\sigma_{23} = \partial^2\psi_2/\partial x_1 \partial x_2 + \partial^2\psi_3/\partial x_1 \partial x_3$, and cyclic permutations of these equations. { mȯ'rer·əz 'stres ˌfəŋk·shənz }

Morgan equation [THERMO] A modification of the Ramsey-Shields equation, in which the expression for the molar surface energy is set equal to a quadratic function of the temperature rather than to a linear one. { 'mȯr·gən iˌkwā·zhən }

morphotropism [CRYSTAL] Similarity of structure, axial ratios, and angles between faces of one or more zones in crystalline substances whose formulas can be derived one from another by substitution. { ¦mȯr·fō'trōˌpiz·əm }

MOS *See* metal oxide semiconductor.

mosaic structure [CRYSTAL] In crystals, a substructure in which neighboring regions are oriented slightly differently. { mō'zā·ik ¦strək·chər }

Moseley's law [SPECT] The law that the square-root of the frequency of an x-ray spectral line belonging to a particular series is proportional to the difference between the atomic number and a constant which depends only on the series. { 'mōz·lēz ˌlȯ }

Mosotti field *See* Lorentz local field. { mȯ'säd·ē ˌfēld }

Mössbauer effect [NUC PHYS] The emission and absorption of gamma rays by certain nuclei, bound in crystals, without loss of energy through nuclear recoil, with the result that radiation emitted by one such nucleus can be absorbed by another. { 'mu̇sˌbau̇·ər iˌfekt }

Mössbauer spectroscopy [SPECT] The study of Mössbauer spectra, for example, for nuclear hyperfine structure, chemical shifts, and chemical analysis. { 'mu̇sˌbau̇·ər spek'träs·kə·pē }

Mössbauer spectrum [SPECT] A plot of the absorption, by nuclei bound in a crystal lattice, of gamma rays emitted by similar nuclei in a second crystal, as a function of the relative velocity of the two crystals. { 'mu̇sˌbau̇·ər ˌspek·trəm }

motion [MECH] A continuous change of position of a body. { 'mō·shən }

motional electromotive force [ELECTROMAG] An electromotive force in a circuit that results from the motion of all or part of the circuit through a magnetic field. { 'mō·shən·əl i¦lek·trə¦mōd·iv 'fȯrs }

motional induction [ELECTROMAG] The production of an electromotive force in a circuit by motion of all or part of the circuit through a magnetic field in such a way that the circuit cuts across the magnetic flux. { 'mō·shən·əl in'dək·shən }

motion picture camera [OPTICS] A camera capable of capturing action by taking a series of still pictures at regular brief intervals on a lengthy strip of film. { 'mō·shən ¦pik·chər ˌkam·rə }

motor effect [ELECTROMAG] The mutually repulsive force exerted by neighboring conductors that carry current in opposite directions. { 'mōd·ər iˌfekt }

motor noise [ACOUS] The noisy sound made by an electric motor. { 'mōd·ər ˌnȯiz }

Mott scattering [QUANT MECH] **1.** The scattering of identical particles due to a Coulomb force. **2.** The scattering of a relativistic electron by a Coulomb field. { 'mät ˌskad·ə·riŋ }

mounce [MECH] A unit of mass, equal to 25 grams. Also known as metric ounce. { mau̇ns }

mount [ELECTROMAG] The flange or other means by which a switching tube, or tube and cavity, is connected to a waveguide. { mau̇nt }

moving-coil instrument [ELEC] Any instrument in which current is sent through one or more coils suspended or pivoted in a magnetic field, and the motion of the coils is used to measure either the current in the coils or the strength of the field. { 'müv·iŋ ¦kȯil 'in·strə·mənt }

moving constraint [MECH] A constraint that changes with time, as in the case of a system on a moving platform. { 'müv·iŋ kən'strānt }

moving load [MECH] A load that can move, such as vehicles or pedestrians. { 'müv·iŋ 'lōd }

mp *See* melting point.

MRTD *See* minimum resolvable temperature difference.

ms *See* millisecond.

Ms *See* megasecond.

msec *See* millisecond.

Msec *See* megasecond.

M shell [ATOM PHYS] The third layer of electrons about the nucleus of an atom, having electrons characterized by the principal quantum number 3. { 'em ˌshel }

MT *See* megaton.

M-theory [PART PHYS] A highly symmetric but only partially understood theory of particles and their interactions that would be a generalization of supergravity and would be related by weak-strong duality to each of the five known superstring theories. { 'em ˌthē·ə·rē }

Mueller matrices [OPTICS] Matrix operators in a calculus used to treat polarized light; in this calculus, the light vector is split into four components one of which is the intensity of the light, and unpolarized light can be treated directly. { 'myül·ər ¦mā·trə,sēz }

muffin-tin potential [SOLID STATE] A potential function used in the augmented plane-wave method and related methods of approximating the energy states of electrons in a crystal lattice, which is spherically symmetric within spheres centered at each atomic nucleus and constant in the region between these spheres. { 'məf·ən ,tin pə,ten·chəl }

mull technique [SPECT] Method for obtaining infrared spectra of materials in the solid state; material to be scanned is first pulverized, then mulled with mineral oil. { 'məl tek,nēk }

multicellular horn [ELECTROMAG] A cluster of horn antennas having mouths that lie in a common surface and that are fed from openings spaced one wavelength apart in one face of a common waveguide. Also known as cellular horn. { ¦məl·tē'sel·yə·lər 'hórn }

multiconfiguration Hartree-Fock approximation [ATOM PHYS] A natural extension of the Hartree-Fock approximation for an atom or molecule in which a number of configurations are chosen and the mixing coefficients, as well as the radial parts of the orbitals, are varied to minimize the expectation value of the energy. Abbreviated MCHF approximation. { ,məl·tē·kən,fig·yə¦rā·shən ¦här,trē 'fäk ə,prok·sə,mā·shən }

multidither COAT *See* multidither coherent adaptive optical techniques. { 'məl·tē,dith·ər ¦sē¦ō¦ā'tē *or* 'kō,at }

multidither coherent adaptive optical techniques [OPTICS] Adaptive optical techniques for concentrating laser radiation into as small an area as possible, involving an array of laser sources, one of which has fixed phase, while the phase of a second element is controlled by a low-amplitude phase modulation or dither, and adjusted to maximize the reflection from a target glint. Abbreviated multidither COAT. { 'məl·tē,dith·ər kō'hir·ənt ə'dap·tiv 'äp·ti·kəl tek,nēks }

multielement array [ELECTROMAG] An antenna array having a large number of antennas. { ¦məl·tē'el·ə·mənt ə'rā }

multielement parasitic array [ELECTROMAG] Antennas consisting of an array of driven dipoles and parasitic elements, arranged to produce a beam of high directivity. { ¦məl·tē'el·ə·mənt ,par·ə'sid·ik ə,rā }

multifocal lens [OPTICS] A lens that has more than one focal length. { ¦məl·tē,fō·kəl 'lenz }

multi-mirror telescope [OPTICS] A telescope in which light from several mirrors of similar shape is brought to a common focus by additional optical elements. Abbreviated MMT. { ,məl·tē ,mir·ər 'tel·ə,scōp }

multipath *See* multipath transmission. { 'məl·tə,path }

multipath transmission [ELECTROMAG] The propagation phenomenon that results in signals reaching a radio receiving antenna by two or more paths, causing distortion in radio and ghost images in television. Also known as multipath. { 'məl·tə,path tranz'mish·ən }

multiphonon emission [SOLID STATE] A process of nonradiative recombination of electrons and holes in which an electron is captured into a deep level near the middle of an energy gap associated with a lattice defect, exciting lattice vibrations, and the trapped electron state captures a hole from the valence band. { ¦məl·tə'fō,nän i'mish·ən }

multiphoton absorption [ATOM PHYS] The excitation of an atom or other microscopic system to a higher quantum state by simultaneous absorption of two or more photons which together provide the necessary energy. { ¦məl·tə¦fō,tän ab'sórp·shən }

multiphoton ionization [ATOM PHYS] The removal of one or more electrons from an atom or other microscopic system as the result of simultaneous absorption of two or more photons. { ¦məl·tə¦fō,tän ,ī·ə·nə'zā·shən }

multiple [ELEC] **1.** Group of terminals arranged to make a circuit or group of circuits accessible at a number of points at any one of which connection can be made. **2.** To connect in parallel. { 'məl·tə·pəl }

multiple-beam antenna [ELECTROMAG] An antenna or antenna array which radiates several beams in different directions. { 'məl·tə·pəl ¦bēm an'ten·ə }

multiple-beam interference [OPTICS] Interference which arises when part of a beam is reflected several times back and forth between a pair of strongly reflecting surfaces before being reflected or transmitted from the pair. { 'məl·tə·pəl ¦bēm ,in·tər'fir·əns }

multiple-beam interferometer [OPTICS] An interferometer in which a beam is reflected several times back and forth between a pair of parallel plane surfaces; examples are the Fizeau interferometer and the Fabry-Perot interferometer. { 'məl·tə·pəl ¦bēm ,in'tər·fə'räm·əd·ər }

multiple isomorphous replacement technique [CRYSTAL] A technique for overcoming the phase problem by growing crystals in three different isomorphic chemical forms and comparing x-ray diffraction data obtained from all three. Abbreviated MIR technique. { ,məl·tə·pəl ,ī·sə,mór·fəs ri'plās·mənt tek,nēk }

multiple-mirror telescope [OPTICS] A type of optical telescope in which images from several complete conventional telescopes that are mounted rigidly on a common frame and coaligned by an active laser and computer system are brought to a common focus by a mirror system. { 'məl·tə·pəl ¦mir·ər 'tel·ə,skōp }

multiple reflection [OPTICS] Reflection of light back and forth several times between a pair of strongly reflecting surfaces. { 'məl·tə·pəl ri 'flek·shən }

multiple reflection echoes [ELECTROMAG] Radar echoes returned from a real target by reflection from some object in the radar beam; such echoes appear at a false bearing and false range. { 'məl·tə·pəl ri'flek·shən 'ek,ōz }

multiple resonance [ELEC] Two or more resonances at different frequencies in a circuit consisting of two or more coupled circuits which are resonant at slightly different frequencies. [QUANT MECH] Two or more resonances at slightly different energies, resulting from the splitting of a single resonance by an interaction that is relatively weak. { 'məl·tə·pəl 'rez·ən·əns }

multiple scattering [PHYS] Process in which a particle undergoes a large number of collisions, and the total change in its momentum is the sum of the many small changes occurring during individual collisions. { 'məl·tə·pəl 'skad·ə·riŋ }

multiplet [QUANT MECH] A collection of relatively closely spaced energy levels which result from the splitting of a single energy level by an interaction which is relatively weak; examples are spin-orbit multiplets and isospin multiplets. [SPECT] A collection of relatively closely spaced spectral lines resulting from transitions to or from the members of a multiplet (as in the quantum-mechanics definition). { 'məl·tə·plət }

multiplet intensity rules [SPECT] Rules for the relative intensities of spectral lines in a spin-orbit multiplet, stating that the sum of the intensities of all lines which start from a common initial level, or end on a common final level, is proportional to $2J+1$, where J is the total angular momentum of the initial level or final level respectively. { 'məl·tə·plət in'ten·səd·ē ,rülz }

multiple-tuned antenna [ELECTROMAG] Low-frequency antenna having a horizontal section with a multiplicity of tuned vertical sections. { 'məl·tə·pəl ¦tünd an'ten·ə }

multiple-unit steerable antenna *See* musa. { 'məl·tə·pəl ¦yü·nət 'stir·ə·bəl an'ten·ə }

multiplex holography [OPTICS] A technique in which a rotating cylindrical hologram is illuminated by a tungsten-filament light bulb on the axis of the cylinder; as the cylinder rotates, or as the viewer moves around, three-dimensional images are seen inside the cylinder. { 'məl·tə,pleks hō'läg·rə·fē }

multiplicative acoustic array [ACOUS] An acoustic array of receiving elements which is divided into two parts, the signal voltages obtained from them being multiplied together. Also known as correlation array. { ,məl·tə¦plik·əd·iv ə¦küs·tik ə'rā }

multiplicity [PHYS] In a system having Russell-Saunders coupling, the quantity $2S+1$, where S is the total spin quantum number. { ,məl·tə'plis·əd·ē }

multipolar [ELECTROMAG] Having more than one pair of magnetic poles. { ¦məl·tə'pō·lər }

multipolar machine [ELECTROMAG] An electric machine that has a field magnet with more than one pair of poles. { ¦məl·tə'pō·lər mə'shēn }

multipole [ELECTROMAG] One of a series of types of static or oscillating distributions of charge or magnetization; namely, an electric multipole or a magnetic multipole. { 'məl·tə,pōl }

multipole fields [ELECTROMAG] The electric and magnetic fields generated by static or oscillating electric or magnetic multipoles. { 'məl·tə,pōl ,fēlz }

multipole radiation [PHYS] **1.** Electromagnetic radiation which has characteristics equivalent to those of radiation generated by an oscillating electric or magnetic multipole, and is made up of photons of well-defined angular momentum and parity. **2.** Internal conversion electrons, or positron-electron pairs having similar characteristics, emitted from an atom when the nucleus makes a transition between two energy states. { 'məl·tə,pōl ,rād·ē'ā·shən }

multipole transition [PHYS] A transition between two energy states of an atom or nucleus in which a quantum of multipole radiation is emitted or absorbed. { 'məl·tə,pōl tran'zish·ən }

multiturn potentiometer [ELEC] A precision wire-wound potentiometer in which the resistance element is formed into a helix, generally having from 2 to 10 turns. { 'məl·tē,tərn pə,ten·chē'äm·əd·ər }

multiwavelength anomalous dispersion [CRYSTAL] A technique for overcoming the phase problem by collecting x-ray diffraction data at several wavelengths around the absorption edge of a strongly absorbing atom. Abbreviated MAD. { ,məl·tē¦wāv,leŋkth ə,näm·ə·ləs di 'spərzh·ən }

mu meson *See* muon. { 'myü ¦mā,sän }

Munsell chroma *See* chroma. { mən'sel ,krō·mə }

Munsell color system [OPTICS] A system for designating colors which employs three perceptually uniform scales (Munsell hue, Munsell value, Munsell chroma) defined in terms of daylight reflectance. { mən'sel 'kəl·ər ,sis·təm }

Munsell hue [OPTICS] The dimension of the Munsell system of color that determines whether a color is blue, green, yellow, red, purple, or the like, without regard to its lightness or saturation. { mən'sel ,hyü }

Munsell value [OPTICS] The dimension, in the Munsell system of object-color specificiation, that indicates the apparent luminous transmittance or reflectance of the object on a scale having approximately equal perceptual steps under the usual conditions of observation. { mən 'sel ,val,yü }

muon [PART PHYS] Collective name for two semistable elementary particles with positive and negative charge, designated μ^+ and μ^- respectively, which are leptons and have a spin of 1/2 and a mass of approximately 105.7 MeV. Also known as mu meson. { 'myü,än }

muon-catalyzed fusion [NUC PHYS] Nuclear fusion that occurs quickly at normal temperatures when the reacting nuclei are bound in an exotic molecule, containing in addition a positively charged muon. { ¦myü,än ,kad·ə¦līzd 'fyü·zhən }

muonic atom [PART PHYS] An atom in which an electron is replaced by a negatively charged muon orbiting close to or within the nucleus. { myü'än·ik ¦ad·əm }

muonium [PART PHYS] An atom consisting of an electron bound to a positively charged muon by their mutual Coulomb attraction, just as an electron is bound to a proton in the hydrogen atom. { myü'ō·nē·əm }

muon lepton number [PART PHYS] The number of muons and muon-associated neutrinos minus the number of antimuons and muon-associated antineutrinos; it is conserved in all known interactions but may not be exactly conserved. { 'myü,än 'lep,tän ,nəm·bər }

muon spin relaxation [PHYS] A technique for studying various phenomena in solids and liquids and chemical reactions of muonium atoms, in which a beam of polarized muons is focused on a sample and the loss of polarization of muons in the sample is monitored by observing the spatial anisotropy of electrons or positrons emitted in the muon decay. Also known as muon spin rotation; muon spin resonance. { 'myü,än ¦spin ri,lak'sā·shən }

muon spin resonance *See* muon spin relaxation. { 'myü,än ¦spin 'rez·ən·əns }

muon spin rotation *See* muon spin relaxation. { 'myü,än ¦spin rō'tā·shən }

musa [ELECTROMAG] An electrically steerable receiving antenna whose directional pattern can be rotated by varying the phases of the contributions of the individual units. Derived from multiple-unit steerable antenna. { 'myü·sə }

musical acoustics [ACOUS] That part of acoustics which is relevant to the composition, performance, and appreciation of music, including the physical characteristics of sounds that may be heard as music, laws governing the action, design, and construction of musical instruments, and the effects of musical sounds upon listeners. { 'myü·zə·kəl ə'kü·stiks }

musical echo [ACOUS] A musical tone produced by the reflection of an impulsive sound from a stepped structure such as a picket fence, when reflections from successive steps reach the observer with suitable frequency. { 'myü·zə·kəl 'ek·ō }

musical quality *See* timbre. { 'myü·zə·kəl 'kwäl·əd·ē }

Muskhelishvili's method [MECH] A method of solving problems concerning the elastic deformation of a planar body that involves using methods from the theory of functions of a complex variable to calculate analytic functions which determine the plane strain of the body. { mə'skel·ish,vil·ēz ,meth·əd }

mutual admittance [ELEC] For two meshes of a network carrying alternating current, the ratio of the complex current in one mesh to the complex voltage in the other, when the voltage in all meshes besides these two is 0. { 'myü·chə·wəl ad'mit·əns }

mutual branch *See* common branch. { 'myü·chə·wəl 'branch }

mutual capacitance [ELEC] The accumulation of charge on the surfaces of conductors of each of two circuits per unit of potential difference between the circuits. { 'myü·chə·wəl kə'pas·əd·əns }

mutual impedance [ELEC] For two meshes of a network carrying alternating current, the ratio of the complex voltage in one mesh to the complex current in the other, when all meshes besides the latter one carry no current. { 'myü·chə·wəl im'pēd·əns }

mutual inductance [ELECTROMAG] Property of two neighboring circuits, equal to the ratio of the electromotive force induced in one circuit to the rate of change of current in the other circuit. { 'myü·chə·wəl in'dək·təns }

mutual induction [ELECTROMAG] The generation of a voltage in one circuit by a varying current in another. { 'myü·chə·wəl in'dək·shən }

mV *See* millivolt.

MV *See* megavolt.

mW *See* milliwatt.

MW *See* megawatt.

Mx *See* maxwell.

myriametric waves [ELECTROMAG] Electromagnetic waves having wavelengths between 10 and 100 kilometers, corresponding to the very low frequency band. { ¦mir·ē·ə¦me·trik 'wāvz }

myriotic field [QUANT MECH] A quantized field that has creation and annihilation operators satisfying specified commutation rules, but no vacuum state. { ,mir·ē'äd·ik 'fēld }

N

N *See* newton.

N.A. *See* numerical aperture.

nacreous [OPTICS] Having an iridescent luster resembling that of mother-of-pearl. Also known as pearly. { 'nā·krē·əs }

nadir point *See* photograph nadir. { 'nā·dər ˌpȯint }

naked singularity [RELAT] A singularity that is not surrounded by an event horizon, and thus gives rise to timelike curves that violate causality. { 'nā·kəd ˌsiŋ·gyə'lar·əd·ē }

nanogram [MECH] One-billionth (10^{-9}) of a gram. Abbreviated ng. { 'nan·əˌgram }

nanometer [MECH] A unit of length equal to one-billionth of a meter, or 10^{-9} meter. Also known as millimicron (μm); nanon. { 'nan·əˌmēd·ər }

nanon *See* nanometer. { 'naˌnän }

nanosecond [MECH] A unit of time equal to one-billionth of a second, or 10^{-9} second. { 'nan·əˌsek·ənd }

nanostructure [SOLID STATE] Something that has a physical dimension smaller than 100 nanometers, ranging from clusters of atoms to dimensional layers. { 'nan·ōˌstrək·chər }

napier *See* neper. { 'nā·pē·ər }

narrow beam [PHYS] In measurements of the attenuation of a beam of ionizing radiation, a beam in which the scattered radiation does not reach the detector. { ¦nar·ō 'bēm }

narrow-beam antenna [ELECTROMAG] An antenna which radiates most of its power in a cone having a radius of only a few degrees. { 'nar·ō ¦bēm an'ten·ə }

narrow cut filter [OPTICS] An optical filter which displays an abrupt change from high transmission to complete absorption over a narrow wavelength region. { ¦nar·ō ˌkət 'film }

natural antenna frequency [ELECTROMAG] Lowest resonant frequency of an antenna without added inductance or capacitance. { 'nach·rəl an'ten·ə ˌfrē·kwən·sē }

natural convection [THERMO] Convection in which fluid motion results entirely from the presence of a hot body in the fluid, causing temperature and hence density gradients to develop, so that the fluid moves under the influence of gravity. Also known as free convection. { 'nach·rəl kən'vek·shən }

natural coordinates [FL MECH] An orthogonal, or mutually perpendicular, system of curvilinear coordinates for the description of fluid motion, consisting of an axis *t* tangent to the instantaneous velocity vector and an axis *n* normal to this velocity vector to the left in the horizontal plane, to which a vertically directed axis *z* may be added for the description of three-dimensional flow; such a coordinate system often permits a concise formulation of atmospheric dynamical problems, especially in the Lagrangian system of hydrodynamics. { 'nach·rəl kō'ȯrd·ən·əts }

natural draft [FL MECH] Unforced gas flow through a chimney or vertical duct, directly related to chimney height and the temperature difference between the ascending gases and the atmosphere, and not dependent upon the use of fans or other mechanical devices. { 'nach·rəl 'draft }

natural frequency [PHYS] The frequency with which a system oscillates in the absence of external forces; or, for a system with more than one degree of freedom, the frequency of one of the normal modes of vibration. { 'nach·rəl 'frē·kwən·sē }

natural linewidth [SPECT] The part of the linewidth of an absorption or emission line that results from the finite lifetimes of one or both of the energy levels between which the transition takes place. { 'nach·rəl 'līnˌwidth }

natural period [PHYS] Period of the free oscillation of a body or system; when the period varies with amplitude, the natural period is the period when the amplitude approaches zero. { 'nach·rəl 'pir·ē·əd }

natural resonance [PHYS] Resonance in which the period or frequency of the applied agency maintaining oscillation is the same as the natural period of oscillation of a system. { 'nach·rəl 'rez·ən·əns }

natural wavelength [ELECTROMAG] Wavelength corresponding to the natural frequency of an antenna or circuit. { 'nach·rəl 'wāvˌleŋkth }

natural width of energy level [PHYS] A measure of the spread in energy of an excited state of a quantized system due to spontaneous transitions to other states; quantitatively, it is the difference between the energies for which the intensity of emission from or absorption by the state, or of the scattering cross section associated with it, is one-half its maximum value, in the absence of any external influence on the system. { 'nach·rəl ¦width əv 'en·ər·jē ,lev·əl }

nautical chain [MECH] A unit of length equal to 15 feet or 4.572 meters. { 'nȯd·ə·kəl 'chān }

Navier's equation [MECH] A vector partial differential equation for the displacement vector of an elastic solid in equilibrium and subjected to a body force. { nä'vyāz i,kwā·zhən }

Navier-Stokes equations [FL MECH] The equations of motion for a viscous fluid which may be written $dV/dt = -(1/\rho)\nabla p + F + \nu\nabla^2 V + (1/3)\nu\nabla(\nabla\cdot V)$, where p is the pressure, ρ the density, F the total external force per unit mass, V the fluid velocity, and ν the kinematic viscosity; for an incompressible fluid, the term in $\nabla\cdot V$ (divergence) vanishes, and the effects of viscosity then play a role analogous to that of temperature in thermal conduction and to that of density in simple diffusion. { nä'vyā 'stōks i,kwā·zhənz }

n-body problem *See* many-body problem. { 'en ¦bad·ē ,präb·ləm }

N center [SOLID STATE] A color center which arises from continued exposure to light in the F band or to x-rays and which produces a faint absorption band on the long-wavelength side of the M band. Also known as G center. { 'en ,sen·tər }

n-component [PART PHYS] Cosmic-ray particles that can take part in nuclear interactions, that is, nucleons, pions, and other baryons and mesons. { 'en kəm,pō·nənt }

nearest neighbors [CRYSTAL] Any pair of atoms in a crystal lattice which are as close to each other, or closer to each other, than any other pair. { 'nir·əst 'nā·bərz }

near field [ACOUS] The acoustic radiation field that is close to an acoustic source such as a loudspeaker. [ELECTROMAG] The electromagnetic field that exists within one wavelength of a source of electromagnetic radiation, such as a transmitting antenna. { 'nir ,fēld }

near-field noise *See* flow noise. { ¦nir ¦fēld ,nȯiz }

near-field scanning optical microscope [OPTICS] An optical microscope in which the intensity of light focused through a pipette with an aperture at its tip is recorded as the tip is moved across the specimen in a raster pattern at a distance of much less than a wavelength. { 'nir ¦fēld 'skan·iŋ 'äp·tə·kəl 'mī·krə,skōp }

near-field scanning optical microscopy [OPTICS] A technique for making optical measurements at dimensions much smaller than the wavelength of light, by scanning a nanometric detector or radiation source in proximity to a sample surface. Also known as scanning near-field optical microscopy. { ¦nir ¦fēld ,skan·iŋ ,äp·tə·kəl mī'kräs·kə·pē }

near-infrared radiation [ELECTROMAG] Infrared radiation having a relatively short wavelength, between 0.75 and about 2.5 micrometers (some scientists place the upper limit from 1.5 to 3 micrometers), at which radiation can be detected by photoelectric cells, and which corresponds in frequency range to the lower electronic energy levels of molecules and semiconductors. Also known as photoelectric infrared radiation. { 'nir ,in·frə'red ,rād·ē'ā·shən }

nearly free electron method [SOLID STATE] A method of approximating the energy levels of electrons in a crystal lattice by considering the potential energy resulting from atomic nuclei and from other electrons in the lattice as a perturbation on free electron states. Abbreviated NFE method. { ¦nir·lē ¦frē i¦lek,trän ,meth·əd }

near-ultraviolet radiation [ELECTROMAG] Ultraviolet radiation having relatively long wavelength, in the approximate range from 300 to 400 nanometers. { 'nir ¦əl·trə¦vī·lət ,rād·ē'ā·shən }

nebulium line [SPECT] An optical emission line in the spectrum of oxygen at a wavelength of 500.7 nanometers, prominent in the spectra of H II regions. { nə'bül·ē·əm ,līn }

Néel ferromagnetism *See* ferrimagnetism. { 'nā·el ,fer·ō'mag·nə,tiz·əm }

Néel point *See* Néel temperature. { 'nā·el ,pȯint }

Néel's theory [SOLID STATE] A theory of the behavior of antiferromagnetic and other ferrimagnetic materials in which the crystal lattice is divided into two or more sublattices; each atom in one sublattice responds to the magnetic field generated by nearest neighbors in other sublattices, with the result that magnetic moments of all the atoms in any sublattice are parallel, but magnetic moments of two different sublattices can be different. { 'nā·elz ,thē·ə·rē }

Néel temperature [SOLID STATE] A temperature, characteristic of certain metals, alloys, and salts, below which spontaneous nonparalleled magnetic ordering takes place so that they become antiferromagnetic, and above which they are paramagnetic. Also known as Néel point. { 'nā·el 'tem·prə·chər }

Néel wall [SOLID STATE] The boundary between two magnetic domains in a thin film in which the magnetization vector remains parallel to the faces of the film in passing through the wall. { 'nā·el ,wȯl }

negative [ELEC] Having a negative charge. { 'neg·əd·iv }

negative acceleration [MECH] Acceleration in a direction opposite to the velocity, or in the direction of the negative axis of a coordinate system. { 'neg·əd·iv ik,sel·ə'rā·shən }

negative charge [ELEC] The type of charge which is possessed by electrons in ordinary matter, and which may be produced in a resin object by rubbing with wool. Also known as negative electricity. { 'neg·əd·iv 'chärj }

negative crystal [CRYSTAL] A crystal containing a cavity, where the form of the cavity is one of the characteristic crystal forms of the mineral in question. [OPTICS] A uniaxial crystal in which the extraordinary wave travels faster than the ordinary wave, such as calcite. { 'neg·əd·iv 'krist·əl }

negative electricity *See* negative charge. { 'neg·əd·iv ,i,lek'tris·əd·ē }

negative electrode *See* negative plate. { 'neg·əd·iv i'lek,trōd }

negative electron *See* electron. { 'neg·əd·iv i'lek,trän }

negative elongation [CRYSTAL] In a section of an anisotropic crystal, a sign of elongation that is parallel to the faster of the two plane-polarized rays. { 'neg·əd·iv ,ē,lȯŋ'gā·shən }

negative g [MECH] In designating the direction of acceleration on a body, the opposite of positive *g*; for example, the effect of flying an outside loop in the upright seated position. { 'neg·əd·iv 'jē }

negative ion [PHYS] An electron or negatively charged subatomic particle. { 'neg·əd·iv 'ī,än }

negative-ion vacancy [CRYSTAL] A point defect in an ionic crystal in which a negative ion is missing from its lattice site. { 'neg·əd·iv ¦ī,än 'vā·kən·sē }

negative lens *See* diverging lens. { 'neg·əd·iv 'lenz }

negative meniscus lens [OPTICS] A lens having one convex and one concave surface, with the radius of curvature of the convex surface greater than that of the concave surface. Also known as diverging meniscus lens. { 'neg·əd·iv mə¦nis·kəs 'lenz }

negative mirror *See* diverging mirror. { 'neg·əd·iv 'mir·ər }

negative nodal points [OPTICS] Two points on the axis of an optical system such that an incident ray passing through one results in an emergent ray passing through the other which makes an angle with the axis having the same magnitude but opposite sign. Also known as antinodal points. { 'neg·əd·iv 'nōd·əl ,pȯins }

negative pion [PART PHYS] A pion having a negative electric charge. { 'neg·əd·iv 'pī,än }

negative pole *See* south pole. { 'neg·əd·iv 'pōl }

negative potential [ELEC] An electrostatic potential which is lower than that of the ground, or of some conductor or point in space that is arbitrarily assigned to have zero potential. { 'neg·əd·iv pə'ten·chəl }

negative pressure [PHYS] A way of expressing vacuum; a pressure less than atmospheric or the standard 760 mmHg (101,325 pascals). { 'neg·əd·iv 'presh·ər }

negative principal planes [OPTICS] Two planes perpendicular to the optical axis such that objects in one plane form images in the other with a lateral magnification of -1. Also known as antiprincipal planes. { 'neg·əd·iv ¦prin·sə·pəl 'plānz }

negative principal point [OPTICS] The intersection of a negative principal plane with the optical axis. Also known as antiprincipal point. { 'neg·əd·iv ¦prin·sə·pəl 'pȯint }

negative temperature [THERMO] The property of a thermally isolated thermodynamic system whose elements are in thermodynamic equilibrium among themselves, whose allowed states have an upper limit on their possible energies, and whose high-energy states are more occupied than the low-energy ones. { 'neg·əd·iv 'tem·prə·chər }

negative temperature coefficient [PHYS] A condition wherein the resistance, length, or some other characteristic of a material decreases when temperature increases. { 'neg·əd·iv 'tem·prə·chər ,kō·i,fish·ənt }

negative terminal [ELEC] The terminal of a battery or other voltage source that has more electrons than normal; electrons flow from the negative terminal through the external circuit to the positive terminal. { 'neg·əd·iv 'tər·mən·əl }

negatron *See* electron. { 'neg·ə,trän }

neighbor [CRYSTAL] One of a pair of atoms or ions in a crystal which are close enough to each other for their interaction to be of significance in the physical problem being studied. { 'nā·bər }

N electron [ATOM PHYS] An electron in the fourth (N) shell of electrons surrounding the atomic nucleus, having the principal quantum number 4. { 'en i'lek,trän }

neodymium glass laser [OPTICS] An amorphous solid laser in which glass is doped with neodymium; characteristics are comparable with those of a pulsed ruby laser, but the wavelength of radiation is outside the visible range. { ,nē·ō'dim·ē·əm ¦glas 'lāz·ər }

neodymium liquid laser *See* inorganic liquid laser. { ,nē·ō'dim·ē·əm ¦lik·wəd 'lāz·ər }

neon-helium laser [OPTICS] A continuous-wave gas laser using a combination of neon and helium gases to obtain a 6328-angstrom (632.8-nanometer) visible red beam. { 'nē,än ¦hē·lē·əm 'lā·zər }

neper [PHYS] Abbreviated Np. Also known as napier. **1.** A unit used for expressing the ratio of two currents, voltages, or analogous quantities; the number of nepers is the natural logarithm of this ratio. **2.** A unit used for expressing the ratio of two powers (even when this ratio is not the square of the corresponding current or voltage ratio); the number of nepers is the natural logarithm of the square root of this ratio; to avoid confusion, this usage should be accompanied by a specific statement. { 'nē·pər }

nephelometer [OPTICS] A type of instrument that measures, at more than one angle, the scattering function of particles suspended in a medium; information obtained may be used to determine the size of the suspended particles and the visual range through the medium. { ,nef·ə'läm·əd·ər }

nephelometry [OPTICS] **1.** The study of suspensoids using the techniques of light scattering. **2.** The study of the scattering properties of small samples of air and its suspensoids. { ,nef·ə'läm·ə·trē }

Nernst approximation formula [THERMO] An equation for the equilibrium constant of a gas reaction based on the Nernst heat theorem and certain simplifying assumptions. { 'nernst ə,präk·sə'mā·shən ,fȯr·myə·lə }

Nernst bridge [ELEC] A four-arm bridge containing capacitors instead of resistors, used for measuring capacitance values at high frequencies. { 'nernst ,brij }

Nernst effect [PHYS] The phenomenon that,

when a conductor is placed in a magnetic field and an electric current flows through the conductor perpendicular to the field, a temperature gradient arises in the direction of the current. { 'nernst i,fekt }

Nernst heat theorem [THERMO] The theorem expressing that the rate of change of free energy of a homogeneous system with temperature, and also the rate of change of enthalpy with temperature, approaches zero as the temperature approaches absolute zero. { 'nernst 'hēt ,thir·əm }

Nernst-Simon statement of the third law of thermodynamics [THERMO] The statement that the change in entropy which occurs when a homogeneous system undergoes an isothermal reversible process approaches zero as the temperature approaches absolute zero. { 'nernst 'sī·mən 'stāt·mənt əv thə 'thərd 'lȯ əv ,thər·mō·dī'nam·iks }

NETD *See* noise equivalent temperature difference.

net head [FL MECH] The difference in elevation between the last free water surface in a power conduit above the waterwheel and the first free water surface in the conduit below the waterwheel, less the friction losses in the conduit. { 'net 'hed }

net heating value *See* low heat value. { 'net 'hēd·iŋ ,val·yü }

net power flow [ELECTROMAG] The difference between the power carried by electromagnetic waves traveling in a given direction along a waveguide and the power carried by waves traveling in the opposite direction. { 'net 'pau̇·ər ,flō }

net ton *See* ton. { 'net 'tən }

network [ELEC] A collection of electric elements, such as resistors, coils, capacitors, and sources of energy, connected together to form several interrelated circuits. Also known as electric network. { 'net,wərk }

network admittance [ELEC] The admittance between two terminals of a network under specified conditions. { 'net,wərk ad'mit·əns }

network analysis [ELEC] Derivation of the electrical properties of a network, from its configuration, element values, and driving forces. { 'net,wərk ə'nal·ə·səs }

network constant [ELEC] One of the resistance, inductance, mutual inductance, or capacitance values involved in a circuit or network; if these values are constant, the network is said to be linear. { 'net,wərk 'kän·stənt }

network flow [ELEC] Flow of current in a network. { 'net,wərk 'flō }

network input impedance [ELEC] The impedance between the input terminals of a network under specified conditions. { 'net,wərk 'in·pu̇t im,pēd·əns }

network synthesis [ELEC] Derivation of the configuration and element values of a network with given electrical properties. { 'net,wərk ,sin·thə·səs }

network theory [ELEC] The systematizing and generalizing of the relations between the currents, voltages, and impedances associated with the elements of an electrical network. { 'net,wərk ,thē·ə·rē }

network transfer admittance [ELEC] The current that would flow through a short circuit between one pair of terminals in a network if a unit voltage were applied across the other pair. { 'net,wərk 'trans·fər ad,mit·əns }

Neugebauer effect [ELEC] A small change in the polarization of an optically isotropic medium in an external electric field, related to the electrooptical Kerr effect. { 'nȯi·gə,bau̇·ər i,fekt }

Neumann-Kopp rule [THERMO] The rule that the heat capacity of 1 mole of a solid substance is approximately equal to the sum over the elements forming the substance of the heat capacity of a gram atom of the element times the number of atoms of the element in a molecule of the substance. { 'nȯi,män 'kȯp ,rül }

Neumann's formula [ELECTROMAG] A formula for the mutual inductance M_{12} between two closed circuits C_1 and C_2; it is

$$M_{12} = \frac{\mu_0}{4\pi} \int_{C_{12}} \int_{C_2} \frac{ds_1 ds_2}{r}$$

where r is the distance between line elements ds_1 and ds_2, and μ_0 is the permeability of the empty space. { 'nȯi,mänz ,fȯr·myə·lə }

Neumann's principle [CRYSTAL] The principle that the symmetry elements of the point group of a crystal are included among the symmetry elements of any property of the crystal. { 'nȯi ,mänz ,prin·sə·pəl }

Neumann's triangle [FL MECH] A triangle whose sides have lengths proportional to the surface tensions of two immiscible liquids and their interfacial tension, and directions parallel to the free surfaces of the liquids and the interface between them at a line where these three surfaces meet, when one liquid is placed on the surface of the other. { 'nȯi·mänz 'trī,aŋ·gəl }

neutral [ELEC] Referring to the absence of a net electric charge. { 'nü·trəl }

neutral atom [ATOM PHYS] An atom in which the number of electrons that surround the nucleus is equal to the number of protons in the nucleus, so that there is no net electric charge. { 'nü·trəl 'ad·əm }

neutral axis [MECH] In a beam bent downward, the line of zero stress below which all fibers are in tension and above which they are in compression. { 'nü·trəl 'ak·səs }

neutral beam [PHYS] A stream of uncharged particles. { 'nü·trəl 'bēm }

neutral current interaction [PART PHYS] A weak interaction in which the charges of the interacting fermions are not changed. { 'nü·trəl 'kə·rənt ,in·tər,ak·shən }

neutral-density filter [OPTICS] An optical filter that reduces the intensity of light without appreciably changing its color; used on a camera when the lens cannot be stopped down sufficiently for use with a given film. Also known as gray filter; neutral filter. { 'nü·trəl ¦den·səd·ē 'fil·tər }

neutral equilibrium [PHYS] A property of the

steady state of a system which exhibits neither instability nor stability according to the particular criterion under consideration; a disturbance introduced into such an equilibrium will thus be neither amplified nor damped. Also known as indifferent equilibrium. { 'nü·trəl ,ē·kwə'lib·rē·əm }

neutral fiber [MECH] A line of zero stress in cross section of a bent beam, separating the region of compressive stress from that of tensile stress. { 'nü·trəl 'fī·bər }

neutral filter *See* neutral-density filter. { 'nü·trəl 'fil·tər }

neutral ground [ELEC] Ground connected to the neutral point or points of an electric circuit, transformer, rotating machine, or system. { 'nü·trəl 'grau̇nd }

neutralize [OPTICS] To place a lens in contact with other lenses of equal and opposite power so that the combination has zero power. { 'nü·trə,līz }

neutralizing power of a lens [OPTICS] The power of a lens, measured by neutralizing it with trial lenses of equal and opposite power; for a spectacle lens, it may differ significantly from back vertex power. { 'nü·trə,līz·iŋ ¦pau̇·ər əv ə 'lenz }

neutral particle [PART PHYS] A particle that carries no electric charge. { 'nü·trəl 'pärd·ə·kəl }

neutral point [ELEC] Point which has the same potential as the point of junction of a group of equal nonreactive resistances connected at their free ends to the appropriate main terminals or lines of the system. [FL MECH] *See* hyperbolic point. [OPTICS] In atmospheric optics, one of several points in the sky for which the degree of polarization of diffuse sky radiation is zero. [PHYS] A point where two fields are equal in magnitude and opposite in direction so that the net field is zero. { 'nü·trəl ,pȯint }

neutral surface [MECH] A surface in a bent beam along which material is neither compressed nor extended. { 'nü·trəl 'sər·fəs }

neutral wave [PHYS] Any wave whose amplitude does not change with time; in most contexts the wave is referred to as a stable wave, the term "neutral wave" being used when it is important to emphasize that the wave is neither damped nor amplified. { 'nü·trəl ¦wāv }

neutrino [PHYS] A neutral particle having zero rest mass and spin 1/2 ($h/2\pi$), where h is Planck's constant; experimentally, there are two such particles known as the e neutrino (ν_e) and the μ neutrino (ν_μ). { nü'trē·nō }

neutrino bremsstrahlung [NUC PHYS] The scattering of an electron from a nucleus with the emission of a neutrino and an antineutrino. { nü'trē·nō 'brem,shträ·lu̇ŋ }

neutrino oscillation [PART PHYS] A phenomenon in which a neutrino in one of the three known flavor states (electron neutrino, mu neutrino, or tau neutrino) becomes a mixture of flavor states that changes back and forth periodically as the neutrino travels through space; it will occur if neutrinos have mass and if each flavor state is a mixture of different mass states. { nü¦trē·nō ,äs·ə'lā·shən }

neutron [PHYS] An elementary particle which has approximately the same mass as the proton but lacks electric charge, and is a constituent of all nuclei having mass number greater than 1. { 'nü,trän }

neutron absorption *See* neutron capture. { 'nü ,trän əb,sȯrp·shən }

neutron-antineutron oscillations [PART PHYS] Hypothetical periodic transitions between the state of a neutron and the state of an antineutron which are predicted by certain unified gauge theories. { 'nü,trän 'ant·i,nü,trän ,äs·ə'lā·shənz }

neutron binding energy [NUC PHYS] The energy required to remove a single neutron from a nucleus. { 'nü,trän ¦bīnd·iŋ ,en·ər·jē }

neutron capture [NUC PHYS] A process in which the collision of a neutron with a nucleus results in the absorption of the neutron into the nucleus with the emission of one or more prompt gamma rays; in certain cases, beta decay or fission of the nucleus results. Also known as neutron absorption; neutron radiative capture. { 'nü,trän ,kap·chər }

neutron-capture cross section [NUC PHYS] The cross section for neutron capture by nuclei in a material; it is a measure of the probability that this reaction will occur. { 'nü,trän ,kap·chər 'krȯs ,sek·shən }

neutron cross section [NUC PHYS] A measure of the probability that an interaction of a given kind will take place between a nucleus and an incident neutron; it is an area such that the number of interactions which occur in a sample exposed to a beam of neutrons is equal to the product of the number of nuclei in the sample and the number of neutrons in the beam that would pass through this area if their velocities were perpendicular to it. { 'nü,trän 'krȯs ,sek·shən }

neutron diffraction [PHYS] The phenomenon associated with the interference processes which occur when neutrons are scattered by the atoms within solids, liquids, and gases. { 'nü,trän di ,frak·shən }

neutron diffraction analysis [PHYS] The study of the atomic structure of solids, liquids, and gases by passing high-flux beams of thermal neutrons through them and measuring the intensity of scattered neutrons in various directions. { ¦nü,trän di'frak·shən ə,nal·ə·səs }

neutron diffractometer [PHYS] A diffractometer in which a beam of neutrons is used for diffraction analysis, and the intensities of the diffracted beams at different angles are measured with an ionization chamber or radiation counter. { 'nü ,trän ,di,frak'täm·əd·ər }

neutron drip-line [NUC PHYS] On a chart of the nuclides, plotting proton number versus neutron number, the boundary beyond which neutron-rich nuclei are unstable against neutron emission. { ¦nü,trän 'drip ,līn }

neutron excess [NUC PHYS] The number of neutrons in a nucleus in excess of the number of protons. Also known as difference number; isotopic number. { 'nü,trän 'ek,ses }

neutronicism [NUC PHYS] In a nuclear reaction, the power carried by the neutrons as a fraction of the total power released in the reaction. { nü'trän·ə,siz·əm }

neutron magnetic moment [NUC PHYS] A vector whose scalar product with the magnetic flux density gives the negative of the energy of interaction of a neutron with a magnetic field. { 'nü ,trän mag¦ned·ik 'mō·mənt }

neutron number [NUC PHYS] The number of neutrons in the nucleus of an atom. { 'nü,trän ,nəm·bər }

neutron optics [PHYS] The study of certain phenomena, for example, crystal diffraction, in which the wave character of neutrons dominates and leads to behavior similar to that of light. { 'nü ,trän 'äp·tiks }

neutron-proton scattering [NUC PHYS] A collision of a neutron with a proton, usually the nucleus of a hydrogen atom. { ¦nü,trän 'prō,tän ,skad·ər·iŋ }

neutron radiative capture *See* neutron capture. { 'nü,trän 'rād·ē,ād·iv 'kap·chər }

neutron reflection [PHYS] Specular reflection of neutrons, either from lattice planes of crystalline substances according to the Bragg law for their de Broglie wavelength, or from highly polished surfaces of certain substances at an angle smaller than their critical angle. { 'nü,trän ri ,flek·shən }

neutron-rich nucleus [NUC PHYS] An atomic nucleus in which the ratio of neutron number to proton number is much larger than that of nuclei found in nature. { ¦nü,trän ,rich 'nü·klē·əs }

neutron spectrometry [NUC PHYS] A method of observing excited states of nuclei in which neutrons are used to bombard a target, causing nuclei to be transmuted into excited states by various nuclear reactions; the resultant excited states are determined by observing resonances in the reaction cross sections or by observing spectra of emitted particles or gamma rays. Also known as neutron spectroscopy. { 'nü,trän spek'träm·ə·trē }

neutron spectroscopy *See* neutron spectrometry. *See* slow-neutron spectroscopy. { 'nü,trän spek 'träs·kə·pē }

neutron spectrum [NUC PHYS] A plot or display of the number of neutrons at various energies, such as the neutrons emitted in a nuclear reaction, or the neutrons in a nuclear reactor. { 'nü ,trän ,spek·trəm }

new achromat [OPTICS] An achromatic lens in which the component lenses are made of glasses chosen from a relatively broad selection, among which refractive index and dispersive power may vary inversely, permitting better correction of optical errors. { ¦nü 'ak·rə,mat }

new candle *See* candela. { 'nü 'kand·əl }

Newman-Penrose formalism [RELAT] A formalism in general relativity, based on spinor analysis, convenient for dealing with gravitational perturbations of space-time. { ¦nü·mən ¦pen,rōz 'fȯr·mə,liz·əm }

newton [MECH] The unit of force in the meter-kilogram-second system, equal to the force which will impart an acceleration of 1 meter per second squared to the International Prototype Kilogram mass. Symbolized N. Formerly known as large dyne. { 'nüt·ən }

Newton formula for the stress *See* Newtonian friction law. { 'nüt·ən 'fȯr·myə·lə fȯr t͟hə 'stres }

Newtonian attraction [MECH] The mutual attraction of any two particles in the universe, as given by Newton's law of gravitation. { nü'tō·nē·ən ə'trak·shən }

Newtonian-Cassegrain telescope [OPTICS] A modification of a Cassegrain telescope in which the light reflected from the hyperboloidal secondary mirror is again reflected from a diagonal plane mirror and focused at a point on the side of the telescope, avoiding the need to pierce the primary mirror and making the eyepiece more accessible. Also known as Cassegrain-Newtonian telescope. { nü'tō·nē·ən 'kas,gran 'tel·ə ,skōp }

Newtonian flow [FL MECH] Flow system in which the fluid performs as a Newtonian fluid, that is, shear stress is proportional to shear rate. { 'nü'tō·nē·ən 'flō }

Newtonian fluid [FL MECH] A simple fluid in which the state of stress at any point is proportional to the time rate of strain at that point; the proportionality factor is the viscosity coefficient. { 'nü'tō·nē·ən 'flü·əd }

Newtonian focus [OPTICS] The position in a Newtonian telescope at which the image is formed, located at the side of the tube near its open end. { 'nü'tō·nē·ən 'fō·kəs }

Newtonian friction law [FL MECH] The law that shear stress in a fluid is proportional to the shear rate; it holds only for some fluids, which are then called Newtonian. Also known as Newton formula for the stress. { 'nü'tō·nē·ən 'frik·shən ,lȯ }

Newtonian mechanics [MECH] The system of mechanics based upon Newton's laws of motion in which mass and energy are considered as separate, conservative, mechanical properties, in contrast to their treatment in relativistic mechanics. { 'nü'tō·nē·ən mi'kan·iks }

Newtonian potential [PHYS] A potential which is associated with an inverse square law of force (such as an electrostatic force), and therefore varies with distance in the same manner as a gravitational potential. { 'nü'tō·nē·ən pə'ten·chəl }

Newtonian reference frame [MECH] One of a set of reference frames with constant relative velocity and within which Newton's laws hold;

the frames have a common time, and coordinates are related by the Galilean transformation rule. { 'nü'tō·nē·ən 'ref·rəns ˌfrām }

Newtonian speed of sound [ACOUS] An approximation to the speed of sound in a perfect gas given by the relation $c^2 = p/\rho$, where p is pressure and ρ the density. { 'nü'tō·nē·ən 'spēd əv 'saủnd }

Newtonian telescope [OPTICS] A reflecting telescope in which the light reflected from a concave mirror is reflected again by a plane mirror making an angle of 45° with the telescope axis, so that it passes through a hole in the side of the telescope containing the eyepiece. { 'nü'tō·nē·ən 'tel·ə ˌskōp }

Newtonian velocity [MECH] The velocity of an object in a Newtonian reference frame, S, which can be determined from the velocity of the object in any other such frame, S′, by taking the vector sum of the velocity of the object in S′ and the velocity of the frame S′relative to S. { 'nü'tō·nē·ən və'läs·əd·ē }

Newtonian viscosity [FL MECH] The viscosity of a Newtonian fluid. { 'nü'tō·nē·ən vi'skäs·əd·ē }

newton-meter of energy *See* joule. { 'nüt·ən ¦mēd·ər əv 'en·ər·jē }

newton-meter of torque [MECH] The unit of torque in the meter-kilogram-second system, equal to the torque produced by 1 newton of force acting at a perpendicular distance of 1 meter from an axis of rotation. Abbreviated N-m. { 'nüt·ən ˌmēd·ər əv 'tȯrk }

Newton's equations of motion [MECH] Newton's laws of motion expressed in the form of mathematical equations. { 'nüt·ənz i'kwā·zhənz əv 'mō·shən }

Newton's first law [MECH] The law that a particle not subjected to external forces remains at rest or moves with constant speed in a straight line. Also known as first law of motion; Galileo's law of inertia. { 'nüt·ənz 'fərst 'lȯ }

Newton's law of cooling [THERMO] The law that the rate of heat flow out of an object by both natural convection and radiation is proportional to the temperature difference between the object and its environment, and to the surface area of the object. { 'nüt·ənz 'lȯ əv 'kül·iŋ }

Newton's law of gravitation [MECH] The law that every two particles of matter in the universe attract each other with a force that acts along the line joining them, and has a magnitude proportional to the product of their masses and inversely proportional to the square of the distance between them. Also known as law of gravitation. { 'nüt·ənz 'lȯ əv ˌgrav·ə'tā·shən }

Newton's law of resistance [FL MECH] The law that the force opposing the motion of an object through a fluid at moderate velocities is proportional to the square of the velocity. { 'nüt·ənz 'lȯ əv ri'zis·təns }

Newton's laws of motion [MECH] Three fundamental principles (called Newton's first, second, and third laws) which form the basis of classical, or Newtonian, mechanics, and have proved valid for all mechanical problems not involving speeds comparable with the speed of light and not involving atomic or subatomic particles. { 'nüt·ənz 'lȯz əv 'mō·shən }

Newton's lens formula [OPTICS] A formula which states that the product of the distances of two conjugate points from the respective principal foci of a lens or mirror is equal to the square of the focal length. { ¦nüt·ənz 'lenz ˌfȯr·myə·lə }

Newton's rings [OPTICS] A series of circular bright and dark bands which appear about the point of contact between a glass plate and a convex lens which is pressed against it and illuminated with monochromatic light. { 'nüt·ənz 'riŋz }

Newton's second law [MECH] The law that the acceleration of a particle is directly proportional to the resultant external force acting on the particle and is inversely proportional to the mass of the particle. Also known as second law of motion. { 'nüt·ənz 'sek·ənd 'lȯ }

Newton's theory of lift [FL MECH] A theory of the forces acting on an airfoil in a fluid current in which these forces are assumed to result from the impact of particles of the fluid on the body. { 'nüt·ənz ˌthē·ə·rē əv 'lift }

Newton's theory of light *See* corpuscular theory of light. { 'nüt·ənz ˌthē·ə·rē əv 'līt }

Newton's third law [MECH] The law that, if two particles interact, the force exerted by the first particle on the second particle (called the action force) is equal in magnitude and opposite in direction to the force exerted by the second particle on the first particle (called the reaction force). Also known as law of action and reaction; third law of motion. { 'nüt·ənz 'thərd 'lȯ }

NFE method *See* nearly free electron method. { ¦en¦ef'ē ˌmeth·əd }

ng *See* nanogram.

nickel-63 [NUC PHYS] Radioactive nickel with beta radiation and 92-year half-life; derived by pile-irradiation of nickel; used in radioactive composition studies and tracer studies. { 'nik·əl ¦sik·stē'thrē }

Nicol prism [OPTICS] A device for producing plane-polarized light, consisting of two pieces of transparent calcite (a birefringent crystal) which together form a parallelogram and are cemented together with Canada balsam. { 'nik·əl ˌpriz·əm }

night-sky camera [OPTICS] A simple camera mounted in a fixed direction to record trails of stars caused by the earth's rotation and the breaks in these trails caused by clouds, in order to determine the reduction, due to clouds, of the effective exposures of astronomical photographs. { 'nīt ˌskī ˌkam·rə }

nighttime visual range *See* night visual range. { 'nīˌtīm 'vizh·ə·wəl 'rānj }

night-vision binoculars [OPTICS] Binoculars that are worn like eyeglasses but use a battery-powered television camera to pick up images;

the images are viewed on tiny television picture tubes built into the binoculars. { 'nīt ˌvizh·ən bə'näk·yə·lərz }

night-vision telescope [OPTICS] A telescope that has sufficient electronic amplification of images to be used at night without artificial illumination; may have television, optoelectronic, or other means of providing the necessary image amplification. { 'nīt ˌvizh·ən 'tel·əˌskōp }

night visual range [OPTICS] The greatest distance at which a point source of light of a given candlepower can be perceived at night by an observer under given atmospheric conditions. Also known as nighttime visual range; penetration range; transmission range. { 'nīt 'vizh·ə·wəl 'rānj }

Nikischov effect [PHYS] The production of electron-positron pairs in the collision of low-energy photons with high-energy photons (gamma rays). { ni'kis·chəf iˌfekt }

nine-j symbol [QUANT MECH] A coefficient used in the general recoupling of four angular momenta, as in a transformation from L-S to *j-j* coupling in a two-electron system. Also known as X coefficient. { 'nīn 'jā ˌsim·bəl }

nit [OPTICS] A unit of luminance, equal to 1 candela per square meter. Abbreviated nt. { nit }

nitrogen cycle *See* carbon-nitrogen cycle. { 'nī·trə·jən ˌsī·kəl }

N line [SPECT] One of the characteristic lines in an atom's x-ray spectrum, produced by excitation of an N electron. { 'en ˌlīn }

N-m *See* newton-meter of torque.

NMR *See* nuclear magnetic resonance.

nodal analysis [ELEC] A method of electrical circuit analysis in which potential differences are taken as independent variables and the sum of the currents flowing into a node is equated to 0. { 'nōd·əl ə'nal·ə·səs }

nodal line [PHYS] **1.** A line or curve in a two-dimensional standing-wave system, such as a vibrating diaphragm, where some specified characteristic of the wave, such as velocity of pressure, does not oscillate. **2.** A line which remains fixed during some deformation or rotation of a body or coordinate system. { 'nōd·əl ˌlīn }

nodal points [ELEC] Junction points in a transmission system; the automatic switches and switching centers are the nodal points in automated systems. [OPTICS] A pair of points on the axis of an optical system such that an incident ray passing through one of them results in a parallel emergent ray passing through the other. { 'nōd·əl ˌpȯins }

node *See* branch point. [NEUROSCI] A point of constriction along a nerve. [PHYS] A point, line, or surface in a standing-wave system where some characteristic of the wave has essentially zero amplitude. { nōd }

node voltage [ELEC] The voltage at a given point in an electric network with respect to that at a node. { 'nōd ˌvōl·tij }

no-hair theorems [RELAT] Popular name for general relativistic theorems proving that black holes are uniquely described by their mass, charge, and angular momentum. { ¦nō ¦hār ¦thir·əmz }

noise [ACOUS] Sound which is unwanted, either because of its effect on humans, its effect on fatigue or malfunction of physical equipment, or its interference with the perception or detection of other sounds. [ELEC] Interfering and unwanted currents or voltages in an electrical device or system. [PHYS] Nonperiodic behavior of a system that results from the presence of random driving forces, such as thermal agitation, as opposed to chaotic behavior. [SPECT] Random fluctuations of electronic signals appearing in a recorded spectrum. { nȯiz }

noise analysis [PHYS] Determination of the frequency components that make up a particular noise being studied. { 'nȯiz əˌnal·ə·səs }

noise control [ACOUS] The process of obtaining an acceptable noise environment for a particular observation point or receiver, involving control of the noise source, transmission path, or receiver, or all three. { 'nȯiz kənˌtrōl }

noise equivalent temperature difference [THERMO] The change in equivalent blackbody temperature that corresponds to a change in radiance which will produce a signal-to-noise ratio of 1 in an infrared imaging device. Abbreviated NETD. { 'nȯiz i¦kwiv·ə·lənt 'tem·prə·chər ˌdif·rəns }

noise immission level [ACOUS] A measure of the cumulative noise energy to which an individual is exposed over time; equal to the average noise level to which the person has been exposed, in decibels, plus 10 times the logarithm of the number of years for which the individual is exposed. { ¦nȯiz i¦mish·ən ˌlev·əl }

noise level [PHYS] The intensity of unwanted sound, or the magnitude of unwanted currents or voltages, averaged over a specified frequency range and time interval, and weighted with frequency in a specified manner; usually expressed in decibels relative to a specified reference. { 'nȯiz ˌlev·əl }

noise measurement [ACOUS] The process of quantitatively determining one or more properties of acoustic noise. { 'nȯiz ˌmezh·ər·mənt }

noise pollution [ACOUS] Excessive noise in the human environment. { 'nȯiz pəˌlü·shən }

noise rating number [ACOUS] The perceived noise level of the noise that can be tolerated under specified conditions; for example, the noise rating number of a bedroom is 25, that of a workshop is 65. { 'nȯiz 'rād·iŋ ˌnəm·bər }

noise-reducing antenna system [ELECTROMAG] Receiving antenna system so designed that only the antenna proper can pick up signals; it is placed high enough to be out of the noise-interference zone, and is connected to the receiver with a shielded cable or twisted transmission line that is incapable of picking up signals. { 'nȯiz ri¦düs·iŋ an'ten·ə ˌsis·təm }

noise reduction coefficient [ACOUS] The average over the logarithm of frequency, in the frequency range from 256 to 2048 hertz inclusive, of the sound absorption coefficient of a material. { 'nȯiz ri‚dək·shən ‚kō·i'fish·ənt }

noise reduction rating [ACOUS] A common method for expressing values of noise reduction or attenuation provided by different types of hearing protectors; values range from 0 to approximately 30, with higher values indicating greater amounts of noise reduction. Abbreviated NRR. { 'nȯiz ri‚dək·shən ‚rād·iŋ }

noise temperature [ELEC] The temperature at which the thermal noise power of a passive system per unit bandwidth would be equal to the actual noise at the actual terminals; the standard reference temperature for noise measurements is 290 K. { 'nȯiz ‚tem·prə·chər }

no-load current [ELEC] The current which flows in a network when the output is open-circuited. { 'nō ¦lōd 'kə·rənt }

no-load voltage *See* open-circuit voltage. { 'nō ¦lōd 'vōl·tij }

Nomarski microscope [OPTICS] A type of interference microscope that is used to study reflecting specimens, such as metallic surfaces or metallized replicas of surfaces, and that gives a true relief image uncomplicated by variations in refractive index. { nə'mär·skē 'mī·krə‚skōp }

nominal value [ELEC] The value of some property (such as resistance, capacitance, or impedance) of a device at which it is supposed to operate, under normal conditions, as opposed to actual value. { 'näm·ə·nəl 'val·yü }

nonabelian gauge theory [PART PHYS] A gauge theory in which the gauge transformations can be represented by a Lie group whose members do not commute. { ‚nän·ə¦bēl·yən ¦gāj ‚thē·ə·rē }

nonabelian quantum Hall state [CRYO] A quantum Hall state that is not amenable to a description in terms of electron dancing steps and that cannot be characterized by a symmetric matrix and a charge vector. { ‚nän·ə¦bēl·yən ¦kwän·təm 'hȯl ‚stāt }

nonadiabatic *See* diabatic. { ‚nän¦ad·ē·ə¦bad·ik }

nonblackbody [THERMO] A body that reflects some fraction of the radiation incident upon it; all real bodies are of this nature. { ¦nän'blak ‚bäd·ē }

noncentral force [PHYS] A force between two particles that is not directed along the line connecting them; for example, a tensor force between two nucleons. { 'nän‚sen·trəl 'fȯrs }

noncoherent scattering [ATOM PHYS] The absorption of a photon and its reemission at a different energy (in the observer's frame of reference) by scattering atoms. { ¦nän·kō'hir·ənt 'skad·ə·riŋ }

nonconservative scattering [PHYS] Scattering that is accompanied by absorption. { ¦nän·kən'sər·vəd·iv 'skad·ə·riŋ }

noncontacting piston *See* choke piston. { ¦nän 'kän ‚tak·tiŋ 'pis·tən }

noncontacting plunger *See* choke piston. { ¦nän 'kän‚tak·tiŋ 'plən·jər }

nondeviated absorption [PHYS] Absorption that occurs without any appreciable slowing up of waves. { ¦nän'dē·vē‚ād·əd əb'sȯrp·shən }

nondirectional antenna *See* omnidirectional antenna. { ¦nän·di'rek·shən·əl an'ten·ə }

nondissipative stub [ELECTROMAG] Nondissipative length of waveguide or transmission line. { ¦nän'dis·ə‚pād·iv 'stəb }

nonelectromagnetic radiation [PHYS] A stream of particles other than photons, such as neutrinos, electrons, positrons, protons, neutrons, or alpha particles (all of which also have wave properties, such as diffraction), or of waves other than electromagnetic waves, such as sound waves (which also have particle properties, for example, as phonons in the case of sound). { ‚nän·i‚lek·trə·mag¦ned·ik ‚rād·ē'ā·shən }

nonequilibrium thermodynamics [THERMO] A quantitative treatment of irreversible processes and of rates at which they occur. Also known as irreversible thermodynamics. { ¦nän‚ē·kwə'lib·rē·əm ‚thər·mō·dī'nam·iks }

nonholonomic system [MECH] A system of particles which is subjected to constraints of such a nature that the system cannot be described by independent coordinates; examples are a rolling hoop, or an ice skate which must point along its path. { ¦nän‚häl·ə'näm·ik 'sis·təm }

nonideal gas [STAT MECH] A gas whose molecules have significant interaction, more than that needed to bring about the equilibrium. { 'nän·ī‚dēl 'gas }

noninductive [ELEC] Having negligible or zero inductance. { ‚nän·in'dək·tiv }

noninductive capacitor [ELEC] A capacitor constructed so it has practically no inductance; foil layers are staggered during winding, so an entire layer of foil projects at either end for contact-making purposes; all currents then flow laterally rather than spirally around the capacitor. { ‚nän·in'dək·tiv kə'pas·əd·ər }

noninductive resistor [ELEC] A wire-wound resistor constructed to have practically no inductance, either by using a hairpin winding or be reversing connections to adjacent sections of the winding. { ‚nän·in'dək·tiv ri'zis·tər }

nonintegrable system [MECH] A dynamical system whose motion is governed by an equation that is not an integrable differential equation. { ‚nän¦int·i·grə·bəl ¦sis·təm }

nonlinear [PHYS] Pertaining to a response which is other than directly or inversely proportional to a given variable. { 'nän‚lin·ē·ər }

nonlinear acoustics [ACOUS] The study of the behavior of sufficiently large sonic and ultrasonic disturbances that nonlinear differential equations are necessary for an adequate mathematical description of the phenomena. { 'nän‚lin·ē·ər ə'kü·stiks }

nonlinear capacitor [ELEC] Capacitor having a mean charge characteristic or a peak charge characteristic that is not linear, or a reversible capacitance that varies with bias voltage. { 'nän‚lin·ē·ər kə'pas·əd·ər }

nonlinear circuit [ELEC] A circuit in which the current and voltage in any element that results from two sources of energy acting together is not equal to the sum of the currents or voltages that result from each of the sources acting alone. { 'nän,lin·ē·ər 'sər·kət }

nonlinear coil [ELECTROMAG] Coil having an easily saturable core, possessing high impedance at low or zero current and low impedance when current flows and saturates the core. { 'nän,lin·ē·ər 'kȯil }

nonlinear crystal [SOLID STATE] A crystal in which some influence (such as stress, electric field, or magnetic field) produces a response (such as strain, electric polarization, or magnetization) which is not proportional to the influence. { 'nän,lin·ē·ər 'krist·əl }

nonlinear damping [PHYS] Damping that is not proportional to velocity. { 'nän,lin·ē·ər 'damp·iŋ }

nonlinear dielectric [ELEC] A dielectric whose polarization is not proportional to the applied electric field. { 'nän,lin·ē·ər ,dī·ə'lek·trik }

nonlinear inductance [ELEC] The behavior of an inductor for which the voltage drop across the inductor is not proportional to the rate of change of current, such as when the inductor has a core of magnetic material in which magnetic induction is not proportional to magnetic field strength. { 'nän,lin·ē·ər in'dək·təns }

nonlinear material [PHYS] A material in which some specified influence (such as stress, electric field, or magnetic field) produces a response (such as strain, electric polarization, or magnetization) which is not proportional to the influence. { 'nän,lin·ē·ər mə'tir·ē·əl }

nonlinear network [ELEC] A network in which the current or voltage in any element that results from two sources of energy acting together is not equal to the sum of the currents or voltages that result from each of the sources acting alone. { 'nän,lin·ē·ər 'net,wərk }

nonlinear optical device [OPTICS] A device based on one of a class of optical effects that result from the interaction of electromagnetic radiation from lasers with nonlinear materials. { 'nän,lin·ē·ər 'äp·tə·kəl di,vīs }

nonlinear optical loop mirror [OPTICS] A fiber-optic device in which a coupler splits the input light into two waves that travel in opposite directions around a fiber-optic loop and acquire different phases because of nonlinearities in the optical medium; the output power is very low for some phase differences and equals the input power for other phase differences. { ¦nän,lin·ē·ər ¦ap·tə·kəl 'lüp 'mir·ər }

nonlinear optics [OPTICS] The study of the interaction of radiation with matter in which certain variables describing the response of the matter (such as electric polarization or power absorption) are not proportional to variables describing the radiation (such as electric field strength or energy flux). { 'nän,lin·ē·ər 'äp ,tiks }

nonlinear physics [PHYS] The study of situations where the measure of an effect is not proportional to the measure of what is considered to be its cause. { ,nän,lin·ē·ər 'fiz·iks }

nonlinear refraction [OPTICS] The phenomenon whereby the refractive index of certain substances varies with light intensity. { 'nän,lin·ē·ər ri'frak·shən }

nonlinear Schrödinger equation [OPTICS] A special form into which the Maxwell equations can be transformed in a medium with an optical nonlinearity that gives rise to self-action effects; this equation resembles the Schrödinger equation of quantum mechanics with the potential term in the latter equation replaced by a nonlinear term proportional to the local intensity of the light field, and it possesses soliton solutions. { ¦nän,lin·ē·ər 'shrād·iŋ·ər i,kwā·zhən }

nonlinear spectroscopy [SPECT] The study of energy levels not normally accessible with optical spectroscopy, through the use of nonlinear effects such as multiphoton absorption and ionization. { 'nän,lin·ē·ər spek'träs·kə·pē }

nonlinear vibration [MECH] A vibration whose amplitude is large enough so that the elastic restoring force on the vibrating object is not proportional to its displacement. { 'nän,lin·ē·ər vī'brā·shən }

nonlinear viscoelasticity [FL MECH] The behavior of a fluid which does not obey a first-order differential equation in stress and strain. { 'nän ,lin·ē·ər ¦vis·gō·i,las'tis·əd·ē }

nonloaded Q [ELEC] Of an electric impedance, the Q value of the impedance without external coupling or connection. Also known as basic Q. { 'nän,lōd·əd 'kyü }

nonlocalized electron [PHYS] An electron whose wave function is not confined to the vicinity of one or two nuclei, but is spread out over a molecule or a crystal lattice. { ¦nän'lō·kə,līzd i'lek,trän }

nonmagnetic [ELECTROMAG] Not magnetizable, and therefore not affected by magnetic fields. { ¦nän·mag'ned·ik }

non-Newtonian fluid [FL MECH] A fluid whose flow behavior departs from that of a Newtonian fluid, so that the rate of shear is not proportional to the corresponding stress. Also known as non-Newtonian system. { ,nän·nü'tō·nē·ən 'flü·əd }

non-Newtonian fluid flow [FL MECH] The flow behavior of non-Newtonian fluids, whose study has applications in many important problems of practical significance such as flow in tubes, extrusion, flow through dies, coating operations, rolling operations, and mixing of fluids. { ,nän·nü'tō·nē·ən 'flü·əd ,flō }

non-Newtonian system *See* non-Newtonian fluid. { ,nän·nü'tō·nē·ən 'sis·təm }

non-Newtonian viscosity [FL MECH] The behavior of a fluid which, when subjected to a constant rate of shear, develops a stress which is not proportional to the shear. Also known as anomalous viscosity. { ,nän·nü'tō·nē·ən vi 'skäs·əd·ē }

non-ohmic [ELEC] Pertaining to a substance or circuit component that does not obey Ohm's law. { ¦nän'ō·mik }

nonpropagating soliton [FL MECH] A stable oscillation in a fluid channel that is stationary and highly localized in the length direction of the channel and can be generated when the floor of the channel is subjected to vertical, sinusoidal oscillations. { ¦nän'präp·ə,gād·iŋ 'säl·ə,tän }

nonquantum mechanics [MECH] The classical mechanics of Newton and Einstein as opposed to the quantum mechanics of Heisenberg, Schrödinger, and Dirac; particles have definite position and velocity, and they move according to Newton's laws. { ,nän¦kwän·təm mi'kan·iks }

nonradioactive reaction [NUC PHYS] A nuclear reaction whose radionuclide content is less than 0.01. { ,nän,rād·ē·ō¦ak·tiv rē'ak·shən }

nonreactive [ELEC] Pertaining to a circuit, component, or load that has no capacitance or impedance, so that an alternating current is in phase with the corresponding voltage. { ,nän·rē'ak·tiv }

nonrelativistic approximation [PHYS] The approximation in which it is assumed that speeds of objects are small compared to the speed of light. { ¦nän,rel·ə·tə'vis·tik ə,präk·sə'mā·shən }

nonrelativistic kinematics [MECH] The study of motions of systems of objects at speeds which are small compared to the speed of light, without reference to the forces which act on the system. { ¦nän,rel·ə·tə'vis·tik ,kin·ə'mad·iks }

nonrelativistic mechanics [MECH] The study of the dynamics of systems in which all speeds are small compared to the speed of light. { ¦nän ,rel·ə·tə'vis·tik mi'kan·iks }

nonrelativistic particle [RELAT] A particle whose velocity is small with respect to that of light. { ¦nän,rel·ə·tə'vis·tik 'pärd·ə·kəl }

nonrelativistic quantum mechanics [QUANT MECH] The modern theory of matter and its interaction with radiation, applicable to systems of material particles which move slowly compared to the speed of light, which are neither created nor destroyed, and whose internal structure (except for spin) either does not change or is irrelevant to the description of the system. { ¦nän,rel·ə·tə'vis·tik 'kwän·təm mi'kan·iks }

nonresonant antenna [ELECTROMAG] A long-wire or traveling-wave antenna which does not have natural frequencies of oscillation, and responds equally well to radiation over a broad range of frequencies. { ¦nän'rez·ən·ənt an 'ten·ə }

nonresonant line [ELECTROMAG] Transmission line having no reflected waves, and neither current nor voltage standing waves. { ¦nän'rez·ən·ənt 'līn }

nonselective radiator *See* graybody. { ¦nän·si'lek·tiv 'rād·ē,ād·ər }

nonsinusoidal waveform [ELEC] The representation of a wave which does not vary in a sinusoidal manner, and which therefore contains harmonics. { ¦nän,sī·nə'sȯid·əl 'wāv,fȯrm }

nonspherical nucleus [NUC PHYS] A nucleus which appears to have a permanent ellipsoidal shape in its ground state, as suggested by a large electric quadrupole moment. { 'nän,sfer·ə·kəl 'nü·klē·əs }

nonsynchronous [ELEC] Not related in phase, frequency, or speed to other quantities in a device or circuit. { ¦nän'siŋ·krə·nəs }

nonthermal decimetric emission [ELECTROMAG] A radio-wave emission above the 4-centimeter wavelength from the planet Jupiter that has a nearly constant flux between 5-centimeter and 1-meter wavelength. Also known as DIM. { 'nän,thər·məl ,des·ə'me·trik i'mish·ən }

nonthermal radiation [PHYS] Electromagnetic radiation emitted by accelerated charged particles that are not in thermal equilibrium; aurora light and fluorescent-lamp light are examples. { 'nän,thər·məl ,rād·ē'ā·shən }

nonuniform flow [FL MECH] Fluid flow which does not have the same velocity at all points in a medium, at a given instant. { ¦nän'yü·nə,fȯrm 'flō }

nonviscous flow *See* inviscid flow. { 'nän,vis·kəs 'flō }

nonviscous fluid *See* inviscid fluid. { 'nän,vis·kəs 'flü·əd }

Nordheim's rule [SOLID STATE] The rule that the residual resistivity of a binary alloy that contains mole fraction x of one element and $1 - x$ of the other is proportional to $x(1 - x)$. { 'nȯrd,hīmz ,rül }

norm [QUANT MECH] **1.** The square of the modulus of a Schrödinger-Pauli wave function, integrated over the space coordinates and summed over the spin coordinates of the particles it describes. **2.** The square root of this quantity. { nȯrm }

normal acceleration [MECH] **1.** The component of the linear acceleration of an aircraft or missile along its normal, or Z, axis. **2.** The usual or typical acceleration. { 'nȯr·məl ak,sel·ə'rā·shən }

normal adjustment [OPTICS] Property of an image formed by an optical system whose viewing position is similar to that of the object, such as an image at infinity formed by a telescope or an image at the viewer's near point formed by a microscope. { 'nȯr·məl ə'jəs·mənt }

normal axis [MECH] The vertical axis of an aircraft or missile. { 'nȯr·məl 'ak·səs }

normal coordinates [MECH] A set of coordinates for a coupled system such that the equations of motion each involve only one of these coordinates. { 'nȯr·məl kō'ȯrd·ən·əts }

normal depth [FL MECH] The depth in an open channel at which a given flow has uniform velocity. { 'nȯr·məl 'depth }

normal dispersion [OPTICS] Dispersion in which the refractive index decreases monotonically and continuously with increasing wavelength. { 'nȯr·məl di'spər·zhən }

normal electrode [ELEC] Standard electrode used for measuring electrode potentials. { 'nȯr·məl i'lek,trōd }

normal fluid [CRYO] The component of liquid

helium II, postulated in the two-fluid theory, that has viscosity and behaves like an ordinary fluid. { 'nór·məl 'flü·əd }

normal frequencies [MECH] The frequencies of the normal modes of vibration of a system. { 'nór·məl 'frē·kwən,sēz }

normal Hall effect [ELECTROMAG] The development, in a current-carrying conductor in a magnetic field, of a transverse voltage in the direction of the deflection of negative charge carriers (electrons) by the Lorentz force. { 'nórm·əl 'hól i,fekt }

normal impact [MECH] **1.** Impact on a plane perpendicular to the trajectory. **2.** Striking of a projectile against a surface that is perpendicular to the line of flight of the projectile. { 'nór·məl 'im,pakt }

normal incidence reflectivity [ELECTROMAG] The ratio of the energy of electromagnetic radiation reflected from the interface between two media to the energy of the incident radiation when the incident radiation travels in a direction perpendicular to the surface. { 'nór·məl ¦in·səd·əns ri,flek'tiv·əd·ē }

normal induction [ELECTROMAG] Limiting induction, either positive or negative, in a magnetic material that is under the influence of a magnetizing force which varies between two specific limits. { 'nór·məl in'dək·shən }

normalize [QUANT MECH] To multiply a wave function by a constant so that its norm is equal to unity. { 'nór·mə,līz }

normalized admittance [ELECTROMAG] The reciprocal of the normalized impedance. { 'nór·mə,līzd ad'mit·əns }

normalized coupling coefficient [ELECTROMAG] Mutual inductance, expressed on a scale running from zero to one. { 'nór·mə,līzd ¦kəp·liŋ ,kō·i,fish·ənt }

normalized current [ELECTROMAG] The current divided by the square root of the characteristic admittance of a waveguide or transmission line. { 'nór·mə,līzd 'kə·rənt }

normalized impedance [ELECTROMAG] An impedance divided by the characteristic impedance of a transmission line or waveguide. { 'nór·mə,līzd im'pēd·əns }

normalized susceptance [ELECTROMAG] The susceptance of an element of a waveguide or transmission line divided by the characteristic admittance. { 'nór·mə,līzd sə'sep·təns }

normalized voltage [ELECTROMAG] The voltage divided by the square root of the characteristic impedance of a waveguide or transmission line. { 'nór·mə,līzd 'vōl·tij }

normally dispersive waves [PHYS] Waves which move more rapidly as their wavelengths increase. { 'nór·mə·lē di'spər·siv 'wāvz }

normal magnetization curve [ELECTROMAG] Curve traced on a graph of magnetic induction versus magnetic field strength in an originally unmagnetized specimen, as the magnetic field strength is increased from zero. Also known as magnetization curve. { 'nór·məl ,mag·nə·tə'zā·shən ,kərv }

normal mass shift [NUC PHYS] The portion of the mass shift that corresponds to the variation of reduced mass, and is thus easily calculated for all transitions. { 'nór·məl 'mas ,shift }

normal-mode helix [ELECTROMAG] A type of helical antenna whose diameter and electrical length are considerably less than a wavelength, and which has a radiation pattern with greatest intensity normal to the helix axis. { 'nór·məl ,mōd 'hē,liks }

normal mode of vibration [MECH] Vibration of a coupled system in which the value of one of the normal coordinates oscillates and the values of all the other coordinates remain stationary. { 'nórməl ¦mōd əv vī'brā·shən }

normal permeability [ELECTROMAG] The permeability of a specimen whose magnetic induction and magnetic field strength lie on the normal magnetization curve. Also known as cyclic permeability. { 'nór·məl ,pər·mē·ə'bil·əd·ē }

normal ray [OPTICS] A ray that is incident perpendicularly on a surface. { 'nór·məl 'rā }

normal reaction [MECH] The force exerted by a surface on an object in contact with it which prevents the object from passing through the surface; the force is perpendicular to the surface, and is the only force that the surface exerts on the object in the absence of frictional forces. { 'nór·məl rē'ak·shən }

normal state [NUC PHYS] A term sometimes used for ground state. { 'nór·məl 'stāt }

normal stress [MECH] The stress component at a point in a structure which is perpendicular to the reference plane. { 'nór·məl 'stres }

normal surface [OPTICS] The surface that is generated by taking, at each point of the ray surface, the intersection of the tangent plane to the ray surface at that point with the perpendicular from the origin to this plane. { 'nór·məl 'sər·fəs }

normal temperature and pressure *See* standard conditions. { 'nór·məl 'tem·prə·chər ən 'presh·ər }

normal twin [CRYSTAL] A twin crystal whose twin axis is perpendicular to the composition surface. { 'nór·məl 'twin }

normal volume *See* standard volume. { 'nór·məl 'väl·yəm }

Norris-Eyring reverberation formula [ACOUS] The reverberation time of a chamber, in seconds, is equal to 0.05 times its volume, in cubic feet, divided by the product of its surface area, in square feet, and the negative of the natural logarithm of 1 minus the absorption coefficient averaged over the surface. { 'när·əs 'ī·riŋ ri,vər·bə'rā·shən ,fór·myə·lə }

north pole [ELECTROMAG] The pole of a magnet at which magnetic lines of force are considered as leaving the magnet; the lines enter the south pole; if the magnet is freely suspended, its north pole points toward the north geomagnetic pole. Also known as positive pole. { 'nórth 'pōl }

Norton equivalent circuit [ELEC] An equivalent circuit that consists of a parallel connection of a current source and a two-terminal circuit, where the current source is usually dependent on the electric signals applied to the input terminals. { ¦nȯrt·ən i'kwiv·ə·lənt ˌsər·kət }

Norton's theorem [ELEC] The theorem that the voltage across an element that is connected to two terminals of a linear network is equal to the short-circuit current between these terminals in the absence of the element, divided by the sum of the admittances between the terminals associated with the element and the network respectively. { 'nȯrt·ənz ˌthir·əm }

nose [FL MECH] The dense, forward part of a turbidity current. { nōz }

Notarys-Mercereau microbridge *See* proximity-effect microbridge. { nō'tar·əs 'mer·səˌrō 'mī·krōˌbrij }

notch antenna [ELECTROMAG] Microwave antenna in which the radiation pattern is determined by the size and shape of a notch or slot in a radiating surface. { 'näch anˌten·ə }

note [ACOUS] **1.** A conventional sign indicating the pitch of a musical sound by its position on a staff, and the duration of the sound by its shape. **2.** The sound indicated by this sign. { nōt }

nox [OPTICS] A unit of illumination, used in measuring low-level illumination, equal to 10^{-3} lux. { näks }

noy [ACOUS] A unit of perceived noisiness equal to the perceived noisiness of random noise occupying the frequency band 910–1090 hertz at a sound pressure level of 40 decibels above 0.0002 microbar; a sound that is *n* times as noisy as this sound has a perceived noisiness of *n* noys, under the assumption that the perceived noisiness of a sound increases with physical intensity at the same rate as the loudness. { nȯi }

nozzle-divergence loss factor [FL MECH] The ratio between the momentum of the gases in a nozzle and the momentum of an ideal nozzle. { 'näz·əl dəˌvər·jəns 'lȯs ˌfak·tər }

NRR *See* noise reduction rating.

NRS *See* nuclear reaction spectrometry.

N shell [ATOM PHYS] The fourth layer of electrons about the nucleus of an atom, having electrons characterized by the principal quantum number 4. { 'en ˌshel }

NSOM *See* near-field scanning optical microscopy. { 'enˌsäm *or* ¦en¦es¦ō'em }

NTP *See* standard conditions.

nuclear [NUC PHYS] Pertaining to the atomic nucleus. { 'nü·klē·ər }

nuclear absorption [NUC PHYS] Absorption of energy by the nucleus of an atom. { 'nü·klē·ər əb'sȯrp·shən }

nuclear adiabatic demagnetization [CRYO] A technique for cooling substances, in which the sample is first cooled to temperatures on the order of 10^{-2} K in an extremely intense magnetic field and is then thermally isolated and removed from the field to reach temperatures on the order of 10^{-6} K. { 'nü·klē·ər ˌad·ē·ə'bad·ik dēˌmag·nə·tə'zā·shən }

nuclear angular momentum *See* nuclear spin. { 'nü·klē·ər 'aŋ·gyə·lər mə'men·təm }

nuclear binding energy [NUC PHYS] The energy required to separate an atom into its constituent protons, neutrons, and electrons. { 'nü·klē·ər 'bīnd·iŋ ˌen·ər·jē }

nuclear capture [NUC PHYS] Any process in which a particle, such as a neutron, proton, electron, muon, or alpha particle, combines with a nucleus. { 'nü·klē·ər 'kap·chər }

nuclear chemistry [ATOM PHYS] Study of the atomic nucleus, including fission and fusion reactions and their products. { 'nü·klē·ər 'kem·ə·strē }

nuclear collision [NUC PHYS] A collision between an atomic nucleus and another nucleus or particle. { 'nü·klē·ər kə'lizh·ən }

nuclear cross section [NUC PHYS] A measure of the probability for a reaction to occur between a nucleus and a particle; it is an area such that the number of reactions which occur in a sample exposed to a beam of particles equals the product of the number of nuclei in the sample and the number of incident particles which would pass through this area if their velocities were perpendicular to it. { 'nü·klē·ər 'krȯs ˌsek·shən }

nuclear decay mode [NUC PHYS] One of the ways in which anucleus can undergo radioactive decay, distinguished from other decay modes by the resulting isotope and the particles emitted. { 'nü·klē·ər di'kā ˌmōd }

nuclear density [NUC PHYS] The mass per unit volume of a nucleus as a function of distance from the center of the nucleus, as determined by a number of different types of experiments which are in reasonably good agreement. { 'nü·klē·ər 'den·səd·ē }

nuclear equation of state [NUC PHYS] Any equation that relates pressure, density, and temperature for nuclear matter. { ¦nü·klē·ər i·kwā·zhən əv 'stāt }

nuclear fission *See* fission. { 'nü·klē·ər 'fish·ən }

nuclear fluid dynamic model [NUC PHYS] A model of high-energy nuclear collisions in which nuclear matter is assumed to behave like a compressible fluid whose pressure, density, and temperature are related by an equation of state. Also known as nuclear hydrodynamic model. { 'nü·klē·ər 'flü·əd dīˌnam·ik 'mäd·əl }

nuclear force [NUC PHYS] That part of the force between nucleons which is not electromagnetic; it is much stronger than electromagnetic forces, but drops off very rapidly at distances greater than about 10^{-13} centimeter; it is responsible for holding the nucleus together. { 'nü·klē·ər 'fȯrs }

nuclear fusion *See* fusion. { 'nü·klē·ər 'fyü·zhən }

nuclear ground state [NUC PHYS] The stationary state of lowest energy of an isotope. { 'nü·klē·ər 'grau̇nd ˌstāt }

nuclear Hanle effect [NUC PHYS] The dependence of the linear polarization of gamma rays emitted from a nucleus on the hyperfine interaction in an external magnetic field. { ¦nü·klē·ər 'han·lē iˌfekt }

nuclear hydrodynamic model *See* nuclear fluid

dynamic model. { 'nü·klē·ər ¦hī·drō·dī¦nam·ik 'mäd·əl }

nuclear induction [PHYS] Magnetic induction originating in the magnetic moments of nuclei; the effect depends on the unequal population of energy states available when the material is placed in a magnetic field. { 'nü·klē·ər in'dək·shən }

nuclear isomer *See* isomer. { 'nü·klē·ər 'ī·sə·mər }

nuclear laser [OPTICS] A gas laser in which the gas molecules are excited by high-energy fission particles produced by a pulsed nuclear reactor. { 'nü·klē·ər 'lā·zər }

nuclear magnetic moment [NUC PHYS] The magnetic dipole moment of an atomic nucleus; a vector whose scalar product with the magnetic flux density gives the negative of the energy of interaction of a nucleus with a magnetic field. { 'nü·klē·ər mag'ned·ik 'mō·mənt }

nuclear magnetic resonance [PHYS] A phenomenon exhibited by a large number of atomic nuclei, in which nuclei in a static magnetic field absorb energy from a radio-frequency field at certain characteristic frequencies. Abbreviated NMR. Also known as magnetic nuclear resonance. { 'nü·klē·ər mag'ned·ik 'rez·ən·əns }

nuclear magnetic resonance spectrometer [SPECT] A spectrometer in which nuclear magnetic resonance is used for the analysis of protons and nuclei and for the study of changes in chemical and physical quantities over wide frequency ranges. { 'nü·klē·ər mag'ned·ik 'rez·ən·əns spek'träm·əd·ər }

nuclear magnetism [PHYS] The phenomena associated with the magnetic dipole, octupole, and higher moments of a nucleus, including the magnetic field generated by the nucleus, the force on the nucleus in an inhomogeneous magnetic field, and the splitting of nuclear energy levels in a magnetic field. { 'nü·klē·ər 'mag·nə,tiz·əm }

nuclear magneton [NUC PHYS] A unit of magnetic dipole moment used to express magnetic moments of nuclei and baryons; equal to the electron charge times Planck's constant divided by the product of 4π, the proton mass, and the speed of light. { 'nü·klē·ər 'mag·nə,tän }

nuclear mass [NUC PHYS] The mass of an atomic nucleus, which is usually measured in atomic mass units; it is less than the sum of the masses of its constituent protons and neutrons by the binding energy of the nucleus divided by the square of the speed of light. { 'nü·klē·ər 'mas }

nuclear molecule [NUC PHYS] A quasistable entity of nuclear dimensions formed in nuclear collisions and comprising two or more nuclei that retain their identities and are bound together by strong nuclear forces. { 'nü·klē·ər 'mäl·ə,kyül }

nuclear moment [NUC PHYS] One of the various static electric or magnetic multipole moments of a nucleus. { 'nü·klē·ər 'mō·mənt }

nuclear number *See* mass number. { 'nü·klē·ər 'nəm·bər }

nuclear orientation [NUC PHYS] The directional ordering of an assembly of nuclear spins with respect to some axis in space. { 'nü·klē·ər ,ȯr·ē·ən'tā·shən }

nuclear paramagnetism [PHYS] Paramagnetism in which a substance develops a net magnetic moment because the magnetic moments of nuclei tend to point in the direction of the field. { 'nü·klē·ər ,par·ə'mag·nə,tiz·əm }

nuclear physics [PHYS] The study of the characteristics, behavior, and internal structures of the atomic nucleus. { 'nü·klē·ər 'fiz·iks }

nuclear polarization [NUC PHYS] For a nucleus in a mixed state, with spin I and probability $p(I_z)$ that the I_z substate is populated, the polarization is the sum over allowed values of I_z of $I_z p(I_z)/I$. { 'nü·klē·ər ,pō·lə·rə'zā·shən }

nuclear potential [NUC PHYS] The potential energy of a nuclear particle as a function of its position in the field of a nucleus or of another nuclear particle. { 'nü·klē·ər pə'ten·chəl }

nuclear potential energy [NUC PHYS] The average total potential energy of all the protons and neutrons in a nucleus due to the nuclear forces between them, excluding the electrostatic potential energy. { 'nü·klē·ər pə'ten·chəl 'en·ər·jē }

nuclear potential scattering [NUC PHYS] That part of elastic scattering of particles by a nucleus which may be treated by studying the scattering of a wave which obeys the Schrödinger equation with a potential determined by the properties of the nucleus. { 'nü·klē·ər pə'ten·chəl 'skad·ə·riŋ }

nuclear quadrupole moment [NUC PHYS] The electric quadrupole moment of an atomic nucleus. { 'nü·klē·ər 'kwäd·rə,pōl ,mō·mənt }

nuclear quadrupole resonance [PHYS] The phenomenon in which certain nuclei in a static, inhomogeneous electric field absorb energy from a radio-frequency field. { 'nü·klē·ər 'kwäd·rə,pōl ,rez·ən·əns }

nuclear radiation [NUC PHYS] A term used to denote alpha particles, neutrons, electrons, photons, and other particles which emanate from the atomic nucleus as a result of radioactive decay and nuclear reactions. { 'nü·klē·ər ,rād·ē'ā·shən }

nuclear radiation spectroscopy [NUC PHYS] Study of the distribution of energies or momenta of particles emitted by nuclei. { 'nü·klē·ər ,rād·ē'ā·shən spek'träs·kə·pē }

nuclear radius [NUC PHYS] The radius of a sphere within which the nuclear density is large, and at the surface of which it falls off sharply. { 'nü·klē·ər 'rād·ē·əs }

nuclear reaction [NUC PHYS] A reaction involving a change in an atomic nucleus, such as fission, fusion, neutron capture, or radioactive decay, as distinct from a chemical reaction, which is limited to changes in the electron structure surrounding the nucleus. Also known as reaction. { 'nü·klē·ər rē'ak·shən }

nuclear reaction spectrometry [SPECT] A method of determining the concentration of a given element as a function of depth beneath the surface of a sample, by measuring the yield of characteristic gamma rays from a resonance reaction occurring when the surface is bombarded by a beam of ions. Abbreviated NRS. { 'nü·klē·ər rē'ak·shən spek'träm·ə·trē }

nuclear recoil [NUC PHYS] The imparting of motion to an atomic nucleus during its emission of particles in radioactive decay, or during its collision with another particle, according to the principle of conservation of momentum. { 'nü·klē·ər 'rē,kȯil }

nuclear relaxation [PHYS] The approach of a system of nuclear spins to a steady-state or equilibrium condition over a period of time, following a change in the applied magnetic field. { 'nü·klē·ər ,rē,lak'sā·shən }

nuclear resonance [NUC PHYS] **1.** An unstable excited state formed in the collision of a nucleus and a bombarding particle, and associated with a peak in a plot of cross section versus energy. **2.** The absorption of energy by nuclei from radio-frequency fields at certain frequencies when these nuclei are also subjected to certain types of static fields, as in magnetic resonance and nuclear quadrupole resonance. { 'nü·klē·ər 'rez·ən·əns }

nuclear response function [NUC PHYS] The probability that a given probing particle with given initial energy is scattered from a nucleus with given energy transfer and given momentum transfer, as a function of the energy transfer and momentum transfer. { 'nü·klē·ər ri¦späns ,fəŋk·shən }

nuclear scattering [NUC PHYS] The change in directions of particles as a result of collisions with nuclei. { 'nü·klē·ər 'skad·ə·riŋ }

nuclear spallation *See* spallation. { 'nü·klē·ər spȯ'lā·shən }

nuclear species *See* nuclide. { 'nü·klē·ər 'spē,shēz }

nuclear spectrum [NUC PHYS] **1.** The relative number of particles emitted by atomic nuclei as a function of energy or momenta of these particles. **2.** The graphical display of data from devices used to measure these quantities. { 'nü·klē·ər 'spek·trəm }

nuclear spin [NUC PHYS] The total angular momentum of an atomic nucleus, resulting from the coupled spin and orbital angular momenta of its constituent nuclei. Also known as nuclear angular momentum. Symbolized I. { 'nü·klē·ər 'spin }

nuclear spontaneous reaction *See* radioactive decay. { 'nü·klē·ər spän'tā·nē·əs rē'ak·shən }

nuclear stability [NUC PHYS] The ability of an isotope to resist decay or fission. { 'nü·klē·ər stə'bil·əd·ē }

nuclear transformation *See* transmutation. { 'nü·klē·ər ,tranz·fər'mā·shən }

nuclear Zeeman effect [SPECT] A splitting of atomic spectral lines resulting from the interaction of the magnetic moment of the nucleus with an applied magnetic field. { 'nü·klē·ər 'zē·mən i,fekt }

nucleon [PHYS] A collective name for a proton or a neutron; these particles are the main constituents of atomic nuclei, have approximately the same mass, have a spin of 1/2, and can transform into each other through the process of beta decay. { 'nü·klē,än }

nucleonium [ATOM PHYS] A bound state of a nucleus and an antinucleus. { ,nü·klē'ō·nē·əm }

nucleon number *See* mass number. { 'nü·klē,än ,nəm·bər }

nucleor [PART PHYS] A hypothetical core of a nucleon, surrounded by a hypothetical cloud of pions. { 'nü·klē,ȯr }

nucleus [NUC PHYS] The central, positively charged, dense portion of an atom. Also known as atomic nucleus. { 'nü·klē·əs }

nuclide [NUC PHYS] A species of atom characterized by the number of protons, number of neutrons, and energy content in the nucleus, or alternatively by the atomic number, mass number, and atomic mass; to be regarded as a distinct nuclide, the atom must be capable of existing for a measurable lifetime, generally greater than 10^{-10} second. Also known as nuclear species; species. { 'nü,klīd }

null cone *See* light cone. { ¦nəl ¦kōn }

null detection [ELEC] Altering of adjustable bridge circuit components, to obtain zero current. { 'nəl di,tek·shən }

null geodesic [RELAT] A curve in space-time which has the property that the infinitesimal interval between any two neighboring points on the curve equals zero; it represents a possible path of a light ray. Also known as zero geodesic. { 'nəl ,jē·ə'des·ik }

nulling interferometry [OPTICS] A technique in which light waves from a bright object such as a star are made to interfere and cancel each other in an optical system, allowing the observation of much fainter nearby objects that would otherwise be invisible in the glare of the bright object. { ¦nəl·iŋ ,in·tə·fə'räm·ə·trē }

null surface [RELAT] A surface in space-time whose normal vector is everywhere null. { 'nəl 'sər·fəs }

null vector [RELAT] In special relativity, a four vector whose spatial part in any Lorentz frame has a magnitude equal to the speed of light multiplied by its time part in that frame; a special case of the mathematics definition. { 'nəl 'vek·tər }

number density [PHYS] The number of particles per unit volume. { 'nəm·bər ,den·səd·ē }

numerical aperture [OPTICS] A measure of the resolving power of a microscope objective, equal to the product of the refractive index of the medium in front of the objective and the sine of the angle between the outermost ray entering the objective and the optical axis. Abbreviated N.A. { nü'mer·i·kəl 'ap·ə·chər }

numerical hydrodynamics [FL MECH] An amalgamation of fluid dynamics, applied mathematics, and numerical analysis, in which numerical algorithms are developed to solve equations of incompressible fluid flows about or within bodies. { nü'mer·i·kəl ,hī·drə·də'nam·iks }

N unit [OPTICS] A unit of index of refraction; a mathematical simplification designed to replace rather awkward numbers involved in the values of the index of refraction n for the atmosphere; it is defined by the relation $N=(n-1)10^6$. { 'en ,yü·nət }

Nusselt equation [THERMO] Dimensionless equation used to calculate convection heat transfer for heating or cooling of fluids outside a bank of 10 or more rows of tubes to which the fluid flow is normal. { 'nu̇s·əlt i,kwā·zhən }

Nusselt number [PHYS] A dimensionless number used in the study of mass transfer, equal to the mass-transfer coefficient times the thickness of a layer through which mass transfer is taking place divided by the moleculor diffusivity. Symbolized Nu_m; N_{Num}. Also known as Sherwood number (N_{Sh}). [THERMO] A dimensionless number used in the study of forced convection which gives a measure of the ratio of the total heat transfer to conductive heat transfer, and is equal to the heat-transfer coefficient times a characteristic length divided by the thermal conductivity. Symbolized N_{Nu}. { 'nu̇s·əlt ,nəm·bər }

nutation [MECH] A bobbing or nodding up-and-down motion of a spinning rigid body, such as a top, as it precesses about its vertical axis. { nü'tā·shən }

nutational scanner [OPTICS] An optical-mechanical system for scanning an image along a series of closely spaced parallel lines, using an oscillating plane mirror and a rotating prism. { nü'tā·shən·əl 'skan·ər }

nu value [OPTICS] The reciprocal of the dispersive power of a medium. Also known as constringence. { 'nü ,val·yü }

N wave [ACOUS] The N-shaped pressure wave that is generated by passage of an aircraft at large distances and has lost all of the fine structure observed closer to the aircraft. { 'en ,wāv }

O

OASM system [PHYS] A system of electrical and mechanical units in which the fundamental quantities are electric resistance, electric current, time, and length, and the base units of these quantities are the ohm, ampere, second, and meter, respectively. { ¦ō¦ā¦es'em ˌsis·təm }

object [OPTICS] A collection of points which may be regarded as a source of light rays in an optical system, whether it actually has this function (as in a real object) or does not (as in a virtual object). { 'äb·jekt }

object contrast [OPTICS] The ratio of the difference between the brightness of an object and of the background to the brightness of the background in an image or reproduction. { 'äb·jekt 'känˌtrast }

object glass *See* objective. { 'äb·jekt ˌglas }

objective [OPTICS] The first lens, lens system, or mirror through which light passes or from which it is reflected in an optical system; many scientists exclude mirrors from the definition. Also known as object glass. { äb'jek·tiv }

objective grating [OPTICS] A series of equally spaced parallel wires placed over the objective lens of a telescope; photographic magnitudes of stars are calculated from the relative brightnesses of images in the resulting diffraction pattern. { äb'jek·tiv 'grād·iŋ }

objective prism [OPTICS] A large prism, usually having a small angle, which is placed in front of the objective of a photographic telescope to make spectroscopic observations. { äb'jek·tiv 'priz·əm }

object lens [OPTICS] The first lens through which light passes in a compound objective. { 'äb·jekt ˌlenz }

object plane [OPTICS] A plane containing the real or virtual object in an optical system; usually perpendicular to the axis of the system. { 'äb·jekt ˌplān }

object space [OPTICS] The region of space where objects are located so that a given optical system can form images of them. { 'äb·jekt ˌspās }

oblique astigmatism *See* radial astigmatism. { ə¦blēk ə'stig·məˌtiz·əm }

oblique extinction *See* inclined extinction. { ə'blēk ik'stiŋk·shən }

oblique-incidence reflectivity [OPTICS] The reflectivity of an interface between two media when the direction of propagation of the incident electromagnetic radiation is not perpendicular to the interface; it differs for the component whose electric vector lies in the plane containing the perpendicular to the surface and the propagation direction, and the component for which this vector is perpendicular to this plane. { ə'blēk ¦in·sə·dəns ˌrēˌflek'tiv·əd·ē }

oblique shock *See* oblique shock wave. { ə'blēk 'shäk }

oblique shock wave [FL MECH] A shock wave inclined at an oblique angle to the direction of flow in a supersonic flow field. Also known as oblique shock. { ə'blēk 'shäk ˌwāv }

oblique visibility *See* oblique visual range. { ə'blēk ˌviz·ə'bil·əd·ē }

oblique visual range [OPTICS] The greatest distance at which a specified target can be perceived when viewed along a line of sight inclined to the horizontal. Also known as oblique visibility; slant visibility. { ə'blēk 'vizh·ə·wəl 'rānj }

obliquity factor [OPTICS] A function which is proportional to the amplitudes of secondary waves propagating in various directions according to Huygens' principle; it is $1 + \cos\theta$, where θ is the angle between the normal to the original wavefront and the normal to the secondary wavefront. { ə'blik·wəd·ē ˌfak·tər }

observable operator [QUANT MECH] A Hermitian operator with a complete, orthonormal set of eigenfunctions on the Hilbert space representing the states of a physical system; such operators are postulated to represent the observable quantities of the system. { əb'zər·və·bəl 'äp·əˌrād·ər }

observable quantity [PHYS] A measurable physical quantity. { əb'zər·və·bəl 'kwän·əd·ē }

obtuse bisectrix [CRYSTAL] The bisectrix of the obtuse angle between the axes of a biaxial crystal. { äb'tüs bī'sek·triks }

occlusion [PHYS] Adhesion of gas or liquid on a solid mass, or the trapping of a gas or liquid within a mass. { ə'klü·zhən }

ocean tomography [ACOUS] A form of acoustic tomography in which an array of acoustic sources and receivers transmits and detects a pulse; the pulse travel times are used to determine temperature distributions in the ocean. { 'ō·shən tō 'mäg·rə·fē }

octahedral cleavage [CRYSTAL] Crystal cleavage in the four planes parallel to the face of the octahedron. { ¦äk·tə¦hē·drəl 'klē·vij }

octahedral normal stress [MECH] The normal component of stress across the faces of a regular octahedron whose vertices lie on the principal axes of stress; it is equal in magnitude to the spherical stress across any surface. Also known as mean stress. { ¦äk·tə¦hē·drəl 'nȯr·məl ˌstres }

octahedral plane [CRYSTAL] The plane in a cubic lattice having three numerically equal Miller indices. { ¦äk·tə¦hē·drəl 'plān }

octahedral shear stress [MECH] The tangential component of stress across the faces of a regular octahedron whose vertices lie on the principal axes of stress; it is a measure of the strength of the deviatoric stress. { ¦äk·tə¦hē·drəl 'shir ˌstres }

octave [ACOUS] The interval in pitch between two tones such that one tone may be regarded as duplicating at the next higher pitch the basic musical import of the other tone; the sounds producing these tones then have a frequency ratio of 2 to 1. [PHYS] The interval between any two frequencies having a ratio of 2 to 1. { 'äk·tiv }

octave frequency band [PHYS] A band of frequencies whose highest frequency is twice its lowest frequency. { 'äk·tiv 'frē·kwən·sē ˌband }

octet [ATOM PHYS] A collection of eight valence electrons in an atom or ion, which form the most stable configuration of the outermost, or valence, electron shell. [PART PHYS] A multiplet of eight elementary particles, corresponding to a representation of the approximate unitary symmetry (SU_3) of the strong interactions. { äk'tet }

octupole [PHYS] **1.** Two electric or magnetic quadrupoles having charge distributions of opposite signs and separated from each other by a small distance. **2.** Any device for controlling beams of electrons or other charged particles, consisting of eight electrodes or magnetic poles arranged in a circular pattern, with alternating polarities; commonly used to correct aberrations of quadrupole systems. { 'äk·təˌpōl }

ocular *See* eyepiece. { 'äk·yə·lər }

ocular prism [OPTICS] The prism employed in a range finder to bend the line of sight through the instrument into the eyepiece. { 'äk·yə·lər 'priz·əm }

OD *See* optical density.

odd-even nucleus [NUC PHYS] A nucleus which has an odd number of protons and an even number of neutrons. { 'äd ˌē·vən 'nü·klē·əs }

odd-odd nucleus [NUC PHYS] A nucleus that has an odd number of protons and an odd number of neutrons. { 'äd 'äd 'nü·klē·əs }

odd parity [QUANT MECH] Property of a system whose state vector is multiplied by −1 under the operation of space inversion, that is, the simultaneous reflection of all spatial coordinates through the origin. { 'äd 'par·əd·ē }

odd term [ATOM PHYS] A term of an atom or molecule for which the sum of the angular-momentum quantum numbers of all the electrons is odd, so that the states have odd parity; designated by a superscript *o* or *u*. { 'äd ˌtərm }

ODMR *See* optical detection of magnetic resonance.

O electron [ATOM PHYS] An electron in the fifth (O) shell of electrons surrounding the atomic nucleus, having the principal quantum number 5. { 'ō iˌlekˌträn }

oersted [ELECTROMAG] The unit of magnetic field strength in the centimeter-gram-second electromagnetic system of units, equal to the field strength at the center of a plane circular coil of one turn and 1-centimeter radius, when there is a current of $1/(2\pi)$ abamp in the coil; 1 oersted corresponds to approximately 79.577 amperes per meter. { 'ər·stəd }

Oersted experiment [ELECTROMAG] An experiment in which the deflection of a magnetic needle is observed when it is placed near a wire carrying an electric current. { 'erˌsted ikˌsper·ə·mənt }

offense against the sine condition [OPTICS] A numerical measure of coma, equal to the sagittal coma divided by the perpendicular distance from the image point to the optical axis. { ə'fens ə'genst t͟hə 'sīn kənˌdish·ən }

offset [MECH] The value of strain between the initial linear portion of the stress-strain curve and a parallel line that intersects the stress-strain curve of an arbitrary value of strain; used as an index of yield stress; a value of 0.2% is common. { 'ȯfˌset }

offset yield strength [MECH] That stress at which the strain surpasses by a specific amount (called the offset) an extension of the initial proportional portion of the stress-strain curve; usually expressed in pounds per square inch. { 'ȯf ˌset 'yēld ˌstreŋkth }

ogdosymmetric class [CRYSTAL] A merohedral crystal class whose general form has one-eighth the number of equivalent faces of the corresponding holohedral form. { ¦äg·dō·si'me·trik 'klas }

ohm [ELEC] The unit of electrical resistance in the rationalized meter-kilogram-second system of units, equal to the resistance through which a current of 1 ampere will flow when there is a potential difference of 1 volt across it. Symbolized Ω. { ōm }

ohmic [ELEC] Pertaining to a substance or circuit component that obeys Ohm's law. { 'ō·mik }

ohmic contact [ELEC] A region where two materials are in contact, which has the property that the current flowing through it is proportional to the potential difference across it. { 'ō·mik 'känˌtakt }

ohmic resistance [ELEC] Property of a substance, circuit, or device for which the current flowing through it is proportional to the potential difference across it. { 'ō·mik ri'zis·təns }

Ohm's law [ELEC] The law that the direct current flowing in an electric circuit is directly proportional to the voltage applied to the circuit; it is valid for metallic circuits and many circuits containing an electrolytic resistance. { 'ōmz ˌlȯ }

old achromat [OPTICS] An achromatic lens in which the component lenses are made of glasses chosen from a limited selection, in which refractive index and dispersive power vary roughly together. { ¦ōld 'a·krəˌmat }

omega hyperon [PART PHYS] A semistable baryon with a mass of approximately 1672 MeV, negative charge, spin of 3/2, and positive parity; constitutes an isotopic spin singlet. Also known as omega particle. Symbolized Ω^-. { ō'meg·ə 'hī·pəˌrän }

omega meson [PART PHYS] An unstable, neutral vector meson having a mass of about 783 MeV, a width of about 8 MeV, and negative charge parity and G parity. Symbolized $\omega(783)$. { ō'meg·ə 'māˌsän }

omega particle *See* omega hyperon. { ō'meg·ə ˌpärd·ə·kəl }

omission solid solution [CRYSTAL] A crystal with certain atomic sites incompletely filled. { ō'mish·ən ¦säl·əd sə'lü·shən }

omnidirectional antenna [ELECTROMAG] An antenna that has an essentially circular radiation pattern in azimuth and a directional pattern in elevation. Also known as nondirectional antenna. { ¦äm·nə·di'rek·shən·əl an'ten·ə }

omnifocal lens [OPTICS] A bifocal eyeglass lens that is shaped to allow a smooth transition from one focus to the other. Also known as progressive lens. { ¦äm·nəˌfō·kəl 'lenz }

onde de choc [ACOUS] The first sound heard as the result of the passage of a high-speed projectile. { ȯn·də 'shȯk }

one-dimensional flow [FL MECH] Fluid flow in which all flow is parallel to some straight line, and characteristics of flow do not change in moving perpendicular to this line. { 'wən diˌmen·chən·əl 'flō }

one-dimensional lattice [CRYSTAL] A simplified model of a crystal lattice consisting of particles lying along a straight line at either equal or periodically repeating distances. { 'wən diˌmen·chən·əl 'lad·əs }

one-face-centered lattice [CRYSTAL] A crystal lattice in which there are lattice points at the centers of one pair of faces in each unit cell as well as at the corners. { 'wən ¦fās ¦sen·tərd 'lad·əs }

1/N-expansion [PHYS] The grouping of scattering processes in a power series in 1/N, where N, assumed to be large, is the degeneracy of some quantum state such as the ground state of an impurity center in a solid, of the number of equivalent colors of quarks. { ¦wən ˌō·vər ¦en ik 'span·chən }

one-particle exchange [PART PHYS] A model for the interaction of two particles in which the interaction results entirely from a single virtual particle being emitted by one interacting particle and absorbed by the other. { 'wən ˌpärd·ə·kəl iks 'chānj }

O network [ELEC] Network composed of four impedance branches connected in series to form a closed circuit, two adjacent junction points serving as input terminals, the remaining two junction points serving as output terminals. { 'ō ˌnetˌwərk }

one-way coupling [FL MECH] The property of a particle flow in which the fluid will affect the particle properties (velocity, temperature, and so forth) but the particles will not influence the fluid properties. { ˌwən ˌwā 'kəp·liŋ }

Onsager reciprocal relations [THERMO] A set of conditions which state that the matrix, whose elements express various fluxes of a system (such as diffusion and heat conduction) as linear functions of the various conjugate affinities (such as mass and temperature gradients) for systems close to equilibrium, is symmetric when certain definitions are chosen for these fluxes and affinities. { 'ȯnˌsäg·ər ri'sip·rə·kəl ri'lā·shənz }

Onsager theory of dielectrics [ELEC] A theory for calculating the dielectric constant of a material with polar molecules in which the local field at a molecule is calculated for an actual spherical cavity of molecular size in the dielectric using Laplace's equation, and the polarization catastrophe of the Lorentz field theory is thereby avoided. { 'ȯnˌsäg·ər ˌthē·ə·rē əv ˌdī·ə'lek·triks }

opacity [OPTICS] The light flux incident upon a medium divided by the light flux transmitted by it. { ō'pas·əd·ē }

opalescence [OPTICS] The milky, iridescent appearance of a dense, transparent medium or colloidal system when it is illuminated by polychromatic radiation in the visible range, such as sunlight. { ˌō·pə'les·əns }

opaque medium [OPTICS] A medium impervious to rays of light, that is, not transparent to the human eye. [PHYS] **1.** A medium which does not transmit electromagnetic radiation of a specified type, such as that in the infrared, x-ray, ultraviolet, and microwave regions. **2.** A medium which prevents the passage of particles of a specified type. { ō'pāk ˌmēd·ē·əm }

opaque projector [OPTICS] A projector designed to project the image of an opaque object, or of graphic material on an opaque support, by reflected light. { ō'pāk prə'jek·tər }

open [ELEC] **1.** Condition in which conductors are separated so that current cannot pass. **2.** Break or discontinuity in a circuit which can normally pass a current. { 'ō·pən }

open circuit [ELEC] An electric circuit that has been broken, so that there is no complete path for current flow. { 'ō·pən 'sər·kət }

open-circuited line [ELECTROMAG] A microwave discontinuity which reflects an infinite impedance. { 'ō·pən ¦sər·kəd·əd 'līn }

open-circuit voltage [ELEC] The voltage at the terminals of a source when no appreciable current is flowing. Also known as no-load voltage. { 'ō·pən ¦sər·kət 'vōl·tij }

open cycle [THERMO] A thermodynamic cycle in which new mass enters the boundaries of the system and spent exhaust leaves it; the automotive engine and the gas turbine illustrate this process. { ¦ō·pən ¦sī·kəl }

open form [CRYSTAL] A crystal form in which the crystal faces do not entirely enclose a space. { 'ō·pən ˌfȯrm }

open-packed structure [CRYSTAL] A crystal structure corresponding to the stacking of spheres in an orthogonal arrangement so that each sphere is in contact with six others. { ¦ō·pən ¦pakt 'strək·chər }

open resonator *See* beam resonator. { 'ō·pən 'rez·ən‚ād·ər }

open system [THERMO] A system across whose boundaries both matter and energy may pass. { 'ō·pən 'sis·təm }

open-window unit *See* sabin. { 'ō·pən ¦win·dō 'yü·nət }

opera glasses [OPTICS] Small binocular telescopes, usually of the Galilean type, adapted for use where magnification and field of view are secondary to compactness and cost. { 'äp·rə ˌglas·əz }

operating power [ELECTROMAG] Power that is actually supplied to a radio transmitter antenna. { 'äp·əˌrād·iŋ ˌpau̇·ər }

operating stress [MECH] The stress to which a structural unit is subjected in service. { 'äp·əˌrād·iŋ ˌstres }

ophthalmometer [OPTICS] **1.** An instrument for measuring refractive errors, especially astigmatism. **2.** An instrument for measuring the capacity of the chamber of the eye. **3.** An instrument for measuring the eye as a whole. { ˌäf·thəl'mäm·əd·ər }

ophthalmoscope [OPTICS] An instrument, consisting essentially of a concave mirror with a hole in it and fitted with lenses of different powers, for examining the interior of the eye through the pupil. { äf'thal·məˌskōp }

Oppenheimer-Phillips reaction [NUC PHYS] A type of stripping reaction which can occur when a deuteron passes near a nucleus, in which the proton in the deuteron experiences Coulomb repulsion from the nucleus while the neutron is attracted to the nucleus by nuclear forces, with the result that the neutron-proton bond in the deuteron is broken, the neutron is absorbed into the nucleus, and the proton is repelled. { 'äp·əˌnau̇·ər 'fil·əps rēˌak·shən }

opponent-colors theory [OPTICS] A theory of color vision according to which various processes in the visual system are capable of responding in two opposite ways; the Hering theory is an example. { ə'pō·nənt 'kəl·ərz ˌthē·ə·rē }

opposition [PHYS] The condition in which the phase difference between two periodic quantities having the same frequency is 180°, corresponding to one half-cycle. { ˌäp·ə'zish·ən }

optic [OPTICS] Pertaining to the lenses, prisms, and mirrors of a camera, microscope, or other conventional optical instrument. { 'äp·tik }

optical [OPTICS] Pertaining to or utilizing visible or near-visible light; the extreme limits of the optical spectrum are about 100 nanometers (0.1 micrometer or 3×10^{15} hertz) in the far ultraviolet and 30,000 nanometers (30 micrometers or 10^{13} hertz) in the far infrared. { 'äp·tə·kəl }

optical aberration [OPTICS] Deviation from perfect image formation by an optical system; examples are spherical aberration, coma, astigmatism, curvature of field, distortion, and chromatic aberration. Also known as aberration. { 'äp·tə·kəl ˌab·ə'rā·shən }

optical achromatism *See* visual achromatism. { 'äp·tə·kəl ā'krō·məˌtiz·əm }

optical activity [OPTICS] The behavior of substances which rotate the plane of polarization of plane-polarized light, as it passes through them. Also known as rotary polarization. { 'äp·tə·kəl ak'tiv·əd·ē }

optical analysis [OPTICS] Study of properties of a substance or medium, such as its chemical composition or the size of particles suspended in it, through observation of effects on transmitted light, such as scattering, absorption, refraction, and polarization. { 'äp·tə·kəl ə'nal·ə·səs }

optical anisotropy [OPTICS] The behavior of a medium, or of a single molecule, whose effect on electromagnetic radiation depends on the direction of propagation of the radiation. { 'äp·tə·kəl ˌan·ə'sä·trə·pē }

optical aspherical surface [OPTICS] An optical surface that does not form part of a sphere, such as a paraboloidal or ellipsoidal surface. { 'äp·tə·kəl ā'sfer·ə·kəl 'sər·fəs }

optical axis [OPTICS] **1.** A line passing through a radially symmetrical optical system such that rotation of the system about this line does not alter it in any detectable way. **2.** *See* optic axis. { 'äp·tə·kəl 'ak·səs }

optical bistability [OPTICS] The property of a substance or device which has two stable states of transmission, high or low, for a single input light intensity. { 'äp·tə·kəl ˌbī·stə'bil·əd·ē }

optical branch [SOLID STATE] The vibrations of an optical mode plotted on a graph of frequency versus wave number; it is separated from, and has higher frequencies than, the acoustic branch. { 'äp·tə·kəl 'branch }

optical center [OPTICS] A point on the axis of a lens so that, for any ray passing through this point, the incident part and the emergent part are parallel. Also known as pole. { 'äp·tə·kəl 'sen·tər }

optical coating [OPTICS] Either a mirror coating, or a film of the proper thickness and refractive index applied to the air-glass surface of a lens to reduce reflection. { 'äp·tə·kəl 'kōd·iŋ }

optical coherence microscopy [OPTICS] A variation of optical coherence tomography which uses a system of high numerical aperture to achieve resolutions comparable to that of confocal microscopy. { ¦äp·tə·kəl kō¦hir·əns mī'kräs·kə·pē }

optical coherence tomography [OPTICS] A noninvasive technique for imaging subsurface tissue structure with micrometer-scale resolution, which is based on a broadband light source and a fiber-optic Michelson interferometer. { ¦äp·tə·kəl kō¦hir·əns tə'mäg·rə·fē }

optical contact [OPTICS] Contact between two surfaces in which the surfaces are separated by a distance much less than a wavelength of light, so that interference fringes are not formed. { 'äp·tə·kəl 'kän,takt }

optical crystal [CRYSTAL] Any natural or synthetic crystal, such as sodium chloride, calcium fluoride, silver chloride, potassium iodide, or stilbene, that is used in infrared and ultraviolet optics and for its piezoelectric effects. { 'äp·tə·kəl 'krist·əl }

optical density [OPTICS] The degree of opacity of a translucent medium expressed by log I_0/I, where I_0 is the intensity of the incident ray, and I is the intensity of the transmitted ray. Abbreviated OD. { 'äp·tə·kəl 'den·səd·ē }

optical detection of magnetic resonance [SPECT] A type of electron paramagnetic resonance (EPR) spectroscopy that takes advantage of the sensitivity of electric dipole transitions, in which paramagnetic states are optically excited and the EPR signal is detected through changes in the optical absorption as the magnetic field is swept through one or more resonances. Abbreviated ODMR. { ¦äp·tə·kəl di ,tek·shən əv mag,ned·ik 'rez·ən·əns }

optical diffraction velocimeter *See* diffraction velocimeter. { 'äp·tə·kəl di'frak·shən ,vel·ə'sim·əd·ər }

optical dispersion [OPTICS] Separation of different colors of light such as occurs when it passes from one medium to another or is reflected from a diffraction grating. { 'äp·tə·kəl di'sper·shən }

optical distance *See* optical path. { 'äp·tə·kəl 'dis·təns }

optical Doppler effect [ELECTROMAG] A change in the observed frequency of light or other electromagnetic radiation caused by relative motion of the source and observer. { 'äp·tə·kəl 'däp·lər i,fekt }

optical Doppler tomography *See* color Doppler optical coherence tomography. { ¦äp·tə·kəl ¦däp·lər tə'mäg·rə·fē }

optical element [OPTICS] A part of an optical instrument which acts upon the light passing through the instrument, such as a lens, prism, or mirror. { 'äp·tə·kəl 'el·ə·mənt }

optical fiber [OPTICS] A long, thin thread of fused silica, or other transparent substance, used to transmit light. Also known as light guide. { 'äp·tə·kəl 'fī·bər }

optical-fiber cable *See* optical waveguide. { 'äp·tə·kəl ¦fī·bər 'kā·bəl }

optical figuring [OPTICS] The final polishing or grinding process used to give glass components of optical instruments their desired shape. { 'äp·tə·kəl 'fig·yə·riŋ }

optical flat [OPTICS] **1.** A disk of high-grade quartz glass approximately 2 centimeters thick, with a deviation in flatness usually not exceeding 0.05 micrometer all over, and a surface quality of 5 microfinish or less; used in determinations of surface contour and in comparison of lineal measurement. **2.** A plane surface, with deviations from a plane surface generally not exceeding one-tenth of a wavelength of light, used to redirect light in a telescope or other optical instrument. { 'äp·tə·kəl 'flat }

optical frequency [PHYS] A frequency comparable to that of electromagnetic waves in the optical region, above about 3×10^{11} hertz. { 'äp·tə·kəl 'frē·kwən·sē }

optical guided wave [ELECTROMAG] An optical-frequency electromagnetic wave confined within an optical waveguide. { ,äp·tə·kəl ,gīd·əd 'wāv }

optical harmonic [SOLID STATE] Light, generated by passing a laser beam with a power density on the order of 10^{10} watts per square centimeter or more through certain transparent materials, which has a frequency which is an integral multiple of that of the incident laser light. { 'äp·tə·kəl här'män·ik }

optical haze *See* terrestrial scintillation. { 'äp·tə·kəl 'hāz }

optical indicatrix *See* index ellipsoid. { 'äp·tə·kəl in'dik·ə,triks }

optical instrument [OPTICS] An optical system which acts on light in some desired way, such as to form a real or virtual image, to form an optical spectrum, or to produce light with a specified polarization or wavelength. { 'äp·tə·kəl 'in·strə·mənt }

optical interference [OPTICS] Interference of light waves. { 'äp·tə·kəl ,in·tər'fir·əns }

optical Kerr effect [OPTICS] An effect in which a very strong linearly polarized light field produces anisotropy in the refractive index of an isotropic medium, usually a liquid. { 'äp·tə·kəl 'kər i,fekt }

optical lattice [OPTICS] A regular pattern of microscopic traps for atoms, formed by the light forces in an interference pattern formed by laser beams. { ,äp·tə·kəl 'lad·əs }

optical length *See* optical path. { 'äp·tə·kəl 'leŋkth }

optical lens *See* lens. { 'äp·tə·kəl 'lenz }

optical lever [OPTICS] A device for measuring small angular displacements of a rotating body in which a narrow fixed beam of light is directed onto a small mirror attached to the body and the reflected beam is directed onto a screen, producing a spot of light whose position is measured. { 'äp·tə·kəl 'lev·ər }

optically pumped laser [OPTICS] A laser that uses absorption of light from an auxiliary light source to excite electrons into an upper energy state. { 'äp·tə·klē ¦pəmpt 'lā·zər }

optical maser *See* laser. { 'äp·tə·kəl 'mā·zər }

optical measurement [OPTICS] Measurement of the intensity, spectral distribution, polarization, or other characteristics of light or of infrared or ultraviolet radiation, which is emitted by or reflected from an object or passes through some medium. { 'äp·tə·kəl 'mezh·ər·mənt }

optical meteor [OPTICS] Any phenomenon of the atmosphere explained in terms of optical laws, such as a mirage or a halo. { 'äp·tə·kəl 'mē·dē·ər }

optical microscope [OPTICS] An instrument used to obtain an enlarged image of a small object, utilizing visible light; in general it consists of a light source, a condenser, an objective lens, and an ocular or eyepiece, which can be replaced by a recording device. Also known as light microscope; photon microscope. { 'äp·tə·kəl 'mī·krə,skōp }

optical mode [SOLID STATE] A type of vibration of a crystal lattice whose frequency varies with wave number only over a limited range, and in which neighboring atoms or molecules in different sublattices move in opposition to each other. { 'äp·tə·kəl ,mōd }

optical model *See* cloudy-crystal-ball model. { 'äp·tə·kəl 'mäd·əl }

optical molasses [OPTICS] **1.** A viscous damping force exerted on neutral atoms by a pair of identical oppositely directed lasers tuned at a frequency below an atomic resonance. **2.** A large number of atoms collected and cooled in a small volume at the intersection of the beams of three orthogonal pairs of such lasers. { 'äp·tə·kəl mə'las·əs }

optical moment [OPTICS] For a ray of light passing through an optical system, the triple product of a vector from an arbitrary origin on the optical axis to a point on the ray, a vector tangent to the ray at that point whose length equals the refractive index, and a unit vector along the optical axis; it does not depend on the point on the ray. { 'äp·tə·kəl 'mō·mənt }

optical monochromator [SPECT] A monochromator used to observe the intensity of radiation at wavelengths in the visible, infrared, or ultraviolet regions. { 'äp·tə·kəl ¦man·ə'kräm·əd·ər }

optical null method [SPECT] In infrared spectrometry, the adjustment of a reference beam's energy transmission to match that of a beam that has been passed through a sample being analyzed. { 'äp·tə·kəl ¦nəl ,meth·əd }

optical parallax [OPTICS] A fault in an optical measuring instrument in which the image being observed does not lie in the plane of the wires or marks used to make the measurement, so that motion of the observer's eye causes displacement of the image relative to these wires or marks. { 'äp·tə·kəl 'par·ə,laks }

optical parametric amplification [OPTICS] A process in which a weak signal beam is amplified at the expense of an intense pump beam simultaneously incident on a nonlinear crystal. { 'äp·tə·kəl ¦par·ə¦me·trik ,am·plə·fə'kā·shən }

optical parametric oscillator [OPTICS] A device, employing a nonlinear dielectric, which when pumped by a laser can generate coherent light whose wavelength can be varied continuously over a wide range. { 'äp·tə·kəl ¦par·ə¦me·trik 'äs·ə,lād·ər }

optical path [OPTICS] For a ray of light traveling along a path between two points, the optical path is the integral, over elements of length along the path, of the refractive index. Also known as optical distance; optical length. { 'äp·tə·kəl 'path }

optical-path difference *See* retardation. { 'äp·tə·kəl ¦path 'dif·rəns }

optical phase conjugation [OPTICS] The use of nonlinear optical effects to precisely reverse the direction of propagation of each plane wave in an arbitrary beam of light, thereby causing the return beam to exactly retrace the path of the incident beam. Also known as time-reversal reflection; wavefront reversal. { 'äp·tə·kəl ¦fāz ,kän·jə'gā·shən }

optical phenomena [ELECTROMAG] Phenomena associated with the generation, transmission, and detection of electromagnetic radiation in the visible, infrared, or ultraviolet regions. { 'äp·tə·kəl fə'näm·ə·nä }

optical phonon [SOLID STATE] A quantum of an optical mode of vibration of a crystal lattice. { 'äp·tə·kəl 'fō,nän }

optical prism *See* prism. { 'äp·tə·kəl 'priz·əm }

optical projection system [OPTICS] An optical system which forms a real image of a suitably illuminated object so that it can be viewed, photographed, or otherwise observed. Also known as optical projector; projector. { 'äp·tə·kəl prə'jek·shən ,sis·təm }

optical projector *See* optical projection system. { 'äp·tə·kəl prə'jek·tər }

optical property [ELECTROMAG] One of the effects of a substance or medium on light or other electromagnetic radiation passing through it, such as absorption, scattering, refraction, and polarization. [OPTICS] Also known as reflection property. **1.** The property of an ellipse whereby rays of light emanating from one focus and reflected from a strip of polished metal at the ellipse come together at the other focus. **2.** The property of a parabola whereby rays of light emanating from the focus and reflected from a strip of polished metal at the parabola are reflected parallel to the axis of the parabola, and likewise rays parallel to the axis of the parabola are reflected and brought together at the focus. **3.** The property of a hyperbola whereby rays emanating from a focus and reflected from a strip of polished metal at the hyperbola appear to emanate from the other focus. { 'äp·tə·kəl 'präp·ərd·ē }

optical pulse [OPTICS] A short flash of light, used to isolate moments of time; pulses as short as 15 femtoseconds have been generated with laser and pulse compression techniques. { 'äp·tə·kəl 'pəls }

optical-pulse compression [OPTICS] The shortening of the duration of an optical pulse by techniques similar to those used in chirp radar, in which a frequency sweep is imposed on the pulse and the pulse is then compressed by using a dispersive delay line. { 'äp·tə·kəl ¦pəls kəm ,presh·ən }

optical pumping [OPTICS] The process of causing strong deviations from thermal equilibrium populations of selected quantized states of different energy in atomic or molecular systems by the use of electromagnetic radiation in or near the visible region. { 'äp·tə·kəl 'pəmp·iŋ }

optical quenching [OPTICS] Reduction in the intensity of luminescent radiation by long-wavelength, visible or infrared radiation. { 'äp·tə·kəl 'kwench·iŋ }

optical rectification [OPTICS] An effect whereby one or more electromagnetic waves propagating in a nonlinear medium produce a second-order polarization that does not oscillate and, in turn, produces an electrical voltage. { 'äp·tə·kəl ,rek·tə·fə'kā·shən }

optical rotation [OPTICS] Rotation of the plane of polarization of plane-polarized light, or of the major axis of the polarization ellipse of elliptically polarized light by transmission through a substance or medium. { 'äp·tə·kəl rō'tā·shən }

optical rotatory dispersion [OPTICS] Specific rotation, considered as a function of wavelength. Abbreviated ORD. { 'äp·tə·kəl 'rōd·ə,tȯr·ē di'spər·zhən }

optical sight [OPTICS] A sight with lenses, prisms, or mirrors that is used in laying weapons, for aerial bombing, or for surveying. { 'äp·tə·kəl 'sīt }

optical spectra [SPECT] Electromagnetic spectra for wavelengths in the ultraviolet, visible and infrared regions, ranging from about 10 nanometers to 1 millimeter, associated with excitations of valence electrons of atoms and molecules, and vibrations and rotations of molecules. { 'äp·tə·kəl 'spek·trə }

optical spectrograph [SPECT] An optical spectroscope provided with a photographic camera or other device for recording the spectrum made by the spectroscope. { 'äp·tə·kəl 'spek·trə,graf }

optical spectrometer [SPECT] An optical spectroscope that is provided with a calibrated scale either for measurement of wavelength or for measurement of refractive indices of transparent prism materials. { 'äp·tə·kəl spek'träm·əd·ər }

optical spectroscope [SPECT] An optical instrument, consisting of a slit, collimator lens, prism or grating, and a telescope or objective lens, which produces an optical spectrum arising from emission or absorption of radiant energy by a substance, for visual observation. { 'äp·tə·kəl 'spek·trə,skōp }

optical spectroscopy [SPECT] The production, measurement, and interpretation of optical spectra arising from either emission or absorption of radiant energy by various substances. { 'äp·tə·kəl spek'träs·kə·pē }

optical spherical surface [OPTICS] An optical surface which forms part of a sphere. { 'äp·tə·kəl 'sfer·ə·kəl 'sər·fəs }

optical staining *See* Rheinberg illumination. { 'äp·tə·kəl 'stān·iŋ }

optical superposition principle [OPTICS] The principle that the optical rotation produced by a compound which is made up of two radicals of opposite optical activity is the algebraic sum of the rotations of each radical alone; not always valid. { 'äp·tə·kəl ¦sü·pər·pə¦zish·ən ,prin·sə·pəl }

optical surface [OPTICS] An interface between two media, such as between air and glass, which is used to reflect or refract light. { 'äp·tə·kəl 'sər·fəs }

optical system [OPTICS] A collection comprising mirrors, lens, prisms, and other devices, placed in some specified configuration, which reflect, refract, disperse, absorb, polarize, or otherwise act on light. { 'äp·tə·kəl ,sis·təm }

optical thickness [OPTICS] The thickness of an optical material times its index of refraction. { 'äp·tə·kəl 'thik·nəs }

optical-to-optical interface device [OPTICS] A device that converts a noncoherently illuminated image into a coherently illuminated object, which can then be used as input to certain types of data processor. Abbreviated OTTO. { 'äp·tə·kəl tü 'äp·tə·kəl 'in·tər,fās di,vīs }

optical train [OPTICS] The series of lenses, mirrors, and prisms of an optical apparatus, such as a microscope or telescope, through which the light rays pass. { 'äp·tə·kəl 'trān }

optical transition [PHYS] A process in which an atom or molecule changes from one energy state to another and emits or absorbs electromagnetic radiation in the visible, infrared, or ultraviolet region. { 'äp·tə·kəl tran'zish·ən }

optical tweezers *See* laser tweezers. { 'äp·tə·kəl 'twēz·ərz }

optical twinning [CRYSTAL] Growing together of two crystals which are the same except that the structure of one is the mirror image of the structure of the other. Also known as chiral twinning. { 'äp·tə·kəl 'twin·iŋ }

optical waveguide [ELECTROMAG] A waveguide in which a light-transmitting material such as a glass or plastic fiber is used for transmitting information from point to point at wavelengths somewhere in the ultraviolet, visible-light, or infrared portions of the spectrum. Also known as fiber waveguide; optical-fiber cable. { 'äp·tə·kəl 'wāv,gīd }

optical window [OPTICS] The spectral region between 300 and 2000 nanometers (0.3 and 2 micrometers in wavelength), in which visible and near-visible radiation will pass through the earth's atmosphere. { 'äp·tə·kəl 'win·dō }

optic angle *See* axial angle. { 'äp·tik ,aŋ·gəl }

optic-axial angle *See* axial angle. { 'äp·tik ¦ak·sē·əl ,aŋ·gəl }

optic axis [OPTICS] The axis in a doubly refracting medium in which the ordinary and extraordinary waves propagate with the same velocity, and double refraction vanishes. Also known as optical axis; principal axis. { 'äp·tik 'ak·səs }

optic ellipse [OPTICS] Any section through the index ellipsoid. { 'äp·tik i'lips }

optic normal [OPTICS] The axis that lies perpendicular to the optic axis. { 'äp·tik 'nȯr·məl }

optics [PHYS] **1.** Narrowly, the science of light

and vision. **2.** Broadly, the study of the phenomena associated with the generation, transmission, and detection of electromagnetic radiation in the spectral range extending from the long-wave edge of the x-ray region to the short-wave edge of the radio region, or in wavelength from about 1 nanometer to about 1 millimeter. { 'äp·tiks }

optimum array current [ELECTROMAG] The current distribution in a broadside antenna array which is such that for a specified side-lobe level the beam width is as narrow as possible, and for a specified first null the side-lobe level is as small as possible. { 'äp·tə·məm ə'rā ,kə·rənt }

optimum coupling *See* critical coupling. { 'äp·tə·məm 'kəp·liŋ }

optimum reverberation time [ACOUS] The reverberation time which is most desirable for a given room size and a given use, such as speech, chamber music, or symphony orchestra. { 'äp·tə·məm ri,vər·bə'rā·shən ,tīm }

optoacoustic effect [PHYS] A phenomenon in which a periodically interrupted beam of light generates sound in a gas through which it is passing; this results from energy in the light beam being transformed first into internal motions of the gas molecules, then into random translational motions of these molecules, or heat, and finally into periodic pressure fluctuations or sound. Also known as thermoacoustic effect. { ¦äp·tō·ə¦küs·tik i,fekt }

optoacoustic modulator *See* acoustooptic modulator. { ¦äp·tō·ə¦küs·tik 'mäj·ə,lād·ər }

optoacoustic spectroscopy *See* photoacoustic spectroscopy. { ¦äp·tō·ə¦kü·stik spek'träs·kə·pē }

optogalvanic effect [PHYS] The alteration of the current through an electrical discharge by light incident on the discharge space. { ¦äp·tō·gal¦van·ik i,fekt }

optogalvanic spectroscopy [SPECT] A method of obtaining absorption spectra of atomic and molecular species in flames and electrical discharges by measuring voltage and current changes upon laser irradiation. { ¦äp·tō·gal¦van·ik spek'träs·kə·pē }

optovoltaic effect [PHYS] The alteration of the potential difference across a discharge by light incident on the discharge space. { ,äp·tō·vōl'tā·ik i,fekt }

OPW method *See* orthogonalized plane-wave method. { ¦ō¦pē'dəb·əl,yü ,meth·əd }

orange [OPTICS] The hue evoked in an average observer by monochromatic radiation having a wavelength in the approximate range from 597 to 622 nanometers; however, the same sensation can be produced in a variety of other ways. { 'är·inj }

orange spectrometer [SPECT] A type of beta-ray spectrometer that consists of a number of modified double-focusing spectrometers employing a common source and a common detector, and has exceptionally high transmission. { 'är·inj spek'träm·əd·ər }

O ray *See* ordinary ray. { 'ō ,rā }

orbit [PHYS] **1.** Any closed path followed by a particle or body, such as the orbit of a celestial body under the influence of gravity, the elliptical path followed by electrons in the Bohr theory, or the paths followed by particles in a circular particle accelerator. **2.** More generally, any path followed by a particle, such as helical paths of particles in a magnetic field, or the parabolic path of a comet. { 'ȯr·bət }

orbital [ATOM PHYS] The space-dependent part of the Schrödinger wave function of an electron in an atom or molecule in an approximation such that each electron has a definite wave function, independent of the other electrons. { 'ȯr·bəd·əl }

orbital angular momentum [MECH] The angular momentum associated with the motion of a particle about an origin, equal to the cross product of the position vector with the linear momentum. Also known as orbital momentum. [QUANT MECH] The angular momentum operator associated with the motion of a particle about an origin, equal to the cross product of the position vector with the linear momentum, as opposed to the intrinsic spin angular momentum. Also known as orbital moment. { 'ȯr·bəd·əl 'aŋ·gyə·lər mə'men·təm }

orbital decay [ATOM PHYS] A change of an atom from one energy state to another of lower energy in which the orbital of one of the electrons changes. { 'ȯr·bəd·əl di'kā }

orbital electron [ATOM PHYS] An electron which has a high probability of being in the vicinity (at distances on the order of 10^{-10} meter or less) of a particular nucleus, but has only a very small probability of being within the nucleus itself. Also known as planetary electron. { 'ȯr·bəd·əl i'lek,trän }

orbital elements [PHYS] A set of seven parameters defining the orbit of a body attracted by a central, inverse-square force. { 'ȯr·bəd·əl 'el·ə·məns }

orbital magnetic moment [QUANT MECH] The magnetic dipole moment associated with the motion of a charged particle about an origin, rather than with its intrinsic spin. { 'ȯr·bəd·əl mag'ned·ik 'mō·mənt }

orbital moment *See* orbital angular momentum. { 'ȯr·bəd·əl 'mō·mənt }

orbital momentum *See* orbital angular momentum. { 'ȯr·bəd·əl mə'men·təm }

orbital motion [PHYS] Continuous motion of a body in a closed path, such as a circle or an ellipse, about some point. { 'ȯr·bəd·əl 'mō·shən }

orbital parity [QUANT MECH] The parity associated with the wave function of a particle, or system of particles, as a function of spatial coordinates; it is opposed to intrinsic parity; if the orbital angular momentum quantum number is l, the orbital parity is $(-1)^l$. { 'ȯr·bəd·əl 'par·əd·ē }

orbital plane [MECH] The plane which contains the orbit of a body or particle in a central force field; it passes through the center of force. { 'ȯr·bəd·əl 'plān }

orbiting collision [ATOM PHYS] An interaction between an ion and an atom in which they approach each other very closely and spend a relatively long time (several orbital periods of the atomic electrons) in proximity. { 'ȯr·bəd·iŋ kə'lizh·ən }

ORD *See* optical rotatory dispersion.

order [PHYS] A range of magnitudes of a quantity (and of all other quantities having the same physical dimensions) extending from some value of the quantity to some small multiple of the quantity (usually 10). Also known as order of magnitude. { 'ȯrd·ər }

order-disorder transition [SOLID STATE] The transition of an alloy or other solid solution between a state in which atoms of one element occupy certain regular positions in the lattice of another element, and a state in which this regularity is not present. { 'ȯrd·ər 'dis,ȯrd·ər tran'zish·ən }

ordering [SOLID STATE] A solid-state transformation in certain solid solutions, in which a random arrangement in the lattice is transformed into a regular ordered arrangement of the atoms with respect to one another; a so-called superlattice is formed. { 'ȯrd·ə·riŋ }

order of aberration [OPTICS] The sum of the powers to which the field height and the pupil coordinates are raised in describing a term in the decomposition of an aberration according to the degree of dependence on these variables. { 'ȯrd·ər əv ,ab·ə'rā·shən }

order of interference [OPTICS] The difference in the number of wavelengths along the paths of two constructively interfering rays of light. { 'ȯrd·ər əv ,in·tər'fir·əns }

order of magnitude *See* order. { 'ȯrd·ər əv 'mag·nə,tüd }

order of phase transition [THERMO] A phase transition in which there is a latent heat and an abrupt change in properties, such as in density, is a first-order transition; if there is not such a change, the order of the transition is one greater than the lowest derivative of such properties with respect to temperature which has a discontinuity. { 'ȯrd·ər əv 'fāz tran,zish·ən }

order parameter [STAT MECH] A measure of the degree of ordering of a system which has value zero above the temperature of a phase transition and acquires some nonzero value below the transition temperature. { 'ȯrd·ər pə,ram·əd·ər }

ordinary index [OPTICS] The index of refraction of the ordinary ray in a crystal. { 'ȯrd·ən,er·ē 'in,deks }

ordinary ray [OPTICS] One of two rays into which a ray incident on an anisotropic uniaxial crystal is split; it obeys the ordinary laws of refraction, in contrast to the extraordinary ray. Also known as O ray. { 'ȯrd·ən,er·ē 'rā }

ordinary-wave component [OPTICS] The component of electromagnetic radiation propagating in an anisotropic uniaxial crystal whose electric displacement vector is perpendicular to the optical axis and the direction normal to the wavefront; gives rise to the ordinary ray. { 'ȯrd·ən,er·ē ¦wāv kəm,pō·nənt }

orient [OPTICS] The play of color upon or just below the surface of a gem-quality pearl. { 'ȯr·ē·ənt }

orientability of sound signal [ACOUS] The property of a sound signal by virtue of which a listener can estimate the direction of the location of the apparatus producing the signal. { ¦ȯr·ē,ent·ə'bil·əd·ē əv 'sau̇nd ,sig·nəl }

orientation [CRYSTAL] The directions of the axes of a crystal lattice relative to the surfaces of the crystal, to applied fields, or to some other planes or directions of interest. [ELECTROMAG] The physical positioning of a directional antenna or other device having directional characteristics. [PHYS] **1.** The direction of some vector or set of vectors, such as the direction of the electric vector and the propagation direction of plane polarized light, or the direction of a preponderance of nuclear spins in a crystal near absolute zero, relative to some other directions of interest. **2.** Any process in which vectors associated with atoms or molecules in the substance are organized relative to some direction, rather than pointed at random; examples include dipole moments of polar molecules in an electric field, and nuclear spins in a crystal in a magnetic field at temperatures near absolute zero. { ,ȯr·ē·ən'tā·shən }

orientation effect [ELEC] Those bulk properties of a material which result from orientation polarization. { ,ȯr·ē·ən'tā·shən i,fekt }

orientation polarization [ELEC] Polarization arising from the orientation of molecules which have permanent dipole moments arising from an asymmetric charge distribution. Also known as dipole polarization. { ,ȯr·ē·ən'tā·shən ,pō·lə·rə,zā·shən }

orifice [ELECTROMAG] Opening or window in a side or end wall of a waveguide or cavity resonator through which energy is transmitted. { 'ȯr·ə·fəs }

orthoaxis [CRYSTAL] The diagonal or lateral axis perpendicular to the vertical axis in the monoclinic system. { ¦ȯr·thō'ak·səs }

orthobaric density [PHYS] The density of a liquid and of a saturated vapor with which it is at equilibrium at a given temperature. { ¦ȯr·thə¦bar·ik 'den·səd·ē }

orthogonal antennas [ELECTROMAG] In radar, a pair of transmitting and receiving antennas, or a single transmitting-receiving antenna, designed for the detection of a difference in polarization between the transmitted energy and the energy returned from the target. { ȯr'thäg·ən·əl an'ten·əz }

orthogonal crystal [CRYSTAL] A crystal whose axes are mutually perpendicular. { ȯr'thäg·ən·əl 'krist·əl }

orthogonalized plane-wave method [SOLID STATE] A method of approximating the energy states of electrons in a crystal lattice: trial wave functions (the orthogonalized plane waves) are constructed which are linear combinations of plane waves and Bloch functions based on core states, and which are orthogonal to the Bloch functions, and linear combinations of these trial functions are then determined by the variational method. Abbreviated OPW method. { ȯr ¦thäg·ən·əl,īzd 'plān ,wāv ,meth·əd }

orthographic projection [CRYSTAL] A projection for displaying the poles of a crystal in which the poles are projected from a reference sphere onto an equatorial plane by dropping perpendiculars from the poles to the plane. { ¦ȯr·thə¦graf·ik prə'jek·shən }

orthohelium [ATOM PHYS] Those states of helium atoms in which the spins of the two electrons are parallel. { ¦ȯr·thō'hē·lē·əm }

orthohexagonal axes [CRYSTAL] A set of crystallographic axes, two of which have a fixed ratio, as in hexagonal or trigonal crystals. { ¦ȯr·thō·hek'säg·ən·əl 'ak,sēz }

orthohydrogen [ATOM PHYS] Those states of hydrogen molecules in which the spins of the two nuclei are parallel. { ¦ȯr·thə'hī·drə·jən }

orthonormal tetrad [RELAT] A collection of four mutually orthogonal unit vectors, three spacelike and one timelike, at a point of space-time, that specify the directions of the four axes of a locally Minkowskian coordinate system. { ¦ȯr·thə¦nȯr·məl 'te,trad }

orthopinacoid *See* front pinacoid. { ¦ȯr·thə'pin·ə,kȯid }

orthopositronium [PART PHYS] The state of positronium in which the positron and electron have parallel spins. { ¦ȯr·thō,päz·ə'trō·nē·əm }

orthorhombic lattice [CRYSTAL] A crystal lattice in which the three axes of a unit cell are mutually perpendicular, and no two have the same length. Also known as rhombic lattice. { ¦ȯr·thə¦räm·bik 'lad·əs }

orthorhombic system [CRYSTAL] A crystal system characterized by three axes of symmetry that are mutually perpendicular and of unequal length. Also known as rhombic system. { ¦ȯr·thə¦räm·bik 'sis·təm }

orthoscope [OPTICS] A polarizing microscope in which light is transmitted by a crystal which is parallel to the microscope axis. { 'ȯr·thə,skōp }

orthoscopic eyepiece [OPTICS] An eyepiece that consists of a single lens, made up of three cemented elements, to which a planoconvex lens is added; designed to minimize distortion and spherical aberration. { ¦ȯr·thə¦skäp·ik 'ī,pēs }

orthoscopic system [OPTICS] An optical system that has been corrected so that distortion and spherical aberration are eliminated. Also known as rectilinear system. { ¦ȯr·thə¦skäp·ik 'sis·təm }

orthosymmetric crystal [CRYSTAL] A crystal that has orthorhombic symmetry. { ¦ȯr·thō·si'me·trik 'krist·əl }

orthotomic system [OPTICS] An optical system in which all the rays may be interesected at right angles by a suitably chosen surface. { ,ȯr·thə'täm·ik ,sis·təm }

orthotropic [MECH] Having elastic properties such as those of timber, that is, with considerable variations of strength in two or more directions perpendicular to one another. { ¦ȯr·thə¦träp·ik }

oscillating magnetic field [ELECTROMAG] A magnetic field which varies periodically in time. { 'äs·ə,lād·iŋ mag'ned·ik 'fēld }

oscillation [PHYS] Any effect that varies periodically back and forth between two values. { ,äs·ə'lā·shən }

oscillation photography [SOLID STATE] A method of x-ray diffraction analysis in which a single crystal is made to oscillate through a small angle about an axis perpendicular to a beam of monochromatic x-rays or particles. { ,äs·ə¦lā·shən fə¦täg·rə·fē }

oscillator [PHYS] Any device (mechanical or electrical) which, in the absence of external forces, can have a periodic back- and-forth motion, the frequency determined by the properties of the oscillator. { 'äs·ə,lād·ər }

oscillator strength [ATOM PHYS] A quantum-mechanical analog of the number of dispersion electrons having a given natural frequency in an atom, used in an equation for the absorption coefficient of a spectral line; it need not be a whole number. Also known as f value; Ladenburg f value. { 'äs·ə,lād·ər ,streŋkth }

oscillatory circuit [ELEC] Circuit containing inductance or capacitance, or both, and resistance, connected so that a voltage impulse will produce an output current which periodically reverses or oscillates. { 'äs·ə·lə,tȯr·ē 'sər·kət }

oscillatory discharge [ELEC] Alternating current of gradually decreasing amplitude which, under certain conditions, flows through a circuit containing inductance, capacitance, and resistance when a voltage is applied. { 'äs·ə·lə,tȯr·ē 'dis,chärj }

oscillatory extinction *See* undulatory extinction. { 'äs·ə·lə,tȯr·ē ik'stiŋk·shən }

oscillatory shear [FL MECH] Application of small-amplitude oscillations to produce shear in viscoelastic fluids for the study of dynamic viscosity. { 'äs·ə·lə,tȯr·ē 'shir }

oscillatory surge [ELEC] Surge which includes both positive and negative polarity values. { 'äs·ə·lə,tȯr·ē 'sərj }

oscillatory twinning [CRYSTAL] Repeated, parallel twinning. { 'äs·ə·lə,tȯr·ē 'twin·iŋ }

oscillatory wave [PHYS] A wave composed of individual particles, each of which oscillates about a point with little, if any, permanent change in position. { 'äs·ə·lə,tȯr·ē 'wāv }

Oseen's flow [FL MECH] Fluid flow in which the velocity of flow is very small but the Reynolds number is greater than 1. { ü'sānz ,flō }

O shell [ATOM PHYS] The fifth layer of electrons about the nucleus of an atom, having electrons characterized by the principal quantum number 5. { 'ō ,shel }

Ostwald's adsorption isotherm [THERMO] An equation stating that at a constant temperature the weight of material adsorbed on an adsorbent dispersed through a gas or solution, per unit weight of adsorbent, is proportional to the concentration of the adsorbent raised to some constant power. { 'ȯst,vālts ad'sörp·shən 'ī·sə ,thərm }

Otto cycle [THERMO] A thermodynamic cycle for the conversion of heat into work, consisting of two isentropic phases interspersed between two constant-volume phases. Also known as spark-ignition combustion cycle. { 'äd·ō ,sī·kəl }

Otto-Lardillon method [MECH] A method of computing trajectories of missiles with low velocities (so that drag is proportional to the velocity squared) and quadrant angles of departure that may be high, in which exact solutions of the equations of motion are arrived at by numerical integration and are then tabulated. { 'äd·ō ,lär·dē'yȯn ,meth·əd }

ounce [MECH] **1.** A unit of mass in avoirdupois measure equal to 1/16 pound or to approximately 0.0283495 kilogram. Abbreviated oz. **2.** A unit of mass in either troy or apothecaries' measure equal to 480 grains or exactly 0.0311034768 kilogram. Also known as apothecaries' ounce or troy ounce (abbreviations are oz ap and oz t in the United States, and oz apoth and oz tr in the United Kingdom). { 'au̇ns }

ouncedal [MECH] A unit of force equal to the force which will impart an acceleration of 1 foot per second per second to a mass of 1 ounce; equal to 0.0086409346485 newton. { 'au̇n·sə,dal }

outer bremsstrahlung [PHYS] Bremsstrahlung involving the acceleration of a charged particle coming from outside the atom whose nucleus produces the acceleration, and in which the energy loss by radiation is much greater than that by ionization, usually seen in electrons with energies greater than about 50 MeV (million electronvolts). { 'au̇d·ər 'brem,shträ·lu̇ŋ }

outer effects [PHYS] Effects on x-ray diffraction that involve neighboring atoms or molecules. { 'au̇d·ər i'feks }

outer-shell electron *See* conduction electron. { 'au̇d·ər ¦shel i'lek,trän }

outer trapped surface [RELAT] A compact, spacelike, two-dimensional surface in a spacetime, such that outgoing light rays perpendicular to the surface are not diverging; whether ingoing light rays are converging or not is immaterial. { ¦au̇d·ər ¦trapt 'sər·fəs }

out of phase [PHYS] Having waveforms that are of the same frequency but do not pass through corresponding values at the same instant. { 'au̇t əv 'fāz }

output winding [ELECTROMAG] Of a saturable reactor, a winding, other than a feedback winding, which is associated with the load, and through which power is delivered to the load. { 'au̇t,pu̇t ,wīnd·iŋ }

overcritical binding [ATOM PHYS] A binding energy for electrons in atoms which is so large that a vacancy in the bound state results in the spontaneous formation of an electron-positron pair; predicted to occur when the atomic number exceeds 173. { 'ō·vər,krid·ə·kəl 'bīnd·iŋ }

overcritical electric field [ATOM PHYS] An electric field so strong that an electron-positron pair is created spontaneously; quantum electrodynamics predicts that this will happen near a nucleus having more than approximately 173 protons. { 'ō·vər,krid·ə·kəl i'lek·trik 'fēld }

overdamping [PHYS] Damping greater than that required for critical damping. { 'ō·vər,damp·iŋ }

overgrowth [CRYSTAL] A crystal growth in optical and crystallographic continuity around another crystal of different composition. { 'ō·vər,grōth }

Overhauser effect [ATOM PHYS] The effect whereby, if a radio frequency field is applied to a substance in an external magnetic field, whose nuclei have spin 1/2 and which has unpaired electrons, at the electron spin resonance frequency, the resulting polarization of the nuclei is as great as if the nuclei had the much larger electron magnetic moment. { 'ō·vər,hau̇z·ər i,fekt }

overheating effect [SOLID STATE] The effect whereby, under certain conditions, a superconductor can be heated above its critical temperature without losing superconductivity. { ,ō·vər'hēd·iŋ i,fekt }

overlap integral [QUANT MECH] The integral over space of the product of the wave function of a particle and the complex conjugate of the wave function of another particle. { 'ō·vər,lap 'int·ə·grəl }

overlapping orbitals [ATOM PHYS] Two orbitals (usually of electrons associated with different atoms in a molecule) for which there is a region of space where both are of appreciable magnitude. { ¦ō·vər¦lap·iŋ 'ȯrb·əd·əlz }

overpressure [FL MECH] The transient pressure, usually expressed in pounds per square inch, exceeding existing atmospheric pressure and manifested in the blast wave from an explosion. { 'ō·vər,presh·ər }

overspin [MECH] In a spin-stabilized projectile, the overstability that results when the rate of spin is too great for the particular design of projectile, so that its nose does not turn downward as it passes the summit of the trajectory and follows the descending branch. Also known as overstabilization. { 'ō·vər,spin }

overstability [PL PHYS] Condition in which the restoring forces acting on an oscillation of a plasma or other conducting fluid drive the fluid back to its equilibrium state at a speed greater than its original outward speed, resulting in continually greater oscillation. { ¦ō·vər·stə'bil·əd·ē }

overstabilization *See* overspin. { ¦ō·vər,stā·bə·lə'zā·shən }

over-the-horizon propagation *See* scatter propagation. { 'ō·vər t͟hə hə'rīz·ən ,präp·ə'gā·shən }

overtone [ACOUS] **1.** A component of a complex sound whose frequency is an integral multiple, greater than 1, of the fundamental frequency.

2. A component of a complex tone having a pitch higher than that of the fundamental pitch. [MECH] One of the normal modes of vibration of a vibrating system whose frequency is greater than that of the fundamental mode. [PHYS] A harmonic other than the fundamental component. { 'ō·vər,tōn }

overtone band [SPECT] The spectral band associated with transitions of a molecule in which the vibrational quantum number changes by 2 or more. { 'ō·vər,tōn ,band }

OW unit *See* sabin. { ¦ō'dəb·əl,yü ,yü·nət }

oxygen burning [NUC PHYS] The synthesis of nuclei in stars through reactions involving the fusion of two oxygen-16 nuclei at temperatures of about 10^9 K. { 'äk·sə·jən ,bərn·iŋ }

oxygen-18 [NUC PHYS] Oxygen isotope with atomic weight 18; found 8 parts to 10,000 of oxygen-16 in water, air, and rocks; used in tracer experiments. Also known as heavy oxygen. { 'äk·sə·jən ā'tēn }

oxygen point [THERMO] The temperature at which liquid oxygen and its vapor are in equilibrium, that is, the boiling point of oxygen, at standard atmospheric pressure; it is taken as a fixed point on the International Practical Temperature Scale of 1968, at −182.962°C. { 'äk·sə·jən ,pȯint }

oz *See* ounce.

oz ap *See* ounce.

oz apoth *See* ounce.

oz t *See* ounce.

oz tr *See* ounce.

P

P *See* poise.

pA *See* picoampere.

Pa *See* pascal.

package power reactor [NUC PHYS] A small nuclear power plant designed to be crated in packages small enough for transportation to remote locations. { 'pak·ij 'pau̇·ər rē,ak·tər }

packet *See* wave packet. { 'pak·ət }

packing [CRYSTAL] Arrangement of atoms or ions in a crystal lattice. { 'pak·iŋ }

packing fraction [NUC PHYS] The quantity (M − A)/A, where M is the mass of an atom in atomic mass units and A is its atomic number. { 'pak·iŋ ,frak·shən }

packing index [CRYSTAL] The volume of ion divided by the volume of the unit cell in a crystal. { 'pak·iŋ ,in,deks }

packing radius [CRYSTAL] One-half the smallest approach distance of atoms or ions. { 'pak·iŋ ,rād·ē·əs }

pair [ELEC] Two like conductors employed to form an electric circuit. { per }

pairing energy [NUC PHYS] An energy associated with extra stability of pairs of nucleons of the same kind, which results in nuclei with odd numbers of neutrons and protons having a lower binding energy and being less stable than nuclei with even numbers of neutrons and protons. { 'per·iŋ ,en·ər·jē }

pairing isomer [NUC PHYS] An excited nuclear state which has an unusually long lifetime because the microscopic motions of its constituent nucleons differ sharply from those of states of lower energy into which it is permitted to decay. { 'per·iŋ ,ī·sə·mər }

pair production [PHYS] The conversion of a photon into an electron and a positron when the photon traverses a strong electric field, such as that surrounding a nucleus or an electron. { 'per prə,dək·shən }

palpable coordinate [MECH] A generalized coordinate that appears explicitly in the Lagrangian of a system. { 'pal·pə·bəl kō'ȯrd·ən·ət }

panchratic eyepiece [OPTICS] A telescope eyepiece whose magnifying power can be varied by moving the erecting lens while keeping the focus at infinity.

panel methods [FL MECH] Methods used in ship design to simplify the computation of the flow of water around a ship by assuming the flow to be frictionless and by taking advantage of the linearity of the Laplace equation, which governs the velocity potential in frictionless flow, to superpose elementary solutions to the problem on panels on the hull and (in most methods) on the free surface of the water. { 'pan·əl ,meth·ədz }

panoramic [OPTICS] Pertaining to a lens or optical instrument that has a wide field of view. { ¦pan·ə¦ram·ik }

paper capacitor [ELEC] A capacitor whose dielectric material consists of oiled paper sandwiched between two layers of metallic foil. { 'pā·pər kə'pas·əd·ər }

parabolic antenna [ELECTROMAG] Antenna with a radiating element and a parabolic reflector that concentrates the radiated power into a beam. { ¦par·ə¦bäl·ik an'ten·ə }

parabolic reflector [ELECTROMAG] An antenna having a concave surface which is generated either by translating a parabola perpendicular to the plane in which it lies (in a cylindrical parabolic reflector), or rotating it about its axis of symmetry (in a paraboloidal reflector). Also known as dish. [OPTICS] *See* paraboloidal reflector. { ¦par·ə¦bäl·ik ri'flek·tər }

paraboloidal antenna *See* paraboloidal reflector. { pə¦rab·ə¦lȯid·əl an'ten·ə }

paraboloidal reflector [ELECTROMAG] An antenna having a concave surface which is a paraboloid of revolution; it concentrates radiation from a source at its focal point into a beam. Also known as paraboloidal antenna. [OPTICS] A concave mirror which is a paraboloid of revolution and produces parallel rays of light from a source located at the focus of the parabola. Also known as parabolic reflector. { pə ¦rab·ə¦lȯid·əl ri'flek·tər }

parachor [PHYS] The molecular weight of a liquid times the fourth root of its surface tension, divided by the difference between the density of the liquid and the density of the vapor in equilibrium with it; essentially constant over wide ranges of temperature. { 'par·ə,kȯr }

parahelium [ATOM PHYS] Those states of helium in which the spins of the two electrons are antiparallel, in contrast to orthohelium. Also spelled parhelium. { ,par·ə'hē·lē·əm }

parahydrogen [ATOM PHYS] Those states of hydrogen molecules in which the spins of the two nuclei are antiparallel; known as spin isomers. { ¦par·ə'hī·drə·jən }

parallax [OPTICS] The change in the apparent relative orientations of objects when viewed from different positions. { 'par·ə,laks }

parallax error [OPTICS] Error in reading an instrument employing a scale and pointer because the observer's eye and pointer are not in a line perpendicular to the plane of the scale. { 'par·ə,laks ,er·ər }

parallel [ELEC] Connected to the same pair of terminals. Also known as multiple; shunt. [PHYS] Of two or more displacements or other vectors, having the same direction. { 'par·ə,lel }

parallel axis theorem [MECH] A theorem which states that the moment of inertia of a body about any given axis is the moment of inertia about a parallel axis through the center of mass, plus the moment of inertia that the body would have about the given axis if all the mass of the body were located at the center of mass. Also known as Steiner's theorem. { 'par·ə,lel ¦ak·səs ,thir·əm }

parallel circuit [ELEC] An electric circuit in which the elements, branches (having elements in series), or components are connected between two points, with one of the two ends of each component connected to each point. { 'par·ə,lel 'sər·kət }

parallel extinction [OPTICS] Nearly total absorption of light that is propagating in an anisotropic crystal in a direction parallel to crystal outlines or traces of cleavage planes. { 'par·ə,lel ik'stiŋk·shən }

parallel flow [ELEC] Also known as loop flow. **1.** The flow of electric current from one point to another in an electric network over multiple paths, in accordance with Kirchhoff's laws. **2.** In particular, the flow of electric current through electric power systems over paths other than the contractual path. { 'par·ə,lel 'flō }

parallel growth *See* parallel intergrowth. { 'par·ə,lel 'grōth }

parallel intergrowth [CRYSTAL] Intergrowth of two or more crystals in such a way that one or more axes in each crystal are approximately parallel. Also known as parallel growth. { 'par·ə,lel 'in·tər,grōth }

parallel-plate laser [OPTICS] A laser which has two small parallel plates facing each other at a distance which is large compared with their diameters; one of them reflects light and the other is partially reflecting, so that light can bounce back and forth between the plates enough to build up a strong pulse. { 'par·ə,lel ¦plāt 'lā·zər }

parallel-plate waveguide [ELECTROMAG] Pair of parallel conducting planes used for propagating uniform circularly cylindrical waves having their axes normal to the plane. { 'par·ə,lel ¦plāt 'wāv,gīd }

parallel resonance [ELEC] Also known as antiresonance. **1.** The frequency at which the inductive and capacitive reactances of a parallel resonant circuit are equal. **2.** The frequency at which the parallel impedance of a parallel resonant circuit is a maximum. **3.** The frequency at which the parallel impedance of a parallel resonant circuit has a power factor of unity. { 'par·ə,lel 'rez·ən·əns }

parallel resonant circuit [ELEC] A circuit in which an alternating-current voltage is applied across a capacitor and a coil in parallel. Also known as antiresonant circuit. { 'par·ə,lel 'rez·ən·ənt ,sər·kət }

parallel series [ELEC] Circuit in which two or more parts are connected together in parallel to form parallel circuits, and in which these circuits are then connected together in series so that both methods of connection appear. { 'par·ə,lel 'sir·ēz }

parallel-slit interferometer [OPTICS] A type of stellar interferometer consisting of a screen with two narrow, parallel slits whose separation is adjustable, placed over the objective of a refracting telescope. { 'par·ə,lel ,slit ,in·tər·fə'räm·əd·ər }

parallel-T network [ELEC] A network used in capacitance measurements at radio frequencies, having two sets of three impedances, each in the form of the letter T, with the arms of the two T's joined to common terminals, and the source and detector each connected between two of these terminals. Also known as twin-T network. { 'par·ə,lel ¦tē 'net,wərk }

parallel-tuned circuit [ELEC] A circuit with two parallel branches, one having an inductance and a resistance in series, the other a capacitance and a resistance in series. { 'par·ə,lel ,tünd 'sər·kət }

parallel twin [CRYSTAL] A twinned crystal whose twin axis is parallel to the composition surface. { 'par·ə,lel 'twin }

paramagnetic [ELECTROMAG] Exhibiting paramagnetism. { ¦par·ə·mag'ned·ik }

paramagnetic amplifier *See* maser. { ¦par·ə·mag'ned·ik 'am·plə,fī·ər }

paramagnetic cooling *See* adiabatic demagnetization. { ¦par·ə·mag'ned·ik 'kül·iŋ }

paramagnetic crystal [ELECTROMAG] A crystal whose permeability is slightly greater than that of vacuum and is independent of the magnetic field strength. { ¦par·ə·mag'ned·ik 'krist·əl }

paramagnetic Faraday effect [OPTICS] The Faraday effect observed in paramagnetic salts at frequencies near an absorption line of the salt which is split due to splitting of the lower energy level. Also known as Becquerel effect. { ¦par·ə·mag'ned·ik 'far·ə,dā i,fekt }

paramagnetic material [ELECTROMAG] A material within which an applied magnetic field is increased by the alignment of electron orbits. { ¦par·ə·mag'ned·ik mə'tir·ē·əl }

paramagnetic relaxation [ELECTROMAG] The approach of a system, which displays paramagnetism because of electronic magnetic moments of atoms or ions, to an equilibrium or steady-state condition over a period of time, following a change in the magnetic field. { ¦par·ə·mag'ned·ik ˌrēˌlak'sā·shən }

paramagnetic resonance *See* electron paramagnetic resonance. { ¦par·ə·mag'ned·ik 'rez·ən·əns }

paramagnetic salt [ELECTROMAG] A salt whose permeability is slightly greater than that of vacuum and is independent of magnetic field strength; used in adiabatic demagnetization. { ¦par·ə·mag'ned·ik 'sȯlt }

paramagnetic spectra [SPECT] Spectra associated with the coupling of the electronic magnetic moments of atoms or ions in paramagnetic substances, or in paramagnetic centers of diamagnetic substances, to the surrounding liquid or crystal environment, generally at microwave frequencies. { ¦par·ə·mag'ned·ik 'spek·trə }

paramagnetic susceptibility [ELECTROMAG] The susceptibility of a paramagnetic substance, which is a positive number and is, in general, much smaller than unity. { ¦par·ə·mag'ned·ik səˌsep·tə'bil·əd·ē }

paramagnetism [ELECTROMAG] A property exhibited by substances which, when placed in a magnetic field, are magnetized parallel to the field to an extent proportional to the field (except at very low temperatures or in extremely large magnetic fields). { ¦par·ə'mag·nəˌtiz·əm }

parameter [CRYSTAL] Any of the axial lengths or interaxial angles that define a unit cell. [ELEC] **1.** The resistance, capacitance, inductance, or impedance of a circuit element. **2.** The value of a transistor or tube characteristic. [PHYS] A quantity which is constant under a given set of conditions, but may be different under other conditions. { pə'ram·əd·ər }

parametric acoustic array [ACOUS] A device for generating very sharp beams of sound devoid of side lobes, consisting of a source of well-collimated high-frequency sound modulated at the frequency of the sound which is to be generated. { ¦par·ə¦me·trik ə'küs·tik ə'rā }

parametric amplifier [OPTICS] A device consisting of an optically nonlinear crystal in which an optical or infrared beam draws power from a laser beam at a higher frequency and is amplified. { ¦par·ə¦me·trik 'am·pləˌfī·ər }

parametric generation [OPTICS] A process in which a single electromagnetic wave propagating in a nonlinear medium is converted to two lower-frequency waves, the sum of whose frequencies equals the frequency of the original wave. { ¦par·ə¦me·trik ˌjen·ə'rā·shən }

parametric mixing [OPTICS] In a medium possessing optical nonlinearities, the mixing of electromagnetic waves to form waves with frequencies linearly related to the frequency of incident radiation. { ¦par·ə¦me·trik 'mikˌsiŋ }

parametric oscillator [OPTICS] A device consisting of an optically nonlinear crystal surrounded by a pair of mirrors to which is applied a relatively high-frequency laser beam and a relatively low-frequency signal, resulting in a low-frequency output whose frequency can be varied, usually by varying the indices of refraction. { ¦par·ə¦me·trik 'äs·əˌlād·ər }

parapositronium [PART PHYS] The state of positronium in which the positron and electron have antiparallel spins. { ¦par·əˌpäz·ə'trō·nē·əm }

parasite [ELEC] Current in a circuit, due to some unintentional cause, such as inequalities of temperature or of composition; particularly troublesome in electrical measurements. { 'par·əˌsīt }

parasite drag [FL MECH] The portion of the total drag of an aircraft exclusive of the induced drag of the wings. { 'par·əˌsīt ˌdrag }

parasitic antenna *See* parasitic element. { ¦par·ə¦sid·ik an'ten·ə }

parasitic element [ELECTROMAG] An antenna element that serves as part of a directional antenna array but has no direct connection to the receiver or transmitter and reflects or reradiates the energy that reaches it, in a phase relationship such as to give the desired radiation pattern. Also known as parasitic antenna; parasitic reflector; passive element. { ¦par·ə¦sid·ik 'el·ə·mənt }

parasitic reflector *See* parasitic element. { ¦par·ə¦sid·ik ri'flek·tər }

parastate [ATOM PHYS] A state of a diatomic molecule in which the spins of the nuclei are antiparallel. { 'par·əˌstāt }

paraxial rays [OPTICS] Rays which are close enough to the opical axis of a system, and thus whose directions are sufficiently close to being parallel to it, so that sines of angles between the rays and the optical axis may be replaced by the angles themselves in calculations. { par'ak·sē·əl 'rāz }

paraxial trajectory [ELEC] A trajectory of a charged particle in an axially symmetric electric or magnetic field in which both the distance of the particle from the axis of symmetry and the angle between this axis and the tangent to the trajectory are small for all points on the trajectory. { par'ak·sē·əl trə'jek·trē }

parent [NUC PHYS] A radionuclide that upon disintegration yields a specified nuclide, the daughter, either directly, or indirectly as a later member of a radioactive series. [QUANT MECH] If an n-electron state is written as a sum of products of 1-electron states and $(n-1)$-electron states, the $(n-1)$-electron states are called the parents of the n-electron states. { 'per·ənt }

parfocal eyepieces [OPTICS] Eyepieces whose lower focal points lie in the same plane, so that they can be interchanged without changing the focus of the instrument with which they are used. { pär'fō·kəl 'īˌpēs·əz }

parhelium *See* parahelium. { pär'hē·lē·əm }

parity [QUANT MECH] A physical property of a wave function which specifies its behavior under an inversion, that is, under simultaneous reflection of all three spatial coordinates through the origin; if the wave function is unchanged by inversion, its parity is 1 (or even); if the function is changed only in sign, its parity is −1 (or odd).

Also known as space reflection symmetry. { 'par·əd·ē }

parity conservation *See* conservation of parity. { 'par·əd·ē ˌkän·sər'vā·shən }

parity selection rules [QUANT MECH] Rules which specify whether or not a change in parity occurs during a given type of transition of an atom, molecule, or nucleus; for example, the Laporte selection rule, or the rule that there is no parity change in an allowed β-decay transition of a nucleus. { 'par·əd·ē si'lek·shən ˌrülz }

parity transformation *See* inversion. { 'par·əd·ē ˌtranz·fər'mā·shən }

Parker-Washburn boundary [SOLID STATE] A surface which separates two regions in a solid in which the crystal axes point in different directions, and which is made up of a single array of dislocations. { 'pär·kər 'wäsh·bərn ˌbaún·drē }

Parry arcs [OPTICS] A class of halos appearing as faintly colored arcs above and below the sun; these refraction phenomena are produced by ice crystals which exhibit a preferred orientation, and are correspondingly more unusual than those associated with randomly oriented crystals. { 'par·ē ˌärks }

partial [ACOUS] Also known as partial tone. **1.** A simple sinusoidal physical component of a complex tone. **2.** A sound sensation component that is distinguishable as a simple tone, cannot be further analyzed by the ear, and contributes to the character of the complex sound; the frequency of a partial may be higher or lower than the basic frequency and may be an integral multiple or submultiple of the basic frequency. { 'pär·shəl }

partial Cauchy surface [RELAT] A spacelike surface S which is intersected only once by each timelike or null curve; "partial" means that only a portion of the future history of the space-time can be predicted from S, that is, there exists a Cauchy horizon. { ¦pär·shəl 'kō·shē ˌsər·fəs }

partial coherence [PHYS] Property of two waves whose relative phase undergoes random fluctuations which are not, however, sufficient to make the wave completely incoherent. { 'pär·shəl kō'hir·əns }

partial dislocation [CRYSTAL] The line at the edge of an extended dislocation where a slip through a fraction of a lattice constant has occurred. { 'pär·shəl ˌdis·lō'kā·shən }

partial node [PHYS] That part (a point, line, or surface) of a standing wave where some characteristic of the wave field has a minimum amplitude other than zero. { 'pär·shəl 'nōd }

partial pressure [PHYS] The pressure that would be exerted by one component of a mixture of gases if it were present alone in a container. { 'pär·shəl 'presh·ər }

partial tone *See* partial. { 'pär·shəl 'tōn }

partial wetting [FL MECH] The situation in which the contact angle between a solid and a liquid is greater than zero but less than 90°. { ¦pär·shəl 'wed·iŋ }

particle [MECH] *See* material particle. [PART PHYS] *See* elementary particle. [PHYS] **1.** Any very small part of matter, such as a molecule, atom, or electron. Also known as fundamental particle. **2.** Any relatively small subdivision of matter, ranging in diameter from a few angstroms (as with gas molecules) to a few millimeters (as with large raindrops). { 'pärd·ə·kəl }

particle beam [PHYS] A concentrated, nearly unidirectional flow of particles. { 'pärd·ə·kəl ˌbēm }

particle derivative [FL MECH] The rate of change of a quantity with respect to time, measured at a point that moves along with a particle of a fluid. { ¦pard·ə·kəl də'riv·ə·div }

particle displacement velocimetry [FL MECH] A method of flow-field measurement in which a two-dimensional sheet of a flow field seeded with fluorescent particles is illuminated by a pulsed laser and particle displacements are recorded by a camera. { ¦pär·də·kəl di¦plās·mənt ˌvel·ə'sim·ə·trē }

particle distribution function [STAT MECH] A function whose value is the number of particles per unit volume of phase space. { 'pärd·ə·kəl ˌdi·strə'byü·shən ˌfəŋk·shən }

particle dynamics [MECH] The study of the dependence of the motion of a single material particle on the external forces acting upon it, particularly electromagnetic and gravitational forces. { 'pärd·ə·kəl dīˌnam·iks }

particle emission [NUC PHYS] The ejection of a particle other than a photon from a nucleus, in contrast to gamma emission. { 'pärd·ə·kəl iˌmish·ən }

particle energy [MECH] For a particle in a potential, the sum of the particle's kinetic energy and potential energy. [RELAT] For a relativistic particle the sum of the particle's potential energy, kinetic energy, and rest energy; the last is equal to the product of the particle's rest mass and the square of the speed of light. { 'pärd·ə·kəl ˌen·ər·jē }

particle flow [FL MECH] The transport of particles in fluids and gases. { 'pärd·i·kəl ˌflō }

particle horizon [RELAT] The spatial boundary beyond which, in certain universe models, it is impossible for an observer at a given time to receive a signal. { 'pärd·ə·kəl həˌrīz·ən }

particle lens [PHYS] An electric or magnetic field, or a combination thereof, which acts upon an electron beam in a manner analogous to that in which an optical lens acts upon a light beam. { 'pärd·ə·kəl ˌlenz }

particle mechanics [MECH] The study of the motion of a single material particle. { 'pärd·ə·kəl miˌkan·iks }

particle multiplet *See* isospin multiplet. { 'pärd·ə·kəl 'məl·tə·plət }

particle physics [PHYS] The branch of physics concerned with understanding the properties and behavior of elementary particles, especially through study of collisions or decays involving energies of hundreds of megaelectronvolts or more. Also known as high-energy physics. { 'pärd·ə·kəl ¦fiz·iks }

particle properties [PART PHYS] The various

quantities which characterize the behavior of an elementary particle, such as mass, charge, baryon number, spin, parity, hypercharge, and isospin. { 'pärd·ə·kəl ˌpräp·ərd·ēz }

particle spectrum *See* mass spectrum. { 'pärd·ə·kəl ˌspek·trəm }

particle track [PHYS] Any visible phenomenon along the path of an ionizing particle, such as a trail of bubbles, water droplets, or sparks in a bubble chamber, cloud chamber, or spark chamber respectively, or of altered material in an emulsion or in glass. { 'pärd·ə·kəl ˌtrak }

particle trap [PHYS] A device used to confine particles, either charged or neutral, in situations where the interaction of the particles with the wall of a container must be avoided. { 'pärd·ə·kəl ˌtrap }

particle velocity [ACOUS] The instantaneous velocity of a given infinitesimal part of a medium, with reference to the medium as a whole, due to the passage of a sound wave. { 'pärd·ə·kəl vəˌläs·əd·ē }

particulate matter [PHYS] matter in the form of small liquid or solid particles. { pär'tik·yə·lət ˌmad·ər }

partition function [STAT MECH] **1.** The integral, over the phase space of a system, of the exponential of $(-E/kT)$, where E is the energy of the system, k is Boltzmann's constant, and T is the temperature; from this function all the thermodynamic properties of the system can be derived. **2.** In quantum statistical mechanics, the sum over allowed states of the exponential of $(-E/kT)$. Also known as sum of states; sum over states. { pär'tish·ən ˌfəŋk·shən }

parton [PART PHYS] One of the very singular (or hard), small charged particles of which hadrons are proposed to be constructed, according to a theory developed to account for the scattering of very-high-energy electrons from protons at large angles and with large momentum transfers. { 'pärˌtän }

parylene capacitor [ELEC] A highly stable fixed capacitor using parylene film as the dielectric; it can be operated at temperatures up to 170°C, as well as at cryogenic temperatures. { 'par·əˌlēn kə'pas·əd·ər }

PAS *See* photoacoustic spectroscopy.

pascal [MECH] A unit of pressure equal to the pressure resulting from a force of 1 newton acting uniformly over an area of 1 square meter. Symbolized Pa. { pa'skal }

Pascal's law [FL MECH] The law that a confined fluid transmits externally applied pressure uniformly in all directions, without change in magnitude. { pa'skalz ˌlȯ }

Paschen-Back effect [SPECT] An effect on spectral lines obtained when the light source is placed in a very strong magnetic field; the anomalous Zeeman effect obtained with weaker fields changes over to what is, in a first approximation, the normal Zeeman effect. { 'päsh·ən 'bäk iˌfekt }

Paschen-Runge mounting [SPECT] A diffraction grating mounting in which the slit and grating are fixed, and photographic plates are clamped to a fixed track running along the corresponding Rowland circle. { 'päsh·ən 'rüŋ·ə ˌmau̇nt·iŋ }

Paschen series [SPECT] A series of lines in the infrared spectrum of atomic hydrogen whose wave numbers are given by $R_H [(1/9) - (1/n^2)]$, where R_H is the Rydberg constant for hydrogen, and n is any integer greater than 3. { 'päsh·ən ˌsir·ēz }

passive antenna [ELECTROMAG] An antenna which influences the directivity of an antenna system but is not directly connected to a transmitter or receiver. { 'pas·iv an'ten·ə }

passive component *See* passive element. { 'pas·iv kəm'pō·nənt }

passive corner reflector [ELECTROMAG] A corner reflector that is energized by a distant transmitting antenna; used chiefly to improve the reflection of radar signals from objects that would not otherwise be good radar targets. { 'pas·iv 'kȯr·nər riˌflek·tər }

passive double reflector [ELECTROMAG] A combination of two passive reflectors positioned to bend a microwave beam over the top of a mountain or ridge, generally without appreciably changing the general direction of the beam. { 'pas·iv 'dəb·əl ri'flek·tər }

passive element [ELEC] An element of an electric circuit that is not a source of energy, such as a resistor, inductor, or capacitor. Also known as passive component. *See* parasitic element. { 'pas·iv 'el·ə·mənt }

passive junction [ELECTROMAG] A waveguide junction that does not have a source of energy. { 'pas·iv 'jəŋk·shən }

passive network [ELEC] A network that has no source of energy. { 'pas·iv 'netˌwərk }

passive reflector [ELECTROMAG] A flat reflector used to change the direction of a microwave or radar beam; often used on microwave relay towers to permit placement of the transmitter, repeater, and receiver equipment on the ground, rather than at the tops of towers. Also known as plane reflector. { 'pas·iv ri'flek·tər }

past [RELAT] For an event in space-time, all events from which a signal could be emitted that could reach the event in question by traveling at speeds less than or equal to the speed of light. { 'past }

past Cauchy development [RELAT] The set of points p relative to a surface S in space-time such that every future-directed timelike or null curve through p intersects S. Symbolized $D^-(S)$. { ¦past 'kō·shē diˌvel·əp·mənt }

past light cone [RELAT] The set of all points in space-time from which signals traveling at the speed of light reach a specified point. { 'past ¦līt ˌkōn }

path integral [QUANT MECH] An integral of a functional over function space; central to a formulation of quantum mechanics developed by R. Feynman. { 'path ˌint·ə·grəl }

Patterson function [SOLID STATE] A function of three spatial coordinates, constructed in the Patterson-Harker method, which has peaks at all vectors between two atoms in a crystal, the heights of the peaks being approximately proportional to the product of the atomic numbers of the corresponding atoms. { 'pad·ər·sən ˌfəŋk·shən }

Patterson-Harker method [SOLID STATE] A method of analyzing the structure of a crystal from x-ray diffraction results; a Fourier series involving squares of the absolute values of the structure factors, which are directly observable, is used to construct a vectorial representation of interatomic distances in the crystal (Patterson map). { 'pad·ər·sən 'här·kər ˌmeth·əd }

Patterson map [SOLID STATE] A contour chart of the Patterson function. { 'pad·ər·sən ˌmap }

Patterson projection [SOLID STATE] A projection of the Patterson function on a section through a crystal. { 'pad·ər·sən prəˌjek·shən }

Patterson vectors [SOLID STATE] In analysis of crystal structure, the vectors of peaks relative to the origin in a Patterson function or Patterson projection. { 'pad·ər·sən ˌvek·tərz }

Pauli anomalous moment term [QUANT MECH] An additional term inserted in the Dirac equation to provide for a *g*-value of the particle different from 2. { 'pȯl·ē ə'näm·ə·ləs 'mō·mənt ˌtərm }

Pauli electron correlation [QUANT MECH] Correlation in space of electrons as a result of the Pauli exclusion principle. { 'pȯl·ē i'lekˌträn ˌkär·əˌlā·shən }

Pauli-Fermi principle [QUANT MECH] The principle that each level of a quantized system can include one, two, or no electrons; if there are two electrons, they must have spins in opposite directions. { 'pȯl·ē 'fer·mē ˌprin·sə·pəl }

Pauli g-permanence rule [ATOM PHYS] For given L, S, and M_J in LS coupling, the sum, over J, of the weak-field *g*-factors is equal to the sum of the strong-field factors. { 'pȯl·ē ¦jē 'pər·mə·nəns ˌrül }

Pauli g-sum rule [ATOM PHYS] For all the states arising from a given electron configuration, the sum of the *g*-factors for levels with the same J value is a constant, independent of the coupling scheme. { 'pȯl·ē ¦jē ˌsəm ˌrül }

Pauling rule [SOLID STATE] A rule governing the number of ions of opposite charge in the neighborhood of a given ion in an ionic crystal, in accordance with the requirement of local electrical neutrality of the structure. { 'pȯl·iŋ ˌrül }

Pauli paramagnetism *See* free-electron paramagnetism. { ¦pau̇·lē ˌpar·ə'mag·nəˌtiz·əm }

Pauli spin matrices [QUANT MECH] Three anticommuting matrices, each having two rows and two columns, which represent the components of the electron spin operator:

$$\sigma_x = \begin{pmatrix} 0 & 1 \\ 1 & 0 \end{pmatrix} - \sigma_y = \begin{pmatrix} 0 & i \\ 1 & 0 \end{pmatrix} - \sigma_z = \begin{pmatrix} 1 & 0 \\ 0 & -1 \end{pmatrix}$$

{ 'pȯl·ē 'spin ˌmā·trə·sēz }

Pauli spin space [QUANT MECH] A two-dimensional vector space over the complex numbers, whose vectors describe orientations of the electron spin. { 'pȯl·ē 'spin 'spās }

Pauli spin susceptibility [SOLID STATE] The susceptibility of free electrons in a metal due to the tendency of their spins to align with a magnetic field. { 'pȯl·ē ¦spin səˌsep·tə'bil·əd·ē }

Pauli-Weisskopf equation [QUANT MECH] The equation resulting from second quantization of the Klein-Gordon equation. { 'pȯl·ē 'vīsˌkȯpf iˌkwā·zhən }

Paul trap [PHYS] A device in which ions and other charged particles can be suspended by radio-frequency electric fields having a quadrupole configuration, for times limited chiefly by collisions with the background gas. { 'pȯl ˌtrap }

Pb-I-Pb junction *See* lead-I-lead junction.

P-branch [SPECT] A series of lines in molecular spectra that correspond, in the case of absorption, to a unit decrease in the rotational quantum number J. { 'pē ˌbranch }

PD *See* potential difference.

pdl-ft *See* foot-poundal.

PDMS *See* plasma desorption mass spectrometry.

peak amplitude [PHYS] The maximum amplitude of an alternating quantity, measured from its zero value. { 'pēk 'am·pləˌtüd }

peak analysis [SPECT] Determination of the relevant peak parameters, such as position or area, from a spectogram. { 'pēk əˌnal·ə·səs }

peak effect [SOLID STATE] In certain hard superconductors, the occurrence of a maximum in the value of the critical current as the external magnetic field is varied, near the critical magnetic field. { 'pēk iˌfekt }

peak factor *See* crest factor. { 'pēk ˌfak·tər }

peak-to-peak amplitude [PHYS] Amplitude of an alternating quantity measured from positive peak to negative peak. { ¦pēk tə 'pēk 'am·pləˌtüd }

peak value [ELEC] The maximum instantaneous value of a varying current, voltage, or power during the time interval under consideration. Also known as crest value. { 'pēk 'val·yü }

pearly *See* nacreous. { 'pər·lē }

peck [MECH] Abbreviated pk. **1.** A unit of volume used in the United States for measurement of solid substances, equal to 8 dry quarts, or 1/4 bushel, or 537.605 cubic inches, or 0.00880976754172 cubic meter. **2.** A unit of volume used in the United Kingdom for measurement of solid and liquid substances, although usually the former, equal to 2 gallons, or 0.00909218 cubic meter. { pek }

pedial class [CRYSTAL] That class in the triclinic system which has no symmetry. { 'ped·ē·əl ˌklas }

pedion [CRYSTAL] A crystal form with only one face; member of the asymmetric class of the triclinic system. { 'ped· ē·ən }

Peierls-Nabarro force [SOLID STATE] The force required to displace a dislocation along its slip plane. { 'pā·ərlz nə'bär·ō ˌfȯrs }

p electron [ATOM PHYS] In the approximation that each electron has a definite central-field wave function, an atomic electron that has an orbital angular momentum quantum number of unity. { 'pē i,lek·trän }

Peltier coefficient [PHYS] The ratio of the rate at which heat is evolved or absorbed at a junction of two metals in the Peltier effect to the current passing through the junction. { pel'tyā ,kō·i,fish·ənt }

Peltier effect [PHYS] The phenomenon in which heat is evolved or absorbed at the junction of two dissimilar metals carrying a small current, depending upon the direction of the current. { pel'tyā i,fekt }

pencil [OPTICS] A bundle of rays that emanate from or converge to a common point. { 'pen·səl }

pencil beam [ELECTROMAG] A beam of radiant energy concentrated in an approximately conical or cylindrical portion of space of relatively small diameter; this type of beam is used for many revolving navigational lights and radar beams. { 'pen·səl ,bēm }

pencil beam antenna [ELECTROMAG] Unidirectional antenna designed so that cross sections of the major lobe formed by planes perpendicular to the direction of maximum radiation are approximately circular. { 'pen·səl ,bēm an ,ten·ə }

pendant-drop method [PHYS] Method for the measurement of liquid surface tension by the elongation of a hanging drop of the liquid. { 'pen·dənt ,dräp ,meth·əd }

pendulous gyroscope [MECH] A gyroscope whose axis of rotation is constrained by a suitable weight to remain horizontal; it is the basis of one type of gyrocompass. { 'pen·jə·ləs 'jī·rə,skōp }

pendulum [PHYS] A rigid body mounted on a fixed horizontal axis, about which it is free to rotate under the influence of gravity. Also known as compound pendulum; gravity pendulum. { 'pen·jə·ləm }

pendulum day [PHYS] The time required for the plane of a freely suspended (Foucault) pendulum to complete an apparent rotation about the local vertical. { 'pen·jə·ləm ,dā }

penetrating shower [NUC PHYS] A cosmic-ray shower, consisting mainly of muons, that can penetrate 6 to 8 inches (15 to 20 centimeters) of lead. { 'pen·ə,trād·iŋ 'shau̇·ər }

penetration ballistics [MECH] A branch of terminal ballistics concerned with the motion and behavior of a missile during and after penetrating a target. { ,pen·ə'trā·shən bə,lis·tiks }

penetration depth [CRYO] The depth beneath the surface of superconductor in a magnetic field at which the magnetic field strength has fallen to $1/e$ of its value at the surface. [ELEC] In induction heating, the thickness of a layer, extending inward from a conductor's surface, whose resistance to direct current equals the resistance of the whole conductor to alternating current of a given frequency. { ,pen·ə'trā·shən ,depth }

penetration probability [QUANT MECH] The probability that a particle will pass through a potential barrier, that is, through a finite region in which the particle's potential energy is greater than its total energy. Also known as transmission coefficient. { ,pen·ə'trā·shən ,präb·ə,bil·əd·ē }

penetration range *See* night visual range. { ,pen·ə'trā·shən ,rānj }

penetration twin *See* interpenetration twin. { ,pen·ə'trā·shən ,twin }

Penning ionization [ATOM PHYS] The ionization of gas atoms or molecules in collisions with metastable atoms. { 'pen·iŋ ,ī·ə·nə'zā·shən }

pennyweight [MECH] A unit of mass equal to 1/20 troy ounce or to 1.55517384 grams; the term is employed in the United States and in England for the valuation of silver, gold, and jewels. Abbreviated dwt; pwt. { 'pen·ē,wāt }

Penrose diagram [RELAT] A diagram of a space-time where the causal and infinity structure is displayed through the use of conformal transformations. Also known as conformal diagram. { 'pen,rōz ,dī·ə,gram }

Penrose-Hawking theorems [RELAT] The general relativistic theorems proving that singularities must occur in space-times, such as the universe, based on reasonable assumptions such as causality and dependent on the existence of a trapped surface. { ¦pen,rōz 'hȯk·iŋ ,thir·əmz }

Penrose process [RELAT] A hypothetical means of extracting energy from a rotating black hole in which a particle spirals into the ergosphere of the black hole in a direction opposite to the black hole's rotation and then breaks up into two fragments, one of which escapes with an energy greater than the energy of the original particle. { 'pen,rōz ,prä·səs }

Penrose theorem [RELAT] A theorem which states that a collapsing object whose radius is smaller than its gravitational radius must collapse into a singularity. { 'pen,rōz ,thir·əm }

pentagonal dodecahedron *See* pyritohedron. { pen'tag·ən·əl dō·dek·ə'hē·drən }

pentane candle [OPTICS] A unit of luminous intensity equal to one-tenth of the luminous intensity of a standard pentane lamp, and approximately equal to 1 candela. { 'pen,tēn 'kand·əl }

penumbra [OPTICS] That portion of a shadow illuminated by only part of a radiating source. { pə'nəm·brə }

perceived noise decibel [ACOUS] A unit of perceived noise level. Abbreviated PNdB. { pər'sēvd ¦nȯiz 'des·ə,bel }

perceived noise level [ACOUS] In perceived noise decibels, the noise level numerically equal to the sound pressure level, in decibels, of a band of random noise of width one-third to one octave centered on a frequency of 1000 hertz which is judged by listeners to be equally noisy. { pər'sēvd ¦nȯiz ,lev·əl }

perch [MECH] Also known as pole; rod. **1.** A unit of length, equal to 5.5 yards, or 16.5 feet, or 5.0292 meters. **2.** A unit of area, equal to 30.25 square yards, or 272.25 square feet, or 25.29285264 square meters. { pərch }

percolation limit [SOLID STATE] In a disordered crystalline alloy having one constituent with a magnetic moment, the concentration of the magnetic element above which the spin-glass phase is replaced by the ferromagnetic state. { pər·kə'lā·shən ,lim·ət }

percussion figure [CRYSTAL] Radiating lines on a crystal section produced by a sharp blow. { pər'kəsh·ən ,fig·yər }

perfect crystal [CRYSTAL] A crystal without lattice defects; it is an unattained ideal or standard. { 'pər·fikt 'krist·əl }

perfect dielectric *See* ideal dielectric. { 'pər·fikt ,dī·ə'lek·trik }

perfect fluid *See* inviscid fluid. { 'pər·fikt 'flü·əd }

perfect gas *See* ideal gas. { 'pər·fikt 'gas }

perfectly diffuse radiator [OPTICS] A body that emits radiant energy in accordance with Lambert's law. { 'pər·fik·lē di'fyüs 'rād·ē,ād·ər }

perfectly diffuse reflector [OPTICS] A body that reflects radiant energy in such a manner that the reflected energy may be treated as if it were being emitted (radiated) in accordance with Lambert's law. { 'pər·fik·lē di'fyüs ri'flek·tər }

perfectly inelastic collision [PHYS] A collision in which as much translational kinetic energy is converted into internal energy of the colliding systems as is consistent with the conservation of momentum. Also known as completely inelastic collision. { 'pər·fik·lē ,in·i'las·tik kə 'lizh·ən }

perfect vacuum *See* absolute vacuum. { 'pər·fikt 'vak·yəm }

pericenter [PHYS] That point on any orbit nearest to the center of attraction. { 'per·ə,sen·tər }

pericline twin law [CRYSTAL] A parallel twin law in triclinic feldspars, in which the *b* axis is the twinning axis and the composition surface is a rhombic section. { 'per·ə,klīn 'twin ,lȯ }

period [PHYS] The duration of a single repetition of a cyclic phenomenon. { 'pir·ē·əd }

period doubling [PHYS] A scenario for the transition of a natural process from regular periodic motion to chaos, in which the time required for the motion of the system to repeat itself doubles again and again as a parameter describing the system is increased. { 'pir·ē·əd ,dəb·liŋ }

periodic antenna [ELECTROMAG] An antenna in which the input impedance varies as the frequency is altered. { ¦pir·ē¦äd·ik an'ten·ə }

periodic attractor [PHYS] A finite sequence of points in phase space such that certain orbits approach each of them in succession, coming closer on each approach. { ,pir·ē,äd·ik ə'frak·tər }

periodic damping [PHYS] Damping which is less than critical damping. { ¦pir·ē¦äd·ik 'damp·iŋ }

periodic lattice *See* lattice. { ¦pir·ē¦äd·ik 'lad·əs }

periodic motion [MECH] Any motion that repeats itself identically at regular intervals. { ¦pir·ē¦äd·ik 'mō·shən }

periodic quantity [PHYS] Oscillating quantity, the values of which recur for equal increments of the independent variable. { ¦pir·ē¦äd·ik 'kwän·əd·ē }

periodic wave [PHYS] A wave whose displacement has a periodic variation with time or distance, or both. { ¦pir·ē¦äd·ik 'wāv }

period of vibration [PHYS] The time for one complete cycle of a vibration. { 'pir·ē·əd əv vī'brā·shən }

periscope [OPTICS] **1.** An optical instrument used to provide a raised line of vision where it may not be practical or possible, as in entrenchments, tanks, or submarines; the raised line of vision is obtained by the use of mirrors or prisms within the structure of the item; it may have single or dual optical systems. **2.** A thin astigmatic lens which approximates a meniscus shape and has a base curve of ± 1.25 diopters. { 'per·ə,skōp }

permanent axis [MECH] The axis of the greatest moment of inertia of a rigid body, about which it can rotate in equilibrium. { 'pər·mə·nənt 'ak·səs }

permanent echo [ELECTROMAG] A signal reflected from an object that is fixed with respect to a radar site. { 'pər·mə·nənt 'ek·ō }

permanent gas [THERMO] A gas at a pressure and temperature far from its liquid state. { 'pər·mə·nənt 'gas }

permanent magnet [ELECTROMAG] A piece of hardened steel or other magnetic material that has been strongly magnetized and retains its magnetism indefinitely. Abbreviated PM. { 'pər·mə·nənt 'mag·nət }

permanent set [MECH] Permanent plastic deformation of a structure or a test piece after removal of the applied load. Also known as set. { 'pər·mə·nənt 'set }

permanent wave [FL MECH] A wave (in a fluid) which moves with no change in streamline pattern, and which, therefore, is a stationary wave relative to a coordinate system moving with the wave. { 'pər·mə·nənt 'wāv }

permeability [ELECTROMAG] A factor, characteristic of a material, that is proportional to the magnetic induction produced in a material divided by the magnetic field strength; it is a tensor when these quantities are not parallel. Also known as magnetic permeability. [FL MECH] **1.** The ability of a membrane or other material to permit a substance to pass through it. **2.** Quantitatively, the amount of substance which passes through the material under given conditions. { ,pər·mē·ə'bil·əd·ē }

permeability coefficient [FL MECH] The rate of water flow in gallons per day through a cross section of 1 square foot under a unit hydraulic gradient, at the prevailing temperature or at 60°F (16°C). Also known as coefficient of permeability; hydraulic conductivity; Meinzer unit. { ,pər·mē·ə'bil·əd·ē ,kō·i,fish·ənt }

permeance [ELECTROMAG] A characteristic of a portion of a magnetic circuit, equal to magnetic flux divided by magnetomotive force; the reciprocal of reluctance. Symbolized P. { 'pər·mē·əns }

permittivity [ELEC] The dielectric constant

multiplied by the permittivity of empty space, where the permittivity of empty space (ϵ_0) is a constant appearing in Coulomb's law, having the value of 1 in centimeter-gram-second electrostatic units, and of 8.854×10^{-12} farad/meter in rationalized meter-kilogram-second units. Symbolized ϵ. { ,pər·mə'tiv·əd·ē }

perpendicular axis theorem [MECH] A theorem which states that the sum of the moments of inertia of a plane lamina about any two perpendicular axes in the plane of the lamina is equal to the moment of inertia about an axis through their intersection perpendicular to the lamina. { ¦pər·pən¦dik·yə·lər 'ak·səs ,thir·əm }

perpetual motion machine of the first kind [PHYS] A mechanism which, once set in motion, continues to do useful work without an input of energy, or which produces more energy than is absorbed in its operation; it violates the principle of conservation of energy. { pər'pech·ə·wəl 'mō·shən mə,shēn əv thə 'fərst ,kīnd }

perpetual motion machine of the second kind [PHYS] A device that extracts heat from a source and then converts this heat completely into other forms of energy; it violates the second law of thermodynamics. { pər'pech·ə·wəl 'mō·shən mə,shēn əv thə 'sek·ənd ,kīnd }

perpetual motion machine of the third kind [PHYS] A device which has a component that can continue moving forever; an example is a superconductor. { pər'pech·ə·wəl 'mō·shən mə,shēn əv thə 'thərd ,kīnd }

persistent current [CRYO] **1.** A magnetically induced electric current that flows undiminished in a superconducting material or circuit. **2.** A superfluid current that flows undiminished around a closed path. { pər'sis·tənt 'kə·rənt }

persistent spectral holeburning [OPTICS] A process in which radiation from a narrow-band laser source sharply reduces the absorption of a solid within a frequency range that is much smaller that the linewidth of an inhomogeneously broadened transition. { pər'sis·tənt 'spek·trəl 'hōl,bərn·iŋ }

perturbation [PHYS] Any effect which makes a small modification in a physical system, especially in case the equations of motion could be solved exactly in the absence of this effect. { ,pər·tər'bā·shən }

perturbation equation [PHYS] Any equation governing the behavior of a perturbation; often this will be a linear differential equation. { ,pər·tər'bā·shən i,kwā·zhən }

perturbation motion [PHYS] The motion of a disturbance (usually but not necessarily assumed infinitesimal), as opposed to the motion of the system on which the perturbation is superimposed. { ,pər·tər'bā·shən ,mō·shən }

perturbation quantity [PHYS] Any characteristic of a system which may be assumed to be a perturbation from an established value. { ,pər·tər'bā·shən ,kwän·əd·ē }

perturbation theory [PHYS] The theory of obtaining approximate solutions to the equations of motion of a physical system when these equations differ by a small amount from equations which can be solved exactly. { ,pər·tər'bā·shən ,thē·ə·rē }

perversion *See* lateral inversion. { pər'vər·zhən }

petrographic microscope [OPTICS] A polarizing microscope used for analysis of petrographic thin sections. { ¦pe·trə¦graf·ik 'mī·krə,skōp }

Petrov classification [RELAT] An algebraic classification of space-times based on eigenvalues of the curvature tensor. { 'pe,träv ,klas·ə·fə,kā·shən }

Petzval condition [OPTICS] The condition where by an optical system will eliminate the aberration of curvature of field only if the Petzval curvature vanishes. { 'pets·väl kən,dish·ən }

Petzval curvature [OPTICS] The axial curvature of the image of a plane object produced by an optical system, equal to the sum over all the optical surfaces in the system of $R(1/n' - 1/n)$, where R is the curvature of the surface, and n and n' are the refractive indices before and after the surface. Also known as Petzval sum. { 'pets·väl 'kər·və·chər }

Petzval lens [OPTICS] A photographic objective which consists of four lenses ordered in two pairs widely separated from each other, with the first pair cemented together and the second usually having a small air space. { 'pets·väl ,lenz }

Petzval sum *See* Petzval curvature. { 'pets·väl ,səm }

Petzval surface [OPTICS] A paraboloidal surface on which point images of point objects are formed by a doublet lens whose separation is such that astigmatism is eliminated. { 'pets·väl ,sər·fəs }

pf *See* power factor.

pF *See* picofarad.

PFE *See* photoferroelectric effect.

Pfund series [SPECT] A series of lines in the infrared spectrum of atomic hydrogen whose wave numbers are given by $R_H[(1/25) - (1/n^2)]$, where R_H is the Rydberg constant for hydrogen, and n is any integer greater than 5. { 'fünt ,sir·ēz }

phantom crystal [CRYSTAL] A crystal containing an earlier stage of crystallization outlined by dust, minute inclusions, or bubbles. Also known as ghost crystal. { 'fan·təm 'krist·əl }

phase [PHYS] **1.** The fractional part of a period through which the time variable of a periodic quantity (alternating electric current, vibration) has moved, as measured at any point in time from an arbitrary time origin; usually expressed in terms of angular measure, with one period being equal to 360° or 2π radians. **2.** For a sinusoidally varying quantity, the phase (first definition) with the time origin located at the last point at which the quantity passed through a zero position from a negative to a positive direction. **3.** The argument of the trigonometric function describing the space and time variation of a sinusoidal disturbance, $y = A \cos [(2\pi/\lambda)(x - vt)]$, where x and t are the space and time coordinates, v is the velocity of propagation, and λ is the wavelength. [THERMO] The type of state of a system, such as solid, liquid, or gas. { fāz }

phase angle [PHYS] The difference between the phase of a sinusoidally varying quantity and the phase of a second quantity which varies sinusoidally at the same frequency. Also known as phase difference. { 'fāz ,aŋ·gəl }

phase boundary [PHYS] The interface between two or more separate phases, such as liquid-gas, liquid-solid, gas-solid, or, for immiscible materials, liquid-liquid or solid-solid. { 'fāz ,bau̇n·drē }

phase change [PHYS] The metamorphosis of a material or mixture from one phase to another, such as gas to liquid, solid to gas. { 'fāz ,chānj }

phase-change coefficient *See* phase constant. { ¦fāz ,chānj ,kō·i,fish·ənt }

phase coherence [PHYS] The existence of a statistical or time coherence between the phases of two or more waves. { 'fāz kō,hir·əns }

phase conjugate system [OPTICS] An adaptive optics system in which the wavefront to be corrected is measured directly, using either a geometric or interferometric test. { 'fāz ¦kän·jə·gət ,sis·təm }

phase constant [ELECTROMAG] A rating for a line or medium through which a plane wave of a given frequency is being transmitted; it is the imaginary part of the propagation constant, and is the space rate of decrease of phase of a field component (or of the voltage or current) in the direction of propagation, in radians per unit length. Also known as phase-change coefficient; wavelength constant. { 'fāz ,kän·stənt }

phase-contrast microscope [OPTICS] A compound microscope that has an annular diaphragm in the front focal plane of the substage condenser and a phase plate at the rear focal plane of the objective, to make visible differences in phase or optical path in transparent or reflecting media. { 'fāz ¦kän,trast 'mī·krə,skōp }

phased array [ELECTROMAG] An array of dipoles on a radar antenna in which the signal feeding each dipole is varied so that antenna beams can be formed in space and scanned very rapidly in azimuth and elevation. { 'fāzd ə'rā }

phase diagram [THERMO] **1.** A graph showing the pressures at which phase transitions between different states of a pure compound occur, as a function of temperature. **2.** A graph showing the temperatures at which transitions between different phases of a binary system occur, as a function of the relative concentrations of its components. { 'fāz ,dī·ə,gram }

phase difference *See* phase angle. { 'fāz ,dif·rəns }

phase factor [ELEC] *See* power factor. [SOLID STATE] The argument (phase) of a structure factor; it cannot be directly observed. { 'fāz ,fak·tər }

phase front [PHYS] A surface of constant phase (or phase angle) of a propagating wave disturbance. { 'fāz ,frənt }

phase function [OPTICS] The angular distribution of light reflected from an object when it is illuminated by light from a specified direction. { 'fāz ,fəŋk·shən }

phase integral *See* action. { 'fāz ¦int·ə·grəl }

phase integral method *See* Wentzel-Kramers-Brillouin method. { 'fāz ¦int·ə·grəl ,meth·əd }

phase lag *See* lag angle. { 'fāz ,lag }

phase lead *See* lead angle. { 'fāz ,lēd }

phase matching [OPTICS] A condition in which the polarization wave produced by two or more beams of incident radiation in a nonlinear medium has the same phase velocity as a freely propagating wave of the same frequency; the amplitude of the polarization wave is then greatly enhanced. { 'fāz ,mach·iŋ }

phase plate [OPTICS] In a polarizing microscope, a plate of doubly refracting material that changes the relative phase of the polarized light's components. { 'fāz ,plāt }

phase problem [CRYSTAL] The problem that arises in determining the electron density function of a crystal from x-ray diffraction data, namely that a complete determination requires knowledge of both the magnitudes and phases of the structure factors, but experimental measurements yield only the magnitudes. { 'fāz ,präb·ləm }

phaser [ELECTROMAG] Microwave ferrite phase shifter employing a longitudinal magnetic field along one or more rods of ferrite in a waveguide. { 'fāz·ər }

phase resonance [PHYS] The frequency at which the angular phase difference between the fundamental components of an oscillation and of the applied agency is 90° ($\pi/2$ radians). Also known as velocity resonance. { 'fāz ,rez·ən·əns }

phase reversal [PHYS] A change of 180°, or one half-cycle, in phase. { 'fāz ri,vər·səl }

phase shift [PHYS] **1.** A change in the phase of a periodic quantity. Also known as phase change. **2.** A change in the phase angle between two periodic quantities. [QUANT MECH] For a partial wave of a particle scattered by a spherically symmetric potential, the phase shift is the difference between the phase of the wave function far from the scatterer and the corresponding phase of a free particle. { 'fāz ,shift }

phase space [STAT MECH] For a system with n degrees of freedom, a euclidean space with $2n$ dimensions, one dimension for each of the generalized coordinates and one for each of the corresponding momenta. { 'fāz ,spās }

phase speed *See* phase velocity. { 'fāz ,spēd }

phase transformation [ELEC] A change of polyphase power from three-phase to six-phase, from three-phase to twelve-phase, and so forth, by use of transformers. *See* phase transition. { 'fāz ,tranz·fər,mā·shən }

phase transition [PHYS] A change of a substance from one phase to another. Also known as phase transformation. { 'fāz tran,zish·ən }

phase velocity [PHYS] The velocity of a point

that moves with a wave at constant phase. Also known as celerity; phase speed; wave celerity; wave speed; wave velocity. { 'fāz və,läs·əd·ē }

phasor [PHYS] **1.** A rotating line used to represent a sinusoidally varying quantity; the length of the line represents the magnitude of the quantity, and its angle with the *x*-axis at any instant represents the phase. **2.** Any quantity (such as impedance or admittance) which is a complex number. [SOLID STATE] A low-energy collective excitation of the conduction electrons in a metal, corresponding to a slowly varying phase modulation of a charge-density wave. { 'fāz·ər }

phi meson [PART PHYS] A neutral vector meson resonance, having a mass of about 1019.4 MeV, a width of about 4.6 MeV, and negative charge parity and G parity. { 'fī 'mā,sän }

phon [ACOUS] A unit of loudness level; the loudness level, in phons, of a sound is numerically equal to the sound pressure level, in decibels, of a 1000-hertz reference tone which is judged by listeners to be equally loud to the sound under evaluation. { fän }

phoneme model [ACOUS] A succinct representation of the acoustic signal that corresponds to a phoneme, usually embedded in an utterance. { 'fō,nēm ,mäd·əl }

phonoatomic effect [ACOUS] An effect that can be observed when a sound source of sufficiently high frequency is placed in liquid helium maintained at a very low temperature, less than 0.1 kelvin above absolute zero, wherein sound quanta, when they arrive at the liquid surface, have sufficient energy to knock helium atoms out of the liquid. { ,fō·nō·ə,täm·ik i'fekt }

phonon [SOLID STATE] A quantum of an acoustic mode of thermal vibration in a crystal lattice. { 'fō,nän }

phonon-drag effect *See* Gurevich effect. { 'fō,nän ,drag i,fekt }

phonon-electron interaction [SOLID STATE] An interaction between an electron and a vibration of a lattice, resulting in a change in both the momentum of the particle and the wave vector of the vibration. { 'fō,nän i'lek,trän ,in·tər,ak·shən }

phonon emission [SOLID STATE] The production of a phonon in a crystal lattice, which may result from the interaction of other phonons via anharmonic lattice forces, from scattering of electrons in the lattice, or from scattering of x-rays or particles which bombard the crystal. { 'fō,nän i,mish·ən }

phonon friction [MECH] Friction that arises when atoms close to a surface are set into motion by the sliding action of atoms in an opposing surface, and the mechanical energy needed to slide one surface over the other is thereby converted to the energy of atomic lattice vibrations (phonons) and is eventually transformed into heat. { 'fō,nän ,frik·shən }

phonon wind [SOLID STATE] A stream of nonthermal phonons that is effective in propelling electron-hole droplets through a crystal. { 'fō ,nän ,wind }

phosphor *See* luminophor. { 'fäs·fər }

phosphorescence [ATOM PHYS] **1.** Luminescence that persists after removal of the exciting source. Also known as afterglow. **2.** Luminescence whose decay, upon removal of the exciting source, is temperature-dependent. { ,fäs·fə 'res·əns }

phosphorogen [PHYS] A substance that promotes phosphorescence in another substance, as manganese does in zinc sulfide. { fä'sfȯr·ə·jən }

phot [OPTICS] A unit of illumination equal to the illumination of a surface, 1 square centimeter in area, on which there is a luminous flux of 1 lumen, or the illumination on a surface all points of which are at a distance of 1 centimeter from a uniform point source of 1 candela. Also known as centimeter-candle (deprecated usage). { fōt }

photino [PART PHYS] A hypothetical counterpart of the photon, postulated to exist in supersymmetry theories. { fō'tē·nō }

photoabsorption [PHYS] A process in which a photon transfers all its energy to an atom, molecule, or nucleus. { ,fōd·ō·əb'sȯrp·shən }

photoacoustic spectroscopy [SPECT] A spectroscopic technique for investigating solid and semisolid materials, in which the sample is placed in a closed chamber filled with a gas such as air and illuminated with monochromatic radiation of any desired wavelength, with intensity modulated at some suitable acoustic frequency; absorption of radiation results in a periodic heat flow from the sample, which generates sound that is detected by a sensitive microphone attached to the chamber. Abbreviated PAS. Also known as optoacoustic spectroscopy. { ¦fōd·ō·ə¦kü·stik spek'träs·kə·pē }

photocapacitative effect [ELEC] A change in the capacitance of a bulk semiconductor or semiconductor surface film upon exposure to light. { ,fōd·ō·kə'pas·ə,tā·tiv i,fekt }

photochromism method [FL MECH] A flow visualization method in which a laser or other light source is used to convert a photochromic compound in the liquid under study from a transparent to an opaque state. { ¦fōd·ō'krō,miz·əm 'meth·əd }

photoconduction [SOLID STATE] An increase in conduction of electricity resulting from absorption of electromagnetic radiation. { ¦fōd·ō·kən'dək·shən }

photoconductivity [SOLID STATE] The increase in electrical conductivity displayed by many nonmetallic solids when they absorb electromagnetic radiation. { ¦fōd·ō,kän,dək'tiv·əd·ē }

photoconductor [SOLID STATE] A nonmetallic solid whose conductivity increases when it is exposed to electromagnetic radiation. { ¦fōd·ō·kən'dək·tər }

photodichroic material [OPTICS] A material which exhibits photoinduced dichroism and birefringence. { ¦fōd·ō·dī'krō·ik mə'tir·ē·əl }

photodisintegration [NUC PHYS] The breakup of an atomic nucleus into two or more fragments as a result of bombardment by gamma radiation. Also known as Chadwick-Goldhaber effect. { ¦fōd·ō·di,sin·tə'grā·shən }

photo echo [OPTICS] A coherent pulse of light generated in a nonlinear medium at a characteristic time after two other pulses, separated by a certain time interval, have entered the medium. { 'fōd·ō ,ek·ō }

photoelastic effect [OPTICS] Changes in optical properties of a transparent dielectric when it is subjected to mechanical stress, such as mechanical birefringence. Also known as photoelasticity. { ¦fōd·ō·i¦las·tik i'fekt }

photoelasticity [OPTICS] **1.** An experimental technique for the measurement of stresses and strains in material objects by means of the phenomenon of mechanical birefringence. **2.** *See* photoelastic effect. { ¦fōd·ō,i,las'tis·əd·ē }

photoelectret [SOLID STATE] An electret produced by the removal of light from an illuminated photoconductor in an electric field. { ¦fōd·ō·i'lek·trət }

photoelectric infrared radiation *See* near-infrared radiation. { ¦fōd·ō·i'lek·trik ¦in·frə¦red ,rā·dē'ā·shən }

photoelectric photometry [OPTICS] In contrast to the methods of visual photometry, an objective approach to the problems of photometry, wherein any of several types of photoelectric devices are used to replace the human eye as the sensing element. { ¦fōd·ō·i'lek·trik fə'täm·ə·trē }

photoelectron holography [ATOM PHYS] A technique for three-dimensional imaging of surface atoms in which electron waves produce holograms that are subjected to numerical image processing to yield computer displays of individual atoms. { ¦fōd·ō·i'lek,trän hō'läg·rə·fē }

photoelectron spectroscopy [SPECT] The branch of electron spectroscopy concerned with the energy analysis of photoelectrons ejected from a substance as the direct result of bombardment by ultraviolet radiation or x-radiation. { ¦fōd·ō·i'lek,trän spek'träs·kə·pē }

photoemitter [SOLID STATE] A material that emits electrons when sufficiently illuminated. { ¦fōd·ō·i¦mid·ər }

photoenlarger *See* enlarger. { ¦fōd·ō·in'lär·jər }

photoferroelectric effect [SOLID STATE] An effect observed in ferroelectric ceramics such as PLZT materials, in which light at or near the band-gap energy of the material has an effect on the electric field in the material created by an applied voltage, and, at a certain value of the voltage, also influences the degree of ferroelectric remanent polarization. Abbreviated PFE. { ¦fōd·ō¦fer·ō·i'lek·trik i,fekt }

photofission [NUC PHYS] Fission of an atomic nucleus that results from absorption by the nucleus of a high-energy photon. { ¦fōd·ō'fish·ən }

photographic field [OPTICS] The area covered or "seen" by the lens of a camera. { ¦fōd·ə¦graf·ik 'fēld }

photographic objective [OPTICS] A camera lens designed to form sharp real images of objects on a photographic film. { ¦fōd·ə¦graf·ik əb'jek·tiv }

photographic photometry [SPECT] The use of a comparator-densitometer to analyze a photographed spectrograph spectrum by emulsion density measurements. { ¦fōd·ə¦graf·ik fə'täm·ə·trē }

photographic zenith tube [OPTICS] A type of zenith telescope in which light is reflected from a pool of mercury, and the photographic plate is held in a carriage just below the objective that is alternately rotated through 180° and moved slowly across the field of view to follow a star image; used for the accurate determination of time. { ¦fōd·ə¦graf·ik 'zēn·əth ,tüb }

photograph nadir [OPTICS] The point at which a vertical line through the perspective center of a camera lens pierces the plane of the photograph. Also known as nadir point. { ¦fōd·ə¦graf·ik 'nā,dir }

photo-Hall effect [PHYS] An effect in which the illumination of a semiconductor in a magnetic field produces a change in its Hall resistance. { ¦fōd·ō 'hȯl i,fekt }

photoheliograph [OPTICS] A refracting telescope specially designed to photograph the sun's disk. { ¦fōd·ō'hē·lē·ə,graf }

photology [OPTICS] The scientific study of light. { fō'täl·ə·jē }

photoluminescence [ATOM PHYS] Luminescence stimulated by visible, infrared, or ultraviolet radiation. { ¦fōd·ō,lü·mə'nes·əns }

photomagnetic effect [PHYS] **1.** The direct effect of light on the magnetic susceptibility of certain substances. **2.** Paramagnetism displayed by certain substances when they are in a phosphorescent state. [NUC PHYS] Photodisintegration that results from the action of the magnetic field component of electromagnetic radiation. { ¦fōd·ō·mag¦ned·ik i'fekt }

photomagnetoelectric effect [ELECTROMAG] The generation of a voltage when a semiconductor material is positioned in a magnetic field and one face is illuminated. { ¦fōd·ō·mag¦ned·ō·i'lek·trik i,fekt }

photometry [OPTICS] The calculation and measurement of quantities describing light, such as luminous intensity, luminous flux, luminous flux density, light distribution, color, absorption factor, spectral distribution, and the reflectance and transmittance of light; sometimes taken to include measurement of near-infrared and near-ultraviolet radiation as well as visible light. { fō'täm·ə·trē }

photon [OPTICS] *See* troland. [QUANT MECH] A massless particle, the quantum of the electromagnetic field, carrying energy, momentum, and angular momentum. Also known as light quantum. { 'fō,tän }

photon antibunching [OPTICS] A quantum phenomenon that occurs in certain types of light emission such as resonance fluorescence, in which the emission of one photon reduces the probability that another photon will be emitted immediately afterward. { 'fō,tän 'an·ti,bənch·iŋ }

photon bunching [OPTICS] The tendency of photoelectric pulses from an illuminated photodetector to occur in bunches rather than at random. { 'fō,tän ,bənch·iŋ }

photon emission spectrum [PHYS] The relative numbers of optical photons emitted by a scintillator material per unit wavelength as a function of wavelength; the emission spectrum may also be given in alternative units such as wave number, photon energy, or frequency. { 'fō,tän i¦mish·ən ,spek·trəm }

photoneutrino [PART PHYS] A member of a neutrino-antineutrino pair that is produced in the collision of a high-energy photon with an electron. { ¦fōd·ō·nü'trē·nō }

photoneutron [NUC PHYS] A neutron released from a nucleus in a photonuclear reaction. { ¦fōd·ō'nü,trän }

photon flux [OPTICS] The number of photons in a light beam reaching a surface, such as the surface of the photocathode of a photomultiplier tube, in a unit of time. { 'fō,tän ,fləks }

photon gas [STAT MECH] An electromagnetic field treated as a collection of photons; it behaves as any other collection of bosons, except that the particles are emitted or absorbed without restriction on their number. { 'fō,tän ,gas }

photonic band-gap material *See* photonic crystal. { fə,täm·ik 'band,gap mə,tir·ē·əl }

photonic crystal [OPTICS] A macroscopic, periodic dielectric structure that possesses spectral gaps (stop bands) for electromagnetic waves, in analogy with the energy bands and gaps in regular semiconductors. Also known as photonic band-gap material. { fə,tän·ik 'krist·əl }

photon microscope *See* optical microscope. { 'fō ,tän ,mī·krə,skōp }

photon theory [QUANT MECH] A theory of photoemission developed by Einstein, according to which a light beam behaves like a stream of particles (called photons) when it delivers energy to a substance displaying photoemission, the particles each having an energy equal to Planck's constant times the frequency of the light. { 'fō ,tän ,thē·ə·rē }

photonuclear reaction [NUC PHYS] A nuclear reaction resulting from the collision of a photon with a nucleus. { ¦fōd·ō'nü·klē·ər rē'ak·shən }

photophoresis [PHYS] Production of unidirectional motion in a collection of very fine particles, suspended in a gas or falling in a vacuum, by a powerful beam of light. { ,fōd·ə·fə'rē·səs }

photoproton [NUC PHYS] A proton released from a nucleus in a photonuclear reaction. { ¦fōd·ō'prō,tän }

photorefractive effect [OPTICS] An effect displayed by many electrooptic materials in which a change in the index of refraction is induced by the presence of light, and this change is retained for a time after the light exposure ceases. { ,fōd·ō·ri'frak·tiv i'fekt }

photostriction [PHYS] The changes in the dimensions of piezoelectric materials that also exhibit one of the photoelectric effects when they are illuminated by light. { ,fōd·ə'strik·shən }

photothermoelasticity [OPTICS] Changes in optical properties of a transparent dielectric when it is subjected to mechanical stress, which is, in turn, induced by temperature gradients. { ¦fōd·ō,thər·mō,i,las'tis·əd·ē }

phototropism [SOLID STATE] A reversible change in the structure of a solid exposed to light or other radiant energy, accompanied by a change in color. Also known as phototropy. { fō'tä·trə,piz·əm }

photoviscoelasticity [OPTICS] Changes in optical properties of a transparent, viscoelastic substance when it is subjected to stress. { ¦fōd·ō,vis·gō,i,las'tis·əd·ē }

phthalocyanine Q switching [OPTICS] Laser Q switching in which a solution of metal-organic compounds known as phthalocyanines is placed in a cell between an uncoated ruby laser crystal and a high-reflectivity mirror; when the incident ruby light reaches a certain level, the solution suddenly becomes almost perfectly transparent to this light, permitting the release of all the energy stored in the ruby as a giant pulse. { ¦thal·ō'sī·ə·nən 'kyü ,swich·iŋ }

physical constant [PHYS] A physical quantity which has a fixed and unchanging numerical value. { 'fiz·ə·kəl 'kän·stənt }

physical law [PHYS] A property of a physical phenomenon, or a relationship between the various quantities or qualities which may be used to describe the phenomenon, that applies to all members of a broad class of such phenomena, without exception. { 'fiz·ə·kəl 'lȯ }

physical measurement [PHYS] Quantitative information on a physical condition, property, or relation, generally in the form of the ratio of the measured quantity to a standard quantity, or to some fixed multiple or fraction thereof. { 'fiz·ə·kəl 'mezh·ər·mənt }

physical optics [OPTICS] The study of the interaction of electromagnetic waves in the optical frequency range with material systems. { 'fiz·ə·kəl 'äp·tiks }

physical theory [PHYS] An attempt to explain a certain class of physical phenomena by deducing them as necessary consequences of some primitive assumptions. { 'fiz·ə·kəl 'thē·ə·rē }

physicist [PHYS] A person who does research in physics. { 'fiz·ə,sist }

physiological acoustics [ACOUS] The study of the responses to acoustic stimuli that take place in the ear or in the associated central neural auditory pathways of humans and animals. { ,fiz·ē·ə,läj·ə·kəl ə'kü·stiks }

Pickering series [SPECT] A series of spectral lines of singly ionized helium, observed in very hot O-type stars, associated with transitions between the level with principal quantum number $n = 4$ and higher energy levels. { 'pik·riŋ ,sir·ēz }

pickup [ELEC] **1.** A device that converts a

sound, scene, measurable quantity, or other form of intelligence into corresponding electric signals, as in a microphone, phonograph pickup, or television camera. **2.** The minimum current, voltage, power, or other value at which a relay will complete its intended function. **3.** Interference from a nearby circuit or system. [NUC PHYS] A type of nuclear reaction in which the incident particle takes a nucleon from the target nucleus and proceeds with this nucleon bound to itself. { 'pik,əp }

picoampere [ELEC] A unit of current equal to 10^{-12} ampere, or one-millionth of a microampere. Abbreviated pA. { ,pē·kō'am,pir }

picofarad [ELEC] A unit of capacitance equal to 10^{-12} farad, or one-millionth of a microfarad. Also known as micromicrofarad (deprecated usage); puff (British usage). Abbreviated pF. { ,pē·kō'far·əd }

picosecond [MECH] A unit of time equal to 10^{-12} second, or one-millionth of a microsecond. Abbreviated ps; psec. { ,pē·kō'sek·ənd }

picowatt [MECH] A unit of power equal to 10^{-12} watt, or one-millionth of a microwatt. Abbreviated pW. { 'pē·kə,wät }

pièze [MECH] A unit of pressure equal to 1 sthène per square meter, or to 1000 pascals. Abbreviated pz. { pē'ez }

piezocaloric effect [SOLID STATE] The production of entropy in a crystal that is subjected to mechanical stress. { pē,ā·zō·kə'lȯr·ik i,fekt }

piezoelectric [SOLID STATE] Having the ability to generate a voltage when mechanical force is applied, or to produce a mechanical force when a voltage is applied, as in a piezoelectric crystal. { pē¦ā·zō·ə'lek·trik }

piezoelectric crystal [SOLID STATE] A crystal which exhibits the piezoelectric effect; used in crystal loudspeakers, crystal microphones, and crystal cartridges for phono pickups. { pē¦ā·zō·ə'lek·trik 'krist·əl }

piezoelectric effect [SOLID STATE] **1.** The generation of electric polarization in certain dielectric crystals as a result of the application of mechanical stress. **2.** The reverse effect, in which application of a voltage between certain faces of the crystal produces a mechanical distortion of the material. { pē¦ā·zō·ə'lek·trik i'fekt }

piezoelectric hysteresis [SOLID STATE] Behavior of a piezoelectric crystal whose electric polarization depends not only on the mechanical stress to which the crystal is subjected, but also on the previous history of this stress. { pē¦ā·zō·ə'lek·trik ,his·tə'rē·səs }

piezoelectricity [SOLID STATE] Electricity or electric polarization resulting from the piezoelectric effect. { pē¦ā·zō·ə,lek'tris·əd·ē }

piezoelectric semiconductor [SOLID STATE] A semiconductor exhibiting the piezoelectric effect, such as quartz, Rochelle salt, and barium titanate. { pē¦ā·zō·ə'lek·trik 'sem·i·kən,dək·tər }

piezoelectric vibrator [SOLID STATE] An element cut from piezoelectric material, usually in the form of a plate, bar, or ring, with electrodes attached to or supported near the element to excite one of its resonant frequencies. { pē¦ā·zō·ə'lek·trik 'vī,brād·ər }

piezomagnetism [SOLID STATE] Stress dependence of magnetic properties. { pē¦ā·zō'mag·nə,tiz·əm }

piezooptical effect [OPTICS] The change produced in the index of refraction of a light-transmitting material by externally applied stress. { pe¦ā·zō'äp·tə·kəl i,fekt }

piezoresistance effect [SOLID STATE] The change in the electrical resistance of a metal or semiconductor that is produced by mechanical stress. { pē,ā·zō·ri'zis·təns i,fekt }

piezotropic [FL MECH] Characterized by piezotropy. { pē¦ā·zō¦träp·ik }

piezotropy [FL MECH] The property of a fluid in which processes are characterized by a functional dependence of the thermodynamic functions of state: $d\rho/dt = b(dp/dt)$, where ρ is the density, p the pressure, and b a function of the thermodynamic variables, called the coefficient of piezotropy. { ,pē·ə'zä·trə·pē }

pile formula [MECH] An equation for the forces acting on a pile at equilibrium: $P = pA + tS + Sn \sin \phi$, where P is the load, A is the area of the pile point, p is the force per unit area on the point, S is the embedded surface of the pile, t is the force per unit area parallel to S, n is the force per unit area normal to S, and ϕ is the taper angle of the pile. { 'pīl ,fȯr·myə·lə }

pill [ELECTROMAG] A microwave stripline termination. { pil }

pillbox antenna [ELECTROMAG] Cylindrical parabolic reflector enclosed by two plates perpendicular to the cylinder, spaced to permit the propagation of only one mode in the desired direction of polarization. { 'pil,bäks an'ten·ə }

pi meson [PART PHYS] **1.** Collective name for three semistable mesons which have charges of +1, 0, and −1 times the proton charge, and form a charge multiplet, with an approximate mass of 138 megaelectronvolts, spin 0, negative parity, negative G parity, and positive charge parity (for the neutral meson). Also known as pion. Symbolized π. **2.** Any meson belonging to an isospin triplet with hypercharge 0, negative G parity, and positive charge parity (for the neutral meson). { 'pī 'mā,sän }

pi-mu atom *See* pionium. { 'pī 'myü 'ad·əm }

pinacoid [CRYSTAL] An open crystal form that comprises two parallel faces. { 'pin·ə,kȯid }

pinacoidal class [CRYSTAL] That crystal class in the triclinic system having only a center of symmetry. { ¦pin·ə¦kȯid·əl ¦klas }

pinacoidal cleavage [CRYSTAL] A type of crystal cleavage that is parallel to one of the crystal's pinacoidal surfaces. { ¦pin·ə¦kȯid·əl 'klē·vij }

pinch effect [ELEC] Manifestation of the magnetic self-attraction of parallel electric currents, such as constriction of ionized gas in a discharge tube, or constriction of molten metal through which a large current is flowing. Also known as cylindrical pinch; magnetic pinch; rheostriction. { 'pinch i,fekt }

pincushion distortion [OPTICS] Aberration in which the magnification produced by an optical system increases with the distance of the object point from the optical axis, so that the image of a square has concave sides. { 'pin,ku̇sh·ən di,stȯr·shən }

pine-tree array [ELECTROMAG] Array of dipole antennas aligned in a vertical plane known as the radiating curtain, behind which is a parallel array of dipole antennas forming a reflecting curtain. { 'pīn ,trē ə,rā }

pi network [ELEC] An electrical network which has three impedance branches connected in series to form a closed circuit, with the three junction points forming an output terminal, an input terminal, and a common output and input terminal. { 'pī ,net,wərk }

pinhole camera [OPTICS] A camera which has no lenses, but consists essentially of a darkened box with a small hole in one side, so that an inverted image of outside objects is projected on the opposite side where it is recorded on photographic film. { 'pin,hōl 'kam·rə }

pink noise [ACOUS] Noise whose intensity is inversely proportional to frequency over a specified range, to give constant energy per octave. { 'piŋk ,nȯiz }

pinning [SOLID STATE] The hindering of motion of dislocations in a solid, and the consequent hardening of the solid, by impurities which collect near the dislocations, resulting in a large energy barrier being imposed against the motion of the dislocations. { 'pin·iŋ }

pint [MECH] Abbreviated pt. **1.** A unit of volume, used in the United States for measurement of liquid substances, equal to 1/8 U.S. gallon, or 231/8 cubic inches, or $4.73176473 \times 10^{-4}$ cubic meter. Also known as liquid pint (liq pt). **2.** A unit of volume used in the United States for measurement of solid substances, equal to 1/64 U.S. bushel, or 107,521/3200 cubic inches, or approximately 5.50610×10^{-4} cubic meter. Also known as dry pint (dry pt). **3.** A unit of volume, used in the United Kingdom for measurement of liquid and solid substances, although usually the former, equal to 1/8 imperial gallon, or 5.6826125×10^{-4} cubic meter. Also known as imperial pint. { pīnt }

pion *See* pi meson. { 'pī,än }

pion bremsstrahlung [NUC PHYS] The emission of pions in the collision of two heavy nuclei in a manner analogous to the radiation that occurs when a charged particle is accelerated or decelerated. { 'pī,än brem'shträ·ləŋ }

pion condensate [NUC PHYS] A state of nuclear matter compressed to abnormally high densities, in which great numbers of pairs of particles, each consisting of a positive pion and a negative pion, are generated, and interact strongly with the nucleons, causing them to form a coherent spin-isospin structure. { 'pī,än 'känd·ən,sāt }

pion double-charge exchange [NUC PHYS] A nuclear reaction in which a positive pion interacts with a nucleus by a two-step process and a negative pion emerges, accompanied by the conversion of two neutrons in the nucleus to two protons, or the inverse reaction. { 'pī,än ¦dəb·əl ¦chärj iks'chānj }

pionium [PART PHYS] **1.** An exotic atom consisting of a muon orbiting about an oppositely charged pion. Also known as pi-mu atom. **2.** An exotic atom consisting of an electron in orbit about an oppositely charged pion. { pī'ō·nē·əm }

pipper [OPTICS] A small hole in the reticle of an optical sight or computing sight. { 'pip·ər }

pipper image [OPTICS] A spot of light projected through the pipper in an optical or computing sight, used in aiming. { 'pip·ər ,im·ij }

Pirani gage [PHYS] A thermal conductivity gage (where the thermal conductivity of a gas heated by a hot wire varies with pressure) connected to a Wheatstone bridge to measure the resistance of the hot wire, thus the gas pressure; used to measure pressure from 1 to 10^{-3} mmHg (133.32 to 0.13332 pascals). { pə'rän·ē ,gāj }

piston [ELECTROMAG] A sliding metal cylinder used in waveguides and cavities for tuning purposes or for reflecting essentially all of the incident energy. Also known as plunger; waveguide plunger. { 'pis·tən }

piston attenuator [ELECTROMAG] A microwave attenuator inserted in a waveguide to introduce an amount of attenuation that can be varied by moving an output coupling device along its longitudinal axis. { 'pis·tən ə'ten·yə,wād·ər }

piston flow [FL MECH] Two-phase (vapor-liquid) flow in which the gas flows as large plugs; occurs for gas superficial velocities from about 2 to 30 feet per second (60 to 900 centimeters per second). Also known as plug flow; slug flow. { 'pis·tən ,flō }

pitch [ACOUS] That psychological property of sound characterized by highness or lowness, depending primarily upon frequency of the sound stimulus, but also upon its sound pressure and waveform. [MECH] **1.** Of an aerospace vehicle, an angular displacement about an axis parallel to the lateral axis of the vehicle. **2.** The rising and falling motion of the bow of a ship or the tail of an airplane as the craft oscillates about a transverse axis. { pich }

pitch acceleration [MECH] The angular acceleration of an aircraft or missile about its lateral, or Y, axis. { 'pich ik,sel·ə,rā·shən }

pitch attitude [MECH] The attitude of an aircraft, rocket, or other flying vehicle, referred to the relationship between the longitudinal body axis and a chosen reference line or plane as seen from the side. { 'pich ,ad·ə,tüd }

pitch axis [MECH] A lateral axis through an aircraft, missile, or similar body, about which the body pitches. Also known as pitching axis. { 'pich ,ak·səs }

pitching axis *See* pitch axis. { 'pich·iŋ ,ak·səs }

pitching moment [MECH] A moment about a lateral axis of an aircraft, rocket, or airfoil. { 'pich·iŋ ,mō·mənt }

pi theorem *See* Buckingham's π theorem. { 'pī ,thir·əm }

pitot pressure [FL MECH] Pressure at the open end of a pitot tube. { pē'tō ˌpresh·ər }

pi-T transformation *See* Y-delta transformation. { ¦pī 'tē ˌtranz·fərˌmā·shən }

pivot [MECH] A short, pointed shaft forming the center and fulcrum on which something turns, balances, or oscillates. { 'piv·ət }

pk *See* peck.

Pl *See* poiseuille.

plagiohedral [CRYSTAL] Pertaining to obliquely arranged spiral faces; in particular, to a member of a group in the isometric system with 13 axes but no center or planes. { ¦plā·jē·ō¦hē·drəl }

planar-array antenna [ELECTROMAG] An array antenna in which the centers of the radiating elements are all in the same plane. { 'plā·nər ə¦rā an'ten·ə }

planar laser-induced fluorescence [FL MECH] A flow visualization technique in which a thin sheet of monochromatic laser radiation is used to excite a particular molecular species in a flow, and the resulting fluorescence emission of this species gives some indication of its number density. Abbreviated PLIF. { ¦plā·nər ¦lā·zər inˌdüst flə'res·əns }

planck [PHYS] A unit of action equal to the product of an energy of 1 joule and a time of 1 second. { plänk }

Planck distribution law *See* Planck radiation formula. { 'plänk ˌdis·trə'byü·shən ˌlò }

Planck function [THERMO] The negative of the Gibbs free energy divided by the absolute temperature. { 'plänk ˌfəŋk·shən }

Planckian locus [OPTICS] The locus of points on a chromaticity diagram that represents blackbody radiators at various temperatures. { 'pläŋ·kē·ən 'lō·kəs }

Planck length [PHYS] The length

$$\sqrt{Gh/2\Pi c^3}),$$

where G is the gravitational constant, h is Planck's constant, and c is the speed of light) at which quantum fluctuations are believed to dominate the geometry of space-time; it is equal to 1.6162×10^{-35} m. { 'plänk ˌleŋkth }

Planck mass [PHYS] The mass

$$\sqrt{hc/2\Pi G},$$

where h is Planck's constant, c is the speed of light, and G is the gravitational constant; equivalently, the mass of a particle whose reduced Compton wavelength equals the Planck length; it is equal to 21.764 micrograms or 1.2209×10^{19} GeV/c^2. { 'plänk ˌmas }

Planck oscillator [QUANT MECH] An oscillator which can absorb or emit energy only in amounts which are integral multiples of Planck's constant times the frequency of the oscillator. Also known as radiation oscillator. { 'plänk ˌäs·əˌlād·ər }

Planck radiation formula [STAT MECH] A formula for the intensity of radiation emitted by a blackbody within a narrow band of frequencies (or wavelengths), as a function of frequency, and of the body's temperature. Also known as Planck distribution law; Planck's law. { 'plänk ˌrād·ē'ā·shən ˌfór·myə·lə }

Planck's constant [QUANT MECH] A fundamental physical constant, the elementary quantum of action; the ratio of the energy of a photon to its frequency, it is equal to 6.62606876 ± $0.00000052 \times 10^{-34}$ joule-second. Symbolized h. Also known as quantum of action. { 'plänks ˌkän·stənt }

Planck's law [QUANT MECH] A fundamental law of quantum theory stating that energy associated with electromagnetic radiation is emitted or absorbed in discrete amounts which are proportional to the frequency of radiation. *See* Planck radiation formula. { 'plänks ˌlò }

Planck time [PHYS] The constant $(Gh/2\pi c^5)^{1/2}$ with dimensions of "time" formed from Planck's constant h, the gravitational constant G, and the speed of light c; approximately 10^{-43} second. { 'plänk ˌtīm }

plane defect [CRYSTAL] A type of crystal defect that occurs along the boundary plane of two regions of a crystal, or between two grains. { 'plān diˌfekt }

plane dendrite *See* plane-dendritic crystal. { 'plān ¦denˌdrīt }

plane-dendritic crystal [CRYSTAL] An ice crystal exhibiting an elaborately branched (dendritic) structure of hexagonal symmetry, with its much larger dimension lying perpendicular to the principal (c-axis) of the crystal. Also known as plane dendrite; stellar crystal. { 'plān den¦drid·ik 'krist·əl }

plane earth [ELECTROMAG] Earth that is considered to be a plane surface as used in ground-wave calculations. { 'plān ˌərth }

plane-earth attenuation [ELECTROMAG] Attenuation of an electromagnetic wave over an imperfectly conducting plane earth in excess of that over a perfectly conducting plane. { 'plān ˌərth əˌten·yə'wā·shən }

plane group [CRYSTAL] The group of operations (rotations, reflections, translations, and combinations of these) which leave a regular, periodic structure in a plane unchanged. { 'plān ˌgrüp }

plane lamina [MECH] A body whose mass is concentrated in a single plane. { 'plān 'lam·ə·nə }

plane lattice [CRYSTAL] A regular, periodic array of points in a plane. { 'plān 'lad·əs }

plane mirror [OPTICS] A mirror whose surface lies in a plane; it forms an image of an object such that the mirror surface is perpendicular to and bisects the line joining all corresponding object-image points. { 'plān 'mir·ər }

plane of departure [MECH] Vertical plane containing the path of a projectile as it leaves the muzzle of the gun. { 'plān əv di'pär·chər }

plane of fire [MECH] Vertical plane containing the gun and the target, or containing a line of site. { 'plān əv 'fīr }

plane of flotation [FL MECH] The plane in which the surface of a liquid intersects a stationary floating body. { 'plān əv flō'tā·shən }

plane of incidence [PHYS] A plane containing

the direction of propagation of a wave striking a surface and a line perpendicular to the surface. Also known as incidence plane. { 'plān əv 'in·səd·əns }

plane of maximum shear stress [MECH] Either of two planes that lie on opposite sides of and at angels of 45° to the maximum principal stress axis and that are parallel to the intermediate principal stress axis. { 'plān əv ¦mak·si·məm 'shir ˌstres }

plane of mirror symmetry [CRYSTAL] In certain crystals, a symmetry element whereby reflection of the crystal through a certain plane leaves the crystal unchanged. Also known as mirror plane of symmetry; plane of symmetry; reflection plane; symmetry plane. { 'plān əv 'mir·ər ˌsim·ə·trē }

plane of polarization [ELECTROMAG] Plane containing the electric vector and the direction of propagation of electromagnetic wave. { 'plān əv ˌpō·lə·rə'zā·shən }

plane of reflection [CRYSTAL] *See* plane of mirror symmetry. [OPTICS] A plane containing the direction of propagation of radiation reflected from a surface, and the normal to the surface. Also known as reflection plane. { 'plān əv ri'flek·shən }

plane of symmetry *See* plane of mirror symmetry. { 'plān əv 'sim·ə·trē }

plane of yaw [MECH] The plane determined by the tangent to the trajectory of a projectile in flight and the axis of the projectile. { 'plān əv 'yȯ }

plane-parallel resonator [OPTICS] A beam resonator that consists of a pair of plane mirrors which are perpendicular to the axis of the beam. { 'plān 'par·əˌlel 'rez·ənˌād·ər }

plane Poiseuille flow [FL MECH] Rheological (viscosity) measurement in which the fluid of interest is propelled through a narrow slot, and the volumetric flow rate and the pressure gradient are measured simultaneously to determine viscosity. { 'plān pwä'zə·ē ˌflō }

plane polarization *See* linear polarization. { 'plān ˌpō·lə·rə'zā·shən }

plane-polarized wave [ELECTROMAG] An electromagnetic wave whose electric field vector at all times lies in a fixed plane that contains the direction of propagation through a homogeneous isotropic medium. { 'plān ¦pō·ləˌrīzd ˌwāv }

plane reflector *See* passive reflector. { 'plān ri ¦flek·tər }

plane strain [MECH] A deformation of a body in which the displacements of all points in the body are parallel to a given plane, and the values of these displacements do not depend on the distance perpendicular to the plane. { 'plān ˌstrān }

plane stress [MECH] A state of stress in which two of the principal stresses are always parallel to a given plane and are constant in the normal direction. { 'plān ˌstres }

plane symmetry group *See* plane group. { 'plān 'sim·ə·trē ˌgrüp }

planetary aberration [OPTICS] The apparent displacement of an object in the solar system that results from the fact that light takes a certain time to travel from the object to earth, during which time the object travels a certain distance in its orbit. { 'plan·əˌter·ē ˌab·ə'rā·shən }

planetary electron *See* orbital electron. { 'plan·əˌter·ē i'lekˌträn }

plane wave [PHYS] Wave in which the wavefront is a plane surface; a wave whose equiphase surfaces form a family of parallel planes. { 'plān ˌwāv }

planoconcave lens [OPTICS] A lens for which one surface is plane and the other is concave. { ¦plā·nō'känˌkāv 'lenz }

planoconvex lens [OPTICS] A lens for which one surface is plane and the other is convex. { ¦plā·nō'känˌveks 'lenz }

planocylindrical lens [OPTICS] A lens, one of whose surfaces is a portion of a plane, while the other is a portion of a cylinder. { ¦plā·nō·sə'lin·drə·kəl 'lenz }

plasma [PL PHYS] **1.** A highly ionized gas which contains equal numbers of ions and electrons in sufficient density so that the Debye shielding length is much smaller than the dimensions of the gas. **2.** A completely ionized gas, composed entirely of a nearly equal number of positive and negative free charges (positive ions and electrons). { 'plaz·mə }

plasma accelerator [PL PHYS] An accelerator that forms a high-velocity jet of plasma by using a magnetic field, an electric arc, a traveling wave, or other similar means. { 'plaz·mə ak'sel·əˌrād·ər }

plasma desorption mass spectrometry [SPECT] A technique for analysis of nonvolatile molecules, particularly heavy molecules with atomic weight over 2000, in which heavy ions with energies on the order of 100 MeV penetrate and deposit energy in thin films, giving rise to chemical reactions that result in the formation of molecular ions and shock waves that result in the ejection of these ions from the surface; the ions are then analyzed in a mass spectrometer. Abbreviated PDMS. { 'plaz·mə dē¦sȯrp·shən 'mas spek 'träm·ə·trē }

plasma frequency *See* Langmuir plasma frequency. { 'plaz·mə ˌfrē·kwən·sē }

plasma gun [ELECTROMAG] An electromagnetic device which creates and accelerates bursts of plasma. { 'plaz·mə ˌgən }

plasma instability [PL PHYS] A sudden change in the quasistatic distribution of positions or velocities of particles constituting a plasma, and a sudden change in the accompanying electromagnetic field. { 'plaz·mə ˌin·stə'bil·əd·ē }

plasma-jet excitation [SPECT] The use of a high-temperature plasma jet to excite an element to provide measurable spectra with many ion lines similar to those from spark-excited spectra. { 'plaz·mə ˌjet ˌek·sə'tā·shən }

plasma oscillations [PL PHYS] Various vibrations and wave motions of the electrons and ions in a plasma. { 'plaz·mə ˌäs·ə'lā·shənz }

plasma physics [PHYS] The study of highly ionized gases. { 'plaz·mə 'fiz·iks }

plasma pinch [PL PHYS] Application of the pinch effect to plasma in attempts to produce controlled nuclear fusion. { 'plaz·mə ,pinch }

plasma radiation [PL PHYS] Electromagnetic radiation emitted from a plasma, primarily by free electrons undergoing transitions to other free states or to bound states of atoms and ions, but also by bound electrons as they undergo transitions to other bound states. { 'plaz·mə ,rād·ē'ā·shən }

plasma wave [PL PHYS] A disturbance of a plasma involving oscillation of its constituent particles and of an electromagnetic field, which propagates from one point in the plasma to another without net motion of the plasma. { 'plaz·mə ,wāv }

plasmoid [PHYS] An isolated collection of electrons, ions, and neutral particles which holds together for a duration many times as long as the collision times between particles. { 'plaz,mȯid }

plasmon [SOLID STATE] A quantum of a collective longitudinal wave in the electron gas of a solid. { 'plaz,män }

plastic [MECH] Displaying, or associated with, plasticity. { 'plas·tik }

plastic collision [MECH] A collision in which one or both of the colliding bodies suffers plastic deformation and mechanical energy is dissipated. { ¦plas·tik kə'lizh·ən }

plastic deformation [MECH] Permanent change in shape or size of a solid body without fracture resulting from the application of sustained stress beyond the elastic limit. { 'plas·tik ,dē,fȯr 'mā·shən }

plastic film capacitor [ELEC] A capacitor constructed by stacking, or forming into a roll, alternate layers of foil and a dielectric which consists of a plastic, such as polystyrene or Mylar, either alone or as a laminate with paper. { 'plas·tik ¦film kə'pas·əd·ər }

plastic flow [PHYS] Rheological phenomenon in which flowing behavior of the material occurs after the applied stress reaches a critical (yield) value, such as with putty. { 'plas·tik 'flō }

plasticity [MECH] The property of a solid body whereby it undergoes a permanent change in shape or size when subjected to a stress exceeding a particular value, called the yield value. { plas'tis·əd·ē }

plasticoviscosity [MECH] Plasticity in which the rate of deformation of a body subjected to stresses greater than the yield stress is a linear function of the stress. { ¦plas·tə·kō·vi'skäs·əd·ē }

plastic viscosity [FL MECH] A measure of the internal resistance to fluid flow of a Bingham plastic, expressed as the tangential shear stress in excess of the yield stress divided by the resulting rate of shear. { 'plas·tik vi'skäs·əd·ē }

plate [ELEC] **1.** One of the conducting surfaces in a capacitor. **2.** One of the electrodes in a storage battery. { plāt }

Plateau's sphere [FL MECH] A small drop of liquid which follows a larger drop that breaks away and falls. { pla¦tōz 'sfir }

plate modulus [MECH] The ratio of the stress component T_{xx} in an isotropic, elastic body obeying a generalized Hooke's law to the corresponding strain component S_{xx}, when the strain components S_{yy} and S_{zz} are 0; the sum of the Poisson ratio and twice the rigidity modulus. { 'plāt ,mäj·ə·ləs }

plate wave [ACOUS] A type of ultrasonic vibration generated in a thin solid, such as a sheet of metal having a thickness of less than one wavelength, and usually consisting of a variety of simultaneous modes having different velocities; it is used in metal inspection. Also known as Lamb wave. { 'plāt ,wāv }

play of color [OPTICS] An optical phenomenon consisting of a rapid succession of flashes of a variety of prismatic colors as certain minerals or cabochon-cut gems are moved about; caused by diffraction of light from spherical particles of amorphous silica stacked in an orderly three-dimensional pattern. Also known as schiller. { 'plā əv 'kəl·ər }

pleochroic halos [OPTICS] Halos of color or color differences that are sometimes observed around inclusions in minerals, resulting from irradiation by alpha particles. { ¦plē·ə¦krō·ik 'hā·lōz }

pleochroism [OPTICS] Phenomenon exhibited by certain transparent crystals in which light viewed through the crystal has different colors when it passes through the crystal in different directions. Also known as polychroism. { plē'äk·rə,wiz·əm }

pleomorphism *See* polymorphism. { ,plē·ō'mȯr ,fiz·əm }

pli [MECH] A unit of line density (mass per unit length) equal to 1 pound per inch, or approximately 17.8580 kilograms per meter. { plē }

PLIF *See* planar laser-induced fluorescence. { ¦pē¦el¦ī'ef *or* plif }

plug flow *See* piston flow. { 'pləg ,flō }

plunger *See* piston. { 'plən·jər }

plural scattering [PHYS] A change in direction of a particle or photon because of a small number of collisions. { ¦plu̇r·əl 'skad·ər·iŋ }

plutonium-238 [NUC PHYS] The first synthetic isotope made of plutonium; similar chemically to uranium and neptunium; atomic number 94; formed by bombardment of uranium with deuterons. { plü'tō·nē·əm ¦tü,thər·dē'āt }

plutonium-239 [NUC PHYS] A synthetic isotope chemically similar to uranium and neptunium; atomic number 94; made by bombardment of uranium-238 with slow neutrons in a nuclear reactor; used as nuclear reactor fuel and an ingredient for nuclear weapons. { plü'tō·nē·əm ¦tü,thər·dē'nīn }

PM *See* permanent magnet.

PMR *See* projection microradiography.

PNdB *See* perceived noise decibel.

pneumatics [FL MECH] Fluid statics and behavior in closed systems when the fluid is a gas. { nu̇'mad·iks }

Pockels cell [OPTICS] A crystal that exhibits the Pockels effect, such as potassium dihydrogen phosphate, which is placed between crossed polarizers and has ring electrodes bonded to two faces to allow application of an electric field; used to modulate light beams, especially laser beams. { 'päk·əlz ˌsel }

Pockels effect [OPTICS] Changes in the refractive properties of certain crystals in an applied electric field, which are proportional to the first power of the electric field strength. { 'päk·əlz iˌfekt }

Poggendorff's first method *See* constant-current dc potentiometer. { 'päg·ənˌdȯrfs 'first ˌmeth·əd }

Poggendorff's second method *See* constant-resistance dc potentiometer. { 'päg·ənˌdȯrfs 'sek·ənd ˌmeth·əd }

Poincaré electron [ELECTROMAG] A classical model of the electron in which nonelectromagnetic forces hold the electron together so that it has zero self-stress; it is unstable and has infinite self-energy in the case of a point electron. { ˌpwänˌkä'rā i'lekˌträn }

Poincaré surface of section [MECH] A method of displaying the character of a particular trajectory without examining its complete time development, in which the trajectory is sampled periodically, and the rate of change of a quantity under study is plotted against the value of that quantity at the beginning of each period. Also known as surface of section. { ˌpwänˌkä'rā 'sər·fəs əv 'sek·shən }

Poinsot ellipsoid *See* inertia ellipsoid. { pwän 'sō ə'lipˌsȯid }

Poinsot motion [MECH] The motion of a rigid body with a point fixed in space and with zero torque or moment acting on the body about the fixed point. { pwän'sō ˌmō·shən }

Poinsot's central axis [MECH] A line through a rigid body which is parallel to the vector sum **F** of a system of forces acting on the body, and which is located so that the system of forces is equivalent to the force **F** applied anywhere along the line, plus a couple whose torque is equal to the component of the total torque **T** exerted by the system in the direction **F**. { ¦pwän·sōz ¦sen·trəl 'ak·səs }

Poinsot's method [MECH] A method of describing Poinsot motion, by means of a geometrical construction in which the inertia ellipsoid rolls on the invariable plane without slipping. { pwän'sōz ¦meth·əd }

point-blank range [MECH] Distance to a target that is so short that the trajectory of a bullet or projectile is practically a straight, rather than a curved, line. { 'pȯint¦blaŋk 'rānj }

point characteristic function [OPTICS] The integral between two points of $n\,ds$ along some path, where ds is the arc length of an infinitesimal piece of the path and n is the refractive index; according to Fermat's principle, it is a maximum or minimum with respect to nearby paths for the actual path of a light ray. { 'pȯint ˌkar·ik·tə'ris·tik ˌfəŋk·shən }

point defect [CRYSTAL] A departure from crystal symmetry which affects only one, or, in some cases, two lattice sites. { 'pȯint diˌfekt }

point function [PHYS] A quantity whose value depends on the location of a point in space, such as an electric field, pressure, temperature, or density. { 'pȯint ˌfəŋk·shən }

point group [CRYSTAL] A group consisting of the symmetry elements of an object having a single fixed point; 32 such groups are possible. { 'pȯint ˌgrüp }

point of contraflexure [MECH] A point at which the direction of bending changes. Also known as point of inflection. { 'pȯint əv ˌkän·trə'flek·shər }

point of fall [MECH] The point in the curved path of a falling projectile that is level with the muzzle of the gun. Also known as level point. { 'pȯint əv 'fȯl }

point of inflection *See* point of contraflexure. { 'pȯint əv in'flek·shən }

point source [PHYS] A source of radiation having definite position but no extension in space; this is an ideal which is a good approximation for distances from the source sufficiently large compared to the dimensions of the source. { 'pȯint ˌsȯrs }

point target [ELECTROMAG] In radar, an object which returns a target signal by reflection from a relatively simple discrete surface; such targets are ships, aircraft, projectiles, missiles, and buildings. { 'pȯint ˌtär·gət }

poise [FL MECH] A unit of dynamic viscosity equal to the dynamic viscosity of a fluid in which there is a tangential force 1 dyne per square centimeter resisting the flow of two parallel fluid layers past each other when their differential velocity is 1 centimeter per second per centimeter of separation. Abbreviated P. { pȯiz }

poiseuille [FL MECH] A unit of dynamic viscosity of a fluid in which there is a tangential force of 1 newton per square meter resisting the flow of two parallel layers past each other when their differential velocity is 1 meter per second per meter of separation; equal to 10 poise; used chiefly in France. Abbreviated Pl. { pwä'zə·ē }

Poiseuille flow [FL MECH] The steady flow of an incompressible fluid parallel to the axis of a circular pipe of infinite length, produced by a pressure gradient along the pipe. { pwä'zə·ē ˌflō }

Poiseuille's law [FL MECH] The law that the volume flow of an incompressible fluid through a circular tube is equal to $\pi/8$ times the pressure differences between the ends of the tube, times the fourth power of the tube's radius divided by the product of the tube's length and the dynamic viscosity of the fluid. { pwä'zə·ēz ˌlȯ }

poison [ATOM PHYS] A substance which reduces the phosphorescence of a luminescent material. { 'pȯiz·ən }

Poisson bracket [MECH] For any two dynamical

variables, X and Y, the sum, over all degrees of freedom of the system, of $(\partial X/\partial q)(\partial Y/\partial p)-(\partial X/\partial p)(\partial Y/\partial q)$, where q is a generalized coordinate and p is the corresponding generalized momentum. { pwä'sōn ,brak·ət }

Poisson constant [PHYS] The ratio k of the gas constant R to the specific heat at constant pressure C_p. { pwä'sōn ,kän·stənt }

Poisson effect [FL MECH] The deflection of a spinning projectile with right-handed spin to the right, and vice versa. Also known as cushion effect. { pwä¦sōn i,fekt }

Poisson number [MECH] The reciprocal of the Poisson ratio. { pwä'sōn ,nəm·bər }

Poisson ratio [MECH] The ratio of the transverse contracting strain to the elongation strain when a rod is stretched by forces which are applied at its ends and which are parallel to the rod's axis. { pwä'sōn ,rā·shō }

polar axis [CRYSTAL] An axis of crystal symmetry which does not have a plane of symmetry perpendicular to it. { 'pō·lər 'ak·səs }

polar crystal *See* ferroelectric crystal. { 'pō·lər 'krist·əl }

polar diagram [PHYS] A diagram employing polar coordinates to show the magnitude of a quantity in some or all directions from a point; examples include directivity patterns and radiation patterns. { 'pō·lər 'dī·ə,gram }

polarimeter [OPTICS] An instrument used to determine the rotation of the plane of polarization of plane polarized light when it passes through a substance; the light is linearly polarized by a polarizer (such as a Nicol prism), passes through the material being analyzed, and then passes through an analyzer (such as another Nicol prism). { ,pō·lə'rim·əd·ər }

polarimetry [OPTICS] The science of determining the polarization state of electromagnetic radiation (x-rays, light, or radio waves). { ,pō·lə'rim·ə·trē }

polariscope [OPTICS] Any of several instruments used to determine the effects of substances on polarized light, in which linearly or elliptically polarized light passes through the substance being studied, and then through an analyzer. { pə'lar·ə,skōp }

polariton [SOLID STATE] A coupled mode of motion in an ionic crystal due to the coupling between the electromagnetic field and transverse optical phonons of long wavelength. { pə'lar·ə,tän }

polarity [PHYS] Property of a physical system which has two points with different (usually opposite) characteristics, such as one which has opposite charges or electric potentials, or opposite magnetic poles. { pə'lar·əd·ē }

polarizability [ELEC] The electric dipole moment induced in a system, such as an atom or molecule, by an electric field of unit strength. { ,pō·lə,rīz·ə'bil·əd·ē }

polarizability catastrophe [ELEC] According to a theory using the Lorentz field concept, the phenomenon where, at a certain temperature, the dielectric constant of a material becomes infinite. { ,pō·lə,rīz·ə'bil·əd·ē kə'tas·trə·fē }

polarizability ellipsoid *See* index ellipsoid. { ,pō·lə,rīz·ə'bil·əd·ē i'lip,sȯid }

polarization [ELEC] **1.** The process of producing a relative displacement of positive and negative bound charges in a body by applying an electric field. **2.** A vector quantity equal to the electric dipole moment per unit volume of a material. Also known as dielectric polarization; electric polarization. **3.** A chemical change occurring in dry cells during use, increasing the internal resistance of the cell and shortening its useful life. [PHYS] **1.** Phenomenon exhibited by certain electromagnetic waves and other transverse waves in which the direction of the electric field or the displacement direction of the vibrations is constant or varies in some definite way. Also known as wave polarization. **2.** The direction of the electric field or the displacement vector of a wave exhibiting polarization (first definition). **3.** The process of bringing about polarization (first definition) in a transverse wave. **4.** Property of a collection of particles with spin, in which the majority have spin components pointing in one direction, rather than at random. { ,pō·lə·rə'zā·shən }

polarization charge *See* bound charge. { ,pō·lə·rə'zā·shən ,chärj }

polarization ellipse [PHYS] The ellipse traced out by the tip of the electric field vector or the displacement vector of a polarized wave at a fixed point in space in the course of time. { ,pō·lə·rə'zā·shən i,lips }

polarization optical coherence tomography [OPTICS] A variation of optical coherence tomography which uses polarization optics in the arms of the fiber-optic Michelson interferometer to determine the sample birefringence from the magnitude of the back-reflected light. { ,pō·lə·rə¦zā·shən ,äp·tə·kəl kō,hir·əns tə'mäg·rə·fē }

polarization potential [ELECTROMAG] One of two vectors from which can be derived, by differentiation, an electric scalar potential and magnetic vector potential satisfying the Lorentz condition. Also known as Hertz vector. { ,pō·lə·rə'zā·shən pə,ten·chəl }

polarization spectroscopy [SPECT] A type of saturation spectroscopy in which a circularly polarized saturating laser beam depletes molecules with a certain orientation preferentially, leaving the remaining ones polarized; the latter are detected through their induction of elliptical polarization in a probe beam, allowing the beam to pass through crossed linear polarizers. { ,pō·lə·rə'zā·shən spek'träs·kə·pē }

polarized electrolytic capacitor [ELEC] An electrolytic capacitor in which the dielectric film is formed adjacent to only one metal electrode; the impedance to the flow of current is then greater in one direction than in the other. { 'pō·lə,rīzd i¦lek·trə¦lid·ik kə'pas·əd·ər }

polarized electromagnetic radiation [ELECTROMAG] Electromagnetic radiation in which the direction of the electric field vector is not random. { 'pō·lə,rīzd i¦lek·trō·mag¦ned·ik ,rād·ē'ā·shən }

polarized light [OPTICS] Polarized electromagnetic radiation whose frequency is in the optical region. { 'pō·lə,rīzd 'līt }

polarized neutrons [PHYS] A collection of neutrons in which the majority have spin pointing in one direction rather than at random. { 'pō·lə,rīzd 'nü,tränz }

polarizer [OPTICS] A device which produces polarized light, such as a Nicol prism or Polaroid sheet. { 'pō·lə,rīz·ər }

polarizing angle *See* Brewster's angle. { 'pō·lə,rīz·iŋ ,aŋ·gəl }

polarizing filter [OPTICS] A device which selectively absorbs components of electromagnetic radiation passing through it, so that light emerging from it is plane-polarized. { 'pō·lə,rīz·iŋ ,fil·tər }

polarizing microscope [OPTICS] A microscope in which an object is viewed in polarized light. { 'pō·lə,rīz·iŋ 'mī·krə,skōp }

polaron [SOLID STATE] An electron in a crystal lattice together with a cloud of phonons that result from the deformation of the lattice produced by the interaction of the electron with ions or atoms in the lattice. { 'pō·lə,rän }

polar radiation pattern [ELECTROMAG] Diagram showing the relative strength of the radiation from an antenna in all directions in a given plane. { 'pō·lər ,rād·ē'ā·shən ,pad·ərn }

polar symmetry [CRYSTAL] A type of crystal symmetry in which the two ends of the central crystallographic axis are not symmetrical. { 'pō·lər 'sim·ə·trē }

polar telescope [OPTICS] A telescope which uses rotating mirrors to enable celestial objects to be observed through a fixed eyepiece. { 'pō·lər 'tel·ə,skōp }

pole [CRYSTAL] **1.** A direction perpendicular to one of the faces of a crystal. **2.** One of the points at which normals to crystal faces or planes intersect a reference sphere at whose center the crystal is located. [ELEC] **1.** One of the electrodes in an electric cell. **2.** An output terminal on a switch; a double-pole switch has two output terminals. [MECH] **1.** A point at which an axis of rotation or of symmetry passes through the surface of a body. **2.** *See* perch. [OPTICS] **1.** The geometric center of a convex or concave mirror. **2.** *See* optical center. { pōl }

pole dominance [PART PHYS] Property of a scattering amplitude, analytically continued to complex values of energy and scattering angle, whose behavior is dominated by the term or terms of negative power in the Laurent series of a nearby pole. { 'pōl ,däm·ə·nəns }

pole face [ELECTROMAG] The end of a magnetic core that faces the air gap in which the magnetic field performs useful work. { 'pōl ,fās }

pole horn [ELECTROMAG] The part of a pole piece or pole shoe in an electrical machine that projects circumferentially beyond the pole core. { 'pōl ,hȯrn }

pole piece [ELECTROMAG] A piece of magnetic material forming one end of an electromagnet or permanent magnet, shaped to control the distribution of magnetic flux in the adjacent air gap. { 'pōl ,pēs }

pole shoe [ELECTROMAG] Portion of a field pole facing the armature of the machine; it may be separable from the body of the pole. { 'pōl ,shü }

pole strength *See* magnetic pole strength. { 'pōl ,streŋkth }

polhode [MECH] For a rotating rigid body not subject to external torque, the closed curve traced out on the inertia ellipsoid by the intersection with this ellipsoid of an axis parallel to the angular velocity vector and through the center. { 'pä,lōd }

polhode cone *See* body cone. { 'pä,lōd ,kōn }

polonium-210 [NUC PHYS] Radioactive isotope of polonium; mass 210, half-life 140 days, α-radiation; used to calibrate radiation counters, and in oil well logging and atomic batteries. Also known as radium F. { pə'lō·nē·əm ¦tü'ten }

polycapillary optics *See* Kumakhov optics. { ,päl·,kap·ə,ler·ē 'äp·tiks }

polychroism *See* pleochroism. { ,päl·i'krō,iz·əm }

polychromatic radiation [ELECTROMAG] Electromagnetic radiation that is spread over a range of frequencies. { ¦päl·i·krō'mad·ik ,rād·ē'ā·shən }

polygonization [SOLID STATE] A phenomenon observed during the annealing of plastically bent crystals in which the edge dislocations created by cold working organize themselves vertically above each other so that polygonal domains are formed. { pə,lig·ə·nə'zā·shən }

polygon wall *See* tilt boundary. { 'päl·i,gän ,wȯl }

polymorph [CRYSTAL] One of the crystal forms of a substance displaying polymorphism. Also known as polymorphic modification. { 'päl·i,mȯrf }

polymorphism [CRYSTAL] The property of a chemical substance crystallizing into two or more forms having different structures, such as diamond and graphite. Also known as pleomorphism. { ,päl·i'mȯr,fiz·əm }

polyrod antenna [ELECTROMAG] End-fire directional dielectric antenna consisting of a polystyrene rod energized by a section of waveguide. { 'päl·i,räd an'ten·ə }

polystyrene capacitor [ELEC] A capacitor that uses film polystyrene as a dielectric between rolled strips of metal foil. { ¦päl·i'stī,rēn kə'pas·əd·ər }

polysynthetic twinning [CRYSTAL] Repeated twinning that involves three or more individual crystals according to the same twin law and on parallel twin planes. { ¦päl·i·sin'thed·ik 'twin·iŋ }

polytropic compression curve [PHYS] Graphical relationship between pressure p and volume V for various values of specific-heat ratios n in the compression formula $pV^n = K$. { ¦päl·i¦träp·ik kəm'presh·ən ˌkərv }

polytropic process [THERMO] An expansion or compression of a gas in which the quantity pV^n is held constant, where p and V are the pressure and volume of the gas, and n is some constant. { ¦päl·i¦träp·ik 'prä·səs }

polytype [CRYSTAL] A type of polymorph whose different forms are due to more than one possible mode of atomic packing. { 'päl·iˌtīp }

polytypism [CRYSTAL] The ability of a mineral to crystallize into more than one form, because of more than one possible mode of atomic packing. { ¦päl·i'tiˌpiz·əm }

Pomeranchuk cooling [CRYO] A method of attaining temperatures as low as 1 millikelvin in which helium-3 is cooled by adiabatic compression at temperatures below 0.3 K. { ¦päm·ə¦ran·chək 'kül·iŋ }

Pomeranchuk pole *See* Pomeron. { ¦päm·ə¦rän·chək ˌpōl }

Pomeranchuk theorem [PART PHYS] The theorem that if the total cross section both for scattering of a particle by a given target particle and for scattering of its antiparticle by the same target particle, approach a limit at high energies, and do so sufficiently rapidly, then these limits must be the same. { ¦päm·ə¦rän·chək ˌthir·əm }

Pomeron [PART PHYS] A Regge pole which is located at + 1 in the angular momentum plane when the momentum transfer in the crossed channel equals zero, corresponding to the fact that total cross sections of reactions are observed to approach constants at high energies. Also known as Pomeranchuk pole. { 'päm·əˌrän }

poncelet [PHYS] A unit of power equal to the power delivered by a force of 100 kilograms-force when the point at which the force is applied is moved at a rate of 1 meter per second in the direction of the force; equal to 980.665 watts. { 'päns·lət }

pond *See* gram-force. { pänd }

ponderomotive force [ATOM PHYS] The part of the interaction of light with atoms that exerts a force on the atoms, rather than coupling with their internal structure. { ˌpän·dərˌmōd·iv 'fȯrs }

Poole-Frenkel effect [ELEC] An increase in the electrical conductivity of insulators and semiconductors in strong electric fields. { ¦pül 'freŋ·kəl iˌfekt }

population inversion [ATOM PHYS] The condition in which a higher energy state in an atomic system is more heavily populated with electrons than a lower energy state of the same system. { ˌpäp·yə'lā·shən inˌvər·zhən }

population of levels [STAT MECH] The number of members of an ensemble which are in each of the allowed energy states of a system. { ˌpäp·yə'lā·shən əv 'lev·əlz }

p orbital [ATOM PHYS] The orbital of an atomic electron with an orbital angular momentum quantum number of unity. { 'pē 'ȯr·bəd·əl }

porcelain capacitor [ELEC] A fixed capacitor in which the dielectric is a high grade of porcelain, molecularly fused to alternate layers of fine silver electrodes to form a monolithic unit that requires no case or hermetic seal. { 'pȯrs·lən kə'pas·əd·ər }

pore diffusion [FL MECH] The movement of fluids (gas or liquid) into the interstices of porous solids or membranes; occurs in membrane separation, zeolite adsorption, dialysis, and reverse osmosis. { 'pȯr diˌfyü·zhən }

porosity [PHYS] **1.** Property of a solid which contains many minute channels or open spaces. **2.** The fraction as a percent of the total volume occupied by these channels or spaces; for example, in petroleum engineering the ratio (expressed in percent) of the void space in a rock to the bulk volume of that rock. { pə'räs·əd·ē }

Porro prism [OPTICS] One of two identical prisms used in the Porro prism erecting system; it is a right-angle prism with the corners rounded to minimize breakage and simplify assembly. { 'pȯ·rō ˌpriz·əm }

Porro-prism erecting system [OPTICS] A compound erecting system, designed by M. Porro, in which there are four reflections to completely erect the image; two Porro prisms are employed; the line of sight is bent through 360°, is displaced, but is not deviated; used in prism binoculars and some telescope systems. { 'pȯ·rō ˌpriz·əm i'rek·tiŋ ˌsis·təm }

port [ELEC] An entrance or exit for a network. [ELECTROMAG] An opening in a waveguide component, through which energy may be fed or withdrawn, or measurements made. { pȯrt }

Portevin-Le Chatelier effect [SOLID STATE] The effect of foreign atoms on the deformation curve of a material, in which steps appear in what was initially a smooth curve. { ¦portˌvan ləˌshat·lē'ā iˌfekt }

position operator [QUANT MECH] The quantum-mechanical operator corresponding to the classical position variable of a particle. { pə'zish·ən ˌäp·əˌrād·ər }

position representation [QUANT MECH] A representation in which the state functions are eigenfunctions of the position operator. Also known as Schrödinger representation. { pə 'zish·ən ˌrep·ri·zən'tā·shən }

positive acceleration [MECH] **1.** Accelerating force in an upward sense or direction, such as from bottom to top, or from seat to head; **2.** The acceleration in the direction that this force is applied. { 'päz·əd·iv akˌsel·ə'rā·shən }

positive birefringence [OPTICS] Birefringence in which the velocity of the ordinary ray is greater than that of the extraordinary ray. { 'päz·əd·iv ˌbi·ri'frin·jəns }

positive charge [ELEC] The type of charge which is possessed by protons in ordinary matter, and which may be produced in a glass object by rubbing with silk. { 'päz·əd·iv 'chärj }

positive crystal [OPTICS] **1.** Uniaxial anisotropic crystal having the ordinary index of refraction greater than the extraordinary index.

2. Biaxial anisotropic crystal having the intermediate index of refraction beta closer in value to alpha, and with Z the acute bisectrix. { 'päz·əd·iv 'krist·əl }

positive electrode *See* anode. { 'päz·əd·iv i'lek,trōd }

positive electron *See* positron. { 'päz·əd·iv i'lek,trän }

positive lens *See* converging lens. { 'päz·əd·iv ,lenz }

positive meniscus lens [OPTICS] A lens having one convex (bulging) and one concave (depressed) surface, with the radius of curvature of the convex surface smaller than that of the concave surface. { 'päz·əd·iv mə'nis·kəs ,lenz }

positive mirror *See* converging mirror. { 'päz·əd·iv 'mir·ər }

positive pole *See* north pole. { 'päz·əd·iv 'pōl }

positive temperature coefficient [THERMO] The condition wherein the resistance, length, or some other characteristic of a substance increases when temperature increases. { 'päz·əd·iv 'tem·prə·chər ,kō·i,fish·ənt }

positive terminal [ELEC] The terminal of a battery or other voltage source toward which electrons flow through the external circuit. { 'päz·əd·iv 'tərm·ən·əl }

positron [PART PHYS] An elementary particle having mass equal to that of the electron, and having the same spin and statistics as the electron, but a positive charge equal in magnitude to the electron's negative charge. Also known as positive electron. { 'päz·ə,trän }

positron depth profiling [SOLID STATE] A technique in which the spread in stopping depths from a low-energy monoenergetic positron beam is measured and used to obtain information on the presence and depth of various crystal defects below the surface. { 'päz·ə,trän 'depth ,prō,fīl·iŋ }

positron emission [NUC PHYS] A β-decay process in which a nucleus ejects a positron and a neutrino. { 'päz·ə,trän i,mish·ən }

positron emission spectroscopy [SPECT] A technique in which a solid surface is bombarded with a low-energy monoenergetic positron beam and the energies of positrons emitted from the surface are measured to determine the amounts of energy lost to molecules adsorbed on the surface. { 'päz·ə,trän i¦mish·ən spek'träs·kə·pē }

positronium [PART PHYS] The bound state of an electron and a positron. { ,päz·ə'trō·nē·əm }

positronium velocity spectroscopy [SPECT] A technique in which a solid surface is bombarded with a low-energy monoenergetic positron beam and the velocities of the emitted positronium atoms are measured to determine the energy and momentum spectrum of the density of electron states near the surface. { ,päz·ə'trō·nē·əm və'läs·əd·ē spek'träs·kə·pē }

postcollision interaction [ATOM PHYS] A phenomenon that arises near a threshold of photoionization in which a slowly receding photoelectron transfers energy to an Auger electron. { ¦pōst·kə¦lizh·ən ,in·tər'ak·shən }

post-Galilean transformation [RELAT] A modification of the Lorentz transformation that includes first-order corrections associated with the effects of general relativity. { ¦pōst ,gal·ə'lē·ən ,tranz·fər'mā·shən }

post-Newtonian effects [RELAT] The first-order corrections of general relativity to classical Newtonian mechanics. { ¦pōst nü'tō·nē·ən i'feks }

pot *See* potentiometer. { pät }

potassium-40 [NUC PHYS] A radioactive isotope of potassium having a mass number of 40, a half-life of approximately 1.31×10^9 years, and an atomic abundance of 0.000122 gram per gram of potassium. { pə'tas·ē·əm ¦fȯr·dē }

potassium-42 [NUC PHYS] Radioactive isotope with mass number of 42; half-life is 12.4 hours, with β- and γ-radiation; radiotoxic; used as radiotracer in medicine. { pə'tas·ē·əm ¦fȯr·dē'tü }

pot core [ELECTROMAG] A ferrite magnetic core that has the shape of a pot, with a magnetic post in the center and a magnetic plate as a cover; the coils for a choke or transformer are wound on the center post. { 'pät ,kȯr }

potential *See* electric potential. [PHYS] A function or set of functions of position in space, from whose first derivatives a vector can be formed, such as that of a static field intensity. { pə'ten·chəl }

potential barrier [PHYS] The potential in a region in a field of force where the force exerted on a particle is such as to oppose the passage of the particle through the region. Also known as barrier; potential hill. { pə'ten·chəl 'bar·ē·ər }

potential density [PHYS] The density that would be reached by a compressible fluid if it were adiabatically compressed or expanded to a standard pressure of the bar. { pə'ten·chəl 'den·səd·ē }

potential difference [ELEC] Between any two points, the work which must be done against electric forces to move a unit charge from one point to the other. Abbreviated PD. { pə'ten·chəl ¦dif·rəns }

potential divider *See* voltage divider. { pə'ten·chəl di'vīd·ər }

potential drop [ELEC] The potential difference between two points in an electric circuit. [FL MECH] The difference in pressure head between one equipotential line and another. { pə'ten·chəl ¦dräp }

potential energy [MECH] The capacity to do work that a body or system has by virtue of its position or configuration. { pə'ten·chəl 'en·ər·jē }

potential flow [FL MECH] Flow in which the velocity of flow is the gradient of a scalar function, known as the velocity potential. { pə'ten·chəl 'flō }

potential gradient [ELEC] Difference in the values of the voltage per unit length along a conductor or through a dielectric. { pə'ten·chəl 'grād·ē·ənt }

potential head *See* elevation head. { pə'ten·chəl 'hed }

potential hill *See* potential barrier. { pə'ten·chəl ˌhil }

potential scattering [QUANT MECH] Scattering of a particle which can be treated as the effect of a potential, representing the particle's potential energy, on the particle's Schrödinger wave function. { pə'ten·chəl 'skad·əˌriŋ }

potential temperature [THERMO] The temperature that would be reached by a compressible fluid if it were adiabatically compressed or expanded to a standard pressure, usually 1 bar. { pə'ten·chəl 'tem·prə·chər }

potential transformer *See* voltage transformer. { pə'ten·chəl tranz'fȯr·mər }

potential transformer phase angle [ELEC] Angle between the primary voltage vector and the secondary voltage vector reversed; this angle is conveniently considered as positive when the reversed, secondary voltage vector leads the primary voltage vector. { pə'ten·chəl tranz'fȯr·mər 'fāz ˌaŋ·gəl }

potential vorticity [FL MECH] The product of the absolute vorticity and the static stability, conservative in adiabatic flow, given by the expression $(\eta/\theta)(\partial\theta/\partial p)$, where η is the absolute vorticity of a fluid parcel, θ the potential temperature, and p the pressure. Also known as absolute potential vorticity. { pə'ten·chəl vȯr'tis·əd·ē }

potential well [PHYS] For an object in a conservative field of force, a region in which the object has a lower potential energy than in all the surrounding regions. { pə'ten·chəl ¦wel }

potentiometer [ELEC] A resistor having a continuously adjusted sliding contact that is generally mounted on a rotating shaft; used chiefly as a voltage divider. Also known as pot (slang). { pəˌten·chē'äm·əd·ər }

potentiometric electrode [ELEC] An electrode that produces a voltage logarithmically dependent on the concentration of a selected ionic substance. { pəˌten·chē·ə¦me·trik i'lekˌtrōd }

potentiometry [ELEC] Use of a potentiometer to measure electromotive forces, and the applications of such measurements. { pəˌten·chē'äm·ə·trē }

pound [MECH] **1.** A unit of mass in the English absolute system of units, equal to 0.45359237 kilogram. Abbreviated lb. Also known as avoirdupois pound; pound mass. **2.** A unit of force in the English gravitational system of units, equal to the gravitational force experienced by a pound mass when the acceleration of gravity has its standard value of 9.80665 meters per second per second (approximately 32.1740 ft/s^2) equal to 4.4482216152605 newtons. Abbreviated lb. Also spelled Pound (Lb). Also known as pound force (lbf). **3.** A unit of mass in the troy and apothecaries' systems, equal to 12 troy or apothecaries' ounces, or 5760 grains, or 5760/7000 avoirdupois pound, or 0.3732417216 kilogram. Also known as apothecaries' pound (abbreviated lb ap in the United States or lb apoth in the United Kingdom); troy pound (abbreviated lb t in the United States, or lb tr or lb in the United Kingdom). { paůnd }

poundal [MECH] A unit of force in the British absolute system of units equal to the force which will impart an acceleration of 1 ft/s^2 to a pound mass, or to 0.138254954376 newton. { 'paůnd·əl }

poundal-foot *See* foot-poundal. { 'paůnd·əl 'fůt }

pound-foot *See* foot-pound. { 'paůnd 'fůt }

pound force *See* pound. { 'paůnd 'fȯrs }

pound mass *See* pound. { 'paůnd 'mas }

pound per square foot [MECH] A unit of pressure equal to the pressure resulting from a force of 1 pound applied uniformly over an area of 1 square foot. Abbreviated psf. { 'paůnd pər ¦skwer 'fůt }

pound per square inch [MECH] A unit of pressure equal to the pressure resulting from a force of 1 pound applied uniformly over an area of 1 square inch. Abbreviated psi. { 'paůnd pər ¦skwer 'inch }

Pound-Rebka experiment [RELAT] A terrestrial experiment demonstrating the gravitational redshift of light. { ¦paůnd 'reb·kə ikˌsper·ə·mənt }

pounds per square inch absolute [MECH] The absolute, thermodynamic pressure, measured by the number of pounds-force exerted on an area of 1 square inch. Abbreviated lbf in.$^{-2}$ abs; psia. { 'paůns pər ¦skwer 'inch 'ab·səˌlüt }

pounds per square inch gage [MECH] The gage pressure, measured by the number of pounds-force exerted on an area of 1 square inch. Abbreviated psig. { 'paůns pər ¦skwer 'inch 'gāj }

pour point [FL MECH] Lowest test temperature at which a liquid will flow. { 'pȯr ˌpȯint }

powder diffraction camera [CRYSTAL] A metal cylinder having a window through which an x-ray beam of known wavelength is sent by an x-ray tube to strike a finely ground powder sample mounted in the center of the cylinder; crystal planes in this powder sample diffract the x-ray beam at different angles to expose a photographic film that lines the inside of the cylinder; used to study crystal structure. Also known as x-ray powder diffractometer. { 'paůd·ər di'frak·shən ˌkam·rə }

powder method [SOLID STATE] A method of x-ray diffraction analysis in which a collimated, monochromatic beam of x-rays is directed at a sample consisting of an enormous number of tiny crystals having random orientation, producing a diffraction pattern that is recorded on film or with a counter tube. Also known as x-ray powder method. { 'paůd·ər ˌmeth·əd }

powder pattern [CRYSTAL] In the powder method of x-ray diffraction analysis, the display of lines made on film by the Debye-Scherrer method or on paper by a recording diffractometer. [ELECTROMAG] The pattern created by very fine powders or colloidal particles, spread over the surface of a magnetic material; reveals the magnetic domains in a single crystal of such material. { 'paůd·ər ˌpad·ərn }

power [OPTICS] *See* focal power. [PHYS] The time rate of doing work. { 'paů·ər }

power component *See* active component. { 'paů·ər kəmˌpō·nənt }

power density [ELECTROMAG] The amount of power per unit area in a radiated microwave or other electromagnetic field, usually expressed in units of watts per square centimeter. { 'pau̇·ər ¦den·səd·ē }

power-density spectrum *See* frequency spectrum. { 'pau̇·ər ¦den·səd·ē ˌspek·trəm }

power divider [ELECTROMAG] A device used to produce a desired distribution of power at a branch point in a waveguide system. { 'pau̇·ər diˌvīd·ər }

power factor [ELEC] The ratio of the average (or active) power to the apparent power (root-mean-square voltage times rms current) of an alternating-current circuit. Abbreviated pf. Also known as phase factor. { 'pau̇·ər ˌfak·tər }

power flow [ELECTROMAG] The rate at which energy is transported across a surface by an electromagnetic field. { 'pau̇·ər ˌflō }

power gain [ELECTROMAG] An antenna ratio equal to 4π (12.57) times the ratio of the radiation intensity in a given direction to the total power delivered to the antenna. { 'pau̇·ər ˌgān }

power-law fluid [FL MECH] A fluid in which the shear stress at any point is proportional to the rate of shear at that point raised to some power. { 'pau̇·ər ˌlȯ ˌflü·əd }

power ratio [ELECTROMAG] The ratio of the maximum power to the minimum power in a waveguide that is improperly terminated. { 'pau̇·ər ˌrā·shō }

power spectrum *See* frequency spectrum. { 'pau̇·ər ˌspek·trəm }

power transfer theorem [ELEC] The theorem that, in an electrical network which carries direct or sinusoidal alternating current, the greatest possible power is transferred from one section to another when the impedance of the section that acts as a load is the complex conjugate of the impedance of the section that acts as a source, where both impedances are measured across the pair of terminals at which the power is transferred, with the other part of the network disconnected. { ¦pau̇·ər 'tranz·fər ˌthir·əm }

Poynting effect [MECH] The effect of torsion of a very long cylindrical rod on its length. { 'pȯin·tiŋ iˌfekt }

Poynting's law [THERMO] A special case of the Clapeyron equation, in which the fluid is removed as fast as it forms, so that its volume may be ignored. { 'pȯint·iŋz ˌlȯ }

Poynting theorem [ELECTROMAG] A theorem, derived from Maxwell's equations, according to which the rate of loss of energy stored in electric and magnetic fields within a region of space is equal to the sum of the rate of dissipation of electrical energy as heat and the rate of flow of electromagnetic energy outward through the surface of the region. { 'pȯint·iŋ ˌthir·əm }

Poynting vector [ELECTROMAG] A vector, equal to the cross product of the electric-field strength and the magnetic-field strength (mks units) whose outward normal component, when integrated over a closed surface, gives the outward flow of electromagnetic energy through that surface. { 'pȯint·iŋ ˌvek·tər }

p-process [NUC PHYS] The synthesis of certain nuclides in stars through capture of protons or ejection of neutrons by gamma rays. { 'pē ˌprä·səs }

PP reaction *See* proton-proton reaction. { ¦pē'pē rēˌak·shən }

practical entropy *See* virtual entropy. { 'prak·ti·kəl 'en·trə·pē }

practical system *See* meter-kilogram-second-ampere system. { 'prak·ti·kəl 'sis·təm }

practical units [ELECTROMAG] The units of the meter-kilogram-second-ampere system. { 'prak·ti·kəl ˌyü·nəts }

Prandtl-Glauert rule [FL MECH] The rule that the pressure coefficient at any point in the subsonic flow of a fluid about a slender body is equal to the pressure coefficient at that point in the corresponding incompressible fluid flow, divided by

$$\sqrt{1 - M^2},$$

where M is the Mach number far from the body. { 'pränt·əl 'glau̇·ərt ˌrül }

Prandtl-Meyer flow [FL MECH] A two-dimensional, supersonic fluid flow in which an initially uniform flow passes a sharp, convex corner in a boundary, resulting in expansion of the fluid. { 'pränt·əl 'mī·ər ˌflō }

Prandtl number [FL MECH] A dimensionless number used in the study of diffusion in flowing systems, equal to the kinematic viscosity divided by the molecular diffusivity. Symbolized Pr_m. Also known as Schmidt number I (N_{Sc}). [THERMO] A dimensionless number used in the study of forced and free convection, equal to the dynamic viscosity times the specific heat at constant pressure divided by the thermal conductivity. Symbolized N_{Pr}. { 'pränt·əl ˌnəm·bər }

precedence effect [ACOUS] The ability of the auditory system to process sound that reaches the ears directly from a source even when significant reflected sounds reach the ears shortly afterward. { prə'sēd·əns iˌfekt *or* 'pre·səd·əns }

precession [MECH] The angular velocity of the axis of spin of a spinning rigid body, which arises as a result of external torques acting on the body. { prē'sesh·ən }

precessional torque [MECH] A torque which causes a rotating body to precess. { prē¦sesh·ən·əl 'tȯrk }

precession camera [CRYSTAL] An x-ray diffraction camera used in the Buerger precession method for recording the diffractions of an individual crystal. { prē'sesh·ən ˌkam·rə }

precipitation attenuation [ELECTROMAG] Loss of radio energy due to the passage through a volume of the atmosphere containing precipitation; part of the energy is lost by scattering, and part by absorption. { prəˌsip·ə'tā·shən əˌten·yə'wā·shən }

preionization *See* autoionization. { ¦prēˌī·ə·nə 'zā·shən }

presence [ACOUS] The impression, as created

by a recording or radio receiver, that the original program source is in the room. { 'prez·əns }

press camera [OPTICS] A folding camera, usually of 4- by 5-inch (10.2- by 12.7-centimeter) format, once widely used in newspaper photography. { 'pres ˌkam·rə }

pressure [MECH] A type of stress which is exerted uniformly in all directions; its measure is the force exerted per unit area. { 'presh·ər }

pressure broadening [SPECT] A spreading of spectral lines when pressure is increased, due to an increase in collision broadening. { 'presh·ər ˌbrȯd·ən·iŋ }

pressure coefficient [THERMO] The ratio of the fractional change in pressure to the change in temperature under specified conditions, usually constant volume. { 'presh·ər ˌkō·iˌfish·ənt }

pressure drag *See* pressure resistance. { 'presh·ər ˌdrag }

pressure drop [FL MECH] The difference in pressure between two points in a flow system, usually caused by frictional resistance to a fluid flowing through a conduit, filter media, or other flow-conducting system. { 'presh·ər ˌdräp }

pressure effect [SPECT] The effect of changes in pressure on spectral lines in the radiation emitted or absorbed by a substance; namely, pressure broadening and pressure shift. { 'presh·ər iˌfekt }

pressure-enthalpy chart [PHYS] A graph of the pressure versus the enthalpy of a substance at various values of temperature, specific volume, and entropy; especially useful in refrigeration calculations. Also known as enthalpy-pressure chart. { 'presh·ər 'enˌthal·pē ˌchärt }

pressure force [FL MECH] The force due to differences of pressure within a fluid mass; the (vector) force per unit volume is equal to the pressure gradient $-\nabla p$, and the force per unit mass (specific force) is equal to the product of the volume force and the specific volume $-\alpha\nabla p$. { 'presh·ər ˌfȯrs }

pressure front *See* shock front. { 'presh·ər ˌfrənt }

pressure gradient [FL MECH] The rate of decrease (that is, the gradient) of pressure in space at a fixed time; sometimes loosely used to denote simply the magnitude of the gradient of the pressure field. Also known as barometric gradient. { 'presh·ər ˌgrād·ē·ənt }

pressure head [FL MECH] Also known as head. **1.** The height of a column of fluid necessary to develop a specific pressure. **2.** The pressure of water at a given point in a pipe arising from the pressure in it. { 'presh·ər ˌhed }

pressure melting [PHYS] The melting of ice due to applied pressure. { 'presh·ər ¦melt·iŋ }

pressure melting temperature [PHYS] The temperature at which ice can melt at a given pressure. { 'presh·ər ¦melt·iŋ ˌtem·prə·chər }

pressure resistance [FL MECH] In fluid dynamics, a normal stress caused by acceleration of the fluid, which results in a decrease in pressure from the upstream to the downstream side of an object acting perpendicular to the boundary. Also known as pressure drag. { 'presh·ər riˌzis·təns }

pressure-sensitive paint [FL MECH] A flow visualization technique in which ultraviolet light is used to excite specific molecules in a special paint affixed to a test surface positioned in a wind tunnel flow. The resulting phosphorescence of these molecules indicates the amount of oxygen in contact with the paint and, thereby, the spatial distribution of surface pressure. { ˌpresh·ər ˌsen·səd·iv 'pānt }

pressure shift [SPECT] An increase in the wavelength at which a spectral line has maximum intensity, which takes place when pressure is increased. { 'presh·ər ˌshift }

pressure tensor [PL PHYS] A tensor which plays a role in magnetohydrodynamics analogous to that of the pressure in ordinary fluid mechanics. { 'presh·ər ˌten·sər }

pressure-travel curve [MECH] Curve showing pressure plotted against the travel of the projectile within the bore of the weapon. { 'presh·ər ¦trav·əl ˌkərv }

pressure viscosity [FL MECH] Property of petroleum lubricating oils to increase in viscosity when subjected to pressure. { 'presh·ər viˌskäs·əd·ē }

pressure wave *See* compressional wave. { 'presh·ər ˌwāv }

Prevost's theory [THERMO] A theory according to which a body is constantly exchanging heat with its surroundings, radiating an amount of energy which is independent of its surroundings, and increasing or decreasing its temperature depending on whether it absorbs more radiation than it emits, or vice versa. { 'prā·vōz ˌthē·ə·rē }

priest [OPTICS] The Z tristimulus value. { prēst }

primary colors [OPTICS] **1.** Three colors, red, yellow, and blue, which can be combined in various proportions to produce any other color. **2.** Any three colors that can be mixed in proper proportions to specify other colors; they need not be physically realizable. { 'prīˌmer·ē 'kəl·ərz }

primary cosmic rays *See* cosmic rays. { 'prīˌmer·ē ¦käz·mik 'rāz }

primary creep [MECH] The initial high strain-rate region in a material subjected to sustained stress. { 'prīˌmer·ē 'krēp }

primary extinction [SOLID STATE] A weakening of the stronger beams produced in x-ray diffraction by a very perfect crystal, as compared with the weaker. { 'prīˌmer·ē ik'stiŋk·shən }

primary focus [OPTICS] In an astigmatic system, a line at which some of the bundle of rays from an off-axis point meet; this line is perpendicular to a plane which contains the point and the optical axis, and has a smaller image distance than the secondary focus. Also known as meridional focus; tangential focus. { 'prīˌmer·ē 'fō·kəs }

primary knocked-on atom [PHYS] An atom in a solid that recoils from a collision with an energetic particle coming from outside the solid, rather than with another knocked-on atom. Abbreviated PKA. { 'prī,mer·ē ¦näkt ,ȯn 'ad·əm }

primary lights [OPTICS] Any three lights used in a system of tristimulus colorimetric analysis of solutions. { 'prī,mer·ē 'līts }

primary optic axis [OPTICS] One of two optic axes in a crystal that are perpendicular to the circular sections of the indicatrix and along which all light rays travel with equal velocity. { 'prī,mer·ē 'äp·tik ,ak·səs }

primary phase [THERMO] The only crystalline phase capable of existing in equilibrium with a given liquid. { 'prī,mer·ē 'fāz }

primary phase region [THERMO] On a phase diagram, the locus of all compositions having a common primary phase. { 'prī,mer·ē 'fāz ,rē·jən }

primary radiation [PHYS] Radiation arriving directly from its source without interaction with matter. { 'prī,mer·ē ,rād·ē'ā·shən }

primary rainbow [OPTICS] The most common of the principal rainbow phenomena, which appears as an arc of angular radius of about 42° about the observer's antisolar point; it is the inner of two rainbows, whose light undergoes only one internal reflection, and which is narrower and brighter than the outer, or secondary, rainbow. { 'prī,mer·ē 'rān,bō }

primary scattering [PHYS] Any scattering process in which radiation is received at a detector, such as the eye, after having been scattered just once; distinguished from multiple scattering. { 'prī,mer·ē 'skad·ə·riŋ }

primary skip zone [ELECTROMAG] Area around a transmitter beyond the ground wave but within the skip distance. { 'prī,mer·ē 'skip ,zōn }

primary stress [MECH] A normal or shear stress component in a solid material which results from an imposed loading and which is under a condition of equilibrium and is not self-limiting. { 'prī,mer·ē 'stres }

prime focus [OPTICS] The position in a reflecting telescope at which light from celestial objects is focused by the main mirror, located on the axis of the mirror near the open end of the tube. { 'prīm 'fō·kəs }

primitive cell [CRYSTAL] A parallelepiped whose edges are defined by the primitive translations of a crystal lattice; it is a unit cell of minimum volume. { 'prim·əd·iv 'sel }

primitive equations [FL MECH] The Eulerian equations of motion of a fluid in which the primary dependent variables are the fluid's velocity components; these equations govern a wide variety of fluid motions and form the basis of most hydrodynamical analysis; in meteorology, these equations are frequently specialized to apply directly to the cyclonic-scale motions by the introduction of filtering approximations. { 'prim·əd·iv i'kwā·zhənz }

primitive lattice [CRYSTAL] A crystal lattice in which there are lattice points only at its corners. Also known as simple lattice. { 'prim·əd·iv 'lad·əs }

primitive translation [CRYSTAL] For a space lattice, one of three translations which can be repeatedly applied to generate any translation which leaves the lattice unchanged. { 'prim·əd·iv tranz'lā·shən }

principal axis [CRYSTAL] The longest axis in a crystal. [MECH] One of three perpendicular axes in a rigid body such that the products of inertia about any two of them vanish. [OPTICS] *See* optic axis. { 'prin·sə·pəl 'ak·səs }

principal axis of strain [MECH] One of the three axes of a body that were mutually perpendicular before deformation. Also known as strain axis. { 'prin·sə·pəl 'ak·səs əv 'strān }

principal axis of stress [MECH] One of the three mutually perpendicular axes of a body that are perpendicular to the principal planes of stress. Also known as stress axis. { 'prin·sə·pəl 'ak·səs əv 'stres }

principal E plane [ELECTROMAG] Plane containing the direction of radiation of electromagnetic waves and arranged so that the electric vector everywhere lies in the plane. { 'prin·sə·pəl 'ē ,plān }

principal focus *See* focal point. { 'prin·sə·pəl 'fō·kəs }

principal function [MECH] The integral of the Lagrangian of a system over time; it is involved in the statement of Hamilton's principle. { 'prin·sə·pəl 'fəŋk·shən }

principal H plane [ELECTROMAG] Plane that contains the direction of radiation and the magnetic vector, and is everywhere perpendicular to the E plane. { 'prin·sə·pəl 'āch ,plān }

principal line [SPECT] That spectral line which is most easily excited or observed. { 'prin·sə·pəl 'līn }

principal lobe [PHYS] The lobe of a radiation pattern or directivity pattern that lies on the axis of symmetry of an acoustic or electromagnetic transmitter or receptor. { 'prin·sə·pəl 'lōb }

principal mode *See* fundamental mode. { 'prin·sə·pəl 'mōd }

principal plane [OPTICS] **1.** Two planes perpendicular to the optical axis such that objects in one plane form images in the other with a lateral magnification of unity. **2.** The vertical plane passing through the internal perspective center and containing the perpendicular from that center to the plane of a tilted photograph. **3.** *See* principal section. { 'prin·sə·pəl 'plān }

principal plane of stress [MECH] For a point in an elastic body, a plane at that point across which the shearing stress vanishes. { 'prin·sə·pəl 'plān əv 'stres }

principal point [OPTICS] The intersection of a principal plane with the optical axis. { 'prin·sə·pəl 'pȯint }

principal quantum number [ATOM PHYS] A quantum number for orbital electrons, which, together with the orbital angular momentum and spin quantum numbers, labels the electron wave function; the energy level and the average distance of an electron from the nucleus depend mainly upon this quantum number. { 'prin·sə·pəl 'kwän·təm ˌnəm·bər }

principal ray [OPTICS] **1.** The one ray within a bundle of incident rays that, upon entering an optical instrument from any given point of the object, passes through the optical center of the lens. **2.** *See* principal visual ray. { 'prin·sə·pəl 'rā }

principal section [OPTICS] A plane in a crystal that contains the crystal's optic axis and the ray of light under consideration. Also known as principal plane. { 'prin·sə·pəl 'sek·shən }

principal series [SPECT] A series occurring in the line spectra of many atoms and ions with one, two, or three electrons in the outer shell, in which the total orbital angular momentum quantum number changes from 1 to 0. { 'prin·sə·pəl 'sir·ēz }

principal strain [MECH] The elongation or compression of one of the principal axes of strain relative to its original length. { 'prin·sə·pəl 'strān }

principal stress [MECH] A stress occurring at right angles to a principal plane of stress. { 'prin·sə·pəl 'stres }

principal visual ray [OPTICS] A perpendicular extending from a station point to a perspective plane and theoretically passing exactly along the visual axis of a viewing eye. Also known as principal ray. { 'prin·sə·pəl 'vish·ə·wəl 'rā }

principle of covariance [RELAT] **1.** In classical physics and in special relativity, the principle that the laws of physics take the same mathematical form in all inertial reference frames. **2.** In general relativity, the principle that the laws of physics take the same mathematical form in all conceivable curvilinear coordinate systems. { 'prin·sə·pəl əv kō'ver·ē·əns }

principle of duality *See* duality principle. { 'prin·sə·pəl əv dü'al·əd·ē }

principle of dynamical similarity [MECH] The principle that two physical systems which are geometrically and kinematically similar at a given instant, and physically similar in constitution, will retain this similarity at later corresponding instants if and only if the Froude number 1 for each independent type of force has identical values in the two systems. Also known as similarity principle. { ¦prin·sə·pəl əv di¦nam·ə·kəl ˌsim·ə'lar·əd·ē }

principle of equivalence *See* equivalence principle. { 'prin·sə·pəl əv i'kwiv·ə·ləns }

principle of inaccessibility *See* Carathéodory's principle. { 'prin·sə·pəl əv ˌin·akˌses·ə'bil·əd·ē }

principle of least action [MECH] The principle that, for a system whose total mechanical energy is conserved, the trajectory of the system in configuration space is that path which makes the value of the action stationary relative to nearby paths between the same configurations and for which the energy has the same constant value. Also known as least-action principle. { 'prin·sə·pəl əv ˌlēst 'ak·shən }

principle of reciprocity *See* reciprocity theorem. { 'prin·sə·pəl əv ˌres·ə'präs·əd·ē }

principle of superposition [ELEC] **1.** The principle that the total electric field at a point due to the combined influence of a distribution of point charges is the vector sum of the electric field intensities which the individual point charges would produce at that point if each acted alone. **2.** The principle that, in a linear electrical network, the voltage or current in any element resulting from several sources acting together is the sum of the voltages or currents resulting from each source acting alone. Also known as superposition theorem. [MECH] The principle that when two or more forces act on a particle at the same time, the resultant force is the vector sum of the two. [PHYS] Also known as superposition principle. **1.** A general principle applying to many physical systems which states that if a number of independent influences act on the system, the resultant influence is the sum of the individual influences acting separately. **2.** In all theories characterized by linear homogeneous differential equations, such as optics, acoustics, and quantum theory, the principle that the sum of any number of solutions to the equations is another solution. { 'prin·sə·pəl əv ˌsü·pər·pə'zish·ən }

principle of virtual work [MECH] The principle that the total work done by all forces acting on a system in static equilibrium is zero for any infinitesimal displacement from equilibrium which is consistent with the constraints of the system. Also known as virtual work principle. { 'prin·sə·pəl əv 'vər·chə·wəl ˌwərk }

prism [CRYSTAL] A crystal which has three, four, six, eight, or twelve faces, with the face intersection edges parallel, and which is open only at the two ends of the axis parallel to the intersection edges. [OPTICS] An optical system consisting of two or more usually plane surfaces of a transparent solid or embedded liquid at an angle with each other. Also known as optical prism. { 'priz·əm }

prismatic binoculars *See* prism binoculars. { priz'mad·ik bə'näk·yə·lərz }

prismatic cleavage [CRYSTAL] A type of crystal cleavage that occurs parallel to the faces of a prism. { priz'mad·ik 'klē·vij }

prismatic error [OPTICS] That error due to lack of parallelism of the two faces of an optical element, such as a mirror or a shade glass. { priz 'mad·ik 'er·ər }

prism binoculars [OPTICS] A type of binoculars, each half of which is a Kepler telescope that employs a Porro prism erecting system both to erect the image and to reduce the length of the instrument. Also known as prismatic binoculars. { 'priz·əm bə'näk·yə·lərz }

prism diopter [OPTICS] A unit used in measuring the deviating power of a prism; this power in prism diopters is 100 times the tangent of the angle of deviation of a ray of light. { 'priz·əm dī'äp·tər }

prism spectrograph [SPECT] Analysis device in

which a prism is used to give two different but simultaneous light wavelengths derived from a common light source; used for the analysis of materials by flame photometry. { 'priz·əm 'spek·trə,graf }

prism transit *See* broken-back transit. { 'priz·əm 'trans·ət }

privileged direction [OPTICS] One of two mutually perpendicular directions for the plane of polarization of a beam of plane-polarized light falling on a plate of anisotropic material such that the light which emerges from the plate is also plane-polarized. { 'priv·ə·lijd də'rek·shən }

probability amplitude *See* Schrödinger wave function. { ,präb·ə'bil·əd·ē 'am·plə,tüd }

probability current density [QUANT MECH] A vector whose component normal to a surface gives the probability that a particle will cross a unit area of the surface during a unit time. { ,präb·ə'bil·əd·ē ¦kə·rənt ,den·səd·ē }

probability density [QUANT MECH] The square of the absolute value of the Schrödinger wave function for a particle at a given point; gives the probability per unit volume of finding the particle at that point. { ,präb·ə'bil·əd·ē ,den·səd·ē }

probe [ELECTROMAG] A metal rod that projects into but is insulated from a waveguide or resonant cavity; used to provide coupling to an external circuit for injection or extraction of energy or to measure the standing-wave ratio. Also known as waveguide probe. [PHYS] A small device which can be brought into contact with or inserted into a system in order to make measurements on the system; ordinarily it is designed so that it does not significantly disturb the system. { prōb }

probe coil [ELECTROMAG] In eddy-current nondestructive tests, a type of test coil which is placed on the surface of an object. { 'prōb ,kȯil }

Proca equations [QUANT MECH] A set of equations, analogous to Maxwell's equations, relating a four-vector potential and a second-rank tensor field describing a particle of spin 1 and nonzero mass. { 'prō·kə i,kwā·zhənz }

process camera [OPTICS] Large camera used to produce materials for reproduction in printing; permits a large range of enlargement and reduction. { 'prä,səs ,kam·rə }

process lens [OPTICS] A highly corrected, apochromatic lens used for precise color-separation work. { 'prä,səs ,lenz }

product of inertia [MECH] Relative to two rectangular axes, the sum of the products formed by multiplying the mass (or, sometimes, the area) of each element of a figure by the product of the coordinates corresponding to those axes. { 'prä·dəkt əv i'nər·shə }

profile chart [ELECTROMAG] A vertical cross-section drawing of a microwave path between two stations, indicating terrain, obstructions, and antenna height requirements. { 'prō,fīl ,chärt }

profile drag [FL MECH] That part of the airfoil drag that results from the skin friction and the shape of the airfoil as indicated by the airfoil profile. { ¦prō,fīl ,drag }

progressive lens *See* omnifocal lens. { prə'gres·iv 'lenz }

progressive wave [PHYS] A wave which transfers energy from one part of a medium to another, in contrast to a standing wave. Also known as free-traveling wave. { prə'gres·iv 'wāv }

progressive-wave antenna *See* traveling-wave antenna. { prə'gres·iv ¦wāv an'ten·ə }

projection microradiography [PHYS] Microradiography in which an electron beam, focused into an extremely fine pencil, generates a point source of x-rays, and enlargement is achieved by placing the sample very near this source, and several centimeters from the recording material. Abbreviated PMR. Also known as shadow microscopy; x-ray projection microscopy. { prə 'jek·shən ¦mī·krō,rād·ē'äg·rə·fē }

projection microscope [PHYS] An x-ray microscope which magnifies by image projection, either in contact microradiography or in projection microradiography. { prə'jek·shən 'mī·krə ,skōp }

projection optics *See* Schmidt system. { prə'jek·shən ,äp·tiks }

projection printer [OPTICS] An optical, image-enlarging device, used in enlarging photographs. { prə'jek·shən ,print·ər }

projector *See* optical projection system. { prə'jek·tər }

promethium-147 [NUC PHYS] Artificially produced isotope with atomic number 61 and mass 147; produced during fission of ^{235}U. Also known as florentium; illinium. { prə'mē·thē·əm ¦wən¦fȯrd·ē'sev·ən }

prompt neutron [NUC PHYS] A neutron released coincident with the fission process, as opposed to neutrons subsequently released. { 'prämpt ,nü,trän }

prompt radiation [NUC PHYS] Radiation emitted within a time too short for measurement, including γ-rays, characteristic x-rays, conversion and Auger electrons, prompt neutrons, and annihilation radiation. { 'prämpt ,rād·ē,ā·shən }

proof plane [ELEC] A small metal plane supported by an insulating handle and used to transfer a small fraction of the electric charge on a body to an electrometer to investigate the charge distribution on the body. { 'prüf ,plān }

proof resilience [MECH] The tensile strength necessary to stretch an elastomer from zero elongation to the breaking point, expressed in foot-pounds per cubic inch of original dimension. { 'prüf ri,zil·yəns }

proof stress [MECH] **1.** The stress that causes a specified amount of permanent deformation in a material. **2.** A specified stress to be applied to a member or structure in order to assess its ability to support service loads. { 'prüf ,stres }

propagation *See* wave motion. { ,präp·ə'gā·shən }

propagation anomaly [PHYS] Change in propagation characteristics due to a resonance in the medium of propagation. { ˌpräp·əˈgā·shən əˌnäm·ə·le }

propagation constant [ELECTROMAG] A rating for a line or medium along or through which a wave of a given frequency is being transmitted; it is a complex quantity; the real part is the attenuation constant in nepers per unit length, and the imaginary part is the phase constant in radians per unit length. { ˌpräp·əˈgā·shən ˌkän·stənt }

propagation mode [ELECTROMAG] A form of propagation of electromagnetic radiation in a periodic beamguide in which the field distributions over cross sections of the beam are identical at positions separated by one period of the guide. { ˈpräp·əˈgā·shən ˌmōd }

propagation velocity [ELECTROMAG] Velocity of electromagnetic wave propagation in the medium under consideration. { ˌpräp·əˈgā·shən vəˌläs·əd·ē }

propagator [QUANT MECH] The probability amplitude for a particle to move or propagate to some new point of space and time when its amplitude at some point of origination is known. { ˈpräp·əˌgād·ər }

propeller cavitation [FL MECH] Formation of vapor-filled and air-filled bubbles or cavities in water at or on the surface of a rotating propeller, occurring when the pressure falls below the vapor pressure of water. { prəˈpel·ər ˌkav·əˌtā·shən }

proper Lorentz transformation [RELAT] A Lorentz transformation which can be represented by a matrix whose determinant is +1. { ˈpräp·ər ˈlȯr·əns ˌtranz·fərˌmā·shən }

proper time [RELAT] The time measured by an ideal clock that is carried along with a specified particle, and is based on the invariant timelike space-time intervals between points along the particle's trajectory. { ˈpräp·ər ˈtīm }

proportional band [ACOUS] One of a series of frequency bands whose members have equal band ratios. { prəˈpȯr·shən·əl ˈband }

proportional elastic limit [MECH] The greatest stress intensity for which stress is still proportional to strain. { prəˈpȯr·shən·əl iˈlas·tik ˌlim·ət }

proportional limit [MECH] The greatest stress a material can sustain without departure from linear proportionality of stress and strain. { prəˈpȯr·shən·əl ˈlim·ət }

proportioning reactor [ELECTROMAG] A saturable-core reactor used for regulation and control; increasing the input control current from zero to rated value makes output current increase in proportion from cutoff up to full load value. { prəˈpȯr·shən·iŋ rēˌak·tər }

propulsion [MECH] The process of causing a body to move by exerting a force against it. { prəˈpəl·shən }

protectoscope [OPTICS] Device in a tank or armored car, similar to the periscope of a submarine; it enables a soldier to see around a shield without exposing himself to enemy gunfire directed at the ports of the vehicle. { prəˈtek·təˌskōp }

protium [NUC PHYS] The lightest hydrogen isotope, having a mass number of 1 and consisting of a single proton and electron. Also known as light hydrogen. { ˈprōd·ē·əm }

proton [PHYS] An elementary particle that is the positively charged constituent of ordinary matter and, together with the neutron, is a building block of all atomic nuclei; its mass is approximately 938 megaelectronvolts and spin 1/2. { ˈprōˌtän }

proton capture [NUC PHYS] A nuclear reaction in which a proton combines with a nucleus. { ˈprōˌtän ˈkap·chər }

proton drip-line [NUC PHYS] On a chart of the nuclides, which plots proton number versus neutron number, the boundary beyond which proton-rich nuclei are unstable against proton emission. { ¦prōˌtän ˈdrip ˌlīn }

proton-electron-proton reaction [NUC PHYS] A nuclear reaction in which two protons and an electron react to form a deuteron and a neutrino; it is an important source of detectable neutrinos from the sun. Abbreviated PeP reaction. { ˈprōˌtän iˈlekˌträn ˈprōˌtän rēˌak·shən }

protonium [ATOM PHYS] A bound state of a proton and an antiproton. { prōˈtō·nē·əm }

proton magnetometer [ELECTROMAG] A highly sensitive magnetometer which measures the frequency of the proton resonance in ordinary water. { ˈprōˌtän ˌmag·nəˈtäm·əd·ər }

proton moment [NUC PHYS] The magnetic dipole moment of the proton, a physical constant equal to $(1.410606633 \pm 0.000000058) \times 10^{-26}$ joule per tesla. { ˈprōˌtän ˌmō·mənt }

proton number *See* atomic number. { ˈprōˌtän ˌnəm·bər }

proton-proton chain [NUC PHYS] An energy-releasing nuclear reaction chain which is believed to be of major importance in energy production in hydrogen-rich stars. Also known as deuterium cycle. { ˈprōˌtän ˈprōˌtän ˌchān }

proton-proton reaction [NUC PHYS] The initiating reaction in the proton-proton chain, in which two protons react to form a deuteron, a positron, and a neutrino. Abbreviated PP reaction. { ˈprōˌtän ˈprōˌtän rēˌak·shən }

proton-proton scattering [NUC PHYS] A collision of a proton with another proton, usually the nucleus of a hydrogen atom. { ˈprōˌtän ˈprōˌtän ˈskad·ər·iŋ }

proton resonance [SPECT] A phenomenon in which protons absorb energy from an alternating magnetic field at certain characteristic frequencies when they are also subjected to a static magnetic field; this phenomenon is used in nuclear magnetic resonance quantitative analysis technique. { ˈprōˌtän ˈrez·ən·əns }

proton-rich nucleus [NUC PHYS] An atomic nucleus in which the ratio of proton number to neutron number is much larger than that of nuclei found in nature. { ¦prōˌtēn ˌrich ˈnü·klē·əs }

proton scattering microscope [SOLID STATE] A

microscope in which protons produced in a cold-cathode discharge are accelerated and focused on a crystal in a vacuum chamber; protons reflected from the crystal strike a fluorescent screen to give a visual and photographable display that is related to the structure of the target crystal. { 'prō,tän ¦skad·ə·riŋ 'mī·krə,skōp }

proton vector magnetometer [ELECTROMAG] A type of proton magnetometer with a system of auxiliary coils that permits measurement of horizontal intensity or vertical intensity as well as total intensity. { 'prō,tän ¦vek·tər ,mag·nə'täm·əd·ər }

proximity effect [ELEC] Redistribution of current in a conductor brought about by the presence of another conductor. { präk'sim·əd·ē i,fekt }

proximity-effect microbridge [CRYO] A Josephson junction formed by overcoating a few micrometers of thin superconducting film with normal metal, thereby weakening the superconductivity in the film beneath the metal. Also known as Notarys-Mercereau microbridge. { präk'sim·əd·ē i¦fekt 'mī·krō,brij }

ps *See* picosecond.

psec *See* picosecond.

pseudocrystal [CRYSTAL] A substance that appears to be crystalline but does not have a true crystalline diffraction pattern. { ¦sü·dō'krist·əl }

pseudo-Goldstone bosons [PART PHYS] Goldstone bosons which accompany the breakdown of approximate, accidental symmetries in certain unified gauge theories of weak and electromagnetic interactions. { ¦sü·dō 'gōl,stōn 'bō,sänz }

pseudoplastic fluid [FL MECH] A fluid whose apparent viscosity or consistency decreases instantaneously with an increase in shear rate. { ¦sü·dō'plas·tik 'flü·əd }

pseudopotential [SOLID STATE] The common effective potential for electrons in a crystal lattice that is calculated in the orthogonalized plane-wave method and in the pseudopotential method, and that is relatively weak (except for diffracted electrons) because the electrons are moving rapidly past the atoms in the lattice. { ,sü·də·pə'ten·chəl }

pseudopotential method [SOLID STATE] A method of approximating the energy states of electrons in a crystal lattice in which the electrons are assumed to move in a common effective potential that is calculated from the experimentally determined energy levels and the effective masses of the electrons. { ,sü·də·pə,ten·chəl 'meth·əd }

pseudoscalar [PHYS] A quantity which has magnitude only, and which acts, under Lorentz transformation, like a scalar but with a sign change under space reflection or time reflection, or both. { ¦sü·dō'skāl·ər }

pseudoscalar coupling [PART PHYS] A type of interaction postulated between a nucleon and a pion in which the interaction energy is a product of the pion's pseudoscalar field and a bilinear pseudoscalar function of the nucleon fields. { ¦sü·dō'skāl·ər 'kəp·liŋ }

pseudoscalar meson [PART PHYS] A meson, such as the pion, which has spin 0 and negative parity, and may be described by a field which is a pseudoscalar quantity. Also known as pseudoscalar particle. { ¦sü·dō'skāl·ər 'mā,sän }

pseudoscalar particle *See* pseudoscalar meson. { ¦sü·dō'skāl·ər 'pärd·ə·kəl }

pseudoscope [OPTICS] A device that produces reversed stereoscopic effects, for example, by transposing the pictures of a stereoscope. { 'süd·ə,skōp }

pseudosymmetry [CRYSTAL] Apparent symmetry of a crystal, resembling that of another system; generally due to twinning. { ¦sü·dō'sim·ə·trē }

pseudotensor [PHYS] **1.** A quantity which transforms as a tensor under space rotations, but which transforms as a tensor, together with a change in sign, under space inversion. **2.** A quantity which transforms as a tensor under Lorentz transformations, but with an additional sign change under space reflection or time reflection or both. { ¦sü·dō'ten·sər }

pseudovector [PHYS] **1.** A quantity which transforms as a vector under space rotations but which transforms as a vector, together with a change in sign, under a space inversion. Also known as axial vector. **2.** A quantity which transforms as a four-vector under Lorentz transformations, but with an additional sign change under space reflection or time reflection or both. { ¦sü·dō'vek·tər }

pseudovector coupling [PART PHYS] A type of interaction postulated between a nucleon and another particle in which the expression for the interaction energy contains a bilinear pseudovector function of nucleon fields. { ¦sü·dō 'vek·tər 'kəp·liŋ }

pseudovector meson [PART PHYS] A meson which has spin quantum number 1 and positive parity, and may be described by a field which is a pseudovector quantity. { ¦sü·dō'vek·tər 'mā,sän }

psf *See* pound per square foot.

P shell [ATOM PHYS] The sixth layer of electrons about the nucleus of an atom, having electrons whose principal quantum number is 6. { 'pē ,shel }

psi *See* pound per square inch.

psia *See* pounds per square inch absolute.

psi function *See* Schrödinger wave function. { 'sī ,fəŋk·shən }

psig *See* pounds per square inch gage.

psi particle *See* J particle. { 'sī ,pärd·ə·kəl }

psi-prime particle [PART PHYS] A neutral meson which has a mass of 3684 megaelectronvolts, spin quantum number 1, and negative parity and charge parity; it has an anomalously long lifetime. Symbolized ψ'. { 'sī ¦prīm ,pärd·ə·kəl }

psychromatic ratio [THERMO] Ratio of the heat-transfer coefficient to the product of the mass-transfer coefficient and humid heat for a gas-vapor system; used in calculation of humidity or saturation relationships. { ,sī·krə'mad·ik 'rā·shō }

psychrometric chart [THERMO] A graph each point of which represents a specific condition of a gas-vapor system (such as air and water vapor) with regard to temperature (horizontal scale) and absolute humidity (vertical scale); other characteristics of the system, such as relative humidity, wet-bulb temperature, and latent heat of vaporization, are indicated by lines on the chart. { ¦sī·krə¦me·trik 'chärt }

psychrometric formula [THERMO] The semiempirical relation giving the vapor pressure in terms of the barometer and psychrometer readings. { ¦sī·krə¦me·trik 'fór·myə·lə }

psychrometric tables [THERMO] Tables prepared from the psychrometric formula and used to obtain vapor pressure, relative humidity, and dew point from values of wet-bulb and dry-bulb temperature. { ¦sī·krə¦me·trik 'tā·bəlz }

pt *See* pint.

puff *See* picofarad. { pəf }

Pulfrich refractometer [OPTICS] A critical angle refractometer in which the material to be tested rests on a prism of material of known higher index of refraction and angle, and the angle of refraction of light which is directed at the interface between the two materials at grazing incidence is observed. { 'pu̇l·frik ˌrē‚frak'täm·əd·ər }

pull strength [MECH] A unit in tensile testing; the bond strength in pounds per square inch. { 'pu̇l ˌstreŋkth }

pulsatance [PHYS] Angular velocity in radians, equal to 2π times frequency in hertz. { 'pəl·səd·əns }

pulsating current [ELEC] Periodic direct current. { 'pəl‚sād·iŋ 'kə·rənt }

pulsating electromotive force [ELEC] Sum of a direct electromotive force and an alternating electromotive force. Also known as pulsating voltage. { 'pəl‚sād·iŋ i¦lek·trə¦mōd·iv 'fórs }

pulsating voltage *See* pulsating electromotive force. { 'pəl‚sād·iŋ 'vōl·tij }

pulse [PHYS] A variation in a quantity which is normally constant; has a finite duration and is usually brief compared to the time scale of interest. { pəls }

pulse amplitude [PHYS] The peak, average, effective, instantaneous, or other magnitude of a pulse, usually with respect to the normal constant value; the exact meaning should be specified when giving a numerical value. { 'pəls ¦am·plə‚tüd }

pulsed laser [OPTICS] A laser in which a pulse of coherent light is produced at fixed time intervals, as required for ranging and tracking applications or to permit higher output power than can be obtained with continuous operation. { 'pəlst 'lā·zər }

pulsed light [OPTICS] A beam of light whose intensity is modulated in some prescribed manner; analogous to a radar pulse. { 'pəlst 'līt }

pulsed ruby laser [OPTICS] A laser in which ruby is used as the active material; the extremely high pumping power required is obtained by discharging a bank of capacitors through a special high-intensity flash tube, giving a coherent beam that lasts for about 0.5 millisecond. { 'pəlst 'rü·bē 'lā·zər }

pulse form [PHYS] The amplitude of a pulse plotted as a function of time. { 'pəls ‚fórm }

pulse-frequency spectrum *See* pulse spectrum. { 'pəls ¦frē·kwən·sē ‚spek·trəm }

pulse group *See* pulse train. { 'pəls ‚grüp }

pulse-height spectrum [PHYS] Distribution of various pulse wavelengths and strengths (heights) developed during activation analysis. { 'pəls ‚hīt 'spek·trəm }

pulse interval *See* pulse spacing. { 'pəls ˌin·tər·vəl }

pulse spacing [PHYS] Time between corresponding points of successive pulses. Also known as pulse interval. { 'pəls ‚spās·iŋ }

pulse spectrum [PHYS] The frequency distribution of the sinusoidal components of a pulse in relative amplitude and in relative phase. Also known as pulse-frequency spectrum. { 'pəls ‚spek·trəm }

pulse train [PHYS] A series of regularly recurrent pulses having similar characteristics. Also known as pulse group. { 'pəls ‚trān }

pulse voltage *See* impulse voltage. { 'pəls ˌvōl·tij }

pumping [FL MECH] Unsteadiness of the mercury in the barometer, caused by fluctuations of the air pressure produced by a gusty wind or due to the motion of a vessel. [PHYS] **1.** The application of optical, infrared, or microwave radiation of appropriate frequency to a laser or maser medium so that absorption of the radiation increases the population of atoms or molecules in higher energy states. Also known as electronic pumping. **2.** The removal of gases and vapors from a vacuum system. { 'pəmp·iŋ }

pumping radiation [PHYS] Electromagnetic radiation applied to a laser or maser in the process of pumping. { 'pəmp·iŋ ˌrād·ē‚ā·shən }

punctum remotum *See* far point. { 'pəŋk·təm ri 'mōd·əm }

puncture [ELEC] Disruptive discharge through insulation involving a sudden and large increase in current through the insulation due to complete failure under electrostatic stress. { 'pəŋk·chər }

puncture voltage [ELEC] The voltage at which a test specimen is electrically punctured. { 'pəŋk·chər ˌvōl·tij }

Pupin coil *See* loading coil. { pyü'pēn ‚kȯil }

pure inverse scattering theory [PHYS] The branch of inverse scattering theory that treats the case in which the data consist of pure, noisefree information about the scattering amplitude. { 'pyu̇r 'in‚vərs ¦skad·ə·riŋ ‚thē·ə·rē }

pure shear [MECH] A particular example of irrotational strain or flattening in which a body is elongated in one direction and shortened at right angles to it as a consequence of differential displacements on two sets of intersecting planes. { 'pyu̇r 'shir }

pure tone *See* simple tone. { 'pyu̇r 'tōn }

purity [OPTICS] The degree to which a primary

color is pure and not mixed with the other two primary colors. { 'pyu̇r·əd·ē }

purity of state [STAT MECH] Property of a system which is definitely in a certain quantum state, rather than having a certain probability of being in any of several quantum states. { 'pyu̇r·əd·ē əv 'stāt }

purple boundary [OPTICS] A straight line connecting the ends of the spectrum locus on the chromaticity diagram. { 'pər·pəl 'bau̇n·drē }

push-pull voltages *See* balanced voltages. { 'pu̇sh ¦pu̇l 'vōl·tij·əz }

pW *See* picowatt.

pwt *See* pennyweight.

pycnometry [PHYS] The determination of liquid density by weighing the liquid in a container (pycnometer) of known volume. { pik'näm·ə·trē }

pyramid [CRYSTAL] An open crystal having three, four, six, eight, or twelve nonparallel faces that meet at a point. { 'pir·ə,mid }

pyramidal cleavage [CRYSTAL] A type of crystal cleavage that occurs parallel to the faces of a pyramid. { ¦pir·ə¦mid·əl 'klē·vij }

pyritohedron [CRYSTAL] A dodecahedral crystal with 12 irregular pentagonal faces; it is characteristic of pyrite. Also known as pentagonal dodecahedron; pyritoid; regular dodecahedron. { pə¦rīd·ō'hē·drən }

pyritoid *See* pyritohedron. { 'pī,rīd,ȯid }

pyroconductivity [SOLID STATE] Electrical conductivity that develops in a material only at high temperature, chiefly at fusion, in solids that are practically nonconductive at atmospheric temperatures. { ¦pī·rō,kän·dək'tiv·əd·ē }

pyroelectric crystal [SOLID STATE] A crystal exhibiting pyroelectricity, such as tourmaline, lithium sulfate monohydrate, cane sugar, and ferroelectric barium titanate. { ¦pī·rō·i¦lek·trik 'krist·əl }

pyroelectricity [SOLID STATE] **1.** The property of certain crystals to produce a state of electrical polarity by a change of temperature. **2.** An electric charge released as the result of a temperature change. { ¦pī·rō,i,lek'tris·əd·ē }

pyrometry [THERMO] The science and technology of measuring high temperatures. { pī'räm·ə·trē }

pyron [PHYS] A unit of area-density of power, equal to the area-density of power resulting from a power of one international table calorie per minute acting uniformly over an area of 1 square centimeter; equal to 697.8 watts per square meter. { 'pī,rän }

Pythagorean scale [ACOUS] A musical scale such that the frequency intervals are represented by the ratios of integral powers of the numbers 2 and 3. { pə,thag·ə'rē·ən 'skāl }

pz *See* pièze.

Q

Q [NUC PHYS] *See* disintegration energy. [PHYS] A measure of the ability of a system with periodic behavior to store energy equal to 2π times the average energy stored in the system divided by the energy dissipated per cycle. Also known as Q factor; quality factor; storage factor. [THERMO] A unit of heat energy, equal to 10^{18} British thermal units, or approximately 1.055×10^{21} joules.

Q branch [SPECT] A series of lines in molecular spectra that correspond to changes in the vibrational quantum number with no change in the rotational quantum number. { 'kyü ,branch }

QCD *See* quantum chromodynamics.

Q factor *See* Q. { 'kyü ,fak·tər }

Q machine [PL PHYS] A device in which a highly ionized, magnetically confined plasma is created by contact ionization of atoms and thermionic emission of electrons. { 'kyü mə,shēn }

Q-machine plasma [PL PHYS] A plasma column in a magnetic field created by surface ionization of a cesium beam on a hot tungsten plate. { 'kyü mə¦shēn 'plaz·mə }

qr *See* quarter.

qr tr *See* quarter.

Q-switched laser [OPTICS] A laser whose Q factor is kept at a low value while an ion population inversion is built up, and then is suddenly switched to a high value just before instability occurs, resulting in a very high rate of stimulated emission. Also known as giant pulse laser. { 'kyü ,swicht 'lā·zər }

qt *See* quart.

quad [ELEC] A series of four separately insulated conductors, generally twisted together in pairs. [THERMO] A unit of heat energy, equal to 10^{15} British thermal units, or approximately 1.055×10^{18} joules. { kwäd }

quadrant [ELECTROMAG] *See* international henry. [OPTICS] A double-reflecting instrument for measuring angles, used primarily for measuring altitudes of celestial bodies; the instrument was replaced by the sextant. { 'kwä·drənt }

quadrant angle of fall [MECH] The vertical acute angle at the level point, between the horizontal and the line of fall of a projectile. { 'kwä·drənt 'aŋ·gəl əv 'fȯl }

quadratic Stark effect [ATOM PHYS] A splitting of spectral lines of atoms in an electric field in which the energy levels shift by an amount proportional to the square of the electric field, and all levels shift to lower energies; observed in lines resulting from the lower energy states of many-electron atoms. { kwä'drad·ik 'stärk i,fekt }

quadratic Zeeman effect [ATOM PHYS] A splitting of spectral lines of atoms in a magnetic field in which the energy levels shift by an amount proportional to the square of the magnetic field. { kwä'drad·ik 'zē·mən i,fekt }

quadrature [PHYS] State of being separated in phase by 90°, or one quarter-cycle. Also known as phase quadrature. { 'kwä·drə·chər }

quadrature component [ELEC] **1.** A vector representing an alternating quantity which is in quadrature (at 90°) with some reference vector. **2.** *See* reactive component. { 'kwä·drə·chər kəm,pō·nənt }

quadrature current *See* reactive current. { 'kwä·drə·chər ,kə·rənt }

quadrupole [ELECTROMAG] A distribution of charge or magnetization which produces an electric or magnetic field equivalent to that produced by two electric or magnetic dipoles whose dipole moments have the same magnitude but point in opposite directions, and which are separated from each other by a small distance. { 'kwä·drə,pōl }

quadrupole field [ELECTROMAG] **1.** An electric or magnetic field equivalent to that produced by two electric or magnetic dipoles whose dipole moments have the same magnitude but point in opposite directions, and which are separated from each other by a small distance. **2.** The field produced by a quadrupole lens. { 'kwä·drə,pōl ,fēld }

quadrupole lens [ELECTROMAG] A device for focusing beams of charged particles which has four electrodes or magnetic poles of alternating sign arranged in a circle about the beam; used in instruments such as electron microscopes and particle accelerators. { 'kwä·drə,pōl ,lenz }

quadrupole moment [ELECTROMAG] A quantity characterizing a distribution of charge or magnetization; it is given by integrating the product of the charge density or divergence of magnetization density, the second power of the distance from the origin, and a spherical harmonic Y^*_{2m}

over the charge or magnetization distribution. { 'kwä·drə,pōl ,mō·mənt }

quality factor *See* Q. { 'kwäl·əd·ē ,fak·tər }

quality of sound *See* timbre. { 'kwäl·əd·ē əv 'sau̇nd }

quantity of electricity *See* charge. { 'kwän·əd·ē əv ,i,lek'tris·əd·ē }

quantization [QUANT MECH] **1.** The restriction of an observable quantity, such as energy or angular momentum, associated with a physical system, such as an atom, molecule, or elementary particle, to a discrete set of values. **2.** The transition from a description of a system of particles or fields in the classical approximation where canonically conjugate variables commute, to a description where these variables are treated as noncommuting operators; quantization (first definition) is a result of this procedure. { ,kwän·tə'zā·shən }

quantized Hall conductance [PHYS] The reciprocal of the von Klitzing constant, equal to e^2/h, where e is the charge of the electron and h is Planck's constant. { ¦kwän,tīzd 'hȯl kən,dək·təns }

quantized Hall resistance *See* von Klitzing constant. { ¦kwän,tīzd 'hȯl ri,zis·təns }

quantized Rabi oscillations [ATOM PHYS] Rabi oscillations that occur when only a small number of photons are present at discrete frequencies determined by the number of photons. { ¦kwän,tīzd 'rä·bē ,äs·ə,lā·shənz }

quantized spin wave *See* magnon. { 'kwän,tīzd 'spin ,wāv }

quantized vortex [CRYO] A circular flow pattern observed in superfluid helium and type II superconductors, in which a superfluid flows about a normal (nonsuperfluid) cylindrical region or core which has the form of a thin line, and either the circulation or the magnetic flux is quantized. { 'kwän,tīzd 'vȯr,teks }

quantum [QUANT MECH] **1.** For certain physical quantities, a unit such that the values of the quantity are restricted to integral multiples of this unit; for example, the quantum of angular momentum is Planck's constant divided by 2π. **2.** An entity resulting from quantization of a field or wave, having particlelike properties such as energy, mass, momentum and angular momentum; for example, the photon is the quantum of an electromagnetic field, and the phonon is the quantum of a lattice vibration. { 'kwän·təm }

quantum acoustics [ACOUS] The study of the properties of propagating sound waves that are directly attributable to the underlying quantum-mechanical nature of the medium. { 'kwän·təm ə'kü·stiks }

quantum anomaly [QUANT MECH] A phenomenon whereby a quantity that vanishes according to the dynamical rules of classical physics acquires a finite value when quantum rules are used. { ,kwän·təm ə'näm·ə·lē }

quantum cascade laser [OPTICS] A semiconductor laser whose light is generated by electronic transitions between bound states created by quantum confinement in alternating ultrathin layers of semiconductor material. { ¦kwänt·əm ,kas,kād 'lā·zər }

quantum chaos [QUANT MECH] The dynamics of quantum systems whose classical counterparts exhibit chaotic behavior. { ,kwänt·əm 'kā,äs }

quantum chromodynamics [PART PHYS] A gauge theory of the strong interactions among quarks; the mathematical structure of the theory resembles that of quantum electrodynamics, with color as the conserved charge. Abbreviated QCD. { 'kwän·təm ¦krō·mō·dī'nam·iks }

quantum defect [ATOM PHYS] The difference between the principal quantum number of an atomic energy level and the effective quantum number obtained by fitting with a Rydberg formula the energy required to ionize the atom from that level. { 'kwän·təm 'dē,fekt }

quantum detector [PHYS] A detector of electromagnetic radiation which converts a quantum of the radiation into a proportionate signal by some process which is insensitive to quanta of less than a certain energy; examples include photographic emulsions, photoelectric cells, and Geiger counters. { 'kwän·təm di,tek·tər }

quantum discontinuity [QUANT MECH] The emission or absorption of a definite amount of energy that accompanies a quantum jump. { 'kwän·təm ,dis,känt·ən'ü·əd·ē }

quantum dot laser [OPTICS] A laser that has a dense array of equal-sized quantum dots in the active region, each with only a few thousand atoms of semiconductor material, and emits light from electronic transitions between the discrete energy levels of these quantum dots. { ¦kwän·təm 'dät ,lä·zər }

quantum electrodynamics [QUANT MECH] The quantum theory of electromagnetic radiation, synthesizing the wave and corpuscular pictures, and of the interaction of radiation with electrically charged matter, in particular with atoms and their constituent electrons. Also known as quantum theory of light; quantum theory of radiation. { 'kwän·təm i¦lek·trō·dī'nam·iks }

quantum entanglement [QUANT MECH] The property of two particles with a common origin whereby a measurement on one of the particles determines not only its quantum state but the quantum state of the other particle as well. { ,kwän·təm in'taŋ·gəl·mənt }

quantum field theory [QUANT MECH] Quantum theory of physical systems possessing an infinite number of degrees of freedom, such as the electromagnetic field, gravitation field, or wave fields in a medium. { 'kwän·təm 'fēld ,thē·ə·rē }

quantum gravitation [QUANT MECH] Also known as quantum gravity. **1.** The quantum theory of the gravitational field. **2.** The study of quantum fields in a curved space-time. { 'kwän·təm ,grav·ə'tā·shən }

quantum gravity *See* quantum gravitation. { 'kwän·təm 'grav·əd·ē }

quantum Hall liquid *See* quantum Hall state.

quantum Hall state [CRYO] A kind of incompressible liquid state obtained by placing a two-dimensional electron gas, confined on the interface of two different semiconductors, in a strong magnetic field at low temperature. Also known as quantum Hall liquid. { ¦kwän·təm 'hȯl ˌstāt }

quantum hydrodynamics [CRYO] The mechanics of a superfluid, such as helium II, investigating phenomena such as the fountain effect and second sound. { 'kwän·təm ˌhī·drō·dī'nam·iks }

quantum hypothesis [QUANT MECH] A hypothesis that some physical quantity can assume only a certain discrete set of values; examples are Planck's law, and the condition in the Bohr-Sommerfeld theory that the action integral of a system must be an integral multiple of Planck's constant. { 'kwän·təm hī'päth·ə·səs }

quantum jump [QUANT MECH] The transition of a quantum system from one stationary state to another, accompanied by emission or absorption of energy. { 'kwän·təm 'jəmp }

quantum limit [SPECT] The shortest wavelength present in a continuous x-ray spectrum. Also known as boundary wavelength; end radiation. { 'kwän·təm 'lim·ət }

quantum measurement paradox [QUANT MECH] A paradox that arises because, at the atomic level where the quantum formalism has been directly tested, the most natural interpretation implies that when two or more different outcomes are possible it is not necessarily true that one or the other is actually realized, whereas at the everyday level such a state of affairs seems to conflict with direct experience. { ¦kwänt·əm 'mezh·ə·mənt ˌpar·əˌdäks }

quantum-mechanical operator [QUANT MECH] A linear, Hermitian operator associated with some physical quantity; for a physical system in any state, the expectation value of the physical quantity equals the integral over configuration space of $\psi^*(A\psi)$, where $A\psi$ is the result of the operator acting on the wave function of the system, and ψ^* is the complex conjugate of the wave function. { 'kwän·təm mi'kan·ə·kəl ˌäp·əˌrād·ər }

quantum mechanics [PHYS] The modern theory of matter, of electromagnetic radiation, and of the interaction between matter and radiation; it differs from classical physics, which it generalizes and supersedes, mainly in the realm of atomic and subatomic phenomena. Also known as quantum theory. { 'kwän·təm mi'kan·iks }

quantum nondemolition measurement [QUANT MECH] A measurement of a physical observable of some system without altering its value. { ¦kwän·təm ˌnänˌdem·ə¦lish·ən 'mezh·ər·mənt }

quantum number [QUANT MECH] One of the quantities, usually discrete with integer or half-integer values, needed to characterize a quantum state of a physical system; they are usually eigenvalues of quantum-mechanical operators or integers sequentially assigned to these eigenvalues. { 'kwän·təm ˌnəm·bər }

quantum of action *See* Planck's constant. { 'kwän·təm əv 'ak·shən }

quantum size effects [SOLID STATE] Unusual properties of extremely small crystals that arise from confinement of electrons to small regions of space in one, two, or three dimensions. { ¦kwänt·əm 'sīz iˌfeks }

quantum solid [SOLID STATE] A solid whose atoms or molecules undergo large zero-point motion even in the quantum ground state (at absolute zero temperature) as a result of their small mass and the weak attractive part of their interaction potential. { 'kwän·təm ˌsäl·əd }

quantum state [QUANT MECH] **1.** The condition of a physical system as described by a wave function; the function may be simultaneously an eigenfunction of one or more quantum-mechanical operators; the eigenvalues are then the quantum numbers that label the state. **2.** *See* energy state. { 'kwän·təm ˌstāt }

quantum statistics [STAT MECH] The statistical description of particles or systems of particles whose behavior must be described by quantum mechanics rather than classical mechanics. { 'kwän·təm stə'tis·tiks }

quantum teleportation [QUANT MECH] The replication of a quantum state at a distant location by utilizing the concept of quantum entanglement. { ˌkwän·təm ˌtel·ə·pȯr'tā·shən }

quantum theory *See* quantum mechanics. { 'kwän·təm ˌthē·ə·rē }

quantum theory of heat capacity [STAT MECH] Application of quantum statistics to calculate heat capacities of various substances; an important result of the theory is the decrease of specific heats at low temperatures to values smaller than their classical values as a result of energy quantization. { 'kwän·təm ˌthē·ə·rē əv 'hēt kəˌpas·əd·ē }

quantum theory of light *See* quantum electrodynamics. { 'kwän·təm ˌthē·ə·rē əv 'līt }

quantum theory of matter [QUANT MECH] The microscopic explanation of the properties of condensed matter, that is, solids and liquids, based on the fundamental laws of quantum mechanics. { ¦kwänt·əm ˌthē·ə·rē əv 'mad·ər }

quantum theory of measurement [QUANT MECH] The attempt to reconcile the counterintuitive features of quantum mechanics with the hypothesis that it is in principle a complete description of the physical world, even at the level of everyday objects. { ¦kwänt·əm ˌthē·ə·rē əv 'mezh·ər·mənt }

quantum theory of radiation [QUANT MECH] **1.** The theory of heat radiation based on Planck's law; its principal result is the Planck radiation formula. **2.** *See* quantum electrodynamics. { 'kwän·təm ˌthē·ə·rē əv ˌrād·ē'ā·shən }

quantum theory of spectra [QUANT MECH] The contemporary theory of spectra, based on the idea that an atom, molecule, or nucleus can exist only in certain allowed energy states, that it emits or absorbs energy as it changes from one state to another, and that the frequency of the associated electromagnetic radiation equals the difference in energies of two states divided by Planck's constant. { 'kwän·təm ˌthē·ə·rē əv 'spek·trə }

quantum turbulence [CRYO] A phenomenon observed in a channel filled with superfluid and subjected to a heat flux which exceeds a certain critical value, in which the superfluid becomes filled with a tangled mass of quantized vortex lines. { 'kwän·təm 'tər·byə·ləns }

quantum-wave equation [QUANT MECH] A partial differential equation which relates the spatial and time dependences of the wave function of a system of one or more atomic or subatomic particles; examples are the Schrödinger equation in nonrelativistic quantum mechanics, and the Klein-Gordon, Dirac, Rarita-Schwinger and Proca equations in relativistic quantum mechanics. { 'kwän·təm ˌwāv iˌkwā·zhən }

quark [PART PHYS] One of the hypothetical basic particles, having charges whose magnitudes are one-third or two-thirds of the electron charge, from which many of the elementary particles may, in theory, be built up; for example, nucleons may be formed from three quarks and mesons from quark-antiquark combinations; no experimental evidence for the actual existence of free quarks has been found. { kwärk }

quark confinement [PART PHYS] The phenomenon wherein quarks can never be removed from the hadrons they compose, even though the interactions between them are relatively weak. { 'kwärk kənˌfīn·mənt }

quark-gluon plasma [NUC PHYS] A state of nuclear matter postulated by quantum chromodynamics to exist at extremely high temperatures and densities in which the neutrons and protons lose their identities and the quarks and gluons form an unstructured collection of particles. { 'kwärk 'glüˌän ˌplaz·mə }

quarkonium [PART PHYS] A meson that is made up of a heavy quark and its antiparticle, an antiquark. { kwär'kō·nē·əm }

quart [MECH] Abbreviated qt. **1.** A unit of volume used for measurement of liquid substances in the United States, equal to 2 pints, or 1/4 gallon, or 573/4 cubic inches, or $9.46352946 \times 10^{-4}$ cubic meter. **2.** A unit of volume used for measurement of solid substances in the United States, equal to 2 dry pints, or $^1/_{32}$; bushel, or 107,521/1600 cubic inches, or approximately 1.10122×10^{-3} cubic meter. **3.** A unit of volume used for measurement of both liquid and solid substances, although mainly the former, in the United Kingdom and Canada, equal to 2 U.K. pints, or 1/4 U.K. gallon, or approximately 1.1365225×10^{-3} cubic meter. { kwȯrt }

quarter [MECH] **1.** A unit of mass in use in the United States, equal to 1/4 short ton, or 500 pounds, or 226.796185 kilograms. **2.** A unit of mass used in troy measure, equal to 1/4 troy hundredweight, or 25 troy pounds, or 9.33104304 kilograms. Abbreviated qr tr. **3.** A unit of mass used in the United Kingdom, equal to 1/4 hundredweight, or 28 pounds, or 12.70058636 kilograms. Abbreviated qr. **4.** A unit of volume used in the United Kingdom for measurement of liquid and solid substances, equal to 8 bushels, or 64 gallons, or approximately 0.29094976 cubic meter. { 'kwȯrd·ər }

quarter-phase *See* two-phase. { 'kwȯrd·ər ˌfāz }

quarter-wave [ELECTROMAG] Having an electrical length of one quarter-wavelength. { 'kwȯrd·ər ˌwāv }

quarter-wave antenna [ELECTROMAG] An antenna whose electrical length is equal to one quarter-wavelength of the signal to be transmitted or received. { 'kwȯrd·ər ˌwāv an'ten·ə }

quarter-wave attenuator [ELECTROMAG] Arrangement of two wire gratings, spaced an odd number of quarter-wavelengths apart in a waveguide, used to attenuate waves traveling through in one direction. { 'kwȯrd·ər ˌwāv ə'ten·yəˌwād·ər }

quarter-wave line *See* quarter-wave stub. { 'kwȯrd·ər ˌwāv ˌlīn }

quarter-wave matching section *See* quarter-wave transformer. { 'kwȯrd·ər ˌwāv 'mach·iŋ ˌsek·shən }

quarter-wave plate [OPTICS] A thin sheet of mica or other doubly refracting crystal material of such thickness as to introduce a phase difference of one quarter-cycle between the ordinary and the extraordinary components of light passing through; such a plate converts circularly polarized light into plane-polarized light. { 'kwȯrd·ər ˌwāv ¦plāt }

quarter-wave stub [ELECTROMAG] A section of transmission line that is one quarter-wavelength long at the fundamental frequency being transmitted; when shorted at the far end, it has a high impedance at the fundamental frequency and all odd harmonics, and a low impedance for all even harmonics. Also known as quarter-wave line; quarter-wave transmission line. { 'kwȯrd·ər ˌwāv ¦stəb }

quarter-wave termination [ELECTROMAG] Metal plate and a wire grating spaced about one-fourth of a wavelength apart in a waveguide, with the plate serving as the termination of the guide; waves reflected from the metal plate are canceled by waves reflected from the grating so that all energy is absorbed (none is reflected) by the quarter-wave termination. { 'kwȯrd·ər ˌwāv tər·mə'nā·shən }

quarter-wave transformer [ELECTROMAG] A section of transmission line approximately one quarter-wavelength long, used for matching a transmission line to an antenna or load. Also known as quarter-wave matching section. { 'kwȯrd·ər ˌwāv tranz'fȯr·mər }

quarter-wave transmission line *See* quarter-wave stub. { 'kwȯrd·ər ˌwāv tranz'mish·ən ˌlīn }

quartz wedge [OPTICS] A very thin wedge of quartz cut parallel to an optic axis; used to determine the sign of double refraction of biaxial crystals, and in other applications involving polarized light and its interaction with matter. { 'kwȯrts 'wej }

quasi-acoustical holography [ACOUS] An optical technique by which images produced by B-mode ultrasonic medical holography are assembled in a single three-dimensional image volume which, when reconstructed with visible light, renders a realistic three-dimensional image of the target under investigation. Also known as holographic multiplexing. { ¦kwä·zē ə'kü·stə·kəl hō 'läg·rə·fē }

quasi-atom [ATOM PHYS] A system formed by two colliding atoms whose nuclei approach each other so closely that, for a very short time, the atomic electrons arrange themselves as if they belonged to a single atom whose atomic number equals the sum of the atomic numbers of the colliding atoms. { ¦kwä·zē 'ad·əm }

quasi-crystal [CRYSTAL] A phase of solid matter that, like a crystal, exhibits long-range orientational order and translational order but whose atoms and clusters repeat in a sequence defined by a sum of periodic functions whose periods are in an irrational ratio. { ¦kwä·zē 'krist·əl }

quasi-fission *See* deep inelastic collision. { ¦kwä·zē 'fish·ən }

quasi-free-electron theory [SOLID STATE] A modification of the free-electron theory of metals to take into account the periodic variation of the potential acting on a conduction electron, in which these electrons are assigned an effective scalar mass which differs from their real mass. { ¦kwä·zē ¦frē i'lek,tran ,thē·ə·rē }

quasi-molecule [ATOM PHYS] The structure formed by two colliding atoms when their nuclei are close enough for the atoms to interact, but not so close as to form a quasi-atom. { ¦kwä·zē 'mäl·ə,kyül }

quasi-particle [PHYS] An entity used in the description of a system of many interacting particles which has particlelike properties such as mass, energy, and momentum, but which does not exist as a free particle; examples are phonons and other elementary excitations in solids, and "dressed" helium-3 atoms in Landau's theory of liquid helium-3. { ¦kwä·zē ¦pard·ə·kəl }

quasi-periodic motion [PHYS] Motion at two frequencies simultaneously, where the ratio of the frequencies is not a rational number. { ¦kwä·zē ,pir·ē'äd·ik 'mō·shən }

quasi-reflection [OPTICS] A term applied to the very strong return of light produced by dust particles and other suspensoids whose diameters are large compared to the wavelength of the incident radiation. { ¦kwä·zē ri'flek·shən }

quasi-stable elementary particle [PART PHYS] A term formerly (before the discovery of charmed particles) used for elementary particles that cannot decay into other particles through strong interactions and that have lifetimes longer than 10^{-20} second. Also known as semistable elementary particle. { ¦kwä·zē 'stā·bəl 'el·ə,men·trē 'pärd·ə·kəl }

quasi-static process *See* reversible process. { ¦kwä·zē 'stad·ik 'prä·səs }

quenching [ATOM PHYS] Phenomenon in which a very strong electric field, such as a crystal field, causes the orbit of an electron in an atom to precess rapidly so that the average magnetic moment associated with its orbital angular momentum is reduced to zero. [SOLID STATE] Reduction in the intensity of sensitized luminescence radiation when energy migrating through a crystal by resonant transfer is dissipated in crystal defects or impurities rather than being reemitted as radiation. { 'kwench·iŋ }

Quincke tube *See* Herschel-Quincke tube. { 'kviŋ·kə ,tüb }

quintal *See* metric centner. { 'kwint·əl }

Q unit [THERMO] A unit of energy, used in measuring the heat energy of fuel reserves, equal to 10^{18} British thermal units, or approximately 1.055×10^{21} joules. { 'kyü ,yü·nət }

Q value *See* disintegration energy. { 'kyü ,val·yü }

R

Rabi oscillation [ATOM PHYS] The periodic exchange of energy between atoms in an electromagnetic field and a single mode of the field. { ¦rä·bē ,äs·ə 'lā·shən }

Racah coefficient [QUANT MECH] A coefficient that appears in the transformation between the modes of coupling eigenfunctions of three angular momenta; they differ only by, at most, a sign from the six-*j* coefficients. Also known as W coefficient. { 'rä·kä ,kō·i'fish·ənt }

radar camera [OPTICS] A special manual or automatic camera used to photograph images on a radarscope. Also known as radarscope camera. { 'rā,där ,kam·rə }

radarscope camera *See* radar camera. { 'rā ,där,skōp ,kam·rə }

radial acceleration *See* centripetal acceleration. { 'rād·ē·əl ak,sel·ə'rā·shən }

radial astigmatism [OPTICS] Astigmatism which affects the imaging of points that lie off the axis of an optical system, due to oblique incidence of rays from these points. Also known as oblique astigmatism. { 'rād·ē·əl ə'stig·mə,tiz·əm }

radial band pressure [MECH] The pressure which is exerted on the rotating band by the walls of the gun tube, and hence against the projectile wall at the band seat, as a result of the engraving of the band by the gun rifling. { 'rād·ē·əl ¦band ,presh·ər }

radial Doppler effect [ELECTROMAG] The part of the optical Doppler effect which depends on the direction of the relative velocity of source and observer, and is analogous to the acoustical Doppler effect, in contrast to the transverse Doppler effect. { 'rād·ē·əl 'däp·lər i,fekt }

radial grating [ELECTROMAG] Conformal wire grating consisting of wires arranged radially in a circular frame, like the spokes of a wagon wheel, and placed inside a circular waveguide to obstruct E waves of zero order while passing the corresponding H waves. { 'rād·ē·əl 'grād·iŋ }

radial heat flow [THERMO] Flow of heat between two coaxial cylinders maintained at different temperatures; used to measure thermal conductivities of gases. { 'rād·ē·əl 'hēt ,flō }

radial motion [MECH] Motion in which a body moves along a line connecting it with an observer or reference point; for example, the motion of stars which move toward or away from the earth without a change in apparent position. { 'rād·ē·əl 'mō·shən }

radial shear interferometer [OPTICS] An interferometer in which a wavefront is interfered with by an expanded version of itself, resulting in fringes along which the radial slope of the wavefront is constant. { 'rād·ē·əl ¦shir ,in·tər·fə'räm·əd·ər }

radial stress [MECH] Tangential stress at the periphery of an opening. { 'rād·ē·əl 'stres }

radial velocity [MECH] The component of the velocity of a body that is parallel to a line from an observer or reference point to the body; the radial velocities of stars are valuable in determining the structure and dynamics of the Galaxy. Also known as line-of-sight velocity. { 'rād·ē·əl və'läs·əd·ē }

radial wave equation [MECH] Solutions to wave equations with spherical symmetry can be found by separation of variables; the ordinary differential equation for the radial part of the wave function is called the radial wave equation. { 'rād·ē·əl ¦wāv i,kwā·zhən }

radiance [OPTICS] The radiant flux per unit solid angle per unit of projected area of the source; the usual unit is the watt per steradian per square meter. Also known as steradiancy. { 'rād·ē·əns }

radiancy *See* radiant emittance. { 'rād·ē·ən·sē }

radian frequency *See* angular frequency. { 'rād·ē·ən 'frē·kwən·sē }

radian length [PHYS] Distance, in a sinusoidal wave, between phases differing by an angle of 1 radian; it is equal to the wavelength divided by 2π. { 'rād·ē·ən 'leŋkth }

radiant [PHYS] **1.** Pertaining to motion of particles or radiation along radii from a common point or a small region. **2.** A point, region, substance, or entity from which particles or radiations are emitted. { 'rād·ē·ənt }

radiant density [PHYS] The instantaneous amount of radiant energy contained in a unit volume of propagation medium. { 'rād·ē·ənt 'den·səd·ē }

radiant efficiency [OPTICS] The ratio of the radiant flux emitted by a radiation source to the power consumed by the source. { 'rād·ē·ənt i'fish·ən·sē }

radiant emittance [ELECTROMAG] The radiant flux per unit area that emerges from a surface.

Also known as radiancy; radiant exitance. { 'rād·ē·ənt i'mit·əns }

radiant exitance *See* radiant emittance. { 'rād·ē·ənt 'ek·sit·əns }

radiant exposure [OPTICS] A measure of the total radiant energy incident on a surface per unit area; equal to the integral over time of the radiant flux density. Also known as exposure. { 'rād·ē·ənt ik'spō·zhər }

radiant flux [OPTICS] The time rate of flow of radiant energy. { 'rād·ē·ənt 'fləks }

radiant flux density [ELECTROMAG] The amount of radiant power per unit area that flows across or onto a surface. Also known as irradiance. { 'rād·ē·ənt ¦fləks 'den·səd·ē }

radiant intensity [ELECTROMAG] The energy emitted per unit time per unit solid angle about the direction considered; usually expressed in watts per steradian. { 'rād·ē·ənt in'ten·səd·ē }

radiant power [ELECTROMAG] The energy carried across or onto a surface by electromagnetic radiation per unit time, or the total radiant energy emitted by a source of electromagnetic radiation per unit time. { 'rād·ē·ənt 'paü·ər }

radiant quantities [OPTICS] Physical quantities used in photometry, such as radiant flux and radiance, which are based on the energy carried by light, and are thus independent of the response of the human eye. { 'rād·ē·ənt 'kwän·əd·ēz }

radiant reflectance [ELECTROMAG] Ratio of reflected radiant power to incident radiant power. { 'rād·ē·ənt ri'flek·təns }

radiant transmittance [ELECTROMAG] Ratio of transmitted radiant power to incident radiant power. { 'rād·ē·ənt tranz'mit·əns }

radiated power [ELECTROMAG] The total power emitted by a transmitting antenna. { 'rād·ē,ād·əd 'paü·ər }

radiating curtain [ELECTROMAG] Array of dipoles in a vertical plane, positioned to reinforce each other; it is usually placed one-fourth wavelength ahead of a reflecting curtain of corresponding half-wave reflecting antennas. { 'rād·ē,ād·iŋ 'kərt·ən }

radiating element [ELECTROMAG] Basic subdivision of an antenna which in itself is capable of radiating or receiving radio-frequency energy. { 'rād·ē,ād·iŋ 'el·ə·mənt }

radiating guide [ELECTROMAG] Waveguide designed to radiate energy into free space; the waves may emerge through slots or gaps in the guide, or through horns inserted in the wall of the guide. { 'rād·ē,ād·iŋ 'gīd }

radiating power *See* emittance. { 'rād·ē,ād·iŋ 'paü·ər }

radiating scattering [PHYS] The diversion of radiation (thermal, electromagnetic, or nuclear) from its orginal path as a result of interactions or collisions with atoms, molecules, or larger particles in the atmosphere or other media between the source of radiation (for example, a nuclear explosion) and a point at some distance away. { 'rād·ē,ād·iŋ 'skad·ə·riŋ }

radiation [PHYS] **1.** The emission and propagation of waves transmitting energy through space or through some medium; for example, the emission and propagation of electromagnetic, sound, or elastic waves. **2.** The energy transmitted by waves through space or some medium; when unqualified, usually refers to electromagnetic radiation. Also known as radiant energy. **3.** A stream of particles, such as electrons, neutrons, protons, α-particles, or high-energy photons, or a mixture of these. { ,rād·ē'ā·shən }

radiation angle [ELECTROMAG] The vertical angle between the line of radiation emitted by a directional antenna and the horizon. { ,rād·ē'ā·shən ,aŋ·gəl }

radiation cooling [PHYS] The cooling of gases to very low temperatures by means of the resonant radiation pressure of intense laser light. { ,rād·ē'ā·shən ,kül·iŋ }

radiation correction *See* cooling correction. { ,rād·ē'ā·shən kə,rek·shən }

radiation damping [ELECTROMAG] Damping of a system which loses energy by electromagnetic radiation. [QUANT MECH] Damping which arises in quantum electrodynamics from the virtual interaction of a particle with its zero point field. { ,rād·ē'ā·shən ,damp·iŋ }

radiation efficiency [ELECTROMAG] Of an antenna, the ratio of the power radiated to the total power supplied to the antenna at a given frequency. { ,rād·ē'ā·shən i,fish·ən·sē }

radiation field [ELECTROMAG] The electromagnetic field that breaks away from a transmitting antenna and radiates outward into space as electromagnetic waves; the other type of electromagnetic field associated with an energized antenna is the induction field. { ,rād·ē'ā·shən ,fēld }

radiation filter [ELECTROMAG] Selectively transparent body, which transmits only certain wavelength ranges. { ,rād·ē'ā·shən ,fil·tər }

radiation impedance *See* radiation resistance. { ,rād·ē'ā·shən im,pēd·əns }

radiation intensity [ELECTROMAG] The power radiated from an antenna per unit solid angle in a given direction. { ,rād·ē'ā·shən in,ten·səd·ē }

radiation ionization [PHYS] Ionization of the atoms or molecules of a gas or vapor by the action of electromagnetic radiation. { ,rād·ē'ā·shən ,ī·ə·nə'zā·shən }

radiation laws [PHYS] **1.** The four physical laws which, together, fundamentally describe the behavior of blackbody radiation, Kirchhoff's law, Planck's law, Stefan-Boltzmann law, and Wien's law. **2.** All of the more inclusive assemblage of empirical and theoretical laws describing all manifestations of radiative phenomena. { ,rād·ē'ā·shən ,lȯz }

radiationless transition [PHYS] A transition of a system between two energy states in which energy is given to or taken up from another system or particle, rather than being emitted or absorbed in electromagnetic radiation; examples include internal conversion, the Auger effect, and excitation or deexcitation of atoms or molecules in collisions with other atoms or molecules. { ,rād·ē'ā·shən·ləs tran'zish·ən }

radiation oscillator *See* Planck oscillator. { ,rād·ē'ā·shən ,äs·ə,lād·ər }

radiation pattern [ELECTROMAG] Directional dependence of the radiation of an antenna. Also known as antenna pattern; directional pattern; field pattern. { ,rād·ē'ā·shən ,pad·ərn }

radiation physics [PHYS] The study of ionizing radiation and its effects on matter. { ,rād·ē'ā·shən 'fiz·iks }

radiation pressure [ACOUS] The average pressure exerted on a surface or interface between two media by a sound wave. [ELECTROMAG] The pressure exerted by electromagnetic radiation on objects on which it impinges. { ,rād·ē'ā·shən ,presh·ər }

radiation quality [PHYS] The spectrum of radiant energy produced by a given radiation source with respect to its penetration or its suitability for a specific application. { ,rād·ē'ā·shən ,kwäl·əd·ē }

radiation resistance [ACOUS] For a medium, the acoustic impedance of a plane wave in that medium. Also known as radiation impedance. [ELECTROMAG] The total radiated power of an antenna divided by the square of the effective antenna current measured at the point where power is supplied to the antenna. { ,rād·ē'ā·shən ri,zis·təns }

radiation scattering [PHYS] The diversion of radiation (thermal, electromagnetic, or nuclear) from its original path as a result of interactions or collisions with atoms, molecules, or larger particles in the atmosphere or other media. { ,rād·ē'ā·shən 'skad·ə·riŋ }

radiation thermocouple [ELEC] An infrared detector consisting of several thermocouples connected in series, arranged so that the radiation falls on half of the junctions, causing their temperature to increase so that a voltage is generated. { ,rād·ē'ā·shən 'thər·mə,kəp·əl }

radiation zone *See* Fraunhofer region. { ,rād·ē'ā·shən ,zōn }

radiative capture [NUC PHYS] A nuclear capture process whose prompt result is the emission of electromagnetic radiation only. { 'rād·ē,ād·iv 'kap·chər }

radiative collision [PHYS] A collision between two charged particles in which part of the kinetic energy of the particles is converted into electromagnetic radiation. { 'rād·ē,ād·iv kə'lizh·ən }

radiative correction [QUANT MECH] The change produced in the value of some physical quantity, such as the mass or charge of a particle, as the result of the particle's interactions with various fields. { 'rād·ē,ād·iv kə'rek·shən }

radiative recombination [PHYS] Recombination of parts of an atom or of an electron and a hole in a semiconductor during which electromagnetic radiation is emitted. { 'rād·ē,ād·iv ri,käm·bə'nā·shən }

radiative transfer [PHYS] The propagation of energy by radiative processes, involving emission, absorption, and scattering of electromagnetic radiation. Also known as radiative transport. { 'rād·ē,ād·iv 'tranz·fər }

radiative transition [QUANT MECH] A change of a quantum-mechanical system from one energy state to another in which electromagnetic radiation is emitted. { 'rād·ē,ād·iv tran'zish·ən }

radiative transport *See* radiative transfer. { 'rād·ē,ād·iv 'tranz,pȯrt }

radiator [ACOUS] A vibrating element of a transducer which radiates sound waves. [ELECTROMAG] **1.** The part of an antenna or transmission line that radiates electromagnetic waves either directly into space or against a reflector for focusing or directing. **2.** A body that emits radiant energy. [PHYS] **1.** In general, a body which emits particles or radiation in any form. **2.** A body placed in a beam of ionizing radiation which, as a result, emits radiation of another kind. { 'rād·ē,ād·ər }

radio- [ELECTROMAG] A prefix denoting the use of radiant energy, particularly radio waves. { 'rād·ē·ō }

radioactinium [NUC PHYS] Conventional name for the isotope of thorium which has mass number 227 and is in the actinium series. Symbolized RdAc. { ¦rād·ē·ō·ak'tin·ē·əm }

radioactive [NUC PHYS] Exhibiting radioactivity or pertaining to radioactivity. { ¦rād·ē·ō'ak·tiv }

radioactive chain *See* radioactive series. { ¦rād·ē·ō'ak·tiv 'chān }

radioactive clock [NUC PHYS] A radioactive isotope such as potassium-40 which spontaneously decays to a stable end product at a constant rate, allowing absolute geologic age to be determined. { ¦rād·ē·ō'ak·tiv 'kläk }

radioactive cobalt [NUC PHYS] Radioactive form of cobalt, such as cobalt-60 with a half-life of 5.3 years. { ¦rād·ē·ō'ak·tiv 'kō,bȯlt }

radioactive collision [NUC PHYS] A nuclear reaction in which a neutron is absorbed by a nucleus and a gamma ray is emitted. { ,rād·ē·ō¦ak·tiv kə'lizh·ən }

radioactive decay [NUC PHYS] The spontaneous transformation of a nuclide into one or more different nuclides, accompanied by either the emission of particles from the nucleus, nuclear capture or ejection of orbital electrons, or fission. Also known as decay; nuclear spontaneous reaction; radioactive disintegration; radioactive transformation; radioactivity. { ¦rād·ē·ō'ak·tiv di'kā }

radioactive decay constant *See* decay constant. { ¦rād·ē·ō'ak·tiv di¦kā ,kän·stənt }

radioactive decay product *See* daughter. { ¦rād·ē·ō'ak·tiv di¦kā ,präd·əkt }

radioactive decay series *See* radioactive series. { ¦rād·ē·ō'ak·tiv di¦kā ,sir·ēz }

radioactive disintegration *See* radioactive decay. { ¦rād·ē·ō'ak·tiv di,sin·tə'grā·shən }

radioactive displacement law [NUC PHYS] The statement of the changes in mass number A and atomic number Z that take place during various nuclear transformations. Also known as displacement law. { ¦rād·ē·ō'ak·tiv di'splās·mənt ˌlȯ }

radioactive element [NUC PHYS] An element all of whose isotopes spontaneously transform into one or more different nuclides, giving off various types of radiation; examples include promethium, radium, thorium, and uranium. { ¦rād·ē·ō'ak·tiv 'el·ə·mənt }

radioactive emanation [NUC PHYS] A radioactive gas given off by certain radioactive elements; all of these gases are isotopes of the element radon. Also known as emanation. { ¦rād·ē·ō'ak·tiv ˌem·ə'nā·shən }

radioactive equilibrium [NUC PHYS] In radioactivity, the condition of equilibrium in which the rate of decay of the parent isotope is exactly matched by the rate of decay of every intermediate daughter isotope. { ¦rād·ē·ō'ak·tiv ˌē·kwə'lib·rē·əm }

radioactive heat [THERMO] Heat produced within a medium as a result of absorption of radiation from decay of radioisotopes in the medium, such as thorium-232, potassium-40, uranium-238, and uranium-235. { ¦rād·ē·ō'ak·tiv 'hēt }

radioactive isotope *See* radioisotope. { ¦rād·ē·ō'ak·tiv 'ī·səˌtōp }

radioactive metal [NUC PHYS] A luminous metallic element, such as actinium, radium, or uranium, that spontaneously and continuously emits its radiation capable in some degree of penetrating matter impervious to ordinary light. { ¦rād·ē·ō'ak·tiv 'med·əl }

radioactive series [NUC PHYS] A succession of nuclides, each of which transforms by radioactive disintegration into the next until a stable nuclide results. Also known as decay chain; decay family; decay series; disintegration chain; disintegration family; disintegration series; radioactive chain; radioactive decay series; series decay; transformation series. { ¦rād·ē·ō'ak·tiv 'sir·ēz }

radioactive transformation *See* radioactive decay. { ¦rād·ē·ō'ak·tiv ˌtranz·fər'mā·shən }

radioactivity [NUC PHYS] **1.** A particular type of radiation emitted by a radioactive substance, such as alpha radioactivity. **2.** *See* radioactive decay. **3.** *See* activity. { ˌrād·ē·ō·ak'tiv·əd·ē }

radioactivity equilibrium [NUC PHYS] A condition which may arise in the decay of a radioactive parent with short-lived descendants, in which the ratio of the activity of a parent to that of a descendant remains constant. { ˌrād·ē·ō·ak'tiv·əd·ē ˌē·kwə'lib·rē·əm }

radio antenna *See* antenna. { 'rād·ē·ō an'ten·ə }

radio attenuation [ELECTROMAG] For one-way propagation, the ratio of the power delivered by the transmitter to the transmission line connecting it with the transmitting antenna to the power delivered to the receiver by the transmission line connecting it with the receiving antenna. { 'rād·ē·ō əˌten·yə'wā·shən }

radio beam [ELECTROMAG] A concentrated stream of radio-frequency energy as used in radio ranges, microwave relays, and radar. { 'rād·ē·ō ˌbēm }

radiocarbon *See* carbon-14. { ¦rad·ē·ō'kär·bən }

radiocesium *See* cesium-137. { ¦rad·ē·ō'sē·zē·əm }

radio element [NUC PHYS] A radioactive isotope of an element, or a sample consisting of one or more radioactive isotopes of an element. { 'rād·ē·ō 'el·ə·mənt }

radio emission [ELECTROMAG] The emission of radio-frequency electromagnetic radiation by oscillating charges or currents. { 'rād·ē·ō iˌmish·ən }

radio energy [ELECTROMAG] The energy carried by radio-frequency electromagnetic radiation. { 'rād·ē·ō ˌen·ər·jē }

radio field intensity [ELECTROMAG] Electric or magnetic field intensity at a given location associated with the passage of radio waves. { 'rād·ē·ō 'fēld inˌten·səd·ē }

radio field-to-noise ratio [ELECTROMAG] Ratio, at a given location, of the radio field intensity of the desired wave to the noise field intensity. { 'rād·ē·ō 'fēld tə 'nȯiz ˌrā·shō }

radio frequency [ELECTROMAG] A frequency at which coherent electromagnetic radiation of energy is useful for communication purposes; roughly the range from 10 kilohertz to 100 gigahertz. Abbreviated rf. { 'rād·ē·ō ˌfrē·kwən·sē }

radio-frequency line *See* radio-frequency transmission line. { 'rād·ē·ō ¦frē·kwən·sē ˌlīn }

radio-frequency resistance *See* high-frequency resistance. { 'rād·ē·ō ¦frē·kwən·sē ri'zis·təns }

radio-frequency shielding [ELECTROMAG] Enclosure of a physical space or an object with a shield that prevents radio-frequency electromagnetic radiation from leaving or entering. { 'rād·ē·ō ¦frē·kwən·sē ˌshēld·iŋ }

radio-frequency spectrometer [SPECT] An instrument which measures the intensity of radiation emitted or absorbed by atoms or molecules as a function of frequency at frequencies from 10^5 to 10^9 hertz; examples include the atomic-beam apparatus, and instruments for detecting magnetic resonance. { 'rād·ē·ō ¦frē·kwən·sē spek'träm·əd·ər }

radio-frequency spectroscopy [SPECT] The branch of spectroscopy concerned with the measurement of the intervals between atomic or molecular energy levels that are separated by frequencies from about 10^5 to 10^9 hertz, as compared to the frequencies that separate optical energy levels of about 6×10^{14} hertz. { 'rād·ē·ō ¦frē·kwən·sē spek'träs·kə·pē }

radio-frequency transformer [ELECTROMAG] A transformer having a tapped winding or two or more windings designed to furnish inductive reactance or to transfer radio-frequency energy from one circuit to another by means of a magnetic field; may have an air core or some form of ferrite core. Also known as radio transformer. { 'rād·ē·ō ¦frē·kwən·sē tranz'fȯr·mər }

radio-frequency transmission line [ELECTROMAG] A transmission line designed primarily to conduct radio-frequency energy, consisting of two or more conductors supported in a fixed spatial relationship along their own length. Also known as radio-frequency line. { 'rād·ē·ō ¦frē·kwən·sē tranz'mish·ən ˌlīn }

radiogenic [NUC PHYS] Pertaining to a material produced by radioactive decay, as the production of lead from uranium decay. { ¦rād·ē·ō¦jen·ik }

radiogenic argon [NUC PHYS] Argon occurring in rocks and minerals that is the result of in-place decay of potassium-40 since the formation of the earth. { ¦rād·ē·ō¦jen·ik 'ärˌgän }

radiogenic isotope [NUC PHYS] An isotope which was produced by the decay of a radioisotope, but which itself may or may not be radioactive. { ¦rād·ē·ō¦jen·ik 'ī·səˌtōp }

radiogenic lead [NUC PHYS] Stable, end-product lead (Pb-206, Pb-207, and Pb-208) occurring in rocks and minerals that is the result of in-place decay of uranium and thorium since the formation of the earth. { ¦rād·ē·ō¦jen·ik 'led }

radiogenic strontium [NUC PHYS] Strontium-87 occurring in rocks and minerals that is the direct result of in-place decay of rubidium-87 since the formation of the earth. { ¦rād·ē·ō¦jen·ik 'strän·chəm }

radioiodine [NUC PHYS] Any radioactive isotope of iodine, especially iodine-131; used as a tracer to determine the activity and size of the thyroid gland, and experimentally, to destroy the thyroid glands of animals. { ¦rād·ē·ō'ī·əˌdīn }

radioisotope [NUC PHYS] An isotope which exhibits radioactivity. Also known as radioactive isotope; unstable isotope. { ¦rād·ē·ō'ī·səˌtōp }

radiolucent [ELECTROMAG] Transparent to x-rays and radio waves. { ¦rād·ē·ō'lüs·ənt }

radioluminescence [PHYS] Luminescence produced by x-rays or γ-rays, or by particles emitted in radioactive decay. { ¦rād·ē·ōˌlü·mə'nes·əns }

radiometer effect [PHYS] The effect of the temperature of a surface on the pressure exerted on it in a low vacuum, due to the effect on the momentum transferred to gas molecules colliding with the surface. { ˌrād·ē'äm·əd·ər iˌfekt }

radiometry [PHYS] The detection and measurement of radiant electromagnetic energy, especially that associated with infrared radiation. { ˌrād·ē'äm·ə·trē }

radionuclide [NUC PHYS] A nuclide that exhibits radioactivity. { ¦rād·ē·ō'nüˌklīd }

radionuclide content [NUC PHYS] The sum of the number of radioactive nuclei before and after a nuclear reaction as a fraction of the total number of nuclei involved (reactant nuclei plus reaction-product nuclei). { ˌrād·ē·ō'nüˌklīd 'känˌtent }

radiopaque [ELECTROMAG] Not appreciably penetrable by x-rays or other forms of radiation. { ¦rād·ē·ō'pāk }

radiophotoluminescence [PHYS] Luminescence exhibited by minerals such as fluorite and kunzite as a result of irradiation with β- and γ-rays followed by exposure to light. { ¦rād·ē·ō¦fōd·ōˌlü·mə'nes·əns }

radio pulse [ELECTROMAG] An intense burst of radio-frequency energy lasting for a fraction of a second. { 'rād·ē·ō ˌpəls }

radio recombination line [SPECT] A radio-frequency spectral line that results from an electron transition between energy levels in an atom or ion having a large principal quantum number *n*, greater than 50. { 'rād·ē·ō rēˌkäm·bə'nā·shən ˌlīn }

radio scattering *See* scattering. { 'rād·ē·ō 'skad·ə·riŋ }

radio sextant [ELECTROMAG] An antenna with a high-resolution beam pattern that measures the angle between local direction references and an astronomical radio signal source such as an artificial satellite, the sun, the moon, or a radio star. { 'rād·ē·ō 'sek·stənt }

radiothorium [NUC PHYS] Conventional name of the isotope of thorium which has mass number 228. Symbolized RdTh. { ¦rād·ē·ō'thȯr·ē·əm }

radio wave [ELECTROMAG] An electromagnetic wave produced by reversal of current in a conductor at a frequency in the range from about 10 kilohertz to about 300,000 megahertz. { 'rād·ē·ō ˌwāv }

radio wavefront distortion [ELECTROMAG] Change in the direction of advance of a radio wave. { 'rād·ē·ō 'wāvˌfrənt diˌstȯr·shən }

radio-wave propagation [ELECTROMAG] The transfer of energy through space by electromagnetic radiation at radio frequencies. { 'rād·ē·ō ¦wāv ˌpräp·əˌgā·shən }

radium F *See* polonium-210. { 'rād·ē·əm 'ef }

radius of gyration [MECH] The square root of the ratio of the moment of inertia of a body about a given axis to its mass. { 'rād·ē·əs əv ji'rā·shən }

radome [ELECTROMAG] A strong, thin shell, made from a dielectric material that is transparent to radio-frequency radiation, and used to house a radar antenna, or a space communications antenna of similar structure. { 'rāˌdōm }

radon [NUC PHYS] The conventional name for radon-222. Symbolized Rn. { 'rāˌdän }

radon-220 [NUC PHYS] The isotope of radon having mass number 220, symbol ^{220}Rn, which is a radioactive member of the thorium series with a half-life of 56 seconds. { 'rāˌdän ˌtü 'twen·tē }

radon-222 [NUC PHYS] The isotope of radon having mass number 222, symbol ^{222}Rn, which is a radioactive member of the uranium series with a half-life of 3.82 days. { 'rāˌdän ˌtü 'twen·tēˌtü }

rainbow [OPTICS] Colored arc seen in the sky when the sun or moon is illuminating large numbers of falling raindrops. { 'rānˌbō }

rainbow scattering [PHYS] The scattering of particles by a potential that contains both attractive and repulsive parts and whose width is much greater than the de Broglie wavelength of the particles; analogous to scattering of light by liquid droplets, which produces a rainbow. { 'rānˌbō ˌskad·ər·iŋ }

Raman effect [OPTICS] A phenomenon observed in the scattering of light as it passes through a transparent medium; the light undergoes a change in frequency and a random alteration in phase due to a change in rotational or vibrational energy of the scattering molecules. Also known as Raman scattering. { 'räm·ən i,fekt }

Raman-induced Kerr effect [OPTICS] The birefringence of an observation beam whose frequency differs from that of a pumping beam by a characteristic frequency of the medium (molecular vibration or rotation). { ¦räm·ən in¦düst 'kər i,fekt }

Raman lidar [OPTICS] A type of lidar that measures the scattered signal at the Raman-shifted wavelength in order to determine atmospheric density, temperature, and water vapor concentration. { 'räm·ən 'lī,där }

Raman-Rayleigh ratio [OPTICS] The ratio of Raman scattering (of a light beam passing through a transparent medium) to Rayleigh scattering (of a light beam horizontal to the medium). { 'räm·ən 'rā,lē ,rā·shō }

Raman scattering *See* Raman effect. { 'räm·ən ,skad·ə·riŋ }

Raman spectroscopy [SPECT] Analysis of the intensity of Raman scattering of monochromatic light as a function of frequency of the scattered light; the information obtained is useful for determining molecular structure. { 'räm·ən spek 'träs·kə·pē }

Raman spectrum [SPECT] A display, record, or graph of the intensity of Raman scattering of monochromatic light as a function of frequency of the scattered light. { 'räm·ən ,spek·trəm }

ram pressure [FL MECH] The pressure that is exerted by a fluid as a result of its motion. { 'ram ,presh·ər }

Ramsauer effect [ATOM PHYS] The vanishing of the scattering cross section of electrons from atoms of a noble gas at some value of the electron energy, always below 25 electronvolts. { 'räm,zaù·ər i,fekt }

Ramsay-Shields-Eötvös equation [THERMO] An elaboration of the Eötvös rule which states that at temperatures not too near the critical temperature, the molar surface energy of a liquid is proportional to $t_c - t - 6$ K, where t is the temperature and t_c is the critical temperature. { 'ram·zē 'shēlz 'ót·vósh i,kwā·zhən }

Ramsay-Young method [THERMO] A method of measuring the vapor pressure of a liquid, in which a thermometer bulb is surrounded by cotton wool soaked in the liquid, and the pressure, measured by a manometer, is reduced until the thermometer reading is steady. { ¦ram·zē 'yəŋ ,meth·əd }

Ramsay-Young rule [THERMO] An empirical relationship which states that the ratio of the absolute temperatures at which two chemically similar liquids have the same vapor pressure is independent of this vapor pressure. { 'ram·zē 'yəŋ ,rül }

Ramsden circle [OPTICS] A sharp, bright circle of light which appears on a sheet of white paper held near the eyepiece of a telescope focused for infinity and pointed at a bright sky. Also known as Ramsden disk. { 'ramz·dən ,sər·kəl }

Ramsden disk *See* Ramsden circle. { 'ramz·dən ,disk }

Ramsden eyepiece [OPTICS] An eyepiece consisting of two planoconvex lenses with their plane sides facing outward, having the same power and focal length, and separated by a distance equal to their common focal length. { 'ramz·dən 'ī,pēs }

Ramsey fringes [PHYS] Oscillations in the number of transitions in a molecular beam that passes through two separated radio-frequency fields, as a function of L/v, where L is the separation of the field and v is the speed of the beam. { 'ram·zē ,frin·jəz }

random diffusion chamber *See* reverberation chamber. { 'ran·dəm di¦fyü·zhən ,chām·bər }

random interstratification [SOLID STATE] A crystalline structure in which two or more types of layers alternate in a random fashion. { 'ran·dəm ,in·tər,strad·i·fə'kā·shən }

random noise [PHYS] Noise characterized by a large number of overlapping transient disturbances occurring at random, such as thermal noise and shot noise. Also known as fluctuation noise. { 'ran·dəm 'nóiz }

random structure [CRYSTAL] A crystal structure in which different types of atoms are associated with the various points in a crystal lattice in a random fashion. { 'ran·dəm 'strək·chər }

random vibration [MECH] A varying force acting on a mechanical system which may be considered to be the sum of a large number of irregularly timed small shocks; induced typically by aerodynamic turbulence, airborne noise from rocket jets, and transportation over road surfaces. { 'ran·dəm vī'brā·shən }

range [MECH] The horizontal component of a projectile displacement at the instant it strikes the ground. [PHYS] The greatest distance between two particles at which a given force between them is appreciable. { rānj }

range attenuation [ELECTROMAG] In radar terminology, the decrease in power density (flux density) caused by the divergence of the flux lines with distance, this decrease being in accordance with the inverse-square law. { 'rānj ə,ten·yə¦wā·shən }

range delay [ELECTROMAG] A control used in radars which permits the operator to present on the radarscope only those echoes from targets which lie beyond a certain distance from the radar; by using range delay, undesired echoes from nearby targets may be eliminated while the indicator range is increased. { 'rānj di,lā }

range deviation [MECH] Distance by which a projectile strikes beyond, or short of, the target; the distance as measured along the gun-target line or along a line parallel to the gun-target line. { 'rānj ,dē·vē¦ā·shən }

range strobe [ELECTROMAG] An index mark which may be displayed on various types of radar indicators to assist in the determination of the exact range of a target. { 'rānj ˌstrōb }

Rankine body [FL MECH] A fluid flow pattern formed by combining a uniform stream with a source and a sink of equal strengths, with the line joining the source and sink along the stream direction. { 'raŋ·kən ˌbäd·ē }

Rankine cycle [THERMO] An ideal thermodynamic cycle consisting of heat addition at constant pressure, isentropic expansion, heat rejection at constant pressure, and isentropic compression; used as an ideal standard for the performance of heat-engine and heat-pump installations operating with a condensable vapor as the working fluid, such as a steam power plant. Also known as steam cycle. { 'raŋ·kən ˌsī·kəl }

Rankine-Hugoniot equations [THERMO] Equations, derived from the laws of conservation of mass, momentum, and energy, which relate the velocity of a shock wave and the pressure, density, and enthalpy of the transmitting fluid before and after the shock wave passes. { 'raŋ·kən yü'gō·nē·ō iˌkwā·zhənz }

Rankine temperature scale [THERMO] A scale of absolute temperature; the temperature in degrees Rankine (°R) is equal to 9/5 of the temperature in kelvins and to the temperature in degrees Fahrenheit plus 459.67. { 'raŋ·kən 'tem·prə·chər ˌskāl }

Rankine vortex [FL MECH] A vortex with a vertical axis and circular motion, in which the motion is that of a rotating solid cylinder inside some fixed radius, and the circulation is constant outside this radius. { 'raŋ·kən ˌvȯrˌteks }

Ranque effect [FL MECH] An effect whereby turbulent flow in a tube supplied with air through a tangential nozzle at high pressure produces warming near the walls of the tube and cooling at the axis. { 'räŋk iˌfekt }

RANS analysis *See* Reynolds-averaged Navier-Stokes analysis. { 'ranz əˌnal·ə·səs }

rapid sequence camera [OPTICS] A conventional camera in most respects except that it is designed to permit a number of photographs to be obtained in rapid succession with one winding of the shutter. { 'rap·əd ¦sē·kwəns ˌkam·rə }

rare-earth chelate laser *See* chelate laser. { 'rer ˌərth 'kēˌlāt 'lā·zər }

rare-earth magnet [ELECTROMAG] Any of several types of magnets made with rare-earth elements, such as rare-earth-cobalt magnets, which have coercive forces up to ten times that of ordinary magnets; used for computers and signaling devices. { 'rer ˌərth 'mag·nət }

rarefaction [ACOUS] The instantaneous, local reduction in density of a gas resulting from passage of a sound wave, or the region in which the density is reduced at some instant. Also known as rarefraction. { ¦rer·ə¦fak·shən }

rarefaction wave [FL MECH] A pressure wave or rush of air or water induced by rarefaction; it travels in the opposite direction to that of a shock wave directly following an explosion. Also known as suction wave. { ¦rer·ə¦fak·shən ˌwāv }

rarefied gas [FL MECH] A gas whose pressure is much less than atmospheric pressure. { 'rer·əˌfīd 'gas }

rarefraction *See* rarefaction. { ˌrer·ə'frak·shən }

Rarita-Schwinger equation [QUANT MECH] A partial differential equation, similar in form to the Dirac equation, relating the spatial and time dependence of a 16-component wave function describing a free relativistic particle with intrinsic spin 3/2, and its antiparticle. { 'rä·rē·tä 'shviŋ·ər iˌkwā·shən }

Rateau formula [FL MECH] A formula, $m = A_2p_1(16.367 - 0.96 \log p_1)/1000$, for determining the discharge m of saturated steam in pounds per second through a well-rounded convergent orifice; A_2 is the area of the orifice in square inches, and p_1 the reservoir pressure in pounds per square inch. { rä'tō ˌfor·myə·lə }

rate of change of acceleration [MECH] Time rate of change of acceleration; this rate is a factor in the design of some items of ammunition that undergo large accelerations. { 'rāt əv 'chānj əv ikˌsel·ə'rā·shən }

rate of flow *See* flow rate. { 'rāt əv 'flō }

rate process [PHYS] Any process in which the derivatives with respect to time of one or more variables, evaluated at any given time t_0, depend on the values of the variables at time t_0 and possibly at times earlier than t_0. { 'rāt ˌpräˌses }

rate receiver [ELECTROMAG] A guidance antenna that receives a signal from a launched missile as to its rate of speed. { 'rāt riˌsē·vər }

ratio arm circuit [ELEC] Two adjacent arms of a Wheatstone bridge, designed so they can be set to provide a variety of indicated resistance ratios. { 'rā·shō ¦ärm ˌsər·kət }

rationalized units [ELEC] A system of electrical units, such as occurs in the International System, in which the factor of 4π is removed from the field equations and appears instead in the explicit expressions for the fields of a point charge and current element. { 'rash·ən·əlˌīzd 'yü·nəts }

ratio of transformer [ELEC] Ratio of the number of turns in one winding of a transformer to the number of turns in the other, unless otherwise specified. { 'rā·shō əv tranz'fȯr·mər }

ratio resistor [ELEC] One of the resistors in a Wheatstone or Kelvin bridge whose resistances appear in a pair of ratios which are equal in a balanced bridge. { 'rā·shō riˌzis·tər }

ray [OPTICS] A curve whose tangent at any point lies in the direction of propagation of a light wave. [PHYS] A moving particle or photon of ionizing radiation. { rā }

ray acoustics [ACOUS] The study of the behavior of sound under the assumption that sound traversing a homogeneous medium travels along straight lines or rays. Also known as geometrical acoustics. { 'rā əˌküs·tiks }

Raychuraduri equation [RELAT] An equation of general relativity theory, useful in proving singularity theorems, that relates the expansion, convergence, and shear of a congruence of time-like or null curves to the amount of matter present. { ˌrā·chu̇r¦ə¦du̇r·ē iˌkwā·zhən }

ray diagram [OPTICS] A diagram showing the paths of selected rays through an optical system. { 'rā ˌdī·əˌgram }

ray ellipsoid *See* Fresnel ellipsoid. { 'rā i'lipˌsȯid }

rayl [ACOUS] A unit of specific acoustical impedance, equal to a sound pressure of 1 dyne per square centimeter divided by a sound particle velocity of 1 centimeter per second. Also known as specific acoustical ohm (Ω_s); unit-area acoustical ohm. { rāl }

rayleigh [OPTICS] A unit of brightness, used to measure the brightness of the night sky and aurorae, equal to $10^{10}/4\pi$ quanta per square meter per second per steradian. { 'rā·lē }

Rayleigh balance [ELECTROMAG] An apparatus for assigning the value of the ampere in which the force exerted on a movable circular coil by larger circular coils above and below, but coaxial with, the movable coil is compared with the gravitational force on a known mass. { 'rā·lē ˌbal·əns }

Rayleigh-Bènard convection [PHYS] Convection of a fluid heated from below, characterized by a regular array of usually hexagonal cells. { 'rā·lē bā'när kənˌvek·shən }

Rayleigh criterion [OPTICS] A criterion for the resolving power of an optical instrument which states that the images of two point objects are resolved when the principal maximum of the diffraction pattern of one falls exactly on the first minimum of the diffraction pattern of the other. { 'rā·lē krīˌtir·ē·ən }

Rayleigh cycle [ELECTROMAG] A cycle of magnetization that does not extend beyond the initial portion of the magnetization curve, between zero and the upward bend. { 'rā·lē ˌsī·kəl }

Rayleigh disk [ACOUS] An acoustic radiometer, used to measure particle velocity, consisting of a thin disk set at an angle of 45° to a sound beam; the particle velocity is calculated from the resulting torque on the disk. { 'rā·lē ˌdisk }

Rayleigh distance [PHYS] For electromagnetic, acoustic, or elastic waves emitted from a uniformly excited planar array transmitting a sinusoidal signal, the distance from the array at which there is a transition from a near-field region, in which the radiated energy is confined to a cylindrical region, to a far-field region, in which the wave field exhibits spherical spreading and the field amplitude varies inversely with range. { 'rā·lē ˌdis·təns }

Rayleigh flow [FL MECH] An idealized type of gas flow in which heat transfer may occur, satisfying the assumptions that the flow takes place in constant-area cross section and is frictionless and steady, that the gas is perfect and has constant specific heat, that the composition of the gas does not change, and that there are no devices in the system which deliver or receive mechanical work. { 'rā·lē ˌflō }

Rayleigh interferometer [OPTICS] An optical interferometer in which two rays of light, emanating from a single slit, are collimated by a lens, pass through separate slits and cells, and are brought to focus by a second lens so that interference fringes become visible. Also known as Rayleigh refractometer. { 'rā·lē ˌin·tər·fə'räm·əd·ər }

Rayleigh-Jansen method [FL MECH] A method for solving equations for compressible fluid flow past a body, in which the velocity potential for the difference between the fluid velocity and the velocity distant from the body (V) is expressed as a power series in the square of the Mach number corresponding to V. { ¦rā·lē 'jan·sən ˌmeth·əd }

Rayleigh-Jeans law [STAT MECH] A law giving the intensity of radiation emitted by a blackbody within a narrow band of wavelengths; it states that this intensity is proportional to the temperature divided by the fourth power of the wavelength; it is a good approximation to the experimentally verified Planck radiation formula only at long wavelengths. { ¦rā·lē 'jēnz ˌlȯ }

Rayleigh law [ELECTROMAG] **1.** For small values of the magnetic field strength H, the normal permeability of a material is approximately by $a + b$H, where a is the initial permeability and b is a constant. **2.** In a magnetic material subject to cyclic magnetization, with maximum magnetic field strength small compared with the coercive force, the hysteresis loss per cycle is proportional to the cube of the maximum value of the magnetic induction. [OPTICS] In Rayleigh scattering, the intensity of light scattered in a direction making an angle θ with the incident direction is proportional to $1 + \cos^2 \theta$ and inversely proportional to the fourth power of the wavelength of the incident radiation. { 'rā·lē ˌlȯ }

Rayleigh lidar [OPTICS] A type of lidar that is designed to measure the Rayleigh scattering of laser light from molecules in the atmosphere, thereby determining atmospheric density. { 'rāˌlē 'līˌdär }

Rayleigh line [MECH] A straight line connecting points corresponding to the initial and final states on a graph of pressure versus specific volume for a substance subjected to a shock wave. [SPECT] Spectrum line in scattered radiation which has the same frequency as the corresponding incident radiation. { 'rā·lē ˌlīn }

Rayleigh loop [ELECTROMAG] A parabolic approximation to a magnetic hysteresis loop. { 'rā·lē ˌlüp }

Rayleigh number 1 [FL MECH] A dimensionless number used in studying the breakup of liquid jets, equal to Weber number 2. Symbolized N_{Ra1}. { 'rā·lē ¦nəm·bər 'wən }

Rayleigh number 2 [THERMO] A dimensionless number used in studying free convection, equal to the product of the Grashof number and the Prandtl number. Symbolized R'_2. { 'rā·lē ¦nəm·bər 'tü }

Rayleigh number 3 [THERMO] A dimensionless number used in the study of combined free and forced convection in vertical tubes, equal to Rayleigh number 2 times the Nusselt number times the tube diameter divided by its entry length. Symbolized Ra_3. { 'rā·lē ¦nəm·bər 'thrē }

Rayleigh prism [OPTICS] A system of prisms used to produce greater dispersion of light than would be produced by a single prism. { 'rā·lē ˌpriz·əm }

Rayleigh ratio [OPTICS] Light-scattering relationship defined by the ratio of intensities of incident and scattered light at a specified distance; used in photometric and refractometric analyses. { 'rā·lē ˌrā·shō }

Rayleigh reciprocity theorem [ELECTROMAG] Reciprocal relationship for an antenna when it is transmitting or receiving; the effective heights, radiation resistance, and the radiation pattern are alike, whether the antenna is transmitting or receiving. { 'rā·lē ˌres·ə'präs·əd·ē ˌthir·əm }

Rayleigh refractometer *See* Rayleigh interferometer. { 'rā·lē ˌrēˌfrak'täm·əd·ər }

Rayleigh scattering [ELECTROMAG] Scattering of electromagnetic radiation by independent particles which are much smaller than the wavelength of the radiation. { 'rā·lē ˌskad·ə·riŋ }

Rayleigh's dissipation function [MECH] A function which enters into the equations of motion of a system undergoing small oscillations and represents frictional forces which are proportional to velocities; given by a positive definite quadratic form in the time derivatives of the coordinates. Also known as dissipation function. { 'rā·lē ˌdis·ə'pā·shən ˌfəŋk·shən }

Rayleigh-Taylor instability [FL MECH] The instability of the interface separating two fluids having different densities when the lighter fluid is accelerated toward the heavier fluid. { ¦rā·lē 'tā·lər ˌin·stə'bil·əd·ē }

Rayleigh wave [MECH] A wave which propagates on the surface of a solid; particle trajectories are ellipses in planes normal to the surface and parallel to the direction of propagation. Also known as surface wave. { 'rā·lē ˌwāv }

ray path [PHYS] An imaginary path along which travels the energy associated with a point on a wavefront. { 'rā ˌpath }

ray surface [OPTICS] The locus of points reached in a unit time in an anisotropic medium by an electromagnetic disturbance that starts from the origin. { 'rā ˌsər·fəs }

ray tracing [OPTICS] Calculation of the paths followed by rays of light through an optical system, using Snell's law and trigonometrical formulas. { 'rā ˌtrās·iŋ }

Razin effect [PL PHYS] An effect whereby electrons in a cool, collisionless plasma strongly reduce the intensity of synchrotron radiation. { 'räz·ən iˌfekt }

R-branch [SPECT] A series of lines in molecular spectra that correspond, in the case of absorption, to a unit increase in the rotational quantum number J. { 'är ˌbranch }

RBS *See* Rutherford backscattering spectrometry.

R-C circuit *See* resistance-capacitance circuit. { ¦är¦sē 'sər·kət }

R-C constant *See* resistance-capacitance constant. { ¦är¦sē 'kän·stənt }

R center [SOLID STATE] A color center whose absorption band lies between the F band and M band, and which is produced by prolonged irradiation with light in the F band or prolonged x-ray exposure at room temperature. Also known as D center; E center. { 'är ˌsen·tər }

R-C network *See* resistance-capacitance network. { ¦är¦sē 'netˌwərk }

RdAc *See* radioactinium.

RdTh *See* radiothorium.

reactance [ELEC] The imaginary part of the impedance of an alternating-current circuit. { rē'ak·təns }

reactance drop [ELEC] The component of the phasor representing the voltage drop across a component or conductor of an alternating-current circuit which is perpendicular to the current. { rē'ak·təns ˌdräp }

reaction [MECH] The equal and opposite force which results when a force is exerted on a body, according to Newton's third law of motion. [NUC PHYS] *See* nuclear reaction. { rē'ak·shən }

reaction energy *See* disintegration energy. { rē'ak·shən ˌen·ər·jē }

reactive [ELEC] Pertaining to either inductive or capacitance reactance; a reactive circuit has a high value of reactance in comparison with resistance. { rē'ak·tiv }

reactive component [ELEC] In the phasor representation of quantities in an alternating-current circuit, the component of current, voltage, or apparent power which does not contribute power, and which results from inductive or capacitive reactance in the circuit, namely, the reactive current, reactive voltage, or reactive power. Also known as idle component; quadrature component; wattless component. { rē'ak·tiv kəm'pō·nənt }

reactive current [ELEC] In the phasor representation of alternating current, the component of the current perpendicular to the voltage, which contributes no power but increases the power losses of the system. Also known as idle current; quadrature current; wattless current. { rē'ak·tiv 'kə·rənt }

reactive factor [ELEC] The ratio of reactive power to apparent power. { rē'ak·tiv ˌfak·tər }

reactive power [ELEC] The power value obtained by multiplying together the effective value of current in amperes, the effective value of voltage in volts, and the sine of the angular phase difference between current and voltage. Also known as wattless power. { rē'ak·tiv 'pau̇·ər }

reactive voltage [ELEC] In the phasor representation of alternating current, the voltage component that is perpendicular to the current. { rē'ak·tiv 'vōl·tij }

reactive volt-ampere *See* volt-ampere reactive. { rē'ak·tiv 'vōlt 'amˌpir }

reactive volt-ampere hour *See* var hour. { rē'ak·tiv 'vōlt 'amˌpir 'au̇·ər }

reactor [ELEC] A device that introduces either inductive or capacitive reactance into a circuit, such as a coil or capacitor. Also known as electric reactor. *See* nuclear reactor. { rē'ak·tər }

reading microscopes [OPTICS] A set of microscopes used to read the division circle of a transit circle in order to precisely determine the inclination of the telescope. { 'rēd·iŋ ˌmī·krəˌskōps }

real crystal [CRYSTAL] A crystal for which the finite extent of the crystal and its various imperfections and defects are taken into account. { 'rēl 'krist·əl }

real fluid flow [FL MECH] The flow in which effects of tangential or shearing forces are taken into account; these forces give rise to fluid friction, because they oppose the sliding of one particle past another. { 'rēl 'flü·əd ˌflō }

real gas [THERMO] A gas, as considered from the viewpoint in which deviations from the ideal gas law, resulting from interactions of gas molecules, are taken into account. Also known as imperfect gas. { 'rēl 'gas }

real image [OPTICS] An optical image such that all the light from a point on an object that passes through an optical system actually passes close to or through a point on the image. { 'rēl 'im·ij }

real object [OPTICS] A collection of points which actually serves as a source of light rays in an optical system. { 'rēl 'äb·jikt }

real power [ELEC] The component of apparent power that represents true work; expressed in watts, it is equal to volt-amperes multiplied by the power factor. { 'rēl ¦paủ·ər }

real source [ACOUS] A source of sound consisting of a macroscopic body that is composed of materials different from those of the medium in which the sound propagates and has sharply delineated physical extent, and which generates sound by executing complex motions while immersed in the medium. { 'rēl 'sȯrs }

real-time holographic interferometry [OPTICS] The study of the interference fringes generated when a hologram is made of an object and is later placed back into its original position relative to the object, now slightly deformed, so that there is interference between the object and its hologram. { 'rēl ˌtīm ¦hō·lə¦graf·ik ˌin·tər·fə'räm·ə·trē }

rearrangement reaction [NUC PHYS] A nuclear reaction in which nucleons are exchanged between nuclei. { ˌrē·ə'rānj·mənt rēˌak·shən }

Réaumur temperature scale [THERMO] Temperature scale where water freezes at 0°R and boils at 80°R. { ¦rā·ō¦myủr 'tem·prə·chər ˌskāl }

received power [ELECTROMAG] **1.** The total power received at an antenna from a signal, such as a radar target signal. **2.** In a mobile communications system, the root-mean-square value of power delivered to a load which properly terminates an isotropic reference antenna. { ri'sēvd 'paủ·ər }

receiver radiation [ELECTROMAG] Radiation of interfering electromagnetic fields by the oscillator of a receiver. { ri'sē·vər ˌrād·ē'ā·shən }

receiving antenna [ELECTROMAG] An antenna used to convert electromagnetic waves to modulated radio-frequency currents. { ri'sēv·iŋ anˌten·ə }

receiving area [ELECTROMAG] The factor by which the power density must be multiplied to obtain the received power of an antenna, equal to the gain of the antenna times the square of the wavelength divided by 4π. { ri'sēv·iŋ ˌer·ē·ə }

recharge [ELEC] To restore a cell or battery to a charged condition by sending a current through it in a direction opposite to that of the discharging current. { rē'chärj }

rechargeable battery *See* storage battery. { rē 'chär·jə·bəl'bad·ə·rē }

reciprocal ellipsoid *See* index ellipsoid. { ri'sip·rə·kəl i'lipˌsȯid }

reciprocal ferrite switch [ELECTROMAG] A ferrite switch that can be inserted in a waveguide to switch an input signal to either of two output waveguides; switching is done by a Faraday rotator when acted on by an external magnetic field. { ri'sip·rə·kəl 'feˌrīt ˌswich }

reciprocal impedance [ELEC] Two impedances Z_1 and Z_2 are said to be reciprocal impedances with respect to an impedance Z (invariably a resistance) if they are so related as to satisfy the equation $Z_1Z_2 = Z^2$. { ri'sip·rə·kəl im'pēd·əns }

reciprocal junction [ELECTROMAG] A waveguide junction in which the transmission coefficient from the *i*th port to the *j*th port is the same as that from the *j*th port to the *i*th port; that is, the S matrix is symmetrical. { ri'sip·rə·kəl 'jəŋk·shən }

reciprocal lattice [CRYSTAL] A lattice array of points formed by drawing perpendiculars to each plane (*hkl*) in a crystal lattice through a common point as origin; the distance from each point to the origin is inversely proportional to spacing of the specific lattice planes; the axes of the reciprocal lattice are perpendicular to those of the crystal lattice. { ri'sip·rə·kəl 'lad·əs }

reciprocal ohm *See* siemens. { ri'sip·rə·kəl 'ōm }

reciprocal ohm centimeter *See* roc. { ri'sip·rə·kəl 'ōm 'sent·iˌmēd·ər }

reciprocal ohm meter *See* rom. { ri'sip·rə·kəl 'ōm ˌmēd·ər }

reciprocal space *See* wave-vector space. { ri'sip·rə·kəl ˌspās }

reciprocal strain ellipsoid [MECH] In elastic theory, an ellipsoid of certain shape and orientation which under homogeneous strain is transformed into a set of orthogonal diameters of the sphere. { ri'sip·rə·kəl ¦strān i'lipˌsȯid }

reciprocal vectors [CRYSTAL] For a set of three vectors forming the primitive translations of a lattice, the vectors that form the primitive translations of the reciprocal lattice. { ri'sip·rə·kəl 'vek·tərz }

reciprocal velocity region [NUC PHYS] The energy region in which the capture cross section for neutrons by a given element is inversely proportional to neutron velocity. { ri'sip·rə·kəl və'läs·əd·ē ˌrē·jən }

reciprocal wavelength *See* wave number. { ri'sip·rə·kəl 'wāv,leŋkth }

reciprocity theorem Also known as principle of reciprocity. [ACOUS] The theorem that, in an acoustic system consisting of a fluid medium with boundary surfaces and subject to no impressed body forces, if p_1 and p_2 are the pressure fields produced respectively by the components of the fluid velocities V_1 and V_2 normal to the boundary surfaces, then the integral over the boundary surfaces of $p_1V_2 - p_2V_1$ vanishes. [ELEC] **1.** The electric potentials V_1 and V_2 produced at some arbitrary point, due to charge distributions having total charges of q_1 and q_2 respectively, are such that $q_1V_2 = q_2V_1$. **2.** In an electric network consisting of linear passive impedances, the ratio of the electromotive force introduced in any branch to the current in any other branch is equal in magnitude and phase to the ratio that results if the positions of electromotive force and current are exchanged. [ELECTROMAG] Given two loop antennas, *a* and *b*, then $I_{ab}/V_a = I_{ba}/V_b$, where I_{ab} denotes the current received in *b* when *a* is used as transmitter, and V_a denotes the voltage applied in *a*; I_{ba} and V_b are the corresponding quantities when *b* is the transmitter, *a* the receiver; it is assumed that the frequency and impedances remain unchanged. [PHYS] In general, any theorem that expresses various reciprocal relations for the behavior of some physical systems, in which input and output can be interchanged without altering the response of the system to a given excitation. { ,res·ə'präs·əd·ē ,thir·əm }

recognition differential [ACOUS] For a specified listener, the amount by which the signal level exceeds the noise level reaching the ear when there is a 50% probability of detection of the signal. { ,rek·ig'nish·ən ,dif·ə'ren·chəl }

recoil *See* gun reaction. { 'rē,kȯil }

recoil electron [PHYS] An electron that has been set into motion by a collision. { 'rē,kȯil i,lek,trän }

recoil ion spectroscopy [ATOM PHYS] A method of studying highly ionized and highly excited atomic states, in which relatively light atoms in a gaseous target are bombarded by highly ionized, fast, heavy projectiles, resulting in single collisions in which the target atoms are raised to very high states of ionization and excitation while incurring relatively small recoil velocities. { 'rē,kȯil 'ī,än ,spek'träs·kə·pē }

recoil particle [PHYS] A particle that has been set into motion by a collision or by a process involving the ejection of another particle. { 'rē ,kȯil ,pärd·ə·kəl }

recombination [PHYS] The combination and resultant neutralization of particles or objects having unlike charges, such as a hole and an electron or a positive ion and a negative ion. { ,rē,käm·bə'nā·shən }

recombination energy [PHYS] The energy released when two oppositely charged portions of an atom or molecule rejoin to form a neutral atom or molecule. { ,rē,käm·bə'nā·shən ,en·ər·jē }

recombination radiation [SOLID STATE] The radiation emitted in semiconductors when electrons in the conduction band recombine with holes in the valence band. { ,rē,käm·bə'nā·shən ,rād·ē,ā·shən }

reconstruction [SOLID STATE] A process in which atoms at the surface of a solid displace and form bands different from those existing in the bulk solid. { ,rē·kən'strək·shən }

reconstructive transformation [CRYSTAL] A type of crystal transformation that involves the breaking of either first- or second-order coordination bonds. { ,rē·kən'strək·tiv ,tranz·fər'mā·shən }

recoupling [QUANT MECH] A transformation between eigenfunctions of total angular momentum resulting from coupling eigenfunctions of three or more angular momenta in some order, and eigenfunctions of total angular momentum resulting from coupling of the same eigenfunctions in a different order. { rē'kəp·liŋ }

recoverable shear [FL MECH] Measure of the elastic content of a fluid, related to elastic recovery (mechanicallike property of elastic recoil); found in unvulcanized, unfilled natural rubber, and certain polymer solutions, soap gels, and biological fluids. { ri'kəv·rə·bəl 'shir }

recovery [MECH] The return of a body to its original dimensions after it has been stressed, possibly over a considerable period of time. { ri'kəv·ə·rē }

recovery temperature *See* adiabatic recovery temperature. { ri'kəv·ə·rē ,tem·prə·chər }

recrystallization [CRYSTAL] A change in the structure of a crystal without a chemical alteration. { rē,krist·əl·ə'zā·shən }

rectangular cavity [ELECTROMAG] A resonant cavity having the shape of a rectangular parallelepiped. { rek'taŋ·gyə·lər 'kav·əd·ē }

rectangular waveguide [ELECTROMAG] A waveguide having a rectangular cross section. { rek 'taŋ·gyə·lər 'wāv,gīd }

rectification [ELEC] The process of converting an alternating current to a unidirectional current. { ,rek·tə·fə'kā·shən }

rectified value [ELEC] For an alternating quantity, the average of all the positive (or negative) values of the quantity during an integral number of periods. { 'rek·tə,fīd 'val·yü }

rectifier [ELEC] A nonlinear circuit component that allows more current to flow in one direction than the other; ideally, it allows current to flow in one direction unimpeded but allows no current to flow in the other direction. { 'rek·tə,fī·ər }

rectilinear motion [MECH] A continuous change of position of a body so that every particle of the body follows a straight-line path. Also known as linear motion. { ¦rek·tə'lin·ē·ər 'mō·shən }

rectilinear system *See* orthoscopic system. { ¦rek·tə'lin·ē·ər 'sis·təm }

rectiliner lens [OPTICS] A lens that is free from

distortion, imaging straight lines onto straight lines regardless of their orientation. { ¦rek·tə'lin·ē·ər 'lenz }

red [OPTICS] The hue evoked in an average observer by monochromatic radiation having a wavelength in the approximate range from 622 to 770 nanometers; however, the same sensation can be produced in a variety of other ways. { red }

reduced Compton wavelength [QUANT MECH] The Compton wavelength of a particle divided by 2π. { ri'düst 'käm·tən 'wāv,leŋkth }

reduced distance [OPTICS] A distance in a medium divided by the medium's index of refraction. { ri'düst 'dis·təns }

reduced equation of state [PHYS] An equation relating the reduced pressure, reduced volume, and reduced temperature of a substance. { ri 'düst i'kwā·zhən əv 'stāt }

reduced frequency *See* Strouhal number. { ri 'düst 'frē·kwən·sē }

reduced mass [MECH] For a system of two particles with masses m_1 and m_2 exerting equal and opposite forces on each other and subject to no external forces, the reduced mass is the mass m such that the motion of either particle, with respect to the other as origin, is the same as the motion with respect to a fixed origin of a single particle with mass m acted on by the same force; it is given by $m = m_1m_2/(m_1 + m_2)$. { ri'düst 'mas }

reduced pressure [THERMO] The ratio of the pressure of a substance to its critical pressure. { ri'düst 'presh·ər }

reduced property *See* reduced value. { ri'düst 'präp·ərd·ē }

reduced temperature [THERMO] The ratio of the temperature of a substance to its critical temperature. { ri'düst 'tem·prə·chər }

reduced value [THERMO] The actual value of a quantity divided by the value of that quantity at the critical point. Also known as reduced property. { ri'düst 'val·yü }

reduced volume [THERMO] The ratio of the specific volume of a substance to its critical volume. { ri'düst 'väl·yəm }

reducing glass [OPTICS] A double-concave lens that reduces the apparent size of objects viewed through it; used by illustrators and painters to create an artificial sense of distance from their work. { ri'düs·iŋ ,glas }

redundancy [MECH] A statically indeterminate structure. { ri'dən·dən·sē }

reed relay [ELECTROMAG] A relay having contacts mounted on magnetic reeds scaled into a length of small glass tubing; an actuating coil is wound around the tubing or wound on an auxiliary ferrite-core structure, to provide the magnetic field required for relay operation. { 'rēd ,rē,lā }

reentrant angle [CRYSTAL] The angle between two plane surfaces on a crystalline solid, in which the external angle is less than 180°. { rē'en·trənt ,aŋ·gəl }

reference acoustic pressure [ACOUS] Magnitude of any complex sound that will produce a sound-level meter reading equal to that produced by a sound pressure of 20 micropascals at 1000 hertz. Also known as reference sound level. { 'ref·rəns ə'küs·tik 'presh·ər }

reference angle [ELECTROMAG] Angle formed between the center line of a radar beam as it strikes a reflecting surface and the perpendicular drawn to that reflecting surface. { 'ref·rəns ,aŋ·gəl }

reference dipole [ELECTROMAG] Straight half-wave dipole tuned and matched for a given frequency, and used as a unit of comparison in antenna measurement work. { 'ref·rəns 'dī,pōl }

reference frame *See* frame of reference. { 'ref·rəns ,frām }

reference sound level *See* reference acoustic pressure. { 'ref·rəns 'saủnd ,lev·əl }

reference volume [ACOUS] The audio volume level that gives a reading of 0 VU (volume units) on a standard volume indicator; the sensitivity of the volume indicator is adjusted so reference volume or 0 is read when the instrument is connected across a 600-ohm resistance to which is delivered a power of 1 milliwatt at 1000 hertz. { 'ref·rəns ,väl·yəm }

reflectance *See* reflection factor; reflectivity. { ri 'flek·təns }

reflectance spectrophotometry [SPECT] Measurement of the ratio of spectral radiant flux reflected from a light-diffusing specimen to that reflected from a light-diffusing standard substituted for the specimen. { ri'flek·təns ,spek·trə·fə'täm·ə·trē }

reflected pressure [PHYS] The pressure from an explosion (especially an airburst bomb), which is reflected from a solid object or surface, rather than dissipated in the air. { ri'flek·təd 'presh·ər }

reflected ray [PHYS] A ray extending outward from a point of reflection. { ri'flek·təd 'rā }

reflected wave [PHYS] A wave reflected from a surface, discontinuity, or junction of two different media, such as the sky wave in radio, the echo wave from a target in radar, or the wave that travels back to the source end of a mismatched transmission line. { ri'flek·təd 'wāv }

reflecting antenna [ELECTROMAG] An antenna used to achieve greater directivity or desired radiation patterns, in which a dipole, slot, or horn radiates toward a larger reflector which shapes the radiated wave to produce the desired pattern; the reflector may consist of one or two plane sheets, a parabolic or paraboloidal sheet, or a paraboloidal horn. { ri'flek·tiŋ an'ten·ə }

reflecting curtain [ELECTROMAG] A vertical array of half-wave reflecting antennas, generally used one quarter-wavelength behind a radiating curtain of dipoles to form a high-gain antenna. { ri'flek·tiŋ 'kərt·ən }

reflecting grating [ELECTROMAG] Arrangement of wires placed in a waveguide to reflect one desired wave while allowing one or more other waves to pass freely. { ri'flek·tiŋ 'grād·iŋ }

reflecting microscope [OPTICS] A microscope whose objective is composed of two mirrors, one convex and the other concave; its imaging properties are independent of the wavelength of light, allowing it to be used even for infrared and ultraviolet radiation. { ri'flek·tiŋ 'mī·krə,skōp }

reflecting prism [OPTICS] A prism used in place of a mirror for deviating light, usually designed so that there is no dispersion of light; the light undergoes at least one internal reflection. { ri 'flek·tiŋ 'priz·əm }

reflecting spectrograph [OPTICS] A solar spectrograph in which the collimator and camera element are long-focus concave mirrors. { ri'flek·tiŋ 'spek·trə,graf }

reflecting telescope [OPTICS] A telescope in which a concave parabolic mirror gathers light and forms a real image of an object. Also known as reflector telescope. { ri'flek·tiŋ 'tel·ə,skōp }

reflection [PHYS] The return of waves or particles from surfaces on which they are incident. { ri'flek·shən }

reflection angle *See* angle of reflection. { ri'flek·shən ,aŋ·gəl }

reflection coefficient [PHYS] The ratio of the amplitude of a wave reflected from a surface to the amplitude of the incident wave. Also known as coefficient of reflection. { ri'flek·shən ,kō·i,fish·ənt }

reflection density [OPTICS] The common logarithm of the ratio of the luminance of a nonabsorbing perfect diffuser to that of the surface under consideration, when both are illuminated at an angle of 45° to the normal and the direction of measurement is perpendicular to the surface. { ri'flek·shən ,den·səd·ē }

reflection diffraction [PHYS] Type of electron diffraction analysis in which the electron beam grazes the sample surface. { ri'flek·shən di ,frak·shən }

reflection factor [ELEC] Ratio of the load current that is delivered to a particular load when the impedances are mismatched to that delivered under conditions of matched impedances. Also known as mismatch factor; reflectance; transition factor. { ri'flek·shən ,fak·tər }

reflection HEED [PHYS] A form of HEED (high-energy electron diffraction) in which electrons are incident at small angles (0.5–4°) to the surface, and are reflected from it. Abbreviated RHEED. { ri'flek·shən ,hēd }

reflection law [PHYS] When a wave, such as electromagnetic radiation or sound, is reflected from a surface in a sharply defined direction, the reflected and incident waves travel in directions that make the same angle with a perpendicular to the surface and lie in a common plane with it. Also known as law of reflection. { ri'flek·shən ,lȯ }

reflection lobes [ELECTROMAG] Three-dimensional sections of the radiation pattern of a directional antenna, such as a radar antenna, which results from reflection of radiation from the earth's surface. { ri'flek·shən ,lōbz }

reflection plane *See* plane of mirror symmetry; plane of reflection. { ri'flek·shən ,plān }

reflection rainbow [OPTICS] A rainbow formed by light rays which have been reflected from an extended water surface; not to be confused with a reflected rainbow whose image may be seen in a still body of water. { ri'flek·shən ,rān,bō }

reflection spectrum [PHYS] The spectrum seen when incident waves are selectively altered by a reflecting substance. { ri'flek·shən ,spek·trəm }

reflection twin [CRYSTAL] A crystal twin whose symmetry is formed by an apparent mirror image across a plane. { ri'flek·shən ,twin }

reflective coating *See* mirror coating. { ri'flek·tiv 'kōd·iŋ }

reflectivity [PHYS] The ratio of the energy carried by a wave which is reflected from a surface to the energy carried by the wave which is incident on the surface. Also known as reflectance. { ,rē,flek'tiv·əd·ē }

reflectometer *See* microwave reflectometer. { ,rē ,flek'täm·əd·ər }

reflector [ELECTROMAG] **1.** A single rod, system of rods, metal screen, or metal sheet used behind an antenna to increase its directivity. **2.** A metal sheet or screen used as a mirror to change the direction of a microwave radio beam. { ri'flek·tər }

reflector plate [OPTICS] A transparent mirror in a computing gunsight, or in some types of optical gunsights and bombsights, that reflects the reticle image or images to the eye. { ri'flek·tər ,plāt }

reflector telescope *See* reflecting telescope. { ri 'flek·tər ,tel·ə,skōp }

reflex camera [OPTICS] A camera in which a mirror is used to reflect a full-size image of a scene on a ground glass so that the composition and focus may be judged. { 'rē,fleks ,kam·rə }

reflex sight [OPTICS] An optical or computing sight that reflects a reticle image or images onto a reflector plate for superimposition on the target by the eye. { 'rē,fleks ,sīt }

refracted ray [PHYS] A ray extending onward from the point of refraction. { ri'frak·təd 'rā }

refracted wave [PHYS] That portion of an incident wave which travels from one medium into a second medium. Also known as transmitted wave. { ri'frak·təd 'wāv }

refracting angle *See* apical angle. { ri'frak·tiŋ ,aŋ·gəl }

refracting edge [OPTICS] The intersection of the two refracting faces of a prism. { ri'frak·tiŋ ,ej }

refracting sphere [OPTICS] A sphere made of a transparent material whose index of refraction differs from the medium surrounding it, so that it refracts light passing through it. { ri'frak·tiŋ ,sfir }

refracting telescope [OPTICS] A telescope in which a lens gathers light and forms a real image of an object. Also known as refractor telescope. { ri'frak·tiŋ 'tel·ə,skōp }

refraction [ELECTROMAG] The change in direction of lines of force of an electric or magnetic field at a boundary between media with different permittivities or permeabilities. [PHYS] The change of direction of propagation of any wave, such as an electromagnetic or sound wave, when it passes from one medium to another in which the wave velocity is different, or when there is a spatial variation in a medium's wave velocity. { ri'frak·shən }

refraction loss [ELECTROMAG] Portion of the transmission loss that is due to refraction resulting from nonuniformity of the medium. { ri 'frak·shən ,lȯs }

refractive constant *See* index of refraction. { ri 'frak·tiv 'kän·stənt }

refractive index *See* index of refraction. { ri'frak·tiv ,in,deks }

refractivity [ELECTROMAG] **1.** Some quantitative measure of refraction, usually a measure of the index of refraction. **2.** The index of refraction minus 1. { ,rē,frak'tiv·əd·ē }

refractometry [OPTICS] The measurement of the index of refraction of a substance; it is an important tool in analytical chemistry. { ,rē,frak 'täm·ə·trē }

refractor telescope *See* refracting telescope. { ri 'frak·tər 'tel·ə,skōp }

refrangible [PHYS] Capable of being refracted. { ri'fran·jə·bəl }

refrigeration cycle [THERMO] A sequence of thermodynamic processes whereby heat is withdrawn from a cold body and expelled to a hot body. { ri,frij·ə'rā·shən ,sī·kəl }

regelation [THERMO] Phenomenon in which ice (or any substance which expands upon freezing) melts under intense pressure and freezes again when this pressure is removed; accounts for phenomena such as the slippery nature of ice and the motion of glaciers. { ¦rē·jə'lā·shən }

regenerative cycle [THERMO] An engine cycle in which low-grade heat that would ordinarily be lost is used to improve the cyclic efficiency. { rē'jen·rəd·iv ,sī·kəl }

Reggeism [PART PHYS] An attempt to account for and correlate hadron resonances and the asymptotic behavior of scattering amplitudes of hadrons at high energies in terms of Regge poles. { 'reg·ə,iz·əm }

Regge pole [PART PHYS] A pole singularity of a scattering amplitude in the complex angular momentum plane; the scattering amplitude is formed by continuing partial wave amplitudes from positive integer values of the angular momentum to the complex plane. { 'reg·ə ,pōl }

Regge recurrence [PART PHYS] One of a sequence of hadrons, with successive hadrons increasing by one in spin and also increasing in mass, but with the same values of other quantum numbers, except for parity, charge parity and G parity, which alternate in sign; it is believed that they are rotationally excited states of a particle, and that they alternate between two Regge trajectories. { 'reg·ə ri,kər·əns }

Regge trajectory [PART PHYS] **1.** The path followed by a Regge pole in the complex angular momentum plane as the center-of-mass energy is varied. **2.** The relationship between the spin and mass of a sequence of hadrons, with successive hadrons increasing by 2 in spin and also increasing in mass, but with the same values of other quantum numbers; the hadrons are thought to correspond to energies at which a Regge pole passes near positive integers (or half integers). { 'reg·ə trə,jek·trē }

regular [ELECTROMAG] In a definite direction; not diffused or scattered, when applied to reflection, refraction, or transmission. { 'reg·yə·lər }

regular dodecahedron [CRYSTAL] *See* pyritohedron. { 'reg·yə·lər dō,dek·ə'hē·drən }

regularization [QUANT MECH] A formal procedure used to eliminate ambiguities which arise in evaluating certain integrals in a quantized field theory; corresponds to adding extra fields whose masses are allowed to approach infinity. { ,reg·yə·lər·ə'zā·shən }

regular reflection *See* specular reflection. { 'reg·yə·lər ri'flek·shən }

regular reflector *See* specular reflector. { 'reg·yə·lər ri'flek·tər }

Rehbinder effect [PHYS] The reduction in the hardness and ductility of a material by a surface-active molecular film. { 'rā,bīn·dər i,fekt }

Rehbock weir formula [FL MECH] Probably the most accurate formula for the rate of flow of water over a rectangular suppressed weir; it includes a correction for the velocity of approach for normal, or fairly uniform, velocity distribution in the upstream channel; the formula is $Q = [3.234 + 5.347/(320h - 3) + 0.428h/d_0]lh^{3/2}$, where Q is the flow rate in cubic feet per second, l is the width of the weir in feet, h is the head of water above the crest of the weir in feet, and d_0 is the height of weir or depth of water at zero head in feet. { 'rā,bäk 'wer ,fȯr·myə·lə }

reheating [THERMO] A process in which the gas or steam is reheated after a partial isentropic expansion to reduce moisture content. Also known as resuperheating. { rē'hēd·iŋ }

Reissner-Nordstrom solution [RELAT] The unique solution of general relativity theory describing a nonrotating, charged black hole. { 'rīs·nər 'nȯrd·strəm sə,lü·shən }

rejection band Also known as stop band. [ELECTROMAG] The band of frequencies below the cutoff frequency in a uniconductor waveguide. [PHYS] A frequency band within which electrical or electromagnetic signals are reduced or eliminated. { ri'jek·shən ,band }

rejector impedance *See* dynamic impedance. { ri 'jek·tər im,pēd·əns }

rel [ELECTROMAG] Unit of reluctance equal to 1 ampere-turn per magnetic line of force. { rel }

relative coordinate system [PHYS] Any coordinate system which is moving with respect to an inertial coordinate system. { 'rel·əd·iv kō'ȯrd·ən·ət ,sis·təm }

relative damping ratio *See* damping ratio. { 'rel·əd·iv 'damp·iŋ ,rā·shō }

relative density *See* specific gravity. { 'rel·əd·iv 'den·səd·ē }

relative dielectric constant *See* dielectric constant. { 'rel·əd·iv ¦dī·i'lek·trik 'kän·stənt }

relative gain [ELECTROMAG] The gain of an antenna in a given direction when the reference antenna is a half-wave, loss-free dipole isolated in space whose equatorial plane contains the given direction. { 'rel·əd·iv ¦gān }

relative index of refraction [OPTICS] The ratio of the velocity of light in one medium to that in another medium. { 'rel·əd·iv 'in,deks əv ri 'frak·shən }

relative Mach number *See* Mach number. { 'rel·əd·iv 'mäk ,nəm·bər }

relative momentum [MECH] The momentum of a body in a reference frame in which another specified body is fixed. { 'rel·əd·iv mə'men·təm }

relative motion [MECH] The continuous change of position of a body with respect to a second body or to a reference point that is fixed. Also known as apparent motion. { 'rel·əd·iv 'mō·shən }

relative permeability [ELECTROMAG] The ratio of the permeability of a substance to the permeability of a vacuum at the same magnetic field strength. { 'rel·əd·iv ,pər·mē·ə'bil·əd·ē }

relative permittivity *See* dielectric constant. { 'rel·əd·iv ,pər·mə'tiv·əd·ē }

relative power gain [ELECTROMAG] Of one transmitting or receiving antenna over another, the measured ratio of the signal power one produces at the receiver input terminals to that produced by the other, the transmitting power level remaining fixed. { 'rel·əd·iv 'paù·ər ,gān }

relative resistance [ELEC] The ratio of the resistance of a piece of a material to the resistance of a piece of specified material, such as annealed copper, having the same dimensions and temperature. { 'rel·əd·iv ri'zis·təns }

relative roughness factor [FL MECH] Roughness of pipe-wall interior (distance from peaks to valleys) divided by pipe internal diameter; used to modify Reynolds number calculations for fluid flow through pipes. { 'rel·əd·iv ¦rəf·nəs ,fak·tər }

relative scatter intensity [OPTICS] For scattering of radiation under any given set of physical conditions, the ratio of the radiant intensity scattered in any given direction to the radiant intensity scattered in the direction of the incident beam. { 'rel·əd·iv 'skad·ər in,ten·səd·ē }

relative-state interpretation *See* Everett-Wheeler interpretation. { ¦rel·əd·iv ,stāt in,tər·prə'tā·shən }

relative velocity [MECH] The velocity of a body with respect to a second body; that is, its velocity in a reference frame where the second body is fixed. { 'rel·əd·iv və'läs·əd·ē }

relativistic beam [RELAT] A beam of particles traveling at a speed comparable with the speed of light. { ,rel·ə·tə'vis·tik 'bēm }

relativistic bremsstrahlung *See* gravitational bremsstrahlung. { ,rel·ə·tə'vis·tik 'brem,shträ·ləŋ }

relativistic electrodynamics [ELECTROMAG] The study of the interaction between charged particles and electric and magnetic fields when the velocities of the particles are comparable with that of light. { ,rel·ə·tə'vis·tik i¦lek·trō·dī'nam·iks }

relativistic kinematics [RELAT] A description of the motion of particles compatible with the special theory of relativity, without reference to the causes of motion. { ,rel·ə·tə'vis·tik ,kin·ə'mad·iks }

relativistic mass [RELAT] The mass of a particle moving at a velocity exceeding about one-tenth the velocity of light; it is significantly larger than the rest mass. { ,rel·ə·tə'vis·tik 'mas }

relativistic mechanics [RELAT] **1.** Any form of mechanics compatible with either the special or the general theory of relativity. **2.** The nonquantum mechanics of a system of particles or of a fluid interacting with an electromagnetic field, in the case when some of the velocities are comparable with the speed of light. { ,rel·ə·tə'vis·tik mi'kan·iks }

relativistic particle [RELAT] A particle moving at a speed comparable with the speed of light. { ,rel·ə·tə'vis·tik 'pärd·ə·kəl }

relativistic quantum theory [QUANT MECH] The quantum theory of particles which is consistent with the special theory of relativity, and thus can describe particles moving arbitrarily close to the speed of light. { ,rel·ə·tə'vis·tik 'kwänt·əm ,thē·ə·rē }

relativistic theory [PHYS] Any theory which is consistent with the special or general theory of relativity. { ,rel·ə·tə'vis·tik 'thē·ə·rē }

relativistic velocity [RELAT] A velocity comparable to the speed of light. { ,rel·ə·tə'vis·tik və'läs·əd·ē }

relativity [PHYS] Theory of physics which recognizes the universal character of the propagation speed of light and the consequent dependence of space, time, and other mechanical measurements on the motion of the observer performing the measurements; it has two main divisions, the special theory and the general theory. { ,rel·ə'tiv·əd·ē }

relaxation [MECH] **1.** Relief of stress in a strained material due to creep. **2.** The lessening of elastic resistance in an elastic medium under an applied stress resulting in permanent deformation. [PHYS] A process in which a physical system approaches a steady state after conditions affecting it have been suddenly changed, and in which the presence of dissipative agents prevents the system from overshooting and then oscillating about this state. { ,rē,lak'sā·shən }

relaxation oscillations [PHYS] Oscillations having a sawtooth waveform in which the displacement increases to a certain value and then drops back to zero, after which the cycle is repeated. { ,rē,lak'sā·shən ,äs·ə,lā·shənz }

relaxation time [PHYS] For many physical systems undergoing relaxation, a time τ such that the displacement of a quantity from its equilibrium value at any instant of time t is the exponential of $-t/\tau$. [SOLID STATE] The travel time of an electron in a metal before it is scattered and loses its momentum. { ˌrēˌlakˈsā·shən ˌtīm }

relaxed peak process *See* deep inelastic collision. { riˈlakst ¦pēk ˌprä·səs }

release adiabat [MECH] A curve or locus of points which defines the succession of states through which a mass that has been shocked to a high-pressure state passes while monotonically returning to zero pressure. { riˈlēs ˈad·ē·əˌbat }

relief [CRYSTAL] The apparent topography exhibited by minerals in thin section as a consequence of refractive index. { riˈlēf }

reluctance [ELECTROMAG] A measure of the opposition presented to magnetic flux in a magnetic circuit, analogous to resistance in an electric circuit; it is equal to magnetomotive force divided by magnetic flux. Also known as magnetic reluctance. { riˈlək·təns }

reluctivity [PHYS] The reciprocal of magnetic permeability; the reluctivity of empty space is unity. Also known as magnetic reluctivity; specific reluctance. { ˌrēˌləkˈtiv·əd·ē }

remaining velocity [MECH] Speed of a projectile at any point along its path of fire. { riˈmān·iŋ vəˈläs·əd·ē }

remanence [ELECTROMAG] The magnetic flux density that remains in a magnetic circuit after the removal of an applied magnetomotive force; if the magnetic circuit has an air gap, the remanence will be less than the residual flux density. { ˈrem·ə·nəns }

Renninger effect [PHYS] A phenomenon observed in the analysis of thick crystals with x-rays or neutrons, in which a strong diffracted beam acts as a primary beam and can undergo further diffraction. { ˈren·iŋ·ər iˌfekt }

renormalizability [QUANT MECH] The property of some quantum field theories whereby all infinite quantities can be absorbed into a renormalization of physical parameters such as mass and charge. { rēˌnȯr·məˌlīz·əˈbil·əd·ē }

renormalization [QUANT MECH] In certain quantum field theories, a procedure in which nonphysical bare values of certain quantities such as mass and charge are eliminated and the corresponding physically observable quantities are introduced. { rēˌnȯr·mə·ləˈzā·shən }

renormalization group methods [STAT MECH] Methods for treating the behavior of substances near critical points, in which the canonical ensemble is generalized by dividing a substance into cells of arbitrary size and forming an ensemble consisting of all microscopic configurations consistent with specified values of the thermodynamic variables in each of these cells. { rēˌnȯr·mə·ləˈzā·shən ¦grüp ˌmeth·ədz }

repeated load [MECH] A force applied repeatedly, causing variation in the magnitude and sometimes in the sense, of the internal forces. { riˈpēd·əd ˈlōd }

repeated twinning [CRYSTAL] Crystal twinning that involves more than two simple crystals. { riˈpēd·əd ˈtwin·iŋ }

repeat glass [OPTICS] Used by textile and wallpaper designers, a device consisting of four lenses formed in one piece of glass; when a single drawing or pattern is viewed, the subject matter is repeated four times. { riˈpēt ˌglas }

replica grating [OPTICS] A diffraction grating made by flowing a plastic solution over an original grating, evaporating the solvent, and removing the resulting film, which has the lines of the original grating impressed on it. { ˈrep·lə·kə ˌgrād·iŋ }

report [ACOUS] Sharp explosive sound, as of a shot, bursting bomb, or projectile. { riˈpȯrt }

representation theory [QUANT MECH] Quantum-mechanical device in which one selects the common eigenfunctions of a complete set of quantum-mechanical operators as a basis of vectors in a Hilbert space, and expresses wave functions and operators in terms of column matrices and square matrices, respectively, which correspond to this basis. { ˌrep·riˌzenˈtā·shən ˌthē·ə·rē }

repulsion [MECH] A force which tends to increase the distance between two bodies having like electric charges, or the force between atoms or molecules at very short distances which keeps them apart. Also known as repulsive force. { riˈpəl·shən }

repulsive force *See* repulsion. { riˈpəl·siv ˈfȯrs }

reserve battery [ELEC] A battery which is inert until an operation is performed which brings all the cell components into the proper state and location to become active. { riˈzərv ˈbad·ə·rē }

residual charge [ELEC] The charge remaining on the plates of a capacitor after initial discharge. { rəˈzij·ə·wəl ˈchärj }

residual error ratio [PHYS] The difference between an optimum result derived from experience or experiment and a supposedly exact result derived from theory. { rəˈzij·ə·wəl ˈer·ər ˌrā·shō }

residual field [ELECTROMAG] The magnetic field left in an iron core after excitation has been removed. { rəˈzij·ə·wəl ¦fēld }

residual flux density [ELECTROMAG] The magnetic flux density at which the magnetizing force is zero when the material is in a symmetrically and cyclically magnetized condition. Also known as residual induction; residual magnetic induction; residual magnetism. { rəˈzij·ə·wəl ˈfləks ˌden·səd·ē }

residual induction *See* residual flux density. { rəˈzij·ə·wəl inˈdək·shən }

residual intensity [SPECT] The intensity of radiation at some wavelength in a spectral line divided by the intensity in the adjacent continuum. { rəˈzij·yə·wəl inˈten·səd·ē }

residual ionization [PHYS] Ionization of air or other gas in a closed chamber, not accounted for by recognizable neighboring agencies; now attributed to cosmic rays. { rəˈzij·ə·wəl ˌī·ə·nəˈzā·shən }

residual magnetic induction *See* residual flux density. { rə'zij·ə·wəl mag'ned·ik in'dək·shən }

residual magnetism *See* residual flux density. { rə'zij·ə·wəl 'mag·nə,tiz·əm }

residual radiation [OPTICS] The nearly monochromatic radiation resulting from several reflections of light or other radiation from polished surfaces of certain substances such as quartz and rock salt, due to high reflectivity of these substances in certain bands of wavelengths. { rə'zij·ə·wəl ,rād·ē'ā·shən }

residual resistance [SOLID STATE] The value to which the electrical resistance of a metal drops as the temperature is lowered to near absolute zero, caused by imperfections and impurities in the metal rather than by lattice vibrations. { rə'zij·ə·wəl ri'zis·təns }

residual stress *See* internal stress. { rə'zij·ə·wəl 'stres }

residual vibration *See* zero-point vibration. { rə'zij·ə·wəl vī'brā·shən }

resilience [MECH] **1.** Ability of a strained body, by virtue of high yield strength and low elastic modulus, to recover its size and form following deformation. **2.** The work done in deforming a body to some predetermined limit, such as its elastic limit or breaking point, divided by the body's volume. { rə'zil·yəns }

resistance [ACOUS] *See* acoustic resistance. [ELEC] **1.** The opposition that a device or material offers to the flow of direct current, equal to the voltage drop across the element divided by the current through the element. Also known as electrical resistance. **2.** In an alternating-current circuit, the real part of the complex impedance. *See* fluid resistance. [MECH] In damped harmonic motion, the ratio of the frictional resistive force to the speed. Also known as damping coefficient; damping constant; mechanical resistance. { ri'zis·təns }

resistance box [ELEC] A box containing a number of precision resistors connected to panel terminals or contacts so that a desired resistance value can be obtained by withdrawing plugs (as in a post-office bridge) or by setting multicontact switches. { ri'zis·təns ,bäks }

resistance bridge *See* Wheatstone bridge. { ri'zis·təns ,brij }

resistance-capacitance circuit [ELEC] A circuit which has a resistance and a capacitance in series, and in which inductance is negligible. Abbreviated R-C circuit. { ri'zis·təns kə'pas·əd·əns ,sər·kət }

resistance-capacitance constant [ELEC] Time constant of a resistive-capacitive circuit, equal in seconds to the resistance value in ohms multiplied by the capacitance value in farads. Abbreviated R-C constant. { ri'zis·təns kə'pas·əd·əns ,kän·stənt }

resistance-capacitance network [ELEC] Circuit containing resistances and capacitances arranged in a particular manner to perform a specific function. Abbreviated R-C network. { ri'zis·təns kə'pas·əd·əns 'net,wərk }

resistance coefficient 1 [FL MECH] A dimensionless number used in the study of flow resistance, equal to the resistance force in flow divided by one-half the product of fluid density, the square of fluid velocity, and the square of a characteristic length. Symbolized c_{ff}. { ri'zis·təns ,kō·i,fish·ənt 'wən }

resistance coefficient 2 *See* Darcy number 1. { ri'zis·təns,kō·i,fish·ənt 'tü }

resistance drop [ELEC] The voltage drop occurring between two points on a conductor due to the flow of current through the resistance of the conductor; multiplying the resistance in ohms by the current in amperes gives the voltage drop in volts. Also known as IR drop. { ri'zis·təns ,dräp }

resistance grounding [ELEC] Electrical grounding in which lines are connected to ground by a resistive (totally dissipative) impedance. { ri'zis·təns ,graủnd·iŋ }

resistance loss [ELEC] Power loss due to current flowing through resistance; its value in watts is equal to the resistance in ohms multiplied by the square of the current in amperes. { ri'zis·təns ,lȯs }

resistance measurement [ELEC] The quantitative determination of that property of an electrically conductive material, component, or circuit called electrical resistance. { ri'zis·təns ,mezh·ər·mənt }

resisting moment [MECH] A moment produced by internal tensile and compressive forces that balances the external bending moment on a beam. { ri'zist·iŋ ,mō·mənt }

resistivity *See* electrical resistivity. { ,rē,zis'tiv·əd·ē }

resistor [ELEC] A device designed to have a definite amount of resistance; used in circuits to limit current flow or to provide a voltage drop. Also known as electrical resistor. { ri'zis·tər }

resistor color code [ELEC] Code adopted by the Electronic Industries Association to mark the values of resistance on resistors in a readily recognizable manner; the first color represents the first significant figure of the resistor value, the second color the second significant figure, and the third color represents the number of zeros following the first two figures; a fourth color is sometimes added to indicate the tolerance of the resistor. { ri'zis·tər 'kəl·ər ,kōd }

resistor core [ELEC] Insulating support on which a resistor element is wound or otherwise placed. { ri'zis·tər ,kȯr }

resistor element [ELEC] That portion of a resistor which possesses the property of electric resistance. { ri'zis·tər ,el·ə·mənt }

resistor network [ELEC] An electrical network consisting entirely of resistances. { ri'zis·tər 'net,wərk }

resolution [ELECTROMAG] In radar, the minimum separation between two targets, in angle or range, at which they can be distinguished on a radar screen. Also known as resolving power. *See* resolving power. [PHYS] **1.** For a measurement of energy or momentum of a collection of particles, the difference between the highest and lowest energies at which the response of an instrument to a beam of monoenergetic particles is at least half its maximum value, divided by the energy of the particles. **2.** The procedure of breaking up a vectorial quantity into its components. *See* resolving power. { ˌrez·ə'lü·shən }

resolution chart [OPTICS] A device to test resolving power; usually alternate black and white lines of equal width arranged in groups of decreasing line width, identified as the number of line pairs per millimeter. { ˌrez·ə'lü·shən ˌchärt }

resolution reading [OPTICS] A number indicating how many lines per millimeter are contained in the finest group which can be distinguished on a resolution chart. { ˌrez·ə'lü·shən ˌrēd·iŋ }

resolving cell [ELECTROMAG] In radar, volume in space whose diameter is the product of slant range and beam width, and whose length is the pulse length. { ri'zälv·iŋ ˌsel }

resolving power [ELECTROMAG] The reciprocal of the beam width of a unidirectional antenna, measured in degrees. [OPTICS] A quantitative measure of the ability of an optical instrument to produce separable images of different points on an object; usually, the smallest angular or linear separation of two object points for which they may be resolved according to the Rayleigh criterion. Also known as resolution. [PHYS] A measure of the ability of a mass spectroscope to separate particles of different masses, equal to the ratio of the average mass of two particles whose mass spectrum lines can just be completely separated, to the difference in their masses. [SPECT] A measure of the ability of a spectroscope or interferometer to separate spectral lines of nearly equal wavelength, equal to the average wavelength of two equally strong spectral lines whose images can barely be separated, divided by the difference in wavelengths; for spectroscopes, the lines must be resolved according to the Rayleigh criterion; for interferometers, the wavelengths at which the lines have half of maximum intensity must be equal. Also known as resolution. { ri'zälv·iŋ ˌpau̇·ər }

resonance [ELEC] A phenomenon exhibited by an alternating-current circuit in which there are relatively large currents near certain frequencies, and a relatively unimpeded oscillation of energy from a potential to a kinetic form; a special case of the physics definition. [PHYS] **1.** A phenomenon exhibited by a physical system acted upon by an external periodic driving force, in which the resulting amplitude of oscillation of the system becomes large when the frequency of the driving force approaches a natural free oscillation frequency of the system. **2.** In general, any phenomenon which is greatly enhanced at frequencies or energies that are at or very close to a given characteristic value. [QUANT MECH] An enhanced coupling between quantum states with the same energy. *See* resonance level. { 'rez·ən·əns }

resonance absorption [QUANT MECH] The absorption of electromagnetic radiation by a quantum-mechanical system at a characteristic frequency satisfying the Bohr frequency condition. Also known as resonance. { 'rez·ən·əns əbˌsȯrp·shən }

resonance bridge [ELEC] A four-arm alternating-current bridge used to measure inductance, capacitance, or frequency; the inductor and the capacitor, which may be either in series or in parallel, are tuned to resonance at the frequency of the source before the bridge is balanced. { 'rez·ən·əns ˌbrij }

resonance capture [NUC PHYS] The combination of an incident particle and a nucleus in a resonance level of the resulting compound nucleus, characterized by having a large cross section at and very near the corresponding resonance energy. { 'rez·ən·əns ˌkap·chər }

resonance curve [ELEC] Graphical representation illustrating the manner in which a tuned circuit responds to the various frequencies in and near the resonant frequency. { 'rez·ən·əns ˌkərv }

resonance energy [PHYS] The characteistic energy at which, or very close to which, the amplitude of a resonance phenomenon is greatly enhanced. { 'rez·ən·əns ˌen·ər·jē }

resonance fluorescence [ATOM PHYS] *See* resonance radiation. [NUC PHYS] Resonant scattering from an atomic nucleus. { 'rez·ən·əns flu̇ˌres·əns }

resonance fluorescence lidar [OPTICS] A type of lidar in which the laser wavelength is tuned to the resonance absorption wavelength of a specific molecular species whose resonant backscatter cross section is measured in order to determine its density in the upper atmosphere. { ¦rez·ən·əns flu̇'res·əns ˌlīˌdär }

resonance frequency [PHYS] A frequency at which some measure of the response of a physical system to an external periodic driving force is a maximum; three types are defined, namely, phase resonance, amplitude resonance, and natural resonance, but they are nearly equal when dissipative effects are small. Also known as resonant frequency. [QUANT MECH] A characteristic frequency, satisfying the Bohr frequency condition, at which a quantum-mechanical system absorbs radiation. { 'rez·ən·əns ˌfrē·kwən·sē }

resonance ionization spectroscopy [SPECT] A technique capable of detecting single atoms or molecules of a given element or compound in a gas, in which an atom or molecule in its ground state is excited to a bound state when a photon is absorbed from a laser beam at a very well-controlled wavelength that is resonant with the excitation energy; a second photon removes the excited electron from the atom or molecule, and this electron is then accelerated by an electric field and collides with the gas molecules, creating additional ionization which is detected by a proportional counter. Abbreviated RIS. { 'rez·ən·əns ˌī·ə·nə'zā·shən spek'träs·kə·pē }

resonance lamp [ATOM PHYS] An evacuated quartz bulb containing mercury, which acts as a source of radiation at the wavelength of the pure resonance line of mercury when irradiated by a mercury-arc lamp. { 'rez·ən·əns ˌlamp }

resonance level [QUANT MECH] An unstable state of a compound system capable of being formed in a collision between two particles, and associated with a peak in a graph of cross section versus energy for the scattering of the particles. Also known as resonance. { 'rez·ən·əns ˌlev·əl }

resonance line [SPECT] The line of longest wavelength associated with a transition between the ground state and an excited state. { 'rez·ən·əns ˌlīn }

resonance luminescence *See* resonance radiation. { 'rez·ən·əns ˌlü·məˌnes·əns }

resonance method [ELEC] A method of determining the impedance of a circuit element, in which resonance frequency of a resonant circuit containing the element is measured. { 'rez·ən·əns ˌmeth·əd }

resonance radiation [ATOM PHYS] The emission of radiation by a gas or vapor as a result of excitation of atoms to higher energy levels by incident photons at the resonance frequency of the gas or vapor; the radiation is characteristic of the particular gas or vapor atom but is not necessarily the same frequency as the absorbed radiation. Also known as resonance fluorescence; resonance luminescence. { 'rez·ən·əns ˌrād·ēˌā·shən }

resonance reaction [NUC PHYS] A nuclear reaction that takes place only when the energy of the incident particles is at or very close to a characteristic value. { 'rez·ən·əns rēˌak·shən }

resonance scattering [NUC PHYS] A peak in the cross section of a nucleus for elastic scattering of neutrons at energies near a resonance level, accompanied by an anomalous phase shift in the scattered neutrons. { 'rez·ən·əns ˌskad·ə·riŋ }

resonance spectrum [SPECT] An emission spectrum resulting from illumination of a substance (usually a molecular gas) by radiation of a definite frequency or definite frequencies. { 'rez·ən·əns ˌspek·trəm }

resonance vibration [MECH] Forced vibration in which the frequency of the disturbing force is very close to the natural frequency of the system, so that the amplitude of vibration is very large. { 'rez·ən·əns vīˌbrā·shən }

resonant antenna [ELECTROMAG] An antenna for which there is a sharp peak in the power radiated or intercepted by the antenna at a certain frequency, at which electric currents in the antenna form a standing-wave pattern. { 'res·ən·ənt an 'ten·ə }

resonant capacitor [ELEC] A tubular capacitor that is wound to have inductance in series with its capacitance. { 'res·ən·ənt kə'pas·əd·ər }

resonant cavity *See* cavity resonator. { 'res·ən·ənt 'kav·əd·ē }

resonant-cavity maser [PHYS] A maser in which the paramagnetic active material is placed in a cavity resonator. { 'res·ən·ənt ¦kav·əd·ē 'mā·zər }

resonant chamber *See* cavity resonator. { 'res·ən·ənt 'chām·bər }

resonant-chamber switch [ELECTROMAG] Waveguide switch in which a tuned cavity in each waveguide branch serves the functions of switch contacts; detuning of a cavity blocks the flow of energy in the associated waveguide. { 'res·ən·ənt ¦chām·bər ˌswich }

resonant circuit [ELEC] A circuit that contains inductance, capacitance, and resistance of such values as to give resonance at an operating frequency. { 'res·ən·ənt 'sər·kət }

resonant coupling [ELEC] Coupling between two circuits that reaches a sharp peak at a certain frequency. { 'res·ən·ənt 'kəp·liŋ }

resonant detector [PHYS] A detector of electromagnetic radiation which is responsive to radiation only at certain frequencies at which resonance is created in the detector. { 'res·ən·ənt di'tek·tər }

resonant diaphragm [ELECTROMAG] Diaphragm, in waveguide technique, so proportioned as to introduce no reactive impedance at the design frequency. { 'res·ən·ənt 'dī·əˌfram }

resonant element *See* cavity resonator. { 'res·ən·ənt 'el·ə·mənt }

resonant frequency *See* resonance frequency. { 'res·ən·ənt 'frē·kwən·sē }

resonant helix [ELECTROMAG] An inner helical conductor in certain types of transmission lines and resonant cavities, which carries currents with the same frequency as the rest of the line or cavity. { 'res·ən·ənt 'hē·liks }

resonant ionization mass spectrometry [SPECT] An instrumental technique for quantitative identification of trace impurities (at or below the part-per-billion level), it begins with laser-induced or ion-induced desorption, followed by resonant laser ionization (usually from two or three lasers), and then analysis by time-of-flight mass spectrometry. Abbreviated RIMS. { ¦rez·ən·ənt ˌī·ə·nəˌzā·shən ¦mas spek'träm·ə·trē }

resonant iris [ELECTROMAG] A resonant window in a circular waveguide; it resembles an optical iris. { 'res·ən·ənt 'ī·rəs }

resonant line [ELECTROMAG] A transmission line having values of distributed inductance and distributed capacitance so as to make the line resonant at the frequency it is handling. { 'res·ən·ənt 'līn }

resonant Raman effect [ATOM PHYS] A process in which a photon whose energy is exactly matched to the transition energy between two atomic energy levels (within the natural linewidth) promotes an atomic electron to an excited state, which decays in the same step. { 'rez·ən·ənt 'räˌmän iˌfekt }

resonant reaction [NUC PHYS] A nuclear reaction whose probability is enhanced at an energy corresponding to an energy level of one of the nuclei. { 'rez·ən·ənt rēˌak·shən }

resonant resistance [ELEC] Resistance value to

which a resonant circuit is equivalent. { 'res·ən·ənt ri'zis·təns }

resonant scattering [QUANT MECH] Scattering of a photon by a quantum-mechanical system (usually an atom or nucleus) in which the system first absorbs the photon by undergoing a transition from one of its energy states to one of higher energy, and subsequently reemits the photon by the exact inverse transition. { 'res·ən·ənt 'skad·ə·riŋ }

resonant ultrasound spectroscopy [ACOUS] An experimental technique for obtaining a complete set of elastic constants of a material in which a sample of the material of well-defined shape, usually a sphere or a rectangular parallelepiped, is held between two transducers, one of which excites the sample while the other measures its response, and the spectrum of free mechanical resonances of the sample is measured. { ¦rez·ən·ənt ˌəl·trə ˌsau̇nd spek'träs·kə·pē }

resonant voltage step-up [ELEC] Ability of an inductor and a capacitor in a series resonant circuit to deliver a voltage several times greater than the input voltage of the circuit. { 'res·ən·ənt ¦vōl·tij 'stepˌəp }

resonant wavelength [ELECTROMAG] The wavelength in free space of electromagnetic radiation having a frequency equal to a natural resonance frequency of a cavity resonator. { 'res·ən·ənt 'wāvˌleŋkth }

resonant window [ELECTROMAG] A parallel combination of inductive and capacitive diaphragms, used in a waveguide structure to provide transmission at the resonant frequency and reflection at other frequencies. { 'res·ən·ənt 'win·dō }

resonate [ELEC] To bring to resonance, as by tuning. { 'rez·ənˌāt }

resonating cavity [ELECTROMAG] Short piece of waveguide of adjustable length, terminated at either or both ends by a metal piston, an iris diaphragm, or some other wave-reflecting device; it is used as a filter, as a means of coupling between guides of different diameters, and as impedance networks corresponding to those used in radio circuits. { 'rez·ənˌād·iŋ 'kav·əd·ē }

resonator [PHYS] A device that exhibits resonance at a particular frequency, such as an acoustic resonator or cavity resonator. { 'rez·ənˌād·ər }

resonator wavemeter [ELECTROMAG] Any resonant circuit used to determine wavelength, such as a cavity-resonator frequency meter. { 'rez·ənˌād·ər 'wāvˌmēd·ər }

resorption [PHYS] Absorption or, less commonly, adsorption of material by a body or system from which the material was previously released. { rē'sȯrp·shən }

response time [ELEC] The time it takes for the pointer of an electrical or electronic instrument to come to rest at a new value, after the quantity it measures has been abruptly changed. { ri 'spä̈ns ˌtīm }

rest density [RELAT] The density of a small portion of a fluid in a Lorentz frame in which that portion of the fluid is at rest. { 'rest ˌden·səd·ē }

rest energy [RELAT] The energy equivalent to the rest mass m_0 of a particle or body; that is, the quantity of m_0c^2, where c is the speed of light; often expressed in electronvolts. { 'rest ˌen·ər·jē }

rest frame [RELAT] The Lorentz frame in which the total momentum of a system equals zero; for an accelerated system, the rest frame varies from instant to instant. { 'rest ˌfrām }

restitution coefficient *See* coefficient of restitution. { ˌres·tə'tü·shən ˌkō·iˌfish·ənt }

rest mass [RELAT] The mass of a particle in a Lorentz reference frame in which it is at rest. { 'rest ˌmas }

resultant of forces [MECH] A system of at most a single force and a single couple whose external effects on a rigid body are identical with the effects of the several actual forces that act on that body. { ri'zəlt·ənt əv 'fȯrs·əz }

resuperheating *See* reheating. { rē¦sü·pər'hēd·iŋ }

retardation [OPTICS] In interference microscopy, the difference in optical path between the light passing through the specimen and the light bypassing the specimen. Also known as optical-path difference. { ˌrēˌtär'dā·shən }

retardation coil [ELECTROMAG] A high-inductance coil used in telephone circuits to permit passage of direct current or low-frequency ringing current while blocking the flow of audio-frequency currents. { ˌrēˌtär'dā·shən ˌkȯil }

retardation plate *See* wave plate. { ˌrēˌtär'dā·shən ˌplāt }

retardation sheet *See* wave plate. { ˌrēˌtär'dā·shən ˌshēt }

retardation theory [OPTICS] General methods of calculating the effect of one or more wave plates on light which is normally incident on the plates and which is initially polarized in some fashion. { ˌrēˌtär'dā·shən ˌthē·ə·rē }

retarded field [ELECTROMAG] An electric or magnetic field strength as found from the retarded potentials. { ri'tärd·əd 'fēld }

retarded potentials [ELECTROMAG] The electromagnetic potentials at an instant in time t and a point in space r as a function of the charges and currents that existed at earlier times at points on the past light cone of the event r,t. { ri'tärd·əd pə'ten·chəlz }

retarding potential [PHYS] A potential which causes the speed of a moving particle to be reduced. { ri'tärd·iŋ pəˌten·chəl }

retentivity [ELECTROMAG] The residual flux density corresponding to the saturation induction of a magnetic material. { ˌrēˌten'tiv·əd·ē }

Retgers' law [SOLID STATE] The law that the properties of crystalline mixtures of isomorphous substances are continuous functions of the percentage composition. { 'ret·gərzˌlȯ }

reticle [OPTICS] A series of intersecting fine lines, wires, or the like which are placed in the focus of the objective of an optical instrument to aid in measurement of angles or distances. { 'red·ə·kəl }

reticle image [OPTICS] A light image of the reticle in a computing gunsight or in certain types of optical gunsights and bombsights, cast on a reflector plate and superimposed on the target. { 'red·ə·kəl ˌim·ij }

retinal illuminance [OPTICS] A psychophysiological quantity which is a measure of the brightness of a visual sensation; it is measured in trolands. { 'ret·ən·əl i'lü·mə·nəns }

retrodirective mirror [OPTICS] **1.** An optical system consisting of two mutually perpendicular plane mirrors; it reflects any beam of light which lies in a plane perpendicular to the mirrors into a direction antiparallel to its original direction. **2.** An optical system consisting of three mutually perpendicular plane mirrors; it reflects any beam of light into a direction antiparallel to its original direction. { ¦re·tro·di'rek·tiv 'mir·ər }

retroreflection [PHYS] Reflection wherein the reflected rays of radiation return along paths parallel to those of their corresponding incident rays. { ¦re·trō·ri'flek·shən }

retroreflector [PHYS] Any instrument used to cause reflected radiation to return along paths parallel to those of their corresponding incident rays; one type, the corner reflector, is an efficient radar target. { ¦re·trō·ri'flek·tər }

reverberant sound [ACOUS] The portion of the room impulse response consisting of sound that arrives at a listener's location more than 150 milliseconds after the first direct sound, and that has been reflected against walls and ceilings many times. { riˌvər·bə·rənt 'saủnd }

reverberation [ACOUS] The prolongation of sound at a given point after direct reception from the source has ceased, due to such causes as reflections from bounding surfaces, scattering from inhomogeneities in a medium, and vibrations excited by the original sound. { riˌvər·bə'rā·shən }

reverberation chamber [ACOUS] An enclosure with heavy surfaces which randomly reflect as great an amount of sound as possible; used in acoustic measurements. Also known as random diffusion chamber. { riˌvər·bə'rā·shən ˌchām·bər }

reverberation time [ACOUS] The time in seconds required for the average sound-energy density at a given frequency to reduce to one-millionth of its initial steady-state value after the sound source has been stopped; this corresponds to a decrease of 60 decibels. { riˌvər·bə'rā·shən ˌtīm }

reversal spectrum [SPECT] A spectrum which may be observed in intense white light which has traversed luminous gas, in which there are dark lines where there were bright lines in the emission spectrum of the gas. { ri'vər·səl ˌspek·trəm }

reversal temperature [SPECT] The temperature of a blackbody source such that, when light from this source is passed through a luminous gas and analyzed in a spectroscope, a given spectral line of the gas disappears, whereas it appears as a bright line at lower blackbody temperatures, and a dark line at higher temperatures. { ri'vər·səl ˌtem·prə·chər }

reverse Brayton cycle [THERMO] A refrigeration cycle using air as the refrigerant but with all system pressures above the ambient. Also known as dense-air refrigeration cycle. { ri'vərs 'brāt·ən ˌsī·kəl }

reverse Carnot cycle [THERMO] An ideal thermodynamic cycle consisting of the processes of the Carnot cycle reversed and in reverse order, namely, isentropic expansion, isothermal expansion, isentropic compression, and isothermal compression. { ri'vərs kär'nō ˌsī·kəl }

reversed image *See* inverted image. { ri'vərst 'im·ij }

reverse voltage [ELEC] In the case of two opposing voltages, voltage of that polarity which produces the smaller current. { ri'vərs 'vōl·tij }

reversibility principle [OPTICS] The principle that if a beam of light is reflected back on itself, it will traverse the same path or paths as it did before reversal. [STAT MECH] *See* microscopic reversibility. { riˌvər·sə'bil·əd·ē ˌprin·sə·pəl }

reversible engine [THERMO] An ideal engine which carries out a cycle of reversible processes. { ri'vər·sə·bəl 'en·jən }

reversible path [THERMO] A path followed by a thermodynamic system such that its direction of motion can be reversed at any point by an infinitesimal change in external conditions; thus the system can be considered to be at equilibrium at all points along the path. { ri'vər·sə·bəl 'path }

reversible process [THERMO] An ideal thermodynamic process which can be exactly reversed by making an indefinitely small change in the external conditions. Also known as quasistatic process. { ri'vər·sə·bəl 'prä·səs }

revolution [MECH] The motion of a body around a closed orbit. { ˌrev·ə'lü·shən }

revolution per minute [MECH] A unit of angular velocity equal to the uniform angular velocity of a body which rotates through an angle of 360° (2π radians), so that every point in the body returns to its original position, in 1 minute. Abbreviated rpm. { ˌrev·ə'lü·shən pər 'min·ət }

revolution per second [MECH] A unit of angular velocity equal to the uniform angular velocity of a body which rotates through an angle of 360° (2π radians), so that every point in the body returns to its original position, in 1 second. Abbreviated rps. { ˌrev·ə'lü·shən pər 'sek·ənd }

reyn [FL MECH] A unit of dynamic viscosity equal to the dynamic viscosity of a fluid in which there is a tangential force of 1 poundal per square foot resisting the flow of two parallel fluid layers past each other when their differential velocity is 1 foot per second per foot of separation; equal to approximately 14.8816 poise. { ren }

Reynolds-averaged Navier-Stokes analysis [FL MECH] The process of determining numerical solutions of the Navier-Stokes equations for a fluid flow, using time averaging of flow variables and modeling of turbulent stresses to simplify the calculations. Abbreviated RANS analysis. { ¦ren·əlz ¦av·rijd ¦näv·ē,ā 'stōks ə,nal·ə·səs }

Reynolds criterion [FL MECH] The principle that the type of fluid motion, that is, laminar flow or turbulent flow, in geometrically similar flow systems depends only on the Reynolds number; for example, in a pipe, laminar flow exists at Reynolds numbers less than 2000, turbulent flow at numbers above about 3000. { 'ren·əlz ,krī,tir·ē·ən }

Reynolds equation [FL MECH] A form of the Navier-Stokes equation which is $\rho \partial u/\partial t = (\partial/\partial x)(p_{xx} - \rho u^2) + (\partial/\partial y)\cdot(p_{xy} - \rho uv) + (\partial/\partial z)(p_{xz} - \rho uw)$ where ρ is the fluid density, u, v, and w are the components of the fluid velocity, and p_{xx}, p_{xy}, and p_{xz} are normal and shearing stresses. { 'ren·əlz i,kwā·zhən }

Reynolds number [FL MECH] A dimensionless number which is significant in the design of a model of any system in which the effect of viscosity is important in controlling the velocities or the flow pattern of a fluid; equal to the density of a fluid, times its velocity, times a characteristic length, divided by the fluid viscosity. Symbolized N_{Re}. Also known as Damköhler number V (*Da*V). { 'ren·əlz ,nəm·bər }

Reynolds stress [FL MECH] The net transfer of momentum across a surface in a turbulent fluid because of fluctuations in fluid velocity. Also known as eddy stress. { 'ren·əlz ,stres }

Reynolds stress tensor [FL MECH] A tensor whose components are the components of the Reynolds stress across three mutually perpendicular surfaces. { 'ren·əlz 'stres ,ten·sər }

rf *See* radio frequency.

rhe [FL MECH] **1.** A unit of dynamic fluidity, equal to the dynamic fluidity of a fluid whose dynamic viscosity is 1 centipoise. **2.** A unit of kinematic fluidity, equal to the kinematic fluidity of a fluid whose kinematic viscosity is 1 centistoke. { rē }

RHEED *See* reflection HEED. { 'är,hēd }

Rheinberg illumination [OPTICS] An illumination technique used in optical microscopes that is a modification of the dark-field method; the central disk is transparent and colored; an annulus of a complementary color fills the remaining condenser aperture; the specimen is seen in the color of the annulus against the background of the central disk. Also known as optical staining. { 'rīn,bərg i,lü·mə,nā·shən }

rheogoniometry [MECH] Rheological tests to determine the various stress and shear actions on Newtonian and non-Newtonian fluids. { ¦rē·ə·gō·nē'äm·ə·trē }

rheology [MECH] The study of the deformation and flow of matter, especially non-Newtonian flow of liquids and plastic flow of solids. { rē'äl·ə·jē }

rheopectic fluid [FL MECH] A fluid for which the structure builds up on shearing; this phenomenon is regarded as the reverse of thixotropy. { ¦rē·ə¦pek·tik 'flü·əd }

rheostat [ELEC] A resistor constructed so that its resistance value may be changed without interrupting the circuit to which it is connected. Also known as variable resistor. { 'rē·ə,stat }

rheostriction *See* pinch effect. { 'rē·ə,strik·shən }

rhomb *See* rhombohedron. { räm }

rhombic antenna [ELECTROMAG] A horizontal antenna having four conductors forming a diamond or rhombus; usually fed at one apex and terminated with a resistance or impedance at the opposite apex. Also known as diamond antenna. { 'räm·bik an'ten·ə }

rhombic dodecahedron [CRYSTAL] A crystal form in the cubic system that is a dodecahedron whose faces are equal rhombuses. { 'räm·bik ,dō·dek·ə'hē·drən }

rhombic lattice *See* orthorhombic lattice. { 'räm·bik 'lad·əs }

rhombic system *See* orthorhombic system. { 'räm·bik 'sis·təm }

rhombohedral [CRYSTAL] **1.** Of or pertaining to the rhombohedral system. **2.** Of or pertaining to crystal cleavage in or a centered lattice of the hexagonal system. { ¦räm·bō¦hē·drəl }

rhombohedral close packing *See* rhombohedral packing. { ¦räm·bō¦hē·drəl 'klōs 'pak·iŋ }

rhombohedral lattice [CRYSTAL] A crystal lattice in which the three axes of a unit cell are of equal length, and the three angles between axes are the same, and are not right angles. Also known as trigonal lattice. { ¦räm·bō¦hē·drəl ¦lad·əs }

rhombohedral packing [CRYSTAL] The tightest manner of systematic arrangement of uniform solid spheres in a clastic sediment or crystal lattice, characterized by a unit cell of six planes passed through eight sphere centers situated at the corners of a regular rhombohedron. Also known as rhombohedral close packing. { ¦räm·bō¦hē·drəl ¦pak·iŋ }

rhombohedral system [CRYSTAL] A division of the trigonal crystal system in which the rhombohedron is the basic unit cell. { ¦räm·bō¦hē·drəl ¦sis·təm }

rhombohedron [CRYSTAL] A trigonal crystal form that is a parallelepiped, the six identical faces being rhombs. Also known as rhomb. { ¦räm·bō¦hē·drən }

rhomboidal prism [OPTICS] A prism with four parallel sides and two slanting, or oblique, parallel ends; it will divert the path of light entering its ends without changing the form of the light. { räm'bȯid·əl 'priz·əm }

rho meson [PART PHYS] Collective name for vector meson resonances belonging to a charge multiplet with total isospin 1, hypercharge 0, negative charge conjugation parity, positive g-parity, mass of about 770 MeV, and width of about 146 MeV. Designated $\rho(770)$. { 'rō 'mā,sän }

rhumbatron *See* cavity resonator. { 'rəm·bə,trän }

ribbon conductor [ELEC] A thin, flat piece of metal suitable for carrying electric current. { 'rib·ən kən,dək·tər }

Riblet coupler *See* three-decibel coupler. { 'rib·lət ˌkəp·lər }

Richardson number [FL MECH] A dimensionless number used in studying the stratified flow of multilayer systems; equal to the acceleration of gravity times the density gradient of a fluid, divided by the product of the fluid's density and the square of its velocity gradient at a wall. Symbolized N_{Ri}. { 'rich·ərd·sən ˌnəm·bər }

ridge waveguide [ELECTROMAG] A circular or rectangular waveguide having one or more longitudinal internal ridges that serve primarily to increase transmission bandwidth by lowering the cutoff frequency. { 'rij 'wāvˌgīd }

Riemannian space-time [RELAT] The space-time of general relativity, having the mathematical structure of a four-dimensional Riemann space. { rē¦män·ē·ən 'spāsˌtīm }

Righi experiment [OPTICS] An experiment in which a rotating Nicol prism, a Fresnel mirror, a quarter-wave plate, and a fixed Nicol prism are used to produce effects in light beams similar to beats between sounds with slightly different frequencies. { 'rē·gē ikˌsper·ə·mənt }

Righi-Leduc effect [PHYS] The phenomenon wherein, if a magnetic field is applied at right angles to the direction of a temperature gradient in a conductor, a new temperature gradient is produced perpendicular to both the direction of the original temperature gradient and to the magnetic field. Also known as Leduc effect. { 'rē·gē lə'du̇k iˌfekt }

right-angle prism [OPTICS] A type of prism used to turn a beam of light through a right angle (90°); it will invert (turn upside-down) or will revert (turn right for left), according to the position of the prism, any light reflected by it. { 'rīt ¦aŋ·gəl 'priz·əm }

right-handed [CRYSTAL] Having a crystal structure with a mirror-image relationship to a left-handed structure. { 'rīt ¦han·dəd }

right-hand helicity [QUANT MECH] Property of a particle whose spin is parallel to its momentum. { 'rīt ¦hand he'lis·əd·ē }

right-hand polarization [ELECTROMAG] In elementary particle discussions, circular or elliptical polarization of an electromagnetic wave in which the electric field vector at a fixed point in space rotates in the right-hand sense about the direction of propagation; in optics, the opposite convention is used; in facing the source of the beam, the electric vector is observed to rotate clockwise. { 'rīt ¦hand ˌpō·lə·rə'zā·shən }

right-hand rule [ELECTROMAG] **1.** For a current-carrying wire, the rule that if the fingers of the right hand are placed around the wire so that the thumb points in the direction of current flow, the fingers will be pointing in the direction of the magnetic field produced by the wire. Also known as hand rule. **2.** For a moving wire in a magnetic field, such as the wire on the armature of a generator, if the thumb, first, and second fingers of the right hand are extended at right angles to one another, with the first finger representing the direction of magnetic lines of force and the second finger representing the direction of current flow induced by the wire's motion, the thumb will be pointing in the direction of motion of the wire. Also known as Fleming's rule. { 'rīt ¦hand 'rül }

righting lever [FL MECH] The horizontal distance from the center of mass of a floating body, slightly displaced from the equilibrium position, to a vertical line passing through the center of buoyancy. { 'rīd·iŋ ˌlev·ər }

rigid body [MECH] An idealized extended solid whose size and shape are definitely fixed and remain unaltered when forces are applied. { 'rij·id 'bäd·ē }

rigid-body dynamics [MECH] The study of the motions of a rigid body under the influence of forces and torques. { 'rij·id ¦bäd·ē dī'nam·iks }

rigid copper coaxial line [ELECTROMAG] A coaxial cable in which the central conductor and outer conductor are formed by joining rigid pieces of copper. { 'rij·id 'käp·ər kō'ak·sē·əl ˌlīn }

rigidity [MECH] The quality or state of resisting change in form. { ri'jid·əd·ē }

rigidity modulus *See* modulus of elasticity in shear. { ri'jid·əd·ē ˌmäj·ə·ləs }

RIMS *See* resonant ionization mass spectrometry. { rimz *or* ¦är¦ī¦em'es }

ring circuit [ELECTROMAG] In waveguide practice, a hybrid T junction having the physical configuration of a ring with radial branches. { 'riŋ ˌsər·kət }

ring micrometer [OPTICS] A flat, thin ring in the focal plane of a telescope; used to measure differences in right ascension and declination. { 'riŋ mi'kräm·əd·ər }

rings and brushes [OPTICS] An interference pattern produced by ordinary and extraordinary rays when a uniaxial crystal is placed between two polarizers. { ¦riŋz ən 'brəsh·əz }

ring vortex *See* vortex ring. { 'riŋ ˌvȯrˌteks }

ripple [ELEC] The alternating-current component in the output of a direct-current power supply, arising within the power supply from incomplete filtering or from commutator action in a dc generator. [FL MECH] *See* capillary wave. { 'rip·əl }

ripple quantity [PHYS] Alternating component of a pulsating quantity when this component is small relative to the constant component. { 'rip·əl ˌkwän·əd·ē }

ripple tank [PHYS] A shallow tray containing a liquid and equipped with means for generating surface waves; used to illustrate several types of wave phenomena, such as interference and diffraction. { 'rip·əl ˌtaŋk }

ripple voltage [ELEC] The alternating component of the unidirectional voltage from a rectifier or generator used as a source of direct-current power. { 'rip·əl ˌvōl·tij }

RIS *See* resonance ionization spectroscopy.

rise time [ELEC] The time for the pointer of an electrical instrument to make 90% of the change to its final value when electric power suddenly is applied from a source whose impedance is high enough that it does not affect damping. { 'rīz ,tīm }

Risley prism system [OPTICS] A type of dispersing prism used to test ocular convergence in ophthalmology; consists of two thin prisms mounted so that they can be rotated simultaneously in opposite directions. { 'riz·lē 'priz·əm ,sis·təm }

Ritchey-Chrétien optics [OPTICS] A modification of the Cassegrain optical system used in large optical telescopes; it has a hyperbolic image-forming primary mirror, no spherical aberration, and no coma; it has a larger usable field than either Newtonian or Cassegrain optical systems. { 'rich·ē 'krā·chən ,äp·tiks }

Ritchie's experiment [THERMO] An experiment that uses a Leslie cube and a differential air thermometer to demonstrate that the emissivity of a surface is proportional to its absorptivity. { 'rich·ēz ik,sper·ə·mənt }

Ritchie wedge [OPTICS] A photometer in which a test source and a standard source of light illuminate two perpendicular white, diffusing surfaces which intersect in a movable wedge, and these surfaces are viewed from a direction perpendicular to a line connecting the sources. { 'rich·ē ,wej }

Ritz formula [ATOM PHYS] A particular expansion of an equation used in studying the spectra of atoms. { 'ritz ,fór·myə·lə }

Ritz's combination principle [SPECT] The empirical rule that sums and differences of the frequencies of spectral lines often equal other observed frequencies. Also known as combination principle. { 'rit·səz ,käm·bə'nā·shən ,prin·sə·pəl }

rms value *See* root-mean-square value. { ¦är ¦em'es ,val·ü }

Robertson-Walker solutions [RELAT] A class of relativistic models for a homogeneous, isotropic universe that are conventionally accepted as describing the real universe. { 'räb·ərt·sən 'wók·ər sə,lü·shənz }

Robin law [PHYS] The law that an increase in pressure on a system in chemical or physical equilibrium favors the system formed with a decrease in volume, and conversely, a change in pressure does not affect a system formed with no change in volume. { 'räb·ən ,ló }

roc [ELEC] A unit of electrical conductivity equal to the conductivity of a material in which an electric field of 1 volt per centimeter gives rise to a current density of 1 ampere per square centimeter. Derived from reciprocal ohm centimeter. { räk }

Rochelle-electric *See* ferroelectric. { rō'shel i,lek·trik }

Roche lobes [MECH] **1.** Regions of space surrounding two massive bodies revolving around each other under their mutual gravitational attraction, such that the gravitational attraction of each body dominates the lobe surrounding it. **2.** In particular, the effective potential energy (referred to a system of coordinates rotating with the bodies) is equal to a constant V_0 over the surface of the lobes, and if a particle is inside one of the lobes and if the sum of its effective potential energy and its kinetic energy is less than V_0, it will remain inside the lobe. { 'rōch ,lōbz }

Rochon polarizing prism [OPTICS] A device for producing linearly polarized beams of light, consisting of two adjacent quartz wedges, the first of which has its optic axis parallel to the beam, while the second has its optic axis perpendicular to the beam; one of the beams (the ordinary ray) is undeviated, and is therefore not spread into a spectrum. { rō'shōn 'pō·lə,rīz·iŋ ,priz·əm }

rod *See* perch. { räd }

roentgen current [ELEC] An electric current arising from the motion of polarization charges, as in the rotation of a dielectric in a charged capacitor. { 'rent·gən ,kər·ənt }

roentgen diffractometry *See* x-ray crystallography. { 'rent·gən ,dē,frak'täm·ə·trē }

roentgenography [PHYS] Radiography by means of x-rays. { ,rent·gə'näg·rə·fē }

roentgenoluminescence [PHYS] Luminescence which can be produced by x-rays. { ,rent·gə·nō,lü·mə'nes·əns }

roentgen optics *See* x-ray optics. { 'rent·gən 'äp·tiks }

roentgen rays *See* x-rays. { 'rent·gən ,rāz }

roentgen spectrometry *See* x-ray spectrometry. { 'rent·gən spek'träm·ə·trē }

Roget's spiral [ELEC] A spiral wire, suspended vertically with the lower end in mercury, that is made to go through a cycle in which an electric current passing through the wire produces mutual attraction between the coils, causing the wire to lift out of the mercury and breaking the current; the spiral then expands under its own weight, so that the lower end drops back into the mercury and the current is reestablished. { rō¦zhāz 'spī·rəl }

roll [MECH] Rotational or oscillatory movement of an aircraft or similar body about a longitudinal axis through the body; it is called roll for any degree of such rotation. { rōl }

roll acceleration [MECH] The angular acceleration of an aircraft or missile about its longitudinal or X axis. { 'rōl ik,sel·ə,rā·shən }

roll axis [MECH] A longitudinal axis through an aircraft, rocket, or similar body, about which the body rolls. { 'rōl ,ak·səs }

Rollin film *See* helium film. { 'räl·ən ,film }

rolling [MECH] Motion of a body across a surface combined with rotational motion of the body so that the point on the body in contact with the surface is instantaneously at rest. { 'rōl·iŋ }

rolling contact [MECH] Contact between bodies such that the relative velocity of the two contacting surfaces at the point of contact is zero. { 'rōl·iŋ 'kän,takt }

rolling friction [MECH] A force which opposes the motion of any body which is rolling over the surface of another. { 'rōl·iŋ 'frik·shən }

rom [ELEC] A unit of electrical conductivity, equal to the conductivity of a material in which an electric field of 1 volt per meter gives rise to a current density of 1 ampere per square meter. Derived from reciprocal ohm meter. { räm }

Römer method [OPTICS] A method of measuring the speed of light, in which apparent changes in the periods of satellites of another planet, such as Jupiter, whose distance from the earth is known, are observed throughout the year. { 'rem·ər ,meth·əd }

Ronchi test [OPTICS] An improvement on the Foucault knife-edge test for testing curved mirrors, in which the knife edge is replaced with a transmission grating with 15–80 lines per centimeter, and the pinhole source is replaced with a slit or a section of the same grating. { 'raŋ·kē ,test }

rood [MECH] A unit of area, equal to $^1/_4$ acre, or 10,890 square feet, or 1011.7141056 square meters. { rüd }

room acoustics [ACOUS] The study of the behavior of sound waves in an enclosed room. { 'rüm ə,küs·tiks }

room impulse response [ACOUS] The sound pressure in a room that results from a very short sound pulse, usually measured as a function of the time after the arrival of the first direct sound from the source to a listener's location. { ¦rüm 'im,pəls ri,späns }

root-mean-square current *See* effective current. { 'rüt ,mēn 'skwer 'kə·rənt }

root-mean-square sound pressure *See* effective sound pressure. { 'rüt ,mēn 'skwer 'saůnd ,presh·ər }

root-mean-square value [PHYS] The square root of the time average of the square of a quantity; for a periodic quantity the average is taken over one complete cycle. Abbreviated rms value. Also known as effective value. { 'rüt ,mēn 'skwer 'val·yü }

root-sum-square value [PHYS] The square root of the sum of the squares of a series of related values; commonly used to express total harmonic distortion. { 'rüt ,səm 'skwər 'val·ü }

Rosa and Dorsey method [ELECTROMAG] A method of measuring the speed of light by comparing the capacitance of a capacitor in electromagnetic units, as measured experimentally, with values of currents determined from a current balance, to the capacitance of the same capacitor in electrostatic units, as determined from its geometrical dimensions. { 'rō·sə ən 'dȯr·sē ,meth·əd }

Rossby diagram [THERMO] A thermodynamic diagram, named after its designer, with mixing ratio as abscissa and potential temperature as ordinate; lines of constant equivalent potential temperature are added. { 'rȯs·bē ,dī·ə,gram }

Rossby number [FL MECH] The nondimensional ratio of the inertial force to the Coriolis force for a given flow of a rotating fluid, given as $R_0 = U/fL$, where U is a characteristic velocity, *f* the Coriolis parameter (or, if the system is cylindrical rather than spherical, twice the system's rotation rate), and L a characteristic length. { 'rȯs·bē ,nəm·bər }

Rossby parameter [FL MECH] The northward variation of the Coriolis parameter, arising from the sphericity of the earth. Also known as Rossby term. { 'rȯs·bē pə,ram·əd·ər }

Rossby regime [FL MECH] A type of flow pattern in a rotating fluid with differential radial heating in which the major radial transport of shear and momentum is effected by horizontal eddies of low wave-number; this regime occurs for low values of the Rossby number (of the order of 0.1). { 'rȯs·bē ri,zhēm }

Rossby term *See* Rossby parameter. { 'rȯs·bē ,tərm }

Rosseland mean absorption coefficient [OPTICS] A coefficient of opacity that is the inverse of a weighted mean of the transmission coefficient over all frequencies. { 'rȯs·lənd 'mēn əb'sȯrp·shən ,kō·i,fish·ənt }

Ross objective [OPTICS] A type of wide-field lens objective in cameras used for astrometric work. { 'rȯs əb,jek·tiv }

rotary beam [ELECTROMAG] Short-wave antenna system highly directional in azimuth and altitude, mounted in such a manner that it can be rotated to any desired position, either manually or by an electric motor drive. { 'rōd·ə·rē 'bēm }

rotary coupler *See* rotating joint. { 'rōd·ə·rē 'kəp·lər }

rotary dispersion [OPTICS] The change in the angle through which an optically active substance rotates the plane of polarization of plane polarized light as the wavelength of the light is varied. Also known as rotatory dispersion. { 'rōd·ə·rē di'spər·zhən }

rotary joint *See* rotating joint. { 'rōd·ə·rē 'jȯint }

rotary polarization *See* optical activity. { 'rōd·ə·rē ,pō·lə·rə'zā·shən }

rotary reflection axis *See* rotoreflection axis. { 'rōd·ə·rē ri'flek·shən ,ak·səs }

rotary solenoid [ELECTROMAG] A solenoid in which the armature is rotated when actuated; the rotary stroke, ranging from 25 to 95°, is usually converted to linear motion to give a longer stroke than is possible with a conventional plunger-type solenoid. { 'rōd·ə·rē 'sō·lə,nȯid }

rotary-vane attenuator [ELECTROMAG] Device designed to introduce attenuation into a waveguide circuit by varying the angular position of a resistive material in the guide. { 'rōd·ə·rē ¦vān ə'ten·yə,wād·ər }

rotating coordinate system [MECH] A coordinate system whose axes as seen in an inertial coordinate system are rotating. { 'rō,tād·iŋ kō'ȯrd·ən·ət ,sis·təm }

rotating crystal method [SOLID STATE] Any method of studying crystalline structures by x-ray or neutron diffraction in which a monochromatic, collimated beam of x-rays or neutrons falls on a single crystal that is rotated about an axis perpendicular to the beam. { 'rō,tād·iŋ ¦krist·əl ,meth·əd }

rotating-cylinder method [FL MECH] A method of measuring the viscosity of a fluid in which the fluid fills the space between two concentric cylinders, and the torque on the stationary inner cylinder is measured when the outer cylinder is rotated at constant speed. { 'rō,tād·iŋ ¦sil·ən·dər ,meth·əd }

rotating joint [ELECTROMAG] A joint that permits one section of a transmission line or waveguide to rotate continuously with respect to another while passing radio-frequency energy. Also known as rotary coupler; rotary joint. { 'rō ,tād·iŋ 'jȯint }

rotating Reynolds number [FL MECH] A nondimensional number arising in problems of a rotating viscous fluid and, in particular, in problems involving the agitation of such a fluid by an impeller, equal to the product of the square of the impeller's diameter and its angular velocity divided by the kinematic viscosity of the fluid. Symbolized Re_r. { 'rō,tād·iŋ 'ren·əlz ,nəm·bər }

rotating wedge [OPTICS] A circular optical wedge mounted to be rotated in the path of light to divert the line of sight to a limited degree. { 'rō,tād·iŋ 'wej }

rotation [MECH] Also known as rotational motion. **1.** Motion of a rigid body in which either one point is fixed, or all the points on a straight line are fixed. **2.** Angular displacement of a rigid body. **3.** The motion of a particle about a fixed point. { rō'tā·shən }

rotational energy [MECH] The kinetic energy of a rigid body due to rotation. { rō'tā·shən·əl 'en·ər·jē }

rotational field [PHYS] A vector field whose curl does not vanish. Also known as circuital field; vortical field. { rō¦tā·shən·əl 'fēld }

rotational flow [FL MECH] Flow of a fluid in which the curl of the fluid velocity is not zero, so that each minute particle of fluid rotates about its own axis. Also known as rotational motion. { rō'tā·shən·əl 'flō }

rotational impedance [MECH] A complex quantity, equal to the phasor representing the alternating torque acting on a system divided by the phasor representing the resulting angular velocity in the direction of the torque at its point of application. Also known as mechanical rotational impedance. { rō'tā·shən·əl im'pēd·əns }

rotational inertia *See* moment of inertia. { rō'tā·shən·əl i'nər·shə }

rotational motion *See* rotational flow. { rō'tā·shən·əl 'mō·shən }

rotational reactance [MECH] The imaginary part of the rotational impedance. Also known as mechanical rotational reactance. { rō'tā·shən·əl rē'ak·təns }

rotational resistance [MECH] The real part of rotational impedance; it is responsible for dissipation of energy. Also known as mechanical rotational resistance. { rō'tā·shən·əl ri'zis·təns }

rotational spectrum [SPECT] The molecular spectrum resulting from transitions between rotational levels of a molecule which behaves as the quantum-mechanical analog of a rotating rigid body. { rō'tā·shən·əl 'spek·trəm }

rotational stability [MECH] Property of a body for which a small angular displacement sets up a restoring torque that tends to return the body to its original position. { rō'tā·shən·əl stə'bil·əd·ē }

rotational strain [MECH] Strain in which the orientation of the axes of strain is changed. { rō'tā·shən·əl 'strān }

rotational sum rule [SPECT] The rule that, for a molecule which behaves as a symmetric top, the sum of the line strengths corresponding to transitions to or from a given rotational level is proportional to the statistical weight of that level, that is, to $2J+1$, where J is the total angular momentum quantum number of the level. { rō'tā·shən·əl 'səm ,rül }

rotational transform [PL PHYS] Property possessed by a magnetic field, in a system used to confine a plasma, in which magnetic lines of force do not close in on themselves after making a circuit around the system, but are rotationally displaced. { rō'tā·shən·əl 'tranz,fȯrm }

rotational transformation [CRYSTAL] A type of crystal transformation that is a change from an ordered phase to a partially disordered phase by rotation of groups of atoms. { rō'tā·shən·əl ,tranz·fər'mā·shən }

rotation axis [CRYSTAL] A symmetry element of certain crystals in which the crystal can be brought into a position physically indistinguishable from its original position by a rotation through an angle of $360°/n$ about the axis, where n is the multiplicity of the axis, equal to 2, 3, 4, or 6. Also known as symmetry axis. { rō'tā·shən ,ak·səs }

rotation camera [SOLID STATE] An instrument for studying crystalline structure by x-ray or neutron diffraction, in which a monochromatic, collimated beam of x-rays or neutrons falls on a single crystal which is rotated about an axis perpendicular to the beam and parallel to one of the crystal axes, and the various diffracted beams are registered on a cylindrical film concentric with the axis of rotation. { rō'tā·shən ,kam·rə }

rotation coefficients [MECH] Factors employed in computing the effects on range and deflection which are caused by the rotation of the earth; they are published only in firing tables involving comparatively long ranges. { rō'tā·shən ,kō·i,fish·əns }

rotation-inversion axis [CRYSTAL] A symmetry element of certain crystals in which a crystal can be brought into a position physically indistinguishable from its original position by a rotation through an angle of $360°/n$ about the axis followed by an inversion, where n is the multiplicity of the axis, equal to 1, 2, 3, 4, or 6. Also known as inversion axis. { rō'tā·shən in'vər·zhən ,ak·səs }

rotation moment *See* torque. { rō'tā·shən ,mō·mənt }

rotation-reflection axis [CRYSTAL] A symmetry element of certain crystals in which a crystal can be brought into a position physically indistinguishable from its original position by a rotation through an angle of 360°/*n* about the axis followed by a reflection in the plane perpendicular to the axis, where *n* is the multiplicity of the axis, equal to 1, 2, 3, 4, or 6. { rō'tā·shən ri'flek·shən ,ak·səs }

rotation Reynolds number *See* rotating Reynolds number. { rō'tā·shən 'ren·əlz ,nəm·bər }

rotation twin [CRYSTAL] A twin crystal in which the parts will coincide if one part is rotated 180° (sometimes 30, 60, or 120°). { rō'tā·shən ,twin }

rotator [ELECTROMAG] A device that rotates the plane of polarization of a plane-polarized electromagnetic wave, such as a twist in a waveguide. [MECH] A rotating rigid body. [QUANT MECH] A molecule or other quantum-mechanical system which behaves as the quantum-mechanical analog of a rotating rigid body. Also known as top. { 'rō,tād·ər }

rotatory [OPTICS] Having the capability to rotate the plane of polarization of polarized electromagnetic radiation. { 'rōd·ə,tȯr·ē }

rotatory dispersion *See* rotary dispersion. { 'rōd·ə,tȯr·ē di'spər·zhən }

rotatory power [OPTICS] A substance's capability to rotate the plane of polarization of polarized electromagnetic radiation. { 'rōd·ə,tȯr·ē ,pau̇·ər }

rotoflector [ELECTROMAG] In radar, elliptically shaped, rotating reflector used to reflect a vertically directed radar beam at right angles so that it radiates in a horizontal direction. { 'rōd·ə,flek·tər }

rotoinversion axis [CRYSTAL] A type of crystal symmetry element that combines a rotation of 60, 90, 120, or 180° with inversion across the center. Also known as symmetry axis of rotary inversion; symmetry axis of rotoinversion. { ¦rōd·ō·in'vər·zhən ,ak·səs }

roton [PHYS] A quantum of rotational motion in a liquid, such as superfluid helium. { 'rō,tän }

rotoreflection axis [CRYSTAL] A type of symmetry element that combines a rotation of 60, 90, 120, or 180° with reflection across the plane perpendicular to the axis. Also known as rotatory reflection axis. { ¦rōd·ō·ri¦flek·shən ,ak·səs }

roughening transition [PHYS] A change in the behavior of the interface between the solid and liquid phases of a substance at a certain critical temperature, below which the surface is flat or sharp and displays distinct terraces and ledges on an atomic level, while above the critical temperature the surface is rough or rounded. { 'rəf·ən·iŋ tran'zish·ən }

roughness [FL MECH] Distance from peaks to valleys in pipe-wall irregularities; used to modify Reynolds number calculations for fluid flow through pipes. { 'rəf·nəs }

roughness factor [FL MECH] A correction factor used in fluid-flow calculations to allow for flow resistance caused by the roughness of the surface over which the fluid must flow. { 'rəf·nəs ,fak·tər }

round-trip echoes [ELECTROMAG] Multiple reflection echoes produced when a radar pulse is reflected from a target strongly enough so that the echo is reflected back to the target where it produces a second echo. { 'rau̇nd ¦trip 'ek·ōz }

Rousseau diagram [OPTICS] A geometric construction used to determine the total luminous flux of a lamp from a number of polar diagrams which give the effective luminous intensity of the lamp in various directions. { rü,sō 'dī·ə,gram }

Routh's procedure [MECH] A procedure for modifying the Lagrangian of a system so that the modified function satisfies a modified form of Lagrange's equations in which ignorable coordinates are eliminated. { 'rüths prə,sē·jər }

Routh's rule of inertia [MECH] The moment of inertia of a body about an axis of symmetry equals $M(a^2 + b^2)/n$, where M is the body's mass, *a* and *b* are the lengths of the body's two other perpendicular semiaxes, and *n* equals 3, 4, or 5 depending on whether the body is a rectangular parallelepiped, elliptic cylinder, or ellipsoid, respectively. { 'rau̇ths 'rül əv i'nər·shə }

rowland [SPECT] A unit of length, formerly used in spectroscopy, equal to 999.81/999.94 angstrom, or approximately 0.99987×10^{-10} meter. { 'rō,lənd }

Rowland circle [SPECT] A circle drawn tangent to the face of a concave diffraction grating at its midpoint, having a diameter equal to the radius of curvature of a grating surface; the slit and camera for the grating should lie on this circle. { 'rō,lənd ,sər·kəl }

Rowland current [ELEC] A convection current that arises when a charged capacitor plate is rotated. { 'rō·lənd ,kər·ənt }

Rowland ghost [SPECT] A false spectral line produced by a diffraction grating, arising from periodic errors in groove position. { 'rō,lənd ,gōst }

Rowland grating *See* concave grating. { 'rō,lənd ,grād·iŋ }

Rowland mounting [SPECT] A mounting for a concave grating spectrograph in which camera and grating are connected by a bar forming a diameter of the Rowland circle, and the two run on perpendicular tracks with the slit placed at their junction. { 'rō,lənd ,mau̇nt·iŋ }

Rowland ring [ELECTROMAG] A ring-shaped sample of magnetic material, generally surrounded by a coil of wire carrying a current. { 'rō,lənd ,riŋ }

rpm *See* revolution per minute.

r-process [NUC PHYS] The synthesis of elements and nuclides in supernovas through rapid captures of neutrons in a matter of seconds, followed by beta decay. { 'är ,prä·səs }

rps *See* revolution per second.

rubidium-vapor frequency standard [PHYS] An atomic frequency standard in which the frequency is established by a gas cell containing rubidium vapor and a neutral buffer gas. { rü'bid·ē·əm ¦vā·pər 'frē·kwən·se ,stan·dərd }

ruby laser [OPTICS] An optically pumped solid-state laser that uses a ruby crystal to produce an intense and extremely narrow beam of coherent red light. { 'rü·bē 'lā·zər }

ruby maser [PHYS] A maser that uses a ruby crystal in the cavity resonator. { 'rü·bē 'mā·zər }

ruling engine [SPECT] A machine operated by a long micrometer screw which rules equally spaced lines on an optical diffraction grating. { 'rül·iŋ ˌen·jən }

runaway effect [PL PHYS] The phenomenon whereby an electric current in a plasma produces heat through the Joule effect, and the resulting temperature increase results in an increase in conductivity and, thereby, an increase in current flow. { 'rən·əˌwā iˌfekt }

Runge vector [MECH] A vector which describes certain unchanging features of a nonrelativistic two-body interaction obeying an inverse-square law, either in classical or quantum mechanics; its constancy is a reflection of the symmetry inherent in the inverse-square interaction. { 'rəŋ·ə ˌvek·tər }

Russell-Saunders coupling [PHYS] A process for building many-electron single-particle eigenfunctions of orbital angular momentum and spin; the orbital functions are combined to make an eigenfunction of the total orbital angular momentum, the spin functions are combined to make an eigenfunction of the total spin angular momentum, and then the results are combined into eigenfunctions of the total angular momentum of the system. Also known as LS coupling. { 'rəs·əl 'sȯn·dərz ˌkəp·liŋ }

Rutherford backscattering spectrometry [SPECT] A method of determining the concentrations of various elements as a function of depth beneath the surface of a sample, by measuring the energy spectrum of ions which are backscattered out of a beam directed at the surface. { 'rəth·ər·fərd 'bakˌskad·ə·riŋ spek'träm·ə·trē }

Rutherford nuclear atom [ATOM PHYS] A theory of atomic structure in which nearly all the mass is concentrated in a small nucleus, electrons surrounding the nucleus fill nearly all the atom's volume, the number of these electrons equals the atomic number, and the positive charge on the nucleus is equal in magnitude to the negative charge of the electrons. { 'rəth·ər·fərd ¦nü·klē·ər 'ad·əm }

Rutherford scattering [ATOM PHYS] Scattering of heavy charged particles by the Coulomb field of an atomic nucleus. { 'rəth·ər·fərd ˌskad·ə·riŋ }

ry *See* rydberg.

rydberg [ATOM PHYS] A unit of energy used in atomic physics, equal to the square of the charge of the electron divided by twice the Bohr radius; equal to 13.605698 ± 0.000004 electronvolts. Symbolized ry. [SPECT] *See* kayser. { 'ridˌbərg }

Rydberg atom [ATOM PHYS] An atom whose outer electron has been excited to very high energy states, far from the nucleus. { 'ridˌbərg ˌad·əm }

Rydberg constant [ATOM PHYS] **1.** The most accurately measured of the fundamental constants, which enters into the formulas for wave numbers of atomic spectra and serves as a universal scaling factor for any spectroscopic transition and as an important cornerstone in the determination of other constants; it is equal to $\alpha^2 mc/(2h)$, or, in International System (SI) units, to $me^4/(8h^3\epsilon_0{}^2c)$, where α is the fine-structure constant, m and e are the electron mass and charge, c is the speed of light, h is Planck's constant, and ϵ_0 is the electric constant; numerically, it is equal to 10,973,731.568 549 ± 0.000 083 inverse meters. Symbolized R_∞. **2.** For any atom, the Rydberg constant (first definition) divided by $1 + m/M$, where m and M are the masses of an electron and of the nucleus. { 'ridˌbərg ˌkän·stənt }

Rydberg correction [ATOM PHYS] A term inserted into a formula for the energy of a single electron in the outermost shell of an atom to take into account the failure of the inner electron shells to screen the nuclear charge completely. { 'ridˌbərg kəˌrek·shən }

Rydberg maser [PHYS] A maser that amplifies microwave radiation by means of stimulated emission of atoms whose outer electrons have been excited to very high energy states. { 'ridˌbərg ˌmā·zər }

Rydberg series formula [SPECT] An empirical formula for the wave numbers of various lines of certain spectral series such as neutral hydrogen and alkali metals; it states that the wave number of the nth member of the series is $\lambda_\infty - R/(n + a)^2$, where λ_∞ is the series limit, R is the Rydberg constant of the atom, and a is an empirical constant. { 'ridˌbərg ¦sir·ēz ˌfȯr·myə·lə }

Rydberg spectrum [SPECT] An ultraviolet absorption spectrum produced by transitions of atoms of a given element from the ground state to states in which a single electron occupies an orbital farther from the nucleus. { 'rīdˌbərg ˌspek·trəm }

S

s *See* second; strange quark.

S *See* siemens; stoke.

sabin [ACOUS] A unit of sound absorption for a surface, equivalent to 1 square foot (0.09290304 square meter) of perfectly absorbing surface. Also known as absorption unit; open window unit (OW unit); square-foot unit of absorption. { 'sā·bən }

Sabine formula [ACOUS] An empirical equation for the reverberation time of sound in a room; its form is identical to that of the Franklin equation. { 'sā,bēn ,fȯr·myə·lə }

Sackur-Tetrode equation [STAT MECH] An equation for the translational entropy of an ideal gas made up of free fermions. { 'säk·ər 'te,trōd i,kwā·zhən }

safe load [MECH] The stress, usually expressed in tons per square foot, which a soil or foundation can safely support. { 'sāf ,lōd }

safety factor [ELEC] The amount of load, above the normal operating rating, that a device can handle without failure. [MECH] *See* factor of safety. { 'sāf·tē ,fak·tər }

sagittal coma [OPTICS] The radius of the circle formed in the focal plane by rays from an off-axis point passing near the edge of a lens that displays coma. { ¦saj·ə·təl ¦kō·mə }

sagittal focus *See* secondary focus. { 'saj·əd·əl 'fō·kəs }

sagittal plane [OPTICS] A plane that is perpendicular to the meridional plane of an optical system and contains a specified ray. Also known as equatorial plane. { ¦saj·ə·təl ¦plān }

sagittal surface [OPTICS] A surface containing the secondary foci of points in a plane perpendicular to the optical axis of an astigmatic system. { ¦saj·ə·təl ¦sər·fəs }

Sagnac effect [OPTICS] The shift in interference fringes from two coherent light beams traveling in opposite directions around a ring when the ring is rotated about an axis perpendicular to the ring. { 'sän·yäk i,fekt }

Saha ionization [STAT MECH] The ionization of a gas which exists when the gas is in thermal equilibrium at a given temperature, in the absence of external influences; it increases with increasing temperature. Also known as thermal ionization. { ,sä,hä ,ī·ə·nə'zā·shən }

Saha's equation [STAT MECH] An equation for Saha ionization of a monatomic gas in terms of the temperature and pressure of the gas, the ionization potential, and statistical weights of ion, electron, and atom. { ,sä,häz i,kwā·zhən }

Saint Elmo's fire [ELEC] A visible electric discharge, sometimes seen on the mast of a ship, on metal towers, and on projecting parts of aircraft, due to concentration of the atmospheric electric field at such projecting parts. { 'sānt 'el·mōz 'fīr }

Saint Venant's compatibility equations [MECH] Equations for the components e_{ij} of the strain tensor that follow from their integrability, namely, $(e_{ij})_{kl} + (e_{kl})_{ij} - (e_{ik})_{jl} - (e_{jl})_{ik} = 0$, where i, j, k, and l can take on any of the values x, y, and z, and subscripts outside the parentheses indicate partial differentiation. { ,sän·və'nänz kəm,pad·ə'bil·əd·ē i,kwā·shənz }

Saint Venant's principle [MECH] The principle that the strains that result from application, to a small part of a body's surface, of a system of forces that are statically equivalent to zero force and zero torque become negligible at distances which are large compared with the dimensions of the part. { ,sän·və'nänz 'prin·sə·pəl }

Sakata-Taketani equation [QUANT MECH] A relativistic wave equation for a particle with spin 1 whose form resembles that of the nonrelativistic Schrödinger equation. { sä'käd·ə ,tä·kə'tä·nē i,kwā·zhən }

Salam-Weinberg theory *See* Weinberg-Salam theory. { sə'läm 'wīn,bərg ,thē·ə·rē }

samarium-cobalt magnet [ELECTROMAG] A rare-earth permanent magnet that is more efficient, has lower leakage and greater resistance to demagnetization, and can be magnetized to higher levels than conventional permanent magnets. { sə'mar·ē·əm 'kō,bȯlt 'mag·nət }

sand heap analogy *See* sand hill analogy. { 'sand ,hēp ə,nal·ə·jē }

sand hill analogy [MECH] A formal identity between the differential equation and boundary conditions for a stress function for torsion of a perfectly plastic prismatic bar, and those for the height of the surface of a granular material, such as dry sand, which has a constant angle of rest. Also known as sand heap analogy. { 'sand ,hil ə,nal·ə·jē }

sand load [ELECTROMAG] An attenuator used as a power-dissipating terminating section for a coaxial line or waveguide; the dielectric space in the line is filled with a mixture of sand and graphite that acts as a matched-impedance load, preventing standing waves. { 'san ˌlōd }

Sargent curve [NUC PHYS] A graph of logarithms of decay constants of radioisotopes subject to beta-decay against logarithms of the corresponding maximum beta-particle energies; most of the points fall on two straight lines. { 'sär·jənt ˌkərv }

Sargent cycle [THERMO] An ideal thermodynamic cycle consisting of four reversible processes: adiabatic compression, heating at constant volume, adiabatic expansion, and isobaric cooling. { 'sär·jənt ˌsī·kəl }

satellite infrared spectrometer [SPECT] A spectrometer carried aboard satellites in the Nimbus series which measures the radiation from carbon dioxide in the atmosphere at several different wavelengths in the infrared region, giving the vertical temperature structure of the atmosphere over a large part of the earth. Abbreviated SIRS. { 'sad·əlˌīt ¦in·frə'red spek'träm·əd·ər }

saturable absorption [OPTICS] A decrease in the absorption coefficient of certain nonlinear materials at high intensities in the incident radiation. { 'sach·rə·bəl ab'sȯrp·shən }

saturated color [OPTICS] A pure color not contaminated by white. { 'sach·əˌrād·əd 'kəl·ər }

saturated interference spectroscopy [SPECT] A version of saturation spectroscopy in which the gas sample is placed inside an interferometer that splits a probe laser beam into parallel components in such a way that they cancel on recombination; intensity changes in the recombined probe beam resulting from changes in absorption or refractive index induced by a laser saturating beam are then measured. { 'sach·əˌrād·əd ˌin·tərˌfir·əns spek'träs·kə·pē }

saturated vapor [THERMO] A vapor whose temperature equals the temperature of boiling at the pressure existing on it. { 'sach·əˌrād·əd 'vā·pər }

saturation [ELECTROMAG] *See* magnetic saturation. [OPTICS] *See* color saturation. [PHYS] **1.** The condition in which a further increase in some cause produces no further increase in the resultant effect. **2.** The property exhibited by certain forces between particles wherein each particle can interact strongly with only a limited number of other particles, as in the forces between atoms in a molecule, and between nucleons in a nucleus. { ˌsach·ə'rā·shən }

saturation flux density *See* saturation induction. { ˌsach·ə'rā·shən 'fləks ˌden·səd·ē }

saturation induction [ELECTROMAG] The maximum intrinsic induction possible in a material. Also known as saturation flux density. { ˌsach·ə'rā·shən in'dək·shən }

saturation magnetization [ELECTROMAG] The maximum possible magnetization of a material. { ˌsach·ə'rā·shən ˌmag·nəd·ə'zā·shən }

saturation scale [OPTICS] A series of colors which appear to have equal differences in color saturation. { ˌsach·ə'rā·shən ¦skāl }

saturation signal [ELECTROMAG] A radio signal (or radar echo) which exceeds a certain power level fixed by the design of the receiver equipment; when a receiver or indicator is "saturated," the limit of its power output has been reached. { ˌsach·ə'rā·shən ¦sig·nəl }

saturation specific humidity [THERMO] A thermodynamic function of state; the value of the specific humidity of saturated air at the given temperature and pressure. { ˌsach·ə'rā·shən spə'sif·ik hyü'mid·əd·ē }

saturation spectroscopy [SPECT] A branch of spectroscopy in which the intense, monochromatic beam produced by a laser is used to alter the energy-level populations of a resonant medium over a narrow range of particle velocities, giving rise to extremely narrow spectral lines that are free from Doppler broadening; used to study atomic, molecular, and nuclear structure, and to establish accurate values for fundamental physical constants. { ˌsach·ə'rā·shən spek 'träs·kə·pē }

saturation vapor pressure [THERMO] The vapor pressure of a thermodynamic system, at a given temperature, wherein the vapor of a substance is in equilibrium with a plane surface of that substance's pure liquid or solid phase. { ˌsach·ə'rā·shən 'vā·pər ˌpresh·ər }

sausage instability *See* kink instability. { 'sȯ·sij ˌin·stə'bil·əd·ē }

savart [ACOUS] A unit of pitch interval, such that the interval between two frequencies measured in savarts is equal to 1000 times the common logarithm of the ratio of the frequencies; one octave equals approximately 301.030 savarts. { sa'vär }

Savart plate [OPTICS] A device consisting of a pair of calcite plates having the same thickness, cut along the natural cleavage faces, and mounted with corresponding faces perpendicular to each other; used to detect polarization of light by means of interference fringes. { sa 'vär ˌplāt }

Savart polariscope [OPTICS] A polariscope consisting of a specially constructed double-plate polarizer and a tourmaline plate analyzer; polarized light passing through the instrument is indicated by the presence of parallel colored fringes, while unpolarized light results in a uniform field. { 'sa'vär pə'lar·əˌskōp }

SAW *See* surface acoustic wave.

saxophone [ELECTROMAG] Vertex-fed linear array antenna giving a cosecant-squared radiation pattern. { 'sak·səˌfōn }

Saybolt chromometer [OPTICS] Device used to measure the color of undyed gasolines, jet fuels, naphthas, kerosines, petroleum waxes, and pharmaceutical white oils. { 'sāˌbōlt krə'mäm·əd·ər }

Saybolt Furol viscosity [FL MECH] The time in seconds for 60 milliliters of fluid to flow through a capillary tube in a Saybolt Furol viscosimeter at specified temperatures between 70 and 210°F (21 and 99°C); used for high-viscosity petroleum oils, such as transmission and gear oils, and heavy fuel oils. { 'sā,bōlt 'fyu̇,ro̊l vi'skäs·əd·ē }

Saybolt Seconds Universal [FL MECH] A unit of measurement for Saybolt Universal viscosity. Abbreviated SSU. { 'sā,bōlt 'sek·ənz ,yü·nə'vər·səl }

Saybolt Universal viscosity [FL MECH] The time in seconds for 60 milliliters of fluid to flow through a capillary tube in a Saybolt Universal viscosimeter at a given temperature. { 'sā,bōlt ,yü·nə'vər·səl vi'skäs·əd·ē }

sb *See* stilb.

scalar [PHYS] **1.** A quantity which has magnitude only and no direction, in contrast to a vector. **2.** A quantity which has magnitude only, and has the same value in every coordinate system. Also known as scalar invariant. { 'skā·lər }

scalar field [PHYS] A field which is characterized by a function of position and time whose value at each point is a scalar. { 'skā·lər 'fēld }

scalar function [PHYS] A function of position and time whose value at each point is a scalar. { 'skā·lər 'fəŋk·shən }

scalar meson [PART PHYS] A meson which has spin 0 and positive parity, and may be described by a scalar field. { 'skā·lər 'mā,sän }

scalar polynomial curvature singularity [RELAT] A singularity in space-time at which a scalar, formed as a polynomial in the curvature tensor, diverges. { ¦skāl·ər ,päl·ə¦nō·mē·əl ¦kərv·ə·chər ,siŋ·gyə'lar·əd·ē }

scalar potential [PHYS] A scalar function whose negative gradient is equal to some vector field, at least when this field is time-independent; for example, the potential energy of a particle in a conservative force field, and the electrostatic potential. { 'skā·lər pə'ten·chəl }

scale [ACOUS] A series of musical notes arranged from low to high by a specified scheme of intervals suitable for musical purposes. [PHYS] **1.** A one-to-one correspondence between numbers and the value of some physical quantity, such as the centigrade or Kelvin temperature scales on the API (American Petroleum Institute) or Baumé scales of specific gravity. **2.** To determine a quantity at some order of magnitude by using data or relationships which are known to be valid at other (usually lower) orders of magnitude. { skāl }

scale effect [FL MECH] An effect in fluid flow that results from changing the scale, but not the shape, of a body around which the flow passes; this effect is relevant to wind tunnel experiments. { 'skāl i,fekt }

scalenohedron [CRYSTAL] A closed crystal form whose faces are scalene triangles. { skə'lē·nō'hē·drən }

scale symmetry [PHYS] The property of certain physical systems whereby their equations of motion in classical mechanics are unchanged when the space and time variables are rescaled; for example, for a point particle moving in a plane in a potential given by a delta function at the origin, when the vector giving the particle's position r is replaced by sr and the time t is replaced by s^2t, where s is any number. { 'skāl ,sim·ə·trē }

scaling [MECH] Expressing the terms in an equation of motion in powers of nondimensional quantities (such as a Reynolds number), so that terms of significant magnitude under conditions specified in the problem can be identified, and terms of insignificant magnitude can be dropped. [NUC PHYS] A property of nuclear collisions whereby the likelihood of a nuclear reaction depends more on the ratio between energy transferred and momentum transferred than on the energy transferred between the colliding particles. { 'skāl·iŋ }

scaling factor [PHYS] A constant of proportionality which appears in a scaling law. { 'skāl·iŋ ,fak·tər }

scaling law [PHYS] A law, stating that two quantities are proportional, which is known to be valid at certain orders of magnitude and is used to calculate the value of one of the quantities at another order of magnitude. { 'skāl·iŋ ,lȯ }

scanning acoustic microscope [ACOUS] A type of acoustic microscope in which a collimated beam of acoustic radiation is focused by a spherical cavity filled with coupling fluid and an object is mechanically scanned by moving it through the focus. { 'skan·iŋ ə¦küs·tik ,mī·krə,skōp }

scanning HEED [PHYS] A form of HEED in which the diffracted electrons are directly measured electronically with a sensitive detector, and the diffraction pattern is recorded either by moving the detector across it or by deflecting the diffracted electrons across a stationary detector. Abbreviated SHEED. { 'skan·iŋ ¦hēd }

scanning loss [ELECTROMAG] In a radar system employing a scanning antenna, the reduction in sensitivity (usually expressed in decibels) due to scanning across the target, compared with that obtained when the beam is directed constantly at the target. { 'skan·iŋ ,lȯs }

scanning near-field optical microscopy *See* near-field scanning optical microscopy. { ¦skan·iŋ ¦nir ,fēld ,äp·tə·kəl mī'kräs·kə·pē }

scatter angle *See* scattering angle. { 'skad·ər ,aŋ·gəl }

scattering [ELECTROMAG] Diffusion of electromagnetic waves in a random manner by air masses in the upper atmosphere, permitting long-range reception, as in scatter propagation. Also known as radio scattering. [PHYS] **1.** The change in direction of a particle or photon because of a collision with another particle or a system. **2.** Diffusion of acoustic or electromagnetic waves caused by inhomogeneity or anisotropy of the transmitting medium. **3.** In general, causing a collection of entities to assume a less orderly arrangement. { 'skad·ə·riŋ }

scattering amplitude [QUANT MECH] A quantity, depending in general on the energy and scattering angle, which specifies the wave function of particles scattered in a collision, and whose squared modulus is proportional to the number of particles scattered in a given direction. { 'skad·ə·riŋ ˌam·pləˌtüd }

scattering angle [PHYS] The angle between the initial and final directions of motion of a scattered particle. Also known as scatter angle. { 'skad·ə·riŋ ˌaŋ·gəl }

scattering coefficient [ELECTROMAG] One of the elements of the scattering matrix of a waveguide junction; that is, a transmission or reflection coefficient of the junction. [PHYS] The fractional decrease in intensity of a beam of electromagnetic radiation or particles per unit distance traversed, which results from scattering rather than absorption. Also known as dissipation coefficient. { 'skad·ə·riŋ ˌkō·iˌfish·ənt }

scattering cross section [ELECTROMAG] The power of electromagnetic radiation scattered by an antenna divided by the incident power. [PHYS] The sum of the cross sections for elastic and inelastic scattering. { 'skad·ə·riŋ 'krȯs ˌsek·shən }

scattering function [ELECTROMAG] The intensity of scattered radiation in a given direction per lumen of flux incident upon the scattering material. { 'skad·ə·riŋ ˌfəŋk·shən }

scattering length [NUC PHYS] A parameter used in analyzing nuclear scattering at low energies; as the energy of the bombarding particle becomes very small, the scattering cross section approaches that of an impenetrable sphere whose radius equals this length. Also known as scattering power. { 'skad·ə·riŋ ˌleŋkth }

scattering loss [ELECTROMAG] The portion of the transmission loss that is due to scattering within the medium or roughness of the reflecting surface. { 'skad·ə·riŋ ˌlȯs }

scattering matrix [ELECTROMAG] A square array of complex numbers consisting of the transmission and reflection coefficients of a waveguide junction. [QUANT MECH] A matrix which expresses the initial state in a scattering experiment in terms of the possible final states. Also known as collision matrix; S matrix. { 'skad·ə·riŋ ˌmā·triks }

scattering-matrix theory *See* S-matrix theory. { 'skad·ə·riŋ ¦mā·triks ˌthē·ə·rē }

scattering operator [PHYS] An operator which acts in the vector space of solutions of a wave equation, transforming solutions representing incoming waves in a domain exterior to a bounded obstacle into solutions representing outgoing waves. { 'skad·ər·iŋ ˌäp·əˌrād·ər }

scattering power *See* scattering length. { 'skad·ə·riŋ ˌpau̇·ər }

scattering theory [PHYS] The discipline that mathematically determines the amplitudes of the scattered fields in a scattering process or collision from the equations of motion of the interacting particles, including the potential energy of the interaction. Also known as direct scattering theory. { 'skad·ə·riŋ ˌthē·ə·rē }

scatter propagation [ELECTROMAG] Transmission of radio waves far beyond line-of-sight distances by using high power and a large transmitting antenna to beam the signal upward into the atmosphere and by using a similar large receiving antenna to pick up the small portion of the signal that is scattered by the atmosphere. Also known as beyond-the-horizon communication; forward-scatter propagation; over-the-horizon propagation. { 'skad·ər ˌpräp·əˌgā·shən }

scatter reflections [ELECTROMAG] Reflections from portions of the ionosphere having different virtual heights, which mutually interfere and cause rapid fading. { 'skad·ər riˌflek·shənz }

scfh [FL MECH] Cubic feet per hour of gas flow at specified standard conditions of temperature and pressure.

scfm [FL MECH] Cubic feet per minute of gas flow at specified standard conditions of temperature and pressure.

schematic circuit diagram *See* circuit diagram. { ski'mad·ik 'sər·kət ˌdī·əˌgram }

Schering bridge [ELEC] A four-arm alternating-current bridge used to measure capacitance and dissipation factor; bridge balance is independent of frequency. { 'sher·iŋ ˌbrij }

schiller *See* play of color. { 'shil·ər }

schillerization [OPTICS] Development of schiller in crystals due to the pattern of inclusions. { ˌshil·ə·rə'zā·shən }

Schleiermacher's method [THERMO] A method of determining the thermal conductivity of a gas, in which the gas is placed in a cylinder with an electrically heated wire along its axis, and the electric energy supplied to the wire and the temperatures of wire and cylinder are measured. { 'shlī·ərˌmäk·ərz ˌmeth·əd }

schlieren [OPTICS] In atmospheric optics, parcels or strata of air having densities sufficiently different from that of their surroundings so that they may be discerned by means of refraction anomalies in transmitted light. { 'shlir·ən }

schlieren method [OPTICS] An optical technique that detects density gradients occurring in a fluid flow; in its simplest form, light from a slit is collimated by a lens and focused onto a knife-edge by a second lens, the flow pattern is placed between these two lenses, and the diffraction pattern that results on a screen or photographic film placed behind the knife-edge is observed. { 'shlir·ən ˌmeth·əd }

Schmidt camera *See* Schmidt system. { 'shmit ˌkam·rə }

Schmidt-Cassegrain telescope [OPTICS] A variant of the Schmidt system which uses a Schmidt corrector plate together with a pair of spheroidal or slightly aspherical mirrors arranged as in a Cassegrain telescope. { 'shmit kas·gran 'tel·əˌskōp }

Schmidt correction plate [OPTICS] In the Schmidt system, a glass plate with one face a plane and the other aspherical and deviating from a plane in such a way that it bends light, traveling to the system's spherical mirror, so as to correct for spherical aberration and coma. Also known as Schmidt lens. { 'shmit kə'rek·shən ˌplāt }

Schmidt lens *See* Schmidt correction plate. { 'shmit ˌlenz }

Schmidt lines [NUC PHYS] Two lines, on a graph of nuclear magnetic moment versus nuclear spin, on which points describing all nuclides should lie, according to the independent particle model; experimentally, however, points describing nuclides are scattered between the lines. { 'shmit ˌlīnz }

Schmidt-Maksutov telescope *See* meniscus-Schmidt telescope. { 'shmit mak'su·täf ¦tel·əˌskōp }

Schmidt number 1 *See* Prandtl number. { 'shmit ˌnəm·bər 'wən }

Schmidt objective *See* Schmidt system. { 'shmit äbˌjek·tiv }

Schmidt optics *See* Schmidt system. { 'shmit ˌäp·tiks }

Schmidt reflector [OPTICS] A telescope employing the Schmidt system. { 'shmit riˌflek·tər }

Schmidt system [OPTICS] An optical system designed to eliminate spherical aberration and coma, which, in its original form, consists of a spherical mirror, a Schmidt correction plate near the focus of the mirror, and usually a curved reflecting plate at the focus of the mirror; used in astronomical telescopes with unusually wide fields of view and in spectroscopes, and to project a television image from a cathode-ray tube onto a screen. Also known as projection optics; Schmidt camera; Schmidt objective; Schmidt optics. { 'shmit ˌsis·təm }

Schoch effect [ACOUS] A shift of a few wavelengths in the position of an ultrasonic sound beam that undergoes total internal reflection from a surface. { 'shäk iˌfekt }

Schönflies crystal symbols [CRYSTAL] Symbols denoting the 32 crystal point groups or symmetry classes; capital letters indicate the general type of class, and subscripts the multiplicity of rotation axes and the existence of additional symmetries. { 'shənˌflēs 'krist·əl ˌsim·bəlz }

Schottky anomaly [SOLID STATE] A contribution to the heat capacity of a solid arising from the thermal population of discrete energy levels as the temperature is raised; the effect is particularly prominent at low temperatures. { 'shät·kē əˌnäm·ə·lē }

Schottky defect [SOLID STATE] **1.** A defect in an ionic crystal in which a single ion is removed from its interior lattice site and relocated in a lattice site at the surface of the crystal. **2.** A defect in an ionic crystal consisting of the smallest number of positive-ion vacancies and negative-ion vacancies which leave the crystal electrically neutral. { 'shät·kē diˌfekt }

Schottky effect [SOLID STATE] The enhancement of the thermionic emission of a conductor resulting from an electric field at the conductor surface. { 'shät·kē iˌfekt }

Schottky line [SOLID STATE] A graph of the logarithm of the saturation current from a thermionic cathode as a function of the square root of anode voltage; it is a straight line according to the Schottky theory. { 'shät·kē ˌlīn }

Schottky theory [SOLID STATE] A theory describing the rectification properties of the junction between a semiconductor and a metal that result from formation of a depletion layer at the surface of contact. { 'shät·kē ˌthē·ə·rē }

Schrödinger equation [QUANT MECH] A partial differential equation governing the Schrödinger wave function ψ of a system of one or more nonrelativistic particles; $\hbar(\partial\psi/\partial t) = H\psi$, where H is a linear operator, the Hamiltonian, which depends on the dynamics of the system, and $\hbar$ is Planck's constant divided by 2π. { 'shrād·iŋ·ər iˌkwā·zhən }

Schrödinger-Klein-Gordon equation *See* Klein-Gordon equation. { 'shrād·iŋ·ər 'klīn 'gȯrd·ən iˌkwā·zhən }

Schrödinger-Pauli equation [QUANT MECH] A modification of the Schrödinger equation to describe a particle with spin of $^1/_2\hbar$, where $\hbar$ is Planck's constant divided by 2π; the wave function has two components, corresponding to the particle's spin pointing in either of two opposite directions. { 'shrād·iŋ·ər 'pau̇·lē iˌkwā·zhən }

Schrödinger picture [QUANT MECH] A mode of description of a quantum-mechanical system in which dynamical states are represented by vectors which evolve in the course of time, and physical quantities are represented by stationary operators, in contrast to the Heisenberg picture. { 'shrād·iŋ·ər ˌpik·chər }

Schrödinger representation *See* position representation. { 'shrād·iŋ·ər ˌrep·ri·zən'tā·shən }

Schrödinger's wave mechanics [QUANT MECH] The version of nonrelativistic quantum mechanics in which a system is characterized by a wave function which is a function of the coordinates of all the particles of the system and time, and obeys a differential equation, the Schrödinger equation; physical quantities are represented by differential operators which may act on the wave function, and expectation values of measurements are equal to integrals involving the corresponding operator and the wave function. Also known as wave mechanics. { 'shrād·iŋ·ərz 'wāv miˌkan·iks }

Schrödinger wave function [QUANT MECH] A function of the coordinates of the particles of a system and of time which is a solution of the Schrödinger equation and which determines the average result of every conceivable experiment on the system. Also known as probability amplitude; psi function; wave function. { 'shrād·iŋ·ər 'wāv ˌfəŋk·shən }

Schuler pendulum [MECH] Any apparatus which swings, because of gravity, with a natural period of 84.4 minutes, that is, with the same period as a hypothetical simple pendulum whose length is the earth's radius; the pendulum arm remains vertical despite any motion of its pivot, and the apparatus is therefore useful in navigation. { 'shü·lər ˌpen·jə·ləm }

Schumann region [OPTICS] The most extreme ultraviolet portion of the electromagnetic spectrum that will affect a photographic plate. { 'shüˌmän ˌrē·jən }

Schuster method [SPECT] A method for focusing a prism spectroscope without using a distant object or a Gauss eyepiece. { 'shu̇s·tər ,meth·əd }

Schwarzschild anastigmat [OPTICS] A Gregorian telescope whose surfaces are altered to reduce astigmatism. { 'shvärts,shilt ,an·ə'stig·mat }

Schwarzschild radius [RELAT] Conventionally taken to be twice the black hole mass appearing in the general relativistic Schwarzschild solution times the gravitational constant divided by the square of the speed of light; the event horizon in a Schwarzschild solution is at the Schwarzschild radius. { 'shvärts,shilt ,rād·ē·əs }

Schwarzschild singularity [RELAT] The coordinate singularity at the event horizon that exists in a certain coordinate system describing a nonrotating black hole in empty space. { ¦shvärts ,shilt ,siŋ·gyə'lar·əd·ē }

Schwarzschild solution [RELAT] The unique solution of general relativity theory describing a nonrotating black hole in empty space. { 'shvärts,shilt sə,lü·shən }

Schwinger critical field [ELEC] That electric field at which an electron is accelerated from rest to a velocity at which its kinetic energy equals its rest energy over a distance of one Compton wavelength. { ¦shviŋ·ər ¦krid·ə·kəl 'fēld }

Schwinger's variational principle [PHYS] A method used in electromagnetic theory, or similar disciplines, to calculate an approximate value or a linear of quadratic functional, such as a scattering amplitude or reflection coefficient, when the function for which the functional is evaluated is the solution of an integral equation. { 'shviŋ·ərz ,ver·ē'ā·shən·əl ,prin·sə·pəl }

scintillation [ELECTROMAG] **1.** A rapid apparent displacement of a target indication from its mean position on a radar display; one cause is shifting of the effective reflection point on the target. Also known as target glint; target scintillation; wander. **2.** Random fluctuation, in radio propagation, of the received field about its mean value, the deviations usually being relatively small. [OPTICS] **1.** Rapid changes of brightness of stars or other distant, celestial objects caused by variations in the density of the air through which the light passes. **2.** Rapid changes in the values of irradiance over the cross section of a laser beam. { ,sint·əl'ā·shən }

scissors mode [NUC PHYS] A mode of nuclear collective motion in permanently deformed nuclei that is of orbital magnetic dipole character and can be explained as a scissorslike vibration of the ellipsoidally deformed bodies of protons and neutrons against each other. { 'siz·ərs ,mōd }

scopometer [OPTICS] An instrument used to measure the absorption or scattering of light in a solution containing solid particles by measuring the contrast between an illuminated line placed behind the solution and a field of constant brightness. { skə'päm·əd·ər }

screen [ELECTROMAG] Metal partition or shield which isolates a device from external magnetic or electric fields. { skrēn }

screening [ATOM PHYS] The reduction of the electric field about a nucleus by the space charge of the surrounding electrons. *See* electric shielding. { 'skrēn·iŋ }

screening constant [ATOM PHYS] The difference between the atomic number of an element and the apparent atomic number for a given process; this difference results from screening. { 'skrēn·iŋ ,kän·stənt }

screening factor [NUC PHYS] The actual rate of a nuclear reaction in a dense plasma divided by the rate that would prevail if there were no free electrons to screen the repulsion between the nuclei. { 'skrēn·iŋ ,fak·tər }

screw axis [CRYSTAL] A symmetry element of some crystal lattices, in which the lattice is unaltered by a rotation about the axis combined with a translation parallel to the axis and equal to a fraction of the unit lattice distance in this direction. { 'skrü ,ak·səs }

screw dislocation [CRYSTAL] A dislocation in which atomic planes form a spiral ramp winding around the line of the dislocation. { 'skrü ,dis·lō,kā·shən }

screw displacement [MECH] A rotation of a rigid body about an axis accompanied by a translation of the body along the same axis. { 'skrü di,splās·mənt }

sea rainbow *See* marine rainbow. { 'sē ¦rān,bō }

search coil *See* exploring coil. { 'sərch ,kȯil }

searchlight [OPTICS] A type of light projector designed to produce a beam of high intensity and minimum divergence; it usually employs a specular paraboloidal reflector to produce parallel rays from a light source located at the focus. { 'sərch,līt }

secohm *See* international henry. { 'sek,ōm }

second [PHYS] The fundamental unit of time equal to 9,192,631,770 periods of the radiation corresponding to the transition between the two hyperfine levels of the ground state of an atom of cesium-133. Abbreviated s; sec. { 'sek·ənd }

secondary battery *See* storage battery. { 'sek·ən,der·ē 'bad·ə·rē }

secondary bow *See* secondary rainbow. { 'sek·ən,der·ē 'bō }

secondary creep [MECH] The change in shape of a substance under a minimum and almost constant differential stress, with the strain-time relationship a constant. Also known as steady-state creep. { 'sek·ən,der·ē 'krēp }

secondary extinction [PHYS] Increased absorption or decreased diffraction of x-rays by a crystal lattice, due to previous reflection of the x-rays from suitably placed crystal planes. { ¦sek·ən,der·ē ik'stiŋk·shən }

secondary flow [FL MECH] A field of fluid motion which can be considered as superposed on a primary field of motion through the action of friction usually in the vicinity of solid boundaries. Also known as frictional secondary flow. { 'sek·ən,der·ē 'flō }

secondary focus [OPTICS] In an astigmatic system, a line at which some of the bundle of rays from an off-axis point meet; this line lies in a plane which contains the point and the optical axis, and has a greater image distance than the primary focus. Also known as sagittal focus. { 'sek·ən,der·ē 'fō·kəs }

secondary lobe *See* minor lobe. { 'sek·ən,der·ē 'lōb }

secondary optic axis [OPTICS] One of two optic axes in a crystal along which all light rays travel with equal velocity. { 'sek·ən,der·ē 'äp·tik 'ak·səs }

secondary radiation [PHYS] Particles or photons produced by the action of primary radiation on matter, such as Compton recoil electrons, delta rays, secondary cosmic rays, and secondary electrons. { 'sek·ən,der·ē ,rād·ē'ā·shən }

secondary rainbow [OPTICS] A faint rainbow of angular radius about 50° which may appear outside the primary rainbow of 42° radius, and which has its colors in reverse order to those of the primary. Also known as secondary bow. { 'sek·ən,der·ē 'rān,bō }

secondary standard [PHYS] **1.** A unit, as of length, capacitance, or weight, used as a standard of comparison in individual countries or localities, but checked against the one primary standard in existence somewhere. **2.** A unit defined as a specified multiple or submultiple of a primary standard, such as the centimeter. { 'sek·ən,der·ē 'stan·dərd }

secondary stress [MECH] A self-limiting normal or shear stress which is caused by the constraint of a structure and which is expected to cause minor distortions that would not result in a failure of the structure. { 'sek·ən,der·ē 'stres }

secondary twinning [CRYSTAL] Twinning of a crystal caused by an external influence, such as pressure in rock. { 'sek·ən,der·ē 'twin·iŋ }

secondary wave [OPTICS] One of the waves that radiate from each point on a wavefront, according to Huygens' principle. { 'sek·ən,der·ē 'wāv }

second law of motion *See* Newton's second law. { 'sek·ənd 'lȯ əv 'mō·shən }

second law of thermodynamics [THERMO] A general statement of the idea that there is a preferred direction for any process; there are many equivalent statements of the law, the best known being those of Clausius and of Kelvin. { 'sek·ənd 'lȯ əv ,thər·mə·dī'nam·iks }

second-moment closure [FL MECH] A model of the Reynolds stresses in turbulent flow which is based upon providing transport equations for the Reynolds stresses themselves, thus abandoning the concept of eddy viscosity. { ¦sek·ənd ,mō·mənt 'klō·zhər }

second-order transition [THERMO] A change of state through which the free energy of a substance and its first derivatives are continuous functions of temperature and pressure, or other corresponding variables. { 'sek·ənd ¦ȯr·dər tran'zish·ən }

second quantization [QUANT MECH] A procedure in which the dependent variables of a classical field or a quantum-mechanical wave function are regarded as operators on which commutation rules are imposed; this produces a formalism in which particles may be created and destroyed. { 'sek·ənd ,kwän·tə'zā·shən }

second radiation constant [STAT MECH] A constant appearing in the Planck radiation formula, equal to the speed of light times Planck's constant divided by Boltzmann's constant, or approximately 1.4388 degree-centimeters. Symbolized c_2; C_2. { 'sek·ənd ,rād·ē'ā·shən ,kän·stənt }

second sound [ACOUS] A transverse sound wave which propagates in smectic liquid crystals, and whose behavior resembles mathematically that of second sound in superfluid helium. [CRYO] A type of wave propagated in the superfluid phase of liquid helium (helium II), in which temperature and entropy variations propagate with no appreciable variation in density or pressure. { 'sek·ənd 'sau̇nd }

sectionalized vertical antenna [ELECTROMAG] Vertical antenna that is insulated at one or more points along its length; the insertion of suitable reactances or applications of a driving voltage across the insulated points results in a modified current distribution giving a more desired radiation pattern in the vertical plane. { 'sek·shən·əl,īzd 'vərd·ə·kəl an,ten·ə }

section modulus [MECH] The ratio of the moment of inertia of the cross section of a beam undergoing flexure to the greatest distance of an element of the beam from the neutral axis. { 'sek·shən 'mäj·ə·ləs }

sector [ELECTROMAG] Coverage of a radar as measured in azimuth. { 'sek·tər }

sectoral horn [ELECTROMAG] Horn with two opposite sides parallel and the two remaining sides which diverge. { 'sek·tə·rəl 'hȯrn }

sector disk [PHYS] A device used to reduce the intensity of a beam of light or other electromagnetic radiation by an accurately known amount; in its simplest form, it consists of a circular, opaque disk with one or more sectors cut out of it, rapidly rotating in the path of the beam. { 'sek·tər ,disk }

seed [SOLID STATE] A small, single crystal of semiconductor material used to start the growth of a large, single crystal for use in cutting semiconductor wafers. { sēd }

seepage [FL MECH] The slow movement of water or other fluid through a porous medium. { 'sēp·ij }

segmented mirror telescope [OPTICS] A telescope consisting of many mirrors, all figured as segments of one parent paraboloidal surface. { 'seg,ment·əd ¦mir·ər 'tel·ə,skōp }

Segrè chart [NUC PHYS] A chart of the nuclides which is laid off in squares, each square displaying data about a nuclide; each column contains nuclides with a given neutron number and each row contains nuclides with a given atomic number; successive columns and rows represent successively higher numbers of neutrons and protons. { sə'grā ˌchärt }

seiche [FL MECH] An oscillation of a fluid body in response to the disturbing force having the same frequency as the natural frequency of the fluid system. { sāsh }

Seidel aberrations [OPTICS] The five types of aberration of monochromatic light deduced from the Seidel theory, namely, spherical aberration, coma, astigmatism, curvature of field, and distortion. { 'zīd·əl ˌab·əˌrā·shənz }

Seidel theory [OPTICS] A theory of aberrations in which the sine of the angle which a ray makes with the optical axis is approximated by the first two terms in the sine's Taylor expansion (the first- and third-order terms), rather than the first term alone, as in a first-order theory. { 'zīd·əl ˌthē·ə·rē }

Seignette-electric *See* ferroelectric. { sen¦yet i'lek·trik }

selection rules [PHYS] Rules summarizing the changes that must take place in the quantum numbers of a quantum-mechanical system for a transition between two states to take place with appreciable probability; transitions that do not agree with the selection rules are called forbidden and have considerably lower probability. { si'lek·shən ˌrülz }

selective absorption [ELECTROMAG] A greater absorption of electromagnetic radiation at some wavelengths (or frequencies) than at others. { si'lek·tiv ab'sȯrp·shən }

selective permeability [PHYS] The property of a membrane or other material that allows some substances to pass through it more easily than others. { si'lek·tiv ˌpər·mē·ə'bil·əd·ē }

selective radiator [PHYS] An object that emits electromagnetic radiation whose spectral energy distribution differs from that of a blackbody with the same temperature. { si'lek·tiv 'rād·ēˌād·ər }

selective reflection [ELECTROMAG] Reflection of electromagnetic radiation more strongly at some wavelengths (or frequencies) than at others. { si'lek·tiv ri'flek·shən }

selective scattering [ELECTROMAG] Scattering of electromagnetic radiation more strongly at some wavelengths than at others. { si'lek·tiv 'skad·ə·riŋ }

s-electron [ATOM PHYS] An atomic electron that is described by a wave function with orbital angular momentum quantum number of zero in the independent particle approximation. { 'es iˌlekˌträn }

self-absorption [SPECT] Reduction of the intensity of the center of an emission line caused by selective absorption by the cooler portions of the source of radiation. Also known as self-reduction; self-reversal. { ¦self əb¦sȯrp·shən }

self-action effect [OPTICS] In a medium with a third-order optical nonlinearity, the modification of the refractive index and absorption coefficient of a light field present in the medium by the strength of the light intensity, so that the light field effectively acts on itself. { ˌself 'ak·shən iˌfekt }

self-charge [QUANT MECH] A contribution to a particle's electric charge arising from the vacuum polarization in the neighborhood of the bare charge. { ¦self ¦chärj }

self-conjugate particle [PART PHYS] An elementary particle which is identical to its antiparticle; it must have zero charge, lepton number, baryon number, and hypercharge. { self 'kän·jə·gət 'pärd·ə·kəl }

self-consistent field method *See* Hartree method. { ¦self kən¦sis·tənt 'fēld ˌmeth·əd }

self-defocusing [OPTICS] The action of a medium whose index of refraction decreases with increasing optical intensity on a laser beam that is more intense in the center than at the edges, whereby the profile of the refractive index corresponds to that of a negative lens, causing the beam to defocus. { ¦self dē¦fō·kə·siŋ }

self-diffusion [SOLID STATE] The spontaneous movement of an atom to a new site in a crystal of its own species. { ¦self di¦fyü·zhən }

self-energy [PHYS] **1.** Classically, the contribution to the energy of a particle that arises from the interaction between different parts of the particle. **2.** In a quantized field theory, the contribution to the energy of a particle due to virtual emission and absorption of other particles, in particular, mesons and photons. { ¦self 'en·ər·jē }

self-excited vibration *See* self-induced vibration. { ¦self ik'sīd·əd vī'brā·shən }

self-fields [ELECTROMAG] The electric and magnetic fields generated by an intense beam of charged particles, which act on the beam itself; they limit the beam intensities which can be achieved in storage rings. { ¦self ¦fēlz }

self-focusing [OPTICS] The action of a medium whose index of refraction increases with increasing optical intensity on a laser beam that is more intense in the center than at the edges, whereby the profile of the refractive index corresponds to that of a positive lens, causing the beam to focus. { ¦self 'fō·kə·siŋ }

self-focusing fiber [OPTICS] A type of optical fiber in which the refractive index decreases continuously along the radius, but progressively more rapidly with distance from the radius, so that light rays which travel longer distances are speeded up, and nearly all light rays travel with the same net axial velocity. { ¦self 'fō·kə·siŋ 'fī·bər }

self-impedance *See* mesh impedance. { ¦self im ¦pēd·əns }

self-induced transparency [OPTICS] A phenomenon in which a pulse of coherent light, with a certain frequency, amplitude, and duration, is transmitted by a normally opaque medium; energy absorbed from the first half of the pulse, whose frequency is at or near the average resonance peak of a band of coherent atomic two-quantum-level optical oscillators, is returned to the last half of the pulse by the medium in the form of coherently emitted light. Abbreviated SIT. { ¦self in¦düst tranz'par·ən·sē }

self-induced vibration [MECH] The vibration of

a mechanical system resulting from conversion, within the system, of nonoscillatory excitation to oscillatory excitation. Also known as self-excited vibration. { ¦self in¦düst vī'brā·shən }

self-inductance [ELECTROMAG] **1.** The property of an electric circuit whereby an electromotive force is produced in the circuit by a change of current in the circuit itself. **2.** Quantitatively, the ratio of the electromotive force produced to the rate of change of current in the circuit. { ¦self in¦dəkt·əns }

self-induction [ELECTROMAG] The production of a voltage in a circuit by a varying current in that same circuit. { ¦self in¦dək·shən }

self-organized criticality [PHYS] Property of a system that persistently operates far from equilibrium, at or near a threshold of instability, having evolved automatically to this critical state independently of external fields. { ˌself ¦ór·gəˌnīzd ˌkrid·ə'kal·əd·ē }

self-phase modulation [OPTICS] The temporal action of a medium whose index of refraction increases with increasing optical intensity on time-varying optical signals or pulses, whereby the rising front edge of the pulse is shifted to lower frequencies and the rear of the pulse is shifted to higher frequencies. { ¦self ˌfāz ˌmäj·ə'lā·shən }

self-reversal *See* self-absorption. { ¦selfri¦vər·səl }

self-running droplet [FL MECH] A droplet that spontaneously coats a surface without the assistance of gravity or any external motion, due to the fact that it contains molecular species which are likely to react with the solid and cause the solvent to cease to wet the solid once it is coated with these molecules. { ˌself ˌrən·iŋ 'dräp·lət }

self-similar flow [FL MECH] A fluid flow whose shape does not change with time, such as a spherical expansion. { ¦self 'sim·ə·lər 'flō }

Sellmeier's equation [ELECTROMAG] An equation for the index of refraction of electromagnetic radiation as a function of wavelength in a medium whose molecules have oscillators of different frequencies. { 'zelˌmī·ərz iˌkwā·zhən }

Semenov number 2 [PHYS] The reciprocal of the Lewis number. { 'se·məˌnóf 'nəm·bər 'tü }

semianechoic room [ACOUS] A room having surfaces which reduce the reflection of sound to less than normal, although not to as great an extent as an anechoic room. { ¦sem·ēˌan·e'kō·ik 'rüm }

semiconducting compound [SOLID STATE] A compound which is a semiconductor, such as copper oxide, mercury indium telluride, zinc sulfide, cadmium selenide, and magnesium iodide. { ¦sem·i·kən¦dək·tiŋ 'kämˌpaůnd }

semiconducting crystal [SOLID STATE] A crystal of a semiconductor, such as silicon, germanium, or gray tin. { ¦sem·i·kən¦dək·tiŋ ¦krist·əl }

semiconductor [SOLID STATE] A solid crystalline material whose electrical conductivity is intermediate between that of a conductor and an insulator, ranging from about 10^5 mhos to 10^{-7} mho per meter, and is usually strongly temperature-dependent. { ¦sem·i·kən¦dək·tər }

semiconductor intrinsic properties [SOLID STATE] Properties of a semiconductor that are characteristic of the ideal crystal. { ¦sem·i·kən¦dək·tər in'trin·sik 'präp·ərd·ēz }

semiconductor laser [OPTICS] A laser in which stimulated emission of coherent light occurs at a *pn* junction when electrons and holes are driven into the junction by carrier injection, electron-beam excitation, impact ionization, optical excitation, or other means. Also known as diode laser; laser diode. { ¦sem·i·kən¦dək·tər 'lā·zər }

semiconvection [FL MECH] A partial convective mixing that causes a region to become convectively stable before complete mixing has been achieved. { ¦sem·i·kən'vek·shən }

semiforbidden line [SPECT] A spectral line associated with a semiforbidden transition. { ¦sem·i·fər'bid·ən 'līn }

semiforbidden transition [ATOM PHYS] An atomic transition whose probability is reduced by selection rules by a factor roughly of the order of 10^6, as compared with 10^9 for a forbidden transition. { ¦sem·i·fər'bid·ən tran'zish·ən }

semipermeable membrane [PHYS] A membrane which allows a solvent to pass through it, but not certain dissolved or colloidal substances. { ¦sem·i'pər·mē·ə·bəl 'memˌbrān }

semistable elementary particle *See* quasi-stable elementary particle. { ¦sem·i'stā·bəl 'el·əˌmen·trē 'pärd·ə·kəl }

semitone [ACOUS] The interval between two sounds whose frequencies have a ratio approximately equal to the twelfth root of 2. Also known as half step. { 'sem·iˌtōn }

Senftleben effect [PHYS] The change in the thermal conductivity of a gas in a magnetic field. { 'senftˌlā·bən iˌfekt }

sensation unit [ACOUS] A unit of loudness, no longer in use; the loudness of a sound is 20 $\log_{10}(p/p_0)$ sensation units above threshold, where p is the pressure level of the sound, and p_0 is the pressure of a sound which can just be detected by the ear. { sen'sā·shən ˌyü·nət }

sensibility [PHYS] The ability of a magnetic compass card to align itself with the magnetic meridian after deflection. { ˌsen·sə'bil·əd·ē }

sensible heat [THERMO] **1.** The heat absorbed or evolved by a substance during a change of temperature that is not accompanied by a change of state. **2.** *See* enthalpy. { 'sen·sə·bəl 'hēt }

sensible-heat factor [THERMO] The ratio of space sensible heat to space total heat; used for air-conditioning calculations. Abbreviated SHF. { 'sen·sə·bəl ¦hēt ˌfak·tər }

sensible-heat flow [THERMO] The heat given up or absorbed by a body upon being cooled or heated, as the result of the body's ability to hold heat; excludes latent heats of fusion and vaporization. { 'sen·sə·bəl ¦hēt 'flō }

separating calorimeter [PHYS] A device for measuring the moisture content of steam. { 'sep·əˌrād·iŋ ˌkal·ə'rim·əd·ər }

separation energy [NUC PHYS] The energy needed to remove a proton, neutron, or alpha particle from a nucleus. { ˌsep·əˈrā·shən ˌen·ər·jē }

separative work unit [NUC PHYS] A fundamental measure of work required to separate a quantity of isotopic mixture into two component parts, one having a higher percentage of concentration of the desired isotope and one having a lower percentage. { ˈsep·rəd·iv ˈwərk ˌyü·nət }

septate coaxial cavity [ELECTROMAG] Coaxial cavity having a vane or septum, added between the inner and outer conductors, so that it acts as a cavity of a rectangular cross section bent transversely. { ˈsepˌtāt kōˈak·sē·əl ˈkav·əd·ē }

septate waveguide [ELECTROMAG] Waveguide with one or more septa placed across it to control microwave power transmission. { ˈsepˌtāt ˈwāvˌgīd }

septum [ELECTROMAG] A metal plate placed across a waveguide and attached to the walls by highly conducting joints; the plate usually has one or more windows, or irises, designed to give inductive, capacitive, or resistive characteristics. { ˈsep·təm }

Serber potential [NUC PHYS] A potential between nucleons, equal to $^{1}/_{2}(1 + M)V(r)$, where $V(r)$ is a function of the distance between the nucleons, and M is an operator which exchanges the spatial coordinates of the particles but not their spins (corresponding to the Majorana force). { ˈsər·bər pəˌten·chəl }

series [ELEC] An arrangement of circuit components end to end to form a single path for current. [SPECT] A collection of spectral lines of an atom or ion for a set of transitions, with the same selection rules, to a single final state; often the frequencies have the general formula $[R/(a + c_1)^2] - [R/(n + c_2)^2]$, where R is the Rydberg constant for the atom, a and c_1 and c_2 are constants, and n takes on the values of the integers greater than a for the various lines in the series. { ˈsir·ēz }

series circuit [ELEC] A circuit in which all parts are connected end to end to provide a single path for current. { ˈsir·ēz ˌsər·kət }

series connection [ELEC] A connection that forms a series circuit. { ˈsir·ēz kəˌnek·shən }

series decay *See* radioactive series. { ˈsir·ēz diˈkā }

series disintegration [NUC PHYS] The successive radioactive transformations in a radioactive series. Also known as chain decay; chain disintegration. { ˈsir·ēz diˌsin·təˈgrā·shən }

series-fed vertical antenna [ELECTROMAG] Vertical antenna which is insulated from the ground and energized at the base. { ˈsir·ēz ¦fed ˈvərd·i·kəl anˈten·ə }

series-parallel circuit [ELEC] A circuit in which some of the components or elements are connected in parallel, and one or more of these parallel combinations are in series with other components of the circuit. { ˈsir·ēz ¦par·əˌlel ¦sər·kət }

series resonance [ELEC] Resonance in a series resonant circuit, wherein the inductive and capacitive reactances are equal at the frequency of the applied voltage; the reactances then cancel each other, reducing the impedance of the circuit to a minimum, purely resistive value. { ˈsir·ēz ˈrez·ən·əns }

series resonant circuit [ELEC] A resonant circuit in which the capacitor and coil are in series with the applied alternating-current voltage. { ˈsir·ēz ¦rez·ən·ənt ˌsər·kət }

series T junction *See* E-plane T junction. { ˈsir·ēz ˈtē ˌjəŋk·shən }

series-tuned circuit [ELEC] A simple resonant circuit consisting of an inductance and a capacitance connected in series. { ˈsir·ēz ¦tünd ˌsər·kət }

sessile bubble method [FL MECH] A method of measuring the surface tension of a liquid that involves measuring the dimensions of a bubble in the liquid resting under a plane or concave-downward surface. { ˈses·əl ¦bəb·əl ˌmeth·əd }

sessile drop method [FL MECH] A method of measuring surface tension in which the depth and mass of a drop resting on a surface that it does not wet are measured; from this, the shape of the drop and, in turn, the surface tension are determined. { ˈses·əl ¦dräp ˌmeth·əd }

set *See* permanent set. { set }

setback [MECH] The relative rearward movement of component parts in a projectile, missile, or fuse undergoing forward acceleration during its launching; these movements, and the setback force which causes them, are used to promote events which participate in the arming and eventual functioning of the fuse. { ˈsetˌbak }

setback force [MECH] The rearward force of inertia which is created by the forward acceleration of a projectile or missile during its launching phase; the forces are directly proportional to the acceleration and mass of the parts being accelerated. { ˈsetˌbak ˌfȯrs }

set forward [MECH] Relative forward movement of component parts which occurs in a projectile, missile, or bomb in flight when impact occurs; the effect is due to inertia and is opposite in direction to setback. { ˈset ˈfȯr·wərd }

set forward force [MECH] The forward force of inertia which is created by the deceleration of a projectile, missile, or bomb when impact occurs; the forces are directly proportional to the deceleration and mass of the parts being decelerated. Also known as impact force. { ˈset ˈfȯr·wərd ˌfȯrs }

set forward point [MECH] A point on the expected course of the target at which it is predicted the target will arrive at the end of the time of flight. { ˈset ˈfȯr·wərd ˌpȯint }

settling velocity [FL MECH] The rate at which suspended solids subside and are deposited. Also known as fall velocity. [MECH] The velocity reached by a particle as it falls through a fluid, dependent on its size and shape, and the difference between its specific gravity and that of the settling medium; used to sort particles by grain size. { ˈset·liŋ vəˌläs·əd·ē }

sextupole magnet [ELECTROMAG] A configuration of six magnets arranged in a circular pattern with alternating polarities, used in the control of electron beams. { 'seks·tə,pōl ,mag·nət }

shade [OPTICS] The color of a mixture of pigments or dyes which has some black pigment or dye in it. { shād }

shade glass [OPTICS] A darkened transparency that can be moved into the line of sight of an optical instrument, such as a sextant, to reduce the intensity of light reaching the eye. { 'shād ,glas }

shadow [OPTICS] A region of darkness caused by the presence of an opaque object interposed between such a region and a source of light. [PHYS] A region which some type of radiation, such as sound or x-rays, does not reach because of the presence of an object, which the radiation cannot penetrate, interposed between the region and the source of radiation. { 'shad·ō }

shadow attenuation [ELECTROMAG] Attenuation of radio waves over a sphere in excess of that over a plane when the distance over the surface and other factors are the same. { 'shad·ō ə,ten·yə'wā·shən }

shadow factor [ELECTROMAG] The ratio of the electric-field strength that would result from propagation of waves over a sphere to that which would result from propagation over a plane under comparable conditions. [OPTICS] A multiplication factor derived from the sun's declination, the latitude of the target and the time of photography, used in determining the heights of objects from shadow length. Also known as tan alt. { 'shad·ō ,fak·tər }

shadowgram [PHYS] A plot or display of a shadow. { 'shad·ō,gram }

shadowgraph [OPTICS] A simple method of making visible the disturbances that occur in fluid flow at high velocity, in which light passing through a flowing fluid is refracted by density gradients in the fluid, resulting in bright and dark areas on a screen placed behind the fluid. { 'shad·ō,graf }

shadow microscopy *See* projection microradiography. { 'shad·ō mī'kräs·kə·pē }

shadow region [ELECTROMAG] Region in which, under normal propagation conditions, the field strength from a given transmitter is reduced by some obstruction which renders effective radio reception of signals or radar detection of objects in this region improbable. { 'shad·ō ,rē·jən }

shadow scattering [QUANT MECH] Scattering that results from the interference of the incident wave and scattered waves. { 'shad·ō 'skad·ər·iŋ }

shadow zone [ACOUS] A region, usually under water or in the atmosphere, which sound waves will not reach, according to ray acoustics. { 'shad·ō ,zōn }

shaped-beam antenna [ELECTROMAG] Antenna with a directional pattern which, over a certain angular range, is of special shape for some particular use. { 'shāpt ¦bēm an 'ten·ə }

shape factor *See* form factor. [FL MECH] The quotient of the area of a sphere equivalent to the volume of a solid particle divided by the actual surface of the particle; used in calculations of gas flow through beds of granular solids. [OPTICS] For a lens, the quantity $(R_2 + R_1)/(R_2 - R_1)$, where R_1 and R_2 are the radii of the first and second surface of the lens. Also known as Coddington shape factor. { 'shāp ,fak·tər }

shape isomer [NUC PHYS] An excited nuclear state which has an unusually long lifetime because of its deformed shape, which differs drastically from that of the lower energy states into which it is permitted to decay. { 'shāp 'ī·sə·mər }

shape resonance [QUANT MECH] A broad resonance or peak in the cross section of a scattering process that reflects the shape of the potential between projectile and target; this shape consists typically of a barrier separating an inner deep well from a shallow, asymptotically vanishing potential at large separation distances. { 'Shāp ,rez·ən·əns }

sharpness function technique [OPTICS] An adaptive optics technique in which a deformable mirror or other corrective element is dithered to maximize a given function, such as the integral of the square of the irradiance in the image plane or the amount of energy within a certain region. { 'shärp·nəs ,fəŋk·shən tek,nēk }

sharpness of resonance [ELEC] The narrowness of the frequency band around the resonance at which the response of an electric circuit exceeds an arbitrary fraction of its maximum response, often 70.7%. { 'shärp·nəs əv 'rez·ən·əns }

sharp series [SPECT] A series occurring in the line spectra of many atoms and ions with one, two, or three electrons in the outer shell, in which the total orbital angular momentum quantum number changes from 0 to 1. { 'shärp 'sir·ēz }

sharp tuning [ELEC] Having high selectivity; responding only to a desired narrow range of frequencies. { 'shärp 'tün·iŋ }

shattering [FL MECH] One theory to explain homogenization or globule fractionation in milk; it is the effect that occurs when whole milk under high velocity strikes a flat surface, such as an impact ring. [MECH] The breaking up into highly irregular, angular blocks of a very hard material that has been subjected to severe stresses. { 'shad·ə·riŋ }

shear *See* shear strain. { shir }

shear center *See* center of twist. { 'shir ,sen·tər }

shear diagram [MECH] A diagram in which the shear at every point along a beam is plotted as an ordinate. { 'shir ,dī·ə,gram }

shear drag *See* shear resistance. { 'shir ,drag }

shear fracture [MECH] A fracture resulting from shear stress. { 'shir ,frak·chər }

shearing field [PL PHYS] A special type of magnetic field, used to confine a plasma whose rotational transform angle changes with distance from the magnetic axis. { 'shir·iŋ ,fēld }

shearing forces [MECH] Two forces that are equal in magnitude, opposite in direction, and act along two distinct parallel lines. { 'shēr·iŋ ˌfȯrs·əz }

shearing instability *See* Helmholtz instability. { 'shir·iŋ ˌin·stəˌbil·əd·ē }

shearing interferometer [OPTICS] An interferometer in which a wavefront is interfered with a version of itself that has been modified in some manner; includes lateral shear interferometers and radial shear interferometers. { 'shir·iŋ ˌin·tər·fə'räm·əd·ər }

shearing strain [MECH] The distortion that results from motion of material on opposite sides of a plane in opposite directions parallel to the plane. { 'shir·iŋ ˌstrān }

shearing stress [MECH] A stress in which the material on one side of a surface pushes on the material on the other side of the surface with a force which is parallel to the surface. Also known as shear stress; tangential stress. { 'shir·iŋ ˌstres }

shear modulus *See* modulus of elasticity in shear. { 'shir ˌmäj·ə·ləs }

shear plane [MECH] A confined zone along which fracture occurs in metal cutting. { 'shir ˌplān }

shear rate [FL MECH] The relative velocities in laminar flow of parallel adjacent layers of a fluid body under shear force. { 'shir ˌrāt }

shear resistance [FL MECH] A tangential stress caused by fluid viscosity and taking place along a boundary of a flow in the tangential direction of local motion. Also known as shear drag. { 'shir riˌzis·təns }

shear strain [MECH] Also known as shear. **1.** A deformation of a solid body in which a plane in the body is displaced parallel to itself relative to parallel planes in the body; quantitatively, it is the displacement of any plane relative to a second plane, divided by the perpendicular distance between planes. **2.** The force causing such deformation. { 'shir ˌstrān }

shear strength [MECH] **1.** The maximum shear stress which a material can withstand without rupture. **2.** The ability of a material to withstand shear stress. { 'shir ˌstreŋkth }

shear stress *See* shearing stress. { 'shir ˌstres }

shear thickening [FL MECH] Viscosity increase of non-Newtonian fluids (for example, complex polymers, proteins, protoplasm) that undergo viscosity increases under conditions of shear stress (that is, viscometric flow). { 'shir ˌthik·ən·iŋ }

shear thinning [FL MECH] Viscosity reduction of non-Newtonian fluids (for example, polymers and their solutions, most slurries and suspensions, lube oils with viscosity-index improvers) that undergo viscosity reductions under conditions of shear stress (that is, viscometric flow). { 'shir ˌthin·iŋ }

shear-viscosity function [FL MECH] The expression of the viscometric flow of a purely viscous, non-Newtonian fluid in terms of velocity gradient and shear stress of the flowing fluid. { 'shir vi'skäs·əd·ē ˌfəŋk·shən }

shear wave [MECH] A wave that causes an element of an elastic medium to change its shape without changing its volume. Also known as rotational wave. { 'shir ˌwāv }

sheath [ELEC] A protective outside covering on a cable. [ELECTROMAG] The metal wall of a waveguide. { shēth }

sheath-reshaping converter [ELECTROMAG] In a waveguide, a mode converter in which the change of wave pattern is achieved by gradual reshaping of the sheath of the waveguide and of conducting metal sheets mounted longitudinally in the guide. { 'shēth rē¦shāp·iŋ kən'vərd·ər }

shed [NUC PHYS] A unit of cross section, used in studying collisions of nuclei and particles, equal to 10^{-24} barn, or 10^{-48} square centimeter. { shed }

SHEED *See* scanning HEED. { shēd *or* 'esˌhēd }

sheen [OPTICS] A subdued and often iridescent or metallic glitter which approaches, but is just short of, optical reflection and which modifies the surface luster of a mineral. { shēn }

sheet cavitation [FL MECH] A type of cavitation in which cavities form on a solid boundary and remain attached as long as the conditions that led to their formation remain unaltered. Also known as steady-state cavitation. { 'shēt ˌkav·ə'tā·shən }

sheet grating [ELECTROMAG] Three-dimensional grating consisting of thin, longitudinal, metal sheets extending along the inside of a waveguide for a distance of about a wavelength, and used to stop all waves except one predetermined wave that passes unimpeded. { 'shēt ˌgrād·iŋ }

sheet polarizer [OPTICS] A mechanism for obtaining linear polarized light; there are several types, one of which is a microcrystalline polarizer in which small crystals of a dichroic material (quinine iodosulfate), oriented parallel to each other in a plastic medium, absorb one polarization and transmit the other. { 'shēt 'pō·lə ˌrīz·ər }

shell [PHYS] A set of energy levels with approximately the same energy in an atom or nucleus. { shel }

shell-form transformer [ELECTROMAG] A transformer in which all of the widings are on the center of three legs. { 'shel ˌfȯrm tranzˌfȯr·mər }

shell model [NUC PHYS] A model of the nucleus in which the shell structure is either postulated or is a consequence of other postulates; especially the model in which the nucleons act as independent particles filling a preassigned set of energy levels as permitted by the quantum numbers and Pauli principle. { 'shel ˌmäd·əl }

shell structure [NUC PHYS] Structure of the nucleus in which nucleons of each kind occupy quantum states which are in groups of approximately the same energy, called shells, the number of nucleons in each shell being limited by the Pauli exclusion principle. { 'shel ˌstrək·chər }

Sherwood number *See* Nusselt number. { 'shərˌwu̇d ˌnəm·bər }

SHF *See* sensible-heat factor.

shielded line [ELECTROMAG] Transmission line, the elements of which confine the propagated waves to an essentially finite space; the external conducting surface is called the sheath. { 'shēl·dəd 'līn }

shielding *See* electric shielding. { 'shēld·iŋ }

shielding distance *See* Debye shielding length. { 'shēld·iŋ ˌdis·təns }

shielding ratio [ELECTROMAG] The ratio of a field in a specified region when electrical shielding is in place to the field in that region when the shielding is removed. { 'shēld·iŋ ˌrā·shō }

shift [SPECT] A small change in the position of a spectral line that is due to a corresponding change in frequency which, in turn, results from one or more of several causes, such as the Doppler effect. { shift }

shimmy [MECH] Excessive vibration of the front wheels of a wheeled vehicle causing a jerking motion of the steering wheel. { 'shim·ē }

SHM *See* harmonic motion.

shock [MECH] A pulse or transient motion or force lasting thousandths to tenths of a second which is capable of exciting mechanical resonances; for example, a blast produced by explosives. { 'shäk }

shock cells [FL MECH] Diamond-shaped regions of alternating high and low pressure in a jet flow, through which the jet exit pressure adjusts to the ambient pressure. { 'shäk ˌselz }

shock diamonds [PHYS] The shock waves that appear in the exhaust stream of a rocket; they are made visible by their luminosity and describe an approximate diamond configuration in side view. { 'shäk ˌdī·mənz }

shock front [PHYS] The outer side of a shock wave whose pressure rises from zero up to its peak value. Also known as pressure front. { 'shäk ˌfrənt }

shock heating [PHYS] The nonisentropic heating of a fluid which takes place when a shock wave passes through it. { 'shäk ˌhēd·iŋ }

Shockley partial dislocation [SOLID STATE] A partial dislocation in which the Burger's vector lies in the fault plane, so that it is able to glide, in contrast to a Frank partial dislocation. Also known as glissile dislocation. { 'shäk·lē 'pär·shəl ˌdis·lō'kā·shən }

shock tube [FL MECH] A long tube divided into two parts by a diaphragm; the volume on one side of the diaphragm constitutes the compression chamber, the other side is the expansion chamber; a high pressure is developed by suitable means in the compression chamber, and the diaphragm ruptured; the shock wave produced in the expansion chamber can be used for the calibration of air blast gages, or the chamber can be instrumented for the study of the characteristics of the shock wave. { 'shäk ˌtüb }

shock wave [PHYS] A fully developed compression wave of large amplitude, across which density, pressure, and particle velocity change drastically. { 'shäk ˌwāv }

shock-wave lip [PHYS] The shock wave obtained from the lip of a free jet nozzle, because of failure to match the stream pressure and the ambient exhaust pressure. { 'shäk ¦wāv ˌlip }

short *See* short circuit. { shȯrt }

short antenna [ELECTROMAG] An antenna shorter than about one-tenth of a wavelength, so that the current may be assumed to have constant magnitude along its length, and the antenna may be treated as an elementary dipole. { 'shȯrt anˌten·ə }

short circuit [ELEC] A low-resistance connection across a voltage source or between both sides of a circuit or line, usually accidental and usually resulting in excessive current flow that may cause damage. Also known as short. { 'shȯrt 'sər·kət }

short-pulse laser [OPTICS] A laser designed to generate a pulse of light lasting on the order of nanoseconds or less, and having very high power, such as by Q switching or mode-locking. { 'shȯrt ¦pəls 'lā·zər }

short-range force [PHYS] A force between two particles which is negligible when the distance between the particles is greater than a certain amount; in particular, nuclear forces whose range is several times 10^{-15} meter. { 'shȯrt ¦rānj 'fȯrs }

short-range order [PHYS] A regularity in the arrangement of atoms in a disordered solid or a liquid in which the probability of a given type of atom having neighbors of a given type is greater than one would expect on a purely random basis. { 'shȯrt ¦rānj 'ȯr·dər }

short-slot coupler *See* three-decibel coupler. { 'shȯrt ¦slät 'kəp·lər }

short ton *See* ton. { 'shȯrt 'tən }

shortwave radiation [ELECTROMAG] A term used loosely to distinguish radiation in the visible and near-visible portions of the electromagnetic spectrum (roughly 0.4 to 1.0 micrometer in wavelength) from long-wave radiation (infrared radiation). { 'shȯrt'wāv ˌrād·ē'ā·shən }

shower *See* cosmic-ray shower. { 'shau̇·ər }

Shpol'skii effect [SPECT] The occurrence of very narrow fluorescent lines in the spectra of certain compounds from molecules frozen at low temperatures. { 'shpōl·skē iˌfekt }

Shubnikov-de Haas effect [SOLID STATE] Oscillations of the resistance or Hall coefficient of a metal or semiconductor as a function of a strong magnetic field, due to the quantization of the electron's energy. { 'shüb·nəˌkȯf də'häs iˌfekt }

Shubnikov groups [SOLID STATE] The point groups and space groups of crystals having magnetic moments. Also known as black-and-white groups; magnetic groups. { 'shüb·nəˌkȯf ˌgrüps }

shunt [ELEC] **1.** A precision low-value resistor placed across the terminals of an ammeter to increase its range by allowing a known fraction of the circuit current to go around the meter. Also known as electric shunt. **2.** To place one part in parallel with another. **3.** *See* parallel. [ELECTROMAG] A piece of iron that provides a parallel path for magnetic flux around an air gap in a magnetic circuit. { shənt }

shunt-excited [ELECTROMAG] Having field windings connected across the armature terminals, as in a direct-current generator. { 'shənt ik¦sīd·əd }

shunt-excited antenna [ELECTROMAG] A tower antenna, not insulated from the ground at the base, whose feeder is connected at a point about one-fifth of the way up the antenna and usually slopes up to this point from a point some distance from the antenna's base. { 'shənt ik¦sīd·əd an'ten·ə }

shunt-fed vertical antenna [ELECTROMAG] Vertical antenna connected to the ground at the base and energized at a point suitably positioned above the grounding point. { 'shənt ¦fed ¦vərd·ə·kəl an'ten·ə }

shunting [ELEC] The act of connecting one device to the terminals of another so that the current is divided between the two devices in proportion to their respective admittances. { 'shənt·iŋ }

shunt T junction *See* H-plane T junction. { 'shənt 'tē ˌjəŋk·shən }

shutter [OPTICS] A mechanical device that cuts off a beam of light by opening and closing at different rates of speed to expose film or plates; used in cameras and motion picture projectors. { 'shəd·ər }

SI *See* International System of Units.

Siacci method [MECH] An accurate and useful method for calculation of trajectories of high-velocity missiles with low quadrant angles of departure; basic assumptions are that the atmospheric density anywhere on the trajectory is approximately constant, and the angle of departure is less than about 15°. { sē'ä·chē ˌmeth·əd }

SIC *See* dielectric constant.

sideband [ELECTROMAG] **1.** The frequency band located either above or below the carrier frequency, within which fall the frequency components of the wave produced by the process of modulation. **2.** The wave components lying within such bands. { 'sīdˌband }

side-centered lattice [CRYSTAL] A type of centered lattice that is centered on the side faces only. { 'sīd ¦sen·tərd 'lad·əs }

side direction [MECH] In stress analysis, the direction perpendicular to the plane of symmetry of an object. { 'sīd diˌrek·shən }

side echo [ELECTROMAG] Echo due to a side lobe of an antenna. { 'sīd ˌek·ō }

side lobe *See* minor lobe. { 'sīd ˌlōb }

side pinacoid [CRYSTAL] A pinacoid with Miller indices (010) in an orthorhombic, monoclinic, or triclinic crystal. { 'sīd 'pin·əˌkȯid }

siderostat [OPTICS] A more precise model of a heliostat; the siderostat uses a modified mirror mounting so that the image of a star is kept steady while the rest of the field is in rotation about the center. { 'sid·ə·rəˌstat }

siegbahn [SPECT] A unit of length, formerly used to express wavelengths of x-rays, equal to 1/3029.45 of the spacing of the (200) planes of calcite at 18°C, or to $(1.00202 \pm 0.00003) \times 10^{-13}$ meter. Also known as x-ray unit; X-unit. Symbolized X; XU. { 'sēgˌbän }

siemens [ELEC] A unit of conductance, admittance, and susceptance, equal to the conductance between two points of a conductor such that a potential difference of 1 volt between these points produces a current of 1 ampere; the conductance of a conductor in siemens is the reciprocal of its resistance in ohms. Formerly known as mho (℧); reciprocal ohm. Symbolized S. { 'sē·mənz }

Siemens' electrodynamometer [ELECTROMAG] An early type of electromagnetic instrument in which current flows through all the coils in series. { 'sē·mənz i¦lek·trōˌdī·nə'mäm·əd·ər }

sight distance [OPTICS] The distance from which an object at eye level remains visible to an observer. { 'sīt ˌdis·təns }

sighting [OPTICS] **1.** The act or procedure of aiming with the aid of a sight. **2.** The action of bringing something into view; the action of seeing something. { 'sīd·iŋ }

sight unit [OPTICS] A compact sighting device composed of an elbow or panoramic telescope, mount, or adapter, usually used for pointing a weapon for direct or indirect fire; it may be attached to, or used in conjunction with, a weapon rocket launcher, or the like. { 'sīt ˌyü·nət }

sigma function [THERMO] A property of a mixture of air and water vapor, equal to the difference between the enthalpy and the product of the specific humidity and the enthalpy of water (liquid) at the thermodynamic wet-bulb temperature; it is constant for constant barometric pressure and thermodynamic wet-bulb temperature. { 'sig·mə ˌfəŋk·shən }

sigma hyperon [PART PHYS] **1.** The collective name for three semistable baryons with charges of +1, 0, and −1 times the proton charge, designated Σ^+, Σ^0, and Σ^-, all having masses of approximately 1193 megaelectronvolts, spin of 1/2, and positive parity; they form an isotopic spin multiplet with a total isotopic spin of 1 and a hypercharge of 0. **2.** Any baryon belonging to an isotopic spin multiplet having a total isotopic spin of 1 and a hypercharge of 0; designated $\Sigma_J{}^P(m)$, where m is the mass of the baryon in megaelectronvolts, and J and P are its spin and parity; the $\Sigma_{3/2}{}^+(1385)$ is sometimes designated Σ^*. { 'sig·mə 'hī·pəˌrän }

sigma-minus hyperonic atom [ATOM PHYS] An atom consisting of a negatively charged sigma hyperon orbiting around an ordinary nucleus. Designated Σ^- hyperonic atom. { 'sig·mə 'mī·nəs ¦hī·pə¦rän·ik 'ad·əm }

sigmoid distortion [OPTICS] A distortion present in line-scan imagery, causing straight lines cut obliquely to appear as sigmoid curves. { 'sigˌmȯid di'stȯr·shən }

signature [QUANT MECH] A quantum number α that characterizes a system with the symmetry of a prolate or oblate spheroid and satisfies the equation $r = \exp(-i\pi\alpha)$, where r is the eigenvalue of the system under a rotation through 180° about an axis perpendicular to the symmetry axis. { 'sig·nə·chər }

sign convention [OPTICS] A convention as to which quantities, such as angles, distances, and radii of curvature, are positive and which are negative in computations involving a lens or a mirror. { 'sīn kən,ven·chən }

SIL *See* speech interference level.

silicon burning [NUC PHYS] The synthesis, in stars, of elements, chiefly in the iron group, resulting from the photodisintegration of silicon-28 and other intermediate-mass nuclei; copious supplies of protons, alpha particles, and neutrons are produced, followed by the capture of these particles by other intermediate-mass nuclei. { 'sil·ə·kən 'bərn·iŋ }

Silsbee effect [CRYO] The ability of an electric current to destroy superconductivity by means of the magnetic field that it generates, without raising the cryogenic temperature. { 'silz·bē i,fekt }

similarity principle *See* principle of dynamical similarity. { ,sim·ə'lar·əd·ē ,prin·sə·pəl }

similitude [PHYS] The use in scientific studies and engineering designs of the corresponding behavior between large and small objects or systems which are of similar nature and, more precisely, have geometrical, kinematic, and dynamical similarity. { si'mil·ə,tüd }

Simon liquefier [CRYO] A device for liquefying helium in which helium is first cooled at high pressure by liquid or solid hydrogen and is then liquefied by a single adiabatic expansion. { ¦sī·mən 'lik·wə,fī·ər }

simple cubic lattice [CRYSTAL] A crystal lattice whose unit cell is a cube, and whose lattice points are located at the vertices of the cube. { 'sim·pəl 'kyü·bik 'lad·əs }

simple harmonic current [ELEC] Alternating current, the instantaneous value of which is equal to the product of a constant, and the cosine of an angle varying linearly with time. Also known as sinusoidal current. { 'sim·pəl här'män·ik 'kə·rənt }

simple harmonic electromotive force [ELEC] An alternating electromotive force which is equal to the product of a constant and the cosine or sine of an angle which varies linearly with time. { 'sim·pəl här'män·ik i¦lek·trə¦mōd·iv 'fȯrs }

simple harmonic motion *See* harmonic motion. { 'sim·pəl här'män·ik 'mō·shən }

simple lattice *See* primitive lattice. { 'sim·pəl 'lad·əs }

simple lens [OPTICS] A lens consisting of a single element. Also known as single lens. { 'sim·pəl 'lenz }

simple metal [SOLID STATE] A metal in which the electrons are basically free to move throughout the volume. { 'sim·pəl 'med·əl }

simple microscope [OPTICS] A diverging lens system, which can form an enlarged image of a small object. Also known as hand lens; magnifier; magnifying glass. { 'sim·pəl 'mī·krə,skōp }

simple pendulum [MECH] A device consisting of a small, massive body suspended by an inextensible object of negligible mass from a fixed horizontal axis about which the body and suspension are free to rotate. { 'sim·pəl 'pen·jə·ləm }

simple sound source [ACOUS] Under free-field conditions, a source that emits sound uniformly in every direction. { 'sim·pəl 'sau̇nd ,sȯrs }

simple tone [ACOUS] Also known as pure tone. **1.** A sound wave whose instantaneous sound pressure is a simple sinusoidal function of time. **2.** A sound sensation characterized by singleness of pitch. { 'sim·pəl 'tōn }

simple twin [CRYSTAL] A twinned crystal composed of only two individuals in twin relation. { 'sim·pəl 'twin }

simplex [QUANT MECH] The eigenvalue of a nucleus or other object with an octupole (pear) shape under an operation consisting of rotation through 180° about an axis perpendicular to the symmetry axis, followed by inversion. { 'sim,pleks }

simultaneity [MECH] Two events have simultaneity, relative to an observer, if they take place at the same time according to a clock which is fixed relative to the observer. { ,sī·məl·tə'nē·əd·ē }

sine wave [PHYS] A wave whose amplitude varies as the sine of a linear function of time. Also known as sinusoidal wave. { 'sīn ,wāv }

single-atom laser [ATOM PHYS] A device in which atoms emit visible-wavelength photons at an increased rate as they pass through a resonant cavity one by one, consistent with a theory of quantized Rabi oscillations. Also known as microlaser. { ,siŋ·gəl ,ad·əm 'lā·zər }

single-atom maser [ATOM PHYS] A device in which atoms emit microwave photons at an increased rate as they pass through a resonant cavity one by one, consistent with a theory of quantized Rabi oscillations. Also known as micromaser. { ,siŋ·gəl ,ad·əm 'mā·zər }

single-carrier theory [SOLID STATE] A theory of the behavior of a rectifying barrier which assumes that conduction is due to the motion of carriers of only one type; it can be applied to the contact between a metal and a semiconductor. { 'siŋ·gəl ¦kar·ē·ər 'thē·ə·rē }

single crystal [CRYSTAL] A crystal, usually grown artificially, in which all parts have the same crystallographic orientation. { 'siŋ·gəl 'krist·əl }

single-degree-of-freedom gyro [MECH] A gyro the spin reference axis of which is free to rotate about only one of the orthogonal axes, such as the input or output axis. { 'siŋ·gəl di¦grē əv ¦frē·dəm 'jī·rō }

single-ended [ELEC] Unbalanced, as when one side of a transmission line or circuit is grounded. { 'siŋ·gəl 'end·əd }

single-ended Q machine [PL PHYS] A Q machine in which the plasma column is generated at a hot tungsten plate at one end and terminated by a cold metal plate at the other. { ¦siŋ·gəl ¦en·dəd 'kyü mə,shēn }

single knock-on [SOLID STATE] A sputtering event in which target atoms are ejected either directly by the bombarding projectiles or after a small number of collisions. { 'siŋ·gəl 'näk·ȯn }

single-layer solenoid [ELECTROMAG] A solenoid which has only one layer of wire, wound in a cylindrical helix. { 'siŋ·gəl ¦lā·ər 'sō·lə,nȯid }

single lens *See* simple lens. { 'siŋ·gəl 'lenz }

single-phase [ELEC] Energized by a single alternating voltage. { 'siŋ·gəl 'fāz }

single-phase circuit [ELEC] Either an alternating-current circuit which has only two points of entry, or one which, having more than two points of entry, is intended to be so energized that the potential differences between all pairs of points of entry are either in phase or differ in phase by 180°. { 'siŋ·gəl ¦fāz 'sər·kət }

single refraction [OPTICS] Any refraction that occurs in an isotropic crystal. { 'siŋ·gəl ri 'frak·shən }

single scattering [PHYS] A change in direction of a particle or photon because of a single collision. { 'siŋ·gəl 'skad·ər·iŋ }

single-shot camera [OPTICS] An underwater camera that takes one picture on each lowering when the camera shutter is tripped by contact with the bottom. { 'siŋ·gəl ¦shät 'kam·rə }

single-stub transformer [ELECTROMAG] Shorted section of a coaxial line that is connected to a main coaxial line near a discontinuity to provide impedance matching at the discontinuity. { 'siŋ·gəl ¦stəb tranz'fȯr·mər }

single-stub tuner [ELECTROMAG] Section of transmission line terminated by a movable short-circuiting plunger or bar, attached to a main transmission line for impedance-matching purposes. { 'siŋ·gəl ¦stəb 'tün·ər }

singlet [QUANT MECH] An energy level that is not split by a relatively weak interaction, and thus is not a multiplet or a member of a multiplet. [SPECT] A spectral line that cannot be resolved into components at even the highest resolution. { 'siŋ·glət }

single-tuned circuit [ELEC] A circuit whose behavior is the same as that of a circuit with a single inductance and a single capacitance, together with associated resistances. { 'siŋ·gəl ¦tünd 'sər·kət }

singularity [RELAT] A region of space-time where one or more components of the Riemann curvature tensor becomes infinite. { ,siŋ·gyə 'lar·əd·ē }

singularity theorems [RELAT] Theorems proving that singularities must develop in certain space-times, such as the universe, given only broad conditions, such as causality, and the existence of a trapped surface. { ,siŋ·gyə'lar·əd·ē ,thir·əmz }

sink [ELECTROMAG] The region of a Rieke diagram where the rate of change of frequency with respect to phase of the reflection coefficient is maximum for an oscillator; operation in this region may lead to unsatisfactory performance by reason of cessation or instability of oscillations. [PHYS] A device or system where some extensive entity is absorbed, such as a heat sink, a sink flow, a load in an electrical circuit, or a region in a nuclear reactor where neutrons are strongly absorbed. { siŋk }

sink flow [FL MECH] **1.** In three-dimensional flow, a point into which fluid flows uniformly from all directions. **2.** In two-dimensional flow, a straight line into which fluid flows uniformly from all directions at right angles to the line. { 'siŋk ,flō }

sinking [OPTICS] In atmospheric optics, a refraction phenomenon, the opposite of looming, in which an object on, or slightly above, the geographic horizon apparently sinks below it. { 'siŋk·iŋ }

sinusoidal current *See* simple harmonic current. { ,sī·nə'sȯid·əl 'kə·rənt }

sinusoidal wave *See* sine wave. { ,sī·nə'sȯid·əl 'wāv }

SIRS *See* satellite infrared spectrometer. { sərz }

SIT *See* self-induced transparency.

six-j-symbol [QUANT MECH] A coefficient that appears in the transformation between various modes of coupling eigenfunctions of three angular momenta; it is equal to the Racah coefficient, except perhaps in sign, and has greater symmetry than the Racah coefficient. { 'siks ¦jā 'sim·bəl }

sixth-power law [FL MECH] A law stating that the size of particles that can be carried by a stream is proportional to the sixth power of its velocity. { 'siksth ¦pau̇·ər 'lȯ }

sixty degrees Fahrenheit British thermal unit *See* British thermal unit. { 'siks·tē di¦grēz 'far·ən,hīt 'brid·ish 'thər·məl ,yü·nət }

six-vector [RELAT] An antisymmetrical, second-rank tensor in Minkowski space; that is, a tensor whose components, $T_{\mu\nu}$, with $\mu,\nu = 1,2,3,4$, satisfy $T_{\mu\nu} = -T_{\nu\mu}$; it has six independent components. { 'siks ,vek·tər }

Sk *See* Stefan number.

skeleton crystal [CRYSTAL] A crystal formed in microscopic outline with incomplete filling in of the faces. { 'skel·ət·ən ,krist·əl }

skiascope [OPTICS] An instrument used to study optical refraction within the eye. { 'skī·ə,skōp }

skin depth [ELECTROMAG] The depth beneath the surface of a conductor, which is carrying current at a given frequency due to electromagnetic waves incident on its surface, at which the current density drops to one neper below the current density at the surface. { 'skin ,depth }

skin effect [ELEC] The tendency of alternating currents to flow near the surface of a conductor thus being restricted to a small part of the total sectional area and producing the effect of increasing the resistance. Also known as conductor skin effect; Kelvin skin effect. { 'skin i,fekt }

skin friction [FL MECH] A type of friction force which exists at the surface of a solid body immersed in a much larger volume of fluid which is in motion relative to the body. { 'skin ˌfrik·shən }

skin resistance [ELEC] For alternating current of a given frequency, the direct-current resistance of a layer at the surface of a conductor whose thickness equals the skin depth. { 'skin riˌzis·təns }

skip distance [ELECTROMAG] The minimum distance that radio waves can be transmitted between two points on the earth by reflection from the ionosphere, at a specified time and frequency. { 'skip ˌdis·təns }

skip fading [ELECTROMAG] Fading due to fluctuations of ionization density at the place in the ionosphere where the wave is reflected which causes the skip distance to increase or decrease. { 'skip ˌfād·iŋ }

skip trajectory [MECH] A trajectory made up of ballistic phases alternating with skipping phases; one of the basic trajectories for the unpowered portion of the flight of a reentry vehicle or spacecraft reentering earth's atmosphere. { 'skip trəˌjek·trē }

skip zone [ACOUS] A region in the air surrounding a source of sound in which no sound is heard, although the sound becomes audible at greater distances. Also known as zone of silence. { 'skip ˌzōn }

skot [OPTICS] A unit of luminance, used particularly to measure low-level luminance, equal to 10^{-3} apostilb, or $10^{-3}/\pi$ nit. { skät }

slant visibility *See* oblique visual range. { 'slant ˌviz·ə'bil·əd·ē }

Slater determinant [QUANT MECH] A quantum-mechanical wave function for n fermions, which is an $n \times n$ determinant whose entries are n different one-particle wave functions depending on the coordinates of each of the particles in the system. { 'slād·ər diˌtər·mə·nənt }

sleeve antenna [ELECTROMAG] A single vertical half-wave radiator, the lower half of which is a metallic sleeve through which the concentric feed line runs; the upper radiating portion, one quarter-wavelength long, connects to the center of the line. { 'slēv anˌten·ə }

sleeve dipole antenna [ELECTROMAG] Dipole antenna surrounded in its central portion by a coaxial cable. { 'slēv 'dīˌpōl an'ten·ə }

slender-body theory [FL MECH] The theory of compressible inviscid fluid flow past bodies which have pointed noses and bases, or flat bases in supersonic flow only, and which satisfy the following conditions: (1) the ratio r of the maximum thickness to the length of the body must be small compared with unity, (2) the angle between the tangent plane to the body and the direction of motion must be small and of order r, and (3) the smoothness conditions. { ¦slen·dər 'bäd·ē ˌthē·ə·rē }

slide-wire bridge [ELEC] A bridge circuit in which the resistance in one or more branches is controlled by the position of a sliding contact on a length of resistance wire stretched along a linear scale. { 'slīd ¦wīr ˌbrij }

slide-wire potentiometer [ELEC] A potentiometer (variable resistor) which employs a movable sliding connection on a length of resistance wire. { 'slīd ¦wīr pəˌten·chē'äm·əd·ər }

sliding friction [MECH] Rubbing of bodies in sliding contact. { 'slīd·iŋ ˌfrik·shən }

sliding vector [MECH] A vector whose direction and line of application are prescribed, but whose point of application is not prescribed. { 'slīd·iŋ 'vek·tər }

slip [CRYSTAL] The movement of one atomic plane over another in a crystal; it is one of the ways that plastic deformation occurs in a solid. Also known as glide. [ELEC] **1.** The difference between synchronous and operating speeds of an induction machine. Also known as slip speed. **2.** Method of interconnecting multiple wiring between switching units by which trunk number 1 becomes the first choice for the first switch, trunk number 2 first choice for the second switch, trunk number 3 first choice for the third switch, and so on. [FL MECH] The difference between the velocity of a solid surface and the mean velocity of a fluid at a point just outside the surface. { slip }

slipband [CRYSTAL] One of the microscopic parallel lines (Lüders' lines) on the surface of a crystalline material stretched beyond its elastic limit, located at the intersection of the surface with intracrystalline slip planes in the grains of the material. Also known as slip line. { 'slip,band }

slip direction [CRYSTAL] The crystallographic direction in which the translation of slip occurs. { 'slip diˌrek·shən }

slip flow [FL MECH] A situation in which the mean free path of a gas is between 1 and 65% of the channel diameter; the gas layer next to the channel wall assumes a velocity of slip past the liquid, known as slip flow. { 'slip 'flō }

slip line *See* slipband. { 'slip ˌlīn }

slip plane *See* glide plane. { 'slip ˌplān }

slip velocity [FL MECH] The difference in velocities between liquids and solids (or gases and liquids) in the vertical flow of two-phase mixtures through a pipe because of the slip between the two phases. { 'slip vəˌläs·əd·ē }

slitless spectrograph [OPTICS] A type of astronomical spectrograph that does not use a slit, sufficient resolution being obtained from the small image sizes of individual stars, and through the use of an objective prism in front of the telescope. { 'slit·ləs 'spek·trəˌgraf }

slitlet mask [SPECT] A metal plate used in astronomical spectroscopy, with several small slits at locations corresponding to astronomical objects of interest. { 'slit·lət ˌmask }

slit spectrograph [OPTICS] A type of astronomical spectrograph that uses a slit to provide resolution. { 'slit 'spek·trəˌgraf }

slope of fall [MECH] Ratio between the drop of a projectile and its horizontal movement; tangent of the angle of fall. { 'slōp əv 'fȯl }

slot antenna [ELECTROMAG] An antenna formed by cutting one or more narrow slots in a large metal surface fed by a coaxial line or waveguide. { 'slät an,ten·ə }

slot coupling [ELECTROMAG] Coupling between a coaxial cable and a waveguide by means of two coincident narrow slots, one in a waveguide wall and the other in the sheath of the coaxial cable. { 'slät ,kəp·liŋ }

slot radiator [ELECTROMAG] Primary radiating element in the form of a slot cut in the walls of a metal waveguide or cavity resonator or in a metal plate. { 'slät ,rād·ē,ād·ər }

slotted line *See* slotted section. { 'släd·əd 'līn }

slotted section [ELECTROMAG] A section of waveguide or shielded transmission line in which the shield is slotted to permit the use of a traveling probe for examination of standing waves. Also known as slotted line; slotted waveguide. { 'släd·əd ,sek·shən }

slotted waveguide *See* slotted section. { 'släd·əd 'wāv,gīd }

slowing of clocks [RELAT] According to the special theory of relativity, a clock appears to tick less rapidly to an observer moving relative to the clock than to an observer who is at rest with respect to the clock. Also known as time dilation effect. { 'slō·iŋ əv 'kläks }

slow neutron [NUC PHYS] A neutron having low kinetic energy, up to about 100 electronvolts. { 'slō 'nü,trän }

slow-neutron spectroscopy [PHYS] The use of beams of slow neutrons, from nuclear reactors or nuclear accelerators, in studies of the structure or structural dynamics of solid, liquid, or gaseous matter, particularly the atomic and magnetic dynamics. Also known as inelastic neutron scattering; neutron spectroscopy. { 'slō ¦nü,trän spek'träs·kə·pē }

slow ray [OPTICS] In crystal optics, that component of light in any birefringent crystal section which travels with the lesser velocity and has the higher index of refraction. { 'slō 'rā }

slow-vibration direction [OPTICS] The direction of the electric field vector of the ray of light that travels with the smallest velocity in an anisotropic crystal and therefore corresponds to the maximum refractive index. { 'slō vī'brā·shən di,rek·shən }

slow wave [ELECTROMAG] A wave having a phase velocity less than the velocity of light, as in a ridge wave guide. { 'slō 'wāv }

slug [ELECTROMAG] **1.** A heavy copper ring placed on the core of a relay to delay operation of the relay. **2.** A movable iron core for a coil. **3.** A movable piece of metal or dielectric material used in a wave guide for tuning or impedance-matching purposes. [MECH] A unit of mass in the British gravitational system of units, equal to the mass which experiences an acceleration of 1 foot per second per second when a force of 1 pound acts on it; equal to approximately 32.1740 pound mass or 14.5939 kilograms. Also known as geepound. { sləg }

slug flow *See* piston flow. { 'sləg ,flō }

SLUG junction [CRYO] A Josephson junction consisting of a drop of lead-tin solder solidified around a niobium wire. { 'sləg ,jəŋk·shən }

slug tuner [ELECTROMAG] Waveguide tuner containing one or more longitudinally adjustable pieces of metal or dielectric. { 'sləg ¦tün·ər }

slug tuning [ELECTROMAG] Means of varying the frequency of a resonant circuit by introducing a slug of material into either the electric field or magnetic field, or both. { 'sləg ¦tün·iŋ }

small-angle scattering [PHYS] Scattering of a beam of electromagnetic or acoustic radiation, or particles, at small angles by particles or cavities whose dimensions are many times as large as the wavelength of the radiation or the de Broglie wavelength of the scattered particles. Also known as low-angle scattering. { ¦smȯl ,aŋ·gəl 'skad·ər·iŋ }

small calorie *See* calorie. { 'smȯl 'kal·ə·rē }

small perturbation [PHYS] A disturbance imposed on a system in steady state, with amplitude assumed small of the first order; that is, the square of the amplitude is negligible in comparison with the amplitude, and the derivatives of the perturbation are assumed to be of the same order of magnitude as the perturbation. { 'smȯl ,pərd·ər'bā·shən }

small polaron [SOLID STATE] A quasiparticle comprising a self-trapped electronic charge localized within a small region of a solid of spatial extent comparable to an interatomic dimension, and the atomic displacement pattern which produces the potential well within which the charge is bound. { 'smȯl 'pō·lə,rän }

S matrix *See* scattering matrix. { 'es ,mā·triks }

S-matrix theory [PART PHYS] A theory of elementary particles based on the scattering matrix, and on its properties such as unitarity and analyticity. Also known as scattering-matrix theory. { 'es ¦mā·triks 'thē·ə·rē }

Smith-Baker microscope [OPTICS] A type of interference microscope in which a beam of polarized light is split by a birefringent calcite plate cemented to the front lens of the condenser, and reunited by another such plate cemented to the objective. { 'smith 'bā·kər 'mī·krə,skōp }

Smith chart [ELECTROMAG] A special polar diagram containing constant-resistance circles, constant-reactance circles, circles of constant standing-wave ratio, and radius lines representing constant line-angle loci; used in solving transmission-line and waveguide problems. { 'smith ,chärt }

Smith-Helmholtz law [OPTICS] For a single refracting surface of sufficiently small aperture, the product of the index of refraction, distance from the optical axis, and the angle which a light ray makes with the optical axis at the object point is equal to the corresponding product at the image point. { 'smith 'helm,hōlts ,lȯ }

Smith-Purcell-Salisbury effect [PHYS] The emission of nearly coherent radiation when a beam of relativistic electrons strikes a metallic diffraction grating at grazing incidence. { ¦smith ¦pər·səl 'sȯlz·brē i,fekt }

smoke technique [FL MECH] A technique used to measure only very-low-speed air velocity; smoke enables the fluid motion to be observed with the eye, and the smoke is timed over a measured distance along an airway of constant cross section to determine the velocity of flow. { 'smōk tek,nēk }

smoke-wire method [FL MECH] A method of flow visualization in which a thin stainless-steel wire coated with a thin film of kerosine is heated electrically to generate smoke, resulting in a series of time lines. { ¦smōk ¦wīr 'meth·əd }

smoothness conditions [FL MECH] Two conditions that must be satisfied by bodies studied in slender body theory: (1) the rate of change of the angle between the tangent plane to the body and the direction of the motion, evaluated along this direction, must be small and of the same order as the ratio of the maximum thickness of the body to its length; (2) the curvature of any section of the body in a plane normal to the direction of motion must be of the same order as the reciprocal of the maximum diameter of the section, at all points where the section is convex outward. { 'smüth·nəs kən,dish·ənz }

Snell laws of refraction [OPTICS] When light travels from one medium into another the incident and refracted rays lie in one plane with the normal to the surface; are on opposite sides of the normal; and make angles with the normal whose sines have a constant ratio to one another. Also known as Descartes laws of refraction; laws of refraction. { 'snel 'lȯz əv ri'frak·shən }

Snoek effect [SOLID STATE] The preferential occupation by carbon impurity atoms of sites on one of the three faces in the cubic lattice of iron. { 'snůk i,fekt }

SNOM *See* near-field scanning optical microscopy.

Soddy's displacement law [NUC PHYS] The atomic number of a nuclide decreases by 2 in alpha decay, increases by 1 in beta negatron decay, and decreases by 1 in beta positron decay and electron capture. { 'säd·ēz di'splās·mənt ,lȯ }

sodium-24 [NUC PHYS] A radioactive isotope of sodium, mass 24, half-life 15.5 hours; formed by deuteron bombardment of sodium; decomposes to magnesium with emission of beta rays. { 'sōd·ē·əm ¦twen·tē¦fȯr }

sodium-line reversal temperature measurement [PHYS] A method of measuring the temperature of a gas containing sodium vapor, in which the gas is placed in the path of a radiator of known temperature, and the temperature of the gas or the radiator is adjusted until the sodium D line disappears against the background of light from the radiator. { ¦sōd·ē·əm ,līn ri¦vər·səl 'tem·prə·chər ,mezh·ər·mənt }

softening point [PHYS] For a substance which does not have a definite melting point, the temperature at which viscous flow changes to plastic flow. { 'sȯf·ən·iŋ ,pȯint }

softening range [PHYS] The temperature range in which material without a melting point goes from a rigid to a soft condition. { 'sȯf·ən·iŋ ,rānj }

soft magnetic material [ELECTROMAG] A magnetic material which is relatively easily magnetized or demagnetized. { 'sȯft mag'ned·ik mə'tir·ē·əl }

soft radiation [PHYS] Radiation whose particles or photons have a low energy, and, as a result, do not penetrate any type of material readily. { 'sȯft ,rād·ē'ā·shən }

soft shower [NUC PHYS] A cosmic-ray shower that cannot penetrate 6 to 8 inches (15 to 20 centimeters) of lead; consists mainly of electrons and positrons. { 'sȯft 'shaů·ər }

soft x-ray [ELECTROMAG] An x-ray having a comparatively long wavelength and poor penetrating power. { 'sȯft 'eks,rā }

soft-x-ray absorption spectroscopy [SPECT] A spectroscopic technique which is used to get information about unoccupied states above the Fermi level in a metal or about empty conduction bands in an inoculator. { 'sȯft ¦eks,ra əb'sȯrp·shən spek'träs·kə·pē }

soft-x-ray appearance potential spectroscopy [SPECT] A branch of electron spectroscopy in which a solid surface is bombarded with monochromatic electrons, and small but abrupt changes in the resulting total x-ray emission intensity are detected as the energy of the electrons is varied. Abbreviated SXAPS. { 'sȯft ¦eks,ra ə'pir·əns pə¦ten·chəl spek'träs·kə·pē }

sogasoid [PHYS] A system of solid particles dispersed in a gas. { 'säg·ə,sȯid }

Sohncke's law [PHYS] The law that the stress per unit area normal to a crystallographic plane needed to produce a fracture in a crystal is a constant characteristic of a crystalline substance. { 'zōŋ·kəz ,lȯ }

solar-excited laser *See* sun-pumped laser. { 'sō·lər ik¦sīd·əd 'lā·zər }

solarization [PHYS] Loss of transparency or coloration of glass exposed to sunlight or ultraviolet radiation. { ,sō·lə·rə'zā·shən }

solar pumping [OPTICS] The use of sunlight focused directly into a laser rod for pumping to induce lasing action. { 'sō·lər 'pəmp·iŋ }

solar telescope [OPTICS] An observational instrument of the solar astronomer; it is designed so that heating effects produced by the sun do not distort the images; the two classes consist of those designed for observations of the brilliant solar disk, and those designed for the study of the much fainter prominences and the still fainter corona through the relatively bright, scattered light of the sky. { 'sō·lər 'tel·ə,skōp }

soleil compensator [OPTICS] A compensator which resembles the Babinet compensator but is constructed so that the phase change is constant over the entire field. { sō'lā 'käm·pən,sād·ər }

solenoid [ELECTROMAG] Also known as electric solenoid. **1.** An electrically energized coil of insulated wire which produces a magnetic field within the coil. **2.** In particular, a coil that surrounds a movable iron core which is pulled to a central position with respect to the coil when the coil is energized by sending current through it. { 'säl·ə,nȯid }

solid [PHYS] **1.** A substance that has a definite volume and shape and resists forces that tend to alter its volume or shape. **2.** A crystalline material, that is, one in which the constituent atoms are arranged in a three-dimensional lattice, periodic in three independent directions. { 'säl·əd }

solid-dielectric capacitor [ELEC] A capacitor whose dielectric is one of several solid materials such as ceramic, mica, glass, plastic film, or paper. { 'säl·əd ¦dī·ə¦lek·trik kə'pas·əd·ər }

solid electrolytic capacitor [ELEC] An electrolytic capacitor in which the dielectric is an anodized coating on one electrode, with a solid semiconductor material filling the rest of the space between the electrodes. { 'säl·əd i¦lek·trə¦lid·ik kə'pas·əd·ər }

solid helium [CRYO] A certain state which is not attained by helium under its own vapor pressure down to absolute zero, but which requires an external pressure of 25 atmospheres at absolute zero. { 'säl·əd 'hē·lē·əm }

solidification [PHYS] The change of a fluid (liquid or gas) into the solid state. { sə'lid·ə·fə'kā·shən }

solid insulator [ELEC] An electric insulator made of a solid substance, such as sulfur, polystyrene, rubber, or porcelain. { 'säl·əd 'in·sə,lād·ər }

solid laser [OPTICS] A laser in which either a crystalline or amorphous solid material, usually in the form of a rod, is excited by optical pumping; the most common crystalline materials are ruby, neodymium-doped ruby, and neodymium-doped yttrium aluminum garnet. { 'säl·əd 'lā·zər }

solid moment of inertia [PHYS] The integral of the products of the mass of each of the infinitesimal elements of the solid with the square of their distance from a given axis. { 'säl·əd 'mō·mənt əv i'nər·shə }

solid Schmidt telescope [OPTICS] A type of Schmidt system which is constructed from a single block of glass, designed to operate at very small aperture ratios. { 'säl·əd 'shmit 'tel·ə,skōp }

solid solution [PHYS] A homogeneous crystalline phase composed of several distinct chemical species, occupying the lattice points at random and existing in a range of concentrations. { 'säl·əd sə'lü·shən }

solid state [PHYS] The condition of a substance in which it is a solid. { 'säl·əd 'stāt }

solid-state battery [ELEC] A battery in which both the electrodes and the electrolyte are solid-state materials. { 'säl·əd ¦stāt 'bad·ə·rē }

solid-state laser [OPTICS] A laser in which a semiconductor material produces the coherent output beam. { 'säl·əd ¦stāt 'lā·zər }

solid-state maser [PHYS] A maser in which a semiconductor material produces the coherent output beam; two input waves are required: one wave, called the pumping source, induces upward energy transitions in the active material, and the second wave, of lower frequency, causes downward transitions and undergoes amplification as it absorbs photons from the active material. { 'säl·əd ¦stāt 'mā·zər }

solid-state physics [PHYS] The branch of physics centering about the physical properties of solid materials. { 'säl·əd ¦stāt 'fiz·iks }

solid tantalum capacitor [ELEC] An electrolytic capacitor in which the anode is a porous pellet of tantalum; the dielectric is an extremely thin layer of tantalum pentoxide formed by anodization of the exterior and interior surfaces of the pellet; the cathode is a layer of semiconducting manganese dioxide that fills the pores of the anode over the dielectric. { 'säl·əd ¦tant·əl·əm kə'pas·əd·ər }

solitary wave [PHYS] A traveling wave in which a single disturbance is neither preceded by nor followed by other such disturbances, but which does not involve unusually large amplitudes or rapid changes in variables, in contrast to a shock wave. { 'säl·ə,ter·ē 'wāv }

soliton [PHYS] An isolated wave that propagates without dispersing its energy over larger and larger regions of space, and whose nature is such that two such objects emerge unchanged from a collision. { 'säl·ə,tän }

Sommerfeld equation *See* Sommerfeld formula. { 'zȯm·ər,felt i,kwā·zhən }

Sommerfeld fine-structure constant *See* fine-structure constant. { 'zȯm·ər,felt 'fīn ¦strək·chər ,kän·stənt }

Sommerfeld formula [ELECTROMAG] An approximate formula for the field strength of electromagnetic radiation generated by an antenna at distances small enough so that the curvature of the earth may be neglected, in terms of radiated power, distance from the antenna, and various constants and parameters. Also known as Sommerfeld equation. { 'zȯm·ər,felt ,fȯr·myə·lə }

Sommerfeld law for doublets [ATOM PHYS] According to the Bohr-Sommerfeld theory, the splitting in frequency of regular or relativistic doublets is $\alpha^2R(Z - \sigma)^4/n^3(l + 1)$, where α is the fine structure constant, R is the Rydberg constant of the atom, Z is the atomic number, σ is a screening constant, n is the principal quantum number, and l is the orbital angular momentum quantum number. { 'zȯm·ər,felt 'lȯ fər 'dəb·ləts }

Sommerfeld model *See* free-electron theory of metals. { 'zȯm·ər,felt ,mäd·əl }

Sommerfeld theory *See* free-electron theory of metals. { 'zȯm·ər,felt ,thē·ə·rē }

Sondhauss tube [ACOUS] A device that converts heat to acoustic energy by heating a small glass bulb that is attached to a cool glass stem whose tip radiates sound. { 'zänd,hau̇s ,tüb }

sone [ACOUS] A unit of loudness, equal to the loudness of a simple 1000-hertz tone with a sound pressure level 40 decibels above 0.0002 microbar; a sound that is judged by listeners to be *n* times as loud as this tone has a loudness of *n* sones. { sōn }

sonic [ACOUS] **1.** Of or pertaining to the speed of sound. **2.** Pertaining to that which moves at acoustic velocity, as in sonic flow. **3.** Designed to operate or perform at the speed of sound, as in sonic leading edge. { 'sän·ik }

sonic boom [ACOUS] A noise caused by a shock wave that emanates from an aircraft or other object traveling at or above sonic velocity. { 'sän·ik 'büm }

sonic radiation [ACOUS] Acoustic radiation with a frequency between about 16 hertz and about 20,000 hertz. { 'sän·ik ˌrād·ē'ā·shən }

sonics [ACOUS] The technology of sound, or elastic wave motion, as applied to problems of measurement, control, and processing. { 'sän·iks }

sonic speed *See* speed of sound. { 'sän·ik ¦spēd }

sonic velocity *See* speed of sound. { 'sän·ik və'läs·əd·ē }

Sonnar lens [OPTICS] A modified triplet lens used as a photographic objective. { 'sōˌnär ˌlenz }

sonoelastography [ACOUS] An ultrasound technique for imaging the relative elastic properties of soft tissue and, in particular, for imaging hard tumors within the human body, in which vibrations (shear waves) with low frequencies (less than 1000 hertz) are propagated through tissue while real-time Doppler techniques are used to image the resulting vibration pattern on an ultrasound scanner. { ˌsō·nōˌi·las'täg·rə·fē }

sonogram [ACOUS] The image produced by ultrasonic imaging. { 'sän·əˌgram }

sonography *See* acoustic imaging. { sə'näg·rə·fē }

sonoluminescence [PHYS] Luminescence produced by high-frequency sound waves or by phonons. { ¦sän·ō·əˌlü·mə'nes·əns }

soot luminosity [OPTICS] The portion of the luminosity of a flame attributable to soot particles in the flame. { 'sůt ˌlü·mə'näs·əd·ē }

Soret coefficient [PHYS] A tabulated value used in binary thermal diffusion calculations in gaseous systems; expressed as $D'/D = \alpha X_1X_2$, where D' is the coefficient of thermal diffusion, D is the coefficient of ordinary diffusion, α is the thermal diffusion constant, X_1 is the mole fraction of the lower-molecular-weight component, and X_2 is the mole fraction of the higher-molecular-weight component. { sȯ'rā ˌkō·iˌfish·ənt }

Soret effect [PHYS] Thermal diffusion in liquids. { sȯ'rā iˌfekt }

sound [ACOUS] An alteration of properties of an elastic medium, such as pressure, particle displacement, or density, that propagates through the medium, or a superposition of such alterations; sound waves having frequencies above the audible (sonic) range are termed ultrasonic waves; those with frequencies below the sonic range are called infrasonic waves. Also known as acoustic wave; sound wave. { saůnd }

sound absorption [ACOUS] A process in which sound energy is reduced when sound waves pass through a medium or strike a surface. Also known as acoustic absorption. { 'saůnd əbˌsȯrp·shən }

sound absorption coefficient [ACOUS] The ratio of sound energy absorbed to that arriving at a surface or medium. Also known as acoustic absorption coefficient; acoustic absorptivity. { 'saůnd abˌsȯrp·shən ˌkō·iˌfish·ənt }

sound attenuation [ACOUS] Diminution of the intensity of sound energy propagating in a medium; caused by absorption, spreading, and scattering. { 'saůnd əˌten·yəˌwā·shən }

sound band pressure level [ACOUS] The effective sound pressure for the sound energy in a given frequency band. { 'saůnd ¦band 'presh·ər ˌlev·əl }

sound channel [ACOUS] A layer of seawater extending from about 2300 feet (700 meters) down to about 4950 feet (1500 meters), in which sound travels at about 1485 feet (450 meters) per second, the slowest it can travel in seawater; below 4950 feet (1500 meters) the speed of sound increases as a result of pressure. { 'saůnd ˌchan·əl }

sound detection [ACOUS] The discrimination of a sound from background noise, either by the ear or by an electronic instrument such as a volume indicator. { 'saůnd diˌtek·shən }

sound energy [ACOUS] The difference between the total energy and the energy which would exist if no sound waves were present. Also known as acoustic energy. { 'saůnd ˌen·ər·jē }

sound-energy density [ACOUS] Sound energy per unit volume; the commonly used unit is the erg per cubic centimeter. { 'saůnd ˌen·ər·jē ˌden·səd·ē }

sound-energy flux [ACOUS] Average over one period of the rate of flow of sound energy through any specified area; the unit is the erg per second. { 'saůnd ˌen·ər·jē ˌfləks }

sound image [ACOUS] The photographic image of a sound, as on a film sound track. { 'saůnd ˌim·ij }

sounding velocity [ACOUS] The vertical velocity of sound in water, usually assumed to be constant at 800 to 820 fathoms (1464 to 1501 meters) per second for sounding measurements. { 'saůnd·iŋ vəˌläs·əd·ē }

sound intensity [ACOUS] For a specified direction and point in space, the average rate at which sound energy is transmitted through a unit area perpendicular to the specified direction. { 'saůnd inˌten·səd·ē }

sound irradiator [ACOUS] A device for focusing sound waves so that sound of high intensity is produced at the focus. { 'saůnd iˌrād·ēˌād·ər }

sound lag [ACOUS] Time necessary for a sound wave to travel from its source to the point of reception. { 'saůnd ˌlag }

sound level [ACOUS] The sound pressure level (in decibels) at a point in a sound field, averaged over the audible frequency range and over a time interval, with a frequency weighting and the time interval as specified by the American National Standards Association. { 'saůnd ˌlev·əl }

sound masking [ACOUS] The ability of one sound to make the ear incapable of perceiving another sound. { 'saůnd ,mask·iŋ }

sound power [ACOUS] The total sound energy radiated by a source per unit time, generally expressed in ergs per second or watts. Also known as acoustic power. { 'saůnd ,paů·ər }

sound pressure *See* effective sound pressure. { 'saůnd ,presh·ər }

sound pressure level [ACOUS] A value in decibels equal to 20 times the logarithm to the base 10 of the ratio of the pressure of the sound under consideration to a reference pressure; reference pressures in common use are 0.0002 microbar and 1 microbar. Abbreviated SPL. { 'saůnd ¦presh·ər ,lev·əl }

sound pressure spectrum level [ACOUS] Ten times the logarithm to base 10 of the ratio of the mean square pressure of the portion of sound within a specified frequency band to the mean square pressure of the portion of a reference sound within the same frequency band. Abbreviated SPSL. { ¦saůnd ,presh·ər 'spek·trəm ,lev·əl }

sound-ray diagram [ACOUS] A plot of the paths taken by sound rays in an acoustical system; analogous to a light-ray diagram in optics. { 'saůnd ¦rā ,dī·ə,gram }

sound reduction factor [ACOUS] A measure of the reduction in the intensity of sound when it crosses an interface, equal to 10 times the common logarithm of the reciprocal of the sound transmission coefficient of the surface. { 'saůnd ri,dək·shən ,fak·tər }

sound reflection coefficient *See* acoustic reflectivity. { 'saůnd ri,flek·shən ,kō·i,fish·ənt }

sound sensation *See* sound. { 'saůnd sen,sā·shən }

sound spectrum [ACOUS] A plot of the strength of a sound at various frequencies. { 'saůnd ,spek·trəm }

sound transmission [ACOUS] Passage of a sound wave through a medium or series of media. { 'saůnd tranz,mish·ən }

sound transmission coefficient [ACOUS] The ratio of transmitted to incident sound energy at an interface in a sound medium; the value depends on the angle of incidence of the sound. Also known as acoustic transmission coefficient; acoustic transmissivity. { 'saůnd tranz,mish·ən ,kō·i,fish·ənt }

sound velocity *See* speed of sound. { 'saůnd və,läs·əd·ē }

sound volume velocity [ACOUS] The rate at which a substance flows through a specified area as a result of a sound wave. { 'saůnd ,väl·yəm və,läs·əd·ē }

sound wave *See* sound. { 'saůnd ,wāv }

sound-wave photography [PHYS] A method of studying propagation, reflection, and refraction of sound waves, in which a sound wave is generated by a spark and is illuminated a fraction of a second later by a second spark, causing the wave to cast a shadow on a photographic plate. { ¦saůnd ,wāv fə'tag·rə·fē }

source [ELEC] The circuit or device that supplies signal power or electric energy or charge to a transducer or load circuit. [PHYS] **1.** In general, a device that supplies some extensive entity, such as energy, matter, particles, or electric charge. **2.** A point, line, or area at which mass or energy is added to a system, either instantaneously or continuously. **3.** A point at which lines of force in a vector field originate, such as a point in an electrostatic field where there is positive charge. [SPECT] The arc or spark that supplies light for a spectroscope. [THERMO] A device that supplies heat. { sȯrs }

source flow [FL MECH] **1.** In three-dimensional flow, a point from which fluid issues at a uniform rate in all directions. **2.** In two-dimensional flow, a line normal to the planes of flow, from which fluid flows uniformly in all directions at right angles to the line. { 'sȯrs ,flō }

source impedance [ELEC] Impedance presented by a source of energy to the input terminals of a device. { 'sȯrs im,pēd·əns }

source level [ACOUS] The sound intensity, in decibels above a reference level, at a point which is a unit distance from a source and on an axis of the source. { 'sȯrs ,lev·əl }

south pole [ELECTROMAG] The pole of a magnet at which magnetic lines of force are assumed to enter. Also known as negative pole. { 'saůth 'pōl }

space attenuation [ACOUS] Loss of energy, expressed in decibels, of a signal in free air; caused by such factors as absorption, reflection, scattering, and dispersion. { 'spās ə,ten·yə,wā·shən }

space centrode [MECH] The path traced by the instantaneous center of a rotating body relative to an inertial frame of reference. { ¦spās 'sen,trōd }

space charge [ELEC] The net electric charge within a given volume. { 'spās ,chärj }

space-charge polarization [ELEC] Polarization of a dielectric which occurs when charge carriers are present which can migrate an appreciable distance through the dielectric but which become trapped or cannot discharge at an electrode. Also known as interfacial polarization. { 'spās ¦chärj ,pō·lə·rə'zā·shən }

space cone [MECH] The cone in space that is swept out by the instantaneous axis of a rigid body during Poinsot motion. Also known as herpolhode cone. { 'spās ,kōn }

spaced antenna [ELECTROMAG] Antenna system consisting of a number of separate antennas spaced a considerable distance apart, used to minimize local effects of fading at short-wave receiving stations. { 'spāst an'ten·ə }

space group [CRYSTAL] A group of operations which leave the infinitely extended, regularly repeating pattern of a crystal unchanged; there are 230 such groups. { 'spās ,grüp }

space-group extinction [CRYSTAL] The absence of certain classes of reflections in the x-ray diffraction pattern of a crystal due to the existence of symmetry elements in the space group of the crystal which are not present in its point group. { 'spās ¦grüp ik'stiŋk·shən }

space inverson See inversion. { 'spās in,vər·zhən }

space lattice See lattice. { 'spās ,lad·əs }

spacelike path [RELAT] A trajectory in space-time such that a vector tangent to any point on the path is a spacelike vector. { 'spās,līk 'path }

spacelike surface [RELAT] A three-dimensional surface in a four-dimensional space-time which has the property that no event on the surface lies in the past or the future of any other event on the surface. { 'spās¦līk 'sər·fəs }

spacelike vector [RELAT] A four vector in Minkowski space whose space component has a magnitude which is greater than the magnitude of its time component multiplied by the speed of light. { 'spās¦līk 'vek·tər }

space permeability [ELECTROMAG] Factor that expresses the ratio of magnetic induction to magnetizing force in a vacuum; in the centimeter-gram-second electromagnetic system of units, the permeability of a vacuum is arbitrarily taken as unity; in the meter-kilogram-second-ampere system, it is $4\pi \times 10^{-7}$. { 'spās ,pər·mē·ə'bil·əd·ē }

space quadrature [PHYS] A difference of a quarter-wavelength in the position of corresponding points of a wave in space. { 'spās ,kwäd·rə·chər }

space quantization [QUANT MECH] The quantization of the component of the angular momentum of a system in some specified direction. { 'spās ,kwän·tə'zā·shən }

space-time [RELAT] A four-dimensional space used to represent the universe in the theory of relativity, with three dimensions corresponding to ordinary space and the fourth to time. Also known as space-time continuum. { 'spās 'tīm }

space-time continuum See space-time. { 'spās 'tīm kən'tin·yə·wəm }

spallation [NUC PHYS] A nuclear reaction in which the energy of the incident particle is so high that more than two or three particles are ejected from the target nucleus and both its mass number and atomic number are changed. Also known as nuclear spallation; spallation reaction. { spȯ'lā·shən }

spallation reaction See spallation. { spȯ'lā·shən rē,ak·shən }

spark [ELEC] A short-duration electric discharge due to a sudden breakdown of air or some other dielectric material separating two terminals, accompanied by a momentary flash of light. Also known as electric spark; spark discharge; sparkover. { spärk }

spark discharge See spark. { 'spärk 'dis,chärj }

spark excitation [SPECT] The use of an electric spark (10,000 to 30,000 volts) to excite spectral line emissions from otherwise hard-to-excite samples; used in emission spectroscopy. { 'spärk ,ek,sī'tā·shən }

spark gap [ELEC] An arrangement of two electrodes between which a spark may occur; the insulation (usually air) between the electrodes is self-restoring after passage of the spark; used as a switching device, for example, to protect equipment against lightning or to switch a radar antenna from receiver to transmitter and vice versa. { 'spärk ,gap }

spark-ignition combustion cycle See Otto cycle. { 'spärk ig¦nish·ən kəm'bəs·chən ,sī·kəl }

sparking potential See breakdown voltage. { 'spärk·iŋ pə,ten·chəl }

sparking voltage See breakdown voltage. { 'spärk·iŋ ,vōl·tij }

sparkover See spark. { 'spärk,ō·vər }

spark photography [OPTICS] Any type of photography in which a spark provides illumination and determines the length of exposure. { 'spk fə,tag·rə·fē }

spark spectrum [SPECT] The spectrum produced by a spark discharging through a gas or vapor; with metal electrodes, a spectrum of the metallic vapor is obtained. { 'spärk ,spek·trəm }

spark-tracing method [FL MECH] A method of flow visualization in which a series of spark discharges produced between two electrodes is photographed with an open shutter to record a system of time lines. { 'spärk ,trās·iŋ ,meth·əd }

Sparrow's criterion [OPTICS] A criterion for the resolution of two light sources, according to which the light sources are resolved if there is some central dip in their combined diffraction pattern. { 'spar·ōz krī,tir·ē·ən }

spatial [PHYS] Of or pertaining to space; occupying space; occurring in, or conditioned by, space; considered with relation to space. { 'spā·shəl }

spatial coherence [PHYS] The existence of a correlation between the phases of waves at points separated in space at a given time. { 'spā·shəl kō'hir·əns }

spatial filter [OPTICS] An optical filter that consists of a very small aperture, such as a pinhole. { 'spā·shəl 'fil·tər }

spatiotemporal [PHYS] Of or pertaining to space time; having extent and duration. { spā·shē·ō'tem·pə·rəl }

special relativity [RELAT] The division of relativity theory which relates the observations of observers moving with constant relative velocities and postulates that natural laws are the same for all such observers. { 'spesh·əl rel·ə'tiv·əd·ē }

species See nuclide. { 'spē·shēz }

specific [PHYS] Indicating the amount of a physical quantity per unit mass, weight, volume, or area, or the ratio of the quantity for the substance under consideration to the same quantity for a standard substance, such as water. { spə'sif·ik }

specific acoustical impedance [ACOUS] The ratio of the pressure phasor associated with a sound wave at any given point in a medium to the velocity phasor at that point. { spə'sif·ik ə'küs·tə·kəl im'pēd·əns }

specific acoustical ohm See rayl. { spə'sif·ik ə'küs·tə·kəl 'ōm }

specific acoustical reactance [ACOUS] The

magnitude of the imaginary part of the specific acoustical impedance. { spə'sif·ik ə'küs·tə·kəl rē'ak·təns }

specific acoustical resistance [ACOUS] The real part of the specific acoustical impedance. { spə'sif·ik ə'küs·tə·kəl ri'zis·təns }

specific charge [ELEC] The ratio of a particle's charge to its mass. { spə'sif·ik 'chärj }

specific conductance *See* conductivity. { spə'sif·ik kən'dək·təns }

specific energy [THERMO] The internal energy of a substance per unit mass. { spə'sif·ik 'en·ər·jē }

specific gravity [MECH] The ratio of the density of a material to the density of some standard material, such as water at a specified temperature, for example, 4°C or 60°F, or (for gases) air at standard conditions of pressure and temperature. Abbreviated sp gr. Also known as relative density. { spə'sif·ik 'grav·əd·ē }

specific heat [THERMO] **1.** The ratio of the amount of heat required to raise a mass of material 1 degree in temperature to the amount of heat required to raise an equal mass of a reference substance, usually water, 1 degree in temperature; both measurements are made at a reference temperature, usually at constant pressure or constant volume. **2.** The quantity of heat required to raise a unit mass of homogeneous material one degree in temperature in a specified way; it is assumed that during the process no phase or chemical change occurs. { spə'sif·ik 'hēt }

specific inductive capacity *See* dielectric constant. { spə'sif·ik in'dək·tiv kə'pas·əd·ē }

specific insulation resistance *See* volume resistivity. { spə'sif·ik ,in·sə'lā·shən ri,zis·təns }

specific mass shift [NUC PHYS] The portion of the mass shift that is produced by the correlated motion of different pairs of atomic electrons and is therefore absent in one-electron systems. { spə'sif·ik ¦mas ,shift }

specific reluctance *See* reluctivity. { spə'sif·ik ri'lək·təns }

specific resistance *See* electrical resistivity. { spə'sif·ik ri'zis·təns }

specific rotation [OPTICS] The calculated rotation of light passing through a solution as related to the solution volume and depth, the amount of solute, and the observed optical rotation at a given wavelength and temperature. { spə'sif·ik rō'tā·shən }

specific viscosity [FL MECH] The specific viscosity of a polymer is the relative viscosity of a polymer solution of known concentration minus 1; usually determined at low concentration of the polymer; for example, 0.5 gram per 100 milliliters of solution, or less. { spə'sif·ik vi'skäs·əd·ē }

specific volume [MECH] The volume of a substance per unit mass; it is the reciprocal of the density. Abbreviated sp vol. { spə'sif·ik 'väl·yəm }

specific weight [MECH] The weight per unit volume of a substance. { spə'sif·ik 'wāt }

speckle [OPTICS] A phenomenon in which the scattering of light from a highly coherent source, such as a laser, by a rough surface or inhomogeneous medium generates a random-intensity distribution of light that gives the surface or medium a granular appearance. { 'spek·əl }

speckle interferometry [OPTICS] The use of speckle patterns in the study of object displacements, vibration, and distortion, and in obtaining diffraction-limited images of stellar objects. { 'spek·əl ,in·tər·fə'räm·ə·trē }

spectral bandwidth [SPECT] The minimum radiant-energy bandwidth to which a spectrophotometer is accurate; that is, 1–5 nanometers for better models. { 'spek·trəl 'band,width }

spectral centroid [OPTICS] An average wavelength; specifically, for a light filter or other light-transmitting device, a weighted average of the spectral energy distribution of the incident light, the transmittance of the device, and the luminosity function. { 'spek·trəl 'sen,tròid }

spectral characteristic [OPTICS] The relation between wavelength and some other variable, such as between wavelength and emitted radiant power of a luminescent screen per unit wavelength interval. { 'spek·trəl ,kar·ik·tə'ris·tik }

spectral color [OPTICS] **1.** A color corresponding to light of a pure frequency; the basic spectral colors are violet, blue-green, yellow, orange, and red. **2.** A color that is represented by a point on the chromaticity diagram that lies on a straight line between some point on the spectral color (first definition) locus and the achromatic points; purple, for example, is not a spectral color. { 'spek·trəl 'kəl·ər }

spectral density *See* spectral energy distribution. { 'spek·trəl 'den·səd·ē }

spectral dimension *See* fracton dimension. { 'spek·trəl di'men·shən }

spectral emissivity [THERMO] The ratio of the radiation emitted by a surface at a specified wavelength to the radiation emitted by a perfect blackbody radiator at the same wavelength and temperature. { 'spek·trəl ,ē,mi'siv·əd·ē }

spectral energy distribution [ELECTROMAG] The power carried by electromagnetic radiation within some small interval of wavelength (of frequency) of fixed amount as a function of wavelength (of frequency). Also known as spectral density. { 'spek·trəl 'en·ər·jē ,dis·trə,byü·shən }

spectral extinction [OPTICS] The selective absorption of different wavelengths of light as a function of depth in water. { 'spek·trəl ik 'stiŋk·shən }

spectral irradiance [OPTICS] The density of the radiant flux that is incident on a surface per unit of wavelength. { 'spek·trəl i'rād·ē·əns }

spectral line [SPECT] A discrete value of a quantity, such as frequency, wavelength, energy, or mass, whose spectrum is being investigated; one may observe a finite spread of values resulting from such factors as level width, Doppler broadening, and instrument imperfections. Also known as spectrum line. { 'spek·trəl ,līn }

spectral locus *See* spectrum locus. { 'spek·trəl 'lō·kəs }

spectral luminous efficacy [OPTICS] The ratio of the luminous flux emitted by a monochromatic light source in lumens to its radiant flux in watts, as a function of the wavelength of the emitted light. { 'spek·trəl 'lü·mə·nəs 'ef·i·kə·sē }

spectral luminous efficiency *See* luminosity fuction. { 'spek·trəl 'lü·mə·nəs ə'fish·ən·sē }

spectral photography [OPTICS] A technique used in airborne surveys for mineral deposits; narrow-band-pass filters and special film are used to accentuate minor color effects caused by mineralization and alteration which would be undetectable by broad-band photography. { 'spek·trəl fə'täg·rə·fē }

spectral radiance [OPTICS] The radiant flux per unit wavelength or frequency interval per unit solid angle per unit of projected area of the source; the usual unit is watt per nanometer per steradian per square meter. { 'spek·trəl 'rād·ē·əns }

spectral regions [SPECT] Arbitrary ranges of wavelength, some of them overlapping, into which the electromagnetic spectrum is divided, according to the types of sources that are required to produce and detect the various wavelengths, such as x-ray, ultraviolet, visible, infrared, or radio-frequency. { 'spek·trəl ˌrē·jənz }

spectral sensitivity [PHYS] The response of a device or material to monochromatic light as a function of wavelength. Also known as spectral response. { 'spek·trəl ˌsen·sə'tiv·əd·ē }

spectral series [SPECT] Spectral lines or groups of lines that occur in sequence. { 'spek·trəl ˌsir·ēz }

spectral temperature [OPTICS] The temperature of a blackbody that produces the same spectral radiance as a given radiation field at a given wavelength or frequency and in a given direction. { 'spek·trəl 'tem·prə·chər }

spectral transmission [OPTICS] The radiant flux which passes through a filter divided by the radiant flux incident upon it, for monochromatic light of a specified wavelength. { 'spek·trəl tranz'mish·ən }

Spectra Pritchard photometer [OPTICS] A photoelectric instrument for measuring the luminance of surfaces; it has a telescopic viewing system for imaging the bright surface to be measured on the cathode of a photoemissive tube, and a separate unit that combines the power supply with the controls and readout meter. { 'spek·trə 'prich·ərd fə'täm·əd·ər }

spectrobolometer [SPECT] An instrument that measures radiation from stars; measurement can be made in a narrow band of wavelengths in the electromagnetic spectrum; the instrument itself is a combination spectrometer and bolometer. { ¦spek·trō·bō'läm·əd·ər }

spectrofluorometer [SPECT] A device used in fluorescence spectroscopy to increase the selectivity of fluorometry by passing emitted fluorescent light through a monochromator to record the fluorescence emission spectrum. { ¦spek·trō·flu̇'räm·əd·ər }

spectrogram [SPECT] The record of a spectrum produced by a spectrograph. Also known as measured spectrum. { 'spek·trəˌgram }

spectrograph [SPECT] A spectroscope provided with a photographic camera or other device for recording the spectrum. { 'spek·trəˌgraf }

spectrography [SPECT] The use of photography to record the electromagnetic spectrum displayed in a spectroscope. { spek'träg·rə·fē }

spectroheliocinematograph [OPTICS] A camera used to make motion pictures of, for example, prominences of the sun; the camera utilizes monochromatic light; it is composed of a camera and a spectrohelioscope. { ¦spek·trō¦hē·lē·ō ˌsin·ə'm ad·əˌgraf }

spectroheliograph [OPTICS] An instrument used to photograph the sun in one spectral band. { ¦spek·trō'hē·lē·əˌgraf }

spectrohelioscope [OPTICS] An instrument based on the principle of the spectroheliograph but used for visual observation, and not for photography. { ¦spek·trō'hē·lē·əˌskōp }

spectrometer [SPECT] **1.** A spectroscope that is provided with a calibrated scale either for measurement of wavelength or for measurement of refractive indices of transparent prism materials. **2.** A spectroscope equipped with a photoelectric photometer to measure radiant intensities at various wavelengths. { spek'träm·əd·ər }

spectrometry [SPECT] The use of spectrographic techniques for deriving the physical constants of materials. { spek'träm·ə·trē }

spectrophotometer [SPECT] An instrument that consists of a radiant-energy source, monochromator, sample holder, and detector, used for measurement of radiant flux as a function of wavelength and for measurement of absorption spectra. { ¦spek·trō·fə'täm·əd·ər }

spectropolarimeter [OPTICS] A device used to measure optical rotation in solutions for different light wavelengths. { ¦spek·trōˌpō·lə'rim·əd·ər }

spectropolarimetry [SPECT] The measurement of the polarization of light that has been dispersed into a continuum or line spectrum as a function of wavelength. { ˌspek·trəˌpō·lə'rim·ə·trē }

spectropyrheliometer [SPECT] An astronomical instrument used to measure distribution of radiant energy from the sun in the ultraviolet and visible wavelengths. { ¦spek·trō¦pīrˌhē·lē'äm·əd·ər }

spectroscope [SPECT] An optical instrument consisting of a slit, collimator lens, prism or grating, and a telescope or objective lens which produces a spectrum for visual observation. { 'spek·trəˌskōp }

spectroscopic displacement law [SPECT] The spectrum of an un-ionized atom resembles that of a singly ionized atom of the element one place higher in the periodic table, and that of a doubly ionized atom two places higher in the table, and so forth. { ¦spek·trə¦skäp·ik di'splās·mənt ˌlȯ }

spectroscopic splitting factor *See* Landé g factor. { ¦spek·trə¦skäp·ik 'splid·iŋ ˌfak·tər }

spectroscopy [PHYS] The branch of physics concerned with the production, measurement, and interpretation of electromagnetic spectra arising from either emission or absorption of radiant energy by various substances. { spek 'träs·kə·pē }

spectrum [PHYS] **1.** A display or plot of intensity of radiation (particles, photons, or acoustic radiation) as a function of mass, momentum, wavelength, frequency, or some related quantity. **2.** The set of frequencies, wavelengths, or related quantities, involved in some process; for example, each element has a characteristic discrete spectrum for emission and absorption of light. **3.** A range of frequencies within which radiation has some specified characteristic, such as audio-frequency spectrum, ultraviolet spectrum, or radio spectrum. { 'spek·trəm }

spectrum analysis [PHYS] The measurement of the amplitude of the components of a complex waveform throughout the frequency range of the waveform. { 'spek·trəm əˌnal·ə·səs }

spectrum line *See* spectral line. { 'spek·trəm ˌlīn }

spectrum locus [OPTICS] The locus of points representing the chromaticities of spectrally pure stimuli in a chromaticity diagram. Also known as spectral locus. { 'spek·trəm ˌlō·kəs }

specular reflection [PHYS] Reflection of electromagnetic, acoustic, or water waves in which the reflected waves travel in a definite direction, and the directions of the incident and reflected waves make equal angles with a line perpendicular to the reflecting surface, and lie in the same plane with it. Also known as direct reflection; mirror reflection; regular reflection. { 'spek·yə·lər ri'flek·shən }

specular reflection factor [OPTICS] The ratio of the specularly reflected light to the incident light. { 'spek·yə·lər ri'flek·shən ˌfak·tər }

specular reflection model [PHYS] A model for the behavior of gas molecules striking the surface of a solid body, in which the molecules are reflected so that the component of velocity tangent to the surface is unchanged while the component of velocity perpendicular to the surface is reversed. { 'spek·yə·lər ri'flek·shən ˌmäd·əl }

specular reflector [OPTICS] A reflecting surface (polished metal or silvered glass) that gives a direct image of the source, with the angle of reflection equal to the angle of incidence. Also known as regular reflector; specular surface. { 'spek·yə·lər ri'flek·tər }

specular surface *See* specular reflector. { 'spek·yə·lər 'sər·fəs }

specular transmittance [ELECTROMAG] The ratio of the power carried by electromagnetic radiation which emerges from a body and is parallel to a beam entering the body, to the power carried by the beam entering the body. { 'spek·yə·lər tranz'mit·əns }

speculum [OPTICS] An optical instrument reflector of polished metal or of glass with a film of metal. { 'spek·yə·ləm }

speech clipping [ACOUS] In tests of the intelligibility of speech signals, the limiting of peak signals to a maximum value, or the reduction of signals of less than a certain value to zero. { 'spēch ˌklip·iŋ }

speech interference level [ACOUS] The average sound pressure, in decibels above 0.0002 microbar, in the frequency range from 600 to 4800 hertz. Abbreviated SIL. { 'spēch ˌin·tər¦fir·əns ˌlev·əl }

speed [MECH] The time rate of change of position of a body without regard to direction; in other words, the magnitude of the velocity vector. [OPTICS] **1.** The light-gathering power of a lens, expressed as the reciprocal of the *f* number. **2.** The time that a camera shutter is open. [PHYS] In general, the rapidity with which a process takes place. { spēd }

speed of light [ELECTROMAG] The speed of propagation of electromagnetic waves in a vacuum, which is a physical constant equal to exactly 299,792.458 kilometers per second. Also known as electromagnetic constant; velocity of light. { 'spēd əv 'līt }

speed of response [PHYS] The time required for a system to react to some signal; for example, the delay time for a photon detector to react to a radiation pulse, or the time needed for a current or voltage in a circuit to reach a definite fraction of its final value as a result of an abrupt change in the electromotive force. { 'spēd əv ri'späns }

speed of sound [ACOUS] The phase velocity of a sound wave. Also known as sonic speed; sonic velocity; sound velocity; velocity of sound. { 'spēd əv 'sau̇nd }

speromagnetic state [SOLID STATE] The condition of a rare-earth glass in which the spins are oriented in fixed directions which are more or less random because of electric fields which exist in the glass. { ¦spir·ə·mag¦ned·ik 'stāt }

sp gr *See* specific gravity.

sphenoid [CRYSTAL] An open crystal, occurring in monoclinic crystals of the sphenoidal class, and characterized by two nonparallel faces symmetrical with an axis of twofold symmetry. { 'sfēˌnȯid }

spherator [PL PHYS] One of the class of low-β, low-density, quasi-steady-state closed devices (like Tokamak) used in studying production of electric power by fusion. { 'sfeˌrād·ər }

sphere gap [ELEC] A spark gap between two equal-diameter spherical electrodes. { 'sfir ˌgap }

sphere photometer *See* integrating-sphere photometer. { 'sfir fə'täm·əd·ər }

spherical aberration [OPTICS] Aberration arising from the fact that rays which are initially at different distances from the optical axis come to a focus at different distances along the axis when they are reflected from a spherical mirror or refracted by a lens with spherical surfaces. { 'sfir·ə·kəl ˌab·ə'rā·shən }

spherical antenna [ELECTROMAG] An antenna having the shape of a sphere, used chiefly in theoretical studies. { 'sfir·ə·kəl an'ten·ə }

spherical capacitor [ELEC] A capacitor made of two concentric metal spheres with a dielectric filling the space between the spheres. { 'sfir·ə·kəl kə'pas·əd·ər }

spherical-earth attenuation [ELECTROMAG] Attenuation over an imperfectly conducting spherical earth in excess of that over a perfectly conducting plane. { 'sfir·ə·kəl ¦ərth əˌten·yəˌwā·shən }

spherical-earth factor [ELECTROMAG] The ratio of the electric field strength that would result from propagation over an imperfectly conducting spherical earth to that which would result from propagation over a perfectly conducting plane. { 'sfir·ə·kəl ¦ərth ˌfak·tər }

spherical lens [OPTICS] A lens whose surfaces form portions of spheres. { 'sfir·ə·kəl 'lenz }

spherical mirror [OPTICS] A mirror, either convex or concave, whose surface forms part of a sphere. { 'sfir·ə·kəl 'mir·ər }

spherical pendulum [MECH] A simple pendulum mounted on a pivot so that its motion is not confined to a plane; the bob moves over a spherical surface. { 'sfir·ə·kəl 'pen·jə·ləm }

spherical stress [MECH] The portion of the total stress that corresponds to an isotropic hydrostatic pressure; its stress tensor is the unit tensor multiplied by one-third the trace of the total stress tensor. { 'sfir·ə·kəl 'stres }

spherical wave [PHYS] A wave whose equiphase surfaces form a family of concentric spheres; the direction of travel is always perpendicular to the surfaces of the spheres. { 'sfir·ə·kəl 'wāv }

spherochromatism [OPTICS] The variation of chromatic aberration with color of light. { ¦sfir·ō'krō·məˌtiz·əm }

spherocylindrical lens [OPTICS] A lens having one surface that is a portion of a sphere, while the other is a portion of a cylinder. { ¦sfir·ō·si¦lin·dri·kəl 'lenz }

spheroidal group [CRYSTAL] A group in the tetragonal symmetry system; the sphenoid is the typical form. { sfir'ȯid·əl 'grüp }

spherotoric lens [OPTICS] A lens having one surface that is a portion of a sphere, while the other is a portion of a torus. { ¦sfir·ə¦tȯr·ik 'lenz }

spiderweb antenna [ELECTROMAG] All-wave receiving antenna having several different lengths of doublets connected somewhat like the web of a spider to give favorable pickup characteristics over a wide range of frequencies. { 'spīd·ərˌweb anˌten·ə }

spike [PHYS] A short-duration transient whose amplitude considerably exceeds the average amplitude of the associated pulse or signal. [SOLID STATE] A sputtering event in which the process from impact of a bombarding projectile to the ejection of target atoms involves motion of a large number of particles in the target, so that collisions between particles become significant. { spīk }

spike antenna *See* monopole antenna. { 'spīk anˌten·ə }

spin [MECH] Rotation of a body about its axis. [QUANT MECH] The intrinsic angular momentum of an elementary particle or nucleus, which exists even when the particle is at rest, as distinguished from orbital angular momentum. { spin }

spin axis [PHYS] The axis of rotation of a gyroscope. { 'spin ˌak·səs }

spin compensation [MECH] Overcoming or reducing the effect of projectile rotation in decreasing the penetrating capacity of the jet in shaped-charge ammunition. { 'spin ˌkäm·pənˌsā·shən }

spin-decelerating moment [MECH] A couple about the axis of the projectile, which diminishes spin. { 'spin di¦sel·əˌrād·iŋ 'mō·mənt }

spin-density wave [SOLID STATE] The ground state of a metal in which the conduction-electron spin density has a sinusoidal variation in space. { 'spin ¦den·səd·ē 'wāv }

spin-dependent force [PHYS] A force between two particles which depends in some way on the spin, possibly on the angle between their spin directions, or on the angles between their spin directions and a line joining the particles. { 'spin di¦pen·dənt 'fȯrs }

spin echo technique [NUC PHYS] A variation of the nuclear magnetic resonance technique in which the radio frequency field is applied in two pulses, separated by a time interval *t*, and a strong nuclear induction signal is observed at a time *t* after the second pulse. { 'spin ¦ek·ō tekˌnēk }

spin-flip laser [OPTICS] A semiconductor laser in which the output wavelength is continuously tunable by a magnetic field; operation is based on exciting conduction-band electrons to a higher energy level by reversing the direction of the electrons as they spin about an axis in the direction of the magnetic field. { 'spin ¦flip 'lā·zər }

spin-flip scattering [QUANT MECH] Scattering of a particle with spin 1/2 in which the direction of the particle's spin is reversed. { 'spin ¦flip 'skad·ə·riŋ }

spin glass [SOLID STATE] A substance in which the atomic spins are oriented in random but fixed directions. { 'spin ˌglas }

spin isomer [NUC PHYS] An excited nuclear state which has an unusually long lifetime because of the large difference between the spin of the state and the spins of the states of lower energy into which it is permitted to decay. { 'spin ˌī·sə·mər }

spin-lattice interaction [SOLID STATE] The state of a solid when the energy of electron spins is being shared with the thermal-vibration energy of the solid as a whole. { 'spin ¦lad·əs ˌin·təˌrak·shən }

spin-lattice relaxation [SOLID STATE] Magnetic relaxation in which the excess potential energy associated with electron spins in a magnetic field is transferred to the lattice. { 'spin ¦lad·əs ,rē ,lak,sā·shən }

spin magnetism [SOLID STATE] Paramagnetism or ferromagnetism that arises from polarization of electron spins in a substance. { 'spin ,mag·nə,tiz·əm }

Spinnbarkeit relaxation [FL MECH] A rheological effect illustrated by the pulling away of liquid threads when an object that has been immersed in a viscoelastic fluid is pulled out. { 'spin ,bär,kīt ,rē,lak,sā·shən }

spinning acoustic modes [ACOUS] Natural acoustic waveforms in a circular duct, consisting of pressure disturbances with wavefronts that follow cylindrical paths. { ¦spin·iŋ ə,kü·stik 'mäd·əlz }

spin-orbit coupling [QUANT MECH] The interaction between a particle's spin and its orbital angular momentum. { 'spin ¦ȯr·bət ,kəp·liŋ }

spin-orbit multiplet [PHYS] A collection of atomic or nuclear states which differ in energy only on account of spin-orbit coupling; the total spin angular momentum quantum number S and total orbital angular momentum quantum number L are the same for all states; the energy levels are labeled by the total angular momentum quantum number J. { 'spin ¦ȯr·bət 'məl·tə·plət }

spin paramagnetism [SOLID STATE] Paramagnetism that arises from the electron spins in a substance. { 'spin ,par·ə'mag·nə,tiz·əm }

spin-parity [PART PHYS] A combined symbol J^P for an elementary particle's spin J, and its intrinsic parity P. { 'spin ,par·əd·ē }

spin-polarized low-energy electron diffraction [SOLID STATE] A version of low-energy electron diffraction in which electrons in the incident beam have their spins aligned in one direction; used in studies of the magnetic properties of atoms near the surface of a material. { 'spin ¦pō·lə,rīzd ¦lō 'en·ər·jē i'lek,trän di,frak·shən }

spin quantum number [QUANT MECH] The ratio of the maximum observable component of a system's spin to Planck's constant divided by 2π; it is an integer or a half-integer. { 'spin 'kwän·təm ,nəm·bər }

spin resonance *See* magnetic resonance. { 'spin ,rez·ən·əns }

spin-spin energy [PHYS] An interaction energy proportional to the dot product of the spin angular momenta of two systems. { 'spin 'spin ,en·ər·jē }

spin-spin relaxation [SOLID STATE] Magnetic relaxation, observed after application of weak magnetic fields, in which the excess potential energy associated with electron spins in a magnetic field is redistributed among the spins, resulting in heating of the spin system. { 'spin 'spin ,rē,lak,sā·shən }

spin state [QUANT MECH] Condition of a particle in which its total spin, and the component of its spin along some specified axis, have definite values; more precisely, the particle's wave function is an eigenfunction of the operators corresponding to these quantities. { 'spin ,stāt }

spin temperature [SOLID STATE] For a system of electron spins in a lattice, a temperature such that the population of the energy levels of the spin system is given by the Boltzmann distribution with this temperature. { 'spin ,tem·prə·chər }

spin wave [SOLID STATE] A sinusoidal variation, propagating through a crystal lattice, of that angular momentum which is associated with magnetism (mostly spin angular momentum of the electrons). { 'spin ,wāv }

SPL *See* sound pressure level.

split cameras [OPTICS] An assembly of two cameras disposed at a fixed overlapping angle relative to each other. { 'split 'kam·rəz }

split interstitial [CRYSTAL] A crystal defect in which a displaced atom forms a bond with a normal atom in such a way that neither atom is on the normal site but the two are symmetrically displaced from it. { 'split ,in·tər'stish·əl }

split-lens interference [OPTICS] Interference produced by a Billet split lens. { 'split ¦lenz ,in·tər'fir·əns }

spoiler [ELECTROMAG] Rod grating mounted on a parabolic reflector to change the pencil-beam pattern of the reflector to a cosecant-squared pattern; rotating the reflector and grating 90° with respect to the feed antenna changes one pattern to the other. { 'spȯi·lər }

spontaneous [PHYS] Occurring without application of an external agency, because of the inherent properties of an object. { spän'tā·nē·əs }

spontaneous emission [ELECTROMAG] The emission of radiation from a system in an excited state at a rate that does not depend on the presence of external fields. { spän'tan·ē·əs i'm·ish·ən }

spontaneous fission [NUC PHYS] Nuclear fission in which no particles or photons enter the nucleus from the outside. { spän'tā·nē·əs 'fi·shən }

spontaneous magnetization [ELECTROMAG] Magnetization which a substance possesses in the absence of an applied magnetic field. { spän'tā·nē·əs ,mag·nə·də'zā·shən }

spontaneous polarization [ELEC] Electric polarization that a substance possesses in the absence of an external electric field. { spän'tā·nē·əs ,pō·lə·rə'zā·shən }

spontaneous process [THERMO] A thermodynamic process which takes place without the application of an external agency, because of the inherent properties of a system. { spän'tā·nē·əs 'prä·səs }

spontaneous symmetry breaking [PHYS] A situation in which the solution of a set of physical equations fails to exhibit a symmetry possessed by the equations themselves; an example is a magnet, in which the underlying equations describing the metal do not distinguish any direction of space from any other, but the magnet certainly does, since it points in some definite direction. { spän'tā·nē·əs 'sim·ə·trē ˌbrāk·iŋ }

spotter *See* dotter. { 'späd·ər }

spray angle [FL MECH] The angle formed by the cone of liquid leaving a nozzle orifice. { 'sprā ˌaŋ·gəl }

spray flow [FL MECH] A two-phase flow in which the liquid phase is the dispersed phase and exists in the form of many droplets, while the gas phase is the continuous phase. { 'sprā ˌflō }

spreading anomaly [PHYS] That part of the propagation anomaly which may be identified with the geometry of the ray paths. { 'spred·iŋ əˌnäm·ə·lē }

spreading coefficient [THERMO] The work done in spreading one liquid over a unit area of another, equal to the surface tension of the stationary liquid, minus the surface tension of the spreading liquid, minus the interfacial tension between the liquids. { 'spred·iŋ ˌkō·iˌfish·ənt }

spreading method [ELEC] A method of calculating the potential due to a set of point charges by replacing them with a continuous distribution of charge or a distribution of charge and polarization. { 'spred·iŋ ˌmeth·əd }

spread reflection [ELECTROMAG] Reflection of electromagnetic radiation from a rough surface with large irregularities. Also known as mixed reflection. { 'spred riˌflek·shən }

spring modulus [MECH] The additional force necessary to deflect a spring an additional unit distance; if a certain spring has a modulus of 100 newtons per centimeter, a 100-newton weight will compress it 1 centimeter, a 200-newton weight 2 centimeters, and so on. { 'spriŋ 'mäj·ə·ləs }

s-process [NUC PHYS] The synthesis of elements, predominantly in the iron group, over long periods of time through the capture of slow neutrons which are produced mainly by the reactions of α-particles with carbon-13 and neon-21. { 'es ˌprä·səs }

SPSL *See* sound pressure spectrum level.

spur [PHYS] A cluster of ionized molecules near the path of an energetic charged particle, consisting of the molecule ionized directly by the charged particle, and secondary ionizations produced by electrons released in the primary ionization; it usually forms a side track from the path of the particle. { spər }

spurious disk [OPTICS] The nearly round image of perceptible diameter of a star as seen through a telescope, due to diffraction of light in the telescope. { 'spyu̇r·ē·əs 'disk }

sp vol *See* specific volume.

sq *See* square.

square [MECH] Denotes a unit of area; if *x* is a unit of length, a square *x* is the area of a square whose sides have a length of 1*x*; for example, a square meter, or a meter squared, is the area of a square whose sides have a length of 1 meter. Also known as monomino. Abbreviated sq. { skwer }

square-foot unit of absorption *See* sabin. { 'skwer 'fu̇t 'yü·nət əv əb'sȯrp·shən }

square-loop ferrite [ELECTROMAG] A ferrite that has an approximately rectangular hysteresis loop. { 'skwer ¦lüp 'feˌrīt }

squareness ratio [ELECTROMAG] **1.** The magnetic induction at zero magnetizing force divided by the maximum magnetic induction, in a symmetric cyclic magnetization of a material. **2.** The magnetic induction when the magnetizing force has changed half-way from zero toward its negative limiting value divided by the maximum magnetic induction in a symmetric cyclic magnetization of a material. { 'skwer·nəs ˌrā·shō }

square-on *See* center. { ¦skwer 'ȯn }

square wave [ELEC] An oscillation the amplitude of which shows periodic discontinuities between two values, remaining constant between jumps. { 'skwer 'wāv }

squeezable waveguide [ELECTROMAG] A waveguide whose dimensions can be altered periodically; used in rapid scanning. { 'skwēz·ə·bəl 'wāvˌgīd }

squeeze [PHYS] Increasing external pressure upon the ears and sinuses in diving. { skwēz }

squeezed state [QUANT MECH] **1.** A quantum state for which one of a pair of conjugate variables, which cannot simultaneously possess definite values according to the Heisenberg uncertainty principle, is specified more accurately than in the vacuum state, at the expense of increasing the uncertainty in the value of the other variable. **2.** A quantum state in which one of a pair of conjugate variables is specified more accurately than in the vacuum state, and the product of the uncertainties of the conjugate variables is the minimum allowed by the Heisenberg uncertainty principle. { 'skwēzd ˌstāt }

squeeze section [ELECTROMAG] Length of waveguide constructed so that alteration of the critical dimension is possible with a corresponding alteration in the electrical length. { 'skwēz ˌsek·shən }

SRMS *See* structure resonance modulation spectroscopy.

s-state [QUANT MECH] A single-particle state whose orbital angular momentum quantum number is zero. { 'es ˌstāt }

SSU *See* Saybolt Seconds Universal.

St *See* stoke.

stability [FL MECH] The resistance to overturning or mixing in the water column, resulting from the presence of a positive (increasing downward) density gradient. [MECH] *See* dynamic stability. [PHYS] **1.** The property of a system which does not undergo any change without the application of an external agency. **2.** The property of a system in which any departure from an equilibrium state gives rise to forces or influences which tend to return the system to equilibrium. Also known as static stability. [PL PHYS] The property of a plasma which maintains its shape against externally applied forces (usually pressure of magnetic fields) and whose constituents can pass through confining fields only by diffusion of individual particles. { stə'bil·əd·ē }

stability matrix *See* stiffness matrix. { stə'bil·əd·ē ˌmā·triks }

stabilization [ELECTROMAG] Treatment of a magnetic material to improve the stability of its magnetic properties. { ˌstā·bə·lə'zā·shən }

stabilizing magnetic field [PL PHYS] A magnetic field which is added to a device confining a plasma in order to increase the plasma's stability. { 'stā·bəˌlīz·iŋ mag'ned·ik 'fēld }

stable [PHYS] Not subject to any change without the application of an external agency, such as radiation; said of a molecule, atom, nucleus, or elementary particle. { 'stā·bəl }

stable isobar [NUC PHYS] One of two or more stable nuclides which have the same mass number but differ in atomic number. { 'stā·bəl 'ī·səˌbär }

stable isotope [NUC PHYS] An isotope which does not spontaneously undergo radioactive decay. { 'stā·bəl 'ī·səˌtōp }

stable nucleus [NUC PHYS] A nucleus which does not spontaneously undergo radioactive decay. { 'stā·bəl 'nü·klē·əs }

stable system [PHYS] A system that returns to a stationary state following sufficiently small perturbations. { ¦stā·bəl ¦sis·təm }

stacked antennas [ELECTROMAG] Two or more identical antennas arranged above each other on a vertical supporting structure and connected in phase to increase the gain. { 'stakt an'ten·əz }

stacked array [ELECTROMAG] An array in which the antenna elements are stacked one above the other and connected in phase to increase the gain. { 'stakt ə'rā }

stacked-dipole antenna [ELECTROMAG] Antenna in which directivity is increased by providing a number of identical dipole elements, excited either directly or parasitically; the resultant radiation pattern depends on the number of dipole elements used, the spacing and phase difference between the elements, and the relative magnitudes of the currents. { 'stakt 'dīˌpōl anˌten·ə }

stacked loops [ELECTROMAG] Two or more loop antennas arranged above each other on a vertical supporting structure and connected in phase to increase the gain. Also known as vertically stacked loops. { 'stakt 'lüps }

stacking [ELECTROMAG] The placing of antennas one above the other, connecting them in phase to increase the gain. { 'stak·iŋ }

stacking fault [CRYSTAL] A defect in a face-centered cubic or hexagonal close-packed crystal in which there is a change from the regular sequence of positions of atomic planes. { 'stak·iŋ ˌfȯlt }

Staebler-Wronski effect [SOLID STATE] A reversible change (usually a reduction) in the dark conductivity and photoconductivity of hydrogenated amorphous silicon during and following illumination by light with sufficient energy to produce electron-hole pairs. { ¦stāb·lər 'rän·skē iˌfekt }

stagnation point [FL MECH] A point in a field of flow about a body where the fluid particles have zero velocity with respect to the body. { stag'nā·shən ˌpȯint }

stagnation pressure *See* dynamic pressure. { stag'nā·shən ˌpresh·ər }

stagnation temperature *See* adiabatic recovery temperature. { stag'nā·shən ˌtem·prə·chər }

standard [PHYS] An accepted reference sample used for establishing a unit for the measurement of a physical quantity. { 'stan·dərd }

standard antenna [ELECTROMAG] An open single-wire antenna (including the lead-in wire) having an effective height of 4 meters. { 'stan·dərd an'ten·ə }

standard atmosphere *See* atmosphere. { 'stan·dərd 'at·məˌsfir }

standard ballistic conditions [MECH] A set of ballistic conditions arbitrarily assumed as standard for the computation of firing tables. { 'stan·dərd bə'lis·tik kən'dish·ənz }

standard candle *See* international candle. { 'stan·dərd 'kand·əl }

standard capacitor [ELEC] A capacitor constructed in such a manner that its capacitance value is not likely to vary with temperature and is known to a high degree of accuracy. Also known as capacitance standard. { 'stan·dərd kə'pas·əd·ər }

standard cell [ELEC] A primary cell whose voltage is accurately known and remains sufficiently constant for instrument calibration purposes; the Weston standard cell has a voltage of 1.018636 volts at 20°C. { 'stan·dərd 'sel }

standard conditions [PHYS] **1.** A temperature of 0°C and a pressure of 1 atmosphere (760 torr). Also known as normal temperature and pressure (NTP); standard temperature and pressure (STP). **2.** According to the American Gas Association, a temperature of 60°F (155/9°C) and a pressure of 762 millimeters (30 inches) of mercury. **3.** According to the Compressed Gas Institute, a temperature of 20°C (68°F) and a pressure of 1 atmosphere. [SOLID STATE] The allotropic form in which a substance most commonly occurs. { 'stan·dərd kən'dish·ənz }

standard free-energy increase [THERMO] The increase in Gibbs free energy in a chemical reaction, when both the reactants and the products of the reaction are in their standard states. { 'stan·dərd 'frē ¦en·ər·jē 'inˌkrēs }

standard gravity [MECH] A value of the acceleration of gravity equal to 9.80665 meters per second per second. { 'stan·dərd 'grav·əd·ē }

standard heat of formation [THERMO] The heat needed to produce one mole of a compound from its elements in their standard state. { 'stan·dərd 'hēt əv fȯr'mā·shən }

standard illuminants [OPTICS] Three standard sources of light, designated A, B, and C, used in specifying the light used when colors are matched; A is light from a filament at a color temperature of 2575°C, and B and C, representing noon sunlight and normal daylight respectively, are obtained by modifying A with rigorously specified filters. { 'stan·dərd i'lü·mə·nəns }

standard inductor [ELECTROMAG] An inductor (coil) having high stability of inductance value, with little variation of inductance with current or frequency and with a low temperature coefficient; it may have an air core or an iron core; used as a primary standard in laboratories and as a precise working standard for impedance measurements. { 'stan·dərd in'dək·tər }

standard lens [OPTICS] Usually the lens provided with a camera as standard equipment; in still cameras, the standard lens is one whose focal length is about equal to the length of the diagonal of the negative area normally provided by the camera; the normal field of view of a standard lens is about 53°. { 'stan·dərd 'lenz }

standard model [PART PHYS] The modern theory of the interactions of elementary particles, comprising the Weinberg-Salam theory and quantum chromodynamics. { 'stan·dərd 'mäd·əl }

standard pitch [ACOUS] A musical pitch based on 440 hertz for tone A; with this standard, the frequency of middle C is 261 hertz. { 'stan·dərd 'pich }

standard plane [CRYSTAL] The crystal plane whose Miller indices are (111), that is, whose intercepts on the crystal axes are proportional to the corresponding sides of a unit cell. { 'stan·dərd 'plān }

standard pressure [PHYS] A pressure of 1 atmosphere (101,325 newtons per square meter), to which measurements of quantities dependent on pressure, such as the volume of a gas, are often referred. Also known as normal pressure. { 'stan·dərd 'presh·ər }

standard propagation [ELECTROMAG] Propagation of radio waves over a smooth spherical earth of specified dielectric constant and conductivity, under conditions of standard refraction in the atmosphere. { 'stan·dərd ˌpräp·ə'gā·shən }

standard refraction [ELECTROMAG] Refraction which would occur in an idealized atmosphere in which the index of refraction decreases uniformly with height at a rate of 39×10^{-6} per kilometer; standard refraction may be included in ground wave calculations by use of an effective earth radius of 8.5×10^{6} meters, or 4/3 the geometrical radius of the earth. { 'stan·dərd ri'frak·shən }

standard state [PHYS] The stable and pure form of a substance at standard pressure and a specified temperature, usually 298 K (77°F). { 'stan·dərd 'stāt }

standard temperature and pressure *See* standard conditions. { 'stan·dərd 'tem·prə·chər ən 'presh·ər }

standard trajectory [MECH] Path through the air that it is calculated a projectile will follow under given conditions of weather, position, and material, including the particular fuse, projectile, and propelling charge that are used; firing tables are based on standard trajectories. { 'stan·dərd trə'jek·trē }

standard volume [PHYS] The volume of 1 mole of a gas at a pressure of 1 atmosphere and a temperature of 0°C. Also known as normal volume. { 'stan·dərd 'väl·yəm }

standing wave [PHYS] A wave in which the ratio of an instantaneous value at one point to that at any other point does not vary with time. Also known as stationary wave. { 'stand·iŋ 'wāv }

standing-wave detector [ELECTROMAG] An electric indicating instrument used for detecting a standing electromagnetic wave along a transmission line or in a waveguide and measuring the resulting standing-wave ratio; it can also be used to measure the wavelength, and hence the frequency, of the wave. Also known as standing-wave indicator; standing-wave meter; standing-wave-ratio meter. { 'stand·iŋ ¦wāv di'tek·tər }

standing-wave indicator *See* standing-wave detector. { 'stand·iŋ ¦wāv 'in·dəˌkād·ər }

standing-wave loss factor [ELECTROMAG] The ratio of the transmission loss in an unmatched waveguide to that in the same waveguide when matched. { 'stand·iŋ ¦wāv 'lȯs ˌfak·tər }

standing-wave meter *See* standing-wave detector. { 'stand·iŋ ¦wāv 'mēd·ər }

standing-wave method [ELECTROMAG] Any method of measuring the wavelength of electromagnetic waves that involves measuring the distance between successive nodes or antinodes of standing waves. { 'stand·iŋ ¦wāv 'meth·əd }

standing-wave producer [ELECTROMAG] A movable probe inserted in a slotted waveguide to produce a desired standing-wave pattern, generally for test purposes. { 'stand·iŋ ¦wāv prə'dü·sər }

standing-wave ratio [PHYS] **1.** The ratio of the maximum amplitude to the minimum amplitude of corresponding components of a wave in a transmission line or waveguide. **2.** The reciprocal of this ratio. { 'stand·iŋ ¦wāv 'rā·shō }

standing-wave-ratio meter *See* standing-wave detector. { 'stand·iŋ ¦wāv 'rā·shō ˌmēd·ər }

Stanhope lens [OPTICS] A thick biconvex lens with front and back surfaces having radii of curvature of two-thirds and one-third the lens thickness; used as a magnifier by placing the object to be viewed in contact with the front surface. { 'stanˌhōp ˌlenz }

Stanton diagram [FL MECH] The plot of the airflow friction coefficient against the Reynolds number. { 'stant·ən ˌdī·əˌgram }

Stanton number [THERMO] A dimensionless number used in the study of forced convection, equal to the heat-transfer coefficient of a fluid divided by the product of the specific heat at constant pressure, the fluid density, and the fluid velocity. Symbolized N_{St}. Also known as Margoulis number (M). { 'stant·ən ˌnəm·bər }

star [OPTICS] A light source that subtends a very small angle at the entrance pupil of an optical instrument and is used to test the instrument. { stär }

Stark effect [SPECT] The effect on spectrum lines of an electric field which is either externally applied or is an internal field caused by the presence of neighboring ions or atoms in a gas, liquid, or solid. Also known as electric field effect. { 'stärk i,fekt }

Stark-Lunelund effect [ELECTROMAG] The polarization of light emitted from a beam of moving atoms in a region where there are no electric or magnetic fields. { 'stärk 'lün·ə,lənd i,fekt }

Stark number *See* Stefan number. { 'stärk ,nəm·bər }

star network [ELEC] A set of three or more branches with one terminal of each connected at a common node to give the form of a star. Also known as star connection. { 'stär ¦net,wərk }

star telescope [OPTICS] An accessory of the marine navigational sextant designed primarily for star observations; it has a large objective to give a greater field of view and increased illumination; it is an erect telescope, that is, the object viewed is seen erect as opposed to the inverting telescope in which the object viewed is inverted. { 'stär ,tel·ə,skōp }

star test [OPTICS] A procedure in which a telescope is directed at a bright star and the in-focus and out-of-focus images and diffraction patterns of the star are examined to detect aberrations and abnormalities in the optical system. { 'stär ,test }

starting friction *See* static friction. { 'stärd·iŋ ,frik·shən }

stat- [ELEC] A prefix indicating an electrical unit in the electrostatic centimeter-gram-second system of units; it is attached to the corresponding SI unit. { stat }

statΩ *See* statmho.

statΩ *See* statohm.

statA *See* statampere. { 'stat¦ā }

statampere [ELEC] The unit of electric current in the electrostatic centimeter-gram-second system of units, equal to a flow of charge of 1 statcoulomb per second; equal to approximately 3.3356×10^{-10} ampere. Abbreviated stat A. { stad'am,pir }

statC *See* statcoulomb. { 'stat,sē }

statcoulomb [ELEC] The unit of charge in the electrostatic centimeter-gram-second system of units, equal to the charge which exerts a force of 1 dyne on an equal charge at a distance of 1 centimeter in a vacuum; equal to approximately 3.3356×10^{-10} coulomb. Abbreviated stat C. Also known as franklin (Fr); unit charge. { 'stat¦kü,läm }

state [PHYS] The condition of a system which is specified as completely as possible by observations of a specified nature, for example, thermodynamic state, energy state. [QUANT MECH] The condition in which a system exists; the state may be pure and describable by a wave function or mixed and describable by a density matrix. { stāt }

state of matter [PHYS] One of three fundamental conditions of matter: the solid, liquid, and gaseous states. { 'stāt əv 'mad·ər }

state of strain [MECH] A complete description, including the six components of strain, of the deformation within a homogeneously deformed volume. { 'stāt əv 'strān }

state of stress [MECH] A complete description, including the six components of stress, of a homogeneously stressed volume. { 'stāt əv 'stres }

state parameter *See* thermodynamic function of state. { 'stāt pə,ram·əd·ər }

state variable *See* thermodynamic function of state. { 'stāt ,ver·ē·ə·bəl }

state vector [QUANT MECH] A vector in Hilbert space which corresponds to the state of a quantum-mechanical system. { 'stāt ,vek·tər }

statF *See* statfarad. { 'stad,ef }

statfarad [ELEC] Unit of capacitance in the electrostatic centimeter-gram-second system of units, equal to the capacitance of a capacitor having a charge of 1 statcoulomb, across the plates of which the charge is 1 statvolt; equal to approximately 1.1126×10^{-12} farad. Abbreviated statF. { 'stat¦fa,rad }

statH *See* stathenry. { 'stad,āch }

stathenry [ELECTROMAG] The unit of inductance in the electrostatic centimeter-gram-second system of units, equal to the self-inductance of a circuit or the mutual inductance between two circuits if there is an induced electromotive force of 1 statvolt when the current is changing at a rate of 1 statampere per second; equal to approximately 8.9876×10^{11} henry. Abbreviated statH. { 'stat¦hen·rē }

static [PHYS] Without motion or change. { 'stad·ik }

statically admissible loads [MECH] Any set of external loads and internal forces which fulfills conditions necessary to maintain the equilibrium of a mechanical system. { 'stad·ik·əl·ē əd'mis·ə·bəl 'lōdz }

static charge [ELEC] An electric charge accumulated on an object. { 'stad·ik 'chärj }

static electricity [ELEC] **1.** The study of the effects of macroscopic charges, including the transfer of a static charge from one object to another by actual contact or by means of a spark that bridges an air gap between the objects. **2.** *See* electrostatics. { 'stad·ik ,i,lek'tris·əd·ē }

static equilibrium *See* equilibrium. { 'stad·ik ,ē·kwə'lib·rē·əm }

static fluid column [FL MECH] An unchanging height of fluid in a vertical pipe, well bore, process vessel, or tank. { 'stad·ik ¦flü·əd ,käl·əm }

static friction [MECH] **1.** The force that resists the initiation of sliding motion of one body over the other with which it is in contact. **2.** The force required to move one of the bodies when they are at rest. Also known as limiting friction; starting friction. { 'stad·ik 'frik·shən }

static gel buildup [FL MECH] A method used to infer the degree of thixotropy of a fluid by viscometric measurement of its gel strength. { 'stad·ik 'jel ,bild,əp }

static head [FL MECH] Pressure of a fluid due to the head of fluid above some reference point. { 'stad·ik 'hed }

static level [FL MECH] Elevation of the water level or a pressure surface at rest. { 'stad·ik 'lev·əl }

static limit *See* stationary limit. { 'stad·ik 'lim·ət }

static load [MECH] A nonvarying load; the basal pressure exerted by the weight of a mass at rest, such as the load imposed on a drill bit by the weight of the drill-stem equipment or the pressure exerted on the rocks around an underground opening by the weight of the superimposed rocks. Also known as dead load. { 'stad·ik 'lōd }

static magnetism *See* magnetostatics. { 'stad·ik 'mag·nə,tiz·əm }

static moment [MECH] **1.** A scalar quantity (such as area or mass) multiplied by the perpendicular distance from a point connected with the quantity (such as the centroid of the area or the center of mass) to a reference axis. **2.** The magnitude of some vector (such as force, momentum, or a directed line segment) multiplied by the length of a perpendicular dropped from the line of action of the vector to a reference point. { 'stad·ik 'mō·mənt }

static pressure [ACOUS] The pressure that would exist at a point in a medium if no sound waves were present. [FL MECH] **1.** The normal component of stress, the force per unit area, exerted across a surface moving with a fluid, especially across a surface which lies in the direction of fluid flow. **2.** The average of the normal components of stress exerted across three mutually perpendicular surfaces moving with a fluid. { 'stad·ik 'presh·ər }

static reaction [MECH] The force exerted on a body by other bodies which are keeping it in equilibrium. { 'stad·ik rē'ak·shən }

statics [MECH] The branch of mechanics which treats of force and force systems abstracted from matter, and of forces which act on bodies in equilibrium. { 'stad·iks }

static stability *See* stability. { 'stad·ik stə'bil·əd·ē }

stationary distribution [PHYS] A time-independent distribution of a scalar quantity. { 'stā·shə,ner·ē ,dis·trə'byü·shən }

stationary field [PHYS] Field which does not change during the time interval under consideration. Also known as constant field. { 'stā·shə,ner·ē 'fēld }

stationary limit [RELAT] In the Kerr solution to Einstein's equations, a surface on which a particle would have to move with the local light velocity in order to appear stationary to a distant observer, and inside which no particle can appear stationary to such an observer. Also known as static limit. { 'stā·shə,ner·ē 'lim·ət }

stationary state *See* energy state. { 'stā·shə,ner·ē 'stāt }

stationary time principle *See* Fermat's principle. { 'stā·shə,ner·ē 'tīm ,prin·sə·pəl }

stationary wave *See* standing wave. { 'stā·shə,ner·ē 'wāv }

statistical mechanics [PHYS] That branch of physics which endeavors to explain and predict the macroscopic properties and behavior of a system on the basis of the known characteristics and interactions of the microscopic constituents of the system, usually when the number of such constituents is very large. Also known as statistical thermodynamics. { stə'tis·tə·kəl mi'kan·iks }

statistical thermodynamics *See* statistical mechanics. { stə'tis·tə·kəl ,thər·mō·dī'nam·iks }

statistical weight [STAT MECH] **1.** The number of microscopic states that correspond to a given macroscopic state. **2.** A multiplicative factor in the expression for the probability of finding a system in a given quantum state, usually equal to the number of degenerate substates contained in the state. { stə'tis·tə·kəl 'wāt }

statmho [ELEC] The unit of conductance, admittance, and susceptance in the electrostatic centimeter-gram-second system of units, equal to the conductance between two points of a conductor when a constant potential difference of 1 statvolt applied between the points produces in this conductor a current of 1 statampere, the conductor not being the source of any electromotive force; equal to approximately 1.1126×10^{-12} mho. Abbreviated statΩ. Also known as statsiemens (statS). { 'stat,mō }

statohm [ELEC] The unit of resistance, reactance, and impedance in the electrostatic centimeter-gram-second system of units, equal to the resistance between two points of a conductor when a constant potential difference of 1 statvolt between these points produces a current of 1 statampere; it is equal to approximately 8.9876×10^{11} ohms. Abbreviated statΩ. { 'stad,ōm }

statS *See* statmho. { ¦stat'es }

statsiemens *See* statmho. { ¦stat'sē·mənz }

statT *See* stattesla. { ¦stat'tē }

stattesla [ELECTROMAG] The unit of magnetic flux density in the electrostatic centimeter-gram-second system of units, equal to one statweber per square centimeter; equal to approximately 2.9979×10^{6} tesla. Abbreviated statT. { ¦stat 'tes·lə }

statute mile *See* mile. { 'stach·üt 'mīl }

statV *See* statvolt.

statvolt [ELEC] The unit of electric potential and electromotive force in the electrostatic centimeter-gram-second system of units, equal to the potential difference between two points such that the work required to transport 1 statcoulomb of electric charge from one to the other is equal to 1 erg; equal to approximately 299.79 volts. Abbreviated statV. { 'stat,vōlt }

statWb *See* statweber.

statweber [ELECTROMAG] The unit of magnetic flux in the electrostatic centimeter-gram-second system of units, equal to the magnetic flux which, linking a circuit of one turn, produces in it an electromotive force of 1 statvolt as it is reduced to zero at a uniform rate in 1 second; equal to approximately 299.79 webers. Abbreviated statWb. { 'stat,web·ər }

steady flow [FL MECH] Fluid flow in which all

the conditions at any one point are constant with respect to time. { 'sted·ē 'flō }

steady state [PHYS] The condition of a body or system in which the conditions at each point do not change with time, that is after initial transients or fluctuations have disappeared. { 'sted·ē 'stāt }

steady-state cavitation *See* sheet cavitation. { 'sted·ē ¦stāt ,kav·ə'tā·shən }

steady-state conduction [THERMO] Heat conduction in which the temperature and heat flow at each point does not change with time. { 'sted·ē ¦stāt kən'dək·shən }

steady-state creep *See* secondary creep. { 'sted·ē ¦stāt 'krēp }

steady-state current [ELEC] An electric current that does not change with time. { 'sted·ē ¦stāt 'kə·rənt }

steady-state vibration [MECH] Vibration in which the velocity of each particle in the system is a continuous periodic quantity. { 'sted·ē ¦stāt vī'brā·shən }

steady-state wave motion [PHYS] Wave motion in which the wave quantities at each point in the region through which the wave is passing repeat themselves periodically. { 'sted·ē ¦stāt 'wāv ,mō·shən }

steam [PHYS] Water vapor, or water in its gaseous state; the most widely used working fluid in external combustion engine cycles. { stēm }

steam cycle *See* Rankine cycle. { 'stēm ,sī·kəl }

steam line [THERMO] A graph of the boiling point of water as a function of pressure. { 'stēm ,līn }

steam point [THERMO] The boiling point of pure water whose isotopic composition is the same as that of sea water at standard atmospheric pressure; it is assigned a value of 100°C on the International Practical Temperature Scale of 1968. { 'stēm ,pȯint }

Stefan-Boltzmann constant [STAT MECH] The energy radiated by a blackbody per unit area per unit time divided by the fourth power of the body's temperature; equal to $(5.6696 \pm 0.0010) \times 10^{-8}$ (watt)(meter)$^{-2}$(kelvin)$^{-4}$. { 'shte,fän 'bōlts,män ,kän·stənt }

Stefan-Boltzmann law [STAT MECH] The total energy radiated from a blackbody is proportional to the fourth power of the temperature of the body. Also known as fourth-power law; Stefan's law of radiation. { 'shte,fän 'bōlts,män ,lȯ }

Stefan number [THERMO] A dimensionless number used in the study of radiant heat transfer, equal to the Stefan-Boltzmann constant times the cube of the temperature times the thickness of a layer divided by the layer's thermal conductivity. Symbolized St. Also known as Stark number (*Sk*). { 'shte,fän ,nəm·bər }

Stefan's law of radiation *See* Stefan-Boltzmann law. { 'shte,fänz 'lȯ əv 'rād·ē,ā·shən }

Steiner's theorem *See* parallel axis theorem. { 'shtīn·ərz ,thir·əm }

Steinheil lens [OPTICS] A type of magnifier lens in which a biconvex crown lens is cemented between a pair of flint lenses. { 'shtīn,hīl ,lenz }

Steinmetz coefficient [ELECTROMAG] The constant of proportionality in Steinmetz's law. { 'stīn,mets ,kō·i,fish·ənt }

Steinmetz's law [ELECTROMAG] The energy converted into heat per unit volume per cycle during a cyclic change of magnetization is proportional to the maximum magnetic induction raised to the 1.6 power, the constant of proportionality depending only on the material. { 'stīn,mets·əz ,lȯ }

stellarator [PL PHYS] A device for confining a high-temperature plasma, consisting of a tube, which closes in on itself in a figure-eight or racetrack configuration, and external coils which generate magnetic fields whose lines of force run parallel to the walls of the tube and prevent the plasma from touching the walls. { 'stel·ə,rād·ər }

stellar crystal *See* plane-dendritic crystal. { 'stel·ər 'krist·əl }

stellar interferometer [OPTICS] An optical interferometer for measuring angular diameters of stars; it is attached to a telescope and measures interference rings at the telescope's focus. { 'stel·ər ,in·tər·fə'räm·əd·ər }

stem correction [THERMO] A correction which must be made in reading a thermometer in which part of the stem, and the thermometric fluid within it, is at a temperature which differs from the temperature being measured. { 'stem kə,rek·shən }

step-and-repeat camera [OPTICS] A type of camera providing a gridlike pattern of latent image frames in a given sequence. { 'step ən ri'pēt ,kam·rə }

steradiancy *See* radiance. { stə'rād·ē·ən·sē }

sterba curtain [ELECTROMAG] Type of stacked dipole antenna array consisting of one or more phased half-wave sections with a quarter-wave section at each end; the array can be oriented for either vertical or horizontal radiation, and can be either center or end fed. { 'stər·bə ,kərt·ən }

stère [MECH] A unit of volume equal to 1 cubic meter; it is used mainly in France, and in measuring timber volumes. { stir }

stereo- [PHYS] A prefix used to designate a three-dimensional characteristic. { 'ste·rē·ō }

stereo camera *See* stereoscopic camera. { 'ster·ē·ō 'kam·rə }

stereocomparagraph [OPTICS] A projection device in which two-dimensional aerial photographs taken at slightly different angles are combined so as to give the appearance of tridimensionality. { ¦ster·ē·ō·kəm'par·ə,graf }

stereo comparator [OPTICS] An instrument that may be used to view two photographs taken of the stars in the same section of sky at different times; viewing the images stereoscopically may reveal stars that have moved between exposures or stars of varying brightness. { 'ster·ē·ō kəm'par·əd·ər }

stereo effect [ACOUS] Reproduction of sound in such a manner that the listener receives the sensation that individual sounds are coming from different locations, just as did the original sounds reaching the stereo microphone system. { 'ster·ē·ō i,fekt }

stereognomogram [CRYSTAL] The projection resulting from the superposition of the projection planes of a stereogram and a gnomogram. { ¦ster·ē·ə'nō·mə,gram }

stereographic projection [CRYSTAL] A method of displaying the positions of the poles of a crystal in which poles are projected through the equatorial plane of the reference sphere by lines joining them with the south pole for poles in the upper hemisphere, and with the north pole for poles in the lower hemisphere. { ¦ster·ē·ə¦graf·ik prə'jek·shən }

stereomicrography [OPTICS] The taking of two microphotographs of the same field at different angles (a stereo pair), then viewing them simultaneously with a stereo viewer. { ¦ster·ē·ə·mī'kräg·rə·fē }

stereo rangefinder *See* stereoscopic rangefinder. { 'ster·ē·ō 'rānj,fīn·dər }

stereoscope [OPTICS] An optical instrument in which each eye views one of two photographs taken with the camera or object of study displaced, or simultaneously with two cameras, or with a stereoscopic camera, so that a sensation of depth is produced. { 'ster·ē·ə,skōp }

stereoscopic camera [OPTICS] A camera which takes photographs simultaneously with two similar lenses a few inches apart, for use in a stereoscope or other optical system which gives a sensation of depth to the viewer. Also known as stereo camera. { ¦ster·ē·ə¦skäp·ik 'kam·rə }

stereoscopic heightfinder *See* stereoscopic rangefinder. { ¦ster·ē·ə¦skäp·ik 'hīt,fīn·dər }

stereoscopic microscope [OPTICS] A microscope with two eyepieces and two objectives, giving the viewer a sensation of depth. { ,ste·rē·ə¦skäp·ik 'mī·krə,skōp }

stereoscopic photography [OPTICS] A technique that simulates stereoscopic vision, in which two photographs are made with the camera or object of study displaced, or simultaneously with two cameras, or with a stereoscopic camera, and each of the photographs is viewed by one eye, using a stereoscope or other optical system. { ¦ster·ē·ə¦skäp·ik fə'täg·rə·fē }

stereoscopic power [OPTICS] The magnifying power of a binoculars or other stereo system multiplied by the ratio of the distance between the objective axes to the distance between the eyepiece axes; it is a measure of the stereoscopic radius. Also known as total relief. { ¦ster·ē·ə¦skäp·ik 'pau̇·ər }

stereoscopic rangefinder [OPTICS] An optical rangefinder which utilizes stereoscopic vision; it is essentially a large binoculars fitted with special reticles which allow a skilled user to superimpose the stereoscopic image formed by the pair of reticles over the images of the target seen in the eyepieces, so that the correct range is obtained when the reticle marks appear to be suspended over the target and at the same apparent distance. Also known as stereo rangefinder; stereoscopic heightfinder. { ¦ster·ē·ə¦skäp·ik 'rānj,fīn·dər }

stereoscopic system [OPTICS] An optical system such as a binoculars or stereoscope that produces two images of the same object viewed from slightly different positions, so that a sensation of depth is created when one image is presented to each eye. { ,ste·rē·ə¦skäp·ik 'sis·təm }

Stern-Gerlach effect [ATOM PHYS] The splitting of a beam of atoms passing through a strong, inhomogeneous magnetic field into several beams. { 'stərn 'ger,läk i,fekt }

Stern-Zartman experiment [STAT MECH] An experiment in which the distribution in speed of atoms or molecules in a beam emitted from an opening in an oven is measured by having the beam impinge on a rotating cylindrical drum, with a slit cut parallel to the drum axis, and measuring the density of atoms or molecules deposited on the inner surface of the drum, roughly opposite the slit, as a function of distance from a point directly opposite the slit; it is used to test the Maxwell-Boltzmann distribution law. { 'stərn 'zärt·mən ik,sper·ə·mənt }

sthène [MECH] The force which, when applied to a body whose mass is 1 metric ton, results in an acceleration of 1 meter per second per second; equal to 1000 newtons. Formerly known as funal. { sthēn }

stick-slip friction [MECH] Friction between two surfaces that are alternately at rest and in motion with respect to each other. { 'stik ,slip ,frik·shən }

stiction [MECH] Friction that tends to prevent relative motion between two movable parts at their null position. { 'stik·shən }

stiffness [ACOUS] *See* acoustic stiffness. [MECH] The ratio of a steady force acting on a deformable elastic medium to the resulting displacement. { 'stif·nəs }

stiffness coefficient [MECH] The ratio of the force acting on a linear mechanical system, such as a spring, to its displacement from equilibrium. { 'stif·nəs ,kō·i,fish·ənt }

stiffness constant [MECH] Any one of the coefficients of the relations in the generalized Hooke's law used to express stress components as linear functions of the strain components. Also known as elastic constant. { 'stif·nəs ,kän·stənt }

stiffness matrix [MECH] A matrix **K** used to express the potential energy V of a mechanical system during small displacements from an equilibrium position, by means of the equation $V = \frac{1}{2}\mathbf{q}^T\mathbf{K}\mathbf{q}$, where **q** is the vector whose components are the generalized components of the system with respect to time and $\mathbf{q}^T$ is the transpose of **q**. Also known as stability matrix. { 'stif·nəs ,mā·triks }

stiffness reactance *See* acoustic stiffness reactance. { 'stif·nəs rē,ak·təns }

stigma [MECH] A unit of length used mainly in

nuclear measurements, equal to 10^{-12} meter. Also known as bicron. { 'stig·mə }

stigmatic [OPTICS] **1.** Property of an optical system whose focal power is the same in all meridians. **2.** *See* homocentric. { stig'mad·ik }

stigmatic concave grating [OPTICS] An optical element with many parallel grooves on a concave optical surface; combines the two functions of light dispersion and focusing in one dispersive element; used in space optics, in food and metal analysis, and as a dispersive element of spectrophotometers and spectrographs. { stig'mad·ik 'kän,kāv 'grād·iŋ }

stilb [OPTICS] A unit of luminance, equal to 1 candela per square centimeter. Abbreviated sb. { stilb }

stimulated emission [ATOM PHYS] Emission of electromagnetic radiation by an atom or molecule as a result of its interaction with incident radiation of the same frequency. Also known as induced emission. [ELECTROMAG] The emission of radiation at a given frequency that results from the presence of an external radiation field of the same frequency. { 'stim·yə,lād·əd i'mish·ən }

stimulated scattering [OPTICS] The amplification, by stimulated emission, of intense radiation from a pulsed laser that has been inelastically scattered; results in exponential growth of the scattered power and, in some cases, almost complete scattering of the incident light. { 'stim·yə,lād·əd 'skad·ə·riŋ }

Stirling cycle [THERMO] A regenerative thermodynamic power cycle using two isothermal and two constant volume phases. { 'stir·liŋ ,sī·kəl }

stirring [PHYS] A turbulent process in which molecular diffusion and molecular heat conduction are speeded up. { 'stər·iŋ }

stirring effect [ELECTROMAG] The circulation in a molten metal carrying electric current as a result of the combined forces of the pinch and motor effects. { 'stər·iŋ i,fekt }

Stodola method [MECH] A method of calculating the deflection of a uniform or nonuniform beam in free transverse vibration at a specified frequency, as a function of distance along the beam, in which one calculates a sequence of deflection curves each of which is the deflection resulting from the loading corresponding to the previous deflection, and these deflections converge to the solution. { 'stō·də·lə ,meth·əd }

stoke [FL MECH] A unit of kinematic viscosity, equal to the kinematic viscosity of a fluid with a dynamic viscosity of 1 poise and a density of 1 gram per cubic centimeter. Symbol St (formerly S). Also known as lentor (deprecated usage); stokes. { stōk }

stokes *See* stoke. { stōks }

Stokes drift [FL MECH] The drift of particles in a gravity wave, which arises from the fact that particle velocities are periodic with a mean which is not zero. { ¦stōks ,drift }

Stokes flow [FL MECH] Fluid flow in which the Reynolds number is very small, so that the nonlinear terms in the Navier-Stokes equations can be neglected. { ¦stōks,flō }

Stokes frequencies [OPTICS] Scattered (secondary) light in the Raman effect (when a high-intensity light beam passes through a transparent medium) that occurs at frequencies smaller than the frequency of the primary beam. { ¦stōks ,frē·kwən·sēz }

Stokes' law [FL MECH] At low velocities, the frictional force on a spherical body moving through a fluid at constant velocity is equal to 6π times the product of the velocity, the fluid viscosity, and the radius of the sphere. [SPECT] The wavelength of luminescence excited by radiation is always greater than that of the exciting radiation. { 'stōks ,lȯ }

Stokes' lens [OPTICS] A variable-power compound lens made up of cylindrical lenses mounted so that the angle between their axes can be varied. { 'stōks ,lenz }

Stokes line [SPECT] A spectrum line in luminescent radiation whose wavelength is greater than that of the radiation which excited the luminescence, and thus obeys Stokes' law. { 'stōks ,līn }

Stokes number 1 [FL MECH] A dimensionless number used in the study of the dynamics of a particle in a fluid, equal to the product of the dynamic viscosity of the fluid and the particle's vibration time, divided by the product of the fluid density and a characteristic length. Symbol St. { 'stōks ¦nəm·bər 'wən }

Stokes parameters [OPTICS] Four quantities that fully describe the polarization of a beam of light. { 'stōks pə,ram·əd·ərz }

Stokes shift [SPECT] The displacement of spectral lines or bands of luminescent radiation toward longer wavelengths than those of the absorption lines or bands. { 'stōks ,shift }

Stokes stream function [FL MECH] A one-component vector potential function used in analyzing and describing a steady, axially symmetric fluid flow; at any point it is equal to $1/2\pi$ times the mass rate of flow inside the surface generated by rotating the streamline on which the point is located about the axis of symmetry. { 'stōks 'strēm ,fəŋk·shən }

stone [MECH] A unit of mass in common use in the United Kingdom, equal to 14 pounds or 6.35029318 kilograms. { stōn }

stop [OPTICS] The aperture or useful opening of a lens, usually adjustable by means of a diaphragm. { stäp }

stop band *See* rejection band. { 'stäp ,band }

stop down [OPTICS] To reduce the size of a lens aperture. { 'stäp'dau̇n }

stop number *See* f number. { 'stäp ,nəm·bər }

stopping range [NUC PHYS] A postulated energy range for extremely high-energy heavy-ion collisions in which the target stops the projectile and the protons and neutrons of target and projectile are mixed together in a fireball forming a quark-gluon plasma. { 'stäp·iŋ ,rānj }

storage battery [ELEC] A connected group of two or more storage cells or a single storage cell. Also known as accumulator; accumulator battery; rechargeable battery; secondary battery. { 'stȯr·ij ˌbad·ə·rē }

storage factor *See* Q. { 'stȯr·ij ˌfak·tər }

storage time *See* decay time. { 'stȯr·ij ˌtīm }

STP *See* standard conditions.

straggling [PHYS] Random variations in some property associated with the passage of ions through matter. { 'strag·liŋ }

straight vertical antenna [ELECTROMAG] An antenna consisting of a straight vertical wire. { 'strāt 'vərd·ə·kəl an'ten·ə }

strain [CELL MOL] A population of cells derived either from a primary culture or from a cell line by the selection or cloning of cells having specific properties or markers. [MECH] Change in length of an object in some direction per unit undistorted length in some direction, not necessarily the same; the nine possible strains form a second-rank tensor. { strān }

strain axis *See* principal axis of strain. { 'strān ˌak·səs }

strain ellipsoid [MECH] A mathematical representation of the strain of a homogeneous body by a strain that is the same at all points or of unequal stress at a particular point. Also known as deformation ellipsoid. { 'strān i'lip,sȯid }

strain energy [MECH] The potential energy stored in a body by virtue of an elastic deformation, equal to the work that must be done to produce this deformation. { 'strān ˌen·ər·jē }

strain figure [PHYS] A series of markings, such as Lüder's lines, that may appear on the surface of a body subjected to stress, indicating its state of deformation. Also known as flow figure. { 'strān ˌfig·yər }

strain rate [MECH] The time rate for the usual tensile test. { 'strān ˌrāt }

strain rosette [MECH] A pattern of intersecting lines on a surface along which linear strains are measured to find stresses at a point. { 'strān rō,zet }

strain shadow *See* undulatory extinction. { 'strān ˌshad·ō }

strain tensor [MECH] A second-rank tensor whose components are the nine possible strains. { 'strān ˌten·sər }

strange attractor [PHYS] A geometrical object in phase space with a fractal structure toward which the trajectory followed by a chaotic system converges in the course of time. { 'strānj ə'trak·tər }

strangeness conservation [PART PHYS] The principle that the sum of the strangeness numbers of the hadrons in an isolated system is constant; it is violated by the weak interactions. { 'strānj·nəs ˌkän·sər'vā·shən }

strangeness number [PART PHYS] A quantum number carried by hadrons, equal to the hypercharge minus the baryon number. Symbol S. { 'strānj·nəs ˌnəm·bər }

strange particle [PART PHYS] A hadron whose strangeness number is not zero, for example, a K-meson or a Σ-hyperon. { 'strānj 'pard·ə·kəl }

strange quark [PART PHYS] A quark with an electric charge of −1/3, baryon number of 1/3, strangeness of −1, and 0 charm. Symbolized *s*. { 'strānj 'kwärk }

stratified flow [FL MECH] A two-phase flow in which a liquid flows along the bottom of a pipe and gas flows separately above it. { 'strad·ə,fīd 'flō }

stratified fluid [FL MECH] A fluid having density variation along the axis of gravity, usually implying upward decrease of density, that is, a stratification characterized by static stability. { 'strad·ə,fīd 'flü·əd }

stratoscope [OPTICS] A balloon-borne astronomical telescope for taking solar or other celestial photographs at high altitudes; subsequently, the photos are transmitted to a ground receiving station. { 'strad·əˌskōp }

Stratton pseudoscope [OPTICS] A type of Wheatstone stereoscope in which the mirrors transpose the right- and left-eye views, producing a reversed stereoscopic effect. { ¦strat·ən 'süd·əˌskōp }

stray emission [PHYS] Emission of radiation that serves no useful purpose. { 'strā i'mish·ən }

stray field [ELECTROMAG] Leakage of magnetic flux that spreads outward from a coil and does no useful work. { 'strā ¦fēld }

streak camera [OPTICS] A special type of high-speed motion picture camera that records an image as a continuous spreadout picture rather than as a sequence of separate frames; special viewing equipment must be used to analyze the image and reconstitute individual pictures. { 'strēk ˌkam·rə }

streak line [FL MECH] A line within a fluid which, at a given instant, is formed by those fluid particles which at some previous instant have passed through a specified fixed point in the fluid; an example is the line of color in a flow into which a dye is continuously introduced through a small tube, all dyed fluid particles having passed the tube's end. { 'strēk ˌlīn }

stream function *See* Lagrange stream function. { 'strēm ˌfəŋk·shən }

streaming current [ELEC] The electric current which is produced when a liquid is forced to flow through a diaphragm, capillary, or porous solid. { 'strēm·iŋ ˌkə·rənt }

streaming potential [ELEC] The difference in electric potential between a diaphragm, capillary, or porous solid and a liquid that is forced to flow through it. { 'strēm·iŋ pə,ten·chəl }

streamline [FL MECH] A line which is every where parallel to the direction of fluid flow at a given instant. { 'strēm,līn }

streamline flow [FL MECH] Flow of a fluid in which there is no turbulence: particles of the fluid follow well-defined continuous paths, and the flow velocity at a fixed point either remains constant or varies in a regular fashion with time. { 'strēm,līn ˌflō }

stream tube [FL MECH] In fluid flow, an imaginary tube whose wall is generated by streamlines passing through a closed curve. { 'strēm ˌtüb }

Strehl ratio [OPTICS] The ratio of the peak field amplitude in the focus of an optical element to the diffraction-limited amplitude. { 'strāl ˌrā·shō }

strength [ACOUS] The maximum instantaneous rate of volume displacement produced by a sound source when emitting a wave with sinusoidal time variation. [MECH] The stress at which material ruptures or fails. { streŋkth }

stress [MECH] The force acting across a unit area in a solid material resisting the separation, compacting, or sliding that tends to be induced by external forces. { stres }

stress analysis [PHYS] The determination of the stresses produced in a solid body when subjected to various external forces. { 'stres əˌnal·ə·səs }

stress axis *See* principal axis of stress. { 'stres ˌak·səs }

stress birefringence *See* mechanical birefringence. { 'stres ¦bī·ri'frin·jəns }

stress concentration [MECH] A condition in which a stress distribution has high localized stresses; usually induced by an abrupt change in the shape of a member; in the vicinity of notches, holes, changes in diameter of a shaft, and so forth, maximum stress is several times greater than where there is no geometrical discontinuity. { 'stres ˌkän·sənˌtrā·shən }

stress concentration factor [MECH] A theoretical factor K_t expressing the ratio of the greatest stress in the region of stress concentration to the corresponding nominal stress. { 'stres ˌkän·sən¦trā·shən ˌfak·tər }

stress crack [MECH] An external or internal crack in a solid body (metal or plastic) caused by tensile, compressive, or shear forces. { 'stres ˌkrak }

stress difference [MECH] The difference between the greatest and the least of the three principal stresses. { 'stres ˌdif·rəns }

stress ellipsoid [MECH] A mathematical representation of the state of stress at a point that is defined by the minimum, intermediate, and maximum stresses and their intensities. { 'stres i'lipˌsȯid }

stress function [MECH] A single function, such as the Airy stress function, or one of two or more functions, such as Maxwell's or Morera's stress functions, that uniquely define the stresses in an elastic body as a function of position. { 'stres ˌfəŋk·shən }

stress intensity [MECH] Stress at a point in a structure due to pressure resulting from combined tension (positive) stresses and compression (negative) stresses. { 'stres inˌten·səd·ē }

stress lines *See* isostatics. { 'stres ˌlīnz }

stress-optic law [OPTICS] In a transparent, isotropic plate subjected to a biaxial stress field, the relative retardation R_t between the two components produced by temporary double refraction is equal to $Ct(p - q)$, which in turn is equal to $n\lambda$; C is the stress-optic coefficient, t the plate thickness, p and q the principal stresses, n the number of fringes which have passed the point during application of the load, and λ the wavelength of the light. { 'stres 'äp·tik ˌlȯ }

stress range [MECH] The algebraic difference between the maximum and minimum stress in one fatigue test cycle. { 'stres ˌrānj }

stress ratio [MECH] The ratio of minimum to maximum stress in fatigue testing, considering tensile stresses as positive and compressive stresses as negative. { 'stres ˌrā·shō }

stress-strain curve *See* deformation curve. { 'stres 'strān ˌkərv }

stress tensor [MECH] A second-rank tensor whose components are stresses exerted across surfaces perpendicular to the coordinate directions. { 'stres ˌten·sər }

stress trajectories *See* isostatics. { 'stres trə ˌjek·trēz }

stress-wave emission *See* acoustic emission. { 'stres ¦wāv iˌmish·ən }

striation technique [ACOUS] A technique for making sound waves visible by using their ability to refract light waves. { strī'ā·shən tekˌnēk }

strich *See* millimeter. { strich }

strike note [ACOUS] The note which is the loudest heard when a bell is struck, and whose pitch is generally assigned to the bell. { 'strīk ˌnōt }

striking velocity *See* impact velocity. { 'strīk·iŋ vəˌläs·əd·ē }

string [MECH] A solid body whose length is many times as large as any of its cross-sectional dimensions, and which has no stiffness. [PART PHYS] A proposed structure for elementary particles, consisting of a one-dimensional curve with zero thickness and length typically of the order of the Planck length, 10^{-35} m. { striŋ }

string duality [PART PHYS] The property of superstring theories that the strongly coupled behavior of each theory appears to be equivalent to that of some other weakly coupled system, and all the seemingly different superstring theories, as well as M-theory, are just different weakly coupled limits of a single theory. { ¦striŋ dü'al·əd·ē }

stripe *See* stripe phase. { strīp }

stripe phase [SOLID STATE] A phase exhibited by the electrons in certain solid systems, such as doped antiferomagnets and quantum Hall systems, in which the behavior of the electrons is similar to that of the molecules in a liquid crystal, exhibiting properties of both a crystal (orientational order and anisotropy) and a liquid (absence of space periodicity). Also known as stripe. { 'strīp ˌfāz }

strip line [ELECTROMAG] A strip transmission line that consists of a flat metal-strip center conductor which is separated from flat metal-strip outer conductors by dielectric strips. { 'strip ˌlīn }

strip-line circuit [ELECTROMAG] A circuit in which one or more strip transmission lines serve as filters or other circuit components. { 'strip ¦līn ˌsər·kət }

stripped atom [ATOM PHYS] An ionized atom which has appreciably fewer electrons than it has protons in the nucleus. { 'stript 'ad·əm }

stripping reaction [NUC PHYS] A nuclear reaction in which part of the incident nucleus combines with the target nucleus, and the other part proceeds with most of its original momentum in practically its original direction; especially the reaction in which the incident nucleus is a deuteron and only a proton emerges from the target. { 'strip·iŋ rē,ak·shən }

strip transmission line [ELECTROMAG] A microwave transmission line consisting of a thin, narrow, rectangular metal strip that is supported above a ground-plane conductor or between two wide ground-plane conductors and is usually separated from them by a dielectric material. { 'strip tranz'mish·ən ,līn }

strong causality condition [RELAT] The strong causality condition is said to hold at a point *p* in a space-time if every neighborhood of *p* contains a neighborhood of *p* which no timelike or null curve intersects more than once. { ¦strȯŋ kȯ'zal·əd·ē kən,dish·ən }

strong interaction [PART PHYS] One of the fundamental interactions of elementary particles, primarily responsible for nuclear forces and other interactions among hadrons. { 'strȯŋ ,in·tər'ak·shən }

strongly damped collision *See* deep inelastic collision. { 'strȯŋ·lē ¦dampt kə'lizh·ən }

strongly future asymptotically predictable space-time [RELAT] A future asymptotically predictable space-time such that a neighborhood of the event horizon is also predictable. { ¦strȯŋ·lē ¦fyü·chər ,ā,sim¦täd·ə·klē prə¦dik·tə·bəl 'spās ,tīm }

strontium-90 [NUC PHYS] A poisonous, radioactive isotope of strontium; 28-year half life with β radiation; derived from reactor-fuel fission products; used in thickness gages, medical treatment, phosphor activation, and atomic batteries. { 'strän·tē·əm 'nīn·tē }

Strouhal frequency [FL MECH] The frequency of vortex shedding from a structure in a uniform flow. { 'strü·əl ,frē·kwən·sē }

Strouhal number [MECH] A dimensionless number used in studying the vibrations of a body past which a fluid is flowing; it is equal to a characteristic dimension of the body times the frequency of vibrations divided by the fluid velocity relative to the body; for a taut wire perpendicular to the fluid flow, with the characteristic dimension taken as the diameter of the wire, it has a value between 0.185 and 0.2 Symbolized S_r. Also known as reduced frequency. { 'strü·əl ,nəm·bər }

structural deflections [MECH] The deformations or movements of a structure and its flexural members from their original positions. { 'strək·chə·rəl di'flek·shənz }

structure amplitude [SOLID STATE] The absolute value of a structure factor. { 'strək·chər ,am·plə,tüd }

structure cell *See* unit cell. { 'strək·chər ,sel }

structured light [OPTICS] Light that is projected in a particular geometrical pattern that is used to aid in computer vision. { 'strək·chərd 'līt }

structure factor [SOLID STATE] A factor which determines the amplitude of the beam reflected from a given atomic plane in the diffraction of an x-ray beam by a crystal, and is equal to the sum of the atomic scattering factors of the atoms in a unit cell, each multiplied by an appropriate phase factor. { 'strək·chər ,fak·tər }

structure resonance [SPECT] An extremely narrow resonance exhibited by a small aerosol particle at a natural electromagnetic frequency at which the dielectric sphere oscillates, observed in the particle's scattered light excitation spectrum. { 'strək·chər ,rez·ən·əns }

structure resonance modulation spectroscopy [SPECT] The infrared modulation of visible scattered light near a structure resonance to determine the absorption spectrum of an aerosol particle. Abbreviated SRMS. { 'strək·chər ,rez·ən·əns ,mäj·ə¦lā·shən spek'träs·kə·pē }

structure-sensitive property [SOLID STATE] A property of a substance that depends on impurities and the imperfections of the crystal structure. { ¦strək·chər ¦sen·sə·tiv 'präp·ərd·ē }

structure type [CRYSTAL] The structural arrangement of a crystal, regardless of the atomic elements present; it corresponds to the crystal's space group. { 'strək·chər ,tīp }

stub [ELECTROMAG] **1.** A short section of transmission line, open or shorted at the far end, connected in parallel with a transmission line to match the impedance of the line to that of an antenna or transmitter. **2.** A solid projection one-quarter-wavelength long, used as an insulating support in a waveguide or cavity. { stəb }

stub angle [ELECTROMAG] Right-angle elbow for a coaxial radio-frequency transmission line which has the inner conductor supported by a quarter-wave stub. { 'stəb ,aŋ·gəl }

stub matching [ELECTROMAG] Use of a stub to match a transmission line to an antenna or load; matching depends on the spacing between the two wires of the stub, the position of the shorting bar, and the point at which the transmission line is connected to the stub. { 'stəb ,mach·iŋ }

stub-supported coaxial [ELECTROMAG] Coaxial whose inner conductor is supported by means of short-circuited coaxial stubs. { 'stəb sə¦pȯrd·əd kō'ak·sē·əl }

stub-supported line [ELECTROMAG] A transmission line that is supported by short-circuited quarter-wave sections of coaxial line; a stub exactly a quarter-wavelength long acts as an insulator because it has infinite reactance. { 'stəb sə¦pȯrd·əd 'līn }

stub tuner [ELECTROMAG] Stub which is terminated by movable short-circuiting means and used for matching impedance in the line to which it is joined as a branch. { 'stəb ,tün·ər }

subatomic particle [PHYS] A particle which is smaller than an atom, namely, an elementary particle or an atomic nucleus. { ¦səb·ə'täm·ik 'pärd·ə·kəl }

subcritical flow *See* subsonic flow. { ¦səb'krid·ə·kəl 'flō }

subharmonic [PHYS] A sinusoidal quantity having a frequency that is an integral submultiple of the frequency of some other sinusoidal quantity to which it is referred; a third subharmonic would be one-third the fundamental or reference frequency. { ¦səb·här'män·ik }

sublevel *See* subshell. { ¦səb'lev·əl }

sublimation [THERMO] The process by which solids are transformed directly to the vapor state or vice versa without passing through the liquid phase. { ˌsəb·lə'mā·shən }

sublimation cooling [THERMO] Cooling caused by the extraction of energy to produce sublimation. { ˌsəb·lə'mā·shən ¦kül·iŋ }

sublimation curve [THERMO] A graph of the vapor pressure of a solid as a function of temperature. { ˌsəb·lə'mā·shən ¦kərv }

sublimation energy [THERMO] The increase in internal energy when a unit mass, or 1 mole, of a solid is converted into a gas, at constant pressure and temperature. { ˌsəb·lə'mā·shən ¦en·ər·jē }

sublimation point [THERMO] The temperature at which the vapor pressure of the solid phase of a compound is equal to the total pressure of the gas phase in contact with it; analogous to the boiling point of a liquid. { ˌsəb·lə'mā·shən ¦pȯint }

sublimation pressure [THERMO] The vapor pressure of a solid. { ˌsəb·lə'mā·shən ¦presh·ər }

sublime [THERMO] To change from the solid to the gaseous state without passing through the liquid phase. { sə'blīm }

submarine sound signal [ACOUS] A sound signal transmitted through water. { ¦səb·mə'rēn 'sau̇n ˌsig·nəl }

submetallic [OPTICS] Referring to a luster intermediate between metallic and nonmetallic, such as exhibited by the mineral chromite. { ¦səb·mə'tal·ik }

submillimeter wave [ELECTROMAG] An electromagnetic wave whose wavelength is less than 1 millimeter, corresponding to frequencies above 300 gigahertz. { ˌsəb¦mil·əˌmēd·ər ˌwāv }

submultiple resonance [PHYS] Resonance at a frequency that is a submultiple of the frequency of the exciting impulses. { ¦səb'məl·tə·pəl 'rez·ən·əns }

subnuclear particle *See* elementary particle. { səb¦nü·klē·ər 'pärd·i·kəl }

subrefraction [ELECTROMAG] Atmospheric refraction which is less than standard refraction. { ¦səb·ri'frak·shən }

subshell [ATOM PHYS] Electrons of an atom within the same shell (energy level) and having the same azimuthal quantum numbers. Also known as sublevel. { 'səbˌshel }

subsonic [ACOUS] *See* infrasonic. [PHYS] Of, pertaining to, or dealing with speeds less than acoustic velocity, as in subsonic aerodynamics. { ¦səb'sän·ik }

subsonic flow [FL MECH] Flow of a fluid at a speed less than that of the speed of sound in the fluid. Also known as subcritical flow. { ¦səb'sän·ik 'flō }

subsonic speed [FL MECH] A speed relative to surrounding fluid less than that of the speed of sound in the same fluid. { ¦səb'sän·ik 'spēd }

substance [PHYS] Tangible material, occurring in macroscopic amounts. { 'səb·stəns }

substandard propagation [ELECTROMAG] The propagation of radio energy under conditions of substandard refraction in the atmosphere; that is, refraction by an atmosphere or section of the atmosphere in which the index of refraction decreases with height at a rate of less than 12 N units (unit of index of refraction) per 1000 feet (304.8 meters). { ¦səb'stan·dərd ˌpräp·ə'gā·shən }

substitutional impurity [SOLID STATE] An atom or ion which is not normally found in a solid, but which resides at the position where an atom or ion would ordinarily be located in the lattice structure, and replaces it. { ˌsəb·stə'tü·shən·əl im'pyu̇r·əd·ē }

substitution method [PHYS] Any method of measurement, such as substitution weighing, in which a quantity is determined by substituting for it a known quantity which produces the same effect. { ˌsəb·stə'tü·shən ˌmeth·əd }

substitution weighing [MECH] A method of weighing to allow for differences in lengths of the balance arms, in which the object to be weighed is first balanced against a counterpoise, and the known weights needed to balance the same counterpoise are then determined. Also known as counterpoise method. { ˌsəb·stə'tü·shən ˌwā·iŋ }

subsurface wave [ELECTROMAG] Electromagnetic wave propagated through water or land; operating frequencies for communications may be limited to approximately 35 kilohertz due to attenuation of high frequencies. { ¦səb'sər·fəs 'wāv }

subtractive primaries [OPTICS] The three colors, usually yellow, magenta, and cyan (greenish-blue), which are mixed together in a subtractive process. { səb'trak·tiv 'prīˌmer·ēz }

subtractive process [OPTICS] The process of producing colors by mixing absorbing media or filters of subtractive primary colors. { səb'trak·tiv 'prä·səs }

suction wave *See* rarefaction wave. { 'sək·shən ˌwāv }

Suhl amplifier [SOLID STATE] A parametric microwave amplifier which utilizes the instability of certain spin waves in a ferromagnetic material subjected to intense microwave fields. { 'sül ˌam·pləˌfī·ər }

sulfur-35 [NUC PHYS] Radioactive sulfur with mass number 35; radiotoxic, with 87.1-day half-life, β radiation; derived from pile irradiation; used as a tracer to study chemical reactions, engine wear, and protein metabolism. { 'səl·fər ˌthərd·ē'fīv }

summation network *See* summing network. { sə'mā·shən ˌnetˌwərk }

summation tone [ACOUS] Combination tone, heard under certain circumstances, whose pitch corresponds to a frequency equal to the sum of the frequencies of the two components. { sə'mā·shən ,tōn }

summing network [ELEC] A passive electric network whose output voltage is proportional to the sum of two or more input voltages. Also known as summation network. { 'səm·iŋ 'net,wərk }

sum of states *See* partition function. { 'səm əv 'stāts }

sum over states *See* partition function. { 'səm ,ō·vər 'stāts }

sum rule [QUANT MECH] A formula for a transition between energy levels in which the sum of the transition strengths is expressed in a simple form. { 'səm ,rül }

sun-pumped laser [OPTICS] A continuous-wave laser in which pumping is achieved by concentrating the energy of the sun on the laser crystal with a parabolic mirror. Also known as solar-excited laser. { 'sən ¦pəmt 'lā·zər }

SU_n symmetry [PART PHYS] A unitary symmetry based on a fundamental multiplet of *n* equivalent particles-in particular, *n* quarks. { ¦es ¦yü 'səb¦en 'sim·ə·trē }

superaerodynamics [FL MECH] That branch of gas dynamics dealing with the flow of gases at such low density that the molecular mean free path is not negligibly small; under these conditions the gas no longer behaves as a continuous fluid. { ¦sü·pər,er·ō·dī'nam·iks }

supercavitation [FL MECH] An extreme form of cavitation in which a single bubble of gas forms around an object moving rapidly through water, enveloping it almost completely so that the water wets very little of the object's surface, thereby drastically reducing viscous drag. { ,sü·pər,kav·ə'ā·shən }

supercompressibility factor *See* compressibility factor. { ¦sü·pər·kəm,pres·ə'bil·əd·ē ,fak·tər }

superconducting ball [CRYO] A ball, typically with a radius of about 0.25 millimeter (0.01 inch), that is formed from the aggregation of several million microscopic superconducting particles in a strong electric field. { ¦sü·pər·kən,duk·tiŋ 'bȯl }

superconducting circuit [CRYO] An electric circuit having elements which are in a superconducting state at least part of the time, such as a cryotron. { ¦sü·pər·kən'dəkt·iŋ 'sər·kət }

superconducting device *See* cryogenic device. { ¦sü·pər·kən'dəkt·iŋ di'vīs }

superconducting magnet [CRYO] An electromagnet whose coils are made of a type II superconductor with a high transition temperature and extremely high critical field, such as niobium tin, Nb_3Sn; it is capable of generating magnetic fields of 100,000 oersteds (8,000,000 amperes per meter) and more with no steady power dissipation. { ¦sü·pər·kən'dəkt·iŋ 'mag·nət }

superconducting material *See* superconductor. { ¦sü·pər·kən'dəkt·iŋ mə'tir·ē·əl }

superconducting memory [CRYO] A computer memory made up of a number of cryotrons, thin-film cryotrons, superconducting thin films, or other superconducting storage devices; these operate only under cryogenic conditions and dissipate power only during the read or write operation, which permits construction of large, dense memories. { ¦sü·pər·kən'dəkt·iŋ 'mem·rē }

superconducting thin film [CRYO] A thin film of indium, tin, or other superconducting element, used as a cryogenic switching or storage device, as in a thin-film cryotron. { ¦sü·pər·kən'dəkt·iŋ 'thin 'film }

superconductivity [SOLID STATE] A property of many metals, alloys, and chemical compounds at temperatures near absolute zero by virtue of which their electrical resistivity vanishes and they become strongly diamagnetic. { ¦sü·pər,kän,dək'tiv·əd·ē }

superconductor [SOLID STATE] Any material capable of exhibiting superconductivity; examples include iridium, lead, mercury, niobium, tin, tantalum, vanadium, and many alloys. Also known as cryogenic conductor; superconducting material. { ¦sü·pər·kən'dək·tər }

supercooling [THERMO] Cooling of a substance below the temperature at which a change of state would ordinarily take place without such a change of state occurring, for example, the cooling of a liquid below its freezing point without freezing taking place; this results in a metastable state. { ¦sü·pər'kül·iŋ }

supercritical [THERMO] Property of a gas which is above its critical pressure and temperature. { ¦sü·pər'krid·ə·kəl }

supercritical field [PHYS] A static field that is strong enough to cause the normal vacuum, which is devoid of real particles, to break down into a new vacuum in which real particles exist. { ¦sü·pər¦krid·ə·kəl 'fēld }

supercritical flow *See* supersonic flow. { ¦sü·pər'krid·ə·kəl 'flō }

supercritical fluid [THERMO] A fluid at a temperature and pressure above its critical point; also, a fluid above its critical temperature regardless of pressure. { ¦sü·pər¦krid·ə·kəl 'flü·əd }

supercurrent [SOLID STATE] In the two-fluid model of superconductivity, the current arising from motion of superconducting electrons, in contrast to the normal current. { ¦sü·pər'kə·rənt }

superdeformed nuclear state [NUC PHYS] A nucleus in a highly excited state, formed in the collision of heavy nuclei, whose shape corresponds to an ellipsoid with an axis ratio approaching 2:1. { ¦sü·pər·di'fȯrmd 'nü·klē·ər 'stāt }

superelastic collision [PHYS] A collision in which potential energy is converted into kinetic energy so that the total kinetic energy of the colliding objects is greater after the collision than before. { ¦sü·pər·i¦las·tik kə'lizh·ən }

superexchange [SOLID STATE] A phenomenon in which two electrons from a double negative ion (such as oxygen) in a solid go to different positive ions and couple with their spins, giving rise to a strong antiferromagnetic coupling between the positive ions, which are too far apart to have a direct exchange interaction. { ¦sü·pər·iks'chānj }

superficial expansivity *See* coefficient of superficial expansion. { ¦sü·pər¦fish·əl ˌikˌspan'siv·əd·ē }

superfluid [PHYS] A collection of particles which obey Bose-Einstein statistics and are all in the lowest energy state allowed by quantum mechanics, having zero entropy and zero resistance to motion; examples are a fraction of the atoms in liquid helium II and a fraction of the pairs of electrons in a superconductor. { ¦sü·pər'flü·əd }

superfluidity [CRYO] The frictionless flow of liquid helium at temperatures very close to absolute zero through holes as small as 10^{-7} centimeter in diameter, and for particle velocities below a few centimeters per second. { ¦sü·pər·flü'id·əd·ē }

superfluorescence [ATOM PHYS] The process of spontaneous emission of electromagnetic radiation from a collection of excited atoms. { ¦sü·pər·flu̇'res·əns }

supergravity [PHYS] A supersymmetry which is used to unify general relativity and quantum theory; it is formed by adding to the Poincaré group, as a symmetry of space-time, four new generators that behave as spinors and vary as the square root of the translations. { ¦sü·pər'grav·əd·ē }

superheat [THERMO] Sensible heat in a gas above the amount needed to maintain the gas phase. { 'sü·pərˌhēt }

superheated vapor [THERMO] A vapor that has been heated above its boiling point. { ¦sü·pər'hēd·əd 'vā·pər }

superheating [THERMO] Heating of a substance above the temperature at which a change of state would ordinarily take place without such a change of state occurring, for example, the heating of a liquid above its boiling point without boiling taking place; this results in a metastable state. { ¦sü·pər'hēd·iŋ }

superheavy boson [PART PHYS] A hypothetical particle, postulated in grand unified theories, which would be responsible for interactions between quarks and leptons in the early universe, and also responsible for proton decay. Also known as X boson. { ¦sü·pər'hev·ē 'bōˌsän }

superionic conduction [SOLID STATE] Extremely fast conduction of ions in certain inorganic crystalline solids, approaching the ionic conductivity of aqueous sodium chloride. { ¦sü·pərˌī'än·ik kən'dək·shən }

superionic conductor [SOLID STATE] An ionic solid whose ionic conductivity is extremely high, on the order of 100 times that normally observed. { ¦sü·pərˌī¦än·ik kən'dək·tər }

superior mirage [OPTICS] A spurious image of an object formed above the object's position by abnormal refraction conditions; opposite to an inferior mirage. { sə'pir·ē·ər mi'räzh }

superlattice [SOLID STATE] An ordered arrangement of atoms in a solid solution which forms a lattice superimposed on the normal solid solution lattice. Also known as artificial crystal; artificially layered structure; superstructure. { ¦sü·pər'lad·əs }

superlayer [FL MECH] A very thin, highly convoluted interface that separates the turbulent from the nonturbulent regions in a flow of high Reynolds number. { 'sü·pərˌlā·ər }

superleak *See* lambda leak. { 'sü·pərˌlēk }

supermode laser [OPTICS] Frequency-modulated laser, the output of which is passed through a second phase modulator driven 180° out of phase and with the same modulation index as the first modulator; all of the energy of the previously existing laser modes is compressed into a single frequency with nearly the full power of the laser concentrated in that signal. { ¦sü·pər'mōd 'lā·zər }

supermultiplet [QUANT MECH] A set of quantum-mechanical states each of which has the same value of some fundamental quantum numbers and differs from the other members of the set by other quantum numbers, which take values from a range of numbers dictated by the fundamental quantum numbers. { ¦sü·pər'məl·tə·plət }

supernumerary rainbow [OPTICS] One of a set of weakly colored rainbow arcs sometimes discernible inside a primary rainbow; they are of smaller angular width, and fade toward the common center. { ¦sü·pər'nü·məˌrer·ē 'rānˌbō }

superparamagnetic particle [SOLID STATE] A crystalline grain in a magnetic medium that is so small that its magnetic properties decrease with time due to thermal fluctuations. { ¦sü·pərˌpar·ə·magˌned·ik 'pärd·i·kəl }

superparamagnetism *See* collective paramagnetism. { ¦sü·pərˌpar·ə'mag·nəˌtiz·əm }

superposition [PHYS] Addition of phenomena when the sum of two physically realizable disturbances is also physically realizable; for example, sound waves are superposable in this sense, but shock waves are not. { ˌsü·pər·pə'zish·ən }

superposition principle *See* principle of superposition. { ˌsü·pər·pə'zish·ən 'prin·sə·pəl }

superposition theorem *See* principle of superposition. { ˌsü·pər·pə'zish·ən 'thir·əm }

superradiant scattering [RELAT] The scattering of radiation from a black hole in such a way that the scattered radiation carries more energy than the incident radiation. { ¦sü·pər'rād·ē·ənt 'skad·ə·riŋ }

super-Schmidt telescope [OPTICS] A type of Schmidt system that has a compound corrector plate consisting of a pair of opposing meniscus lenses and an achromatic doublet. { 'sü·pər 'shmit 'tel·əˌskōp }

supersonic [ACOUS] *See* ultrasonic. [PHYS] Of, pertaining to, or dealing with speeds greater than the speed of sound. { ¦sü·pər¦sän·ik }

supersonic aerodynamics [FL MECH] The study of aerodynamics of supersonic speeds. { ¦sü·pər¦sän·ik ¦er·ō·dī'nam·iks }

supersonic flow [FL MECH] Flow of a fluid over a body at speeds greater than the speed of sound in the fluid, and in which the shock waves start at the surface of the body. Also known as supercritical flow. { ¦sü·pər¦sän·ik 'flō }

superspace [PHYS] The space of all three-geometries on a three-manifold, used in discussions of quantum gravity. { 'sü·pər,spās }

superstandard propagation [ELECTROMAG] The propagation of radio waves under conditions of superstandard refraction in the atmosphere, that is, refraction by an atmosphere or section of the atmosphere in which the index of refraction decreases with height at a rate of greater than 12 N units (unit of index of refraction) per 1000 feet (304.8 meters). { ¦sü·pər'stan·dərd ,präp·ə'gā·shən }

superstring theory [PART PHYS] A theory of elementary particles which obeys supersymmetry and in which the particles are one-dimensional, closed curves with zero thickness and length of the order of the Planck length, 10^{-35} m. { 'sü·pər,striŋ ,thē·ə·rē }

superstructure *See* superlattice. { 'sü·pər,strək·chər }

supersymmetry [PART PHYS] A generalization of previously known symmetries of elementary particles to new kinds of supermultiplets that include both bosons and fermions; it is based on graded Lie algebras rather than on Lie algebras. { ¦sü·pər'sim·ə·trē }

superturbulent flow [FL MECH] The flow of water in which the energy loss by friction is so great that Reynolds criterion for the transition of laminar to turbulent flow does not apply. { ¦sü·pər'tər·byə·lənt 'flō }

supplementary condition [QUANT MECH] In a quantized field theory, an auxiliary condition required of a state vector to make it correspond to an actual state. { ¦səp·lə¦men·trē kən'dish·ən }

supplementary units [PHYS] Dimensionless units that are used along with base units to form derived units in the International System; this class contains only two, purely geometrical units: the radian and the steradian. { ¦səp·lə,men·trē 'yü·nəts }

supported end [MECH] An end of a structure, such as a beam, whose position is fixed but whose orientation may vary; for example, an end supported on a knife-edge. { sə'pȯrd·əd ,end }

suppressor [ELEC] **1.** In general, a device used to reduce or eliminate noise or other signals that interfere with the operation of a communication system, usually at the noise source. **2.** Specifically, a resistor used in series with a spark plug or distributor of an automobile engine or other internal combustion engine to suppress spark noise that might otherwise interfere with radio reception. [SPECT] In an analytical procedure, a substance added to the analyte to reduce the extraneous emission, absorption, or light scattering caused by the presence of an impurity. { sə'pres·ər }

supraaural cushion [ACOUS] An earphone cushion that fits against the auricle. { ,sü·prə,ȯr·əl 'ku̇sh·ən }

surface acoustic wave [ACOUS] A sound wave that propagates along and is bound to the surface of a solid; ordinarily it contains both compressional and shear components. Abbreviated SAW. { 'sər·fəs ə'kü·stik 'wāv }

surface-coated mirror [OPTICS] A mirror produced by depositing a thin film of highly reflective material on a glass surface. { 'sər·fəs ¦kōd·əd 'mir·ər }

surface color [OPTICS] The color of light reflected from the surface of a body; in contrast to the color of light that is reflected after penetrating some distance into the body. { 'sər·fəs ,kəl·ər }

surface density [PHYS] The quantity of anything distributed over a surface per unit area of surface. { 'sər·fəs ,den·səd·ē }

surface drag [FL MECH] That portion of drag which is caused by skin friction. { 'sər·fəs ,drag }

surface energy [FL MECH] The energy per unit area of an exposed surface of a liquid; generally greater than the surface tension, which equals the free energy per unit surface. { 'sər·fəs ,en·ər·jē }

surface force [MECH] An external force which acts only on the surface of a body; an example is the force exerted by another object with which the body is in contact. { 'sər·fəs ,fȯrs }

surface leakage [ELEC] The passage of current over the surface of an insulator. { 'sər·fəs ,lē·kij }

surface magnetic wave [ELECTROMAG] A magnetostatic wave that can be propagated on the surface of a magnetic material, as on a slab of yttrium iron garnet. { 'sər·fəs mag'ned·ik 'wāv }

surface of discontinuity [FL MECH] A surface within a fluid across which there is a discontinuity in fluid velocity; often generated in the wake of a body moving relative to the fluid. { ¦sər·fəs əv ,dis,känt·ən'ü·əd·ē }

surface of section *See* Poincaré surface of section. { ¦sər·fəs əv 'sek·shən }

surface oil-film technique [FL MECH] A method of flow visualization in which a solid surface is coated with a mixture of oil and powdered pigment, and the airstream carries the oil away, leaving a streaky deposit of pigment that provides information on airflow. { ¦sər·fəs 'ȯil ,film ,tek,nēk }

surface physics [SOLID STATE] The study of the structure and dynamics of atoms and their associated electron clouds in the vicinity of a surface, usually at the boundary between a solid and a low-density gas. { 'sər·fəs 'fiz·iks }

surface plasmon [SOLID STATE] A quantum of a collective oscillation of charges on the surface of a solid induced by a time-varying electric field. { 'sər·fəs 'plaz,män }

surface pressure *See* film pressure. { 'sər·fəs ,presh·ər }

surface recombination rate [SOLID STATE] The rate at which free electrons and holes at the surface of a semiconductor recombine, thus neutralizing each other. { ¦sər·fəs rē,käm·bə'nā·shən ,rāt }

surface recombination velocity [SOLID STATE] A measure of the rate of recombination between electrons and holes at the surface of a semiconductor, equal to the component of the electron or hole current density normal to the surface divided by the excess electron or hole volume charge density close to the surface. { 'sər·fəs rē,käm·bə'nā·shən və,läs·əd·ē }

surface resistivity [ELEC] The electric resistance of the surface of an insulator, measured between the opposite sides of a square on the surface; the value in ohms is independent of the size of the square and the thickness of the surface film. { 'sər·fəs ,rē,zis'tiv·əd·ē }

surface state [SOLID STATE] An electron state in a semiconductor whose wave function is restricted to a layer near the surface. { 'sər·fəs ,stāt }

surface tension [FL MECH] The force acting on the surface of a liquid, tending to minimize the area of the surface; quantitatively, the force that appears to act across a line of unit length on the surface. Also known as interfacial force; interfacial tension; surface tensity. { 'sər·fəs ,ten·chən }

surface tension number [FL MECH] A dimensionless number used in studying mass transfer in packed columns equal to the square of the dynamic viscosity of a fluid times the length of the perimeter of a packing element, divided by the product of the surface area of the packing element, the surface tension, and the density of the liquid. Symbol T_s. { 'sər·fəs ,ten·chən ,nəm·bər }

surface tensity *See* surface tension. { 'sər·fəs ,ten·səd·ē }

surface wave [ELECTROMAG] A wave that can travel along an interface between two different mediums without radiation; the interface must be essentially straight in the direction of propagation; the commonest interface used is that between air and the surface of a circular wire. [FL MECH] A wave that distorts the free surface that separates two fluid phases, usually a liquid and a gas. *See* Rayleigh wave. { 'sər·fəs ,wāv }

surface-wave transmission line [ELECTROMAG] A single conductor transmission line energized in such a way that a surface wave is propagated along the line with satisfactorily low attenuation. { 'sər·fəs ¦wāv tranz'mish·ən ,līn }

surge [ELEC] A momentary large increase in the current or voltage in an electric circuit. [FL MECH] A wave at the free surface of a liquid generated by the motion of a vertical wall, having a change in the height of the surface across the wavefront and violent eddy motion at the wavefront. { sərj }

surge stress [MECH] The physical stress on process equipment or systems resulting from a sudden surge in fluid (gas or liquid) flow rate or pressure. { 'sərj ,stres }

survey foot [MECH] A unit of length, used by the U.S. Coast and Geodetic Survey, equal to 12/39.37 meter, or approximately 1.000002 feet. { 'sər,vā 'fu̇t }

susceptance [ELEC] The imaginary component of admittance. { sə'sep·təns }

susceptibility *See* electric susceptibility; magnetic susceptibility. { sə,sep·tə'bil·əd·ē }

suspended transformation [THERMO] The cessation of change before true equilibrium is reached, or the failure of a system to change immediately after a change in conditions, such as in supercooling and other forms of metastable equilibrium. { sə'spen·dəd ,tranz·fər 'mā·shən }

sustained oscillation [PHYS] Oscillation in which forces outside the system, but controlled by the system, maintain a periodic oscillation of the system at a period or frequency that is nearly the natural period of the system. { sə'stānd ,äs·ə'lā·shən }

SU_3 symmetry [PART PHYS] A unitary symmetry based on a fundamental multiplet of three equivalent particles—in particular, the approximate symmetry based on the up, down, and strange quarks, and the exact symmetry based on the three differently colored quarks of a given flavor. { ¦es ¦yü ,səb¦thrē 'sim·ə·trē }

Sutherland's formula [STAT MECH] **1.** The absolute viscosity of a gas is proportional to $T^{3/2}/(C + T)$, where T is the absolute temperature and C is a constant for a given gas. **2.** The mean free path of a molecule in a gas is proportional to $1/[nd\sqrt{1 + (C/T)}]$, where n is the number of molecules per unit volume, d is the diameter of a molecule, T is the absolute temperature, and C is a constant. { 'səth·ər·lənz ,fȯr·myə·lə }

Svedberg equation [STAT MECH] An equation which states that the amplitude of vibration of a particle which exhibits Brownian motion is proportional to its period. { 'sfed,bərg i,kwā·zhən }

sverdrup [FL MECH] A unit of volume transport equal to 1,000,000 cubic meters per second. { 'sfər·drəp }

sweep-frequency reflectometer [ELECTROMAG] A reflectometer that measures standing-wave ratio and insertion loss in decibels over a wide range of frequencies, in either single- or sweep-frequency operation. { 'swēp ¦frē·kwən·sē ,rē,flek'täm·əd·ər }

swept-frequency reflectometer [ELECTROMAG] A microwave reflectometer with a swept-signal source and an oscilloscope for displaying the output as a function of frequency. { 'swept ¦frē·kwən·sē ,rē,flek'täm·əd·ər }

swing [ELEC] Variation in frequency or amplitude of an electrical quantity. { swiŋ }

SXAPS *See* soft-x-ray appearance potential spectroscopy.

symmetrical achromat lens [OPTICS] An older

type of camera lens in which two positive achromatic meniscus lenses are symmetrically arranged about the stop. { sə'me·trə·kəl 'ak·rə,mat 'lenz }

symmetrical alternating quantity [PHYS] Alternating quantity of which all values separated by a half period have the same magnitude but opposite sign. { sə'me·trə·kəl 'òl·tər,nād·iŋ 'kwän·əd·ē }

symmetrical inductive diaphragm [ELECTROMAG] A waveguide diaphragm which consists of two plates that leave a space at the center of the waveguide, and which introduces an inductance in the waveguide. { sə'me·trə·kəl in'dək·tiv 'dī·ə,fram }

symmetrical lens [OPTICS] A lens system consisting of two parts, each of which is the mirror image of the other. { sə'me·trə·kəl 'lenz }

symmetry *See* invariance. { 'sim·ə,trē }

symmetry axis *See* axis of symmetry; rotation axis. { 'sim·ə,trē ,ak·səs }

symmetry axis of rotary inversion *See* rotoinversion axis. { 'sim·ə,trē ,ak·səs əv 'rōd·ə·rē in'vər·zhən }

symmetry axis of rotoinversion *See* rotoinversion axis. { 'sim·ə,trē ,ak·səs əv ¦rōd·ō·in'vər·zhən }

symmetry breaking [PHYS] The deviation from exact symmetry exhibited by many physical systems; it encompasses explicit symmetry breaking and spontaneous symmetry breaking. { 'sim·ə·trē ,brāk·iŋ }

symmetry class *See* crystal class. { 'sim·ə,trē ,klas }

symmetry element [CRYSTAL] **1.** Some combination of rotations and reflections and translations which brings a crystal into a position that cannot be distinguished from its original position. Also known as symmetry operation; symmetry transformation. **2.** The rotational axes, mirror planes, and center of symmetry characteristic of a given crystal. { 'sim·ə,trē ,el·ə·mənt }

symmetry law *See* invariance principle. { 'sim·ə,trē ,lȯ }

symmetry operation *See* symmetry element. { 'sim·ə,trē ,äp·ə,rā·shən }

symmetry operation of the second kind [CRYSTAL] A combination of rotations, reflections, and translations that brings a crystal into a position that is a mirror image of its original position. { ¦sim·ə·trē ,äp·ə,rā·shən əv thə 'sek·ənd ,kīnd }

symmetry plane *See* plane of mirror symmetry. { 'sim·ə,trē ,plān }

symmetry principle *See* invariance principle. { 'sim·ə,trē ,prin·sə·pəl }

symmetry transformation *See* symmetry element. { 'sim·ə,trē ,tranz·fər'mā·shən }

sympathetic vibration [PHYS] The driving of a mechanical or acoustical system at its resonant frequency by energy from an adjacent system vibrating at the same frequency. { ,sim·pə'thed·ik vī'brā·shən }

synchronism [ELEC] Of a synchronous motor, the condition under which the motor runs at a speed which is directly related to the frequency of the power applied to the motor and is not dependent upon variables. [PHYS] Condition of two periodic quantities which have the same frequency, and whose phase difference is either constant or varies around a constant average value. { 'siŋ·krə,niz·əm }

synchrotron process [ELECTROMAG] The emission of electromagnetic radiation by relativistic electrons orbiting in a magnetic field. { 'siŋ·krə,trän ,prä·səs }

synchrotron radiation [ELECTROMAG] Electromagnetic radiation generated by the acceleration of charged relativistic particles, usually electrons, in a magnetic field. { 'siŋ·krə,trän ,rād·ē,ā·shən }

synoptic correlation *See* Eulerian correlation. { sə'näp·tik ,kär·ə'lā·shən }

syntony [ELEC] Condition in which two oscillating circuits have the same resonant frequency. { 'sin·tə·nē }

system [PHYS] A region in space or a portion of matter that has a certain amount of one or more substances, ordered in one or more phases. { 'sis·təm }

Système International d'Unités *See* International System of Units. { si'stem ,in·tər,näs·ē·ə'näl dyün·i'tāz }

T

t *See* troy system.

T *See* tesla.

TΩ *See* teraohm.

tablespoonful [MECH] A unit of volume used particularly in cookery, equal to 4 fluid drams or 1/2 fluid ounce; in the United States this is equal to approximately 14.7868 cubic centimeters, in the United Kingdom to approximately 14.2065 cubic centimeters. Abbreviated tbsp. { 'tā·bəl¦spün‚fu̇l }

tabular crystal [CRYSTAL] A crystal that appears broad and flat due to two prominent parallel faces. { 'tab·yə·lər 'krist·əl }

tachyon [PART PHYS] A hypothetical particle that travels faster than light, consistent with the theory of relativity. { 'tak·ē‚än }

talbot [OPTICS] A unit of luminous energy equal to the luminous energy carried by a luminous flux of 1 lumen during a period of 1 second. { 'tal·bət }

Talbot's bands [OPTICS] A series of dark bands that appear in the spectrum of white light when a glass plate of the proper thickness is placed across one half of the aperture of a spectroscope on the side of the blue end of the spectrum. { 'tal·bəts ‚banz }

Talbot's law [OPTICS] The law that apparent brightness of an object flashing at a frequency greater than about 10 hertz is equal to its actual brightness times the ratio of the exposure time to the total time. { 'tal·bəts ‚lȯ }

Tamm-Dancoff method [QUANT MECH] A method of forming an approximate wave function of a system of interacting particles, particularly nucleons and mesons, by describing it as an algebraic sum of several possible states, the number of such states determining the order of the approximation. { 'tam 'dan‚kȯf ‚meth·əd }

tan alt *See* shadow factor. { 'tan 'ȯlt }

tandem [ELEC] Two-terminal pair networks are in tandem when the output terminals of one network are directly connected to the input terminals of the other network. { 'tan·dəm }

tangential acceleration [MECH] The component of linear acceleration tangent to the path of a particle moving in a circular path. { tan'jen·chəl ak‚sel·ə'rā·shən }

tangential coma [OPTICS] For a lens that displays coma, the length of a tangent from the vertex of the patch of light formed in the focal plane by rays from an off-axis point to the comatic circle of rays from this point that pass near the edge of the lens. { tan'jen·chəl 'kō·mə }

tangential focus *See* primary focus. { tan'jen·chəl 'fō·kəs }

tangential plane *See* meridional plane. { tan'jen·chəl 'plān }

tangential stress *See* shearing stress. { tan'jen·chəl 'stres }

tangential surface [OPTICS] A surface containing the primary foci of points in a plane perpendicular to the optical axis of an astigmatic system. { tan'jen·chəl 'sər·fəs }

tangential velocity [MECH] **1.** The instantaneous linear velocity of a body moving in a circular path; its direction is tangential to the circular path at the point in question. **2.** The component of the velocity of a body that is perpendicular to a line from an observer or reference point to the body. { tan'jen·chəl və'läs·əd·ē }

tangential wave path [ELECTROMAG] In radio propagation over the earth, a path of propagation of a direct wave which is tangential to the surface of the earth; the tangential wave path is curved by atmospheric refraction. { tan'jen·chəl 'wāv ‚path }

tangling [SOLID STATE] The reduction of motion of dislocations in a substance by increasing the total number of dislocations, so that they tangle and interfere with each other's motions. { 'taŋ·gliŋ }

tank periscope [OPTICS] A periscope permitting a tank occupant to observe without being exposed to bullet fire; employs a pair of plane, parallel, reflecting surfaces (either mirrors or prisms) so arranged in a mount that the path of light through the instrument forms a letter Z. { 'taŋk ¦per·ə‚skōp }

tantalum capacitor [ELEC] An electrolytic capacitor in which the anode is some form of tantalum; examples include solid tantalum, tantalum-foil electrolytic, and tantalum-slug electrolytic capacitors. { 'tant·əl·əm kə'pas·əd·ər }

tantalum-foil electrolytic capacitor [ELEC] An electrolytic capacitor that uses plain or etched tantalum foil for both electrodes, with a weak acid electrolyte. { 'tant·əl·əm ¦fȯil i¦lek·trə¦lid·ik kə'pas·əd·ər }

tantalum-slug electrolytic capacitor [ELEC] An electrolytic capacitor that uses a sintered slug of tantalum as the anode, in a highly conductive acid electrolyte. { 'tant·əl·əm ¦sləg i¦lek·trə¦lid·ik kə'pas·əd·ər }

T antenna [ELECTROMAG] An antenna consisting of one or more horizontal wires, with a lead-in connection being made at the approximate center of each wire. { 'tē an,ten·ə }

Taos hum [ACOUS] An irritating low-frequency sound of mysterious origin perceived by some individuals in and around Taos, New Mexico. { ¦tä,ōs 'həm }

tap [ELEC] A connection made at some point other than the ends of a resistor or coil. { tap }

tapered transmission line *See* tapered waveguide. { 'tā·pərd tranz'mish·ən ,līn }

tapered waveguide [ELECTROMAG] A waveguide in which a physical or electrical characteristic changes continuously with distance along the axis of the waveguide. Also known as tapered transmission line. { 'tā·pərd 'wāv,gīd }

tapped resistor [ELEC] A wire-wound fixed resistor having one or more additional terminals along its length, generally for voltage-divider applications. { 'tapt ri'zis·tər }

tare [MECH] The weight of an empty vehicle or container; subtracted from gross weight to ascertain net weight. { ter }

tare effect [FL MECH] In wind tunnel testing, the forces and moments due to support assembly and mutual interference between support assembly and model. { 'ter i,fekt }

target [ATOM PHYS] The atom or nucleus in an atomic or nuclear reaction which is initially stationary. [PHYS] An object or substance subjected to bombardment or irradiation by particles or electromagnetic radiation. { 'tär·gət }

target strength [ACOUS] A measure of the reflecting power of a sonar target, which is expressed in decibels by the equation E + 2L − S, where E is the echo level, L is the total transmission loss, and S is the source level. { 'tär·gət ,streŋkth }

Taub NUT space [RELAT] An exact homogeneous, aniso-tropic solution of Einstein's equations of general relativity that contains many of the pathologies (such as closed timelike lines) possible in a space-time. { ¦taủb 'nət ,spās }

tau meson [PART PHYS] Former name for the K meson, especially one which decays into three pions. { 'taủ 'mā,sän }

tauon *See* tau particle. { 'taủ,än }

tau particle [PART PHYS] A heavy, charged lepton with a mass of approximately 1777 megaelectronvolts, observed as a resonance in electron-positron collisions. Also known as tauon. { 'taủ ,pärd·ə·kəl }

Taylor effect [FL MECH] A phenomenon in which the relative motion of a homogeneous rotating liquid tends to be the same in all planes perpendicular to the axis of rotation. { 'tā·lər i,fekt }

Taylor instability [FL MECH] An instability in Couette flow between rotating cylinders that arises at a critical angular velocity, in which periodic eddies along the cylinder axis form a regular pattern. { 'tā·lər ,in·stə'bil·əd·ē }

Taylor number [FL MECH] A nondimensional number arising in problems of a rotating viscous fluid, written as $T = (f^2h^4)/v^2$, where f is the Coriolis parameter (or, for a cylindrical system, twice the rate of rotation of the system), h the depth of the fluid, and v the kinematic viscosity; the square root of the Taylor number is a rotating Reynolds number, and the fourth root is proportional to the ratio of the depth h to the depth of the Ekman layer. { 'tā·lər ,nəm·bər }

Taylor-Orowan dislocation *See* edge dislocation. { 'tā·lər ō'rō·wən ,dis·lō,kā·shən }

TBE *See* binding energy.

tbsp *See* tablespoonful.

T circulator [ELECTROMAG] A circulator in which three identical rectangular waveguides are joined asymmetrically to form a T-shaped structure, with a ferrite post or wedge at its center; power entering any waveguide emerges from only one adjacent waveguide. { 'tē ,sər·kyə ,lād·ər }

TEA laser [OPTICS] A gas laser in which a glow discharge is maintained without arc formation at atmospheric pressure (which is relatively high for a gas laser) by using a discharge which is transverse rather than parallel to the optic axis. Derived from transversely excited atmospheric pressure laser. { ,tē,ē'ā 'lā·zər }

tear strength [MECH] The force needed to initiate or to continue tearing a sheet or fabric. { 'ter ,streŋkth }

teaspoonful [MECH] A unit of volume used particularly in cookery and pharmacy, equal to $1^1/_3$ fluid drams, or 1/3 tablespoonful; in the United States this is equal to approximately 4.9289 cubic centimeters, in the United Kingdom to approximately 4.7355 cubic centimeters. Abbreviated tsp; tspn. { 'tē,spün,fül }

technical atmosphere [MECH] A unit of pressure in the metric technical system equal to one kilogram-force per square centimeter. Abbreviated at. { 'tek·nə·kəl 'at·mə,sfir }

technicolor [PART PHYS] A hypothetical fifth force proposed to explain the breakdown of symmetry in the unified electroweak interaction. { 'tek·nə,kəl·ər }

technihadron [PART PHYS] Any of a class of hypothetical particles that would be bound together by technicolor forces in the same manner that the ordinary color force binds quarks, antiquarks, and gluons into hadrons. { ¦tek·ni'had,rän }

technipion [PART PHYS] A technihadron that would be anomalously light compared to the characteristic technicolor energy scale, analogous to pions which are light compared to the strong interaction scale. { ¦tek·ni'pī,än }

telecentric system [OPTICS] A telescopic system whose aperture stop is located at one of the foci of the objective lens. { ¦tel·ə¦sen·trik 'sis·təm }

telephoto lens [OPTICS] A lens for photographing distant objects; it is designed in a compact manner so that the distance from the front of the lens to the film plane is less than the focal length of the lens. { ¦tel·ə¦fōd·ō 'lenz }

telephotometry [OPTICS] The body of principles and techniques concerned with measuring atmospheric extinction by using various types of telephotometers. { ¦tel·ə·fə'täm·ə·trē }

telescope [OPTICS] An assemblage of lenses or mirrors, or both, that enhances the ability of the eye either to see objects with greater resolution or to see fainter objects. { 'tel·ə,skōp }

telescope effect [PHYS] An effect in which stretching a high-polymer fiber increases hardness and reduces cross-sectional areas of the fibers. { ¦tel·ə,skäp·ik i¦fekt }

telluric line [SPECT] Any of the spectral bands and lines in the spectrum of the sun and stars produced by the absorption of their light in the atmosphere of the earth. { tə'lu̇r·ik 'līn }

tellurium method [FL MECH] A method of flow visualization in which a cloud of tellurium is generated in a liquid through an irreversible electrolytic reaction, forming a tracer. { tə'lu̇r·ē·əm ,meth·əd }

TEM mode *See* transverse electromagnetic mode. { ,tē,ē'em ,mōd }

TE mode *See* transverse electric mode. { ¦tē'ē ,mōd }

temperament [ACOUS] The adjustment of the pitch of the notes of a keyboard instrument so that the diatonic scale in all keys approximates just tuning; this permits modulation to any key. { 'tem·prə·mənt }

temperature [THERMO] A property of an object which determines the direction of heat flow when the object is placed in thermal contact with another object: heat flows from a region of higher temperature to one of lower temperature; it is measured either by an empirical temperature scale, based on some convenient property of a material or instrument, or by a scale of absolute temperature, for example, the Kelvin scale. { 'tem·prə·chər }

temperature bath [THERMO] A relatively large volume of a homogeneous substance held at constant temperature, so that an object placed in thermal contact with it is maintained at the same temperature. { 'tem·prə·chər ,bath }

temperature coefficient [PHYS] The rate of change of some physical quantity (such as resistance of a conductor or voltage drop across a vacuum tube) with respect to temperature. { 'tem·prə·chər ,kō·i,fish·ənt }

temperature color scale [THERMO] The relation between an incandescent substance's temperature and the color of the light it emits. { 'tem·prə·chər 'kəl·ər ,skāl }

temperature gradient [THERMO] For a given point, a vector whose direction is perpendicular to an isothermal surface at the point, and whose magnitude equals the rate of change of temperature in this direction. { 'tem·prə·chər ,grād·ē·ənt }

temperature resistance coefficient [ELEC] The ratio of the change of electrical resistance in a wire caused by a change in its temperature of 1°C as related to its resistance at 0°C. { 'tem·prə·chər ri'zis·təns ,kō·i,fish·ənt }

temperature scale [THERMO] An assignment of numbers to temperatures in a continuous manner, such that the resulting function is single valued; it is either an empirical temperature scale, based on some convenient property of a substance or object, or it measures the absolute temperature. { 'tem·prə·chər ,skāl }

temperature wave [CRYO] A disturbance in which a variation in temperature propagates through a medium; the chief example of this is second sound. Also known as thermal wave. { 'tem·prə·chər ,wāv }

tempon [PHYS] A unit of time equal to the time required for light to traverse the classical radius of an electron. { 'tem,pän }

temporal coherence [PHYS] The existence of a correlation in time between the phases of waves at a point in space. { 'tem·prəl kō'hir·əns }

TEM wave *See* transverse electromagnetic wave. { ,tē,ē'em ,wāv }

Ten Broecke chart [THERMO] A graphical plot of heat transfer and temperature differences used to calculate the thermal efficiency of a countercurrent cool-fluid-warm-fluid heat-exchange system. { 'ten ,brü·kə ,chärt }

tenebrescence [PHYS] Darkening and bleaching under suitable irradiation; materials having this property are called scotophors; darkening may be produced by primary x-rays or cathode rays, while bleaching may be produced by heat or by photons of appropriate wavelength. { ,ten·ə'bres·əns }

tensile modulus [MECH] The tangent or secant modulus of elasticity of a material in tension. { 'ten·səl ,mäj·ə·ləs }

tensile strength [MECH] The maximum stress a material subjected to a stretching load can withstand without tearing. Also known as hot strength. { 'ten·səl ,streŋkth }

tensile stress [MECH] Stress developed by a material bearing a tensile load. { 'ten·səl ,stres }

tensiometer method [FL MECH] A method of determining the surface tension of a liquid that involves measuring the force necessary to remove a ring of known radius from the liquid surface, usually by means of a torsion balance. { ,ten·sē'äm·əd·ər ,meth·əd }

tension [MECH] **1.** The condition of a string, wire, or rod that is stretched between two points. **2.** The force exerted by the stretched object on a support. { 'ten·chən }

tensor force [NUC PHYS] A spin-dependent force between nucleons, having the same form as the interaction between magnetic dipoles; it is introduced to account for the observed values of the magnetic dipole moment and electric quadrupole moment of the deuteron. { 'ten·sər ,fȯrs }

tenthmeter *See* angstrom. { 'tenth,mēd·ər }

terahertz [PHYS] A unit of frequency, equal to 10^{12} hertz, or 1,000,000 megahertz. Abbreviated THz. { 'ter·ə‚hərtz }

terahertz gap [ELECTROMAG] A region of the electromagnetic spectrum, roughly in the frequency range from 0.05 to 20 terahertz, that lies at the boundary between the microwave region where electronic devices such as high-speed transistors operate and the infrared and visible regions where photonic devices such as lasers operate, and whose exploitation has therefore suffered from a lack of bright sources or coherent sensitive detectors. { 'ter·ə‚hərts ‚gap }

teraohm [ELEC] A unit of electrical resistance, equal to 10^{12} ohms, or 1,000,000 megohms. Abbreviated TΩ. { 'ter·ə‚ōm }

terawatt [PHYS] A unit of power, equal to 10^{12} watts, or 1,000,000 megawatts. Abbreviated TW. { 'ter·ə‚wät }

tercentesimal thermometric scale *See* approximate absolute temperature. { ¦tər·sen'tes·ə·məl ¦thər·mə¦me·trik 'skāl }

term [SPECT] A set of (2S+1)(2L+1) atomic states belonging to a definite configuration and to definite spin and orbital angular momentum quantum numbers S and L. { tərm }

terminal indecomposable future [RELAT] A means of attaching a causal boundary to a space-time; it is the future of a past inextendible time-like curve. Abbreviated TIF. { ¦tər·mən·əl ‚in·di·kəm¦pōz·ə·bəl 'fyü·chər }

terminal indecomposable past [RELAT] A means of attaching a causal boundary to a space-time; it is the past of a future inextendible time-like curve. Abbreviated TIP. { ¦tər·mən·əl ‚in·di·kəm¦pōz·ə·bəl 'past }

terminal speed *See* terminal velocity. { 'tər·mən·əl ‚spēd }

terminal velocity [FL MECH] The velocity with which a body moves relative to a fluid when the resultant force acting on it (due to friction, gravity, and so forth) is zero. [PHYS] The maximum velocity attainable, especially by a freely falling body, under given conditions. Also known as terminal speed. { 'tər·mən·əl və'läs·əd·ē }

termination [ELECTROMAG] **1.** Load connected to a transmission line or other device; to avoid wave reflections, it must match the characteristic of the line or device. **2.** In waveguide technique, the point at which energy flowing along a waveguide continues in a nonwaveguide mode of propagation. { ‚tər·mə'nā·shən }

term splitting [QUANT MECH] The separation of the energies of the states in a term; in the Russell-Saunders case this is produced by the spin-orbit interaction. { 'tərm ‚splid·iŋ }

terrestrial refraction [OPTICS] Any refraction phenomenon observed in the light originating from a source lying within the earth's atmosphere; this is applied only to refraction caused by inhomogeneities of the atmosphere itself, not, for example, to that caused by ice crystals suspended in the atmosphere. { tə'res·trē·əl ri'frak·shən }

terrestrial scintillation [OPTICS] A generic term for scintillation phenomena observed in light that reaches the eye from sources lying within the earth's atmosphere. Also known as atmospheric boil; atmospheric shimmer; optical haze. { tə'res·trē·əl ‚sint·əl'ā·shən }

terrestrial telescope [OPTICS] Any telescope which produces an erect image. { tə'res·trē·əl 'tel·ə‚skōp }

tertiary pyroelectricity [SOLID STATE] The polarization due to temperature and gradients and corresponding nonuniform stresses and strains when the crystal is heated nonuniformly; found in pyroelectric and nonpyroelectric crystals, that is, crystals which have no polar directions. Also known as false pyroelectricity. { 'tər·shē‚er·ē ¦pī·rō‚i‚lek'tris·əd·ē }

tesla [ELECTROMAG] The International System unit of magnetic flux density, equal to one weber per square meter. Symbolized T. { 'tes·lə }

Tesla coil [ELECTROMAG] An air-core transformer used with a spark gap and capacitor to produce a high voltage at a high frequency. { 'tes·lə ‚kȯil }

Tessar lens [OPTICS] An anastigmatic lens made up of a negative lens at the aperture stop with two positive lenses, one in front and the other in back; the last positive lens is a cemented doublet. { 'te‚sär ‚lenz }

tetrachord [ACOUS] The basis of a variety of ancient musical scales, consisting of four notes, with an interval of a perfect fourth between the highest and lowest notes. { 'te·trə‚kȯrd }

tetrad axis [CRYSTAL] A rotation axis whose multiplicity is equal to 4. { 'te‚trad ‚ak·səs }

tetragonal lattice [CRYSTAL] A crystal lattice in which the axes of a unit cell are perpendicular, and two of them are equal in length to each other, but not to the third axis. { te'trag·ən·əl 'lad·əs }

tetragonal trisoctahedron *See* trapezohedron. { te'trag·ən·əl ‚tris¦äk·tə'hē·drən }

tetragonal tristetrahedron *See* deltohedron. { te'trag·ən·əl ‚tris¦te·trə'hē·drən }

tetrahedral symmetry [PHYS] Having the same rotation symmetries as a regular tetrahedron. { 'te·trə¦hē·drəl 'sim·ə·trē }

tetrahedron [CRYSTAL] An isometric crystal form in cubic crystals, in the shape of a four-faced polyhedron, each face of which is a triangle. { ‚te·trə'hē·drən }

tetrahexahedron [CRYSTAL] A form of regular crystal system with four triangular isosceles faces on each side of a cube; there are altogether 24 congruent faces. { ‚te·trə'hek·sə‚drän }

tetratohedral crystal [CRYSTAL] A crystal which has one quarter of the maximum number of faces allowed by the crystal system to which the crystal belongs. { 'te·trəd·ō'hē·drəl 'krist·əl }

TE wave *See* transverse electric wave. { ¦tē'ē ‚wāv }

texture [CRYSTAL] The nature of the orientation, shape, and size of the small crystals in a polycrystalline solid. { 'teks·chər }

theodolite [OPTICS] An optical instrument used

in surveying which consists of a sighting telescope mounted so that it is free to rotate around horizontal and vertical axes, and graduated scales so that the angles of rotation may be measured; the telescope is usually fitted with a right-angle prism so that the observer continues to look horizontally into the eyepiece, whatever the variation of the elevation angle; in meteorology, it is used principally to observe the motion of a pilot balloon. { thē'äd·əl,īt }

theorem of corresponding states [STAT MECH] A theorem stating that two substances which have the same reduced temperature and the same reduced pressure have the same reduced volume. { 'thir·əm əv ,kär·ə'spänd·iŋ 'stāts }

theoretical cutoff frequency [ELEC] Of an electric structure, a frequency at which, disregarding the effects of dissipation, the attenuation constant changes from zero to a positive value or vice versa. { ,thē·ə'red·ə·kəl 'kəd,ȯf ,frē·kwən·sē }

theoretical draft [FL MECH] Draft in a ducted space, neglecting flow losses due to fluid friction. { ,thē·ə'red·ə·kəl 'draft }

theoretical physics [PHYS] The description of natural phenomena in mathematical form. { ,thē·ə'red·ə·kəl 'fiz·iks }

therm [THERMO] A unit of heat energy, equal to 100,000 international table British thermal units, or approximately 1.055×10^8 joules. { thərm }

thermal [THERMO] Of or concerning heat. { 'thər·məl }

thermal agitation [SOLID STATE] Random movements of the free electrons in a conductor, producing noise signals that may become noticeable when they occur at the input of a high-gain amplifier. Also known as thermal effect. { 'thər·məl ,aj·ə'tā·shən }

thermal blooming [OPTICS] The phenomenon of self-defocusing in certain weakly absorbing media; it is prominent, for example, in the propagation of high-power infrared laser beams through the atmosphere. { 'thər·məl 'blüm·iŋ }

thermal bremsstrahlung [PL PHYS] Radiation emitted by electrons in a hot plasma when they are accelerated by positive ions. { 'thər·məl 'brem,shträ·ləŋ }

thermal capacitance [THERMO] The ratio of the entropy added to a body to the resulting rise in temperature. { 'thər·məl kə'pas·əd·əns }

thermal capacity *See* heat capacity. { 'thər·məl kə'pas·əd·ē }

thermal conductance [THERMO] The amount of heat transmitted by a material divided by the difference in temperature of the surfaces of the material. Also known as conductance. { 'thər·məl kən'dək·təns }

thermal conductimetry [THERMO] Measurement of thermal conductivities. { 'thər·məl ,kän,dək'tim·ə·trē }

thermal conductivity [THERMO] The heat flow across a surface per unit area per unit time, divided by the negative of the rate of change of temperature with distance in a direction perpendicular to the surface. Also known as coefficient of conductivity; heat conductivity. { 'thər·məl ,kan,dək'tiv·əd·ē }

thermal conductor [THERMO] A substance with a relatively high thermal conductivity. { 'thər·məl kən'dək·tər }

thermal convection *See* heat convection. { 'thər·məl kən'vek·shən }

thermal coulomb [THERMO] A unit of entropy equal to 1 joule per kelvin. { 'thər·məl 'kü,läm }

thermal diffusivity *See* diffusivity. { 'thər·məl ,di·fyü'siv·əd·ē }

thermal effect *See* thermal agitation. { 'thər·məl i,fekt }

thermal efficiency *See* efficiency. { 'thər·məl i'fish·ən·sē }

thermal effusion *See* thermal transpiration. { 'thər·məl e'fyü·zhən }

thermal electromotive force [PHYS] An electromotive force arising from a difference in temperature at two points along a circuit, as in the Seebeck effect. { 'thər·məl i¦lek·trə¦mōd·iv 'fȯrs }

thermal emissivity *See* emissivity. { 'thər·məl ,ē·mi'siv·əd·ē }

thermal equilibrium [THERMO] Property of a system all parts of which have attained a uniform temperature which is the same as that of the system's surroundings. { 'thər·məl ,ē·kwə'lib·rē·əm }

thermal excitation [ATOM PHYS] The process in which atoms or molecules acquire internal energy in collisions with other particles. { 'thər·məl ,ek·sī'tā·shən }

thermal expansion [PHYS] The dimensional changes exhibited by solids, liquids, and gases for changes in temperature while pressure is held constant. { 'thər·məl ik'span·chən }

thermal expansion coefficient [PHYS] The fractional change in length or volume of a material for a unit change in temperature. { 'thər·məl ik¦span·chən ,kō·i,fish·ənt }

thermal farad [THERMO] A unit of thermal capacitance equal to the thermal capacitance of a body for which an increase in entropy of 1 joule per kelvin results in a temperature rise of 1 kelvin. { 'thər·məl 'far,ad }

thermal flux *See* heat flux. { 'thər·məl 'fləks }

thermal henry [THERMO] A unit of thermal inductance equal to the product of a temperature difference of 1 kelvin and a time of 1 second divided by a rate of flow of entropy of 1 watt per kelvin. { 'thər·məl 'hen·rē }

thermal hysteresis [THERMO] A phenomenon sometimes observed in the behavior of a temperature-dependent property of a body; it is said to occur if the behavior of such a property is different when the body is heated through a given temperature range from when it is cooled through the same temperature range. { 'thər·məl ,his·tə'rē·səs }

thermal inductance [THERMO] The product of temperature difference and time divided by entropy flow. { 'thər·məl in'dək·təns }

thermal instability [FL MECH] The instability resulting in free convection in a fluid heated at a boundary. { 'thər·məl ˌin·stə'bil·əd·ē }

thermal ionization *See* Saha ionization. { 'thər·məl ˌī·ə·nə'zā·shən }

thermalize [NUC PHYS] To bring neutrons into thermal equilibrium with the surroundings. { 'thər·məˌlīz }

thermal magnon [SOLID STATE] A magnon with a relatively short wavelength, on the order of 10^{-6} centimeter. { 'thər·məl 'magˌnän }

thermal ohm [THERMO] A unit of thermal resistance equal to the thermal resistance for which a temperature difference of 1 kelvin produces a flow of entropy of 1 watt per kelvin. Also known as fourier. { 'thər·məl 'ōm }

thermal potential difference [THERMO] The difference between the thermodynamic temperatures of two points. { 'thər·məl pə¦ten·chəl 'dif·rəns }

thermal pulse method [SOLID STATE] A method of measuring properties of insulating and conducting crystals, in which a heat pulse of known duration is measured after propagating through a crystal; the pulse can be generated by directing a laser pulse at an absorbing film evaporated onto one face of the crystal, and detected by a thin-film circuit on the other face. { 'thər·məl ¦pəls ˌmeth·əd }

thermal radiation *See* heat radiation. { 'thər·məl ˌrād·ē'ā·shən }

thermal resistance [THERMO] A measure of a body's ability to prevent heat from flowing through it, equal to the difference between the temperatures of opposite faces of the body divided by the rate of heat flow. Also known as heat resistance. { 'thər·məl ri'zis·təns }

thermal resistivity [THERMO] The reciprocal of the thermal conductivity. { 'thər·məl rēˌzis'tiv·əd·ē }

thermal resistor [ELEC] A resistor designed so its resistance varies in a known manner with changes in ambient temperature. { 'thər·məl ri'zis·tər }

thermal Rossby number [FL MECH] The nondimensional ratio of the inertial force due to the thermal wind and the Coriolis force in the flow of a fluid which is heated from below. { 'thər·məl 'rȯs·bē ˌnəm·bər }

thermal scattering [SOLID STATE] Scattering of electrons, neutrons, or x-rays passing through a solid due to thermal motion of the atoms in the crystal lattice. { 'thər·məl 'skad·ə·riŋ }

thermal shock [MECH] Stress produced in a body or in a material as a result of undergoing a sudden change in temperature. { 'thər·məl 'shäk }

thermal stress [MECH] Mechanical stress induced in a body when some or all of its parts are not free to expand or contract in response to changes in temperature. { 'thər·məl 'stres }

thermal stress cracking [MECH] Crazing or cracking of materials (plastics or metals) by overexposure to elevated temperatures and sudden temperature changes or large temperature differentials. { 'thər·məl ¦stres 'krak·iŋ }

thermal transpiration [THERMO] The formation of a pressure gradient in gas inside a tube when there is a temperature gradient in the gas and when the mean free path of molecules in the gas is a significant fraction of the tube diameter. Also known as thermal effusion. { 'thər·məl ˌtranz·pə'rā·shən }

thermal value [THERMO] Heat produced by combustion, usually expressed in calories per gram or British thermal units per pound. { 'thər·məl ˌval·yü }

thermal volt *See* kelvin. { 'thər·məl 'vōlt }

thermal wave [CRYO] *See* temperature wave. [SOLID STATE] A sound wave in a solid which has a short wavelength. { 'thər·məl 'wāv }

thermal x-rays [ELECTROMAG] The electromagnetic radiation, mainly in the soft (low-energy) x-ray region. { 'thər·məl 'eksˌrāz }

thermie [THERMO] A unit of heat energy equal to the heat energy needed to raise 1 tonne of water from 14.5°C to 15.5°C at a constant pressure of 1 standard atmosphere; equal to 10^6 fifteen-degrees calories or $(4.1855 \pm 0.0005) \times 10^6$ joules. Abbreviated th. { 'thər·mē }

thermoacoustic array [ACOUS] A sound source consisting of a light beam (usually a laser beam) modulated at the frequency of the sound to be generated; the resulting sound has its maximum value in a direction perpendicular to the axis of the light beam, and its directivity pattern has no side lobes. { ¦thər·mō·ə¦kü·stik ə'rā }

thermoacoustic effect [PHYS] **1.** Any effect that arises from the combination of the pressure oscillations of a sound wave with the accompanying adiabatic temperature oscillations. **2.** *See* optoacoustic effect. { ¦thər·mō·ə¦kü·stik i'fekt }

thermoacoustics [PHYS] The study of phenomena that involve both thermodynamics and acoustics. { ˌthər·mō·ə'kü·stiks }

thermochemical calorie *See* calorie. { ¦thər·mō 'kem·ə·kəl 'kal·ə·rē }

thermochromism [PHYS] A reversible change in the color of a substance as its temperature is varied. { ˌthər·mə'krōˌmiz·əm }

thermodynamic cycle [THERMO] A procedure or arrangement in which some material goes through a cyclic process and one form of energy, such as heat at an elevated temperature from combustion of a fuel, is in part converted to another form, such as mechanical energy of a shaft, the remainder being rejected to a lower temperature sink. Also known as heat cycle. { ¦thər·mō·dī'nam·ik 'sī·kəl }

thermodynamic equation of state [THERMO] An equation that relates the reversible change in energy of a thermodynamic system to the pressure, volume, and temperature. { ¦thər·mō·dī'nam·ik i'kwā·zhən əv 'stāt }

thermodynamic equilibrium [THERMO] Property of a system which is in mechanical, chemical, and thermal equilibrium. { ¦thər·mō·dī'nam·ik ˌē·kwə'lib·rē·əm }

thermodynamic function of state [THERMO] Any of the quantities defining the thermodynamic state of a substance in thermodynamic equilibrium; for a perfect gas, the pressure, temperature, and density are the fundamental thermodynamic variables, any two of which are, by the equation of state, sufficient to specify the state. Also known as state parameter; state variable; thermodynamic variable. { ¦thər·mō·dī'nam·ik 'fəŋk·shən əv 'stāt }

thermodynamic potential [THERMO] One of several extensive quantities which are determined by the instantaneous state of a thermodynamic system, independent of its previous history, and which are at a minimum when the system is in thermodynamic equilibrium under specified conditions. { ¦thər·mō·dī'nam·ik pə'ten·chəl }

thermodynamic potential at constant volume *See* free energy. { ¦thər·mō·dī'nam·ik pe¦ten·chəl at 'kän·stənt 'väl·yəm }

thermodynamic principles [THERMO] Laws governing the conversion of energy from one form to another. { ¦thər·mō·dī'nam·ik 'prin·sə·pəlz }

thermodynamic probability [THERMO] Under specified conditions, the number of equally likely states in which a substance may exist; the thermodynamic probability Ω is related to the entropy S by $S = k \ln \Omega$, where k is Boltzmann's constant. { ¦thər·mō·dī'nam·ik ˌpräb·ə'bil·əd·ē }

thermodynamic process [THERMO] A change of any property of an aggregation of matter and energy, accompanied by thermal effects. { ¦thər·mō·dī'nam·ik 'prä·səs }

thermodynamic property [THERMO] A quantity which is either an attribute of an entire system or is a function of position which is continuous and does not vary rapidly over microscopic distances, except possibly for abrupt changes at boundaries between phases of the system; examples are temperature, pressure, volume, concentration, surface tension, and viscosity. Also known as macroscopic property. { ¦thər·mō·dī'nam·ik 'präp·ərd·ē }

thermodynamics [PHYS] The branch of physics which seeks to derive, from a few basic postulates, relationships between properties of matter, especially those which are affected by changes in temperature, and a description of the conversion of energy from one form to another. { ¦thər·mō·dī'nam·iks }

thermodynamic system [THERMO] A part of the physical world as described by its thermodynamic properties. { ¦thər·mō·dī'nam·ik 'sis·təm }

thermodynamic temperature scale [THERMO] Any temperature scale in which the ratio of the temperatures of two reservoirs is equal to the ratio of the amount of heat absorbed from one of them by a heat engine operating in a Carnot cycle to the amount of heat rejected by this engine to the other reservoir; the Kelvin scale and the Rankine scale are examples of this type. { ¦thər·mō·dī'nam·ik 'tem·prə·chər ˌskāl }

thermodynamic variable *See* thermodynamic function of state. { ¦thər·mō·dī'nam·ik 'ver·ē·ə·bəl }

thermoelasticity [PHYS] Dependence of the stress distribution of an elastic solid on its thermal state, or of its thermal conductivity on the stress distribution. { ¦thər·mō·iˌlas'tis·əd·ē }

thermoelectric effect *See* thermoelectricity. { ¦thər·mō·i'lek·trik i¦fekt }

thermoelectricity [PHYS] The direct conversion of heat into electrical energy, or the reverse; it encompasses the Seebeck, Peltier, and Thomson effects but, by convention, excludes other electrothermal phenomena, such as thermionic emission. Also known as thermoelectric effect. { ¦thər·mōˌiˌlek'tris·əd·ē }

thermoelectric properties [PHYS] Properties of materials associated with thermoelectricity, namely, the electromotive force generated in the Seebeck effect, the heat generated or absorbed in the Peltier and Thomson effects, and the influence of magnetic fields upon these quantities. { ¦thər·mō·i'lek·trik 'präp·ərd·ēz }

thermoelectromotive force [ELEC] Voltage developed due to differences in temperature between parts of a circuit containing two or more different metals. { ¦thər·mō·i¦lek·trə¦mōd·iv 'fȯrs }

thermograph [OPTICS] A far-infrared image-forming device that provides a thermal photograph by scanning a far-infrared image of an object or scene. { 'thər·məˌgraf }

thermoluminescence [ATOM PHYS] **1.** Broadly, any luminescence appearing in a material due to application of heat. **2.** Specifically, the luminescence appearing as the temperature of a material is steadily increased; it is usually caused by a process in which electrons receiving increasing amounts of thermal energy escape from a center in a solid where they have been trapped and go over to a luminescent center, giving it energy and causing it to luminesce. { ¦thər·mōˌlü·mə'nes·əns }

thermomagnetic effect [PHYS] An electrical or thermal phenomenon occurring when a conductor or semiconductor is placed simultaneously in a temperature gradient and a magnetic field; examples are the Ettingshausen-Nernst effect and the Righi-Leduc effect. { ¦thər·mō·mag'ned·ik i¦fekt }

thermomechanical effect *See* fountain effect. { ¦thər·mō·mi'kan·ə·kəl i¦fekt }

thermometric conductivity *See* diffusivity. { ¦thər·mə¦me·trik ˌkänˌdəkˌtiv·əd·ē }

thermometric fluid [THERMO] A fluid that has properties, such as a large and uniform thermal expansion coefficient, good thermal conductivity, and chemical stability, that make it suitable for use in a thermometer. { ˌthər·mə¦me·trik 'flü·əd }

thermometric property [THERMO] A physical property that changes in a known way with temperature, and can therefore be used to measure temperature. { ¦thər·mə¦me·trik 'präp·ərd·ē }

thermometry [THERMO] The science and technology of measuring temperature, and the establishment of standards of temperature measurement. { thər'mäm·ə·trē }

thermonuclear reaction [NUC PHYS] A nuclear fusion reaction which occurs between various nuclei of light elements when they are constituents of a gas at very high temperature. { ¦thər·mō'nü·klē·ər rē'ak·shən }

thermooptic effect [PHYS] The change in optical properties of a material because of heat radiation. { ¦thər·mō'äp·tik ə'fekt }

thermophoresis [THERMO] The movement of particles in a thermal gradient from high to low temperatures. { ˌthər·mə·fə'rē·səs }

thermopower [ELEC] A measure of the temperature-induced voltage in a conductor. { 'thər·məˌpaù·ər }

thermoviscous attenuation *See* classical attenuation. { ˌthər·mōˌvis·kəs əˌten·yə'wā·shən }

thetagram [THERMO] A thermodynamic diagram with coordinates of pressure and temperature, both on a linear scale. { 'thād·əˌgram }

theta pinch [PL PHYS] A device for producing a controlled nuclear fusion reaction, in which plasma in a long torus or skinny tube is confined by a magnetic field produced by current-carrying coils, and is shock-heated and compressed by pulses in this field to produce the high temperatures at which fusion reactions take place; the magnetic field is then sustained in order to maintain the plasma confinement. { 'thād·ə ˌpinch }

theta polarization [ELECTROMAG] State of a wave in which the E vector is tangential to the meridian lines of some given spherical frame of reference. { 'thād·ə ˌpō·lə·rə'zā·shən }

Thévenin equivalent circuit [ELEC] An equivalent circuit that consists of a series connection of a voltage source and a two-terminal circuit, where the voltage source is usually dependent on the electric signals applied to the input terminals. { tā·vō¦na iˌkwiv·ə·lənt 'sər·kət }

Thévenin generator [ELEC] The voltage generator in the equivalent circuit of Thévenin's theorem. { tā·vō'na ˌjen·əˌrād·ər }

Thévenin's theorem [ELEC] A theorem in network problems which allows calculation of the performance of a device from its terminal properties only: the theorem states that at any given frequency the current flowing in any impedance, connected to two terminals of a linear bilateral network containing generators of the same frequency, is equal to the current flowing in the same impedance when it is connected to a voltage generator whose generated voltage is the voltage at the terminals in question with the impedance removed, and whose series impedance is the impedance of the network looking back from the terminals into the network with all generators replaced by their internal impedances. Also known as Helmholtz's theorem. { tā·vō'naz ˌthir·əm }

thick-film capacitor [ELEC] A capacitor in a thick-film circuit, made by successive screen-printing and firing processes. { 'thik ¦film kə'pas·əd·ər }

thick-film resistor [ELEC] Fixed resistor whose resistance element is a film well over 0.001 inch (25 micrometers) thick. { 'thik ¦film ri'zis·tər }

thick lens [OPTICS] A lens in which the separation between the two surfaces is too great to be ignored in calculations of such quantities as focal length and magnification. { ¦thik 'lenz }

thickness noise [ACOUS] The component of propeller noise that is caused by the displacement of the air by the rotating propeller blade. { 'thik·nəs ˌnòiz }

thin-film capacitor [ELEC] A capacitor that can be constructed by evaporation of conductor and dielectric films in sequence on a substrate; silicon monoxide is generally used as the dielectric. { 'thin ¦film kə'pas·əd·ər }

thin-film ferrite coil [ELECTROMAG] An inductor made by depositing a thin flat spiral of gold or other conducting metal on a ferrite substrate. { 'thin ¦film 'feˌrīt 'kòil }

thin-film resistor [ELEC] A fixed resistor whose resistance element is a metal, alloy, carbon, or other film having a thickness of about 0.000001 inch (25 nanometers). { 'thin ¦film ri'zis·tər }

thin-film transducer [SOLID STATE] A film a few molecules thick, usually consisting of cadmium sulfide, evaporated on a crystal substrate, used to convert microwave radiation into hypersonic sound waves in the crystal. { 'thin ¦film tran'zdü·sər }

thin lens [OPTICS] A lens whose thickness is small enough to be neglected in calculations of such quantities as object distance, image distance, and magnification. { 'thin 'lenz }

third harmonic [PHYS] A sine-wave component having three times the fundamental frequency of a complex signal. { 'thərd här'män·ik }

third law of motion *See* Newton's third law. { 'thərd 'lò əv 'mō·shən }

third law of thermodynamics [THERMO] The entropy of all perfect crystalline solids is zero at absolute zero temperature. { 'thərd 'lò əv ¦thər·mō·də'nam·iks }

third sound [CRYO] A type of wave propagated in thin films of superfluid helium (helium II), consisting of variations in film thickness and temperature. { 'thərd ˌsaùnd }

13.0 temperature *See* annealing point. { ¦thərˌtēn 'tem·prə·chər }

Thomas-Fermi atom model [ATOM PHYS] A method of approximating the electrostatic potential and the electron density in an atom in its ground state, in which these two quantities are related by the Poisson equation on the one hand, and on the other hand by a semiclassical formula for the density of quantum states in phase space. { 'täm·əs 'fer·mē 'ad·əm ˌmäd·əl }

Thomas-Fermi equation [ATOM PHYS] The differential equation $x^{1/2}(d^2y/dx^2) = y^{3/2}$ that arises in calculating the potential in the Thomas-Fermi atom model; the physically meaningful solution satisfies the boundary conditions $y(0) = 1$ and $y(\infty) = 0$. { 'täm·əs 'fer·mē i,kwā·zhən }

Thomas precession [RELAT] The precession of a vector in an accelerated system, relative to an observer for whom the system has a given velocity and acceleration, when this vector appears to be constant to an observer attached to the system; this precession is the kinematical basis of one type of spin-orbit coupling. { 'täm·əs prē'sesh·ən }

Thomas-Reiche-Kuhn sum rule *See* f-sum rule. { 'täm·əs 'rīk·ə 'kyün 'səm ,rül }

Thomson bridge *See* Kelvin bridge. { 'täm·sən ,brij }

Thomson coefficient [PHYS] The ratio of the voltage existing between two points on a metallic conductor to the difference in temperature of those points. { 'täm·sən ,kō·i,fish·ənt }

Thomson cross section [ELECTROMAG] The total scattering cross section for Thomson scattering, equal to $(8/3)\pi(e^2/mc^2)^2$, where e and m are the charge (in electrostatic units) and mass of the scattering particle, and c is the speed of light. { 'täm·sən 'krȯs ,sek·shən }

Thomson effect [PHYS] A thermoelectric effect in which heat flows into or out of a homogeneous conductor when an electric current flows between two points in the conductor at different temperatures, the direction of heat flow depending upon whether the current flows from colder to warmer metal or from warmer to colder. { 'täm·sən i,fekt }

Thomson formula [ELECTROMAG] **1.** The formula for the intensity of scattered electromagnetic radiation in Thomson scattering as a function of the scattering angle ϕ; the intensity is proportional to $1 + \cos^2 \phi$. **2.** A formula for the period of oscillation of a current when a capacitor is discharged. Also known as Kelvin's formula. { 'täm·sən ,fȯr·myə·lə }

Thomson heat [PHYS] The heat generated or absorbed in the Thomson effect in a reversible manner when a current passes through a conductor in which there is a temperature gradient; it is proportional to the product of the current and the temperature gradient. { 'täm·sən ,hēt }

Thomson parabolas [ELECTROMAG] A pattern of parabolas which appear on a photographic plate exposed to a beam of ions of an element which has passed through electric and magnetic fields applied in the same direction normal to the path of the ions; each parabola corresponds to a different charge-to-mass ratio, and thus to a different isotope. { 'täm·sən pə'rab·ə·ləz }

Thomson relations [PHYS] Equations in the study of thermoelectricity, relating the Peltier coefficient and the Thomson coefficient to the Seebeck voltage; they are derived by thermodynamics. Also known as Kelvin relations. { 'täm·sən ri,lā·shənz }

Thomson scattering [ELECTROMAG] Scattering of electromagnetic radiation by free (or very loosely bound) charged particles, computed according to a classical nonrelativistic theory: energy is taken away from the primary radiation as the charged particles accelerated by the transverse electric field of the radiation, radiate in all directions. { 'täm·sən ,skad·ə·riŋ }

Thomson voltage [PHYS] The voltage that exists between two points that are at different temperatures in a conductor. { 'täm·sən ,vōl·tij }

thoron [NUC PHYS] The conventional name for radon-220. Symbolized Tn. { 'thȯr,än }

thou *See* mil. { thau̇ }

thought experiment *See* Gedanken experiment. { 'thȯt ik'sper·ə·mənt }

thousandth mass unit [PHYS] A unit of energy equal to the energy equivalent of a mass of 10^{-3} atomic mass unit according to the Einstein mass-energy relation, that is, to the product of 10^{-3} atomic mass unit and the square of the speed of light; equal to approximately 1.49176×10^{-13} joule. { 'thau̇z·ənth 'mas 'yü·nət }

three-body problem [MECH] The problem of predicting the motions of three objects obeying Newton's laws of motion and attracting each other according to Newton's law of gravitation. { 'thrē ¦bäd·ē ,präb·ləm }

three-body recombination [ATOM PHYS] The combination of an electron with a positive ion in a gas in such a way that the incoming free electron transfers energy and momentum to another free electron in the neighborhood of the ion. { ¦thrē ,bäd·ē rē,käm·bə'nā·shən }

three-decibel coupler [ELECTROMAG] Junction of two waveguides having a common H wall; the two guides are coupled together by H-type aperture coupling; the coupling is such that 50% of the power from either channel will be fed into the other. Also known as Riblet coupler; short-slot coupler. { 'thrē ¦des·ə·bəl 'kəp·lər }

three-dimensional flow [FL MECH] Any fluid flow which is not a two-dimensional flow. { 'thrē di¦men·chən·əl 'flō }

three-j number [QUANT MECH] A coefficient used in coupling eigenfunctions of two commuting angular momenta to form eigenfunctions of the total angular momentum; closely related to the Clebsch-Gordan coefficients. Also known as Wigner 3-*j* symbol. { 'thrē ¦jā 'nəm·bər }

three-level laser [OPTICS] A laser involving three energy levels, one of which is the ground state; laser action usually occurs between the intermediate and ground states. { 'thrē ¦lev·əl 'lā·zər }

three-level maser [PHYS] A solid-state maser in which three energy levels are used; successful operation has been obtained with crystals of gadolinium ethyl sulfate and crystals of potassium chromecyanide at the temperature of liquid helium. { 'thrē ¦lev·əl 'mā·zər }

threeling *See* trilling. { 'thrēl·iŋ }

three-phase system [PHYS] Any physical system in which three distinct phases coexist; phases can be liquid, solid, vapor (gas), or three mutually insoluble liquids, or any combination thereof. { 'thrē ¦fāz 'sis·təm }

threshold [PHYS] The minimum level of some input quantity needed for some process to take place, such as a threshold energy for a reaction, or the minimum level of pumping at which a laser can go into self-excited oscillation. { 'thresh,hōld }

threshold contrast [OPTICS] The smallest contrast of luminance (or brightness) that is perceptible to the human eye under specified conditions of adaptation luminance and target visual angle. Also known as contrast sensitivity; contrast threshold; liminal contrast. { 'thresh ,hōld ¦kän,trast }

threshold detector [NUC PHYS] An element or isotope in which radioactivity is induced only by the capture of neutrons having energies in excess of a certain characteristic threshold value; used to determine the neutron spectrum from a nuclear explosion. { 'thresh,hōld di,tek·tər }

threshold illuminance [OPTICS] The lowest value of illuminance which the eye is capable of detecting under specified conditions of background luminance and degree of dark adaptation of the eye. Also known as flux density threshold. { 'thresh,hōld i'lü·mə·nəns }

threshold of reaction [PHYS] The minimum energy, for an incident particle or photon, below which a particular reaction does not occur. { 'thresh,hōld əv rē'ak·shən }

throttled flow [FL MECH] Flow which is forced to pass through a restricted area, where the velocity must increase. { 'thräd·əld 'flō }

throttling [THERMO] An adiabatic, irreversible process in which a gas expands by passing from one chamber to another chamber which is at a lower pressure than the first chamber. { 'thräd·əl·iŋ }

thrust [MECH] **1.** The force exerted in any direction by a fluid jet or by a powered screw. **2.** Force applied to an object to move it in a desired direction. { thrəst }

thulium-170 [NUC PHYS] The radioactive isotope of thulium, with mass number 170; used as a portable x-ray source. { 'thü·lē·əm ,wən'sev·ən·tē }

thymol blue method [FL MECH] A flow-visualization method in which the pH indicator thymol blue is added to the liquid under study and made to change color by electrolysis near a wire inserted in the liquid. { ¦thī,mōl ¦blü ¦meth·əd }

THz *See* terahertz.

tier array [ELECTROMAG] Array of antenna elements, one above the other. { 'tir ə,rā }

TIF *See* terminal indecomposable future.

tight binding approximation [SOLID STATE] A method of calculating energy states and wave functions of electrons in a solid in which the wave function is assumed to be a sum of pure atomic wave functions centered about each of the atoms in the lattice, each multiplied by a phase factor; it is suitable for deep-lying energy levels. { 'tīt ¦bīnd·iŋ ə,präk·sə'mā·shən }

tight coupling *See* close coupling. { 'tīt 'kəp·liŋ }

tilt [ELECTROMAG] **1.** Angle which an antenna forms with the horizontal. **2.** In radar, the angle between the axis of radiation in the vertical plane and a reference axis which is normally the horizontal. [OPTICS] The angle between the plane of a photograph from a downward-pointing camera and the horizontal plane. { tilt }

tilt boundary [SOLID STATE] A boundary between two crystals that differ in orientation by only a few degrees, consisting of a series of edge dislocations; it is formed during polygonization. Also known as bend plane; polygon wall. { 'tilt ,baún·dre }

timbre [ACOUS] That attribute of auditory sensation in terms of which a listener can judge that two sounds similarly presented and having the same loudness and pitch are dissimilar. Also known as musical quality; quality of sound. { 'tam·bər }

time [PHYS] **1.** The dimension of the physical universe which, at a given place, orders the sequence of events. **2.** A designated instant in this sequence, as the time of day. Also known as epoch. { tīm }

time-average holographic interferometry [OPTICS] The study of holograms of a vibrating surface which have been averaged over time; illumination of such a hologram yields an image of the surface on which are superimposed interference fringes which are contour lines of equal displacement of the surface, enabling vibrational amplitudes of the surface to be measured with precision. { 'tīm ¦av·rij ¦häl·ə¦graf·ik ,in·tər·fə'räm·ə·trē }

time constant [PHYS] **1.** The time required for a physical quantity to rise from zero to $1-1/e$ (that is, 63.2%) of its final steady value when it varies with time t as $1 - e^{-kt}$. **2.** The time required for a physical quantity to fall to $1/e$ (that is, 36.8%) of its initial value when it varies with time t as e^{-kt}. **3.** Generally, the time required for an instrument to indicate a given percentage of the final reading resulting from an input signal. Also known as lag coefficient. { 'tīm ,kän·stənt }

time delay [PHYS] The time required for a signal to travel between two points in a circuit or for a wave to travel between two points in space. { 'tīm di,lā }

time dilation effect *See* slowing of clocks. { 'tīm də,lā·shən i,fekt }

time lag [PHYS] The time between a cause and a resultant effect, as between occurrence of a primary ionizing event and its count by a counter. { 'tīm ,lag }

timelike path [RELAT] A trajectory in space-time such that a vector tangent to any point on the path is a timelike vector. { 'tīm,līk 'path }

timelike surface [RELAT] A surface in space-time whose normal vector is everywhere space-like. { 'tīm,līk 'sər·fəs }

timelike vector [RELAT] A four vector in Minkowski space whose space component has a magnitude which is less than the magnitude of its time component multiplied by the speed of light. { 'tīm,līk 'vek·tər }

time line [FL MECH] A line of tracers in a fluid, produced by a voltage or laser-beam pulse, whose deformation follows the contour of the local velocity profile. { 'tīm ,līn }

time of flight [MECH] Elapsed time in seconds from the instant a projectile or other missile leaves a gun or launcher until the instant it strikes or bursts. [PHYS] The elapsed time from the instant a particle leaves a source to the instant it reaches a detector. { 'tīm əv 'flīt }

time-of-flight mass spectrometer [SPECT] A mass spectrometer in which all the positive ions of the material being analyzed are ejected into the drift region of the spectrometer tube with essentially the same energies, and spread out in accordance with their masses as they reach the cathode of a magnetic electron multiplier at the other end of the tube. { ¦tīm əv ¦flīt 'mas spek 'träm·əd·ər }

time phase [PHYS] Two disturbances are in time phase if they reach corresponding peak values at the same instants of time, though not necessarily at the same points in space. { 'tīm ˌfāz }

time quadrature [PHYS] **1.** Differing by a time interval corresponding to one-fourth the time of one cycle of the frequency in question. **2.** An integration over time. { 'tīm ˌkwäd·rə·chər }

time-resolved laser spectroscopy [SPECT] A method of studying transient phenomena in the interaction of light with matter through the exposure of samples to extremely short and intense pulses of laser light, down to subnanosecond or subpicosecond duration. { 'tīm ri¦zälvd 'lā·zər spek'träs·kə·pē }

time reversal [PHYS] The replacement of the time coordinate t by its negative $-t$ in the equations of motion of a dynamical system; the time reversal operator, a symmetry operator for a quantum-mechanical system, contains also the complex conjugation operator and a matrix operating on the spin coordinate. { 'tīm riˌvər·səl }

time-reversal invariance [PHYS] A symmetry of the fundamental (microscopic) equations of a system such that, if it holds, the time reversal of any motion of the system is also a motion of the system. { 'tīm ri¦vər·səl in'ver·ē·əns }

time-reversal reflection *See* optical phase conjugation. { 'tīm ri¦vər·səl riˌflek·shən }

tint [OPTICS] The mixture of a pure color with white. { 'tint }

tint of passage [OPTICS] The color produced when a plate which is colorless, but which rotates the plane of polarization of polarized light passing through it by an amount which depends on the wavelength of the light, is placed between crossed polarizers. { 'tint əv 'pas·ij }

tintometer [OPTICS] A device used to estimate the intensity of a colored solution by comparing it with standard solutions or colored glass slides, as with the Lovibond tintometer. { tin'täm·əd·ər }

TIP *See* terminal indecomposable past.

TME *See* metric-technical unit of mass.

TM mode *See* transverse magnetic mode. { ¦tē'em ˌmōd }

TM wave *See* transverse magnetic wave. { ¦tē'em ˌwāv }

Tn *See* thoron.

T network [ELEC] A network composed of three branches, with one end of each branch connected to a common junction point, and with the three remaining ends connected to an input terminal, an output terminal, and a common input and output terminal, respectively. { 'tē ˌnetˌwərk }

tokamak [PL PHYS] A device for confining a plasma within a toroidal chamber, which produces plasma temperatures, densities, and confinement times greater than that of any other such device; confinement is effected by a very strong externally applied toroidal field, plus a weaker poloidal field produced by a toroidally directed plasma current, and this current causes ohmic heating of the plasma. { 'täk·əˌmak }

Tolman and Stewart effect [ELEC] The development of negative charge at the forward end of a metal rod which is suddenly stopped after rapid longitudinal motion. { ¦täl·mən ən 'stü·ərt iˌfekt }

ton [MECH] **1.** A unit of weight in common use in the United States, equal to 2000 pounds or 907.18474 kilogram-force. Also known as just ton; net ton; short ton. **2.** A unit of mass in common use in the United Kingdom equal to 2240 pounds, or to 1016.0469088 kilogram-force. Also known as gross ton; long ton. **3.** A unit of weight in troy measure, equal to 2000 troy pounds, or to 746.4834432 kilogram-force. **4.** *See* tonne. { tən }

tondal [MECH] A unit of force equal to the force which will impart an acceleration of 1 foot per second to a mass of 1 long ton; equal to approximately 309.6911 newtons. { 'tənd·əl }

tone [ACOUS] **1.** A sound oscillation capable of exciting an auditory sensation having pitch. **2.** An auditory sensation having pitch. { 'tōn }

tonne [MECH] A unit of mass in the metric system, equal to 1000 kilograms or to approximately 2204.62 pound mass. Also known as metric ton; millier; ton; tonneau. { tən }

tonneau *See* tonne. { tə'nō }

top [MECH] A rigid body, one point of which is held fixed in an inertial reference frame, and which usually has an axis of symmetry passing through this point; its motion is usually studied when it is spinning rapidly about the axis of symmetry. [PART PHYS] The new quantum number associated with the top quark. Also known as truth. *See* rotator. { täp }

top-loaded vertical antenna [ELECTROMAG] Vertical antenna constructed so that, because of its greater size at the top, there results modified current distribution, giving a more desirable radiation pattern in the vertical plane. { ¦täp ¦lōd·əd 'vərd·ə·kəl an'ten·ə }

Topogon lens [OPTICS] A periscopic lens with supplementary thick menisci to permit the correction of aperture aberrations for a moderate aperture and a large field; one or two plane-parallel plates are sometimes added to correct distortion. { 'täp·əˌgän ¦lenz }

topological order [CRYO] An internal order in a

quantum Hall state which describes the quantum motions of the electrons with respect to one another; it is different from any other known order in not being associated with any symmetries or breaking of symmetries. { ˌtäp·əˌläj·ə·kəl 'ȯrd·ər }

topological soliton [PHYS] A soliton whose confinement can be related to the existence of multiple states of minimum energy whose arrangement around the wave has special geometrical properties. { ¦täp·ə¦läj·ə·kəl 'säl·əˌtän }

topology of circuits [ELEC] The study of electric networks in terms of the geometry of their connections only; used in finding such properties of circuits as equivalence and duality, and in analyzing and synthesizing complex circuits. { tə¦päl·ə·jē əv 'sər·kəts }

toponium [PART PHYS] A hypothetical meson that is made up of the top quark *t* and its antiquark $\bar{t}$. { tä'pō·nē·əm }

topple [MECH] In gyroscopes for marine or aeronautical use, the condition of a sudden upset gyroscope or a gyroscope platform evidenced by a sudden and rapid precession of the spin axis due to large torque disturbances such as the spin axis striking the mechanical stops. Also known as tumble. { 'täp·əl }

topple axis [MECH] Of a gyroscope, the horizontal axis, perpendicular to the horizontal spin axis, around which topple occurs. Also known as tumble axis. { 'täp·əl ˌak·səs }

top quark [PART PHYS] A hypothetical quark that is postulated to have a mass of approximately 30–50 GeV, electric charge of +2/3, baryon number of 1/3, zero isotopic spin, strangeness, and charm, and a new quantum number associated with it. Also known as *t* quark; truth quark. Symbolized *t*. { 'täp ˌkwärk }

tor *See* pascal. { tȯr }

toric lens [OPTICS] A lens whose surfaces form portions of toric surfaces. Also known as toroidal lens. { 'tȯr·ik ˌlenz }

toroid *See* toroidal magnetic circuit. { 'tȯrˌȯid }

toroidal coil *See* toroidal magnetic circuit. { tə'rȯid·əl 'kȯil }

toroidal core [ELECTROMAG] The doughnut-shaped piece of magnetic material in a toroidal magnetic circuit. { tə'rȯid·əl 'kȯr }

toroidal lens *See* toric lens. { tə'rȯid·əl ˌlenz }

toroidal magnetic circuit [ELECTROMAG] Doughnut-shaped piece of magnetic material, together with one or more coils of current-carrying wire wound about the doughnut, with the permeability of the magnetic material high enough so that the magnetic flux is almost completely confined within it. Also known as toroid; toroidal coil. { tə'rȯid·əl mag'ned·ik 'sər·kət }

torque [MECH] **1.** For a single force, the cross product of a vector from some reference point to the point of application of the force with the force itself. Also known as moment of force; rotation moment. **2.** For several forces, the vector sum of the torques (first definition) associated with each of the forces. { tȯrk }

torr [MECH] A unit of pressure, equal to 1/760 atmosphere; it differs from 1 millimeter of mercury by less than one part in seven million; approximately equal to 133.3224 pascals. { tȯr }

Torricellian vacuum [FL MECH] The space enclosed above a column of mercury when a tube, closed at one end, is filled with mercury and then placed, open end downward, in a well of mercury; this space is evacuated except for mercury vapor. { ¦tȯr·ə¦chel·ē·ən 'vak·yəm }

Torricelli's law of efflux [FL MECH] The velocity of efflux of liquid from an orifice in a container is equal to that which would be attained by a body falling freely from rest at the free surface of the liquid to the orifice. { ¦tȯr·ə¦chel·ēz 'lȯ əv 'ēˌfləks }

torsion [MECH] A twisting deformation of a solid body about an axis in which lines that were initially parallel to the axis become helices. { 'tȯr·shən }

torsional angle [MECH] The total relative rotation of the ends of a straight cylindrical bar when subjected to a torque. { 'tȯr·shən·əl 'aŋ·gəl }

torsional compliance [MECH] The reciprocal of the torsional rigidity. { ¦tȯr·shə·nəl kəm'pli·əns }

torsional hysteresis [MECH] Dependence of the torques in a twisted wire or rod not only on the present torsion of the object but on its previous history of torsion. { ¦tȯr·shə·nəl ˌhis·tə'rē·səs }

torsional modulus [MECH] The ratio of the torsional rigidity of a bar to its length. Also known as modulus of torsion. { 'tȯr·shən·əl 'mäj·ə·ləs }

torsional pendulum [MECH] A device consisting of a disk or other body of large moment of inertia mounted on one end of a torsionally flexible elastic rod whose other end is held fixed; if the disk is twisted and released, it will undergo simple harmonic motion, provided the torque in the rod is proportional to the angle of twist. Also known as torsion pendulum. { 'tȯr·shən·əl 'pen·jə·ləm }

torsional rigidity [MECH] The ratio of the torque applied about the centroidal axis of a bar at one end of the bar to the resulting torsional angle, when the other end is held fixed. { 'tȯr·shən·əl ri'jid·əd·ē }

torsional vibration [MECH] A periodic motion of a shaft in which the shaft is twisted about its axis first in one direction and then in the other; this motion may be superimposed on rotational or other motion. { 'tȯr·shən·əl vī'brā·shən }

torsional wave [PHYS] A wave motion in which the vibrations of the medium are periodic rotational motions around the direction of propagation. { 'tȯr·shən·əl 'wāv }

torsion function [MECH] A harmonic function, $\phi(x,y) = w/\tau$, expressing the warping of a cylinder undergoing torsion, where the x, y, and z coordinates are chosen so that the axis of torsion lies along the z axis, w is the z component of the displacement, and τ is the torsion angle. Also known as warping function. { 'tȯr·shən ˌfəŋk·shən }

torsion pendulum *See* torsional pendulum. { 'tȯr·shən 'pen·jə·ləm }

total binding energy *See* binding energy. { 'tōd·əl 'bīnd·iŋ 'en·ər·jē }

total curvature [OPTICS] The difference between the reciprocals of the radii of curvature of the two surfaces of a lens. { 'tōd·əl 'kər·və·chər }

total head [FL MECH] The sum of the velocity head and the pressure head corresponding to the static pressure. { ¦tōd·əl ¦hed }

total heat *See* enthalpy. { 'tōd·əl 'hēt }

total internal reflection [OPTICS] A phenomenon in which electromagnetic radiation in a given medium which is incident on the boundary with a less-dense medium (one having a lower index of refraction) at an angle less than the critical angle is completely reflected from the boundary. { 'tōd·əl in'tərn·əl ri'flek·shən }

totally stable system [PHYS] A system that returns to a stationary state following arbitrarily large perturbations. { ¦tōd·əl·ē 'stā·bəl ˌsis·təm }

total pressure [FL MECH] *See* dynamic pressure. [MECH] The gross load applied on a given surface. { 'tōd·əl 'presh·ər }

total relief *See* stereoscopic power. { 'tōd·əl ri'lēf }

total stability [PHYS] The property of a system that returns to an equilibrium configuration after the system is subjected to arbitrarily large perturbations. { ¦tōd·əl stə'bil·əd·ē }

total vorticity [FL MECH] Usually, the magnitude of the vorticity vector, all components included, as opposed to the vertical (component of the) vorticity. { 'tōd·əl vȯr'tis·əd·ē }

total wetting [FL MECH] The situation in which a liquid surface meets a solid surface with zero contact angle. { ˌtōd·əl 'wed·iŋ }

toughness [MECH] A property of a material capable of absorbing energy by plastic deformation; intermediate between softness and brittleness. { 'təf·nəs }

Touschek effect [PHYS] In electron storage rings, an effect in which the maximum particle concentration in the circulating electron bunches is restricted at low energies by the loss of electrons in Møller scattering. { 'tüs,chek i,fekt }

towed load [MECH] The weight of a carriage, trailer, or other equipment towed by a prime mover. { 'tōd 'lōd }

t quark *See* top quark. { 'tē ˌkwärk }

track adaptation effect [ACOUS] A drop in the sound insulation of a wall at the frequency at which bending oscillations excited by obliquely incident sound waves correspond to the free bending oscillations of the wall. Also known as coincidence effect. { ¦trak a,dap'tā·shən i,fekt }

tracking telescope [OPTICS] A long-focal-length telescope mounted to track missiles in flight precisely while collecting missile performance data. { 'trak·iŋ ˌtel·ə,skōp }

traction [MECH] Pulling friction of a moving body on the surface on which it moves. { 'trak·shən }

tractional force [FL MECH] The force exerted on particles under flowing water by the current; it is proportional to the square of the velocity. { 'trak·shən·əl 'fȯrs }

trajectory [MECH] The curve described by an object moving through space, as of a meteor through the atmosphere, a planet around the sun, a projectile fired from a gun, or a rocket in flight. { trə'jek·trē }

transfer impedance [ELEC] The ratio of the voltage applied at one pair of terminals of a network to the resultant current at another pair of terminals, all terminals being terminated in a specified manner. { 'tranz·fər im,pēd·əns }

transfer-matrix method [MECH] A method of analyzing vibrations of complex systems, in which the system is approximated by a finite number of elements connected in a chainlike manner, and matrices are constructed which can be used to determine the configuration and forces acting on one element in terms of those on another. { 'tranz·fər ˌmā·triks ˌmeth·əd }

transfer reaction [NUC PHYS] A nuclear reaction in which one or more nucleons are exchanged between the target nucleus and an incident projectile. { 'tranz·fər rē,ak·shən }

transferred-electron effect [SOLID STATE] The variation in the effective drift mobility of charge carriers in a semiconductor when significant numbers of electrons are transferred from a low-mobility valley of the conduction band in a zone to a high-mobility valley, or vice versa. { 'tranz'fərd i¦lek,trän i,fekt }

transformation [CRYSTAL] *See* inversion. [ELEC] For two networks which are equivalent as far as conditions at the terminals are concerned, a set of equations giving the admittances or impedances of the branches of one circuit in terms of the admittances or impedances of the other. { ˌtranz·fər'mā·shən }

transformation constant *See* decay constant. { ˌtranz·fər'mā·shən ˌkän·stənt }

transformation matrix [ELECTROMAG] A two-by-two matrix which relates the amplitudes of the traveling waves on one side of a waveguide junction to those on the other. { ˌtranz·fər'mā·shən ˌmā·triks }

transformation series *See* radioactive series. { ˌtranz·fər'mā·shən ˌsir·ēz }

transformation theory [QUANT MECH] The study of coordinate and other transformations in quantum mechanics, especially those which leave some properties of the system invariant. { ˌtranz·fər'mā·shən ˌthē·ə·rē }

transformation twin [CRYSTAL] A crystal twin developed by a growth transformation from a higher to a lower symmetry. { ˌtranz·fər'mā·shən ˌtwin }

transformer [ELECTROMAG] An electrical component consisting of two or more multiturn coils of wire placed in close proximity to cause the magnetic field of one to link the other; used to transfer electric energy from one or more alternating-current circuits to one or more other circuits by magnetic induction. { tranz'fȯr·mər }

transformer coupling *See* inductive coupling. { tranz'fȯr·mər ,kəp·liŋ }

transformer voltage ratio [ELEC] Ratio of the root-mean-square primary terminal voltage to the root-mean-square secondary terminal voltage under specified conditions of load. { tranz 'fȯr·mər 'vōl·tij ,rā·shō }

transforming section [ELECTROMAG] Length of waveguide or transmission line of modified cross section, or with a metallic or dielectric insert, used for impedance transformation. { tranz 'fȯrm·iŋ ,sek·shən }

transient [PHYS] A pulse, damped oscillation, or other temporary phenomenon occurring in a system prior to reaching a steady-state condition. { 'tranch·ənt }

transient motion [PHYS] An oscillatory or other irregular motion occurring while a quantity is changing to a new steady-state value. { 'tranch·ənt 'mō·shən }

transient overshoot [PHYS] The maximum value of the overshoot of a quantity as a result of a sudden change in conditions. { 'tranch·ənt 'ō·vər,shüt }

transient phenomena [ELEC] Rapidly changing actions occurring in a circuit during the interval between closing of a switch and settling to a steady-state condition, or any other temporary actions occurring after some change in a circuit. { 'tranch·ənt fə,näm·ə·nä }

transient problem *See* initial-value problem. { 'tranch·ənt 'präb·ləm }

transient RANS modeling *See* TRANS modeling. { 'tranch·ənt 'ranz ,mäd·əl·iŋ }

transient response [PHYS] The behavior of a system following a sudden change in its input. { 'tranch·ənt ri'späns }

transillumination [OPTICS] Illumination of translucent material by light being transmitted through the material from the rear. { ,tranz·ə,lü·mə'nā·shən }

transit instrument *See* transit telescope. { 'trans·ət ,in·strə·mənt }

transition [QUANT MECH] The change of a quantum-mechanical system from one energy state to another. [THERMO] A change of a substance from one of the three states of matter to another. { tran'zish·ən }

transitional flow [FL MECH] A flow in which the viscous and Reynolds stresses are of approximately equal magnitude; it is transitional between laminar flow and turbulent flow. { tran 'zish·ən·əl 'flō }

transition element [ELECTROMAG] An element used to couple one type of transmission system to another, as for coupling a coaxial line to a waveguide. { tran'zish·ən ,el·ə·mənt }

transition factor *See* reflection factor. { tran'zish·ən ,fak·tər }

transition frequency [QUANT MECH] The characteristic frequency of radiation emitted or absorbed by a quantum-mechanical system as it changes from one energy state to another; equal to the energy difference between the states divided by Planck's constant. { tran'zish·ən ,frē·kwən·sē }

transition moment [QUANT MECH] Any type of multipole moment which determines radiative transitions between states; it consists of an integral of the product of the conjugate of the final state wave function, a multipole moment operator, and the initial state wave function. { tran 'zish·ən ,mō·mənt }

transition point [ELECTROMAG] A point at which the constants of a circuit change in such a way as to cause reflection of a wave being propagated along the circuit. [THERMO] Either the temperature at which a substance changes from one state of aggregation to another (a first-order transition), or the temperature of culmination of a gradual change, such as the lambda point, or Curie point (a second-order transition). Also known as transition temperature. { tran'zish·ən ,pȯint }

transition probability [QUANT MECH] The probability per unit time that a quantum-mechanical system will make a transition from a given initial state to a given final state. { tran'zish·ən ,präb·ə'bil·əd·ē }

transition radiation detector [PART PHYS] A detector of energetic charged particles that makes use of the radiation emitted as the particle crosses boundaries between regions with different indices of refraction. { tran'zish·ən ,rād·ē'ā·shən di,tek·tər }

transition region [SOLID STATE] The region between two homogeneous semiconductors in which the impurity concentration changes. { tran'zish·ən ,rē·jən }

transition temperature *See* transition point. { tran'zish·ən ,tem·prə·chər }

transition to chaos [PHYS] The process by which a system evolves from periodic toward chaotic behavior as one or more parameters governing the behavior of the system are varied. { tran'zish·ən tə 'kā,äs }

transition zone [FL MECH] Those conditions of fluid flow in which the nature of the flow is changing from laminar to turbulent. { tran'zish·ən ,zōn }

transit telescope [OPTICS] A telescopic instrument adapted to the observation of the passage, or transit, of an astronomical object across the meridian of an observer; consists of a telescope mounted on a single fixed horizontal axis of rotation which has a central hollow cube (sometimes a sphere) and two conical semiaxes ending in cylindrical pivots; the objective and eyepiece halves of the instrument are also fastened to the cube of the instrument, perpendicular to the horizontal axis. Also known as transit instrument. { 'trans·ət 'tel·ə,skōp }

translation [MECH] The linear movement of a point in space without any rotation. { tran'slā·shən }

translational motion [MECH] Motion of a rigid

body in such a way that any line which is imagined rigidly attached to the body remains parallel to its original direction. { tran'slā·shən·əl 'mō·shən }

translation gliding *See* crystal gliding. { tran'slā·shən ,glīd·iŋ }

translation group [CRYSTAL] The collection of all translation operations which carry a crystal lattice into itself. { tran'slā·shən ,grüp }

translation operation [PHYS] The process of moving an object along a straight line in such a way that any line which is fixed with respect to the object remains parallel to its original direction. { tran'slā·shən ,äp·ə,rā·shən }

translucent medium [OPTICS] A medium which transmits rays of light so diffused that objects cannot be seen distinctly; examples are various forms of glass which admit considerable light but impede vision. { tran'slüs·əns 'mēd·ē·əm }

transmissibility [MECH] A measure of the ability of a system either to amplify or to suppress an input vibration, equal to the ratio of the response amplitude of the system in steady-state forced vibration to the excitation amplitude; the ratio may be in forces, displacements, velocities, or accelerations. { tranz,mis·ə'bil·əd·ē }

transmission *See* transmittance. { tranz'mish·ən }

transmission anomaly [ACOUS] The ratio of the transmission loss of underwater sound at a given distance from the source to the inverse square of this distance, usually expressed in decibels. { tranz'mish·ən ə,näm·ə·lē }

transmission band [ELECTROMAG] Frequency range above the cutoff frequency in a waveguide, or the comparable useful frequency range for any other transmission line, system, or device. { tranz'mish·ən ,band }

transmission coefficient [PHYS] **1.** The value of some quantity associated with the resultant field produced by incident and reflected waves at a given point in a transmission medium divided by the corresponding quantity in the incident wave. **2.** The ratio of transmitted to incident energy flux or flux of some other quantity at a discontinuity in a transmission medium; for sound waves, it is called the sound transmission coefficient. **3.** The ratio of the transmitted flux of some quantity to the incident flux for a substance of unit thickness. *See* penetration probability. { tranz'mish·ən ,kō·i,fish·ənt }

transmission factor [PHYS] The ratio of the flux of some quantity transmitted through a body to the incident flux. { tranz'mish·ən ,fak·tər }

transmission grating [OPTICS] A diffraction grating produced on a transparent base so radiation is transmitted through the grating instead of being reflected from it. { tranz'mish·ən ,grād·iŋ }

transmission line [ELEC] A system of conductors, such as wires, waveguides, or coaxial cables, suitable for conducting electric power or signals efficiently between two or more terminals. { tranz'mish·ən ,līn }

transmission plane [OPTICS] The plane of vibration of polarized light that will pass through a Nicol prism or other polarizer. { tranz'mish·ən ,plān }

transmission range *See* night visual range. { tranz'mish·ən ,rānj }

transmissivity [ELECTROMAG] The ratio of the transmitted radiation to the radiation arriving perpendicular to the boundary between two mediums. { ,tranz·mə'siv·əd·ē }

transmissometry [OPTICS] The technique of determining the extinction characteristics of a medium by measuring the transmission of a light beam of known initial intensity directed into that medium. { ,tranz·mə'säm·ə·trē }

transmittance [ELECTROMAG] The radiant power transmitted by a body divided by the total radiant power incident upon the body. Also known as transmission. { tranz'mid·əns }

transmittancy [ELECTROMAG] The transmittance of a solution divided by that of the pure solvent of the same thickness. { tranz'mid·ən·sē }

transmitted wave *See* refracted wave. { tranz'mid·əd 'wāv }

transmittivity [ELECTROMAG] The internal transmittance of a piece of nondiffusing substance of unit thickness. { ,trans·mə'tiv·əd·ē }

TRANS modeling [FL MECH] A type of turbulence modeling which is based on solving the Reynolds-averaged Navier-Stokes equations in a time-dependent mode. Derived from transient RANS modeling. { 'tranz ,mäd·əl·iŋ or 'tē,ranz }

transmutation [NUC PHYS] A nuclear process in which one nuclide is transformed into the nuclide of a different element. Also known as nuclear transformation. { ,trans·myü'tā·shən }

transonic [PHYS] That which occurs or is occurring within the range of speed in which flow patterns change from subsonic to supersonic (or vice versa), about Mach 0.8 to 1.2, as in transonic flight or transonic flutter. { tran'sän·ik }

transonic flow [FL MECH] Flow of a fluid over a body in the range just above and just below the acoustic velocity. { tran'sän·ik 'flō }

transonic range [FL MECH] The range of speeds between the speed at which one point on a body reaches supersonic speed, and the speed at which all points reach supersonic speed. { tran'sän·ik 'rānj }

transonic speed [FL MECH] The speed of a body relative to the surrounding fluid at which the flow is in some places on the body subsonic and in other places supersonic. { tran'sän·ik 'spēd }

transparency [OPTICS] The ability of a substance to transmit light of different wavelengths, sometimes measured in percent of radiation which penetrates a distance of 1 meter. { tranz 'par·ən·sē }

transparency range [NUC PHYS] A postulated energy range for extremely high-energy heavy-ion collisions in which the projectile passes through the target and emerges with its temperature and density raised to the point at which a quark-gluon plasma forms. { tranz'par·ən·sē ,rānj }

transparent [PHYS] Permitting passage of radiation or particles. { tranz'par·ənt }

transparent medium [OPTICS] **1.** A medium which has the property of transmitting rays of light in such a way that the human eye may see through the medium distinctly. **2.** A medium transparent to other regions of the electromagnetic spectrum, such as x-rays and microwaves. { tranz'par·ənt 'mēd·ē·əm }

transphasor [OPTICS] A nonlinear optical device that uses one light beam to modulate another, in a manner analogous to an electronic transistor, and that operates through the transference of a phase shift from one beam to the other. { ¦tranz'fāz·ər }

transport cross section [PHYS] The product of the total scattering cross section and the average value of $1 - \cos\theta$, where θ is the laboratory scattering angle. { ¦tranz,pȯrt 'krȯs ,sek·shən }

transport effect [PHYS] Any phenomenon, such as diffusion, thermal conductivity, or electrical conductivity, that involves the movement of some entity, such as matter, energy, or electric charge. { 'tranz,pȯrt i,fekt }

transport properties [PHYS] Properties of a compound or material associated with mass or heat transport; for example, viscosity and thermal conductivity of liquids, gases, or solids. { 'tranz,pȯrt ,präp·ərd·ēz }

transrectification [ELEC] Rectification that occurs in one circuit when an alternating voltage is applied to another circuit. { tranz,rek·tə·fə'kā·shən }

transresistance [ELEC] The ratio of the voltage between any two connections of a four-terminal junction to the current passing between the other two connections. { ,tranz·ri'zis·təns }

transverse chromatic aberration [OPTICS] A departure of an optical image-forming system from ideal behavior in which different colors have conjugate image planes which are separated transversely. { trans¦vərs krō'mad·ik ,ab·ə'rā·shən }

transverse Doppler effect [ELECTROMAG] An aspect of the optical Doppler effect, occurring when the direction of motion of the source relative to an observer is perpendicular to the direction of the light received by the observer; the observed frequency is smaller than the source frequency by the factor $[1-(v/c)^2]^{1/2}$, where v is the speed of the source and c is the speed of light. { trans¦vərs 'däp·lər i,fekt }

transverse electric mode [ELECTROMAG] A mode in which a particular transverse electric wave is propagated in a waveguide or cavity. Abbreviated TE mode. Also known as H mode (British usage). { trans¦vərs i¦lek·trik ,mōd }

transverse electric wave [ELECTROMAG] An electromagnetic wave in which the electric field vector is everywhere perpendicular to the direction of propagation. Abbreviated TE wave. Also known as H wave (British usage). { trans¦vərs i'lek·trik 'wāv }

transverse electromagnetic mode [ELECTROMAG] A mode in which a particular transverse electromagnetic wave is propagated in a waveguide or cavity. Abbreviated TEM mode. { trans¦vərs i¦lek·trō·mag'ned·ik 'mōd }

transverse electromagnetic wave [ELECTROMAG] An electromagnetic wave in which both the electric and magnetic field vectors are everywhere perpendicular to the direction of propagation. Abbreviated TEM wave. { trans¦vərs i¦lek·trō·mag'ned·ik 'wāv }

transversely excited atmospheric pressure laser *See* TEA laser. { trans¦vərs·lē ek¦sīd·əd ¦at·mə¦sfir·ik 'presh·ər 'lā·zər }

transverse magnetic mode [ELECTROMAG] A mode in which a particular transverse magnetic wave is propagated in a waveguide or cavity. Abbreviated TM mode. Also known as E mode (British usage). { trans¦vərs mag'ned·ik 'mōd }

transverse magnetic wave [ELECTROMAG] An electromagnetic wave in which the magnetic field vector is everywhere perpendicular to the direction of propagation. Abbreviated TM wave. Also known as E wave (British usage). { trans¦vərs mag'ned·ik 'mōd 'wāv }

transverse magnetoresistance [ELECTROMAG] One of the galvanomagnetic effects, in which a magnetic field perpendicular to an electric current gives rise to an electrical potential change in the direction of the current. { trans¦vərs ¦mag·ned·ō·ri'zis·təns }

transverse mass [RELAT] The ratio of a force acting on a relativistic particle in a direction perpendicular to its velocity to the resulting acceleration; equal to $m_0\,(1 - v^2/c^2)^{-1/2}$, where m_0 is the particle's rest mass, v is its speed, and c is the speed of light. { trans¦vərs 'mas }

transverse piezoelectric effect [SOLID STATE] The manifestation of the piezoelectric effect in which the applied stress is perpendicular to the direction of the resultant electric field, or in which the applied electric field is perpendicular to the direction of the resultant stress. { ¦tranz,vərs pē,āt·sō·i'lek·trik i,fekt }

transverse ray aberration [OPTICS] The transverse displacement from the ideal image point to the ray intersection with the ideal image plane; a measure of monochromatic aberration. { 'tranz,vərs ¦rā ,ab·ə'rā·shən }

transverse vibration [MECH] Vibration of a rod in which elements of the rod move at right angles to the axis of the rod. { trans¦vərs vī'brā·shən }

transverse wave [PHYS] A wave in which the direction of the disturbance at each point of the medium is perpendicular to the wave vector and parallel to surfaces of constant phase. { trans¦vərs 'wāv }

trap [SOLID STATE] Any irregularity, such as a vacancy, in a semiconductor at which an electron or hole in the conduction band can be caught and trapped until released by thermal agitation. Also known as semiconductor trap. { trap }

trapezohedron [CRYSTAL] An isometric crystal form of 24 faces, each face of which is an irregular four-sided figure. Also known as icositetrahedron; leucitohedron; tetragonal trisoctahedron. { trə¦pē·zō¦hē·drən }

traveling microscope [OPTICS] A low-power microscope equipped with a graticule and rails enabling it to move horizontally or vertically, used to make accurate length determinations. { 'trav·əl·iŋ 'mī·krə,skōp }

traveling wave [PHYS] A wave in which energy is transported from one part of a medium to another, in contrast to a standing wave. { 'trav·əl·iŋ 'wāv }

traveling-wave antenna [ELECTROMAG] An antenna in which the current distributions are produced by waves of charges propagated in only one direction in the conductors. Also known as progressive-wave antenna. { 'trav·əl·iŋ ¦wāvan'ten·ə }

traveling-wave maser [PHYS] A ruby maser used with a comblike slow-wave structure and a number of yttrium iron garnet isolators to give L-band amplification (390 to 1550 megahertz); operation is at the temperature of liquid helium (4.2 K). { 'trav·əl·iŋ ¦wāv 'mā·zər }

treble [ACOUS] High audio frequencies, such as those handled by a tweeter in a sound system. { 'treb·əl }

Tresca criterion [MECH] The assumption that plastic deformation of a material begins when the difference between the maximum and minimum principal stresses equals twice the yield stress in shear. { 'tres·kə krī,tir·ē·ən }

Trevelyan rocker [PHYS] A prismatic metal block having one edge grooved to form two ridges; it vibrates when heated and placed on the grooved edge, providing a simple example of heat-maintained vibrations. { trə'vel·yən ,räk·ər }

triad axis [CRYSTAL] A rotation axis whose multiplicity is equal to 3. { 'trī,ad ,ak·səs }

triangle of forces [MECH] A triangle, two of whose sides represent forces acting on a particle, while the third represents the combined effect of these forces. { 'trī,aŋ·gəl əv 'fȯr·səs }

triaxial pinch [PL PHYS] A device for heating a confined plasma, in which a discharge in an annular space between two concentric cylindrical conductors forms a cylindrical sheet of plasma, and this plasma is then confined and compressed by magnetic fields produced by currents flowing in the axial direction in the discharge itself and in the two conductors. { trī'ak·sē·əl 'pinch }

tribo- [PHYS] A prefix meaning pertaining to or resulting from friction. { 'trī·bo }

triboelectricity *See* frictional electricity. { ¦trī·bō,i,lek'tris·əd·ē }

triboelectric series [ELEC] A list of materials that produce an electrostatic charge when rubbed together, arranged in such an order that a material has a positive charge when rubbed with a material below it in the list, and has a negative charge when rubbed with a material above it in the list. { ¦trī·bō·i¦lek·trik 'sir·ēz }

triboelectrification [ELEC] The production of electrostatic charges by friction. { ¦trī·bō·i,lek·trə·fə'kā·shən }

tribology [PHYS] The study of the phenomena and mechanisms of friction, lubrication, and wear of surfaces in relative motion. { trī'bäl·ə·jē }

triboluminescence [ATOM PHYS] Luminescence produced by friction between two materials. { ¦trī·bō·ə,lü·mə'nes·əns }

trichroism [OPTICS] Phenomenon exhibited by certain optically anisotropic transparent crystals when subjected to white light, in which a cube of the material is found to transmit a different color through each of the three pairs of parallel faces. { 'trī,krō,iz·əm }

trichromatic theory [OPTICS] A theory of color vision which states that three primary colors may be chosen in such a way that, combined in various proportions, they can match any color. { ¦trī·krō'mad·ik 'thē·ə·rē }

triclinic crystal [CRYSTAL] A crystal whose unit cell has axes which are not at right angles, and are unequal. Also known as anorthic crystal. { trī'klin·ik 'krist·əl }

triclinic system [CRYSTAL] The most general and least symmetric crystal system, referred to by three axes of different length which are not at right angles to one another. { trī'klin·ik 'sis·təm }

trigger pull [MECH] Resistance offered by the trigger of a rifle or other weapon; force which must be exerted to pull the trigger. { 'trig·ər ,pu̇l }

trigonal lattice *See* rhombohedral lattice. { trī'gōn·əl 'lad·əs }

trigonal system [CRYSTAL] A crystal system which is characterized by threefold symmetry, and which is usually considered as part of the hexagonal system since the lattice may be either hexagonal or rhombohedral. { trī'gōn·əl 'sis·təm }

trill *See* trilling. { tril }

trilling [CRYSTAL] A cyclic crystal twin consisting of three individual crystals. Also known as threeling; trill. { 'tril·iŋ }

trimmer capacitor [ELEC] A relatively small variable capacitor used in parallel with a larger variable or fixed capacitor to permit exact adjustment of the capacitance of the parallel combination. { 'tim·ər kə,pas·əd·ər }

trimuon event [PART PHYS] An inelastic collision of a neutrino or antineutrino with a nucleus in which there are three muons among the products of the collision. { trī'myü,än i,vent }

triple collision [PHYS] A process in which three particles collide simultaneously. { 'trip·əl kə'lizh·ən }

triple harmonic [PHYS] A harmonic whose frequency is three times the fundamental frequency. { ¦trip·əl här'män·ik }

triple-stub transformer [ELECTROMAG] Microwave transformer in which three stubs are placed a quarter-wavelength apart on a coaxial line and adjusted in length to compensate for impedance mismatch. { 'trip·əl ¦stəb tranz'fȯr·mər }

triplet [OPTICS] A compound lens made up of three components. { 'trip·lət }
triplet state [ATOM PHYS] Electronic state of an atom or molecule whose total spin angular momentum quantum number is equal to 1. [QUANT MECH] Any multiplet having three states. { 'trip·lət ˌstāt }
tristimulus colorimeter [OPTICS] A colorimeter that measures a color stimulus in terms of tristimulus values. { trī'stim·yə·ləs ˌkə·lə'rim·əd·ər }
tristimulus values [OPTICS] The magnitudes of three standard stimuli needed to match a given sample of light. { trī'stim·yə·ləs ˌval·yüz }
tritium [NUC PHYS] The hydrogen isotope having mass number 3; it is one form of heavy hydrogen, the other being deuterium. Symbolized ^{3}H; T. { 'trid·ē·əm }
triton [NUC PHYS] The nucleus of tritium. { 'trīˌtän }
trochoidal mass analyzer [PHYS] A mass spectrometer in which the ion beams traverse trochoidal paths within mutually perpendicular electric and magnetic fields. { trə'kȯid·əl 'mas 'an·əˌlīz·ər }
trochoidal wave [FL MECH] A progressive oscillatory wave whose form is that of a prolate cycloid or trochoid; it is approximated by waves of small amplitude. { trə'kȯid·əl 'wāv }
troland [OPTICS] A unit of retinal illuminance, equal to the retinal illuminance produced by a surface whose luminance is one nit when the apparent area of the entrance pupil of the eye is 1 square millimeter. Also known as luxon; photon. { 'trō·lənd }
trombone [ELECTROMAG] U-shaped, adjustable, coaxial-line matching assembly. { träm'bōn }
Trouton-Noble experiment [ELECTROMAG] An experiment to detect ether drift by measuring the deflection of a charged parallel plate capacitor which is suspended so that it is free to turn. [RELAT] An experiment to measure the torque acting on a suspended, charged parallel-plate capacitor; the observed absence of the torque supports the special theory of relativity. { ¦trü·tən 'nō·bəl ik'sper·ə·mənt }
Trouton's rule [THERMO] The rule that, for a nonassociated liquid, the latent heat of vaporization in calories is equal to approximately 22 times the normal boiling point on the Kelvin scale. { 'trau̇t·ənz ˌrül }
troy ounce *See* ounce. { 'trȯi 'au̇ns }
troy pound *See* pound. { 'trȯi 'pau̇nd }
troy system [MECH] A system of mass units used primarily to measure gold and silver; the ounce is the same as that in the apothecaries' system, being equal to 480 grains or 31.1034768 grams. Abbreviated t. Also known as troy weight. { 'trȯi ˌsis·təm }
troy weight *See* troy system. { 'trȯi ˌwāt }
true horizon [OPTICS] The boundary of a horizontal plane passing through a point of vision, or in photogrammetry, the perspective center of a lens system. { 'trü hə'rīz·ən }
truncated paraboloid [ELECTROMAG] Paraboloid antenna in which a portion of the top and bottom have been cut away to broaden the main lobe in the vertical plane. { 'trəŋˌkād·əd pə'rab·əˌlȯid }
truth *See* top. { 'trüth }
truth quark *See* top quark. { 'trüth ˌkwärk }
tsi [MECH] A unit of force equal to 1 ton-force per square inch; equal to approximately 1.54444 × 10^7 pascals. { sī *or* ˌtēˌes'ī }
tsp *See* teaspoonful.
tspn *See* teaspoonful.
Tsytovich effect [ELECTROMAG] An effect wherein the index of refraction of a medium is much less than unity so that the phase velocity of electromagnetic waves in the medium exceeds the speed of light. { 'sīd·əˌvich iˌfekt }
tube of flux *See* tube of force. { 'tüb əv 'fləks }
tube of force [ELEC] A region of space bounded by a tubular surface consisting of the lines of force which pass through a given closed curve. Also known as tube of flux. { 'tüb əv 'fȯrs }
tubular capacitor [ELEC] A paper or electrolytic capacitor having the form of a cylinder, with leads usually projecting axially from the ends; the capacitor plates are long strips of metal foil separated by insulating strips, rolled into a compact tubular shape. { 'tü·byə·lər kə'pas·əd·ər }
tuft method [FL MECH] A technique of surface flow visualization in which an array of short pieces of flexible string or yarn are attached to a surface in such a way that they can move freely under the influence of a flow. { 'təft ˌset }
tumble *See* topple. { 'təm·bəl }
tumble axis *See* topple axis. { 'təm·bəl ˌak·səs }
tunable laser [OPTICS] A laser in which the frequency of the output radiation can be tuned over part or all of the ultraviolet, visible, and infrared regions of the spectrum. { 'tü·nə·bəl 'lā·zər }
tuned cavity *See* cavity resonator. { ¦tünd 'kav·əd·ē }
tuned resonating cavity [ELECTROMAG] Resonating cavity half a wavelength long or some multiple of a half wavelength, used in connection with a waveguide to produce a resultant wave with the amplitude in the cavity greatly exceeding that of the wave in the waveguide. { ¦tünd 'rez·ənˌād·iŋ ˌkav·əd·ē }
tuning core [ELECTROMAG] A ferrite core that is designed to be moved in and out of a coil or transformer to vary the inductance. { 'tün·iŋ ˌkȯr }
tuning screw [ELECTROMAG] A screw that is inserted into the top or bottom wall of a waveguide and adjusted as to depth of penetration inside for tuning or impedance-matching purposes. { 'tün·iŋ ˌskrü }
tuning stub [ELECTROMAG] Short length of transmission line, usually shorted at its free end, connected to a transmission line for impedance-matching purposes. { 'tün·iŋ ˌstəb }
tunnel effect [QUANT MECH] The ability of a particle to pass through a region of finite extent in which the particle's potential energy is greater

than its total energy; this is a quantum-mechanical phenomenon which would be impossible according to classical mechanics. Also known as tunneling. { 'tən·əl i,fekt }

tunneling *See* tunnel effect. { 'tən·əl·iŋ }

tunneling magnetoresistance [SOLID STATE] A type of magnetoresistance displayed by a trilayer thin-film structure consisting of two metallic ferromagnetic thin films sandwiching an insulating film that is thin enough (less than about 2 nanometers) that electrons can pass through it via quantum-mechanical tunneling. Also known as junction magnetoresistance. { ,tən·əl·iŋ mag,ned·ō·ri'zis·təns }

turbidimeter [OPTICS] A device that measures the loss in intensity of a light beam as it passes through a solution with particles large enough to scatter the light. { ,tər·bə'dim·əd·ər }

turbidity coefficient [OPTICS] A factor in the absorption (light) law equation that describes the extinction of the incident light beam. { tər'bid·əd·ē ,kō·i,fish·ənt }

turbulence *See* turbulent flow. { 'tər·byə·ləns }

turbulence energy *See* eddy kinetic energy. { 'tər·byə·ləns ,en·ər·jē }

turbulence modeling [FL MECH] The construction of models of the Reynolds stresses in turbulent flow. { 'tər·byə·ləns ,mäd·əl·iŋ }

turbulent boundary layer [FL MECH] The layer in which the Reynolds stresses are much larger than the viscous stresses. { 'tər·byə·lənt 'baün·drē ,lā·ər }

turbulent diffusion *See* eddy diffusion. { 'tər·byə·lənt di'fyü·zhən }

turbulent flow [FL MECH] Motion of fluids in which local velocities and pressures fluctuate irregularly, in a random manner. Also known as turbulence. { 'tər·byə·lənt 'flō }

turbulent flux *See* eddy flux. { 'tər·byə·lənt 'fləks }

turbulent Lewis number [PHYS] A dimensionless number used in the study of combined turbulent heat and mass transfer, equal to the ratio of the eddy mass diffusivity to the eddy thermal diffusivity. Symbolized Le_T. { 'tər·byə·lənt 'lü·əs ,nəm·bər }

turbulent Prandtl number [PHYS] A dimensionless number used in the study of heat transfer in turbulent flow, equal to the ratio of the eddy viscosity to the eddy thermal diffusivity. Symbolized Pr_T. { 'tər·byə·lənt 'pränt·əl ,nəm·bər }

turbulent Schmidt number [FL MECH] A dimensionless number used in the study of mass transfer in turbulent flow, equal to the ratio of the eddy viscosity to the eddy mass diffusivity. Symbolized Sc_T. { 'tər·byə·lənt 'shmit ,nəm·bər }

turbulent shear force [FL MECH] A shear force in a fluid which arises from turbulent flow. { 'tər·byə·lənt 'shir ,fȯrs }

turn [ELEC] One complete loop of wire. { 'tərn }

turns ratio [ELEC] The ratio of the number of turns in a secondary winding of a transformer to the number of turns in the primary winding. { 'tərnz ,rā·shō }

turnstile antenna [ELECTROMAG] An antenna consisting of one or more layers of crossed horizontal dipoles on a mast, usually energized so the currents in the two dipoles of a pair are equal and in quadrature; used with television, frequency modulation, and other very-high-frequency or ultra-high-frequency transmitters to obtain an essentially omnidirectional radiation pattern. { 'tərn,stīl an,ten·ə }

TW *See* terawatt.

21-centimeter line [SPECT] A radio-frequency spectral line of neutral atomic hydrogen at a wavelength of approximately 21 centimeters and a frequency of approximately 1420 megahertz, that results from hyperfine transitions between states in which the spins of the electron and proton are parallel and antiparallel. { ¦twen·tē¦wən 'sen·tə,mēd·ər ,līn }

twilight phenomena [OPTICS] Those meteorological optical phenomena which occur during twilight, including such effects as the antitwilight arch, dark segment, bright segment, green flash, purple light, and crepuscular rays. { 'twī,līt fə,näm·ə·nä }

twin *See* twin crystal. { 'twin }

twin axis [CRYSTAL] The crystal axis about which one individual of a twin crystal may be rotated (usually 180°) to bring it into coincidence with the other individual. { 'twin 'ak·səs }

twin boundary [CRYSTAL] A grain boundary whose lattice structures are mirror images of each other in the plane of the boundary. { 'twin 'baün·drē }

twin crystal [CRYSTAL] A compound crystal which has one or more parts whose lattice structure is the mirror image of that in the other parts of the crystal. Also known as crystal twin; twin. { 'twin 'krist·əl }

twin law [CRYSTAL] A statement relating two or more individuals of a twin to one another in terms of their crystallography (twin plane, twin axis, and so on). { 'twin ,lȯ }

twinning [CRYSTAL] The development of a twin crystal by growth, translation, or gliding. { 'twin·iŋ }

twinning plane *See* twin plane. { 'twin·iŋ ,plān }

twin paradox *See* clock paradox. { 'twin ,par·ə,däks }

twin plane [CRYSTAL] The plane common to and across which the individual crystals or components of a crystal twin are symmetrically arranged or reflected. Also known as twinning plane. { 'twin ,plān }

twin-T network *See* parallel-T network. { 'twin 'tē ,net,wərk }

twist [ELECTROMAG] A waveguide section in which there is a progressive rotation of the cross section about the longitudinal axis of the waveguide. { twist }

twist boundary [SOLID STATE] A boundary between two crystals that differ in orientation by only a few degrees, consisting of a series of screw dislocations. { 'twist ,baün·drē }

twister [SOLID STATE] A piezoelectric crystal

that generates a voltage when twisted. { 'twis·tər }

two-beam interference [PHYS] Interference between two waves. { 'tü ¦bēm ,in·tər'fir·əns }

two-body force [PHYS] A force between two particles which is not affected by the existence of other particles in the vicinity, such as a gravitational force or a Coulomb force between charged particles. { 'tü ¦bäd·ē 'fȯrs }

two-body problem [MECH] The problem of predicting the motions of two objects obeying Newton's laws of motion and exerting forces on each other according to some specified law such as Newton's law of gravitation, given their masses and their positions and velocities at some initial time. { 'tü ¦bäd·ē 'präb·ləm }

two-carrier theory [SOLID STATE] A theory of the conduction properties of a material in bulk or in a rectifying barrier which takes into account the motion of both electrons and holes. { 'tü ¦kar·ē·ər 'thē·ə·rē }

two-component model [FL MECH] A dynamical model of a gas flow in which the gas has two components of differing temperature. { 'tü kəm¦pō·nənt 'mäd·əl }

two-component neutrino theory [PART PHYS] A theory according to which the neutrino and antineutrino have exactly zero rest mass, and the neutrino spin is always antiparallel to its motion, while the antineutrino spin is parallel to its motion. { 'tü ¦kəm¦pō·nənt nü'trē·nō ,thē·ə·re }

two-degrees-of-freedom gyro [MECH] A gyro whose spin axis is free to rotate about two orthogonal axes, not counting the spin axis. { 'tü ¦di¦grēz əv ¦frē·dəm 'jī·rō }

two-dimensional electron gas [SOLID STATE] A system of electrons that are confined by opposing forces to a thin planar region adjacent to an interface or within a thin layer of material, but are free to move along the plane scattering off each other. { ,tü di'men·shən·əl i¦lek,trän 'gas }

two-dimensional flow [FL MECH] Fluid flow in which all flow occurs in a set of parallel planes with no flow normal to them, and the flow is identical in each of these parallel planes. { 'tü ¦di¦men·shən·əl 'flō }

two-fluid model [CRYO] A theoretical model of helium II which assumes that it consists of two interpenetrating components, a normal fluid and a superfluid with zero entropy, viscosity, and thermal conductivity. { 'tü,flü·əd ,mäd·əl }

two-phase [PHYS] Having a phase difference of one quarter-cycle or 90°. Also known as quarter-phase. { 'tü ¦fāz }

two-phase flow [CRYO] Flow of helium II, or of electrons in a superconductor, thought of as consisting of two interpenetrating, noninteracting fluids, a superfluid component which exhibits no resistance to flow and is responsible for superconducting properties, and a normal component, which behaves as does an ordinary fluid or as conduction electrons in a nonsuperconducting metal. [FL MECH] Cocurrent movement of two phases (for example, gas and liquid) through a closed conduit or duct (for example, a pipe). { 'tü ¦fāz 'flō }

two-photon coherent state [QUANT MECH] A quantum state of an electromagnetic field in which the product of the uncertainties of two quadrature components of the field is the minimum allowed by the Heisenberg uncertainty principle. { ¦tü ¦fō,tän kō¦hir·ənt 'stāt }

two-port junction [ELECTROMAG] A waveguide junctionwith two openings; it can consist either of a discontinuity or obstacle in a waveguide, or of two essentially different waveguides connected together. { 'tü ¦pȯrt 'jəŋk·shən }

two-slit experiment [QUANT MECH] A thought experiment that demonstrates the essence of the wave-particle duality, in which radiation (either light or massive particles) passes a diaphragm with two openings, and interference fringes can be observed behind the diaphragm even when the intensity of radiation is so low that the photons or massive particles can be detected one by one. { ¦tü ,slit ik'sper·ə·mənt }

two-slit interference *See* Young's two-slit interference. { 'tü ¦slit ,in·tər'fir·əns }

two-way coupling [FL MECH] The property of a particle flow in which there is a mutual interaction between the particles and the fluid. { ,tü ,wā 'kəp·liŋ }

Twyman-Green interferometer [OPTICS] An interferometer similar to the Michelson interferometer except that it is illuminated with a point source of light instead of an extended source. { 'twī·mən 'grēn ,in·tər·fə'räm·əd·ər }

Tyndall cone [OPTICS] The luminous path of a beam of light resulting from the Tyndall effect. { 'tind·əl ,kōn }

Tyndall effect [OPTICS] Visible scattering of light along the path of a beam of light as it passes through a system containing discontinuities, such as the surfaces of colloidal particles in a colloidal solution. { 'tind·əl i,fekt }

type A wave *See* continuous wave. { 'tīp ¦ā ,wāv }

type I superconductor [CRYO] A superconductor for which there is a single critical magnetic field; magnetic flux is completely excluded from the interior of the material at field strengths below this critical field, while at field strengths above the critical field, magnetic flux penetrates the superconductor completely and it reverts to the normal state. { 'tīp ¦wən ¦sü·pər·kən'dək·tər }

type II superconductor [CRYO] A superconductor for which there are two critical magnetic fields; magnetic flux is completely excluded from the interior of the material only at field strengths below the smaller critical field, and at field strengths between the two critical fields the magnetic flux consists of flux vortices in the form of filaments embedded in the superconducting material. Also known as high-field superconductor (HFS). { 'tīp ¦tü ¦sü·pər·kən'dək·tər }

U

u *See* up quark.

u-band [OPTICS] The absorption band in the ultraviolet resulting from a U-center type of point lattice defect. { 'yü ˌband }

Übler effect [FL MECH] An effect in which secondary normal stresses cause bubbles or suspended bodies to lag behind the fluid in the accelerated flow of a viscoelastic fluid through a narrowing tube. { ē·blər iˌfekt }

U center [CRYSTAL] The color-center type of point lattice defect in ionic crystals created by the incorporation of an impurity such as hydrogen into alkali halides. { 'yü ˌsen·tər }

U coefficient [QUANT MECH] A coefficient that appears in the transformation between modes of coupling eigenfunctions of three angular momenta; it is equal to the product of the Racah coefficient and $[(2j_{12} + 1)(2j_{23} + 1)]^{1/2}$, where j_{12} and j_{23} are the intermediate angular momenta in the respective modes. { 'yü ˌkō·iˌfish·ənt }

Uda antenna *See* Yagi-Uda antenna. { 'ü·də anˌten·ə }

Uehling effect [ATOM PHYS] The departure of the electrostatic potential of a point charge from a Coulomb potential, due to vacuum polarization. { 'ēl·iŋ iˌfekt }

Uhlbricht sphere [OPTICS] A sphere whose inside surface has a diffusely reflecting white finish, used in an integrating sphere photometer. { ůlˌbrik̲t ˌsfir }

ultimate load *See* breaking load. { 'əl·tə·mət ˌlōd }

ultimate strength [MECH] The tensile stress, per unit of the original surface area, at which a body will fracture, or continue to deform under a decreasing load. { 'əl·tə·mət 'streŋkth }

ultracold atom [PHYS] One of a collection of atoms cooled to a thermodynamic temperature below 1 millikelvin. { ¦əl·trəˌkōld 'ad·əm }

ultracold molecules [PHYS] A collection of molecules cooled to a thermodynamic temperature below 1 millikelvin. { ¦əl·trəˌkōld 'mäl·əˌkyülz }

ultracold neutron [PHYS] A neutron whose energy is of the order of 10^{-7} electronvolt or less, so that it is totally reflected from various materials and suitably constructed magnetic fields, regardless of the angle of incidence, and can be stored in suitably constructed bottles. { ¦əl·trə¦kōld 'nüˌträn }

ultragravity waves [FL MECH] Gravity waves which are characterized by periods in the 0.1–1.0 second range. { ¦əl·trə'grav·əd·ē ˌwāvz }

ultrahigh vacuum [PHYS] A vacuum in which the pressure is of the order of 10^{-10} millimeter of mercury or less. { ¦əl·trə'hī 'vak·yəm }

ultramicroscope [OPTICS] An instrument for investigating particles of submicroscopic dimensions; it consists of a high-intensity illumination system for producing a Tyndall cone in a colloidal system, coupled with a compound microscope to examine the points of light scattered from the individual particles. { ¦əl·trə'mī·krəˌskōp }

ultraphotic rays [ELECTROMAG] Rays outside the visible part of the spectrum, including infrared and ultraviolet rays. { ¦əl·trə¦fäd·ik 'rāz }

ultrarelativistic [RELAT] Having a speed that is nearly equal to the speed of light. { ¦əl·trəˌrel·ə·tə'vis·tik }

ultrasonic [ACOUS] Pertaining to signals, equipment, or phenomena involving frequencies just above the range of human hearing, hence above about 20,000 hertz. Also known as supersonic (deprecated usage). { ¦əl·trə'sän·ik }

ultrasonic coagulation [PHYS] The bonding of small particles into large aggregates by the action of ultrasonic waves. { ¦əl·trə'sän·ik kōˌag·yə'lā·shən }

ultrasonic echo [ACOUS] An ultrasonic wave that has been reflected or otherwise returned with sufficient magnitude and time delay to be perceived in some manner as a wave distinct from that directly transmitted. { ¦əl·trə'sän·ik 'ek·ō }

ultrasonic imaging *See* acoustic imaging. { ¦əl·trə'sän·ik 'im·ij·iŋ }

ultrasonic light modulator [OPTICS] Device containing a fluid which, by action of ultrasonic waves passing through the fluid, modulates a beam of light passed transversely through the fluid. { ¦əl·trə'sän·ik 'līt ˌmäj·əˌlād·ər }

ultrasonic medical tomography [ACOUS] A form of acoustic tomography in which ultrasonic pulses are emitted by an acoustoelectric transducer and echoes are received from acoustic impedance discontinuities along the assumed line-of-sight propagation path. { ¦əl·trə'sän·ik 'med·ə·kəl tō'mäg·rə·fē }

ultrasonic microscope [OPTICS] A special type of microscope which employs ultrasonic radiation. { ¦əl·trə'sän·ik 'mī·krəˌskōp }

ultrasonic radiation [ACOUS] Ultrasonic waves propagating through a solid, liquid, or gaseous medium. { ¦əl·trə'sän·ik ˌrād·ē'ā·shən }

ultrasonics [ACOUS] The science of ultrasonic sound waves. { ¦əl·trə'sän·iks }

ultrasonic wave [ACOUS] A sound wave that has a frequency above about 20,000 hertz. { ¦əl·trə'sän·ik 'wāv }

ultrasonography *See* acoustic imaging.

ultrasound [ACOUS] Sound with a frequency above about 20,000 Hz, the upper limit of human hearing. { 'əl·trəˌsau̇nd }

ultraviolet [PHYS] Pertaining to ultraviolet radiation. Abbreviated UV. { ¦əl·trə'vī·lət }

ultraviolet absorber [OPTICS] Any substance that absorbs ultraviolet radiant energy, then dissipates the energy in a harmless form; used in plastics and rubbers to decrease light sensitivity. { ¦əl·trə'vī·lət əb'sȯr·bər }

ultraviolet absorption [OPTICS] Absorption of specific ultraviolet radiation wavelengths by a material; for example, by a sample solution during spectroscopic analysis. { ¦əl·trə'vī·lət əb 'sȯrp·shən }

ultraviolet absorption spectrophotometry [SPECT] The study of the spectra produced by the absorption of ultraviolet radiant energy during the transformation of an electron from the ground state to an excited state as a function of the wavelength causing the transformation. { ¦əl·trə'vī·lət əb'sȯrp·shən ˌspek·trō·fə'täm·ə·trē }

ultraviolet catastrophe [STAT MECH] The prediction of the Rayleigh-Jeans law that the energy radiated by a blackbody at extremely short wavelengths is extremely large, and the total energy radiated is infinite, whereas in reality it must be finite. { ¦əl·trə'vī·lət kə'tas·trə·fē }

ultraviolet densitometry [SPECT] An ultraviolet-spectrophotometry technique for measurement of the colors on thin-layer chromatography absorbents following elution. { ¦əl·trə'vī·lət ˌden·sə'täm·ə·trē }

ultraviolet light *See* ultraviolet radiation. { ¦əl·trə'vī·lət 'līt }

ultraviolet microscope [OPTICS] A special type of microscope which uses electromagnetic radiation in the range 180–400 nanometers; it requires reflecting optics or special quartz and crystal objectives. { ¦əl·trə'vī·lət 'mī·krəˌskōp }

ultraviolet photoemission spectroscopy [SPECT] A spectroscopic technique in which photons in the energy range 10–200 electronvolts bombard a surface and the energy spectrum of the emitted electrons gives information about the states of electrons in atoms and chemical bonding. Abbreviated UPS. { ¦əl·trə'vī·lət ¦fōd·ō·i'mish·ən spek'träs·kə·pē }

ultraviolet radiation [ELECTROMAG] Electromagnetic radiation in the wavelength range 4–400 nanometers; this range begins at the short-wavelength limit of visible light and overlaps the wavelengths of long x-rays (some scientists place the lower limit at higher values, up to 40 nanometers). Also known as ultraviolet light. { ¦əl·trə'vī·lət ˌrād·ē'ā·shən }

ultraviolet spectrometer [SPECT] A device which produces a spectrum of ultraviolet light and is provided with a calibrated scale for measurement of wavelength. { ¦əl·trə'vī·lət spek 'träm·əd·ər }

ultraviolet spectrophotometry [SPECT] A method of chemical analysis based on the absorption of electromagnetic radiation between 200 and 400 nanometers, used extensively for quantitative analysis of simple inorganic ions and their complexes, as well as organic molecules that have at least one double bond. { ¦əl·trə'vī·lət ¦spek·trō·fə'täm·ə·trē }

ultraviolet spectroscopy [SPECT] Absorption spectroscopy involving electromagnetic wavelengths in the range 4–400 nanometers. { ¦əl·trə'vī·lət spek'träs·kə·pē }

ultraviolet spectrum [ELECTROMAG] **1.** The range of wavelengths of ultraviolet radiation, covering 4–400 nanometers. **2.** A display or graph of the intensity of ultraviolet radiation emitted or absorbed by a material as a function of wavelength or some related parameter. { ¦əl·trə'vī·lət 'spek·trəm }

ultraviolet telescope [OPTICS] An assemblage of mirrors, with special coatings imparting high ultraviolet reflectivity, which forms magnified ultraviolet images of objects in the same manner as an optical telescope forms images in visible light. { ¦əl·trə'vī·lət 'tel·əˌskōp }

umbra [OPTICS] That portion of a shadow which is screened from light rays emanating from any part of an extended source. { 'əm·brə }

umbrella antenna [ELECTROMAG] Antenna in which the wires are guyed downward in all directions from a central pole or tower to the ground, somewhat like the ribs of an open umbrella. { əm'brel·ə anˌten·ə }

Umkehr effect [OPTICS] An anomaly of the relative zenith intensities of scattered sunlight at certain wavelengths in the ultraviolet as the sun approaches the horizon; it is due to the presence of the ozone layer. { 'u̇mˌker iˌfekt }

Umklapp process [SOLID STATE] The interaction of three or more waves in a solid, such as lattice waves or electron waves, in which the sum of the wave vectors is not equal to zero but, rather, is equal to a vector in the reciprocal lattice. Also known as flip-over process. { 'u̇m ˌkläp ˌprä·səs }

unavailable energy [THERMO] That part of the energy which, when an irreversible process takes place, is initially in a form completely available for work and is converted to a form completely unavailable for work. { ¦ən·ə¦vāl·ə·bəl 'en·ər·jē }

unbounded wave [PHYS] A wave which propagates through a nondissipative, homogeneous medium which is infinite in extent, without any boundaries. { ¦ən'bau̇n·dəd 'wāv }

uncertainty principle [QUANT MECH] The precept that the accurate measurement of an observable quantity necessarily produces uncertainties in one's knowledge of the values of other observables. Also known as Heisenberg uncertainty principle; indeterminacy principle. { ¦ən'sərt·ən·tē ˌprin·sə·pəl }

uncertainty relation [QUANT MECH] The relation whereby, if one simultaneously measures values of two canonically conjugate variables, such as position and momentum, the product of the uncertainties of their measured values cannot be less than approximately Planck's constant divided by 2π. Also known as Heisenberg uncertainty relation. { ¦ən'sərt·ən·tē riˌlā·shən }

uncharged [ELEC] Having no electric charge. { ¦ən'chärjd }

uncoupling phenomena [SPECT] Deviations of observed spectra from those predicted in a diatomic molecule as the magnitude of the angular momentum increases, caused by interactions which could be neglected at low angular momenta. { ¦ən'kəp·liŋ fəˌnäm·ə·nä }

undamped wave [PHYS] A continuous wave produced by oscillations having constant amplitude. { ¦ən'damt 'wāv }

undercooling effect [SOLID STATE] The effect whereby a superconductor can be cooled below its critical temperature without the onset of superconductivity. { ¦ən·dər¦kül·iŋ iˌfekt }

underdamping [PHYS] Condition of a system when the amount of damping is sufficiently small so that, when the system is subjected to a single disturbance, either constant or instantaneous, one or more oscillations are executed by the system. { ¦ən·dər¦dam·piŋ }

underspin [MECH] Property of a projectile having insufficient rate of spin to give proper stabilization. { 'ən·dərˌspin }

underwater acoustics [ACOUS] Study of the propagation of sound waves in water, especially in the oceans, and of phenomena produced by these sound waves. Also known as hydroacoustics. { ¦ən·dər¦wȯd·ər ə'küs·tiks }

underwater camera [OPTICS] A camera designed for use under the surface of the water; it is usually a conventional type enclosed in a casing to withstand water pressure, preferably with a correction lens to compensate for aberrations caused by the water. { ¦ən·dər¦wȯd·ər 'kam·rə }

underwater sound [ACOUS] The production, transmission, and reception of sounds in the ocean; used for locating submarines and other submerged objects, and to determine the physical structure of the ocean and its bottom, and to study organisms found in the sea. { ¦ən·dər¦wȯd·ər 'sau̇nd }

undisturbed motion [PHYS] The steady state of a system before perturbations are introduced. { ¦ən·di'stərbd 'mō·shən }

undulatory extinction [OPTICS] Extinction that occurs successively in adjacent areas as the microscope stage is turned. Also known as oscillatory extinction; strain shadow; wavy extinction. { 'ən·jə·ləˌtȯr·ē ik'stiŋk·shən }

uniaxial crystal [OPTICS] A doubly refracting crystal which has a single axis along which light can propagate without exhibiting double refraction. { ¦yü·nē'ak·sē·əl 'krist·əl }

uniaxial stress [MECH] A state of stress in which two of the three principal stresses are zero. { ¦yü·nē'ak·sē·əl 'stres }

unidirectional [PHYS] **1.** Flowing in only one direction, such as direct current. **2.** Radiating in only one direction. { ¦yü·nə·də'rek·shən·əl }

unidirectional antenna [ELECTROMAG] An antenna that has a single well-defined direction of maximum gain. { ¦yü·nə·də'rek·shən·əl an'ten·ə }

unidirectional log-periodic antenna [ELECTROMAG] A broad-band antenna in which the cut-out portions of a log-periodic antenna are mounted at an angle to each other, to give a unidirectional radiation pattern in which the major radiation is in the backward direction, off the apex of the antenna; impedance is essentially constant for all frequencies, as is the radiation pattern. { ¦yü·nə·də'rek·shən·əl 'läg ˌpir·ē'äd·ik an'ten·ə }

unified field theory [RELAT] Any theory which attempts to express gravitational theory and electromagnetic theory within a single unified framework; usually, an attempt to generalize Einstein's general theory of relativity from a theory of gravitation alone to a theory of gravitation and classical electromagnetism. { 'yü·nəˌfīd 'fēld ˌthē·ə·rē }

uniform circular motion [MECH] Circular motion in which the angular velocity remains constant. { 'yü·nəˌfȯrm 'sər·kyə·lər 'mō·shən }

uniform field [PHYS] A field which, at the instant under consideration, has the same value at every point in the region under consideration. { 'yü·nəˌfȯrm 'fēld }

uniform load [MECH] A load distributed uniformly over a portion or over the entire length of a beam; measured in pounds per foot. { 'yü·nəˌfȯrm 'lōd }

uniform luminance [OPTICS] Property of a surface for which the luminous intensity of any area of the surface is proportional to the area. { 'yü·nəˌfȯrm 'lü·mə·nəns }

uniform plane wave [ELECTROMAG] Plane wave in which the electric and magnetic intensities have constant amplitude over the equiphase surfaces; such a wave can only be found in free space at an infinite distance from the source. { 'yü·nəˌfȯrm 'plān 'wāv }

unipole [ELECTROMAG] A hypothetical antenna that radiates or receives signals equally well in all directions. Also known as isotropic antenna. { 'yü·nəˌpōl }

unit [PHYS] A quantity adopted as a standard of measurement. { 'yü·nət }

unit-area acoustical ohm *See* rayl. { 'yü·nət 'er·ē·ə ə'küs·tə·kəl 'ōm }

unitarity condition [PART PHYS] The condition that the scattering matrix for any process be unitary, as a result of the fact that the probability for the system to end in some final state must be unity. { ˌyü·nə'tar·əd·ē kənˌdish·ən }

unitary decuplet [PART PHYS] A collection of 10 hadrons whose isospin and hypercharge values form a symmetrical pattern, and which are related by unitary symmetry operations. { 'yü·nəˌter·ē dē'kəp·lət }

unitary octet [PART PHYS] A collection of eight hadrons whose isospin and hypercharge values form a symmetrical pattern, and which are related by unitary symmetry operations. { 'yü·nəˌter·ē äk'tet }

unitary spin [PART PHYS] A quantum number associated with SU_3 symmetry and which determines the SU_3 supermultiplet to which a particle belongs, such as singlet, octet, or decuplet. { 'yü·nəˌter·ē 'spin }

unitary symmetry [PART PHYS] An approximate internal symmetry law obeyed by the strong interactions of elementary particles; a system of particles has such a symmetry if all the particles can be described as compounds of a fundamental multiplet of particles, and if all physical properties of the system are unchanged by an arbitrary unitary transformation of this fundamental multiplet. { 'yü·nəˌter·ē 'sim·ə·trē }

unit cell [CRYSTAL] A parallelepiped which will fill all space under the action of translations which leave the crystal lattice unchanged. Also known as structure cell. { 'yü·nət 'sel }

unit charge *See* statcoulomb. { 'yü·nət 'chärj }

unit magnetic pole [ELECTROMAG] Two equal magnetic poles of the same sign have unit value when they repel each other with a force of 1 dyne if placed 1 centimeter apart in a vacuum. { 'yü·nət mag'ned·ik 'pōl }

unit strain [MECH] **1.** For tensile strain, the elongation per unit length. **2.** For compressive strain, the shortening per unit length. **3.** For shear strain, the change in angle between two lines originally perpendicular to each other. { 'yü·nət 'strān }

unit stress [MECH] The load per unit of area. { 'yü·nət 'stres }

unit systems [OPTICS] Optical systems that have a lateral magnification of +1 or −1. [PHYS] Groups of units suitable for use in measurement of physical quantities and in the convenient statement of physical laws relating physical quantities. { 'yü·nət ˌsis·təmz }

unity coupling [ELECTROMAG] Perfect magnetic coupling between two coils, so that all magnetic flux produced by the primary winding passes through the entire secondary winding. { 'yü·nəd·ē ˌkəp·liŋ }

unity power factor [ELEC] Power factor of 1.0, obtained when current and voltage are in phase, as in a circuit containing only resistance or in a reactive circuit at resonance. { 'yü·nəd·ē 'paù·ər ˌfak·tər }

univariant system [THERMO] A system which has only one degree of freedom according to the phase rule. { ¦yü·nə¦ver·ē·ənt 'sis·təm }

universal constants *See* fundamental constants. { ¦yü·nə¦vər·səl 'kän·stəns }

universal gas constant *See* gas constant. { ¦yü·nə¦vər·səl 'gas ˌkän·stənt }

universality [STAT MECH] The hypothesis that the critical exponents of a substance are the same within broad classes of substances of widely varying characteristics, and depend only on the microscopic symmetry properties of the substance. { ˌyü·nə·vər'sal·əd·ē }

universality class [STAT MECH] A class of substances which have the same critical exponents according to the universality hypothesis. { ˌyü·nə·vər'sal·əd·ē ˌklas }

universal resonance curve [ELEC] A plot of Y/Y_0 against $Q_0\delta$ for a series-resonant circuit, or of Z/Z_0 against $Q_0\delta$ for a parallel-resonant circuit, where Y and Z are the admittance and impedance of a circuit, Y_0 and Z_0 are the values of these quantities at resonance, Q_0 is the Q value of the circuit at resonance, and δ is the deviation of the frequency from resonance divided by the resonant frequency; it can be applied to all resonant circuits. { ¦yü·nə¦vər·səl 'rez·ən·əns ˌkərv }

universal sequence [PHYS] A sequence of periodic states which reappear as a parameter governing a mapping is varied after the mapping has undergone a transition to chaos by a sequence of period-doubling bifurcations, and which have a particular order that does not depend on the details of the mapping, within certain limits. { ¦yü·nə¦vər·səl 'sē·kwəns }

universal stage [OPTICS] A stage attached to the rotating stage of a polarizing microscope that has three, four, or five axes and thin sections of low-symmetry minerals to be tilted about two mutually perpendicular horizontal axes. Also known as Fedorov stage; U stage. { ¦yü·nə¦vər·səl 'stāj }

universal wavelength function [OPTICS] One of four functions which enable one to compute easily, with reasonable accuracy, the refractive index of glass or other transparent material when this index is known for four standard wavelengths. { ¦yü·nə¦vər·səl 'wāvˌleŋkth ˌfəŋk·shən }

unpaired electron [ATOM PHYS] An orbital electron for which there is no other electron in the same atom with the same energy but opposite spin. { ¦ənˌperd i'lekˌträn }

unpitched sound [ACOUS] Sound that includes a wide range of frequencies and thus does not have a definite pitch. { ¦ənˌpichd 'saùnd }

unpolarized light [OPTICS] Light in which the electric vector is oriented in a random, unpredictable fashion. { ¦ən'pō·ləˌrīzd 'līt }

unpolarized particle beam [PHYS] A beam of particles with spin in which the directions of the spins are random. { ¦ən'pō·ləˌrīzd 'pard·ə·kəl ˌbēm }

unsaturated standard cell [ELEC] One of two

types of Weston standard cells (batteries); used for voltage calibration work not requiring an accuracy greater than 0.01%. { ¦ən'sach·ə,rād·əd 'stan·dərd 'sel }

unstable [PHYS] Capable of undergoing spontaneous change, as in a radioactive nuclide or an excited nuclear system. { ¦ən'stā·bəl }

unstable equilibrium [PHYS] An equilibrium state of a system in which any departure of the system from equilibrium gives rise to forces or tendencies moving the system further away from equilibrium; for example, mechanical equilibrium in which the potential energy is a maximum, as a sphere sitting on top of a hill. { ¦ən'stā·bəl ē·kwə'lib·rē·əm }

unstable isotope *See* radioisotope. { ¦ən'stā·bəl 'ī·sə,tōp }

unstable particle [PART PHYS] **1.** Any elementary particle that spontaneously decays into other particles. **2.** An elementary particle that can decay through the strong interactions, as opposed to a semistable particle; it has a lifetime on the order of 10^{-23} second. { ¦ən'stā·bəl 'pärd·ə·kəl }

unstable wave [PHYS] A wave motion whose amplitude increases with time or whose total energy increases at the expense of its environment. { ¦ən'stā·bəl 'wāv }

unsteady flow [FL MECH] Fluid flow in which properties of the flow change with respect to time. { ¦ən'sted·ē 'flō }

unsteady-state flow [FL MECH] A condition of fluid flow in which the volumetric ratios of two or more phases (liquid-gas, liquid-liquid, and so on) vary along the course of flow; can be the result of changes in temperature, pressure, or composition. { ¦ən'sted·ē ¦stāt 'flō }

untuned [ELEC] Not resonant at any of the frequencies being handled. { ¦ən'tünd }

upper consolute temperature *See* consolute temperature. { 'əp·ər 'kän·sə,lüt 'tem·prə·chər }

upper critical field [SOLID STATE] The magnetic field strength above which a type II superconductor is completely normal. Symbolized H_{c2}. { ¦əp·ər ¦krid·i·kəl 'fēld }

upper critical solution temperature *See* consolute temperature. { 'əp·ər ¦krid·ə·kəl sə¦lü·shən 'tem·prə·chər }

up quark [PART PHYS] A quark with an electric charge of +2/3, baryon number of 1/3, and 0 strangeness and charm. Symbolized u. { 'əp ,kwärk }

UPS *See* ultraviolet photoemission spectroscopy.

upsilon particle [PART PHYS] One of a family of elementary particles having about 10 times the mass of the proton and consisting of an atomlike combination of a bottom quark with its antiquark. { 'əp·sə·lən ,pärd·ə·kəl }

uranium decay series *See* uranium series. { yə'rā·nē·əm di'kā ,sir·ēz }

uranium-radium series *See* uranium series. { yə'rā·nē·əm 'rād·ē·əm ,sir·ēz }

uranium series [NUC PHYS] The series of nuclides resulting from the decay of uranium-238, including isotopes of uranium, thorium, protactinium, radium, radon, pdonium, lead, bismuth, and thallium with mass number $4n+2$, where n is an integer. Also known as uranium decay series; uranium-radium series. { yə'rā·nē·əm ,sir·ēz }

U stage *See* universal stage. { 'yü ,stāj }

UV *See* ultraviolet.

V

V *See* electric potential; volt.

VA *See* volt-ampere.

vac *See* millibar.

vacancy [SOLID STATE] A defect in the form of an unoccupied lattice position in a crystal. { 'vā·kən·sē }

vacuum [PHYS] **1.** Theoretically, a space in which there is no matter. **2.** Practically, a space in which the pressure is far below normal atmospheric pressure so that the remaining gases do not affect processes being carried on in the space. [QUANT MECH] The lowest possible energy state of a system, conceived of as a polarizable gas of virtual particles, fluctuating randomly. { 'vak·yəm }

vacuum capacitor [ELEC] A capacitor with separated metal plates or cylinders mounted in an evacuated glass envelope to obtain a high breakdown voltage rating. { 'vak·yəm kə'pas·əd·ər }

vacuum correction [PHYS] The correction to the reading of a mercury barometer required by the imperfections in the vacuum above the mercury column, due to the presence of water vapor and air; this correction is a function of both temperature and pressure. { 'vak·yəm kə,rek·shən }

vacuum polarization [QUANT MECH] A process in which an electromagnetic field gives rise to virtual electron-positron pairs that effectively alter the distribution of charges and currents that generated the original electromagnetic field. { 'vak·yəm ,pō·lə·rə'zā·shən }

vacuum Rabi oscillation [ATOM PHYS] A process in which an excited atom placed in a very small cavity emits a photon much more quickly than it would in free space if the cavity is resonant with the radiation emitted by the photon. { ¦vak·yəm ¦rä·bē ,äs·ə'lā·shən }

vacuum ultraviolet radiation [ELECTROMAG] Ultraviolet radiation with a wavelength of less than 200 nanometers; absorption of radiation in this region by air and other gases requires the use of evacuated apparatus for transmission. Abbreviated VUV radiation. Also known as extreme ultraviolet radiation (EUV radiation). { 'vak·yəm ¦əl·trə'vī·lət ,rād·ē'ā·shən }

vacuum ultraviolet spectroscopy [SPECT] Absorption spectroscopy involving electromagnetic wavelengths shorter than 200 nanometers; so called because the interference of the high absorption of most gases necessitates work with evacuated equipment. { 'vak·yəm ¦əl·trə'vī·lət spek'träs·kə·pē }

Vaisala comparator [OPTICS] An interferometer measuring distances on the order of 100 meters with accuracies on the order of 1 part in 10^7 { ¦vī·sə·lə kəm'par·əd·ər }

Väisälä period [ACOUS] The natural period of oscillation of a small parcel of air which is displaced adiabatically in a vertical direction, in an isothermal atmosphere horizontally stratified by gravity; equal to 337 seconds for a sound velocity of 333 meters per second. { 'väl·sä·lä ,pir·ē·əd }

valence band [SOLID STATE] The highest electronic energy band in a semiconductor or insulator which can be filled with electrons. { 'vā·ləns ,band }

valence crystal *See* covalent crystal. { 'vā·ləns ,krist·əl }

valence electron [ATOM PHYS] An electron that belongs to the outermost shell of an atom. [SOLID STATE] *See* conduction electron. { 'vā·ləns i,lek,trän }

valence shell [ATOM PHYS] The electrons that form the outermost shell of an atom. { 'vā·ləns ,shel }

valency effect [SOLID STATE] The dependence of the transition temperature of a superconductor on the concentration of doping atoms. { 'vā·lən·sē i,fekt }

Van Atta array [ELECTROMAG] Antenna array in which pairs of corner reflectors or other elements equidistant from the center of the array are connected together by a low-loss transmission line in such a way that the received signal is reflected back to its source in a narrow beam to give signal enhancement without amplification. { va'nad·ə ə,rā }

Van der Pol oscillator [PHYS] A vibrating system that is governed by an equation of the form $\ddot{x} + \epsilon(-\dot{x} + 1/3\dot{x}^3) + x = 0$. { 'van dər ,pōl ,äs·ə,lād·ər }

van der Waals structure [CRYSTAL] The structure of a molecular crystal. { 'van dər ,wölz ,strək·chər }

van der Waals surface tension formula [THERMO] An empirical formula for the dependence of the surface tension on temperature: $\gamma = Kp_c^{2/3}T_c^{1/3}(1 - T/T_c)^n$, where γ is the surface tension, T is the temperature, T_c and p_c are the critical temperature and pressure, K is a constant, and n is a constant equal to approximately 1.23. { 'van dər ,wölz 'sər·fəs ,ten·chən ,fór·myə·lə }

vane attenuator *See* flap attenuator. { 'vān ə'ten·yə,wād·ər }

V antenna [ELECTROMAG] An antenna having a V-shaped arrangement of conductors fed by a balanced line at the apex; the included angle, length, and elevation of the conductors are proportioned to give the desired directivity. Also spelled vee antenna. { 'vē an,ten·ə }

van't Hoff factor [PHYS] The ratio of the observed osmotic pressure of a solution to that predicted by van't Hoff's law. { van'tóf ,fak·tər }

van't Hoff's law [PHYS] The law that the osmotic pressure of a dissolved substance equals the gas pressure it would exert if it were an ideal gas that occupied the same volume as that of the solution. { van'tófs ,ló }

Van Vleck equation [QUANT MECH] An equation based on quantum theory for the molar paramagnetism of a magnetically susceptible material from magnetic moment, absolute temperature, and various constants. { van 'vlek i,kwā·zhən }

Van Vleck paramagnetism [QUANT MECH] The paramagnetism of a collection of atoms, ions, or molecules, as computed by quantum theory; the atoms, ions, or molecules in a magnetic field are distributed among the various allowed energy levels according to a Boltzmann distribution, and the magnetization of the system is computed by finding the average component of angular momentum parallel to the field. { van 'vlek ¦par·ə'mag·nə,tiz·əm }

vapor [THERMO] A gas at a temperature below the critical temperature, so that it can be liquefied by compression, without lowering the temperature. { 'vā·pər }

vapor cycle [THERMO] A thermodynamic cycle, operating as a heat engine or a heat pump, during which the working substance is in, or passes through, the vapor state. { 'vā·pər ,sī·kəl }

vaporization *See* volatilization. { ,vā·pə·rə'zā·shən }

vaporization coefficient [THERMO] The ratio of the rate of vaporization of a solid or liquid at a given temperature and corresponding vapor pressure to the rate of vaporization that would be necessary to produce the same vapor pressure at this temperature if every vapor molecule striking the solid or liquid were absorbed there. { ,vā·pə·rə'zā·shən ,kō·ə·fish·ənt }

vapor lock [FL MECH] Interruption of the flow of fuel in a gasoline engine caused by formation of vapor or gas bubbles in the fuel-feeding system. { 'vā·pər ,läk }

vapor-phase epitaxy [SOLID STATE] The use of chemical vapor deposition to grow epitaxial layers. Abbreviated VPE. { 'vā·pər ¦fāz 'ep·ə ,tak·sē }

vapor pressure [THERMO] For a liquid or solid, the pressure of the vapor in equilibrium with the liquid or solid. { 'vā·pər ,presh·ər }

var *See* volt-ampere reactive.

var hour [ELEC] A unit of the integral of reactive power over time, equal to a reactive power of 1 var integrated over 1 hour; equal in magnitude to 3600 joules. Also known as reactive volt-ampere hour; volt-ampere-hour reactive. { 'vär ,aür }

variable capacitor [ELEC] A capacitor whose capacitance can be varied continuously by moving one set of metal plates with respect to another. { 'ver·ē·ə·bəl kə'pas·əd·ər }

variable coupling [ELEC] Inductive coupling that can be varied by moving one coil with respect to another. { 'ver·ē·ə·bəl 'kəp·liŋ }

variable field [PHYS] Field which changes during the time under consideration. { 'ver·ē·ə·bəl 'fēld }

variable flow [FL MECH] Fluid flow in which the velocity changes both with time and from point to point. { 'ver·ē·ə·bəl 'flō }

variable-focal-length lens *See* zoom lens. { 'ver·ē·ə·bəl ¦fō·kəl ¦leŋkth 'lenz }

variable-focus condenser [OPTICS] A condenser that is used to obtain a large illuminated field area, and has two lenses, the first of which can be adjusted to bring light to a focus between the lenses. { 'ver·ē·ə·bəl ¦fō·kəs kən'den·sər }

variable force [MECH] A force whose direction or magnitude or both change with time. { 'ver·ē·ə·bəl 'fórs }

variable inductance *See* variable inductor. { 'ver·ē·ə·bəl in'dək·təns }

variable inductor [ELECTROMAG] A coil whose effective inductance can be changed. Also known as variable inductance. { 'ver·ē·ə·bəl in'dək·tər }

variable parameter [PHYS] A parameter which may be varied to assume any value in some range. { 'ver·ē·ə·bəl pə'ram·əd·ər }

variable resistor *See* rheostat. { 'ver·ē·ə·bəl ri'zis·tər }

variable-thickness microbridge [CRYO] A Josephson junction formed by a short, narrow constriction in a thin superconducting film, which is thinner than the rest of the film. { 'ver·ē·ə·bəl ¦thik·nəs 'mī·krō,brij }

variable waveguide attenuator [ELECTROMAG] Device designed to introduce attenuation into a waveguide circuit by moving a lossy vane either sideways across the waveguide or into the waveguide through a longitudinal slot. { 'ver·ē·ə·bəl 'wāv,gīd ə,ten·yə,wād·ər }

variational method [QUANT MECH] A method of calculating an upper bound on the lowest energy level of a quantum-mechanical system and an approximation for the corresponding wave function; in the integral representing the expectation value of the Hamiltonian operator, one substitutes a trial function for the true wave function, and varies parameters in the trial function to minimize the integral. { ,ver·ē'ā·shən·əl ,meth·əd }

varifocal lens *See* zoom lens. { ¦ver·ə¦fō·kəl 'lenz }

Varignon's theorem [MECH] The theorem that the moment of a force is the algebraic sum of the moments of its vector components acting at a common point on the line of action of the force. { var·ən'yōnz ˌthir·əm }

varindor [ELECTROMAG] Inductor in which the inductance varies markedly with the current in the winding. { 'var·ənˌdȯr }

variometer [ELECTROMAG] A variable inductance having two coils in series, one mounted inside the other, with provisions for rotating the inner coil in order to vary the total inductance of the unit over a wide range. { ˌver·ē'äm·əd·ər }

var measurement [ELEC] The measurement of reactive power in a circuit. { 'vär ˌmezh·ər·mənt }

V band [ELECTROMAG] A radio-frequency band of 46.0 to 56.0 gigahertz. [SPECT] Absorption bands that appear in the ultraviolet part of the spectrum due to color centers produced in potassium bromide by exposure of the crystal at temperature of liquid nitrogen (81 K) to intense penetrating x-rays. { 'vē ˌband }

V coefficient [QUANT MECH] Either of two coefficients used in the coupling of eigenfunctions of two angular momenta, differing from the Wigner 3-*j* symbol by at most a sign. Symbolized V and $\bar{V}$. { 'vē ˌkō·iˌfish·ənt }

VCSEL *See* vertical-cavity surface-emitting laser. { ¦vē¦sē¦es¦ē'el *or* ¦vē'sēˌsel }

vector [PHYS] A quantity which has both magnitude and direction, and whose components transform from one coordinate system to another in the same manner as the components of a displacement. Also known as polar vector. { 'vek·tər }

vector coupling coefficient [QUANT MECH] One of the coefficients used to express an eigenfunction of the sum of two angular momenta in terms of sums of products of eigenfunctions of the original two angular momenta. Also known as Clebsch-Gordan coefficient; Wigner coefficient. { 'vek·tər ¦kəp·liŋ ˌkō·iˌfish·ənt }

vector current [PART PHYS] A current which behaves as a vector under Lorentz transformations, rather than as an axial vector. { 'vek·tər ˌkə·rənt }

vector field [PHYS] A field which is characterized by a vector function. { 'vek·tər ˌfēld }

vector function [PHYS] A function of position and time whose value at each point is a vector. Also known as vector point function. { 'vek·tər ˌfəŋk·shən }

vector meson [PART PHYS] A meson which has spin quantum number 1 and negative parity, and may be described by a vector field; examples include the ω, ρ, φ, and K* mesons. { 'vek·tər ˌmāˌsän }

vector model of atomic structure [ATOM PHYS] A model of atomic structure in which spin and orbital angular momenta of the electrons are represented by vectors, with special rules for their addition imposed by underlying quantum-mechanical considerations. { 'vek·tər ¦mäd·əl əv ə'täm·ik 'strək·chər }

vector momentum *See* momentum. { 'vek·tər mə'men·təm }

vector point function *See* vector function. { 'vek·tər 'pȯint ˌfəŋk·shən }

vector potential [ELECTROMAG] A vector function whose curl is equal to the magnetic induction. Symbolized **A**. Also known as magnetic vector potential. [PHYS] Any vector function whose curl is equal to some solenoidal vector field. { 'vek·tər pəˌten·chəl }

vector power [ELEC] Vector quantity equal in magnitude to the square root of the sum of the squares of the active power and the reactive power. { 'vek·tər ˌpaù·ər }

vector-power factor [ELEC] Ratio of the active power to the vector power; it is the same as power factor in the case of simple sinusoidal quantities. { 'vek·tər ¦paù·ər ˌfak·tər }

vee antenna *See* V antenna. { 'vē anˌten·ə }

veiling glare [OPTICS] The reduction in contrast of an optical image caused by superposition of scattered light. { 'vāl·iŋ ˌglār }

velocity [MECH] **1.** The time rate of change of position of a body; it is a vector quantity having direction as well as magnitude. Also known as linear velocity. **2.** The speed at which the detonating wave passes through a column of explosives, expressed in meters or feet per second. { və'läs·əd·ē }

velocity analysis [MECH] A graphical technique for the determination of the velocities of the parts of a mechanical device, especially those of a plane mechanism with rigid component links. { və'läs·əd·ē əˌnal·ə·səs }

velocity coefficient [FL MECH] The ratio of the actual velocity of gas emerging from a nozzle to the velocity calculated under ideal conditions; it is less than 1 because of friction losses. Also known as coefficient of velocity. { və'läs·əd·ē ˌkō·iˌfish·ənt }

velocity dispersion [PHYS] The root-mean-square value of the magnitudes of the random velocities of particles about their mean velocity. { və'läs·əd·ē di'spər·zhən }

velocity distribution [STAT MECH] For the molecules of a gas, a function of velocity whose value at any velocity *v* is proportional to the number of molecules with velocities in an infinitesimal range about *v*, per unit velocity range. { və'läs·əd·ē ˌdi·strəˌbyü·shən }

velocity-focusing mass spectrograph *See* velocity spectrograph. { və'läs·əd·ē ¦fō·kəs·iŋ 'mas 'spek·trəˌgraf }

velocity gradient [FL MECH] The rate of change of velocity of propagation with distance normal to the direction of flow. { və'läs·əd·ē ˌgrād·ē·ənt }

velocity head [FL MECH] The square of the speed of flow of a fluid divided by twice the acceleration of gravity; it is equal to the static pressure head corresponding to a pressure equal to the kinetic energy of the fluid per unit volume. { və'läs·əd·ē ˌhed }

velocity level [ACOUS] A sound rating in decibels, equal to 20 times the logarithm to the base 10 of the ratio of the particle velocity of the sound to a specified reference particle velocity. { və'läs·əd·ē ˌlev·əl }

velocity of light *See* speed of light. { və'läs·əd·ē əv 'līt }

velocity of sound *See* speed of sound. { və'läs·əd·ē əv 'sau̇nd }

velocity potential [FL MECH] For a fluid flow, a scalar function whose gradient is equal to the velocity of the fluid. { və'läs·əd·ē pəˌten·chəl }

velocity pressure *See* wind pressure. { və'läs·əd·ē ˌpresh·ər }

velocity profile [FL MECH] A graph of the speed of a fluid flow as a function of distance perpendicular to the direction of flow. { və'läs·əd·ē ˌprōˌfīl }

velocity resonance *See* phase resonance. { və'läs·əd·ē ˌrez·ən·əns }

velocity spectrograph [PHYS] A mass spectrograph in which only positive ions having a certain velocity pass through all three slits and enter a chamber where they are deflected by a magnetic field in proportion to their charge-to-mass ratio. Also known as velocity-focusing mass spectrograph. { və'läs·əd·ē 'spek·trəˌgraf }

vena contracta [FL MECH] The contraction of a jet of liquid which comes out of an opening in a container to a cross section smaller than the opening. { ˌvē·nə kən'trak·tə }

Verdet constant [OPTICS] A constant of proportionality in the equation of the Faraday effect; it is equal to the angle of rotation of plane-polarized light in a magnetized substance divided by the product of the length of the light path in the substance and the strength of the magnetic field. { ˌvər'dā ˌkän·stənt }

vernier capacitor [ELEC] Variable capacitor placed in parallel with a larger tuning capacitor to provide a finer adjustment after the larger unit has been set approximately to the desired position. { 'vər·nē·ər kə'pas·əd·ər }

vertex [OPTICS] One of the points where the surface of a lens intersects the optical axis. { 'vərˌteks }

vertex power [OPTICS] The reciprocal of the back focal length of a lens. { 'vərˌteks ˌpau̇·ər }

vertical antenna [ELECTROMAG] A vertical metal tower, rod, or suspended wire used as an antenna. { 'vərd·ə·kəl an'ten·ə }

vertical-cavity surface-emitting laser [OPTICS] A very small semiconductor laser in which stacks of dielectric mirrors above and below the optically active region form the optical cavity, the active gain medium consists of one or more semiconductor quantum wells placed parallel to the mirrors at an antinode of the cavity resonance, and lasing light emission is from the surface of the semiconductor substrate, normal to the plane of the gain medium. Abbreviated VCSEL. { ¦vərd·i·kəl ˌkav·əd·ē ¦sər·fəs iˌmid·iŋ 'lā·zər }

vertical component effect *See* antenna effect. { 'vərd·ə·kəl kəm'pō·nənt iˌfekt }

vertical drop [MECH] The drop of an object in trajectory or along a plumb line, measured vertically from its line of departure to the object. { 'vərd·ə·kəl 'dräp }

vertical field-strength diagram [ELECTROMAG] Representation of the field strength at a constant distance from an antenna and in a vertical plane passing through the antenna. { 'vərd·ə·kəl 'fēld ¦streŋkth ˌdī·əˌgram }

vertical illuminator [OPTICS] A microscope designed for observing surfaces of opaque substances such as metals, which has a mechanism for passing light down through the objective lens in order to illuminate the surface to be observed with a beam perpendicular to the surface. { 'vərd·ə·kəl i'lü·məˌnād·ər }

vertical-incidence transmission [ELECTROMAG] Transmission of a radio wave vertically to the ionosphere and back. { 'vərd·ə·kəl ¦in·sə·dəns tranz'mish·ən }

vertically stacked loops *See* stacked loops. { 'vərd·ə·klē ¦stakt 'lüps }

vertical vorticity [FL MECH] The vertical component of the vorticity vector. { 'vərd·ə·kəl vȯr'tis·əd·ē }

vibration [MECH] A continuing periodic change in a displacement with respect to a fixed reference. { vī'brā·shən }

vibrational spectrum [SPECT] The molecular spectrum resulting from transitions between vibrational levels of a molecule which behaves like the quantum-mechanical harmonic oscillator. { vī'brā·shən·əl 'spek·trəm }

vibrational sum rule [SPECT] **1.** The rule that the sums of the band strengths of all emission bands with the same upper state is proportional to the number of molecules in the upper state, where the band strength is the emission intensity divided by the fourth power of the frequency. **2.** The sums of the band strengths of all absorption bands with the same lower state is proportional to the number of molecules in the lower state, where the band strength is the absorption intensity divided by the frequency. { vī'brā·shən·əl 'səm ˌrül }

vibrato [ACOUS] A musical embellishment that depends primarily on periodic variations of frequency which are often accompanied by variations in amplitude and waveform. { vi'bräd·ō }

vicinal faces [CRYSTAL] Macroscopic crystal faces which are inclined only a few minutes of arc to crystal faces with low Miller indices, and which therefore must have high Miller indices themselves. { 'vis·ən·əl ˌfās·əz }

videomicroscopy [OPTICS] The use of television cameras to brighten magnified images that are otherwise too dark to be seen with the naked eye. { ¦vid·ē·ō·mī'kräs·kə·pē }

view camera [OPTICS] A camera that can be focused at both front and back, with adjustments for tilts, swings, shifts, and rise and fall, to control the shape of the subject in the image; it has a groundglass on the back which enables the photographer to view the image to be recorded. { 'vyü ˌkam·rə }

viewfinder [OPTICS] A device which provides

the user of a camera with the view of the subject that is focused by the lens. { 'vyü,fīn·dər }

vignetting [OPTICS] Reduction in intensity of illumination near the edges of an optical instrument's field of view caused by obstruction of light rays by the edge of the aperture. { vin'yed·iŋ }

Villari effect [PHYS] A change of magnetic induction within a ferromagnetic substance in a magnetic field when the substance is subjected to mechanical stress. { və'lär·ē i,fekt }

Villari reversal [PHYS] A change in the sign of the Villari effect which occurs with some ferromagnetic materials when the magnetic field strength reaches a certain value. { və'lär·ē ri,vər·səl }

violet [OPTICS] The hue evoked in an average observer by monochromatic radiation having a wavelength in the approximate range from 390 to 455 nanometers; however, the same sensation can be produced in a variety of other ways. { 'vī·ə·lət }

violle [OPTICS] A unit of luminous intensity, equal to the luminous intensity of 1 square centimeter of platinum at its temperature of solidification; it is found experimentally to be equal to 20.17 candelas. { vyȯl }

virial coefficients [THERMO] For a given temperature T, one of the coefficients in the expansion of P/RT in inverse powers of the molar volume, where P is the pressure and R is the gas constant. { 'vir·ē·əl ,kō·i'fish·əns }

virial of a system [STAT MECH] The average over a long period of time of $-1/2$ the sum over the particles in the system of the scalar product of the total force acting on the particle and its radius vector. { 'vir·ē·əl əv ə 'sis·təm }

virial theorem *See* Clausius virial theorem. { 'vir·ē·əl ,thir·əm }

virtual displacement [MECH] **1.** Any change in the positions of the particles forming a mechanical system. **2.** An infinitesimal change in the positions of the particles forming a mechanical system, which is consistent with the geometrical constraints on the system. { 'vər·chə·wəl di 'splās·mənt }

virtual entropy [THERMO] The entropy of a system, excluding that due to nuclear spin. Also known as practical entropy. { 'vər·chə·wəl 'en·trə·pē }

virtual image [OPTICS] An optical image from which rays of light only appear to diverge, without actually being focused there. { 'vər·chə·wəl 'im·ij }

virtual level [NUC PHYS] The energy of a virtual state. { 'vər·chə·wəl ,lev·əl }

virtual object [OPTICS] A collection of points which may be regarded as a source of light rays for a portion of an optical system but which does not actually have this function. { 'vər·chə·wəl 'äb·jekt }

virtual particle *See* virtual quantum. { 'vər·chə·wəl 'pard·ə·kəl }

virtual process [QUANT MECH] A process which contributes in a stage of a theoretical model but is not, by itself, physically realizable. { 'vər·chə·wəl ,prä·səs }

virtual quantum [QUANT MECH] A photon or other particle in an intermediate state which appears in matrix elements connecting initial and final states in second-and higher-order perturbation theory; energy is not conserved in the transitions to or from the intermediate state. Also known as virtual particle. { 'vər·chə·wəl 'kwän·təm }

virtual source [ACOUS] A source of sound which is composed of the same material as that in which the sound propagates and which does not have sharply delineated boundaries; such sources include thermal sources (such as lightning and thermoacoustic arrays), turbulence (as in rocket and jet engines), and sound itself (as in a parametric acoustic array). { 'vər·chə·wəl 'sȯrs }

virtual state [NUC PHYS] An unstable state of a compound nucleus which has a lifetime many times longer than the time it takes a nucleon, with the same energy as it has in the virtual state, to cross the nucleus. { 'vər·chə·wəl 'stāt }

virtual work [MECH] The work done on a system during any displacement which is consistent with the constraints on the system. { 'vər·chə·wəl 'wərk }

virtual work principle *See* principle of virtual work. { 'vər·chə·wəl ¦wərk ,prin·sə·pəl }

viscoelastic fluid [FL MECH] A fluid that displays viscoelasticity. { ¦vis·kō·i¦las·tik 'flü·əd }

viscoelasticity [MECH] Property of a material which is viscous but which also exhibits certain elastic properties such as the ability to store energy of deformation, and in which the application of a stress gives rise to a strain that approaches its equilibrium value slowly. { ¦vis·kō,i,las'tis·əd·ē }

viscoelastic theory [MECH] The theory which attempts to specify the relationship between stress and strain in a material displaying viscoelasticity. { ¦vis·kō·i¦las·tik 'thē·ə·rē }

viscometric analysis [FL MECH] Measurement of the flow properties of substances by viscometry. { ¦vis·kə¦me·trik ə'nal·ə·səs }

viscosity [FL MECH] The resistance that a gaseous or liquid system offers to flow when it is subjected to a shear stress. Also known as flow resistance; internal friction. { vi'skäs·əd·ē }

viscosity coefficient [FL MECH] An empirical number used in equations of fluid mechanics to account for the effects of viscosity. { vi'skäs·əd·ē ,kō·i,fish·ənt }

viscosity curve [FL MECH] A graph showing the viscosity of a liquid or gaseous material as a function of temperature. { vi'skäs·əd·ē ,kərv }

viscous dissipation function [FL MECH] A quadratic function of spatial derivatives of components of fluid velocity which gives the rate at which mechanical energy is converted into heat in a viscous fluid per unit volume. Also known as dissipation function. { 'vis·kəs ,dis·ə'pā·shən ,fəŋk·shən }

viscous drag [FL MECH] That part of the rearward force on an aircraft that results from the aircraft carrying air forward with it through viscous adherence. { 'vis·kəs 'drag }

viscous flow [FL MECH] **1.** The flow of a viscous fluid. **2.** The flow of a fluid through a duct under conditions such that the mean free path is small in comparison with the smallest, transverse section of the duct. { 'vis·kəs 'flō }

viscous fluid [FL MECH] A fluid whose viscosity is sufficiently large to make the viscous forces a significant part of the total force field in the fluid. { 'vis·kəs 'flü·əd }

viscous force [FL MECH] The force per unit volume or per unit mass arising from viscous effects in fluid flow. { 'vis·kəs 'fȯrs }

viscous sublayer [FL MECH] In a turbulent flow, a very thin region next to a wall, typically only 1% of the boundary layer thickness, where turbulent mixing is impeded and transport occurs partly or, if the limit as the wall is approached, entirely by viscous diffusion. { ˌvis·kəs 'səbˌlā·ər }

visibility function *See* luminous function. { 'viz·ə'bil·əd·ē ˌfəŋk·shən }

visibility meter [OPTICS] A type of photometer that operates on the principle of artificially reducing the visibility of objects to threshold values (borderline of seeing and not seeing) and measuring the amount of the reduction on an appropriate scale. { ˌviz·ə'bil·əd·ē ˌmēd·ər }

visible absorption spectrophotometry [SPECT] Study of the spectra produced by the absorption of visible-light energy during the transformation of an electron from the ground state to an excited state as a function of the wavelength causing the transformation. { 'viz·ə·bəl əb'sȯrp·shən ¦spek·trō·fə'täm·ə·trē }

visible radiation *See* light. { 'viz·ə·bəl ˌrād·ē'ā·shən }

visible spectrophotometry [SPECT] In spectrophotometric analysis, the use of a spectrophotometer with a tungsten lamp that has an electromagnetic spectrum of 380–780 nanometers as a light source, glass or quartz prisms or gratings in the monochromator, and a photomultiplier cell as a detector. { 'viz·ə·bəl ¦spek·trō·fə'täm·ə·trē }

visible spectrum [SPECT] **1.** The range of wavelengths of visible radiation. **2.** A display or graph of the intensity of visible radiation emitted or absorbed by a material as a function of wavelength or some related parameter. { 'viz·ə·bəl 'spek·trəm }

visual achromatism [OPTICS] In an optical system, the removal of chromatic aberration or chromatic differences of magnification between light at the wavelength of the Fraunhofer C line at 656.3 nanometers and the F line at 486.1 nanometers in order to minimize these defects at wavelengths at which the human eye is most sensitive. Also known as optical achromatism. { 'vizh·ə·wəl ā'krō·məˌtiz·əm }

visual angle [OPTICS] The angle which an object subtends at the nodal point of the eye of an observer. { 'vizh·ə·wəl ˌaŋ·gəl }

visual photometer [OPTICS] A photometer in which the luminance of two surfaces is compared by human vision; it usually utilizes the Lummer-Brodhun sightbox or some adaptation of its principles. { 'vizh·ə·wəl fə'täm·əd·ər }

vitreous luster [OPTICS] A type of luster resembling that of glass. { 'vi·trē·əs ˌləs·tər }

vitreous state [SOLID STATE] A solid state in which the atoms or molecules are not arranged in any regular order, as in a crystal, and which crystallizes only after an extremely long time. Also known as glassy state. { 'vi·trē·əs ˌstāt }

Vlasov equation [PL PHYS] A modification of the Boltzmann transport equation for the study of a plasma, in which particles interact only through the mutually induced space-charge field, and collisions are assumed to be negligible. Also known as collisionless Boltzmann equation. { 'vla·sȯf iˌkwā·zhən }

Vlasov-Maxwell equations [PL PHYS] Equations for the propagation of electromagnetic radiation in a hot, collisionless plasma. { 'vlä·sȯf 'maksˌwel iˌkwā·zhənz }

Voigt body *See* Kelvin body. { 'fȯit ˌbäd·ē }

Voigt effect [OPTICS] Double refraction of light passing through a substance that is placed in a magnetic field perpendicular to the direction of light propagation. { 'fȯit iˌfekt }

Voigt notation [MECH] A notation employed in the theory of elasticity in which elastic constants and elastic moduli are labeled by replacing the pairs of letters *xx*, *yy*, *zz*, *yz*, *zx*, and *xy* by the number 1, 2, 3, 4, 5, and 6 respectively. { 'fȯit nōˌtā·shən }

volatility [THERMO] The quality of having a low boiling point or subliming temperature at ordinary pressure or, equivalently, of having a high vapor pressure at ordinary temperatures. { ˌväl·ə'til·əd·ē }

volatilization [THERMO] The conversion of a chemical substance from a liquid or solid state to a gaseous or vapor state by the application of heat, by reducing pressure, or by a combination of these processes. Also known as vaporization. { ˌväl·əd·əl·ə'zā·shən }

volt [ELEC] The unit of potential difference or electromotive force in the meter-kilogram-second system, equal to the potential difference between two points for which 1 coulomb of electricity will do 1 joule of work in going from one point to the other. Symbolized V. { vōlt }

Volta effect *See* contact potential difference. { 'vōl·tə iˌfekt }

voltage [ELEC] Potential difference or electromotive force measured in volts. { 'vōl·tij }

voltage coefficient [ELEC] For a resistor whose resistance varies with voltage, the ratio of the fractional change in resistance to the change in voltage. { 'vōl·tij ˌkō·iˌfish·ənt }

voltage-current dual [ELEC] A pair of circuits in which the elements of one circuit are replaced by their dual elements in the other circuit according to the duality principle; for example, currents are replaced by voltages, capacitances by resistances. { 'vōl·tij 'kə·rənt 'dül }

voltage divider [ELEC] A tapped resistor, adjustable resistor, potentiometer, or a series arrangement of two or more fixed resistors connected across a voltage source; a desired fraction of the total voltage is obtained from the intermediate tap, movable contact, or resistor junction. Also known as potential divider. { 'vōl·tij di,vīd·ər }

voltage drop [ELEC] The voltage developed across a component or conductor by the flow of current through the resistance or impedance of that component or conductor. { 'vōl·tij ,dräp }

voltage feed [ELECTROMAG] Excitation of a transmitting antenna by applying voltage at a point of maximum potential (at a voltage loop or antinode). { 'vōl·tij ,fēd }

voltage gradient [ELEC] The voltage per unit length along a resistor or other conductive path. { 'vōl·tij ,grād·ē·ənt }

voltage measurement [ELEC] Determination of the difference in electrostatic potential between two points. { 'vōl·tij ,mezh·ər·mənt }

voltage node [ELECTROMAG] Point having zero voltage in a stationary wave system, as in an antenna or transmission line; for example, a voltage node exists at the center of a half-wave antenna. { 'vōl·tij ,nōd }

voltage phasor [ELEC] A line whose length represents the magnitude of a sinusoidally varying voltage and whose angle with the positive *x*-axis represents its phase. { 'vōl·tij ,fā·zər }

voltage ratio [ELEC] The root-mean-square primary terminal voltage of a transformer divided by the root-mean-square secondary terminal voltage under a specified load. { 'vōl·tij ,rā·shō }

voltage reflection coefficient [ELECTROMAG] The ratio of the phasor representing the magnitude and phase of the electric field of the backward-traveling wave at a specified cross section of a waveguide to the phasor representing the forward-traveling wave at the same cross section. { 'vōl·tij ri¦flek·shən ,kō·i,fish·ənt }

voltage transformer [ELEC] An instrument transformer whose primary winding is connected in parallel with a circuit in which the voltage is to be measured or controlled. Also known as potential transformer. { 'vōl·tij tranz,fȯr·mər }

voltaic cell [ELEC] A primary cell consisting of two dissimilar metal electrodes in a solution that acts chemically on one or both of them to produce a voltage. { vōl'tā·ik 'sel }

volt-ampere [ELEC] The unit of apparent power in the International System; it is equal to the apparent power in a circuit when the product of the root-mean-square value of the voltage, expressed in volts, and the root-mean-square value of the current, expressed in amperes, equals 1. Abbreviated VA. { 'vōlt 'am,pir }

volt-ampere hour [ELEC] A unit for expressing the integral of apparent power over time, equal to the product of 1 volt-ampere and 1 hour, or to 3600 joules. { 'vōlt 'am,pir 'aủr }

volt-ampere-hour reactive *See* var hour. { 'vōlt 'am,pir 'aủr rē'ak·tiv }

volt-ampere reactive [ELEC] The unit of reactive power in the International System; it is equal to the reactive power in a circuit carrying a sinusoidal current when the product of the root-mean-square value of the voltage, expressed in volts, by the root-mean-square value of the current, expressed in amperes, and by the sine of the phase angle between the voltage and the current, equals 1. Abbreviated var. Also known as reactive volt-ampere. { 'vōlt 'am,pir rē'ak·tiv }

volt box [ELEC] A series of resistors arranged so that a desired fraction of a voltage can be measured, and the voltage thereby computed. { 'vōlt ,bäks }

Volterra dislocation [SOLID STATE] A model of a dislocation which is formed in a ring of crystalline material by cutting the ring, moving the cut surfaces over each other, and then rejoining them. { vol'ter·ə ,dis·lō'kā·shən }

volume [ACOUS] The intensity of a sound. { 'väl·yəm }

volume acoustic wave *See* bulk acoustic wave. { 'väl·yəm ə'küs·tik 'wāv }

volume flow rate [FL MECH] The volume of the fluid that passes through a given surface in a unit time. { 'väl·yəm 'flō ,rāt }

volume lifetime [SOLID STATE] Average time interval between the generation and recombination of minority carriers in a homogeneous semiconductor. { 'väl·yəm 'līf,tīm }

volume recombination rate [SOLID STATE] The rate at which free electrons and holes within the volume of a semiconductor recombine and thus neutralize each other.

volume resistivity [ELEC] Electrical resistance between opposite faces of a 1-centimeter cube of insulating material, commonly expressed in ohm-centimeters. Also known as specific insulation resistance. { 'väl·yəm ,rē,zis'tiv·əd·ē }

volume shift *See* field shift. { 'väl·yəm ,shift }

volumetric strain [MECH] One measure of deformation; the change of volume per unit of volume. { ¦väl·yə¦me·trik 'strān }

volume velocity [ACOUS] The rate of flow of a medium through a specified area due to a sound wave. { 'väl·yəm və,läs·əd·ē }

von Kármán *See* Kármán. { fōn 'kär·män }

von Klitzing constant [PHYS] **1.** The quantity $RK = h/e^2$, where h is Planck's constant and e is the charge of the electron; materials that exhibit the quantum Hall effect have a Hall resistance equal to RK/n, where n is either an integer or a rational fraction. Also known as quantized Hall resistance. **2.** The conventional value of this quantity adopted by international agreement on January 1, 1990, to establish a standard for the ohm, $RK - 90 = 25{,}812.807$ ohms. { fōn 'klit·siŋ ,kän·stənt }

von Mises yield criterion [MECH] The assumption that plastic deformation of a material begins when the sum of the squares of the principal components of the deviatoric stress reaches a certain critical value. { fōn ¦mēz·əz 'yēld ,krī,tir·ē·ən }

vortex [FL MECH] **1.** Any flow possessing vorticity; for example, an eddy, whirlpool, or other rotary motion. **2.** A flow with closed streamlines, such as a free vortex or line vortex. **3.** *See* vortex tube. [SOLID STATE] *See* fluxoid. { 'vȯr,teks }

vortex breakdown [FL MECH] An abrupt change in the structure of the core of a swirling flow. { ,vȯr,teks 'brāk,dau̇n }

vortex distribution method [FL MECH] An analytic method used in ideal aerodynamics which ignores the thickness of the profile of the aerodynamic figure being studied. { 'vȯr,teks ,di·strə'byü·shən ,meth·əd }

vortex filament [FL MECH] The line of concentrated vorticity in a line vortex. Also known as vortex line. { 'vȯr,teks ,fil·ə·mənt }

vortex line [FL MECH] **1.** A line drawn through a fluid such that it is everywhere tangent to the vorticity. **2.** *See* vortex filament. { 'vȯr,teks ,līn }

vortex ring [FL MECH] A line vortex in which the line of concentrated vorticity is a closed curve. Also known as collar vortex; ring vortex. { 'vȯr ,teks ,riŋ }

vortex shedding [FL MECH] In the flow of fluids past objects, the shedding of fluid vortices periodically downstream from the restricting object (for example, smokestacks, pipelines, or orifices). { 'vȯr,teks ,shed·iŋ }

vortex sheet [FL MECH] A surface across which there is a discontinuity in fluid velocity, such as in slippage of one layer of fluid over another; the surface may be regarded as being composed of vortex filaments. { 'vȯr,teks ,shēt }

vortex street [FL MECH] A series of vortices which are systematically shed from the downstream side of a body around which fluid is flowing rapidly. Also known as vortex trail; vortex train. { 'vȯr,teks ,strēt }

vortex trail *See* vortex street. { 'vȯr,teks ,trāl }

vortex train *See* vortex street. { 'vȯr,teks ,trān }

vortex tube [FL MECH] A tubular surface consisting of the collection of vortex lines which pass through a small closed curve. Also known as vortex. { 'vȯr,teks ,tüb }

vortical field *See* rotational field. { ¦vȯrd·ə·kəl 'fēld }

vorticity [FL MECH] For a fluid flow, a vector equal to the curl of the velocity of flow. { vȯr'tis·əd·ē }

vorticity equation [FL MECH] An equation of fluid mechanics describing horizontal circulation in the motion of particles around a vertical axis: $(d/dt)\ (S + f) = -\ (S + f)\ \mathrm{div}_H c$, where $(S + f)$ is the absolute vorticity (S is the relative vorticity and f is the Coriolis parameter) and $\mathrm{div}_H c$ is the horizontal divergence of the fluid velocity. { vȯr'tis·əd·ē i,kwā·zhən }

vorticity-transport hypothesis [FL MECH] The hypothesis that, owing to the existence of pressure fluctuations, vorticity, and not momentum, is conservative in turbulent eddy flux. { vȯr'tis·əd·ē ¦tranz,pȯrt hī,päth·ə·səs }

V particle [PART PHYS] The name first used for the unstable particles whose decay is responsible for the production of characteristic V-shaped tracks observed in cloud chambers exposed to cosmic radiation; they are neutral semistable particles such as neutral K mesons or lambda hyperons. { 'vē ,pärd·ə·kəl }

VPE *See* vapor-phase epitaxy.

v-process [NUC PHYS] The synthesis of certain elements and nuclides in type II supernovas; in this process, the inelastic scattering of neutrinos emitted from the core of the supernova excites states that then decay via single or multiple nucleon emission. { 'vē ,prä·səs }

VU *See* volume unit. { vyü *or* ¦vē'yü }

VUV radiation *See* vacuum ultraviolet radiation. { ¦vē¦yü'vē ,rād·ē,ā·shən }

W

W *See* watt.

Wadsworth mounting [OPTICS] **1.** A device in which light passes through a prism and is then reflected from a plane mirror; it has the effect of a constant-deviation prism. **2.** A mounting for a diffraction grating in which the slit is placed at the principal focus of a concave mirror, so that the light falling on the grating is in a parallel beam; it greatly reduces astigmatism. { 'wädz,wərth ,maůnt·iŋ }

Wagner earth connection *See* Wagner ground. { 'wag·nər 'ərth kə,nek·shən }

Wagner ground [ELEC] A ground connection used with an alternating-current bridge to minimize stray capacitance errors when measuring high impedances; a potentiometer is connected across the bridge supply oscillator, with its movable tap grounded. Also known as Wagner earth connection. { 'wag·nər ,graůnd }

Waidner-Burgess standard [OPTICS] A unit of luminous intensity equal to the luminous intensity of 1 square centimeter of a blackbody at the melting point of platinum, or to 60 candelas. { 'wīd·nər 'bər·jəs ,stan·dərd }

wake [FL MECH] The region behind a body moving relative to a fluid in which the effects of the body on the fluid's motion are concentrated. { wāk }

wake flow [FL MECH] Turbulent eddying flow that occurs downstream from bluff bodies. { 'wāk ,flō }

wake-induced flutter *See* buffeting flutter. { ¦wāk in,düst 'fləd·ər }

wake-induced galloping *See* buffeting flutter. { ¦wāk in,düst 'gal·əp·iŋ }

wall-attachment amplifier [FL MECH] A bistable fluidic device utilizing two walls set back from the supply jet port, control ports, and channels to define two downstream outputs. Also known as flip-flop amplifier. { 'wȯl ə¦tach·mənt 'am·plə,fī·ər }

wall energy [SOLID STATE] The energy per unit area of the boundary between two ferromagnetic domains which are oriented in different directions. { 'wȯl ¦en·ər·jē }

wall friction [FL MECH] The drag created in the flow of a liquid or gas because of contact with the wall surfaces of its conductor, such as the inside surfaces of a pipe. { 'wȯl ,frik·shən }

wall superheat [THERMO] The difference between the temperature of a surface and the saturation temperature (boiling point at the ambient pressure) of an adjacent liquid that is heated by the surface. { ¦wȯl 'sü·pər,hēt }

Wannier function [SOLID STATE] The Fourier transform of a Bloch function defined for an entire band, regarded as a function of the wave vector. { vän'yā ,fəŋk·shən }

warble tone [ACOUS] A tone whose frequency varies periodically several times per second over a small range; used to prevent standing-wave patterns from forming in reverberation chambers. { 'wȯr·bəl ,tōn }

warpage [MECH] The action, process, or result of twisting or turning out of shape. { 'wȯr·pij }

warping function *See* torsion function. { 'wȯrp·iŋ ,fəŋk·shən }

wash [FL MECH] The surge of disturbed air or other fluid resulting from the passage of something through the fluid. { wäsh }

water dropper [ELEC] A simple electrostatic generator in which each of two series of water drops falls through cylindrical metal cans into lower cans with funnels, and the cans are electrically connected in such a way that charge accumulates on them, energy being supplied by the gravitational force on the water drops. { 'wȯd·ər ,dräp·ər }

water hammer [FL MECH] Pressure rise in a pipeline caused by a sudden change in the rate of flow or stoppage of flow in the line. { 'wȯd·ər ,ham·ər }

water load [ELECTROMAG] A matched waveguide termination in which the electromagnetic energy is absorbed in water; the resulting rise in the temperature of the water is a measure of the output power. { 'wȯd·ər ,lōd }

water noise [ACOUS] Underwater acoustic energy resulting primarily from the movement of the water itself. { 'wȯd·ər ,nȯiz }

waterpower [MECH] Power, usually electric, generated from an elevated water supply by the use of hydraulic turbines. { 'wȯd·ər,paů·ər }

water vapor [PHYS] Water in the form of a vapor, especially when below the boiling point and diffused. { 'wȯd·ər ,vā·pər }

water-vapor laser [OPTICS] A laser whose active substance is water vapor, and which emits infrared radiation at wavelengths of 27.97, 47.7, 78.46, and 118.6 micrometers. { 'wȯd·ər ¦vā·pər 'lā·zər }

watt [PHYS] The unit of power in the meter-kilogram-second system of units, equal to 1 joule per second. Symbolized W. { wät }

watt balance [PHYS] A device for making a highly accurate comparison of electrical and mechanical powers, in which the force on a current-carrying coil in a constant magnetic field is balanced by the gravitational force on an accurately measured mass, and then, with the current source removed, the coil is moved at constant speed in the same magnetic field and the induced potential difference is measured. { 'wät ˌbal·əns }

watt current *See* active current. { 'wät ˌkə·rənt }

watt-hour [ELEC] A unit of energy used in electrical measurements, equal to the energy converted or consumed at a rate of 1 watt during a period of 1 hour, or to 3600 joules. Abbreviated Wh. { 'wät ¦au̇r }

wattless component *See* reactive component. { 'wätˌləs kəm'pō·nənt }

wattless current *See* reactive current. { 'wätˌləs 'kə·rənt }

wattless power *See* reactive power. { 'wätˌləs 'pau̇·ər }

watt-second [PHYS] Amount of energy corresponding to 1 watt acting for 1 second; 1 watt-second is equal to 1 joule. { 'wät ¦sek·ənd }

Watt's law [THERMO] A law which states that the sum of the latent heat of steam at any temperature of generation and the heat required to raise water from 0°C to that temperature is constant; it has been shown to be substantially in error. { 'wäts ˌlȯ }

wave [FL MECH] A disturbance which moves through or over the surface of a liquid, as of a sea. [PHYS] A disturbance which propagates from one point in a medium to other points without giving the medium as a whole any permanent displacement. { wāv }

wave aberration [OPTICS] The departure of the geometrical wavefront from a reference sphere with its vertex at the center of the exit pupil and its center of curvature located at the ideal image point; a measure of monochromatic aberration. { 'wāv ˌab·əˌrā·shən }

wave acoustics [ACOUS] The study of the propagation of sound based on its wave properties. { 'wāv əˌküs·tiks }

wave amplitude [PHYS] The magnitude of the greatest departure from equilibrium of the wave disturbance. { 'wāv ˌam·pləˌtüd }

wave angle [ELECTROMAG] The angle, either in bearing or elevation, at which a radio wave leaves a transmitting antenna or arrives at a receiving antenna. { 'wāv ˌaŋ·gəl }

wave antenna [ELECTROMAG] Directional antenna composed of a system of parallel, horizontal conductors, varying from a half to several wavelengths long, terminated to ground at the far end in its characteristic impedance. { 'wāv anˌten·ə }

wave celerity *See* phase velocity. { 'wāv səˌler·əd·ē }

wave converter [ELECTROMAG] Device for changing a wave of a given pattern into a wave of another pattern, for example, baffle-plate converters, grating converters, and sheath-reshaping converters for waveguides. { 'wāv kən ˌvərd·ər }

wave-corpuscle duality *See* wave-particle duality. { 'wāv ¦kȯr·pə·səl dü'al·əd·ē }

wave crest [PHYS] The position at which the disturbance of a progressive wave attains its maximum positive value. { 'wāv ˌkrest }

wave duct [ELECTROMAG] **1.** Waveguide, with tubular boundaries, capable of concentrating the propagation of waves within its boundaries. **2.** Natural duct, formed in air by atmospheric conditions, through which waves of certain frequencies travel with more than average efficiency. { 'wāv ˌdəkt }

wave equation [PHYS] **1.** In classical physics, a special equation governing waves that suffer no dissipative attenuation; it states that the second partial derivative with respect to time of the function characterizing the wave is equal to the square of the wave velocity times the Laplacian of this function. Also known as classical wave equation; d'Alembert's wave equation. **2.** Any of several equations which relate the spatial and time dependence of a function characterizing some physical entity which can propagate as a wave, including quantum-wave equations for particles. { 'wāv iˌkwā·zhən }

waveform [PHYS] The pictorial representation of the form or shape of a wave, obtained by plotting the displacement of the wave as a function of time, at a fixed point in space. { 'wāvˌfȯrm }

waveform analysis [PHYS] The determination of the amplitude and phase of the components of a complex waveform, either mathematically or by means of electronic instruments. { 'wāv ˌfȯrm əˌnal·ə·səs }

wavefront [PHYS] **1.** A surface of constant phase. **2.** The portion of a wave envelope that is between the beginning zero point and the point at which the wave reaches its crest value, as measured either in time or distance. { 'wāvˌfrənt }

wavefront reversal *See* optical phase conjugation. { 'wāvˌfrənt ri'vər·səl }

wavefront splitting [OPTICS] Any method of producing interference in which light from a single source is split into two parts which can then be recombined; examples include Young's two-slit experiment, the Fresnel double mirror, and the Fresnel biprism. { 'wāvˌfrənt 'splid·iŋ }

wave function *See* Schrödinger wave function. { 'wāv ˌfəŋk·shən }

wave group [PHYS] A series of waves in which the wave direction, length, and height vary only slightly. { 'wāv ˌgrüp }

waveguide [ELECTROMAG] **1.** Broadly, a device which constrains or guides the propagation of electromagnetic waves along a path defined by the physical construction of the waveguide; includes ducts, a pair of parallel wires, and a coaxial cable. Also known as microwave waveguide. **2.** More specifically, a metallic tube which can confine and guide the propagation of electromagnetic waves in the lengthwise direction of the tube. { 'wāv,gīd }

waveguide assembly [ELECTROMAG] An item consisting of one or more definite lengths of straight or formed, flexible or rigid, prefabricated hollow tubing of conductive material; the tubing has a predetermined cross-section, and is designed to guide or conduct high-frequency electromagnetic energy through its interior; one or more ends are terminated. { 'wāv,gīd ə¦sem·blē }

waveguide attenuation [ELECTROMAG] The decrease from one point of a waveguide to another, in the power carried by an electromagnetic wave in the waveguide. { 'wāv,gīd ə,ten·yə'wā·shən }

waveguide bend [ELECTROMAG] A section of waveguide in which the direction of the longitudinal axis is changed; an **E**-plane bend in a rectangular waveguide is bent along the narrow dimension, while an **H**-plane bend is bent along the wide dimension. Also known as waveguide elbow. { 'wāv,gīd ¦bend }

waveguide cavity [ELECTROMAG] A cavity resonator formed by enclosing a section of waveguide between a pair of waveguide windows which form shunt susceptances. { 'wāv,gīd ¦kav·əd·ē }

waveguide connector [ELECTROMAG] A mechanical device for electrically joining and locking together separable mating parts of a waveguide system. Also known as waveguide coupler. { 'wāv,gīd kə¦nek·tər }

waveguide coupler *See* waveguide connector. { 'wāv,gīd ¦kəp·lər }

waveguide critical dimension [ELECTROMAG] Dimension of waveguide cross section which determines the cutoff frequency. { 'wāv,gīd 'krid·ə·kəl də'men·shən }

waveguide cutoff frequency [ELECTROMAG] Frequency limit of propagation along a waveguide for waves of a given field configuration. { 'wāv,gīd 'kəd,ȯf ,frē·kwən·sē }

waveguide discontinuity *See* discontinuity. { 'wāv,gīd dis,känt·ən'ü·əd·ē }

waveguide elbow *See* waveguide bend. { 'wāv,gīd ¦el·bō }

waveguide filter [ELECTROMAG] A filter made up of waveguide components, used to change the amplitude-frequency response characteristic of a waveguide system. { 'wāv,gīd ¦fil·tər }

waveguide hybrid [ELECTROMAG] A waveguide circuit that has four arms so arranged that a signal entering through one arm will divide and emerge from the two adjacent arms, but will be unable to reach the opposite arm. { 'wāv,gīd 'hī·brəd }

waveguide plunger *See* piston. { 'wāv,gīd ¦plən·jər }

waveguide resonator *See* cavity resonator. { 'wāv,gīd ¦rez·ən,ād·ər }

waveguide shim [ELECTROMAG] Thin resilient metal sheet inserted between waveguide components to ensure electrical contact. { 'wāv,gīd ,shim }

waveguide slot [ELECTROMAG] A slot in a waveguide wall, either for coupling with a coaxial cable or another waveguide, or to permit the insertion of a traveling probe for examination of standing waves. { 'wāv,gīd ,slät }

waveguide window *See* iris. { ,wāv,gīd ¦win·dō }

wave height [PHYS] Twice the wave amplitude. { 'wāv ,hīt }

wave impedance [ELECTROMAG] The ratio, at every point in a specified plane of a waveguide, of the transverse component of the electric field to the transverse component of the magnetic field. { 'wāv im,pēd·əns }

wave intensity [PHYS] The average amount of energy transported by a wave in the direction of wave propagation, per unit area per unit time. { 'wāv in,ten·səd·ē }

wave interference *See* interference. { ,wāv ,in·tər'fir·əns }

wavelength [PHYS] The distance between two points having the same phase in two consecutive cycles of a periodic wave, along a line in the direction of propagation. { 'wāv,leŋkth }

wavelength constant *See* phase constant. { 'wāv ,leŋkth ,kän·stənt }

wavelength standards [SPECT] Accurately measured lengths of waves emitted by specified light sources for the purpose of obtaining the wavelengths in other spectra by interpolating between the standards. { 'wāv,leŋkth ,stan·dərdz }

wave-making resistance *See* wave resistance. { ¦wāv ¦māk·iŋ ri,zis·təns }

wave mechanics *See* Schrödinger's wave mechanics. { 'wāv mi,kan·iks }

wave motion [PHYS] The process by which a disturbance at one point is propagated to another point more remote from the source with no net transport of the material of the medium itself; examples include the motion of electromagnetic waves, sound waves, hydrodynamic waves in liquids, and vibration waves in solids. Also known as propagation; wave propagation. { 'wāv ,mō·shən }

wave normal [PHYS] **1.** A unit vector which is perpendicular to an equiphase surface of a wave, and has its positive direction on the same side of the surface as the direction of propagation. **2.** One of a family of curves which are everywhere perpendicular to the equiphase surfaces of a wave. { 'wāv 'nȯr·məl }

wave number [PHYS] The reciprocal of the wavelength of a wave, or sometimes 2π divided by the wavelength. Also known as reciprocal wavelength. { 'wāv ,nəm·bər }

wave optics [OPTICS] The branch of optics which treats of light (or electromagnetic radiation in general) with explicit recognition of its wave nature. { 'wāv ,äp·tiks }

wave packet [PHYS] In wave phenomena, a superposition of waves of differing lengths, so phased that the resultant amplitude is negligibly small except in a limited portion of space whose dimensions are the dimensions of the packet. Also known as packet. { 'wāv ˌpak·ət }

wave-particle duality [QUANT MECH] The principle that both matter and electromagnetic radiation exhibit phenomena in which they behave as waves and other phenomena in which they behave as particles, the two aspects being associated by the de Broglie relations. Also known as duality principle; wave-corpuscle duality. { 'wāv 'pärd·ə·kəl dü'al·əd·ē }

wave period [PHYS] The time between the attainment of successive maxima, at a fixed point, of a quantity characterizing a wave. { 'wāv ˌpir·ē·əd }

wave plate [OPTICS] A plate of material which is linearly birefringent. Also known as retardation plate; retardation sheet. { 'wāv ˌplāt }

wave polarization *See* polarization. { 'wāv ˌpō·lə·rəˌzā·shən }

wave propagation *See* wave motion. { 'wāv ˌpräp·əˌgā·shən }

wave refraction [PHYS] The process by which the direction of a wave train moving in shallow water at an angle to the contours is changed. { 'wāv riˌfrak·shən }

wave resistance [FL MECH] The portion of fluid resistance to a body moving on the surface of a liquid that results from energy dissipation in the formation of waves on the liquid surface. Also known as wave-making resistance. { 'wāv riˌzis·təns }

wave speed *See* phase velocity. { 'wāv ˌspēd }

wave theory of light [OPTICS] A theory which assumes that light is a wave motion, rather than a stream of particles. { 'wāv 'thē·ə·rē əv 'līt }

wave train [PHYS] A series of waves produced by the same disturbance. { 'wāv ˌtrān }

wave trough [PHYS] The lowest part of a wave form between successive wave crests. { 'wāv ˌtrȯf }

wave vector [PHYS] A vector whose direction is the direction of phase propagation of a wave at each point in space, and whose magnitude is sometimes set at $2\pi/\lambda$ and sometimes at $1/\lambda$, where λ is the wavelength. { 'wāv ˌvek·tər }

wave-vector space [SOLID STATE] The space of the wave vectors of the state functions of some system; this would be used, for example, for electron wave functions in a crystal and thermal vibrations of a lattice. Also known as **k**-space; reciprocal space. { 'wāv ¦vek·tər ˌspās }

wave velocity *See* phase velocity. { 'wāv vəˌläs·əd·ē }

wavy extinction *See* undulatory extinction. { 'wāv·ē ik'stiŋk·shən }

wax-block photometer *See* Joly photometer. { ¦waks ¦bläk fō'tam·əd·ər }

Wb *See* weber.

W boson [PART PHYS] An intermediate vector boson with positive or negative electric charge that mediates the charged-current weak interactions. Also known as W particle. { 'dəb·əlˌyü ˌbōˌsän }

W coefficient *See* Racah coefficient. { 'dəb·əlˌyü ˌkō·iˌfish·ənt }

weak coupling [PART PHYS] The coupling of four fermion fields in the weak interaction, having a strength many orders of magntiude weaker than that of the strong or electromagnetic interactions. { 'wēk 'kəp·liŋ }

weak energy condition [RELAT] The condition in general relativity theory that all observers see a nonnegative energy density. { ¦wēk 'en·ər·jē kənˌdish·ən }

weak interaction [PART PHYS] One of the fundamental interactions among elementary particles, responsible for beta decay of nuclei, and for the decay of elementary particles with lifetimes greater than about 10^{-10} second, such as muons, K mesons, and lambda hyperons; it is several orders of magnitude weaker than the strong and electromagnetic interactions, and fails to conserve strangeness or parity. Also known as beta interaction. { 'wēk ˌin·tər'ak·shən }

weakly interacting massive particle [PART PHYS] A hypothetical massive elementary particle, interacting only through gravity and the weak nuclear interaction. Abbreviated WIMP. { ¦wēk·lē ˌin·tər¦ak·tiŋ ¦mas·iv 'pärd·ə·kəl }

weak-strong duality [PHYS] A property of some physical systems that have a dual description such that, when a coupling constant governing the strengths of interactions in the original system is large, the coupling constant in the dual description is small, so that the system can be accurately described by means of perturbation theory. { ¦wēk ¦strȯŋ dü'al·əd·ē }

web *See* wire. { web }

weber [ELECTROMAG] The unit of magnetic flux in the meter-kilogram-second system, equal to the magnetic flux which, linking a circuit of one turn, produces in it an electromotive force of 1 volt as it is reduced to zero at a uniform rate in 1 second. Symbolized Wb. { 'vā·bər }

Weber number 1 [FL MECH] A dimensionless number used in the study of surface tension waves and bubble formation, equal to the product of the square of the velocity of the wave or the fluid velocity, the density of the fluid, and a characteristic length, divided by the surface tension. Symbolized N_{We1}, *We*. { 'vā·bər ¦nəm·bər 'wən }

Weber number 2 [FL MECH] A dimensionless number, equal to the square root of Weber number 1. Symbolized N_{We2}. { 'vā·bər ¦nəm·bər 'tü }

wedge [ELECTROMAG] A waveguide termination consisting of a tapered length of dissipative material introduced into the guide, such as carbon. [OPTICS] **1.** An optical filter in which the transmission decreases continuously or in steps from one end to the other. **2.** A refracting prism of very small angle, inserted in an optical train to introduce a bend in the ray path. { wej }

wedge spectrograph [SPECT] A spectrograph in which the intensity of the radiation passing through the entrance slit is varied by moving an optical wedge. { 'wej 'spek·trəˌgraf }

Weigert effect [OPTICS] Dichroism introduced

in a silver-silver chloride photographic emulsion by a beam of linearly polarized light. { 'vī·gərt i,fekt }

weight [MECH] **1.** The gravitational force with which the earth attracts a body. **2.** By extension, the gravitational force with which a star, planet, or satellite attracts a nearby body. { wāt }

weight density [PHYS] The weight of a body or portion of a body divided by its volume. { 'wāt ,den·səd·ē }

weighted oscillator strength *See* gf-value. { 'wād·əd 'äs·ə,lād·ər ,strəŋkth }

weight factor [STAT MECH] The number of microstates that correspond to a given macrostate. { 'wāt ,fak·tər }

weightlessness [MECH] A condition in which no acceleration, whether of gravity or other force, can be detected by an observer within the system in question. Also known as zero gravity. { 'wāt·ləs·nəs }

Weinberg-Salam theory [PART PHYS] A gage theory in which the electromagnetic and weak nuclear interactions are described by a single unifying framework in which both have a characteristic coupling paramenter equal to the fine-structure constant; it predicts the existence of intermediate vector bosons and neutral current interactions. Also known as Salam-Weinberg theory. { 'wīn,bərg sə'läm ,thē·ə·rē }

Weissenberg effect [FL MECH] An alteration of the normal stresses in a non-Newtonian fluid on account of elasticity, so that such a fluid, when placed between two concentric, rotating cylinders, can rise on the inner cylinder in spite of centrifugal forces. { 'vīs·ən,bərg i,fekt }

Weissenberg method [SOLID STATE] A method of studying crystal structure by x-ray diffraction in which the crystal is rotated in a beam of x-rays, and a photographic film is moved parallel to the axis of rotation; the crystal is surrounded by a sleeve which has a slot that passes only diffraction spots from a single layer of the reciprocal lattice, permitting positive identification of each spot in the pattern. { 'vīs·ən,berk ,meth·əd }

Weiss magneton [ATOM PHYS] A unit of magnetic moment, equal to 1.853×10^{-24} joule/tesla, about one-fifth of the Bohr magneton; it is experimentally derived, the magnetic moments of certain molecules being close to integral multiples of this quantity. { 'ves 'mag·nə,tän }

Weiss molecular field [SOLID STATE] The effective magnetic field postulated in the Weiss theory of ferromagnetism, which acts on atomic magnetic moments within a domain, tending to align them, and is in turn generated by these magnetic moments. { 'ves mə'lek·yə·lər 'fēld }

Weiss theory [SOLID STATE] A theory of ferromagnetism based on the hypotheses that below the Curie point a ferromagnetic substance is composed of small, spontaneously magnetized regions called domains, and that each domain is spontaneously magnetized because a strong molecular magnetic field tends to align the individual atomic magnetic moments within the domain. Also known as molecular field theory. { 'ves ,thē·ə·rē }

Weizäcker-Williams method [QUANT MECH] A method of calculating the bremsstrahlung emitted when two particles, whose relative kinetic energies are much larger than their rest energies, collide; in the rest frame of one of the particles, the field of the other is equivalent to a set of virtual photons, and Compton scattering of these photons by the particle at rest is computed. { 'vīt,sek·ər 'wil·yəms ,meth·əd }

Wentzel-Kramers-Brillouin method [QUANT MECH] Method of approximating quantum-mechanical wave functions and energy levels, in which the logarithm of the wave function is expanded in powers of Planck's constant, and all except the first two terms are neglected. Also known as phase integral method; WKB method. { 'vent·səl 'krä·mərz brē'wan ,meth·əd }

Werner band [SPECT] A band in the ultraviolet spectrum of molecular hydrogen extending from 116 to 125 nanometers. { 'ver·nər ,band }

Wertheim effect [ELECTROMAG] A potential difference that appears between the ends of a wire twisted in a longitudinal magnetic field. { 'vert,hīm i,fekt }

Weston standard cell [ELEC] A standard cell used as a highly accurate voltage source for calibrating purposes; the positive electrode is mercury, the negative electrode is cadmium, and the electrolyte is a saturated cadmium sulfate solution; the Weston standard cell has a voltage of 1.018636 volts at 20°C. { 'wes·tən 'stan·dərd 'sel }

wet [PHYS] A liquid is said to wet a solid if the contact angle between the solid and the liquid, measured through the liquid, lies between 0 and 90°, and not to wet the solid if the contact angle lies between 90 and 180°. { wet }

wet cell [ELEC] A primary cell in which there is a substantial amount of free electrolyte in liquid form. { 'wet ,sel }

wet electrolytic capacitor [ELEC] An electrolytic capacitor employing a liquid electrolyte. { 'wet i¦lek·trə¦lid·ik kə'pas·əd·ər }

wetting angle [FL MECH] A contact angle which lies between 0 and 90°. { 'wed·iŋ ,aŋ·gəl }

Weyl equations [QUANT MECH] Two sets of relativistic wave equations into which the Dirac equation decomposes for a massless, spin-1/2 particle. { 'vīl i,kwā·zhənz }

Weyl tensor [RELAT] A tensor with the symmetries of the curvature tensor such that all contractions on its indices vanish; the curvature tensor is decomposable in terms of the metric, the scalar curvature, and the Weyl tensor. { 'wīl ,ten·sər }

Wheatstone bridge [ELEC] A four-arm bridge circuit, all arms of which are predominately resistive; used to measure the electrical resistance of an unknown resistor by comparing it with a known standard resistance. Also known as resistance bridge; Wheatstone network. { 'wēt ˌstōn 'brij }

Wheatstone network *See* Wheatstone bridge. { 'wētˌstōn 'netˌwərk }

Wheatstone stereoscope [OPTICS] A type of stereoscope that uses plane mirrors to enable the eyes to form a fused image of two pictures whose separation is greater than the interocular distance. { ¦wētˌstōn 'ster·ē·ōˌskōp }

Wheeler-Feynman theory [RELAT] A relativistic action-at-a-distance theory in which it is assumed that there are enough absorbers in the universe to serve as sinks for all actions that emanate from any charged particle; radiation damping is a consequence of the theory. { 'wēl·ər 'fīn·mən ˌthē·ə·rē }

whip antenna [ELECTROMAG] A flexible vertical rod antenna, used chiefly on vehicles. Also known as fishpole antenna. { 'wip anˌten·ə }

whisker *See* crystal whisker. { 'wis·kər }

whispering gallery [ACOUS] A domed gallery in which weak sounds can be heard at great distances. { 'wis·pər·iŋ ˌgal·rē }

whispering-gallery resonance [PHYS] A resonance that rises in the propagation of waves around the circumference of a circular structure when an integral number of wavelengths can fit into the circumference. { ¦wis·pər·iŋ ˌgal·rē ¦rez·ən·əns }

whistler wave *See* electron cyclotron wave. { 'wis·lər ˌwāv }

white body [PHYS] A hypothetical substance whose surface absorbs no electromagnetic radiation of any wavelength, that is, one which exhibits zero absorptivity for all wavelengths. { 'wīt ˌbäd·ē }

white light [OPTICS] Any radiation producing the same color sensation as average noon sunlight. { 'wīt 'līt }

white light hologram [OPTICS] A reflection hologram which can be viewed with an ordinary light source. { 'wīt ¦līt 'häl·əˌgram }

white noise [PHYS] Random noise that has a constant energy per unit bandwidth at every frequency in the range of interest. { 'wīt ˌnȯiz }

white object [OPTICS] An object that reflects all wavelengths of light with substantially equal high efficiencies and with considerable diffusion. { 'wīt 'äb·jekt }

white radiation *See* continuous radiation. { ¦wīt ˌrād·ē¦ā·shən }

white rainbow *See* fogbow. { 'wīt 'rānˌbō }

whole step *See* whole tone. { 'hōl 'step }

whole tone [ACOUS] The interval between two sounds whose basic frequency ratio is approximately equal to the sixth root of 2. Also known as whole step. { 'hōl 'tōn }

whr *See* watt-hour.

wide-angle lens [OPTICS] An optical lens having a large angular field, generally greater than 80°. { 'wīd ¦aŋ·gəl 'lenz }

Wiedemann effect [ELECTROMAG] The twist produced in a current-carrying wire when placed in a longitudinal magnetic field. Also known as circular magnetostriction. { 'vēd·əˌmän iˌfekt }

Wiedemann-Franz law [SOLID STATE] The law that the ratio of the thermal conductivity of a metal to its electrical conductivity is a constant, independent of the metal, times the absolute temperature. Also known as Lorentz relation. { 'vēd·əˌmän 'fränts ˌlȯ }

Wiegand effect [ELEC] The generation of an electrical pulse in a coil wrapped around or located near a Wiegand wire subjected to a changing magnetic field. { 'vē·gänt iˌfekt }

Wiegand module [ELEC] The apparatus for generating an electrical pulse by means of the Wiegand effect, consisting of a Wiegand wire, two small magnets, and a pickup coil. { 'vē·gänt ˌmäj·əl }

Wiegand wire [ELEC] A work-hardened wire whose magnetic permeability is much greater near its surface than at its center. { 'vē·gänt ˌwīr }

Wien capacitance bridge [ELEC] A four-arm alternating-current bridge used to measure capacitance in terms of resistance and frequency; two adjacent arms contain capacitors respectively in parallel and in series with resistors, while the other two arms are nonreactive resistors; bridge balance depends on frequency. { 'vēn kə'pas·əd·əns ˌbrij }

Wien constant [STAT MECH] The product of the temperature and the wavelength at which the intensity of radiation from a blackbody reaches its maximum; it is equal to approximately 2898 micrometer-kelvins. { 'vēn ˌkän·stənt }

Wien-DeSauty bridge *See* DeSauty's bridge. { 'vēn də·sō'tē ˌbrij }

Wiener experiment [OPTICS] An experiment in which a front-faced mirror is covered with a thick photographic emulsion which is then exposed to light incident perpendicular to the surface; upon development, it is found that standing waves are set up in the emulsion whose nodes coincide with those of the electric vector, rather than those of the magnetic vector. { 'vē·nər ikˌsper·ə·mənt }

Wien frequency bridge [ELEC] A modification of the Wien capacitance bridge, used to measure frequencies. { 'vēn 'frē·kwən·sē ˌbrij }

Wien inductance bridge [ELEC] A four-arm alternating-current bridge used to measure inductance in terms of resistance and frequency; two adjacent arms contain inductors respectively in parallel and in series with resistors, while the other two arms are nonreactive resistors; bridge balance depends on frequency. { 'vēn in'dək·təns ˌbrij }

Wien-Maxwell bridge *See* Maxwell bridge. { 'vēn 'maksˌwel ˌbrij }

Wien's displacement law [STAT MECH] A law for blackbody radiation which states that the wavelength at which the maximum amount of radiation occurs is a constant equal to approximately 2898 times the product of 1 micrometer and 1 kelvin. Also known as displacement law; Wien's radiation law. { 'vēnz di'splās·mənt ˌlȯ }

Wien's distribution law [STAT MECH] A formula for the spectral distribution of radiation from a blackbody, which is a good approximation to the Planck radiation formula at sufficiently low temperatures or wavelengths, for example, in the visible region of the spectrum below 3000 K. Also known as Wien's radiation law. { 'vēnz ,di·strə'byü·shən ,lȯ }

Wien's radiation law [STAT MECH] **1.** The law that the intensity of radiation emitted by a blackbody per unit wavelength, at that wavelength at which this intensity reaches a maximum, is proportional to the fifth power of the temperature. **2.** *See* Wien's displacement law; Wien's distribution law. { 'vēnz ,rād·ē'ā·shən ,lȯ }

Wigner coefficient *See* vector coupling coefficient. { 'wig·nər ,kō·i,fish·ənt }

Wigner-Eckart theorem [QUANT MECH] A theorem in the quantum theory of angular momentum which states that the matrix elements of a tensor operator can be factored into two quantities, the first of which is a vector-coupling coefficient, and the second of which contains the information about the physical properties of the particular states and operator, and is completely independent of the magnetic quantum numbers. { 'wig·nər 'ek·ərt ,thir·əm }

Wigner nuclides [NUC PHYS] The most important class of mirror nuclides, comprising pairs of odd-mass-number isobars for which the atomic number and the neutron number differ by 1. { 'wig·nər 'nü,klīdz }

Wigner-Seitz cell [CRYSTAL] A polyhedron about an atom in a face-centered cubic structure, made by drawing planes which perpendicularly bisect the lines to the nearest neighbors; in a body-centered cubic structure, bisecting planes of lines to nearest neighbors and next-nearest neighbors are used; such polyhedra fill space. { 'wig·nər 'zīts ,sel }

Wigner-Seitz method [SOLID STATE] A method of approximating the band structure of a solid: Wigner-Seitz cells surrounding atoms in the solid are approximated by spheres, and band solutions of the Schrödinger equation for one electron are estimated by using the assumption that an electronic wave function is the product of a plane wave function and a function whose gradient has a vanishing radial component at the sphere's surface. { 'wig·nər 'zīts ,meth·əd }

Wigner's theorem [QUANT MECH] **1.** The theorem that, if ψ is an eigenfunction of the Hamiltonian operator and R is a symmetry element of the Hamiltonian, then $R\psi$ is an eigenfunction of the Hamiltonian having the same eigenvalue as ψ. **2.** Angular momentum of the electron spin is conserved in a collision of the second kind. { 'wig·nərz ,thir·əm }

Wigner supermultiplet [NUC PHYS] A set of quantum-mechanical states of a collection of nucleons which form the basis of a representation of SU(4), especially appropriate when spin and isospin dependence of the nuclear interaction may be disregarded; several combinations of spin and isospin multiplets may occur in a supermultiplet. { 'wig·nər ¦sü·pər'məl·tə·plət }

Wigner three-j symbol *See* three-*j* number. { 'wig·nər 'thrē 'jā ,sim·bəl }

Williams-Hazen formula [FL MECH] In a liquid-flow system, a method for calculation of head loss due to the friction in a pipeline. { 'wil·yəmz 'hāz·ən ,fȯr·myə·lə }

Williams refractometer [OPTICS] A refractometer in which light from a single slit is divided into two beams by a pentagonal prism. { 'wil·yəmz ,rē,frak,täm·əd·ər }

Wilson electroscope [ELEC] An electroscope that has a single gold leaf which, when charged, is attracted to a grounded metal plate inclined at an angle that maximizes the instrument's sensitivity. { 'wil·sən i'lek·trə,skōp }

Wilson experiment [ELECTROMAG] An experiment that tests the validity of electromagnetic theory; a hollow cylinder of dielectric material, having layers of metal on its outer and inner cylindrical surfaces, is rotated about its axis in a magnetic field parallel to the axis; a sensitive electrometer, connected to the metal layers, indicates a charge that has the magnitude and sign predicted by theory. { 'wil·sən ik,sper·ə·mənt }

WIMP *See* weakly interacting massive particle. { wimp }

Wimshurst machine [ELEC] An electrostatic generator consisting of two glass disks rotating in opposite directions, having sectors of tinfoil and collecting combs so arranged that static electricity is produced for charging Leyden jars or discharging across a gap. { 'wimz,hərst mə,shēn }

windage [MECH] **1.** The deflection of a bullet or other projectile due to wind. **2.** The correction made for such deflection. { 'win·dij }

wind deflection [MECH] Deflection caused by the influence of wind on the course of a projectile in flight. { 'win di,flek·shən }

wind drift [ACOUS] Shift in the apparent position of a sound source or target observed by sound apparatus; it is caused by the effect of wind on sound waves, which changes their direction and increases or decreases sound lag. { 'win ,drift }

wind noise [ACOUS] Noise caused by turbulent airflow over and around an object. { 'win ,nȯiz }

window [ELECTROMAG] A hole in a partition between two cavities or waveguides, used for coupling. { 'win·dō }

wind pressure [MECH] The total force exerted upon a structure by wind. Also known as velocity pressure. { 'win ,presh·ər }

wire [ELEC] A single bare or insulated metallic conductor having solid, stranded, or tinsel construction, designed to carry current in an electric circuit. Also known as electric wire. [OPTICS] A filament, usually consisting of a stretched strand of spider's web or a fine metal wire, mounted in the field of view of a telescope eyepiece to serve as a reference or for measurements. Also known as web. { wīr }

wiregrating [ELECTROMAG] A series of wires placed in a waveguide that allow one or more types of waves to pass and block all others. { 'wīr,grād·iŋ }

wire-wound cryotron [CRYO] A cryotron that consists of a central insulated wire surrounded by a control coil; it is designed so that a relatively small current passed through the control coil produces a magnetic field which makes the gate resistive. { 'wīr ¦waủnd 'krī·ə,trän }

wire-wound potentiometer [ELEC] A potentiometer which is similar to a slide-wire potentiometer, except that the resistance wire is wound on a form and contact is made by a slider which moves along an edge from turn to turn. { 'wīr ¦waủnd pə,ten·chē'äm·əd·ər }

wire-wound resistor [ELEC] A resistor employing as the resistance element a length of high-resistance wire or ribbon, usually Nichrome, wound on an insulating form. { 'wīr ¦waủnd ri'zis·tər }

wire-wound rheostat [ELEC] A rheostat in which a sliding or rolling contact moves over resistance wire that has been wound on an insulating core. { 'wīr ¦waủnd 'rē·ə,stat }

wiring diagram *See* circuit diagram. { 'wīr·iŋ ,dī·ə,gram }

WKB method *See* Wentzel-Kramers-Brillouin method. { ,dəb·əl·yü,kā'bē ,meth·əd }

Wobbe index [THERMO] A measure of the amount of heat released by a gas burner with a constant orifice, equal to the gross calorific value of the gas in British thermal units per cubic foot at standard temperature and pressure divided by the square root of the specific gravity of the gas. { 'wä·bə ,in,deks }

wolf [ACOUS] A dissonant interval which appears when the meantone scale is extended to include chromatic notes. { wủlf }

Wollaston polarizing prism [OPTICS] A device for producing linearly polarized beams of light, consisting of two adjacent quartz wedges with their optic axes perpendicular to each other and to the direction of incident light. { 'wủl·ə·stən 'pō·lə,rīz·iŋ ,priz·əm }

Wood effect [OPTICS] Transparence of alkali metals to ultraviolet light. { 'wủd i,fekt }

work *See* load. [MECH] The transference of energy that occurs when a force is applied to a body that is moving in such a way that the force has a component in the direction of the body's motion; it is equal to the line integral of the force over the path taken by the body. { wərk }

work function [SOLID STATE] The minimum energy needed to remove an electron from the Fermi level of a metal to infinity; usually expressed in electronvolts. *See* free energy. { 'wərk ,fəŋk·shən }

working Q *See* loaded Q. { 'wərk·iŋ 'kyü }

work-kinetic energy theorem [MECH] The theorem that the change in the kinetic energy of a particle during a displacement is equal to the work done by the resultant force on the particle during this displacement. { 'wərk ki'ned·ik ¦en·ər·jē ,thir·əm }

work of adhesion *See* adhesional work. { 'wərk əv ad'hē·zhən }

world [RELAT] Pertaining to Lorentz transformations and four-dimensional space-time, rather than rotations and three-dimensional space, as in world scalar, world vector, world line. { wərld }

world line [RELAT] A path in four-dimensional space-time that represents a continuous sequence of events relating to a given particle. { 'wərld ,līn }

W particle *See* W boson. { 'dəb·əl,yü ,pärd·ə·kəl }

wrench [MECH] The combination of a couple and a force which is parallel to the torque exerted by the couple. { rench }

Wright telescope [OPTICS] A modification of the Schmidt system in which the spherical primary mirror is replaced by an ellipsoidal mirror, and the corrector plate is modified accordingly. { 'rīt ¦tel·ə,skōp }

Wullenweber antenna [ELECTROMAG] An antenna array consisting of two concentric circles of masts, connected to be electronically steerable; used for ground-to-air communication at Strategic Air Command bases. { 'wủl·ən,web·ər an,ten·ə }

wye connection *See* Y network. { 'wī kə,nek·shən }

X *See* siegbahn.

XAFS *See* x-ray absorption fine structure.

XANES *See* x-ray absorption near-edge structure.

x axis [CRYSTAL] A reference axis within a quartz crystal. { 'eks ˌak·səs }

X boson *See* superheavy boson. { 'eks 'bōˌsän }

X coefficient *See* nine-*j* symbol. { 'eks ˌkō·iˌfish·ənt }

X cut [CRYSTAL] A quartz-crystal cut made in such a manner that the *x* axis is perpendicular to the faces of the resulting slab. { 'eks ˌkət }

xenocryst [CRYSTAL] A crystal in igneous rock that resembles a phenocryst and is foreign to the enclosing body of rock. Also known as chadacryst. { 'zēn·əˌkrist }

xenon-135 [NUC PHYS] A radioactive isotope of xenon produced in nuclear reactors; readily absorbs neutrons; half-life is 9.2 hours. { 'zēˌnän ¦wən¦thərd·ē'fīv }

xi hyperon [PART PHYS] Also known as xi particle. **1.** Collective name for the xi-minus and xi-zero particles, which form an isotopic-spin multiplet of quasi-stable baryons, designated Ξ, having a hypercharge of -1, a total isotopic spin of 1/2, a spin of 1/2, positive parity, and an average mass of approximately 1318 megaelectronvolts. Also known as cascade hyperon; cascade particle. **2.** A baryon belonging to any isotopic-spin multiplet having a hypercharge of -1 and a total isotopic spin of 1/2; designated by $\Xi_{JP}(m)$, where m is the mass of the baryon in MeV, and J and P are its spin and parity (if known); the $\Xi_{3/2}+(1530)$ is sometimes designated Ξ^*. { 'zī 'hī·pəˌrän }

xi-minus particle [PART PHYS] A negatively charged xi hyperon, designated Ξ^-. Also known as cascade particle. { 'zī ¦mī·nəs ˌpärd·ə·kəl }

xi particle *See* xi hyperon. { 'zī ˌpärd·ə·kəl }

xi-zero particle [PART PHYS] An uncharged xi hyperon, designated Ξ^0. { 'zī ¦zir·ō ˌpard·ə·kəl }

Y-process [NUC PHYS] The synthesis of certain nuclides in stars through nuclear reactions in which gamma rays remove neutrons, protons, or alpha particles from the nucleus; it is one mode of the *p*-process. { 'wī ˌprä·səs }

XPS *See* x-ray photoelectron spectroscopy.

x-radiation *See* x-rays. { ¦eks ˌrād·ē'ā·shən }

x-ray absorption [ELECTROMAG] The taking up of energy from an x-ray beam by a medium through which the beam is passing. { 'eks ˌrā əb'sórp·shən }

x-ray absorption fine structure [SPECT] The structure in the x-ray absorption spectrum of a substance at energies above the absorption edge, including both the x-ray absorption near-edge structure and the extended x-ray absorption fine structure. Abbreviated XAFS. { ˌeks ˌrā əb¦sorp·shən ¦fīn 'strək·chər }

x-ray absorption near-edge structure [PHYS] A ripplelike structure in the x-ray absorption spectrum of a substance, at energies just above the absorption edge associated with liberation of core electrons into the continuum, and much closer to the absorption edge than the energies associated with extended x-ray absorption fine structure. Abbreviated XANES. { 'eks ˌrā əb'sórp·shən ¦nir ˌej 'strək·chər }

x-ray analysis [PHYS] The use of x-ray radiations to detect heavy elements in the presence of lighter ones, to give critical-edge absorption to identify elemental composition, and to identify crystal structures by diffraction patterns. { 'eks ˌrā ə'nal·ə·səs }

x-ray crystallography [CRYSTAL] The study of crystal structure by x-ray diffraction techniques. Also known as roentgen diffractometry. { 'eks ˌrā ˌkrist·əl'äg·rə·fē }

x-ray crystal spectrometer [SPECT] An instrument designed to produce an x-ray spectrum and measure the wavelengths of its components, by diffracting x-rays from a crystal with known lattice spacing. { 'eks ˌrā 'krist·əl spek'träm·əd·ər }

x-ray diffraction [PHYS] The scattering of x-rays by matter, especially crystals, with accompanying variation in intensity due to interference effects. Also known as x-ray microdiffraction. { 'eks ˌrā di'frak·shən }

x-ray diffraction analysis [CRYSTAL] Analysis of the crystal structure of materials by passing x-rays through them and registering the diffraction (scattering) image of the rays. { 'eks ˌrā di'frak·shən əˌnal·ə·səs }

x-ray emission *See* x-ray fluorescence. { 'eks ˌrā i'mish·ən }

x-ray fluorescence [ATOM PHYS] Emission by a substance of its characteristic x-ray line spectrum upon exposure to x-rays. Also known as x-ray emission. { 'eks ˌrā flu̇'res·əns }

x-ray fluorescence analysis [SPECT] A nondestructive physical method used for chemical elemental analysis in which a material is irradiated by photons or charged particles of sufficient energy to cause its elements to emit (fluoresce) characteristic x-ray line spectra. { 'eks ˌrā flu̇'res·əns əˌnal·ə·səs }

x-ray fluorescent emission spectrometer [SPECT] An x-ray crystal spectrometer used to measure wavelengths of x-ray fluorescence; in order to concentrate beams of low intensity, it has bent reflecting or transmitting crystals arranged so that the theoretical curvature required can be varied with the diffraction angle of a spectrum line. { 'eks ˌrā flu̇'res·ənt i¦mish·ən spek 'träm·əd·ər }

x-ray hardness [ELECTROMAG] The penetrating ability of x-rays; it is an inverse function of the wavelength. { 'eks ˌrā 'härd·nəs }

x-ray holography [ELECTROMAG] The use of holographic techniques to image objects beyond the reach of optical microscopes by using high-intensity coherent sources of electromagnetic radiation with wavelengths between 0.1 and 10 nanometers. Also known as microholography. { 'eks ˌrā hō'läg·rə·fē }

x-ray image spectrography [SPECT] A modification of x-ray fluorescence analysis in which x-rays irradiate a cylindrically bent crystal, and Bragg diffraction of the resulting emissions produces a slightly enlarged image with a resolution of about 50 micrometers. { 'eks ˌrā ¦im·ij spek 'träg·rə·fē }

x-ray irradiation [PHYS] Subjection of a material, object, or patient to x-rays. { 'eks ˌrā iˌrād·ē'ā·shən }

x-ray laser [ELECTROMAG] A device that uses the principle of amplification by stimulated emission of radiation to produce an intense beam of coherent x-rays. { 'eks ˌrā 'lā·zər }

x-ray microdiffraction *See* x-ray diffraction. { 'eks ˌrā ¦mī·krō·di'frak·shən }

x-ray microprobe *See* microprobe. { 'eks ˌrā 'mī·krəˌprōb }

x-ray optics [ELECTROMAG] A title-by-analogy of those phases of x-ray physics in which x-rays demonstrate properties similar to those of light waves. Also known as roentgen optics. { 'eks ˌrā 'äp·tiks }

x-ray photoelectron spectroscopy [SPECT] A form of electron spectroscopy in which a sample is irradiated with a beam of monochromatic x-rays and the energies of the resulting photoelectrons are measured. Abbreviated XPS. Also known as electron spectroscopy for chemical analysis (ESCA). { 'eks ˌrā ¦fōd·ō·i¦lekˌträn spek'träs·kə·pē }

x-ray powder diffractometer *See* powder diffraction camera. { 'eks ˌrā 'pau̇d·ər ˌdiˌfrak'täm·əd·ər }

x-ray powder method *See* powder method. { 'eks ˌrā 'pau̇d·ər ˌmeth·əd }

x-ray projection microscopy *See* projection microradiography. { 'eks ˌrā prə¦jek·shən mī'kräs·kə·pē }

x-rays [PHYS] A penetrating electromagnetic radiation, usually generated by accelerating electrons to high velocity and suddenly stopping them by collision with a solid body, or by inner-shell transitions of atoms with atomic number greater than 10; their wavelengths range from about 10^{-5} angstrom to 10^{3} angstroms, the average wavelength used in research being about 1 angstrom. Also known as roentgen rays; x-radiation. { 'eks ˌrāz }

x-ray spectrograph [SPECT] An x-ray spectrometer equipped with photographic or other recording apparatus; one application is fluorescence analysis. { 'eks ˌrā 'spek·trəˌgraf }

x-ray spectrometer [SPECT] An instrument for producing the x-ray spectrum of a material and measuring the wavelengths of the various components. { 'eks ˌrā spek'träm·əd·ər }

x-ray spectrometry [SPECT] A technique for quantitative analysis of the elemental composition of specimens. Irradiation of a sample by high-energy electrons, protons, or photons ionizes some of the atoms, which then emit characteristic *x*-rays whose wavelength depends on the atomic number of the element and whose intensity is related to the concentration of that element. { 'eks ˌrā spek'träm·ə·trē }

x-ray spectroscopy *See* x-ray spectrometry. { 'eks ˌrā spek'träs·kə·pē }

x-ray spectrum [SPECT] A display or graph of the intensity of x-rays, produced when electrons strike a solid object, as a function of wavelengths or some related parameter; it consists of a continuous bremsstrahlung spectrum on which are superimposed groups of sharp lines characteristic of the elements in the target. { 'eks ˌrā ˌspek·trəm }

x-ray unit *See* siegbahn. { 'eks ˌrā ˌyü·nət }

XU *See* siegbahn.

X unit *See* siegbahn. { 'eks ˌyü·nət }

x,y chromaticity diagram *See* Maxwell triangle. { ¦eks¦wī ˌkrō·mə'tis·əd·ē ˌdī·əˌgram }

Y

Yagi antenna *See* Yagi-Uda antenna. { 'yäg·ē an,ten·ə }

Yagi-Uda antenna [ELECTROMAG] An end-fire antenna array having maximum radiation in the direction of the array line; it has one dipole connected to the transmission line and a number of equally spaced unconnected dipoles mounted parallel to the first in the same horizontal plane to serve as directors and reflectors. Also known as Uda antenna; Yagi antenna. { 'yäg·ē 'üd·ə an,ten·ə }

yag laser *See* yttrium-aluminum-garnet laser. { 'yag ,lā·zər }

Yang-Mills theory [PART PHYS] A theory of nuclear forces between nucleons based on the hypothesis that they can be derived by imposing local isospin invariance; this implies that the interaction must occur through the exchange of three massless vector bosons. { 'yaŋ 'milz ,thē·ə·rē }

yard [MECH] A unit of length in common use in the United States and United Kingdom, equal to 0.9144 meter, or 3 feet. Abbreviated yd. { yärd }

yardage [MECH] An amount expressed in yards. { 'yärd·ij }

yaw [MECH] **1.** The rotational or oscillatory movement of a ship, aircraft, rocket, or the like about a vertical axis. Also known as yawing. **2.** The amount of this movement, that is, the angle of yaw. **3.** To rotate or oscillate about a vertical axis. { yȯ }

yaw acceleration [MECH] The angular acceleration of an aircraft or missile about its normal or Z axis. { 'yȯ ak,sel·ə'rā·shən }

yaw axis [MECH] A vertical axis through an aircraft, rocket, or similar body, about which the body yaws; it may be a body, wind, or stability axis. Also known as yawing axis. { 'yȯ ,ak·səs }

yawing *See* yaw. { 'yȯ·iŋ }

yawing axis *See* yaw axis. { 'yȯ·iŋ ,ak·səs }

y axis [CRYSTAL] A line perpendicular to two opposite parallel faces of a quartz crystal. { 'wī ,ak·səs }

Y circulator [ELECTROMAG] Circulator in which three identical rectangular waveguides are joined to form a symmetrical Y-shaped configuration, with a ferrite post or wedge at its center; power entering any waveguide will emerge from only one adjacent waveguide. { 'wī 'sər·kyə ,lād·ər }

Y connection *See* Y network. { 'wī kə,nek·shən }

Y cut [CRYSTAL] A quartz-crystal cut such that the *y* axis is perpendicular to the faces of the resulting slab. { 'wī ,kət }

Y-delta transformation [ELEC] One of two electrically equivalent networks with three terminals, one being connected internally by a Y configuration and the other being connected internally by a delta transformation. Also known as delta-Y transformation; pi-T transformation. { 'wī 'del·tə ,tranz·fər,mā·shən }

yellow [OPTICS] The hue evoked in an average observer by monochromatic radiation having a wavelength in the approximate range from 577 to 597 nanometers; however, the same sensation can be produced in a variety of other ways. { yel·ō }

yield [MECH] That stress in a material at which plastic deformation occurs. { yēld }

yield point [MECH] The lowest stress at which strain increases without increase in stress. { 'yēld ,pȯint }

yield strength [MECH] The stress at which a material exhibits a specified deviation from proportionality of stress and strain. { 'yēld ,streŋkth }

yield stress [FL MECH] The minimum stress needed to cause a Bingham plastic to flow. [MECH] The lowest stress at which extension of the tensile test piece increases without increase in load. { 'yēld ,stres }

Y junction [ELECTROMAG] A waveguide in which the longitudinal axes of the waveguide form a Y. { 'wī ,jəŋk·shən }

Y network [ELEC] A star network having three branches. Also known as wye connection' Y connection. { 'wī ,net,wərk }

yoke [ELECTROMAG] Piece of ferromagnetic material without windings, which permanently connects two or more magnet cores. { yōk }

Young construction [OPTICS] A graphical procedure for tracing a light ray through a boundary between two media having different refractive indices. { 'yəŋ kən,strək·shən }

Young-Helmholtz laws [MECH] Two laws describing the motion of bowed strings; the first states that no overtone with a node at the point of excitation can be present; the second states that when the string is bowed at a distance of $1/n$ times the string's length from one of the ends, where n is an integer, the string moves back and forth with two constant velocities, one of which has the same direction as that of the bow and is equal to it, while the other has the opposite direction and is $n - 1$ times as large. { ¦yəŋ 'helm,hōlts ,lȯz }

Young's modulus [MECH] The ratio of a simple tension stress applied to a material to the resulting strain parallel to the tension. Also known as modulus of elasticity. { 'yəŋz ,mäj·ə·ləs }

Young's two-slit interference [OPTICS] Interference of light from two parallel slits which are illuminated by light from a single slit, which in turn is illuminated by a source; the interference can be seen by letting the light fall on a screen, which then shows a series of parallel fringes. Also known as double-slit interference; two-slit interference. { 'yəŋz ¦tü ¦slit ,in·tər'fir·əns }

y-process [NUC PHYS] The synthesis of certain nuclides in stars through nuclear reactions in which gamma rays remove neutrons, protons, or alpha particles from the nucleus; it is one mode of the p-process. { 'wī ,prä·səs }

yrast state [NUC PHYS] An energy state of a nucleus whose energy is less than that of any other state with the same spin. { ē'rast ,stāt }

yttrium-aluminum-garnet laser [OPTICS] A four-level infrared laser in which the active material is neodymium ions in an yttrium-aluminum-garnet crystal; it can provide a continuous output power of several watts. Abbreviated yag laser. { 'i·trē·əm ə¦lü·mə·nəm ¦gär·nət 'lā·zər }

Yukawa force [NUC PHYS] The strong, short-range force between nucleons, as calculated on the assumption that this force is due to the exchange of a particle of finite mass (Yukawa meson), just as electrostatic forces are interpreted in quantum electrodynamics as being due to the exchange of photons. { yü'kä·wä ,fȯrs }

Yukawa meson [PART PHYS] A particle, having a finite rest mass, whose exchange between nucleons is postulated to account for the strong, short-range forces between them; such a contributor is the pi meson. { yü'kä·wä ,mā,sän }

Yukawa potential [NUC PHYS] The potential function that is associated with the Yukawa force, with the form $V(r) = -V_0(b/r) \exp(-r/b)$, where r is the distance between the nucleons and V_0 and b are constants, giving measures of the strength and range of the force respectively. { yü'kä·wä pə,ten·chəl }

Z

ZAA spectrometry *See* Zeeman-effect atomic absorption spectrometry. { ¦zē¦ā¦ā spek'träm·ə·trē }

z axis [CRYSTAL] The optical axis of a quartz crystal, perpendicular to both the *x* and *y* axes. { 'zē ˌak·səs }

Z^0 boson [PART PHYS] An intermediate vector boson which has zero electric charge and mediates the neutral current weak interactions. Also known as Z^0 particle. { 'zē¦zir·ō 'bōˌsän }

Z coefficient [QUANT MECH] A coefficient used in the transformation between modes of coupling eigenfunctions of three angular momenta, and especially in calculating matrix elements in beta decay and similar problems. { 'zē ˌkō·iˌfish·ənt }

Zeeman displacement [SPECT] The separation, in wave numbers, of adjacent spectral lines in the normal Zeeman effect in a unit magnetic field, equal (in centimeter-gram-second Gaussian units) to $e/4\pi mc^2$, where e and m are the charge and mass of the electron, or to approximately 4.67×10^{-5} (centimeter)$^{-1}$ (gauss)$^{-1}$. { 'zā·mən diˌsplās·mənt }

Zeeman effect [SPECT] A splitting of spectral lines in the radiation emitted by atoms or molecules in a static magnetic field. { 'zā·mən iˌfekt }

Zeeman-effect atomic absorption spectrometry [SPECT] A type of atomic absorption spectrometry in which either the light source or the sample is placed in a magnetic field, splitting the spectral lines under observation into polarized components, and a rotating polarizer is placed between the source and the sample, enabling the absorption caused by the element under analysis to be separated from background absorption. Abbreviated ZAA spectrometry. { ¦zē·mən iˌfekt ə¦täm·ik əp¦sórp·shən spek'träm·ə·trē }

Zeeman energy [ATOM PHYS] The energy of interaction between an atomic or molecular magnetic moment and an applied magnetic field. { 'zā·mən ˌen·ər·jē }

Zener voltage *See* breakdown voltage. { 'zē·nər ˌvōl·tij }

zenith telescope [OPTICS] A type of telescope that is fixed in the vertical or moves only a small amount from the vertical; it is used to get positional measurement of stars moving near the zenith. { 'zē·nəth ˌtel·əˌskōp }

Zepp antenna [ELECTROMAG] Horizontal antenna which is a multiple of a half-wavelength long and is fed at one end by one lead of a two-wire transmission line that is some multiple of a quarter-wavelength long. { 'zep anˌten·ə }

zero branch [SPECT] A spectral band whose Fortrat parabola lies between two other Fortrat parabolas, with its vertex almost on the wave number axis. { 'zir·ō 'branch }

zero gravity *See* weightlessness. { 'zir·ō 'grav·əd·ē }

zero-point energy [STAT MECH] The kinetic energy retained by the molecules of a substance at a temperature of absolute zero. { 'zir·ō ¦póint 'en·ər·jē }

zero-point entropy [STAT MECH] The entropy that a substance such as glass, which is not in thermodynamic equilibrium, retains at a temperature of absolute zero. { 'zir·ō ¦póint 'en·trə·pē }

zero-point pressure [STAT MECH] The pressure exerted by a degenerate electron gas, whose electrons are in motion even at absolute zero. { 'zir·ō ¦póint 'presh·ər }

zero-point vibration [STAT MECH] The vibrational motion which molecules in a crystal lattice, or particles in any oscillator potential, retain at a temperature of absolute zero; it is quantum-mechanical in origin. Also known as residual vibration. { 'zir·ō ¦póint vī'brā·shən }

zero potential [ELEC] Expression usually applied to the potential of the earth, as a convenient reference for comparison. { 'zir·ō pə'ten·chəl }

zeroth law of thermodynamics [THERMO] A law that if two systems are separately found to be in thermal equilibrium with a third system, the first two systems are in thermal equilibrium with each other, that is, all three systems are at the same temperature. { ¦zirˌōth ˌló əv ˌthər·mō·dī'nam·iks }

zeta potential [PHYS] The electrical potential that exists across the interface of all solids and liquids. Also known as electrokinetic potential. { 'zād·ə pəˌten·chəl }

zigzag reflections [ELECTROMAG] From a layer of the ionosphere, high-order multiple reflections which may be of abnormal intensity; they occur in waves which travel by multihop ionosphere reflections and finally turn back toward their starting point by repeated reflections from a slightly curved or sloping portion of an ionized layer. { 'zig,zag ri,flek·shənz }

zinc-65 [NUC PHYS] A radioactive isotope of zinc, which has a 250-day half-life with beta and gamma radiation; used in alloy-wear tracer studies and body metabolism studies. { 'ziŋk ¦siks·tē'fīv }

zirconium-95 [NUC PHYS] A radioactive isotope of zirconium; half-life of 63 days with beta and gamma radiation; used to trace petroleum-pipeline flows and in the circulation of a catalyst in a cracking plant. { ,zər'kō·nē·əm ¦nīn·tē'fīv }

zitterbewegung [QUANT MECH] An oscillatory motion of an electron suggested in some interpretations of the Dirac electron theory, having a frequency greater than $4\pi mc^2/h$, where m is the electron's mass, c is the speed of light, and h is Planck's constant, or approximately 1.5×10^{21} hertz. { ¦tsid·ər·be'vā,gu̇ŋ }

zone [CRYSTAL] A set of crystal faces which intersect (or would intersect, if extended) along edges which are all parallel. { zōn }

zone axis [CRYSTAL] A line through the center of a crystal which is parallel to all the faces of a zone. { 'zōn 'ak·səs }

zone indices [CRYSTAL] Three integers identifying a zone of a crystal; they are the crystallographic coordinates of a point joined to the origin by a line parallel to the zone axis. { 'zōn 'in·də,sēz }

zone law [CRYSTAL] A law which states that the Miller indices (h, k, l) of any crystal plane lying in a zone with zone indices (u, v, w) satisfy the equation $hu + lv + kw = 0$. { 'zōn ,lȯ }

zone of silence *See* skip zone. { 'zōn əv 'sī·ləns }

zone plate [OPTICS] A plate with alternate transparent and opaque rings, designed to block off every other Fresnel half-period zone; light from a point source passing through the plate produces an intense point image much like that produced by a lens. { 'zōn ,plāt }

zoning [CRYSTAL] A variation in the composition of a crystal from core to margin due to a separation of the crystal phases during its growth by loss of equilibrium in a continuous reaction series. [ELECTROMAG] The displacement of various portions of the lens or surface of a microwave reflector so the resulting phase front in the near field remains unchanged. Also known as stepping. { 'zōn·iŋ }

zoom lens [OPTICS] A system of lenses in which two or more parts are moved with respect to each other to obtain a continuously variable focal length and hence magnification, while the image is kept in the same image plane. Also known as variable-focal-length lens; varifocal lens. { 'züm ¦lenz }

Z^0 particle *See* Z^0 boson. { 'zē ,zir·ō ,pärd·ə·kəl }

Appendix

Equivalents of commonly used units for the U.S. Customary System and the metric system

1 inch = 2.5 centimeters (25 millimeters) 1 foot = 0.3 meter (30 centimeters) 1 yard = 0.9 meter 1 mile = 1.6 kilometers	1 centimeter = 0.4 inch 1 meter = 3.3 feet 1 meter = 1.1 yards 1 kilometer = 0.62 mile	1 inch = 0.083 foot 1 foot = 0.33 yard (12 inches) 1 yard = 3 feet (36 inches) 1 mile = 5280 feet (1760 yards)
1 acre = 0.4 hectare 1 acre = 4047 square meters	1 hectare = 2.47 acres 1 square meter = 0.00025 acre	
1 gallon = 3.8 liters 1 fluid ounce = 29.6 milliliters 32 fluid ounces = 946.4 milliliters	1 liter = 1.06 quarts = 0.26 gallon 1 milliliter = 0.034 fluid ounce	1 quart = 0.25 gallon (32 ounces; 2 pints) 1 pint = 0.125 gallon (16 ounces) 1 gallon = 4 quarts (8 pints)
1 quart = 0.95 liter 1 ounce = 28.35 grams 1 pound = 0.45 kilogram 1 ton = 907.18 kilograms	1 gram = 0.035 ounce 1 kilogram = 2.2 pounds 1 kilogram = 1.1 × 10^{-3} ton	1 ounce = 0.0625 pound 1 pound = 16 ounces 1 ton = 2000 pounds
°F = (1.8 × °C) + 32	°C = (°F − 32) ÷ 1.8	

Appendix

Conversion factors for the U.S. Customary System, metric system, and International System

A. Units of length

Units	*cm*	*m*	*in.*	*ft*	*yd*	*mi*
1 cm	= 1	0.01	0.3937008	0.03280840	0.01093613	6.213712×10^{-6}
1 m	= 100.	1	39.37008	3.280840	1.093613	6.213712×10^{-4}
1 in.	= 2.54	0.0254	1	0.08333333...	0.02777777...	1.578283×10^{-5}
1 ft	= 30.48	0.3048	12.	1	0.3333333...	$1.893939... \times 10^{-4}$
1 yd	= 91.44	0.9144	36.	3.	1	$5.681818... \times 10^{-4}$
1 mi	= 1.609344×10^{5}	1.609344×10^{3}	6.336×10^{4}	5280.	1760.	1

B. Units of area

Units	cm^2	m^2	$in.^2$	ft^2	yd^2	mi^2
1 cm^2	= 1	10^{-4}	0.1550003	1.076391×10^{-3}	1.195990×10^{-4}	3.861022×10^{-11}
1 m^2	= 10^4	1	1550.003	10.76391	1.195990	3.861022×10^{-7}
1 $in.^2$	= 6.4516	6.4516×10^{-4}	1	$6.944444... \times 10^{-3}$	7.716049×10^{-4}	2.490977×10^{-10}
1 ft^2	= 929.0304	0.09290304	144.	1	0.1111111...	3.587007×10^{-8}
1 yd^2	= 8361.273	0.8361273	1296.	9.	1	3.228306×10^{-7}
1 mi^2	= 2.589988×10^{10}	2.589988×10^{6}	4.014490×10^{9}	2.78784×10^{7}	3.0976×10^{6}	1

C. Units of volume

Units	m^3	cm^3	liter	$in.^3$	ft^3	qt	gal
1 m^3	= 1	10^6	10^3	6.102374×10^4	35.31467×10^{-3}	1.056688	264.1721
1 cm^3	= 10^{-6}	1	10^{-3}	0.06102374	3.531467×10^{-5}	1.056688×10^{-3}	2.641721×10^{-4}
1 liter	= 10^{-3}	1000.	1	61.02374	0.03531467	1.056688	0.2641721
1 $in.^3$	= 1.638706×10^{-5}	16.38706	0.01638706	1	5.787037×10^{-4}	0.01731602	4.329004×10^{-3}
1 ft^3	= 2.831685×10^{-2}	28316.85	28.31685	1728.	1	2.992208	7.480520
1 qt	= 9.46352×10^{-4}	946.359	0.946351	57.75	0.03342014	1	0.25
1 gal (U.S.)	= 3.785412×10^{-3}	3785.412	3.785412	231.	0.1336806	4.	1

D. Units of mass

Units	g	kg	oz	lb	metric ton	ton
1 g	= 1	10^{-3}	0.03527396	2.204623×10^{-3}	10^{-6}	1.102311×10^{-6}
1 kg	= 1000.	1	35.27396	2.204623	10^{-3}	1.102311×10^{-3}
1 oz (avdp)	= 28.34952	0.02834952	1	0.0625	2.834952×10^{-5}	3.125×10^{-5}
1 lb (avdp)	= 453.5924	0.4535924	16.	1	4.535924×10^{-4}	$5. \times 10^{-4}$
1 metric ton	= 10^8	1000.	35273.96	2204.623	1	1.102311
1 ton	= 907184.7	907.1847	32000.	2000.	0.9071847	1

Conversion factors for the U.S. Customary System, metric system, and International System (*cont.*)

E. Units of density

Units	$g \cdot cm^{-3}$	$g \cdot L^{-1}$, $kg \cdot m^{-3}$	$oz \cdot in.^{-3}$	$lb \cdot in.^{-3}$	$lb \cdot ft^{-3}$	$lb \cdot gal^{-1}$
1 $g \cdot cm^{-3}$	= 1	1000.	0.5780365	0.03612728	62.42795	8.345403
1 $g \cdot L^{-1}$, $kg \cdot m^{-3}$	= 10^{-3}	1	5.780365×10^{-4}	3.612728×10^{-5}	0.06242795	8.345403×10^{-3}
1 $oz \cdot in.^{-3}$	= 1.729994	1729.994	1	0.0625	108.	14.4375
1 $lb \cdot in.^{-3}$	= 27.67991	27679.91	16.	1	1728.	231.
1 $lb \cdot ft^{-3}$	= 0.01601847	16.01847	9.259259×10^{-3}	5.787037×10^{-4}	1	0.1336806
1 $lb \cdot gal^{-1}$	= 0.1198264	119.8264	4.749536×10^{-3}	4.329004×10^{-3}	7.480519	1

F. Units of pressure

Units	Pa, $N \cdot m^{-2}$	$dyn \cdot cm^{-2}$	bar	atm	$kgf \cdot cm^{-2}$	mmHg (torr)	in. Hg	$lbf \cdot in.^{-2}$
1 Pa, 1 $N \cdot m^{-2}$	= 1	10	10^{-5}	9.869233×10^{-6}	1.019716×10^{-5}	7.500617×10^{-3}	2.952999×10^{-4}	1.450377×10^{-4}
1 $dyn \cdot cm^{-2}$	= 0.1	1	10^{-6}	9.869233×10^{-7}	1.019716×10^{-6}	7.500617×10^{-4}	2.952999×10^{-5}	1.450377×10^{-5}
1 bar	= 10^5	10^6	1	0.9869233	1.019716	750.0617	29.52999	14.50377
1 atm	= 101325	101325.0	1.01325	1	1.033227	760.	29.92126	14.69595
1 $kgf \cdot cm^{-2}$	= 98066.5	980665	0.980665	0.9678411	1	735.5592	28.95903	14.22334
1 mmHg (torr)	= 133.3224	1333.224	1.333224×10^3	1.315789×10^{-3}	1.359510×10^{-3}	1	0.03937008	0.01933678
1 in. Hg	= 3386.388	33863.88	0.03386388	0.03342105	0.03453155	25.4	1	0.4911541
1 $lbf \cdot in.^{-2}$	= 6894.757	68947.57	0.06894757	0.06804596	0.07030696	51.71493	2.036021	1

G. Units of energy

Units	g mass (energy equiv.)	J	eV	cal	cal_{IT}	Btu_{IT}	kWh	hp-h	ft-lbf	$ft^3 \cdot lbf \cdot in.^{-2}$	liter-atm
1 g mass (energy equiv.)	= 1	8.987552×10^{13}	5.609589×10^{32}	2.148076×10^{3}	2.146640×10^{13}	8.518555×10^{10}	2.496542×10^{7}	3.347918×10^{7}	6.628878×10^{13}	4.603388×10^{11}	8.870024×10^{11}
1 J	= 1.112650×10^{-14}	1	6.241510×10^{18}	0.2390057	0.2388459	9.478172×10^{-4}	$2.777777... \times 10^{-7}$	3.725062	0.7375622	5.121960×10^{-3}	9.869233×10^{-3}
1 eV	= 1.782662×10^{-33}	1.602176×10^{-19}	1	3.829293×10^{-20}	3.826733×10^{-20}	1.518570×10^{-22}	4.450490×10^{-26}	5.968206×10^{-26}	1.181705×10^{-19}	8.206283×10^{-22}	1.581225×10^{-21}
1 cal	= 4.655328×10^{-14}	4.184	2.611448×10^{19}	1	0.9993312	3.965667×10^{-3}	$1.1622222... \times 10^{-6}$	1.558562×10^{-6}	3.085960	2.143028×10^{-2}	0.04129287
1 cal_{IT}	= 4.658443×10^{-14}	4.1868	2.613195×10^{19}	1.000669	1	3.968321×10^{-3}	1.163×10^{-6}	1.559609×10^{-6}	3.088025	2.144462×10^{-2}	0.04132050
1 Btu_{IT}	= 1.173908×10^{-11}	1055.056	6.585141×10^{21}	252.1644	251.9958	1	2.930711×10^{-4}	3.930148×10^{-4}	778.1693	5.403953	10.41259
1 kWh	= 4.005540×10^{-8}	3600000.	2.246944×10^{25}	860420.7	859845.2	3412.142	1	1.341022	2655224.	18349.06	35529.24
1 hp-h	= 2.986931×10^{-8}	2384519.	1.675545×10^{25}	641615.6	641186.5	2544.33	0.7456998	1	1980000.	13750.	26494.15
1 ft-lbf	= 1.508551×10^{-14}	1.355818	8.462351×10^{18}	0.3240483	0.3238315	1.285067×10^{-3}	3.766161×10^{-7}	$5.050505... \times 10^{-7}$	1	$6.944444... \times 10^{-3}$	0.01338088
1 ft^3 lbf · $in.^{-2}$	= 2.172313×10^{-12}	195.2378	1.218579×10^{21}	46.66295.	46.63174	0.1850497	5.423272×10^{-5}	$7.272727... \times 10^{-5}$	144.	1	1.926847
1 liter-atm	= 1.127393×10^{-12}	101.325	6.324210×10^{20}	24.21726	24.20106	0.09603757	2.814583×10^{-5}	3.774419×10^{-5}	74.73349	0.5189825	1

Appendix

Internal energy and generalized work

Type of energy	*Intensive factor*	*Extensive factor*	*Element of work*
Mechanical			
Expansion	Pressure (P)	Volume (V)	$-PdV$
Stretching	Surface tension (γ)	Area (A)	γdA
Extension	Tensile stretch (F)	Length (l)	Fdl
Thermal	Temperature (T)	Entropy (S)	TdS
Chemical	Chemical potential (μ)	Amount (n)	μdn
Electrical	Electric potential (E)	Charge (Q)	EdQ
Gravitational	Gravitational field strength (mg)	Height (h)	$mgdh$
Polarization			
Electrostatic	Electric field strength ($\mathcal{E}$)	Total electric polarization (dP)	$\mathcal{E}dP$
Magnetic	Magnetic field strength ($\mathcal{H}$)	Total magnetic polarization (M)	$\mathcal{H}dM$

Appendix

The fundamental particles[a]

Gauge bosons $J^P_C = 1^-_-$ Self-conjugate except $\overline{W^+} = W^-$.

Name	Symbol	Charge[b]	Mass and width, GeV	Couplings
Photon	γ	0	0	$A \Rightarrow \gamma A$
Gluon[c]	g	0	0	$A \Rightarrow gA'$
Weak bosons				
Charged	$W^\pm$	± 1	80.4, 2.1	$U \Rightarrow W + D$
Neutral	Z^0	0	91.2, 2.5	$A \Rightarrow Z^0A$

Fermions[d] $J = 1/2$ All have distinct antiparticles, except perhaps the neutrinos.

Name	Charge[b]	Symbol and mass, GeV	Symbol and mass, GeV	Symbol and mass, GeV
Leptons				
Neutrinos	0	ν_e $<3 \times 10^{-9}$	ν_μ $<3 \times 10^{-9}$	ν_τ $<3 \times 10^{-9}$
Charged leptons	-1	e 0.00051	μ 0.106[e]	τ 1.78[e]
Quarks[c]				
Up type	$2/3$	u 0.0015–0.0045	c 1.0–1.4	t 170–180[f]
Down type	$-1/3$	d 0.005–0.0085	s 0.08–0.155	b 4.0–4.5

[a]The graviton, with $J^P_C = 2^+_+$, has been omitted, since it plays no role in high-energy particle physics.
[b]In units of the proton charge.
[c]The gluon is a color SU_3 octet (8); each quark is a color triplet (3). These colored particles are confined constituents of hadrons; they do not appear as free particles.
[d]The three known families (generations) of fermions are displayed in three columns.
[e]The μ and τ leptons are unstable.
[f]The t quark has a width $\approx$ 2 GeV, with dominant decay to Wb.

Appendix

List of frequently occurring dimensionless groups*

Symbol and name	*Definition*	*Notation*†	*Physical significance*
Ar; Archimedes number	$d_p^3 g \rho_f(\rho_s - \rho_f)/\mu^2$		(Inertia force) (gravity force)/(Viscous force)^2
Bi; Biot number	hl/k_s	k_s = thermal conductivity of solid, $ML/T^3\theta$	(Internal thermal resistance)/(Surface thermal resistance)
Bi_m; mass transport Biot number	$k_m L/D_i$	L = Layer thickness, L D_i = interface diffusivity, L^2/T	(Mass transport conductivity at solid/fluid interface)/(Internal transport conductivity of solid wall of thickness L)
Bo; Bond number (also Eo, Eötvös number)	$d^2 g(\rho - \rho_f)/\gamma$	d = bubble or droplet diameter, L	(Gravity force)/(Surface tension force)
Bq; Boussinesq number	$v/(2gm)^{1/2}$	m = mean hydraulic depth of open channel, L	(Inertia force)$^{1/2}$/(Gravity force)$^{1/2}$
Dn; Dean number	$(d/2R)^{1/2}(Re)$	d = pipe diameter, L R = radius of curvature of channel centerline, L Re = Reynolds number	Effect of centrifugal force on flow in a curved pipe
Ec; Eckert number	$v^2/C_p\Delta\theta$	$\Delta\theta$ = temperature difference, θ	Used in study of variable-temperature flow
Ek; Ekman number	$(\nu/2\omega l^2)^{1/2}$		(Viscous force)$^{1/2}$/(Coriolis force)$^{1/2}$
f; Fanning friction factor	$2\tau_w/\rho v^2 = D\Delta p/2L\rho v^2$	τ_w = wall shear stress, M/LT^2 D = pipe diameter, L L = pipe length, L	(Wall shear stress)/(Velocity head)
Fo_f, Fourier flow number	$\nu t/l^2$	—	Used in undimensionalization
Fo; Fourier flow number	$\alpha t/l^2$	—	Indicates the extent of thermal penetration in unsteady-state heat transport
Fo_m; mass transport Fourier number	$k_m t/l$	—	Indicates the extent of substance penetration in unsteady-state mass transport
Fr; Froude number	$v/(gl)^{1/2}$	—	(Inertia force)$^{1/2}$/(Gravity force)$^{1/2}$
Ga; Galileo number	$l^3 g/\nu^2$	—	(Inertia force)(gravity force)/(Viscous force)^2
Gz; Graetz number	$\dot{m}C_p/k_f l$	$\dot{m}$ = mass flow rate, M/T C_p = specific heat capacity (constant pressure), $L^2/T^2\theta$	(Fluid thermal capacity)/(Thermal energy transferred by conduction)
Gr; Grashof number	$l^3 g\Delta\rho/\rho_f\nu^2$	$\Delta\rho$ = density driving force, M/L^3	(Inertia force)(buoyancy force)/(Viscous force)^2
Gr_m; mass transport Grashof number	$l^3 g\beta_c\Delta c/\nu^2$	β_c = volumetric expansion coefficient Δc = concentration driving force	(Inertia force)(buoyancy force)/(Viscous force)^2
Ha; Hartmann number (M)	$lB(\sigma/\mu)^{1/2}$	B = magnetic flux density, M/QT σ = electrical conductivity, Q^2T/ML^3	(Magnetically induced stress)$^{1/2}$/(Viscous shear stress)$^{1/2}$
Kn; Knudsen number	λ/l	λ = length of mean free path	Used in study of low-pressure gas flow
Le; Lewis number	D/α	—	(Molecular diffusivity)/(Thermal diffusivity)
Lu; Luikov number	$k_m l/\alpha$	—	(Mass diffusivity)/(Thermal diffusivity)
Ly; Lykoudis number	—	—	(Hartman number)^2/(Grashof number)$^{1/2}$
Ma; Mach number	v/v_s	v_s = velocity of sound in fluid, L/T	(Linear velocity)/(Velocity of sound)

Ne; Newton number	$F/\rho f v^2 l^2$	F = hydrodynamic drag force, ML/T^2	(Resistance force)/(Inertia force)
Nu; Nusselt number	hl/k_f	—	(Thermal energy transport in forced convection)/(Thermal energy transport if it occurred by conduction)
Pe; Peclet number	lv/α	—	(Reynolds number)(Prandtl number); (Bulk thermal energy transport in forced convection)/(Thermal energy transport by conduction)
Pe_m; mass transport Peclet number	lv/D	—	(Bulk mass transport)/(Diffusional mass transport)
Ps; Poiseuille number	$v\mu/gd_p^2(\rho_s - \rho f)$	—	(Viscous force)/(Gravity force)
Pr; Prandtl number	ν/α	—	(Momentum diffusivity)/Thermal diffusivity)
Ra; Rayleigh number			(Grashof number) (Prandtl number)
Ra_m; mass transport Rayleigh number			(Mass transport Grashof number) (Schmidt number)
Re; Reynolds number	vl/ν	—	
Re_R; rotational Reynolds number	l^2N/ν	N = rate of rotation (revolution per time), $1/T$}	(Inertia force)/(Viscous force)
Sc; Schmidt number	ν/D	—	(Momentum diffusivity)/(Molecular diffusivity)
Sh; Sherwood number	$k_m l/D$	—	(Mass diffusivity)/(Molecular diffusivity)
St; Stanton number	$h/C_p\rho v$	—	(Nusselt number)/(Reynolds number) (Prandtl number); (Thermal energy transferred)/(Fluid thermal capacity)
St_m; mass transport Station number	k_m/v	—	(Sherwood number)/(Reynolds number) (Schmidt number)
Sk; Stokes number	$l^2\Delta p/\mu\nu$	—	(Pressure force)/(Viscous force)
Sr; Strouhal number	fl/v	f = oscillation frequency, $1/T$	Used in study of unsteady flow
Su; Suratnam number	$\rho l/\gamma/\mu_f^2$	—	(Inertia force)(surface tension force)/(Viscous force)^2
Ta; Taylor number	$\omega^2\gamma\bar{r}(r_o - r_i)^3/\nu^2$	r_o = outer radius, L; r_i = inner radius, L; $\bar{r}$ = mean radius, L	Criterion for Taylor vortex stability in rotating concentric cylinder systems
We; Weber number	$v(\rho_f 1/\gamma)^{1/2}$	—	(Inertia force)$^{1/2}$/(Surface tension force)$^{1/2}$

*In many cases, roots of, or power functions of, the groups listed here may be designated by the same names in the technical literature.

†Names of quantities are followed by their dimensions. Fundamental dimensions are taken to be length [L], mass [M], time [T], temperature [θ], and electrical charge [Q]. Notation for quantities that appear only once is given in the table. The following quantities appear more than once:

C_p = specific heat at constant pressure, $L^2/T^2\theta$
d_p = particle diameter, L
D = molecular diffusivity, L^2/T
g = acceleration due to gravity = weight/mass ratio, L/T^2
h = heat transfer coefficient, $M/T^3\theta$
k_f = thermal conductivity of fluid, $ML/T^3\theta$
k_m = mass transfer coefficient, L/T

l = characteristic length, L
$\Delta\rho$ = pressure drop, M/LT^2
t = time, T
v = characteristic velocity, L/T
α = thermal diffusivity = $k_f/\rho_f C_p$, L^2/T
γ = surface tension, M/T^2

μ = dynamic or absolute viscosity, M/LT
ν = kinematic viscosity = μ/ρ_f, L^2/T
ρ = bubble or droplet density, M/L^3
ρ_f = fluid density, M/L^3
ρ_s = solid density, M/L^3
ω = angular velocity of fluid, 1/T

SOURCE: After N. P. Cheremisinoff (ed.), *Encyclopedia of Fluid Mechanics*, vol. 1: *Flow Phenomena and Measurement*, Gulf Publishing Co., Houston, copyright © 1986. Used with permission.

Appendix

Dimensional formulas of common quantities

Quantity	*Definition*	*Dimensional formula*
Mass	Fundamental	M
Length	Fundamental	L
Time	Fundamental	T
Velocity	Distance/time	LT^{-1}
Acceleration	Velocity/time	LT^{-2}
Force	Mass × acceleration	MLT^{-2}
Momentum	Mass × velocity	MLT^{-1}
Energy	Force × distance	ML^2T^{-2}
Angle	Arc/radius	1
Angular velocity	Angle/time	T^{-1}
Angular acceleration	Angular velocity/time	T^{-2}
Torque	Force × lever arm	ML^2T^{-2}
Angular momentum	Momentum × lever arm	ML^2T^{-1}
Moment of inertia	Mass × radius squared	ML^2
Area	Length squared	L^2
Volume	Length cubed	L^3
Density	Mass/volume	ML^{-3}
Pressure	Force/area	$ML^{-1}T^{-2}$
Action	Energy × time	ML^2T^{-1}
Viscosity	Force per unit area per unit velocity gradient	$ML^{-1}T^{-1}$

Defining fixed points of the International Temperature Scale of 1990 (ITS-90)

Equilibrium state	*Temperature*		
	K	°C	°F
Vapor pressure equation of helium	3 to 5	−270.15 to −268.15	−454.27 to 450.67
Triple point of equilibrium hydrogen	13.80	−259.35	−434.82
Vapor pressure point of equilibrium hydrogen (or constant volume gas thermometer point of helium)	≈17 ≈20.3	≈−256.15 ≈−252.85	−429.07 −423.13
Triple point of neon	24.56	−248.59	−415.47
Triple point of oxygen	54.36	−218.79	−361.82
Triple point of argon	83.81	−189.34	−308.82
Triple point of mercury	234.32	−38.83	−37.90
Triple point of water	273.16	0.01	32.02
Melting point of gallium	302.91	29.76	85.58
Freezing point of indium	429.75	156.60	313.88
Freezing point of tin	505.08	231.93	449.47
Freezing point of zinc	692.68	419.53	787.15
Freezing point of aluminum	933.47	660.32	1220.58
Freezing point of silver	1234.93	961.78	1763.20
Freezing point of gold	1337.33	1064.18	1947.52
Freezing point of copper	1357.77	1084.62	1984.32

Appendix

Primary thermometry methods

Method	Approximate useful range of T, K	Principal measured variables	Relation of measured variables to T	Remarks
Gas thermometry	1.3–950	Pressure P and volume V	Ideal gas law plus correction; $PV \propto k_B T$ plus corrections	Careful determination of corrections necessary, but capable of high accuracy
Acoustic interferometry	1.5–3000	Speed of sound W	$W^2 \propto k_B T$ plus corrections	(as above)
Magnetic thermometry 1. Electron paramagnetism 2. Nuclear paramagnetism	 0.001–35 0.000001–1	Magnetic susceptibility	Curie's law plus corrections: $\chi \propto 1/k_B T$ plus corrections	(as above)
Gamma-ray anisotropy or nuclear orientation thermometry	0.01–1	Spatial distribution of gamma-ray emission	Spatial distribution related to Boltzmann factor for nuclear spin states	Useful standard for $T < 1$ K
Thermal electric noise thermometry 1. Josephson junction point contact 2. Conventional amplifier	 0.001–1 4–1400	Mean square voltage fluctuation $\overline{V}^2$	Nyquist's law: $\overline{V}^2 \propto k_B T$	Other sources of noise serious problem for $T > 4$ K
Radiation thermometry (visual, photoelectric, or photodiode)	500–50,000	Spectral intensity J at wavelength λ	Planck's radiation law, related to Boltzmann factor for radiation quanta	Needs blackbody conditions or well-defined emittance
Infrared spectroscopy	100–1500	Intensity I of rotational lines of light molecules	Boltzmann factor for rotational levels related to I	Also Doppler line broadening ($\propto \sqrt{k_B T}$) useful; principal applications to plasmas and astrophysical observations; proper sampling, lack of equilibrium, atmospheric absorption often problems
Ultraviolet and x-ray spectroscopy	5000–2,000,000	Emission spectra from ionized atoms—H, He, Fe, Ca, and so on	Boltzmann factor for electron states related to band structure and line density	(as above)

Appendix

Principal spectral regions and fields of spectroscopy

Spectral region	*Approx. wave-length range*	*Typical source*	*Typical detector*	*Energy transitions studied in matter*
Gamma	1–100 pm	Radioactive nuclei	Geiger counter; scintillation counter	Nuclear transitions and disintegrations
X-rays	6 pm–100 nm	X-ray tube (electron bombardment of metals)	Geiger counter	Ionization by inner electron removal
Vacuum ultraviolet	10–200 nm	High-violet discharge; high-vacuum spark	Photomultiplier	Ionization by outer electron removal
Ultraviolet	200–400 nm	Hydrogen-discharge lamp	Photomultiplier	Excitation of valence electrons
Visible	400–800 nm	Tungsten lamp	Phototubes	Excitation of valence electrons
Near-infrared	0.8–2.5 μm	Tungsten lamp	Photocells	Excitation of valence electrons; molecular vibrational overtones
Infrared	2.5–50 μm	Nernst glower; Globar lamp	Thermocouple; bolometer	Molecular vibrations; stretching, bending, and rocking
Far-infrared	50–1000 μm	Mercury lamp (high-pressure)	Thermocouple; bolometer	Molecular rotations
Microwave	0.1–30 cm	Klystrons; magnetrons	Silicon-tungsten crystal; bolometer	Molecular rotations; electron spin resonance
Radio-frequency	10^{-1}–10^{3} m	Radio transmitter	Radio receiver	Molecular rotations; nuclear magnetic resonance

Appendix

Types of particulate systems

System			Hydrosol	Aerosol	Powder
Continuous phase		Solid	Liquid	Gas	None (or gas)
Dispersed or particulate phase	Gas	Sponge	Foam	—	—
	Liquid	Gel	Emulsion	Mist Spray Fog Rain	—
	Solid	Alloy	Slurry Suspension	Fume Dust Snow Hail	Single phase
					Multiphase (ores, flour)

Density and speed of sound in selected materials*

Gases	*Density* (ρ_0), *kg/m*^{3}	*Speed of sound* (c), *m/s*
Air	1.21	343
Oxygen (0°C; 32°F)	1.43	317
Hydrogen (0°C; 32°F)	0.09	1,270

Liquids	*Density* (ρ_0), *kg/m*^{3}	*Speed of sound* (c), *m/s*
Water	998	1,481
Seawater (13°C; 55°F)	1,026	1,500
Ethyl alcohol	790	1,150
Mercury	13,600	1,450
Glycerin	1,260	1,980

Solids	*Density* (ρ_0), *kg/m*^{3}	*Speed of sound* (c), *m/s*	
		bar	*bulk*
Aluminum	2,700	5,150	6,300
Brass	8,500	3,500	4,700
Lead	11,300	1,200	2,050
Steel	7,700	5,050	6,100
Glass	2,300	5,200	5,600
Lucite	1,200	1,800	2,650
Concrete	2,600	—	3,100

*Temperature = 20°C = 68°F unless otherwise indicated.
Pressure = 1 atm = 101.325 kPa.

Appendix

Recommended values (1998) of selected fundamental physical constants

Quantity	*Symbol**	*Numerical value†*	*Units†*	*Relative uncertainty (standard deviation)*
Speed of light in vacuum	c	299792458	m/s	(defined)
Permeability of vacuum	μ_0	$4\pi \times 10^{-7}$	N/A^2	(defined)
Permittivity of vacuum	ε_0	8.854187817. . .	10^{-12} F/m	(defined)
Constant of gravitation	G	6.673 (10)	10^{-11} $m^3/(kg \cdot s^2)$	1.5×10^{-3}
Planck constant	h	6.62606876 (52)	10^{-34} J · s	7.8×10^{-8}
Elementary charge	e	1.602176462 (63)	10^{-19} C	3.9×10^{-8}
Magnetic flux quantum, $h/(2e)$	Φ_0	2.067833636 (81)	10^{-15} Wb	3.9×10^{-8}
Fine-structure constant,	α	7.297352533 (27)	10^{-3}	3.7×10^{-9}
$\mu_0 c e^2/(2h)$	α^{-1}	137.03599976 (50)		3.7×10^{-9}
Electron mass	m_e	9.10938188 (72)	10^{-31} kg	7.9×10^{-8}
Proton mass	m_p	1.67262158 (13)	10^{-27} kg	7.9×10^{-8}
Neutron mass	m_n	1.67492716 (13)	10^{-27} kg	7.9×10^{-8}
Proton-electron mass ratio	m_p/m_e	1836.1526675 (39)		2.1×10^{-9}
Rydberg constant, $m_e c\alpha^2/(2h)$	R_∞	10973731.568549 (83)	m^{-1}	7.6×10^{-12}
Bohr radius, $\alpha/(4\pi R_\infty)$	a_0	5.291772083 (19)	10^{-11} m	3.7×10^{-9}
Compton wavelength of the electron, $h/(m_e c) = \alpha^2/(2R_\infty)$	λ_c	2.426310215 (18)	10^{-12} m	7.3×10^{-9}
Classical electron radius, $\mu_0 e^2/(4\pi m_e) = \alpha^3/(4\pi R_\infty)$	r_e	2.817940285 (31)	10^{-15} m	1.1×10^{-8}
Bohr magneton, $eh/(4\pi m_e)$	μ_B	9.27400899 (37)	10^{-24} J/T	4.0×10^{-8}
Electron magnetic moment μ_e		−9.28476362 (37)	10^{-24} J/T	4.0×10^{-8}
Electron magnetic moment/Bohr magneton ratio	μ_e/μ_B	−1.0011596521869 (41)		4.1×10^{-12}
Nuclear magneton, $eh/(4\pi m_p)$	μ_N	5.05078317 (20)	10^{-27} J/T	4.0×10^{-8}
Proton magnetic moment/nuclear magneton ratio	μ_p/μ_N	2.792847337 (29)		1.0×10^{-8}
Avogadro constant	N_A	6.02214199 (47)	10^{23}	7.9×10^{-8}
Faraday constant, $N_A e$	F	96485.3415 (39)	C/mol	4.0×10^{-8}
Molar gas constant	R	8.314472 (15)	J/(mol · K)	1.7×10^{-6}
Boltzmann constant, R/N_A	k	1.3806503 (24)	10^{-23} J/K	1.7×10^{-6}

*A = ampere, C = coulomb, F = farad, J = joule, kg = kilogram, K = kelvin, m = meter, mol = mole, N = newton, s = second, T = tesla, Wb = weber.

†Recommended by CODATA Task Group on Fundamental Constants. Digits in parentheses represent one-standard-deviation uncertainties in final two digits of quoted value.

Appendix

Electromagnetic spectrum

Frequency, Hz	*Wavelength, m*	*Nomenclature*	*Typical source*
10^{23}	3×10^{-15}	Cosmic photons	Astronomical
10^{22}	3×10^{-14}	γ-rays	Radioactive nuclei
10^{21}	3×10^{-13}	γ-rays, x-rays	
10^{20}	3×10^{-12}	x-rays	Atomic inner shell, positron-electron annihilation
10^{19}	3×10^{-11}	Soft x-rays	Electron impact on a solid
10^{18}	3×10^{-10}	Ultraviolet, x-rays	Atoms in sparks
10^{17}	3×10^{-9}	Ultraviolet	Atoms in sparks and arcs
10^{16}	3×10^{-8}	Ultraviolet	Atoms in sparks and arcs
10^{15}	3×10^{-7}	Visible spectrum	Atoms, hot bodies, molecules
10^{14}	3×10^{-6}	Infrared	Hot bodies, molecules
10^{13}	3×10^{-5}	Infrared	Hot bodies, molecules
10^{12}	3×10^{-4}	Far-infrared	Hot bodies, molecules
10^{11}	3×10^{-3}	Microwaves	Electronic devices
10^{10}	3×10^{-2}	Microwaves, radar	Electronic devices
10^{9}	3×10^{-1}	Radar	Electronic devices, interstellar hydrogen
10^{8}	3	Television, FM radio	Electronic devices
10^{7}	30	Short-wave radio	Electronic devices
10^{6}	300	AM radio	Electronic devices
10^{5}	3000	Long-wave radio	Electronic devices
10^{4}	3×10^{4}	Induction heating	Electronic devices
10^{3}	3×10^{5}		Electronic devices
100	3×10^{6}	Power	Rotating machinery
10	3×10^{7}	Power	Rotating machinery
1	3×10^{8}		Commutated direct current
0	Infinity	Direct current	Batteries

Types of radioactivity

Type	*Symbol*	*Particles emitted*	*Change in atomic number, ΔZ*	Change in atomic mass number, ΔA
Alpha	α	Helium nucleus	−2	−4
Beta negatron	β^-	Negative electron and antineutrino[a]	+1	0
Beta positron	β^+	Positive electron and neutrino[a]	−1	0
Electron capture	EC	Neutrino[a]	−1	0
Isomeric transition[b]	IT	Gamma rays or conversion electrons or both (and positive-negative electron pair)[c]	0	0
Proton	p	Proton	−1	−1
Spontaneous fission (hot)	SF	Two intermediate-mass nuclei and 1–10 neutrons	Various	Various
Spontaneous fission (cold)	SF	Two intermediate-mass nuclei (zero neutrons)	Various	Various
Ternary spontaneous fission (hot)	TSF	Two intermediate-mass nuclei, a light particle (^{2}H, α, up to ^{10}Be), and neutrons	Various	Various
Ternary spontaneous fission (cold)	TSF	Two intermediate-mass nuclei and a light particle	Various	Various
Isomeric spontaneous fission	ISF	Heavy fragments and neutrons	Various	Various
Beta-delayed spontaneous fission	$(EC + \beta^+)SF$	Positive electron, neutrino, heavy fragments, and neutrons	Various	Various
	β^-SF	Negative electron, antineutrino heavy fragments, and neutrons	Various	Various
Beta-delayed neutron	$\beta^- n$	Negative electron, and antineutrino, neutron	+1	−1
Beta-delayed two-neutron (three-, four-neutron)	$\beta^- 2n(3n, 4n)$	Negative electron, antineutrino, and two (three, four) neutrons	+1	−2(−3,−4)
Beta-delayed proton	β^+p or $(\beta^+ + EC)p$	Positive electron, neutrino, and proton	−2	−1

Beta-delayed two-proton	$\beta^+ 2p$	Positive electron, neutrino, and two protons	−3	−2
Beta-delayed triton	β_1^{-3}H	Negative electron, antineutrino, and triton	0	−3
Beta-delayed alpha	$\beta^+\alpha$	Positive electron, neutrino, and alpha	−3	−4
	$\beta^-\alpha$	Negative electron, antineutrino, and alpha	−1	−4
Beta-delayed alpha-neutron	$\beta^-\alpha,n$	Negative electron, antineutrino, alpha, and neutron	−1	−5
Double beta decay	$\beta^-\beta^-$	Two negative electrons and two antineutrinos	+2	0
	$\beta^+\beta^+$	Two positive electrons and two neutrinos	−2	0
Double electron capture[d]	EC EC	Two neutrinos	−2	0
Neutrinoless double beta decay[e]	$\beta^-\beta^-$	Two negative electrons	+2	0
Two-proton[d]	$2p$	Two protons	−2	−2
Neutron[d]	n	Neutron	0	−1
Two-neutron	$2n$	Two neutrons	0	−2
Heavy clusters[f]	${}^{14}_{6}$C	${}^{14}_{6}$C nucleus	−6	−14
	${}^{20}_{8}$O	${}^{20}_{8}$O nucleus	−8	−20
	${}^{24}_{10}$Ne	${}^{24}_{10}$Ne nucleus	−10	−24

[a]The neutrinos and antineutrinos emitted in beta decay are electron neutrinos and antineutrinos.
[b]Excited states with relatively long measured half-lives are called isomeric.
[c]Pair emission occurs as an additional competing decay mode when the decay energy exceeds 1.022 MeV.
[d]Theoretically predicted but not established experimentally.
[e]Theoretically possible and predicted in some grand unified theories.
[f]There are other possible clusters in addition to those shown. Cold spontaneous fission is also a type of cluster radioactivity.

Appendix

Physical properties of crystalline optical materials

Material	Symbol	Refractive index (wavelength = 500 nm)	Density, g/cm^3	Melting or softening temperature, K
Germanium	Ge	4	5.33	1210
Lithium fluoride	LiF	1.394	3.5	1140
Magnesium fluoride	MgF_2	1.39	3.18	1528
Sodium chloride	NaCl	1.53	2.17	1070
Zinc sulfide	ZnS	2.42	4.08	2100
Zinc selenide	ZnSe	2.43	5.42	1790
Barium titanate	$BaTiO_3$	—	5.9	1870
Cesium iodide	CsI	1.75	4.51	894
Diamond	C	2.4	3.51	3770
Lanthanum fluoride	LaF_3	1.6	5.94	1766
Magnesia	MgO	1.74	3.585	3053
Potassium chloride	KCl	1.49	1.98	1050
Sapphire	Al_2O_3	1.77	3.98	2300
Crystal quartz	SiO_2	1.55	2.65	1740
Barium fluoride	BaF_2	1.47	4.89	1550
Calcium fluoride	CaF_2	1.43	3.18	1630
Cadmium telluride	CdTe	2.69	6.2	1320
Calcite	$CaCO_3$	n_o = 1.665, n_e = 1.490	2.710	1612
Cuprous chloride	CuCl	2.0	4.14	695
Gallium phosphide	GaP	3.65	4.13	1623
Indium arsenide	InAs	4.5	—	1215
Lead fluoride	PbF_2	1.78	8.24	1100
Lead sulfide	PbS	4.3 at 3 μm	7.5	1387
Silicon carbide	SiC	2.68	3.217	3000
Selenium	Se	2.83	4.82	490
Silicon	Si	3.45	2.329	1690

The 14 Bravais lattices, derived by centering of the seven crystal classes (P and R) defined by symmetry operators

Bravais lattice cells	*Examples*
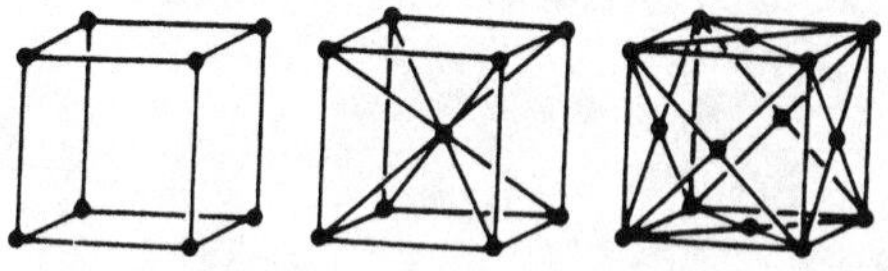 Cubic P Cubic I Cubic F	Copper (Cu), silver (Ag), sodium chloride (NaCl)
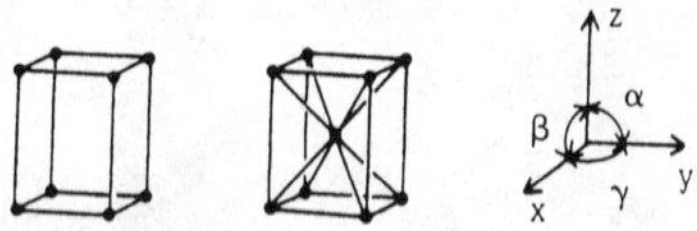 Tetragonal P Tetragonal I	While tin (Sn), rutile (TiO_2), β-spodumene ($LiAlSi_2O_6$)
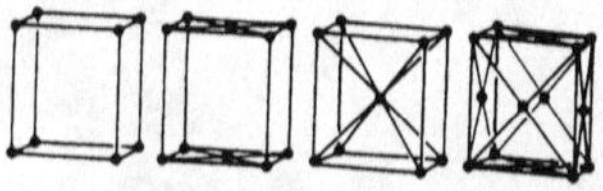 P C I F Orthorhombic	Gallium (Ga), perovskite ($CaTiO_3$)
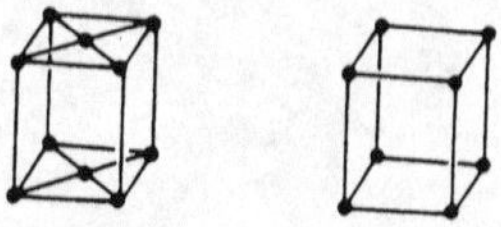 Monoclinic P Monoclinc C	Gypsum ($CaSO_4 \cdot 2H_2O$)
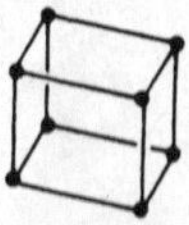 Triclinic P	Potassium chromate (K_2CrO_7)
 Trigonal R (rhombohedral)	Calcite ($CaCO_3$), arsenic (As), bismuth (Bi)
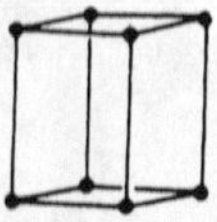 Trigonal and hexagonal C (or P)	Zinc (Zn), cadmium (Cd), quartz (SiO_2) [P]